Wastewater and Waste Treatment: Overview, Challenges and Current Trends

Wastewater and Waste Treatment: Overview, Challenges and Current Trends

Guest Editors

Dimitris Zagklis
Georgios Bampos

MDPI

Basel • Beijing • Wuhan • Barcelona • Belgrade • Novi Sad • Cluj • Manchester

Guest Editors
Dimitris Zagklis
Industrial Engineering and Management
International Hellenic University
Thessaloniki
Greece

Georgios Bampos
Department of Chemical Engineering
University of Patras
Patras
Greece

Editorial Office
MDPI AG
Grosspeteranlage 5
4052 Basel, Switzerland

This is a reprint of the Special Issue, published open access by the journal *Processes* (ISSN 2227-9717), freely accessible at: www.mdpi.com/journal/processes/special_issues/Wastewater_Waste_Treatment.

For citation purposes, cite each article independently as indicated on the article page online and using the guide below:

Lastname, A.A.; Lastname, B.B. Article Title. *Journal Name* **Year**, *Volume Number*, Page Range.

ISBN 978-3-7258-2958-3 (Hbk)
ISBN 978-3-7258-2957-6 (PDF)
https://doi.org/10.3390/books978-3-7258-2957-6

Contents

processes

Editorial

Editorial for the Special Issue "Wastewater and Waste Treatment: Overview, Challenges and Current Trends"

Dimitris P. Zagklis [1,*] and Georgios Bampos [2,*]

1 Department of Industrial Engineering and Management, International Hellenic University (IHU), GR-57400 Thessaloniki, Greece
2 Department of Chemical Engineering, University of Patras, GR-26504 Patras, Greece
* Correspondence: dimitriszag@chemeng.upatras.gr (D.P.Z.); geoba@chemeng.upatras.gr (G.B.)

1. Introduction

Today's environmental challenges, marked by escalating pollution levels, climate change and diminishing natural resources, urgently require innovative solutions, particularly in waste and wastewater management. The existing conventional treatment methods are in need of improved efficiency, reduced energy demands and the ability to effectively tackle complex environmental problems. These shortcomings have been highlighted in the literature, indicating the serious need for more sustainable and environmentally friendly approaches [1,2]. In response to this pressing aim, recent environmental research has turned towards discovering and developing more effective, sustainable and innovative strategies for handling waste and wastewater. This includes leveraging cutting-edge technologies and exploring new ways to improve the efficiency of pollutant removal whilst also reducing the environmental impact of these processes.

This Special Issue has compiled relevant research addressing the aforementioned issues. An overview of the papers published in this Special Issue will follow, highlighting some emerging trends in the research carried out in the field and the interesting solutions that have been proposed to target more specific environmental concerns.

2. Overview of Contributions

The range of topics covered in this Special Issue reflects the dynamic nature of the wastewater and waste treatment research field, offering a broad overview characterized by a diverse array of research into waste streams and treatment processes. The focus on waste streams is notably varied, with urban waste leading in the number of publications, comprising 41% of the total papers in this Special Issue, reflecting the urgent need to address pollution in densely populated areas. Industrial and agro-industrial wastes also received significant attention, constituting 21% and 15%, respectively, highlighting the environmental impact of these sectors. On the treatment front, physicochemical methods are predominant, representing 35% of the studies, indicative of their widespread applicability and effectiveness. Biological treatment methods, integral for their sustainability and eco-friendliness [3,4], account for 32%, while advanced oxidation processes, crucial for recalcitrant contaminants [5], make up 12%. The composition of the Special Issue, with 24% of the works being review papers, provides a solid foundation of knowledge and context, while the 76% original research papers drive the field forward with new insights and innovations, collectively painting a comprehensive picture of the current state of the field and future directions in wastewater and waste treatment research. Figure 1 summarizes the distribution of papers published in this Special Issue.

check for updates

Citation: Zagklis, D.P.; Bampos, G. Editorial for the Special Issue "Wastewater and Waste Treatment: Overview, Challenges and Current Trends". *Processes* **2024**, *12*, 853. https://doi.org/10.3390/pr12050853

Received: 18 April 2024
Accepted: 22 April 2024
Published: 24 April 2024

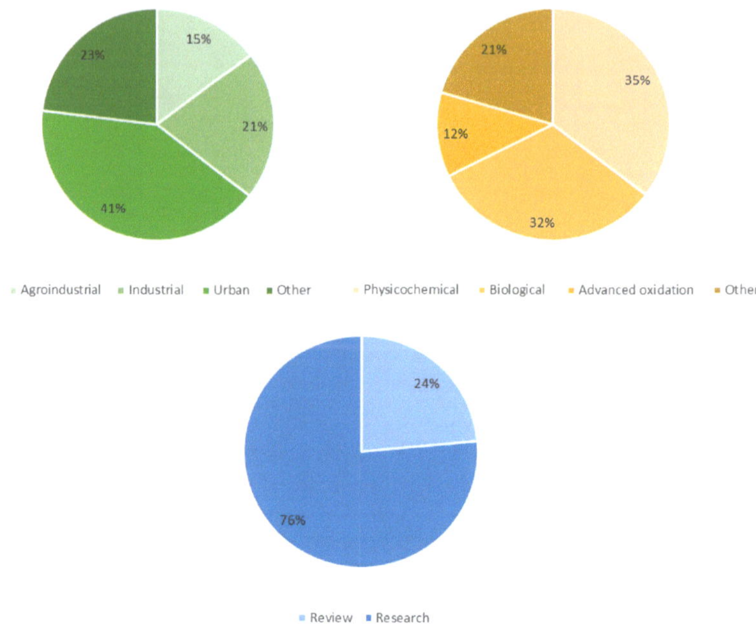

Figure 1. Distribution of topics and nature of the published papers in this Special Issue.

It would be impossible to discuss all of the high-quality contributions in the Editorial of this Special Issue, but a brief discussion will follow to highlight some of them.

As previously mentioned, most of the contributions present innovative research work describing treatment technologies. In the work entitled "Characterization of Sludge Resulting from Chemical Coagulation and Electrocoagulation of Pumping Water from Fishmeal Factories", the authors present a physicochemical treatment method for the management of an industrial waste stream (namely from a fishmeal factory). Through the use of chemical coagulation and electrocoagulation, the authors managed to recover some of the valuable constituents of the waste stream. In the work entitled "Preparation and Phosphorus Removal Performance of Zr–La–Fe Ternary Composite Adsorbent Embedded with Sodium Alginate", the authors discuss the preparation of a novel composite material that can be used as an adsorbent for phosphorous removal. The material was tested on both synthetic and real urban wastewater, achieving a removal of phosphorous of more than 99%. Papers such as the one entitled "Development of a Continuous Hydrothermal Treatment Process for Efficient Dewatering of Industrial Wastewater Sludge" showcase novel treatment technologies that promise enhanced efficiency. This study demonstrates the benefits of hydrothermal carbonization for the dewatering of biological paper sludge without the use of dewatering aids such as fiber sludge or polyelectrolytes.

In addition to the works that present novel treatment methods, several of the contributions examine ways to optimize existing treatment processes; for example, "Optimization of the Biotreatment of GTL Process Water Using Pseudomonas aeruginosa Immobilized in PVA Hydrogel" and "Improving the Treatment Efficiency and Lowering the Operating Costs of Electrochemical Advanced Oxidation Processes" examine the optimization of a biological and an advanced oxidation process, respectively. In the work entitled "Increase by Substitution of Galvanized Steel for Aluminum Mirrors in the UV Solar Radiation in Canal with Fins and Side Panels That Disinfect Wastewater", the authors examine the addition of mirrors in water disinfection canals, intensifying solar radiation by 8%.

Emerging contaminants are an ever-growing environmental problem, posing unique challenges to existing waste and wastewater treatment systems [6]. The authors of the work entitled "Occurrence of Selected Emerging Contaminants in Southern Europe WWTPs: Comparison of Simulations and Real Data" examine the occurrence of six targeted emerging contaminants in the outflow of an urban wastewater treatment plant over nine months. In the review work entitled "Emerging Pollutants in Wastewater, Advanced Oxidation Processes as an Alternative Treatment and Perspectives", the authors provide an overview of the most significant groups of emerging contaminants that could potentially be found in wastewater and the environment.

Papers such as "Fermentation Wastes from *Chrypthecodinium cohnii* Lipid Production for Energy Recovery by Anaerobic Digestion", "Circular Economy Indicators for the Assessment of Waste and By-Products from the Palm Oil Sector" and "Recovery of Phosphorus in Wastewater in the Form of Polyphosphates: A Review" examine the recovery of valuable materials and energy from waste streams, a concept that promotes the incorporation of circular economy principles and sustainable practices in waste treatment. Such methodologies not only address waste treatment but also underscore the importance of resource conservation and recovery in achieving environmental sustainability.

3. Challenges and Future Directions

Despite the significant advances in the field of waste and wastewater treatment, the field faces persistent challenges due to the ever-growing complexity of industrial effluents and the scalability of novel treatment methods. In most cases, new processes, and in this case wastewater treatment processes, need to have a positive fiscal impact on the waste producer in order to be adopted or imposed by authorities. More work is needed to improve the economic sustainability of novel treatment methods, at least to minimize the treatment cost, and the authorities that have oversight regarding environmental issues need to be updated on recently developed solutions to existing environmental concerns. These solutions should exhibit their positive impact on the environment through environmental assessment studies, as this will facilitate their adoption. Future research must continue to foster interdisciplinary collaborations, integrating insights from engineering, microbiology, chemistry and environmental science to develop holistic and adaptable solutions.

The varied collection of research presented in this Special Issue not only advances our understanding of wastewater and waste treatment but also catalyzes action towards more sustainable practices, offering both a reflection of our current standing and an aim of a cleaner, more sustainable future.

Conflicts of Interest: The authors declare no conflict of interest.

References

1. Dutta, D.; Arya, S.; Kumar, S. Industrial wastewater treatment: Current trends, bottlenecks, and best practices. *Chemosphere* **2021**, *285*, 131245. [CrossRef] [PubMed]
2. Rashid, R.; Shafiq, I.; Akhter, P.; Iqbal, M.J.; Hussain, M. A state-of-the-art review on wastewater treatment techniques: The effectiveness of adsorption method. *Environ. Sci. Pollut. Res.* **2021**, *28*, 9050–9066. [CrossRef]
3. Hussain, F.; Shah, S.Z.; Ahmad, H.; Abubshait, S.A.; Abubshait, H.A.; Laref, A.; Manikandan, A.; Kusuma, H.S.; Iqbal, M. Microalgae an ecofriendly and sustainable wastewater treatment option: Biomass application in biofuel and bio-fertilizer production. A review. *Renew. Sustain. Energy Rev.* **2021**, *137*, 110603. [CrossRef]
4. Balasubramani, R.; Awasthi, M.K.; Varjani, S.; Karmegam, N. Aerobic and anaerobic digestion of agro-industrial and livestock wastes: A green and sustainable way toward the future. *Agronomy* **2023**, *13*, 2607. [CrossRef]
5. Ghime, D.; Ghosh, P. Advanced oxidation processes: A powerful treatment option for the removal of recalcitrant organic compounds. In *Advanced Oxidation Processes—Applications, Trends, and Prospects*; IntechOpen: Rijeka, Croatia, 2020; ISBN 1789848911.
6. Sol, D.; Menéndez-Manjón, A.; Arias-García, P.; Laca, A.; Laca, A.; Rancaño, A.; Díaz, M. Occurrence of Selected Emerging Contaminants in Southern Europe WWTPs: Comparison of Simulations and Real Data. *Processes* **2022**, *10*, 2491. [CrossRef]

processes

Article

Characterization of Sludge Resulting from Chemical Coagulation and Electrocoagulation of Pumping Water from Fishmeal Factories

Edwar Aguilar-Ascón [1,*], Armando Solari-Godiño [2], Miguel Cueva-Martínez [3], Walter Neyra-Ascón [4] and Miguel Albrecht-Ruíz [3]

1 Programa de Estudios Generales, Instituto de Investigación Científica (IDIC), Universidad de Lima, Av. Javier Prado 4600, Surco, Lima 15023, Peru
2 Pesquera Diamante S.A., Amador Merino Reyna 307, Edificio Nacional, piso 12 y 13, San Isidro, Lima 15036, Peru
3 Laboratorio de Bioquímica, Dirección de Investigación (DIDITT), Instituto Tecnológico de la Producción, Carretera a Ventanilla Km. 5,2, Callao 07061, Peru
4 Instituto de Investigación Científica (IDIC), Universidad de Lima, Av. Javier Prado 4600, Surco, Lima 15023, Peru
* Correspondence: eaguilaa@ulima.edu.pe

Abstract: In the fishmeal industry, seawater is used to transport fish to the factories. Due to this, "pumping water" with high concentrations of organic matter is generated that is treated via chemical coagulation before letting it into the sea. The objective of this study is to characterize and compare the sludge obtained from conventional chemical coagulation and the sludge from the electrocoagulation process. A pilot electrocoagulation plant was built next to a chemical coagulation plant. The sludge obtained from both methodologies was analyzed for its proximal composition, its iron and aluminum content, and the fatty acid profile in its contained fat. Electrocoagulation was found to produce sludge with a higher concentration of lipids and ash, which indirectly confirmed that it removes more organic pollutants and salts than chemical coagulation. The contents of aluminum and iron in the sludge obtained by electrocoagulation were 4.2% and 0.025%, respectively, while those in the sludge obtained from chemical coagulation were 0.01% and 4.8%, respectively. Aluminum comes from the sacrificial electrode of the electrocoagulation tank, while iron comes from the salts used in chemical coagulation. The sum of w-3 fatty acid values (EPA + DHA) was 12.5% and 18.8% for sludges from the electrocoagulation and chemical coagulation processes, respectively, so we can assume that electrocoagulation is a more oxidizing process than chemical coagulation. Due to their high organic load, both sludges must be assessed as an alternative feed ingredient.

Keywords: electrocoagulation; pumping water fishmeal industry; sludge from pumping water; chemical coagulation

Citation: Aguilar-Ascón, E.; Solari-Godiño, A.; Cueva-Martínez, M.; Neyra-Ascón, W.; Albrecht-Ruíz, M. Characterization of Sludge Resulting from Chemical Coagulation and Electrocoagulation of Pumping Water from Fishmeal Factories. *Processes* **2023**, *11*, 567. https://doi.org/10.3390/pr11020567

Academic Editors: Dimitris Zagklis and Georgios Bampos

Received: 29 November 2022
Revised: 9 January 2023
Accepted: 10 January 2023
Published: 13 February 2023

1. Introduction

In the fishmeal industry, ~30% of the world's production comes from Peru and primarily comprises anchovy (*Engraulis ringens*) [1]. The transport and discharge of water fish from the boats to the processing plants, after sieving, is called "pumping water" (PW) [2]. On average, the process uses two volumes of seawater for every volume of fish transported [3,4]. However, updated reports show this ratio has turned out to be 1. Depending on the capture time and the subsequent treatment of fish, PW can contain between 9000 and 30,000 mg/L of Total Suspended Solids (TSS) and between 400 and 650 mg/L of Fats, Oils and Grease (FOG) [5,6]. Given that the annual anchovy catch had an average annual landing of 4.21 million tons during the last decade [7], these are significant statistics to consider.

Before returning to the sea, PW must be treated to reduce its organic material content, and thus, avoid pollution. The treatment of PW includes several consecutive processes (Figure 1). Initially, the fish is unloaded from the boat and taken to the processing plant. Once inside the plant, water is removed via rotating filters that recover scales and larger solids incorporated into fishmeal production. Then, water is taken to inclined plane pools, "grease traps", where grease is recovered by flotation at a water flow rate of 200–250 m^3/h. The water that comes from the grease traps is stored in an "equalizer tank", from where it can be diverted to discharge by recirculation, depending on its physicochemical parameters, making efficient use of the water resource. In these instances, the water passes to a Dissolved Air Flotation "DAF" system to continue the recovery of lipophilic foams. Then, the PW is sent to the clarifier, where chemical agents are added to allow the coagulation and flocculation of the solids present in the water, called the "Chemical Coagulation Process" (CCP), which, by the action of injected air microbubbles, allow flocs to float and be collected and taken to centrifugal decanters where the solids are separated to form the sludge [8]. Once clarified, the water is returned to the sea, complying with government regulations that establish values that should not exceed the Maximum Permissible Limit (MPL): 700 mg/L of TSS and 350 mg/L of FOG [9].

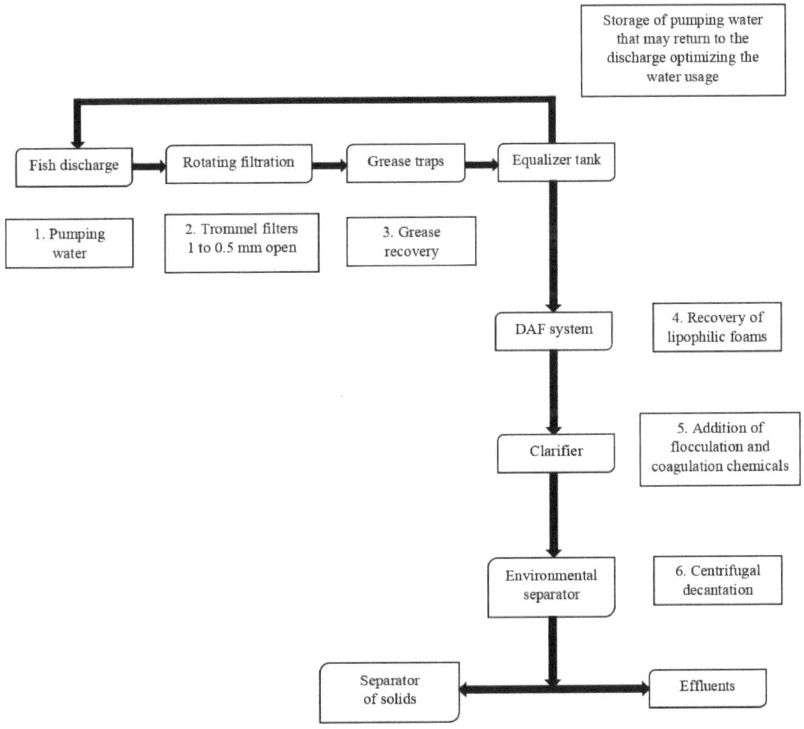

Figure 1. Scheme for the fish PW treatment system. Source: Pesquera Diamante.

The most studied coagulation/flocculation chemical agents in the fishmeal industry are iron salts, which also reduce the pH of the medium and bring most proteins to its isoelectric point, increasing their precipitation [10]. Currently, for the chemical treatment of PW, ferric sulfate and polyelectrolytes of an anionic or cationic nature are added for the destabilization and flocculation of proteins [11]. Chemical agent application increases treatment costs and encourages the recovery of iron-rich sludge, which has only a limited value.

An alternative method to the CCP is electrocoagulation (EC), where the electric current destabilizes the polluting particles in the water through metal plates used as electrodes, with iron and aluminum being the most used elements due to their environmental and economic advantages [12,13]. In a sacrificial electrode, the metal ions are released and dispersed in the liquid medium. They tend to form metal hydroxides that electrically attract the polluting solids that have been destabilized.

Flocs larger than those observed in the CCP are formed [14,15]. Because of this, they precipitate or float, forming sludges which contain the polluting material and are easier to dehydrate with other secondary processes [16,17]. In addition, hydrogen and oxygen bubbles are generated on the electrodes, which help lift the coagulated particles via electroflotation [18,19].

Compared to CCP, EC provides a compact and robust treatment alternative where electrochemical reactions are started at the anode that provides active metal cations for coagulation and flocculation and where the resulting use of high-cost chemicals and secondary pollution is avoided [13].

In the search for alternative systems for treating PW from fishmeal factories, the EC process has been proposed as a more economical system [14]. Previous studies have shown that it reduces the organic levels in residues from fish processing. However, on a pilot scale, this treatment did not turn out to be as effective as initially anticipated Beck et al. 1974, mentioned by [8]. Recent improvements to the EC processes are again making this technology an alternative to treating this type of water [12,13].

Pumping water in fishmeal factories usually comes from the sea, where NaCl is higher than other industrial wastewater. These ionic strength conditions favor conductivity in the EC process and reduce the formation of calcium carbonate-type anions that can precipitate on the cathode surface and increase ohmic resistance [20]. The conductivity of the PW varies between 20,000 and 50,000 uS/cm. For this reason, this technology must be adapted to avoid high demands for electric current and excessive electrode consumption in the electrocoagulation process. Recent studies [21] removed organic material in PW using a laboratory-level EC system. They sufficiently complied with the limits proposed by the Peruvian legal regulations for treated water [9].

As environmental restrictions become stringent, there is a growing interest in researching wastewater sludge from water treatment to assess its potential use and recovery within a circular economy model. This would result in cheaper costs and sufficient final waste disposal [22].

In this regard, high concentrations of silica, aluminum, ferric oxide, lime, and heavy metals have been found in the sludge from CCP drinking water plants. Thus, a safe disposal route for this sludge would be manufacturing bricks or their use as a substitute for building materials [23]. Alternatively, a product has been obtained from sludge treated with an alum; the said product can remove colloidal suspensions and apply them to river water [24]. Regarding the use and reassessment of sludge obtained by EC, it has been proposed as a partial substitute for cement [25], used for the adsorption of dyes in aqueous solutions [26] and the manufacture of alumina microcapsules for corrosion protection [27].

PW has a high conductivity due to dissolved salts and high concentrations of organic substances. In previous studies at the laboratory level [1], EC technology managed to remove 99% of TSS under conditions of $I = 13$ A and $t = 30$ min, generating sludge that contained concentrations of 0.3% of iron and 1.7% aluminum, which can be considered as a technology for the purification of this type of water and generating sludge that can serve as a feed component. Thus, to use EC technology in PW, a pilot electrocoagulation system was designed and built to evaluate the efficiency of solids removal in PW while characterizing the sludge obtained to evaluate its possible use as feed and compare them with sludges obtained by the CCP.

2. Materials and Methods

2.1. Electrocoagulation Pilot System

The system consists of an equalizer tank with a 0.6 m^3 capacity, where the PW is received and homogenized. This water is sent to the reaction tanks, where aluminum electrodes are located (three tanks connected in series). The electric current used is continuous and was supplied to the electrodes with a power source with a capacity of up to 100 A. After passing through the electrocoagulation process of the third tank, the water is diverted to a clarifying tank for the final sedimentation process and then sent to the discharge point by pumping. The start-up of the loading and unloading pumps, as well as the retention time in the reactor tanks, are monitored with a control panel. The sludge (recovered solids) formed managed to float as foam in the three tanks until overflowed by the action of the microbubbles generated in the electrodes. These foams are collected to obtain the sludge (Figure 2A,B).

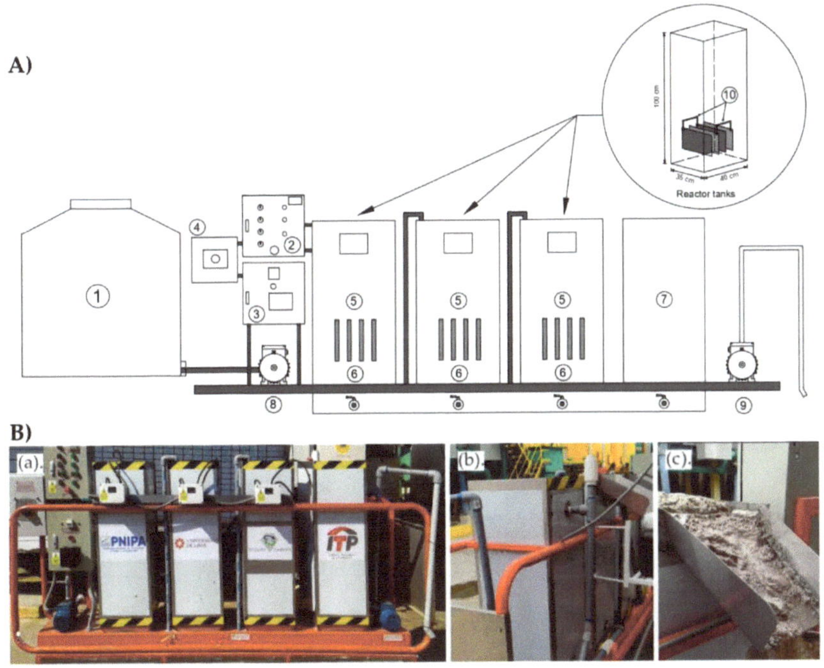

Figure 2. (**A**): Scheme of the electrocoagulation pilot plant. (1) Equalizer tank. (2) Control board. (3) Power source. (4) Voltage variator. (5) Reactor tanks. (6) Aluminum electrodes. (7) Clarifying tank. (8) Charging pump. (9) Discharge pump. (10) Cable connection. (**B**): Electrocoagulation pilot plant images. (**a**) Image of front. (**b**) Back image. (**c**) Sludge collector.

2.2. Obtaining and Preparing Samples

The samples were obtained during the first week of October 2021. To obtain sludge by EC, the PW was placed in the equalizer tank, then sent to the reactor tanks at a flow of 125 L/h and a current of 26 A (Figure 2) so that the PW in the successive reactor tanks had an increasingly lower concentration of suspended solids to comply with MPL [9]. The time to treat 600 L of PW contained in the equalizing tank was 3 h. Simultaneously, for each batch, the sludge from the CCP was obtained from the separator centrifuge PW treatment plant (Figure 1). The sludge generated by the EC floats in the form of foam due to the action of the microbubbles generated by the electrodes. These foams increased until they overflowed from the reactor tanks and were collected in polyethylene bags and placed in thermal boxes at 4 °C, where they returned to a liquid state. These samples were sent to the laboratory and centrifuged at 2000 rpm for 15 min. The supernatant liquid was discarded and the precipitated sludge was collected.

The sludge from both processes was spread on nylon meshes placed on aluminum trays and dried at 70 °C for 18 h in an oven with forced air convection (VENTICELL®). The dried samples were ground and sieved through a mesh of <1 mm, bagged in polyethylene bags and stored at room temperature (20 °C ± 2) for 30 days before being analyzed The samples were called LCQ (chemical coagulation sludge) and LEC (electrocoagulation sludge).

2.3. Physicochemical Tests

The proximate composition of LCQ and LEC was conducted according to the methodologies recommended by the FAO [28]: moisture by oven drying at 104 °C (VENTICELL®) and ash content using a muffle furnace at 600 °C (Barnstead/thermolyne 48000®), both until constant weight. The protein was obtained using the Kjeldahl method using the kjeldatherm/Turbosog/Vapodest® equipment system (C. Gerhardt GmbH & Co. KG Cäsariusstraße 97, 53639 Königswinter, Germany), using a total nitrogen to protein conversi on factor of 6.25. The crude fat was realized using hexane as an extraction solvent in Soxtherm® equipment (C. Gerhardt GmbH & Co. KG). All tests were duplicated and expressed in g 100 g^{-1} of sample.

Iron (Fe) was quantified by atomic absorption spectrometry (AA) following the 999.11 AOAC methodologies [29]. Aluminum quantification was performed by AA on the sample calcined at 600 °C and dissolved in 1N HCl, which was read at 396.2 nm, using a nitrous oxide-acetylene flame following the recommendations of the team [30].

The fatty acid profile was obtained by gas chromatography (Perkin Elmer Autosystem XL GC with FID detector) using a Carbowax 20 M column following the Ce 1f-96 AOCS methodology. The sample analyzed was the fat extracted from the sludge by the Bligh and Dyer methodology [31] before drying it in the oven to avoid oxidation.

The Tukey test verified the differences at 95% confidence intervals using the Minitab® 21 Statistical Software (Chicago, IL, USA).

3. Results and Discussion

The chemical coagulation process (CCP) and electrocoagulation (EC) have been studied and compared with variable results depending on the type of water to be treated, ionic charge, electrical conditions, and pH [13,23,32–35]. A quantitative comparison is not feasible because both operations are different; in CCP, the coagulant is administered discretely, whereas in EC, the addition of the coagulant is a function of time and current [36].

The content of solids and fats in PW varies greatly. It depends on factors after the fish catch, such as time, type and storage temperature, discharge speed, type of pumps used in the discharge, and flow rate of water used in the discharge, among others. The water samples analyzed presented values of about 4800 mg/L for TSS, 265.4 mg/L for FOG, 6.4 for pH value, and 20,500 uS/cm for conductivity.

At the flow rate and current intensity conditions analyzed (125 L/h and 26 A), the treated water obtained in the clarification tank sufficiently met the maximum permissible limits (MPL) required by the regulations, as it presented values of 41 mg/L in SST (MPL < 700 mg/L), 2 mg/L FOG (MPL < 350 mg/L), and a pH value of 7.08 (MPL pH 5–9). Even though CCP was accomplished with MPL, reaching values of 417.4 mg/L TSS, 30 mg/L A&G, and a pH value of 5.5, these values were more elevated than EC. These results indicate that EC more effectively removed the contaminants from PW than CCP.

The estimated cost of the EC process was USD 1.24/m^3 PW, including electrode consumption and energy cost USD/KWh. The fishing industry works with CCP and reports a cost of around USD 1.5/m^3 PW by chemical consumption, energy, and operations.

3.1. Proximate Composition of Sludge

The proximate composition included moisture and served to convert the protein, fat, and ash on a dry basis. The LEC sample's moisture presented lower values than their LCQ pairs (2.33 ± 0.6 vs. 5.32 ± 1.3, respectively). The higher fat content in LEC proved to be a barrier to the balance with ambient humidity, given that the samples were stored for similar amounts of time. Pronounced differences were observed in the composition of the LCQ and LEC samples (Table 1), and it was confirmed that both types of coagulation operate differently. Under the conditions analyzed, LEC presented a higher fat and ash than LCQ. Furthermore, EC was more effective than CCP in removal of pumping water contaminants, so it is deduced that EC has a better performance than CCP.

Table 1. Proximate composition (dry basis) of sludge flour from PW.

	Batch	% Fat	% Ash	% Proteins
Chemical coagulation sludge (LCQ)	1	13.75 ± 0.40 ^{a}	14.14 ± 0.04 ^{a}	41.26 ± 1.44 ^{a,b}
	2	10.23 ± 0.33 ^{b}	16.94 ± 0.08 ^{b}	38.72 ± 0.96 ^{b,c}
	3	21.46 ± 0.44 ^{c}	13.95 ± 0.09 ^{c}	44.17 ± 2.21 ^{a}
	4	19.76 ± 0.71 ^{d}	13.76 ± 0.03 ^{c}	35.98 ± 0.96 ^{c}
Electrocoagulation sludge (LEC)	1	24.71 ± 0.54 ^{a}	22.89 ± 0.31 ^{b}	35.43 ± 0.52 ^{b,c}
	2	25.07 ± 2.21 ^{a}	22.85 ± 0.36 ^{b}	34.89 ± 0.61 ^{c}
	3	26.19 ± 0.60 ^{a}	27.11 ± 0.14 ^{a}	39.84 ± 0.70 ^{a}
	4	25.41 ± 0.67 ^{a}	27.19 ± 0.24 ^{a}	37.59 ± 1.66 ^{a,b}

Different letters in LCQ and LEC columns indicate significant differences ($p < 0.05$). Each batch number worked in LEC had its corresponding sampling in LCQ.

Proximate composition on a dry basis in LCQ samples presented values of 10.2–21.5% fat, 13.8–16.9% ash, and 35.9–44.2% protein. Meanwhile, LEC samples showed 24.7–26.2% fat, 22.8–27.2% ash, and 34.9–39.8% protein. LCQ samples presented variable results and greater significant differences between batches because they represented larger processed volumes, even though they came from the same work period (Table 2: LCQ part). Meanwhile, LEC samples showed more homogeneous results: 24.7–26.2% fat, 22.8–27.2% ash, and 34.9–39.8% protein. They represent 600 L of PW only. We must remember that the chemical coagulation process worked at 150–250 m^3/h, while the electrocoagulation process worked at a flow of 0.125 m^3/h.

Table 2. Composition of Fe and Al (dry base) in sludge flour from PW.

	Batch	Fe %	Al %
Chemical coagulation sludge (LCQ)	1	4.60 ± 0.08 ^{b}	0.01
	2	4.64 ± 0.06 ^{b}	0.01
	3	5.31 ± 0.03 ^{a}	0.01
	4	4.69 ± 0.13 ^{b}	0.01
Electrocoagulation sludge (LEC)	1	0.03	4.12 ± 0.11
	2	0.03	4.31 ± 0.17
	3	0.03	4.36 ± 0.10
	4	0.03	4.02 ± 0.02

Different letters in LCQ and LEC columns indicate significant differences ($p < 0.05$). Each batch analyzed in the electrocoagulation had its corresponding sampling in the chemical coagulation.

There is evidence of a higher percentage of fat and ash content in the LEC samples; consequently, there is a percentage decrease in the protein content (Figure 3). The average results of LCQ versus LEC were 16.3 ± 4.7% and 25.3 ± 1.2% in fat; 14.7 ± 1.4% and 25.01 ± 2.2% in ash; and 40.0 ± 3.4% and 36.9 ± 2.2% in protein.

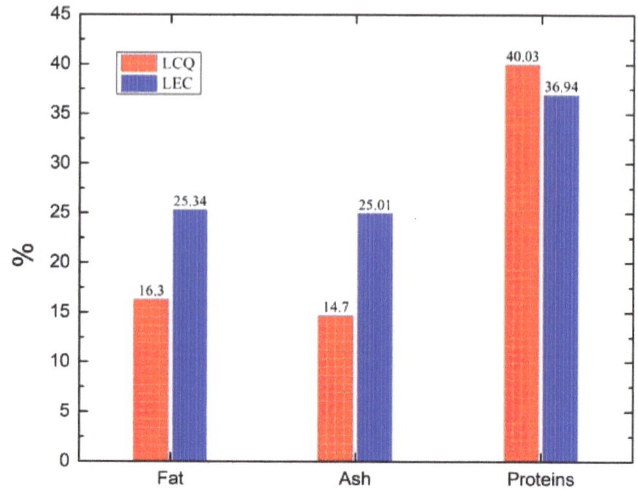

Figure 3. Proximal composition on a dry basis of the sludge obtained by LCQ and LEC. The bar values represent the average of all the batches worked.

3.2. Iron and Aluminum Content in Sludge

The treatment of PW in fishmeal factories is carried out by chemical coagulation by adding mainly iron salts. In contrast, in the tests carried out by electrocoagulation, aluminum electrodes were used, so the formation of aluminum salts from the sacrificial electrode was expected. Consequently, the resulting sludge's iron and aluminum content differed (Table 2).

The Fe content in the LCQ was around 4.8% $\pm$ 0.3% and mainly came from ferric sulfate added during the treatment of PW. Previous tests on the iron content in this type of sludge showed iron values ranging from 2.5% to 3.5% on a dry basis. These values show a notable difference from those found in this study. Chemical coagulation processes in the plant are complicated due to the large volumes treated in short times (250 m^3/hour); thus, over an hour, PW can show changes of up to 1500 mg/L of SST, depending on the raw material quality, according to data reported by the company. Although the jar test in the laboratory suggests adding precise coagulant contents, the answer to treating these large volumes in adequate time depends in part on technical experience, so, oftentimes, the addition of coagulants is performed empirically, resulting in varying concentrations of coagulants in the sludge obtained. Alternatively, Fe content in the LEC had an average value of 278 $\pm$ 16 mg/kg, higher than that found in fishmeal, possibly due to some pollution in the line and due to the recovery of Fe associated with proteins and iron salts present in seawater.

Aluminum content in LCQ had an average value of 106 $\pm$ 12 mg/kg, an average found in marine foods [36], while aluminum content in LEC was around 4.2 $\pm$ 0.2%, mainly from the sacrificial anode. The oxidation and reduction reactions that occur in the aluminum electrodes of the electrocoagulation reactor depend on the pH of the medium. The reactions at the anode and cathode are observed below:

Acidic pH

$$Al \rightarrow Al^{3+} + 3e^{-} \quad (1)$$

$$Anode \quad Al^{3+}_{(ac)} + 3H_2O \rightarrow Al(OH)_{3(S)} + 3H^{+}_{(ac)} \quad (2)$$

$$nAl(OH)_3 \rightarrow Al_n(OH)_{3n} \quad (3)$$

$$O_2 + 2H^{+} + 2e^{-} \rightarrow H_2O \quad (4)$$

Cathode

$$3H_2O + 3e^- \rightarrow 3/2H_2 + 3OH^- \quad (5)$$

Neutral/Alkaline pH

$$Anode \quad Al_{(S)} + 4OH^- \rightarrow AlO_2^- + 2H_2O + 4e^- \quad (6)$$

$$Cathode \quad H_2O + O_{2(g)} + 2e^- \rightarrow 2OH^- \quad (7)$$

The PW in fishmeal factories has pH values of 6.1–6.5. Under these conditions, the sacrificial electrode (anode) releases aluminum ions that, in contact with the hydroxyls formed by electrolysis, favor the formation of hydroxides or polyhydroxides, which have a great affinity for dispersed particles and ions and allow their coagulation. This explains the high concentration of aluminum in the recovered sludge (Figure 4).

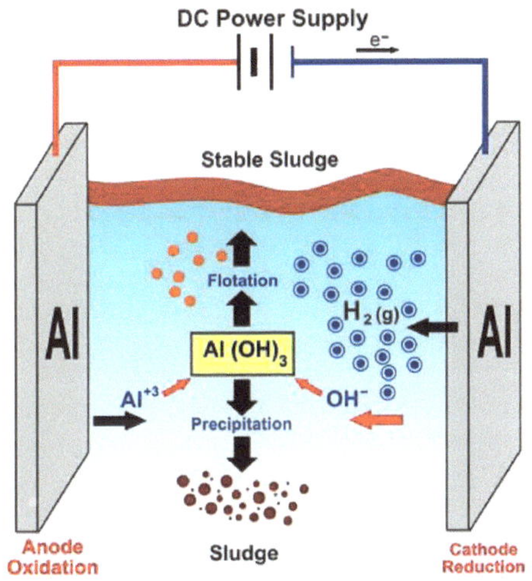

Figure 4. Simplified diagram of the electrocoagulation process with an aluminum electrode.

3.3. Fatty Acid Profile in the Fat Extracted from the Sludge

The initial tests of the fatty acid profile in LCQ and LEC showed values in the content of omega-3 fatty acids (W-3) [eicosapentaenoic (EPA) + docosahexaenoic (DHA)], well below the expected value in the anchovy fat (E. ringens), whose W-3 values are around 30% [37]. It is possible that the drying process during the preparation of the flour samples oxidized the polyunsaturated fatty acids and greatly decreased them. Thus, the fat was extracted from the newly arrived samples at the laboratory using the Bligh and Dyer methodology to avoid oxidation; then, its methylation was performed, and the fatty acid profile was determined.

The ions formed by electrolysis are responsible for the oxidation of polyunsaturated fatty acids. The fatty acid profile of the fat extracted from the sludge (Table 3) shows that the EC was more susceptible to oxidation than the CCP. The EPA + DHA content in LEC was 12.46%, while it was 18.78% in LCQ. The average percentage content of EPA was $9.98 \pm 0.89\%$ and $4.38 \pm 0.09\%$, while DHA was $8.80 \pm 1.1\%$ and $8.08 \pm 0.34\%$ in LCQ and LEC, respectively. The unsaturation of fatty acids directly influences their oxidation, so it is striking that DHA (22:6 W-3) has not decreased as much as or more than EPA (20:5 W-3). The DHA: EPA ratio in anchovy is higher in phospholipids than in triglycerides [38]. The increased EPA concentration may have been the target of the oxidative assault, because it is possible that the lipids recovered in EC are primarily derived from triglycerides, even if there is no more information on the lipids present in the sludge.

Table 3. Fatty acid profile of the fat extracted from the sludge from PW. The values are expressed in percentage.

	Chemical Coagulation Sludge (LCQ)				Electrocoagulation Sludge (LEC)			
Ac. Fatty/Batch	**1**	**2**	**3**	**4**	**1**	**2**	**3**	**4**
C 14:0 (Myristic)	10.65	9.73	10.89	10.44	10.72	10.62	10.63	10.63
C 16:0 (Palmitic)	24.88	23.69	25.8	25.03	29.85	30.11	30.29	31.1
C 16:1 (Palmitoleic)	10.93	8.96	11.24	11.01	9.37	9	9.13	9.01
C 18:0 (Stearic)	5.19	5.04	5.37	5.26	7.28	7.37	7.46	7.62
C 18:1 w-9 (Oleic)	10.11	11.67	10.28	10.07	9.87	9.28	9.34	9.58
C 18:1 w-7 (Vaccenic)	3.72	3.33	4.11	4.08	4.4	4.28	4.34	4.49
C 18:2 w-6 (Linoleic)	1.24	1.04	1.24	1.28	1.05	0.96	1.03	1.05
C 20:5 w-3 (EPA)	9.69	11.12	9	10.11	4.5	4.4	4.35	4.28
C 22:5 w-3 (Cupladonic)	1.06	1.43	1.02	1.13	1.61	1.63	1.73	1.78
C 22:6 w-3 (DHA)	8.59	1.43	7.64	8.66	8.08	8.41	8.21	7.62

4. Conclusions

This study compared sludges obtained from the treatment of PW treated with a 125 L/h continuous system electrocoagulation pilot plant and conventional chemical coagulation used in the fishing industry. EC was found to produce sludge with a higher concentration of lipids and ash, which indirectly confirmed that it performs a greater removal of biomolecules and salts compared to CCP. The aluminum content in LEC was 4.2%, while the iron content in LCQ was 4.8%. The fatty acid profile of the lipids extracted from the sludge indirectly showed that the electrocoagulation process, under the test conditions, turned out to be more oxidizing than the chemical coagulation process, which was verified by a lower concentration of W-3 fatty acids (EPA + DHA).

Although it should be noted that, due to the large volumes of water handled by this type of industry, electrocoagulation tanks must have a continuous flow to meet the water purification criteria and simultaneously obtain sludge that serves as a feed ingredient under a circular economy concept. Electrocoagulation has shown to be an up-and-coming technology for treating PW in fishmeal factories.

Author Contributions: Conceptualization/visualization: E.A.-A. and M.A.-R.; Methodology: E.A.-A., M.A.-R., A.S.-G. and M.C.-M.; Draft preparation: E.A.-A., M.A.-R., M.C.-M., A.S.-G. and W.N.-A.; Manuscript review, editing, and supervision: E.A.-A., A.S.-G., W.N.-A. and M.A.-R. All authors have read and agreed to the published version of the manuscript.

Funding: From the Instituto de Investigación Científica (IDIC) of the Universidad de Lima, Pesquera Diamante S. A., Instituto Tecnológico de la Producción (ITP) and the Programa Nacional de Innovación en Pesca y Acuicultura (PNIPA)-PESSIA-PP-000001, Contract: No. 141–2018, who fully supported the development of this study.

Data Availability Statement: Not applicable.

Acknowledgments: Engineers Miguel Gallo and Alberto Salas for their support in the project.

Conflicts of Interest: The authors declare no conflict of interest.

References

1. Indexmundi, S.F. Producción de Harina de Pescado Por País en Miles de Toneladas. R. Available online: https://www.indexmundi.com/agriculture/?producto=harina-de-pescado&variable=produccion&l=es (accessed on 14 July 2022).
2. Martí, C.; Roeckel, M.; Aspe, E.; Kanda, H. Recovery of proteins from fishmeal factory wastewaters. *Process Biochem.* **1994**, *29*, 39–46. [CrossRef]
3. Figueroa, J.G.; Jung, H.Y.; Jeong, G.T.; Kim, J.K. The high reutilization value potential of high-salinity anchovy fishmeal wastewater through microbial degradation. *World J. Microbiol. Biotechnol.* **2015**, *31*, 1575–1586. [CrossRef]
4. Centurión, R.I.; Ganoza, R.E.; Torres, R.M. *Propuesta de Mejora en EL Control de Efluentes de Una Planta Pesquera*; Universidad Peruana de Ciencias Aplicadas (UPC): Lima, Perú, 2007. [CrossRef]

5. Aguilar, E.M.; Yupanqui, J.A. *Optimización de la Dosis de Coagulante Y Floculante en El Tratamiento Químico Del Agua de Bombeo en Una Empresa Pesquera de Chimbote*; Universidad Nacional de Trujillo: Trujillo, Perú, 2018; Available online: https://www.unitru.edu.pe/bitstream/handle/UNITRU/AguilarCruz (accessed on 12 July 2022).
6. Espinoza, V.M.I. *Propuesta de Adición de Lodos Recuperados Del Agua de Bombeo Para Mejorar El Rendimiento de Harina de Pescado en Una Empresa Pesquera*; Universidad Nacional de Trujillo: Trujillo, Perú, 2016; Available online: https://www.dspace.unitru.edu.pe/bitstream/handle/UNITRU/7853 (accessed on 10 July 2022).
7. Ministerio de la Producción. Anuario Estadístico Pesquero y Acuícola 2020. Lima 27, Perú. Available online: https://ogeiee.produce.gob.pe/index.php/en/shortcode/oee-documentos-publicaciones/publicaciones-anuales/item/1001-anuario-estadisticoo-pesquero-y-acuicola-2020 (accessed on 20 June 2022).
8. Tay, J.H.; Show, K.Y.; Hung, Y.T. Seafood Processing Wastewater Treatment. ChemInform 37, 29–65. Available online: https://www.taylorfrancis.com/chapters/edit/10.1201/9780203026519-20/seafood-processing-wastewater-treatment-joo-hwa-tay-kuan-yeow-show-yung-tse-hung (accessed on 12 June 2022).
9. Ministerio del Ambiente. Decreto Supremo N° 010-2018-MINAM. Lima, Perú. 2018. Available online: https://sinia.minam.gob.pe/dowload/file/fid/63554 (accessed on 12 March 2022).
10. Brandt, M.J.; Johnson, K.M.; Elphinston, A.J.; Ratnayaka, D.D. Storage, clarification and chemical treatment. In *Twort's Water Supply*; Elsevier: Amsterdam, The Netherlands, 2017.
11. Jarvis, P.; Banks, J.; Molinder, R.; Stephenson, T.; Parsons, S.A.; Jefferson, B. Processes for enhanced NOM removal: Beyond Fe and Al coagulation. *Water Sci. Technol. Water Supply* **2008**, *8*, 709–716. [CrossRef]
12. Mollah, M.Y.; Morkovsky, P.; Gomes, J.A.; Kesmez, M.; Parga, J.; Cocke, D.L. Fundamentals, present and future perspectives of electrocoagulation. *J. Hazard. Mater.* **2004**, *114*, 199–210. [CrossRef] [PubMed]
13. Hutcherson, J.R. A Comparison of Electrocoagulation and Chemical Coagulation Treatment Effectiveness on Frac Flowback and Produced Water. Ph.D. Thesis, Colorado State University, Fort Collins, CO, USA, 2015.
14. Piña, M.; Martín, A.; González, C.; Prieto, F.; Guevara, A.; García, J. Revisión de variables de diseño y condiciones de operación en la electrocoagulación. *Rev. Mex. De Ing. Química* **2011**, *10*, 257–271. Available online: http://www.scielo.org.mx/scielo.php?script=sci_arttext&pid=S1665--27382011000200010 (accessed on 15 April 2022).
15. Restrepo, A.; Arango, A.; Garcés, L. La Electrocoagulación: Retos y oportunidades en el tratamiento de aguas. *Prod. Limpia* **2006**, *1*, 58–77.
16. Bensadok, K.; Belkacem, M.; Nezzal, G. Treatment of cutting oil/water emulsion by coupling coagulation and dissolved air flotation. *Desalination* **2007**, *206*, 440–448. [CrossRef]
17. Mella, B.; Glanert, A.C.C.; Gutterres, M. Removal of chromium from tanning wastewater by chemical precipitation and electrocoagulation. In Proceedings of the XXXII Congress of the IULTCS, Istanbul, Türkiye, 29–31 May 2013.
18. Apshankar, K.R.; Goel, S. Review and analysis of defluoridation of drinking water by electrocoagulation. *J. Water Supply Res. Technol.—AQUA* **2018**, *67*, 97–316. [CrossRef]
19. Pranjal, P.D.; Mukesh, S.; Mihir, K.P. Recent progress on electrocoagulation process for wastewater treatment: A review. *Sep. Purif. Technol.* **2022**, *292*, 121058. [CrossRef]
20. Kabdaşlı, I.; Arslan-Alaton, I.; Ölmez-Hancı, T.; Tünay, O. Electrocoagulation applications for industrial wastewaters: A Critical review. *Environ. Technol. Rev.* **2012**, *1*, 2–45. [CrossRef]
21. Aguilar-Ascón, E.; Neyra-Ascón, W.; Albrecht-Ruiz, M.; Ibarra-Basurto, A. Optimizing the Reduction of Total Suspended Solids in Pump Water from Fish Factories Through Electrocoagulation using Response Surface Methodology. *J. Ecol. Eng.* **2021**, *22*, 269–277. [CrossRef]
22. Rajaniemi, K.; Tuomikoski, S.; Lassi, U. Electrocoagulation Sludge Valorization—A Review. *Resources* **2021**, *10*, 127. [CrossRef]
23. Ahmad, T.; Ahmad, K.; Ahad, A.; Alam, M. Characterization of water treatment sludge and its reuse as coagulant. *J. Environ. Manag.* **2016**, *182*, 606–611. [CrossRef] [PubMed]
24. Ahmad, T.; Ahmad, K.; Alam, M. Characterization of water treatment plant's sludge and its safe disposal options. *Procedia Environ. Sci.* **2016**, *35*, 950–955. [CrossRef]
25. Sharma, P.; Joshi, H. Utilization of electrocoagulation-treated spent wash sludge in making building blocks. *Int. J. Environ. Sci. Technol.* **2016**, *13*, 349–358. [CrossRef]
26. Golder, A.K.; Samanta, A.N.; Ray, S. Anionic reactive dye removal from aqueous solution using a new adsorbent: Sludge generated in removal of heavy metal by electrocoagulation. *Chem. Eng. J.* **2002**, *122*, 107–115. [CrossRef]
27. Santana, J.J.; Mena, V.F.; Betancor-Abreu, A.; Rodriguez-Raposo, R.; Izquierdo, J.; Souto, R.M. Use of alumina sludge arising from an electrocoagulation process as functional mesoporous microcapsules for active corrosion protection of aluminum. *Prog. Org. Coat.* **2021**, *151*, 106044. [CrossRef]
28. FAO. *Food and Nutrition Paper 14/7. Manuals of Food Quality Control. 7 Food Analisys: General Techniques, Additives, Contaminants and Composition*; Food and Agriculture Organization of the United Nations: Rome, Italy, 1986.
29. Horwitz, W. *Official Methods of Analysis of AOAC International. Volume I, Agricultural Chemicals, Contaminants, Drugs/Edited by William Horwitz, 2010*; AOAC International: Gaithersburg, MD, USA, 1997.
30. Williams, T.R. *Analytical Methods for Atomic Absorption Spectrophotometry*; Perkin-Elmer Corp: Nolwalk, CT, USA, 1972.
31. Sündermann, A.; Eggers, L.F.; Schwudke, D. Liquid Extraction: Bligh and Dyer. In *Encyclopedia of Lipidomics*; Wenk, M., Ed.; Springer: Dordrecht, The Netherlands, 2016. [CrossRef]

32. Haan, T.Y.; Fen, C.S.; Radzi, M.F.; Ganasen, U. Comparative study for lake water remediation: Chemical coagulation and electrocoagulation. *J. Kejuruter. SI* **2018**, *1*, 81–87.
33. Tchamango, S.R.; Kamdoum, O.; Donfack, D.; Babale, D. Comparison of electrocoagulation and chemical coagulation processes in the treatment of an effluent of a textile factory. *J. Appl. Sci. Environ. Manag.* **2017**, *21*, 1317–1322. [CrossRef]
34. Tlaiaa, Y.S.; Naser, Z.A.R.; Ali, A.H. Comparison between coagulation and electrocoagulation processes for the removal of reactive black dye RB-5 and COD reduction. *Desalin. Water Treat.* **2020**, *195*, 154–161. [CrossRef]
35. Holt, P.K.; Barton, G.W.; Wark, M.; Mitchell, C.A. A quantitative comparison between chemical dosing and electrocoagulation. *Colloids Surf. A Physicochem. Eng. Asp.* **2002**, *211*, 233–248. [CrossRef]
36. Hardisson, A.; Revert, C.; Gonzales-Weler, D.; Rubio, C. Aluminium exposure through the diet. *Food Sci.* **2017**, *3*, 19. [CrossRef]
37. Salas-Maldonado, A.; Ayala-Galdós, M.E.; Albrecht-Ruiz, M. Contenido de EPA y DHA en aceite crudo de pescado producido en el perú durante el periodo 1996–2000. *CYTA-J. Food* **2002**, *3*, 283–287. [CrossRef]
38. Albrecht-Ruiz, M.; Salas-Maldonado, A. Chemical composition of light and dark muscle of Peruvian anchovy (Engraulis ringens) and its seasonal variation. *J. Aquat. Food Prod. Technol.* **2015**, *24*, 191–196. [CrossRef]

processes

Article

Assessment of Natural Zeolite Clinoptilolite for Remediation of Mercury-Contaminated Environment

Ivona Nuić [1,*], Mateja Gosar [2], Marin Ugrina [1] and Marina Trgo [1]

1 Faculty of Chemistry and Technology, Department of Environmental Engineering, University of Split, Ruđera Boškovića 35, 21 000 Split, Croatia; mugrin@ktf-split.hr (M.U.); marina.trgo@ktf-split.hr (M.T.)
2 Geological Survey of Slovenia, Dimičeva ulica 14, 1000 Ljubljana, Slovenia; mateja.gosar@geo-zs.si
* Correspondence: ivona@ktf-split.hr; Tel.: +385-21-329-454

Abstract: The soil at ancient roasting sites in the surroundings of the Idrija mine (Slovenia) is highly contaminated with mercury. To assess the impact of mercury on groundwater by infiltration and find an eco-friendly remediation method, the leaching of mercury from the soil containing 1347 mg Hg/kg, followed by sorption of the total leached mercury on cost-effective natural zeolite (NZ) clinoptilolite, was performed. The leaching of soil in ultrapure water of pH_o = 3.00–11.46 after 24 h resulted in the total leached mercury concentration in the range 0.33–17.43 μg/L. Much higher concentrations (136.9–488.0 μg/L) were determined after the first few hours of leaching and were high above the maximum permissible level in water for human consumption. The NZ showed very good sorption of the total leached mercury, with a maximum removal efficiency of 94.2%. The leaching of mercury in presence of the NZ resulted in a significant decrease of the total leached mercury (1.9–20.3 μg/L compared to 12.8–42.2 μg/L), with removal efficiencies up to 90.5%, indicating immobilization of mercury species. The NZ has a great potential for economically viable remediation of mercury-contaminated environment. However, efforts should be made in the further study of mercury leachability to reduce the mercury concentration in water to acceptable levels.

Keywords: Idrija mine; mercury-contaminated soil and water; leaching; natural zeolite clinoptilolite; remediation

Citation: Nuić, I.; Gosar, M.; Ugrina, M.; Trgo, M. Assessment of Natural Zeolite Clinoptilolite for Remediation of Mercury-Contaminated Environment. *Processes* **2022**, *10*, 639. https://doi.org/10.3390/pr10040639

Academic Editor: Raja Ghosh

Received: 17 February 2022
Accepted: 22 March 2022
Published: 24 March 2022

1. Introduction

Mercury (Hg) and its compounds are identified as priority hazardous substances in the field of water policy and are listed among the 10 chemicals of major public health concern [1,2]. It is well-known that mining and ore processing areas are known to be major sources of mercury pollution, especially mercury-enriched ore deposits that have been exploited for centuries by inadequate techniques [3]. The other anthropogenic sources of mercury are small-scale gold mining, coal combustion, metal smelting and cement production [4,5]. Despite the awareness of the negative mercury impact on both human health and global environment, some of the mercury compounds are still used in the production of vinyl chloride, chlorine-based chemicals, in dentistry and in skin lightening products, but also in traditional medicines and some ritualistic and religious practices [6,7].

Once emitted in the atmosphere, mercury can be transmitted long distances, becoming a global concern and must be considered at international levels. For example, 40–80% of the annual amount of mercury deposited in Europe originates from mercury emissions from outside its boundaries [2]. Even if all anthropogenic emissions cease immediately, the significant amount of mercury in the environment will lag behind for centuries [4]. Mercury toxicity is very dependent on its speciation and transformation, and the process of mercury methylation is perhaps the most significant process in the global circulation of mercury in the environment [4,8]. Among all mercury compounds, organic mercury is considered to be the most dangerous due to rapid and extensive absorption by the human body through the food chain. For preventing or reducing the exposure to these pollutants, the remediation of mercury-polluted environments is of primary interest [5,9]. In a selection of remediation technology, the main approaches include immobilization of the contaminant or its removal from polluted media, using ex situ or in situ technologies. Ex situ technologies are quite expensive, while in situ technologies using thermal, biological or physical and chemical methods are more economically viable [9–12]. Alternative photocatalytic and electrocatalytic approaches for eco-friendly mercury removal are also being developed [13–16]. There is no single technology appropriate for all contaminants; thus, a combination of different technologies is needed for achieving satisfying results. So far, different materials have been studied for mercury removal, from waste to modified and novel materials [17–20]. The usage of novel materials is very effective but quite expensive, while waste materials are cost effective but sometimes limited due to the possible content of toxic ingredients. Among natural materials, zeolites seem adequate for remediation due to the prevalence of their natural deposits, simple and economically acceptable exploitation, high removal capacity and great reusability [21–23].

The present study aimed to investigate the potential of natural zeolite (NZ) for mercury capturing in order to remediate the extremely mercury-contaminated sites of Idrija district in Slovenia, without causing secondary pollution. Namely, in the Idrija mine, which is the second largest mercury mine in the world, during 500 years (1490–1995) of mercury exploitation from ore containing cinnabar (HgS) and/or native mercury (Hg^0), around 37,500 t of mercury were lost into the environment [24–30]. Extremely high concentrations of mercury (10 to even 10,000 mg/kg) found in the soil represent a potential source of mercury in aquatic systems by runoff and migration through the porous media [26,28]. The geochemical studies estimated that mercury will leach from these sites into groundwater for at least the next 100 years [26–28,30]. Therefore, the plan for remediation is extremely necessary in order to protect sources of drinking water.

A promising approach to remediation of contaminated groundwater is a permeable reactive barrier, where natural zeolite can be used as a filler material for immobilization of mercury by sorption process [31]. To the best of our knowledge, this is the first attempt at using natural zeolite clinoptilolite for possible remediation of the mercury-contaminated environment at these sites. It is expected that this research will expand the knowledge on the possible application of natural zeolites in environmentally friendly and economically viable in situ remediation technologies of the mercury-contaminated environment and contribute to the reduction of mercury migration. Ultimately, the results obtained could contribute to solving serious environmental problems caused by the presence of mercury and other heavy metals in the environment, especially in waters, with the final goal of protecting human health.

2. Materials and Methods

2.1. Samples Preparation

The experiments were carried out on the contaminated soil taken within the first 30 cm of the soil profile F1 at the Ferbejžene trate roasting site in Idrija district in Slovenia (location of F1 profile is given in Gosar and Čar [32], for more details see also Teršič et al. [26]). The air-dried sample of soil was gently disaggregated in a ceramic mortar, sieved through a 2 mm sieve, then pulverized in an agate mill to a fine-grain size of <0.075 mm and stored in polyethylene bags.

The natural zeolite (NZ) used in this research originates from the Zlatokop deposit in Vranjska Banja (Serbia). The sample was milled and sieved (Retsch AS200 basic device) in order to obtain two particle sizes (<0.04 mm and 0.6–0.8 mm), washed several times with ultrapure water in order to remove impurities, dried at 60 °C and finally stored in a desiccator until use.

2.2. Physicochemical Characterization of Contaminated Soil

Elemental analysis of mercury-contaminated soil (3 parallel analysis/3 sample replicates) was performed by simultaneous multielement analysis of 53 elements. Elemental concentrations were analyzed at AcmeLabs, Vancouver, Canada (accredited under ISO 9001:2008) with inductively coupled plasma (ICP) mass spectrometry (MS) after digestion of an aliquot of 15 g sample material with aqua regia (1:1:1 $HCl:HNO_3:H_2O$) for 3 h at 160 °C. Samples exceeding the upper detection limit for Hg (100 mg/kg) were analyzed using ICP emission spectrometry (ICP-ES).

The soil pH value was determined in a 1:5 soil:suspensions ratio in water (pH-H_2O), as well as in electrolyte solutions, 1 M potassium chloride (pH-KCl) and 0.01 M calcium chloride (pH-$CaCl_2$) following the DIN ISO standard method [33].

The soil buffering capacity (BC) was determined by the addition of dilute 0.1 M HCl or 0.1 M NaOH solutions (0, 5, 10, 20, 40 and 80 mmol/kg of soil) to 1:5 soil:water suspensions. The suspensions were shaken occasionally, and the pH values were monitored after 30 min, 1 day and 4 days, in order to establish pH equilibration [34,35].

The cation exchange capacity (CEC) of the soil was determined according to HRN EN ISO standard method [36] using barium chloride (barium chloride dihydrate for analysis EMSURE® ACS, ISO, Reag. Ph Eur, Merck, Darmstadt, Germany).

The rough estimation of soil organic matter (SOM) content was performed by a loss on ignition (LOI) procedure at 400 °C to accomplish the destruction of organic matter while avoiding overestimation of its content due to destruction of present mineral matter through the loss of structural water [37–39]. After oven-drying at 90 °C, the weighed soil sample was heated at 400 °C to constant mass. The SOM percentage was calculated as follows:

$$SOM = \frac{m_{90^\circ C} - m_{400^\circ C}}{m_{90^\circ C}} \cdot 100, \ (\%) \tag{1}$$

where SOM is the estimated soil organic matter content (%), $m_{90^\circ C}$ is the soil weight after oven-drying at 90 °C (g) and $m_{400^\circ C}$ is the soil weight after heating at 400 °C (g).

The morphological characterization of the soil sample and zeolite samples after sorption were analyzed by scanning electron microscopy (SEM), while the semi-quantitative elemental composition by energy dispersive spectrometry (EDS) technique on Jeol Scanning Microscope—6610LV. The SEM analysis through the secondary electron image (SEI) and backscattered electrons (BEC) at magnifications from 300× to 1400× was performed. The BEC view provided insight into the phases of different chemical compositions present on the surface of the soil particle. Imaging was also performed in selected sites (spectra) on the particle surface or over the entire surface (mapping analysis).

For identification and quantification of the mineralogical composition of the soil, the X-ray powder diffraction (XRPD) technique was used. The X-ray diffraction patterns were collected in the Bragg–Brentano geometry using a Panalytical Aeris bench-top diffractometer with Ni-filtered CuKα radiation obtained from an X-ray tube operating at 7.5 mA and 40 kV, in the 2θ range of 5–80°. Samples were prepared in a thin film on a zero-background silicon holder, which was spinning during data collection. The powder X-ray diffraction patterns were analyzed by the Rietveld method using the Topas program (Bruker-AXS, Karlsruhe, Germany).

2.3. Leaching and Sorption Experiments

Leaching experiments provide insight into the changes in solubility of soil compounds and enable the environmental impact assessment of soil pollutants. For investigation of mercury leaching from the soil and its possible capturing onto natural zeolite, two different approaches were used. The first one was performed in two steps, by leaching the total mercury from the contaminated soil and its subsequent sorption onto natural zeolite. On the contrary, the second one involved simultaneous leaching and sorption by mixing the soil with ultrapure water without or with the addition of the natural zeolite to the soil.

2.3.1. Leaching of Mercury from the Contaminated Soil Followed by Sorption from the Obtained Leachates onto Natural Zeolite

The leaching experiments were performed in accordance with the standard batch method DIN 38414 S4 [40]. Nine portions of ultrapure water of different initial pH_o values in the range from 3.00 to 11.46, adjusted by addition of 1 M HNO_3 or 1 M KOH, were mixed with mercury-contaminated soil in the solid/liquid ratio, S/L = 1/10, in an incubator shaker at 23 ± 2 °C, at 25 rpm for 24 h. The nine obtained leachates were separated from the solid phase by double centrifugation at 4000 rpm for 15 min followed by filtration through 0.45 μm filter, and the concentration of the total mercury in supernatants was determined by atomic absorption spectrometry on a direct thermal decomposition mercury analyzer (Mercury Analyzer MA—3000, NIC). During the experiment, the pH and electrical conductivity were monitored.

Afterwards, obtained leachates were subjected to sorption on NZ of two particle sizes (<0.04 mm or 0.6–0.8 mm) by batch method. The NZ and each leachate were shaken in the S/L ratio of 1/100 at 23 ± 2 °C, at 25 rpm for 24 h. The experiment was monitored by measuring the pH and electrical conductivity at defined time intervals within 24 h, while the residual total mercury concentration was determined after 24 h. The experiments were performed in duplicate.

2.3.2. Simultaneous Leaching of Mercury from the Contaminated Soil and Its Sorption onto Natural Zeolite

Further leaching experiments were carried out in batch mode at three different initial pH_o values of ultrapure water of 4.11, 5.09 and 6.04, chosen based on previously obtained results. The ultrapure water and the contaminated soil were stirred at ≈800 rpm in S/L = 1/10 (300 g/3000 mL) at 23 ± 2 °C for 24 h, simultaneously in two 5-liter batch reactors, without and with the addition of 10% of NZ (0.6–0.8 mm) to the soil. The stirring speed was chosen in such a way to keep the soil in suspension. For monitoring the concentration of the total mercury leached over time, a certain volume of suspension was taken at each time interval (1, 2, 3, 4, 5, 7 and 24 h), where the total sampled volume did not exceed 5% of the total suspension volume. In the obtained supernatants, the residual concentration of the total mercury was analyzed on a mercury analyzer. The pH and electrical conductivity were measured in the suspensions at the same time intervals.

3. Results and Discussion

3.1. Physicochemical Characterization of the Mercury-Contaminated Soil

The elemental analysis of mercury-contaminated soil is presented in Table 1. Element content in the soil represents the mean value of three measurements. As it can be seen the total mercury content of 1347 mg/kg indicates extremely contaminated soil. For comparison, agricultural land is considered contaminated when mercury content is higher than the threshold value of 0.5 mg/kg [41–43]. According to the previous research, the average percentage of non-cinnabar compounds at roasting sites of the Idrija district was 35–40%, showing that soil contains considerable amounts of potentially bioavailable mercury that could leach into the environment [26]. Among others, the major elements in the soil are Fe and Al. There are also substantial amounts of Ca, Mg, K and Mn followed by S and P. In significantly lower amounts, Ce, Zn, Li, Ba, V, Pb, La, Rb, Cr and As are present, while other elements are present only in traces.

Table 1. Elemental content of mercury-contaminated soil sample.

Element	Element Content, mg/kg		
	Min.	Max.	Average
Ag	0.06	0.12	0.08
Al	21,500.00	22,700.00	22,200.00
As	22.20	24.90	23.60
B	1.00	3.00	2.00
Ba	59.70	61.00	60.47
Be	3.10	6.10	4.60
Bi	LDL	0.15	0.15
Ca	700.00	700.00	700.00
Cd	0.43	0.89	0.60
Ce	72.80	77.60	75.10
Co	11.20	12.10	11.73
Cr	23.80	25.00	24.53
Cs	2.34	2.66	2.53
Cu	18.37	19.01	18.69
Fe	24,800.00	25,300.00	25,100.00
Ga	4.10	7.50	5.40
Ge	LDL	0.40	0.40
Hf	0.11	0.45	0.28
Hg	1290.00	1410.00	1347.00
K	2100.00	2400.00	2300.00
La	29.40	31.20	30.37
Li	58.80	69.70	63.37
Mg	4900.00	5900.00	5500.00
Mn	1062.00	1087.00	1074.67
Mo	4.07	6.22	5.23
Na	LDL	100.00	100.00
Nb	0.18	0.18	0.18
Ni	21.30	22.10	21.77
P	300.00	300.00	300.00
Pb	35.04	43.43	39.87
Rb	25.80	29.50	27.37
S	400.00	400.00	400.00
Sb	0.65	3.05	1.85
Sc	3.70	7.30	4.97
Se	7.00	8.90	7.70
Sn	0.20	1.00	0.60
Sr	6.50	6.80	6.63
Te	LDL	1.44	1.44
Th	5.00	6.10	5.67
Ti	LDL	20.00	20.00
Tl	0.37	0.65	0.51
U	2.50	2.80	2.67
V	47.00	48.00	47.67
Y	18.02	19.92	18.87
Zn	63.70	66.30	65.27
Zr	5.00	5.60	5.20
W	0.30	0.30	0.30

LDL—low detection limit; the elements (Au, In, Pd, Pt, Re and Ta) below the detectable limits were excluded from the table.

The goal of SEM-EDS analysis was to examine the surface morphology and to identify the main elements and their proportions, especially mercury. Figure 1 represents the SEM image of mercury-contaminated soil with corresponding EDS results and mapping analysis. It can be seen that mercury is uniformly distributed over entire selected area, but among the most abundant elements (O, Si and Al), its content (in weight %) was too low to be determined.

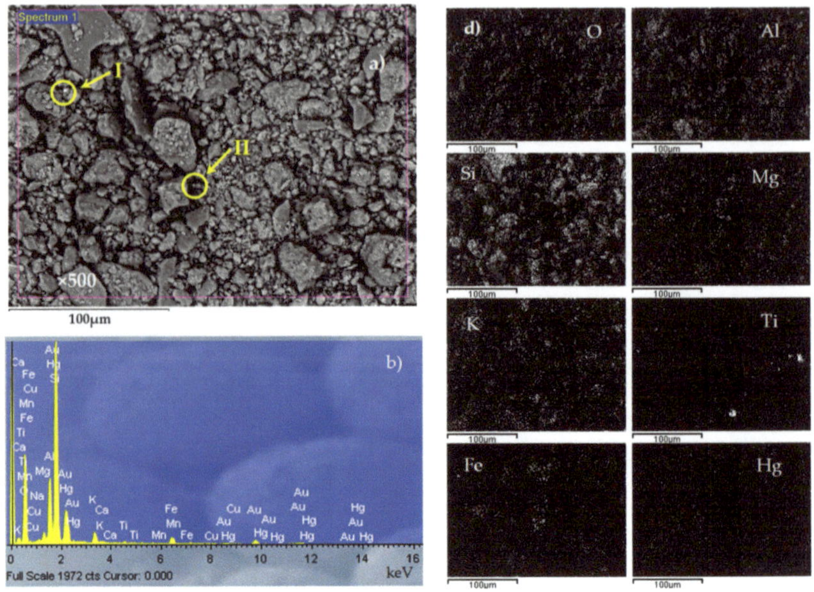

c)					Element, wt. (%)						
O	Na	Mg	Al	Si	K	Ca	Ti	Mn	Fe	Cu	Hg
57.73	0.34	1.44	8.04	25.07	2.13	0.00	0.66	0.00	3.88	0.73	0.00

Figure 1. (**a**) SEM image, (**b**) EDS peaks of elements over the entire surface (Spectrum 1), (**c**) element content in wt. (%) over the entire surface and (**d**) distribution of main elements and mercury (Hg) over the entire surface.

Nevertheless, two bright agglomerates were observed on SEM image in Figure 1 and marked as I and II. Therefore, additional SEM-EDS analysis was performed at magnification of 1400× as shown in Figure 2, where Spectrum 1 represents agglomerate I and Spectrum 3 represents agglomerate II.

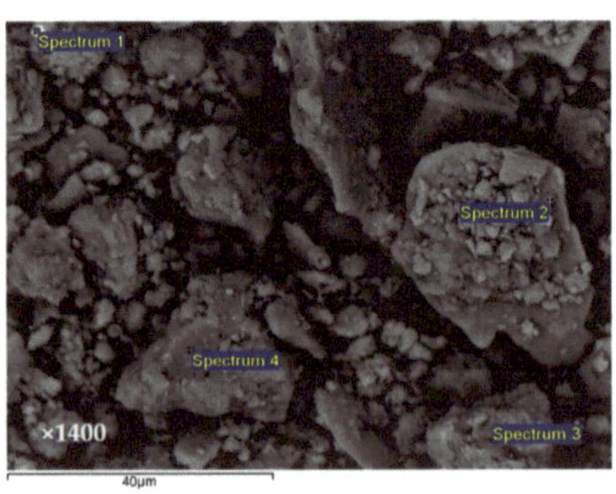

Element	Selected sites on the particle surface			
	Sp 1	Sp 2	Sp 3	Sp 4
	wt. (%)			
O	30.67	29.59	26.68	59.44
Mg	0.66	1.99	0.89	3.10
Al	2.33	6.95	3.43	3.82
Si	5.07	14.39	6.17	32.79
P	0.00	5.28	0.00	0.00
S	8.52	0.00	0.00	0.00
K	0.47	1.17	0.77	0.00
Ti	0.00	0.00	5.21	0.00
Fe	0.91	3.61	28.18	0.87
Cu	1.04	1.82	0.00	0.00
La	0.00	9.65	0.00	0.00
Ce	0.00	18.03	0.00	0.00
Nd	0.00	7.51	0.00	0.00
Hg	50.32	0.00	28.66	0.00

Figure 2. SEM image of mercury-contaminated soil particles with four selected sites (spectra, Sp) for EDS analysis and corresponding EDS results.

According to EDS analysis, the analyzed sites (Spectra 1 and 3) in Figure 2 contain mercury in large quantities probably in the form of HgS and HgO. In the analyzed site (Spectrum 3), besides mercury, iron oxide is also found, which is in agreement with the high content of iron in the soil obtained by elemental analysis given in Table 1. A particle in Figure 3 is another confirmation of the cinnabar (HgS) presence in the analyzed soil.

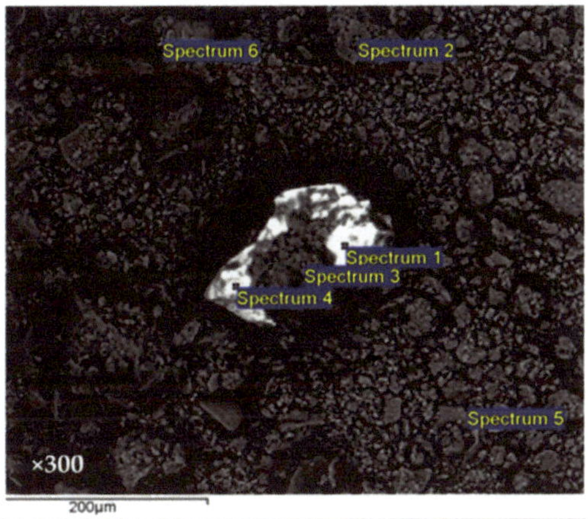

Element	Selected sites on the particle surface					
	Sp 1	Sp 2	Sp 3	Sp 4	Sp 5	Sp 6
	wt. (%)					
O	17.59	35.87	44.13	9.43	55.98	48.06
Na	0.00	0.00	0.00	0.00	0.00	0.74
Mg	0.44	18.76	1.99	0.00	1.48	1.23
Al	1.84	1.11	15.17	1.16	7.50	6.55
Si	2.55	36.63	26.68	1.66	30.19	13.24
S	9.00	0.00	0.00	12.07	0.00	0.00
K	0.74	0.00	5.11	0.00	1.62	0.83
Ti	0.00	0.00	0.83	0.00	0.00	0.00
Cr	0.00	0.00	0.00	0.00	0.00	7.28
Fe	0.00	6.69	6.09	0.68	3.23	21.53
Cu	0.00	0.94	0.00	0.00	0.00	0.53
Hg	67.84	0.00	0.00	74.99	0.00	0.00

Figure 3. SEM image of mercury-contaminated soil particle magnified 300× with six selected sites (spectra, Sp) analyzed, with corresponding EDS results.

It can be seen that the bright areas (Spectra 1 and 4) contain higher amounts of sulfur (9.00 and 12.07 wt. %), oxygen (17.59 and 9.43 wt. %) and mercury (67.84 and 74.99 wt. %), indicating the presence of HgS and HgO. The other spectra in Figure 3 contain neither sulfur nor mercury, and represent the common soil composition, with elevated content of O, Si, Al, Mg and Fe as major elements, which are commonly expressed as oxides [44]. According to the obtained SEM-EDS results, it can be observed that at all sites without mercury content, the amount of silicon and aluminum components are elevated. Elemental analysis from Figures 1 and 3 confirm silicon content on the marked spectra, which is in correspondence to SiO_2 determined by XRD analysis of the soil and presented in Figure 4. The Rietveld analysis shows that the soil sample mostly consists of alpha-quartz [45]. There are also additional small diffraction peaks in the patterns, which remain unidentified. There is no amorphous halo in the pattern background indicating that the sample is largely crystalline.

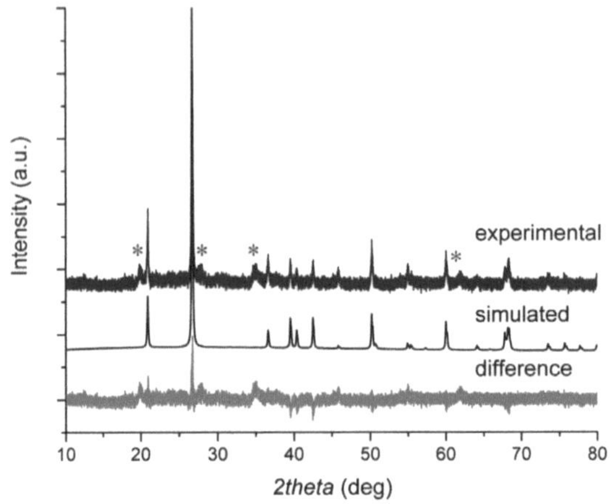

Figure 4. Rietveld fit for the XRD spectrum of the studied mercury-contaminated soil using the crystal structure of alpha-quartz ("simulated"). "Difference" curve represents the difference between "experimental" and "simulated" using alpha-quartz as the only phase. Some additional unidentified reflections are marked with "*".

It was found that SOM content in the analyzed soil was 5.16%, indicating the soil has low organic content [46,47]. In general, soil is composed of mineral fraction and organic matter, where the total amount of organic matter usually ranges from 0.5% to 10%. However, despite these usually small proportions of the soil organic components, their impact on soil properties is often much greater than would be expected, since, unlike minerals, organic matter in the soil is transformed much faster [48,49]. Organic matter significantly increases the soil capacity for ion sorption by regulating the balance between ions in the soil solution and those exchangeable on colloidal soil particles. Thus, the soil organic matter greatly affects the mobility and availability of metals in soil. Organic matter has a large capacity to bind mercury, creating very stable complexes [50,51]. Previous analyses of mercury species in highly Hg-contaminated soil from ancient small-scale ore roasting sites in the Idrija area (Pšenk and Ferbejžene trate) have shown that more than 90% of soluble mercury occurs in non-reactive organic complexes, confirming the preferential binding of mercury to organic matter. The amount of easily reactive mercury is generally low and ranges from 0.038% to 1.7% of the total soluble mercury [52].

3.2. *Chemical Behavior of the Mercury-Contaminated Soil*

The metal mobility in the soil directly depends on soil pH, BC and CEC. By knowing BC, it is possible to evaluate the soil's ability to resist the pH changes and the capability of immobilization of the soil contaminants. The soils with higher BC can absorb more acid or base with no significant change in pH [53]. Knowing the pH reaction of the soil provides insight into soil processes, where at higher pH values, metal precipitation occurs, while at lower pH values, cation exchange becomes dominant. Additionally, the higher content of soil organic matter or clays improves BC due to the large numbers of surface sites able to adsorb hydrogen ions [34,54]. Thus, its estimation could be one of the parameters that is unavoidable in remediation studies and in environmental risk assessment. In order to elucidate the changes happening in such a real soil system and for better interpretation of the leaching and sorption results, the soil pH, BC and CEC were determined prior to the leaching experiments.

The soil pH was determined in the suspension of the contaminated soil in water (pH–H_2O), KCl (pH–KCl) and $CaCl_2$ (pH–$CaCl_2$) solutions. The acidity measured in water refers to the acidity of the soil solution, while acidity in KCl or $CaCl_2$ solutions also includes the potential exchangeable acidity of the soil adsorption complex, and thus usually yields lower pH values than in water. The obtained pH values of the soil of 5.22 in water, 4.13 in $CaCl_2$ and 3.89 in KCl classify the mercury-contaminated soil into moderate to strong acid soils [55]. The lower pH values in the $CaCl_2$ and KCl solutions indicate that mercury-contaminated soil contains a considerable amount of potential exchangeable acidity. In addition to free hydrogen ions in soil solution, acidity in electrolyte solutions is also caused by exchangeable acidic cations, H^+, Al^{3+} and Fe^{3+}/Fe^{2+}, that are weakly bound to the soil surface exchange complex. Namely, the H^+ can be displaced upon the addition of K^+ and Ca^{2+}, which leads to an additional decrease in pH.

The BC of the soil was estimated by the addition of either dilute HCl or NaOH solutions to the contaminated soil suspended in water. The obtained results are compared in Figure 5.

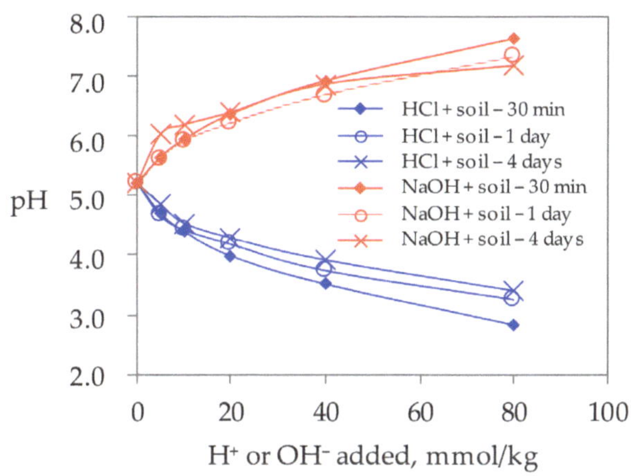

Figure 5. Change in pH of soil suspension after the addition of different amounts of HCl or NaOH.

The data show that mercury-contaminated soil resists well to pH change, upon the addition of both an acid or base, and a period of 30 min seems sufficient to reach equilibrium pH. In general, flatter curves correspond to a smaller change in pH, indicating better BC. On the contrary, the steeper curves are characterized by low BC. Maximum buffering is defined as the minimal slope of pH in soil suspension vs. the amount of added acid or base [56–59]. The BC is calculated from the slope of the curves when the amount of added H^+ or OH^- (mmol/kg) was plotted against the pH, and the mean value equals 3.89 cmol/kg·pH. The comparison of obtained soil BC with other soils is very difficult. Soils vary considerably in their BC because it is greatly affected by soil properties. For example, the BC of 18 acid tropical and subtropical soils in China ranged from 0.91 up to 3.21 cmol/kg·pH, depending on their properties [35]. The BC of organic soils is mostly higher than 80 cmol/kg·pH [60]. Usually, the BC lower than 1 cmol/kg·pH is interpreted as low, between 1 and 2 cmol/kg·pH as moderate, while higher than 2 cmol/kg·pH is interpreted as high [61]. If soil BC is not high enough, metals will not retain in the soil, but rather will be leached [34,54]. Some soil components are effective in buffering in a wider pH range, while some of them are effective only in a limited range of the pH values. The buffer reactions at low pH are most controlled by aluminum compounds, while at high pH they are controlled by carbonates [49]. The oxides and hydroxides of Fe and Al are effective in the whole pH range due to the sorption/desorption of H^+ on hydroxyl surface sites. Thus, due to their high content (Table 1) in mercury-contaminated soil, they probably greatly contribute to its BC [57,58].

The value of CEC for mercury-contaminated soil is 15.4 cmol/kg, with the amounts of individual non-acid exchangeable cations as follows: Ca^{2+} at 2.49 cmol/kg, Mg^{2+} at 1.89 cmol/kg, K^+ at 0.35 cmol/kg and Na^+ at 0.084 cmol/kg. As it can be seen, the Ca^{2+} is the dominant non-acid cation in this soil sample. The amount of acid cations H^+ and Al^{3+} equals 10.586 cmol/kg, which leads to 68.74% of soil acid saturation (the ratio of the amount of exchangeable H^+ and Al^{3+} and CEC multiplied with 100) [49,62]. This is in line with the obtained values of the soil pH and corresponds to elemental soil analysis. Since the CEC was determined by the standard method using barium as the exchanger cation (pH = 8.2), the overestimation of CEC in this case was very likely since mercury-contaminated soil is acidic. Thus, the obtained value represents the maximum cation exchange capacity of the soil. If we consider that for soil humus the CEC is higher than 150 cmol/kg, then the obtained soil CEC is quite low [49]. However, even high organic soils will have low effective CEC if soil pH is low. In soils with pH lower than 6.0, the negative exchange sites will be occupied by tightly held acid-forming cations, mostly H^+ and Al^{3+}, as well as by Mn^{2+} or Fe^{3+}/Fe^{2+}, enabling nutrient binding and exchanging [49,62–65]. Some studies reported a high correlation of CEC with BC, where for acid soils the main buffering mechanism was cation exchange of H^+ with non-acid cations [35]. The CEC is a key parameter describing the sorption/desorption of both nutrients and contaminants by soil, and the obtained CEC points onto soil vulnerability in mercury losses through the leaching.

3.3. Mercury Leaching from Contaminated Soil and Its Subsequent Sorption from the Obtained Leachates onto Natural Zeolite

The results of mercury leaching from contaminated soil in the ultrapure water of different initial pH_o values in the range 3.00–11.46 according to the standard leaching method are presented in Figure 6 as the concentration of total mercury (c_{TM}) leached after 24 h vs. pH_o.

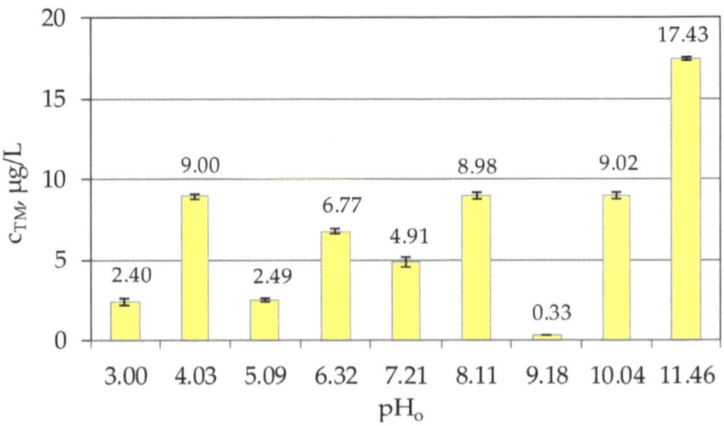

Figure 6. The concentration (mean ± SD) of total mercury leached in ultrapure water of different pH_o values in the range 3.00–11.46 after 24 h.

In Figure 6 it can be observed that there is no regularity in the concentration of the total mercury leached in the examined pH range. The concentration of the total mercury after 24 h was in the range 0.33–17.43 μg/L, suggesting a strong affinity of mercury toward soil constituents. However, these low concentrations still exceeded the maximum permissible mercury concentration of 1.0 μg/L in water for human consumption [66]. As it can be seen, the highest concentration of the total mercury was obtained under extremely alkaline conditions at pH_o = 11.46. Namely, in the soil system, humic and fulvic acids along with other substances form a dissolved organic carbon (DOC) that controls the binding of heavy metals. The higher solubility of DOC is expected with increasing pH, thus affecting the release of mercury [67]. Fulvic acids are soluble in acidic and alkaline conditions, while humic acids are soluble only in alkaline conditions. The mercury in the soil will be mobile depending on the mobility of its complexes. In general, dissolving or even degradation of the structure of organo-mineral mercury complexes is considered to have the most dominant influence on mercury mobility at highly acidic or alkaline conditions. Since the mercury-contaminated soil is a very complex system, the synergistic interaction of numerous factors on the mercury speciation and transformations finally affects the rate and extent of its transportation. Mercury in the soil in an oxidized state as mercuric (Hg^{2+}) tends to formation of inorganic compounds ($HgCl_2$, HgO or HgS), or binds to the organic matter forming different organo-mercury compounds. Additionally, it was found that soluble mercury complexes, as well as nano-sized mercury particles, have a major impact on mercury mobility, methylation in soil and availability [68]. However, the variation in the concentration of leached mercury at examined pH_o range indicates that besides pH, some other factors also might be important for leaching. Although the standard leaching method prescribes a contact time of 24 h, the concentration of the total mercury was monitored in time intervals within 24 h during the same experiment, in order to get insight into mercury behavior over time. The results are presented in Figure 7.

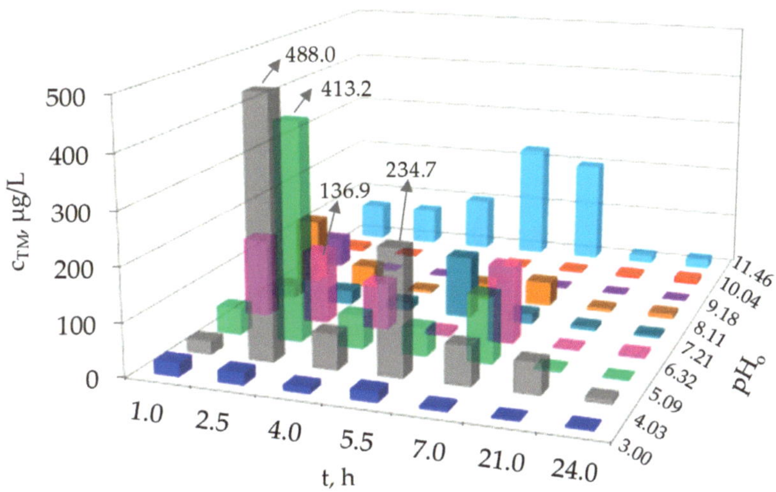

Figure 7. The concentration of the total mercury leached in soil suspensions at different pH_o values in time intervals within 24 h. Note: the results after 24 h correspond to those given in Figure 6.

As can be seen in Figure 7, at each pH_o value, there is an alternating increase and decrease in the concentration of total mercury. In the first few hours of the experiment, the concentration of total mercury is considerably higher than after 24 h. It reaches even 488.0 µg/L after 2.5 h from the beginning of the experiment, which is more than 50 times higher compared to the same sample in Figure 6. High mercury concentrations were obtained at neutral or slightly acidic pH values, which more likely occur in nature. High concentrations of leached mercury cannot only be attributed to the release from organo-mineral complexes, but also to mercury speciation and soil properties. In addition, the influence of soil inhomogeneity cannot be ignored. The observed irregularity in the concentration of the total mercury is most likely due to changes in mercury mobility, which significantly depends on its chemical form and bonds established. In the soil solution, organic compounds and water are the most abundant ligands, so hydrolysis and organic complexation are the most common reactions. The mobility of metals in the environment depends on the many different processes—dissolution, sorption, complex formation, migration, deposition, diffusion (into minerals), binding to organic matter, adsorption, desorption, evaporation, etc. Each metal can be precipitated and/or adsorbed even with a small change in equilibrium conditions, which alter very easily in the soil depending on the location [69]. Mercury is therefore most likely released from one phase of the soil and re-binds to another phase in the soil, causing a variation in the concentration of the total mercury in the soil suspension over 24 h.

To better understand this complex soil system, the pH and electrical conductivity (k) in suspensions during the leaching experiment were monitored. The results are given in Figures 8 and 9. The electrical conductivity is presented in two separate graphs in Figure 9 in order to get better visibility of the obtained results.

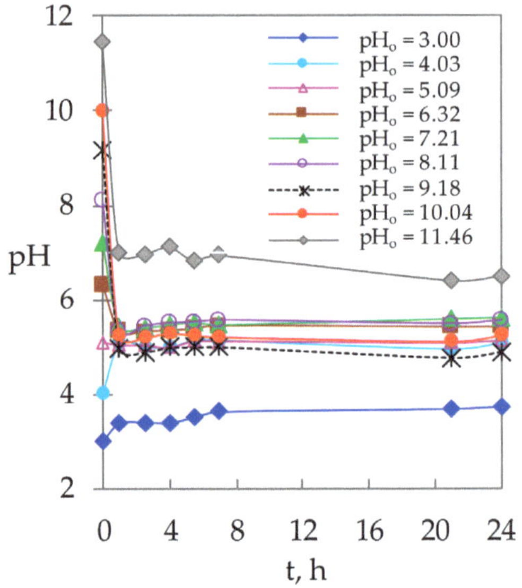

Figure 8. Change of pH values in suspensions during the leaching.

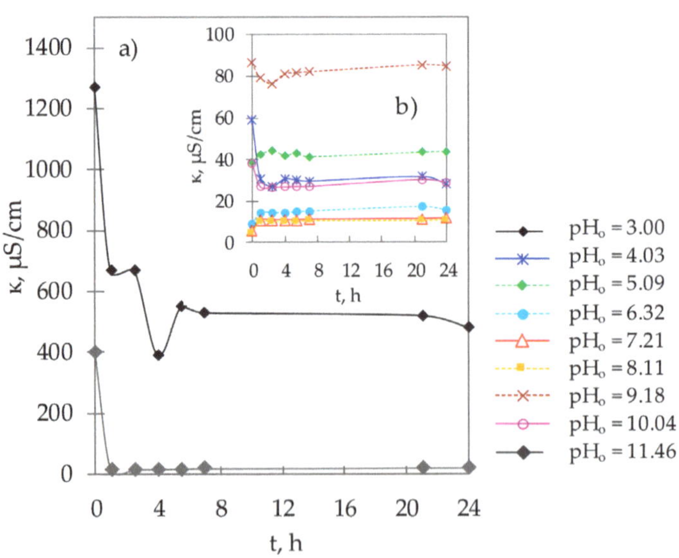

Figure 9. Change in electrical conductivity during the leaching for: (**a**) pH_o = 3.00 and pH_o = 11.46, and (**b**) pH_o = 4.03–10.04.

According to Figure 8, the highest change in pH can be observed in the first two hours of the experiment. The constant value of pH is established very rapidly, and soil suspensions tend to be slightly acidic or neutral pH range, except for $pH_o = 3.00$. This can be attributed to the buffering capacity of the soil, which is least expressed for the lowest pH_o due to the extremely high concentration of H_3O^+. The electrical conductivity in Figure 9 shows a similar trend for all pH_o values. The highest electrical conductivity in Figure 9a is measured for $pH_o = 3.00$ and $pH_o = 11.46$ due to the high concentration of H_3O^+ or OH^-. The specific electrical conductivity of both H^+, $\lambda_o(H^+) = 350$ S·cm^2/mol and OH^-, $\lambda_o(OH^-) \approx 200$ S·cm^2/mol at 25 °C, compared to other ionic species, is extremely high [70]. For all other pH_o values in Figure 9b, the electrical conductivity is lower than 100 µS/cm.

The changes in pH and electrical conductivity values were negligible after ≈5 h for all examined pH_o values, indicating that variations in total mercury concentrations in Figure 7 are not in function of pH and k.

Leachates obtained at different pH_o values in Figure 6 were mixed with natural zeolite of two particle sizes. Based on the experimental results after 24 h of contact with natural zeolite, the mercury uptake and removal efficiency of both zeolite particle sizes were calculated according to the following equations and presented graphically in Figures 10 and 11:

$$q = [c_o\,(Hg) - c_e\,(Hg)] \cdot \frac{m}{V} \quad (2)$$

$$\alpha = \frac{c_o\,(Hg) - c_e\,(Hg)}{c_o(Hg)} \cdot 100 \quad (3)$$

where q is mercury uptake per unit mass of zeolite in equilibrium (µg/g), α is the removal efficiency in equilibrium (%), $c_o(Hg)$ is the concentration of total mercury in leachates (µg/L), $c_e(Hg)$ is the equilibrium concentration of total mercury in eluates after sorption (µg/L), m is mass of the zeolite (g) and V is the volume of leachate before sorption (L).

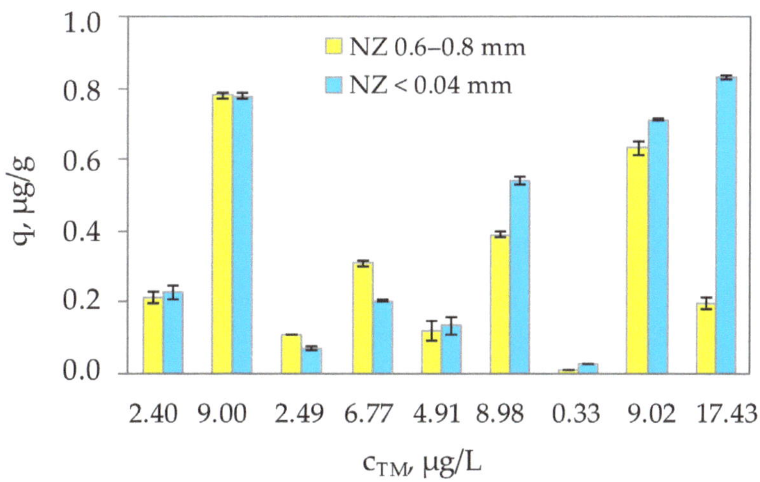

Figure 10. Mercury uptake (mean ± SD) onto natural zeolite of two particle sizes in dependence on the initial concentration of total mercury leached.

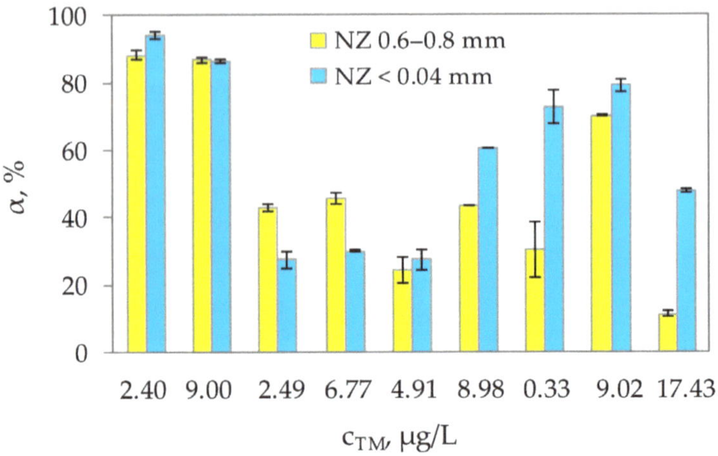

Figure 11. Removal efficiency (mean ± SD) of natural zeolite of two particle sizes toward total mercury in dependence on the initial concentration of total mercury leached.

According to the results presented in Figures 10 and 11, it is noticeable that there is no direct relationship between mercury uptake and removal efficiency on the initial total mercury concentration in leachates. The obtained values of mercury uptake (q) were in the range 0.010–0.781 μg/g for the particle size of 0.6–0.8 mm, and 0.024–0.834 μg/g for the particle size < 0.04 mm, respectively. The highest values of mercury uptake were mainly achieved for the highest initial concentrations of total mercury. The visible oscillations are most likely due to differences in the pH_o values, which affected the type of mercury speciation in leachates, and consequently its sorption onto natural zeolite. A slightly larger amount of total mercury was uptake by the powdered zeolite, due to the larger specific surface area. This is especially noticeable for the highest initial concentration of the total mercury of 17.43 μg/L. The removal efficiency (α) was in the range 11.2–88.3% for the particle size of 0.6–0.8 mm, and 27.3–94.2% for the particle size < 0.04 mm, respectively. The maximum removal efficiency of 94.2% for the NZ of a smaller particle size was obtained for initial mercury concentration of 2.40 μg/L and $pH_o = 3.00$. Interestingly, the removal efficiencies higher than 60% were obtained for different initial concentrations of total mercury of 2.4, 9.0, 8.98, 0.33 and 9.02 μg/L, which is quite unexpected. This indicates that regardless of mercury concentration, some other factors greatly affect mercury uptake on the zeolite, most likely the speciation of mercury in the leachates.

In literature reports, higher but inconsistent sorption capacities and removal efficiencies of mercury can be found, but sorption is mostly performed from modal solutions with significantly higher initial mercury concentrations [71,72]. Higher concentration usually leads to improved sorption capacity due to the greater driving force [73]. Furthermore, compared to modal solutions, obtained leachates comprise of various organic and inorganic mercury species, which significantly affect mercury uptake. For instance, it was found that mercury uptake onto natural zeolite was hindered by the presence of humic acids, due to its partial sorption to organic matter [74]. Removal of mercury from industrial effluents (0.0028–0.0246 mg Hg/kg) resulted in lower capacity of natural zeolite in the range 0.0039–0.021 mg/kg, due to the competition of other present cations, which also undergo sorption on zeolite [75]. Moreover, the use of unmodified natural low-cost sorbents commonly leads to lower sorption capacities and removal efficiencies, compared to synthesized and modified sorbents. As in the case of low-cost sugarcane bagasse, the removal efficiency was only 20%, while its activation increased the efficiency to 85–95%, with a capacity of up to 4.8 mg/g, for an initial mercury concentration of 1 mg/L [76]. Non-modified magnetic iron oxide nanoparticles removed mercury with ≈44% efficiency, while modification of the adsorbent increased the efficiency to 98.6%, with a capacity of 590 μg/g, for an initial mercury concentration of 50 μg/L [77]. Modification improves sorption capacity but can cause higher costs and secondary pollution. Generally, natural zeolites have a great ion-exchange capacity, but their adsorption properties need improvement [78]. In order to increase mercury uptake, further research of mercury leachability from the contaminated soil should be focused on the modification of the natural zeolite surface by sulfurization, since the mercury tendency for sulfur-containing species has been long established.

In Figure 12, a comparison of pH values before leaching, prior to sorption and after sorption on the natural zeolite can be seen.

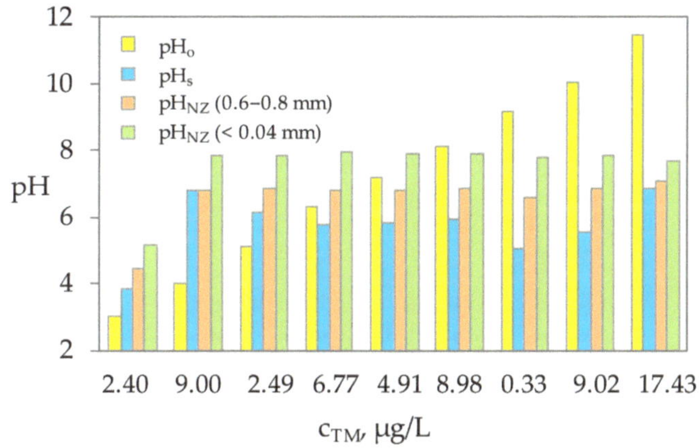

Figure 12. The comparison of pH values before leaching (pH_o), prior to sorption (pH_S) and after sorption (pH_{NZ}) on two particle sizes of natural zeolite.

Despite a wide range of initial pH_o values prior to leaching, the neutral or slightly acidic range of pH_S values indicates good buffering capacity of the soil. The pH_{NZ} values are slightly higher than pH_S and are mostly in the neutral pH range.

3.4. SEM-EDS Analysis of Zeolite Samples before and after Sorption of Total Mercury

Figure 13 presents SEM image of natural zeolite surface before sorption with the corresponding EDS analysis.

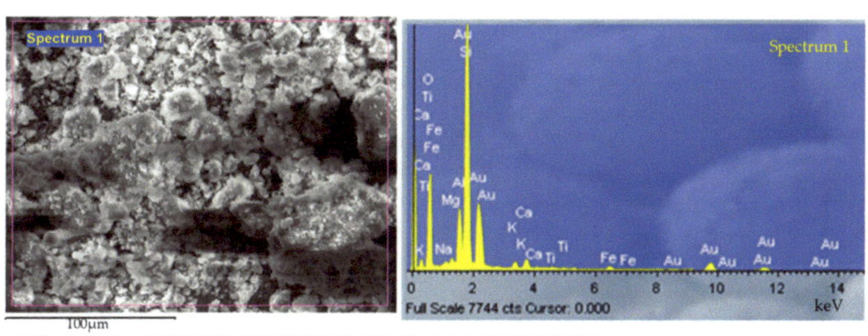

Element, wt. (%)							
Na	K	Ca	Mg	Fe	Si	Al	O
0.86	1.27	2.72	0.76	1.78	35.87	6.56	49.84

Figure 13. SEM image and EDS analysis with mass percentage values of the detected elements on the marked (Spectrum 1) surface for raw natural zeolite.

According to standard chemical analysis of aluminosilicates [79], zeolite contains (in mass%) the following: SiO_2, 66.36; Al_2O_3, 13.81; Fe_2O_3, 1.69; MgO, 1.03; CaO, 3.65; K_2O, 0.99; and Na_2O, 1.03, with a loss on ignition 13.40, while the major mineralogical component according to the semi-quantitative mineralogical analysis is clinoptilolite (up to 80%) with quartz as an impurity [80].

The morphology of the zeolite surface of two particle sizes after sorption of total mercury with corresponding EDS results is given in Figures 14 and 15.

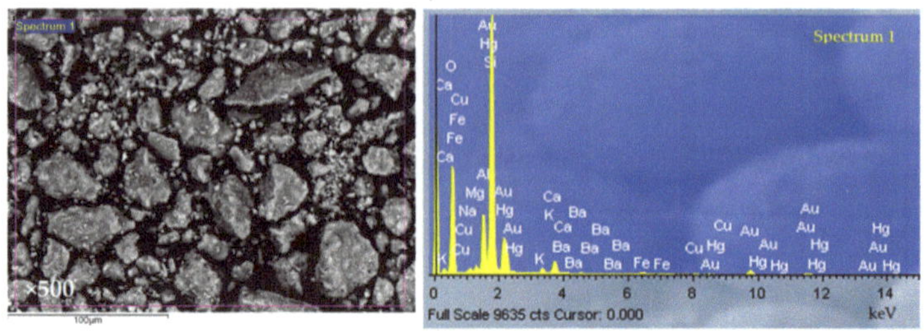

Element, wt. (%)										
O	Na	Mg	Al	Si	K	Ca	Fe	Cu	Ba	Hg
60.49	0.62	0.73	6.23	26.60	0.93	2.28	0.87	0.85	0.41	0.00

Figure 14. SEM image and EDS results for NZ of particle size < 0.04 mm after sorption of total mercury leached for pH_o of ultrapure water of 6.32.

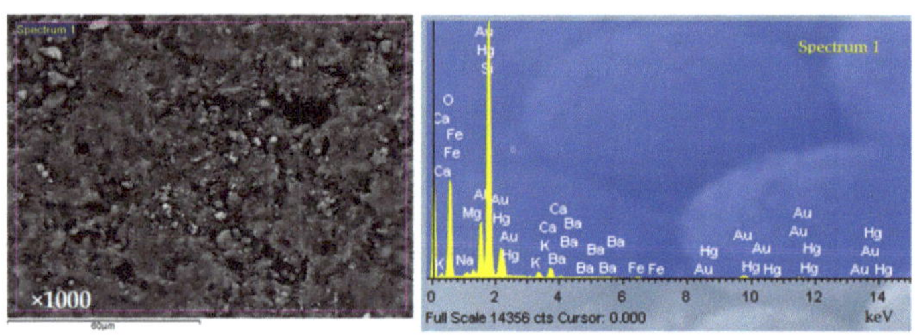

Element, wt. (%)									
O	Na	Mg	Al	Si	K	Ca	Fe	Ba	Hg
60.45	0.52	0.72	6.05	28.09	0.89	2.06	0.87	0.35	0.00

Figure 15. SEM image and EDS results for NZ of particle size 0.6–0.8 mm after sorption of total mercury leached for pH_o of ultrapure water of 6.32.

According to the obtained results, the presence of mercury onto the surface of both zeolite samples can be observed in given EDS spectra, but there were no visible mercury agglomerates. The content of mercury in weight % again was not possible to determine, as it was in the case of SEM-EDS results obtained for a sample of mercury-contaminated soil. This is due to the low initial mercury concentration in leachates. Thus, the significant difference in the element composition on the zeolite surface before and after sorption was not observed. A slight decrease in the Na and K content on the surface of zeolite after sorption was noted, which could be attributed to ion exchange with mercury species.

3.5. Simultaneous Leaching of Mercury from the Contaminated Soil and Its Sorption onto Natural Zeolite

The leaching experiments were continued with an emphasis on the first few hours of the experiment for pH_o values of ultrapure water of 4.11, 5.09 and 6.04, according to the earlier obtained results (Figure 7). Leaching was performed simultaneously in two 5-liter batch reactors, without and with the addition of natural zeolite to the soil, in order to evaluate its effectiveness in reducing mercury leaching. Larger reactors enabled getting more representative suspension samples and avoiding or minimizing possible inhomogeneity of the soil.

3.5.1. Concentration of the Total Mercury Leached

The concentrations of the total mercury leached at time intervals within 24 h without and with the addition of natural zeolite to the soil in ultrapure water of three different initial pH_o values are graphically shown in Figure 16. For all three pH_o values, the same trend in the increase of the total mercury concentration without natural zeolite addition can be observed. The mercury concentration did not exhibit a clear pattern as a function of time, but these variations were not so considerable as in Figure 6, where they are probably a consequence of soil inhomogeneity and experimental performance in nine batch reactors. In all leachates the concentration of the total mercury was higher than the maximum allowed value of 1 μg/L in water for human consumption [66]. The concentration of the total mercury in suspensions without natural zeolite ranged from 12.8 μg/L to 42.2 μg/L, with the maximum concentration achieved after 24 h for all pH_o values. The lower concentration of the total mercury from 1.9 to 20.3 μg/L was determined in all leachates in the presence of natural zeolite for all examined pH_o values, implying uptake of mercury species on the zeolite. These three pH_o values correspond to real conditions in nature, and mercury is most likely present in a form that is much more readily available for binding to zeolite.

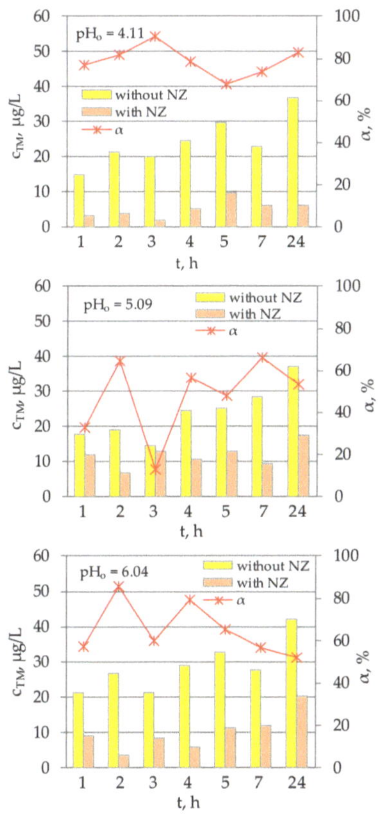

Figure 16. The concentration of total mercury leached from the soil and removal efficiency of natural zeolite at time intervals within 24 h without and with the addition of natural zeolite for three different pH_o values of ultrapure water.

For determining the efficiency α (%) of natural zeolite in reducing the concentration of leached mercury, Equation (3) was used, where c_o(Hg) in μg/L represents the concentration of the total mercury leached in soil suspension without natural zeolite, while c_e(Hg) in μg/L represents the concentration of the total mercury leached in soil suspension with the presence of natural zeolite. The calculated removal efficiencies are also shown in Figure 16. It can be seen that the removal efficiency of natural zeolite toward the total mercury under given conditions was from 12.9% to as much as 90.5%. The best removal on the natural zeolite was achieved at the lowest initial pH_o of 4.11, where α was in the range 67.7–90.5%. Moreover, at this pH_o, the highest removal efficiency was achieved up to the first 3 h. The same tendency, although less pronounced, was observed for pH_o of 5.09 and 6.04 up to the first two hours, which is in correspondence with previously obtained results. At pH_o = 6.04, the removal efficiency was in the range 51.9–85.7%, while the lowest removal was achieved at pH_o of 5.09, where it was in the range 12.9–66.4%.

3.5.2. pH and Electrical Conductivity

Changes in the pH value in soil–ultrapure water and soil with natural zeolite–ultrapure water suspensions at different initial pH_o values of 4.11, 5.09 and 6.04 are shown graphically in Figure 17. All suspensions tend to have a slightly acidic pH, which can be attributed to the buffering capacity of the soil. The largest change in pH in Figure 17a–c in all suspensions occurs within the first two hours of the experiment after which it becomes almost constant. The pH value in suspensions depends on the exchange of H_3O^+ and OH^- with ionic species from the soil, as well as with cations from the zeolite structure. The difference in pH values in the case of suspensions without and with the presence of natural zeolite is almost negligible, indicating the strong buffering capacity of the soil. This is in line with several similar field studies [81,82].

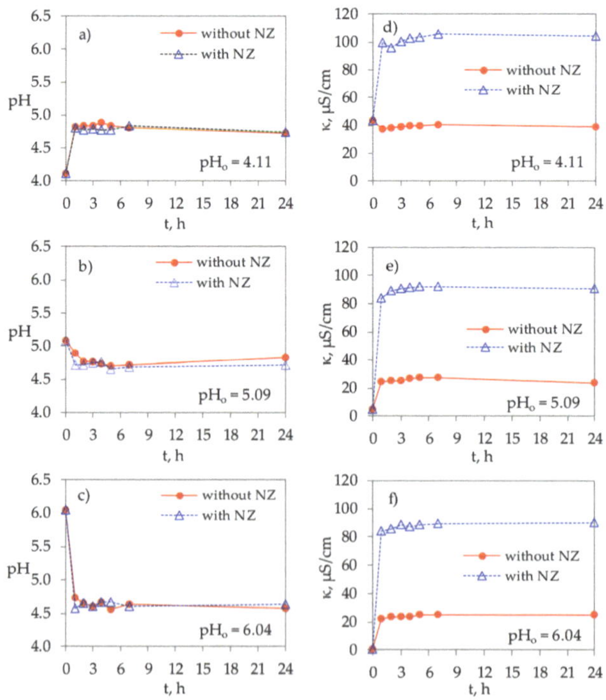

Figure 17. Change in: (**a–c**) pH and (**d–f**) electrical conductivity in suspensions during soil leaching without and with the addition of natural zeolite for three different pH_o values of ultrapure water.

The largest change in electrical conductivity in Figure 17d–f in all suspensions occurs within the first two hours and remains almost constant until the end of the experiment. The electrical conductivity decreases with increasing in pH_o, both in suspensions without and with the presence of natural zeolite. The difference in electrical conductivity of ≈60 µS/cm in soil suspensions with and without the NZ (Figure 17d–f), as a consequence of the zeolite–soil interaction is negligible.

4. Conclusions

In the surroundings of the Idrija mine (Slovenia) extremely mercury-contaminated ancient small-scale cinnabar roasting sites were identified. Soil from one of these locations (Ferbejžene trate) containing 1347 mg Hg/kg was tested. Leaching of this soil in ultrapure water of pH_o values in the range 3.00–11.46 resulted in equilibrium concentrations of the total leached mercury up to 17.43 µg/L, while maximal concentrations during leaching were from 136.9 to 488.0 µg/L, for the pH_o range 4.03–6.32. These concentrations are quite high if we consider that the maximum permissible mercury concentration in water for human consumption is 1.0 µg/L, and mercury and its compounds have been identified as priority hazardous substances in the field of water policy. The pH significantly not only affected the mobility and solubility of mercury species in the soil, but also very likely the mercury speciation in leachates.

The natural zeolite clinoptilolite of two particle sizes showed very good sorption of leached mercury, where the maximum removal efficiency of 94.2% was achieved. The removal efficiencies are not in correlation with total mercury concentrations in leachates, indicating that presumably mercury speciation is the dominant factor affecting mercury uptake on the natural zeolite. Furthermore, by leaching of mercury in the presence of the natural zeolite, a significant decrease of the total leached mercury is observed (1.9–20.3 µg/L compared to 12.8–42.2 µg/L), for all examined pH_o values, pointing to immobilization of mercury species, with obtained removal efficiencies up to 90.5%. In both, the soil–ultrapure water and soil–zeolite–ultrapure water suspensions, the tendency toward neutral or weak acidic equilibrium stages indicates the high buffering capacity of these systems.

The obtained preliminary results indicate the potential of natural zeolite clinoptilolite for remediation of mercury-contaminated environments. However, additional efforts should be made to maintain the mercury concentration in the water below the maximum permissible limits. Natural zeolites are proven as great ion exchangers, but their adsorption properties need to be improved. Due to the well-known mercury tendency toward sulfur-containing species, the sulfurization of the zeolite surface might improve its effectiveness in mercury capturing. Moreover, the performance of multistage sorption could reduce the mercury concentration in water up to satisfying levels.

The further studying of mercury leachability at ancient roasting sites of the Idrija district in Slovenia should be continued by estimation of the efficiency of modified natural materials in their mercury capture capability, especially in column mode. Additionally, in some part of further research the attention should be addressed to an adequate method of disposal of heavy metal-bearing zeolite to prevent secondary pollution. The solidification/stabilization process by cementitious binders seems very promising in reducing the release of toxic species into the environment, especially since many researchers found that the addition of low content of zeolitic tuff improves the properties of the concrete.

Author Contributions: Conceptualization, investigation, methodology, visualization, writing—original draft preparation, review and editing, I.N.; formal analysis, visualization, writing—review and editing, M.G.; visualization, writing—review and editing, M.U.; visualization, writing—review and editing, M.T. All authors have read and agreed to the published version of the manuscript.

Funding: This research was fully financially supported by the bilateral Croatian–Slovenian project "Application of natural zeolite for remediation of mercury-contaminated soil". The study was partly financed by the Slovenian Research Agency (ARRS) through the research program Groundwater and Geochemistry (P1-0020). The APC was funded by Geological Survey of Slovenia (P1-0020).

Institutional Review Board Statement: Not applicable.

Informed Consent Statement: Not applicable.

Data Availability Statement: Not applicable.

Acknowledgments: Many thanks to colleague Mariusz Macherzyński from AGH University of Science and Technology, Faculty of Energy and Fuels, Department of Coal Chemistry and Environmental Sciences in Krakow, Poland and the AGH UST (grant No. 11.11.210.374).

Conflicts of Interest: The authors declare no conflict of interest. The funders had no role in the design of the study; in the collection, analyses, or interpretation of data; in the writing of the manuscript, or in the decision to publish the results.

References

1. WHO. *Exposure to Mercury: A Major Public Health Concern*, 2nd ed.; Preventing Disease through Healthy Environments; World Health Organization: Geneva, Switzerland, 2021; ISBN 9789240023567. Available online: https://www.who.int/publications/i/item/9789240023567 (accessed on 14 December 2021).
2. EEA Report No 11/2018, TH-AL-18-011-EN-N, Mercury in Europe's Environment, A Priority for European and Global Action, European Environment Agency. Available online: https://www.eea.europa.eu/publications/mercury-in-europe-s-environment/download (accessed on 14 December 2021).
3. Gray, J.E.; Pribil, M.J.; Higueras, P.L. Mercury isotope fractionation during ore retorting in the Almadén mining district, Spain. *Chem. Geol.* **2013**, *357*, 150–157. [CrossRef]
4. Swartzendruber, P.; Jaffe, D. Sources and transport a global issue. In *Mercury in the Environment: Pattern and Process*, 1st ed.; Bank, M.S., Ed.; University of California Press: Berkeley, CA, USA; Los Angeles, CA, USA; London, UK, 2012; Chapter 1; pp. 3–18.
5. Calao-Ramos, C.; Bravo, A.G.; Paternina-Uribe, R.; Marrugo-Negrete, J.; Díez, S. Occupational human exposure to mercury in artisanal small-scale gold mining communities of Colombia. *Environ Int.* **2021**, *146*, 106216. [CrossRef] [PubMed]
6. Masur, L.C. A review of the use of mercury in historic and current ritualistic and spiritual practices. *Altern. Med. Rev.* **2011**, *16*, 314–320. [PubMed]
7. WHO. Preventing Disease through Healthy Environments, Mercury in Skin Lightening Products, World Health Organization, WHO/CED/PHE/EPE/19.13. 2019. Available online: https://apps.who.int/iris/bitstream/handle/10665/330015/WHO-CED-PHE-EPE-19.13-eng.pdf (accessed on 16 December 2021).
8. Shiwei, Z.; Dongmei, Z.; Yuan, X.; Ying, M.; Liu, S.; Huiying, L. Study on mercury methylation in phragmites australis soil and its influencing factors. *Water Air Soil Pollut.* **2020**, *231*, 426. [CrossRef]
9. He, F.; Gao, J.; Pierce, E.; Strong, P.J.; Wang, H.; Liang, L. In situ remediation technologies for mercury-contaminated soil. *Sci Pollut Res.* **2015**, *22*, 8124–8147. [CrossRef]
10. McCarthy, D.; Edwards, G.C.; Gustin, M.S.; Care, A.; Miller, M.B.; Sunna, A. An innovative approach to bioremediation of mercury contaminated soils from industrial mining operations. *Chemosphere* **2017**, *184*, 694–699. [CrossRef]
11. Moharem, M.; Elkhatib, E.; Mesalem, M. Remediation of chromium and mercury polluted calcareous soils using nanoparticles: Sorption-desorption kinetics, speciation and fractionation. *Environ. Res.* **2019**, *170*, 366–373. [CrossRef]
12. Eckley, C.S.; Gilmour, C.C.; Janssen, S.; Luxton, T.P.; Randall, P.M.; Whalin, L.; Austin, C. The assessment and remediation of mercury contaminated sites: A review of current approaches. *Sci Total Environ.* **2020**, *707*, 136031. [CrossRef]
13. Spanu, D.; Bestetti, A.; Hildebrand, H.; Schmuki, P.; Altomare, M.; Recchia, S. Photocatalytic reduction and scavenging of Hg(ii) over templated-dewetted Au on TiO2 nanotubes. *Photochem. Photobiol. Sci.* **2019**, *18*, 1046–1055. [CrossRef]
14. Wang, T.; He, S.; Zhang, Y.; Wang, J.; Pan, W.-P. Photocatalytic removal of elemental mercury on TiO2-BiOIO3 heterostructures: Mercury transformation, sulfur tolerance and SO2/SO3 conversion. *Chem. Eng. J.* **2020**, *388*, 124390. [CrossRef]
15. Tunsu, C.; Wickman, B. Effective removal of mercury from aqueous streams via electrochemical alloy formation on platinum. *Nat. Commun.* **2018**, *9*, 4876. [CrossRef] [PubMed]
16. Bengtsson, M.K.O.; Tunsu, C.; Wickman, B. Decontamination of mercury-containing aqueous streams by electrochemical alloy formation on copper. *Ind. Eng. Chem. Res.* **2019**, *58*, 9166–9172. [CrossRef]
17. O'Connor, D.; Peng, T.; Li, G.; Wang, S.; Duan, L.; Mulder, J.; Cornelissen, G.; Cheng, Z.; Yang, S.; Hou, D. Sulfur-modified rice husk biochar: A green method for the remediation of mercury contaminated soil. *Sci. Total Environ.* **2018**, *621*, 819–826. [CrossRef] [PubMed]
18. Al-Ghouti, M.A.; Da'ana, D.; Abu-Dieyeh, M.; Khraisheh, M. Adsorptive removal of mercury from water by adsorbents derived from date pits. *Sci. Rep.* **2019**, *9*, 1–15. [CrossRef]
19. Abdillah, A.M.; Abidin, Z.; Widiatmaka, W.; Johan, E.; Matsue, N. Adsorptive Removal of Mercury by Zeolites and Montmorillonite. *J. Pengelolaan Sumberd. Alam Lingk.* **2019**, *9*, 797–801. [CrossRef]
20. Fayazi, M. Removal of mercury(II) from wastewater using a new and effective composite: Sulfur-coated magnetic carbon nanotubes. *Environ. Sci. Pollut. Res. Int.* **2020**, *27*, 12270–12279.

21. Favvas, E.; Tsanaktsidis, C.; Sapalidis, A.; Tzilantonis, G.; Papageorgiou, S.; Mitropoulos, A. Clinoptilolite, a natural zeolite material: Structural characterization and performance evaluation on its dehydration properties of hydrocarbon-based fuels. *Microporous Mesoporous Mater.* **2016**, *225*, 385–391. [CrossRef]
22. Nuić, I.; Trgo, M.; Vukojević Medvidović, N. The application of the packed bed reactor theory to Pb and Zn uptake from the binary solution onto the fixed bed of natural zeolite. *Chem. Eng. J.* **2016**, *295*, 347–357. [CrossRef]
23. Nuić, I.; Trgo, M.; Vukojević Medvidović, N.; Ugrina, M. A mass transfer analysis of competitive binding of Pb, Cd, and Zn from binary systems onto a fixed zeolite bed. *Int. J. Environ. Res. Public Health* **2019**, *16*, 426. [CrossRef]
24. Cigale, M. Rudnik živega srebra Idrija od 1490 do 2006. In *2. Slovenski Geološki Kongres, Zbornik Povzetkov. Rudnik ŽIvega Srebra v Zapiranju d.o.o., Idrija*; Režun, B., Eržen, U., Petrič, M., Gantar, I., Eds.; Rudnik Živega srebra v Zapiranju: Idrija, Slovenia, 2006; pp. 13–16. (In Slovenian)
25. Kavčič, I. *History of Smelting in Idrija*; Založba Bogataj: Idrija, Slovenia, 2008; pp. 1–352. (In Slovenian)
26. Teršič, T.; Gosar, M.; Biester, H. Distribution and speciation of mercury in soil in the area of an ancient mercury ore roasting site, Frbejžene trate (Idrija area, Slovenia). *J. Geochem. Explor.* **2011**, *110*, 136–145. [CrossRef]
27. Teršič, T.; Gosar, M.; Biester, H. Environmental impact of ancient small-scale mercury ore processing at Pšenk on soil (Idrija area, Slovenia). *Appl. Geochem.* **2011**, *26*, 1867–1876. [CrossRef]
28. Gosar, M.; Teršič, T. Environmental geochemistry studies in the area of Idrija mercury mine, Slovenia. *Environ. Geochem. Health* **2012**, *34*, 27–41. [CrossRef] [PubMed]
29. Gosar, M.; Teršič, T. Contaminated sediment loads from ancient mercury ore roasting sites, Idrija area, Slovenia. *J. Geochem. Explor.* **2015**, *149*, 97–105. [CrossRef]
30. Gosar, M.; Šajn, R.; Teršič, T. Distribution pattern of mercury in the Slovenian soil: Geochemical mapping based on multiple geochemical datasets. *J. Geochem. Explor.* **2016**, *167*, 38–48. [CrossRef]
31. Vukojević Medvidović, N.; Nuić, I.; Ugrina, M.; Trgo, M. Evaluation of natural zeolite as a material for permeable reactive barrier for remediation of zinc contaminated groundwater based on column study. *Water Air Soil Pollut.* **2018**, *229*, 1–14. [CrossRef]
32. Gosar, M.; Čar, J. Influence of mercury ore roasting sites from 16th and 17th century on the mercury dispersion in surroundings of Idrija. *Geologija* **2006**, *49*, 91–101. [CrossRef]
33. *DIN ISO 10390:2005*; Soil Quality-Determination of pH. International Organization for Standardization. ISO: Geneva, Switzerland, 2005.
34. Nelson, P.; Su, N. Soil pH buffering capacity: A descriptive function and its application to some acidic tropical soils. *Aust. J. Soil Res.* **2010**, *48*, 201–207. [CrossRef]
35. Xu, R.-K.; Zhao, A.-Z.; Yuan, J.; Jiang, J. pH buffering capacity of acid soils from tropical and subtropical regions of China as influenced by incorporation of crop straw biochars. *J. Soils Sediments* **2012**, *12*, 494–502. [CrossRef]
36. *HRN EN ISO 11260:2012*; Soil Quality-Determination of Effective Cation Exchange Capacity and Base Saturation Level Using Barium Chloride Solution, (ISO 11260:1994+Cor 1:1996; EN ISO 11260:2011). Croatian Standards Institute: Zagreb, Croatia, 2012; p. 14. Available online: https://repozitorij.hzn.hr/norm/HRN+EN+ISO+11260%3A2012(accessed on 16 February 2022).
37. Zhang, H.; Wang, J.J. Loss on ignition method. In *Soil Test Methods from the Southeasten United States. Southern Extension and Research Activity Information Exchange Group 6*; Sikora, F.J., Moore, K.P., Eds.; University of Georgia: Athens, Greek, 2014; pp. 155–157. Available online: https://aesl.ces.uga.edu/sera6/PUB/MethodsManualFinalSERA6.asp (accessed on 16 December 2021).
38. Roper, W.; Robarge, W.; Osmond, D.; Heitman, J.L. Comparing four methods of measuring soil organic matter in North Carolina soils. *Soil Sci. Soc. Am. J.* **2019**, *83*, 466–474. [CrossRef]
39. Vogel, S.; Bönecke, E.; Kling, C.; Kramer, E.; Lück, K.; Nagel, A.; Philipp, G.; Rühlmann, R.; Schröter, I.; Gebbers, R. Base neutralizing capacity of agricultural soils in a quaternary landscape of north-east Germany and its relationship to best management practices in lime requirement determination. *Agronomy* **2020**, *10*, 877. [CrossRef]
40. *DIN 38414 S4*; German Standard Procedure for Water, Wastewater and Sediment Testing–Sludge and Sediment. Determination of Leachability. Institut für Normung: Berlin, Germany, 1984.
41. The Environmental Management (Soil Quality Standards) Regulations. 2007. Available online: https://extwprlegs1.fao.org/docs/pdf/tan151538.pdf (accessed on 18 June 2021).
42. Government Decree on the Assessment of Soil Contamination and Remediation Needs (214/2007), Ministry of the Environment, Finland. 2007. Available online: https://www.finlex.fi/en/laki/kaannokset/2007/en20070214.pdf (accessed on 16 February 2022).
43. Vodyanitskii, Y.N. Standards for the contents of heavy metals in soils of some states. *Ann. Agrar. Sci.* **2016**, *14*, 257–263. [CrossRef]
44. Krupenikov, I.A.; Boincean, B.P.; Dent, D. Soil mineralogy and elemental composition. In *The Black Earth*; International Year of Planet Earth; Springer: Dordrecht, The Netherlands, 2011; pp. 27–32. [CrossRef]
45. D'Amour, H.; Denner, W.; Schulz, H. Structure determination of [alpha]-quartz up to 68 x 108 Pa. *Acta Cryst.* **1979**, *B35*, 550–555. [CrossRef]
46. *ISO 14688-2:2004*; Geotechnical Investigation and Testing-Identification and Classification of Soil-Part 2: Principles for a Classification. International Organization for Standardization, ISO: Geneva, Switzerland, 2004.

47. Huang, P.-T.; Patel, M.; Santagata, M.C.; Bobet, A. *FHWA/IN/JTRP-2008/2, Classification of Organic Soils*; Final Report; Indiana Department of Transportation and Federal Highway Administration: Indianapolis, IN, USA, 2009; pp. 46–62. Available online: https://www.geostructures.com/library/technical-bulletins/pdf/Classification-of-Organic-Soils-FHWA-IN-JTRP-2008-2.pdf (accessed on 16 February 2022).
48. Brümmer, G.; Herms, U. Influence of soil reaction and organic matter on the solubility of heavy metals in soils. In *Effects of Accumulation of Air Pollutants in Forest Ecosystems*; Urlich, B., Pankrath, J., Eds.; D. Reidel Publishing Company: Dordrecht, The Netherlands, 1983; pp. 233–243.
49. Weil, R.R.; Brady, N.C. *The Nature and Properties of Soils*, 15th ed.; Pearson Education Limited, Edinburgh Gate: Harlow, UK, 2017; pp. 20–51, 345–391, 392–437.
50. Schuster, E. The behaviour of mercury in the soil with special emphasis on complexation and adsorption processes-a review of the literature. *Water Air Soil Pollut.* **1991**, *56*, 667–680. [CrossRef]
51. Xu, J.; Kleja, D.B.; Biester, H.; Lagerkvist, A.; Kumpiene, J. Influence of particle size distribution, organic carbon, pH and chlorides on washing of mercury contaminated soil. *Chemosphere* **2014**, *109*, 99–105. [CrossRef] [PubMed]
52. Teršič, T.; Biester, H.; Gosar, M. Leaching of mercury from soils at extremely contaminated historical roasting sites (Idrija area, Slovenia). *Geoderma* **2014**, *226–227*, 213–222. [CrossRef]
53. Sharpley, A.N. Effect of soil pH on cation and anion solubility. *Commun. Soil Sci. Plant Anal.* **1991**, *22*, 827–841. [CrossRef]
54. Jiang, J.; Wang, Y.-P.; Yu, M.; Cao, N.; Yan, J. Soil organic matter is important for acid buffering and reducing aluminum leaching from acidic forest soils. *Chem. Geol.* **2018**, *501*, 86–94. [CrossRef]
55. Tan, K.H. *Principles of Soil Chemistry*, 4th ed.; CRC Press Taylor & Francis Group: Boca Raton, FL, USA, 2011; pp. 243–266.
56. Matsusaka, Y.; Sherman, G.D. *Titration Curves and Buffering Capacities of Hawaiian Soils*; Technical Bulletin No 11; University of Hawaii, College of Agriculture, Agricultural Experiment Station: Honolulu, HA, USA, 1950; pp. 5–36.
57. Jansen van Rensburg, H.G.; Claassens, A.S.; Beukes, D.J. Relationships between soil buffer capacity and selected soil properties in a resource-poor farming area in the Mpumalanga Province of South Africa. *S. Afr. J. Plant Soil.* **2009**, *26*, 237–243. [CrossRef]
58. Bloom, P.R.; Skyllberg, U. Soil pH and pH buffering. In *Handbook of Soil Sciences, Properties and Processes*, 2nd ed.; Huang, P.M., Li, Y., Sumner, M.E., Eds.; CRC Press: Boca Raton, FL, USA, 2012; pp. 1–14.
59. Pagani, A.; Sawyer, J.E.; Mallarino, A.P.; Moody, L.; Davis, J.; Phillips, S. *Site-Specific Nutrient Management: For Nutrient Management Planning to Improve Crop Production, Environmental Quality, and Economic Return*; Extension and Outreach Publications: Washington, DC, USA, 2013; Volume 116, pp. 103–113. Available online: https://lib.dr.iastate.edu/extension_pubs/116 (accessed on 17 December 2021).
60. Curtin, D.; Trolove, S. Predicting pH buffering capacity of New Zealand soils from organic matter content and mineral characteristics. *Soil Res.* **2013**, *51*, 494–502. [CrossRef]
61. Moore, G.; Dolling, P.; Porter, B.; Leonard, L. Chemical factors affecting plant growth. In *Soil Guide: A Handbook for Understanding and Managing Agricultural Soils*; No. 4343; Moore, G.A., Ed.; Department of Agriculture, Western Australia: Bulletin, Australia, 2001; pp. 127–140.
62. Reganold, J.P.; Harsh, J.B. Expressing cation exchange capacity in milliequivalents per 100 grams and in SI units. *J. Agric. Educ.* **1985**, *14*, 84–90. [CrossRef]
63. Johnson, C.E. Cation exchange properties of acid forest soils of the north eastern USA. *Eur. J. Soil Sci.* **2002**, *53*, 271–282. [CrossRef]
64. Nešić, L.; Vasin, J.; Belic, M.; Ćirić, V.; Gligorijevic, J.; Milunovic, K.; Sekulić, P. The colloid fraction and cation-exchange capacity in the soils of Vojvodina, Serbia. *Ratar. Povrt.* **2015**, *52*, 18–23. [CrossRef]
65. Gumbara, R.H.; Darmawan; Sumawinata, B. A comparison of cation exchange capacity of organic soils determined by ammonium acetate solutions buffered at some pHs ranging between around field pH and 7.0. *IOP Conf Ser. Earth Environ Sci.* **2019**, *393*, 012015. [CrossRef]
66. Directive (EU) 2020/2184 of the European Parliament and of the Council of 16 December 2020 on the Quality of Water Intended for Human Consumption. ELI. Available online: http://data.europa.eu/eli/dir/2020/2184/oj (accessed on 17 December 2021).
67. Cappuyns, V.; Swennen, R. The Application of pHstat leaching tests to assess the pH-dependent release of trace metals from soils, sediments and waste materials. *J. Hazard. Mater.* **2008**, *158*, 185–195. [CrossRef]
68. O'Connor, D.; Hou, D.; Ok, Y.S.; Mulder, J.; Duan, L.; Wu, Q.; Wang, S.; Tack, F.M.G.; Rinklebe, J. Mercury speciation, transformation, and transportation in soils, atmospheric flux, and implications for risk management: A critical review. *Environ. Int.* **2019**, *126*, 747–761. [CrossRef]
69. Clifford, M.J.; Hilson, G.M.; Hodson, M.E. Characteristics and behaviour of individual elements, Tin and Mercury. In *Trace Elements in Soils*; Hooda, P.S., Ed.; John Wiley & Sons Ltd.: Chichester, UK, 2010; pp. 497–513.
70. Miller, R.L.; Bradford, W.L.; Peters, N.E. *Specific Conductance; Theoretical Considerations and Application to Analytical Quality Control*; U.S. Geological survey Water Supply Paper 2311; United States Government Printing Office: Washington, DC, USA, 1988. [CrossRef]
71. Zhang, D.; Crini, G.; Lichtfouse, E.; Rhimi, B.; Wang, C. Removal of mercury ions from aqueous solutions by crosslinked chitosan-based adsorbents: A mini review. *Chem. Rec.* **2020**, *20*, 1–16. [CrossRef]
72. Sharma, A.; Sharma, A.; Arya, R.K. Removal of mercury(II) from aqueous solution: A review of recent work. *Sep. Sci. Technol.* **2015**, *50*, 1310–1320. [CrossRef]

73. Liu, Z.; Sun, Y.; Xu, X.; Qu, J.; Qu, B. Adsorption of Hg(II) in an aqueous solution by activated carbon prepared from rice husk using KOH activation. *ACS Omega* **2020**, *5*, 29231–29242.
74. Prasakti, L.; Utami, F.N.; Prasetya, A. Removal of mercury ion from wastewater using natural zeolite: Effect of humic acid to adsorption isotherm. *IOP Conf. Ser. Earth Environ. Sci.* **2021**, *882*, 012068.
75. Chojnacki, A.; Chojnacka, K.; Hoffmann, J.; Gorecki, H. The application of natural zeolites for mercury removal: From laboratory tests to industrial scale. *Miner. Eng.* **2004**, *17*, 933–937. [CrossRef]
76. Giraldo, S.; Robles, I.; Ramirez, A.; Flórez, E.; Acelas, N. Mercury removal from wastewater using agroindustrial waste adsorbents. *SN Appl. Sci.* **2020**, *2*, 1029. [CrossRef]
77. Parham, H.; Zargar, B.; Shiralipour, R. Fast and efficient removal of mercury from water samples using magnetic iron oxide nanoparticles modified with 2-mercaptobenzothiazole. *J. Hazard. Mater.* **2012**, *205–206*, 94–100. [CrossRef]
78. Wang, S.; Peng, Y. Natural zeolites as effective adsorbents in water and wastewater treatment. *Chem. Eng. J.* **2010**, *156*, 11–24. [CrossRef]
79. Voinovitch, I.; Debrad-Guedon, J.; Louvrier, J. *The Analysis of Silicates*; Israel Program for Scientific Translations: Jerusalem, Israel, 1966; pp. 127–129.
80. Vukojević Medvidović, N.; Perić, J.; Trgo, M. Column performance in lead removal from aqueous solutions by fixed bed of natural zeolite-clinoptilolite. *Sep. Purif. Technol.* **2006**, *49*, 237–244.
81. Litaor, M.I.; Katz, L.; Shenker, M. The influence of compost and zeolite co-addition on the nutrients status and plant growth in intensively cultivated Mediterranean Soils. *Soil Use Manag.* **2017**, *33*, 72–80. [CrossRef]
82. Tallai, M.; Olah Zsuposne, A.; Sandor, Z.; Katai, J. The effect of using zeolite on some characteristics of sandy soil and on the amount of the test plant biomass. *Ann. Ser. Agric. Silvic Vet.* **2017**, *6*, 115–120.

processes

Article

Increase by Substitution of Galvanized Steel for Aluminum Mirrors in the UV Solar Radiation in Canal with Fins and Side Panels That Disinfect Wastewater

Pedro Cisterna-Osorio *, María Galvez-Gonzalez, Miguel Moraga-Chaura and Sergio Quijada-Vera

Department of Civil and Environmental Engineering, University of Bío Bío, Concepción 378000, Chile
* Correspondence: pcisterna@ubiobio.cl; Tel.: +56-9-90007339

Abstract: The need arises to seek new depuration technological responses aimed at the reuse of wastewater, which requires the development and promotion of economically and environmentally sustainable technologies. In this paper, it studies an improvement to a disinfection system sustainable, low-cost, patented in 2019, and based on solar energy. The water passes through a canal of reflective material in the continuous regime, and in the batch regime, the water remains in the canal. The panels are located parallel to the lateral faces of the canal. The fraction of the radiation reflected outside the canal reaches the reflective side panels that return the radiation to the canal. These panels concentrate the radiation in the canal through reflection. The disinfectant canal with fins and side panels uses ultraviolet radiation to eliminate the bacterial load carried by treated wastewater. For this reason, the present work analyzes the incidence in the area of influence of the disinfectant canal. When reflective aluminum mirrors were installed on the sloping walls of the canal, global radiation increased by 4%, when they were used on the side panels, it increased 3%, and when the aluminum mirrors were used on the canal walls and side panels, it increased 8%. The important thing about this work is that it opens windows for improving the system through materiality so the new challenge is the search for the optimal material considering the impact on global radiation and consequently on the bacteriological elimination.

Keywords: solar radiation; disinfection; wastewater

Citation: Cisterna-Osorio, P.; Galvez-Gonzalez, M.; Moraga-Chaura, M.; Quijada-Vera, S. Increase by Substitution of Galvanized Steel for Aluminum Mirrors in the UV Solar Radiation in Canal with Fins and Side Panels That Disinfect Wastewater. *Processes* **2023**, *11*, 84. https://doi.org/10.3390/pr11010084

Academic Editors: Dimitris Zagklis and Georgios Bampos

Received: 24 October 2022
Revised: 6 December 2022
Accepted: 7 December 2022
Published: 28 December 2022

1. Introduction

Climate change is one of the most complex and important problems and externalities in the environmental paradigm [1]. The speed of climate change will make survival and adaptation more difficult. This being a global dilemma, it will require mitigation measures at all scales, from the local as well as the international community [2].

In the global context, the water resource fulfills three main functions in sustainable development: it plays an important role for human health, it allows economic development, and it allows the different ecosystems worldwide to be viable [3]. The future will present hydrological uncertainty, which is accentuated by the climate change so it is necessary to implement measures to manage water resources [4]. Depending on how greenhouse gas emissions are managed, the temperature in southern Chile could increase by 0.68 °C to 1.51 °C between the years 2050–2065 [5].

A water resource management policy that strives to ensure the water availability for future generations is the reuse of wastewater [6]. This reuse of water is highly desirable since it promotes the conservation of water resources and allows control of water pollution [7]. This management policy allows adaptation to climate change and is also sustainable; thus providing an alternative source of water for use in irrigation, having great potential in rural areas with water scarcity. Since agriculture tends to be one of the most important economic activities [8] it is vital to develop technologies that guarantee optimal treatment and low construction, operation, and maintenance costs [9]. The energy costs and the environmental impact it produces should be determinate to optimize the wastewater treatment [10].

The industry and domestic activities, in their operation, produce wastewater that contains contaminants, so the protection of the water resource is required. To eliminate these pollutants (soluble and insoluble), the treatment methods include physical, chemical, and biological processes [11]. These mainly seeks to regulate water parameters such as turbidity, pH, alkalinity, biological oxygen demand (BOD), chemical oxygen demand (COD), total solids (TS), total organic carbon (TOC), fecal coliforms, and others specified by local regulations [12].

Microbial activity is a contaminant that affects the water quality, both for consumption and for recreation; therefore, it represents a danger to human health [13]. Disinfectants such as ultraviolet radiation (UV), chlorine, and ozone are the most widely used in wastewater treatment and water disinfection [14]. The generation of toxic by-products in chlorination must be considered. Tt has also been identified that these chlorine by-products could be carcinogenic, posing a risk to human health [15,16]. Even regulatory bodies encourage the use of ecologically sustainable technologies [17]. Efforts are also being made to eliminate the use of chlorine in other fields of water treatment [18].

UV radiation is a viable option to chlorination. This technology has been extensively studied as it does not generate harmful by-products and is effective against many of the microbes present in water [16]. Hartman and Eisenstark in 1978 reported the first positive evidence of this process for water disinfection. They demonstrated the bactericidal effect at very low concentration of H_2O_2 ($6{\cdot}10^{-3}$–$6{\cdot}10^{-1}$ M) via a synergistic effect with near UV-radiation (300–400 nm) over Escherichia coli cells in water [19].

This manuscript presents the impact on solar radiation around the disinfectant canal with fins and side panels when galvanized steel is replaced by aluminum mirrors as reflecting material. This new technology, based on the use of solar radiation to disinfect treated wastewater, with a low cost of investment and operation, was patented in 2019 [20].

A comparative analysis of scenarios that progressively shows the differential impact caused by aluminum mirrors using the radiation rate was obtained by measuring the solar radiation in the disinfection canal, the atmospheric solar radiation, and the elimination of total coliforms as the principal quantitative indicators

This system, the disinfectant canal with fins and side panels as well as the Solar disinfection systems (SODIS) strive to inactivate microorganisms through the application of UV radiation, damaging the nucleic acids of the microorganism, making it unable to replicate. There are also photosensitive molecules in the water that manage to absorb visible light, causing oxidizing activities that manage to damage the structure of the cells [21].

Equally the disinfectant canal with fins and side panels and SODIS systems mainly uses ultraviolet radiation from the sun (UVA 320–400 nm and UVB—280–320 nm). In these systems the water circulates through a reactor under the incidence of solar radiation. These reactors are permeable to ultraviolet rays so that the radiation can inactivate the microorganisms present in the water [22], the higher the radiation intensity, the higher the cell damage; however, the inactivation of the microorganisms also depends strongly on the wavelength of the radiation applied [23]. UV radiation (200–400 nm) has three different classifications: UVA (320–400 nm), UVB (280–320 nm), and UVC (200–280 nm); however, UVC radiation and part of UVB are absorbed by the ozone layer while UVA is the main component of UV radiation that, coming from the sun, manages to reach the earth's surface [24].

Different advances made in recent decades have shown that SODIS method of supplying drinking water is effective and economical, being able to reduce prevalence and mortality for waterborne diseases [25]. Usually, SODIS is oriented to developing countries since the system should be able to withstand adverse weather conditions and its maintenance requirements should be minimal [26]. The conventional SODIS use transparent water containers (usually polyethylene terephthalate (PET) plastic bottles) to expose it directly to sunlight for a minimum of 6 h. The pathogens present in the water are inactivated, thus producing safe drinking water for human consumption [27].

The efficiency and time of the SODIS treatment is affected by many factors [28]. In fact, it has been shown that when the turbidity and total suspended solids (TSS) of the water increases, the inactivation of microorganisms is less, preventing UV radiation from achieving disinfection [29].

There are treatments, such as constructed wetlands, that can be designed to remove more than 90% of TSS [30], only requiring good maintenance constructed wetlands [31]. An advantage of artificial wetlands is that they can perform well as a biofilter because they have different physical, chemical, and biological factors that participate in reducing of the number of bacteria of human origin [32,33]. More research efforts must be done in these UV technologies and other sustainable technologies capable of obtaining reclaimed Urban Waste Water for reuse in agriculture with an affordable cost [34]. Constructed wetlands are more cost-effective and low-energy consumption compared to conventional wastewater treatment [35]. Regarding the performance of the wetland in the removal of all pathogens, efficiency of 90% can be achieved when the hydraulic residence time is 1 to 2 days with adequate hydrodynamics [36,37].

In 2019, a new and innovative Solar Disinfection Technology for Treated Wastewater was patented that Integrates Materiality, Geometry, and Reflective Panels. The objective of this technology is the elimination of pathogens by means of UV radiation [38], which is in an optimization process. In this manuscript, the development of one of the improvement ways is exposed, which consists of a change in the materiality using mirrors on the walls and panels that are part of the disinfectant channel. The aim is to increase the UV radiation in the disinfectant canal and consequently increase the elimination of pathogens. We analyzed four different configurations linked to the area where the mirrors are positioned.

2. Methodology

The experimental work shown below searches to define how much the values of the solar radiation rate vary when aluminum mirrors are included as a material in the solar disinfection system since it seeks to increase the amount of UV radiation that manages to reach the canal in order to increase the efficiency of the treatment of fecal coliforms within the wastewater treatment.

Four different configurations were compared and analyzed focusing on the variations of the materiality of the canal inclined walls and that of the panels arranged on the sides. The proposed cases were:

- Condition 1: Canal and side panels of galvanized steel.
- Condition 2: Galvanized steel side panels and installation of aluminum mirrors on sloped canal walls.
- Condition 3: Galvanized steel canal and side panels with aluminum mirror installation.
- Condition 4: Aluminum mirrors on sloping canal walls and on side panels.

For the measurement of solar radiation, different points along and across the canal were considered. The analyzed system is located in the dependencies of the Canteras wastewater treatment plant, Quilleco commune, Bío-Bío region, Chile.

a. *Proposed disinfection system*

The design of the proposed canal and panels, both in geometry and materiality, has the aim of increasing the radiation inside the canal, and finally, in the water [38]. The canal consists of a trapezoidal cross section (see Figure 1) that allows redirecting the sun's rays, in order to increase the radiation in the water. In addition, it has side panels that manage to redirect the outgoing radiation back to the canal which allows for optimizing the disinfection process.

Figure 1. Canal and panels installed. The left figure shows the arrangement of the aluminum mirrors inside the canal, which are installed in the required conditions.

The canal and the installed panels are made of galvanized steel allowing it to maintain the radiation in the residual water and achieve disinfection. Galvanized steel was used as a reflector, due to the ease of obtaining it, its constructability, and the cost reduction in the project.

The supporting structure of the side panels is made up of 4 × 2 radiata pine wood boards. The reflective mirrors are made of aluminum and are 2 mm thick, 1 m long, and 0.5 m wide. This configuration is chosen since there is a decrease in reflectance as the thickness of the aluminum mirror increases, caused by the greater attenuation of the glass, which must be traversed twice [39].

The details and dimensions of the canal can be seen in Figure 2.

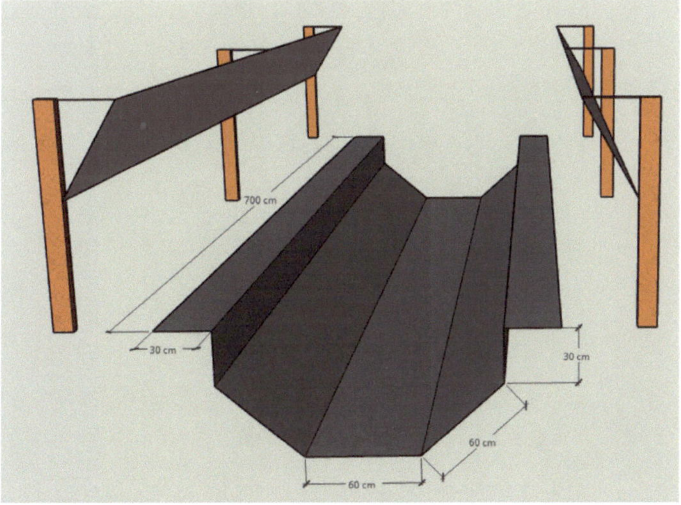

Figure 2. Dimensions of the disinfection canal.

b. *Measurement equipment*

For the measurements, a UV sensor was used, ultraviolet radiation meter (UVA-UVB), model PUV-360, whose technical specifications are detailed below in Table 1.

Table 1. Technical specifications of meter PUV-360.

Monitor/UV Sensor Spectrum	**LCD Size: 28 × 19 mm/Bandpass from 290 nm to 390 nm**
Measuring/Resolution Ranges	range 1: 2000 uW/cm^2: 1999 uW/cm^2 × 1 uW/cm^2 range 2: 20 mW/cm^2: 19.99 mW/cm^2 × 1 mW/cm^2
Precision	(4% of the read + 2 dig.). Calibration was performed under UVA light and compared with a standard reference field light meter less than 3 V/M and frequency less than 30 MHz
Sensor structure/Sample time	Photo UV sensor with cosine correction filter/1 s approx..
Off/Weight	Auto power off saves battery life/190 g/0.2 LB
Humidity (HR)/Temperature Operation	Less than 85%HR/0 °C to 50 °C
Dimensions/Power Supply	210 × 49 × 40 mm/4 AAA batteries

Likewise, a level was also used, since the meter must be as level as possible to avoid measurement errors.

c. *Measurement distribution*

The measurements made in the similar disinfection canal are distributed in the same way as proposed and worked by Pedro Cisterna et al. [38]. Likewise, to determine the radiation present in the environment, four measurements are considered, one for each side. The distribution of the measurements made can be seen graphically in Figure 3. In addition, four measurements were made at the following times: 9.55; 11:30, 12:15; 13.00. and by the instrument PUV-360.

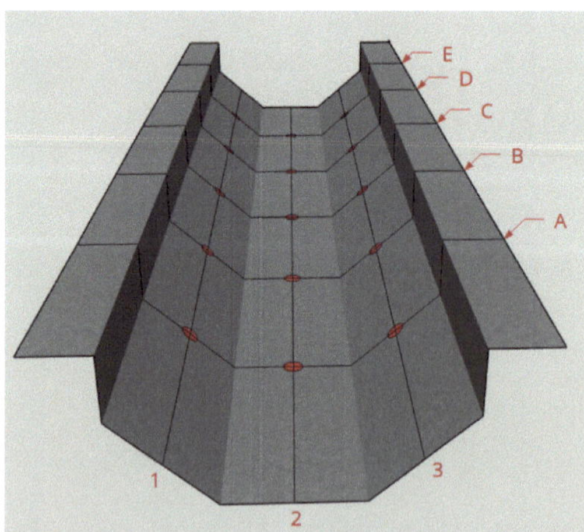

Figure 3. Mesh of measurement points.

d. *Radiation rate as control parameter*

Due to the impossibility of comparing the measurements between them, due to the environmental radiation conditions where the radiation is obtained in the measurement matrix, a dimensionless indicator is proposed that allows this comparison of the measurements of different days in a consistent way. Since there are daily different values of solar radiation that are random and cannot be controlled or reproduced due to the natural variable, a parameter is defined as: Radiation Ratio. This allows comparing the radiation measurements Rij corresponding to each point of the radiation matrix, linking the radiation measured in the matrix in a position, with respect to the Environmental Radiation (Ra) [38].

If Rij = Ra, Rij/Ra, RR = 1, for Rij > Ra, RR > 1 and for Rij < Ra, RR will be less to 1.

Hence: $RR = Rij/Ra, i = A,,,D,E; j = 1,2,3.$

For this case, the environmental solar radiation (Ra) is considered the average of the four measurements made around the canal.

e. *Determination of the Total Coliform*

Measurements of total coliform elimination were also carried out in a disinfectant canal with similar characteristics at different radiation ratios, which are used as references in order to correlate them with those obtained in this investigation.

The treated sewage was deposited in the canal and subjected to disinfection using the technology based on exposure to solar radiation. A sample was collected to measure the total coliforms. This activity was performed in a time range of 225 min, with samples taken each 45 min.

The wastewater treated without disinfection corresponded to time 0. After 225 min of exposure to UV radiation, the canal was emptied and washed. A sample of the residual treated water was taken at the beginning of the experiment using a 250 mL closed glass with a top. This closed glass was submerged between 5 and 10 cm. Once submerged, the glass was opened, allowing the entrance of water. Then, the glass was pulled out of the water and introduced to a cooler. This process was repeated for the following samples and according to the predefined residence times. Finally, the samples were sent to the laboratory for analysis. The samples collected for microbiological analysis were kept at a low temperature in a cooler and analyzed on the same day they were collected. The multiple tube technique was used for the total coliform counts and most probable number (MPN) determination [40].

The treated wastewater was disinfected under two situations. (A) In the disinfection canal and (B) in the disinfection canal with reflective side panels. Two experiments of coliform abatement were conducted considering a residence time range from 0 to 225 min. In both cases, two measurements were made.

The initial coliform condition, TCo, in the wastewater that was subjected to disinfection was measured at different hydraulic residence times, and, with this data, we estimated the efficiency of the coliform removal.

$$\text{Elimination} = (\text{TCt} - \text{TCo})/\text{TCo}$$

Such that,

$$\text{TCt} = \text{MPN to time t}$$

$$\text{TCo} = \text{MPN to time 0.}$$

3. Results and Analysis

The radiation measurements were made according to Figure 2. The positions of the respective UV radiation measurements are indicated, with the letters A, B, C, D, E corresponding to the abscissas and the ordinates with the numbers 1, 2, 3.

These results were obtained in real atmospheric conditions, and due to its variability, the data shown are normalized and compared by using the Radiation Ratio. This was done for each of the proposed configurations since it seeks to establish the inference in solar radiation when aluminum mirrors are used.

Stainless steel shows a low spectral reflectance of around 50–60% in the visible wavelength range, like galvanized steel, less than other material, such as aluminum and copper. Therefore, the results can improve using others materials [41].

a. *Results*

Next, the results obtained with the normalization are shown, applying the radiation ratio, already carried out.

The values obtained in condition 1 (Table 2), that is, with canal and galvanized steel side panels, showed that 28 data out of 60, which represent 47% of the total sample, presented values greater than or equal to 1; that is, values of radiation equal to or greater than the environmental radiation (Ra).

Table 2. Radiation ratios obtained in canal and galvanized steel side panels.

	Data	A	B	C	D	E	Time
Measurement 1	1 2 3	1.7 0.98 1.01	1.02 1.00 1.03	1.07 0.93 1.00	1.06 1.01 1.03	1.10 1.01 1.05	11:30
Measurement 2	1 2 3	0.99 0.99 1.05	0.99 0.98 1.04	0.97 1.00 1.00	0.99 1.00 0.96	0.97 0.98 1.01	12:30
Measurement 3	1 2 3	0.95 0.97 1.02	0.94 0.93 1.02	0.95 0.95 1.00	0.94 0.95 1.01	0.92 0.93 1.02	13:30
Measurement 4	1 2 3	0.92 0.92 1.05	0.93 0.93 1.01	0.96 0.93 1.02	0.93 0.95 1.05	0.95 0.97 1.05	14:30

For condition 2 (Table 3), it was obtained that 48 of 60 data are greater than or equal to the environmental radiation (Ra). This increase from 47% to 80% of radiation rate values greater than or equal to 1 can be attributed to the presence of aluminum mirrors on the slopes of the canal, despite the fact that the side panels are made of galvanized steel.

Table 3. Radiation ratios with galvanized steel side panels and installation of aluminum mirrors on inclined canal walls.

	Data	A	B	C	D	E	Time
Measurement 1	1 2 3	0.98 1.02 1.05	1.00 1.01 1.03	0.94 0.97 1.00	1.00 1.01 1.00	0.98 1.01 0.98	11:30
Measurement 2	1 2 3	1.05 1.10 1.12	1.05 1.08 1.07	1.05 1.03 1.02	1.05 1.08 1.13	1.05 1.08 1.06	12:30
Measurement 3	1 2 3	1.01 1.02 1.04	1.02 1.01 1.05	1.01 1.01 0.97	1.01 1.04 1.00	1.01 1.00 0.98	13:30
Measurement 4	1 2 3	0.98 1.00 1.04	0.98 1.03 1.08	0.98 1.03 1.01	0.97 1.02 1.07	0.97 1.03 1.15	14:30

Regarding the radiation measured in the canal, when aluminum mirrors are available on the side panels and the canal retains its materiality of galvanized steel corresponding condition 3, (Table 4), 29 of 60 radiation rate values greater than or equal to the environmental radiation are obtained; in other words, 48% of the total sample. This amount is similar to condition 1 and much lower than the values obtained in condition 2.

Finally, the condition 4, regarding the condition in which aluminum mirrors are present on the inclined walls canal and on the side panels (Table 5), the highest percentage of radiation rate is observed, which exceeds the value measured in the environment. Measurements above environment radiation represent 82% of the sample, obtaining 49 of 60 values that fulfill this condition.

Table 4. Radiation ratios with galvanized steel canal and side panels with aluminum mirror installation.

	Data	A	B	C	D	E	Time
Measurement 1	1	0.97	0.97	0.97	0.97	0.97	11:30
	2	1.00	1.00	1.01	1.01	1.00	
	3	1.00	0.99	0.99	1.05	1.02	
Measurement 2	1	0.99	0.99	0.99	0.99	0.99	12:30
	2	0.99	0.99	0.99	0.99	0.98	
	3	1.12	1.13	1.02	1.12	1.15	
Measurement 3	1	0.98	0.98	0.98	0.98	0.98	13:30
	2	1.00	1.00	1.00	1.00	1.00	
	3	1.09	1.11	1.15	1.20	1.19	
Measurement 4	1	0.98	0.96	0.98	0.96	0.97	14:30
	2	1.02	0.99	0.99	0.99	0.98	
	3	1.15	1.10	1.11	1.13	1.16	

Table 5. Radiation ratios with galvanized steel canal and side panels with aluminum mirror installation.

	Data	A	B	C	D	E	Time
Measurement 1	1	0.99	1.02	1	1.05	1.09	11:30
	2	1.01	1.04	1.02	1.04	1.03	
	3	1.26	1.26	1.22	1.24	1.25	
Measurement 2	1	0.97	0.98	0.96	1	1.02	12:30
	2	1	1.02	1.01	1.03	1.03	
	3	1.1	1.11	1.08	1.17	1.25	
Measurement 3	1	0.97	0.99	0.96	1.01	1.04	13:30
	2	1.01	1.03	1.01	1.04	1.05	
	3	1.1	1.11	1.08	1.17	1.24	
Measurement 4	1	0.94	0.97	0.96	0.99	1	14:30
	2	1.02	1.03	1	1.02	1.01	
	3	1.2	1.23	1.22	1.18	1.12	

From the above and Figure 4, a hierarchical correlation can be established, in terms of the raw values of the radiation rates. In comparison the value that the environmental radiation presented during the measurements, from the lowest value to the highest, would be: condition 1, condition 3, condition 2 and finally, condition 4. Respect to the data variability, it can be seen that some cases are more variable than others; however, this is analyzed later.

b. *Analysis and results comparison*

According to the measurements made, the results were processed and analyzed statistically, this is shown in Figure 5 (average of each point of the canal, for this calculation the four measurements made for each point were considered).

From the analysis of the results for condition 1, it is revealed that this is the sample with the lowest average values of radiation ratio, obtaining a maximum value of 1.10 and a minimum of 0.92. Likewise with all 60 data obtained, a global average of 0.99 is calculated for this condition. Although this value may be reliable, a confidence interval was also estimated for the global average (95% confidence). It has a lower limit of 0.98 and greater than 1.00, thus having an error of ±0.01. Figure 6 shows that this global average is the lowest of all conditions. From this it could be inferred that, by adding aluminum mirrors to the side walls of the canal and/or to the side panels, the system's ability to concentrate solar radiation is improved.

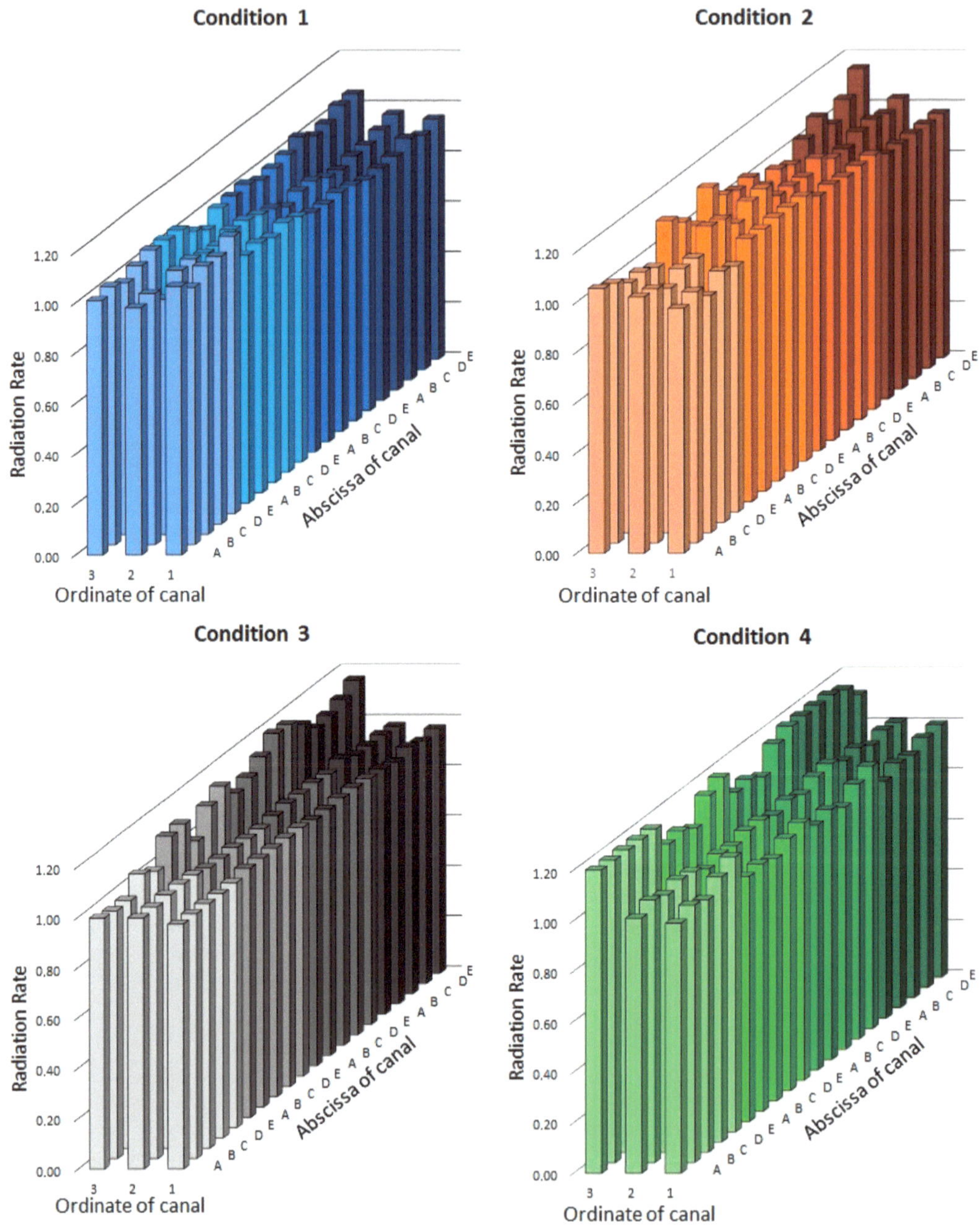

Figure 4. Graphic representation of estimated radiation ratios.

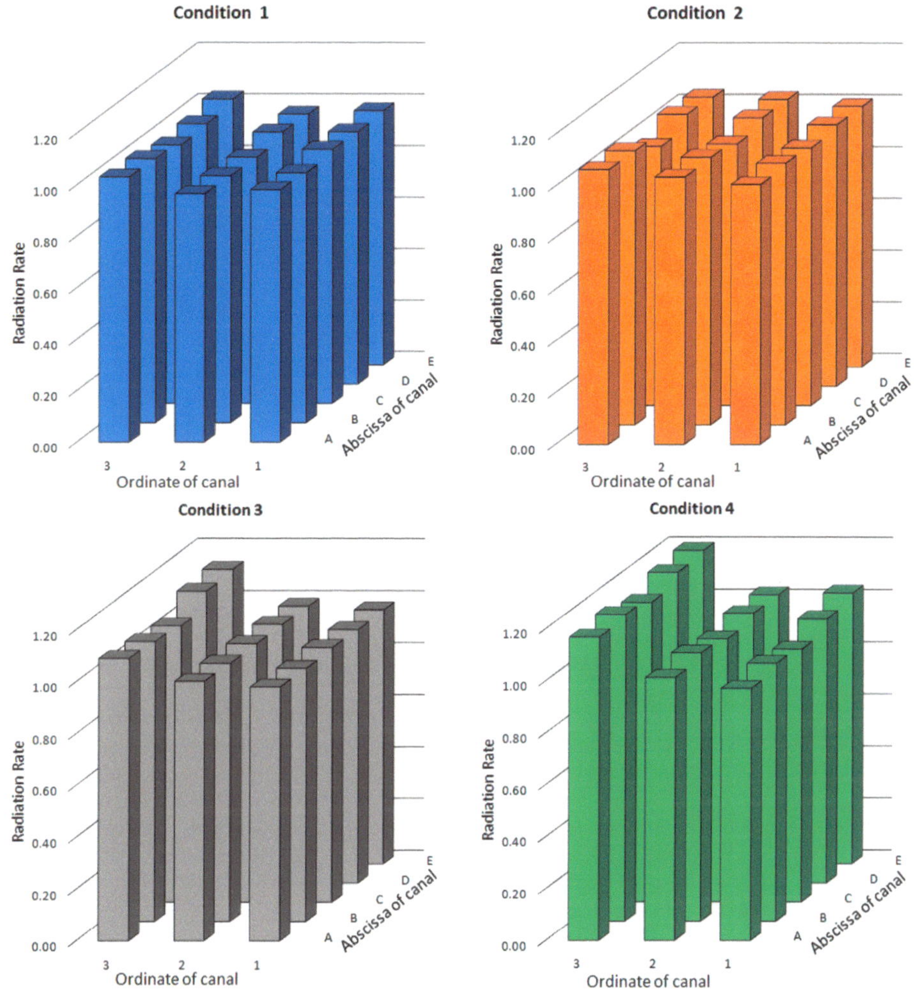

Figure 5. Graphic representation of the means of the Radiation Ratios indices obtained, for each condition.

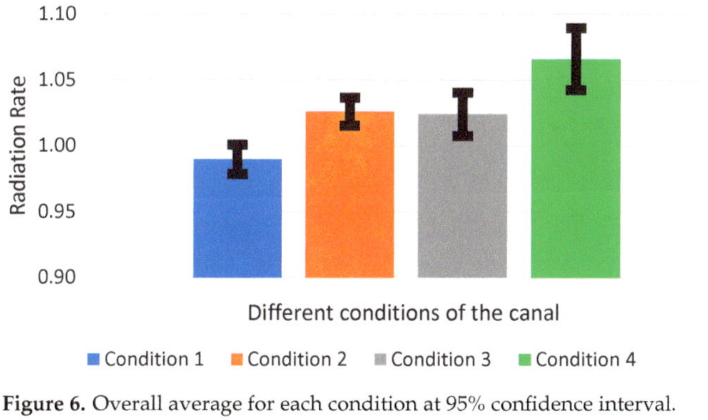

Figure 6. Overall average for each condition at 95% confidence interval.

Regarding conditions 2 and 3, related to the global average of the radiation ratio, these are similar, showing values of 1.03 and 1.02, respectively. Likewise, it can be seen in Figure 4, that the global average of radiation ratio presents greater variability for condition 3, with an error of ±0.02, while for condition 2, it is ±0.01 at a confidence of 95%. In addition, for condition 2, the data obtained show a maximum average radiation index of 1.15 and a minimum of 0.94. On the other hand, the radiation index values for condition 3 reach a maximum of 1.20 and a minimum of 0.96. These data are extracted directly from the canal-averaged data and are plotted in Figure 3.

This similarity in the results obtained for conditions 2 and 3 could be due to the presence of aluminum mirrors, but only in one of the components of the system.

According to condition 4, which presents aluminum mirrors both on the lateral walls of the canal and on the side panels, the highest radiation index results are obtained with a global average radiation index of 1.07 and a variability of ± 0.02. Relative to the mean values observed in Figure 3, a maximum of 1.26 and a minimum of 0.94 were determined.

Regarding the overall averages, it clearly shows in Figure 6 that the results obtained in condition 1 are the most unfavorable, being the one with the lowest general average radiation ratio. Subsequently, conditions 2 and 3 are the most similar, with only a difference of 1%, condition 2 being superior; however, the presence of aluminum mirrors in one part of the system, the canal or the side panels, allows obtaining better results than when the material is only galvanized steel. Finally, when conditions 2 and 3 overlap, condition 4 is obtained, which consists of the presence of aluminum mirrors on the side walls and on the side panels of the canal. It is the condition that has been the one of the best results, as to the general performance of the radiation ratio, being 0.08 higher than the base condition, that is, condition 1.

The variability of the data presented for each condition is different. According to Figures 4 and 6, it can be established that the data with the greatest variability are those presented in condition 4, and this decreases until reaching condition 1, which is the most uniform. This is shown by having the smallest confidence interval and thus the smallest standard deviation and variance.

For this reason, an analysis was also carried out regarding how the radiations are distributed throughout the complete canal, thus obtaining Figure 7.

Concerning the cross section, the zone that presents the highest values of radiation ratio is zone 3, and from there it decreases to zone 1. This is the case for all conditions except for 1, which has its lowest value in zone 2. Figure 7 shows that the radiation, with respect to the cross section, is highly variable, presenting grid differences between zone 3 and 1; however, condition 2 is the one that achieves greater uniformity in the results of the radiation ratio, which may be mainly due to the presence of aluminum mirrors, but only on the canal walls. This large difference in the values of the radiation ratio may be due to the plan orientation of the system, which causes the radiation to be concentrated in a single zone of the canal.

This can be a problem in terms of coliform removal as the process is not guaranteed to be homogeneous.

For the longitudinal section, it can be mentioned that the distribution of the radiation ratios is uniform, different to conditions 2 and 4, which present a great variability; however, this does not mean that it affects the performance of the system, since the water crosses the complete canal, thus achieving the disinfection of the fluid.

In general, the distribution of radiation follows the same behavior regarding the general average, since the values observed are greater for condition 4, except in the cross section, and being so variable there are values that are less than condition 2. Therefore the latter is more uniform. Conditions 2 and 3 are an intermediate condition between conditions 1 and 4; however, they differ more in terms of cross section since condition 2 is more uniform than condition 3. Finally, condition 1 is the one with the lowest values, so it serves as a standard situation for the other cases for which a comparison of the partial averages was made and can be seen in Figure 8.

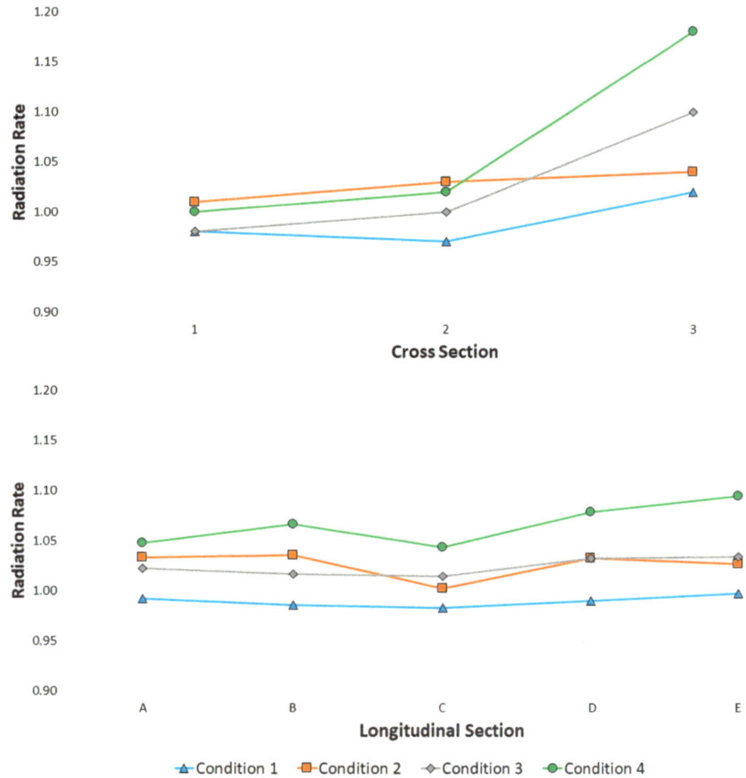

Figure 7. Plot of average radiation ratio distributions.

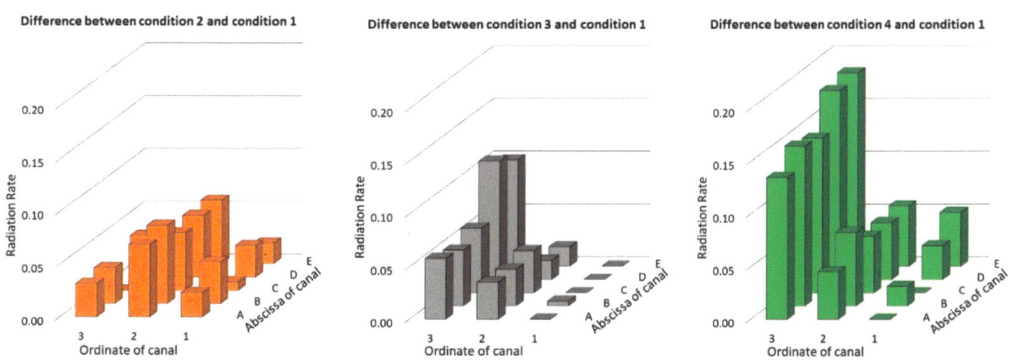

Figure 8. Difference of the partial means between conditions 2, 3 and 4, with respect to the standard condition 1.

Figure 8 reaffirms, in accord to the magnitude, distribution, and variability, that condition 4 is the highest value, although they are also the most variable and their distribution concentrates the highest radiation in zone 1. Concerning to conditions 2 and 3, these present similar differences with respect to the magnitudes of radiation, being lower than condition 4 but greater than condition 1; however, the difference between these last two cases is the distribution of radiation, being that condition 2 is the one that presents more uniform radiations with respect to the number, which could be a difference at the moment of the elimination of coliforms.

c. *Incidence of the radiation ratio on the elimination of coliforms.*

According to the prototype proposed in the study "First Results: Innovative Solar Disinfection Technology for Treated Wastewater that Integrates Materiality, Geometry, and Reflective Panels" [38], a direct relationship can be established between the total elimination of coliforms and the radiation ratio. To find this relationship, the radiation and coliform elimination data presented in the aforementioned study were used and a graph was made in which the cases of the present study were extrapolated as related to the efficiency of the prototype to remove coliforms. This relationship of total elimination of coliforms against radiation ratio was worked with three hydraulic residence times: 135, 180 and 225 min. The graph made is represented in Figure 8. It is worth mentioning that the extrapolated points are the ones with white filling; instead, the real data are A and B. Also, since point B and condition 1 have the same radiation ratio, these have the same efficiency and therefore overlap.

When comparing the results shown in Figure 9 with similar research, their consistency is verified. The maximum removal of total coliform was found to be 92.95% at 10 cm depth for sunny conditions, at 110 NTU, white container, 240 min [42].

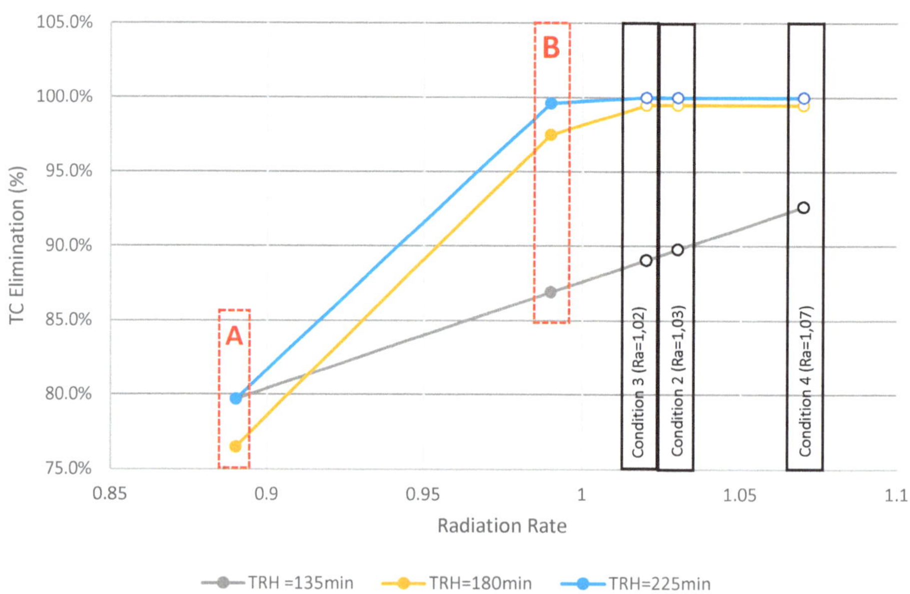

Figure 9. Theoretical relationship between the radiation ratio and the total removal of coliforms.

Figure 9 shows that total removal of coliform increases as the average radiation ratio increases for each hydraulic residence time. This considering that the same climatic and radiation conditions are developed for all cases. This could mean that, to the extent that aluminum mirrors are included in the materiality of the prototype, it could increase its total removal of coliforms. In other research, the high reflection coatings significantly improve the surface reflectivity in a specific wavelength range. A coating can be designed as a long or short pass filter, or an aluminum mirror with a specific reflectivity [43]. The increase in efficiency is reflected for different hydraulic residence times so this increase is applied to the different cases that may be found.

In sewage treatment experiences with a treatment system consisting of a constructed wetland and a disinfectant canal with flaps-panels, an excellent result of MPN/100 mL = 23 was reached [38]. Previous work of the constructed wetland is substantive for this achievement. The higher removal of fecal and pathogenic bacteria in constructed wetlands depends on certain factors, such as: the hydraulic regime, wastewater characteristics, and even the local climate [44]. The constructed wetlands with hydraulic improvement can also be proposed as a sustainable solution for pathogen removal [45].

Noteworthy, photochemical processes for the treatment of water and wastewater have been receiving increased attention for being sustainable treatments. Even different oxidative agents such as peroxymonosulfate, persulfate, and peracetic acid, have been recently investigated to enhance the action of the solar photons for water disinfection, aiming for their future implementation at full scale [34]; thus validating this alternative technology developed in Chile, since it is very easy to implement at different scales due to the continuous flow condition that characterizes it.

4. Conclusions

When only the aluminum mirrors are applied on the sloping walls of canal (condition 2) or on the side panels (condition 3), a similar increase in radiation (3 and 4% increase, respectively) is evident in the treatment zone. However, the best performance of the system, with the 8% increase in radiation compared to the base case, is achieved when using aluminum mirrors simultaneously on the sloped canal walls and on the side panels (condition 4).

It is concluded that the use of aluminum mirrors increases UV solar radiation in the area of influence of the canal, thus obtaining better results with respect to galvanized steel, and consequently increasing the elimination of pathogens.

The results obtained with the use of aluminum mirrors in the disinfectant channel, confirm the validity of this new technological solution to replace chlorine treatment as a disinfectant for treated wastewater.

This technology can be improved and, with further research directed towards the geometry, materiality, and fins-panels, it will be possible to reduce residence times, equipment size, and associated investment costs.

It should be noted that this equipment is technologically very simple, with a low investment and operating cost. In addition, as a goal, it is intended to obtain a competitive disinfectant equipment that is attractive to communities of people who need to reuse treated wastewater in the context of water and economic crisis through research and technological optimization.

Author Contributions: Conceptualization, P.C.-O.; Methodology, P.C.-O., M.G.-G., M.M.-C. and S.Q.-V.; Formal analysis, P.C.-O., M.G.-G., M.M.-C. and S.Q.-V.; Investigation, P.C.-O., M.G.-G. and M.M.-C.; Resources, M.G.-G. and S.Q.-V.; Data curation, P.C.-O.; Writing—original draft, P.C.-O. and M.M.-C.; Writing—review & editing, P.C.-O.; Visualization, P.C.-O. and S.Q.-V.; Supervision, P.C.-O., M.G.-G. and M.M.-C.; Project administration, M.G.-G. All authors have read and agreed to the published version of the manuscript.

Funding: This research received no external funding.

Data Availability Statement: Not applicable.

Conflicts of Interest: The authors declare no conflict of interest.

References

1. Tol, R.S.J. The Economic Effects of Climate Change. *J. Econ. Perspect.* **2009**, *23*, 29–51. [CrossRef]
2. Abbass, K.; Qasim, M.Z.; Song, H.; Murshed, M.; Mahmood, H.; Younis, I. A Review of the Global Climate Change Impacts, Adaptation, and Sustainable Mitigation Measures. *Environ. Sci. Pollut. Res.* **2022**, *29*, 42539–42559. [CrossRef] [PubMed]
3. Oficina de Estudios y Políticas Agrarias. *Agricultura Chilena: Reflexiones y Desafíos al 2030*; ODEPA: Santiago, Chile, 2017; pp. 147–178. ISBN 9789567244300.
4. Valdés-Pineda, R.; Pizarro, R.; García-Chevesich, P.; Valdés, J.B.; Olivares, C.; Vera, M.; Balocchi, F.; Pérez, F.; Vallejos, C.; Fuentes, R.; et al. Water Governance in Chile: Availability, Management and Climate Change. *J. Hydrol.* **2014**, *519*, 2538–2567. [CrossRef]

5. Verichev, K.; Zamorano, M.; Carpio, M. Effects of Climate Change on Variations in Climatic Zones and Heating Energy Consumption of Residential Buildings in the Southern Chile. *Energy Build.* **2020**, *215*, 109874. [CrossRef]
6. Akpan, V.E.; Omole, D.O.; Bassey, D.E. Assessing the Public Perceptions of Treated Wastewater Reuse: Opportunities and Implications for Urban Communities in Developing Countries. *Heliyon* **2020**, *6*, e05246. [CrossRef]
7. Li, P.; Wang, D.; Li, W.; Liu, L. Sustainable Water Resources Development and Management in Large River Basins: An Introduction. *Environ. Earth Sci.* **2022**, *81*, 179. [CrossRef]
8. Livia, S.; María, M.S.; Marco, B.; Marco, R. Assessment of Wastewater Reuse Potential for Irrigation in Rural Semi-Arid Areas: The Case Study of Punitaqui, Chile. *Clean Technol. Environ. Policy* **2020**, *22*, 1325–1338. [CrossRef]
9. Torres, P. Tendencias En El Tratamiento de Aguas Residuales Domésticas. *Ingeniería y Competitividad* **2001**, *3*, 35–42. [CrossRef]
10. Rashidi, J.; Rhee, G.H.; Kim, M.; Nam, K.J.; Heo, S.; Yoo, C.K.; Karbassi, A. Life Cycle and Economic Assessments of Key Emerging Energy Efficient Wastewater Treatment Processes for Climate Change Adaptation. *Int. J. Environ. Res.* **2018**, *12*, 815–827. [CrossRef]
11. Crini, G.; Lichtfouse, E. Advantages and Disadvantages of Techniques Used for Wastewater Treatment. *Environ. Chem. Lett.* **2019**, *17*, 145–155. [CrossRef]
12. Patel, A.; Arkatkar, A.; Singh, S.; Rabbani, A.; Solorza Medina, J.D.; Ong, E.S.; Habashy, M.M.; Jadhav, D.A.; Rene, E.R.; Mungray, A.A.; et al. Physico-Chemical and Biological Treatment Strategies for Converting Municipal Wastewater and Its Residue to Resources. *Chemosphere* **2021**, *282*, 130881. [CrossRef] [PubMed]
13. Some, S.; Mondal, R.; Mitra, D.; Jain, D.; Verma, D.; Das, S. Microbial Pollution of Water with Special Reference to Coliform Bacteria and Their Nexus with Environment. *Energy Nexus* **2021**, *1*, 100008. [CrossRef]
14. Macauley, J.J.; Qiang, Z.; Adams, C.D.; Surampalli, R.; Mormile, M.R. Disinfection of Swine Wastewater Using Chlorine, Ultraviolet Light and Ozone. *Water Res.* **2006**, *40*, 2017–2026. [CrossRef] [PubMed]
15. Hrudey, S.E. Chlorination Disinfection By-Products, Public Health Risk Tradeoffs and Me. *Water Res.* **2009**, *43*, 2057–2092. [CrossRef]
16. Oparaku, N.F.; Mgbenka, B.O.; Ibeto, C.N. Waste Water Disinfection Utilizing Ultraviolet Light. *J. Environ. Sci. Technol.* **2011**, *4*, 73–78. [CrossRef]
17. Cisterna-Osorio, P.; Calabran-Caceres, C.; Tiznado-Bustamante, G.; Bastias-Toro, N. Influent with Particulate Substrate, Clean, Innocuous and Sustainable Solution for Bulking Control and Mitigation in Activated Sludge Process. *Water* **2021**, *13*, 984. [CrossRef]
18. Al-Abri, M.; Al-Ghafri, B.; Bora, T.; Dobretsov, S.; Dutta, J.; Castelletto, S.; Rosa, L.; Boretti, A. Chlorination Disadvantages and Alternative Routes for Biofouling Control in Reverse Osmosis Desalination. *NPJ Clean Water* **2019**, *2*, 2. [CrossRef]
19. Hartman, P.S.; Eisenstark, A. Synergistic Killing of Escherichia Coli by Near-UV Radiation and Hydrogen Peroxide: Distinction between RecA-Repairable and RecA-Nonrepairable Damage. *J. Bacteriol.* **1978**, *133*, 769–774. [CrossRef]
20. National Institute of Industrial Property, INAPI, Chile. 2019. Available online: https://www.inapi.cl/ (accessed on 23 October 2022).
21. World Health Organization. *Results of Round II of the WHO International Scheme to Evaluate Household Water Treatment Technologies*; World Health Organization: Geneva, Switzerland, 2019.
22. Chaúque, B.J.M.; Brandão, F.G.; Rott, M.B. Development of Solar Water Disinfection Systems for Large-Scale Public Supply, State of the Art, Improvements and Paths to the Future—A Systematic Review. *J. Environ. Chem. Eng.* **2022**, *10*, 107887. [CrossRef]
23. García-Gil, Á.; García-Muñoz, R.A.; McGuigan, K.G.; Marugán, J. Solar Water Disinfection to Produce Safe Drinking Water. A Review of Parameters, Enhancements, and Modelling Approaches to Make SODIS Faster and Safer. *Molecules* **2021**, *26*, 3431. [CrossRef]
24. Byrne, J.A.; Fernandez-Ibañez, P.A.; Dunlop, P.S.M.; Alrousan, D.M.A.; Hamilton, J.W.J. Photocatalytic Enhancement for Solar Disinfection of Water: A Review. *Int. J. Photoenergy* **2011**, *2011*, 798051. [CrossRef]
25. Chaúque, B.J.M.; Rott, M.B. Solar Disinfection (SODIS) Technologies as Alternative for Large-Scale Public Drinking Water Supply: Advances and Challenges. *Chemosphere* **2021**, *281*, 130754. [CrossRef] [PubMed]
26. Gómez-Couso, H.; Fontán-Sainz, M.; Navntoft, C.; Fernández-Ibáñez, P.; Ares-Mazás, E. Comparison of Different Solar Reactors for Household Disinfection of Drinking Water in Developing Countries: Evaluation of Their Efficacy in Relation to the Waterborne Enteropathogen Cryptosporidium Parvum. *Trans. R. Soc. Trop. Med. Hyg.* **2012**, *106*, 645–652. [CrossRef]
27. Asiimwe, J.K.; Quilty, B.; Muyanja, C.K.; McGuigan, K.G. Field Comparison of Solar Water Disinfection (SODIS) Efficacy between Glass and Polyethylene Terephalate (PET) Plastic Bottles under Sub-Saharan Weather Conditions. *J. Water Health* **2013**, *11*, 729–737. [CrossRef]
28. Loo, S.L.; Fane, A.G.; Krantz, W.B.; Lim, T.T. Emergency Water Supply: A Review of Potential Technologies and Selection Criteria. *Water Res.* **2012**, *46*, 3125–3151. [CrossRef] [PubMed]
29. Acosta Castellanos, P.M.; Caro Camargo, C.A.; Perico Granados, N.R. Análisis de Interferencia de Parámetros Físicos Del Agua, En Desinfección Por Radiación UV. *Rev. Technol.* **2015**, *14*, 2. [CrossRef]
30. Verhoeven, J.T.A.; Meuleman, A.F.M. Wetlands for Wastewater Treatment: Opportunities and Limitations. *Ecol. Eng.* **1999**, *12*, 5–12. [CrossRef]
31. Vymazal, J. Is Removal of Organics and Suspended Solids in Horizontal Sub-Surface Flow Constructed Wetlands Sustainable for Twenty and More Years? *Chem. Eng. J.* **2019**, *378*. [CrossRef]

32. Vymazal, J. Removal of Enteric Bacteria in Constructed Treatment Wetlands with Emergent Macrophytes: A Review. *J. Environ. Sci. Health* **2005**, *40*, 1355–1367. [CrossRef]
33. Alufasi, R.; Gere, J.; Chakauya, E.; Lebea, P.; Parawira, W.; Chingwaru, W. Mechanisms of Pathogen Removal by Macrophytes in Constructed Wetlands. *Environ. Technol. Rev.* **2017**, *6*, 135–144. [CrossRef]
34. Berruti, I.; Nahim-Granados, S.; Abeledo-Lameiro, M.J.; Oller, I.; Polo-López, M.I. Recent Advances in Solar Photochemical Processes for Water and Wastewater Disinfection. *Chem. Eng. J. Adv.* **2022**, *10*, 100248. [CrossRef]
35. Hunter, R.G.; Day, J.W.; Wiegman, A.R.; Lane, R.R. Municipal Wastewater Treatment Costs with an Emphasis on Assimilation Wetlands in the Louisiana Coastal Zone. *Ecol. Eng.* **2019**, *137*, 21–25. [CrossRef]
36. Wu, S.; Carvalho, P.N.; Müller, J.A.; Manoj, V.R.; Dong, R. Sanitation in Constructed Wetlands: A Review on the Removal of Human Pathogens and Fecal Indicators. *Sci. Total Environ.* **2016**, *541*, 8–22. [CrossRef]
37. Quiñónez-Díaz, M.D.J.; Karpiscak, M.M.; Ellman, E.D.; Gerba, C.P. Removal of Pathogenic and Indicator Microorganisms by a Constructed Wetland Receiving Untreated Domestic Wastewater. *J. Environ. Sci. Health A Tox Hazard Subst. Environ. Eng.* **2001**, *36*, 1311–1320. [CrossRef] [PubMed]
38. Cisterna-Osorio, P.; Quijada-Vera, S.; Ruiz-Duran, D.; Peirano-Cuevas, R.; Ortiz-Briones, P. First Results: Innovative Solar Disinfection Technology for Treated Wastewater That Integrates Materiality, Geometry, and Reflective Panels. *Int. J. Environ. Res. Public Health* **2020**, *17*, 6523. [CrossRef]
39. Echazú, R.; Cadena, C.; Saravia, L. Estudio de Materiales Reflectivos Para Concentradores Solares. *Av. En Energías Renov. Y Medio Ambiente (AVERMA)* **2000**, *4*, 11–16.
40. Instituto Nacional de Normalización. *Norma Chilena Oficial NCh 2313/22. Of. 95*; Aguas Residuales—Métodos de análisis—Parte 22: Determinación de coliformes fecales en medio EC; Instituto Nacional de Normalización: Santiago, Chile, 1995.
41. Grätzel, M. Dye-Sensitized Solar Cells. *J. Photochem. Photobiol. C Photochem. Rev.* **2003**, *4*, 145–153. [CrossRef]
42. Abbassi, B.; Dababneh, B.; Shquirat, W. Coliform-specific solar disinfection of treated wastewater. *Pol. J.Environ. Stud.* **2012**, *21*, 1577–1581.
43. Jeong, S.-H.; Kim, J.-K.; Kim, B.-S.; Shim, S.-H.; Lee, B.-T. Characterization of SiO2 and TiO2 Films Prepared Using Rf Magnetron Sputtering and Their Application to Anti-Reflection Coating. *Vacuum* **2004**, *76*, 507–515. [CrossRef]
44. Alexandros, S.I.; Akratos, C.S. Removal of Pathogenic Bacteria in Constructed Wetlands: Mechanisms and Efficiency. In *Phytoremediation: Management of Environmental Contaminants*; Springer International Publishing: Berlin/Heidelberg, Germany, 2016; Volume 4, pp. 327–346, ISBN 9783319418117.
45. Cisterna-Osorio, P.; Lazcano-Castro, V.; Silva-Vasquez, G.; Llanos-Baeza, M.; Fuentes-Ortega, I. Innovative effluent capture and evacuation device that increases COD removal efficiency in subsurface flow wetlands. *Processes* **2019**, *7*, 418. [CrossRef]

processes

Article

Development of a Continuous Hydrothermal Treatment Process for Efficient Dewatering of Industrial Wastewater Sludge

Douwe S. Zijlstra [1], Esther Cobussen-Pool [1], Dennis J. Slort [1], Mark Visser [1], Pavlina Nanou [2], Jan R. Pels [2] and Heather E. Wray [1,*]

[1] TNO, Energy Transition, Biobased and Circular Technologies, Westerduinweg 3, 1755 LE Petten, The Netherlands
[2] TORWASH BV, Welgelegen 1, 1754 JK Burgerbrug, The Netherlands
* Correspondence: heather.wray@tno.nl

Abstract: Sludges from the papermaking industry represent a challenging residue stream that is difficult to dewater using conventional processes. The successful development and scale-up of innovative processes from lab- to pilot- to industrial-scale are required to tackle challenges for waste treatment, including paper sludges. Biological paper sludge was treated via a mild hydrothermal carbonization process (TORWASH®) to improve dewaterability of the sludge, including long-duration, continuous testing. Initial lab-scale experiments indicated the optimal treatment temperature for sludge dewatering was 190 °C. Dewaterability improved with increasing temperature, but the obtained solid yield decreased. Scaling-up to a continuous flow pilot plant required a temperature of 200 °C to achieve optimum dewatering. Pilot-scale hydrothermal treatment and dewatering resulted in solid cakes with an average dry matter content of 38% and a solid yield of 39%. This study demonstrates the benefits of hydrothermal carbonization for the dewatering of biological paper sludge without the use of dewatering aids such as fiber sludge or polyelectrolytes. The results also demonstrate the successful adaptation of a lab-scale batch process to a pilot-scale continuous flow process for hydrothermal carbonization of industrial wastewater sludge.

Keywords: hydrothermal carbonization; industrial wastewater; biological paper sludge; dewatering; TORWASH

check for updates

Citation: Zijlstra, D.S.; Cobussen-Pool, E.; Slort, D.J.; Visser, M.; Nanou, P.; Pels, J.R.; Wray, H.E. Development of a Continuous Hydrothermal Treatment Process for Efficient Dewatering of Industrial Wastewater Sludge. *Processes* **2022**, *10*, 2702. https://doi.org/10.3390/pr10122702

Academic Editors: Dimitris Zagklis and Georgios Bampos

Received: 25 October 2022
Accepted: 11 December 2022
Published: 14 December 2022

1. Introduction

The paper industry is one of the largest global industries with an estimated production of >400 Mton of board and paper produced per year. During this production, approximately 400 Mton of wet sludges are generated annually as a residue stream of wastewater treatment at paper mills [1,2]. Of these, the most problematic for the paper mill is biological (bio-) sludge, so named because it is a product of biological wastewater treatment (typically aeration followed by clarification). The composition of biological paper sludge is complex and can contain high concentrations of harmful substances, including heavy metals, inorganics and pathogenic microorganisms [3,4]. Typically, these sludges are disposed of via landfilling or incineration. Landfilling of bio-sludge has been banned in many countries due to associated negative environmental impacts. Incineration is still a common disposal pathway but is difficult and expensive due to the high concentration of inorganic particles in the paper sludges combined with their poor dewaterability [5–7]. Dewatering, the separation of solid and liquid components, of biological paper sludge in order to improve the calorific value at incineration, can be so challenging that it can be responsible for over half of the total wastewater treatment costs at a paper mill [8], often requiring the use of costly dewatering aids such as coagulants or polyelectrolytes. A widely applied method to improve the dewaterability of the bio-sludge is the addition of fiber sludge, a by-product of paper making. Fiber sludge is composed of short cellulose fibers and is obtained from primary clarification in the wastewater treatment process. The addition of fiber sludge to bio-sludge to form a mixed sludge enhances the overall dewaterability via mechanical dewatering, which is highly preferred over thermal dewatering [8,9]. However, the use of fiber sludge as a dewatering aid is not ideal as it is rich in cellulose fibers and organics and therefore has the potential for reuse as a raw material in the papermaking process [5]. Therefore, alternative methods to improve the dewaterability of bio-sludge should be explored.

A variety of thermochemical processes have been developed for the upgrading of biomass. The majority of these processes use dry biomass as a feedstock, requiring an energy- and cost-intensive upstream drying step [10]. For hydrothermal carbonization this is not required, as water is the applied reaction medium. Therefore, it is an excellent upgrading process for wet biogenic feedstocks, including sludges [6]. Hydrothermal carbonization (HTC) is typically applied at temperatures in the range of 180–250 °C under autogenous pressure (1–5 MPa). Historically, the higher temperatures in this temperature range in combination with long residence times were targeted, but the focus has been shifting towards the lower temperatures of this temperature range in combination with shorter residence times [11]. At TNO, the TORWASH® technology has been developed as a reactor concept in which a mild HTC process, with relatively low temperatures (150–210 °C) and an adjustable short residence time, is applied for the upgrading of wet low-grade biomass feedstocks towards high-grade solid fuel of a quality suitable for co-firing [12,13].

At the applied conditions, a wide range of chemical reactions take place, including hydrolysis, dehydration, decarboxylation, recondensation and aromatization. These reactions result in lower O/C and H/C ratios for the obtained product compared to the original biomass, with values similar to conventional solid fuels [14,15]. During this carbonization process, organic acids (acetic acid, lactic acid, formic acid, levulinic acid) are generated, which decreases the pH of the reaction media. The decrease in pH further enhances the hydrolysis and decomposition of the polymers, making HTC an autocatalytic process [16,17]. As such, a drop in pH of the product slurry is a good indication for an efficient carbonization reaction, and the resulting decrease in the O/C ratio can therefore be linked to physicochemical properties such as enhanced dewaterability and higher heating value (HHV) [18].

The main product of HTC is a solid fraction (hydrochar), with a carbon-rich aqueous solution and a small amount of gas (70–90% of which is CO_2) as side products. The distribution between these components is influenced by the reaction conditions and the feedstock [17,19]. The most influential reaction parameter is the applied temperature. Moderately low temperatures are applied in HTC, as higher temperatures (>250 °C) favor the formation of liquids and gases. With increasing temperature, the atomic ratios of H/C and O/C gradually decrease [20,21], but this comes at the expense of the hydrochar yield [10,22]. Another important parameter is the reaction time. Longer reaction times result in a higher hydrochar yield with better properties such as enhanced porosity, pore volume and surface area and a higher calorific value due to a decrease in the O/C ratio [23,24].

The generated hydrochar has many improved properties compared to the initial biomass, including higher mass and energy density, and improved grindability, hydrophobicity and stability to microbial degradation, and better combustion performance as a solid. The increased hydrophobicity results in better dewaterability of the hydrochar and removal of part of the inorganics, which significantly improves the combustion performance [19,23,25]. The hydrochar has many applications, including as a solid bio-fuel, adsorbent, catalyst, soil improver and carbon sequestration medium [19,21].

Hydrothermal carbonization of paper sludge has been widely reported in recent years [26–31]. Studies on the effect of temperature on different paper sludges show similar trends. With increasing temperature, a decrease in oxygen content relative to the carbon content was observed, resulting in a higher heating value (HHV) for the obtained hydrochar. Additionally, a clear decrease in solid yield was observed for both a higher reaction temperature and increased residence time [26–28].

One of the current gaps in hydrothermal carbonization research is the lack of reported data for pilot- and full-scale processes. Most of the existing research focuses on bench-scale operation and model simulations [11,32,33]. Recent work by Zaccariello et al. [34] demonstrated that the reaction conditions are scalable from a 3 L to a 100 L batch reactor but that the thermal efficiency decreases for the larger reactor. Moving from batch to (semi-)continuous reactors for hydrothermal carbonization is also highly relevant from an industrial point of view. The use of liter-scale semi-continuous reactors has been reported in a few studies and the results tend to be in agreement with small batch experiments [35,36]. Higher degrees of carbonization are reported for the hydrochar obtained by semi-continuous operation, which also lead to a higher hydrophobicity and thermal stability [35]. These results are promising, but additional research is needed on continuous HTC at a larger/industrially relevant (multi-kilogram to ton) scale for longer (days/weeks/months) time periods.

The aim of this project is the scale-up of a mild continuous hydrothermal carbonization of paper sludge from batch lab-scale to a continuous pilot-scale (>25 kg/h) in an industrial setting for a long duration (>400 h). A special focus is given to raw bio-sludge, as this is the problematic waste stream of the paper industry. Improving dewatering of this stream following hydrothermal treatment can forego the use of dewatering aids such as polyelectrolytes and fiber sludge. A reaction condition screening is performed at the lab-scale to optimize the HTC reaction conditions for both good dewaterability (reported as dry matter content of the obtained press cakes) and solid yield, and these conditions are applied at a continuous pilot-scale.

2. Materials and Methods

2.1. Raw Feedstock and Reference Dewatering Process

Biological paper sludge (bio-sludge) and mixed sludge (bio-sludge and fiber sludge) were provided by the Smurfit Kappa paper mill in Piteå, Sweden. The bio-sludge is a by-product of the biological wastewater treatment process of the paper mill. The mixed sludge is bio-sludge (33%) mixed with fiber sludge (67%). Fiber sludge is a by-product of paper production and is a wet stream containing solids in the form of very short cellulose fibers. During the initial lab testing, two different batches of bio-sludge were applied. The first batch had a low dry matter content (1.0%), which was a sample obtained directly after a maintenance stop. The second batch was obtained during normal operation, which resulted in a significantly higher dry matter content (3.4%)

Currently, the Smurfit Kappa paper mill in Piteå treats (dewaters) the mixed sludge as a residue stream, where the addition of fiber sludge aims to improve dewaterability of the bio-sludge. The mixed sludge is first treated with the addition of a polyelectrolyte as a chemical dewatering aid and this mixture is dewatered via a screw press to achieve solids with approximately 30–35% dry matter content. The solids are used on-site in a biomass boiler for heat production.

2.2. TORWASH® Process

This study uses the TORWASH® hydrothermal treatment process for short and mild HTC of bio-sludge. For this specific feedstock a temperature of 200 °C and autogenous pressure are applied. TORWASH® is a patented process that applies hydrothermal carbonization on wet organic residues via TORWASH® reactor technology. This process yields a product slurry that can be dewatered mechanically, whilst part of the inorganic ions are removed. Organic material that is dissolved in the filtrate is biodegradable and can be converted to biogas. Figure 1 shows a schematic representation of the overall process. The TORWASH® reactor technology includes flow operation on a small angle. During the carbonization reaction that takes place in the TORWASH® section (Figure 2), CO_2 is generated as a result of dehydration and decarboxylation reactions. The angle of the reactor pipe retains the CO_2 in the reactor. This results in a decrease in pH of the reaction media, catalyzing the hydrothermal carbonization [12].

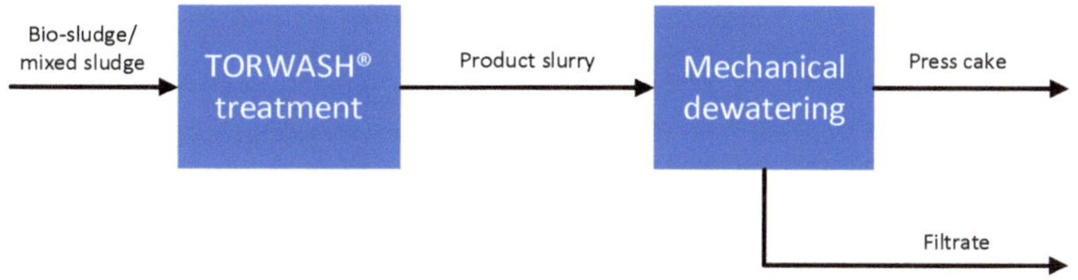

Figure 1. Schematic representation of the TORWASH® process.

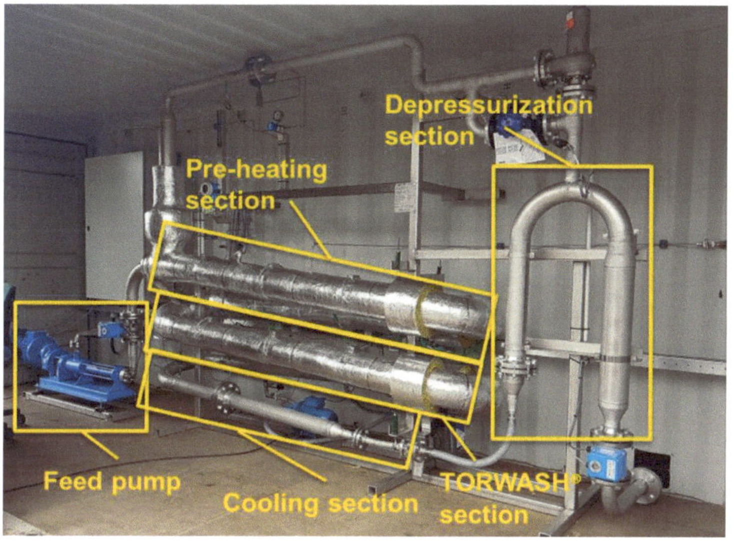

Figure 2. Containerized pilot plant applied for the continuous flow hydrothermal experiments.

2.3. Lab-Scale Batch Experiments

Initial screening tests were performed in small microclave vessels (125 mL) to determine the optimal temperature for hydrothermal treatment. Six different temperatures (160–210 °C) were selected to test the effect of the reaction temperature on the properties and the dewaterability of the obtained product slurry. In total, 80 g of feedstock was added to the microclave vessel, which was then tightly sealed and inserted into a heating block and heated to a target temperature. A typical test was performed with a stirring speed of 500 rpm and a run time at setpoint of 30 min.

Autoclave experiments were performed in a stainless steel vessel (20 L), which was indirectly heated. These experiments were performed to scale up optimized conditions as determined by the initial screening experiments. A total of 15 kg of feedstock was added to the vessel and the vessel was tightly sealed. The mixture was stirred by a stirrer equipped inside the vessel. Heating to the set temperature required approximately three hours. Upon reaching the temperature set point, the reactor was kept at this temperature for 30 min. Afterwards the reactor was left to cool to room temperature.

2.4. Pilot-Scale Continuous Flow Experiments

The TORWASH® continuous flow pilot plant (Figure 2) was designed for a maximum throughput of 50 kg/h, maximum temperature of 250 °C and a maximum pressure of 25 bar. The feedstock passes through three distinct temperature regimes in the pilot plant. The first section is the preheat section (~30 min residence time), in which the feed is gradually heated from ambient temperature to approximately 180 °C. Hydrolysis reactions occur in this section, but dehydration reactions are limited. In the TORWASH® section the temperature is increased to the desired temperature (200 °C for bio-sludge, ~1 h residence time) to achieve hydrothermal carbonization. In the cooling section the mixture is rapidly cooled to 40 °C to stop the reaction. The product slurry is depressurized and afterwards collected in external tanks. For the heating sections, electrical heating is applied, whereas the cooling section applies cooling water.

The typical throughput during the pilot tests was 25 kg/h, which corresponds to an overall residence time of ~2.5 h. The pilot plant is containerized and mobile and was used for long-duration (>400 h) experiments performed on-site at the Smurfit Kappa paper mill in Piteå, Sweden.

2.5. Dewatering Experiments

For lab-scale experiments and initial testing of the TORWASH® reactor, the obtained product slurry after hydrothermal carbonization was filtered (Whatman GF/D glass microfiber filter, 2.7 µm pore size) to remove the bulk of the liquid from the product slurry, yielding a transparent orange–brown filtrate and a solid filter cake. The filter cake was further mechanically dewatered with a carver die unit. For the carver die (58 mm diameter), nylon filters (20 µm pore size) were used at the top and bottom of the die. The loaded die was pressed unidirectionally in a hydraulic press, operated at 65 bar for 1 min to generate a solid press cake.

For on-site pilot-scale testing, dewatering of the large quantities of product slurry obtained during the long-duration test was performed by a pilot-scale membrane filter press of Limburg Filter, which employs their Leaktite technology. Briefly, the product slurry was pumped into the chambers of the membrane filter press, which was set at a (adjustable) pressure. The pressure slowly increased, which removed the bulk of the water from the solids. Next, the membranes were squeezed to dewater the solids as much as possible, yielding the solid press cake.

2.6. Analysis

The dry matter content of the feedstocks, TORWASH® product slurry, press cakes and filtrate was determined by drying overnight in an oven at 105 °C at atmospheric pressure. The dry matter content of press cakes was used to evaluate the dewaterability of the solids following the hydrothermal treatment.

Elemental analysis was performed on all streams (dried) with the Element Analyser Flash 2000 (Thermo Fisher Scientific, Lelystad, The Netherlands) and the ash content was determined using a Nabertherm LV5/11/B180 oven. CHN content was determined according to EN 15104 and the ash content according to EN 14775. Volatile matter was measured according to EN 15148. The higher heating value (HHV) was determined according to EN 14918. Inorganics were measured by Inductively Coupled Plasma with an ICAP6300 ICP-OES destruction according to NEN 6963 standard and ICP measurements according to NEN 6966. F, Cl and Br were analyzed according to NEN-EN-ISO-10304-1. Mercury was analyzed by Cold Vapour Atomic Fluorescence Spectroscopy. Dissolved chemical oxygen demand (COD) and dissolved phosphorus and nitrogen in the liquid fractions were measured with the Hach Lange cuvette test method.

3. Results and Discussion

3.1. Feedstock Analysis

Analysis of the two feedstocks showed that both streams are similar in composition (Table 1). Both feedstocks have close to neutral pH and are very dilute (<3% dry solids) with approximately 42% carbon content in the solids. The oxygen content of the mixed sludge is higher (38%) than in the bio-sludge (31%) due to the presence of cellulose fibers in the mixed sludge. The ash content of both feedstocks is high (up to 16%) due to the presence of a wide range of metals. Heavy metals are common in paper mill wastewater, mainly from the black liquor by-product of papermaking. These metals are concentrated in the sludge during wastewater treatment. The sulfur content in both sludges is particularly high, which is a result of the use of sulfur in the pulping process. The obtained values are within the range of other reported sludges derived from various paper mills [26].

Table 1. Characteristics of biological sludge and mixed sludge (33% biological sludge and 67% fiber sludge) as received. [1] db = dry basis; analysis on solids that were dried at 105 °C.

Parameter	Bio-Sludge	Mixed Sludge
pH	7.48	6.41
Conductivity (mS/cm)	1.87	2.01
Dry matter content (%)	1.0	2.2
Moisture content (%)	99.0	97.8
Solids Analysis:		
Ash content, 815 °C (% db [1])	15.8	12.4
Volatile matter (% db)	71.8	72.8
HHV (MJ/kg)	18.6	17.8
C (% db)	42.4	41.8
H (% db)	5.9	5.8
N (% db)	7.1	2.7
O (% db)	30.8	38.3
S (mg/kg)	11,000	7400
F (mg/kg)	37	39
Cl (mg/kg)	580	150
Br (mg/kg)	150	76
Hg (mg/kg)	0.059	0.055

3.2. Lab-Scale Experiments

3.2.1. Microclave Tests

A good indication of an efficient hydrothermal carbonization process, and thus dewaterability, is the decrease in pH of the resulting product slurry [17]. For both mixed sludge and bio-sludge, a clear decrease in pH was observed with hydrothermal treatment in the screening tests (Tables 2 and 3). For bio-sludge, the optimum temperature for pH reduction was 180 °C. For mixed sludge, the pH decreased with increasing temperature to 180 °C and then did not further decrease with higher temperatures. The chemical oxygen demand (COD) of the product slurry increased with increasing reaction temperature, confirming the increase in organics in the liquid fraction by hydrolysis reactions. Hydrothermally treated mixed sludge had increased concentrations of dissolved ammonium, total nitrogen and orthophosphate compared to the feedstock, and a strong relationship between the reaction temperature and the total dissolved matter was observed.

Table 2. Characteristics of the mixed sludge feedstock before and after processing, and the effect of hydrothermal treatment reaction temperature. Reported values for COD, NH_4^+, total nitrogen and PO_4^{3-} correspond to the dissolved fraction.

Temperature (°C)	pH	Conductivity (mS/cm)	COD (g/L)	NH_4^+ (mg/L)	Total Nitrogen (mg/L)	PO_4^{3-} (mg/L)
Mixed sludge feed	6.4	2.0	0.8	8	22	7
160	6.7	1.7	4.7	11	247	24
170	5.8	2.5	6.0	63	312	24
180	5.4	2.6	7.5	87	367	24
190	5.5	2.8	7.8	101	417	20
200	5.3	3.1	8.1	154	437	37
210	5.2	3.3	8.3	164	445	37

Table 3. Characteristics of two bio-sludge feedstocks before and after processing, and the effect of hydrothermal treatment reaction temperature. * Second bio-sludge batch used, n.d. = not determined. Reported values for COD, NH_4^+, total nitrogen and PO_4^{3-} correspond to the dissolved fraction.

Temperature (°C)	pH	Conductivity (mS/cm)	COD (g/L)	NH_4^+ (mg/L)	Total Nitrogen (mg/L)	PO_4^{3-} (mg/L)
Bio-sludge feed	7.5	1.9	0.2	n.d.	n.d.	n.d.
160	6.7	2.0	5.0	n.d.	n.d.	n.d.
170	6.4	2.2	5.1	n.d.	n.d.	n.d.
180	5.6	2.3	6.6	n.d.	n.d.	n.d.
190	5.9	2.8	6.0	n.d.	n.d.	n.d.
200	5.9	2.9	6.5	n.d.	n.d.	n.d.
210	5.9	3.1	7.2	n.d.	n.d.	n.d.
Second bio-sludge feed	6.4	2.5	2.7	200	223	42
180 *	5.4	4.4	24.8	548	2258	112
200 *	5.3	5.0	23.7	675	2153	134

Hydrothermally treated bio-sludge and mixed sludge from the screening tests were pressed with the carver die to test the dewaterability of the obtained product slurries. Sludges that were treated at higher reaction temperatures were more easily pressable, which is closely related to a drop in pH (Tables 2 and 3). For mixed sludge, the dewaterability was consistent as a good press cake could be obtained in combination with a clear liquid. The dry matter content of the obtained press cake was in the range of 59–63% at treatment temperatures between 160–190 °C (Table S1). The dry matter content in the press cake did show a small (and unexpected) decrease at higher temperatures, to 54% at 200 and 210 °C, as dewaterability is expected to increase for hydrochar obtained at higher temperatures [23]. For bio-sludge, no reliable data for the dry matter content could be obtained due to the small quantity of solids obtained in each experiment. The product slurry obtained at 160 °C was not dewaterable, whereas all the other samples yielded a solid and a clear liquid after mechanical dewatering.

Determination of the total solid yield did show a clear trend for both feedstocks (Figure 3) as a decrease in the solid yield (as a percentage of the total initial solids) was observed with increasing reaction temperature. For mixed sludge, this decrease was gradual from 160 °C (83% solid yield) to 190 °C (69% solid yield). Increasing the temperature to 200 °C resulted in a sharp drop to 56% solid yield. For bio-sludge treated at 170 °C and 180 °C, the solid yield was comparable (42% and 38%, respectively). Higher temperatures resulted in a sharp decrease in solid yield, ranging from 26% at 190 °C to 10% at 210 °C, as was also observed in comparable studies [26,27]. The very low values obtained for bio-sludge are also a result of the small quantities of solids obtained (less than 0.4 g), which makes any product loss during work-up result in a significant yield loss percentage-wise.

The differences between both feedstocks confirm the positive effect of the addition of fiber sludge to bio-sludge for both the dewaterability and the solid yield of the obtained product. Based on these results the optimal reaction temperature for mixed sludge was determined to be 190 °C. It should be noted that the results (and trends) observed for the experiments at the higher temperatures are likely an actual overestimation of the temperature effect. This is due to the relatively slow heating and cooling profile of the microclave, causing a longer exposure to reactive conditions compared to the experiments performed at lower temperatures.

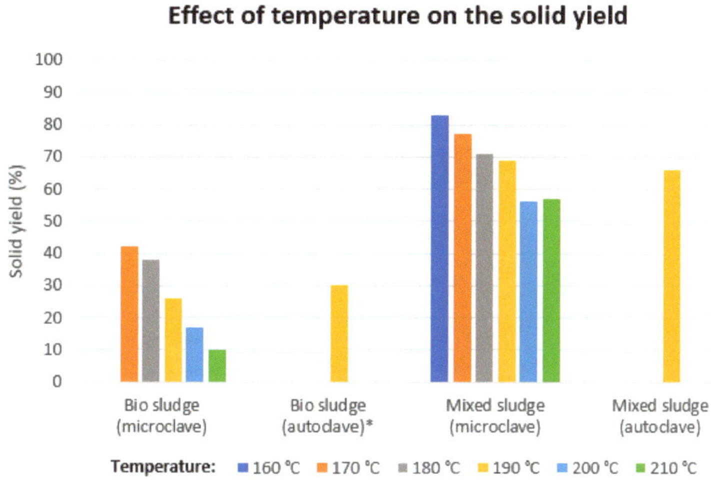

Figure 3. Solid yield of the press cakes from the obtained product slurry from bio-sludge and mixed sludge after mild hydrothermal carbonization at different temperatures and carver die dewatering. * Second batch of bio-sludge.

Some follow-up experiments were performed with a different (second) batch of bio-sludge from Smurfit Kappa, which had different properties compared to the first batch (Table 3). In particular the dry matter content (3.4% vs. 1%) and the dissolved organics (COD of 2.7 g/L vs. 0.2 g/L) were significantly higher than those of the original bio-sludge sample. Two microclave experiments were performed for this new bio-sludge, at 180 °C and 200 °C, and confirmed the trends observed with the previous feedstock. The result for the total solid yield was similar at 180 °C (40% solid yield), but a significant improvement was observed at 200 °C (28% solid yield vs. 17% for the first batch). This is likely a result of the higher amount of dry matter content in the second batch and the lower pH of both the feedstock and the resulting product slurry, resulting in more efficient carbonization.

3.2.2. Autoclave Tests

Based on the results obtained with the microclave experiments, a reaction temperature of 190 °C was determined as a setpoint for the autoclave tests for mixed sludge and the second batch of bio-sludge. The dewaterability of both product slurries was very good, with consistent quality of the solid press cake and a clear liquid. For mixed sludge, the obtained results were similar to the microclave experiment, although the pH of the resulting mixture was slightly more acidic (pH 5.4) and the COD slightly higher (9.2 g/L) (Table 4). Due to the high volume of product slurry created, 18 press cakes were pressed in order to obtain good average values. A total solid yield of 66% and an average dry matter content of 61% were obtained for the press cakes, which is in correspondence with the results obtained with the microclave experiments.

Table 4. Characteristics of the product slurry obtained with the hydrothermal treatment at 190 °C with the autoclave. Reported values for COD, NH_4^+, total nitrogen and PO_4^{3-} correspond to the dissolved fraction.

Feedstock	pH	Conductivity (mS/cm)	COD (g/L)	NH_4^+ (mg/L)	Total Nitrogen (mg/L)	PO_4^{3-} (mg/L)
Mixed sludge	5.4	3.2	9.2	172	535	44
Bio-sludge	5.3	5.1	24.0	760	2489	194

For bio-sludge, the results were also in line with the microclave experiments. The characteristics of the obtained product slurry are similar to the results with the microclave at 200 °C, with a slightly higher amount of dissolved components. This is likely a result of the longer heating and cooling time of the autoclave compared to the microclave, effectively resulting in a longer reaction time. A total solid yield of 30% was obtained and an average dry matter content of 61% was obtained for the press cakes (average of six press cakes). These results are also comparable to the results obtained with the microclave at 200 °C.

3.3. Pilot-Scale Continuous Flow Experiments

The first experiments in the pilot plant were performed at TNO in an effort to prepare for and de-risk the long-duration experiments on-site. The first experiment was performed with mixed sludge as the feedstock, during which a range of temperatures were tested. A clear improvement in dewaterability was observed with increasing temperature. The liquid obtained after pressing the product slurry obtained at 185 °C contained solids, whilst at higher reaction temperatures, a clear liquid was obtained. The press cake obtained from the mixed sludge treated at 185 °C showed large fluctuations in the dry matter content, ranging from 20–46%. At temperatures of 195 °C and 205 °C, more consistent results for the press cakes were obtained with an average dry matter content of 61% and 63%, respectively (Table 5), which is in correspondence with the results obtained with the microclave and autoclave at 190 °C. The improved dewaterabilty at elevated temperatures is in accordance with the reported literature [23]. The solid yields at 195 °C and 205 °C were 69% and 68%, respectively, which was similar to the result with the microclave at 190 °C (69%). pH analysis of the product slurries confirmed the trend of improved dewaterability with decreasing pH. The product slurry obtained at 185 °C had a pH of 6.0, compared to a pH of 5.8 for treatment at 195 °C and 5.6 at 205 °C. The pH of the obtained product slurries was less acidic compared to the corresponding results with the microclave, indicating a less efficient carbonization reaction. Overall, it was noticed that a higher temperature (~10 °C) was required for the pilot plant to obtain similar results to those for the microclave and autoclave tests. This could be a result of a low degree of mixing in the pilot plant, which limits the heat transfer in the reaction mixture. The requirement for higher temperatures could also be a result of the shorter residence time in the pilot plant compared to the autoclave and microclave. Therefore, the carbonization reaction is not as effective in the pilot plant compared to the batch systems at the same temperature, and a higher temperature is required in the pilot reactor to obtain similar results to the batch system.

Table 5. Results obtained for the press cake and the pH of the obtained product slurry.

	Dry Matter Content Press Cake (%)	Solid Yield (%)	pH
Mixed sludge			6.4
185 °C	34 ± 9	70	6.0
195 °C	61 ± 3	69	5.8
205 °C	63 ± 2	68	5.6
Bio-sludge			6.4
185 °C	Not pressable		5.9
195 °C	48 ± 7	49	5.7
205 °C	50 ± 5	44	5.7

Following 8.5 h of continuous pilot plant operation with mixed sludge, the flow profile became irregular and the experiment was stopped. Subsequent inspection of the pilot plant showed the accumulation of fibrous material inside the reactor. This would be problematic for continuous flow operation for an extended time period.

Initial pilot experiments with bio-sludge were performed at the same temperatures as mixed sludge. As with the mixed sludge, a clear increase in dewaterability was observed with increasing temperature. Bio-sludge treated at 185 °C was not dewaterable. A temperature increase to 195 °C yielded a product slurry that was pressable, albeit with difficulty. The product slurry obtained at 205 °C was significantly easier and more consistent to press. The obtained press cakes from the product slurry generated at 195 °C and 205 °C had average dry matter contents of 48% and 50%, respectively, which were lower than the 61% obtained from the autoclave experiment at 190 °C. The solid yields at 195 °C and 205 °C were 49% and 44%, respectively, which were markedly higher compared to the microclave and autoclave experiments, which had a solid yield of around 30%, which could be an effect of the shorter residence time in the flow setup [26–28]. The pH of the product slurry followed the same trend of decreasing with increasing reaction temperature. Inspection of the pilot plant following the experiments did not show any accumulation of solids, demonstrating the suitability of bio-sludge for long duration experiments. The improved dewaterability with increasing temperature is in good accordance with literature reports. Enhanced dewaterability is directly linked to the polarity index ((O+N)/C), where a low value (due to a high carbon content) corresponds to hydrophobicity [37,38]. At higher temperatures, the carbon content of the obtained hydrochar is higher, resulting in enhanced dewaterability.

Based on the results of the initial pilot plant experiments with mixed sludge and bio-sludge, bio-sludge was selected as the feedstock for long duration (>400 h) testing on-site at the Smurfit Kappa paper mill. The preliminary experiments showed that hydrothermal treatment of mixed sludge does not offer many benefits in terms of dewaterability over the bio-sludge and, in fact, is detrimental at the pilot-scale due to the fibers settling in the reactor. The use of bio-sludge as the sole feedstock for hydrothermal treatment is also beneficial from an industrial perspective, as this material is the problematic waste stream for the paper industry. Fiber sludge is typically added as a dewatering aid, but it could be used instead for higher value applications [8].

The long-duration pilot-scale experiment was performed on the site of the wastewater treatment plant at the paper mill of Smurfit Kappa in Piteå, Sweden. This allowed for pilot testing on the actual location where the residue stream was generated and ensured a constant supply of fresh bio-sludge for the entire duration of the test, providing a better approximation of actual scale-up conditions. The long-duration experiment was performed over a period of 24 days, with a pilot plant uptime of 75% (410 h). The reactor down-time was mainly caused by the unusual flow behavior of heated bio-sludge and fouling within the reactor, more details of which can be found in the Supporting Information.

During the long-duration experiment, the target temperature was set at 200 °C, and throughout the experiment the temperature ranged between 195 and 205 °C. Nearly 11,000 kg of product slurry was produced in this experiment. The obtained product slurry was easily dewaterable with a membrane filter press, yielding a clear filtrate. In total, 142 kg of press cake was obtained from the product slurry with the membrane filter press (Figure 4). The press cakes had an average dry matter content of 38% (54.4 kg dry solid) and a total solid yield of 39%. The dry matter content of the press cake was consistent throughout the entire test with values ranging from 33–42% (9 filter press runs). The solid yield is comparable but slightly lower than those obtained in the preliminary pilot plant experiments. The values for the dry matter content of the press cakes obtained during continuous operation were approximately 10% lower compared to the values obtained during the lab and pilot tests. This is due to the difference in operating pressure of the applied filter press. The carver die operates at a higher pressure compared to the membrane filter press, resulting in a more compressed cake. The dry matter content of the obtained press cakes was higher than the values obtained with the currently applied dewatering method at the paper mill, showing the potential of the hydrothermal treatment process.

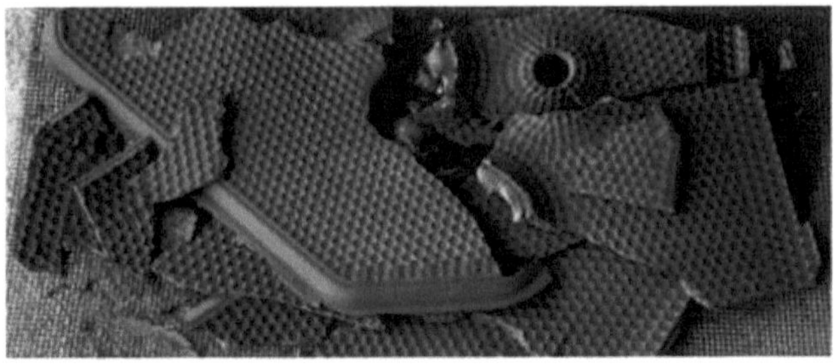

Figure 4. Press cake obtained with the membrane filter press during the long duration experiment.

Analysis of the bio-sludge feedstock on-site showed values that were an intermediary of the two bio-sludge samples used in the lab-scale testing. This demonstrates the variability within this feedstock and the requirement for a robust process to treat it. Hydrothermally treated bio-sludge had a slightly smaller dry matter content compared to untreated bio-sludge (Table 6). This loss is likely a combined effect of loss due to the formation of gases (~5% of the initial organic content) and some settling of solids in both the pilot plant and the collection tanks. The high dry matter content of the filtrate (53% of the feed) indicates that a large fraction of the initial feed is dissolved in the liquid fraction as a result of the hydrothermal treatment and valorization of this stream is essential. The decrease in pH of the initial feedstock compared to both the obtained product slurry and filtrate and the marked increase in dissolved COD are indicators for efficient hydrothermal carbonization.

Table 6. Characteristics of the bio-sludge feed, product slurry and filtrate obtained with the pilot plant operated at 200 °C. Reported values for COD, NH_4^+, total nitrogen and PO_4^{3-} correspond to the dissolved fraction. Early stage denotes the first two weeks of long duration testing, late stage denotes the final two weeks of long duration testing.

	Dry Matter Content (%)	pH	Conductivity (mS/cm)	COD (g/L)	NH_4^+ (mg/L)	Total Nitrogen (mg/L)	PO_4^{3-} (mg/L)
Early stage							
Feed	1.59	7.0	2.2	0.4	36	84	17
Product slurry	1.45	6.1	3.3	11.2	93	885	34
Filtrate	0.86	5.9	3.4	11.0	103	947	43
Late stage							
Feed	1.57	6.9	2.0	0.6	32	47	34
Product slurry	1.46	6.0	3.0	10.6	93	866	110
Filtrate	0.83	6.1	3.1	10.8	97	802	98

In order to obtain more insight into the chemical changes during the process and the final composition of the obtained products, an extensive analysis was performed on the feedstock, hydrothermal carbonization product slurry, the obtained press cake from the membrane filter press and the resulting filtrate of the membrane filter press (Table 7). One of the main observations is the high ash content (24.9%) of the obtained press cake. More specifically, the press cake has a high affinity for metals such as aluminum, barium, cadmium, chromium, copper, iron, lead, manganese, mercury, silicon, strontium, titanium, vanadium and zinc compared to the obtained filtrate. This makes the hydrothermal carbonization of bio-sludge an efficient method to concentrate and trap metals in a wastewater stream [39]. Calcium, magnesium, phosphorus and sulfur were also detected in the press cake, but these were more evenly distributed between the press cake and filtrate. The filtrate has high affinity for boron, bromine, chlorine, potassium and sodium and is also rich in calcium and sulfur. Another trend is the increase in the higher heating value (HHV) for the obtained product slurry and press cake (19.2 and 20.8 MJ/kg db) compared to the feedstock (18.1 MJ/kg db). This is associated with an increase in the carbon content relative to oxygen in the product slurry and press cake, a direct effect of the carbonization of the feedstock. The increase in the HHV and the removal of most of the water (99.2% of the initial feedstock) could make this a suitable process to produce solid fuel. Further investigation to see if the press cake can be used directly as a solid fuel (e.g., in the steelmaking industry) should be performed [40], whereas the effluent could be used for nutrient recovery [41] and biogas formation via anaerobic digestion [42]. The effect of the high ash content of the hydrochar obtained from this specific feedstock should be carefully evaluated as this will exclude the use as, e.g., soil remediation. Even if no direct applications are possible for this feedstock, this process is still of high interest for the paper industry as the enhanced dewaterability of bio-sludge can lead to improved energy efficiency of the sludge and cost savings. Preliminary techno-economic analysis indicates the proposed process can save three times the primary fuel and reduce treatment costs by 28% when compared to the conventional dewatering and incineration process at the mill. Detailed techno-economic and environmental assessments are ongoing and represent topics for future research.

Table 7. Analysis results of bio-sludge, the hydrothermal carbonization product slurry, the press cake obtained with the membrane press and the resulting filtrate.

Parameter	Unit	Bio-Sludge	Product Slurry	Press Cake	Filtrate
Ash content (550 °C)	% db	19.7	20.3	24.9	18.4
Ash content (815 °C)	% db	19.0	19.9	23.4	18.0
Volatile matter	% db	70.0	67.7	59.9	74.4
Moisture	% ar	5.6	4.6	1.6	7.9
HHV	MJ/kg db	18.1	19.2	20.8	17.3
Element					
C	% db	41.3	43.7	46.0	40.7
N	% db	6.3	7.2	4.2	9.0
H	% db	5.8	5.5	5.8	5.5
O	% db	32.7	29.2	22.2	33.1
Br	mg/kg	120	130	30	140
Cl	mg/kg	820	960	120	790
F	mg/kg	18	18	64	<10
Al	mg/kg db	7000	6900	18,000	210
As	mg/kg db	<4	<4	<4	<4
B	mg/kg db	170	180	9.8	220
Ba	mg/kg db	360	360	820	5.6
Ca	mg/kg db	18,000	14,000	21,000	9000
Cd	mg/kg db	4.8	5.0	12	<0.3
Co	mg/kg db	1.4	1.4	1.7	<0.9
Cr	mg/kg db	18	18	40	4.1
Cu	mg/kg db	27	27	69	<5
Fe	mg/kg db	3100	3000	6300	730
K	mg/kg db	6100	7000	1300	11,000
Li	mg/kg db	3.4	3.8	3.1	4.4
Mg	mg/kg db	2500	2900	3200	3100
Mn	mg/kg db	7600	7300	12,000	1900
Mo	mg/kg db	7.4	7.9	7.9	3.9
Na	mg/kg db	21,000	24,000	1400	44,000
Ni	mg/kg db	8.3	8.8	18	4.6
P	mg/kg db	13,000	13,000	22,000	6500
Pb	mg/kg db	17	17	43	<2
S	mg/kg db	22,000	23,000	9300	33,000
Sb	mg/kg db	<10	<10	<10	<10
Se	mg/kg db	<4	<4	<4	<4
Si	mg/kg db	13,000	14,000	35,000	3800
Sn	mg/kg db	<3	<3	<3	<3
Sr	mg/kg db	59	51	110	15
Ti	mg/kg db	250	390	860	2.2
V	mg/kg db	14	15	29	1.7
W	mg/kg db	<4	<4	<4	<4
Zn	mg/kg db	530	550	1200	24
Hg	mg/kg ar	0.062	0.065	0.14	<0.004

4. Conclusions

This study reports the successful scale-up from a batch lab-scale process to a continuous flow pilot plant for the mild hydrothermal carbonization of biological sludge from the wastewater treatment plant of a paper mill. Initial experiments were performed at lab-scale on bio-sludge and mixed sludge (mix of bio- and fiber sludge, with the latter as a dewatering aid) to determine the optimal reaction temperature for dewaterability of the sludge whilst maintaining a good solid yield. At lab-scale, a temperature of around 190 °C was established as the optimal temperature as determined by the overall solid yield (26% for bio-sludge and 69% for mixed sludge) and ease of mechanical dewaterability (pressing). The experiments could be increased in scale from a 100 mL microclave to a 20 L autoclave with good and comparable results for both bio-sludge (30% solid yield, 61% dry matter content) and mixed sludge (66% solid yield, 61% dry matter content) at 190 °C. In short tests at the pilot-scale, it was possible to achieve the same results at a small increase in temperature to approximately 200 °C. This increase in temperature was likely required due to poorer mixing and uneven heating in the pilot reactor. For mixed sludge, the pilot-scale test showed that the accumulation of cellulose fibers in the reactor would be problematic. For bio-sludge, a significant improvement in solid yield to over 40% was achieved by hydrothermal treatment in combination with a dry matter content close to 50%, which is higher than the current applied methods for enhancing bio-sludge dewaterability. Therefore, bio-sludge was selected as feedstock for the long duration testing. A long duration experiment, 24 days, 410 h for a total uptime of 75%, was performed on-site at a paper mill with bio-sludge as the feedstock. The results were in good correspondence with the lab tests, with a solid yield of 39% and a dry matter content of 38% for the obtained press cakes (solids). Compared to the current practice at the paper mill, which uses valuable fiber sludge and polymer as dewatering aids to achieve 30% dry matter content, this study highlights the potential of hydrothermal treatment to improve the dewaterability of wastewater sludges without the use of dewatering aids. The obtained press cake shows potential as a means to concentrate metals present in wastewater, and due to the enhanced dewaterability of bio-sludge, it should have enhanced potential as a solid fuel. Furthermore, 99.2% of the initial water content was removed from the bio-sludge during the process of hydrothermal carbonization and dewatering. Applying a more concentrated feed and valorization of the filtrate are topics that should be addressed in follow-up research.

Supplementary Materials: The following supporting information can be downloaded at: https://www.mdpi.com/article/10.3390/pr10122702/s1, Figure S1: Build-up and blockages of sludge in the reactor (left) and unexpected phase separation of floating sludge upon heating of bio-sludge (right) to explain the blockages at the top of the reactor tubes; Figure S2: Fouling on the inside of the TORWASH pilot plant upon inspection and cleaning at the end of the long-duration tests; Table S1: Results for the microclave and autoclave experiments at different reaction temperatures on bio-sludge and mixed sludge as feedstock.

Author Contributions: Conceptualization., P.N., J.R.P. and H.E.W.; methodology, E.C.-P., M.V. and D.J.S.; validation. P.N., J.R.P. and H.E.W.; investigation. E.C.-P., M.V. and D.J.S.; resources, P.N. and H.E.W.; data curation. D.S.Z.; writing—original draft preparation. D.S.Z.; writing—review and editing. H.E.W., J.R.P. and P.N.; supervision, H.E.W. and P.N.; project administration. H.E.W.; funding acquisition, P.N. All authors have read and agreed to the published version of the manuscript.

Funding: This project has received funding from the European Union's Horizon 2020 research and innovation program under grant agreement No. 884226.

Data Availability Statement: Not applicable.

Acknowledgments: The authors acknowledge Stefan Lundqvist, Ingemar Lundstrom, Frank Kruip, Martin Viklund, Ludvig Ånnhagen and Peter Skemark for assistance during the on-site trials in Piteå. We also acknowledge Sayujya Shah and Jan Wilco Dijkstra for preliminary techno-economic information.

Conflicts of Interest: The authors declare no conflict of interest.

References

1. Turner, T.; Wheeler, R.; Oliver, I.W. Evaluating land application of pulp and paper mill sludge: A review. *J. Environ. Manag.* **2022**, *317*, 115439. [CrossRef] [PubMed]
2. Oumabady, S.; Selvaraj, P.S.; Periasamy, K.; Veeraswamy, D.; Ramesh, P.T.; Palanisami, T.; Ramasamy, S.P. Kinetic and isotherm insights of Diclofenac removal by sludge derived hydrochar. *Sci. Rep.* **2022**, *12*, 2184. [CrossRef] [PubMed]
3. Su, B.; Wang, G.; Li, R.; Xu, K.; Wu, J.; Li, D.; Liu, J. Co-Combustion Behavior of Paper Sludge Hydrochar and Pulverized Coal: Low Rank Coal and Its Product by Hydrothermal Carbonization. *Energies* **2022**, *15*, 5619. [CrossRef]
4. Lin, Y.; Ma, X.; Peng, X.; Hu, S.; Yu, Z.; Fang, S. Effect of hydrothermal carbonization temperature on combustion behavior of hydrochar fuel from paper sludge. *Appl. Therm. Eng.* **2015**, *91*, 574–582. [CrossRef]
5. Nguyen, L.H.; Van, H.T.; Chu, T.H.H.; Nguyen, T.H.V.; Nguyen, T.D.; Hoang, L.P.; Hoang, V.H. Paper waste sludge-derived hydrochar modified by iron (III) chloride for enhancement of ammonium adsorption: An adsorption mechanism study. *Environ. Technol. Innov.* **2021**, *21*, 101223. [CrossRef]
6. Makela, M.; Benavente, V.; Fullana, A. Hydrothermal carbonization of industrial mixed sludge from a pulp and paper mill. *Bioresour. Technol.* **2016**, *200*, 444–450. [CrossRef] [PubMed]
7. Makela, M.; Forsberg, J.; Soderberg, C.; Larsson, S.H.; Dahl, O. Process water properties from hydrothermal carbonization of chemical sludge from a pulp and board mill. *Bioresour. Technol.* **2018**, *263*, 654–659. [CrossRef]
8. Meyer, T.; Amin, P.; Allen, D.G.; Tran, H. Dewatering of pulp and paper mill biosludge and primary sludge. *J. Environ. Chem. Eng.* **2018**, *6*, 6317–6321. [CrossRef]
9. Jarnerud, T.; Karasev, A.V.; Wang, C.; Bäck, F.; Jönsson, P.G. Utilization of Organic Mixed Biosludge from Pulp and Paper Industries and Green Waste as Carbon Sources in Blast Furnace Hot Metal Production. *Sustainability* **2021**, *13*, 7706. [CrossRef]
10. Khan, T.A.; Saud, A.S.; Jamari, S.S.; Rahim, M.H.A.; Park, J.-W.; Kim, H.-J. Hydrothermal carbonization of lignocellulosic biomass for carbon rich material preparation: A review. *Biomass Bioenergy* **2019**, *130*, 105384. [CrossRef]
11. Román, S.; Libra, J.; Berge, N.; Sabio, E.; Ro, K.; Li, L.; Ledesma, B.; Álvarez, A.; Bae, S. Hydrothermal Carbonization: Modeling, Final Properties Design and Applications: A Review. *Energies* **2018**, *11*, 216. [CrossRef]
12. Meijden, C.M.V.D.; Pels, J.R. Reactor for the Hydrothermal Treatment of Biomass. WIPO (PCT) Patent WO 2021/032842 A1, 25 February 2021.
13. Pels, J.R.; Blijendaal, L.P.J.; Nijman, M.N.W.; Zandvoort, M.H.; Cieplik, M.K. Conversion of water plants to biomass fuel using torwash. Proceedings of 22nd European Biomass Conference and Exhibitions, Hamburg, Germany, 23–26 June 2014.
14. Ischia, G.; Fiori, L. Hydrothermal Carbonization of Organic Waste and Biomass: A Review on Process, Reactor, and Plant Modeling. *Waste Biomass Valorization* **2020**, *12*, 2797–2824. [CrossRef]
15. Pauline, A.L.; Joseph, K. Hydrothermal carbonization of organic wastes to carbonaceous solid fuel—A review of mechanisms and process parameters. *Fuel* **2020**, *279*, 118472. [CrossRef]
16. Sharma, H.B.; Sarmah, A.K.; Dubey, B. Hydrothermal carbonization of renewable waste biomass for solid biofuel production: A discussion on process mechanism, the influence of process parameters, environmental performance and fuel properties of hydrochar. *Renew. Sustain. Energy Rev.* **2020**, *123*, 109761. [CrossRef]
17. Wang, T.; Zhai, Y.; Zhu, Y.; Li, C.; Zeng, G. A review of the hydrothermal carbonization of biomass waste for hydrochar formation: Process conditions, fundamentals, and physicochemical properties. *Renew. Sustain. Energy Rev.* **2018**, *90*, 223–247. [CrossRef]
18. Li, L.; Wang, Y.; Xu, J.; Flora, J.R.V.; Hoque, S.; Berge, N.D. Quantifying the sensitivity of feedstock properties and process conditions on hydrochar yield, carbon content, and energy content. *Bioresour. Technol.* **2018**, *262*, 284–293. [CrossRef]
19. Aragón-Briceño, C.I.; Pozarlik, A.K.; Bramer, E.A.; Niedzwiecki, L.; Pawlak-Kruczek, H.; Brem, G. Hydrothermal carbonization of wet biomass from nitrogen and phosphorus approach: A review. *Renew. Energy* **2021**, *171*, 401–415. [CrossRef]
20. Jamari, S.S.; Howse, J.R. The effect of the hydrothermal carbonization process on palm oil empty fruit bunch. *Biomass Bioenergy* **2012**, *47*, 82–90. [CrossRef]
21. Maniscalco, M.P.; Volpe, M.; Messineo, A. Hydrothermal Carbonization as a Valuable Tool for Energy and Environmental Applications: A Review. *Energies* **2020**, *13*, 4098. [CrossRef]
22. Saqib, N.U.; Sharma, H.B.; Baroutian, S.; Dubey, B.; Sarmah, A.K. Valorisation of food waste via hydrothermal carbonisation and techno-economic feasibility assessment. *Sci. Total Environ.* **2019**, *690*, 261–276. [CrossRef]
23. Heidari, M.; Dutta, A.; Acharya, B.; Mahmud, S. A review of the current knowledge and challenges of hydrothermal carbonization for biomass conversion. *J. Energy Inst.* **2019**, *92*, 1779–1799. [CrossRef]
24. Nizamuddin, S.; Baloch, H.A.; Griffin, G.J.; Mubarak, N.M.; Bhutto, A.W.; Abro, R.; Mazari, S.A.; Ali, B.S. An overview of effect of process parameters on hydrothermal carbonization of biomass. *Renew. Sustain. Energy Rev.* **2017**, *73*, 1289–1299. [CrossRef]
25. Tasca, A.L.; Puccini, M.; Gori, R.; Corsi, I.; Galletti, A.M.R.; Vitolo, S. Hydrothermal carbonization of sewage sludge: A critical analysis of process severity, hydrochar properties and environmental implications. *Waste Manag.* **2019**, *93*, 1–13. [CrossRef] [PubMed]
26. Saha, N.; Saba, A.; Saha, P.; McGaughy, K.; Franqui-Villanueva, D.; Orts, W.; Hart-Cooper, W.; Reza, M. Hydrothermal Carbonization of Various Paper Mill Sludges: An Observation of Solid Fuel Properties. *Energies* **2019**, *12*, 858. [CrossRef]
27. Hamalainen, A.; Kokko, M.; Kinnunen, V.; Hilli, T.; Rintala, J. Hydrothermal carbonization of pulp and paper industry wastewater treatment sludges—Characterization and potential use of hydrochars and filtrates. *Bioresour. Technol.* **2022**, *355*, 127258. [CrossRef]

28. Sadish, O.; Paul Sebastian, S.; Banu, K.S.P.; Mahendran, R. Hydrochar as an Energy Alternative to Coal: Effect of Temperature on Hydrothermal Carbonization of Paper Board Mill Sludge. *Int. J. Curr. Microbiol. Appl. Sci.* **2019**, *8*, 1668–1675. [CrossRef]
29. Ejilane, J.; Balasubramaniam, P.; Maheswari, M.; Sebastian, S.P.; Sabarish, K. Production and Characterization of Paper Board Mill ETP Sludge Derived Hydrochar. *Int. J. Environ. Clim. Change* **2021**, *11*, 1–8. [CrossRef]
30. Mäkelä, M.; Benavente, V.; Fullana, A. Hydrothermal carbonization of lignocellulosic biomass: Effect of process conditions on hydrochar properties. *Appl. Energy* **2015**, *155*, 576–584. [CrossRef]
31. Alatalo, S.M.; Repo, E.; Makila, E.; Salonen, J.; Vakkilainen, E.; Sillanpaa, M. Adsorption behavior of hydrothermally treated municipal sludge & pulp and paper industry sludge. *Bioresour. Technol.* **2013**, *147*, 71–76. [CrossRef]
32. Briongos, J.V.; Taramona, S.; Gómez-Hernández, J.; Mulone, V.; Santana, D. Solar and biomass hybridization through hydrothermal carbonization. *Renew. Energy* **2021**, *177*, 268–279. [CrossRef]
33. Danso-Boateng, E.; Holdich, R.G.; Martin, S.J.; Shama, G.; Wheatley, A.D. Process energetics for the hydrothermal carbonisation of human faecal wastes. *Energy Convers. Manag.* **2015**, *105*, 1115–1124. [CrossRef]
34. Zaccariello, L.; Battaglia, D.; Morrone, B.; Mastellone, M.L. Hydrothermal Carbonization: A Pilot-Scale Reactor Design for Bio-waste and Sludge Pre-treatment. *Waste Biomass Valorization* **2022**, *13*, 3865–3876. [CrossRef]
35. Heidari, M.; Norouzi, O.; MacDermid-Watts, K.; Acharya, B.; Zhang, Y.; Dutta, A. Product evaluation of hydrothermal carbonization of biomass: Semi-continuous vs. batch feeding. *Biomass Convers. Biorefinery* **2020**, *12*, 15–25. [CrossRef]
36. Güleç, F.; Riesco, L.M.G.; Williams, O.; Kostas, E.T.; Samson, A.; Lester, E. Hydrothermal conversion of different lignocellulosic biomass feedstocks—Effect of the process conditions on hydrochar structures. *Fuel* **2021**, *302*, 121166. [CrossRef]
37. Sliz, M.; Tuci, F.; Czerwinska, K.; Fabrizi, S.; Lombardi, L.; Wilk, M. Hydrothermal carbonization of the wet fraction from mixed municipal solid waste: Hydrochar characteristics and energy balance. *Waste Manag.* **2022**, *151*, 39–48. [CrossRef] [PubMed]
38. Lu, L.; Namioka, T.; Yoshikawa, K. Effects of hydrothermal treatment on characteristics and combustion behaviors of municipal solid wastes. *Appl. Energy* **2011**, *88*, 3659–3664. [CrossRef]
39. Masoumi, S.; Borugadda, V.B.; Nanda, S.; Dalai, A.K. Hydrochar: A Review on Its Production Technologies and Applications. *Catalysts* **2021**, *11*, 939. [CrossRef]
40. Liang, W.; Nanou, P.; Wray, H.; Zhang, J.; Lundstrom, I.; Lundqvist, S.; Wang, C. Feasibility Study of Bio-Sludge Hydrochar as Blast Furnace Injectant. *Sustainability* **2022**, *14*, 5510. [CrossRef]
41. Gerner, G.; Meyer, L.; Wanner, R.; Keller, T.; Krebs, R. Sewage Sludge Treatment by Hydrothermal Carbonization: Feasibility Study for Sustainable Nutrient Recovery and Fuel Production. *Energies* **2021**, *14*, 2697. [CrossRef]
42. Lucian, M.; Volpe, M.; Merzari, F.; Wust, D.; Kruse, A.; Andreottola, G.; Fiori, L. Hydrothermal carbonization coupled with anaerobic digestion for the valorization of the organic fraction of municipal solid waste. *Bioresour. Technol.* **2020**, *314*, 123734. [CrossRef]

processes

Article

Optimization of the Biotreatment of GTL Process Water Using *Pseudomonas aeruginosa* Immobilized in PVA Hydrogel

Somaya A. Ahmed, Riham Surkatti, Muneer M. Ba-Abbad and Muftah H. El-Naas *

Gas Processing Center, College of Engineering, Qatar University, Doha P.O. Box 2703, Qatar
* Correspondence: muftah@qu.edu.qa

Abstract: The COD reduction in gas to liquid (GTL) process water was optimized using response surface methodology (RSM). The biodegradation process was carried out in a spouted bed bioreactor (SBBR) using *Pseudomonas aeruginosa* immobilized in polyvinyl alcohol (PVA) gel. Different factors affecting the biological treatment of GTL process water (PW) were investigated. Three variables including PVA volume fraction, initial COD, and pH were investigated in the batch experiments. The biodegradation experiments were carried out by varying the initial COD values from 1000 to 3000 mg/L, pH from 5 to 8, and PVA v% from 20 to 30%. The maximum COD reduction was estimated to occur at an initial COD of 2595 mg/L, PVA v% of 27%, and pH of 7.3. At optimum conditions, the bioreactor system was able to achieve a maximum COD reduction of 89%, which is quite close to the RSM prediction value of 90%. The optimum operating conditions were used to carry out continuous biodegradation, and the results indicated that the COD reduction increased from 60% to 62% with an increase in the air flow rate from 2 to 3.3 L_a/L_r.min. However, by increasing the liquid flow rate from 2.1 to 4.2 mL/min and back to 2.1 mL/min, the COD reduction decreased from 66% to 39%. The system responded quickly to the change in liquid flow rate and returned to the initial COD level. This indicates that the system is highly stable and can easily recover.

Keywords: GTL process water; biodegradation; spouted bed bioreactor (SBBR); *Pseudomonas aeruginosa*; response surface methodology (RSM)

Citation: Ahmed, S.A.; Surkatti, R.; Ba-Abbad, M.M.; El-Naas, M.H. Optimization of the Biotreatment of GTL Process Water Using *Pseudomonas aeruginosa* Immobilized in PVA Hydrogel. *Processes* **2022**, *10*, 2568. https://doi.org/10.3390/pr10122568

Academic Editors: Dimitris Zagklis and Georgios Bampos

Received: 25 October 2022
Accepted: 24 November 2022
Published: 2 December 2022

1. Introduction

The need for environmentally friendly fuels for energy has recently been steadily increasing. Gas-to-liquid (GTL) is a technological breakthrough that uses the Fischer–Tropsch (FT) process to transform natural gas (NG) into high-performance, ultra-clean liquid fuels. This innovative and rising technology is likely to contribute to a higher proportion of global gas processing in the future. The development of GTL technology has advanced technologically over the past ten years, and many commercial-scale plants have been constructed across the world [1]. GTL process water (PW) is a by-product of the Fischer–Tropsch (FT) reaction that produces water in significant quantities (~25%), more than other hydrocarbon products on a weight basis. Process water produced by GTL industries must be treated to meet the regulatory agency standards for both safe releases into water bodies and effective reuse [2]. The majority of COD in the GTL PW water stream comes from alcohols, aldehydes, and ketones, which can be effectively treated biologically.

Biological approaches are popular in the wastewater treatment field because of their ease of use, environmental friendliness, low cost, and long-term application [3]. Biological approaches involve using organisms such as bacteria, fungi, and algae to mineralize or break down contaminants into less toxic compounds [4]. The use of bacteria is beneficial since these microorganisms are flexible representatives and contain a variety of Actinomycetes [5]. Furthermore, these bacteria can produce spores, resist different contaminants, and survive in several environments. The organic pollutants in the GTL PW can be effectively treated biologically under aerobic and anaerobic conditions. The course of treatment may also combine anaerobic and aerobic systems. Several studies on GTL PW, namely FT wastewater, have been investigated on the bench and laboratory scales using both synthetic and actual wastewater, and biological treatment has been deliberated under anaerobic conditions [6]. GTL PW is usually treated using a well-known biological process in conventional activated sludge (CAS) systems. This method of treatment is based on biomass, which is held by a settler, aerobically degrading organic contaminants. Three steps of GTL PW treatment, including chemical, biological, and physical treatment approaches, were described by Pon Saravanan and Van Vuuren [2]. The three integrated processes of the GTL PW treatment plant used chemical treatment in the first stage to remove free oil and suspended hydrocarbons, followed by biotreatment in an aeration tank to eliminate carbonic and nitrogenous compounds, and finally physical treatment such as sand filtration in the third stage to eliminate suspended solids, oil, chemical oxygen demand, and related biological oxygen demand [2]. The CAS process was combined with ultrafiltration (CAS-UF) and used in the treatment of GTL PW. The CAS effluent can undergo post-treatment ultrafiltration (UF) to remove unsettled particles and further lower the COD in the effluent, allowing for the reuse of the treated water [7]. Additionally, Fischer–Tropsch (FT) reaction water from gas-to-liquid (GTL) industries was also treated using a membrane bioreactor (MBR) system and compared to the current treatment system [7]. Majone et al. [8] investigated the anaerobic biodegradation of synthetic FT wastewater with a high concentration of COD (~28 g/L) generated by long-chain alcohols utilizing a continuous flow-packed bed biofilm reactor (FPBBR) on a laboratory scale. They steadily increased the COD content in tests to evaluate the inhibitory effect of long-chain alcohol concentrations and achieved about 96% of COD reduction. Aerobic degradation of FT wastewater was studied by Chain et al. [9] to decrease the high COD specifically from short-chain alcohols and volatile fatty acids as they represented around 87% of the given wastewater. The FT wastewater was synthesized from SCAs and VFAs in a mineral salt solution and a COD of 67.9 g/L and biodegraded using *Bacillus* sp. Within 3 days, the strains reduced COD by up to 90% and effectively degraded the organic pollutants. Other studies [6,10] have suggested a combination of anaerobic biological treatment and chemical strategies to address the issue of incomplete degradation for long-chain alcohol under high organic load. Bio-electrochemical systems (BES) have been used in the treatment of FT wastewater by utilizing electrochemically active bacteria as a catalyst for oxidation and/or reduction processes at the anode and/or the cathode [6]. By combining an anaerobic digester with a BES, the treatment performance was improved as well as the production of biogas [11].

In most biological treatment processes of GTL PW, activated sludge is used as a biomass source where the biological community, considered as a mixed culture, contains a wide range of bacteria, fungi, and algae. Despite the biodegradability of these microorganisms, several types of pure cultures have been used over the years for the biodegradation of organic contaminants. *Pseudomonas* was first identified as a Gram-negative, polar-flagellated, and rod-shaped bacteria by Migula in 1894, during the 19th century [12]. It is one of the prokaryote genera that has undergone the most research (bacteria). *Pseudomonas aeruginosa* is a Gram-negative bacterium that can be found in almost any environment. Their metabolic capacity is vast, as evidenced by their ability to synthesize a wide range of secondary metabolites and polymers as well as their ability to employ a wide range of carbon sources and electron acceptors [13]. Many researchers have studied the ability of *P. aeruginosa* to degrade various chemicals in the batch and continuous bioreactors under various conditions. These include the degradation of 0.02% naphthalene [14], the biodegradation of volatile organic compounds [15], the biodegradation of petroleum compounds [16], the degradation of octamethylcyclotetrasiloxane (D4) [17], and the bioremediation of heavy metals [18]. However, the application of *P. aeruginosa* for the biotreatment of GTL PW has never been previously documented. Hence, this study aimed to investigate the biodegradation of GTL water by *P. aeruginosa* bacteria using a laboratory-scale spouted bed bioreactor system (SBBS). Moreover, the interaction between the operating parameters and the optimal conditions for the maximum reduction efficiencies by varying test parameters such as the initial COD concentration, PVA v%, and pH using RSM was investigated.

2. Materials and Methods

2.1. Chemicals and Materials

Most of the chemicals including NaOH pellets for pH adjustment and mineral salts including $MgSO_4{\cdot}7H_2O$, K_2HPO_4, $CaCl_2{\cdot}2H_2O$, $(NH_4)_2CO_3$, $FeSO_4{\cdot}7H_2O$, $ZnSO_4{\cdot}7H_2O$, $MnCl_2{\cdot}4H_2O$, $CuSO_4{\cdot}5H_2O$, $CoCl_2{\cdot}6H_2O$, and $Na_2MoO_4{\cdot}2H_2O$ were obtained from Sigma Aldrich, St. Louis, MO, USA. However, PVA powder was obtained from BDH, London, UK.

2.2. GLT PW Samples

The water samples were obtained from a local GTL plant in Qatar and pretreated using air stripping to remove volatile organic pollutants. GTL PW is characterized by high acidity and COD content. Table 1 shows the physical and chemical properties of the original and pretreated GTL process.

Table 1. The physical and chemical characteristics of GTL.

Characteristic	GTL PW	Pretreated GTL PW
COD (mg/L)	5000–7000	2000 to 4000
TOC (mg/L)	1500–1700	700–1400
pH	2.9	3.3

2.3. Isolation and Immobilization of Bacterial Culture

Pseudomonas aeruginosa was isolated with other hydrocarbon-degrading bacteria from different highly contaminated soils in Qatar. According to Al Disi et al. [19], at a tillage depth of 1–2 cm, random sampling was taken with a sterile spatula from various spots. The soil samples were collected and placed in clean glass bottles, which were then securely sealed, labeled, and twisted with foil to prevent contamination and prevent any further light reactions. The technique for enrichment, isolation, identification, and studying the activation of hydrocarbon-degrading bacteria was described thoroughly by Al Disi et al. [19]. The isolated *Pseudomonas aeruginosa* bacterial cells were collected in a glass jar and kept refrigerated until they were immobilized in PVA gel.

To prepare PVA with 10 wt%, a homogenous PVA solution was made by mixing 100 g of PVA powder with 850 mL of distilled water and 50 mL of a suspension of the isolated bacteria cells at roughly 70–80 °C. The solution was then mixed using a glass rod to ensure its homogeneity. PVA solution was poured into ice molds and cross-linked using freezing–thawing cycles, by freezing at −20 °C for 20 h, and thawing at +20 °C for 4 h. The FT process was repeated four times. This prepared PVA gel is known to produce a high-porosity polymeric matrix with good mechanical strength and stability [20].

2.4. Biomass Acclimatization

The immobilized bacteria were slowly acclimatized to GTL PW by placing them in a batch SBBR with GTL PW and a mineral nutrient solution. The acclimatization process was performed by a gradual increase in the GTL PW COD in the range from 500 to 2000 mg/L over a period of two weeks. After this step, the bacteria were fully acclimatized to GTL-PW and were ready for the biodegradation process [20]. Table 2 shows the composition of mineral salts added to the GTL PW solution.

Table 2. Composition of the mineral salt medium [21].

Component	Concentration (mg/L)
$MgSO_4 \cdot 7H_2O$	300
K_2HPO_4	250
$CaCl_2 \cdot 2H_2O$	150
$(NH_4)_2CO_3$	120
$FeSO_4 \cdot 7H_2O$	3.5
$ZnSO_4 \cdot 7H_2O$	1.3
$MnCl_2 \cdot 4H_2O$	0.13
$CuSO_4 \cdot 5H_2O$	0.018
$CoCl_2 \cdot 6H_2O$	0.015
$Na_2MoO_4 \cdot 2H_2O$	0.013
Total	824.98

2.5. Spouted Bed Bioreactor System (SBBR)

The SBBR utilized in this experiment was made of Plexiglas and had a total volume of 1.5 L. The SBBR is characterized by systematic intensive mixing caused by the cyclic motion of particles within the bed, which is caused by a single air jet injected through an aperture at the bottom of the reactor [22]. The temperature (at 32 °C) of the bioreactor was controlled through the circular movement of the water in the water jacket surrounding the reactor. Air was continuously supplied into the reactor at a certain flow rate to improve mixing and to also provide oxygen to maintain aerobic conditions [23,24]. A schematic diagram of the SBBR is shown in Figure 1.

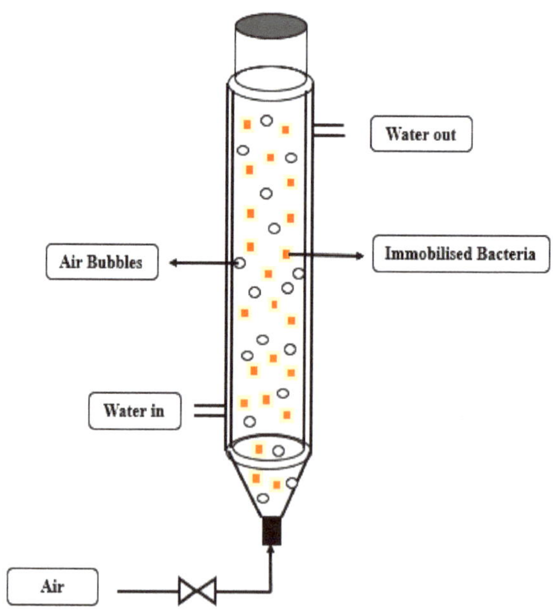

Figure 1. A schematic diagram of the spouted bed bioreactor (SBBR).

2.6. Batch Biological Treatment of GTL PW

In batch experiments of GTL PW, the reactor was initially filled with the standard nutrient medium solution as well as the biocatalyst, which is the PVA gel with the immobilized bacteria, which occupied 20 to 30% of the reactor volume based on previous studies [22]. The reactor was filled with a total volume of 1.5 L (GTL PW and immobilized bacteria). Air was injected into the conical bottom of the bioreactor to ensure proper mixing and interaction between the substrate and the bacterial cells within the bioreactor as well as to provide the oxygen required for biodegradation. The air flow rates, and liquid flow rates were adjusted to 3.3 L_a/L_r.min and 2.4 mL/min, respectively. The reactor temperature was held at almost 32 °C. This temperature was found to be optimum in a past study [23], and is in good agreement with the values mentioned in the literature [24]. The initial pH solution was kept at pH 7 using NaOH pellets as these conditions are regarded to be the best for organic biodegradation according to El-Naas et al. [22].

2.7. Continuous Biological Treatment of GTL PW

Continuous experiments were performed to study the effects of air flow rates at 2, and 3.3 L_a/L_r.min, and liquid flow rates at 2.1 and 4.2 mL/min on the biotreatment of GTL PW under optimum conditions selected based on RSM. To study the response of the SBBR system to other operating factors including air and liquid flow rates, continuous experiments were carried out at constant COD of 2200 mg/L, pH of 7.29, PVA v% of 27%, and temperature of 32 °C. The stripped GTL PW mineral solution was constantly introduced to the reactor during all experimental runs, utilizing a peristaltic pump at a stable liquid flow rate with an accuracy of ±1 mL/min. The samples were collected and analyzed using the COD at various periods. All continuous experiments were carried out in replicates and the average was plotted. The standard error ranged from 5 to 10% of the reported average.

2.8. Analytical Methods

A HAC-UV spectrophotometer with COD reagents was used to conduct the COD analysis. Two mL of the water sample was added to the HAC LCK514 cuvettes and heated for 2 h to complete the reaction between the reagent and the water sample. The COD content in mg/L was measured using the HAC DR 3900.

2.9. Statistical Analysis and Optimization Using RSM

The current study aimed to investigate the biotreatment of GTL PW by *Pseudomonas aeruginosa* bacteria as well as the interaction between operating parameters and the optimum conditions for the maximum COD reduction efficiencies by varying the test parameters such as the initial COD concentration, immobilized bacteria volume percentage, and pH using RSM. The RSM was carried out for the biotreatment of the two types of wastewater under specific experimental conditions. The parameters were selected based on the screening experiments. Using RSM, a large amount of data can be obtained with a small number of experimental operations, and there are many important advantages such as the ability to determine not only the effects of individual parameters, but also their relative importance in a given and the interactive effects of two or more variables. Table 3 lists the three variables used in this study, each of which has two levels (upper and lower).

Table 3. Variables and levels applied for the removal of COD in GTL PW.

Factor	Units	$-\alpha$	Lower Limit (−1)	0	Upper Limit (+1)	$+\alpha$
Concentration	mg/L	318.21	1000	2000	3000	3681.79
pH	-	3.98	5	6.5	8	9.02
PVA v%	-	16.59	20	25	30	33.41

The central composite method (CCD) experimental design of RSM was chosen to find the relationship between the response functions and variables using the statistical software tool MINITAB 20. In the central composite method, a total number of 20 experiments including center points was augmented with a group of axial points (also called star points) to estimate the curvature.

The reduction rate of COD was determined according to the following equation:

$$\text{COD Reduction \%} = \frac{C_0 - C}{C_0} \tag{1}$$

where C (mg/L) is the final COD concentration of the sample, and C_0 (mg/L) represents the initial COD of the water sample.

Using the optimal design factor (RSM), the coefficients were also predicted in a quadratic polynomial mathematical model to predict the COD reduction efficiencies. Table 4 lists the several parameter combinations used in the central composite design by response surface methodology using CCD.

Table 4. The several parameter combinations used in the design of the GTL removal experiments.

RunOrder	PtType	Blocks	COD	pH	PVA Vol. %
1	0	1	2000	6.5	25
2	1	1	3000	8	20
3	1	1	1000	5	30
4	−1	1	2000	3.98	25
5	−1	1	2000	6.5	16.59
6	1	1	3000	5	20
7	1	1	3000	5	30
8	0	1	2000	6.5	25
9	−1	1	2000	9.02	25
10	−1	1	3681.79	6.5	25
11	−1	1	318.20	6.5	25
12	1	1	1000	8	20
13	0	1	2000	6.5	25
14	0	1	2000	6.5	25
15	0	1	2000	6.5	25
16	1	1	1000	8	30
17	−1	1	2000	6.5	33.40
18	1	1	1000	5	20
19	0	1	2000	6.5	25
20	1	1	3000	8	30

3. Results and Discussion

3.1. Statistical Analysis

All statistical combinations of the variables were evaluated using the DOE software, which yielded the design outcomes from the experiments. *P. aeruginosa* had high coefficients of determination R2 of 0.8140 for the biotreatment of GTL PW. RSM was used to optimize the degradation conditions. The CCD method was used to investigate the effects of the initial COD concentration, PVA v%, and pH on COD reduction. Equation (2) shows the quadratic polynomial mathematical model to predict the COD reduction efficiencies.

Regression Equation in Uncoded Units:

$$\begin{aligned}\text{COD Reduction \%} = {} & -304.5 + 0.0198\,\text{COD} + 37.0\,\text{pH} + 17.29\,\text{PVA v\%} - 0.000004\,\text{COD} * \text{COD} - 1.654\,\text{pH} * \text{pH} \\ & -0.2728\,\text{PVA v\%} * \text{PVA v\%} - 0.00077\,\text{COD} * \text{pH} + 0.000151\,\text{COD} * \text{PVA v\%} - 0.403\,\text{pH} \\ & *\text{PVA v\%}\end{aligned} \tag{2}$$

3.2. Effect of Initial COD

The contaminant concentration is one of the most important factors that affect the biological treatment of wastewater. The increase in the organic concentration may inhibit the biodegradation process or increase the degradation efficiency [25]. Figure 2a,b shows the effect of interaction between the initial COD and other two process parameters, namely, pH, and PVA volume fraction on the removal of COD. At a constant pH of 5, a PVA v% of 20, and an initial COD of 0f 1000 mg/L, the COD reduction % was 58. However, under the same conditions, increasing the initial COD to 3000 mg/L resulted in a COD reduction % of 69.89 (around a 12% increase). Similar behavior was also noticed at a constant pH of 5, and PVA v% of 30, where an increase of the initial COD from 1000 mg/L to 3000 mg/L resulted in a COD reduction from 62.8% to 83.27%, respectively (approximately 21% increase). This behavior reveals that when the COD levels are higher, there is a greater oxygen demand. This implies that water with high COD levels likely contains more oxidizable organic material, and the microorganism can adapt to the higher organic concentration, resulting in high removal efficiency [26]. Moreover, one can observe that the biodegradation increased as COD increased from 500 to ~2600 mg/L and then decreased. It is anticipated that exposure to a high concentration of contaminants might suddenly have detrimental effects on the bacterial enzymes that are often responsible for the major stages in the biodegradation process [27]. This also implies that wastewater with high COD levels has lower dissolved oxygen (DO) values. Life forms need oxygen to survive, hence a low quantity of dissolved oxygen is dangerous. Therefore, it is advisable to increase DO concentrations by reducing COD levels in wastewater before releasing it [26].

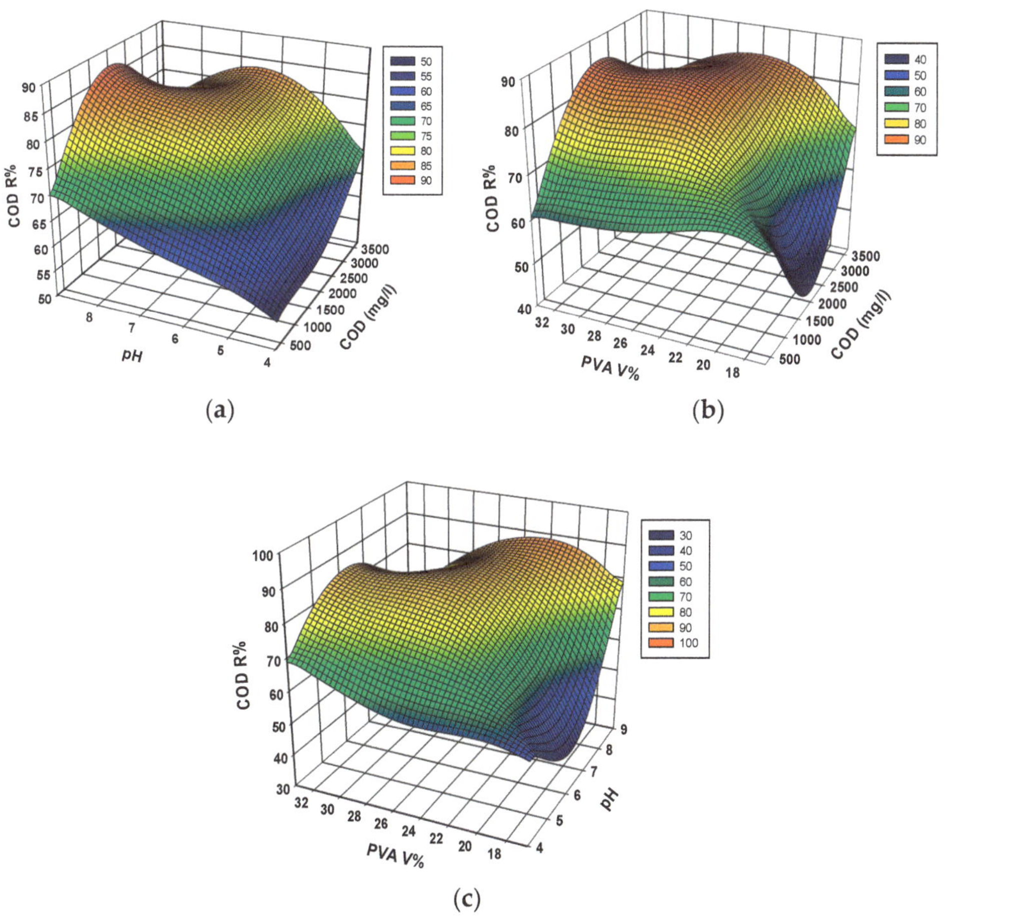

Figure 2. The interaction between the experimental parameters in the GTL experiment. (**a**) COD reduction % vs. pH and initial COD (mg/L); (**b**) COD reduction % vs. initial COD concentration (mg/L) and PVA volume fraction; (**c**) COD reduction % vs. pH and PVA volume fraction.

3.3. Effect of pH

The initial pH of the solution has a big impact on the microbial growth and enzyme activity, therefore, it is significant when it comes to developing biological treatment methods. Similarly, the effects of the interaction relation between pH with the initial COD and PVA volume fraction on the COD reduction rate were determined and the results are shown in Figure 2a,c. At a constant initial COD of 2000 mg/L, a PVA v% of 20, and an initial pH of 3.9, the COD reduction % was 58.15. However, under the same conditions, increasing the pH to 6.5 resulted in a COD reduction % of 86.77 (around a 29% increase). Because almost all biological species seemed to have optimal pH conditions, pH has a physiological effect on microbial activity, similar to temperature [28]. Although some species need a restricted pH range, others are more tolerant of a larger pH range. Some species thrive in high-pH environments, whereas others thrive in low-pH environments. Research of *Pseudomonas aeruginosa* biotreatment of GTL-produced water mixtures found that biomass activity was entirely reduced at pH 5, 9, and 10, with pH 6–8 being the optimal [28]. Because most bacteria are neutrophils, the optimal pH for the most efficient biotreatment of organic substances is usually about 7.5 [29].

3.4. Effect of PVA Volume %

The rate of biodegradation of organic pollutants is significantly influenced by the volume fraction of PVA pellets, which is directly related to the quantity of active biomass cells in the bioreactor [30]. Figure 2b,c show the effect of interactions between PVA v% with initial COD concentration, and pH respectively. The number of active biomass cells in the bioreactor is directly proportional to the volume fraction of PVA pellets, and so plays an essential role in determining the rate of COD reduction. At a constant initial COD of 2000 mg/L, pH of 6.5, and PVA v% of 16.9, the COD reduction % was 42.85. However, under the same conditions, increasing the PVA v% to 30 resulted in a COD reduction % of 86.77 (around a 44% increase). The COD reduction increased as the PVA volume fraction increased. However, the removal effectiveness of PVA v% 25 and 33% appeared to be nearly the same. This could be because 33% had a greater biomass and 25% had better mixing. It is believed that increasing the volume fraction of PVA above 30% is likely to result in reduced mixing and, as a result, lower degradation performance [30].

The optimization of the biodegradation of the GTL PW was evaluated. Figure 3 shows that the optimum conditions obtained from RSM to achieve a maximum COD reduction efficiency of 90% was predicted to occur at an initial COD of 2595 mg/L, PVA volume fraction of 27%, and pH of 7.3.

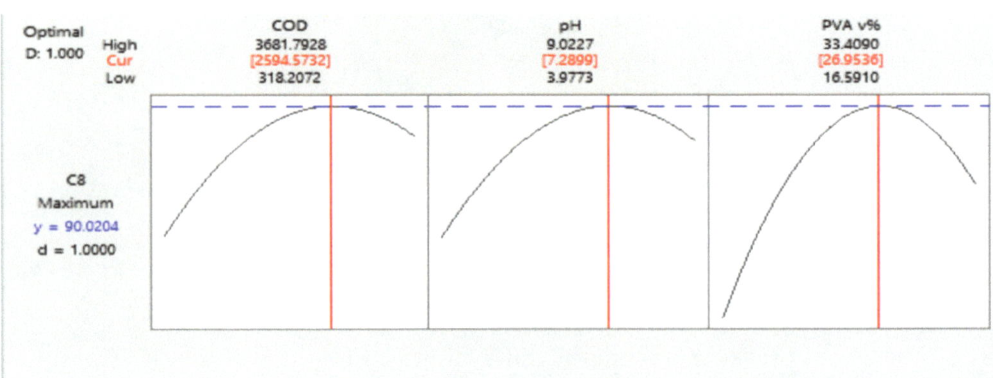

Figure 3. The optimum conditions for COD reduction % were obtained using the response optimizer.

3.5. Model Validation

The experimental response (i.e., COD reduction %) was compared to the predicted response to validate the accuracy of RSM, as shown in Figure 4. The collected experimental values, illustrated by scattered points, showed a linear relationship between the experimental and predicted values. The R^2 value of 0.8481 indicates a close fit between the modeled and collected data. Moreover, the biodegradation experiment was carried out at optimum conditions obtained from the optimization of the COD reduction to the maximum value. Results showed that the predicted optimal COD from RSM had a good match to the experimental value (90% and 89%, respectively). This means that the model for the biotreatment of GTL PW was valid.

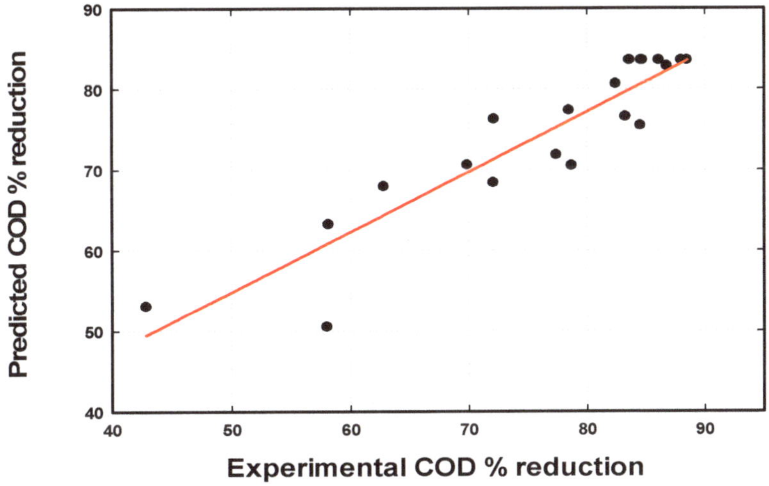

Figure 4. Predicted versus experimental COD reduction values for GTL PW.

3.6. Continues Biological Treatment of GTL PW

3.6.1. Effect of Air Flow Rate

The bioreactor's air flow rate is crucial in ensuring that there is enough oxygen for biodegradation and adequate mixing via particle movement. Different air flow rates of 1, 2, and 3.3 L_a/L_r.min (liter of air per liter of the reactor per min) were operated for 24 h to evaluate the impact of air flow rate on the continuous biological treatment of GTL PW. However, at 1 L_a/L_r.min, the aeration and mixing of the immobilized bacteria were limited, thus, only 2, and 3.3 L_a/L_r.min were studied. The initial COD concentration, liquid flow rate, and temperature were kept at the optimum conditions of ~2200 mg/L, 2.1 mL/min, and 32 °C, respectively. Samples were collected from the effluent for the COD analysis. Figure 5 shows COD reduction with increasing time for two different air flow rates 2, 3.3 L_a/L_r.min. When comparing the airflow rates of 2 and 3.3 L_a/L_r.min, the difference in DO was found to be insignificant (5.2, and 5.5 mg/L, respectively). Moreover, there was a slight difference in the mixing of immobilized bacteria between the two flow rates, which could be attributed to the lower COD reduction at 2 L_a/L_r.min (60.6%) compared to the air flow rate of 3.3 L_a/L_r.min (62%). The findings demonstrate that the air flow rate affects the biodegradation rate of organic pollutants. In general, at a specific range, the higher the airflow rate, the better the biodegradation rate, which are two key factors for how the air flow rate affects the biotreatment of organic compounds by feeding the necessary amount of oxygen [21]. Moreover, Gopalakrishnan et al. [31] reported that the excess supply of oxygen led to the higher degradation activity of the biomass, therefore, the increase in airflow rate will increase the reduction in COD. However, at a lower range, the biodegradation rate will decrease due to the fact that, after a certain limit, critical velocity occurs, causing the biomass support particles to settle down at the bottom of the reactor. As a result, there are fewer interactions between the substrate, biomass, and air, which leads to less reduction in COD. For the air flow rate at a lower range, oxygen is the growth-limiting factor [31].

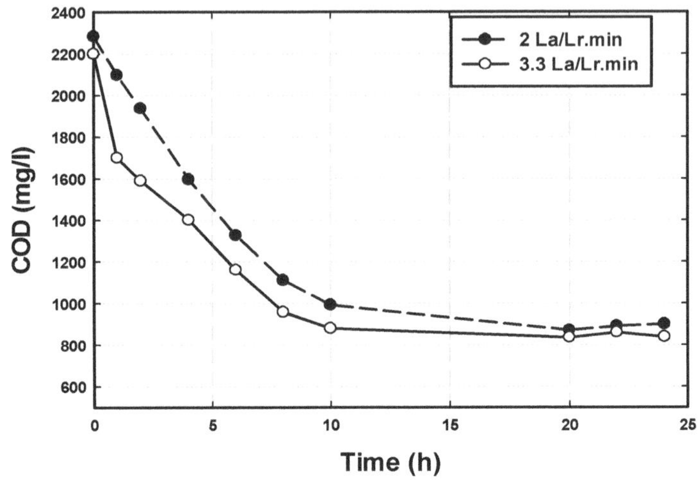

Figure 5. The concentration of GTL PW in SBBR vs. time for two different air flow rates (2, and 3.3 L_a/L_r). Initial COD = ~2200 mg/L; PVA volume % = 27; reactor temperature = 32 °C; liquid flow rate = 2.1 mL/min.

3.6.2. Effect of Liquid Flow Rate

To assess the effect of liquid flow rate step change on the biotreatment of GTL PW, a continuous experiment was carried out at the optimum conditions obtained from batch experiments in Section 3.4 (initial COD of ~2200 mg/L; PVA v% of 27; reactor temperature at 32 °C; and an airflow rate of 2.1 L/min and HRT of 12 h). Figure 6 shows the decrease in COD reduction through these step changes in the liquid flow rate. First, the experiment was carried out for around 22 h to reach a steady state and a COD reduction of 65.69% was achieved. Then, a step change of liquid flow rate from 2.1 to 4.2 mL/min was performed, reducing the HRT to 6 h. A second steady state condition with a liquid flow rate of 4.2 mL/min was reached, where the COD reduction declined to 38.61%. After this time, the liquid flow rate was returned to its previous value to reach the third steady state condition with a 63.68% COD reduction. This is to be expected because a higher liquid flow rate reduces the residence time in the bioreactor HRT from 12 h to 6 h at liquid flow rates of 2.1 and 4.2 mL/min, respectively. Thus, the contact time between immobilized bacteria and organic pollutants PW reduced and resulted in lower COD reduction [21]. It is generally accepted that the inlet feed flow rate has the greatest influence on determining the efficiency of bioreactors by governing the retention time of the contaminants within the bioreactor [20,32]. The more the HRT, the greater the chance for the efficient decomposition of the organic contaminant. A very low liquid flow rate, however, may also result in mass transfer constraints, which could lead to a decrease in pollutant removal [33]. By increasing the liquid flow rate, more organic pollutants are introduced to the bioreactor, thus the substrate may serve as a nutrient for the microbial biomass until a certain initial concentration, however, an increase in substrate concentration may impart the toxic effect of pollutants on the metabolic activity of microbial biomass, resulting in lower removal efficiency [33].

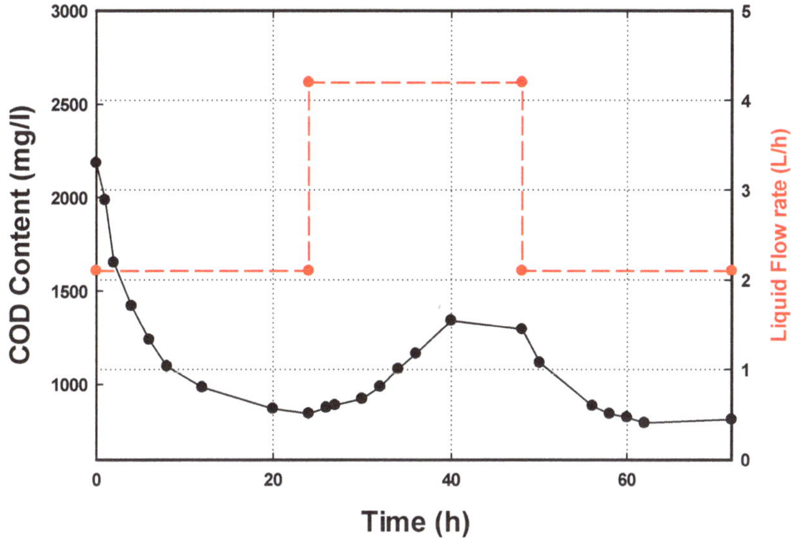

Figure 6. Liquid flow rate step-change of (2.1–4.5 and back to 2.1 mL/min). Initial COD = 2200 mg/L; temperature = 32 °C; air flow rate 3.3 L_a/L_r.min.

Kawan et al. [34] reported that the water flow or movement in the reactor caused a high water velocity in the reactor, a short duration of contact between the organic matter, and less time for biofilm formation when the reactor was run at a high flow rate (short retention time (HRT). As a result, the capability of the organisms to metabolize organic materials was diminished. Additionally, Chen et al. noted that minimizing the hydraulic retention time (HRT) and lowering the pH level in the system are two ways to minimize the excessive accumulation of butyric acid and propionic acid, which is a critical component of successful FT wastewater treatment [35].

3.6.3. Dynamic Behavior

The variation in the load, the hydraulic characteristics of the reactor (transport and mixing), and the transformation processes are all factors that affect the dynamic, time-dependent behavior of reactors. Many effects are unreasonable and cannot be fully understood without thoughtful deliberation and analysis. A set of time constants that can be assessed depending on the rate of specific processes governs the time-dependent behavior of systems. Additionally, comparing these time constants enables the assessment of which variable has a significant or minor impact on the system behavior. The reactor time constant (τ_p), which represents the reactor response to a change in liquid flowrate, and the dead time (t_d), which measures the interval between a step change and the first response of the measured COD, were calculated using Equations (3) and (4), respectively [36].

$$t_d = t_1 - t_2 \tag{3}$$

$$\tau_p = t_{63.2\%} - t_2 \tag{4}$$

where t_1 is the time when the step change is made; t_2 represents the time when the measured COD first responds to the step change; $t_{63.2\%}$ is the time when the measured COD reaches 63.2% of its total final change. The results given in Table 5 and Figure 7 demonstrate that $t_d < \tau_p$ in the experiments, which indicates a tight and easy overall process control [36]. It was found that the dead time was connected to the step change direction in the liquid flow rate. For instance, when the liquid flow rate was reduced, the dead time decreased. These findings are quite encouraging in terms of the adaptation and suitability of the reactor system for large-scale processes.

Table 5. Step-change in the liquid flow rate and HRT; dead time and time constant.

Condition	Step Change	t_1 (h)	t_2 (h)	$t_{63.2\%}$ (h)	t_d (h)	τ_p (h)
Liquid flow rate	2.1–4.2	22	24	33.376	2	9.376
Liquid flow rate	4.2–2.1	48	48.5	56.848	0.5	6.848

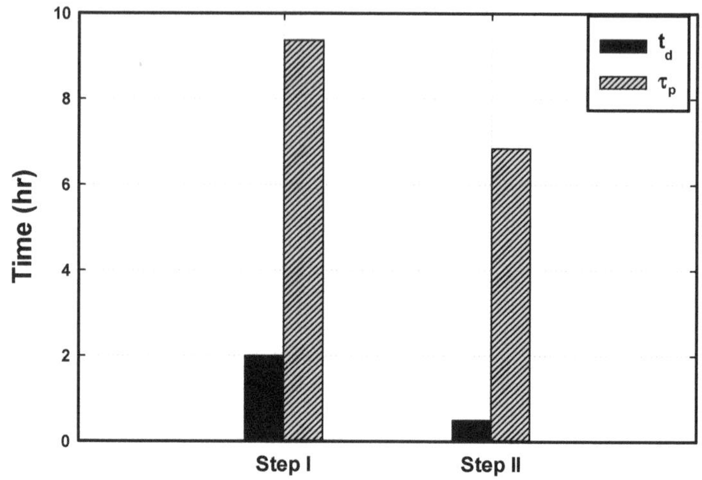

Figure 7. The dead time and reactor time constants for a step change experiment.

Although these continuous experiments were limited to studying only the effect of the liquid and air flow rates, other effects such as the initial COD and immobilized PVA particle size can be investigated in the future. A study of the continuous biotreatment of phenol in SBBR was performed by El-Naas et al. [21] who investigated the effect of the initial COD and PVA immobilized particle size on the biotreatment of phenol. It revealed that the rate of phenol reduction is stable for a longer period at high initial COD concentrations than at low initial COD concentrations, as it appears to decrease over time. Bacteria immobilized within the PVA particles may only have limited access to organic compounds at low COD concentrations, which can be explained by mass transfer limitations. In addition, the particle size of PVA has a significant effect on the continuous biotreatment of phenol, which is increased by particle size reduction, especially for high air flow rates.

The average COD for GTL process water treated in this study was found to be ~441 mg/L, which is considered to be purified water, as reported by Luis et al. [37]. Therefore, further pre/post-treatment is required. Table 6 shows GTL PW characteristics after treatment. Table 6 shows the treated GTL PW characteristics.

Table 6. The physical and chemical characteristics of GTL.

Characteristic	GTL PW	Pretreated GTL PW	Treated GTL PW (Batch Experiments)
COD (mg/L)	5000–7000	2000 to 4000	~441
TOC (mg/L)	1500–1700	700–1400	~154
pH	2.9	3.3	7.2

4. Conclusions

In a specially designed SBBR system, the biotreatment of organic contaminants in the GTL process was investigated. Using RSM, the second-order polynomial was found to be effective in forecasting COD reduction when three independent variables were used: initial COD concentration, pH, and PVA v%. All three factors had a considerable impact on the GTL PW biotreatment, and the maximum COD reduction for GTL PW was found to be 89% in batch experiments. Additionally, when applying the RSM optimizer for GTL PW, the optimal COD reduction was predicted to be 90% at the optimum condition of an initial COD of 2595 mg/L, PVA v% of 27%, and pH of 7.3. A continuous study was carried out to investigate the effect of the liquid flow rate and air flow rate on the organic removal from GTL PW. The results showed that when increasing the air flow rate from 2 to 3.3 L_a/L_r.min, the COD reduction and DO of GTL PW increased from 60.6% to 62%, 5.2 and 5.5 mg/L, respectively. This demonstrates that the difference between the two flow rates was insignificant. A step change experiment of the liquid flow rate from 2.1 to 4.2 mL/min and then back to 2.1 mL/min was performed. The system responded quickly to the change in liquid flow rate and returned to the initial COD level. This indicates that the system is highly stable and can easily recover. These findings are quite encouraging in terms of the adaptation and suitability of the reactor system for large-scale processes.

Author Contributions: Conceptualization, S.A.A. and M.H.E.-N.; Methodology S.A.A., R.S., M.M.B.-A. and M.H.E.-N.; Software, S.A.A.; Investigation, S.A.A., R.S., M.M.B.-A. and M.H.E.-N.; Data curation, S.A.A. and R.S.; Writing—original draft preparation, S.A.A.; Writing—review and editing, S.A.A., R.S., M.M.B.-A. and M.H.E.-N.; Supervision, M.M.B.-A. and M.H.E.-N.; Project administration, M.H.E.-N.; Funding acquisition, M.H.E.-N. All authors have read and agreed to the published version of the manuscript.

Funding: The authors would like to acknowledge the support of Qatar National Research Fund (a member of Qatar Foundation) through grant #NPRP100129170278.

Data Availability Statement: Not applicable.

Acknowledgments: The authors would like to acknowledge the help of Zulfa Al Disi from the biological science lab at Qatar University and Eng. Dan Cortes for his valuable technical assistance.

Conflicts of Interest: The authors declare no conflict of interest.

References

1. Elbashir, N.O.; Eljack, F.T. A Method to Design an Advanced Gas-to-Liquid Technology Reactor for Fischer-Tropsch Synthesis. In Proceedings of the 2nd Annual Gas Processing Symposium, Doha, Qatar, 10–14 January 2010; pp. 369–377.
2. Saravanan, N.P.; van Vuuren, M.J. Process wastewater treatment and management in gas-to-liquids industries. In Proceedings of the SPE Oil and Gas India Conference and Exhibition, Mumbai, India, 20–22 January 2010; Volume 1, pp. 176–184.
3. Surkatti, R.; El-Naas, M.H.; van Loosdrecht, M.C.M.; Benamor, A.; Al-Naemi, F.; Onwusogh, U. Biotechnology for gas-to-liquid (GTL) wastewater treatment: A review. *Water* **2020**, *12*, 2126. [CrossRef]
4. Ding, P.; Chu, L.; Wang, J. Biological treatment of actual petrochemical wastewater using anaerobic/anoxic/oxic process and the microbial diversity analysis. *Appl. Microbiol. Biotechnol.* **2016**, *100*, 10193–10202. [CrossRef] [PubMed]
5. Satish, G.P.; Ashokrao, D.M.; Arun, S.K. Microbial degradation of pesticide: A review. *Afr. J. Microbiol. Res.* **2017**, *11*, 992–1012. [CrossRef]
6. Wang, D.; Ma, W.; Han, H.; Li, K.; Hao, X. Enhanced treatment of Fischer–Tropsch (F-T) wastewater by novel anaerobic biofilm system with scrap zero valent iron (SZVI) assisted. *Biochem. Eng. J.* **2017**, *117*, 66–76. [CrossRef]
7. Laurinonyte, J.; Meulepas, R.J.W.; van den Brink, P.; Temmink, H. Membrane Bioreactor (MBR) as Alternative to a Conventional Activated Sludge System Followed by Ultrafiltration (CAS-UF) for the Treatment of Fischer-Tropsch Reaction Water from Gas-to-Liquids Industries. *Water Air Soil Pollut.* **2017**, *228*, 1–13. [CrossRef]
8. Majone, M.; Aulenta, F.; Dionisi, D.; D'Addario, E.N.; Sbardellati, R.; Bolzonella, D.; Beccari, M. High-rate anaerobic treatment of Fischer–Tropsch wastewater in a packed-bed biofilm reactor. *Water Res.* **2010**, *44*, 2745–2752. [CrossRef]
9. Moreroa, M.; Hildebrandt, D.; Matambo, T. Aerobic bioremediation of fischer-tropsch effluent—Short chain alcohols and volatile fatty acids. *Lect. Notes Eng. Comput.* **2018**, *2238*, 528–531.
10. Enyi, G.C.; Nasr, G.G.; Burby, M. Economics of wastewater treatment in GTL plant using spray technique. *Int. J. Energy Environ.* **2013**, *4*, 2076–2909.

11. Vashi, H.; Iorhemen, O.T.; Tay, J.H. Aerobic granulation: A recent development on the biological treatment of pulp and paper wastewater. *Environ. Technol. Innov.* **2018**, *9*, 265–274. [CrossRef]
12. Anayo, O.F.; Scholastica, E.C.; Peter, O.C.; Nneji, U.G.; Obinna, A.; Mistura, L.O. The Beneficial Roles of Pseudomonas in Medicine, Industries, and Environment: A Review. In *Pseudomonas aeruginosa: An Armory Within*; IntechOpen: Lagos, Nigeria, 2019.
13. Moradali, M.F.; Ghods, S.; Rehm, B.H.A. *Pseudomonas aeruginosa* Lifestyle: A Paradigm for Adaptation, Survival, and Persistence. *Front. Cell. Infect. Microbiol.* **2017**, *7*, 39. [CrossRef]
14. Retmana, B.P.; Sitaresmi, S.; Yasman, Y.; Oetari, A. Biodegradation of 0.02% Naphthalene by *Pseudomonas aeruginosa* DRK 9.1 Isolated from Volcanic Mud in Renokenongo Village, Sidoarjo. *AIP Conf. Proc.* **2020**, *2242*, 50025.
15. Yoshikawa, M.; Zhang, M.; Toyota, K. Biodegradation of Volatile Organic Compounds and Their Effects on Biodegradability under Co-Existing Conditions. *Microbes Environ.* **2017**, *32*, 188–200. [CrossRef] [PubMed]
16. Darsa, K.V.; Thatheyus, A.J. Biodegradation of Petroleum Compound Using *Pseudomonas aeruginosa*. *OALib* **2014**, *1*, 1–9. [CrossRef]
17. He, S.; Ni, Y.; Lu, L.; Chai, Q.; Liu, H.; Yang, C. Enhanced biodegradation of n-hexane by *Pseudomonas* sp. strain NEE2. *Sci. Rep.* **2019**, *9*, 16615. [CrossRef] [PubMed]
18. Kaur, R.; Mavi, G.K.; Raghav, S. Bioremediation of Sludge using *Pseudomonas aeruginosa*. *Int. J. Curr. Microbiol. Appl. Sci.* **2019**, *8*, 69–79. [CrossRef]
19. Al Disi, Z.; Jaoua, S.; Al-Thani, D.; Al-Meer, S.; Zouari, N. Considering the Specific Impact of Harsh Conditions and Oil Weathering on Diversity, Adaptation, and Activity of Hydrocarbon-Degrading Bacteria in Strategies of Bioremediation of Harsh Oily-Polluted Soils. *BioMed Res. Int.* **2017**, *2017*, 8649350. [CrossRef]
20. El-Naas, M.H.; Mourad, A.-H.I.; Surkatti, R. Evaluation of the characteristics of polyvinyl alcohol (PVA) as matrices for the immobilization of *Pseudomonas putida*. *Int. Biodeterior. Biodegrad.* **2013**, *85*, 413–420. [CrossRef]
21. El-Naas, M.H.; Al-Zuhair, S.; Makhlouf, S. Continuous biodegradation of phenol in a spouted bed bioreactor (SBBR). *Chem. Eng. J.* **2010**, *160*, 565–570. [CrossRef]
22. El-Naas, M.H.; Al-Zuhair, S.; Makhlouf, S. Batch degradation of phenol in a spouted bed bioreactor system. *J. Ind. Eng. Chem.* **2010**, *16*, 267–272. [CrossRef]
23. El-Naas, M.H.; Al-Zuhair, S.; Alhaija, M.A. Removal of phenol from petroleum refinery wastewater through adsorption on date-pit activated carbon. *Chem. Eng. J.* **2010**, *162*, 997–1005. [CrossRef]
24. Kumar, A.; Kumar, S.; Kumar, S. Biodegradation kinetics of phenol and catechol using *Pseudomonas putida* MTCC 1194. *Biochem. Eng. J.* **2005**, *22*, 151–159. [CrossRef]
25. El-Naas, M.H.; Al-Muhtaseb, S.A.; Makhlouf, S. Biodegradation of phenol by *Pseudomonas putida* immobilized in polyvinyl alcohol (PVA) gel. *J. Hazard. Mater.* **2009**, *164*, 720–725. [CrossRef] [PubMed]
26. Burns, T. How to Reduce COD in Water | Bioprocess H_2O. 2021. Available online: https://www.bioprocessh2o.com/blog/cod (accessed on 11 September 2022).
27. El-Naas, M.H.; Al-Zuhair, S.; Al-Lobaney, A.; Makhlouf, S. Assessment of electrocoagulation for the treatment of petroleum refinery wastewater. *J. Environ. Manag.* **2009**, *91*, 180–185. [CrossRef] [PubMed]
28. You, C.; Okano, H.; Hui, S.; Zhang, Z.; Kim, M.; Gunderson, C.W.; Wang, Y.-P.; Lenz, P.; Yan, D.; Hwa, T. Coordination of bacterial proteome with metabolism by cyclic AMP signalling. *Nature* **2013**, *500*, 301–306. [CrossRef]
29. Achrem, M.; Skuza, L.; Kalinka, A.; Szućko, I.; Filip, E.; Słominska-Walkowiak, R.; Rogalska, S.M. Role of Epigenetic Mechanisms in Plant Response to Low Temperature. *Acta Biol. Crac. Bot.* **2012**, *54*, 7–15. [CrossRef]
30. El Telib, A.E.; El-Naas, M.H.; Acio, J.A. Biodegradation of BTEX: Optimization through Response Surface Methodology. *Am. J. Eng. Appl. Sci.* **2017**, *10*, 20–31. [CrossRef]
31. Gopalakrishnan, B.; Muthukumarapandian, A.; Sujatha, S.; Raja, S.; Rajamohan, N.; Rajasimman, M. Statistical modeling and optimization of tannery wastewater treatment in a fluidized bed bioreactor with low density biomass support. *Model. Earth Syst. Environ.* **2021**, *8*, 1099–1107. [CrossRef]
32. Azimi, N.; Hassani, A.H.; Darzi, G.N.; Borghei, S.M. Biodegradation of Wastewater Containing High Concentration of Sulfamethoxazole by Antibiotic Adopted Biofilm in Attached Growth Bioreactor. *Pol. J. Environ. Stud.* **2017**, *26*, 2463–2469. [CrossRef]
33. Mohanty, S.S.; Jena, H.M. Biodegradation of Herbicide by the Immobilized Microbial Consortium SMC1 in Continuous Packed-Bed Biofilm Reactor. *Front. Chem. Eng.* **2022**, *4*, 44. [CrossRef]
34. Kawan, J.A.; Suja', F.; Pramanik, S.K.; Yusof, A.; Rahman, R.A.; Hasan, H.A. Effect of Hydraulic Retention Time on the Performance of a Compact Moving Bed Biofilm Reactor for Effluent Polishing of Treated Sewage. *Water* **2022**, *14*, 81. [CrossRef]
35. Chen, Y.; Jay, J.; Cheng, J.J.; Creamer, K.S. Inhibition of anaerobic digestion process: A review. *Bioresour. Technol.* **2008**, *99*, 4044–4064. [CrossRef] [PubMed]
36. El-Naas, M.H.; Mohammad, A.F.; Suleiman, M.I.; Al Musharfy, M.; Al-Marzouqi, A.H. Evaluation of a novel gas-liquid contactor/reactor system for natural gas applications. *J. Nat. Gas Sci. Eng.* **2017**, *39*, 133–142. [CrossRef]
37. Luis, K.; Francois, D.; Trevor, P. Method for Purifying Fischer-Tropsch Derived Water. 2003, pp. 1–25. Available online: https://patents.google.com/patent/US20050131086A1/en (accessed on 15 September 2022).

processes

MDPI

Article

Occurrence of Selected Emerging Contaminants in Southern Europe WWTPs: Comparison of Simulations and Real Data

Daniel Sol [1], Andrea Menéndez-Manjón [1], Paula Arias-García [1], Amanda Laca [1], Adriana Laca [1,*], Amador Rancaño [2] and Mario Díaz [1]

[1] Department of Chemical and Environmental Engineering, University of Oviedo, C/Julián Clavería s/n, 33006 Oviedo, Spain
[2] ACCIONA Agua S.A., 28108 Alcobendas, Spain
* Correspondence: lacaadriana@uniovi.es; Tel.: +34-985-10-29-74

Abstract: Emerging contaminants (ECs) include a diverse group of compounds not commonly monitored in wastewaters, which have become a global concern due to their potential harmful effects on aquatic ecosystems and human health. In the present work, six ECs (ibuprofen, diclofenac, erythromycin, triclosan, imidacloprid and 17α-ethinylestradiol) were monitored for nine months in influents and effluents taken from four wastewater treatment plants (WWTPs). Except for the case of ibuprofen, which was in all cases in lower concentrations than those usually found in previous works, results found in this work were within the ranges normally reported. Global removal efficiencies were calculated, in each case being very variable, even when the same EC and facility were considered. In addition, the SimpleTreat model was tested by comparing simulated and real ibuprofen, diclofenac and erythromycin data. The best agreement was obtained for ibuprofen which was the EC with the highest removal efficiencies.

Keywords: emerging contaminants; removal; wastewater; WWTP; SimpleTreat

Citation: Sol, D.; Menéndez-Manjón, A.; Arias-García, P.; Laca, A.; Laca, A.; Rancaño, A.; Díaz, M. Occurrence of Selected Emerging Contaminants in Southern Europe WWTPs: Comparison of Simulations and Real Data. *Processes* **2022**, *10*, 2491. https://doi.org/10.3390/pr10122491

Academic Editors: Dimitris Zagklis and Georgios Bampos

Received: 7 November 2022
Accepted: 21 November 2022
Published: 23 November 2022

1. Introduction

The progressive population growth, closely related to industrial and technological development, has led to a considerable increase in the production of consumer goods [1]. Processes associated with the different stages of the life cycle of these products entail the emission of different pollutants into the environment. Among the numerous pollutants of anthropogenic origin, emerging contaminants (ECs) should be highlighted [2]. The United States Geological Survey (USGS) defined these contaminants as "any compound of engineered or normal root or any microorganism that is not usually observed in the surrounding, however it can possibly cause unfriendly environmental and human wellbeing impacts" [3].

The concentration of emerging pollutants in wastewater varies from ng·L^{-1} to μg·L^{-1} depending on the specific compound analysed [4,5]. Additionally, this group includes diverse contaminants with different origins, chemical characteristics and potential harmful effects [6,7]. Certainly, endocrine disrupting substances (e.g., hormones), drugs, hygiene and personal care products, nanoparticles, perfluorinated substances, fire retardants, fertilizers, pesticides, and oily contaminants, among others, are considered ECs [4,8,9]. It is very difficult to control the spread of ECs as they can be found in many everyday products. In addition, most of these compounds are toxic or chemical substances capable of mimicking the hormones of living organisms, which alters their proper development and behaviour [10]. Mostly, biologically active forms of emerging pollutants, with or without limited treatment, are intentionally or unintentionally dumped into marine ecosystems causing health hazards and directly affecting the aquatic environments [11].

One of the main obstacles to controlling EC dispersion is the absence of specific regulations and quality standards that establish limits for these pollutants [10–13]. Therefore, ECs are not usually analysed in the environment, so their harmful potential is underestimated. Due to the lack of knowledge about these substances and their hazardousness, organizations such as the Environmental Protection Agency (EPA), the World Health Organization (WHO) and the European Commission are prioritizing the study of their effects on human health and on the environment with the final aim of reducing its presence in water [10,13].

One of the main routes of entry of these compounds in aquatic ecosystems is through wastewater. In most cases, their elimination in wastewater treatment plants (WWTPs) is not complete due to the use of conventional methods, which are ineffective or insufficient for EC removal, so an efficient and profitable additional water treatment would be required [14–17]. In general, the concentrations of ECs in the effluents of WWTPs are usually low, but their presence implies potential harmful effects on the environment and human health due to their structure and characteristics, such as their persistence and their ability to bioaccumulate even at low concentrations [18]. In addition, it should be noted that, even in the same facility, removal efficiencies in WWTPs are highly variable depending on the type of treatment, season, compound nature, etc. [15,17,19].

In the present work, a follow-up was conducted for nine months of influents and effluents of four WWTPs sited in the southwest of Spain to analyze the incidence of six selected ECs (ibuprofen, diclofenac, erythromycin, triclosan, imidacloprid and 17α-ethinylestradiol). The occurrence and the effectiveness of EC removal of the different facilities were evaluated. In addition, the SimpleTreat 4.0 software was employed to compare simulated and real data with the aim to obtain results that could be extrapolated at European level.

2. Materials and Methods

2.1. Selected ECs

The following six emerging contaminants were analysed: ibuprofen, diclofenac, erythromycin, triclosan, imidacloprid and 17α-ethinylestradiol. These compounds were chosen since they are representative of diverse groups of ECs widely reported in aquatic and land environments. Ibuprofen and diclofenac are non-steroidal anti-inflammatory drugs that, due to their high consumption, are included within the top ten high priority pharmaceuticals identified in a European assessment of pharmaceuticals and personal care products (PPCPs) [20–22]. Erythromycin is the most toxic antibiotic, which inhibits protein synthesis in bacteria [23–25]. Triclosan is a pharmaceutical drug (antiseptic) used in hospital and personal hygiene products [13,26]. Imidacloprid is a neuroactive insecticide [27,28] and 17α-ethinylestradiol is an estrogen derived from estradiol, used in the formulation of contraceptive pills [29].

2.2. Characteristics of Wastewater Treatment Plants (WWTPs)

Four WWTPs located in the southwest of Spain were studied. The flow diagram of each WWTP is shown in Figure 1.

WWTP 1 treats approximately 3000 m^3/day (11,375 population equivalent, PE) and it consists of: screening systems, a grit and grease removal system, an activated sludge treatment (6800 m^3) composed of three anaerobic chambers and two oxic biological reactors followed by a secondary settling, a tertiary treatment based on a coagulation/flocculation process, a lamellar settling, a filtration employing disc filters and, finally, UV disinfection equipment. The secondary sludge is removed from the secondary settler and driven towards the thickener, together with the sludge from the lamella decanter. After that, the mixed sludge is pumped up to two centrifuges for dehydration.

WWTP 2 treats around 2000 m^3/day (13,586 PE) and consists of two roughing channels equipped with automatic screens and automatic cleaning sieves, and a third channel equipped with a manual cleaning screen, two grit and grease removal systems, a primary decantation, a secondary treatment based on an activated sludge treatment (1191 m^3), followed by a secondary settling. The tertiary treatment is comprised a homogenization tank, a coagulation/flocculation chamber, a lamella settler, a sand filter (RSF) and a disinfection

UV system. The primary, biological sludge and homogenization-tank sludge are sent to the aerobic digester. Once digested, the mixed sludge is driven to the thickener and after that is dewatered in two centrifuges.

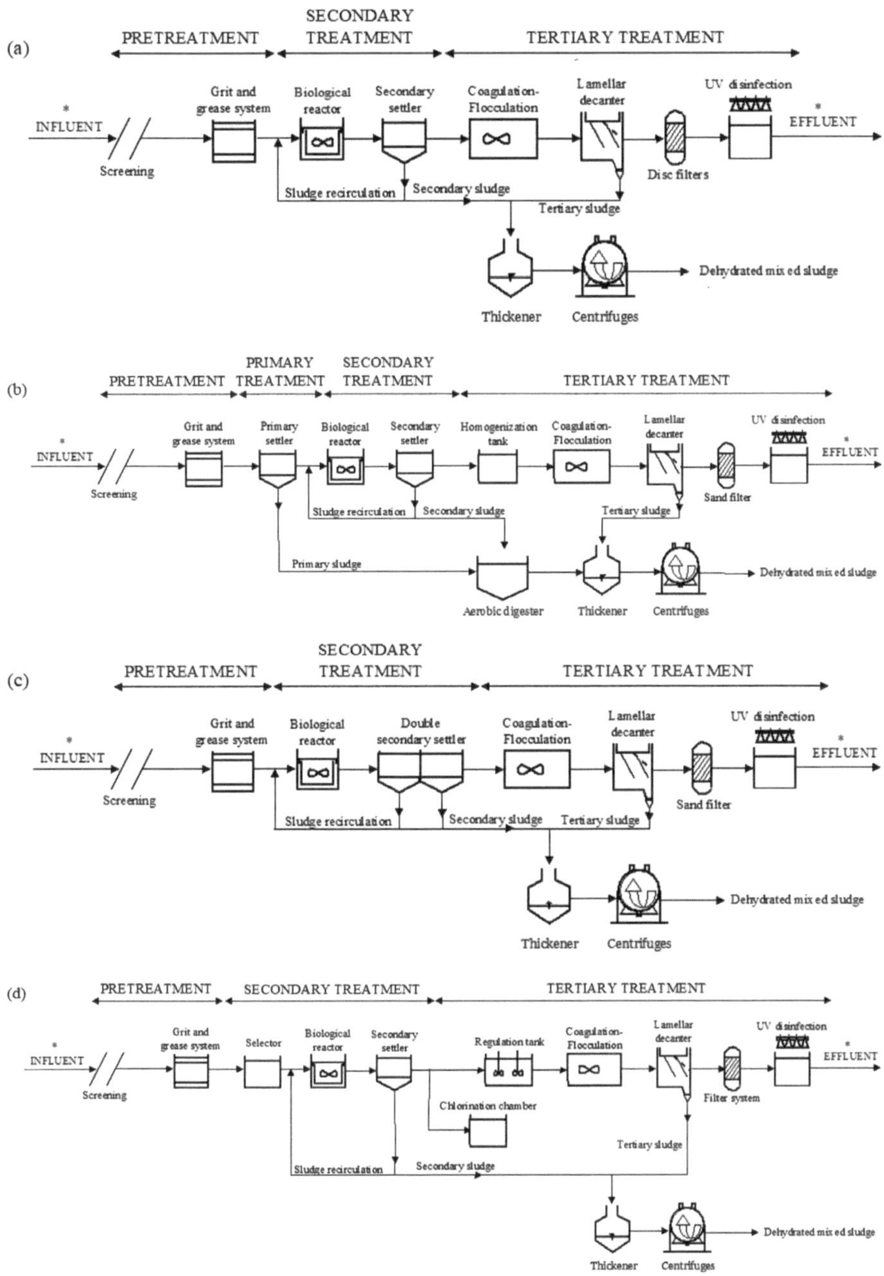

Figure 1. Scheme of the WWTPs studied in this work: (**a**) WWTP 1, (**b**) WWTP 2, (**c**) WWTP 3, (**d**) WWTP 4. Asterisks indicate the sampling points.

The wastewater that arrives at WWTP 3 (4800 m^3/day; 24,866 PE) is collected in channels with automatic and manual systems for wastewater screening. This facility has two grit and grease removal systems, a secondary treatment constituted by an anoxic chamber and two oxic biological reactors (19,000 m^3), followed by a double settler. Finally, a tertiary treatment consisting of coagulation/flocculation chambers, lamella settler equipment, a pair of sand filters and a UV disinfection channel. To treat the mixed sludge, a gravity thickener is employed and, subsequently, the sludge is driven to the centrifuges for dewatering.

The treatment carried out at the WWTP 4 (1500 m^3/day; 5423 PE) entails the following systems: two channels equipped with automatic grates and sieves for coarse roughing, two grit and grease removal systems, two biological treatment lines that have two carousel-type reactors (12,036 m^3) and a secondary settler. Finally, a tertiary treatment consisting of a chlorination chamber, a regulation tank, a coagulation/flocculation chamber followed by a lamella settler, a filter system and a UV disinfection system. Sludge collected from secondary and lamella settling is sent to a thickener, and after that, is centrifuged to obtain the dehydrated mixed sludge.

2.3. Sample Collection

Wastewater was sampled from the influent and effluent flows of the four WWTPs after the roughing processes. Grab samples were collected once a month for nine months, using a sample device consisting of a plastic bottle attached to a stick. After the collection, the samples were transferred to 2.5 L amber glass bottles and transported at 4 °C to the laboratory where 3% (v/v) of methanol (Sigma-Aldrich, San Luis, MO, USA) was added to them, avoiding the degradation of analytes until sample processing.

2.4. Analytical Method

Influent samples were centrifuged at 10,000 G for 30 min (Kubota 6500) and the supernatant was filtered under vacuum through 0.45 µm (Ahlstrom-Munksjö, Helsinki, Finland). The effluent samples were directly filtered. Once filtered, the pH of the samples was adjusted to 2 with H_2SO_4 (98%, Sigma-Aldrich, San Luis, MO, USA) and a solid-phase extraction (SPE) was carried out to concentrate the emerging pollutants using OASIS HLB cartridges (Waters, Milford, MA, USA) and a vacuum pump at 300 mbars as described below. First, acetone (2 × 5 mL), 5 mL of acetic acid and 5 mL of distilled water were successively flowed through the cartridge to prepare it. Then, 200 mL of sample was filtered through the cartridge and the retained compounds were desorbed employing firstly 6 mL of 40:60 acetone and buffer solution (0.1 M $NaHCO_3$ at pH 10 adjusted with 1 M NaOH), obtaining in this way the acidic fraction. After that, 6 mL of acetone was flowed through the cartridge, obtaining the acetonic fraction. The acidic fraction, after being adjusted again to pH 2, was subjected to three consecutive extractions employing each time 2 mL of ethyl acetate by centrifugation for 10 min at 3500 rpm (Sorvall ST-16R, Thermo Fisher, Waltham, MA, USA), and approximately 6 mL of supernatant was recovered. The acidic fraction (now in ethyl acetate) and the acetonic fraction were dried under a stream of nitrogen. The internal standards (2,4-dichlorophenoxyacetic acid for acidic fraction, 17β-estradiol d5 and Imidacloprid d4 for acetonic fraction) were added at this moment and the samples were again dried under a nitrogen stream. Next, the derivatization agents, 50 µL MTBSTFA + 1% TBDMCS (N-methyl-N-(tert-butyldimethylsilyl) trifluoroacetamide with 1% tert-butyldimethylchlorosilane) for acidic fraction and 50 µL BSTFA + 1% TCMS (N,O-Bis(trimethylsilyl)trifluoroacetamide with 1% Trimethylchlorosilane) for acetonic fraction, and the solvent (ethyl acetate for the acid fraction and pyridine for the acetone fraction) were added and the samples were placed in a water bath at 60 °C for 30 min. Once the samples and standards had been tempered, they are brought to a final volume of 1 mL with ethyl acetate. All reagents were supplied by Merck KGaA (Darmstadt, Germany).

The samples were analysed using the GC-MS technique, using a 7890A GC Gas Chromatograph coupled to a 5975C Inert XL MSD Mass Spectrometer (Agilent Technologies,

Santa Clara, CA, USA), fitted with a column DB5MS (Agilent Technologies, Santa Clara, CA, USA). Ultrapure helium was used as carrier gas at a constant flow of 1.3 mL/min. The oven temperature was held at 100 °C for 1 min, then programmed at 15 °C/min to 250 °C, and finally at 20 °C/min to 300 °C, with the final temperature being held for 10 min. 1 μL of sample was injected in the splitless mode. The transfer line and ion source were set at 280 °C and 230 °C, respectively. Calibration curves were obtained employing standards of each EC, and limits of quantification (LOQ) are shown in Table 1.

Table 1. LOQ for each compound.

	LOQ (ppb)
Ibuprofen	2
Diclofenac	3
Erythromycin	2
Triclosan	3
Imidacloprid	3
17α-ethinylestradiol	1

2.5. Kinetic Constants Estimation

SimpleTreat 4.0 is a basic tool used to predict exposure in environmental risk assessment and includes the most important processes such as sorption, volatilization, dissolution, and biodegradation [30]. The program requires several parameters to model the evolution of the six ECs analysed (ibuprofen, diclofenac, erythromycin, triclosan, imidacloprid and 17α-ethinylestradiol). The physicochemical parameters employed for the modellization, i.e., type of compound, molecular weight, solubility, vapor pressure, octanol-water partition coefficient (K_{ow}), solubility, pKa and Henry coefficient, are available in Table S1. Some sorption parameters not found in the literature, such as K_{oc}, Kp_{s} and Kp_{as}, were estimated by the program [30]. In addition, physicochemical and biological parameters related to the specific wastewater treatment have been provided by the WWTPs and are shown in Tables S2–S5.

Firstly, the SimpleTreat 4.0 program was used to estimate the theoretical biodegradation constant (k) as a function of the experimental removal rate for each compound and sample. As samples were taken in different months, which implies different ambient temperatures, the kinetic constants were used to obtain Arrhenius parameters (only data with deviations lower than 10% from mean values were considered). Using these parameters, the theoretical k was obtained for each month, considering the daily temperature. This value was implemented in the SimpleTreat 4.0 program and predicted, and real ECs removal efficacies were compared.

3. Results and Discussion

3.1. Occurrence and Removal Efficiencies

The occurrence and removal efficiency of six emerging contaminants (ibuprofen, diclofenac, erythromycin, triclosan, imidacloprid and 17α-ethinylestradiol) were studied in four WWTPs. The results are shown in Figure 2 and Tables S6–S11. The removal efficiency was calculated in each case, except when the effluent concentration was higher than the influent concentration or the EC was "not detected".

3.1.1. Ibuprofen

Ibuprofen is the third most prescribed drug worldwide and due to its high demand more than 30 kilotons are synthesized annually. The main route of emission is through the excretion of non-metabolized and metabolized drug in the urine of humans and animals after consumption. It has been detected in wastewater, sewage sludge, hospital wastewaters, surface waters and drinking water. Potential effects of this carcinogenic and endocrine disrupting drug have been described for microorganisms, algae, fish species, and even human health [31,32].

As can be observed in Figure 2a and Table S6, the concentration of ibuprofen in influent samples of the four WWTPs generally varies between 1 and 20 ppb with average values between 3 and 18 ppb. The highest values were observed in October (32–59 ppb in WWTP 2, WWTP 3 and WWTP 4). After wastewater treatment, the concentration of ibuprofen was notably reduced, with efficacies above 80%, in most cases (Figure 3a and Table S6). So, ibuprofen concentrations in the effluent usually ranged between "not detected" and 0.8 ppb. The only exceptions were in October and November with ibuprofen concentrations in the WWTP 2 and WWTP 4 effluents, respectively, of around 2.8 ppb. In the case of WWTP 2, this higher level was due to the poor elimination efficacy observed for ibuprofen in this month (46.2%). In general, in the literature, the ibuprofen concentrations in influents and effluents of urban WWTPs were within the ranges of 9–17,500 ppb and 1–3777 ppb, respectively. The ibuprofen concentrations observed in this work in the influents and effluents were lower than those usually obtained in previous works [5,33]. However, removal efficiencies reported for this micropollutant varied between 25% and 100% [30,34,35], in agreement with percentages obtained here.

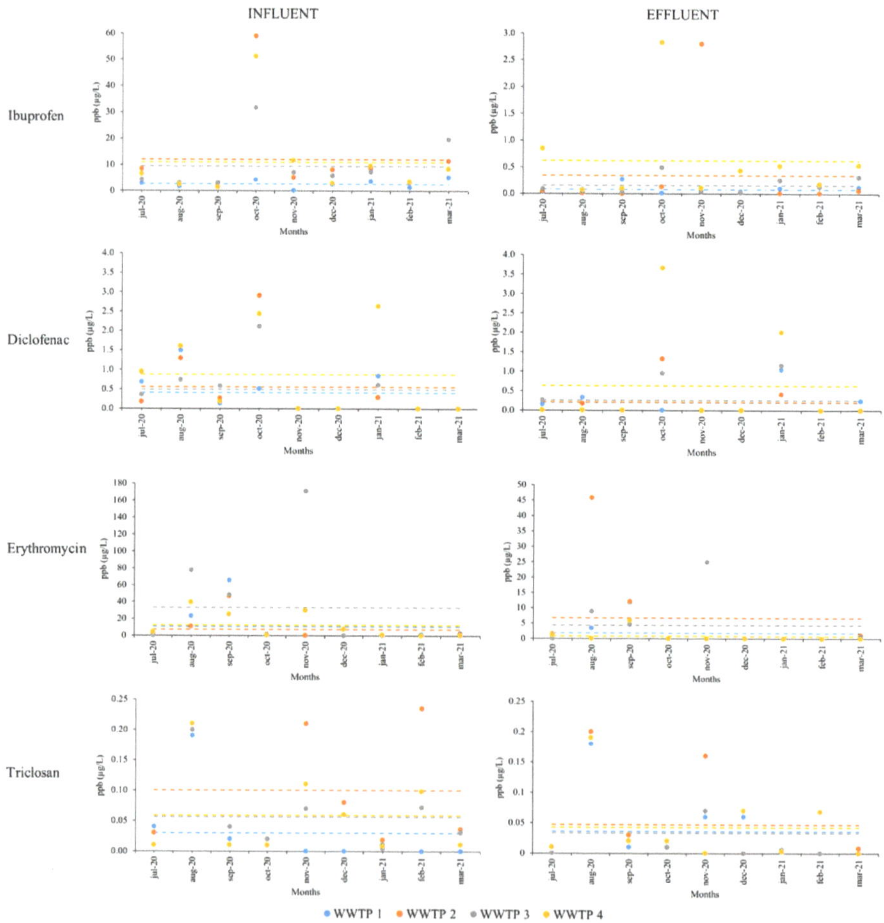

(a)

Figure 2. *Cont.*

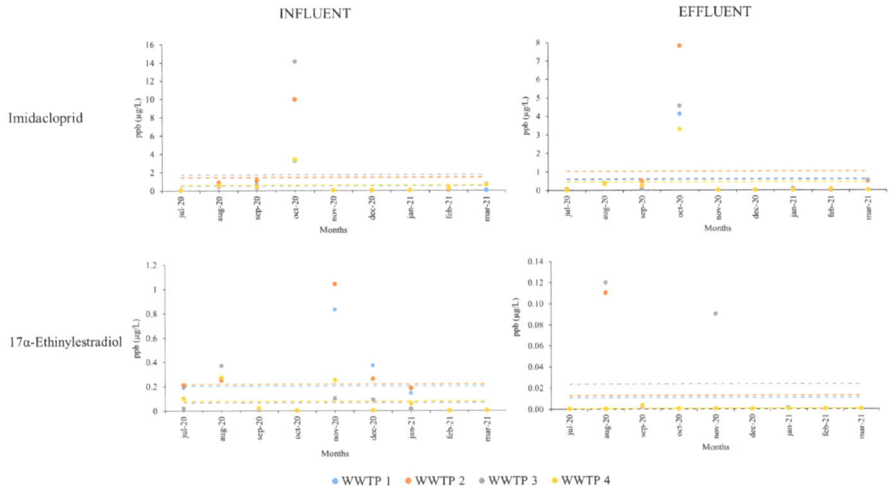

(b)

Figure 2. (**a**) Evolution of ibuprofen, diclofenac, erythromycin and triclosan from July 2020 to March 2021 in influent and effluent samples of the WWTPs studied. (**b**) Evolution of imidacloprid and 17α-ethinylestradiol in influent and effluent samples in the four WWTPs analysed, from July 2020 to March 2021. Dashed lines show the mean concentration for each WWTP.

3.1.2. Diclofenac

Diclofenac is one of the pharmaceutical wastes included in the watch list of the Commission Implementing Decision (EU) published by the European Commission in 2015. It is an oral tablet or topical gel non-steroidal anti-inflammatory drug (NSAID), and its global consumption is around 940 tons per year. A total of 65% is usually released through urine into wastewater [36,37].

Diclofenac was observed in concentrations even lower than those obtained for ibuprofen, with similar profiles in the four WWTPs (Figure 2a and Table S7). In influent samples, diclofenac exhibited values that were in most cases below 1.5 ppb, with average concentrations of 0.41 ± 0.52 ppb, 0.55 ± 0.98 ppb, 0.49 ± 0.68 ppb and 0.87 ± 1.10 ppb for WWTP 1, WWTP 2, WWTP 3 and WWTP 4, respectively. The highest concentrations were measured in October in WWTP 2, WWTP 3 and WWTP 4, and in January in WWTP 4, with values between 2 and 3 ppb. After wastewater treatment, the concentration of diclofenac was notably reduced, with removal efficiencies usually above 50% (Figure 3b and Table S7) in most cases. These percentages are higher than those reported in the literature, typically below 40% [38]. In accordance with the literature data, poor removal efficiencies were obtained for WWTP 3 in July (28%) and WWTP 4 in January (24%). In effluents, most diclofenac concentrations were below 1 ppb, with slightly higher values observed in certain samples. Particularly, WWTP 4 in October and January exhibited concentrations between 2 and 4 ppb in both influent and effluent. The diclofenac concentrations found in this work are within the wide ranges reported for different WWTPs, including from "not detected" to 7100 ppb [5,39]. It is remarkable that, in a few cases, the concentrations of diclofenac detected in the effluent were higher than those found in the influent for the same date (October for WWTP 4 and January for WWTP 1, WWTP 2 and WWTP 3). This behaviour has been previously described and many studies reported the difficulty in removing this persistent analgesic from wastewaters [33]. In addition, it has to be considered that influent and effluent samples were taken almost simultaneously, and the water sample taken at the end of the treatment might correspond to a higher concentration with respect to the raw wastewater.

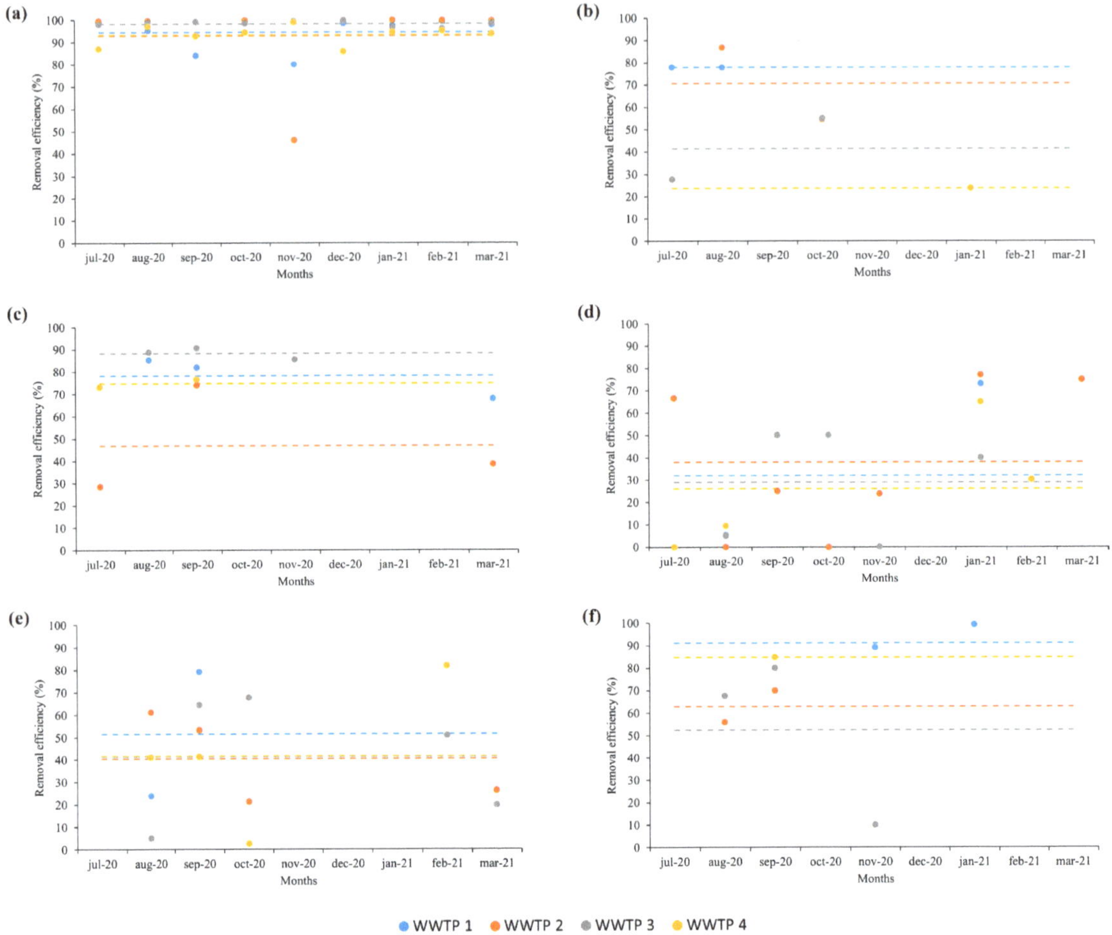

Figure 3. Removal efficiencies of the WWTPs studied obtained for: (**a**) ibuprofen, (**b**) diclofenac, (**c**) erythromycin, (**d**) triclosan, (**e**) imidacloprid and (**f**) 17α-ethinylestradiol from July 2020 to March 2021.

3.1.3. Erythromycin

Erythromycin is an antibiotic, also included in the watch list of the Commission Implementing Decision (EU) published by the European Commission in 2015. It has been found in different environments, such as rivers, surface water, wastewater, seawater, sediments, etc. Low concentrations of this compound can contribute to the proliferation of antibiotic resistant bacteria [40,41].

Regarding erythromycin (Figure 2a and Table S8), the influent samples showed higher concentrations in the warmest months (2–78 ppb, July–September) compared to the coldest months ("not detected"–8 ppb, October-March), with the exception of WWTP 3 and WWTP 4, which showed in November a peak in concentration of 171 ppb and 30 ppb, respectively. In general, in the effluent samples the concentrations of erythromycin were also higher during the warmest months ("not detected"–46 ppb) than in the coldest months, with concentrations at levels close to 0 ppb. Again, the only exception was a value of 25 ppb detected in November in WWTP 3. In the literature, a wide variation of erythromycin values can be found both in influent and effluent samples ("not detected"–14,700 ppb) [42], having reported concentrations much higher than those obtained here. The concentration of erythromycin was reduced in the WWTPs studied with efficacies above 70%, except in WWTP 2 with 47% (Figure 3c and Table S8). WWTP 2 showed low efficacies of 29% and 39% in July and March, respectively, and, in addition, in August the concentration detected in the effluent was higher than that of the influent. In this work, removal efficiencies higher than those obtained by other authors were observed, typically below 59%, in conventional WWTPs [43].

3.1.4. Triclosan

Triclosan is an antimicrobial agent widely used in PPCPs. It has persistent characteristics and bioaccumulation properties that make it a major concern. In 2016, the US Food and Drugs Administration (FDA) regulated the use of this product in commercial products. Although it is a very stable chemical, it can be decomposed by light, ozone, chlorine and some microorganisms, into compounds harmful to living beings, such as chlorinated phenols and dibenzo-p-dioxins [44].

The emerging contaminant triclosan showed, in general terms, a similar behaviour in the influent and effluent samples of the four WWTPs, being in all cases in concentrations close to 0 ppb (Figure 2a and Table S9). Specifically, these values ranged between "not detected" and 0.24 ppb. These concentrations of triclosan were lower than those usually found in the literature (between 1 ppb and 33 ppm). It has been reported that triclosan is largely eliminated from wastewater by conventional treatments, with removal efficiencies between 85% and 98%. In this work, the removal efficiencies were very variable, even for the same WWTP, but always with values below those described in the literature [45,46] (Figure 3d and Table S9). The reason may be that the measured concentrations are very close to the quantification limit, which implies higher uncertainty associated with the values (Table 1).

3.1.5. Imidacloprid

Imidacloprid is a neonicotinoid insecticide that has been widely used as a pesticide, biocide and veterinary medicinal product. This compound presents a high toxicity, a long persistence, and is easily and flexibly applied. The most sensitive organisms to this EC are aquatic invertebrates [47].

The profile of imidacloprid concentrations in the influent and effluent samples are remarkably similar for the four WWTPs (Figure 2b and Table S10). Low concentrations, ranging between "not detected" and 1 ppb in influent and below 0.6 ppb in effluent, were detected. Oddly, higher concentrations were found in October in all the WWTPs, varying between 3 and 14 ppb in influent and effluent samples. In general, imidacloprid concentrations in the literature range between 20 and 387 ppb, and its removal during wastewater treatment was quite low (<20–30%) [48]. In the present work, imidacloprid showed concentrations much lower, which is in agreement with the amount of this EC detected in the effluents of 16 Spanish WWTPs sampled in 2019 (<0.4 ppb) [49]. The reduction in the concentrations of imidacloprid in European wastewaters and water courses is expectable due to its use as a plant protection product is no longer authorized in the EU (Regulation EU 2020/1643), and since 2020 no emission is expected from this use. Regarding the removal efficiencies here again the results were very variable (Figure 3e and Table S10) since the concentrations used to calculate them were very close to the detection limit.

3.1.6. 17α-ethinylestradiol

In 2013, Directive 2013/39/EU extended the list of Priority Pollutants, including the 17α-ethinylestradiol. WWTPs are considered the main source of emission of this EC into the aquatic environment. In 2018, this emerging contaminant was included in the watch list of the Commission Implementing Decision 2018/840/EU as a substance to be monitored in water in the field [50,51].

The concentration of 17α-ethinylestradiol in the four facilities was between "not detected" and 0.37 ppb, with the exception of samples taken in November from WWTP 2 and WWTP 3 that exhibited concentrations around 1 ppb (Figure 2b and Table S11). In the effluent samples, the concentration varied between "not detected" and 0.12 ppb in all cases. These results were quite low but in agreement with other concentrations published, usually below 78 ppb. The reported removal efficiencies of this hormone were between 47.5–83.6% [29,52]. In concordance with these data, the removal efficiencies obtained here were usually above 56% (Figure 3f and Table S11).

3.2. *Simulation of Emerging Contaminants Removal*

SimpleTreat is the recommended sewage treatment plant model in Europe for environmental risk assessment of industrial chemicals (REACH), chemicals covered under the Biocidal Products Regulation (BPR) and active pharmaceutical ingredients regulated by the European Medicines Agency (EMA) [30]. Using a software that implements this model (SimpleTreat 4.0), a first-order biodegradation constant (k_i) was determined as a function of the percentage of removal obtained for each compound analysed in each WWTP each month. Samples where an EC was "not detected" in the effluent have not been considered for the estimation, as considering 100% efficacies could lead to a large error. So, using data shown in Tables S1–S11, k_i values were obtained (Tables S12 and S13). It should be borne in mind that the software does not allows the consideration of all the treatment processes employed in the WWTPs. For example, most parts of tertiary treatments cannot be considered, and it is evident that these can contribute to reduce the concentrations of ECs [53–55]. So, the kinetic constants calculated could be overestimated to a certain extent. Another factor to be considered is the existence of errors due to the estimation of sorption parameters by the program, leading to a deviation in the accuracy of the model results [56,57].

In the case of ibuprofen, the majority of the apparent degradation constants for WWTP 1, WWTP 3 and WWTP 4 were between 0.4 and 3 h^{-1}, values in general higher than those obtained in WWTPs which did not include tertiary treatments (0.06–0.7 h^{-1}) [5], whereas the simulation of ibuprofen in SimpleTreat usually employs a default value of 1 h^{-1} [55,58]. Specifically, Alvarino et al. [59] indicated a degradation rate for ibuprofen of 0.3 h^{-1} in an UASB reactor coupled to a hybrid aerobic membrane bioreactor for municipal wastewater treatment. Furthermore, values above 30 h^{-1} were obtained in WWTP 2, which was the only facility that includes primary settling. Ibuprofen is a fairly easily biodegradable compound and can be removed by sorption on several materials [60], so this settling step contributed to ibuprofen removal by being adsorbed on primary sludge or even biodegradation.

For diclofenac, erythromycin, triclosan, imidacloprid and 17α-ethinylestradiol, kinetic constants were quite similar, most of them being within the range 0.02 to 0.3 h^{-1}. These values were lower than those obtained in this work for ibuprofen. This is in agreement with other authors who reported slow biodegradations for these compounds, being 0.75, 0.03 and 0.08 h^{-1} for diclofenac, erythromycin and 17α-ethinylestradiol, respectively, in an UASB reactor employed at pilot scale to treat municipal wastewater [59], 0.002–0.005 h^{-1} for imidacloprid degraded by pure bacterial isolates in the laboratory [61] and 0.1–0.2 h^{-1} for triclosan when bioaugmentation treatment for wastewater was applied [62]. Furthermore, it should be noted that default values employed in previous works to simulate the evolution of these contaminants by means of the SimpleTreat model were, in all cases, 0.3 h^{-1} for diclofenac, erythromycin and 17α-ethinylestradiol, and 0.1 h^{-1} for triclosan [55,58]. However, no data for imidacloprid has been reported until now, since, to the best of our knowledge, this is the first time that the SimpleTreat software has been used to model the behaviour of this EC in WWTPs. It is noteworthy that few constants could be estimated in the cases of diclofenac, erythromycin, triclosan and 17α-ethinylestradiol because effluent concentrations were below the LOQ or even slightly higher than the influent concentrations. This behaviour has already been described in previous works for several ECs [5] and indicates that conventional wastewater treatment processes are not effective enough to remove these complex compounds. In addition, some products of human metabolism may act as a reservoir from which a later yield of the parent EC can occur [63]. Additionally, measured concentrations were very low, near to the LOQ, which implies higher uncertainty of the values quantified.

As wastewater samples were taken on different dates with different ambient temperatures, the calculated constants correspond to different temperatures. It would be expected that ki values were higher for higher temperatures, following the Arrhenius expression $k_i = Ae^{-E_a/RT}$. Nevertheless, a lack of tendency with temperature was observed for triclosan, imidacloprid and 17α-ethinylestradiol kinetic constants. This can be explained because the concentrations measured, in particular in the effluent, were very low, near LOQ, which means a larger uncertainty in the data and, therefore, in the calculated kinetic constants. So only the kinetic constants obtained for ibuprofen, diclofenac and erythromycin were used to estimate the parameters A and E_a (apparent activation energy) of a type of Arrhenius equation. The next line equations were obtained, $\ln k_{ib} = 11.292 - 3281.3\frac{1}{T}$, $\ln k_{di} = 31.767 - 9971.3\frac{1}{T}$ and $\ln k_{er} = 6.7050 - 2526.2\frac{1}{T}$ for ibuprofen, diclofenac, and erythromycin (k in h^{-1} and T in K), and the apparent activation energies were 27.3, 82.9 and 21.0 kJ/mol, respectively. These values were similar to those reported for the apparent activation energies in the biodegradation of other ECs, i.e., caffeine, paracetamol, naproxen and ibuprofen (21–86 kJ/mol) [5]. Using the obtained equations, new kinetic constants were calculated for these ECs for each sampling temperature. In Tables S14–S16 these kinetic constants (h^{-1}) for each month are shown. These values have been used in SimpleTreat 4.0 to obtain the theoretical percentage of removal for ibuprofen, diclofenac and erythromycin for each month and WWTP. In Figure 4, the theoretical percentages are represented versus those obtained experimentally. Considering a 10% error interval (dashed green lines), it can be seen that the majority of the points fall within this interval for ibuprofen in WWTPs 1, 3 and 4, indicating that the simulation of ibuprofen is acceptable for these cases (Figure 4a). These plants have in common that they do not have primary treatment, whereas WWTP 2 does. It seems that the presence of this settling step makes the software predict lower removal efficacies than those obtained from the real data.

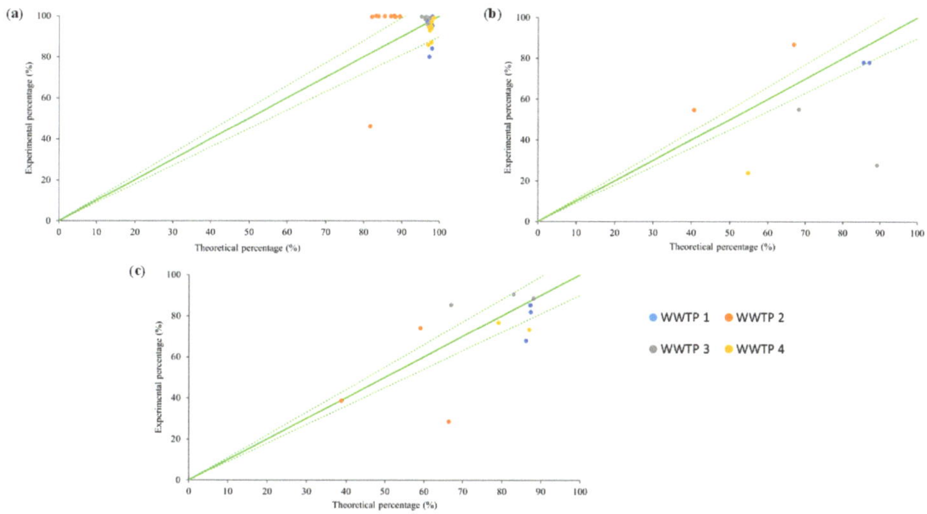

Figure 4. Representation of the experimental percentage (%) obtained by simulation in SimpleTreat 4.0 and theoretical percentage (%) of the WWTPs studied for (**a**) ibuprofen, (**b**) diclofenac and (**c**) erythromycin. The green dashed lines indicate a 10% error.

On the opposite, the simulations carried out for diclofenac (Figure 4b) and erythromycin (Figure 4c) led to worse results. Approximately half the erythromycin estimated removal efficacies are within the 10% error interval, whereas only one third of diclofenac predicted efficacies could be considered as acceptable under this criterion. Lautz et al. [55] suggested that, when using SimpleTreat software, the most influentual parameters are biodegradation and the hydraulic retention time and, in addition, model performance is highly dependent on the nature and quality of the data, i.e., the degree of uncertainty.

Table S1 shows the chemical character of the compounds analysed, i.e., acidic (ibuprofen and diclofenac), basic (erythromycin, imidacloprid and 17α-ethinylestradiol) and neutral (triclosan). Although an exhaustive analysis has not been conducted in this sense, the results seem to indicate more reliable simulations for those acidic compounds. Nevertheless, it is remarkable that Lautz et al. [55], who evaluated the SimpleTreat performance to predict the removal of 43 pharmaceuticals from wastewater, indicated that there were no significant differences depending on the nature of the compound (acidic, base or neutral).

4. Conclusions

After having evaluated the incidence of six selected ECs (ibuprofen, diclofenac, erythromycin, triclosan, imidacloprid and 17α-ethinylestradiol) for nine months in four WWTPs sited in the southwest of Spain, the most remarkable fact is the low concentrations (at ppb level) and the high variability found, regarding not only the occurrence, but also the removal efficiencies. In general, these fluctuations were in accordance with the literature data, which show wide ranges of concentrations and elimination percentages for the ECs studied.

By using the SimpleTreat 4.0 software, apparent first kinetic constants were obtained for each sample and an Arrhenius-type equation was employed to determine the relation between the constants and the temperature in the cases of ibuprofen, diclofenac and erythromycin (it was not possible for the rest of the ECs due to the high dispersion of data). These equations were employed to obtain new kinetic constants, which were used to predict the removal efficacies for each case. The agreement between theoretical and experimental efficacies was not very good for diclofenac and erythromycin. However, the software could reasonably predict, with an error lower than 10%, the behaviour of ibuprofen in three of the four WWTPs studies. This could be due to the fact that the presence of a primary clarifier made the software underestimate the removal efficacy in the WWTP 2.

The present work seems to indicate that SimpleTreat is suitable to predict the behaviour of emerging pollutants with a high degree of biodegradability, such as ibuprofen, in simple-configuration WWTPs. Nevertheless, in order to transform this software into a model for general application, even for ECs present in very low concentrations, it should be updated to better reflect the real processes that takes place in WWTPs.

Supplementary Materials: The following supporting information can be downloaded at: https://www.mdpi.com/article/10.3390/pr10122491/s1, Table S1: Physicochemical characteristics of selected emerging compounds analysed in the present work; Table S2: Physicochemical and biological parameters provided by WWTP 1. BOD, biological oxygen demand; TTS, total suspended solids (INF: influent water after screening; TER: influent tertiary treatment); MLSS, mixed liquor suspended solids suspended within the mixed liquor; Table S3: Physicochemical and biological parameters provided by WWTP 2. BOD, biological oxygen demand; TTS, total suspended solids (INF: influent water after screening; TER: influent tertiary treatment); MLSS, mixed liquor suspended solids suspended within the mixed liquor; Table S4: Physicochemical and biological parameters provided by WWTP 3. BOD, biological oxygen demand; TTS, total suspended solids (INF: influent water after screening; TER: influent tertiary treatment); MLSS, mixed liquor suspended solids suspended within the mixed liquor; Table S5: Physicochemical and biological parameters provided by WWTP 4. BOD, biological oxygen demand; TTS, total suspended solids (INF: influent water after screening; TER: influent tertiary treatment); MLSS, mixed liquor suspended solids suspended within the mixed liquor; Table S6: Ibuprofen concentrations (μg/L) and removal efficiencies (%) in influent and effluent in the four WWTPs analysed; n.d. is not detected; Table S7: Diclofenac concentrations (μg/L) and removal efficiencies (%) in influent and effluent in the four WWTPs analysed; n.d. is not detected; Table S8: Erythromycin concentrations (µg/L) and removal efficiencies (%) in influent and effluent in the four WWTPs analysed; n.d. is not detected; Table S9: Triclosan concentrations (µg/L) and removal efficiencies (%) in influent and effluent in the four WWTPs analysed; n.d. is not detected; Table S10: Imidacloprid concentrations (µg/L) and removal efficiencies (%) in influent and effluent in the four WWTPs analysed; n.d. is not detected; Table S11: 17α-ethinylestradiol concentrations (µg/L) and removal efficiencies (%) in influent and effluent in the four WWTPs analysed; n.d. is not detected; Table S12: Biodegradation constant (ki) for ibuprofen, diclofenac and erythromycin in the four WWTPs analysed; Table S13: Biodegradation constant (ki) for triclosan, imidacloprid and 17α-ethinylestradiol in the four WWTPs analysed; Table S14: Biodegradation constant (kib) obtained using the Arrhenius equation for ibuprofen in the four WWTPs analysed; Table S15: Biodegradation constant (kdi) obtained using the Arrhenius equation for diclofenac in the four WWTPs analysed; Table S16: Biodegradation constant (ker) obtained using the Arrhenius equation for erythromycin in the four WWTPs analysed.

Author Contributions: Conceptualization, A.M.-M., P.A.-G. and D.S.; methodology, A.M.-M., P.A.-G. and D.S.; validation, A.L. (Amanda Laca), A.L. (Adriana Laca) and M.D.; formal analysis, A.M.-M., D.S. and A.L. (Amanda Laca); investigation, A.M.-M., P.A.-G. and D.S.; resources, A.M.-M. and D.S.; data curation, D.S., A.L. (Amanda Laca) and A.L. (Adriana Laca); writing—original draft preparation, A.M.-M. and D.S.; writing—review and editing, A.M.-M., D.S. and A.L. (Amanda Laca); visualization, A.L. (Amanda Laca) and A.L. (Adriana Laca); supervision, A.L. (Amanda Laca), A.L. (Adriana Laca) and M.D.; project administration, M.D. and A.R.; funding acquisition, A.R. All authors have read and agreed to the published version of the manuscript.

Funding: This work was financed by University of Oviedo Foundation: research projects FUO-396-19 and FUO-397-19.

Data Availability Statement: Not applicable.

Acknowledgments: Authors gratefully acknowledge the cooperation of ACCIONA Agua, S.A. and ESAMUR (Regional Entity for Sanitation and Wastewater Treatment of the Region of Murcia) with wastewater samples collection, especially to the workers of the WWTPs.

Conflicts of Interest: The authors declare no conflict of interest.

Abbreviations

BPR	Biocidal products regulation
BSTFA	N,O-Bis(trimethylsilyl)trifluoroacetamide
EC	Emerging contaminant
EMA	European Medicines Agency
EPA	Environmental Protection Agency
FDA	US Food and Drugs Administration
GC-MS	Gas chromatography–mass spectrometry
k	Biodegradation rate
K_{oc}	Organic carbon partition coefficient
K_{ow}	N-octanol-water partition coefficient
Kp_{as}	Activated sludge solids-water equilibrium partition constant
Kp_{s}	Sewage solids–water equilibrium partition constant
LOQ	Limit of quantification
MTBSTFA	N-Methyl-N-(tert-butyldimethylsilyl)trifluoroacetamide
NSAID	Non-steroidal anti-inflammatory drugs
PE	Population equivalent
PPCP	Pharmaceutical and personal care product
RSF	Rapid sand filter
SPE	Solid-phase extraction
TBDMCS	tert-butyldimethylchlorosilane
TCMS	Trimethylchlorosilane
USGS	United States Geological Survey

UV	Ultraviolet
WHO	World Health Organization
WWTP	Wastewater treatment plant

References

1. Ahammad, N.A.; Ahmad, M.A.; Hameed, B.H.; Din, A.T.M.D. A mini review of recent progress in the removal of emerging contaminants from pharmaceutical waste using various adsorbents. *Environ. Sci. Pollut. Res.* **2022**. [CrossRef] [PubMed]
2. Rout, P.R.; Zhang, T.C.; Bhunia, P.; Surampalli, R.Y. Treatment technologies for emerging contaminants in wastewater treatment plants: A review. *Sci. Total Environ.* **2021**, *753*, 141990. [CrossRef] [PubMed]
3. USGS. Contaminants of Emerging Concern in the Environment. Environmental Health—Toxic Substances Hydrology Program. U.S. Geological Survey. 2017. Available online: https://toxics.usgs.gov/investigations/cec/index.php (accessed on 15 September 2022).
4. Olasupo, A.; Suah, F.B.M. Recent advances in the removal of pharmaceuticals and endocrine-disrupting compounds in the aquatic system: A case of polymer inclusion membranes. *J. Hazard. Mater.* **2021**, *406*, 124317. [CrossRef] [PubMed]
5. Fernández, M.; Fernández, M.; Laca, A.; Laca, A.; Díaz, M. Seasonal occurrence and removal of pharmaceutical products in municipal wastewaters. *J. Environ. Chem. Eng.* **2014**, *2*, 495–502. [CrossRef]
6. Gogoi, A.; Mazumder, P.; Tyagi, V.K.; Chaminda, G.G.T.; An, A.K.; Kumar, M. Occurrence and fate of emerging contaminants in water environment: A review. *Groundw. Sustain. Dev.* **2018**, *6*, 169–180. [CrossRef]
7. Bhagat, C.; Kumar, M.; Tyagi, V.K.; Mohapatra, P.K. Proclivities for prevalence and treatment of antibiotics in the ambient water: A review. *npj Clean Water* **2020**, *3*, 42. [CrossRef]
8. Tran, N.H.; Reinhard, M.; Gin, K.Y.H. Occurrence and fate of emerging contaminants in municipal wastewater treatment plants from different geographical regions-a review. *Water Res.* **2018**, *133*, 182–207. [CrossRef]
9. McCance, W.; Jones, O.A.H.; Edwards, M.; Surapaneni, A.; Chadalavada, S.; Currell, M. Contaminants of Emerging Concern as novel groundwater tracers for delineating wastewater impacts in urban and peri-urban areas. *Water Res.* **2018**, *146*, 118–133. [CrossRef]
10. Ahmad, A.; Kurniawan, S.B.; Abdullah, S.R.S.; Othman, A.R.; Hasan, H.A. Contaminants of emerging concern (CECs) in aquaculture effluent: Insight into breeding and rearing activities, alarming impacts, regulations, performance of wastewater treatment unit and future approaches. *Chemosphere* **2022**, *290*, 133319. [CrossRef]
11. Wilkinson, J.; Hooda, P.S.; Barker, J.; Barton, S.; Swinden, J. Occurrence, fate and transformation of emerging contaminants in water: An overarching review of the field. *Environ. Pollut.* **2017**, *231*, 954–970. [CrossRef]
12. Vale, F.; Sousa, C.A.; Sousa, H.; Santos, L.; Simões, M. Parabens as emerging contaminants: Environmental persistence, current practices and treatment processes. *J. Clean. Prod.* **2022**, *347*, 131244. [CrossRef]
13. Surana, D.; Gupta, J.; Sharma, S.; Kumar, S.; Ghosh, P. A review on advances in removal of endocrine disrupting compounds from aquatic matrices: Future perspectives on utilization of agri-waste based adsorbents. *Sci. Total Environ.* **2022**, *826*, 154129. [CrossRef] [PubMed]
14. Rizzo, L.; Malato, S.; Antakyali, D.; Beretsou, V.G.; Đolić, M.B.; Gernjak, W.; Heath, E.; Ivancev-Tumbas, I.; Karaolia, P.; Ribeiro, A.R.L.; et al. Consolidated vs. new advanced treatment methods for the removal of contaminants of emerging concern from urban wastewater (Review). *Sci. Total Environ.* **2019**, *655*, 986–1008. [CrossRef] [PubMed]
15. Lozano, I.; Pérez-Guzmán, J.; Mora, A.; Mahlknecht, J.; Aguilar, C.L.; Cervantes-Avilés, P. Pharmaceuticals and personal care products in water streams: Occurrence, detection, and removal by electrochemical advanced oxidation processes. *Sci. Total Environ.* **2022**, *827*, 154348. [CrossRef]
16. Chen, Y.; Lin, M.; Zhuang, D. Wastewater treatment and emerging contaminants: Bibliometric analysis. *Chemosphere* **2022**, *297*, 133932. [CrossRef]
17. Kumar, M.; Ngasepam, J.; Dhangar, K.; Mahlknecht, J.; Manna, S. Critical review on negative emerging contaminant removal efficiency of wastewater treatment systems: Concept, consistency and consequences. *Bioresour. Technol.* **2022**, *352*, 127054. [CrossRef] [PubMed]
18. Yang, Y.; Ok, Y.S.; Kim, K.H.; Kwon, E.E.; Tsang, Y.F. Occurrences and removal of pharmaceuticals and personal care products (PPCPs) in drinking water and water/sewage treatment plants: A review. *Sci. Total Environ.* **2017**, *596–597*, 303–320. [CrossRef]
19. Mainardis, M.; Cecconet, D.; Moretti, A.; Callegari, A.; Goi, D.; Freguia, S.; Capodaglio, A.G. Wastewater fertigation in agriculture: Issues and opportunities for improved water management and circular economy. *Environ. Pollut.* **2022**, *296*, 118755. [CrossRef]
20. Sousa, J.C.G.; Ribeiro, A.R.; Barbosa, M.O.; Pereira, M.F.R.; Silva, A.M.T. A review on environmental monitoring of water organic pollutants identified by EU guidelines. *J. Hazard. Mater.* **2018**, *344*, 146–162. [CrossRef]
21. Rastogi, A.; Tiwari, M.K.; Ghangrekar, M.M. A review on environmental occurrence, toxicity and microbial degradation of Non-Steroidal Anti-Inflammatory Drugs (NSAIDs). *J. Environ. Manag.* **2021**, *300*, 113694. [CrossRef]
22. Maculewicz, J.; Kowalska, D.; Świacka, K.; Toński, M.; Stepnowski, P.; Bialk-Bielińska, A.; Dołżonek, J. Transformation products of pharmaceuticals in the environment: Their fate, (eco)toxicity and bioaccumulation potential. *Sci. Total Environ.* **2022**, *802*, 149916. [CrossRef] [PubMed]
23. Dos Santos, C.R.; Arcanjo, G.S.; Santos, L.V.S.; Koch, K.; Amaral, M.C.S. Aquatic concentration and risk assessment of pharmaceutically active compounds in the environment. *Environ. Pollut.* **2021**, *290*, 118049. [CrossRef]

24. De Ilurdoz, M.S.; Sadhwani, J.J.; Reboso, J.V. Antibiotic removal processes from water & wastewater for the protection of the aquatic environment—A review. *J. Water Process Eng.* **2022**, *45*, 102474. [CrossRef]
25. Hu, Y.; Zhu, Q.; Wang, Y.; Liao, C.; Jiang, G. A short review of human exposure to antibiotics based on urinary biomonitoring. *Sci. Total Environ.* **2022**, *830*, 154775. [CrossRef]
26. Dar, O.I.; Aslam, R.; Pan, D.; Sharma, S.; Andotra, M.; Kaur, A.; Jia, A.Q.; Faggio, C. Source, bioaccumulation, degradability and toxicity of triclosan in aquatic environments: A review. *Environ. Technol. Innov.* **2022**, *25*, 102122. [CrossRef]
27. Rezende-Teixeira, P.; Dusi, R.G.; Jimenez, P.C.; Espindola, L.S.; Costa-Lotufo, L.V. What can we learn from commercial insecticides? Efficacy, toxicity, environmental impacts, and future developments. *Environ. Pollut.* **2022**, *300*, 118983. [CrossRef]
28. Zhang, D.; Lu, S. Human exposure to neonicotinoids and the associated health risks: A review. *Environ. Int.* **2022**, *163*, 107201. [CrossRef]
29. Wan, Y.P.; Chai, B.W.; Wei, Q.; Hayat, W.; Dang, Z.; Liu, Z.H. 17α-ethynylestradiol and its two main conjugates in seven municipal wastewater treatment plants: Analytical method, their occurrence, removal and risk evaluation. *Sci. Total Environ.* **2022**, *812*, 152489. [CrossRef]
30. Strujis, J. *Application of SimpleTreat 4.0 in European Substance Regulations, Texte 13/2015*; German Environment Agency (Umweltbundesamt, UBA): Dessau-Roßlau, Germany, 2015; Available online: https://www.umweltbundesamt.de/sites/default/files/medien/378/publikationen/texte_13_2015_application_of_simple_treat_4.0.pdf (accessed on 1 October 2022).
31. Chopra, S.; Kumar, D. Ibuprofen as an emerging organic contaminant in environment, distribution and remediation. *Heliyon* **2020**, *6*, e04087. [CrossRef]
32. Brillas, E. A critical review on ibuprofen removal from synthetic waters, natural waters, and real wastewaters by advanced oxidation processes. *Chemosphere* **2022**, *286*, 131849. [CrossRef]
33. Adeleye, A.S.; Xue, J.; Zhao, Y.; Taylor, A.A.; Zenobio, J.E.; Sun, Y.; Han, Z.; Salawu, O.A.; Zhu, Y. Abundance, fate, and effects of pharmaceuticals and personal care products in aquatic environments. *J. Hazard. Mater.* **2022**, *424*, 127284. [CrossRef] [PubMed]
34. Archer, E.; Petrie, B.; Kasprzyk-Hordern, B.; Wolfaardt, G.M. The Fate of Pharmaceuticals and Personal Care Products (PPCPs), Endocrine Disrupting Contaminants (EDCs), Metabolites and Illicit Drugs in a WWTW and Environmental Waters. *Chemosphere* **2017**, *174*, 437–446. [CrossRef] [PubMed]
35. Kermia, A.E.B.; Fouial-Djebbar, D.; Trari, M. Occurrence, Fate and Removal Efficiencies of Pharmaceuticals in Wastewater Treatment Plants (WWTPs) Discharging in the Coastal Environment of Algiers. *C. R. Chim.* **2016**, *19*, 963–970. [CrossRef]
36. Elshikh, M.S.; Hussein, D.S.; Al-khattaf, F.S.; El-Naggar, R.A.R.; Almaary, K.S. Diclofenac removal from the wastewater using activated sludge and analysis of multidrug resistant bacteria from the sludge. *Environ. Res.* **2022**, *208*, 112723. [CrossRef] [PubMed]
37. Alessandretti, I.; Rigueto, C.V.T.; Nazari, M.T.; Rosseto, M.; Dettmer, A. Removal of diclofenac from wastewater: A comprehensive review of detection, characteristics and tertiary treatment techniques. *J. Environ. Chem. Eng.* **2021**, *9*, 106743. [CrossRef]
38. Kasonga, T.K.; Coetzee, M.A.A.; Kamika, I.; Ngole-Jeme, V.M.; Momba, M.N.B. Endocrine-disruptive chemicals as contaminants of emerging concern in wastewater and surface water: A review. *J. Environ. Manag.* **2021**, *277*, 11485. [CrossRef]
39. Shamsudin, M.S.; Azha, S.F.; Ismail, S. A review of diclofenac occurrences, toxicology, and potential adsorption of clay-based materials with surfactant modifier. *J. Environ. Chem. Eng.* **2022**, *10*, 107541. [CrossRef]
40. Shang, Q.; Chi, W.; Zhang, P.; Ling, Y.; Liu, X.; Cui, G.; Liu, W.; Shi, X.; Tang, B. Optimization of Bi_2O_3/TS-1 preparation and photocatalytic reaction conditions for low concentration Erythromycin wastewater treatment based on artificial neural network. *Process Saf. Environ. Prot.* **2022**, *157*, 297–305. [CrossRef]
41. Da Silva Rodrigues, D.A.; da Cunha, C.C.R.F.; do Espirito Santo, D.R.; de Barros, A.L.C.; Pereira, A.R.; de Queiroz Silva, S.; da Fonseca Santiago, A.; de Cássia Franco Afonso, R.J. Removal of cephalexin and erythromycin antibiotics, and their resistance genes, by microalgae-bacteria consortium from wastewater treatment plant secondary effluents. *Environ. Sci. Pollut. Res.* **2021**, *28*, 67822–67832. [CrossRef]
42. Omuferen, L.O.; Maseko, B.; Olowoyo, J.O. Occurrence of antibiotics in wastewater from hospital and convectional wastewater treatment plants and their impact on the effluent receiving rivers: Current knowledge between 2010 and 2019. *Environ. Monit. Assess.* **2022**, *194*, 306. [CrossRef]
43. Aziz, A.; Sengar, A.; Basheer, F.; Farooqi, I.H.; Isa, M.H. Anaerobic digestion in the elimination of antibiotics and antibiotic-resistant genes from the environment—A comprehensive review. *J. Environ. Chem. Eng.* **2022**, *10*, 106423. [CrossRef]
44. Nandikes, G.; Pathak, P.; Razak, A.S.; Narayanamurthy, V.; Singh, L. Occurrence, environmental risks and biological remediation mechanisms of Triclosan in wastewaters: Challenges and perspectives. *J. Water Process Eng.* **2022**, *49*, 103078. [CrossRef]
45. Abbott, T.; Kor-Bicakci, G.; Islam, M.S.; Eskicioglu, C. A Review on the Fate of Legacy and Alternative Antimicrobials and Their Metabolites during Wastewater and Sludge Treatment. *Int. J. Mol. Sci.* **2020**, *21*, 9241. [CrossRef] [PubMed]
46. Montes-Grajales, D.; Fennix-Agudelo, M.; Miranda-Castro, W. Occurrence of personal care products as emerging chemicals of concern in water resources: A review. *Sci. Total Environ.* **2017**, *595*, 601–614. [CrossRef] [PubMed]
47. Mundhe, S.; Birajdar, S.; Chavan, S.; Pawar, N. Imidacloprid poisoning: An emerging cause of potentially fatal poisoning. *Indian J. Crit. Care Med.* **2017**, *21*, 786–788. [CrossRef] [PubMed]
48. Webb, D.T.; Zhi, H.; Kolpin, D.W.; Klaper, R.D.; Iwanowicz, L.R.; LeFevre, G.H. Emerging investigator series: Municipal wastewater as a year-round point source of neonicotinoid insecticides that persist in an effluent-dominated stream. *Environ. Sci. Process. Impacts* **2021**, *23*, 678–688. [CrossRef]

49. De la Casa-Resino, I.; Villegas, J.M.; Puig, A.; Haro, A.; Bartolomé, A.; Herrera, E.M.; Flores, C.; Caixach, J.; Carapeto, R. *Residues of Imidacloprid in Surface Water Treatment Plants and Watercourses in Spain: Are Veterinary Medicinal Products (VMPs) a Significant Source of Emission?* Spanish Agency of Medicines and Medical Devices (AEMPS): Madrid, Spain, 2022. Available online: https://www.aemps.gob.es/publicaciones/publica/docs/Imidacloprid.pdf (accessed on 1 October 2022).
50. Wang, L.; Jiang, X.; Li, A. Ammonia oxidation intermediates-induced abiotic transformation of 17α-ethinylestradiol in nitrification process of wastewater biological treatment. *J. Environ. Chem. Eng.* **2022**, *10*, 108171. [CrossRef]
51. Klaic, M.; Jirsa, F. 17α-Ethinylestradiol (EE2): Concentrations in the environment and methods for wastewater treatment—An update. *RSC Adv.* **2022**, *12*, 12794–12805. [CrossRef]
52. Tang, Z.; Liu, Z.H.; Wang, H.; Dang, Z.; Liu, Y. Occurrence and removal of 17α-ethynylestradiol (EE2) in municipal wastewater treatment plants: Current status and challenges. *Chemosphere* **2021**, *271*, 129551. [CrossRef]
53. Ahmed, M.B.; Zhou, J.L.; Ngo, H.H.; Guo, W.; Thomaidis, N.S.; Xu, J. Progress in the biological and chemical treatment technologies for emerging contaminant removal from wastewater: A critical review. *J. Hazard. Mater.* **2017**, *323*, 274–298. [CrossRef]
54. Rivera-Utrilla, J.; Sánchez-Polo, M.; Ferro-García, M.A.; Prados-Joya, G.; Ocampo-Pérez, R. Pharmaceuticals as emerging contaminants and their removal from water. A review. *Chemosphere* **2013**, *93*, 1268–1287. [CrossRef] [PubMed]
55. Lautz, L.S.; Strujis, J.; Nolte, T.M.; Breure, A.M.; van der Grinten, E.; van de Meent, D.; van Zelm, R. Evaluation of SimpleTreat 4.0: Simulations of pharmaceutical removal in wastewater treatment plant facilities. *Chemosphere* **2017**, *168*, 870–876. [CrossRef] [PubMed]
56. Li, W.L.; Zhang, Z.F.; Kilgallon, J.; Sparham, C.; Li, Y.F.; Yuan, Y.X. Fate of household and personal care chemicals in typical urban wastewater treatment plants indicate different seasonal patterns and removal mechanisms. *Environ. Pollut.* **2022**, *294*, 118618. [CrossRef] [PubMed]
57. Schulte, R.A.A.; Link, A.; Siegert, M.W.; Finkbeiner, M. Modeling the use and end-of-life phase of pharmaceuticals in support of a life cycle inventory analysis—Case study on different antibiotics in Germany. *Sustain. Chem. Pharm.* **2022**, *25*, 100589. [CrossRef]
58. Comber, S.; Gardner, M.; Sörme, P.; Ellor, B. The removal of pharmaceuticals during wastewater treatment: Can it be predicted accurately? *Sci. Total Environ.* **2019**, *676*, 222–230. [CrossRef]
59. Alvarino, T.; Suárez, S.; Garrido, M.; Lema, J.M.; Omil, F. A UASB reactor coupled to a hybrid aerobic MBR as innovative plant configuration to enhance the removal of organic micropollutants. *Chemosphere* **2016**, *144*, 452–458. [CrossRef]
60. Show, S.; Chakraborty, P.; Karmakar, B.; Halder, G. Sorptive and microbial riddance of micro-pollutant ibuprofen from contaminated water: A state of the art review. *Sci. Total Environ.* **2021**, *786*, 147327. [CrossRef]
61. Gautam, P.; Dubey, S.K. Biodegradation of imidacloprid: Molecular and kinetic analysis. *Bioresour. Technol.* **2022**, *350*, 126915. [CrossRef]
62. Balakrishnan, P.; Mohan, S. Treatment of triclosan through enhanced microbial biodegradation. *J. Hazard. Mater.* **2021**, *420*, 126430. [CrossRef]
63. Kosma, C.; Lambropoulou, D.; Albanis, T. Occurrence and removal of PPCPs in municipal and hospital wastewaters in Greece. *J. Hazard. Mater.* **2010**, *179*, 804–817. [CrossRef]

processes

MDPI

Communication

Bioremediation of Crude Oil by Haematococcus Pluvialis: A Preliminary Study

Rosa Paola Radice [1,2,*], Maria Sansone [3], Gabriele D'Arienzo [3], Antonio Scopa [1] and Giuseppe Martelli [1]

1 Department of Science, University of Basilicata, Viale dell'Ateneo Lucano, 85100 Potenza, PZ, Italy
2 Bioinnova srls, Via Ponte Nove Luci, 85100 Potenza, PZ, Italy
3 SCA Lab Service, Zona Pip, 75015 Pisticci, MT, Italy
* Correspondence: rosapaolaradice@gmail.com

Abstract: Nowadays, oil pollution is one of the main environmental problems. The current methods for recovering spills mainly involve chemical agents, but scientific research has focused on more natural and less harmful techniques for the environment, including a consortium of bacteria and microalgae to clean up water contaminated by hydrocarbons. The purpose of this preliminary study was to evaluate the ability of a microalga belonging to Chlorophyceae to grow in the presence of crude oil and remove the principal contaminants. *H. pluvialis*, which is usually used for nutraceutical purposes, thanks to the production of astaxanthin, was able to grow in anaerobic conditions, varying its metabolism from autotrophic to heterotrophic, exploiting the carbon present in the solution deriving from the presence of 1% of crude oil. Furthermore, the results of bioremediation showed a relevant reduction in chemical pollutants such as nitrate, fluoride, sulfate, and phosphate. The most important aspect of the study was the reduction after 160 days in the hydrocarbon concentration inside not only the culture medium (−32%) but also the algal biomass (−80.25%), demonstrating an optimized degradation rather than a simple absorption inside the alga.

Keywords: bioremediation; crude oil; Haematococcus pluvialis; microalgae; anaerobic growth; heterotrophic metabolism

Citation: Radice, R.P.; Sansone, M.; D'Arienzo, G.; Scopa, A.; Martelli, G. Bioremediation of Crude Oil by Haematococcus Pluvialis: A Preliminary Study. *Processes* **2022**, *10*, 2472. https://doi.org/10.3390/pr10122472

Academic Editors: Dimitris Zagklis and Georgios Bampos

Received: 3 November 2022
Accepted: 19 November 2022
Published: 22 November 2022

1. Introduction

The world economy is based on fossil fuels for obtaining energy and their use is a problem for the entire world's ecosystem [1]. In fact, the spill and loss of oil and its by-products are the main causes of oil pollution [2]. Today, oil recovery techniques involve the use of chemical dispersants that break down the oil molecules, making them more available to attack by atmospheric agents and for microorganisms that participate in bioremediation [3,4]. The ability to degrade hydrocarbons is found in some bacteria, fungi, and yeasts [5]. Until now, researchers focused mostly on the bacteria that are able to remove various oil pollutants, including polycyclic aromatic hydrocarbons (PAHs) [6]. Even if bacteria can act efficiently again oil pollution, they are not able to remove each contaminant. To avoid this, it is possible to use bacteria engineered with specific enzymes to degrade petroleum pollutants. However, engineered bacteria could not live naturally in the areas around the spills and could not adapt to the "natural environment", dying before acting against pollution [7].

Other microorganisms belonging to phytoplankton help bacteria during the process. Researchers have often focused on a synergistic mechanism of bioremediation between oil-degrading bacteria and microalgae thanks to their ability to produce oxygen used by bacteria for their metabolism. Still, microalgae could represent the principal actors [8]. They are unicellular and photosynthetic organisms belonging to the prokaryotic (cyanobacteria) and eukaryotic (microalgae) divisions [9]. Microalgae are used in different fields for different purposes. In the nutraceuticals field, they are valuable for the nutrients produced such as proteins, polysaccharides, antioxidants, polyunsaturated fatty acids, lipids, vitamins, and pigments [10,11]. Furthermore, they are used for wastewater treatment [12,13] and biofuel production thanks to the high quality and quantity of lipids contained [14]. One of the most significant aspects related to the use of microalgae is their high ability to adapt to different environmental stimuli (nutrients, temperature, pH), which allows them to work under several conditions. Many species can change their metabolism by using organic carbon instead of carbon dioxide. Thanks to their adaptability, several studies have been conducted on different species of microalgae, mainly *Chlorophyceae*, to test their bioremediation capacities, for example, *Chlorella vulgaris* and *Scenedesmus obliquus* [15,16].

It has been reported that some species can significantly reduce aliphatic compounds, such as 3-methyl decane, heptadecane, and octadecane [16], and that the ability of *Chlorella* spp. to reduce emulsified oil reached 80 % [17]. Further studies have shown that *Chlorella* spp. can remove phosphorus from petroleum effluent in just 13 days, also reducing nitrogen by 78% and reducing the Chemical Oxygen Demand (COD) from 504 mg L^{-1} to 144 mg L^{-1} [18]. Ashwaniy et al. found that the microalgae grown in petroleum refinery effluent (PRE) can reduce the concentration of COD by 70%, the Biochemical Oxygen Demand (BOD) by 81%, sulfide by 61%, total suspended solids (TSS) by 622%, and phosphorous and Total Dissolved Solids (TDS) by 67% and can act as a substrate for bacterial growth in a microbial desalination cell (MDC) to produce clean energy [19]. *Haematococcus pluvialis* is a green microalga capable of colonizing different environments. *H. pluvialis* is a special microalga as its life cycle consists of different phases characterized by different morphologies. Initially, the microalga, called macrozooids, has two flagella; subsequently, through mitotic division, daughter cells called microzooids originate, which grow up to form coccoid cells [20]. Its main applications are in the nutraceutical and pharmaceutical fields, thanks to the production in stressful conditions of several carotenoids and pigments. First, astaxanthin, a powerful antioxidant, is useful for the treatment of many diseases, reducing adipose tissue weight, glucose serum levels, and systolic blood pressure in an animal model of obesity and reducing liver and body weight in an animal model of NAFLD [21,22]. *H. pluvialis* is usually studied for other metabolites of interest. For example, in optimal conditions, it can produce essential amino acids such as leucine, glutaminic acid, and alanine. Furthermore, by varying the growth conditions, it is also possible to improve the lipid content, which can represent 30% of the total weight. In addition, the nature of the lipids makes *H. pluvialis* very interesting, with a high content of polyunsaturated lipids (PUFA) within cells, and in the red encystic stage, h pluvialis dramatically increases triglyceride (TAG) levels [23]. All these characteristics lead to the belief that *H. pluvialis* is a microalga with a high nutritional value that can be used for metabolites of nutraceutical and pharmaceutical interest. However, the high lipid (especially the intermediate fractions C16-C18) and carbohydrate contents could be used for biofuels such as biodiesel and bioethanol production [24]. For this reason, our research also focused on the industrial applications of *H. pluvialis*, especially its role in bioremediation, hypothesizing a mechanism for removing contaminants and reusing algal biomass. However, based on our research, there are no studies about *H. pluvialis* and crude oil bioremediation. Only Suresh Kumar et al. hypothesized how the degradation mechanism of PAHs in *H. pluvialis* could be similar to prokaryotes. However, several parameters were analyzed to create a model that could simulate the link between PAHs and CYP450 of *Haematococcus pluvialis*. Thirty-eight PAHs formed from one to six benzene rings were involved in the analysis and the results showed that hydrogen, hydrophobic, electrostatic, π-π, and Van der Waals interactions occurred in the active site of CYP450. Specifically, 18 PAHs interacted with Threonine282 (Thr282), Alanine337 (Ala337), Serine404 (Ser404), and Lisyne407 (Lys407) via hydrogen bonds. However, in this study, it is evident that only LMW-PAHs were able to bind CYP450, whereas HMW-PAHs were not [25]. Based on this, this preliminary study aimed to study the potential of *H. pluvialis* for crude oil bioremediation, analyzing (i) the ability to grow in the presence of such a toxic agent; (ii) the quantitative concentration of hydrocarbons degraded in the aqueous phase and biomass, and (iii) the ability of *H. pluvialis* to grow in heterotrophic conditions.

2. Materials and Methods

2.1. Algae Growth

H. pluvialis UTEX 2505 was grown in a self-produced media produced by dissolving 0.3 gr of Greenhouse Special 20–20–20 (BIOGARD) powder in 1 L of distilled water (Greenhouse Special powder contains HNO_3 6% (*w*/*v*), NH_4^+ 5.2% (*w*/*v*), CH_4N_2O 8.8% (*w*/*v*), P_2O_5 20% (*w*/*v*), K_2O 20% (*w*/*v*), B 0.05% (*w*/*v*), Cu 0.01% (*w*/*v*), Fe 0.2% (*w*/*v*), Mn 0.1%(*w*/*v*), Mo 0.005% (*w*/*v*), Zn 0.01%(*w*/*v*), chelating agent EDTA) under a light intensity of 120 mmol photons $m^{-2}s^{-1}$ in a 12 h: 12 h light/ dark cycle at 25 °C. Cultures were not supplied with an extra source of CO_2 and were shaken by a mechanical agitator (g24 environmental incubator shaker, American Laboratory Trading) at 70 rpm.

Cultivation with Crude Oil

Five ml (1%) of crude oil was sonicated in 500 mL of standard medium, and then 25 mL of culture containing approximately 1.5 × 107 cells/mL was inoculated in an anaerobiosis tank. The culture was placed under a light intensity of 120 mmol photons $m^{-2}s^{-1}$ in a 12 h: 12 h light/dark cycle at 25 °C. Cultures were shaken by a mechanical agitator (g24 environmental incubator shaker, American Laboratory Trading) at 70 rpm. In another tank, the same *H. pluvialis* cell concentration was inoculated without crude oil as a control. Algal growth was assessed by measuring optical density at 750 nm (SPEC-TROstar® Nano, BMG Labtech) and cell counts by light microscopy (Zeiss Axioplan) using the Burker chamber (BLAUBRAND). The experiment was conducted in triplicate.

2.2. Pigment Estimation

On days 1, 5, 7, 12, 20, and 25, from the *H. pluvialis* culture with 1% of crude oil and the control culture, one ml of culture was centrifuged for 5 min at 14,000 rpm. The pellet was suspended in 1 mL of DMSO and incubated for 10 min at 70 °C. The protocol was repeated until the pellet was white. The supernatant was collected and absorbance was registered at 480 nm, 649 nm, and 665 nm. The protocol used to quantify chlorophyll and carotenoid was the one reported by Wellburn et al. [26].

2.3. Biodegradation Activity

Several analyses were carried out to evaluate the ability of *H. pluvialis* to remove and degrade the oil. On day 0 and day 10, some parameters were preliminarily analyzed including the TOC, fluorides, chlorides, nitrates, sulfates, and phosphates in the supernatant. On day 20 and day 160, two samples were taken to conduct a quantitative analysis using GC-FID. An amount of 150 mL of culture was placed in a separating funnel to separate and collect the microalgae biomass and aqueous phase separately. Crude oil biodegradation was analyzed using a GC-FID Varian 3900 gas carrier helium (1 mL min^{-1}). The mineral column was a 15 m × 0.32 mm × 0.1 μm film, and the temperature programming was 40 °C (hold 1 min), −80 °C rate 50 °C/min (hold 1 min), −350 °C rate 20 °C/min (hold 8 min).

3. Results

As mentioned above, there are no studies regarding the possibility of using *H. pluvialis* as a bioremediator. To evaluate whether the metabolism of the cells grown under anaerobic conditions varied, the cell growth was followed for 25 days. The growth of *H. pluvialis* in the presence of 1% pure crude oil and anaerobiosis demonstrated how the microalga could metabolize the organic carbon present in the medium, reaching the exponential phase after 12 days, unlike the control grown in anaerobiosis but in the absence of carbon in the medium, which showed difficulties in reaching the exponential stage (Figure 1).

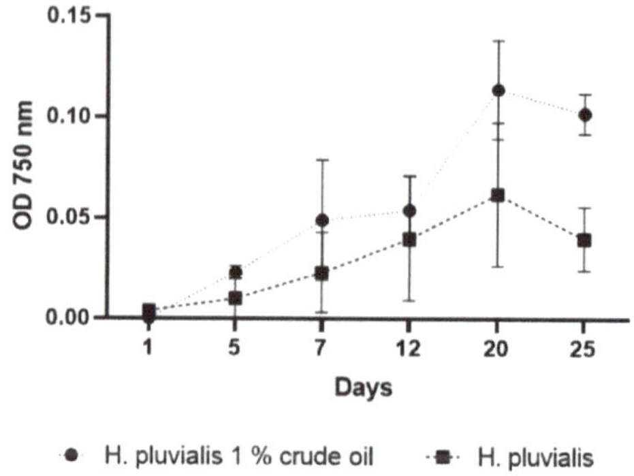

Figure 1. Cell growth. Values are reported as mean ± SEM.

The concentrations of the chlorophylls and pigments were analyzed to confirm the transition of the culture metabolism. The results showed a reduction in the chlorophylls in the culture grown with crude oil, with organic carbon available for their metabolism. However, in the control culture, the production of chlorophylls increased due to the lack of organic carbon in the media and the photosynthetic reaction was optimized (Figure 2a) [20,27]. In addition, the extracted pigments seen in Figure 2b showed a lower production in the mixotrophic culture including those predisposed to the defense of the cells against a stressful agent (e.g., astaxanthin).

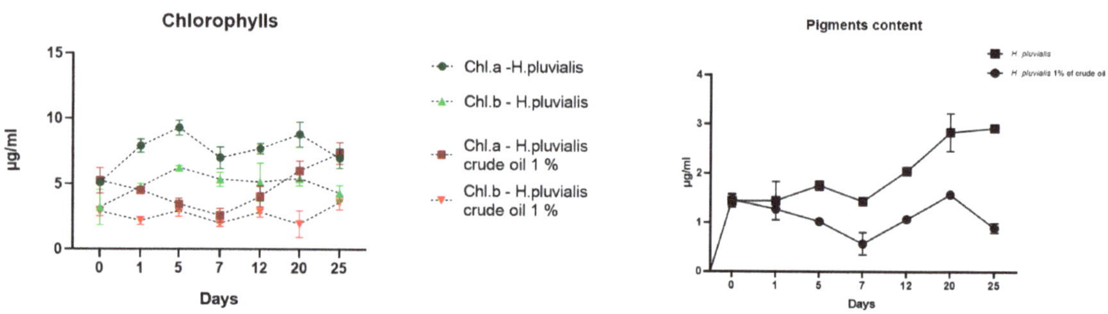

(a) (b)

Figure 2. (**a**) Chlorophyll concentrations and (**b**) pigment concentrations under anaerobiosis conditions with and without crude oil in the media. Values are reported as mean ± SEM.

The chemical analyses carried out on the culture media and the algal biomass showed that after ten days, some parameters and contaminants had already changed compared to day 0 (Table 1).

Table 1. Preliminary data after 10 days of bioremediation, expressed in mg mL^{-1} (mean ± SEM).

	TOC	Nitrates	Fluorides	Chlorides	Sulfates	Phosphates
DAY 0	23.90 ± 0.78	0.95 ± 0.18	0.40 ± 0.14	10.75 ± 0.18	11.50 ± 1.06	70.20 ± 1.27
DAY 10	44.50 ± 1.06	0.24 ± 0.04	0.20 ± 0.07	12.50 ± 0.35	9.00 ± 0.85	53.50 ± 1.06

The TOC values doubled after 10 days, showing how *H. pluvialis* acted against hydrocarbons, making them more available for metabolism and using them for cell reproduction [28]. Nitrates, fluorides, sulfates, and phosphates were reduced, confirming the results obtained by numerous studies carried out on different microalgal species used for wastewater bioremediation (Figure 3) [8,29,30].

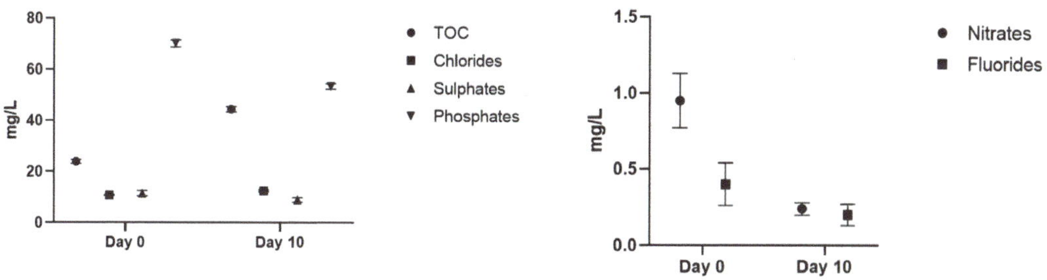

Figure 3. Preliminary data on bioremediation. Values are reported as mean ± SEM.

However, the most significant results were derived from the GC-FID analysis, which was directed at the hydrocarbons from crude oil. The analysis was performed on the aqueous fraction and the algal biomass after 20 and 160 days of treatment. As evident in Figure 4, there was a quantitative reduction in the hydrocarbons. In water, the reduction was 32% (9.24 ppm at day 20; 6.28 ppm at day 160), whereas in the algal biomass, it was 80.25% (4725 mg/Kg at day 20; 933 mg/Kg at day 160). However, the GC-FID spectra showed a quantitative but not qualitative reduction in the water, whereas in the microalgal biomass, the low-molecular-weight hydrocarbon (LMW-HC) fraction (C6–C12) was significantly reduced compared to day 20 (Figure 5).

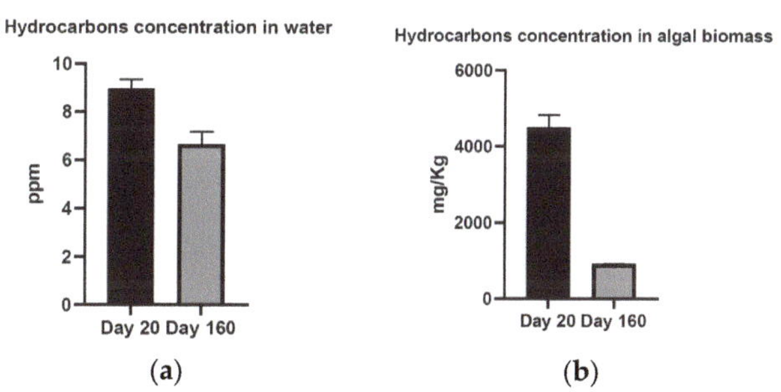

Figure 4. Hydrocarbon concentration in (**a**) water and (**b**) algal biomass. Data are reported as mean ± SEM.

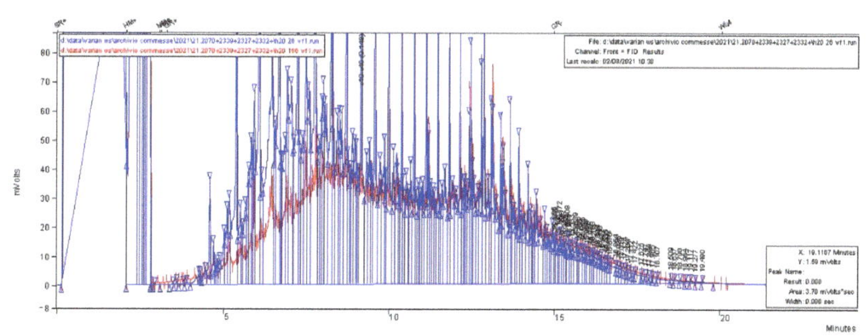

(**a**)

Figure 5. *Cont.*

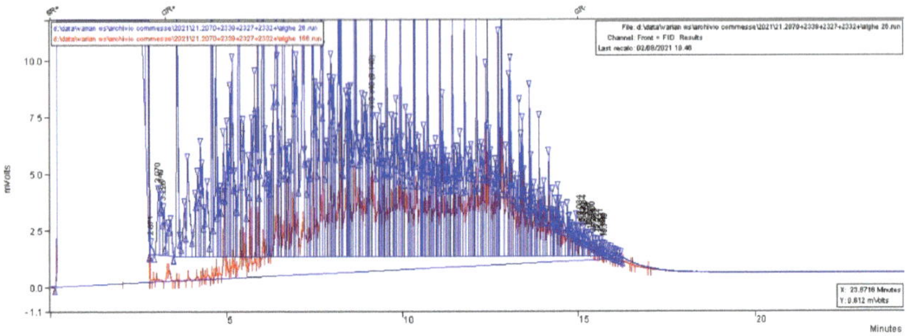

(b)

Figure 5. GC-FID spectra. (**a**) Water and (**b**) microalgal biomass hydrocarbon concentrations (blue = day 20; red = day 160).

4. Discussion

These preliminary results show how the action of *H. pluvialis* in the bioremediation of oil-contaminated water is not only feasible but also optimal. The adaptation of the microalgae to a hostile environment (anaerobic and crude oil) demonstrates how *H. pluvialis* can vary the enzymatic and metabolic expression rapidly and efficiently, supporting the hypothesis that the microalgae genome is highly reactive. In the model alga *Chlamydomonas reinhartdtii*, it was recently discovered that the membrane protein HLA3 is involved in the activation of inorganic carbon uptake in low CO_2 situations; this could also be valid for *H. pluvialis*, which increased growth using the carbon from the hydrocarbons. [31]. The reduction in the nitrogen inside the media is also significant because nitrogen is an abundant contaminant in different processing wastewater [32,33] and *H. pluvialis* could be used to remove it in other bioremediation processes. Considering the amount of crude oil used in this study is crucial because other studies on microalgae bioremediation reported that the optimal crude-oil concentration was 0.5%. In the study by El-Sheekh et al., *S. obliquus* and *C. vulgaris* under heterotrophic conditions at 0.5% of crude oil removed compounds such as n-alkanes and Polycyclic aromatic hydrocarbons (PAHs) [15]. *H.pluvialis*, indeed, grows and degrades higher concentrations, assuming a more efficient remediation process in water where the percentage of oil is lower than that tested.

5. Conclusions

This preliminary study for the biodegradative activity of *H. pluvialis* shows for the first time how this green alga can grow in the presence of pure crude oil and the biodegradable part of the oil in the solution. It will be necessary to extend our analyses by evaluating the biochemical processes and mechanisms involved and performing a mass balance and stoichiometry analysis to identify, in detail, the affected and reduced classes of hydrocarbons. Furthermore, once all the biochemical processes involved are understood, it will be possible to optimize them through high selective pressure to obtain more performance strains, which will be useful for this type of preliminary study. These results are fundamental to not only improving the knowledge about *H. pluvialis*, which is used principally in pharmaceutical fields and not in bioengineering, but also because they could represent the first step in using innovative technology to combine bioremediation and astaxanthin production. In this way, the applications of *H. pluvialis* could expand in environmental and industrial areas, where *H. pluvialis* could initially grow on contaminated substrates and then be used for the extraction of interest compounds. It must be highlighted that microalgae are excellent candidates for biofuel production thanks to the high amount and purity of lipids, and *H. pluvialis* could place itself in a circular green economy by eliminating fossil fuel contaminants, producing biofuels, and reducing emissions and contaminations.

Author Contributions: G.M.: Project administration, Conceptualization, Supervision; R.P.R.: Methodology, Writing—Original draft; A.S.: Visualization; M.S.: Resources; G.D.: Investigation. All authors have read and agreed to the published version of the manuscript.

Funding: This research received no external funding.

Conflicts of Interest: The authors declare no conflict of interest.

References

1. Hassan, A.; Ilyas, S.Z.; Jalil, A.; Ullah, Z. Monetization of the Environmental Damage Caused by Fossil Fuels. *Environ. Sci. Pollut. Res.* **2021**, *28*, 21204–21211. [CrossRef] [PubMed]
2. Chilvers, B.L.; Morgan, K.J.; White, B.J. Sources and Reporting of Oil Spills and Impacts on Wildlife 1970–2018. *Environ. Sci. Pollut. Res.* **2021**, *28*, 754–762. [CrossRef] [PubMed]
3. Thomas, G.E.; Brant, J.L.; Campo, P.; Clark, D.R.; Coulon, F.; Gregson, B.H.; McGenity, T.J.; McKew, B.A. Effects of Dispersants and Biosurfactants on Crude-Oil Biodegradation and Bacterial Community Succession. *Microorganisms* **2021**, *9*, 1200. [CrossRef] [PubMed]
4. Faisal, W.; Almomani, F. A Critical Review of the Development and Demulsification Processes Applied for Oil Recovery from Oil in Water Emulsions. *Chemosphere* **2022**, *291*, 133099. [CrossRef] [PubMed]
5. Dell' Anno, F.; Rastelli, E.; Sansone, C.; Brunet, C.; Ianora, A.; Dell' Anno, A. Bacteria, Fungi and Microalgae for the Bioremediation of Marine Sediments Contaminated by Petroleum Hydrocarbons in the Omics Era. *Microorganisms* **2021**, *9*, 1695. [CrossRef] [PubMed]
6. Chernikova, T.N.; Bargiela, R.; Toshchakov, S.V.; Shivaraman, V.; Lunev, E.A.; Yakimov, M.M.; Thomas, D.N.; Golyshin, P.N. Hydrocarbon-Degrading Bacteria Alcanivorax and Marinobacter Associated With Microalgae Pavlova Lutheri and Nannochloropsis Oculata. *Front. Microbiol.* **2020**, *11*, 572931. [CrossRef]
7. French, K.E.; Zhou, Z.; Terry, N. Horizontal 'Gene Drives' Harness Indigenous Bacteria for Bioremediation. *Sci. Rep.* **2020**, *10*, 15091. [CrossRef]
8. Yadav, G.; Dash, S.K.; Sen, R. A Biorefinery for Valorization of Industrial Waste-Water and Flue Gas by Microalgae for Waste Mitigation, Carbon-Dioxide Sequestration and Algal Biomass Production. *Sci. Total Environ.* **2019**, *688*, 129–135. [CrossRef]
9. Kato, Y.; Hasunuma, T. Metabolic Engineering for Carotenoid Production Using Eukaryotic Microalgae and Prokaryotic Cyanobacteria. *Adv. Exp. Med. Biol.* **2021**, *1261*, 121–135. [CrossRef]
10. Amorim, M.L.; Soares, J.; dos Reis Coimbra, J.S.; de Leite, M.O.; Albino, L.F.T.; Martins, M.A. Microalgae Proteins: Production, Separation, Isolation, Quantification, and Application in Food and Feed. *Crit. Rev. Food Sci. Nutr.* **2021**, *61*, 1976–2002. [CrossRef]
11. Dolganyuk, V.; Belova, D.; Babich, O.; Prosekov, A.; Ivanova, S.; Katserov, D.; Patyukov, N.; Sukhikh, S. Microalgae: A Promising Source of Valuable Bioproducts. *Biomolecules* **2020**, *10*, 1153. [CrossRef] [PubMed]
12. Plöhn, M.; Spain, O.; Sirin, S.; Silva, M.; Escudero-Oñate, C.; Ferrando-Climent, L.; Allahverdiyeva, Y.; Funk, C. Wastewater Treatment by Microalgae. *Physiol. Plant* **2021**, *173*, 568–578. [CrossRef] [PubMed]
13. Kumar Patel, A.; Tseng, Y.-S.; Rani Singhania, R.; Chen, C.-W.; Chang, J.-S.; di Dong, C. Novel Application of Microalgae Platform for Biodesalination Process: A Review. *Bioresour. Technol.* **2021**, *337*, 125343. [CrossRef] [PubMed]
14. Ma, X.; Mi, Y.; Zhao, C.; Wei, Q. A Comprehensive Review on Carbon Source Effect of Microalgae Lipid Accumulation for Biofuel Production. *Sci. Total Environ.* **2022**, *806*, 151387. [CrossRef]
15. El-Sheekh, M.M.; Hamouda, R.A.; Nizam, A.A. Biodegradation of Crude Oil by Scenedesmus Obliquus and Chlorella Vulgaris Growing under Heterotrophic Conditions. *Int. Biodeterior. Biodegrad.* **2013**, *82*, 67–72. [CrossRef]
16. Hamouda, R.A.E.F.; Sorour, N.M.; Yeheia, D.S. Biodegradation of Crude Oil by Anabaena Oryzae, Chlorella Kessleri and Its Consortium under Mixotrophic Conditions. *Int. Biodeterior. Biodegrad.* **2016**, *112*, 128–134. [CrossRef]
17. Kuttiyathil, M.S.; Mohamed, M.M.; Al-Zuhair, S. Using Microalgae for Remediation of Crude Petroleum Oil-Water Emulsion. *Biotechnol. Prog.* **2020**, *37*, e3098. [CrossRef]
18. Znad, H.; al Ketife, A.M.D.; Judd, S.; AlMomani, F.; Vuthaluru, H.B. Bioremediation and Nutrient Removal from Wastewater by Chlorella Vulgaris. *Ecol. Eng.* **2018**, *110*, 1–7. [CrossRef]
19. Ashwaniy, V.R.V.; Perumalsamy, M.; Pandian, S. Enhancing the Synergistic Interaction of Microalgae and Bacteria for the Reduction of Organic Compounds in Petroleum Refinery Effluent. *Environ. Technol. Innov.* **2020**, *19*, 100926. [CrossRef]
20. Bauer, A.; Minceva, M. Examination of Photo-, Mixo-, and Heterotrophic Cultivation Conditions on Haematococcus Pluvialis Cyst Cell Germination. *Appl. Sci.* **2021**, *16*, 7201. [CrossRef]
21. Radice, R.P.; Limongi, A.R.; Viviano, E.; Padula, M.C.; Martelli, G.; Bermano, G. Effects of Astaxanthin in Animal Models of Obesity-Associated Diseases: A Systematic Review and Meta-Analysis. *Free Radic. Biol. Med.* **2021**, *171*, 156–168. [CrossRef] [PubMed]
22. Chen, F.; Xu, N.; Liu, K.; Lv, R.; Shi, J.; Liu, J.; Sun, X.; Hu, C. Increasing Production and Bio-Accessibility of Natural Astaxanthin in Haematococcus Pluvialis by Screening and Culturing Red Motile Cells under High Light Condition. *Bioresour. Technol.* **2022**, *364*, 128067. [CrossRef] [PubMed]
23. Shah, M.M.R.; Liang, Y.; Cheng, J.J.; Daroch, M. Astaxanthin-Producing Green Microalga Haematococcus Pluvialis: From Single Cell to High Value Commercial Products. *Front. Plant Sci.* **2016**, *7*, 531. [CrossRef] [PubMed]

24. Nishshanka, G.K.S.H.; Liyanaarachchi, V.C.; Nimarshana, P.H.V.; Ariyadasa, T.U.; Chang, J.-S. Haematococcus Pluvialis: A Potential Feedstock for Multiple-Product Biorefining. *J. Clean Prod.* **2022**, *344*, 131103. [CrossRef]
25. SureshKumar, P.; Thomas, J.; Poornima, V. Structural Insights on Bioremediation of Polycyclic Aromatic Hydrocarbons Using Microalgae: A Modelling-Based Computational Study. *Environ. Monit. Assess* **2018**, *190*, 92. [CrossRef] [PubMed]
26. Wellburn, A.R. The Spectral Determination of Chlorophylls a and b, as Well as Total Carotenoids, Using Various Solvents with Spectrophotometers of Different Resolution. *J. Plant Physiol.* **1994**, *144*, 307–313. [CrossRef]
27. Kobayashi, M.; Kakizono, T.; Yamaguchi, K.; Nishio, N.; Nagai, S. Growth and Astaxanthin Formation of Haematococcus Pluvialis in Heterotrophic and Mixotrophic Conditions. *J. Ferment. Bioeng.* **1992**, *74*, 17–20. [CrossRef]
28. Do, C.V.T.; Pham, M.H.T.; Pham, T.Y.T.; Dinh, C.T.; Bui, T.U.T.; Tran, T.D.; Nguyen, V.T. Microalgae and Bioremediation of Domestic Wastewater. *Curr. Opin. Green Sustain. Chem.* **2022**, *34*, 100595. [CrossRef]
29. Bhatt, N.C.; Panwar, A.; Bisht, T.S.; Tamta, S. Coupling of Algal Biofuel Production with Wastewater. *Sci. World J.* **2014**, *2014*, 210504. [CrossRef]
30. Das, P.; AbdulQuadir, M.; Thaher, M.; Khan, S.; Chaudhary, A.K.; Alghasal, G.; Al-Jabri, H.M.S.J. Microalgal Bioremediation of Petroleum-Derived Low Salinity and Low PH Produced Water. *J. Appl. Phycol.* **2019**, *31*, 435–444. [CrossRef]
31. Gao, H.; Wang, Y.; Fei, X.; Wright, D.A.; Spalding, M.H. Expression Activation and Functional Analysis of HLA3, a Putative Inorganic Carbon Transporter in Chlamydomonas Reinhardtii. *Plant J.* **2015**, *82*, 1–11. [CrossRef] [PubMed]
32. Chen, H.; Wang, Q. Microalgae-Based Nitrogen Bioremediation. *Algal Res.* **2020**, *46*, 101775. [CrossRef]
33. Su, Y. Revisiting Carbon, Nitrogen, and Phosphorus Metabolisms in Microalgae for Wastewater Treatment. *Sci. Total Environ.* **2021**, *762*, 144590. [CrossRef] [PubMed]

processes

MDPI

Article

Fermentation Wastes from *Chrypthecodinium cohnii* Lipid Production for Energy Recovery by Anaerobic Digestion

Ana Eusébio *, Patrícia Moniz, Teresa Lopes da Silva and Isabel Paula Marques

LNEG—National Laboratory for the Energy and Geology, I.P.—Bioenergy and Biorefineries Unit, Estrada do Paço do Lumiar, 22, Ed. F–R/C, 1649-038 Lisbon, Portugal

* Correspondence: ana.eusebio@lneg.pt

Abstract: Wastes generated during the cultivation of marine microalga *Crypthecodinium cohnii* and after the lipid extraction process, were energetically valorized into biogas production through anaerobic digestion (AD). The tested wastes were extracted microalgae (Ae) with hexane (AeH) using supercritical extraction methods (AeS) and the supernatant obtained after culture medium centrifugation (M). The digestion of the algae biomass in the admixture with the supernatant medium (AeH+M+I and AeS+M+I) provided a higher methane content and a higher methane yield (582 and 440 L CH_4/kg VS) than the substrates Ae and M, individually digested (155 and 96 L CH_4/kg VS, respectively). Flow cytometry monitoring processes during AD indicated that the yield of the accumulated biogas was influenced by the operating conditions. The mixture of AeH+M+I was the only assay with a proportion of cells with less damaged membranes after AD, providing the highest methane yield and productivity (582 L CH_4/kg VS and 31 L CH_4/kg VS.d, respectively) and the highest energetic potential of 5.8 KWh/kg VS of all the substrates. From the results, AD integration to lipid production by *C. cohnii* to recover energy from the generated wastes enhanced the sustainability of the entire process and promoted the practice of zero waste.

Keywords: marine microalgae; *Crypthecodinium cohnii*; wastes; flow cytometry; anaerobic digestion; biogas/methane

Citation: Eusébio, A.; Moniz, P.; Lopes da Silva, T.; Marques, I.P. Fermentation Wastes from *Chrypthecodinium cohnii* Lipid Production for Energy Recovery by Anaerobic Digestion. *Processes* **2022**, *10*, 2463. https://doi.org/10.3390/pr10112463

Academic Editors: Dimitris Zagklis and Georgios Bampos

Received: 13 September 2022
Accepted: 16 November 2022
Published: 21 November 2022

1. Introduction

The heterotrophic marine dinoflagellate microalga *Crypthecodinium cohnii* produces significant amounts of lipids (20–50% cell dry weight) with a high proportion of docosahexaenoic acid (DHA), a ω-3 polyunsaturated fatty acid, necessary for brain development during pregnancy and childhood [1]. Furthermore, this compound has been recognized benefits on human health, improving vision, psoriasis, cancer prevention, heart health and inflammatory response reduction [2]. Currently, DHA has several applications in the nutritional, pharmaceutical and cosmetic industries, foreseeing a growing market size [3]. A few companies have commercialized DHA-rich oil obtained from heterotrophic microalgal oil, such as DSM and Lonza [3].

Several methods can be used to extract the intracellular microalgal lipids, namely, the Soxhlet lipid extraction method that is widely applied in the food industry as it uses hexane. Supercritical extraction (SCE), by resorting to supercritical carbon dioxide (CO_2), is a cleaner method of lipid extraction by way of no residues being present in the extracted lipids.

In addition to obtaining lipids, the microalgal oil production process generates wastes such as fermentation broth supernatant and microalgal biomass leftovers, obtained after lipid extraction which are usually neglected. Because they contain carbon, nitrogen, and other nutrients, they could be useful as substrates for the biogas/methane production. The anaerobic digestion process (AD) transforms organic matter into biogas (mainly methane and carbon dioxide) and preserves nutrients (N and P) in the digestate, through the action of a microbial community, under anaerobic conditions. Biogas can be applied to electric and thermal energy supply or converted into biomethane and injected into the natural gas network [4]. In this way, it can be regarded as an energy-carrying gas and useful for the *C. cohnii* lipid production process, reducing the energy demand of the overall process and fulfilling the circular economy principles [4]. A solid stream of digestate can be stabilized (composted) and converted into a product for agriculture or animal feed while the liquid fraction can be recovered as water irrigation or, even, as "water" directed towards the production of e-methane through the electrolysis process, taking advantage of the power-to-methane concept [5]. AD is a well-known process with recognized advantages associated with mitigating climate change, economic benefits as well as the diversification of opportunities [6]. It has been suggested as the most environmentally sustainable option [7]. Diverse studies have been carried out regarding the energy recovery of algal biomass, highlighting the inherent advantages of the joint digestion of algal biomass and other organic effluents [8–11].

During the bioprocess evolution, microorganisms are often exposed to adverse conditions such as nutrient limitation, pH gradients, osmotic stress, inhibitor presence, etc., which may affect cellular functions and viability, and thus the process efficiency. Usually, these conditions damage the cell membrane, thus compromising the cellular integrity, which may result in cell dormancy or death. If a high proportion of dormant or dead cells are present during the process development, this will inevitably lead to a reduction in bioprocess yield. Therefore, it is essential to monitor the cell viability throughout any bioprocess, to evaluate the cell physiological status throughout the process development [12].

The main objective of the present research was to take advantage of all wastes from *C. cohnii* microalgae fermentation and lipid extraction process, foreseeing zero wastes during the OmegaFuel project (PTDC/EAM-AMB/30169/2017) [13,14]. In an innovative approach, it was intended to evaluate the integration of microalgae lipid production with the energetic and agricultural valorization of the generated wastes by AD. The definite goal was to understand the benefits of digesting the different wastes simultaneously, taking advantage of the substrate complementarity concept, to make the overall system simpler and cheaper and, consequently, more sustainable. Since the wastes (biomass leftovers containing solvents and fermentation broths with salt) used as substrates for biogas production contain compounds that may inhibit the anaerobic microbial cells of the inoculum, flow cytometry was used to evaluate and understand the impact of these substrates on the consortium of cells.

This is the first work that reports the valorization of wastes generated during the *C. cohnii* lipid production process, for biogas production through AD, accomplishing the circular economy principles, which represents an important improvement of the microalgal lipid production process.

2. Materials and Methods

C. cohnii was the focus of the study in the OmegaFuel project regarding its ability to produce DHA and biofuels. *C. cohnii* biomass was produced in a 7 L fed-batch bioreactor. At the end of fermentation, the culture medium was centrifuged, and the biomass was taken for further intracellular lipid extraction [14].

2.1. Substrates and Inoculum

Several wastes were generated during the *C. cohnii* lipid production step, which were used as substrates for AD: microalgae (Ae) after lipid extraction using hexane (AeH) and supercritical extraction methods (AeS), and the supernatant obtained after culture medium centrifugation (M). This medium was composed of yeast extract (corresponding to 0.11 g/L nitrogen), sea salt (25 g/L), and glycerol (19.8 g/L) [14]. The proportion used to prepare both mixtures (AeH+M and AeS+M) was based on the quantities of each residue obtained at the end of the cultivation process (M) and after the lipid extraction steps (AeH or AeS): 8 g Ae/L M (ratio of 0.008:1, *m*/*v*).

Biological solids (1.3 ± 0.0 g VSS/L) were collected in an anaerobic digester plant at Quinta do Conde (Portugal) and used as the inoculum (I, 30% v/v) in the AD process at a substrate/inoculum ratio value of 2.3 expressed in volatile solids.

2.2. Anaerobic Digestion Experimental Set-Up

The experiment was carried out in triplicate under batch conditions, using 70 mL glass units, with useful volume of 40 mL and leaving 30 mL of headspace. The reactors were sealed, and the headspace of each unit was flushed with nitrogen at the beginning of the assay to ensure anaerobic conditions. The test units were incubated within the constant mesophilic range of temperature (37 ± 1 °C) and maintained until it was possible to positively confirm the initial assumptions, that is, to verify the possibility of using AD to valorize the remaining wastes from the microalgae fermentation and lipid extraction processes, as well as to assess the benefits of digesting the different wastes together. According to this, the experiments lasted about a month.

2.3. Analytical and Chromatograph Methods

Performance of the process was monitored by analytical characterizations of all samples and by the volume and quality of the obtained biogas. Chemical oxygen demand (COD), volatile solids (VS), volatile suspended solids (VSS), total nitrogen (Kjeldahl, TN), ammonium (NH_4^+-N), and pH, were assayed according to standard methods [15].

Analytical measures of the described parameters were performed at the beginning ($_{in}$) and at the end of the assay experimental time (data not shown). Results are presented as a percentage of the removal of each parameter. The biogas production was monitored daily with a pressure transducer, expressed under standard conditions of temperature and pressure (STP: 0 °C, 1 bar) defined by IUPAC (International Union of Pure Applied Chemistry). The methane content of the biogas collected in each unit headspace was measured by injecting 0.25 mL of the gas sample into a gas chromatograph (GC Thermo Electron Corporation Trace GC Ultra), equipped with a thermal conductivity detector and a Carboxen®-1010 PLOT Capillary GC Column (L $\times$ I.D. 30 m $\times$ 0.32 mm, average thickness 15 µm). The injector and detector temperatures were 200 and 230 °C, respectively. The column temperature profile was: isot 35 °C for 7.5 min, ramp 24 °C/min, 5 min isot. Helium was utilized as the carrier gas (1.5 mL/min). Quantification of each gas produced was performed by comparing the obtained graphical peak areas with the pattern of an injected gas mix at the beginning of each analysis.

All values of the biogas and methane yields are presented under STP conditions and divided by the mass of volatile solids (L CH4/kg SV_{in}) or by the organic matter content (L CH4/kg COD_{in}) of the substrate fed at the beginning of the assay. The primary energy yield (kWh/kg VS_{in} and kWh/kg COD_{in}) of the tested mixtures was calculated using the lower methane heating value (LHV) of 9.97 kWh/m^3 CH4 [16].

2.4. Flow Cytometric Analysis

Flow cytometry was used to evaluate the physiological status of the anaerobic microbial consortium of cells, used to inoculate all samples, to understand the response of these cells to the different operational conditions of the AD experiment, using membrane integrity as a cell viability marker [12]. This information complements data given by traditional methods used to monitor anaerobic digestion cultivations, such as biogas and methane productions, which does not provide any information on cell physiological status. Two fluorescent dyes were used: SYBR Green I (SYBR) and propidium iodide (PI) [17]. SYBR is a nucleic acid-specific stain that penetrates all cells, and PI is a stain indicator of membrane integrity, as intact membranes exclude this dye. In this way, it was possible to distinguish the microbial consortium of cells from the medium particles.

Samples were sonicated for 15 s to disintegrate cellular aggregates and ensure the analysis of individual cells. For staining, 2 µL of SYBR Green I solution (1:30 dilution of SYBR Green I commercial stock (from Invitrogen) solution made in dimethyl sulfoxide) and 2 μL of PI (from Invitrogen, stock solution of 1 mg mL^{-1} prepared with water) were added simultaneously to 500 μL of sample and incubated for 30 min in the dark.

Samples taken from the vials were analyzed at the beginning and the end of the experiment using a Cytoflex Beckman-Coulter flow cytometer, equipped with a blue laser, FSC/SSC light scattering detectors and five fluorescence detectors.

All analyses were carried out in duplicate, and each analysis was stopped after more than 5000 events were detected. The data were analyzed using the CytExpert 2.5 software.

The proportion of cells with damaged membrane (permeabilized cells) was calculated according to Equation (1):

$$\Delta(\text{SYBR/PI}) = \%\,(\text{SYBR/PI})_f - \%\,(\text{SYBR/PI})_0 \quad (1)$$

where % $(\text{SYBR/PI})_f$ is the percentage of permeabilized cells detected at the end of the AD process; % $(\text{SYBR/PI})_0$ is the percentage of permeabilized cells detected at the beginning of the AD process; and Δ(SYBR/PI) is the difference between the above two parcels.

In this way, an increase in the number permeabilized cells, reflecting the cells' stress response to each assay condition, was assessed throughout the AD process.

3. Results

3.1. Chemical Composition of Substrates and Inoculum

There was a considerable difference between the chemical composition of AeH and M (Table 1). Algae clearly had a higher content than the substrate M, which was practically devoid of nitrogen. Given these characteristics, M could complement and somehow work as a liquid medium when both substrates AeH and M are digested together. Biological solids, used as inoculum for the assays, held the highest concentration of nitrogen present in its ionized form (NH_4^+) suggesting the occurrence of organic material degradation.

Table 1. Substrates and inoculum: chemical composition.

Substrates and Inoculum	COD (g/L)	VS (g/L)	TN (g/L)	NH_4^+-N (mg/L)
AeH+M	96.6 ± 5.6	37.3 ± 0.7	0.17 ± 0.0	0.56 ± 0.0
AeS+M	103.6 ± 2.2	37.6 ± 1.2	0.15 ± 0.0	0.56 ± 0.0
AeH	1638.3 ± 90	413.6 ± 30	17.4 ± 0.4	14.0 ± 0.0
M	96.5 ± 0.0	37.3 ± 0.2	0.17 ± 0.0	0.28 ± 0.0
I	20.0 ± 0.6	5.1 ± 0.2	0.45 ± 0.0	462.0 ± 4.0

COD—chemical oxygen demand; VS—volatile solids; TN—total nitrogen.

The mixtures containing the defatted microalgae biomass—AeH+M and AeS+M—showed an identical chemical composition in terms of COD, VS, and nitrogen, indicating that the extraction process did not seem to influence the energy potential of the remaining algal biomass, as shown in Table 1. Comparatively, the medium (M) with similar concentrations constituted a very interesting potential in terms of energy recovery. Therefore, organic flows above 96 g COD/L that came from *C. cohnii* lipid production as by-products were available and susceptible to energy valorization.

3.2. Anaerobic Digestion of Microalgae Fermentation Wastes

Biogas production was observed in all the operational conditions tested, with no lag phase evidenced in any of them, as shown in Figure 1. The experiments can be described in three distinct phases: an initial phase (the first 7 experimental days), with a steeper slope than the others; an intermediate phase (the following 12 days) where stabilization of gas production was noticed and, finally, a terminus phase of the experiment in which an increase in biogas production was observed, mainly in the case of AeH+M+I and AeS+M+I. The comparison between these two units showed that the first had supremacy in terms of the ability to generate more accentuated increases in gas production than the other and this can be associated with a faster microbial consortium development that in the final stage is more adapted to the substrate in digestion. Effectively, in the case of AeH+M+I, it was possible to verify the similarity between gas production volume obtained in the first seven days and that generated in the last seven experimental days.

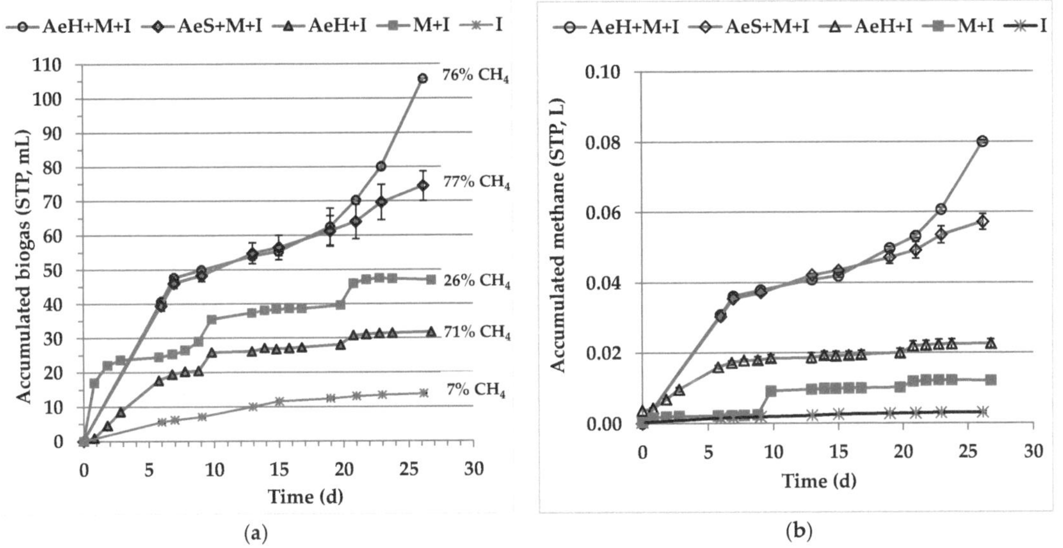

Figure 1. Anaerobic digestion of wastes from fermentation: (**a**) cumulative biogas production; (**b**) cumulative methane production. STP—standard temperature and pressure. The values presented for the percentage of methane refer to the last day of the experiment.

The highest gas productions were recorded in the units digesting the defatted microalgae biomass mixed with the culture medium—AeH+M+I and AeS+M+I—followed by the medium alone, M+I, with accumulated biogas volumes of 106, 74, and 48 mL, respectively. Regarding the quality of the produced biogas at the end of the experiment, it was noted that both mixtures provided a crucial concentration of methane (76–77%), while the collected biogas from the digestion of medium alone was very poor in methane (26%).

Both mixtures, AeH+M+I and AeS+M+I, with COD_{in} concentrations of 27 and 22 g/L, respectively, showed the highest removals in proportions of 68–61% (Table 2), resulting in a methane yield of 75 and 65 L CH_4/kg COD_{in}, respectively (Figure 2a). The highest yields of methane production, in terms of volatile solids, were also obtained with the mixtures AeH+M+I and AeS+M+I, resulting in 582 and 440 L CH_4/kg VS_{in}, respectively (Figure 2b). It was interesting to observe that despite the high production of ammonium levels in these mixtures, an inhibitory effect on the anaerobic digestion process was not observed. Under these conditions, this fact can be explained by the low ammonium concentrations determined, 0.06 g/L at the beginning, increasing to 0.18 g/L (data not shown) at the end of the process (Table 2). On the other hand, M+I with 30 g COD_{in}/L and 63% of COD removal, showed a low methane yield of 0.02 L CH_4/kg COD_{in}. In fact, the reduced biogas volume of M+I and its low quality may be related to the nitrogen involved, whose content of 0.60 g/L and 0.17 g/L NH_4^+-N_{in} at the beginning of the process may have caused an imbalance and thus inhibition of the process.

Table 2. Performance of anaerobic digestion of microalgae fermentation wastes.

Mixture	pH		COD		Volatile Solids		Total Nitrogen		Ammonia Nitrogen	
	pH_{in}	pH_f	COD_{in} (g/L)	COD_r (%)	VS_{in} (g/L)	VS_r (%)	TN_{in} (g/L)	TN_r (%)	NH_4^+-N_{in} (g/L)	NH_4^+-N_r (%)
AeH+M+I	9.4	7.25	26.7 ± 0.0	68	3.4 ± 0	42	0.15 ± 0.000	*−97*	0.06 ± 0.002	*−181*
AeS+M+I	9.4	7.26	22.0 ± 0.0	61	3.2 ± 0	40	0.16 ± 0.010	*−90*	0.06 ± 0.002	*−99*
AeH+I	7.63	7.75	12.9 ± 0.0	*−8*	5.1 ± 1	*−59*	0.59 ± 0.004	25	0.23 ± 0.000	N.d.
M+I	7.46	7.75	29.8 ± 0.4	63	12.3 ± 1	48	0.60 ± 0.000	25	0.17 ± 0.000	N.d.
I	7.68	7.19	6.4 ± 1.8	27	1.6 ± 0	75	0.26 ± 0.004	67	0.04 ± 0.001	−22

COD—chemical oxygen demand; VS—volatile solids; TN—total nitrogen. N.d.—Not determined.

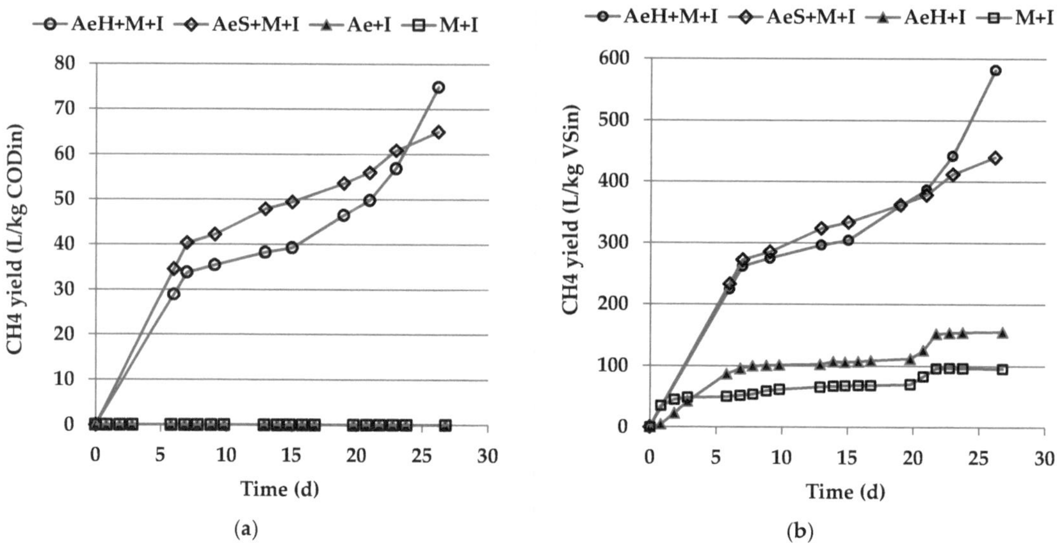

Figure 2. Anaerobic digestion of wastes from fermentation: (**a**) cumulative methane yield expressed as L CH_4/g COD added at the beginning of the AD; (**b**) cumulative methane yield expressed as L CH_4/g VS added at the beginning of the AD.

Despite the relevant proportion of methane (71%) present in the biogas provided by the digestion of extracted algae with hexane AeH+I, the volume of gas obtained in this assay was the lowest of all experiments performed (Figure 1). It becomes evident, accordingly, that the joint digestion of the algae with the culture medium, taking advantage of the substrate complementarity concept [8], is much more interesting as it allows doubling/tripling of the production of biogas and further promoting the gas quality. Biogas volumes of 35 mL (71% CH_4) and about 100 mL (76% CH_4) were accumulated in AeH+I and AeH+M+I anaerobic units, respectively. In comparison, this can be justified by the difference between the organic material initially available in AeH+I which was about half that of the existing organic material in AeH+M+I (13 vs. 27 g COD/L, Table 2) and, conversely, with the initial nitrogen levels which were about four times higher than those recorded in the AeH+M+I.

As expected, the digestion mixture AeH+M+I from the fermentation wastes displayed the highest methane productivity of the whole experiment, 31 L CH_4/kg VS_{in}.d and 4.0 L CH_4/kg COD_{in}.d, followed by the digestion mixture AeS+M+I. These data indicate that the sustainability of the lipids production, from the heterotrophic marine microalga *C. cohnii*, could be promoted by energetic valorization of the remaining materials through integration of the anaerobic digestion techniques. An energetic potential of 4.4–5.8 KWh/Kg VS_{in} and

0.65–0.75 KWh/Kg COD_{in} (Table 3) can be retrieved and applied to the lipid production process itself.

Table 3. Methane productivity and energy potential estimated at the end of the anaerobic digestion for microalgae fermentation wastes and mixtures used as feedstock.

Mixture	Methane Productivity		Energy Content	
	(L CH_4/kg VS.d)	(L CH_4/kg COD.d)	(KWh/Kg VS)	(KWh/Kg COD)
AeH+M+I	31.4	4.04	5.80 ± 0.00	0.75 ± 0.00
AeS+M+I	11.2	1.66	4.39 ± 0.26	0.65 ± 0.04
AeH+I	9.53	0.006	1.10 ± 0.06	0.0007 ± 0.00
M+I	3.32	0.002	0.24 ± 0.01	0.0002 ± 0.00

3.3. Flow Citometry

Figure 3 shows the evolution of permeabilized cells of the anaerobic microbial consortium during AD, shown as Δ(SYBR/PI), calculated according to Equation (1). Therefore, the lower the difference values of Δ(SYBR/PI), the lower the percentage of permeabilized cells at the end of the anaerobic process. A negative Δ(SYBR/PI) value means that the percentage of permeabilized cells at the end of the assay (t_f) was lower than that at the beginning of the experiment (t_0), which indicates that the microbial consortium of cells was exposed to favorable conditions during the AD process, such that the proportion of permeabilized cells decreased during the assay, relatively to the percentage of permeabilized cells detected at the beginning of the assay. This fact happened for the assay which used AeH+M+I as substrate (−8.3%).

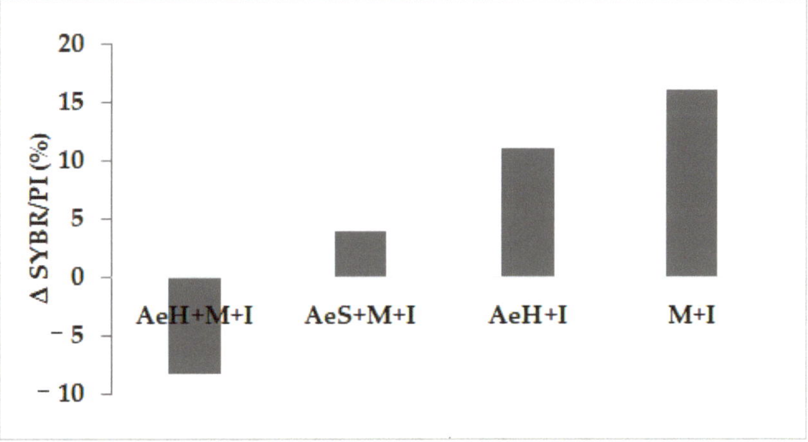

Figure 3. ΔSYBR/PI calculated for all the assays.

In contrast, a Δ(SYBR/PI) positive signal indicates that the percentage of permeabilized cells increased throughout the AD process, relatively to the beginning of the assay. This means that the microbial consortium cells were exposed to adverse conditions that induced the cell membrane permeabilization. This was observed for assays AeS+M+I (+4%), AeH+I (+11%) and M+I (+16%).

4. Discussion

Cultivation of marine microalgae requires a high sodium chloride content (0.5–1 M); however, substrates with salinity values of 0.4 M can already affect methane production during anaerobic digestion, and above 0.5 M are toxic [18,19]. The culture medium supernatant (M) used in this work came from the fermentation of the marine microalgae *C. cohnii*, whose culture medium contains sea salt (25 g/L) [14], that is, the medium has a salinity of about 0.4 M. Indeed, as shown in Figure 1, the digestion of the culture medium (M) produced some biogas volume, 48 mL, but with low quality (26% CH_4), and all yields and methane productivity values obtained from the anaerobic digestion of this residue were also the lowest of all the tests, probably due to the concentration of sea salt. The option of mixing of microalgae biomass leftovers was a good approach to overcome the negative impact of this parameter, making it possible to obtain a biogas production of 74 and 106 mL (76–77% CH_4) during the anaerobic digestion of the global wastes.

In this work, AeH+I or M+I showed the highest content of nitrogen and ammonia (Table 2), thus producing the lowest biogas volumes of all the substrates tested by anaerobic digestion, 32 and 48 mL, respectively. The mono-digestion of microalgae has been shown to be difficult [6]. The high content in nitrogen, usually as proteins, leads to low C:N ratios. During the anaerobic digestion process, acidogenic bacteria and methanogens can be inhibited by high levels of ammonia released by the degradation of these proteins [20]. On the other hand, as proposed by some authors [18,20], the co-digestion of microalgae with carbon-rich feedstock can be a cost-effective and efficient approach to avoid ammonia inhibition. In this case, we used the remaining culture medium (M) as a carbon-rich feedstock, in terms of COD and VS, for the AD of the two extracted microalgae (AeH and AeS), and consequently, biogas production was highly increased by up to 106 mL (76% methane) for AeH+M+I and 74 mL (77% methane) for AeS+M+I. Although at lower percentages, up to 62%, other authors also recorded significant increases in methane production using this approach, the co-digestion of microalgae with high-carbon wastes [21–23].

The results from flow cytometric analysis were in accordance with the results obtained for methane production and can be explained by the presence of available nutrients in the media resulting from the extracted alga residues, which contain proteins and carbohydrates that can be readily uptaken by the microbial consortium during anaerobic digestion. Therefore, it seems that the assayed AeH+M+I (Δ(SYBR/PI) = −8.3%) contained a higher nutrient availability in the culture medium than the other assays, which not only favored biogas production, but also protected the cells against nutrient limitation, a condition that often induces cell membrane damage [24]. It should be noted that no hexane or hexane traces were present in the extracted algal biomass (AeH), in the assayed AeH+M+I and AeH+I since it was previously evaporated. The lower Δ(SYBR/PI) value observed for AeS+M+I (Δ(SYBR/PI) = +4%), relatively to AeH+M+I (Δ(SYBR/PI) = +11%) may be explained by the less efficient alga extraction with hexane (AeH), compared to the supercritical alga extraction (AeS), resulting in biomass leftovers with a higher nutrient content that could be consumed by the microbial consortium, resulting in the highest methane yield and productivity (Figure 2 and Table 3, respectively). Indeed, it is well known that supercritical extraction is much more efficient than traditional solvent extraction methods [25]. This means that less nutrients remain in the microalgal biomass leftovers, after supercritical extraction, compared to the biomass leftovers resulting from hexane extraction. The highest difference Δ(SYBR/PI) observed for the M+I assay (16%), was attributed to lower nutrient availability due to the fermentation broth (M), since it corresponded with the exhausted medium collected at the end of the microalgal biomass production step. Additionally, the high fermentation broth salinity value of 0.4 M (25 g/L) might have had a negative impact on the consortium of cells involved in the M+I assay, resulting in the cell permeabilization of 16% of the total cell population. The presence of microalgal biomass residues in the remaining assays (AeH+M+I, AeS+M+I, AeH+I) might have mitigated the negative effect of the salt present in M since the salt present in the microalgal biomass was removed during the extraction step.

There have been some works reporting anaerobic co-digestion of microalgal biomass- and organic-rich feedstocks in batch condition tests. The mesophilic co-digestion of a mixture of 75% OMSW (olive mill solid waste) and 25% *Dunaliella* salina generated a methane yield of 330 L CH_4/kg VS added [22]. Hermann et al. [6] reported that the co-digestion of 15% *Arthrospira platensis* with 85% seaweed achieved a biochemical methane yield of 311 L CH_4/kg VS and a maximum specific methane production rate of 29 L CH_4/kg VS.d. The co-digestion of mixtures of algae bloom (*Microcystis* spp.) and lake water with corn straw achieved 325 L CH_4/kg VS [21]. *Tetraselmis suecica* biomass co-digested with glycerol increased the methane yield rate from 174 to 438 L CH_4/kg VS in relation to its mono-digestion [26]. The synergetic effect was observed in a mixture containing 63% undigested sewage sludge and 37% wet algae slurry, in which a CH_4 yield of 408 L/Kg VS was reached under mesophilic conditions [23]. All these values were considerably lower than those obtained in the present work for the digestion of the AeH+M+I mixture, 582 L CH_4/kg VS, and lower than those obtained with the AeS+M+I mixture, 440 L CH_4/kg VS (Figure 2).

The selection of biomass for the AD process among other operational conditions decides the composition of the biogas. To be economically feasible, the overall process should guarantee the maximal concentration of methane in the biogas content, so the calorific value of the biogas is increased. The extraction of lipids from microalgae biomass prior to the processing of gaseous bioenergy will results in an increase in the production yield and the biorefinery approach of microalgae biomass has a zero-waste discharge [27]. The concept of the *C. conhii* biorefinery achieved with a high-energy content, 4.4–5.8 KWh/Kg VS, revealed in the biogas produced (Table 3) after anaerobic digestion of both mixtures containing the lipids extracted *C. cohnii* biomass leftovers and the supernatant of the culture medium (AeS+M+I and AeH+M+I). The overall approach offers solutions that address societal needs both in terms of products and processes [28].

5. Conclusions

The anaerobic digestion process can be successfully applied to the energetic valorization of *C. cohnii* defatted biomass and the culture medium supernatant, the main wastes generated during the microalgal lipid production process. AD integration with the lipid production step avoids wastes, using the biomass leftovers as AD feedstocks.

The results indicate that the process used for lipid extraction had an influence on the energy recovery through anaerobic digestion. Concerning the substrates used for AD, the joint digestion of algae with the culture medium was more interesting than the individual digestion of each one, insofar as the combination of the two wastes, taking advantage of the substrate complementarity concept, promoted a zero-waste practice and provided a greater volume of biogas of better quality.

Given the interest of the results presented here, the authors consider that the proposal of this work deserves larger scale trials. Thus, based on the achieved data, they intend to proceed with the research on the energetic valorization of the tested wastes generated from *C. cohnii* microalgae fermentation and the lipid extraction process, and evaluate the impact on the microbial consortium of cells by the anaerobic digestion process. It will be important to design AD assays on a larger scale to consolidate all the results obtained in the present work.

Author Contributions: Conceptualization, all authors; methodology, A.E. and I.P.M.; validation, all authors.; formal analysis, all authors; investigation, A.E. and I.P.M.; resources, all authors; data curation, A.E. and I.P.M.; writing—all authors.; writing—review and editing, A.E.; supervision, A.E. and I.P.M.; project administration, T.L.d.S.; funding acquisition, T.L.d.S. All authors have read and agreed to the published version of the manuscript.

Funding: This research has been funded by the project PTDC/EAM-AMB/30169/2017, entitled "OmegaFuel—New platform for biofuels and omega-3 compounds production, from the marine microalga *Crypthecodinium cohnii* sustainable biorefinery" supported by FCT (Fundação para a Ciência e a Tecnologia), and was developed in the Biomass and Bioenergy Research Infrastructure (BBRI)-LISBOA-01-0145-FEDER-022059, supported by Operational Programme for Competitiveness and Internationalization (PORTUGAL2020), by Lisbon Portugal Regional Operational Programme (Lisboa 2020) and by North Portugal Regional Operational Programme (Norte 2020) under the Portugal 2020 Partnership agreement, through the European Regional Development Fund (ERDF).

Data Availability Statement: Not applicable.

Acknowledgments: Authors acknowledge SIMARSUL (Quinta do Conde, Portugal) for biologic solids. Special thanks to Natércia Santos for technical laboratory assistance.

Conflicts of Interest: The authors declare no conflict of interest. The funders had no role in the design of the study; in the collection, analyses, or interpretation of data; in the writing of the manuscript; or in the decision to publish the results.

References

1. Song, P.; Kuryatov, A.; Axelsen, P.H. A new synthetic medium for the optimization of docosahexaenoic acid production in Crypthecodinium cohnii. *PLoS ONE* **2020**, *15*, e0229556. [CrossRef] [PubMed]
2. Diao, J.; Song, X.; Zhang, X.; Chen, L.; Zhang, W. Genetic Engineering of Crypthecodinium cohnii to Increase Growth and Lipid Accumulation. *Front. Microbiol.* **2018**, *9*, 492. [CrossRef] [PubMed]
3. da Silva, T.L.; Moniz, P.; Silva, C.; Reis, A. The Dark Side of Microalgae Biotechnology: A Heterotrophic Biorefinery Platform Directed to ω-3 Rich Lipid Production. *Microorganisms* **2019**, *7*, 670. [CrossRef] [PubMed]
4. Cabrita, I.; Silva, L.; Marques, I.P.; Gírio, F.; Di Berardino, S. *Avaliação do Potencial e Impacto do Biometano em Portugal: Sumário Executivo*, 1st ed.; LNEG: Lisbon, Portugal, 2015; p. 17.
5. Ghaib, K.; Ben-Fares, F.-Z. Power-to-Methane: A state-of-the-art review. *Renew. Sustain. Energy Rev.* **2018**, *81*, 433–446. [CrossRef]
6. Herrmann, C.; Kalita, N.; Wall, D.; Xia, A.; Murphy, J.D. Optimised biogas production from microalgae through co-digestion with carbon-rich co-substrates. *Bioresour. Technol.* **2016**, *214*, 328–337. [CrossRef]
7. Slorach, P.C.; Jeswani, H.K.; Cuéllar-Franca, R.; Azapagic, A. Environmental sustainability in the food-energy-water-health nexus: A new methodology and an application to food waste in a circular economy. *Waste Manag.* **2020**, *113*, 359–368. [CrossRef]
8. Assemany, P.; Marques, I.D.P.; Calijuri, M.L.; da Silva, T.L.; Reis, A. Energetic valorization of algal biomass in a hybrid anaerobic reactor. *J. Environ. Manag.* **2018**, *209*, 308–315. [CrossRef]
9. Neves, A.; Lopes da Silva, T.; Reis, A.; Ramalho, L.; Eusebio, A.; Marques, I.P. Anaerobic digestion of pre-treated microalgae biomass. *Chem. Eng. Trans.* **2018**, *64*, 169–174. [CrossRef]
10. Saratale, R.G.; Kumar, G.; Banu, R.; Xia, A.; Periyasamy, S.; Saratale, G.D. A critical review on anaerobic digestion of microalgae and macroalgae and co-digestion of biomass for enhanced methane generation. *Bioresour. Technol.* **2018**, *262*, 319–332. [CrossRef]
11. Veerabadhran, M.; Gnanasekaran, D.; Wei, J.; Yang, F. Anaerobic digestion of microalgal biomass for bioenergy production, removal of nutrients and microcystin: Current status. *J. Appl. Microbiol.* **2021**, *131*, 1639–1651. [CrossRef]
12. da Silva, T.L.; Roseiro, J.C.; Reis, A. Applications and perspectives of multi-parameter flow cytometry to microbial biofuels production processes. *Trends Biotechnol.* **2012**, *30*, 225–232. [CrossRef]
13. Moniz, P.; Silva, C.; Oliveira, A.C.; Reis, A.; da Silva, T.L. Raw Glycerol Based Medium for DHA and Lipids Production, Using the Marine Heterotrophic Microalga *Crypthecodinium cohnii*. *Processes* **2021**, *9*, 2005. [CrossRef]
14. Moniz, P.; Martins, D.; Oliveira, A.C.; Reis, A.; Lopes da Silva, T. The Biorefinery of the Marine Microalga Crypthecodinium cohnii as a Strategy to Valorize Microalgal Oil Fractions. *Fermentation* **2022**, *8*, 502. [CrossRef]
15. APHA; AWWA; EWF. *Standard Methods for Examination of Water and Wastewater*, 22nd ed.; American Public Health Association: Washington, DC, USA; American Water Works Association: Denver, CO, USA; Water Environment Federation: Washington, DC, USA, 2012.
16. Heat Values of Various Fuels. Available online: https://world-nuclear.org/information-library/facts-and-figures/heat-values-of-various-fuels.aspx (accessed on 5 July 2022).
17. Lelong, A.; Hégaret, H.; Soudant, P. Cell-based measurements to assess physiological status of Pseudo-nitzschia multiseries, a toxic diatom. *Res. Microbiol.* **2011**, *162*, 969–981. [CrossRef]
18. Sialve, B.; Bernet, N.; Bernard, O. Anaerobic digestion of microalgae as a necessary step to make microalgal biodiesel sustainable. *Biotechnol. Adv.* **2009**, *27*, 409–416. [CrossRef]
19. Mottet, A.; Habouzit, F.; Steyer, J.P. Anaerobic digestion of marine microalgae in different salinity levels. *Bioresour. Technol.* **2014**, *158*, 300–306. [CrossRef]
20. Ward, A.; Lewis, D.; Green, F. Anaerobic digestion of algae biomass: A review. *Algal Res.* **2014**, *5*, 204–214. [CrossRef]
21. Zhong, W.; Zhang, Z.; Luo, Y.; Qiao, W.; Xiao, M.; Zhang, M. Biogas productivity by co-digesting Taihu blue algae with corn straw as an external carbon source. *Bioresour. Technol.* **2012**, *114*, 281–286. [CrossRef] [PubMed]
22. Fernández-, M.J.; Rincón, B.; Fermoso, F.G.; Jiménez-Rodríguez, A.-M.; Borja, R. Assessment of two-phase olive mill solid waste and microalgae co-digestion to improve methane production and process kinetics. *Bioresour. Technol.* **2014**, *157*, 263–269. [CrossRef] [PubMed]
23. Olsson, J.; Feng, X.M.; Ascue, J.; Gentili, F.G.; Shabiimam, M.; Nehrenheim, E.; Thorin, E. Co-digestion of cultivated microalgae and sewage sludge from municipal waste water treatment. *Bioresour. Technol.* **2014**, *171*, 203–210. [CrossRef] [PubMed]
24. Pagán, R.; Mackey, B. Relationship between Membrane Damage and Cell Death in Pressure-Treated *Escherichia coli* Cells: Differences between Exponential- and Stationary-Phase Cells and Variation among Strains. *Appl. Environ. Microbiol.* **2000**, *66*, 2829–2834. [CrossRef] [PubMed]

25. Santana, A.; Jesus, S.; Larrayoz, M.; Filho, R. Supercritical Carbon Dioxide Extraction of Algal Lipids for the Biodiesel Production. *Procedia Eng.* **2012**, *42*, 1755–1761. [CrossRef]
26. Santos-Ballardo, D.U.; Font-Segura, X.; Ferrer, A.S.; Barrena, R.; Rossi, S.; Valdez-Ortiz, A. Valorisation of biodiesel production wastes: Anaerobic digestion of residual *Tetraselmis suecica* biomass and co-digestion with glycerol. *Waste Manag. Res.* **2015**, *33*, 250–257. [CrossRef]
27. Kannah, R.Y.; Kavitha, S.; Karthikeyan, O.P.; Rene, E.R.; Kumar, G.; Banu, J.R. A review on anaerobic digestion of energy and cost effective microalgae pretreatment for biogas production. *Bioresour. Technol.* **2021**, *332*, 125055. [CrossRef] [PubMed]
28. Silva, C.; Moniz, P.; Oliveira, A.C.; Vercelli, S.; Reis, A.; da Silva, T.L. Cascading *Crypthecodinium cohnii* Biorefinery: Global Warming Potential and Techno-Economic Assessment. *Energies* **2022**, *15*, 3784. [CrossRef]

processes

Article

Simultaneous Anaerobic Ammonium Oxidation and Electricity Generation in Microbial Fuel Cell: Performance and Electrochemical Characteristics

Jiqiang Zhang [1,2], Zaiwang Zhang [1], Kun Rong [1], Haiying Guo [1], Jing Cai [2,3], Yajuan Xing [2,4], Lili Ren [1], Jiayun Ren [1], Tao Wu [1], Jialiang Li [1,*] and Ping Zheng [2,*]

1 Shandong Engineering and Technology Research Center for Ecological Fragile Belt of Yellow River Delta, School of Biological & Environmental Engineering, Binzhou University, Binzhou 256600, China
2 Department of Environmental Engineering, Zhejiang University, Hangzhou 310058, China
3 College of Environmental Science and Engineering, Zhejiang Gongshang University, Hangzhou 310058, China
4 School of Ceomatics and Municipal Engineering, Zhejiang University of Water Resources and Electric Power, Hangzhou 310058, China
* Correspondence: ljliang@bzu.edu.cn (J.L.); pzheng@zju.edu.cn (P.Z.)

Abstract: In this study, a microbial fuel cell (MFC) that can achieve simultaneous anode anaerobic ammonium oxidation (anammox) and electricity generation (anode anammox MFC) by high-effective anammox bacteria fed with purely inorganic nitrogen media was constructed. As the influent concentrations of ammonium (NH_4^+-N) and nitrite (NO_2^--N) gradually increased from 25 to 250 mg/L and 33–330 mg/L, the removal efficiencies of NH_4^+-N, NO_2^--N and TN were over 90%, 90% and 80%, respectively, and the maximum volumetric nitrogen removal rate reached 3.01 ± 0.27 kgN/(m^3·d). The maximum voltage and maximum power density were 225.48 ± 10.71 mV and 1308.23 ± 40.38 mW/m^3, respectively. Substrate inhibition took place at high nitrogen concentrations (NH_4^+-N = 300 mg/L, NO_2^--N = 396 mg/L). Electricity production performance significantly depended upon the nitrogen removal rate under different nitrogen concentrations. The reported low coulombic efficiency (CE, 4.09–5.99%) may be due to severe anodic polarization. The anode charge transfer resistance accounted for about 90% of the anode resistance. The anode process was the bottleneck for energy recovery and should be further optimized in anode anammox MFCs. The high nitrogen removal efficiency with certain electricity recovery potential in the MFCs suggested that anode anammox MFCs may be used in energy sustainable nitrogen-containing wastewater treatment.

Keywords: microbial fuel cell; anode anaerobic ammonium oxidation; electricity generation; electrochemical characteristics

Citation: Zhang, J.; Zhang, Z.; Rong, K.; Guo, H.; Cai, J.; Xing, Y.; Ren, L.; Ren, J.; Wu, T.; Li, J.; et al. Simultaneous Anaerobic Ammonium Oxidation and Electricity Generation in Microbial Fuel Cell: Performance and Electrochemical Characteristics. *Processes* **2022**, *10*, 2379. https://doi.org/10.3390/pr10112379

Academic Editors: Dimitris Zagklis and Georgios Bampos

Received: 20 October 2022
Accepted: 9 November 2022
Published: 12 November 2022

1. Introduction

Traditional wastewater treatment is an energy-intensive process and has received considerable attention due to the pressure of carbon neutralization and energy shortage, so it is of great necessity to reduce energy consumption in this process [1,2]. As a matter of fact, substantial internal chemical energy is contained in water pollutants, and the wastewater may contain 9.3 times as much energy as is used to treat it [3]. As a promising technology for sustainable wastewater treatment, microbial fuel cells (MFC) can use these pollutants as fuels and directly convert the chemical energy into electricity with microorganisms as catalysts [4]. In theory, it is feasible to turn wastewater treatment into a self-sustaining, or even a net energy-producing, process by MFCs [1,5,6].

MFCs generate electricity by harvesting the electrons donated to the anode from fuel oxidation [7]. Various pollutants from wastewater and pure chemical compounds have been examined as fuel in MFCs [8,9]. A large number of studies have been conducted on the oxidation of most ubiquitous organic pollutants in wastewater to produce electric power (organic MFCs). However, nutrient removal remains a significant challenge in efficient wastewater treatment with MFCs and emerging studies have mainly focused on energy-containing nitrogen pollutants [10–12]. As one of the most common inorganic water pollutants, ammonium can also be used as the fuel in MFCs [13]. Anode nitrification MFCs represent one type of MFCs based on the nitrification process, involving aerobic oxidation of ammonium as the main reaction for anode ammonium oxidation [13,14]. Electricity can be acquired from anode nitrification MFCs, but the oxidation of ammonium by nitrifying bacteria requires oxygen supply. The presence of oxygen would hamper extracellular electron transfer and greatly decrease power generation in the MFCs [13–15]. If oxygen can be removed from the anode ammonium oxidation process, more electric energy might be recovered via the harvesting of the internal chemical energy stored in ammonium.

Anaerobic ammonium oxidation (anammox), a process in which ammonium is oxidized by anammox bacteria that take nitrite as an electron acceptor ($NH_4^+ + NO_2^- \rightarrow N_2 + 2H_2O$), is known as a low-cost and eco-friendly nitrogen removal method [16,17]. As a result, the combination of the anammox process and MFCs (anode anammox MFCs) might have several advantages over the anode nitrification MFCs. First, ammonium can be oxidized without oxygen, which leads to less electron loss. On the other hand, the anammox process has a significantly higher ammonium oxidation rate than that of the nitrification process [18,19], implying that it involves faster electron transfer in electricity generation. The possibility of MFCs with simultaneous anode anammox and electricity generation abilities has been verified in several studies [20,21]. However, little evidence is available for the ways in which anammox bacteria, as exoelectrogens, can limit the development of anode anammox MFCs. A recent report concluded that anammox bacteria are a newly discovered type of electrochemically active bacteria in microbial electrolysis cells (MECs) and that anode anammox MFCs have broad prospects for application in this field [22]. Electrogenesis bacteria are the core of the MFC system [1,5]. In addition, highly effective electrochemically active bacteria would be the key for the improvement of these promising anode anammox MFCs with great functions in pollutant removal and electricity generation. In a previous study, a high-loaded anammox reactor reached an anammox rate of 76.7 kg $N/(m^3 \cdot d)$ [18]. Exploring the potential of nitrogen removal and electricity generation using anammox bacteria with such high activity in MFCs would improve our understanding of the bioelectrochemical processes in anode anammox MFCs.

In the present study, anammox bacteria from the highly loaded anammox reactor were collected as the anode inoculum to construct the anode anammox MFC, and the MFC was fed with purely inorganic nitrogen media. The objectives of the present study were to investigate the nitrogen removal and the electricity generation performance of the anode anammox MFC, and develop a better understanding of the electrochemical characteristics of this bioelectrochemical process.

2. Materials and Methods

2.1. Consruction of the Anode Anammox MFCs

The anode anammox MFC was a dual chamber MFC (Figure 1). The anode and cathode chamber both had a net volume of 400 mL with graphite felt (8 cm × 4 cm × 0.5 cm, Liaoyang Jingu Carbon Fiber Technology Co., Ltd., Liaoyang, China) as electrodes, separated by cation exchange membranes (CEM, Φ 6 cm, Membranes International, Ringwood, NJ, USA). The electrodes were fixed in the center of the anode and cathode chamber by stainless steel wires. A 1000 Ω resistor was connected to the electrodes with copper wires to form a closed circuit. The output voltage was monitored by a data acquisition system (DAS, Agilent 34970A, Agilent Technologies, Santa Clara, CA, USA).

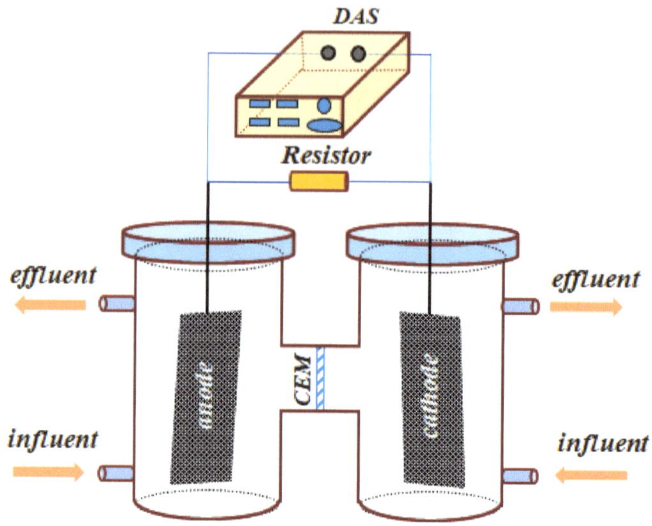

Figure 1. Schematic diagram of the anode anammox MFC.

2.2. Operating Conditions

The anode inoculum (highly effective anammox bacteria) was taken from a highly loaded anammox reactor in our lab [18]. The synthetic wastewater contained $(NH_4)_2SO_4$ and $NaNO_2$ and other trace components. The trace element solution was the same as in our previous study [23]. The ammonium (NH_4^+-N) and nitrite (NO_2^--N) concentrations were administered as $(NH_4)_2SO_4$, and $NaNO_2$, respectively. The cathode chamber was filled with a mixture of potassium permanganate solution (10 Mm) and phosphate buffer (50 mM, pH 7.0).

The anode chamber was continuously supplied with the wastewater using a peristaltic pump, and the hydraulic retention time (HRT) was set to 4 h. The ratio of NH_4^+-N to NO_2^--N was fixed at 1:1.32, and the influent concentrations of NH_4^+-N and NO_2^--N were varied along with the different operating periods. When the output voltage reached and remained at a relatively stable value for at least 6 HRT, it was considered that the MFC reached a steady state, then the influent concentration was raised to the next value. The catholyte was supplied using a peristaltic pump (5 mL/min). The anode and cathode electrolytes were bubbled with argon to remove oxygen before being mixed by a magnetic stirrer (HJ-2, Jin Tan Co., Ltd., Jintan, China) during the test. The anode anammox MFC was kept at 30 ± 1 °C in a biochemical incubator.

2.3. Analyses and Calculations

The ammonium–nitrogen (NH_4^+-N), nitrite–nitrogen (NO_2^--N), nitrate–nitrogen (NO_3^--N) and total nitrogen (TN) values were measured using the APHA Standard Methods [24]. The pH was determined by an S20 K pH meter (Mettler Toledo, Zurich, Switzerland). The conductivity was measured by an FE30 conductivity meter (Mettler Toledo, Zurich, Switzerland). The output voltages were recorded every 2 min by the DAS. The power density and the coulombic efficiency (CE) were calculated according to the methods outlined in previous studies [25]. The electrode potential was measured using an Ag/AgCl electrode (+0.195 V vs. SHE) as the reference electrode.

Polarization tests were performed with the gradual change (80,000 Ω −10 Ω) in the external resistance and the quasi-stable voltage, the anode potential and the cathode potential were recorded, and the current density was calculated. Polarization curves were plotted by taking the current density, the voltage, the anode potential, and the cathode potential as the ordinate and the current density as the abscissa [26]. A three-electrode system and a potentiostat (CHI660D, Beijing Huake Putian Technology Co., Ltd., Beijing, China) were used to perform electrochemical impedance spectroscopy (EIS) [27]. The sinusoidal disturbance voltage was 10 mV and the frequency range was 10^{-3}–10^5 Hz. The EIS equivalent circuit was fitted by ZSimpWin3.10 software.

3. Results and Discussion

3.1. Nitrogen Removal

The nitrogen removal performance of the anode anammox MFC under different nitrogen loads is shown in Figure 2. As the influent concentrations of NH_4^+-N and NO_2^--N gradually elevated from 25 to 250 mg/L and 33–330 mg/L, respectively, the removal efficiencies of NH_4^+-N, NO_2^--N and TN were over 90%, 90% and 80%, respectively (Figure 2A,B). Meanwhile, the volumetric nitrogen removal rate rose from 0.32 ± 0.05 kg N/(m^3·d) to a maximum of 3.01 ± 0.27 kg N/(m^3·d) (Figure 2C). However, when the influent NH_4^+-N and NO_2^--N concentrations further increased to 300 mg/L and 396 mg/L, the nitrogen removal efficiencies suddenly declined and the effluent concentrations of NH_4^+-N and NO_2^--N rose drastically to values almost equal to those of the influent (Figure 2A). Under these circumstances, the activities of anammox bacteria were fully inhibited [28]. The decrease in nitrogen removal in the anode anammox MFC indicated that the nitrogen load or nitrogen concentration had exceeded the maximum threshold value that the MFC could withstand.

Ammonium and nitrite are the two main substrates used by anammox bacteria, but they are potential inhibitors when their concentrations surpass the threshold inhibition values [29,30]. In order to reverse the self-inhibition at high nitrogen concentrations and increase microbial activity in the anode anammox MFC, the residual ammonium and nitrite were firstly washed out of the anode chamber, followed by a progressive increase in influent N concentrations, according to the remedial method of Tang et al. [28]. The nitrogen removal performance recovered after the remedial measures (Figure 2A,B).

When the anode anammox MFC ran stably, the ratio of the consumed NO_2^--N to the consumed NH_4^+-N ranged from 1.25 to 1.40, and the ratio of the produced NO_3^--N to the consumed NH_4^+-N was 0.15–0.23 (Figure 2D), which was close to the theoretical stoichiometry for the following anammox reaction (Equation (1)) [17]:

$$NH_4^+ + 1.32NO_2^- + 0.066HCO_3^- + 0.13H^+ \rightarrow 1.02N_2 + 0.26NO_3^- + 0.066CH_2O_{0.5}N_{0.15} + 2.03H_2O \quad (1)$$

In addition, nitrate (NO_3^-) was produced in the effluent. These results indicated that the anode nitrogen conversion in the MFC was in accordance with the typical anammox reaction. Owing to the faster oxidation rate of ammonium in the anammox process than in the nitrification process, the maximum volumetric nitrogen removal rate of the anode anammox MFC was 3.01 ± 0.27 kg N/(m^3·d), which was much higher than that of anode nitrification MFCs (0.240 kg N/(m^3·d)) [13,14] and previously reported anode anammox MFCs (0.417 kg N/(m^3·d)) [20,21]. Considering that the inoculated anammox bacteria demonstrated very high nitrogen removal abilities, the potential nitrogen removal efficiency of the anode anammox MFC might be further improved under optimized operating conditions [18,31,32].

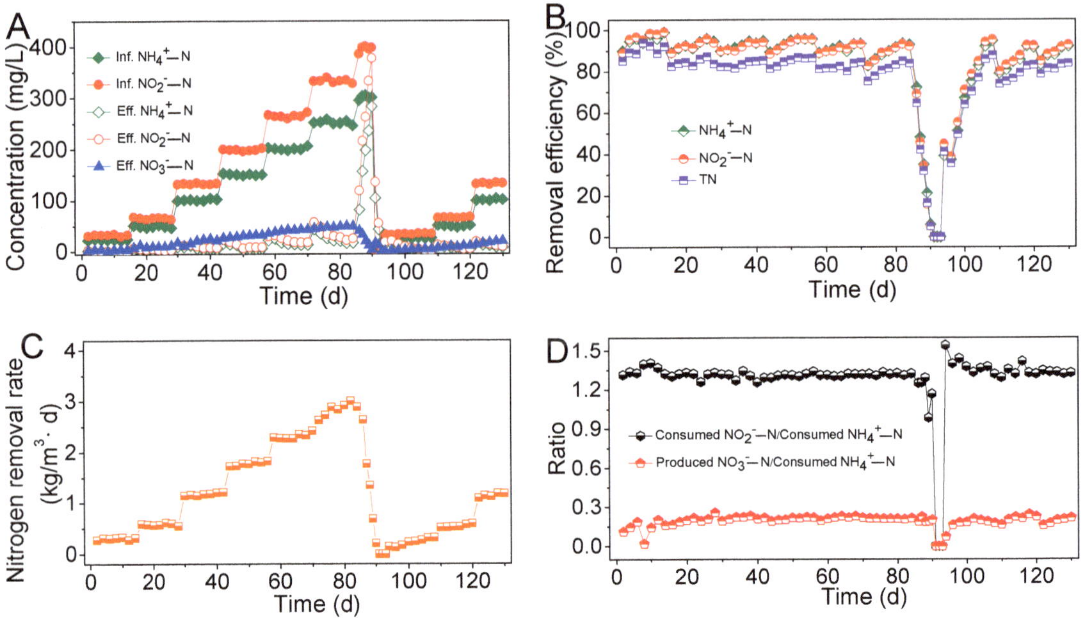

Figure 2. Nitrogen removal in the anode anammox MFC: (**A**) the concentrations of NH_4^+-N, NO_2^--Nand NO_3^--N in the influent and effluent; (**B**) removal efficiency changes in NH_4^+-N, NO_2^--N and TN; (**C**) variation in volumetric nitrogen removal rate over time and (**D**) ratio of consumed NO_2^--N to consumed NH_4^+-N and ratio of produced NO_3^--N to consumed NH_4^+-N.

3.2. Electricity Generation

The electricity generation performance of the anode anammox MFC under different nitrogen loads is depicted in Figure 3. The output voltage gradually increased from 35.13 ± 4.96 mV to 225.48 ± 10.71 mV as the influent nitrogen concentrations increased (Figure 3A). The power density of the anode anammox MFC also rose with escalating influent nitrogen concentrations and the maximum value reached 1308.23 ± 40.38 mW/m^3 (Figure 3B). The output voltage suffered a great loss following substrate self-inhibition and the electricity generation performance increased after the high nitrogen inhibition rate was counteracted (Figure 3A). Simultaneous change was observed between electricity generation and nitrogen removal, and Pearson correlation analysis was conducted to explore the relationship between them. Significant correlation between the maximum voltage and the volumetric nitrogen removal rate at each concentration was observed (R^2 = 0.9862, $p < 0.05$), which showed that electricity generation depends on ammonium oxidation in the anode anammox MFCs. The results corresponded well with reports that electrogenesis from MFCs showed a meaningful relationship with substrate degradation [33,34].

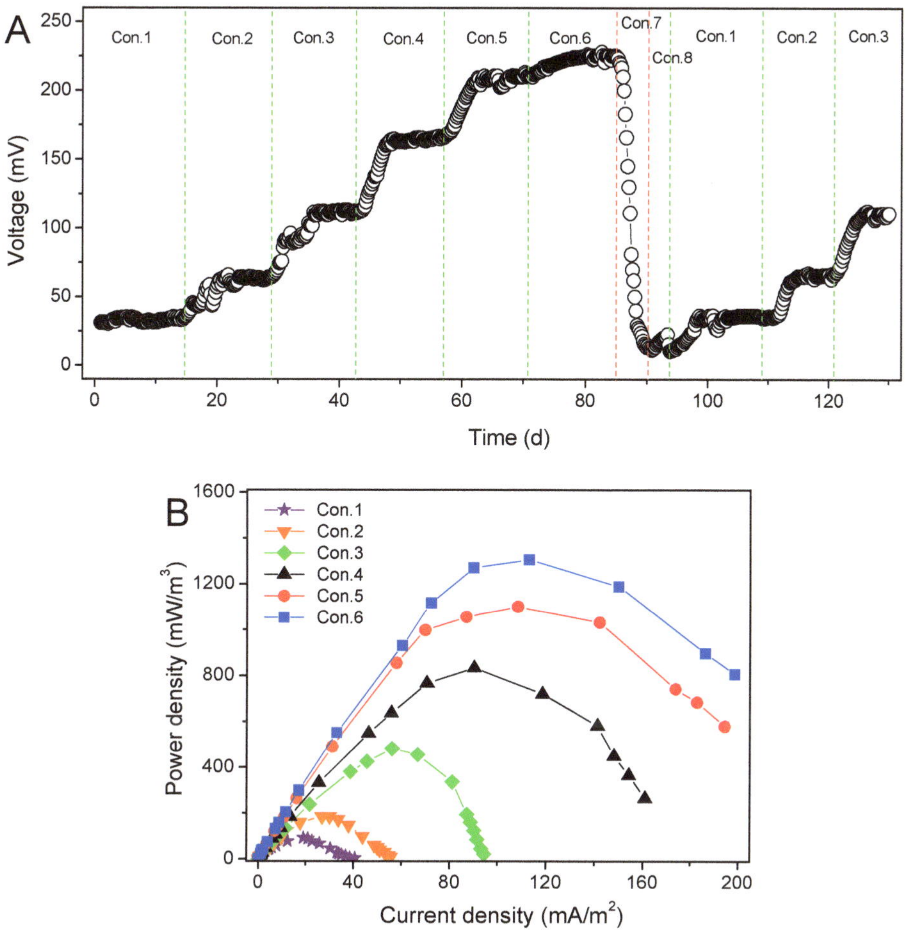

Figure 3. Profiles of (**A**) output voltage and (**B**) power density as a function of current output under different influent nitrogen concentrations (NH_4^+-N concentrations: Con.1: 25 mg/L, Con.2: 50 mg/L, Con.3: 100 mg/L, Con.4: 150 mg/L, Con.5: 200 mg/L, Con.6: 250 mg/L, Con.7: 300 mg/L; Con.8: 0 mg/L) in the anode anammox MFC.

By measuring the electrode potentials of the anode anammox MFC across different influent nitrogen concentrations, it was found that the cathode potential remained nearly stable (about 865 mV vs. SHE), while the anode potential gradually reduced with the rising nitrogen concentrations (Figure 4). The output voltage was the potential difference between the cathode and the anode, so the change in the output voltage was mainly caused by the evolution of the anode potential [35,36]. Coulomb efficiency (CE) is the ratio of electrons captured by the MFC for electricity production and the electrons derived from the oxidation of fuel, reflecting the utilization efficiency of fuel by the MFC [25]. The CE of the anode anammox MFC reduced from 5.99 $\pm$ 0.12% to 4.09 $\pm$ 0.46% with the increasing nitrogen concentrations (Figure 5). The utilization efficiency of fuel by the anode anammox MFC declined with the increase in substrate concentration, which was consistent with other reported studies [37,38].

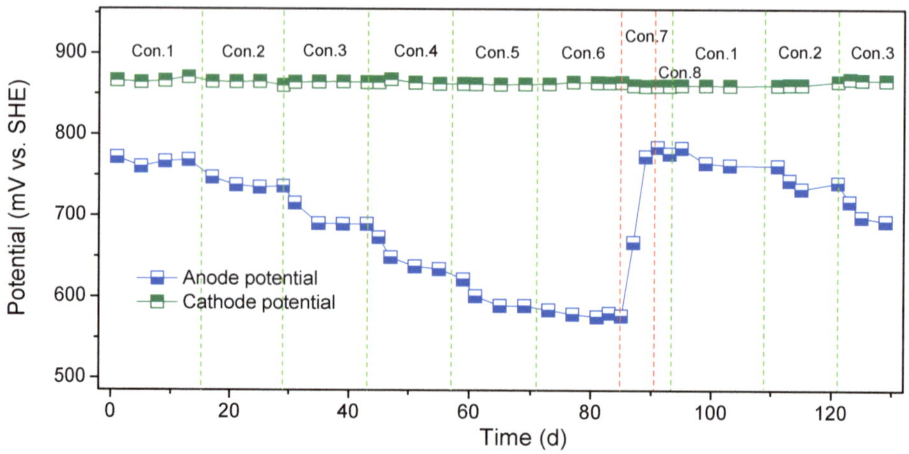

Figure 4. Variation in electrode potentials in the anode anammox MFC.

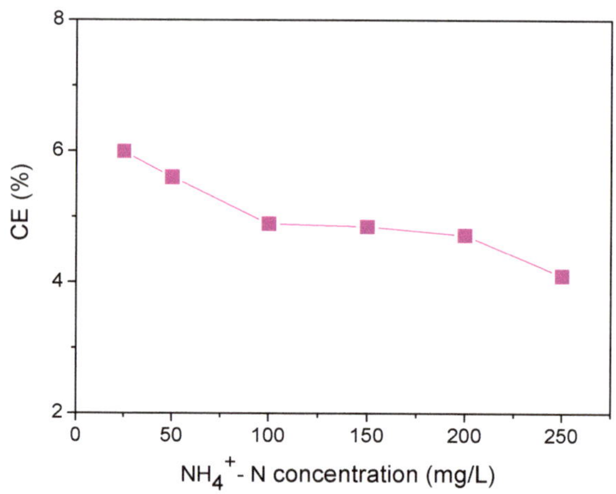

Figure 5. Change in CE with increasing influent concentrations.

The electricity generation performance of the anode anammox MFC was compared with other types of MFCs (Table 1). The anode anammox MFC had clear advantages over anode nitrification MFCs and previous anode anammox MFCs, while the electricity production parameters were lower than those for organic MFCs, especially for CE. In organic MFCs, the organic compounds are primarily consumed by electrogenic/non-electrogenic fermentation without exogenous electron acceptors and the anode potentials usually remain negative [39–42]. Therefore, the potential difference between the two electrodes is large, and the organic MFCs can produce more electric power [8,43–45]. With regard to ammonium-based bioelectrochemical systems, the oxidation of ammonium can be realized by two ways. The first method, in MFCs, is to add exogenous electron acceptors (oxygen or nitrite), which might compete with the anode for electrons, leading to low CE [13,14,20,21]. The second method, in MECs, is to place the anode at oxidative potentials (0.6 V vs. Ag/AgCl) [22,46,47].

Table 1. Comparison of electricity generation in the anode anammox MFC, anode nitrification MFC and Organic MFC.

MFC Types	Maximum Output Voltage (mV)	Maximum Power Density (mW/m³)	CE (%)	Anode Reactions and Potentials (vs. SHE)	References
Anode anammox MFC	225.48	1308.23 ± 40.38	4.09~5.99	① $NO_2^- + 2H^+ + e^- \rightarrow NO + H_2O$ ($E^{\ominus'}$ = +0.38 V) ② $NO + NH_4^+ + 2H^+ + 3e^- \rightarrow N_2H_4 + H_2O$ ($E^{\ominus'}$ = +0.06 V) ③ $N_2H_4 \rightarrow N_2 + 4H^+ + 4e^-$ ($E^{\ominus'}$ = −0.75 V) ④ $NO_2^- + H_2O \rightarrow NO_3^- + 2e^- + 2H^+$ ($E^{\ominus'}$ = −0.43 V)	This research
	48	-	-		[20]
	201.6	90.3	-		[21]
Anode nitrification MFC	98.5	2.43 ± 0.07	0.31~1.1	① $NH_4^+ + O_2 + H^+ + 2e^- \rightarrow NH_2OH + H_2O$ ($E^{\ominus'}$ = +0.73 V) ② $NH_2OH + H_2O \rightarrow NO_2^- + 5H^+ + 4e^-$ ($E^{\ominus'}$ = −0.06 V) ③ $NO_2^- + H_2O \rightarrow NO_3^- + 2e^- + 2H^+$ ($E^{\ominus'}$ = −0.43 V)	[13]
	82	93.3	-		[14]
Organic MFC	340	902.8	72.3	$CH_3COO^- + 2H_2O \rightarrow 2CO_2 + 7H^+ + 8e^-$ ($E^{\ominus'}$ = −0.29 V)	[8]
	466	7250	71		[43]
	540	66,000	43	$C_6H_{12}O_6 + 6H_2O \rightarrow 6CO_2 + 24H^+ + 24e^-$ ($E^{\ominus'}$ = −0.428 V)	[44]
	198	885 ± 346	59 ± 4		[45]

Both the nitrification process and the anammox process were multistage reactions, including positive potential and negative potential reaction steps (Table 1). Higher anode potentials are essential for ammonium oxidation but would be unfavorable for electricity generation. The anammox reactions demonstrated lower positive anode potential (Equation (2)) than that of the aerobic ammonia oxidation reactions (Equation (3)) [48], which might be the reason why the anode anammox MFC exhibited better performance in power generation than the anode nitrification MFCs.

$$NO_2^- + 2H^+ + e^- \rightarrow NO + H_2O,\ E^{\ominus'} = +0.38\ V \tag{2}$$

$$NH_4^+ + O_2 + H^+ + 2e^- \rightarrow NH_2OH + H_2O,\ E^{\ominus'} = +0.73\ V \tag{3}$$

3.3. Electrochemical Characteristics

The electrochemical characteristics of the anode anammox MFC were investigated in order to understand why those electrons released from the anammox process were not sufficiently utilized for electricity generation and led to losses of power output. The internal resistance of an MFC (mainly including activation loss, ohmic loss and mass transport loss) is closely related to the power output losses and polarization curves are useful tools for determining the resistance [25,26]. The polarization curves of the anode anammox MFC plotted at various influent concentrations are illustrated in Figure 6A. The slopes of the polarization curves in the low current density region were steep, indicating that the activation loss was significant while slow drops of the curves represented much a smaller ohmic loss and mass transport loss in the medium and high current density regions [25]. The polarization curves for different electrodes are depicted in Figure 6B. The cathode potentials reduced slowly with the increase in current density, and the cathodic polarization was not significant. By contrast, the variation in anode potentials was more visible than for the cathode potentials, especially in the low current density region, indicating that anodic polarization was more significant than cathodic polarization and activation loss was noted in the anodic electrochemical reaction [49,50]. Based on the above tests, the activation loss of the anode might be the main limiting factor for electricity generation in the anode anammox MFC.

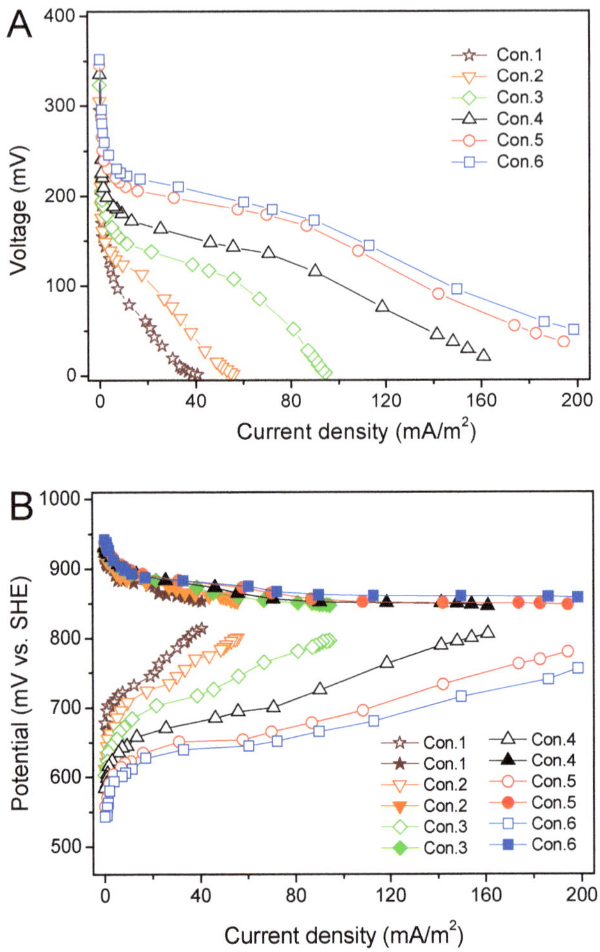

Figure 6. Polarization curves of (**A**) whole MFCs and (**B**) electrodes (cathode: filled symbols; anode: open symbols) under different influent concentrations (Con.1 to Con.6 represent NH_4^+-N concentrations from 25 mg/L to 250 mg/L).

Electrochemical impedance spectroscopy (EIS) causes very little disturbance to the test system and is a non-destructive electrochemical test technology that can obtain data on MFC resistance and other important parameters under actual operating conditions [25,27]. In order to analyze in detail the anode process during ammonium oxidation, EIS was performed to quantify the different components of anode internal resistance (R_{in}), including charge transfer resistance (R_{ct}), ohmic resistance (R_o) and diffusion resistance (R_d) [27]. It can be observed that the plots were mainly composed of semicircles at the high frequencies (Figure 7), indicating that R_{ct} is the main type of anode resistance. The diameter of the semicircles decreased as the influent concentrations increased, showing that the R_{ct} reduced as the biochemical reaction rate increased [27,51]. The accurate results in Table 2 were obtained by fitting the Nyquist plot data to an appropriate equivalent circuit (Figure 8) using ZSimpWin3.10 software [52,53]. At each influent concentration, the resistance distribution of the anode was as follows: R_{ct} (90%) $\gg R_o > R_d$. The results indicated that the anode charge transfer resistance was mostly responsible for the anode anammox MFC internal resistance.

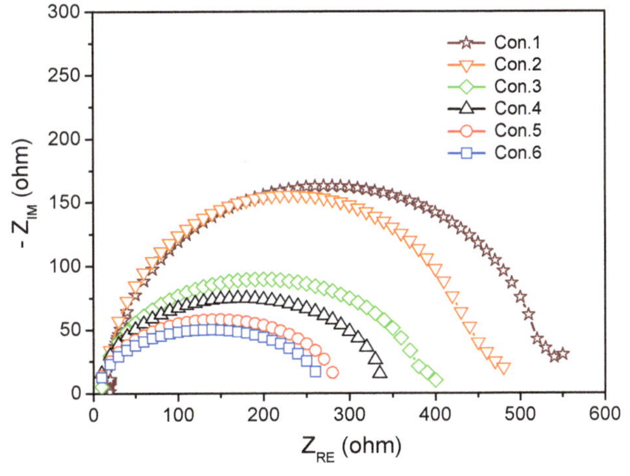

Figure 7. Nyquist plots of the anode under different influent concentrations.

Table 2. Resistance distribution of the anode obtained from the equivalent electrical circuit.

NH_4^+-N Concentration (mg/L)	R_{ct} (Ω)	R_o (Ω)	R_d (Ω)	R_{ct}/R_{in} (%)
25	407.4 ± 15.3	19.4 ± 2.3	5.5 ± 0.9	68.2 ± 3.6
50	346.3 ± 6.4	14.8 ± 1.5	3.5 ± 1.1	66.1 ± 4.7
100	298.8 ± 7.2	9.7 ± 2.9	2.1 ± 0.8	65.2 ± 3.3
150	267.2 ± 2.8	6.3 ± 1.8	1.8 ± 0.6	68.1 ± 9.5
200	223.4 ± 9.1	6.0 ± 2.6	0.9 ± 0.2	59.7 ± 4.8
250	209.2 ± 3.4	5.9 ± 1.2	0.6 ± 0.1	58.6 ± 8.1

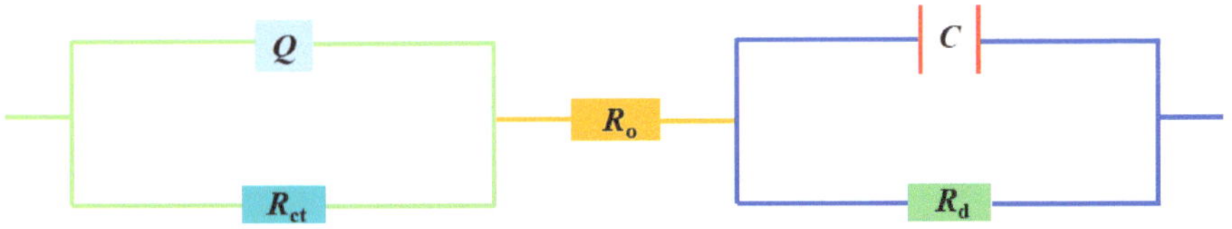

Figure 8. Equivalent electrical circuit for the electrochemical impedance spectra data simulation (*Q* is the constant phase element and C stands for double-layer capacitor).

The anode charge transfer resistance (or active resistance) is caused by the energy loss in the process of electron transfer from microbial cells to the electrode surface, and is affected by the kinetics of the electrode reactions [54,55]. Large anode charge transfer resistance generally means a low electron transfer rate or low electrode reaction rate [51,52]. Anammox bacteria were the main functional bacteria in the anode anammox MFC, whose cells possess a unique compartment structure called anammoxosome. The anammox reaction takes place inside the anammoxosome bounded by a low-permeability ladderane lipid-containing membrane, decreasing its permeability to protons and hydrazine. Additionally, the anammoxosome was also enclosed by the intracytopplasmic membrane and cytoplasmic membrane [17]. Compared with other reported electrochemically active bacteria [56], an additional electron transfer barrier (the anammoxosome compartment) must be crossed by anammox bacteria to transfer the electrons and the unique cell structure might hinder extracellular electron transfer [22], which was in accordance with the large anode charge transfer resistance in our study. The performance of the anode might be improved by selecting materials with better electrical properties (such as coated carbon nanotubes), adding artificial electron mediators in the electrolyte or even editing bacterial genes related to the electron transport chain to minimize the electron transfer resistance and reinforce the electricity generation performance of the anode anammox MFCs [57–59]. In fact, the anode electricity generation process in the anode anammox MFC is a new complex microbial and bioelectrochemical reaction. More in-depth study of the underlying interaction between microorganisms and electrodes during the electricity generating process, and the reaction kinetics or the effect of bacterial metabolism on electrodes, in future can help to further improve the process.

4. Conclusions

The emergence of ammonium-based MFCs provided an alternative approach for sustainable nitrogen-containing wastewater treatment. An anode anammox MFC developed by highly effective anammox bacteria showed stable nitrogen removal and electricity generation under different nitrogen concentrations during a long-term operation. It showed excellent nitrogen removal performance with a maximum volumetric nitrogen removal rate of 3.01 ± 0.27 kg $N/(m^3 \cdot d)$. It also presented certain electricity recovery potential with a maximum power density of 1308.23 ± 40.38 mW/m^3. Compared with the previous anode nitrification MFCs, the anode anammox MFC achieved much higher nitrogen removal and electricity generation performance. The performance of the anode anammox MFC was significantly influenced by nitrogen concentration. The main challenge for the anode anammox MFC was low energy recovery efficiency. Substantial power losses were due to the anodic overpotential and the anode charge transfer resistance was the main reason for the internal resistance. To further reduce the anode charge transfer resistance in the anode anammox MFC, a better understanding of the whole process of the extracellular electron transfer between microorganisms and electrodes is essential in future studies.

Author Contributions: J.L. and P.Z.: conceptualization, obtained research funding, and editing; J.Z. and J.R.: software, data analysis, and writing—original draft preparation; Z.Z. and H.G.: review and editing; K.R., J.C., Y.X., L.R. and T.W.: investigation and validation. All authors have read and agreed to the published version of the manuscript.

Funding: This work was supported by the Natural Science Foundation of Shandong Province (ZR2019QEE039) and the National Natural Science Foundation of China (52070163 and 41977124).

Data Availability Statement: Not applicable.

Conflicts of Interest: The authors declare no conflict of interest.

References

1. Logan, B.E.; Rabaey, K. Conversion of Wastes into Bioelectricity and Chemicals by Using Microbial Electrochemical Technologies. *Science* **2012**, *337*, 686–690. [CrossRef]
2. McCarty, P.L.; Bae, J.; Kim, J. Domestic Wastewater Treatment as a Net Energy Producer–Can This be Achieved? *Environ. Sci. Technol.* **2011**, *45*, 7100–7106. [CrossRef] [PubMed]
3. Heidrich, E.S.; Curtis, T.P.; Dolfing, J. Determination of the Internal Chemical Energy of Wastewater. *Environ. Sci. Technol.* **2011**, *45*, 827–832. [CrossRef] [PubMed]
4. Rozendal, R.A.; Hamelers, H.; Rabaey, K.; Keller, J.; Buisman, C.J. Towards practical implementation of bioelectrochemical wastewater treatment. *Trends Biotechnol.* **2008**, *26*, 450–459. [CrossRef]
5. Logan, B.E.; Regan, J.M. Microbial Fuel Cells—Challenges and Applications. *Environ. Sci. Technol.* **2006**, *40*, 5172–5180. [CrossRef]
6. Li, W.-W.; Yu, H.-Q.; He, Z. Towards sustainable wastewater treatment by using microbial fuel cells-centered technologies. *Energy Environ. Sci.* **2014**, *7*, 911–924. [CrossRef]
7. Bond, D.R.; Holmes, D.E.; Tender, L.M.; Lovley, D.R. Electrode-Reducing Microorganisms That Harvest Energy from Marine Sediments. *Science* **2002**, *295*, 483–485. [CrossRef]
8. Chae, K.-J.; Choi, M.-J.; Lee, J.-W.; Kim, K.-Y.; Kim, I.S. Effect of different substrates on the performance, bacterial diversity, and bacterial viability in microbial fuel cells. *Bioresour. Technol.* **2009**, *100*, 3518–3525. [CrossRef]
9. Pandey, P.; Shinde, V.N.; Deopurkar, R.L.; Kale, S.P.; Patil, S.A.; Pant, D. Recent advances in the use of different substrates in microbial fuel cells toward wastewater treatment and simultaneous energy recovery. *Appl. Energy* **2016**, *168*, 706–723. [CrossRef]

10. Kelly, P.T.; He, Z. Nutrients removal and recovery in bioelectrochemical systems: A review. *Bioresour. Technol.* **2014**, *153*, 351–360. [CrossRef]
11. Virdis, B.; Rabaey, K.; Rozendal, R.A.; Yuan, Z.; Keller, J. Simultaneous nitrification, denitrification and carbon removal in microbial fuel cells. *Water Res.* **2010**, *44*, 2970–2980. [CrossRef] [PubMed]
12. Park, Y.; Park, S.; Nguyen, V.K.; Yu, J.; Torres, C.I.; Rittmann, B.E.; Lee, T. Complete nitrogen removal by simultaneous nitrification and denitrification in flat-panel air-cathode microbial fuel cells treating domestic wastewater. *Chem. Eng. J.* **2017**, *316*, 673–679. [CrossRef]
13. He, Z.; Kan, J.; Wang, Y.; Huang, Y.; Mansfeld, F.; Nealson, K.H. Electricity Production Coupled to Ammonium in a Microbial Fuel Cell. *Environ. Sci. Technol.* **2009**, *43*, 3391–3397. [CrossRef]
14. Xie, Z.; Chen, H.; Zheng, P.; Zhang, J.; Cai, J.; Abbas, G. Influence and mechanism of dissolved oxygen on the performance of Ammonia-Oxidation Microbial Fuel Cell. *Int. J. Hydrogen Energy* **2013**, *38*, 10607–10615. [CrossRef]
15. Quan, X.-C.; Quan, Y.-P.; Tao, K. Effect of anode aeration on the performance and microbial community of an air–cathode microbial fuel cell. *Chem. Eng. J.* **2012**, *210*, 150–156. [CrossRef]
16. Kuypers, M.M.M.; Marchant, H.K.; Kartal, B. The microbial nitrogen-cycling network. *Nat. Rev. Microbiol.* **2018**, *16*, 263–276. [CrossRef] [PubMed]
17. Kartal, B.; Maalcke, W.J.; de Almeida, N.M.; Cirpus, I.; Gloerich, J.; Geerts, W.; Op den Camp, H.J.M.; Harhangi, H.R.; Janssen-Megens, E.M.; Francoijs, K.-J.; et al. Molecular mechanism of anaerobic ammonium oxidation. *Nature* **2011**, *479*, 127–130. [CrossRef] [PubMed]
18. Tang, C.-J.; Zheng, P.; Wang, C.-H.; Mahmood, Q.; Zhang, J.-Q.; Chen, X.-G.; Zhang, L.; Chen, J.-W. Performance of high-loaded ANAMMOX UASB reactors containing granular sludge. *Water Res.* **2011**, *45*, 135–144. [CrossRef]
19. Du, R.; Cao, S.; Li, B.; Niu, M.; Wang, S.; Peng, Y. Performance and microbial community analysis of a novel DEAMOX based on partial-denitrification and anammox treating ammonia and nitrate wastewaters. *Water Res.* **2017**, *108*, 46–56. [CrossRef]
20. Xu, M.Y.; Zhou, S.Q.; Liu, Z.J.; Wang, J.P.; Ma, F.Z. Study on performance of dual-chamber MFC coupled with Anammox process in a high nitrogen load circumstance. *Acta Sci. Circumstant.* **2017**, *37*, 154–161. (In Chinese)
21. Zu, B.; Ma, L.; Liu, B.; Lu, P.-L.; Xu, J. Effects of Organic Substrates on ANAMMOX-MFC Denitrification Electrogenesis Performance. *Environ. Sci.* **2018**, *39*, 3937–3945.
22. Shaw, D.R.; Ali, M.; Katuri, K.P.; Gralnick, J.A.; Reimann, J.; Mesman, R.; van Niftrik, L.; Jetten, M.S.M.; Saikaly, P.E. Extracellular electron transfer-dependent anaerobic oxidation of ammonium by anammox bacteria. *Nat. Commun.* **2020**, *11*, 2058. [CrossRef]
23. Zhang, J.; Zheng, P.; Zhang, M.; Chen, H.; Chen, T.; Xie, Z.; Cai, J.; Abbas, G. Kinetics of substrate degradation and electricity generation in anodic denitrification microbial fuel cell (AD-MFC). *Bioresour. Technol.* **2013**, *149*, 44–50. [CrossRef] [PubMed]
24. American Public Health Association (APHA). *Standard Methods for the Examination of Water and Wastewater*, 21st ed.; American Public Health Association: Washington, DC, USA, 2005; ISBN 087-553-047-8.
25. Logan, B.E.; Hamelers, B.; Rozendal, R.; Schröder, U.; Keller, J.; Freguia, S.; Aelterman, P.; Verstraete, W.; Rabaey, K. Microbial Fuel Cells: Methodology and Technology. *Environ. Sci. Technol.* **2006**, *40*, 5181–5192. [CrossRef] [PubMed]
26. Wang, B.; Han, J.-I. A single chamber stackable microbial fuel cell with air cathode. *Biotechnol. Lett.* **2009**, *31*, 387–393. [CrossRef] [PubMed]
27. Manohar, A.K.; Bretschger, O.; Nealson, K.H.; Mansfeld, F. The use of electrochemical impedance spectroscopy (EIS) in the evaluation of the electrochemical properties of a microbial fuel cell. *Bioelectrochemistry* **2008**, *72*, 149–154. [CrossRef] [PubMed]
28. Tang, C.-J.; Zheng, P.; Mahmood, Q.; Chen, J.-W. Start-up and inhibition analysis of the Anammox process seeded with anaerobic granular sludge. *J. Ind. Microbiol. Biotechnol.* **2009**, *36*, 1093–1100. [CrossRef] [PubMed]
29. Jin, R.-C.; Yang, G.-F.; Yu, J.-J.; Zheng, P. The inhibition of the Anammox process: A review. *Chem. Eng. J.* **2012**, *197*, 67–79. [CrossRef]
30. Hong, S.; De Clippeleir, H.; Goel, R. Response of mixed community anammox biomass against sulfide, nitrite and recalcitrant carbon in terms of inhibition coefficients and functional gene expressions. *Chemosphere* **2022**, *308*, 136232. [CrossRef]
31. Li, Y.-Y.; Huang, X.-W.; Li, X.-Y. Use of a packed-bed biofilm reactor to achieve rapid formation of anammox biofilms for high-rate nitrogen removal. *J. Clean. Prod.* **2021**, *321*, 128999. [CrossRef]
32. Li, B.; Wang, Y.; Wang, W.; Huang, X.; Kou, X.; Wu, S.; Shao, T. High-rate nitrogen removal in a continuous biofilter anammox reactor for treating low-concentration nitrogen wastewater at moderate temperature. *Bioresour. Technol.* **2021**, *337*, 125496. [CrossRef] [PubMed]
33. Juang, D.-F.; Yang, P.-C.; Chou, H.-Y.; Chiu, L.-J. Effects of microbial species, organic loading and substrate degradation rate on the power generation capability of microbial fuel cells. *Biotechnol. Lett.* **2011**, *33*, 2147–2160. [CrossRef] [PubMed]
34. Cai, J.; Qaisar, M.; Sun, Y.; Wang, K.; Lou, J.; Wang, R. Coupled substrate removal and electricity generation in microbial fuel cells simultaneously treating sulfide and nitrate at various influent sulfide to nitrate ratios. *Bioresour. Technol.* **2020**, *306*, 123174. [CrossRef]
35. Wei, J.; Liang, P.; Cao, X.; Huang, X. A New Insight into Potential Regulation on Growth and Power Generation of *Geobacter sulfurreducens* in Microbial Fuel Cells Based on Energy Viewpoint. *Environ. Sci. Technol.* **2010**, *44*, 3187–3191. [CrossRef] [PubMed]
36. Lu, M.; Qian, Y.; Huang, L.; Xie, X.; Huang, W. Improving the performance of microbial fuel cells through anode manipulation. *ChemPlusChem* **2015**, *80*, 1216–1225. [CrossRef] [PubMed]

37. Gao, Y.; Yin, F.; Ma, W.; Wang, S.; Liu, Y.; Liu, H. Rapid detection of biodegradable organic matter in polluted water with microbial fuel cell sensor: Method of partial coulombic yield. *Bioelectrochemistry* **2020**, *133*, 107488. [CrossRef]
38. Jadhav, G.; Ghangrekar, M. Performance of microbial fuel cell subjected to variation in pH, temperature, external load and substrate concentration. *Bioresour. Technol.* **2009**, *100*, 717–723. [CrossRef]
39. Logan, B.E. Exoelectrogenic bacteria that power microbial fuel cells. *Nat. Rev. Microbiol.* **2009**, *7*, 375–381. [CrossRef]
40. de los Ángeles Fernandez, M.; de los Ángeles Sanromán, M.; Marks, S.; Makinia, J.; Del Campo, A.G.; Rodrigo, M.; Fernandez, F.J. A grey box model of glucose fermentation and syntrophic oxidation in microbial fuel cells. *Bioresour. Technol.* **2016**, *200*, 396–404. [CrossRef]
41. Mei, X.; Lu, B.; Yan, C.; Gu, J.; Ren, N.; Ren, Z.J.; Xing, D. The interplay of active energy harvesting and wastewater organic loading regulates fermentation products and microbiomes in microbial fuel cells. *Resour. Conserv. Recycl.* **2022**, *183*, 106366. [CrossRef]
42. Esquivel, D.Y.A.; Guo, Y.; Brown, R.K.; Müller, S.; Schröder, U.; Harnisch, F. Investigating Community Dynamics and Performance During Microbial Electrochemical Degradation of Whey. *ChemElectroChem* **2020**, *7*, 989–997. [CrossRef]
43. Lee, H.-S.; Parameswaran, P.; Kato-Marcus, A.; Torres, C.I.; Rittmann, B.E. Evaluation of energy-conversion efficiencies in microbial fuel cells (MFCs) utilizing fermentable and non-fermentable substrates. *Water Res.* **2008**, *42*, 1501–1510. [CrossRef] [PubMed]
44. Rabaey, K.; Clauwaert, P.; Aelterman, P.; Verstraete, W. Tubular Microbial Fuel Cells for Efficient Electricity Generation. *Environ. Sci. Technol.* **2005**, *39*, 8077–8082. [CrossRef] [PubMed]
45. Kim, K.-Y.; Chae, K.-J.; Choi, M.-J.; Ajayi, F.F.; Jang, A.; Kim, C.-W.; Kim, I.S. Enhanced Coulombic efficiency in glucose-fed microbial fuel cells by reducing metabolite electron losses using dual-anode electrodes. *Bioresour. Technol.* **2011**, *102*, 4144–4149. [CrossRef]
46. Vilajeliu-Pons, A.; Koch, C.; Balaguer, M.D.; Colprim, J.; Harnisch, F.; Puig, S. Microbial electricity driven anoxic ammonium removal. *Water Res.* **2018**, *130*, 168–175. [CrossRef]
47. Qu, B.; Fan, B.; Zhu, S.; Zheng, Y. Anaerobic ammonium oxidation with an anode as the electron acceptor. *Environ. Microbiol. Rep.* **2014**, *6*, 100–105. [CrossRef]
48. Jetten, M.S.M.; Van Niftrik, L.; van Strous, M.; Kartal, B.; Keltjens, J.T.; op den Camp, H.J.M. Biochemistry and molecular biology of anammox bacteria. *Crit. Rev. Biochem. Mol. Biol.* **2009**, *44*, 65–84. [CrossRef]
49. Manohar, A.K.; Bretschger, O.; Nealson, K.H.; Mansfeld, F. The polarization behavior of the anode in a microbial fuel cell. *Electrochim. Acta* **2008**, *53*, 3508–3513. [CrossRef]
50. Zhang, Y.; Mo, G.; Li, X.; Zhang, W.; Zhang, J.; Ye, J.; Huang, X.; Yu, C. A graphene modified anode to improve the performance of microbial fuel cells. *J. Power Sources* **2011**, *196*, 5402–5407. [CrossRef]
51. He, Z.; Mansfeld, F. Exploring the use of electrochemical impedance spectroscopy (EIS) in microbial fuel cell studies. *Energy Environ. Sci.* **2009**, *2*, 215–219. [CrossRef]
52. Wang, H.; Long, X.; Sun, Y.; Wang, D.; Wang, Z.; Meng, H.; Jiang, C.; Dong, W.; Lu, N. Electrochemical impedance spectroscopy applied to microbial fuel cells: A review. *Front. Microbiol.* **2022**, *13*, 973501. [CrossRef]
53. Sindhuja, M.; Kumar, N.S.; Sudha, V.; Harinipriya, S. Equivalent circuit modeling of microbial fuel cells using impedance spectroscopy. *J. Energy Storage* **2016**, *7*, 136–146. [CrossRef]
54. Shi, L.; Dong, H.; Reguera, G.; Beyenal, H.; Haluk, B.; Liu, J.; Yu, H.-Q.; Fredrickson, J.K. Extracellular electron transfer mechanisms between microorganisms and minerals. *Nat. Rev. Microbiol.* **2016**, *14*, 651–662. [CrossRef] [PubMed]
55. Light, S.H.; Su, L.; Rivera-Lugo, R.; Cornejo, J.A.; Louie, A.; Iavarone, A.T.; Ajo-Franklin, C.M.; Portnoy, D.A. A flavin-based extracellular electron transfer mechanism in diverse Gram-positive bacteria. *Nature* **2018**, *562*, 140–144. [CrossRef] [PubMed]
56. Hirose, A.; Kouzuma, A.; Watanabe, K. Towards development of electrogenetics using electrochemically active bacteria. *Biotechnol. Adv.* **2019**, *37*, 107351. [CrossRef] [PubMed]
57. Sonawane, J.M.; Yadav, A.; Ghosh, P.C.; Adeloju, S.B. Recent advances in the development and utilization of modern anode materials for high performance microbial fuel cells. *Biosens. Bioelectron.* **2017**, *90*, 558–576. [CrossRef]
58. Kumar, R.; Singh, L.; Zularisam, A. Exoelectrogens: Recent advances in molecular drivers involved in extracellular electron transfer and strategies used to improve it for microbial fuel cell applications. *Renew. Sustain. Energy Rev.* **2016**, *56*, 1322–1336. [CrossRef]
59. Xie, R.; Wang, S.; Wang, K.; Wang, M.; Chen, B.; Wang, Z.; Tan, T. Improved energy efficiency in microbial fuel cells by bioethanol and electricity co-generation. *Biotechnol. Biofuels* **2022**, *15*, 84. [CrossRef]

processes

MDPI

Review

Tertiary Wastewater Treatment Technologies: A Review of Technical, Economic, and Life Cycle Aspects

Dimitris P. Zagklis * and Georgios Bampos *

Department of Chemical Engineering, University of Patras, GR-26504 Patras, Greece
* Correspondence: dimitriszag@chemeng.upatras.gr (D.P.Z.); geoba@chemeng.upatras.gr (G.B.); Tel.: +30-2610997223 (G.B.)

Abstract: The activated sludge process is the most widespread sewage treatment method. It typically consists of a pretreatment step, followed by a primary settling tank, an aerobic degradation process, and, finally, a secondary settling tank. The secondary effluent is then usually chlorinated and discharged to a water body. Tertiary treatment aims at improving the characteristics of the secondary effluent to facilitate its reuse. In this work, through a literature review of the most prominent tertiary treatment methods, a benchmarking of their technical efficiency, economic feasibility, and environmental impact was carried out. The photo-Fenton method proved to be the most technically efficient process, significantly reducing the microbial load and pharmaceutical content (by 4.9 log and 84%, respectively) of the secondary effluent. Chlorination and UV irradiation exhibited the lowest treatment costs (0.004 EUR/m^{-3}) and the lowest global warming potential (0.04 and 0.09 kg CO_2eq. m^{-3}, respectively). After all the data were aggregated, a decision-making tool was constructed in the form of a ternary diagram, which indicates the most appropriate tertiary treatment method according to the weight-per-process aspect (technical, economic, and environmental) selected by the user, with chlorination, UV irradiation, ozonation, microalgae cultivation, and constructed wetlands prevailing in the final results.

Keywords: tertiary wastewater treatment; chlorination; constructed wetlands; microalgae; ozonation; life cycle assessment; technoeconomic

Citation: Zagklis, D.P.; Bampos, G. Tertiary Wastewater Treatment Technologies: A Review of Technical, Economic, and Life Cycle Aspects. *Processes* **2022**, *10*, 2304. https://doi.org/10.3390/pr10112304

Academic Editor: Monika Wawrzkiewicz

Received: 28 September 2022
Accepted: 3 November 2022
Published: 5 November 2022

1. Introduction

Global population growth, as well as modern lifestyles, has resulted in an increase in the amount of produced sewage, thus making wastewater management urgent. According to United Nations estimations, the global population will reach ca. 11 billion by 2100, whereas other researchers estimate that the population will reach a lower peak earlier than 2100 [1]. In any case, human activity is expected to dramatically increase, followed by increased urban wastewater production.

Typically, wastewater can be classified as sewage when derived from domestic activities. Further categorization can divide sewage into blackwater, greywater, and yellow water, with the first corresponding to wastewater contaminated with human waste, the second to sewage derived from non-toilet units (e.g., showers) without being contaminated with human feces or urine, and the latter being polluted by urine collected from sewer channels [2–4]. Apart from domestic and industrial wastewater, hospitals, agricultural activity, and livestock farming can lead to aquatic environmental pollution, including contamination by emerging contaminants (ECs) (e.g., pharmaceuticals), posing a threat to both the environment and humans [5].

The most widespread sewage treatment method involves the activated sludge process. A typical wastewater treatment plant consists of a pretreatment step where grease and large solids are removed from the wastewater, followed by a primary settling tank. After most of the suspended solids are removed, the wastewater is led to the aerobic degradation process that implements the activated sludge. At this stage, most of the organic content is converted into carbon dioxide and microbial biomass. The effluent of the aerobic process is treated with a secondary settling tank that removes and partially recirculates the produced biomass. These are the steps that are typically included in such a process, with some also implementing a nitrification/denitrification process in the activated sludge process. The effluent of the secondary settling tank (secondary effluent) is then typically chlorinated and discharged into a water body.

Despite the fact that the aforementioned steps are enough to remove most of the organic content from wastewater, residual pollutants such as inorganic nutrients (mostly nitrogen and phosphorus), increased microbial content, and ECs can have a significant impact on the final treated wastewater recipients. Moreover, strict quality criteria need to be met before the treated water can be reused instead of ending up in a water recipient [6]. In this scope, several secondary effluent treatment methods have been developed and implemented, typically called tertiary treatment processes. Among the tertiary treatment methods, chlorination, ultraviolet irradiation, membrane filtration, constructed wetlands, microalgae cultivation, ozonation, and photo-Fenton processes have attracted significant research interest due to their low operational costs, as well as their high efficiency. Some indicative works related to tertiary treatment techniques are demonstrated in the present paper.

An effort was made by the authors to present the most important technical efficacy parameters, namely the ability of each tertiary process to reduce pathogens and remove inorganic nutrients and EC. Furthermore, to put the technical efficacy results into a sustainability perspective, data regarding the treatment costs and global warming potential were also collected. Most of the published literature assesses the technical efficacy of these methods, with a much smaller number of publications examining their economic and life cycle aspects, a fact that unfortunately causes a disparity between the data available for each examined aspect.

2. Tertiary Wastewater Treatment Technologies

2.1. Chlorination

Chlorination is a widely used disinfection method applied in the last stages of sewage treatment. As a strong oxidizer, chlorine reacts with organic compounds, but can also lead to the formation of harmful chlorinated byproducts with negative effects on human health and the environment [7]. Several studies have reported the potency of chlorination in disinfection. In the work of Decol et al. [8], a 2.5-log reduction in *E. coli* was achieved, whereas Francy et al. [9] reported a 0.7- to 2.6-log reduction in a number of microbiological indicators such as *E. coli* (highest log reduction) and viruses. Chlorination does not affect the nutrient content of secondary effluent (in terms of N and P), making this method unsuitable when the treated wastewater is led to recipients prone to eutrophication. Pharmaceutical compounds, on the other hand, are prone to oxidation from chlorine, with Li and Zhang [10] testing the effects of the chlorination of compounds, i.e., sulfamethoxazole, ciprofloxacin, norfloxacin, tetracycline, trimethoprim, and erythromycin, reporting a reduction of 43 to 73%. Regarding the economic aspect of chlorination, it is considered one of the cheapest tertiary methods, with a treatment cost of 0.0003 to 0.006 EUR/m^3 [11–13]. Finally, due to its minimum energy needs, chlorination leads to very low greenhouse gas (GHG) emissions, with Walsh and Mellor [14] reporting GHG as low as 0.004 kg CO_2 eq. m^{-3}, whereas Pasqualino et al. [15] reported a slightly higher value of 0.07 kg CO_2 eq. m^{-3}.

2.2. Ultraviolet Irradiation

Ultraviolet technology, alone or combined with other degradation processes, has been extensively investigated as a tertiary treatment process. Zhang et al. examined the UV process coupled with H_2O_2 for secondary effluent treatment, achieving an ARG reduction equal to 3.48 log at pH 3.0 and 2.32 log at pH 7.0, whereas the cost was estimated at 0.296 USD/m^3 [16]. The coupling of UV with H_2O_2 technology was studied in the removal of microcystin-RR by Qiao et al., with the reduction percentage of this pharmaceutical being 94.83% [17]. The combination of UV and the ozonation process by Chin and Bérubé in the treatment of surface water resulted in 50%, 80%, and 70% reductions in TOC, trihalomethane formation, and haloacetic acid formation, respectively [18].

A reduction of 100% for both diclofenac and bezafibrate was achieved when coupling UV irradiation with H_2O_2, as reported by De la Cruz et al., whereas the combination of three technologies UV/Fe^{2+}/H_2O_2 resulted in a 100% elimination of diclofenac [19]. Moreover, Guo et al. examined the UV disinfection process over ARB and ARGs present in municipal sewage [20]. The reduction in bacteria resistant to erythromycin and tetracycline was found to be equal to 1.4 and 1.1 log, respectively.

2.3. *Membrane Filtration*

Membrane filtration methods include a diverse group of processes, with the most common ones being pressure-driven membranes. During pressure-driven membrane filtration, a pressure difference is imposed on the two sides of a semi-permeable membrane, with the kinds of solutes permeating the membrane, further defining the membrane types. Membranes with a pore size on a scale of 1 μm (microfiltration, MF) typically reject suspended solids. Ultrafiltration (UF) has a smaller pore diameter and can reject larger dissolved molecules. Membranes with a pore size on a scale of 1 nm fall within the nanofiltration (NF) group and can reject smaller dissolved molecules (typically up to 200 Da) and divalent ions. Finally, reverse osmosis (RO) has no pores and separation occurs through the different diffusion rates of the solutes in the polymer of the membrane. RO membranes can even reject monovalent ions [21].

An integrated pilot unit combining UF, RO, and electrooxidation to manage municipal sewage was introduced by Urtiaga et al. [22]. All the target compounds, namely naproxen, ofloxacin, furosemide, bezafibrate, and fenofibric acid, were rejected with a percentage higher than 99%. Cheng et al. studied the efficiency of an anaerobic MF system in the removal of some antibiotic resistance bacteria (ARB) and their associated ARGs found in municipal sewage and the microbial load reduction was equal to 2–3 log units [23]. The elimination of ARB and ARGs present in secondary wastewater effluent by employing a TiO_2-modified polyvinylidene fluoride (PVDF) UF membrane was investigated by Ren et al. [24]. ARGs were removed at a rate of 98%.

Dolar et al. examined performance as it concerned the degradation of selected veterinary Phs present in pharmaceutical sewage using a laboratory and pilot scale RO/NF membrane treatment process [25]. The removal of TOC and COD was 70.8% and 35.4%, respectively, whereas the degradation percentage of the selected Phs ranged from 94% to 100% for the NF and RO membranes, respectively. Furthermore, Ho et al. examined palm oil mill effluent treatment by implementing graphene oxide (GO)/multi-walled carbon nanotube (MWCNTs) conductive membranes [26].

Treatment costs using membrane filtration can range from 0.4 to 1 EUR/m^3 [27] and the produced emissions can range from 0.2 to 2.3 kg CO_2 eq. m^{-3}, with the total emissions depending on the number of membrane steps and the required transmembrane pressure [28,29].

2.4. *Constructed Wetlands*

Constructed wetlands (CW) have been used since the 1950s as a waste management option [30]. The premise of this wastewater treatment method is based on the naturally occurring processes involving vegetation, soil, and microorganisms in a controlled environment [30].

Breitholtz et al. [31] achieved a BOD reduction of 40%, whereas reductions in 92 pharmaceuticals (Phs) ranged between 42 and 52%. The treatment of wastewater derived from residential areas was investigated over horizontal flow constructed wetlands (HFCW), vertical flow constructed wetlands (VFCW), and biofilters by Adrados et al. [32]. According to their study, the TN reduction ranged from 21 to 85%. Younger and Henderson [33] reported

41%, 59%, and 66% reductions in BOD, P-PO_4, and N-NH_4, respectively, employing an innovative full-scale mine water/sewage cotreatment CW for polluted mine waters. The yield of a vertical up-flow CW for swine wastewater was studied by Huang et al. [34] with remarkable results. Reductions in COD, TN, N-NH_4, and TP were 92.2%, 92.7%, 94.4%, and 97.8%, respectively, whereas the degradation of Phs ranged between 98.3% and 99.9%. The removal of various ECs (77.2%) and antibiotic resistance genes (ARGs) derived from landfill leachate was investigated by Yi et al. [35] by employing a full-scale horizontal subsurface flow CW. An integrated surface flow CW was employed over a 10-year period for the removal of ARGs found in domestic wastewater, as reported by Fang et al. [36]. The COD, BOD, TN, TP, and N-NH_4 degradations were 70.8%, 75.2%, 60.2%, 55.6%, and 61.3%, respectively. Chen et al. examined the ARG (85.8%) and antibiotic eliminations from wastewater derived from residential areas using mesocosm-scale horizontal subsurface flow CWs [37]. VFCWs were applied for the removal of ciprofloxacin HCl, oxytetracycline HCl, and sulfamethazine from swine sewage achieving elimination rates of 85%, 95%, and 73% for each of the aforementioned antibiotics, as found by Liu et al. [38]. The elimination of ARGs from municipal wastewater was studied by Nõlvak et al. [39] by employing horizontal subsurface flow CWs, achieving a 92% and 25% removal of BOD and TN, respectively. Chen et al. studied the effect on the degradation of antibiotics and ARGs of domestic sewage by employing six mesocosm-scale CWs [40]. The removal rates of COD, TOC, TN, and N-NH_3 were 80.2%, 80.3%, 54.7%, and 44%, respectively, whereas the total removal of all detected antibiotics was 98.6%. A 62% reduction in the rate of tet genes and a 90% average total removal rate of oxytetracycline and difloxacin antibiotics from swine sewage were achieved by applying a VFCW by Huang et al. [41].

Ledón et al. found a 90% reduction in the BOD rate from domestic wastewater by employing horizontal subsurface flow (HSSF) CW with the HSSF pretreatment cost being equal to 1903 USD/p.e, whereas the GHG emissions ranged between 3.8 and 4.7 kg $CO_{2\text{-eq}}$/kg for BOD_5 [42]. The coupling of microbial fuel cell (MFCs) technology with a conventional HSSF CW was investigated for municipal wastewater treatment by Corbella et al., achieving a remarkable 85% BOD reduction, where the cost of a conventional CW was estimated at 430 EUR/p.e [43]. Winery sewage treatment employing various scenarios of CW operations was examined by Flores et al., with the GHG emissions derived from the LCA being 1.3 kg $CO_{2\text{-eq}}/m^3$ [44]. Significant variations in costs were observed via life cycle costing (LCC) analysis of two different small-scale WWTPs coupled with CW technology for the treatment of wastewater produced by a student residential building and its coffee shop located in Brazil [45]. The cost of the scenario implementing a mobile CW (2.42×10^4 kg $CO_{2\text{-eq}}$) was found to be higher than that of the scenario using a decentralized VFCW (1.03×10^4 kg $CO_{2\text{-eq}}$) for wastewater treatment by Lakho et al. [46]. Pan et al. investigated the treatment of wastewater produced by the residential area of Changzhou in China using a vertical subsurface flow CW system, achieving 96% and 83% reductions in BOD and N-NH_4, respectively, whereas the total GHG emissions were estimated at 38.83 kg $CO_{2\text{-eq}}$/d [47]. A subsurface flow CW system was implemented for blackwater and greywater treatment coming from a rural area in Southern Brazil and showed an impressive COD, BOD, TKN, N—NH_3, and total P reduction, as described by Lutterbeck et al. [48]. The LCA revealed GHG emissions of 1.33×10^3 kg $CO_{2\ eq}$. Finally, Garfí et al. examined the performance of a combined VFCW and HFCW system for the treatment of sewage produced by small rural areas [49]. The BOD reduction was 89%, the capital cost was 210.36 EUR/p.e., the operational and maintenance costs were 0.4 EUR/m^3, and the GHG emissions were ca. 990 g CO_2/m^3_{water}.

2.5. Microalgae

WWTPs based on microalgae have gained significant attention since they combine environmentally friendly tertiary treatment technology with enhanced biomass production [50]. Two different microalgae-based WWTPs were compared considering their performance in agro-industrial sewage treatment, as reported by Magalhães and co-workers [50]. Both proposed WWTPs were examined for wastewater treatment, as well as microalgae biomass production. The first system was a bubble column photobioreactor (PBR), whereas the second was a high-rate pond (HRP). The COD reduction of the former was 54.3%, whereas the reduction percentage of the latter was 47.7%. The removal rates of N-NH_4 and P were found to be complete in the case of the PBR and were 59.5% and 100%, respectively, in the case of the HPR, highlighting the superior performance of the PBR. Silambarasan et al. studied the performance of coupling microalgae (*Scenedesmus* sp. and *Chlorella* sp.) with lipid augmentation in order to remove nutrients from domestic sewage [51]. The reduction percentages of COD, TOC, TN, N-NH_4^+, N-NO_3^-, and PO_4^{3-} were estimated at 83%, 86%, 94%, 98%, 96%, and 95%, respectively.

Marangon et al. compared the environmental impact of two different scenarios for domestic sewage treatment [52]. According to the first scenario, a high-rate algal pond (HRAP) treatment system resulted in 0.1 kg CO_{2eq} GHG emissions, whereas following the second scenario based on a hybrid reactor formed by an HRAP and a BR, the emissions corresponded to 0.19 kg CO_{2eq}. Li and co-workers investigated the treatment of municipal sewage by a non-separated nutrient resource derived from municipal wastewater and integrated in order to facilitate biofuel production from microalgae [53]. The GHG emitted from the proposed process were 20,881 kg CO_{2eq}/y.

2.6. Ozonation

Ozonation was used for wastewater treatment purposes in 1906 in Paris, France [54]. In ozonation processes, ozone reacts with organic contaminants and degrades them but can form intermediary toxic products, and its low water solubility leads to low process efficiency [54]. This process has been examined in the literature as a tertiary treatment method, producing great results for the microbial load and the pharmaceuticals removed, but did not significantly affect the nutrient content of the secondary effluent. More specifically, Lamba and Ahammad [55], reported a 4 log reduction in *coliforms*, whereas Nasuhoglu et al. [56], Shi et al. [57], and Maniakova et al. [58], reported a 2.2- to 5.3-log reduction in *coliforms*, *E. coli*, *Salmonella*, and *Enterococcus*. Regarding pharmaceutical removal, Liu et al. [59] found that ozonation was capable of removing a range of pharmaceuticals by 5 to 80%. Antoniou et al. [60] also reported a removal of 70 to 100% for carbamazepine, naproxen, beclomethasone, and memantine. Regarding the treatment costs, 0.03 EUR/m^3 has been reported in the literature [61], with a global warming potential of 0.025 to 0.3 kg CO_2 eq. m^{-3}.

2.7. Photo-Fenton

Among the various advanced oxidation processes (AOPs), the Fenton oxidation process has gained significant attention, mostly because of its low operational costs [62]. The photo-Fenton process was employed in order to be studied for the removal of sulfamethazine, resulting in complete degradation of this antibiotic, whereas the TOC reduction was 56%, as reported by Pérez-Moya et al. [62]. A 100% elimination of amoxicillin and an 81% TOC removal were obtained by implementing the photo-Fenton process as presented by Trovó et al. [63]. The degradation of 22 micropollutants present in municipal wastewater was investigated by employing a photo-Fenton (UV254/H_2O_2/Fe) process, as shown by De la Cruz et al. [19]. The average percentage concerning the removal of all 22 pollutants was 80%.

The solar photo-Fenton process was quite effective for ARB and ARG removal from urban sewage, as mentioned by Giannakis et al. [64]. Various Phs, namely ofloxacin, carbamazepine, flumequine, ibuprofen, and sulfamethoxazole, were found to be almost completely removed from municipal sewage when coupling NF and solar photo-Fenton technology, as reported by Miralles-Cuevas et al. [65]. Quite interesting is the work of Elmolla and Chaudhuri, who combined the photo-Fenton process with a sequencing batch reactor (SBR) to study performance concerning the treatment of antibiotic wastewater [66]. They achieved an 89% reduction in soluble COD.

Reductions in the Phs present in wastewater produced by a Spanish pharmaceutical industrial unit were investigated by considering the environmental impact of employing an LCA over heterogeneous and homogeneous Fenton processes, as reported by Rodríguez and co-workers [67]. The heterogeneous Fenton process exhibited lower GHG emissions of 0.04 kg CO_{2eq}. An LCA study was performed by Pesqueira et al. of various solar-based treatments including solar circumneutral photo-Fenton (SPF) [68]. The GHG emissions of the latter were estimated at 0.331 kg CO_{2eq}. The emitted GHG of 554 kg $CO_{2eq}/1000$ m^3 as it concerns the operation of a solar photo-Fenton process at acidic pH for municipal wastewater treatment were found to be lower compared to systems operating at neutral pH, as presented by Gallego-Schmid and co-workers [69].

A semi-industrial solar photo-Fenton reactor was investigated by Foteinis et al. concerning the environmental sustainability of the proposed process over real wastewater effluent derived from a pharmaceutical laboratory [70]. The TOC reduction was found to be 79%, whereas the GHG emissions obtained from the LCA corresponded to 2.71 kg CO_{2eq} m^{-3}. Photo-Fenton processes in compound parabolic concentrator-type solar reactors have attracted significant interest recently, which is mostly attributed to their enhanced performance over the degradation of recalcitrant pollutants [71]. As mentioned in the analysis performed by Belalcázar-Saldarriaga and co-workers of the above-mentioned process employed for the degradation of acid orange 52 dye (AO52), the obtained COD and TOC reduction percentages were 55% and 35%, respectively, and the GHG emitted from the system operation were 0.762 kg CO_{2eq}/m^3 wastewater [71].

3. Benchmarking of Tertiary Wastewater Treatment Technologies

The data collected from the literature regarding the aforementioned methods are presented in Appendix A and summarized in Table 1. As can be observed in Table 1, the different tertiary treatment methods exhibited different strengths and weaknesses in the examined aspects. The physicochemical and advanced oxidation methods had a significant impact on the microbial load of the secondary effluent, whereas the biological methods did not exhibit microbial load reductions (or this was not reported at all in the examined literature). On the other hand, the biological processes were the only ones with significant nutrient removal through their assimilation with the produced biomass. Regarding the presence of pharmaceuticals, all the examined tertiary methods showed significant reductions. Regarding the economic aspect, the simpler (and more widespread) methods, such as chlorination, UN irradiation, and ozonation, had the lowest treatment costs. Finally, the GWP of the examined methods seems to follow a similar trend to their energy consumption, with the energy-intensive processes, i.e., membrane filtration, exhibiting the highest GHG emissions.

Table 1. Average technical economic and life cycle performances of the examined tertiary treatment methods.

Category	Treatment Method	Technical			Economic (Cost EUR/m^3)	Life Cycle (kg CO_2 eq. m^{-3})
		Microbial (Log Reduction)	Nutrients (% Reduction)	Pharmaceuticals (% Reduction)		
Physicochemical	Chlorination	2.14	0	42	0.004	0.040
	UV	2.92	0	53	0.004	0.086
	Membrane filtration	3.50	6	70	0.614	0.754
Biological	Constructed wetlands	0.87	53	57	0.784	0.511
	Microalgae	0.00	77	73	0.291	0.468
Advanced oxidation	Ozonation	3.18	0	63	0.030	0.219
	Photo-Fenton	4.93	0	84	0.405	0.549

In order to facilitate a comparison of the different tertiary treatment methods investigated, a scoring methodology was implemented, with a score given for the three aspects (technical, economic, and life cycle) for each treatment method. In this way, the performance of each tertiary treatment method could be put into perspective and easily compared with that of the other examined processes. The scoring system ranged from 0% (worst score) to 100% (best score), with different normalization steps being needed for each aspect. Here, it should be noted that despite making the results of the literature review more comprehensible, the necessary assumptions for the scoring step, such as using data for the removal of specific pathogens or pharmaceuticals for each tertiary treatment method, increased the uncertainty. For the highest accuracy, the reader should refer to the raw data extracted from the literature.

Regarding the technical aspect, the three subcategories of microbial load, nutrient, and pharmaceutical reductions were considered to contribute equally to the average technical aspect score. After normalizing the microbial reduction data by dividing by the highest value, an average reduction was calculated for each method using the microbial, nutrient, and pharmaceutical reductions. Finally, in order to have the technical aspect scores ranging from zero to one hundred, the average reduction values were normalized a second time by subtracting the lowest average reduction value and dividing by the highest average reduction value.

Because the economic and life cycle aspects had a negative impact on the evaluation of the processes, the values used for scoring were the comparison of the methods and the worst results for each criterion. For example, constructed wetlands had the highest treatment cost amongst the processes examined, whereas chlorination and UV had the lowest. The value used for scoring the economic aspect of microalgae was 0.784 − 0.291 = 0.493 EUR/m^3, which was then normalized by dividing by the difference between the highest value and the minimum value among the processes. This way, the worst-performing method had a score of zero and the best-performing method had a score of one hundred after normalization. The three obtained scores for each aspect examined are presented in Figure 1.

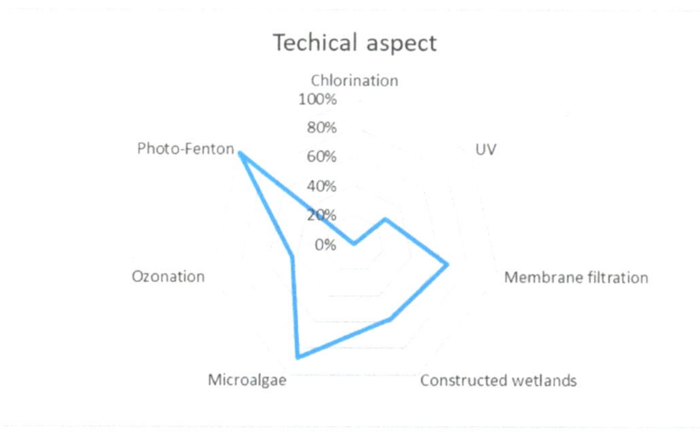

(a)

Figure 1. *Cont.*

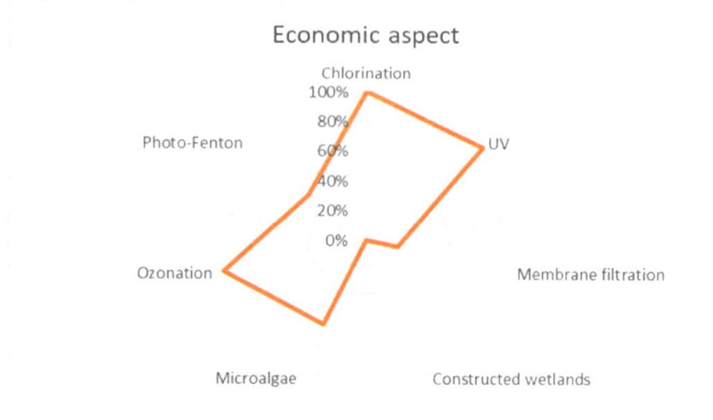

(**b**)

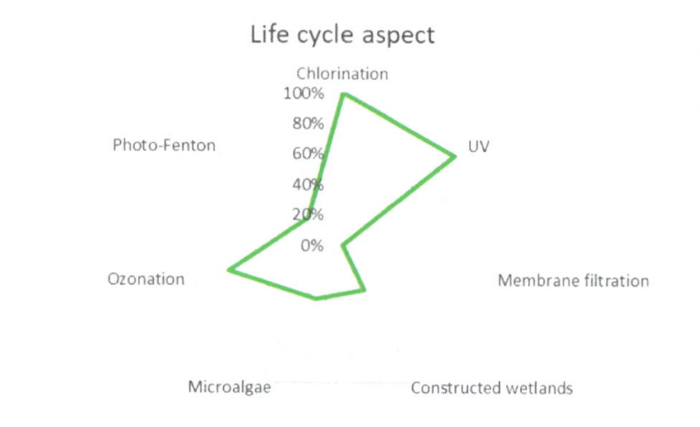

(**c**)

Figure 1. Normalized scores for the three aspects examined for each tertiary treatment method; (**a**) technical aspect, (**b**) economic aspect, (**c**) life cycle aspect.

Because of the complex nature of choosing the most appropriate tertiary treatment method, since each method had certain strengths and weaknesses, the weight given to each of the three examined aspects was decided. In order to facilitate this process, a ternary diagram was constructed, with its three axes corresponding to the different weights to be given to each aspect (Figure 2). Each point in Figure 2 denotes the method with the highest weighted score for the three corresponding weight values. The ternary diagram was constructed by calculating the overall score of all the examined methods at each point and indicating the tertiary method with the highest overall score. The overall score for each method at each point was calculated using the three weights for the three examined aspects (corresponding to the coordinates of the points on the diagram) and the score for each aspect (as depicted in Figure 1). For example, for the points 0.2, 0.6, 0.2 (technical, economic, and life cycle) in the ternary plot, the overall score of all the methods was calculated using the formula 0.2 $\times$ Technical aspect score + 0.6 $\times$ Economic aspect score + 0.2 $\times$ Life cycle aspect score. The resulting scores for all the methods at this point in the plot were compared and the UV method was proven to have the highest overall score. This process was repeated for all the points in Figure 2 in increments of 0.025, calculating a total of 860 sets of overall scores. In this way, the best-performing (in terms of overall score) methods were determined and are depicted in Figure 2. The fact that not all the examined tertiary treatment methods appear in the final plot corresponds to the fact that the missing methods did not exhibit the highest overall score amongst the examined methods at any point in the plot. The final plot can be used to indicate the tertiary method with the highest overall score for a set of weights determined by the user, corresponding to the importance of the technical, economic, and life cycle aspects of the application under investigation.

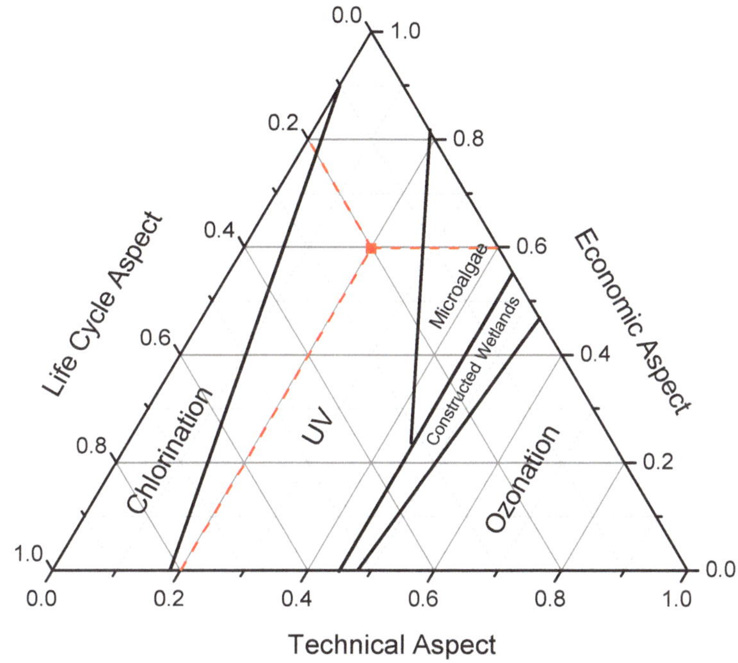

Figure 2. Ternary diagram for the evaluation of tertiary treatment methods, with an example of 0.2, 0.6, and 0.2 weights for the technical, economic, and life cycle aspects, respectively.

4. Conclusions

Many options for tertiary treatment have been examined in the literature, with each treatment process exhibiting different positive and negative points. The first step of the analysis presented herein was the aggregation of data regarding the technical, economic, and environmental impact of each process. This step had the fewest assumptions, as the data were reported as extracted from the literature, but a further aggregation of the results could facilitate a comparison of the tertiary treatment methods examined. As a next step, the data were normalized (assuming the equivalent significance of microbial load, nutrient, and pharmaceutical reductions) and scored. Despite the fact that this process facilitated a comparison of the examined methods, some applications may have attributed higher importance to one technical aspect, for example, when the water recipient was prone to eutrophication or if the sewage derived from a hospital with a high pharmaceutical concentration. The final step was the construction of the ternary diagram illustrating the tertiary treatment method with the highest weighted score for each weight set chosen by the user. This tool can be used to quickly illustrate the strengths and weaknesses of the examined methods and also for decision making in a typical tertiary treatment process, but as previously stated, when specific challenges need to be overcome, directly examining the data aggregated from the literature would be more suitable. UV, ozonation, and chlorination dominated the optimal tertiary treatment methods according to Figure 2, followed by microalgae cultivation and constructed wetlands. The most important advantage of UV and chlorination appears to be their lower treatment costs and low GHG emissions, whereas ozonation exhibited increased efficiency in microbial load and pharmaceutical reductions with moderate treatment costs. The increased treatment costs and GHG emissions of membrane filtration and photo-Fenton did not allow for their assessment as optimal tertiary methods under any set of weights, as seen in Figure 2. It should be noted that these results were dependent on the implemented methodology, and the high efficiency of both the membrane filtration and photo-Fenton methods in the removal of pathogens and pharmaceuticals can be of high value in cases where increased treatment efficiency is required. Through the analysis presented herein, the complementarity of different tertiary treatment methods is illustrated, facilitating their combination in multi-step processes for the removal of all contaminants (i.e., combining biological and physicochemical or advanced oxidation methods).

Author Contributions: Conceptualization, D.P.Z. and G.B.; methodology, D.P.Z.; validation, G.B.; investigation, D.P.Z. and G.B.; data curation, D.P.Z. and G.B.; writing—original draft preparation, D.P.Z. and G.B.; writing—review and editing, D.P.Z. and G.B.; All authors have read and agreed to the published version of the manuscript.

Funding: This research received no external funding.

Institutional Review Board Statement: Not applicable.

Data Availability Statement: Data are contained within the article in Appendix A.

Conflicts of Interest: The authors declare no conflict of interest.

Abbreviations

AO52	acid orange 52
AOPs	advance oxidation processes
ARB	anaerobic resistant bacteria
ARG	antibiotic resistance genes
BOD	biochemical oxygen demand
BR	biofilm reactor
COD	chemical oxygen demand
CW	constructed wetlands
EC	emerging contaminants
GHG	greenhouse gases
GO	graphene oxide
HFCW	horizontal flow constructed wetlands
HRAP	high-rate algal pond
HRP	high-rate pond
HSSF	horizontal subsurface flow
LCC	life cycle costing
MF	microfiltration
MFCs	microbial fuel cells
MWCNTs	multi-walled carbon nanotubes
NF	nanofiltration
PBR	photobioreactor
PVDF	polyvinylidene fluoride
RO	reverse osmosis
SBR	sequencing batch reactor
SPF	solar photo-Fenton
TOC	total organic carbon
UF	ultrafiltration
UV	ultraviolet
VFCW	vertical flow constructed wetlands

Appendix A

Table A1. Literature data used for scoring the chlorination method.

Process Information	Technical			Economic (Cost EUR/m^3)	Life Cycle (kg CO_2 eq./m^3)	Ref.
	Microbial (Log Reduction)	Nutrients (% Reduction)	Pharmaceuticals (% Reduction)			
Sulfamethoxazole 220 ng/L			73			[10]
Ciprofloxacin 153 ng/L			66			[10]
Norfloxacin 92 ng/L			50			[10]
Tetracycline 86 ng/L			39			[10]
Trimethoprim 155 ng/L			65			[10]
Erythromycin 273 ng/L			43			[10]
Diclofenac 40 μg/L			97			[72]
Ibuprofen 40 μg/L			0			[72]
Clofibric acid 40 μg/L			5			[72]
Naproxen 40 μg/L			11			[72]
Gemfibrozil 40 μg/L			45			[72]
Mefenamic acid 40 μg/L			12			[72]
E. coli 3.7 Log CFU/100 mL	2.5					[8]
E. coli 4.34 Log CFU/100 mL	2.57					[9]
Enterococci 3.46 Log CFU/100 mL	1.18					[9]
Fecal coliforms 4.57 Log CFU/100 mL	2.34					[9]
F-specific coliphage 2.33 Log CFU/100 mL	0.71					[9]
Somatic coliphage 3.92 Log CFU/100 mL	1.68					[9]
Adenovirus 0.97 Log CFU/100 mL	0.81					[9]
Norovirus 0.74 Log CFU/100 mL	0.74					[9]
Coliforms 4 Log CFU/100 mL	4					[55]
Antib. Resist. Genes 6 Log	1.97					[13]
E. coli 7 Log CFU/mL	5					[57]
				0.0003		[11]
				0.005		[12]
				0.006		[13]
					0.046	[73]
					0.007	[15]
					0.004	[14]

Table A2. Literature data used for scoring the UV method.

Process Information	Technical			Economic (Cost EUR/m^3)	Life Cycle (kg CO_2 eq./m^3)	Ref.
	Microbial (Log Reduction)	Nutrients (% Reduction)	Pharmaceuticals (% Reduction)			
E. coli 5 × 10^5 CFU/100 mL	4					[74]
E. coli 7.7 Log × 10 CFU/L	3.82					[9]
Enterococci 8.56 Log × 10 CFU/L	3.38					[9]
Fecal coliforms 8.26 Log × 10 CFU/L	3.89					[9]
F-specific coliphage 6.4 Log × 10 CFU/L	1.17					[9]
Somatic coliphage 7.36 Log × 10 CFU/L	2.98					[9]
Adenovirus 2.73 Log gc/L	0.24					[9]
Coliforms 5 Log CFU/mL	4					[55]
Antib. Resist. Genes 6 Log	1					[13]
Antib. Resist. Genes 5 Log copies/L	2.5					[20]
E. coli 2 × 10^7 CFU/mL	5.1					[57]
Sulfamethoxazole 250 ng/L			100			[75]
Trimethoprim 90 ng/L			100			[75]
Erythromycin 200 ng/L			100			[75]
Acetaminophen 0.1 mM, caffeine 0.12 mM, antipyrine 0.05 mM, doxycycline 0.03 mM, ketorolac 0.05 mM			100			[76]

Table A2. *Cont.*

Process Information	Technical			Economic (Cost EUR/m^3)	Life Cycle (kg CO_2 eq./m^3)	Ref.
	Microbial (Log Reduction)	Nutrients (% Reduction)	Pharmaceuticals (% Reduction)			
Atrazine diuron, alachlor, pentachlorophenol 1 mg/L			72			[76]
Boldenone 6.57 µM			98			[76]
BPA 60 µM			22			[76]
Butylparaben 8 × 10^{-5} M			97			[76]
Carbamazepine 3 µM			52			[76]
Chlorfenvinphos			91			[76]
Ciprofloxacin			99			[76]
Chloromycetin 10 mg/L			80			[76]
Clofibric acid 10 mg/L			98			[76]
Cyclophosphamide 10 µg/L			28			[76]
Cytarabine 10 mg/L			10			[76]
Diatrizoate 50 µM			97			[76]
Diclofenac 20 mg/L			74			[76]
Diphenhydramine 5 µM			26			[76]
Doxycycline 5 × 10^{-5} M			27			[76]
E1 20 mg/L			69			[76]
E2 20 mg/L			59			[76]
EE2			37			[76]
Hydrochlorothiazide 1 µM			59			[76]
Ibuprofen 10^{-4} M			74			[76]
Iopromide			53			[76]
Iohexol 3 µM			12			[76]
Irinotecan 10 µg/L			18			[76]
Isoproturon 1 mg/L			12			[76]
Ketoprofen 50 µM			99			[76]
Mefenamic acid 5.5 Log M			56			[76]
Melatonin 20 mg/L			32			[76]
Metoprolol 5 × 10^{-4} M			69			[76]
Metronidazole 6 µM			55			[76]
Naproxen 3 µM			65			[76]
NDMA 1 mM			100			[76]
Norfloxacin 5 × 10^{-5} M			55			[76]
Oxtetracycline			93			[76]
Phenazone 5 µM			96			[76]
Phenytion 5 µM			88			[76]
Primidone 50 µM			9			[76]
Propranolol 100 mg/L			61			[76]
Sulfadimethoxine 3.2 mM			99			[76]
Sulfamethoxazole 10 mg/L			83			[76]
Tamoxifen 10 µg/L			43			[76]
TCE 8.14 × 10^{-3} mol/L			95			[76]
Tibetene 0.03 mM			87			[76]
Bezafibrate 112 ng/L			0			[19]
Metformin 1736 ng/L			27			[19]
Carbamazepine 333 ng/L			48			[19]
Gabapentin 1508 ng/L			0			[19]
Diclofenac 925 ng/L			96			[19]
Ketoprofen 40 ng/L			97			[19]
Naproxen 372 ng/L			70			[19]
Primidone 65 ng/L			3			[19]
Atenolol 320 ng/L			0			[19]
Metoprolol 255 ng/L			0			[19]
Ciprofloxacin 72 ng/L			56			[19]
Clarithromycin 187 ng/L			10			[19]
Sulfamethoxazole 355 ng/L			3			[19]
Trimethoprim 31 ng/L			0			[19]
Iohexol 4313 ng/L			16			[19]
Iomeprol 5806 ng/L			0			[19]
Benzotriazole 6736 ng/L			18			[19]
Atrazin 25 ng/L			58			[19]
Isoproturon 4 ng/L			0			[19]
Mecoprop 365 ng/L			0			[19]
Terbutryn 23 ng/L			39			[19]
				0.00001		[11]

Table A2. *Cont.*

Process Information	Technical			Economic (Cost EUR/m^3)	Life Cycle (kg CO_2 eq./m^3)	Ref.
	Microbial (Log Reduction)	Nutrients (% Reduction)	Pharmaceuticals (% Reduction)			
				0.00644		[13]
				0.0063		[13]
					0.013	[73]
					0.026	[15]
					0.22	[77]

Table A3. Literature data used for scoring the membrane filtration method.

Process Information	Technical Aspect			Economic Aspect (Cost EUR/m^3)	Life Cycle (kg CO_2 eq./m^3)	Ref.
	Microbial (Log Reduction)	Nutrients (% Reduction)	Pharmaceuticals (% Reduction)			
UF enterococcus 1.87×10^5 CFU/100 mL	5					[6]
UF other coliforms 5.05×10^4 CFU/100 mL	2					[6]
UF N 3.62 mg/L		10				[6]
UF P 1.86 mg/L		9				[6]
UF K 16.15 mg/L		0				[6]
NF-90 2 μg/L			73			[78]
NF-200 neutral PhACs 65 μg/L			70			[78]
NF-200 ionic PhACs 65 μg/L			94			[78]
NF-90 neutral 65 μg/L			97			[78]
NF-90 ionic 65 μg/L			99			[78]
NF-90 65 μg/L			73			[78]
UF 2000 0.5 mg/L			70			[78]
UF-NF90 750 μg/L			50			[78]
NF 270 RO 2 μg/L			95			[78]
NF 150-RO 100 ng/L			95			[78]
NF 200 100 ng/L			80			[78]
UF 8000-NF 600 10 ng/L			60			[78]
UF 8000 10 ng/L			30			[78]
NF 90-RO 10 mg/L			99			[78]
NF 200 21 ng/L			100			[78]
NF 270 10 mg L			60			[78]
NF270 800 μg/L			58			[78]
NF90 750 μg/L			97			[78]
RO 0.55 mg/L			100			[78]
NF90 10 mg/L			90			[78]
NF270 10 mg/L			61			[78]
NF90 0.5 mg/L			98			[78]
NF270 0.5 mg/L			71			[78]
RO 0.5 mg/L			89			[78]
NF90 5400 μg/L			77			[78]
NF270 5400 μg/L			58			[78]
RO 5400 μg/L			93			[78]
UF-Atenolol 778 ng/L			0			[22]
UF-Bezafibrate 208 ng/L			21			[22]
UF-Caffeine 17,725 ng/L			0			[22]
UF-Fenofibric acid 139 ng/L			0			[22]
UF-Furosemide 1302 ng/L			17			[22]
UF-Gemfibrozil 18,504 ng/L			71			[22]
UF-Hydrochlorothiazide 16,628 ng/L			90			[22]
UF-Ibuprofen 2514 ng/L			1			[22]
UF-4-AAA 7364 ng/L			0			[22]
UF-Naproxen 2672 ng/L			12			[22]
UF-Nicotine 10,954 ng/L			63			[22]
UF-Ofloxacin 94 ng/L			0			[22]
RO-Atenolol 1044 ng/L			100			[22]
RO-Bezafibrate 164 ng/L			100			[22]
RO-Caffeine 6288 ng/L			99			[22]
RO-Fenofibric acid 194 ng/L			100			[22]

Table A3. *Cont.*

Process Information	Technical Aspect			Economic Aspect (Cost EUR/m^3)	Life Cycle (kg CO_2 eq./m^3)	Ref.
	Microbial (Log Reduction)	Nutrients (% Reduction)	Pharmaceuticals (% Reduction)			
RO-Furosemide 811 ng/L			100			[22]
RO-Gemfibrozil 1035 ng/L			99			[22]
RO-Hydrochlorothiazide 239 ng/L			95			[22]
RO-Ibuprofen 574 ng/L			97			[22]
RO-4-AAA 4472 ng/L			99			[22]
RO-Naproxen 2583 ng/L			98			[22]
RO-Nicotine 75 ng/L			76			[22]
RO-Ofloxacin 87 ng/L			95			[22]
Including RO				0.46		[27]
FO-NF				0.96		[27]
UF-RO				0.4		[27]
UF				0.45		[27]
UF-RO				0.8		[27]
UF-RO					2.32	[29]
NF					0.2	[28]
UF					0.25	[77]
UF					0.40	[73]
MF-RO					0.89	[73]
UF-RO					0.91	[73]

Table A4. Literature data used for scoring the constructed wetlands method.

Process Information	Technical Aspect			Economic Aspect (Cost EUR/m^3)	Life Cycle (kg CO_2 eq./m^3)	Ref.
	Microbial (Log Reduction)	Nutrients (% Reduction)	Pharmaceuticals (% Reduction)			
16 s rDNA, intI1, and tet genes	1.78					[34]
14 antibiotic resistance genes	0.50					[36]
ARGs 8–9 Log of copies/mL	0.49					[40]
N 23.4 mg/L		5				[31]
P 0.2 mg/L		0				[31]
N 29.3 mg/L		35				[31]
P 0.2 mg/L		50				[31]
N 17.3 mg/L		39				[31]
P 0.2 mg/L		50				[31]
N 1.39 mg/L		66				[33]
P 3 mg/L		46				[33]
N 84.4 mg/L		63				[34]
P 28.2 mg/L		92				[34]
N 35 mg/L		46				[40]
N 72 mg/L		99				[79]
P 11.7 mg/L		97				[79]
65 pharmaceuticals 4.3 µg/L			64			[31]
55 pharmaceuticals 300 ng/L			43			[31]
53 pharmaceuticals			50			[31]
56 pharmaceuticals 190 ng/L			32			[31]
6 pharmaceuticals 7.6–150 µg/L			93			[80]
Antibiotics 300 ng/L			58			[40]
Pharmaceuticals 50–200 ng/L			59			[79]
				1.224		[45]
				0.729		[45]
				0.4		[49]
					0.129	[73]
					0.432	[42]
					0.646	[42]
					0.911	[42]
					0.5	[42]
					0.26	[47]
					0.7	[49]

Table A5. Literature data used for scoring the microalgae.

Process Information	Technical			Economic (Cost EUR/m³)	Life Cycle (kg CO_2 eq./m³)	Ref.
	Microbial (Log Reduction)	Nutrients (% Reduction)	Pharmaceuticals (% Reduction)			
N 40 mg/L		55				[81]
P 80 mg/L		15				[81]
N 52 mg/L		82				[82]
P 8.5 mg/L		95				[82]
N 18 mg/L		100				[83]
P 1.4 mg/L		84				[83]
N 46 mg/L		94				[51]
P 5.5 mg/L		95				[51]
Metronidazole 5 µM			100			[84]
Florfenicol 46 mg/L			97			[84]
Enrofloxacin 1 mg/L			23			[84]
Tetracycline 100 µg/L			99			[84]
Methyl parathion 20 mg/L			80			[84]
Trimethoprim, Sulfamethoxazole, Triclosan 1.6 ng/L, 360 ng/L, 8 ng/L			44			[84]
7-amino cephalosporanic acid 100 mg/L			70			[84]
Cefradine 100 mg/L			94			[84]
β-estradiol			93			[84]
17 α-estradiol, 17 β-estradiol, Estrone, Estriol 5 µg/L			90			[84]
Sulfathiazole, Sulfapyridine, Sulfamethazine, Sulfamethoxazole, Tetracycline, Oxytetracycline 200 µg/L			47			[84]
Norfloxacin mg/L			37			[84]
				0.42		[49]
				0.162		[85]
					0.6	[49]
					0.336	[86]

Table A6. Literature data used for scoring the ozonation method.

Process Information	Technical			Economic (Cost EUR/m³)	Life Cycle (kg CO_2 eq./m³)	Ref.
	Microbial (Log Reduction)	Nutrients (% Reduction)	Pharmaceuticals (% Reduction)			
Coliforms 4 Log CFU/100 mL	4					[55]
Coliforms 5 Log MPN/100 mL	2.5					[56]
Coliforms 5 Log MPN/100 mL	2.2					[56]
Coliforms 7 Log MPN/100 mL	4.8					[56]
E. coli 7.3 Log CFU/mL	5.3					[57]
E. coli 4 Log CFU/mL	2.2					[58]
Salmonella 2.9 Log CFU/mL	2.2					[58]
Enterococcus 3 Log CFU/mL	2.2					[58]
Carbamazepine			75			[59]
Alachlor			20			[59]
Bisphenol A			60			[59]
Atrazine			5			[59]
Pentachlorophenol			35			[59]
17-α thinylestradiol			80			[59]
Carbamazepine 1 µg/L			100			[60]
Naproxen 1 µg/L			100			[60]
Beclomethasone 1 µg/L			70			[60]
Memantine 1 µg/L			80			[60]
				0.03		[61]
				0.03		[61]
					0.025	[87]
					0.25	[88]
					0.3	[28]
					0.3	[89]

Table A7. Literature data used for scoring the Photo-Fenton method.

Process Information	Technical			Economic (Cost EUR/m^3)	Life Cycle (kg CO_2 eq./m^3)	Ref.
	Microbial (Log Reduction)	Nutrients (% Reduction)	Pharmaceuticals (% Reduction)			
S. aureus 6 Log CFU/mL	6					[64]
MRSA ATCC 29213 6 Log CFU/mL	6					[64]
E. coli 6 Log CFU/mL	6					[64]
K. pneumoniae 6 Log CFU/mL	6					[64]
MSSA 1112 6 Log CFU/mL	6					[64]
MSSA 1112 RifR 6 Log CFU/mL	6					[64]
MSSA 1112 CipR 6 Log CFU/mL	6					[64]
MSSA 133 6 Log CFU/mL	6					[64]
MSSA 133 CipR 6 Log CFU/mL	6					[64]
MRSA PC1 6 Log CFU/mL	6					[64]
VISA PC# 6 Log CFU/mL	6					[64]
E. coli 4 Log CFU/mL	1					[58]
Salmonella 2.9 Log CFU/mL	2					[58]
Enterococcus 3 Log CFU/mL	0					[58]
Sulfamethazine 50 mg/L			100			[62]
Amoxicillin 50 mg/L			100			[63]
Bezafibrate 112 ng/L			0			[19]
Gemfibrozil 9 ng/L			96			[19]
Metformin 1736 ng/L			63			[19]
Carbamazepine 333 ng/L			94			[19]
Gabapentin 1508 ng/L			77			[19]
Diclofenac 925 ng/L			100			[19]
Ketoprofen 40 ng/L			97			[19]
Naproxen372 ng/L			97			[19]
Primidone 65 ng/L			77			[19]
Atenolol 320 ng/L			87			[19]
Metoprolol 255 ng/L			90			[19]
Ciprofloxacin 72 ng/L			61			[19]
Clarithromycin 187 ng/L			76			[19]
Sulfamethoxazole 355 ng/L			82			[19]
Trimethoprim 31 ng/L			88			[19]
Iohexol 4313 ng/L			94			[19]
Iomeprol 5806 ng/L			87			[19]
Benzotriazole 6736 ng/L			95			[19]
Atrazin 25 ng/L			82			[19]
Isoproturon 4 ng/L			32			[19]
Mecoprop 365 ng/L			93			[19]
Terbutryn 23 ng/L			83			[19]
Ofloxacin 110 µg/L			100			[65]
Carbamazepine130 µg/L			96			[65]
Flumequine 145 µg/L			98			[65]
Ibuprofen 130 µg/L			95			[65]
Sulfamethoxazole 140 µg/L			90			[65]
				0.25		[90]
				0.56		[90]
					0.331	[68]
					0.554	[69]
					0.762	[71]

References

1. Adam, D. How far will global population rise? Researchers can't agree. *Nature* **2021**, *597*, 462–465. [CrossRef] [PubMed]
2. Deaver, J.A.; Kerr, C.A.; Popat, S.C. Primary sludge-based blackwater favors electrical current over methane production in microbial electrochemical cells. *J. Water Process Eng.* **2022**, *47*, 102848. [CrossRef]
3. Wallin, J.; Knutsson, J.; Karpouzoglou, T. A multi-criteria analysis of building level graywater reuse for personal hygiene. *Resour. Conserv. Recycl. Adv.* **2021**, *12*, 200054. [CrossRef]
4. Saliu, T.D.; Ali, J.; Ololade, I.A.; Oladoja, N.A. Preparation and characterization of a decentralized modular yellow water nutrient recovery system. *J. Environ. Manag.* **2020**, *276*, 111345. [CrossRef] [PubMed]
5. Tran, N.H.; Reinhard, M.; Gin, K.Y.-H. Occurrence and fate of emerging contaminants in municipal wastewater treatment plants from different geographical regions-a review. *Water Res.* **2018**, *133*, 182–207. [CrossRef]

6. Zagklis, D.; Katrivesis, F.K.; Sygouni, V.; Tsarouchi, L.; Tsigkou, K.; Kornaros, M.; Paraskeva, C.A. Recovery of Water from Secondary Effluent through Pilot Scale Ultrafiltration Membranes: Implementation at Patras' Wastewater Treatment Plant. *Membranes* **2021**, *11*, 663. [CrossRef]
7. Albolafio, S.; Marín, A.; Allende, A.; García, F.; Simón-Andreu, P.J.; Soler, M.A.; Gil, M.I. Strategies for mitigating chlorinated disinfection byproducts in wastewater treatment plants. *Chemosphere* **2022**, *288*, 132583. [CrossRef]
8. Tombini Decol, L.; López-Gálvez, F.; Truchado, P.; Tondo, E.C.; Gil, M.I.; Allende, A. Suitability of chlorine dioxide as a tertiary treatment for municipal wastewater and use of reclaimed water for overhead irrigation of baby lettuce. *Food Control* **2019**, *96*, 186–193. [CrossRef]
9. Francy, D.S.; Stelzer, E.A.; Bushon, R.N.; Brady, A.M.G.; Williston, A.G.; Riddell, K.R.; Borchardt, M.A.; Spencer, S.K.; Gellner, T.M. Comparative effectiveness of membrane bioreactors, conventional secondary treatment, and chlorine and UV disinfection to remove microorganisms from municipal wastewaters. *Water Res.* **2012**, *46*, 4164–4178. [CrossRef]
10. Li, B.; Zhang, T. Mass flows and removal of antibiotics in two municipal wastewater treatment plants. *Chemosphere* **2011**, *83*, 1284–1289. [CrossRef]
11. Tak, S.; Kumar, A. Chlorination disinfection by-products and comparative cost analysis of chlorination and UV disinfection in sewage treatment plants: Indian scenario. *Environ. Sci. Pollut. Res.* **2017**, *24*, 26269–26278. [CrossRef]
12. Gallego Valero, L.; Moral Pajares, E.; Román Sánchez, I.M.; Sánchez Pérez, J.A. Analysis of Environmental Taxes to Finance Wastewater Treatment in Spain: An Opportunity for Regeneration? *Water* **2018**, *10*, 226. [CrossRef]
13. Zhuang, Y.; Ren, H.; Geng, J.; Zhang, Y.; Zhang, Y.; Ding, L.; Xu, K. Inactivation of antibiotic resistance genes in municipal wastewater by chlorination, ultraviolet, and ozonation disinfection. *Environ. Sci. Pollut. Res.* **2015**, *22*, 7037–7044. [CrossRef] [PubMed]
14. Walsh, T.; Mellor, J. Comparative life cycle assessment of four commonly used point-of-use water treatment technologies. *J. Water Sanit. Hyg. Dev.* **2020**, *10*, 862–873. [CrossRef]
15. Pasqualino, J.C.; Meneses, M.; Castells, F. Life Cycle Assessment of Urban Wastewater Reclamation and Reuse Alternatives. *J. Ind. Ecol.* **2011**, *15*, 49–63. [CrossRef]
16. Zhang, Y.; Zhuang, Y.; Geng, J.; Ren, H.; Xu, K.; Ding, L. Reduction of antibiotic resistance genes in municipal wastewater effluent by advanced oxidation processes. *Sci. Total Environ.* **2016**, *550*, 184–191. [CrossRef]
17. Qiao, R.-P.; Li, N.; Qi, X.-H.; Wang, Q.-S.; Zhuang, Y.-Y. Degradation of microcystin-RR by UV radiation in the presence of hydrogen peroxide. *Toxicon* **2005**, *45*, 745–752. [CrossRef] [PubMed]
18. Chin, A.; Bérubé, P.R. Removal of disinfection by-product precursors with ozone-UV advanced oxidation process. *Water Res.* **2005**, *39*, 2136–2144. [CrossRef] [PubMed]
19. De la Cruz, N.; Esquius, L.; Grandjean, D.; Magnet, A.; Tungler, A.; de Alencastro, L.F.; Pulgarín, C. Degradation of emergent contaminants by UV, UV/H2O2 and neutral photo-Fenton at pilot scale in a domestic wastewater treatment plant. *Water Res.* **2013**, *47*, 5836–5845. [CrossRef]
20. Guo, M.-T.; Yuan, Q.-B.; Yang, J. Ultraviolet reduction of erythromycin and tetracycline resistant heterotrophic bacteria and their resistance genes in municipal wastewater. *Chemosphere* **2013**, *93*, 2864–2868. [CrossRef]
21. Zagklis, D.P.; Papageorgiou, C.S.; Paraskeva, C.A. 18—Valorization of phenolic extracts from *Olea europaea* L. by membrane operations. In *Membrane Engineering in the Circular Economy: Renewable Sources Valorization in Energy and Downstream Processing in Agro-Food Industry*; Iulianelli, A., Cassano, A., Conidi, C., Petrotos, K., Eds.; Elsevier: Amsterdam, The Netherlands, 2022; pp. 495–524. ISBN 978-0-323-85253-1.
22. Urtiaga, A.M.; Pérez, G.; Ibáñez, R.; Ortiz, I. Removal of pharmaceuticals from a WWTP secondary effluent by ultrafiltration/reverse osmosis followed by electrochemical oxidation of the RO concentrate. *Desalination* **2013**, *331*, 26–34. [CrossRef]
23. Cheng, H.; Hong, P.-Y. Removal of Antibiotic-Resistant Bacteria and Antibiotic Resistance Genes Affected by Varying Degrees of Fouling on Anaerobic Microfiltration Membranes. *Environ. Sci. Technol.* **2017**, *51*, 12200–12209. [CrossRef] [PubMed]
24. Ren, S.; Boo, C.; Guo, N.; Wang, S.; Elimelech, M.; Wang, Y. Photocatalytic Reactive Ultrafiltration Membrane for Removal of Antibiotic Resistant Bacteria and Antibiotic Resistance Genes from Wastewater Effluent. *Environ. Sci. Technol.* **2018**, *52*, 8666–8673. [CrossRef] [PubMed]
25. Dolar, D.; Ignjatić Zokić, T.; Košutić, K.; Ašperger, D.; Mutavdžić Pavlović, D. RO/NF membrane treatment of veterinary pharmaceutical wastewater: Comparison of results obtained on a laboratory and a pilot scale. *Environ. Sci. Pollut. Res.* **2012**, *19*, 1033–1042. [CrossRef] [PubMed]
26. Ho, K.C.; Teoh, Y.X.; Teow, Y.H.; Mohammad, A.W. Life cycle assessment (LCA) of electrically-enhanced POME filtration: Environmental impacts of conductive-membrane formulation and process operating parameters. *J. Environ. Manag.* **2021**, *277*, 111434. [CrossRef]
27. Pérez, G.; Gómez, P.; Ortiz, I.; Urtiaga, A. Techno-economic assessment of a membrane-based wastewater reclamation process. *Desalination* **2022**, *522*, 115409. [CrossRef]
28. Zepon Tarpani, R.R.; Azapagic, A. Life cycle environmental impacts of advanced wastewater treatment techniques for removal of pharmaceuticals and personal care products (PPCPs). *J. Environ. Manag.* **2018**, *215*, 258–272. [CrossRef]
29. Teow, Y.H.; Chong, M.T.; Ho, K.C.; Mohammad, A.W. Comparative environmental impact evaluation using life cycle assessment approach: A case study of integrated membrane-filtration system for the treatment of aerobically-digested palm oil mill effluent. *Sustain. Environ. Res.* **2021**, *31*, 15. [CrossRef]

30. Vymazal, J. Constructed Wetlands for Wastewater Treatment. *Water* **2010**, *2*, 530–549. [CrossRef]
31. Breitholtz, M.; Näslund, M.; Stråe, D.; Borg, H.; Grabic, R.; Fick, J. An evaluation of free water surface wetlands as tertiary sewage water treatment of micro-pollutants. *Ecotoxicol. Environ. Saf.* **2012**, *78*, 63–71. [CrossRef]
32. Adrados, B.; Sánchez, O.; Arias, C.A.; Becares, E.; Garrido, L.; Mas, J.; Brix, H.; Morató, J. Microbial communities from different types of natural wastewater treatment systems: Vertical and horizontal flow constructed wetlands and biofilters. *Water Res.* **2014**, *55*, 304–312. [CrossRef] [PubMed]
33. Younger, P.L.; Henderson, R. Synergistic wetland treatment of sewage and mine water: Pollutant removal performance of the first full-scale system. *Water Res.* **2014**, *55*, 74–82. [CrossRef] [PubMed]
34. Huang, X.; Liu, C.; Li, K.; Su, J.; Zhu, G.; Liu, L. Performance of vertical up-flow constructed wetlands on swine wastewater containing tetracyclines and tet genes. *Water Res.* **2015**, *70*, 109–117. [CrossRef] [PubMed]
35. Yi, X.; Tran, N.H.; Yin, T.; He, Y.; Gin, K.Y.-H. Removal of selected PPCPs, EDCs, and antibiotic resistance genes in landfill leachate by a full-scale constructed wetlands system. *Water Res.* **2017**, *121*, 46–60. [CrossRef]
36. Fang, H.; Zhang, Q.; Nie, X.; Chen, B.; Xiao, Y.; Zhou, Q.; Liao, W.; Liang, X. Occurrence and elimination of antibiotic resistance genes in a long-term operation integrated surface flow constructed wetland. *Chemosphere* **2017**, *173*, 99–106. [CrossRef]
37. Chen, J.; Ying, G.-G.; Wei, X.-D.; Liu, Y.-S.; Liu, S.-S.; Hu, L.-X.; He, L.-Y.; Chen, Z.-F.; Chen, F.-R.; Yang, Y.-Q. Removal of antibiotics and antibiotic resistance genes from domestic sewage by constructed wetlands: Effect of flow configuration and plant species. *Sci. Total Environ.* **2016**, *571*, 974–982. [CrossRef]
38. Liu, L.; Liu, C.; Zheng, J.; Huang, X.; Wang, Z.; Liu, Y.; Zhu, G. Elimination of veterinary antibiotics and antibiotic resistance genes from swine wastewater in the vertical flow constructed wetlands. *Chemosphere* **2013**, *91*, 1088–1093. [CrossRef]
39. Nõlvak, H.; Truu, M.; Tiirik, K.; Oopkaup, K.; Sildvee, T.; Kaasik, A.; Mander, Ü.; Truu, J. Dynamics of antibiotic resistance genes and their relationships with system treatment efficiency in a horizontal subsurface flow constructed wetland. *Sci. Total Environ.* **2013**, *461–462*, 636–644. [CrossRef]
40. Chen, J.; Wei, X.-D.; Liu, Y.-S.; Ying, G.-G.; Liu, S.-S.; He, L.-Y.; Su, H.-C.; Hu, L.-X.; Chen, F.-R.; Yang, Y.-Q. Removal of antibiotics and antibiotic resistance genes from domestic sewage by constructed wetlands: Optimization of wetland substrates and hydraulic loading. *Sci. Total Environ.* **2016**, *565*, 240–248. [CrossRef]
41. Huang, X.; Zheng, J.; Liu, C.; Liu, L.; Liu, Y.; Fan, H. Removal of antibiotics and resistance genes from swine wastewater using vertical flow constructed wetlands: Effect of hydraulic flow direction and substrate type. *Chem. Eng. J.* **2017**, *308*, 692–699. [CrossRef]
42. Casas Ledón, Y.; Rivas, A.; López, D.; Vidal, G. Life-cycle greenhouse gas emissions assessment and extended exergy accounting of a horizontal-flow constructed wetland for municipal wastewater treatment: A case study in Chile. *Ecol. Indic.* **2017**, *74*, 130–139. [CrossRef]
43. Corbella, C.; Puigagut, J.; Garfí, M. Life cycle assessment of constructed wetland systems for wastewater treatment coupled with microbial fuel cells. *Sci. Total Environ.* **2017**, *584–585*, 355–362. [CrossRef] [PubMed]
44. Flores, L.; García, J.; Pena, R.; Garfí, M. Constructed wetlands for winery wastewater treatment: A comparative Life Cycle Assessment. *Sci. Total Environ.* **2019**, *659*, 1567–1576. [CrossRef] [PubMed]
45. Resende, J.D.; Nolasco, M.A.; Pacca, S.A. Life cycle assessment and costing of wastewater treatment systems coupled to constructed wetlands. *Resour. Conserv. Recycl.* **2019**, *148*, 170–177. [CrossRef]
46. Lakho, F.H.; Qureshi, A.; Igodt, W.; Le, H.Q.; Depuydt, V.; Rousseau, D.P.L.; Van Hulle, S.W.H. Life cycle assessment of two decentralized water treatment systems combining a constructed wetland and a membrane based drinking water production system. *Resour. Conserv. Recycl.* **2022**, *178*, 106104. [CrossRef]
47. Pan, T.; Zhu, X.-D.; Ye, Y.-P. Estimate of life-cycle greenhouse gas emissions from a vertical subsurface flow constructed wetland and conventional wastewater treatment plants: A case study in China. *Ecol. Eng.* **2011**, *37*, 248–254. [CrossRef]
48. Lutterbeck, C.A.; Kist, L.T.; Lopez, D.R.; Zerwes, F.V.; Machado, Ê.L. Life cycle assessment of integrated wastewater treatment systems with constructed wetlands in rural areas. *J. Clean. Prod.* **2017**, *148*, 527–536. [CrossRef]
49. Garfí, M.; Flores, L.; Ferrer, I. Life Cycle Assessment of wastewater treatment systems for small communities: Activated sludge, constructed wetlands and high rate algal ponds. *J. Clean. Prod.* **2017**, *161*, 211–219. [CrossRef]
50. Magalhães, I.B.; Ferreira, J.; de Castro, J.S.; de Assis, L.R.; Calijuri, M.L. Agro-industrial wastewater-grown microalgae: A techno-environmental assessment of open and closed systems. *Sci. Total Environ.* **2022**, *834*, 155282. [CrossRef]
51. Silambarasan, S.; Logeswari, P.; Sivaramakrishnan, R.; Incharoensakdi, A.; Cornejo, P.; Kamaraj, B.; Chi, N.T.L. Removal of nutrients from domestic wastewater by microalgae coupled to lipid augmentation for biodiesel production and influence of deoiled algal biomass as biofertilizer for Solanum lycopersicum cultivation. *Chemosphere* **2021**, *268*, 129323. [CrossRef]
52. Marangon, B.B.; Calijuri, M.L.; de Siqueira Castro, J.; Assemany, P.P. A life cycle assessment of energy recovery using briquette from wastewater grown microalgae biomass. *J. Environ. Manag.* **2021**, *285*, 112171. [CrossRef] [PubMed]
53. Li, P.; Luo, Y.; Yuan, X. Life cycle and techno-economic assessment of source-separated wastewater-integrated microalgae biofuel production plant: A nutrient organization approach. *Bioresour. Technol.* **2022**, *344*, 126230. [CrossRef] [PubMed]
54. Wang, J.; Chen, H. Catalytic ozonation for water and wastewater treatment: Recent advances and perspective. *Sci. Total Environ.* **2020**, *704*, 135249. [CrossRef] [PubMed]
55. Lamba, M.; Ahammad, S.Z. Performance comparison of secondary and tertiary treatment systems for treating antibiotic resistance. *Water Res.* **2017**, *127*, 172–182. [CrossRef]

56. Nasuhoglu, D.; Isazadeh, S.; Westlund, P.; Neamatallah, S.; Yargeau, V. Chemical, microbial and toxicological assessment of wastewater treatment plant effluents during disinfection by ozonation. *Chem. Eng. J.* **2018**, *346*, 466–476. [CrossRef]
57. Shi, Q.; Chen, Z.; Liu, H.; Lu, Y.; Li, K.; Shi, Y.; Mao, Y.; Hu, H.-Y. Efficient synergistic disinfection by ozone, ultraviolet irradiation and chlorine in secondary effluents. *Sci. Total Environ.* **2021**, *758*, 143641. [CrossRef]
58. Maniakova, G.; Salmerón, I.; Nahim-Granados, S.; Malato, S.; Oller, I.; Rizzo, L.; Polo-López, M.I. Sunlight advanced oxidation processes vs ozonation for wastewater disinfection and safe reclamation. *Sci. Total Environ.* **2021**, *787*, 147531. [CrossRef]
59. Liu, Z.; Chys, M.; Yang, Y.; Demeestere, K.; Van Hulle, S. Oxidation of Trace Organic Contaminants (TrOCs) in Wastewater Effluent with Different Ozone-Based AOPs: Comparison of Ozone Exposure and •OH Formation. *Ind. Eng. Chem. Res.* **2019**, *58*, 8896–8902. [CrossRef]
60. Antoniou, M.G.; Hey, G.; Rodríguez Vega, S.; Spiliotopoulou, A.; Fick, J.; Tysklind, M.; la Cour Jansen, J.; Andersen, H.R. Required ozone doses for removing pharmaceuticals from wastewater effluents. *Sci. Total Environ.* **2013**, *456–457*, 42–49. [CrossRef]
61. Chys, M.; Audenaert, W.T.M.; Stapel, H.; Ried, A.; Wieland, A.; Weemaes, M.; Van Langenhove, H.; Nopens, I.; Demeestere, K.; Van Hulle, S.W.H. Techno-economic assessment of surrogate-based real-time control and monitoring of secondary effluent ozonation at pilot scale. *Chem. Eng. J.* **2018**, *352*, 431–440. [CrossRef]
62. Pérez-Moya, M.; Graells, M.; Castells, G.; Amigó, J.; Ortega, E.; Buhigas, G.; Pérez, L.M.; Mansilla, H.D. Characterization of the degradation performance of the sulfamethazine antibiotic by photo-Fenton process. *Water Res.* **2010**, *44*, 2533–2540. [CrossRef]
63. Trovó, A.G.; Pupo Nogueira, R.F.; Agüera, A.; Fernandez-Alba, A.R.; Malato, S. Degradation of the antibiotic amoxicillin by photo-Fenton process—Chemical and toxicological assessment. *Water Res.* **2011**, *45*, 1394–1402. [CrossRef] [PubMed]
64. Giannakis, S.; Le, T.-T.M.; Entenza, J.M.; Pulgarin, C. Solar photo-Fenton disinfection of 11 antibiotic-resistant bacteria (ARB) and elimination of representative AR genes. Evidence that antibiotic resistance does not imply resistance to oxidative treatment. *Water Res.* **2018**, *143*, 334–345. [CrossRef] [PubMed]
65. Miralles-Cuevas, S.; Oller, I.; Ruiz Aguirre, A.; Sánchez Pérez, J.A.; Malato Rodríguez, S. Removal of pharmaceuticals at microg L−1 by combined nanofiltration and mild solar photo-Fenton. *Chem. Eng. J.* **2014**, *239*, 68–74. [CrossRef]
66. Elmolla, E.S.; Chaudhuri, M. Combined photo-Fenton–SBR process for antibiotic wastewater treatment. *J. Hazard. Mater.* **2011**, *192*, 1418–1426. [CrossRef]
67. Rodríguez, R.; Espada, J.J.; Pariente, M.I.; Melero, J.A.; Martínez, F.; Molina, R. Comparative life cycle assessment (LCA) study of heterogeneous and homogenous Fenton processes for the treatment of pharmaceutical wastewater. *J. Clean. Prod.* **2016**, *124*, 21–29. [CrossRef]
68. Pesqueira, J.F.J.R.; Pereira, M.F.R.; Silva, A.M.T. A life cycle assessment of solar-based treatments (H_2O_2, TiO_2 photocatalysis, circumneutral photo-Fenton) for the removal of organic micropollutants. *Sci. Total Environ.* **2021**, *761*, 143258. [CrossRef]
69. Gallego-Schmid, A.; Tarpani, R.R.Z.; Miralles-Cuevas, S.; Cabrera-Reina, A.; Malato, S.; Azapagic, A. Environmental assessment of solar photo-Fenton processes in combination with nanofiltration for the removal of micro-contaminants from real wastewaters. *Sci. Total Environ.* **2019**, *650*, 2210–2220. [CrossRef]
70. Foteinis, S.; Monteagudo, J.M.; Durán, A.; Chatzisymeon, E. Environmental sustainability of the solar photo-Fenton process for wastewater treatment and pharmaceuticals mineralization at semi-industrial scale. *Sci. Total Environ.* **2018**, *612*, 605–612. [CrossRef]
71. Belalcázar-Saldarriaga, A.; Prato-Garcia, D.; Vasquez-Medrano, R. Photo-Fenton processes in raceway reactors: Technical, economic, and environmental implications during treatment of colored wastewaters. *J. Clean. Prod.* **2018**, *182*, 818–829. [CrossRef]
72. Hey, G.; Ledin, A.; la Cour Jansen, J.; Andersen, H.R. Removal of pharmaceuticals in biologically treated wastewater by chlorine dioxide or peracetic acid. *Environ. Technol.* **2012**, *33*, 1041–1047. [CrossRef] [PubMed]
73. Akhoundi, A.; Nazif, S. Life-cycle assessment of tertiary treatment technologies to treat secondary municipal wastewater for reuse in agricultural irrigation, artificial recharge of groundwater, and industrial usages. *J. Environ. Eng.* **2020**, *146*, 4020031. [CrossRef]
74. Licciardello, F.; Milani, M.; Consoli, S.; Pappalardo, N.; Barbagallo, S.; Cirelli, G. Wastewater tertiary treatment options to match reuse standards in agriculture. *Agric. Water Manag.* **2018**, *210*, 232–242. [CrossRef]
75. Kim, I.; Yamashita, N.; Tanaka, H. Performance of UV and UV/H_2O_2 processes for the removal of pharmaceuticals detected in secondary effluent of a sewage treatment plant in Japan. *J. Hazard. Mater.* **2009**, *166*, 1134–1140. [CrossRef]
76. Yang, W.; Zhou, H.; Cicek, N. Treatment of Organic Micropollutants in Water and Wastewater by UV-Based Processes: A Literature Review. *Crit. Rev. Environ. Sci. Technol.* **2014**, *44*, 1443–1476. [CrossRef]
77. Carré, E.; Beigbeder, J.; Jauzein, V.; Junqua, G.; Lopez-Ferber, M. Life cycle assessment case study: Tertiary treatment process options for wastewater reuse. *Integr. Environ. Assess. Manag.* **2017**, *13*, 1113–1121. [CrossRef]
78. Ganiyu, S.O.; van Hullebusch, E.D.; Cretin, M.; Esposito, G.; Oturan, M.A. Coupling of membrane filtration and advanced oxidation processes for removal of pharmaceutical residues: A critical review. *Sep. Purif. Technol.* **2015**, *156*, 891–914. [CrossRef]
79. Chen, J.; Liu, Y.-S.; Su, H.-C.; Ying, G.-G.; Liu, F.; Liu, S.-S.; He, L.-Y.; Chen, Z.-F.; Yang, Y.-Q.; Chen, F.-R. Removal of antibiotics and antibiotic resistance genes in rural wastewater by an integrated constructed wetland. *Environ. Sci. Pollut. Res.* **2015**, *22*, 1794–1803. [CrossRef]
80. Cardinal, P.; Anderson, J.C.; Carlson, J.C.; Low, J.E.; Challis, J.K.; Beattie, S.A.; Bartel, C.N.; Elliott, A.D.; Montero, O.F.; Lokesh, S.; et al. Macrophytes may not contribute significantly to removal of nutrients, pharmaceuticals, and antibiotic resistance in model surface constructed wetlands. *Sci. Total Environ.* **2014**, *482–483*, 294–304. [CrossRef]

81. Kong, Q.; Li, L.; Martinez, B.; Chen, P.; Ruan, R. Culture of Microalgae Chlamydomonas reinhardtii in Wastewater for Biomass Feedstock Production. *Appl. Biochem. Biotechnol.* **2009**, *160*, 9. [CrossRef]
82. Ansari, F.A.; Ravindran, B.; Gupta, S.K.; Nasr, M.; Rawat, I.; Bux, F. Techno-economic estimation of wastewater phycoremediation and environmental benefits using Scenedesmus obliquus microalgae. *J. Environ. Manag.* **2019**, *240*, 293–302. [CrossRef] [PubMed]
83. Salama, E.-S.; Abou-Shanab, R.A.I.; Kim, J.R.; Lee, S.; Kim, S.-H.; Oh, S.-E.; Kim, H.-C.; Roh, H.-S.; Jeon, B.-H. The effects of salinity on the growth and biochemical properties of Chlamydomonas mexicana GU732420 cultivated in municipal wastewater. *Environ. Technol.* **2014**, *35*, 1491–1498. [CrossRef] [PubMed]
84. Hena, S.; Gutierrez, L.; Croué, J.-P. Removal of pharmaceutical and personal care products (PPCPs) from wastewater using microalgae: A review. *J. Hazard. Mater.* **2021**, *403*, 124041. [CrossRef] [PubMed]
85. Chaudry, S. Integrating Microalgae Cultivation with Wastewater Treatment: A Peek into Economics. *Appl. Biochem. Biotechnol.* **2021**, *193*, 3395–3406. [CrossRef]
86. De Schneider, R.C.S.; de Moura Lima, M.; Hoeltz, M.; de Farias Neves, F.; John, D.K.; de Azevedo, A. Life cycle assessment of microalgae production in a raceway pond with alternative culture media. *Algal Res.* **2018**, *32*, 280–292. [CrossRef]
87. Igos, E.; Mailler, R.; Guillossou, R.; Rocher, V.; Gasperi, J. Life cycle assessment of powder and micro-grain activated carbon in a fluidized bed to remove micropollutants from wastewater and their comparison with ozonation. *J. Clean. Prod.* **2021**, *287*, 125067. [CrossRef]
88. Arzate, S.; Pfister, S.; Oberschelp, C.; Sánchez-Pérez, J.A. Environmental impacts of an advanced oxidation process as tertiary treatment in a wastewater treatment plant. *Sci. Total Environ.* **2019**, *694*, 133572. [CrossRef]
89. Muñoz, I.; Rodríguez, A.; Rosal, R.; Fernández-Alba, A.R. Life Cycle Assessment of urban wastewater reuse with ozonation as tertiary treatment: A focus on toxicity-related impacts. *Sci. Total Environ.* **2009**, *407*, 1245–1256. [CrossRef]
90. Sánchez Pérez, J.A.; Arzate, S.; Soriano-Molina, P.; García Sánchez, J.L.; Casas López, J.L.; Plaza-Bolaños, P. Neutral or acidic pH for the removal of contaminants of emerging concern in wastewater by solar photo-Fenton? A techno-economic assessment of continuous raceway pond reactors. *Sci. Total Environ.* **2020**, *736*, 139681. [CrossRef]

processes

Article

Degradation of Polyvinyl Chloride (PVC) Waste with Supercritical Water

Maja Čolnik [1], Petra Kotnik [1,2], Željko Knez [1,2] and Mojca Škerget [1,*]

1 Laboratory for Separation Processes and Product Design, Faculty of Chemistry and Chemical Engineering, University of Maribor, Smetanova 17, SI-2000 Maribor, Slovenia
2 Department for Chemistry, Faculty of Medicine, University of Maribor, Taborska 8, SI-2000 Maribor, Slovenia
* Correspondence: mojca.skerget@um.si

Abstract: The chemical degradation of PVC waste in SCW between 400 and 425 °C and reaction times from 30 to 60 min was studied. The PVC waste in SCW decomposed into the gas, oil, water soluble, and solid phases. The highest yield of the gas and oil phases was achieved at the temperature of 425 °C after 60 min. By increasing the reaction time at 400 °C, the yield of chloride ions in the aqueous phase increased and reached the maximum at 60 min. The gas and oil phases contained many valuable compounds similar to crude oil. Alkanes and chloroalkanes; alkenes, alicyclic, and aromatic hydrocarbons; as well as alcohols were the main groups of hydrocarbons in the oil phase, while the gas phase contained only light hydrocarbons (C_1–C_6), CO_2, and small amounts of H_2. This confirmed that the largest chlorine content remains in the aqueous phase and does not pass into the gas phase. It can be concluded that SCW presents effective decomposition media for plastic waste.

Keywords: polyvinyl chloride; supercritical water; chemical recycling; plastic waste

check for updates

Citation: Čolnik, M.; Kotnik, P.; Knez, Ž.; Škerget, M. Degradation of Polyvinyl Chloride (PVC) Waste with Supercritical Water. *Processes* **2022**, *10*, 1940. https://doi.org/10.3390/pr10101940

Academic Editors: Dimitris Zagklis and Georgios Bampos

Received: 28 August 2022
Accepted: 22 September 2022
Published: 26 September 2022

1. Introduction

Plastic is one of the most important materials in our daily lives. Synthetic materials are very useful and convenient, with many positive properties (good chemical resistance, technically sophisticated, lightweight, innovative, and inexpensive) [1]. Global world production of plastics and plastic products has increased twenty times since 1960, and by 2036 this number is expected to double [2]. On the other hand, the environmental impacts resulting from the accumulation of plastic waste are increasingly huge. The data show that by the end of 2020, about 585 million tons of plastic waste were generated by over 7.7 billion people globally [3]. The information about plastic waste is alarming. Additionally, plastic debris affects wildlife, human health, and the environment [4]. Recently, society has been increasingly inclined toward cleaner production and the reuse of industrial and municipal waste. Chemical recycling, which is becoming increasingly popular, breaks down plastics into valuable products (crude oil, fuel) or depolymerizes them back into monomers or chemicals. Such processes are still in the early stages, although several projects are already underway, which provides hope that dependence on naphtha can be reduced in this way. Since 2007, the company APChemi has been focusing on developing sustainable pyrolysis technology for the disposal of plastic waste and tires. It has established more than 35 pyrolysis plants in India, Kenya, Indonesia, the UK, etc. [5]. In 2018, the Austrian energy company OMV started to turn plastic waste back into crude oil. The ReOil pilot plant successfully recycled 200,000 kg of plastic waste in two years [6]. The tire manufacturer Michelin is establishing the first tire pyrolysis plant in collaboration with Enviro, a Swedish company that has developed a patented tire pyrolysis technology to recover carbon black, oil, steel, and gas from end-of-life tires [7]. Exxon Mobil and TotalEnergies opened two recycling centers in France. Loop Industries (Canada) and Eastman Chemicals (USA) plan to open two plants in France to process PET by 2025 [8].

Hydrothermal processing is one of the attractive chemical degradation methods in which sub- or supercritical water (SubCW and SCW) is used and can act as a solvent, reactant, or catalyst. Hydrothermal processes can be divided into four main processes according to the type of product: hydrothermal carbonization [9], water reforming, hydrothermal liquefaction, and hydrothermal gasification [10]. The main hydrothermal processes for breaking down biomass and plastic into syngas, bio-oil, or crude oil are supercritical water gasification [11–15] and hydrothermal liquefaction [16–18]. These processes are more energy efficient than thermal processes that produce the same products (pyrolysis and thermal gasification). The drying step of feedstock is not required [19]. This leads to higher energy efficiency, especially for biomass with high moisture content. Temperature, pressure, feedstock concentration (plastic/biomass to water ratio), and reaction time are the important parameters that affect the desired products of the SCW processes [12,20,21]. The operating temperature ranges of supercritical liquefaction and gasification are from 350 °C to 800 °C and depend on the feedstock type, reactor configuration, catalyst, and desired products [10,11,14,16,18,22–25]. The use of SCW techniques exhibits many advantages, such as high yields of the selected products, very clean gas products that contain very low concentrations of NOx and SOx, reduced levels of CO, and inhibited tar and char formation. However, the method has some disadvantages, especially for biomass processing: the problems include plugging, corrosion, catalyst deactivation, and high production costs [14,26].

Polyvinyl chloride (PVC) is a synthetic polymer prepared from vinyl chloride monomer [27]. PVC is a universal polymer, and it is the third most commonly produced synthetic polymer in the world. It can be processed into a wide range of short-life products, as well as for longer-life products [28]. PVC polymer is composed of 56.8% chlorine [27], which causes the biggest problem in environmental pollution and recycling. PVC can be recycled using various recycling techniques (material and energy recovery methods). Secondary (mechanical) recycling has been used for years for PVC waste. Here, the recycling of PVC is preferred when pure PVC with a known composition and history is available. Incineration with energy recovery is an undesirable recycling technology used to remove PVC waste. Due to the high chlorine content, PVC is quite resistant to combustion, and toxic pollutants are also formed during recycling [28,29]. In general, by thermal decomposition of PVC, high amounts of HCl are produced, which can cause corrosion in furnaces and release harmful gases, which in turn can cause serious environmental and health problems [30]. In addition, PVC is not used as a pure polymer, but many additives (plasticizers, stabilizers, lubricants, and fillers) are added to improve its properties [31]. Huge problems arise in landfills, where PVC slowly decomposes (oxidative degradation of PVC), and heavy metals and phthalates from additives can contaminate the groundwater [32].

In recent years, the chemical recycling of PVC materials has attracted increased attention from researchers. One of the common approaches to the chemical recycling of PVC waste is pyrolysis, including hydrocracking, thermal cracking, and catalytic cracking [28]. The thermal degradation of PVC waste was performed in a batch (vessel, autoclave) and continuous process (fluidized bed reactor) [33–37]. The thermal degradation of PVC also encompasses catalytic and noncatalytic processes [38–40]. In general, during the thermal decomposition (250–525 °C) of PVC, the mass loss and dechlorination reaction occur in the temperature range between 250 and 350 °C, depending on the PVC material, the stabilizers, and the additives. Under these conditions, up to 99.5% of chlorine was removed from the PVC molecule [33]. Moreover, organohalogen compounds, chlorinated dioxins, furans, volatile components (benzene, toluene, and hydrocarbons), and char were also formed during the pyrolysis reaction [28], while the main intermediate product of the thermal degradation is a polyene. At higher temperatures (500–900 °C), PVC waste pyrolyzed to hydrocarbons (oil), HCl, and chlorinated hydrocarbons, where the HCl needs to be removed from the gas products. The main product of PVC pyrolysis is oil for the oil industry [29]. Despite the fact that the thermal degradation of PVC has been studied for years, the mechanism of the PVC thermal degradation is still not fully developed [37].

Compared to thermal degradation, during hydrothermal degradation of virgin PVC, 100% of chlorine was removed at a temperature higher than 300 °C [30,41]. Numerous different catalysts have also been used for more efficient dechlorination. The NaOH solution accelerated the dechlorination of flexible PVC between 150 and 250 °C. The degree of dehydrochlorination of flexible PVC increased with an increasing reaction temperature, and was of about 100% at 250 °C and after 5 h [42,43]. In some other studies, several other organic solvents (ethylene glycol [44], dimethylsulfoxide [43]) were added, which also improved the efficiency. Most of these studies focused on the successful hydrothermal dechlorination of PVC and on the hydrochar characteristics. However, the use of cosolvents and catalysts leads to problems such as solvent recycling and the use of large amounts of inorganic salts, which increase process costs and damage the equipment [30]. The degradation of virgin PVC in SubCW and SCW was studied by Takeshita et al. (2004) and Sato et al. (1998). The decomposition of PVC in SubCW produced a mixture of aromatic and aliphatic compounds (CO_2, benzene, acetone, benzoic acid) in the water and gas phases [31,45]. At a higher temperature of over 350 °C, the light hydrocarbons (C_1–C_5), CO_2, acetone, benzene, benzoic acid [31,45], phenol, naphthalene, acetic acid (Sato et al., 1998) in liquid and gas phases were obtained during the degradation of virgin PVC. The obtained residual solid after hydrolysis at 300 °C has a high combustion enthalpy and consequently exhibits good potential for use as a fuel ingredient [31,45]. However, despite a very detailed investigation, Takeshita et al. [31] did not report the yield of the obtained oil phase and its chemical composition, which is very important for understanding the recycling of PVC by hydrothermal processes.

According to the literature review, previous studies addressing the hydrothermal decomposition process of PVC have been conducted on virgin PVC. The aim of this research work was to study the decomposition of the real PVC waste (colorless PVC packaging bags) in SCW at different temperatures (400 and 425 °C), reaction times (30 min and 60 min), and without any catalyst. The main goal was to determine the yield and chemical composition of the obtained products. The chemical components generated in the oil and gas phases during the degradation of PVC in SCW have the potential to be used as fuel blends or feedstock for chemicals. The presented degradation mechanism plays an important role and will contribute significantly to a better understanding of the degradation of PVC in SCW and the simultaneous recycling of several types of plastic waste by this method.

2. Materials and Methods

2.1. Materials

Postconsumer colorless PVC packaging bags were collected from the local market (Maribor, Slovenia). Nitrogen (N_2) (99.5%) as an inert gas was supplied by Messer (Ruše, Slovenia). Potassium chromate (99.0%), silver nitrate solution, and n-hexane (97%) were obtained from Merck (Darmstadt, Germany) and Sigma-Aldrich (Steinheim, Germany). All standards and solvents were of analytical grades.

2.2. Degradation of PVC Waste in SCW

First, the colorless PVC waste (packaging for clothes) was cut into equal-sized particles of 1 × 1 cm. The hydrothermal degradation experiments were performed in a high-temperature and high-pressure batch reactor (Parr Instruments, Moline, IL, USA) [17], presented in Figure 1 at 400 °C and 425 °C for a reaction time of 30 min and 60 min. The reactor was filled with material (~4.5 g) and water (~22.5 mL), so that the water/PVC waste ratio was 5/1 (mL/g). The filled reactor was sealed, and then it was purged with nitrogen (N_2) to avoid the oxidation of the products. In order to enter the supercritical region when heating to the desired temperature (400 or 425 °C), the initial pressure in the reactor was set to 5 bar (N_2). The reactor was wrapped with an electrical wire and covered with glass wool in order to prevent heat losses. Afterwards, the reactor was first heated to the desired temperature and then maintained at that temperature for a certain time. Temperature, pressure, and reaction time are very important reaction parameters for hydrothermal degradation processes. For this reason, pressure and temperature changes during reactor temperature increases were precisely monitored. The heating rate in the temperature range from room temperature to 160 °C was 28 °C/min ± 0.5 °C/min, followed by periods of slightly slower heating rate of 18 °C/min ± 0.5 °C/min (to 250 °C), 16 °C/min ± 0.5 °C/min (to 350 °C), 14 °C/min ± 0.5 °C/min (to 400 °C), and 8.3 °C/min ± 0.5 °C/min (to 425 °C), and the average heating rate was 17 °C/min. At the same time, the pressure increased during the heating of reactor from initial 5 bar to 10 bar at 160 °C, to 45 bar at 250 °C, to 145 bar at 350 °C, to 255 bar at 400 °C, and to 320 bar at 425 °C. While the reaction proceeded at a constant temperature, the pressure still increased slightly due to the decomposition of PVC and was 260 bar after 30 min and 270 bar after 60 min at 400 °C, while at 425 °C it increased to 325 bar after 60 min. After the hydrothermal degradation of PVC waste, the reactor was immediately quenched in a water bath to cool down to room temperature. The generated gases were collected through a control valve into a Tedlar sampling bag (Supelco, Bellefonte, PA, USA). The total volume of the gas phase was measured with a volume of water disposed in a measuring chamber, and then the yield of the gas phase was estimated based on the ideal gas law equations. After collecting the gas phase, the reactor was opened, and the post-reaction mixture was transferred into the beaker. The reactor was washed with n-hexane to remove any solid particles and prevent possible losses. The reaction mixture from the reactor consisted of three phases: solid particles, aqueous phase, and oil phase. The solid particles in the aqueous phase were separated by filtration, while the n-hexane was evaporated from the oily phase. The yield of oil and solid phase was calculated by Equations (1) and (2) and it was expressed in weight percentage (%). The experiments were conducted in duplicate, and the error bars in Figure 2 indicate the standard deviation.

$$\gamma_{\text{oil}} = \frac{m_{\text{oil}}}{m_{\text{PVC waste}}} \times 100\% \quad (1)$$

$$\gamma_{\text{SP}} = \frac{m_{\text{SP}}}{m_{\text{PVC waste}}} \times 100\% \quad (2)$$

where, γ_{oil}, γ_{SP} present yields of oil and solid phases, respectively. m_{oil} and m_{SP} are masses of oil and solid phases obtained by SCW degradation of PVC waste plastics. $m_{\text{PVC waste}}$ is the initial mass of colorless PVC waste loaded in the reactor.

2.3. *Decomposition Products Analysis*

The oily and gaseous products were analyzed by GC–MS while the content of chloride ions in aqueous phase was determined by the titration method. Solid phase was analyzed by Fourier-transform infrared spectroscopy (FTIR). All analyses of oil, gas, and aqueous phases were performed in triplicate and the results were expressed as their mean value ± standard deviation.

2.3.1. FTIR Analysis of PVC Waste and Solid Residue

Functional groups contained in solid residue after degradation of PVC waste in SCW were determined by ATR-FTIR (attenuated reflection method) spectrometry and compared with specific functional groups of non-degraded PVC waste. The spectra of individual samples were recorded in the range of wavenumbers between 4000 and 400 cm^{-1}, at a resolution of 16 cm^{-1} and a total of 30 co-added scans. The background measurement against air was conducted with the same settings. The obtained data were analyzed with high-performance IRsolution software.

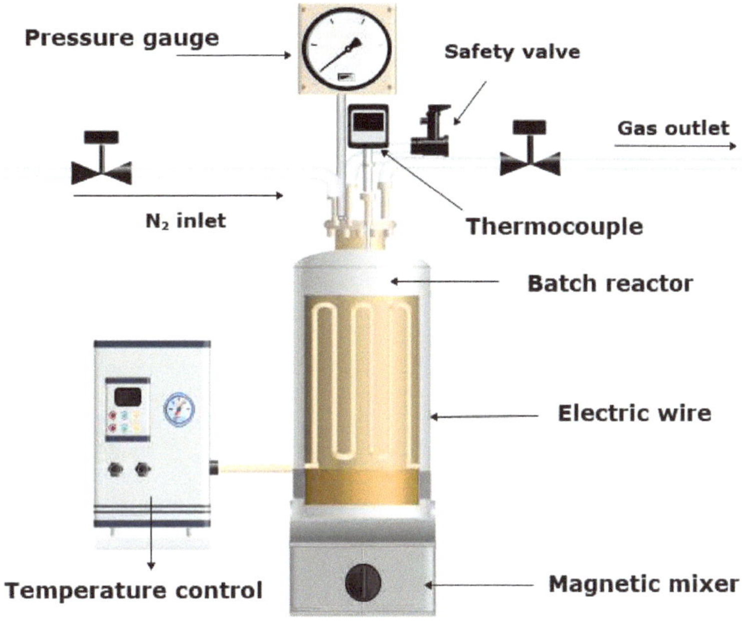

Figure 1. SCW batch reactor system for the chemical degradation of PVC waste in SCW.

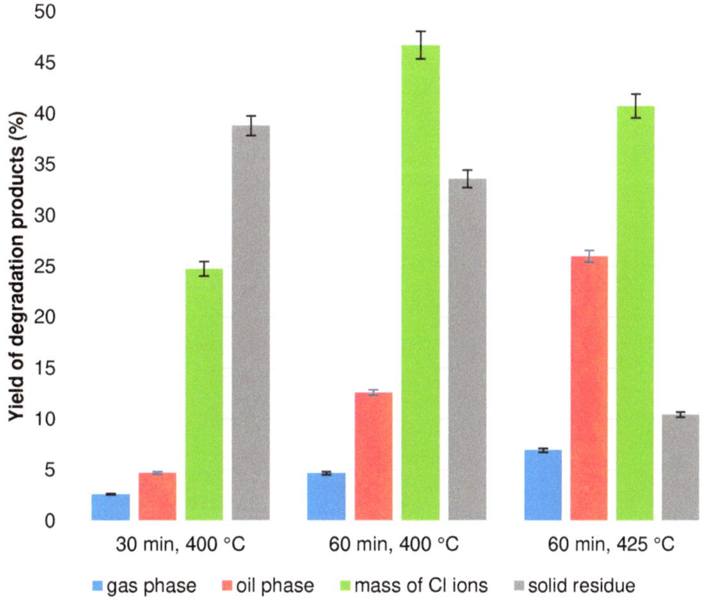

Figure 2. Yield of generated products after degradation of colorless PVC waste in SCW. Error bars in the figure indicate the standard deviation among the duplicate measurements.

2.3.2. Analysis of the Oil and Gas Phase Composition after SCW Degradation

A Shimadzu GCMS-QP2010 Ultra system with a mass spectrometer was used to determine the chemical composition of the gas and oil phase (Shimadzu, Tokyo, Japan). The GC–MS analysis of the oil phase was performed on HP-1 column (30 m × 0.25 mm × 0.25 μm) using temperature gradient from 40 °C to 230 °C in 27.5 min, while the analysis of the gas phase was performed on HP-PLOT/Q column (30 m × 0.32 mm × 20 μm) using temperature gradient from 35 °C to 200 °C in 30 min, where helium was used as a carrier gas. All compounds were identified using NIST library. A detailed description of the methods is presented in our previous study [17].

2.3.3. Determination of Chloride Ions in the Aqueous Phase

The mass of chloride ions in the aqueous phase was determined using the silver nitrate volumetric method [46]. Titrations were performed by adding potassium chromate indicator solution to the diluted aqueous phase sample and titrated with 0.1 M $AgNO_3$. In this case, silver ions reacted with chloride ions, and thus a white precipitate (AgCl) was first formed. In the following procedure, a reddish-brown precipitate (Ag_2CrO_4) was formed between the indicator and the standard solution. The mass of chloride ions ($m_{\text{Cl ion}}$) in the aqueous phase was calculated from the consumption of $AgNO_3$. The yield of chloride ions in the aqueous phase ($\gamma_{\text{Cl ion}}$) after degradation of PVC waste in SCW was determined by Equation (3).

$$\gamma_{\text{Cl ion}} = \frac{m_{\text{Cl ion}}}{m_{\text{PVC waste}}} \times 100\% \quad (3)$$

2.3.4. The Energetic Aspect of the Laboratory Process

For a laboratory batch reactor, energy consumption for the heat transfer ($Q = P \times t$) was roughly estimated. Q (kWh) is energy consumption, P (kW) is the power consumption of heater (0.5 kW for laboratory heater) and t (h) heating time. Heat losses during heating were neglected.

3. Results and Discussion

3.1. The Yield of Degradation Products after Hydrothermal Decomposition of Colorless PVC Waste

After the hydrothermal degradation in SCW at 400 and 425 °C and reaction times of 30 and 60 min, the PVC waste material decomposed into four phases: oil, gas, water soluble phase, and solid residue (Figure 2). In the study of Takeshita et al. [31], in which virgin PVC was degraded in sub- and supercritical water at various temperatures (200–600 °C) for one hour, only three phases (solid, liquid, and gas) were found and there are no data about the yield of the oil phase. In the present work, it was observed that with an increasing temperature (from 400 to 425 °C) and reaction time (from 30 to 60 min), the yield of gas and oil phases increased, while the yield of solid residue decreased from 39.0 ± 0.9% to 10.0 ± 0.3%. Most likely, the PVC molecule was dechlorinated during the heating of the reaction mixture to the desired temperature. At temperatures between 250 and 300 °C, PVC dechlorination occurred, in which the Cl ion was cleaved from one monomeric unit of two carbons, forming the polyene [47]. Polyene is an organic compound containing alternating double and single carbon bonds. By further decomposition of polyene in SCW at 400 and 425 °C, various hydrocarbons were likely formed in the oil phases via hydrogen abstraction, cyclization, aromatization, and hydration reaction, and consequently, the yield of the oil phase increased. Similarly, with the degradation, gasification, and decarboxylation reactions, gas components were formed in the gas mixture, and as a result, the yield of the gas phase increased. Namely, the physical properties of water at temperatures higher than 300 °C change significantly with temperature and pressure; therefore, under these conditions, water has a low ionic product, low dielectric constant, low viscosity, and high diffusivity. The highest yield of the oil and gas phases was found at 425 °C and 60 min, where the yield of the oil phase was 26.0 ± 0.6% and the yield of the gas phase was 7.0 ± 0.2%. An increase in the yield of the gas phase and a decrease in the yield of the solid phase with increasing temperature (200–450 °C) was also found in the work of Takeshita et al. [31]. In our work, the yield of chloride ions in the aqueous phase at a temperature of 400 °C increased with time, which is consistent with the decreasing pH of the post-reaction mixture. The pH of the aqueous phases after the PVC degradation in SCW decreased from 3.14 at 400 °C and 30 min to 2.38 at 400 °C and 60 min. In these conditions (400 °C and 60 min), the maximum concentration of chloride ions in the aqueous phase was determined as 47.0 ± 1.4% (Figure 2). As the temperature increased to 425 °C, the mass of chloride ions in the aqueous phase began to decrease slightly, as the halogenation of alkanes in the oil phase was most likely to occur. The pH of the aqueous phase was the lowest in this case and amounted to 2.1.

3.2. Chemical Composition of the Gas Phase

The chemical composition of the gas phase was determined using the GC–MS method (depending on the peak area (%)). During the degradation of PVC waste H_2, CO_2 and C_1–C_6 saturated and unsaturated aliphatic hydrocarbons were mainly formed (Table 1), most likely due to the decomposition of shorter hydrocarbons into gases. CO_2 is also the most represented component in all cases. The concentration of CO_2 in the gas phase decreased with increasing time and temperature. CO_2 was formed due to decarboxylation, where intramolecular and intermolecular Diels–Alder polyene reactions occurred during the benzene formation [31,45]. Similar to the degradation of polyolefins (PE) in SCW [17], the alkane concentration increased, while the alkene concentration decreased when the temperature and reaction time increased. Among the light hydrocarbons formed at 425 °C and 60 min, propane is the most abundant component in the gas phase. During the pyrolysis (530 °C and 25 min) of a plastic mixture of PVC, HDPE, PP, and PS, similar light hydrocarbons (C_1–C_4) were determined, among which the main compounds in the gas mixture were propene and propane. In contrast to hydrolysis, pyrolysis also produced chlorinated compounds in the gas mixture, of which HCl was the dominant component and its concentration increased with the PVC content in the plastic mixture [48]. The gaseous hydrocarbons C_2, C_3, and C_4 are versatile products which may be used as a fuel, in electric power generation, for the production of platform chemicals, for polyolefin synthesis, etc. [16]. Similarly, CO_2, which was the main component in the gas phase after the degradation of PVC in SCW, is widely used for food processing, in the chemical industry, agriculture, for methanol production, etc. [49].

Table 1. Composition of the gas phase (peak area (%)) at different temperatures and reaction times.

Components	400 °C, 30 min Peak Area (%)	400 °C, 60 min Peak Area (%)	425 °C, 60 min Peak Area (%)
Hydrogen	2.82 ± 0.08	1.00 ± 0.01	0.82 ± 0.01
CO_2	40.52 ± 1.10	36.71 ± 0.78	17.93 ± 0.45
Methane	2.24 ± 0.06	3.00 ± 0.01	3.10 ± 0.05
Ethene	1.50 ± 0.04	1.50 ± 0.02	0.52 ± 0.01
Ethane	5.80 ± 0.12	5.40 ± 0.13	10.71 ± 0.28
Propene	5.01 ± 0.12	3.10 ± 0.06	1.01 ± 0.01
Propane	7.43 ± 0.11	6.60 ± 0.17	17.41 ± 0.43
Isobutane	5.42 ± 0.16	1.10 ± 0.01	14.94 ± 0.21
2-Methyl-1-propene	4.41 ± 0.07	6.02 ± 0.12	1.40 ± 0.02
Butane	5.60 ± 0.12	5.60 ± 0.07	11.21 ± 0.22
2-Butene	7.10 ± 0.18	0.0 ± 0.0	0.0 ± 0.0
2-Methylbutane	0.90 ± 0.02	1.10 ± 0.01	8.40 ± 0.09
2-Pentene	2.50 ± 0.01	1.10 ± 0.02	0.221 ± 0.006
Pentane	0.0 ± 0.0	0.0 ± 0.0	4.21 ± 0.06
1-Methylcyclopentene	0.0 ± 0.0	12.52 ± 0.21	5.3 ± 0.11
Benzene	2.40 ± 0.05	6.54 ± 0.06	1.72 ± 0.04

3.3. Chemical Components in the Oil Phase

Components in the oil phase were determined by the GC–MS method and quantified according to the peak area (%). Six main groups of compounds were detected, i.e., saturated and unsaturated aliphatic hydrocarbons, alicyclic hydrocarbons, aromatic hydrocarbons, halogenated hydrocarbons, and alcohols (Table 2). At the temperature of 425 °C and at 60 min, the oil phase is mainly composed of aromatic hydrocarbons (39.31 ± 0.93%). This indicates that some alkenes and alicyclic hydrocarbons are probably converted via the aromatization reaction to aromatic compounds. Consequently, the concentration of alkenes (3.80 ± 0.11%) and alicyclic hydrocarbons (0.61 ± 0.013%) was quite low. The most represented aromatic components in the oil phase at 425 °C and 60 min are benzene derivatives (18.70 ± 0.39%) and naphthalenes (12.40 ± 0.19%), as well as small amounts of polycyclic aromatic hydrocarbons (2.61 ± 0.06%) (fluorenes and anthracenes). In addition, quite high concentrations of alkanes (24.41 ± 0.51%) and chloroalkanes (15.33 ± 0.38%) were detected in the oil phase. Pentadecane (6.04 ± 0.05%) and 1-chloro-undecane (5.40 ± 0.15%) were the most represented components in this group. Alkanes were produced during the hydrogen abstraction of polyene, after which the chlorine was eliminated from the PVC molecule into the aqueous phase. During the same period, the halogenation of alkanes took place, where radical substitution occurred, and hydrogen was replaced by chlorine. In the following, alcohols were formed during the alkene hydration [17]. Among alcohols (5.23 ± 0.12%), 3,7-dimethyl-1-octanol (1.20 ± 0.02%) is the most common component. The results obtained in this study differ from the results obtained by Sato et al. [50], who studied the decomposition of virgin PVC in SCW at higher temperatures, i.e., at 450 °C and 600 °C, and only aromatic compounds and acetic acid were detected in the oil phase. In comparison, the thermal degradation of PVC (at 500 °C) caused the PVC to be decomposed to 30% of the liquid fraction which was rich with aromatic compounds (88%) of which the major part was benzene (80%), while alkanes, alkenes, and alicyclic hydrocarbons which represented 2.47%, 1.52%, and 1.20% of the liquid, respectively. The content of the chlorinated hydrocarbons in the liquid fraction was 1.75% [36]. At a lower degradation temperature (300 °C), only HCl, benzene, and anthracene were detected after the thermal degradation of PVC due to the dehydrochlorination reaction [37]. The obtained oil phase after the degradation of colorless PVC waste in SCW contained valuable chemical components which are very similar to the components in the crude oil. Namely, crude oil is mainly composed of three types of hydrocarbon components (paraffins, cycloparaffins, and aromatics) [51].

Table 2. Composition of the oil phase (peak area (%)) at 425 °C and 60 min.

Main Groups	Most Represented Components (Peak Area (%))
Saturated aliphatic hydrocarbons (24.41 ± 0.51%)	Octanes (methyl, ethyl, dimethyl) (5.3 ± 0.09%) Decane (2.1 ± 0.03%) Tridecane (4.2 ± 0.11%) Pentadecane (6.04 ± 0.05%) Hexadecane (1.3 ± 0.01%) Octadecane (1.7 ± 0.03%) Heneicosane (1.2 ± 0.01%)
Halogenated hydrocarbons (chloroalkanes) (15.33 ± 0.38%)	1-chloro-nonane (4.8 ± 0.12%) 1-chloro-decane (4.9 ± 0.09%) 1-chloro-undecane (5.4 ± 0.15%)
Unsaturated aliphatic hydrocarbons (3.80 ± 0.11%)	3,3,5-trimethyl-1-hexene (1.8 ± 0.04%) 2,4-dimethyl-1-heptene (1.9 ± 0.01%)
Alicyclic hydrocarbons (0.609 ± 0.013%)	Cyclopentane (0.32 ± 0.011%)

Table 2. *Cont.*

Main Groups	Most Represented Components (Peak Area (%))
Aromatic hydrocarbons (39.31 ± 0.93%)	Benzene derivatives (18.7 ± 0.39%) Naphthalene derivatives (12.4 ± 0.19%) Polycyclic aromatic hydrocarbons (2.6 ± 0.06%)
Alcohols (5.23 ± 0.12%)	2-methyl-1-butanol (0.7 ± 0.012%) 6-methyl-1-heptanol (1.1 ± 0.014%) 3,7-dimethyl1-octanol (1.2 ± 0.02%) 1-nonanol (1.1 ± 0.019%)

3.4. FTIR Analysis of Colorless PVC and Solid Residue after Hydrolysis

FTIR analysis of PVC waste was performed before and after the degradation in SCW (Figure 3). The FTIR spectrum for PVC waste shows strong absorption bands at 2924 cm^{-1} and 2858 cm^{-1} [52], which were attributed to the asymmetric and symmetric elongation of the –CH_2 group, which confirmed the existence of the methylene group (–CH_2–). In comparison, after the hydrothermal decomposition of the PVC waste at 400 °C for 30 and 60 min, the absorption bands for solid residues at 2924 cm^{-1} and 2858 cm^{-1} were visibly weakened, while at 425 °C and 60 min these absorption bands completely disappeared. Furthermore, the C–H band on the adjacent CHCl groups at 2969 cm^{-1} and the absorption bands at 1429 cm^{-1} and 1330 cm^{-1}, which were attributed to the oscillation and bending of C–H vibrations, also disappeared after the hydrothermal process in all cases. The band for undegraded colorless PVC waste at 1247 cm^{-1} corresponded to skeletal vibrations (Cl–C–H), while for solid residues after hydrothermal degradation, this band disappeared in all three cases. The bands at 1097 cm^{-1} and 964 cm^{-1} were attributed to C–C elongation and –CH_2 oscillation. In addition, the bands at 611 cm^{-1}, 688 cm^{-1}, and 837 cm^{-1} can be attributed to C–Cl tensile vibrations. For the solid residues obtained after SCW degradation of PVC, these absorption bands were not present anymore. The presence of an additive (plasticizer) in PVC waste may be attributed to the absorption bands at 495 cm^{-1} and 420 cm^{-1} [30], and it was also present in solid residues after hydrolysis at 400 °C and 30 min. With an increase in temperature to 425 °C and a reaction time of 60 min, the absorption bands for the plasticizer were weakened or disappeared. It can be concluded that after hydrothermal decomposition of the PVC waste in all three cases, the absorption bands characteristic for Cl ions disappeared. It was confirmed that all chlorine was removed from the polymer molecule during the hydrothermal degradation reaction, and the remaining solid carbonaceous material could find many potential applications, e.g., as a solid fuel, as adsorbent, in catalysis, energy storage, sensor industry, etc.

3.5. Degradation Mechanism of PVC Waste in SCW

Based on the obtained results, the decomposition of PVC waste in SCW was proposed (Figure 4). During the heating to the supercritical conditions, HCl was cleaved from the PVC molecule by the "zipper" mechanism. The loss of each HCl molecule from the polymer led to the formation of a double bond, which activated the adjacent chlorine and caused the propagation of the degradation reaction along the polymer chain [53]. As is known from the literature, the dechlorination of PVC waste occurs in the SCW area at about 300 °C and 30 min [30,41]. By further heating the reaction mixture to the supercritical region, the polyenes were broken down into lower molecular weight molecules by the β-cleavage reaction of the C–C bond and by the transformation of hydrogen, resulting in the formation of saturated and unsaturated aliphatic hydrocarbons. Most alkenes were converted by the cyclization reaction to alicyclic hydrocarbons, by the aromatization reaction to aromatic components, and by the hydration reaction to alcohols [16,17] at the reaction temperature of 425 °C. In the same period, the polycyclic aromatic hydrocarbons formed via the polymerization reaction of aromatic hydrocarbons. After the separation of HCl from the PVC molecule, the degradation pathway of PVC waste is very similar to that of polyolefins [17], except that in the case of PVC, a large amount of chloroalkanes (15.33 ± 0.38%) appears in the oil phase, while the chlorine components do not pass into the gas phase. The halogenation of alkanes takes place at high temperatures, where radical substitution occurs, and hydrogen is replaced by chlorine. Gases are also formed during the decomposition of PVC waste in SCW. Namely, with increasing temperature and time, long hydrocarbons are decomposed into shorter ones (C_1–C_6). The decarboxylation reaction of benzene produces CO_2, which is the most abundant component in the gas phase.

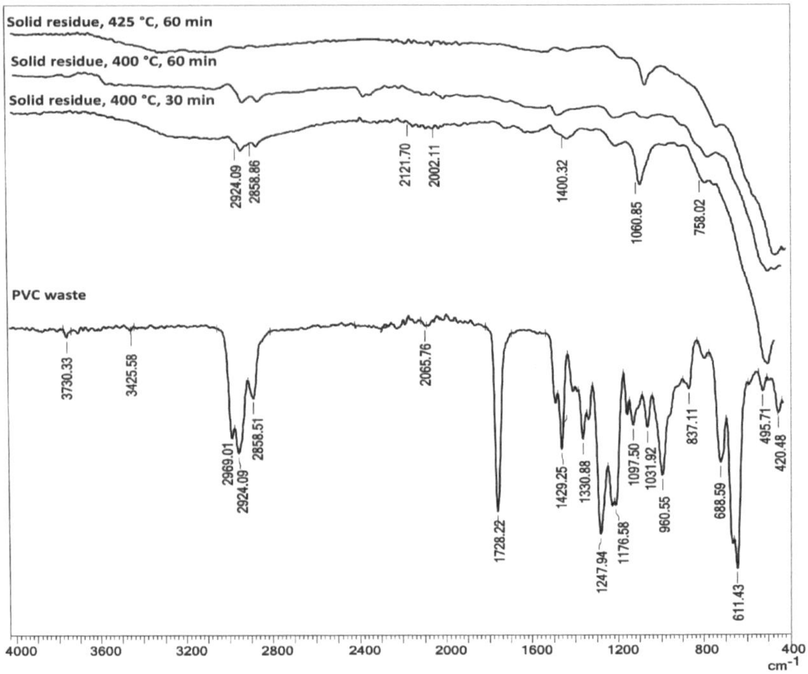

Figure 3. FTIR spectra of colorless PVC waste before and after hydrothermal degradation under different reaction conditions.

3.6. Electricity Cost for Laboratory Scale Hydrothermal Process

For the laboratory-scale hydrothermal decomposition of 1 kg of PVC waste (75 mL) at 400 °C for 30 min, the electricity cost was estimated to be 92.6 kWh. By extending the reaction time and increasing the temperature, the energy consumption would increase to 148.1 kWh per kg of PVC at 400 °C and 60 min and to 153.7 kWh per kg of PVC at 425 °C and 60 min. The energy consumption was calculated based on the actual material consumption for the reaction. Energy costs could also be reduced by optimizing the PVC and water masses in the reactor and by increasing the capacity of the reactor. Chemical recycling using SCW has several advantages compared to other chemical recycling techniques, e.g., faster process rates and higher process efficiency, no required material drying, it is more environmentally acceptable, the reaction temperature is much lower than in the pyrolysis process, but its implementation requires a higher investment and high operating costs due to the considerable energy consumption to heat the water to the reaction temperature, as can also be seen from our study. Thus, the implementation still depends on the economics of the process. However, the advanced recycling company Mura Technology has developed an innovative recycling process, Hydro-PRT^{SM} (Hydrothermal Plastic Recycling Technology). HydroPRS^{TM} utilizes SCW to convert waste plastics, including flexible and multi-layered materials, into high yields of stable hydrocarbon products for use in manufacturing new plastics and other materials [54].

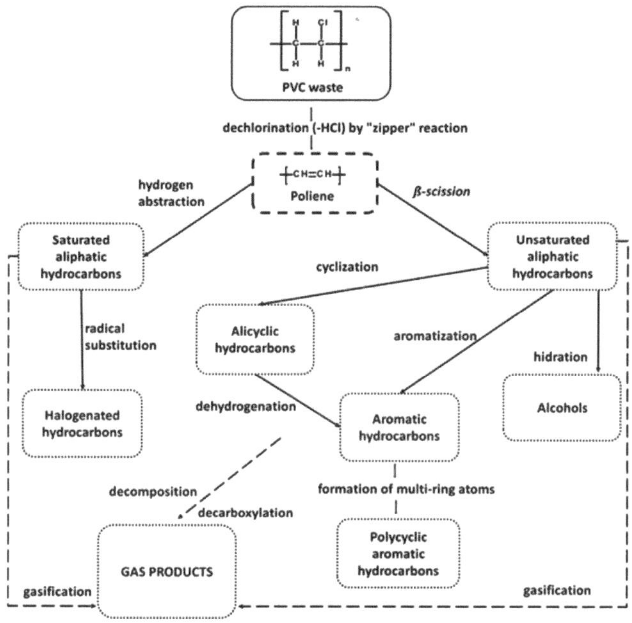

Figure 4. Degradation mechanisms of PVC waste in SCW.

Figure 4. Degradation mechanisms of PVC waste in SCW.

4. Conclusions

The decomposition of colorless PVC waste in SCW at temperatures from 400 to 425 °C and reaction times from 30 to 60 min was investigated. It was found that chlorine was first released from the PVC molecule during degradation, and HCl was formed in the aqueous phase. After the degradation of PVC waste in SCW, an oil, gas, aqueous, and solid phase were obtained. The highest chlorine content (47.0 ± 1.4%) in the aqueous phase was achieved at 400 °C and at a reaction time of 60 min. As the temperature and reaction time increased, the amount of solid residue decreased, while the yield of the oil and gas phases increased up to 425 °C in 60 min. The highest yield was achieved under these conditions, namely 7.0 ± 0.2% for the gas phase and 26.0 ± 0.6% for the oil phase. GC–MS analysis was used to determine the chemical composition of the oil and gas phases. PVC waste in SCW at 425 °C and 60 min decomposes into six main groups of compounds, i.e., saturated and unsaturated aliphatic hydrocarbons, alicyclic and aromatic hydrocarbons, alcohols, and halogenated hydrocarbons (chloroalkanes), of which the highest concentration was detected for aromatic hydrocarbons (39.31 ± 0.93%) where benzene derivatives (18.7 ± 0.39%) were the most common. The presence of polycyclic aromatic hydrocarbons (2.61 ± 0.06%) in the oil phase was also detected. In the gas phase, CO_2, alkanes, and alkenes of light hydrocarbons (C_1–C_6) were determined, among which the most common component was propane (6.60 ± 0.17%–17.41 ± 0.43%). It was found that oil and gas phases obtained from colorless PVC waste after hydrothermal degradation in SCW are composed of many valuable chemical components, and that the composition of the oil phase is similar to the composition of crude oil. Additionally, the advantage of this recycling method was also that most of the chloride ions remained in the aqueous phase and did not pass into the gas phase. As a result, however, the pH of the aqueous phase also dropped significantly, which may cause damage to the equipment. The effect of corrosion depends on the behavior of the material and its ability to resist acid. Metals such as aluminum, cast iron, steel, copper, and titanium will be rapidly attacked by HCl at all concentrations and temperatures. Most types of stainless steel will be exposed to corrosion because their chromium content is not sufficient to form a passive protective layer. Nickel alloys have better resistance than stainless steel but are still not completely resistant to corrosion at different concentrations of HCl. Nickel molybdenum alloys have good resistance at all concentrations and temperatures. However, oxidizing contaminants can change nickel molybdenum alloys from very resistant to very susceptible to attack, and it is difficult to eliminate these oxidizing contaminants from practice. Reactive metals (zirconium and tantalum) offer the best resistance to HCl [55,56]. Considering the problem of corrosivity of the material, the selection of appropriate equipment would be very important if PVC were to be recycled into SCW on a larger scale.

The energy consumption of the process performed in the laboratory autoclave (75 mL) was also analyzed in this study. It was found that at the reaction conditions (425 °C, 60 min) where the highest oil (26.0 ± 0.6%) and gas yield (7.0 ± 0.2%) was detected, the cost of electricity would be 153.7 kWh per 1 kg of PVC waste. The cost of electricity could be reduced by increasing the capacities of the reactors as well as by optimizing the PVC waste and water amounts in the reactor.

The decomposition pathway of PVC waste in SCW provides a significant contribution to the eventual recycling of PVC waste in SCW. Based on the results, it can be concluded that the innovative SCW treatment process is suitable for the chemical degradation of plastic waste into valuable chemicals.

Author Contributions: M.Č.: Formal analysis, Investigation, Conceptualization, Writing—original draft, Writing—review and editing. P.K.: Formal analysis, Writing—review and editing, Ž.K.: Funding acquisition, Review of draft, M.Š.: Investigation, Conceptualization, Writing—review and editing, Supervision. All authors have read and agreed to the published version of the manuscript.

Funding: This research was funded by Slovenian Research Agency (No. P2-0421, No. P2-0046, and No. J7-3149).

Institutional Review Board Statement: Not applicable.

Informed Consent Statement: Not applicable.

Data Availability Statement: Not applicable.

Acknowledgments: The authors would like to acknowledge the Slovenian Research Agency (research core funding No. P2-0421, No. P2-0046, and project No. J7-3149) for financing this research.

Conflicts of Interest: The authors declare no conflict of interest.

References

1. Klemeš, J.J.; Fan, Y.V.; Jiang, P. Plastics: Friends or Foes? The Circularity and Plastic Waste Footprint. *Energy Sources Part Recovery Util. Environ. Eff.* **2021**, *43*, 1549–1565. [CrossRef]
2. Hahladakis, J.; Iacovidou, E.; Gerasimidou, S. Plastic Waste in a Circular Economy. In *Plastic Waste and Recycling*; Letcher, T.M., Ed.; Academic Press/Elsevier: Cambridge, MA, USA, 2020; pp. 481–512. ISBN 978-0-12-817880-5.
3. Benson, N.U.; Bassey, D.E.; Palanisami, T. COVID Pollution: Impact of COVID-19 Pandemic on Global Plastic Waste Footprint. *Heliyon* **2021**, *7*, e06343. [CrossRef] [PubMed]
4. Helmer Pedersen, T.; Conti, F. Improving the Circular Economy via Hydrothermal Processing of High-Density Waste Plastics. *Waste Manag.* **2017**, *68*, 24–31. [CrossRef] [PubMed]
5. Pyrolysis & PyOil Purification Plants. Available online: https://www.apchemi.com (accessed on 16 September 2022).
6. ReOil: OMV's Circular Economy Pilot Project. Available online: https://www.omv.com/en/news/reoil-200-000-kg-of-plastic-waste-recycled-with-omv-s-circular-economy-pilot-project- (accessed on 15 September 2022).
7. Kononov, S. Michelin Launches Construction of Its First Tire Pyrolysis Plant | Weibold–Tire Recycling & Pyrolysis Consulting. Available online: https://www.michelin.com/en/press-releases/michelin-launches-construction-on-its-first-tire-recycling-plant-in-the-world/ (accessed on 15 September 2022).
8. Recycled Plastic: Important Things to Know. Available online: https://theaseanpost.com/article/recycled-plastic-important-things-know (accessed on 15 September 2022).

9. Pavlovič, I.; Knez, Ž.; Škerget, M. Hydrothermal Reactions of Agricultural and Food Processing Wastes in Sub- and Supercritical Water: A Review of Fundamentals, Mechanisms, and State of Research. *J. Agric. Food Chem.* **2013**, *61*, 8003–8025. [CrossRef] [PubMed]
10. Kotnik, P.; Čolnik, M.; Finšgar, M.; Knez, Ž.; Škerget, M. Determination of C1C6 Hydrocarbons in Gaseous Plastic Degradation Products by GC–MS Method. *Polym. Degrad. Stab.* **2020**, *182*, 109386. [CrossRef]
11. Bai, B.; Liu, Y.; Wang, Q.; Zou, J.; Zhang, H.; Jin, H.; Li, X. Experimental Investigation on Gasification Characteristics of Plastic Wastes in Supercritical Water. *Renew. Energy* **2019**, *135*, 32–40. [CrossRef]
12. Bai, B.; Liu, Y.; Meng, X.; Liu, C.; Zhang, H.; Zhang, W.; Jin, H. Experimental Investigation on Gasification Characteristics of Polycarbonate (PC) Microplastics in Supercritical Water. *J. Energy Inst.* **2020**, *93*, 624–633. [CrossRef]
13. Onwudili, J.A.; Williams, P.T. Catalytic Supercritical Water Gasification of Plastics with Supported RuO_2: A Potential Solution to Hydrocarbons–Water Pollution Problem. *Process Saf. Environ. Prot.* **2016**, *102*, 140–149. [CrossRef]
14. Özdenkçi, K.; Prestipino, M.; Björklund-Sänkiaho, M.; Galvagno, A.; De Blasio, C. Alternative Energy Valorization Routes of Black Liquor by Stepwise Supercritical Water Gasification: Effect of Process Parameters on Hydrogen Yield and Energy Efficiency. *Renew. Sustain. Energy Rev.* **2020**, *134*, 110146. [CrossRef]
15. Ghavami, N.; Özdenkçi, K.; Salierno, G.; Björklund-Sänkiaho, M.; De Blasio, C. Analysis of Operational Issues in Hydrothermal Liquefaction and Supercritical Water Gasification Processes: A Review. *Biomass Convers. Biorefinery* **2021**, 1–28. [CrossRef]
16. Chen, W.-T.; Jin, K.; Linda Wang, N.-H. Use of Supercritical Water for the Liquefaction of Polypropylene into Oil. *ACS Sustain. Chem. Eng.* **2019**, *7*, 3749–3758. [CrossRef]
17. Čolnik, M.; Kotnik, P.; Knez, Ž.; Škerget, M. Hydrothermal Decomposition of Polyethylene Waste to Hydrocarbons Rich Oil. *J. Supercrit. Fluids* **2021**, *169*, 105136. [CrossRef]
18. Jin, K.; Vozka, P.; Kilaz, G.; Chen, W.-T.; Wang, N.-H.L. Conversion of Polyethylene Waste into Clean Fuels and Waxes via Hydrothermal Processing (HTP). *Fuel* **2020**, *273*, 117726. [CrossRef]
19. Knez, Ž.; Markočič, E.; Hrnčič, M.K.; Ravber, M.; Škerget, M. High Pressure Water Reforming of Biomass for Energy and Chemicals: A Short Review. *J. Supercrit. Fluids* **2015**, *96*, 46–52. [CrossRef]
20. Knez, Ž.; Hrnčič, M.K.; Čolnik, M.; Škerget, M. Chemicals and Value Added Compounds from Biomass Using Sub- and Supercritical Water. *J. Supercrit. Fluids* **2018**, *133*, 591–602. [CrossRef]
21. Adschiri, T. Chapter 4—Biomass Conversion in Supercritical Water. In *Supercritical Fluid Technology for Energy and Environmental Applications*; Anikeev, V., Fan, M., Eds.; Elsevier: Boston, MA, USA, 2014; pp. 89–98. ISBN 978-0-444-62696-7.
22. Bai, B.; Jin, H.; Fan, C.; Cao, C.; Wei, W.; Cao, W. Experimental Investigation on Liquefaction of Plastic Waste to Oil in Supercritical Water. *Waste Manag.* **2019**, *89*, 247–253. [CrossRef]
23. Gemechu, E.D.; Kumar, A. Chapter 12—The Environmental Performance of Hydrogen Production Pathways Based on Renewable Sources. In *Renewable-Energy-Driven Future*; Ren, J., Ed.; Academic Press: Cambridge, MA, USA, 2021; pp. 375–406. ISBN 978-0-12-820539-6.
24. Sawai, O.; Nunoura, T.; Yamamoto, K. Supercritical Water Gasification of Sewage Sludge Using Bench-Scale Batch Reactor: Advantages and Drawbacks. *J. Mater. Cycles Waste Manag.* **2014**, *16*, 82–92. [CrossRef]
25. Cao, C.; Bian, C.; Wang, G.; Bai, B.; Xie, Y.; Jin, H. Co-Gasification of Plastic Wastes and Soda Lignin in Supercritical Water. *Chem. Eng. J.* **2020**, *388*, 124277. [CrossRef]
26. Elliott, D.C. Catalytic Hydrothermal Gasification of Biomass. *Biofuels Bioprod. Biorefining* **2008**, *2*, 254–265. [CrossRef]
27. Akovali, G. 2—Plastic Materials: Polyvinyl Chloride (PVC). In *Toxicity of Building Materials*; Pacheco-Torgal, F., Jalali, S., Fucic, A., Eds.; Woodhead Publishing Series in Civil and Structural Engineering; Woodhead Publishing: Sawston, UK, 2012; pp. 23–53. ISBN 978-0-85709-122-2.
28. Yu, J.; Sun, L.; Ma, C.; Qiao, Y.; Yao, H. Thermal Degradation of PVC: A Review. *Waste Manag.* **2016**, *48*, 300–314. [CrossRef]
29. Sadat-Shojai, M.; Bakhshandeh, G.-R. Recycling of PVC Wastes. *Polym. Degrad. Stab.* **2011**, *96*, 404–415. [CrossRef]
30. Li, T.; Zhao, P.; Lei, M.; Li, Z. Understanding Hydrothermal Dechlorination of PVC by Focusing on the Operating Conditions and Hydrochar Characteristics. *Appl. Sci.* **2017**, *7*, 256. [CrossRef]
31. Takeshita, Y.; Kato, K.; Takahashi, K.; Sato, Y.; Nishi, S. Basic Study on Treatment of Waste Polyvinyl Chloride Plastics by Hydrothermal Decomposition in Subcritical and Supercritical Regions. *J. Supercrit. Fluids* **2004**, *31*, 185–193. [CrossRef]
32. Yousif, E.; Hasan, A. Photostabilization of Poly(Vinyl Chloride)—Still on the Run. *J. Taibah Univ. Sci.* **2015**, *9*, 421–448. [CrossRef]
33. Yuan, G.; Chen, D.; Yin, L.; Wang, Z.; Zhao, L.; Wang, J.Y. High Efficiency Chlorine Removal from Polyvinyl Chloride (PVC) Pyrolysis with a Gas-Liquid Fluidized Bed Reactor. *Waste Manag.* **2014**, *34*, 1045–1050. [CrossRef] [PubMed]
34. Cho, M.-H.; Jung, S.-H.; Kim, J.-S. Pyrolysis of Mixed Plastic Wastes for the Recovery of Benzene, Toluene, and Xylene (BTX) Aromatics in a Fluidized Bed and Chlorine Removal by Applying Various Additives. *Energy Fuels* **2010**, *24*, 1389–1395. [CrossRef]
35. Borgianni, C.; De Filippis, P.; Pochetti, F.; Paolucci, M. Gasification Process of Wastes Containing PVC. *Fuel* **2002**, *81*, 1827–1833. [CrossRef]
36. McNeill, I.C.; Memetea, L.; Cole, W.J. A Study of the Products of PVC Thermal Degradation. *Polym. Degrad. Stab.* **1995**, *49*, 181–191. [CrossRef]
37. Xu, Z.; Kolapkar, S.S.; Zinchik, S.; Bar-Ziv, E.; McDonald, A.G. Comprehensive Kinetic Study of Thermal Degradation of Polyvinylchloride (PVC). *Polym. Degrad. Stab.* **2020**, *176*, 109148. [CrossRef]

38. Keane, M.A. Catalytic Conversion of Waste Plastics: Focus on Waste PVC. *J. Chem. Technol. Biotechnol.* **2007**, *82*, 787–795. [CrossRef]
39. Keane, M.A. Catalytic Transformation of Waste Polymers to Fuel Oil. *ChemSusChem* **2009**, *2*, 207–214. [CrossRef]
40. Ali, M.F.; Siddiqui, M.N. Thermal and Catalytic Decomposition Behavior of PVC Mixed Plastic Waste with Petroleum Residue. *J. Anal. Appl. Pyrolysis* **2005**, *74*, 282–289. [CrossRef]
41. Endo, K.; Emori, N. Dechlorination of Poly(Vinyl Chloride) without Anomalous Units under High Pressure and at High Temperature in Water. *Polym. Degrad. Stab.* **2001**, *74*, 113–117. [CrossRef]
42. Shin, S.-M.; Yoshioka, T.; Okuwaki, A. Dehydrochlorination Behavior of Flexible PVC Pellets in NaOH Solutions at Elevated Temperature. *J. Appl. Polym. Sci.* **1998**, *67*, 2171–2177. [CrossRef]
43. Yoshinaga, T.; Yamaye, M.; Kito, T.; Ichiki, T.; Ogata, M.; Chen, J.; Fujino, H.; Tanimura, T.; Yamanobe, T. Alkaline Dechlorination of Poly(Vinyl Chloride) in Organic Solvents under Mild Conditions. *Polym. Degrad. Stab.* **2004**, *86*, 541–547. [CrossRef]
44. Yoshioka, T.; Kameda, T.; Imai, S.; Noritsune, M.; Okuwaki, A. Dechlorination of Poly(Vinylidene Chloride) in NaOH/Ethylene Glycol as a Function of NaOH Concentration, Temperature, and Solvent. *Polym. Degrad. Stab.* **2008**, *93*, 1979–1984. [CrossRef]
45. Kershaw, J.; Edmond, L.; Rizzardo, E. Diels-Alder Reactions of Partially Dehydrochlorinated PVC in Solution and in the Melt. *Polymer* **1989**, *30*, 360–363. [CrossRef]
46. Wu, D.; Hu, Y.; Liu, Y.; Zhang, R. Review of Chloride Ion Detection Technology in Water. *Appl. Sci.* **2021**, *11*, 11137. [CrossRef]
47. Gandon-Ros, G.; Soler, A.; Aracil, I.; Gómez-Rico, M.F. Dechlorination of Polyvinyl Chloride Electric Wires by Hydrothermal Treatment Using K_2CO_3 in Subcritical Water. *Waste Manag.* **2020**, *102*, 204–211. [CrossRef]
48. Miskolczi, N.; Bartha, L.; Angyal, A. Pyrolysis of Polyvinyl Chloride (PVC)-Containing Mixed Plastic Wastes for Recovery of Hydrocarbons. *Energy Fuels* **2009**, *23*, 2743–2749. [CrossRef]
49. Vansant, J.; Koziel, P.-W. Technical and Industrial Applications of CO_2. In *An Economy Based on Carbon Dioxide and Water: Potential of Large Scale Carbon Dioxide Utilization*; Aresta, M., Karimi, I., Kawi, S., Eds.; Springer International Publishing: Cham, Switzerland, 2019; pp. 73–103. ISBN 978-3-030-15868-2.
50. Sato, Y.; Kato, K.; Takeshita, Y.; Takahashi, K.; Nishi, S. Decomposition of Ployvinylchloride Using Supercritical Water. *Jpn. J. Appl. Phys.* **1998**, *37*, 6270. [CrossRef]
51. Lois, E.; Keating, E.L.; Gupta, A.K. Fuels. In *Encyclopedia of Physical Science and Technology*, 3rd ed.; Meyers, R.A., Ed.; Academic Press: New York, NY, USA, 2003; pp. 275–314. ISBN 978-0-12-227410-7.
52. Casazza, A.; Spennati, E.; Converti, A.; Busca, G. Study on the Thermal Decomposition of Plastic Residues. *Chem. Eng. Trans.* **2019**, *74*, 1141–1146. [CrossRef]
53. Baum, B.; Wartman, L.H. Structure and Mechanism of Dehydrochlorination of Polyvinyl Chloride. *J. Polym. Sci.* **1958**, *28*, 537–546. [CrossRef]
54. HydroPRSTM. Available online: https://muratechnology.com/hydroprs/ (accessed on 16 September 2022).
55. Hydrochloric Acid—Rolled Alloys, Inc. Available online: https://www.rolledalloys.com/technical-resources/environments/hydrochloric-acid/ (accessed on 15 September 2022).
56. C-276 Hastelloy Alloy Coil-Tisco Shanxi Stainless Steel Co., Ltd. Available online: https://www.tiscoalloy.com/products/hastelloy/hastelloy_alloy_coil/242750.html (accessed on 15 September 2022).

processes

Review

Biotreatment Potential and Microbial Communities in Aerobic Bioreactor Systems Treating Agro-Industrial Wastewaters

Nikolaos Remmas

Laboratory of Wastewater Management and Treatment Technologies, Department of Environmental Engineering, Democritus University of Thrace, Vas. Sofias 12, 67132 Xanthi, Greece; nremmas@env.duth.gr

Abstract: The thriving agro-industry sector accounts for an essential part of the global gross domestic product, as the need for food and feed production is rising. However, the industrial processing of agricultural products requires the use of water at all stages, which consequently leads to the production of vast amounts of effluents with diverse characteristics, which contain a significantly elevated organic content. This fact reinforces the need for action to control and minimize the environmental impact of the produced wastewater, and activated sludge systems constitute a highly reliable solution for its treatment. The current review offers novel insights on the efficiency of aerobic biosystems in the treatment of agro-industrial wastewaters and their ecology, with an additional focus on the biotechnological potential of the activated sludge of such wastewater treatment plants.

Keywords: aerobic wastewater treatment systems; agro-industrial effluents; activated sludge; microbial ecology; nutrients removal efficiency

Citation: Remmas, N. Biotreatment Potential and Microbial Communities in Aerobic Bioreactor Systems Treating Agro-Industrial Wastewaters. *Processes* **2022**, *10*, 1913. https://doi.org/10.3390/pr10101913

Academic Editors: Dimitris Zagklis and Georgios Bampos

Received: 26 August 2022
Accepted: 19 September 2022
Published: 21 September 2022

1. Introduction

For more than a century, activated sludge in batch setups was used to treat sewage samples under continuous cycles of operation. Since then, the activated sludge process has rapidly expanded across the whole world, as industrialization and urbanization have entailed the need for the treatment of produced wastewaters. Nowadays, in the European Union (E.U.), the treatment of municipal wastewater complies with Directive 91/271/EEC, which, through the setting of rules regarding treatment efficiency and discharge, aims to protect the receiving water bodies from the vast amounts of wastewater produced.

The process involves the mixing and aeration of the activated sludge; the oxidation of organic carbon and inorganic nitrogen, wherein additional biomass is produced; the separation of the liquid and solid phases in order to control the concentration of the total suspended solids (TSS); the recirculation and retention of the biomass and possible disposal of the excess activated sludge [1]. During this step, the removal of phosphorus nutrients is also feasible, whereas the implementation of an anoxic stage results in the reduction and removal of the oxidized nitrogen compounds.

Apart from municipal wastewater, activated sludge systems have gained ground in the treatment of a range of agro-industrial effluents, which, among others, include winery, brewery, dairy, citrus, olive mill, vinasse, cassava, pepper, caper, and even coffee processing wastewater [2–5]. Several full-scale treatment plants have been installed and operated under moderate organic loading rates and aerobic or oxic–anoxic conditions to effectively remove nutrients from agro-industrial effluents [2], mainly carbon and nitrogen but even phosphorus [6].

The present review provides new insights on the efficiency of aerobic wastewater treatment plants processing agro-industrial wastewaters and their ecology, with additional reference to the biotechnological potential of activated sludge in such biosystems.

2. Agricultural By-Products and Wastewaters

The agriculture and food sectors globally face significant challenges in the 21st century. Food waste occupies an increasing section of waste treatment facilities and landfills. Today, remaining residues and losses throughout the food supply chain are drawing attention due to the vast waste of valuable resources, constituting a complex environmental problem. In low-income countries, food waste causes serious socio-economic consequences, while, on the other hand, consumer attitudes and the mass consumption of goods and products cause the production of huge amounts of household waste in middle and high-income countries [7].

Food waste can be classified into animal and agricultural origins. In the first, the main sources of waste are those from the dairy, meat, and fishery industries. In the latter, it is possible to classify a variety of residues according to the source, which may include cereals, roots and tubers, oil seeds and legumes, fruits, and vegetables. However, spoiled foodstuffs, along with heterogeneity, make them difficult to exploit, while a comprehensive characterization process is required in order to determine their composition. Nevertheless, both animal and agricultural wastes are often characterized by high organic matter [8].

As reported by Leite et al. [9], in the European Union alone in 2018, more than 21 million tons of waste were generated, which derived from the agriculture, fishery, and forestry sectors. Ravindran et al. [10] pointed out that one third of the food produced worldwide which is intended for human consumption is wasted every year, corresponding to losses of about 1.3 billion tons, while 40 to 50% (520–650 million tons) of global food waste per year derives from fruits, vegetables, and roots. In the E.U., food waste is estimated to reach 89 million tons per year, about half of which is generated during production, while the total production of agricultural residues (crop residues or parts of plants that are not consumed as food) amounts to 367 million tons per year, although some of these residues are commonly used at the farm level as bedding and fodder [11].

Recently, the Food and Agriculture Organization (FAO) in a report under the title "The Future of Food and Agriculture—Alternative Pathways to 2050" stated that the planet's population is expected to increase to about 10 billion by 2050, with an expected increase in agricultural demand, in a scenario of moderate economic growth, of about 50% compared to 2013. Furthermore, an increase in income in low- and middle-income countries would hasten the dietary transition to a greater consumption of meat, fruits, and vegetables over cereals, necessitating subsequent production changes that are expected to put additional strain on natural resources. In fact, although investment in agriculture and available technological tools and innovations boost productivity, final output growth is less profound, as food losses account for a significant proportion of agricultural production; therefore, tackling the crucial parameters of food loss would in itself reduce the need to increase production, which is already hampered by the degradation of natural resources and the loss of biodiversity, as well as the cross-border spread of pests and diseases of plants and animals which are highly resistant to applied antimicrobials [12].

Recently, the European Commission (E.C.) presented the action plan for the circular economy, which sets a coordinated strategy for a climate-neutral, resource-efficient, and competitive economy, as annual waste production until 2050 is expected to increase by 70%, given that "50% of total greenhouse gas emissions and more than 90% of biodiversity loss and pressure from water scarcity are due to resource extraction and processing" [13]. Moreover, regarding the economic aspect of this issue, the FAO stated in a report published in 2016 [14] that food losses worldwide correspond to economic losses of 490 billion US dollars per year, while the amount of water used corresponds to about 1/4 of all fresh water resources used in agriculture worldwide. These, in addition to the economic impact, are responsible for an estimated 8% of global greenhouse gas emissions [14], as well as 23% of the global use of fertilizers [15].

3. Quantities, Composition, and General Valorization Aspects of Agro-Industrial Wastewaters

Agro-industrial waste consists of agricultural and industrial residues, with the by-products of agricultural production being further distinguished into on-site residues (produced on the field) and those deriving from treatment processes [16]. Usually, these residues, which are agricultural by-products, are used in the manufacture of animal feed, as soil conditioners and fertilizers, and even in some construction applications, although significant quantities remain unexploited due to the heterogeneity of their characteristics. Their composition, which is also reflected in the wastewater produced as a consequence of the processing procedure, usually comprises mainly cellulose in a percentage approaching 60% [17], followed by hemicellulose, lignin, and minerals [18]. In the food industry, significant amounts of waste are generated in semi-solid and liquid forms due to the industrial processing of juices, potato products, and confectionery, which also mainly consist of cellulosic and hemicellulosic compounds [19].

Agro-industrial wastewaters are produced throughout the whole multistep processing procedure due to the washing, sterilization, centrifugation, distillation, sanitation, chemical treatment, and cleaning that take place, depending on the initial raw material to be processed [20–24], and are expected to contain high biological oxygen demand (BOD) and chemical oxygen demand (COD) concentrations, as well as suspended solids (SS) (Table 1). In any case, physicochemical and biological processes are applied until final disposal becomes necessary. Ideally, prior to disposal, important components of the wastewater could be recovered, while the reuse of the treated effluent in agriculture could also be considered. Components of the produced wastewater, such as sugars, organic acids, aromatic compounds, pigments, and proteins, may be immediately recovered or upgraded by biotransformation, although the presence of detergents, pesticide residues, antibiotics, and essential oils complicates this process [25,26].

Table 1. Physicochemical composition of the key agro-industrial effluents treated in aerobic wastewater treatment systems.

Agro-Industrial Wastewater	pH	BOD (g/L)	COD (g/L)	TS (g/L)	SS (g/L)	VSS (g/L)	TN/TP (g/L)	Phenols (g/L)	Oil and Grease/ Total Carbohydrates (g/L)	Reference
Brewery	5.1–10	2.0–5.0	3.6–49.5	1.7–38.9	0.7–5.7	38.9	0.07–0.17/ 0.07–0.06			[27–30]
Caper processing	6.8–7	1.5–4.0	2.2–8.4	-	-	-	0.12–0.35	-	-	[4]
Cassava processing	4.02–4.6	1.4–12.3	6.0–38.2	6.6–60.5	-	5.1–49.4	0.19–0.4/0.08–0.24	-	0.6/-	[31–34]
Cheese whey	3.92–6.6	90.1	45.0–91.6	47.6–73.9	9.4	8.3–59.9	0.15–3.2 (TKN)/0.12–0.70	0.27	-	[35–39]
Citrus processing	3.21–3.8	4.7–6.6	2.3–32.1	16.5	1.2	15.1	0.15/0.013	-	E.O. 0.04–1.0	[32,40–43]
Coffee-processing	3.50–4.4	4.3–37.9	7.6–45.9	3.8–19.5	2.9–8.6	1.9–8.2	0.27–0.7/0.01–0.04	0.05–0.28	-	[37,44–47]
Olive mill	4.0–5.7	10.2	36.7–299.0	38.9–94.9	24.7–42.8	23.2–83.2	0.1–0.6/0.06–0.2	0.54–11.0	-	[35,36,38, 39,48–50]
Palm oil mill	4.11–4.8	0.32–15.6	25.5–86.2	20.0–60.1	12.3–35.3	17.5–50.2	0.5–1.1/0.09–0.35	0.26–0.46	2.0–42.8	[31,51–54]
Pepper processing	4.4	1.4	1.9	-	-	-	0.05 (TKN) 0.02(NH_4^+)/0.01	-	-	[5]
Potato processing	4.6–7.1	0.9–5.0	1.5–37.0	4.8–42.0	2.0–3.8	3.5–4.4	0.03–0.62/0.04–0.1	-	-	[55–59]
Soyamilk	4.1	-	7.3	4.7	-	4.1	0.3	-	0.8	[31]
Vinasse	3.6–4.5	14.4–54.8	44.7–131.0	52.2–64.9	4.5–29.0	48.5–49.8	0.2–1.6 /0.1–0.6	0.35–0.7	-/25.1–47.30	[37,60–62]
Winery	3.6–4.9	0.15–8.0	0.5–15.9	-	0.08–2.4	0.07–1.6	0.06–0.2/ 0.01–0.055	0.03–0.2	-	[48,63–67]

Indeed, the processing of agricultural products is responsible for the generation of vast amounts of wastewater all around the world. According to Martinez-Burgos et al. [2], 0.085 million m^3 of wastewater was produced from cassava processing; 1.65 million m^3 was produced from vinasse processing; and palm oil processing wastewater amounted to

0.256 million m^3; whereas milk and cheese whey processing formed 143 and 158 billion m^3 of wastewater worldwide. Zema et al. [24] reported the production of 500 million m^3 of wastewater from the processing of citrus fruits, while this vast volume of processed effluent is expected to further increase and reach 750 million m^3. Interestingly, Brazil is the leading country regarding the processing of citrus fruits (12.11 million MT processed, 80% oranges), covering almost 50% of global citrus processing, followed by the United States of America, which processes 20% of citrus fruits. Moreover, 2.45 and 1.65 billion m^3 of effluents were produced due to slaughterhouse activity [2]. The produced effluents from each process are characterized by intense heterogeneity, even for raw materials of the same origin [2]. Due to the environmental conditions, the type of process, the technology, the seasonality, and the variety of plants (when plants are used as raw materials), the COD in the effluents of the winery and coffee processing industries may range between 0.5 and 16 g/L and 7 to 48 g/L, respectively; for palm oil effluent, from 25 to 86 g/L; and for cheese whey wastewater, from 45 to 91 g/L; whereas the COD of olive mill wastewater lies between 37 and 300 g/L and that of vinasse wastewater between 44 and 131 g/L. Similarly, cassava and citrus wastewaters present similar concentrations of organics, as the effluent COD of the processed material can range from 6 to 38 and 2 to 98 g/L, respectively. The remarkably high organic fraction of these effluents may be attributed to the presence of carbohydrates and proteins, as well as fats and grease. Carbohydrate concentrations in vinasse wastewater and palm oil effluent can reach 47 g/L and 43 g/L, respectively. Furthermore, the strongly acidic pH of the majority of agro-industrial wastewaters (average pH value of 4.5), as well as total and volatile suspended solids, contribute to the uniqueness of these effluents in terms of treatment and exploitation for the recovery of value-added products (Table 1).

In general, the criteria and possible approach strategies for developing valorization processes are based on a number of key pillars, which are described in detail by Castro-Muñoz et al. [68]. In brief, the following are required:

i. Case-by-case analysis of the production process and the characteristics of agro-industrial waste and residues, taking into account organic load, seasonality, chemical stability, and volume of produced wastewater
ii. Selection of one or more objectives, such as recycling of compounds and/or water, and recovery of molecules and components, as well as energy production.
iii. Assessment of possible technological and economic advantages in order to select appropriate disposal and exploitation alternatives.
iv. Identification of possible biotechnological approaches to achieve the initial goal.

4. Main Bioreactor Types, Nutrient Removal Processes, and Factors Affecting Systems Performance under Aerobic Conditions

Various aerobic bioreactor systems performing nutrient removal have been employed to treat agro-industrial wastewaters. The majority are membrane bioreactor (MBR) and sequencing batch reactor (SBR) systems due to their high efficiency and/or simplicity in operation. For instance, Tatoulis et al. [69] reported the high removal efficiency (80.3–96.5%) of an SBR system treating table-olive processing wastewater. Moreover, Vergine et al. [70] treated winery and vegetable canning factory wastewater in a microfiltration module, achieving removal efficiencies steadily above 90%. Most of these focus on the removal of carbonaceous compounds, although nitrification–denitrification schemes are also used to remove both the carbon and nitrogen load of the effluents [71].

The main factors affecting the performance of aerobic wastewater treatment systems are the oxygen supply, the temperature, the organic loading rate (OLR), the hydraulic retention time (HRT), the solids retention time (SRT), and the nature and composition of the wastewater [1,72]. In nitrification–denitrification treatment systems, the concentration of easily biodegradable organic matter is also crucial for enhancing nitrogen removal efficiency.

The initial microbial population and its adaptation to the applied operating conditions are the driving forces of the activated sludge processes [73,74]. A specialized microbial community often proliferates based on the specific composition of the treated agro-industrial effluent [75].

Conventional activated sludge systems are effective in removing macronutrients and solids during the treatment of agro-industrial wastewaters. However, micropollutants are often detected in the effluents of such bioreactor systems, a fact that induces toxicity to certain organisms. Nowadays, the use of membrane reactors, e.g., at nanofiltration scale, can improve micropollutants' removal efficiency during the biological treatment of such effluents [76]. Recently, the application of advanced oxidation processes in the effluents of aerobic treatment systems has improved the quality of the treated wastewater and minimized toxicity [77].

5. Biotreatment of Agro-Industrial Wastewaters in Aerobic Bioreactor Systems

Aerobic biological treatment systems, including nitrification–denitrification plants, are commonly applied for the biotreatment of domestic wastewater and a range of agro-industrial wastewaters, due to their simplicity in operation, low cost of installation, high efficiency, and ability to biologically remove nitrogen through nitrification–denitrification. During the activated sludge process, organic matter is oxidized using air, mainly to carbon dioxide and water, and the microbial flocs formed are separated in a sedimentation tank [78]. Despite the fact that effluents of high organic content could be subjected to anaerobic digestion, the inability to biologically remove nitrogen in an efficient and simple manner and the high cost of installation often make the activated sludge process attractive for the biotreatment of certain agro-industrial effluents, especially those in which the COD concentration is low or moderate, or those that can be co-processed with municipal wastewater or washings. Even though the anaerobic treatment of agro-industrial wastewaters has the benefit of the production of biogas, this is balanced by the high HRT required, increasing the volume of the required digesters and resulting in specific space requirements, as well as the instability of the process, which provides no assurance of stable and satisfactory energy production [79]. Moreover, aerobic treatment enables the effective removal of nutrients, which is considered a strong benefit of the process, as high quality effluents are produced, capable of satisfying the stricter standards for disposal, which are not met in the case of anaerobic treatment systems [80].

Thus, there are several examples of using aerobic biological treatment systems for the depuration of agro-industrial effluents. For instance, activated sludge was immobilized on polyurethane particles in an aerobic bench-scale bioreactor for the treatment of winery wastewater under a maximum organic loading rate of 8.8 kg COD/m^3·d and a hydraulic retention time of 0.8 d. Even at an OLR of 3 kg COD/m^3·d, the ability of the aerobic immobilized cell bioreactor to remove COD was high, recording a COD removal efficiency of 87% [81]. Moreover, Roveroto et al. [82] treated brewery wastewater in a fixed-bed batch reactor, which operated under an intermittent aeration of 3 h aeration in 4 h cycle and a hydraulic retention time (HRT) of 0.83 d. The COD and BOD of the raw brewery wastewater ranged between 2 and 10 g/L and 1.2 to 3.6 g/L, respectively, while the total nitrogen reached up to 0.08 g/L. The highest removal efficiency, 92%, was recorded in the bioreactor when the influent COD was 2.7 g/L and the COD/N ratio was 107. Under these conditions, the nitrification efficiency was 88% and the total nitrogen (TN) removal was 85%.

Antiloro et al. [40] investigated the biotreatment of citrus processing wastewater with a high organic content and essential oils concentration, i.e., between 20 and 30 g/L and 0.6 to 1.0 g/L, respectively, in an aerated lagoon system, reporting COD removal efficiencies from 59 to 97% and the establishment of a microbial community capable of coping with the increased concentration of essential oils. In addition, two aerobic granular sludge bench-scale SBRs operating under a sludge retention time (SRT) of 10 d and organic loading rates (OLRs) ranging from 3 to 15 kg COD/m^3·d were used for the biotreatment of a citrus processing effluent of 5.5 g/L COD. At a neutral pH, the biosystem could remove COD by 90% regardless of the organic loading rate applied, although the reactor's efficiency under acidic conditions declined to 75% when the OLR exceeded 7 kg COD/m^3·d. Furthermore, Zema et al. [83] treated citrus processing wastewater of 5.0g/L COD and an essential oils concentration of 0.5 g/L under aerobic conditions in a full-scale treatment plant, reporting reasonable COD and essential oils removal efficiencies.

Moore at al. [84] treated wastewater deriving from mixtures of fruits and vegetables in an aerobic pilot-scale ultrafiltration membrane bioreactor (MBR), for potential water reuse. Lettuce, beets, carrots, and cassava were processed to produce the first wastewater mixture, while potatoes, carrots, apples, onions, lettuce, beets, and bananas constituted the raw materials for the production of the second mixture of wastewater. The COD and total Kjeldahl nitrogen (TKN) content of the first mixture were 1.5 g/L and 0.01 g/L, respectively, whereas the respective concentrations in the second mixture were 7.1 g/L and 0.23 g/L. The HRT in the two experimental schemes examined varied from 24 to 52 h, whereas the OLR ranged from 0.82 to 2.7 kg COD/m^3·d in the first and from 2.9 to 6.5 kg COD/m^3·d in the second experimental setup. For both fruit- and vegetable-derived effluents treated in the MBR, high COD removal efficiencies of 97–98% were recorded, whereas the TKN removal efficiencies exceeded 91% for both wastewater mixtures. In this case, the activated sludge system, in combination with UV disinfection and the implementation of activated carbon for color removal, could produce an effluent of enhanced quality, which could be used in the agri-food sector.

More than 10.5 million tons of coffee were exported by its producing countries in 2020 [85], a process that leads to the production of significant amounts of wastewater, since up to 45 kg of wastewater is generated during the pulping and washing of 1 kg of green coffee. Villa-Montoya et al. [3] treated coffee processing wastewater of a high organic content (COD of 7 to 15 g/L) and a TN concentration between 0.03 and 0.04 g/L in a sequencing batch reactor (SBR) under an OLR of 9 g COD/L.d, reporting that the intermittently aerated biological system achieved a COD removal efficiency of 92%. Coffee processing wastewater of a high COD concentration (17 g/L) was also treated in a constructed wetland system by Rossmann et al. [86], in order to achieve the efficient removal of nutrients and phenolic content. At an HRT of 11.8 d, the biosystem could remove total nitrogen (TN), total phosphorus, and total phenolic compounds by 69.1, 72.1, and 72.2%, respectively.

6. Biomass Valorization of Aerobic Biosystems Treating Agro-Industrial Wastewaters

Microorganisms are an important source of enzymes, as they grow rapidly in a short period of time. In addition, a wide variety of agro-industrial residues and wastes can be used as substrate, thus reducing overall production costs and the use of natural resources while value-added products are produced. Enzymes of microbial origin can find a variety of applications in industry, such as in the production of food and beverages, as well as in the manufacture of chemicals and pharmaceuticals. The properties and activities of an enzyme are considered to be directly dependent on the strain that is capable of inducing them, while their effectiveness in biotechnological applications is being constantly and increasingly evaluated. Therefore, there is a strong scientific interest and a wide scientific field in the search for new strains capable of producing high-activity enzymes at a reduced cost with potential uses in industry [87]. Moreover, aerobic bioreactor systems treating agro-industrial wastewater can be considered as microbial cell factories producing a wide range of industrial enzymes, such as cellullases, xylanases, glycosidases, lipases, and proteases.

For instance, Zerva et al. [4] assessed the hydrolytic potential of an immobilized cell bioreactor treating caper wastewater at an elevated salinity (3.12 to 101 g/L). The non-halotolerant microbiota of the immobilized cells at a salinity of up to 20 g/L were able to highly hydrolyse celluloses, hemicelluloses, starch, fats, and proteins. Increased endo-1,4-β-xylanase activity above 1785 U/g protein was recorded throughout the experimental period. Endo-1,4-β-D-glucanase activity of 250 U/g protein was also reported, even though it was highly affected by the elevated salinity. Regarding polygalacturonase, its activity exceeded 533 U/g protein and further increased to 959 U/g protein under the highest salinity. Furthermore, β-1,4-D-glucosidase activity was above 510 U/g protein, while the increase in the organic loading rate and low salinity resulted in the elevation of α-1,4-D-glucosidase activity up to 905 U/g protein. Initial lipase activity was above 352 U/g protein but was affected by a salinity concentration of 1% w/v and decreased to 130 U/g protein. Moreover, Zerva et al. [5] treated pepper processing wastewater in an aerobic immobilized cell bioreactor and monitored the hydrolytic potential of bacteria isolated from the immobilized biomass of the biosystem, reporting a high endo-1,4-β-xylanase activity of 107,000, 72,000 and 70,000 U/g protein for three bacterial isolates belonging to the genera *Nocardia* and *Gordonia*. Bacterial isolates related to Aquincola, *Microbacterium*, *Planococcus*, *Sphigopyxis*, and *Xanthobacter* were also found to exert endo-1,4-β-xylanase activity from 29,700 to 37,400 U/mg protein.

In addition, several white-rot fungi can be used for the biotreatment of various agro-industrial enzymes and produce ligninolytic enzymes. For instance, a *Phanerochaete chrysosporium* strain was immobilized by Sharari et al. [88] on polyurethane foam for the treatment of bagasse wastewater and the simultaneous production of ligninolytic enzymes, reporting peroxidase activity of 260 U/L and laccase activity of 131 U/L, whereas xylanase activity of 74 U/L was also detected.

Moreover, Mafakher et al. [89] isolated lipase-and citric acid-producing yeasts from agro-industrial wastewater treatment plants. Among the 300 yeast isolates examined, 6 exhibited a high lipase activity, which were identified as *Yarrowia lypolitica* isolates.

7. Microbial Communities' Structure in Aerobic Biosystems Treating Agro-Industrial Wastewaters

The recent development and application of high-throughput sequencing techniques have led to a better understanding of microbial communities' structure and functions in bioengineering systems. In the last decade, the implementation of molecular methods, such as next generation sequencing techniques, has permitted the elucidation of the microbial ecology and biotechnological potential of certain aerobic bioreactor systems treating agro-industrial wastewaters (Table 2).

In that direction, by implementing high-throughput sequencing techniques, Fang et al. [90] stated the dominance of *Zoogloea* in the activated sludge of an SBR treating rice winery wastewater under an OLR of 2.4 g COD/L.d. Apart from the presence of *Zoogloea* species, *Rhodobacter* and *Rubrivax* were also detected in high abundances. The dominance of *Zoogloea* spp. in the activated sludge of this aerobic bioreactor system can find a biotechnological application potential, since this genus is considered an important PHA accumulating microorganism [91]. Bacteria of the genus *Amaricoccus*, *Zoogloea*, and *Azoarcus* were also identified in winery wastewater using FISH, whereas *Amaricoccus* species dominated the constructed clone library [92].

Moreover, in meat processing wastewater treated in an SBR, the activated sludge microbial community was dominated by the class *Alphaproteobacteria*, which are frequently identified in similar samples [93], where *Amaricoccus* spp. covered 11% of the microbial diversity in the SBR. Furthermore, the biotreatment of dairy wastewater in a full-scale aerobic SBR under an OLR of 2.5 kg COD/m^3·d revealed the predominance of the genera *Proteiniphilum*, *Byssovorax*, *Acidobacterium*, and *Zoogloea*, which covered 35.9, 14.5, 10.1, and 8.3% of the total relative abundance [94], despite the fact that *Proteiniphilum* and *Byssovorax* bacteria are rarely reported as microbial constituents of activated sludge. The same authors also reported that *Thiothrix* and *Leptothrix* spp. were the main filamentous bacteria of the activated sludge system, with the presence of *Thauera* being involved with the formation of granular structures and the cohesion of the activated sludge due to the release of extracellular polymeric substances (EPS).

Other major inhabitants of activated sludge systems treating agricultural wastewater are members of the genera *Bacillus*, *Pseudomonas*, *Thauera*, *Xanthomonas*, *Spingobacterium*, and *Comamonas*, such as in aerobic biosystems treating olive mill [95], winery [96], and dairy [97] wastewaters.

In addition, Pires et al. [98] isolated bacterial and fungal strains from a coffee processing wastewater treatment plant. The bacterial isolates were mainly members of the phyla *Proteobacteria*, e.g., *Acetobacter, Serratia,* and *Enterobacter* spp.; *Actinobacteria*, e.g., *Corynebacterium* and *Arthrobacter*; and *Bacteroidetes*, e.g., *Chrysobacterium*. Regarding the fungal community structure, the majority of isolates were identified as yeasts of the order Sacharomycetales, such as *Wickerhamomyces, Torulaspora, Kazachstania, Saturnispora, Meyerozyma, Hanseniaspora,* and *Pichia* spp., which have been often detected in municipal wastewater treatment plants and in other biosystems treating agro-industrial wastewater, e.g., palm oil effluent [99,100]. Pires et al. [98] also detected filamentous fungi, such as *Alternaria alternata* and *Fusarium oxysporum*. Petruccioli et al. [101] also isolated yeasts from the activated sludge of an aerobic jet loop reactor treating winery wastewater, identifying microbiota such as *Saccharomyces, Candida,* and *Trichosporum*, reporting a link between the presence of *Saccharomyces* and the formation of biofilm.

Regarding the bacterial community structure in aerobic immobilized cell bioreactor systems, Zerva et al. [5] treated pepper processing wastewater under two different loading rates (0.31 and 0.70 g COD/L.d), revealing the composition of microbial communities in the aerobic immobilized biomass. The microbial community structure in the immobilized cell bioreactor was affected by the shift in the OLR, where the relative abundance of the genera *Pirellula, Nakamurella, Nitrospira,* and *Planctomyces* decreased from 54 to 19%. Moreover, the recorded total reads of the genera *Denitratisoma, Blastopirellula,* and *Holophaga* decreased from 3.8 to 0.9%. In the meantime, the OLR increase favored the presence of the genera *Gemmata, Nitrosospira,* and *Chitinophaga*. A thriving nitrifying population was also reported (near 10% of the total relative abundance at the OLR examined), even though bacteria of the genus *Nitrosospira* dominated the *Nitrospira* population after the OLR increase.

Zerva et al. [4] also examined the succession in microbial communities during the treatment of caper processing wastewater in an immobilized cell bioreactor under elevated salinity (from 0.3 to 10% *w/v*). A diverse ecology was revealed in the immobilized activated sludge at the lower salinities examined, while the increase in the salt concentration resulted in the restriction of the diversification of the microbial community. At low salinities, members of the genera *Pirellula, Amaricoccus, Planctomyces,* and *Arenibacter* accounted for 31.5 and 35.8% of total reads, respectively, while, at moderate salinities, representatives of the genera *Defluviimonas, Formosa, Muricauda, Arenibacter, Rhodobacter,* and *Roseovarius* were the predominant taxa. Remarkably, at the highest salinity examined, *Halomonas*, which represented 45% of the total relative abundance, followed by members of the genera *Roseovarius* (18%), *Idiomarina* (7%), and *Cyclobacterium* (5.7%) dominated hypersaline immobilized sludge.

Table 2. Microbial communities in aerobic biosystems treating agro-industrial effluents.

System	Effluent /Method	HRT (d)	OLR (g COD/ L.d)	COD (g/L)	NH_4^+-N (g/L)	CODrem (%)	NH_4^+-Nrem (%)	Microbial Community	Reference
Activated sludge reactor	Winery/ Isolates	2.1–4.4	0.4–5.9	0.8–12.8	0.001–2.0	90–98	100	Bacteria: *Bacillus* sp., *Pseudomons paucimobilis, Pseudomonas* sp., *Agrobacterium radiobacter, Acinetobacter*. Fungi: *Saccharomyces cerevisiae, Candida* sp., *Candida humicola, Candida kefyr, Trichosporum capitatum, Geotrichum penicillatum*.	[101]
Aerobic batch reactors	Rice winery/ Illumina	0.5	1.2, 2.4, 3.6	170	2.3 (as TN)	91.8–93.2	-	PHA-accumulating microorganisms: *Zoogloea* (5–41.1%), *Rhodobacter* (0.6–3.2%), *Rubrivivax* (0.1–2.6%), *Leptothrix* (0.1–2.5%), *Burkholderiaceae, Comamonas, Haliscomenobacter, Rhodobacteraceae, Amaricoccus and Plasticicumulans*. Under OLR of 2.4 g COD/L.d PHA-accumulating microorganisms covered 29.6% of the relative abundance. Phylum level: *Proteobacteria, Bacteroidetes, Verrucomicrobia, Patescibacteria, Acidpbacteria, Chloroflexi*.	[90]

Table 2. *Cont.*

System	Effluent /Method	HRT (d)	OLR (g COD/ L.d)	COD (g/L)	NH_4^+-N (g/L)	CODrem (%)	NH_4^+-Nrem (%)	Microbial Community	Reference
Full-scale A/O	Dairy/ Clone library	7	-	0.4–2.9 (as BOD)	10.1	98 (as BOD)	87	*Lactococcus, Veillonela, Atopobium, Olsenella, Zoogloea* spp., *Dechloromonas* spp., *Leptothrix* spp.	[102]
Activated sludge system	Dairy/ Clone library	5	-	14.9	0.8	77 (as BOD)	88	Phylum: *Proteobacteria* (55.1%), *Bacteroidetes* (15.4%), *Actinobacteria* (10.9%), *Firmicutes* (9.5%), *TM7* (4.5%) Species Aerobic effluent: *Thuera terpenica, Aeromicrobium marinum, Pseudomonas* sp., *Dyella japonica, Roseobacter* sp., *Sphingobacterium thalpophilum, Xanthomonas axonopodis, Dyella koreensis*, UBA318142, AF507866. Aerobic storage: *Thauera terpenica, Pirellula* sp., *Roseobacter* sp., *Rhodobacter praeacuta*, AY570630, *Xanthomonas axonopodis, Tissierella praeacuta*, AY438740, *Thermononas hydrothermalis, Petrimonas sulfuriphila*	[34]
SBR	Dairy/ Illumina	4	-	10.1	0.3	89	99	Phyla: *Proteobacteria* (73–75%) *Bacteroidetes* (15–18%) *Verrucomicrobia* (0.6–3%) *Planctomycetes* (0.5–3%), *Actinobacteria* (1.2–1.8%), *Chloroflexi* (0.7–1%). Class: *Alphaproteobacteria* (43–68%), *Betaproteobacteria* (5–21%), *Deltaroteobacteria* (1.5–9%), *Gammaproteobacteria* (1–7%). *Sphingobacteriia* (10–11%), *Flavobacteriia* (3–6%). Genus: *Terrimonas* (2.6–4.6%), *Thauera* (2.2–12%), *Nannocystis* (2–6%), *Flavitalea* (2.7–6%), *Hyphomonas* (2.7%).	[103]
Immobilized cell bioreactor	Pepper process-ing/ Illumina-Isolates	2.75	0.31–0.70	1.92	0.01	81.0–92.2	83.3–95.0	AOB: *Nitrosospira* (1.45–9.21%), NOB: *Nitrospira* (1.73–7.53%), *Nitrobacter* (0.04–0.13%), AOB abundance (2.00–9.77%), NOB abundance (71.77–7.66%), *Pirellula, Nakamurella, Nitrospira, Planctomyces* (54.03 to 19.10%), *Gemmata, Nitrosospira, Chitinophaga* (from 4.40 to 24.20%). Fungal taxa: *Rhizopus, Paramicrosporidium, Candida, Acaulospora, Neobulgaria*, (94.93 and 87.64%). Isolated microorganisms: *Paracoccus laeviglucosivorans, P. lutimaris, Microbacterium lacus, Microbacterium aurum, Sphingopyxis soli, Pseudoxanthomonas japonensis* and *P. mexicana, Gordonia hongkongensis, G. terrae, Oleiharenicola* sp.	[5]
Immobilized cell bioreactor	Capper process-ing/ Illumina	2.75–13.75	0.22–0.59	0.8–8.5	0.05–0.4 (as TKN)	70	37–70	Periods I and II: *Pirellula, Amaricoccus and Planctomyces, Arenibacter* (31.50 and 35.76%). Period III: *Amaricoccus* (16.14%), *Planctomyces* (12.39%), *Defluviimonas* (10.04%), *Formosa* (6.13%), *Arenibacter* (5.96%). Periods IV and V: *Defluviimonas, Formosa, Muricauda, Arenibacter, Rhodobacter, Roseovarius* (56.57 and 55.61%). Period VI: *Halomonas* (45.16%), *Roseovarius* (18.12%), *Idiomarina* (7.01%), *Cyclobacterium* (5.68%)	[4]
Full scale SBRs	Dairy/ DGGE-Pyrosequencing	-	2.5	1.8	0.7 (as TN)	94	95 (as TN)	Bacteria: *Proteiniphilum* (35.9%), *Byssovorax* (14.5%), *Acidobacterium* (10.1%), *Zoogloea* (8.3%), *Rhodomicrobium* (3,8%), *Roseomonas* (3.8%), *Comamonas* (3.8%), Leptothrix (2.1%), *Hydrogenophaga* (1.7%), *Ingella* (1.7%), *Thiothrix* (1.1%), *Gemmatimonas* (1.1%) DGGE: *Candidatus* Accumulibacter sp., *Lysobacter brunescens, Thauera* sp., *Saprospiraceae, Xanthomonadaceae, Caldilineaceae, Micrococcineae* (order *Actinomycetales*), *Zoogloea caeni, Amaricoccus* sp., *Chiayiivirga flava, Candidatus* Competibacter sp.	[94]

Table 2. *Cont.*

System	Effluent /Method	HRT (d)	OLR (g COD/ L.d)	COD (g/L)	NH_4^+-N (g/L)	CODrem (%)	NH_4^+-Nrem (%)	Microbial Community	Reference
Full-scale WWTP	Coffee/ Isolates	-	-	13.3	0.1 (as TN)	-	-	Bacteria: *Acetobacter indonesiensis, Chrysobacterium bovls, Corynebacterium flavescens, Serratia marcescens, Enterobacter* sp., *Corynebacterium callunae, Moxarela osloensis, Arthrobacter woluwenis* Fungi: *Wickerhamomyces anomalus, Torulaspora delbrueckii, Kazachstania gamospora, Saturnispora gosingensis, Meyerozyma caribbica, Kazachstania exigua, Hanseniaspora uvarum, Pichia fermentans* Filamentous fungi: *Fusarium oxysporum, Geotrichum silvicola, Geotrichum candidum, Alternaria alternata*	[98]
Full scale SBR	Winery/FISH- Clone library	0.5	-	5–16	-	-	-	FISH: *Amaricoccus* spp., *Defluviicoccus* cluster I and II, *Candidatus* 'Alysiosphaera europaea', *Zoogloea* sp., *Azoarcus* sp., type 0803, type 1851, *Haliscomenobacter hydrosis.* Clone library: *Amaricoccus kaplicensis, A. capsulatus, A. veronensis, A. tamworthnensis, Rhodocyclus tenuis, Geminococcus roseus, Tetracoccus cechii.*	[92]
Jet loop reactor	Winery/ Isolates	0.8–1	-	3.1–27.2	0.02–0.06 (as TKN)	63–84	-	Bacteria: *Acinetobacter* sp., *Bacillus* sp., *Pseudomonas* sp., *Sphingomonas paucimobilis, Agrobacerium radiobacter* Fungi: *Blastoschizomyces capitalis, Candida* sp., *Saccharomyces cerevisiae*	[96]
SBR	Olive mill/ Isolates	30	-	75.1	0.6 (as TKN)	60	-	Phyla: *Firmicutes* (57.1%), *Proteobacteria* (35.2%), *Actinobacteria* (7.7%) Genera: *Bacillus, Lysinibacillus, Brevibacillus, Paenibacillus, Roseomonas, Ochrobactrum, Pseudomonas, Klebsiella, Rhodococcus* Species: *Bacillus amyloliquefaciens* (11 isolates), *B. cereus* (8 isolates), *B. nealsonii* (7 isolates), *B. thioparans* (4 isolates), *B. thuringiensis* (3 isolates), *B. subtilis* (1 isolate), *Lysinibacillus macroids* (6 isolates), *Brevibacillus laterosporus* (6 isolates), *Paenibacillus xylanilyticus* (6 isolates), *Klebsiella oxytoca* (6 isolates), *Pseudomonas aeruginosa* (1 isolate), *Kocuria rosea* (1 isolate), *Cellulosimicrobium cellulans* (1 isolate)	[95]

8. Conclusions

Aerobic systems constitute a particularly reliable solution for the treatment of agro-industrial wastewater due to their stability and high nutrient removal efficiencies, particularly in cases where bioreactor systems operate under intermediate organic loading. Moreover, the ability to remove both carbon and nitrogen makes them attractive. In addition, activated sludge systems, in the case of bacterial communities, are dominated by members of the phylum *Proteobacteria*, while due to their nature, a significant diversity of yeasts is also observed. Most importantly, biological activated sludge systems that treat agro-industrial wastewater can be considered as a valuable source for the isolation of novel microbial strains, while the biotechnological potential of such wastewaters can be exploited through the production of hydrolytic enzymes and the recovery of added-value products. Future explorations of aerobic bioreactor systems, e.g., through the use of immobilized cell bioreactors and highly aerated reactors, could allow the biotreatment of agro-industrial wastewaters at higher organic loading rates, a modification that will expand the practical applicability of such biosystems in the treatment of highly polluted agro-industrial effluents.

Funding: This research received no external funding.

Data Availability Statement: Not applicable.

Conflicts of Interest: The author declares no conflict of interest.

References

1. Jenkins, D.; Wanner, J. (Eds.) *Activated Sludge-100 Years and Counting*; IWA Publishing: London, UK, 2014.
2. Martinez-Burgos, W.J.; Bittencourt Sydney, E.; Bianchi Pedroni Medeiros, A.; Magalhães, A.I.; de Carvalho, J.C.; Karp, S.G.; Soccol, C.R. Agro-industrial wastewater in a circular economy: Characteristics, impacts and applications for bioenergy and biochemicals. *Bioresour. Technol.* **2021**, *341*, 125795. [CrossRef] [PubMed]
3. Villa-Montoya, A.C.; Ferro, M.I.T.; de Oliveira, R.A. Removal of phenols and methane production with coffee processing wastewater supplemented with phosphorous. *Int. J. Environ. Sci. Technol.* **2017**, *14*, 61–74. [CrossRef]
4. Zerva, I.; Remmas, N.; Melidis, P.; Ntougias, S. Biotreatment efficiency, hydrolytic potential and bacterial community dynamics in an immobilized cell bioreactor treating caper processing wastewater under highly saline conditions. *Bioresour. Technol.* **2021**, *325*, 124694. [CrossRef]
5. Zerva, I.; Remmas, N.; Melidis, P.; Sylaios, G.; Stathopoulou, P.; Tsiamis, G.; Ntougias, S. Biotreatment, microbial community structure and valorization potential of pepper processing wastewater in an immobilized cell bioreactor. *Waste Biomass Valorization* **2022**, *13*, 1431–1447. [CrossRef]
6. Tanaka, Y.; Yatagai, A.; Masujima, H.; Waki, M.; Yokoyama, H. Autotrophic denitrification and chemical phosphate removal of agro-industrial wastewater by filtration with granular medium. *Bioresour. Technol.* **2017**, *98*, 787–791. [CrossRef]
7. Usubiaga, A.; Butnar, I.; Schepelmann, P. Wasting food, wasting resources: Potential environmental savings through food waste reductions. *J. Ind. Ecol.* **2018**, *22*, 574–584. [CrossRef]
8. Kosseva, M.R. Processing of food wastes. *Adv. Food Nutr. Res.* **2009**, *58*, 57–136.
9. Leite, P.; Sousa, D.; Fernandes, H.; Ferreira, M.; Costa, A.R.; Filipe, D.; Salgado, J.M. Recent advances in production of lignocellulolytic enzymes by solid-state fermentation of agro-industrial wastes. *Curr. Opin. Green Sustain. Chem.* **2021**, *27*, 100407. [CrossRef]
10. Ravindran, R.; Hassan, S.S.; Williams, G.A.; Jaiswal, A.K. A review on bioconversion of agro-industrial wastes to industrially important enzymes. *Bioengineering* **2018**, *5*, 93. [CrossRef]
11. European Union. Available online: https://food.ec.europa.eu/safety/food-waste_en (accessed on 1 August 2022).
12. Food and Agriculture Organization. Available online: http://www.fao.org/faostat/en/#data/QC (accessed on 1 August 2022).
13. European Commission. Available online: https://environment.ec.europa.eu/strategy/circular-economy-action-plan_en (accessed on 1 August 2022).
14. Food Loss and Waste Protocol. Available online: https://flwprotocol.org/ (accessed on 1 August 2022).
15. Blakeney, M. *Food Loss and Food Waste: Causes and Solutions*; Edward Elgar Publishing: Cheltenham, UK, 2019; pp. 1–204.
16. Sadh, P.K.; Duhan, S.; Duhan, J.S. Agro-industrial wastes and their utilization using solid state fermentation: A review. *Bioresour. Bioprocess.* **2018**, *5*, 35–42. [CrossRef]
17. El-Tayeb, T.S.; Abdelhafez, A.A.; Ali, S.H.; Ramadan, E.M. Effect of acid hydrolysis and fungal biotreatment on agro-industrial wastes for obtainment of free sugars for bioethanol production. *Braz. J. Microbiol.* **2012**, *43*, 1523–1535. [CrossRef] [PubMed]
18. Motte, J.C.; Trably, E.; Escudié, R.; Hamelin, J.; Steyer, J.P.; Bernet, N.; Delgenes, J.P.; Dumas, C. Total solids content: A key parameter of metabolic pathways in dry anaerobic digestion. *Biotechnol. Biofuels* **2013**, *6*, 164. [CrossRef] [PubMed]
19. Alayu, E.; Yirgu, Z. Advanced technologies for the treatment of wastewaters from agro-processing industries and cogeneration of by-products: A case of slaughterhouse, dairy and beverage industries. *Int. J. Environ. Sci. Technol.* **2018**, *15*, 1581–1596. [CrossRef]
20. Rezzadori, K.; Benedetti, S.; Amante, E.R. Proposals for the residues recovery: Orange waste as raw material for new products. *Food Bioprod. Processing* **2012**, *90*, 606–614. [CrossRef]
21. Christofoletti, C.A.; Escher, J.P.; Correia, J.E.; Marinho, J.F.U.; Fontanetti, C.S. Sugarcane vinasse: Environmental implications of its use. *Waste Manag.* **2013**, *33*, 2752–2761. [CrossRef]
22. Ahmed, Y.; Yaakob, Z.; Akhtar, P.; Sopian, K. Production of biogas and performance evaluation of existing treatment processes in palm oil mill effluent (POME). *Renew. Sustain. Energy Rev.* **2015**, *42*, 1260–1278. [CrossRef]
23. Tran, T.; Da, G.; Moreno-Santander, M.A.; Vélez-Hernández, G.A.; Giraldo-Toro, A.; Piyachomkwan, K.; Dufour, D. A comparison of energy use, water use and carbon footprint of cassava starch production in Thailand, Vietnam and Colombia. *Resour. Conserv. Recycl.* **2015**, *100*, 31–40. [CrossRef]
24. Zema, D.A.; Calabro, P.S.; Folino, A.; Tamburino, V.; Zappia, G.; Zimbone, S.M. Wastewater management in citrus processing industries: An overview of advantages and limits. *Water* **2019**, *11*, 2481. [CrossRef]
25. Akpomie, O.O.; Buzugbe, H.S.; Eze, P.M. Effect of brewery effluent on the microbiological quality of Ikpoba River and surrounding borehole waters in Benin City, Nigeria. *Microbiol. Res. J. Int.* **2014**, *5*, 76–82. [CrossRef]
26. Vergine, P.; Salerno, C.; Libutti, A.; Beneduce, L.; Gatta, G.; Berardi, G.; Pollice, A. Closing the water cycle in the agro-industrial sector by reusing treated wastewater for irrigation. *J. Clean. Prod.* **2017**, *164*, 587–596. [CrossRef]
27. Choi, H. Effect of mg-sericite flocculant for treatment of brewery wastewater. *Appl. Clay Sci.* **2015**, *115*, 145–149. [CrossRef]
28. Tonhato Junior, A.; Hasan, S.D.M.; Sebastien, N.Y. Optimization of Coagulation/Flocculation treatment of brewery wastewater employing organic flocculant based of vegetable tannin. *Water Air Soil Pollut.* **2019**, *230*, 202. [CrossRef]
29. Xu, S.; Zhu, J.; Meng, Z.; Li, W.; Ren, S.; Wang, T. Hydrogen and methane production by co-digesting liquid swine manure and brewery wastewater in a two-phase system. *Bioresour. Technol.* **2019**, *293*, 122041. [CrossRef] [PubMed]
30. Chen, J.; Liu, Y.; Liu, K.; Hu, L.; Yang, J.; Wang, X.; Wang, R. Bacterial community composition of internal circulation reactor at different heights for large-scale brewery wastewater treatment. *Bioresour. Technol.* **2021**, *331*, 125027. [CrossRef]

31. Auphimai, C.; Rukruam, W.; Chaiprasert, P. Efficacies of various inoculum sources on methane production from agro-industrial wastewaters. *Energy Procedia* **2014**, *52*, 167–172. [CrossRef]
32. Lucas, S.D.M.; Peixoto, G.; Mockaitis, G.; Zaiat, M.; Gomes, S.D. Energy recovery from agro-industrial wastewaters through biohydrogen production: Kinetic evaluation and technological feasibility. *Renew. Energy* **2015**, *75*, 496–504. [CrossRef]
33. Fleck, L.; Tavares, M.H.; Eyng, E.; Andrade, M.A.D.M.; Frare, L.M. Optimization of anaerobic treatment of cassava processing wastewater. *Eng. Agrícola* **2017**, *37*, 574–590. [CrossRef]
34. Jiang, Q.; Xin, Y.; Jiang, Y.; Huang, L.; Shen, P. Improving the efficiency of anaerobic digestion of molasses alcohol wastewater using cassava alcohol wastewater as a mixed feedstock. *Bioresour. Technol.* **2022**, *344*, 126179. [CrossRef]
35. Demirer, G.N.; Duran, M.; Güven, E.; Ugurlu, O.; Tezel, U.; Ergüder, T.H. Anaerobic treatability and biogas production potential studies of different agro-industrial wastewaters in Turkey. *Biodegradation* **2000**, *11*, 401–405. [CrossRef]
36. Maragkaki, A.E.; Fountoulakis, M.; Gypakis, A.; Kyriakou, A.; Lasaridi, K.; Manios, T. Pilot-scale anaerobic co-digestion of sewage sludge with agro-industrial by-products for increased biogas production of existing digesters at wastewater treatment plants. *Waste Manag.* **2017**, *59*, 362–370. [CrossRef]
37. Cruz-Salomón, A.; Ríos-Valdovinos, E.; Pola-Albores, F.; Meza-Gordillo, R.; Lagunas-Rivera, S.; Ruíz-Valdiviezo, V.M. Anaerobic treatment of agro-industrial wastewaters for COD removal in expanded granular sludge bed bioreactor. *Biofuel Res. J.* **2017**, *4*, 715–720. [CrossRef]
38. Thanos, D.; Maragkaki, A.; Venieri, D.; Fountoulakis, M.; Manios, T. Enhanced Biogas Production in Pilot Digesters Treating a Mixture of Olive Mill Wastewater and Agro-industrial or Agro-livestock By-Products in Greece. *Waste Biomass Valorization* **2021**, *12*, 135–143. [CrossRef]
39. Sounni, F.; Elgnaoui, Y.; El Bari, H.; Merzouki, M.; Benlemlih, M. Effect of mixture ratio and organic loading rate during anaerobic co-digestion of olive mill wastewater and agro-industrial wastes. *Biomass Convers. Biorefinery* **2021**, *11*, 1–7. [CrossRef]
40. Andiloro, S.; Bombino, G.; Tamburino, V.; Zema, D.A.; Zimbone, S.M. Aerated lagooning of agro-industrial wastewater: Depuration performance and energy requirements. *J. Agric. Eng.* **2013**, *44*, 827–832. [CrossRef]
41. Guzmán, J.; Mosteo, R.; Ormad, M.P.; Ovelleiro, J.L. Combined photo-fenton-SBR processes for the treatment of wastewater from the citrus processing industry. *J. Agric. Food Chem.* **2015**, *63*, 391–397. [CrossRef]
42. Guzmán, J.; Mosteo, R.; Sarasa, J.; Alba, J.A.; Ovelleiro, J.L. Evaluation of solar photo-fenton and ozone based processes as citrus wastewater pre-treatments. *Sep. Purif. Technol.* **2016**, *164*, 155–162. [CrossRef]
43. Corsino, S.F.; Di Trapani, D.; Torregrossa, M.; Viviani, G. Aerobic granular sludge treating high strength citrus wastewater: Analysis of pH and organic loading rate effect on kinetics, performance and stability. *J. Environ. Manag.* **2018**, *214*, 23–35. [CrossRef]
44. Kulandaivelu, V.; Bhat, R. Changes in the physico-chemical and biological quality attributes of soil following amendment with untreated coffee processing wastewater. *Eur. J. Soil Biol.* **2012**, *50*, 39–43. [CrossRef]
45. Alemayehu, Y.A.; Asfaw, S.L.; Tirfie, T.A. Management options for coffee processing wastewater. A review. *J. Mater. Cycles Waste Manag.* **2020**, *22*, 454–469. [CrossRef]
46. Samuel, Z.A. Treatment of combined coffee processing wastewater using constructed wetland/*Cyperus ustulatus* and *typha latifolia* plants process. *Glob. Nest J.* **2021**, *23*, 429–433.
47. Sujatha, G.; Shanthakumar, S.; Chiampo, F. UV light-irradiated photocatalytic degradation of coffee processing wastewater using TiO_2 as a catalyst. *Environments* **2020**, *7*, 47. [CrossRef]
48. Fountoulakis, M.S.; Drakopoulou, S.; Terzakis, S.; Georgaki, E.; Manios, T. Potential for methane production from typical Mediterranean agro-industrial by-products. *Biomass Bioenergy* **2008**, *32*, 155–161. [CrossRef]
49. Lucas, M.S.; Peres, J.A. Treatment of olive mill wastewater by a combined process: Fenton's reagent and chemical coagulation. *J. Environ. Sci. Health-Part A Toxic/Hazard. Subst. Environ. Eng.* **2009**, *44*, 198–205. [CrossRef] [PubMed]
50. Dias, B.; Lopes, M.; Ramõa, R.; Pereira, A.S.; Belo, I. *Candida tropicalis* as a promising oleaginous yeast for olive mill waste water bioconversion. *Energies* **2021**, *14*, 640. [CrossRef]
51. Kamyab, H.; Md Din, M.F.; Ponraj, M.; Keyvanfar, A.; Rezania, S.; Taib, S.M.; Abd Majid, M.Z. Isolation and screening of microalgae from agro-industrial wastewater (POME) for biomass and biodiesel sources. *Desalin. Water Treat.* **2016**, *57*, 29118–29125. [CrossRef]
52. Shak, K.P.Y.; Wu, T.Y. Synthesis and characterization of a plant-based seed gum via etherification for effective treatment of high-strength agro-industrial wastewater. *Chem. Eng. J.* **2017**, *307*, 928–938. [CrossRef]
53. Aka, B.E.Z.; Djeni, T.N.; Amoikon, S.L.T.; Kannengiesser, J.; Ouazzani, N.; Dje, M.K. High-throughput 16S rRNA gene sequencing of the microbial community associated with palm oil mill effluents of two oil processing systems. *Sci. Rep.* **2021**, *11*, 13232. [CrossRef]
54. Nasrullah, M.; Ansar, S.; Krishnan, S.; Singh, L.; Peera, S.G.; Zularisam, A.W. Electrocoagulation treatment of raw palm oil mill effluent: Optimization process using high current application. *Chemosphere* **2022**, *299*, 134387. [CrossRef]
55. Hernández, D.; Riaño, B.; Coca, M.; García-González, M.C. Treatment of agro-industrial wastewater using microalgae-bacteria consortium combined with anaerobic digestion of the produced biomass. *Bioresour. Technol.* **2013**, *135*, 598–603. [CrossRef]
56. Muniraj, I.K.; Xiao, L.; Hu, Z.; Zhan, X.; Shi, J. Microbial lipid production from potato processing wastewater using oleaginous filamentous fungi *Aspergillus oryzae*. *Water Res.* **2013**, *47*, 3477–3483. [CrossRef]

57. Antwi, P.; Li, J.; Boadi, P.O.; Meng, J.; Shi, E.; Deng, K.; Bondinuba, F.K. Estimation of biogas and methane yields in an UASB treating potato starch processing wastewater with backpropagation artificial neural network. *Bioresour. Technol.* **2017**, *228*, 106–115. [CrossRef] [PubMed]
58. İnce, E.; İnce, M.; Önkal Engin, G. Comparison of thermophilic and mesophilic anaerobic treatments for potato processing wastewater using a contact reactor. *Glob. Nest J.* **2017**, *19*, 318–326.
59. Bouchareb, R.; Bilici, Z.; Dizge, N. Potato processing wastewater treatment using a combined process of chemical coagulation and membrane filtration. *Clean Soil Air Water* **2021**, *49*, 2100017. [CrossRef]
60. Daskalaki, V.M.; Marakas, H.; Mantzavinos, D.; Katsaounis, A.; Gikas, P. Use of seawater for the boron-doped diamond electrochemical treatment of diluted vinasse wastewater. *Water Sci. Technol.* **2013**, *68*, 2344–2350. [CrossRef]
61. Petta, L.; De Gisi, S.; Casella, P.; Farina, R.; Notarnicola, M. Evaluation of the treatability of a winery distillery (vinasse) wastewater by UASB, anoxic-aerobic UF-MBR and chemical precipitation/adsorption. *J. Environ. Manag.* **2017**, *201*, 177–189. [CrossRef]
62. García-Depraect, O.; Valdez-Vázquez, I.; Rene, E.R.; Gómez-Romero, J.; López-López, A.; León-Becerril, E. Lactate- and acetate-based biohydrogen production through dark co-fermentation of tequila vinasse and nixtamalization wastewater: Metabolic and microbial community dynamics. *Bioresour. Technol.* **2019**, *282*, 236–244. [CrossRef]
63. Santos, C.; Lucas, M.S.; Dias, A.A.; Bezerra, R.M.F.; Peres, J.A.; Sampaio, A. Winery wastewater treatment by combination of *Cryptococcus laurentii* and fenton's reagent. *Chemosphere* **2014**, *117*, 53–58. [CrossRef]
64. Moreira, F.C.; Boaventura, R.A.R.; Brillas, E.; Vilar, V.J.P. Remediation of a winery wastewater combining aerobic biological oxidation and electrochemical advanced oxidation processes. *Water Res.* **2015**, *75*, 95–108. [CrossRef]
65. Rodríguez-Chueca, J.; Amor, C.; Silva, T.; Dionysiou, D.D.; Li Puma, G.; Lucas, M.S.; Peres, J.A. Treatment of winery wastewater by sulphate radicals: HSO5−/transition metal/UV-A LEDs. *Chem. Eng. J.* **2017**, *310*, 473–483. [CrossRef]
66. Amor, C.; Rodríguez-Chueca, J.; Fernandes, J.L.; Domínguez, J.R.; Lucas, M.S.; Peres, J.A. Winery wastewater treatment by sulphate radical based-advanced oxidation processes (SR-AOP): Thermally vs UV-assisted persulphate activation. *Process Saf. Environ. Prot.* **2019**, *122*, 94–101. [CrossRef]
67. Jorge, N.; Teixeira, A.R.; Matos, C.C.; Lucas, M.S.; Peres, J.A. Combination of coagulation–flocculation–decantation and ozonation processes for winery wastewater treatment. *Int. J. Environ. Res. Public Health* **2021**, *18*, 8882. [CrossRef]
68. Castro-Muñoz, R.; Barragán-Huerta, B.E.; Fíla, V.; Denis, P.C.; Ruby-Figueroa, R. Current role of membrane technology: From the treatment of agro-industrial by-products up to the valorization of valuable compounds. *Waste Biomass Valorization* **2018**, *9*, 513–529. [CrossRef]
69. Tatoulis, T.I.; Zapantiotis, S.; Frontistis, Z.; Akratos, C.S.; Tekerlekopoulou, A.G.; Pavlou, S.; Vayenas, D.V. A hybrid system comprising an aerobic biological process and electrochemical oxidation for the treatment of black table olive processing wastewaters. *Int. Biodeterior. Biodegrad.* **2016**, *109*, 104–112. [CrossRef]
70. Vergine, P.; Salerno, C.; Berardi, G.; Pappagallo, G.; Pollice, A. The self-forming dynamic membrane BioReactor (SFD MBR) as a suitable technology for agro-industrial wastewater treatment. *New Biotechnol.* **2020**, *56*, 87–95. [CrossRef] [PubMed]
71. Carucci, A.; Cappai, G.; Erby, G.; Milia, S. Aerobic granular sludge formation in a sequencing batch reactor treating agro-industrial digestate. *Environ. Technol.* **2021**, *42*, 3932–3941. [CrossRef]
72. Levett, K.J.; Vanderzalm, J.L.; Page, D.W.; Dillon, P.J. Factors affecting the performance and risks to human health of on-site wastewater treatment systems. *Water Sci. Technol.* **2010**, *62*, 1499–1509. [CrossRef]
73. Seviour, R.; Nielsen, P.H. (Eds.) *Microbial Ecology of Activated Sludge*; IWA Publishing: London, UK, 2010.
74. Ibarbalz, F.M.; Figuerola, E.L.; Erijman, L. Industrial activated sludge exhibit unique bacterial community composition at high taxonomic ranks. *Water Res.* **2013**, *47*, 3854–3864. [CrossRef]
75. Eusébio, A.; Mateus, M.; Baeta-Hall, L.; Almeida-Vara, E.; Duarte, J.C. Microflora evaluation of two agro-industrial effluents treated by the JACTO jet-loop type reactor system. *Water Sci. Technol.* **2005**, *51*, 107–112. [CrossRef]
76. Ochando-Pulido, J.M.; Stoller, M.; Martinez-Ferez, A. Boundary flux modelling for purification optimization of differently-pretreated agro-industrial wastewater with nanofiltration. *Sep. Purif. Technol.* **2018**, *193*, 147–154. [CrossRef]
77. Davididou, K.; Frontistis, Z. Advanced oxidation processes for the treatment of winery wastewater: A review and future perspectives. *J. Chem. Technol. Biotechnol.* **2021**, *96*, 2436–2450. [CrossRef]
78. Riffat, R.; Husnain, T. *Fundamentals of Wastewater Treatment and Engineering*; CRC Press: Boca Raton, FL, USA, 2013; pp. 1–324.
79. Rajagopal, R.; Saady, N.M.C.; Torrijos, M.; Thanikal, J.V.; Hung, Y. Sustainable agro-food industrial wastewater treatment using high rate anaerobic process. *Water* **2013**, *5*, 292–311. [CrossRef]
80. Meneses-Jácome, A.; Diaz-Chavez, R.; Velásquez-Arredondo, H.I.; Cárdenas-Chávez, D.L.; Parra, R.; Ruiz-Colorado, A.A. Sustainable energy from agro-industrial wastewaters in latin-america. *Renew. Sustain. Energy Rev.* **2016**, *56*, 1249–1262. [CrossRef]
81. Petruccioli, M.; Duarte, J.C.; Federici, F. High-rate aerobic treatment of winery wastewater using bioreactors with free and immobilized activated sludge. *J. Biosci. Bioeng.* **2000**, *90*, 381–386. [CrossRef]
82. Roveroto, G.P.; Teles, J.C.; Vuitik, G.A.; Batista, J.S.D.S.; Barana, A.C. Craft brewery wastewater treatment: A fixed-bed single-batch reactor with intermittent aeration to remove COD and TN. *Braz. Arch. Biol. Technol.* **2021**, *64*, 1–14. [CrossRef]
83. Zema, D.A.; Andiloro, S.; Bombino, G.; Tamburino, V.; Sidari, R.; Caridi, A. Depuration in aerated ponds of citrus processing wastewater with a high concentration of essential oils. *Environ. Technol.* **2012**, *33*, 1255–1260. [CrossRef] [PubMed]
84. Moore, A.W.; Zytner, R.G.; Chang, S. Potential water reuse for high strength fruit and vegetable processor wastewater with an MBR. *Water Environ. Res.* **2016**, *88*, 852–870. [CrossRef]

85. International Coffee Organization. Available online: https://www.ico.org/new_historical.asp?section=Statistics (accessed on 10 August 2022).
86. Rossmann, M.; de Matos, A.T.; Abreu, E.C.; e Silva, F.F.; Borges, A.C. Performance of constructed wetlands in the treatment of aerated coffee processing wastewater: Removal of nutrients and phenolic compounds. *Ecol. Eng.* **2012**, *49*, 264–269. [CrossRef]
87. Coelho, A.L.S.; Orlandelli, R.C. Immobilized microbial lipases in the food industry: A systematic literature review. *Crit. Rev. Food Sci. Nutr.* **2020**, *61*, 1689–1703. [CrossRef]
88. Sharari, M.; Roohani, M.; Jahan Latibari, A.; Guillet, A.; Aurousseau, M.; Sharari, A. Treatment of bagasse preparation effluent by *phanerochaete chrysosporium* immobilized on polyurethane foam: Enzyme production versus pollution removal. *Ind. Crops Prod.* **2013**, *46*, 226–233. [CrossRef]
89. Mafakher, L.; Mirbagheri, M.; Darvishi, F.; Nahvi, I.; Zarkesh-Esfahani, H.; Emtiazi, G. Isolation of lipase and citric acid producing yeasts from agro-industrial wastewater. *N. Biotechnol.* **2010**, *27*, 337–340. [CrossRef]
90. Fang, F.; Xu, R.; Huang, Y.; Wang, S.; Zhang, L.; Dong, J.; Cao, J. Production of polyhydroxyalkanoates and enrichment of associated microbes in bioreactors fed with rice winery wastewater at various organic loading rates. *Bioresour. Technol.* **2019**, *292*, 121978. [CrossRef] [PubMed]
91. Inoue, D.; Fukuyama, A.; Ren, Y.; Ike, M. Optimization of aerobic dynamic discharge process for very rapid enrichment of polyhydroxyalkanoates-accumulating bacteria from activated sludge. *Bioresour. Technol.* **2021**, *336*, 125314. [CrossRef] [PubMed]
92. McIlroy, S.J.; Speirs, L.B.M.; Tucci, J.; Seviour, R.J. In situ profiling of microbial communities in full-scale aerobic sequencing batch reactors treating winery waste in Australia. *Environ. Sci. Technol.* **2011**, *45*, 8794–8803. [CrossRef] [PubMed]
93. Jachimowicz, P.; Cydzik-Kwiatkowska, A.; Szklarz, P. Effect of aeration mode on microbial structure and efficiency of treatment of TSS-rich wastewater from meat processing. *Appl. Sci.* **2020**, *10*, 7414. [CrossRef]
94. Meunier, C.; Henriet, O.; Schroonbroodt, B.; Boeur, J.; Mahillon, J.; Henry, P. Influence of feeding pattern and hydraulic selection pressure to control filamentous bulking in biological treatment of dairy wastewaters. *Bioresour. Technol.* **2016**, *221*, 300–309. [CrossRef]
95. Arous, F.; Jamdi, C.; Kmiha, S.; Khammassi, N.; Ayari, A.; Neifar, M.; Mechichi, T.; Jaouani, A. Treatment of olive mill wastewater through employing sequencing batch reactor: Performance and microbial diversity assessment. *3 Biotech* **2018**, *8*, 481. [CrossRef]
96. Eusébio, A.; Petruccioli, M.; Lageiro, M.; Federici, F.; Duarte, J.C. Microbial characterization of activated sludge in jet-loop bioreactors treating winery wastewaters. *J. Ind. Microbiol. Biotechnol.* **2004**, *31*, 29–34. [CrossRef]
97. McGarvey, J.A.; Miller, W.G.; Zhang, R.; Ma, Y.; Mitloehner, F. Bacterial population dynamics in dairy waste during aerobic and anaerobic treatment and subsequent storage. *Appl. Environ. Microbiol.* **2007**, *73*, 193–202. [CrossRef]
98. Pires, J.F.; Cardoso, L.S.; Schwan, R.F.; Silva, C.F. Diversity of microbiota found in coffee processing wastewater treatment plant. *World J. Microbiol. Biotechnol.* **2017**, *33*, 211. [CrossRef]
99. Ganapathy, B.; Yahya, A.; Ibrahim, N. Bioremediation of palm oil mill effluent (POME) using indigenous *meyerozyma guilliermondii*. *Environ. Sci. Pollut. Res.* **2019**, *26*, 11113–11125. [CrossRef]
100. Buratti, S.; Girometta, C.E.; Baiguera, R.M.; Barucco, B.; Bernardi, M.; De Girolamo, G.; Savino, E. Fungal diversity in two wastewater treatment plants in North Italy. *Microorganisms* **2022**, *10*, 1096. [CrossRef] [PubMed]
101. Petruccioli, M.; Cardoso Duarte, J.; Eusebio, A.; Federici, F. Aerobic treatment of winery wastewater using a jet-loop activated sludge reactor. *Process Biochem.* **2002**, *37*, 821–829. [CrossRef]
102. Hirota, K.; Yokota, Y.; Sekimura, T.; Uchiumi, H.; Guo, Y.; Ohta, H.; Yumoto, I. Bacterial communities in different locations, seasons and segments of a dairy wastewater treatment system consisting of six segments. *J. Environ. Sci.* **2016**, *46*, 109–115. [CrossRef] [PubMed]
103. Hanson, A.J.; Guho, N.M.; Paszczynski, A.J.; Coats, E.R. Community proteomics provides functional insight into polyhydroxyalkanoate production by a mixed microbial culture cultivated on fermented dairy manure. *Appl. Microbiol. Biotechnol.* **2016**, *100*, 7957–7976. [CrossRef]

processes

Article

Preparation and Phosphorus Removal Performance of Zr–La–Fe Ternary Composite Adsorbent Embedded with Sodium Alginate

Xiuling Li, Hanyu Liang, Yanling Mo and Yansong Wei *

College of Chemical and Biological Engineering, Hechi University, Yizhou 546300, China
* Correspondence: 06020@hcnu.edu.cn; Tel.: +86-13607786565

Abstract: Using single metal salts of zirconium, lanthanum, and iron as raw materials and sodium alginate as a cross-linking agent, a new composite adsorbent was prepared via the co-precipitation method and embedding immobilization technology, and its phosphorus adsorption performance in wastewater was evaluated. Scanning electron microscopy (SEM), Fourier transform infrared spectroscopy (FT-IR), X-ray diffraction (XRD), and X-ray photoelectron spectroscopy (XPS) were used for characterization, and a 0.5 mol·L^{-1} sodium hydroxide solution was used to regenerate the adsorbent. The experimental results demonstrated that the adsorption rate reached 99.88% when the wastewater volume was 50 mL, the initial concentration of phosphorus-containing wastewater was 5 mg·L^{-1}, the pH was 5, the dosage of composite adsorbent was 0.2 g, and the adsorption time was 200 min. The prepared adsorbent could reduce the initial phosphorus concentration of 5 mg·L^{-1} to 0.006 mg·L^{-1} in simulated wastewater, and from 4.17 mg·L^{-1} in urban sewage to undetected (<0.01 mg·L^{-1}), thus meeting the discharge requirements of the grade A standard of the Urban Sewage Treatment Plant Pollutant Discharge Standard (GB18918-2002). The adsorption process conformed to the Freundlich adsorption isothermal equation and quasi-second-order kinetic equation, and the adsorption reaction was exothermic and spontaneous. More importantly, after three lye regeneration tests, the removal rate of phosphorus in water remained above 68%, that is, the composite adsorbent could be reproducibly fabricated and recycled. The characterization results showed that the surface of the composite adsorbent was rough, with a complex pore structure. After phosphorus removal, the surface morphology of the composite adsorbent showed a similar honeycomb structure, with a P-H, P-O stretching vibration peak and a characteristic P2p peak. At the same time, the proportion of hydroxyl groups (M-OH) on the metal surface decreased after adsorption. Our findings thus demonstrate that the mechanism of phosphorus removal is mainly based on the coordination exchange reaction between phosphate and metal active sites and surface hydroxyl groups, resulting in the formation of granular phosphate deposits.

Keywords: sodium alginate; Zr–La–Fe; composite adsorbent; phosphorus removal; regeneration

Citation: Li, X.; Liang, H.; Mo, Y.; Wei, Y. Preparation and Phosphorus Removal Performance of Zr–La–Fe Ternary Composite Adsorbent Embedded with Sodium Alginate. *Processes* **2022**, *10*, 1761. https://doi.org/10.3390/pr10091761

Academic Editors: Dimitris Zagklis and Georgios Bampos

Received: 29 July 2022
Accepted: 29 August 2022
Published: 2 September 2022

1. Introduction

Due to the increase in industrial output and human activities, such as the excessive use of phosphorus fertilizer and pesticides in rural areas, and the arbitrary discharge of livestock and poultry breeding wastewater, large amounts of phosphorus are discharged into aquatic environments and cause eutrophication, which leads to water hypoxia, a large amount death to aquatic life, human gastrointestinal disease, and so on. Urban sewage treatment plants are responsible for treating sewage water. However, urban sewage treatment plants must be urgently upgraded to further improve sewage treatment capacity and treatment standards, meet the requirements of China's key river basin environmental governance, and alleviate water eutrophication. This would facilitate compliance with the "Urban Sewage Treatment Plant Pollutant Discharge Standard" (GB18918-2002), in addition to reducing phosphorus discharge through in-depth treatment. Therefore, research on phosphorus removal from wastewater has recently garnered increasing attention.

Current phosphorus removal methods include chemical precipitation [1–3], biological methods [4–6], and adsorption [7–10]. Among these, adsorption can be easily and stably implemented and delivers good effluent quality [11]. This approach can not only be used to supplement biological phosphorus removal for the treatment of wastewater with low phosphorus concentrations, but can also be used as a stand-alone method for the treatment of wastewater with high phosphorus concentrations. Moreover, phosphorus removal by adsorption does not cause secondary pollution, and the adsorbed phosphate can be desorbed and recovered. Therefore, the adsorption method has become a widely used treatment technology in the field of phosphorus wastewater treatment.

Studies have found that some metal elements such as Zr, Fe, La, and Mn have strong affinities for phosphorus, and single or binary composite oxides of these metal elements can be synthesized as adsorbents to remove excess phosphorus in water [12–19]. According to the existing literature, we find that lanthanum is a kind of benign rare-earth material that is rich in reserves, environmentally friendly, and non-toxic. Due to its strong binding ability to phosphate ($LaPO_4$ pKsp = 26.16), it can exhibit a strong affinity for phosphorus. Meanwhile, iron (Fe) and zirconium (Zr) compounds are widely used for phosphate removal in wastewater treatment due to their higher surface charge, lower cost, and higher resistance to acid, alkali, oxidant and reducing agent compared to La, which could stably adsorb phosphate in a wider pH range [20]. Multi-metal oxide composites usually exhibit significantly different physical and chemical properties from their single-component oxides and can combine the advantages of single metals, which improves phosphate adsorption performance to a certain extent. Thereby, we chose the single metal salts of zirconium, lanthanum, and iron as raw materials to prepare a new composite adsorbent. However, recent studies have demonstrated that ternary composite adsorbents show greater advantages in phosphorus removal from water compared to other conventionally used materials. Yu et al. [21] synthesized an Fe–Mg–La ternary metal complex for phosphorus removal in water, which reached its maximum adsorption of 415.2 $mg \cdot g^{-1}$ at a pH of 6.0. Yang et al. [22] prepared an Fe–Mg–La ternary composite adsorbent and its maximum adsorption capacity was 30.16 $mg \cdot g^{-1}$ at a pH of 3.02. Lu et al. [23] synthesized an Fe–Al–Mn ternary composite adsorbent by adding alumina to Fe–Mn to improve adsorption and phosphorus removal performance.

Although the aforementioned adsorbents can effectively remove phosphate from water, most of them occur in powder form and, therefore, cannot be easily separated and recycled. Sodium alginate is a kind of biopolymer that contains large amounts of reactive functional groups, such as -COOH and -OH. The presence of calcium ions in aqueous solution can induce the formation of a three-dimensional mesh structure of salt alginate gel particles via ion exchange-mediated polymerization. In turn, this immobilizes powdery adsorbents into a granular material, which not only improves adsorbent settlement but also facilitates its recovery and utilization. Additionally, sodium alginate is safe, stable, non-toxic, environmentally friendly, and does not cause secondary water pollution.

Therefore, based on the findings of previous studies, our study employs three metal salts of zirconium, lanthanum, and iron as raw materials, and sodium alginate was used as a crosslinking agent to prepare a new type of ternary composite adsorbent via the co-precipitation method coupled with embedding immobilization technology. The adsorption performance of the composite adsorbent on phosphorus in water was characterized by XRD, SEM, XPS, and FT-IR. Moreover, the phosphorus removal mechanism of the composite adsorbent was studied to establish a novel approach for the treatment of phosphorus-containing wastewater.

2. Experimental

2.1. Materials

$LaCl_3 \cdot 7H_2O$, $Fe(NO_3)_3 \cdot 9H_2O$, $ZrOCl_2 \cdot 8H_2O$, KH_2PO_4, H_2SO_4, HCl, NaOH, $K_2S_2O_8$, $C_6H_8O_6$, $(NH_4)_6MoO_{24} \cdot 4H_2O$, and $K(SbO)C_4H_4O_6 \cdot 5H_2O$ were purchased from Tianjin Kemi Ou Chemical Co., LTD. (Tianjin, China) All reagents were of analytical grade.

Potassium dihydrogen phosphate (KH_2PO_4) was used to prepare a phosphorus stock solution (1000 mg·L^{-1}), and simulated wastewater with different phosphorus levels was prepared by diluting this stock solution to the desired concentrations. Wastewater samples were collected from the outlet of the secondary sedimentation tank of the First Sewage Treatment Plant in Yizhou, Hechi, Guangxi.

2.2. Preparation of Zr–La–Fe Ternary Adsorbent Embedded with Sodium Alginate

Certain masses of $ZrOCl_2·8H_2O$, $LaCl_3·7H_2O$, and $Fe(NO_3)_3·9H_2O$ were weighed, dissolved and shaken in 250 mL of deionized water, and the three dissolved materials were poured into a 1 L beaker, stirred, and shaken. The pH value of the mixture was then adjusted to 8.00 by slowly adding a 2 mol·L^{-1} NaOH solution, and a suspension gradually appeared when adding the alkali. Next, the mixture was transferred to a blender and stirred for more than 1 hour. The finished mixture was transferred to a round-bottom flask, the temperature of the water bath thermostat was adjusted to 25 °C, and the solution was allowed to age for 24–30 h. The solution was then removed, and the solids were washed with deionized water while measuring the pH value of the solution. Once the pH was neutral, the mixture was placed in a suction filter machine for extraction and filtration, after which the adsorption material was transferred to a 70 °C oven for drying. After drying, the resulting Zr–La–Fe ternary composite material was ground and sieved.

Next, the prepared Zr–La–Fe ternary composite adsorbent was immobilized with sodium alginate. A certain amount of sodium alginate was first weighed and dissolved in water, after which the mixture was transferred to a volumetric flask. A certain proportion of sodium alginate was then added to the Zr–La–Fe terpolymer adsorbent, after which the preparation was stirred, evenly dispersed, and slowly added to an aqueous solution with a certain concentration of calcium chloride. After soaking and cross-linking, the solids were washed 2–3 times with deionized water. After drying in an oven at 85 °C, the composite adsorbent required for phosphorus removal in this experiment was obtained.

2.3. Sample Characterization

The composition and internal morphology of the composite adsorbent were analyzed by XRD. The groups and structures of the composite adsorbent were analyzed by characterizing its FT-IR spectra. The morphology features of the composite adsorbent were analyzed by SEM. The elemental composition of the composite adsorbent was analyzed based on its XPS spectra. Point-of-zero-charge measurements were conducted to evaluate the surface charge distribution of the composite adsorbent under different pH conditions.

2.4. Experimental Methods

The wastewater volume in the adsorption experiment was fixed at 50 mL, the rotating speed of the shaker was 180 r·min^{-1}, and the adsorption experiment was carried out in a 250 mL conical flask. Different experimental conditions were changed to study the influence of various factors on phosphorus removal. After adsorption, the wastewater was filtered with a 0.45 μm drainage membrane and the residual phosphorus concentration was determined by spectrophotometry. The phosphorus removal rate (E) and adsorption capacity (Q_e) were used as evaluation indices, and the calculation formula was as follows:

$$E = \frac{(C_0 - C_e)}{C_0} \times 100\% \tag{1}$$

$$Q_e = \frac{(C_0 - C_e) \cdot V}{m \times 1000} \tag{2}$$

where C_0 is the initial concentration of total phosphorus, mg·L^{-1}; C_e is the equilibrium concentration of total phosphorus, mg·L^{-1}; V is the volume of simulated wastewater containing phosphorus, mL; m is the dosage of adsorbent, g; and Q_e is the equilibrium adsorption capacity, mg·g^{-1}.

2.5. Thermodynamic and Kinetic Fitting of Adsorption

2.5.1. Thermodynamic Fitting of Adsorption

Langmuir adsorption isotherm [24]:

$$\frac{1}{Q_e} = \frac{K_L}{Q_m C_e} + \frac{1}{Q_m} \tag{3}$$

where C_e and Q_e represent the same as in Equations (1) and (2); Q_m is the maximum adsorption capacity, mg·g^{-1}; K_L is the equilibrium constant.

Freundlich sorption isotherm [25]:

$$lgQ_e = lgK_F + \frac{1}{n} lgC_e \tag{4}$$

where C_e and Q_e represent the same as in Equations (1) and (2); K_F and n are both empirical constants.

Van't Hoff thermodynamic fitting equations [26]:

$$lnK = \frac{\Delta ST - \Delta H}{RT} \tag{5}$$

where K is the adsorption equilibrium constant; T is the absolute temperature, K; ΔS is the entropy change, J·$\left(\text{mol·K}^{-1}\right)^{-1}$; and ΔH is the enthalpy change, J·mol^{-1}.

Gibbs free energy change [26]:

$$\Delta G = -\text{R}TlnK \tag{6}$$

where K is the adsorption equilibrium constant; T is the absolute temperature, K; and ΔG is the Gibbs free energy change, kJ·mol^{-1}.

2.5.2. Adsorption Kinetics Fitting

Quasi first-order adsorption kinetic model [27]:

$$ln(Q_e - Q_t) = lnQ_e - k_1 t \tag{7}$$

where Q_t is the adsorption capacity at any time, mg·g^{-1}; Q_e represents the same as in Equation (2); t is the adsorption time, min; and k_1 is the first-order rate constant, h^{-1}.

Quasi-second-order adsorption kinetic model [27]:

$$\frac{t}{Q_e} = \frac{1}{k_2 Q_e^2} + \frac{t}{Q_e} \tag{8}$$

where Q_e represents the same as in Equation (2); t is the adsorption time, min; and k_2 is the second-order rate constant, h^{-1}.

3. Results and Discussion

3.1. XRD and FT-IR Characterization

XRD patterns of composite adsorbents are shown in Figure 1.

As shown in Figure 1, the characteristic peaks of the Zr–La–Fe ternary composite adsorbent were observed at diffraction angles of 31.56°, 45.43°, 56.57°, 67.97°, and 74.95°. Compared with the standard card, the characteristic diffraction peaks of ZrO were located at 31.56° and 67.97°. The characteristic diffraction peaks of La_2O_3 at 45.43° and 74.95° corresponded to the characteristic diffraction peaks of Fe_3O_4 at 56.57°, indicating that the preparation of the ternary composite adsorbent was successful.

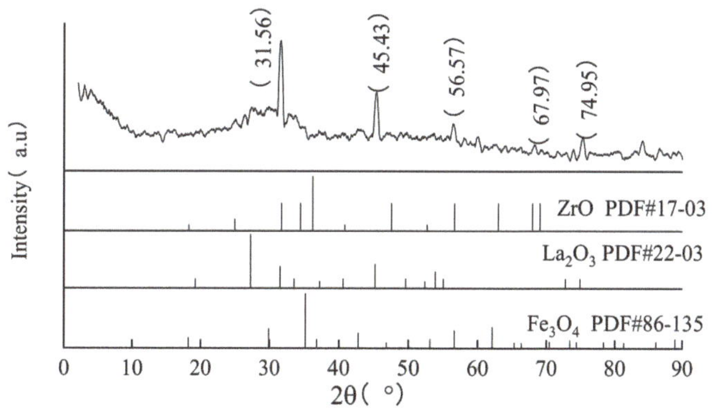

Figure 1. XRD pattern of composite adsorbent and standard card.

3.2. FT-IR Characterization Results and Analysis

The FT-IR spectra of composite adsorbents are shown in Figure 2. As shown in Figure 2, before phosphorus removal, the samples exhibited an O-H deformation vibration peak and C-O stretching vibration peak of sodium alginate at 1400 cm^{-1} and 1090 cm^{-1}, as well as an O-H stretching vibration peak of sodium alginate at 3500 cm^{-1} and H-O-H bending vibration peak of water molecules at 1650 cm^{-1}. After phosphorus removal by adsorption, two new peaks, the P-H stretching vibration peak and P-O stretching vibration peak, were observed at 2350 cm^{-1} and 1080 cm^{-1}, respectively, thus confirming that phosphate had a coordination exchange reaction with surface hydroxyl groups, which enabled phosphorus removal.

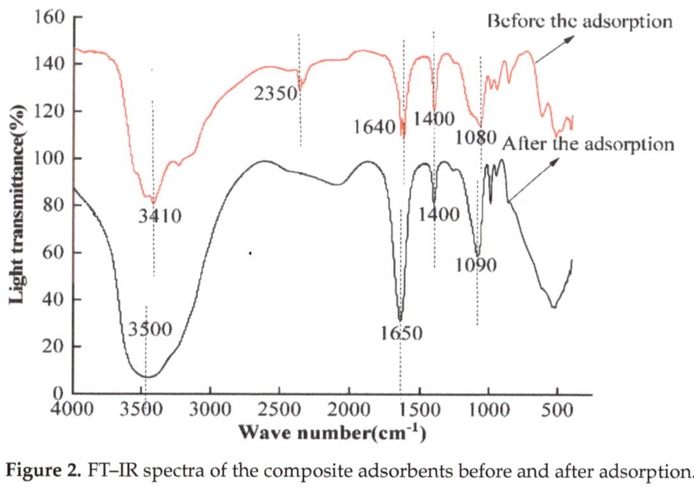

Figure 2. FT–IR spectra of the composite adsorbents before and after adsorption.

3.3. SEM Characterization Results and Analysis

Figure 3 illustrates the SEM images before and after phosphorus adsorption by the composite adsorbent.

As shown in Figure 3, before phosphorus removal by the composite adsorbent, its surface was rough and irregular, and its pore structure was complex. After adsorption and phosphorus removal, some obvious changes occurred in the surface structure, such as a honeycomb structure, increased pore connection, and the formation of granular deposits.

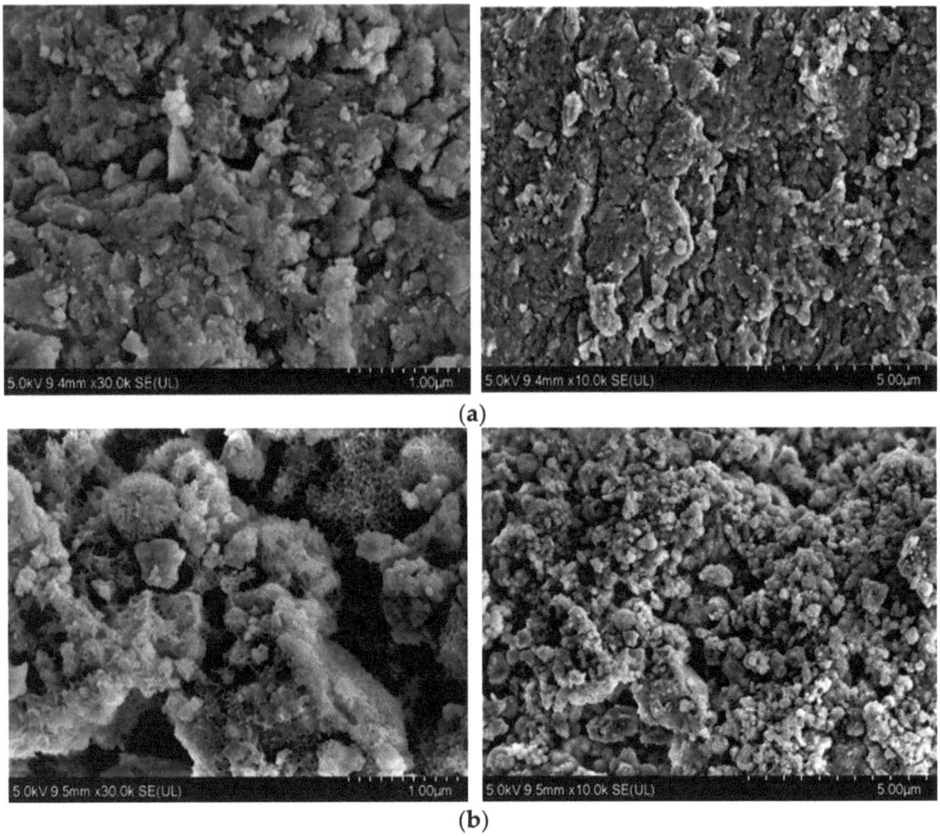

Figure 3. SEM image of composite adsorbents. (**a**) Before adsorption 30,000×; before adsorption 10,000×. (**b**) After adsorption 30,000×; after adsorption 10,000×.

3.4. XPS Characterization Results and Analysis

The full XPS spectrum of the composite adsorbent is shown in Figure 4. As shown in Figure 4, the binding energies of the composite adsorbents correspond to the Zr3d characteristic peaks at 181 eV; the La4d, La3d, and La3p3 characteristic peaks are at 106, 836, and 1129 eV, respectively; and the binding energies of 711 eV correspond to the Fe-OOH characteristic peaks, demonstrating that the preparation of ternary composite adsorbents was successful. The binding energy of 131 eV corresponds to the P2p characteristic peak, confirming the phosphate adsorption reaction.

Figure 5 illustrates the O1s spectrum before and after phosphorus removal by the composite adsorbent. As shown in Figure 5, the proportion of M-OH decreased from 90.10% to 89.11% before and after the adsorption, and the proportion of M-O increased from 9.90% to 10.89%, indicating that hydroxyl M-OH on the metal surface is involved in the removal of phosphorus via adsorption.

3.5. Point-of-Zero-Charge Test

The point of zero charge on the surface of the adsorbed material was determined by potentiometric titration, and the results are shown in Figure 6. A certain mass of composite adsorbent was added into a 0.1 mol·L^{-1} NaCl solution, and the initial pH was adjusted to 5, 7, 9, and 10 with HCl and NaOH. Then, the samples were transferred to a 298 K temperature-controlled shaker for 48 h. The samples were then centrifuged, and the final pH of the supernatant liquid was determined using a pH meter. Finally, DpH (i.e., the difference between the initial and final pH) was plotted against pH.

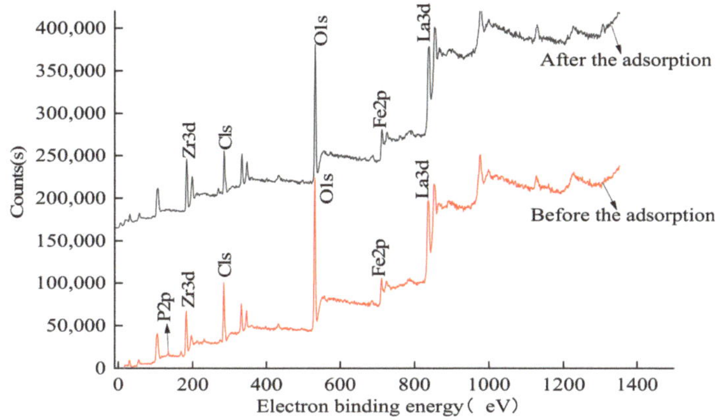

Figure 4. XPS spectra of composite adsorbents.

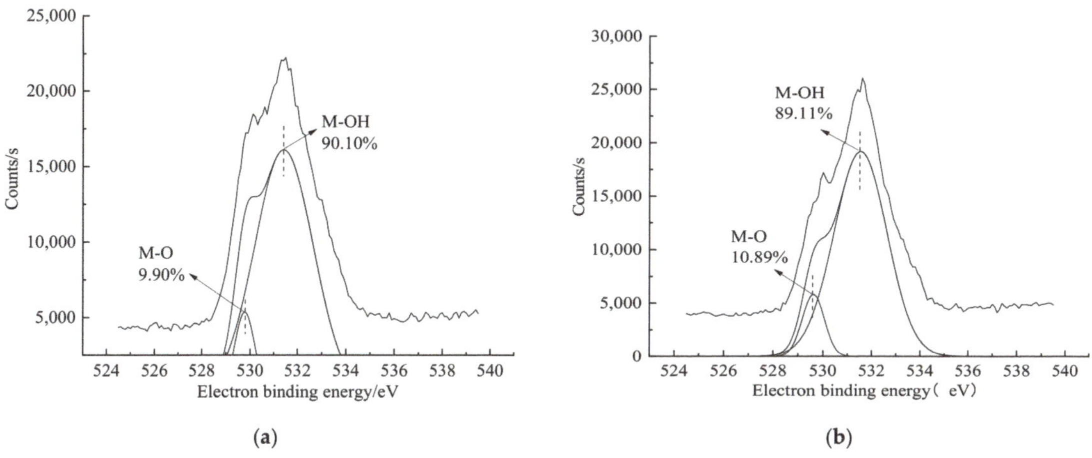

Figure 5. XPS spectra of O1s before (**a**) and after (**b**) phosphorus removal by composite adsorbent.

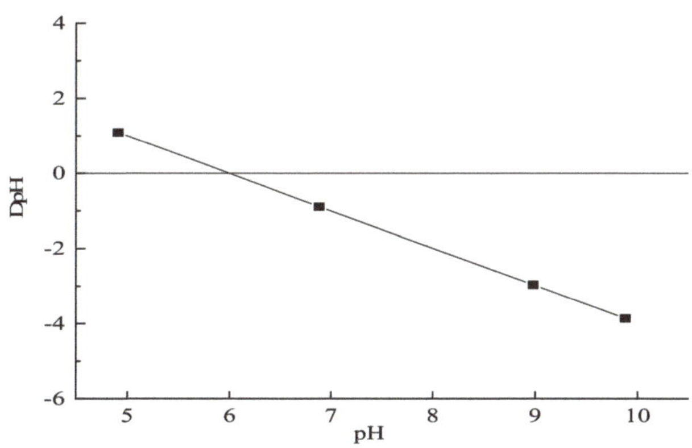

Figure 6. Test chart of surface zero-potential point.

As shown in Figure 6, the point of zero charge of the Zr–La–Fe ternary composite adsorbent embedded with sodium alginate was pH = 6. When pH < 6, the surface of the adsorbent was positively charged due to the proportion of the hydroxyl groups on the surface, which can promote the adsorption of phosphate through electrostatic attraction. When pH > 6, the surface of the adsorbent was electronegative, which was not conducive to the adsorption of phosphate, resulting in a reduced phosphorus removal rate. These results are consistent with the analysis of the influence of pH on phosphorus removal performance in subsequent single-factor experiments.

3.6. Influence of Adsorbent Dosage on Phosphorus Removal Performance

Test conditions: wastewater volume, 50 mL; initial concentration of phosphorus containing wastewater, 10 $mg \cdot L^{-1}$; pH of wastewater, 6; temperature, 25 °C; reaction time, 4 h; shaking speed, 180 $r \cdot min^{-1}$. The test results were obtained by varying the amount of composite adsorbent, as shown in Figure 7.

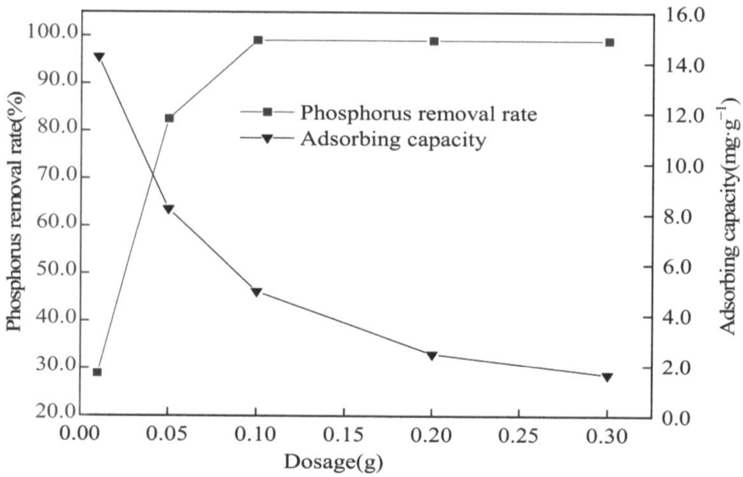

Figure 7. Effect of adsorbent dosage on phosphorus removal performance.

As shown in Figure 7, with the increase in the amount of adsorbent, phosphorus removal first increased rapidly and then gradually stabilized, but the adsorbing capacity decreased slowly. The main reason was that when the initial concentration of phosphorus-containing wastewater is fixed, more active sites are used when the amount of adsorbent increases, thus increasing the removal rate. However, when more adsorbent is added, many active sites are not fully utilized, resulting in a gradual stabilization of the removal rate. The adsorbing capacity is the adsorption quantity of a unit mass of adsorbent when the amount of adsorbate phosphorus is fixed, and increasing the amount of adsorbent is bound to reduce the adsorption capacity. Our findings indicate that the optimal adsorbent dose to meet the "Urban Sewage Treatment Plant Pollutant Discharge Standard" (GB18918-2002) level A standard is 0.1 g.

3.7. Influence of Reaction Time on Phosphorus Removal Performance

Test conditions: all tests were conducted as described in Section 3.6. The adsorption material (0.1 g) was added to the reaction and the other conditions remained unchanged. Test results were obtained by changing the adsorption time, as shown in Figure 8. With increased adsorption time, the phosphorus removal rate of the composite adsorbent increased rapidly and then stabilized. Upon reaching 160 min, the effluent concentration of phosphorus reached the level A standard of the Urban Sewage Treatment Plant Pollutant

Discharge Standard (GB18918-2002). Therefore, the optimal adsorption time was 160 min, and the subsequent single-factor experiment was conducted.

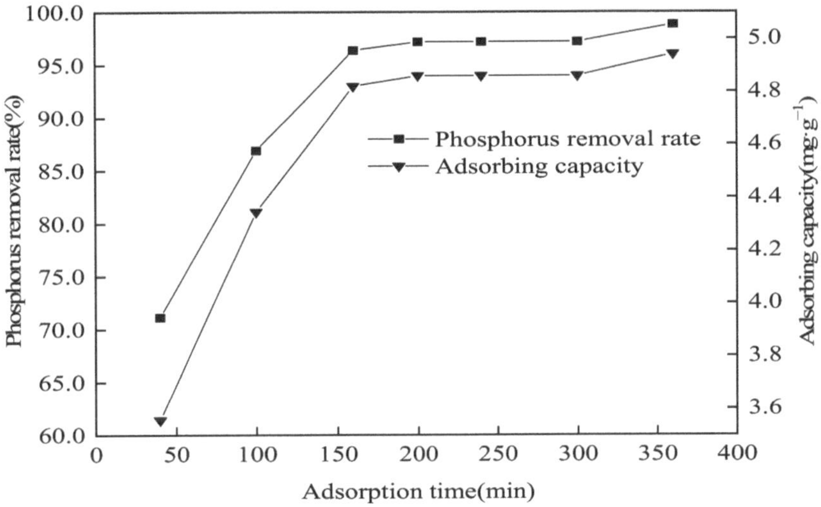

Figure 8. Effect of adsorption time on phosphorus removal performance.

3.8. Influence of pH on Phosphorus Removal Performance of Phosphorus-Containing Wastewater

Test conditions: the test was carried out according to Section 3.7. The reaction time was 160 min, and other conditions remained unchanged. The test results were obtained by changing the pH of wastewater, as shown in Figure 9. The removal rate and adsorption capacity of wastewater decreased when the pH value increased. When the wastewater pH ranged between 4 and 6, the removal rate was above 90%, and the removal rate remained above 80% when the wastewater pH was lower than 7. This change in trend could be explained by the zero-charge point of the material. These findings indicated that the surfaces of the ternary composite adsorbent were protonated and carried positive charges in an acidic medium. However, under alkaline conditions, the surface of the adsorbent is electronegative, and OH^- competes with PO_4^{3-}, thus reducing the removal rate and adsorption capacity of phosphorus. These observations were consistent with those of a previous study [26].

3.9. Influence of Initial Phosphorus Concentration of Wastewater on Phosphorus Removal Performance

Test conditions: All tests were conducted as described in Section 3.8. Other conditions remained unchanged, and the test results were obtained by changing the initial concentration of wastewater, as shown in Figure 10. With the increase in phosphorus concentration in wastewater, the removal rate decreased and the adsorption capacity increased. This is because when the concentration of phosphorus in wastewater is higher, the amount of adsorbent increases when the mass of adsorbent is fixed, and the active sites in wastewater are limited, thus reducing the removal rate. Increasing the adsorbent concentration accelerates its diffusion, as well as phosphorus adsorption capacity.

3.10. Orthogonal Adsorption Experiment

Following an orthogonal experimental design, four factors including time, dosage, initial concentration, and pH were selected, and the results are summarized in Table 1. As shown in Table 1, the optimal phosphorus removal conditions of ternary composite adsorbent embedded with sodium alginate are as follows: initial concentration, 5 mg·L^{-1}; pH, 5; adsorbent dosage, 0.2 g; adsorption time, 200 min. According to the R value range, we concluded that the factors affecting the adsorption effect were the following: dosage of composite adsorbent > initial concentration of phosphorus-containing wastewater > pH of simulated wastewater > adsorption time.

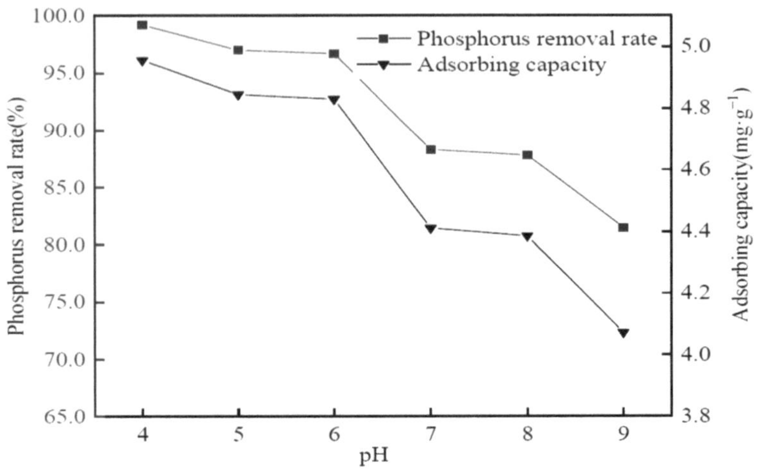

Figure 9. Effect of pH of phosphorus-containing wastewater on phosphorus removal performance.

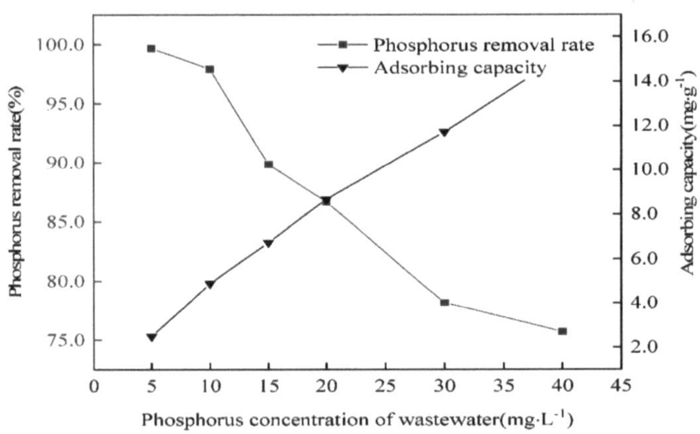

Figure 10. Effect of initial concentration of phosphorus-containing wastewater on phosphorus removal performance.

Under the optimal test conditions, five tests were repeated to verify the repeatability of the results. The results showed that the average phosphorus removal rate of the five tests was 99.88%, and the relative standard deviation was 0.04%. This demonstrated that the experimental results were relatively stable.

3.11. Orthogonal Verification of the Treatment Effect of Composite Adsorbent on Actual Urban Domestic Sewage

Water samples were taken from the inlet of the Yizhou No.1 Sewage Treatment Plant in Hechi City, Guangxi Province, and the optimal process conditions for phosphorus removal were obtained via orthogonal tests to verify the treatment effect of the composite adsorbent prepared in this study on actual domestic sewage. First, the characteristics of the water samples were determined, after which the test conditions were adjusted to conduct three parallel adsorption tests. After the adsorption test, a certain volume of water sample was removed for digestion and the residual phosphorus concentration was determined by spectrophotometry. Table 2 lists the characteristics of actual urban domestic sewage.

Table 1. Orthogonal experiment results and range analysis.

Test Number	pH	*Initial Concentration* (mg·L^{-1})	*Adsorbent Dosage* (g)	*Adsorption Time* (min)	*Removal Rate* (%)
1	5	5	0.05	100	77.80
2	5	10	0.1	160	86.50
3	5	15	0.2	200	99.60
4	6	5	0.1	200	98.20
5	6	10	0.2	100	98.50
6	6	15	0.05	160	64.13
7	7	5	0.2	160	98.00
8	7	10	0.05	200	68.60
9	7	15	0.1	100	69.67
K_1	263.90	274.00	210.53	245.97	
K_2	260.83	253.60	254.37	248.63	
K_3	236.27	233.40	296.10	266.40	
k_1	87.97	91.33	70.18	81.99	
k_2	86.94	84.53	84.79	82.88	
k_3	78.76	77.80	98.70	88.80	
R	9.21	13.53	28.52	6.81	

Table 2. Characteristics of sewage.

Sewage Quality Parameters	Concentration	Unit
COD	321 ± 0.45	mg·L^{-1}
BOD	180 ± 0.16	mg·L^{-1}
NH_3-N	3.26 ± 0.06	mg·L^{-1}
TP	4.17 ± 0.04	mg·L^{-1}
ORP	365 ± 0.78	mV
pH	6.87 ± 0.05	/
Conductivity	985 ± 0.13	μS/cm

The results showed that the mean concentration of total phosphorus in water samples at the intake of urban sewage treatment plant was 4.17 mg·L^{-1}, and the residual concentration of total phosphorus was not detected after adsorption; that is, the concentration was lower than 0.01 mg·L^{-1}. Therefore, the composite adsorbent prepared in this experiment had a significant effect on the treatment of low-concentration municipal wastewater, and could reduce the concentration of total phosphorus in municipal wastewater treatment plant influents to close to 0 mg·L^{-1}. These characteristics highlight the applicability of the adsorbent.

3.12. Results of Adsorption Isotherm Fitting

Experimental conditions: the volume of wastewater was 50 mL; the amount of adsorbent was 0.1 g; the initial concentrations of phosphorus-containing wastewater were 5, 10, 15, 20, 30, and 40 mg·L^{-1}; pH was 6; the experiments were conducted at three temperatures (25 °C, 40 °C and 55 °C); and the shock adsorption was 160 min. The isotherm fitting results are shown in Figure 11 and Table 3.

As shown in Table 3, under three different temperature conditions, the linear correlation coefficient of the Freundlich model was higher, indicating that the reaction process was more consistent with the Freundlich adsorption isotherm equation. The adsorption index 1/n was between 0.1 and 0.5, indicating that phosphorus was readily removed via adsorption.

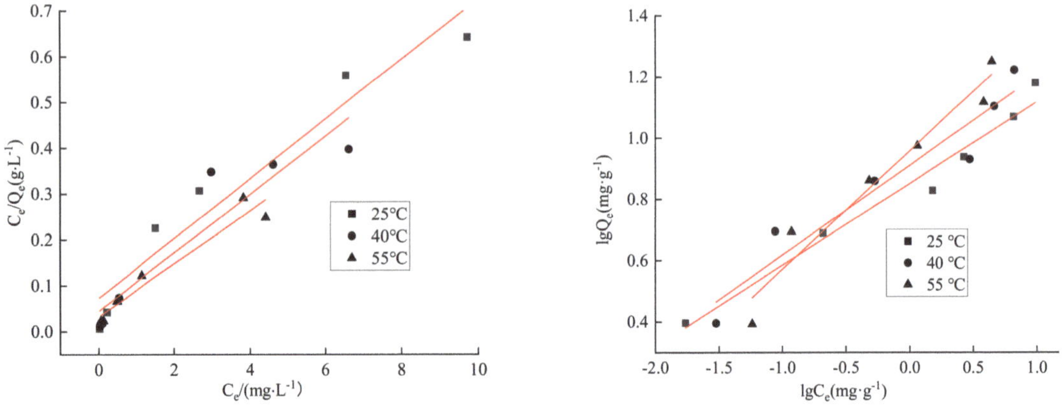

(a) (b)

Figure 11. Langmuir (**a**) and Freundlich adsorption isotherm (**b**).

Table 3. Fitting results of two isothermal lines.

	Langmuir Model			Freundlich Model		
Temperature (°C)	**a**	**$Q_m/(mg{\cdot}g^{-1})$**	**R^2**	**k**	**1/n**	**R^2**
25	0.07	15.3	0.9394	7.11	0.27	0.9729
40	0.04	15.7	0.8792	8.15	0.29	0.9234
55	0.03	17.3	0.9406	9.02	0.39	0.9513

According to the adsorption isotherm data, the calculation results are shown in Figure 12 and Table 4.

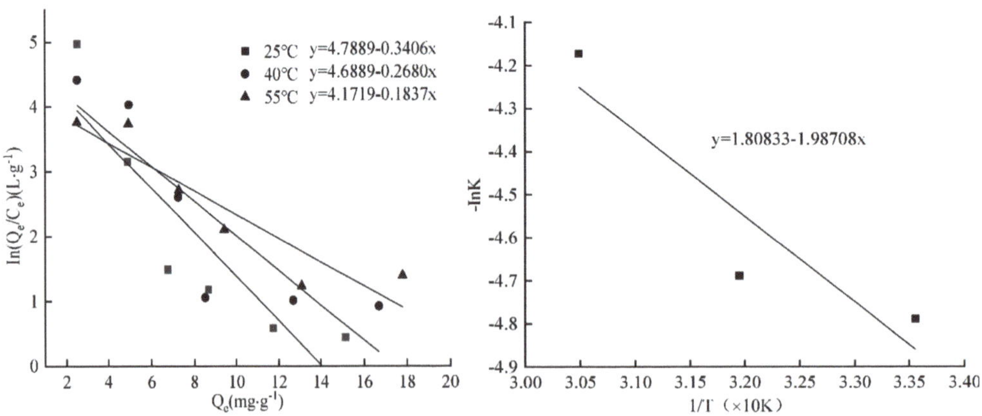

Figure 12. Thermodynamic calculation fitting of adsorption.

Table 4. Thermodynamic constants of phosphorus adsorption on composite adsorbents at different temperatures.

Thermodynamic Constant	298 K	313 K	328 K
$\Delta G/(kJ{\cdot}mol^{-1})$	−11.9	−12.2	−11.4
$\Delta H/(kJ{\cdot}mol^{-1})$	−16.5	−16.5	−16.5
$\Delta S/(J{\cdot}mol^{-1}{\cdot}K^{-1})$	15.0	15.0	15.0

As shown in Table 4, the ΔG values of the three Gibbs free energy variations at different temperatures were all negative, indicating that the adsorption reaction could occur spontaneously. The ΔH enthalpy changes were all negative, indicating that the reaction was exothermic. The ΔS entropy changes were all positive, indicating that adsorption can increase the confusion degree of the solid–liquid interface, and also demonstrating that adsorption can occur spontaneously.

3.13. Results of Adsorption Kinetics Fitting

The fitting results of adsorption kinetics are shown in Figure 13 and Table 5. It can be seen from Table 5 that the adsorption process is more consistent with the quasi-second-order kinetic equation by comparing the linear correlation coefficients. The results demonstrate that the adsorption process is based on the compound effect of multiple adsorption mechanisms, meaning that the adsorption rate is affected by multiple factors such as wastewater pH, concentration, and electron sharing or electron transfer between adsorbent and adsorbate. The chemisorption is the primary controlling step because that is the basic assumption of the quasi-second-order kinetic model.

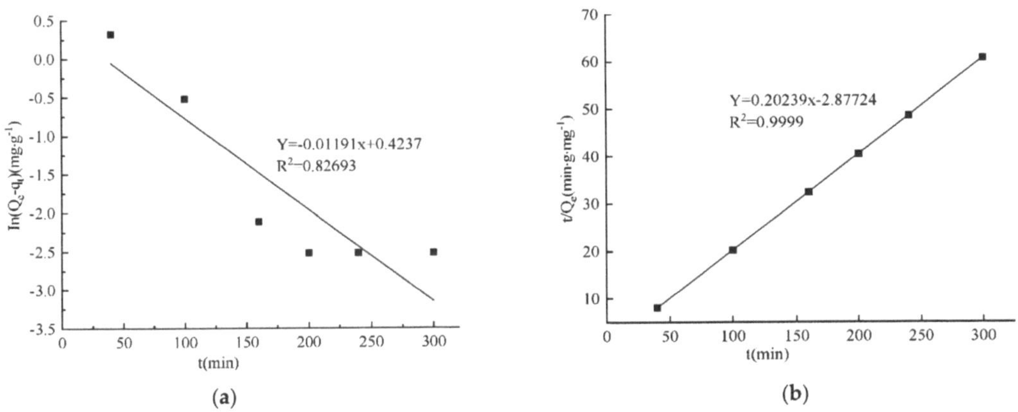

Figure 13. Quasi-first-order kinetic model (**a**) and quasi-second-order kinetic model (**b**).

Table 5. Quasi-first-order and quasi-second-order adsorption kinetics fitting parameters.

Quasi First Order Dynamics Model			Quasi Second Order Dynamics Model		
$Q_e/(mg \cdot g^{-1})$	k_1/h^{-1}	R^2	$Q_e/(mg \cdot g^{-1})$	k_2/h^{-1}	R^2
1.53	0.0119	0.8269	4.94	0.0142	0.9999

3.14. Regeneration Performance of Composite Adsorbent

The first adsorption test was carried out under optimal adsorption conditions, after which the adsorbed samples were centrifuged and dried in a drying oven at 50 °C. After cooling, a 0.5 mol·L^{-1} sodium hydroxide solution was used for desorption for 3 h. The samples were then centrifuged, and the desorption materials were recovered after drying in a drying oven at 50 °C. The recovered material was then used for a second adsorption, and the adsorption–desorption cycle was repeated three times. Through this test, the regeneration capacity of the adsorbed material was determined, and the test results are shown in Figure 14.

As illustrated in Figure 14, after two regeneration cycles, the removal rate of 5 mg·L^{-1} phosphorous wastewater by the embedded composite adsorbent decreased to 83.53%. After three cycles of desorption and adsorption, the removal rate of phosphorus by the adsorbent remained above 68%, indicating that the embedded composite adsorbent prepared in this experiment could be effectively used up to three times, after which its absorbance capacity decreased significantly.

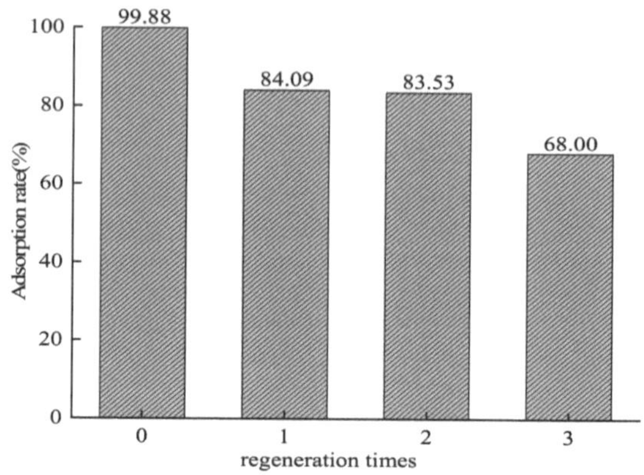

Figure 14. Effect of continuous regeneration cycles on the adsorption rates of the adsorbent.

4. Conclusions

The mechanism of phosphorus removal by the sodium alginate-embedded Zr–La–Fe ternary composite adsorbent prepared in this experiment is mainly based on the coordination of phosphate at the metal active site and the coordination exchange reaction with the surface hydroxyl groups, resulting in granular phosphate deposits. The prepared adsorbent not only exhibited a high adsorption performance on laboratory-simulated wastewater, but was also effective on actual domestic wastewater, thus providing a promising new approach for the treatment of phosphorus-containing wastewater. The adsorption process was more consistent with the Freundlich adsorption isotherm equation and quasi-second-order kinetic model, and the adsorption phosphorus removal reaction was exothermic and could be spontaneous. After three lye regeneration tests, the removal rate of phosphorus in water remained above 68%, and the granulation of the adsorbent was achieved via embedding immobilization technology, which overcomes the problem of the easy loss of powder adsorbent, thus enabling the recycling of the adsorbent.

Author Contributions: Data collation and analysis, writing—original draft, X.L.; adsorption single-factor and orthogonal experiment, H.L.; characterization of materials and regeneration performance of adsorbent, Y.M.; funding acquisition, project administration, supervision, writing—review and editing, Y.W.; All authors have read and agreed to the published version of the manuscript.

Funding: This work was financially supported by the 2021 Young and Middle-aged Teachers' Scientific Research Basic Ability Promotion Project of Guangxi University (No. 2021KY0616), as well as the 2022 Autonomous Region-level College Students' Innovation and Entrepreneurship Training Plan (No. S202210605052).

Institutional Review Board Statement: Not applicable.

Informed Consent Statement: Not applicable.

Data Availability Statement: Not applicable.

Acknowledgments: The authors would like to thank all the reviewers who participated in the review.

Conflicts of Interest: The authors declare no conflict of interest.

References

1. Kim, J.-O.; Chung, J. Implementing chemical precipitation as a pretreatment for phosphorus removal in membrane bioreactor-based municipal wastewater treatment plants. *KSCE J. Civ. Eng.* **2014**, *18*, 956–963. [CrossRef]
2. Wang, H.; Tian, Z.; Wang, H.; Yan, Q. Optimization and reaction kinetics analysis for phosphorus removal in struvite precipitation process. *Water Environ. Res.* **2020**, *92*, 1162–1172. [CrossRef] [PubMed]

3. Ochoa, C.; Hernández, M.A.; Bayona, O.L.; Camargo, H.A.; Cabeza, I.O.; Candela, A.M. Phosphorus recovery by struvite from anaerobic co-digestion effluents during residual biomass treatment. *Biomass Convers. Biorefin.* **2020**, *11*, 261–274. [CrossRef]
4. Izadi, P.; Eldyasti, A. Design, operation and technology configurations for enhanced biological phosphorus removal (EBPR) process: A review. *Rev. Environ. Sci. Bio/Technol.* **2020**, *19*, 561–593. [CrossRef]
5. Matsuura, N.; Masakke, Y.; Karthikeyan, S.; Kanazawa, S.; Honda, R.; Yamamoto-Ikemoto, R.; Konstantinidis, K.T. Metagenomic insights into the effect of sulfate on enhanced biological phosphorus removal. *Appl. Microbiol. Biotechnol.* **2021**, *105*, 2181–2193. [CrossRef]
6. Banu, J.R.; Merrylin, J.; Kavitha, S.; Kannah, R.Y.; Selvakumar, P.; Gopikumar, S.; Sivashanmugam, P.; Do, K.-U.; Kumar, G. Trends in Biological Nutrient Removal for the Treatment of Low Strength Organic Wastewaters. *Curr. Pollut. Rep.* **2021**, *7*, 1–30. [CrossRef]
7. Khalil, A.K.A.; Dweiri, F.; Almanassra, I.W.; Chatla, A.; Atieh, M.A. Mg-Al Layered Double Hydroxide Doped Activated Carbon Composites for Phosphate Removal from Synthetic Water: Adsorption and Thermodynamics Studies. *Sustainability* **2022**, *14*, 6991. [CrossRef]
8. Huang, C.; Tang, C.; Wu, Q.; Zhu, Q. Magnetic $MnFe_2O_4$/ZnFe-LDH for Enhanced Phosphate and Cr (VI) Removal from Water. *Environ. Sci. Pollut. Res.* **2022**, *29*, 59224–59234. [CrossRef]
9. Feng, L.; Zhang, Q.; Ji, F.; Jiang, L.; Liu, C.; Shen, Q.; Liu, Q. Phosphate removal performances of layered double hydroxides (LDH) embedded polyvinyl alcohol/lanthanum alginate hydrogels. *Chem. Eng. J.* **2021**, *430*, 132754. [CrossRef]
10. Alagha, O.; Manzar, M.S.; Zubair, M.; Anil, I.; Mu'Azu, N.D.; Qureshi, A. Comparative Adsorptive Removal of Phosphate and Nitrate from Wastewater Using Biochar-MgAl LDH Nanocomposites: Coexisting Anions Effect and Mechanistic Studies. *Nanomaterials* **2020**, *10*, 336. [CrossRef]
11. Parasana, N.; Shah, M.; Unnarkat, A. Recent advances in developing innovative sorbents for phosphorus removal—Perspective and opportunities. *Environ. Sci. Pollut. Res.* **2022**, *29*, 38985–39016. [CrossRef]
12. Xu, Z.; Zhong, Y.; Wang, Y.; Song, X.; Huang, W. Removal performance and mechanism of phosphorus by different Fe-based layered double hydroxides. *Environ. Sci. Pollut. Res.* **2022**, 1–11. [CrossRef]
13. Yuan, J.; Zhu, Y.; Wang, J.; Liu, Z.; He, M.; Zhang, T.; Li, P.; Qiu, F. Facile Modification of Biochar Derived from Agricultural Straw Waste with Effective Adsorption and Removal of Phosphorus from Domestic Sewage. *J. Inorg. Organomet. Polym. Mater.* **2021**, *31*, 3867–3879. [CrossRef]
14. Duan, H.; Zhang, L.; Wang, Y.; Liu, Y.; Wang, Y. Phosphate removal from aqueous solution by Fe–La binary (hydr)oxides: Characterizations and mechanisms. *Environ. Sci. Pollut. Res.* **2021**, *28*, 62662–62676. [CrossRef]
15. Zhang, X.; Ma, C.; Wen, K.; Han, R. Adsorption of phosphate from aqueous solution by lanthanum modified macroporous chelating resin. *Korean J. Chem. Eng.* **2020**, *37*, 766–775. [CrossRef]
16. Park, Y.; Gorman, C.; Ford, E. Lanthanum carbonate nanofibers for phosphorus removal from water. *J. Mater. Sci.* **2020**, *55*, 5008–5020. [CrossRef]
17. Nakarmi, A.; Moreira, R.; Bourdo, S.E.; Watanabe, F.; Toland, A.; Viswanathan, T. Removal and Recovery of Phosphorus from Contaminated Water Using Novel, Reusable, Renewable Resource-Based Aluminum/Cerium Oxide Nanocomposite. *Water Air Soil Pollut.* **2020**, *231*, 559. [CrossRef]
18. Qiu, L.; Zhang, M.; Yu, X.; Zheng, P. A novel Fe(II)-Ca synergistic phosphorus removal process: Process optimization and phosphorus recovery. *Environ. Sci. Pollut. Res.* **2017**, *25*, 1543–1550. [CrossRef]
19. Banu, R.J.; Do, K.U.; Yeom, I.T. Phosphorus removal in low alkalinity secondary effluent using alum. *Int. J. Environ. Sci. Technol.* **2007**, *5*, 93–98. [CrossRef]
20. Liu, B.; Lou, S.; Zeng, Y.; Qin, Y.; Zhang, W.; Zhang, L.; Liu, X. High-efficiency adsorption of phosphate by Fe-Zr-La tri-metal oxide composite from aqueous media: Performance and mechanism. *Adv. Powder Technol.* **2021**, *32*, 4587–4598. [CrossRef]
21. Yu, Y.; Chen, J.P. Key factors for optimum performance in phosphate removal from contaminated water by a Fe–Mg–La tri-metal composite sorbent. *J. Colloid Interface Sci.* **2015**, *445*, 303–311. [CrossRef] [PubMed]
22. Yu, Y.; Yu, L.; Chen, J.P. Adsorption of fluoride by Fe–Mg–La triple-metal composite: Adsorbent preparation, illustration of performance and study of mechanisms. *Chem. Eng. J.* **2015**, *262*, 839–846. [CrossRef]
23. Lu, J.; Liu, H.; Liu, R.; Zhao, X.; Sun, L.; Qu, J. Adsorptive removal of phosphate by a nanostructured Fe–Al–Mn trimetal oxide adsorbent. *Powder Technol.* **2013**, *233*, 146–154. [CrossRef]
24. Langmuir, I. The constitution and fundamental properties of solids and liquids. *J. Am. Chem. Soc.* **1916**, *38*, 2221–2295. [CrossRef]
25. Freundlich, H. Uber die adsorption in lasungen. *J. Z. Phys. Chem.* **1906**, *57*, 385–470. [CrossRef]
26. Yang, H.; Luo, B.; Zhang, Y.; Zhou, B.; Ahmed, S.M.; Liu, H.; Liu, X.; He, Y.; Xia, S. Study of Humic Acid Adsorption Character on Natural Maifan Stone: Characterization, Kinetics, Adsorption Isotherm, and Thermodynamics. *ACS Omega* **2020**, *5*, 7683–7692. [CrossRef]
27. Zubair, M.; Manzar, M.S.; Suleiman, M.A.; Fernandes, D.P.; Meili, L.; Al Bin Essa, W.; Al-Adam, H.; AlGhamdi, J.M.; Mu'Azu, N.D.; Haladu, S.A.; et al. Production of magnetic biochar-steel dust composites for enhanced phosphate adsorption. *J. Water Process Eng.* **2022**, *47*, 102793. [CrossRef]

processes MDPI

Article

Comparison between Regression Models, Support Vector Machine (SVM), and Artificial Neural Network (ANN) in River Water Quality Prediction

Nur Najwa Mohd Rizal [1], Gasim Hayder [2,*], Mohammed Mnzool [3], Bushra M. E. Elnaim [4], Adil Omer Yousif Mohammed [5] and Manal M. Khayyat [6]

1 College of Graduate Studies, Universiti Tenaga Nasional (UNITEN), Kajang 43000, Malaysia
2 Department of Civil Engineering, College of Engineering, Universiti Tenaga Nasional (UNITEN), Kajang 43000, Malaysia
3 Department of Civil Engineering, College of Engineering, Taif University, P.O. Box 11099, Taif 21944, Saudi Arabia
4 Department of Computer Science, College of Science and Humanities in Al-Sulail, Prince Sattam bin Abdulaziz University, Al-Kharj 16278, Saudi Arabia
5 Department of Computer Science, College of Science and Arts, Qassim University, P.O. Box 1162, Al-Bukairiyah 51941, Saudi Arabia
6 Department of Information Systems, College of Computers and Information Systems, Umm Al-Qura University, P.O. Box 7607, Makkah 24382, Saudi Arabia
* Correspondence: gasim@uniten.edu.my

Abstract: Both anthropogenic and natural sources of pollution are regionally significant. Therefore, in order to monitor and protect the quality of Langat River from deterioration, we use Artificial Intelligence (AI) to model the river water quality. This study has applied several machine learning models (two support vector machines (SVMs), six regression models, and artificial neural network (ANN)) to predict total suspended solids (TSS), total solids (TS), and dissolved solids (DS)) in Langat River, Malaysia. All of the models have been assessed using root mean square error (RMSE), mean square error (MSE) as well as the determination of coefficient (R^2). Based on the model performance metrics, the ANN model outperformed all models, while the GPR and SVM models exhibited the characteristic of over-fitting. The remaining machine learning models exhibited fair to poor performances. Although there are a few researches conducted to predict TDS using ANN, however, there are less to no research conducted to predict TS and TSS in Langat River. Therefore, this is the first study to evaluate the water quality (TSS, TS, and DS) of Langat River using the aforementioned models (especially SVM and the six regression models).

Keywords: river; water quality parameters; regression models; ANN; SVM

Citation: Najwa Mohd Rizal, N.; Hayder, G.; Mnzool, M.; Elnaim, B.M.E.; Mohammed, A.O.Y.; Khayyat, M.M. Comparison between Regression Models, Support Vector Machine (SVM), and Artificial Neural Network (ANN) in River Water Quality Prediction. *Processes* **2022**, *10*, 1652. https://doi.org/10.3390/pr10081652

Academic Editor: George Z. Kyzas

Received: 20 July 2022
Accepted: 18 August 2022
Published: 20 August 2022

1. Introduction

River water is one of the prime natural resources that are vital for living beings, especially humans. River water is also considered one of the resources that is at risk even though it is essential for one's life [1]. This is due to the fact that it is also being exploited for several purposes such as electric power generation, agriculture, irrigation, industrialization, and recreation [2–4]. Consequently, the quality of the river will deteriorate and bring harm to the people as well as to the surrounding environment because the contamination occurred either from human activities such as the discharge of effluents from chemical, toxic, and human waste [5] or natural sources of pollution, namely flood and landslides. Therefore, to guarantee that high-quality water is available to be utilized for sundry purposes, it is requisite to have the river water quality is controlled [6].

In the past decades, conventional and statistical techniques that involved in the collection and evaluation of raw data manually have been applied to assess water quality [7].

However, the approaches are time-consuming, expensive, and require a labor-intensive procedure that needed a specialized measuring tool [8,9]. Recently, the usage of Artificial Intelligence (AI) in predicting the river water quality is no more outlandish as AI able to help monitor the condition of the river efficiently compared to statistical and conventional techniques of lab testing [10]. Moreover, AI approaches have been explored by researchers and scholars all around the world and AI has shown great capability in monitoring and forecasting the water quality [11]. Machine learning, as a part of AI has been widely applied in the various fields, especially in hydrology. According to Moubayed [12], there are four types of machine learning algorithms viz. reinforcement learning, unsupervised learning, semi-supervised learning, and supervised learning algorithms. In addition, a supervised learning algorithm is known when a machine learning algorithm obtains the target pattern and the feature vector as an input to develop a model. The developed model can be applied to determine the latest patterns and set output to the model [13].

There are several studies that have utilized machine learning models to predict water quality parameters. For instance, a study conducted by Niroobakhsh et al. [14] have developed two different artificial neural network (ANN) models to predict total dissolved solids (TDS) for Jajrood River in Iran. The type of ANN models used were radial basis function (RBF) and a multilayer perceptron (MLP). Two evaluation metrics (i.e., R^2 and RMSE) were used to evaluate the performance of the model. Based on the study conducted, the authors concluded that RBF model performed better than the developed MLP model, where the values of R^2 were 0.9362 compared to MLP model that achieved $R^2 = 0.8968$. Furthermore, Talib and Amat [15] proposed an ANN model to forecast chemical oxygen demand (COD) concentration for Dondang River in Penang, Malaysia. A total of nine water quality parameters i.e., phosphate, temperature, nitrate, biochemical oxygen demand (BOD), total solids (TS), dissolved solids (DO), suspended solids, pH, and ammonia were used as input parameters for the modeling. From the results obtained, the developed ANN model has obtained R^2 and R values of 0.83 and 0.94, respectively.

On the other hand, Najah et al. [16] have developed several machine learning models such as ensemble ANN, MLP-ANN, and also support vector machine (SVM) models to predict three water quality parameters (i.e., BOD, DO, and COD) in Johor River, Malaysia. From the study, the authors concluded that SVM model with five input parameters outperformed the other developed models by having the mean square error (MSE) and correlation of efficiency (CE) values of 0.07 and 0.95, 0.07 and 0.91, and also 0.12 and 0.93 for COD, BOD, and DO predictions, respectively. Moreover, Zhou et al. [17] predict the sulphate content of lakes in China using different Kernel functions of Gaussian process regression (GPR) model, such as Matern 5/2, rational quadratic, squared exponential, and exponential functions. The authors also tested several machine learning models (i.e., support vector regression models, bagging tree model, boosted tree model, and decision tree model) and compare the mentioned models with the GPR models. According to the study, the authors concluded that exponential GPR model outperformed other models, where the values of RMSE, R^2, and mean absolute error (MAE) obtained were 7.269, 0.72, and 5.046, respectively.

Other previous studies such as [9] and [18] also applied machine learning models and have successfully predicted the water quality parameters with high accuracy. Thus, it has shown that machine learning models are suitable in predicting river water quality parameters with high degree of robustness and accuracy. Therefore, this study aims to predict DS, TS, and TSS of Langat River, Malaysia by using nine different machine learning models i.e., six different regression models, two SVMs, and an ANN model. In order to determine the best and optimum machine learning model in predicting the aforementioned water quality parameters, the outcomes of the models were compared. This study also able to contribute in monitoring the water quality of Langat River as there are less to none research that have predicted TS and TSS of Langat River. Hence, this is the first study to evaluate the water quality (TSS, TS, and DS) of Langat River using machine learning models (especially using SVM and the six regression models).

The study area and the methodology are explained in the next section. The results of the study and the discussion regarding the results of the models are discussed in Section 3. Lastly, the conclusion of the study is provided in Section 4.

2. Methodology

2.1. Study Area

The study area chosen for the study was the Langat River in Malaysia. Langat River is located in the State of Selangor and it originated from the peak of Mount Nuang (Gunung Nuang). From there, the river flows southward towards the Straits of Malacca [19,20]. In addition, The Langat River basin is the second biggest basin in the State of Selangor, where it has an approximate catchment area of 1815 km^2 and is 141 km long [21,22].

The daily historical data of 24 water quality parameters (Station No. 2917601) which spans from January 1891 until March 2019 have been obtained from the Department of Irrigation and Drainage (DID), Malaysia [23]. The historical data yields 161 available data points for each parameters. BOD_5, potassium, manganese, iron, phosphate, sulphate, silica, chemicals, magnesium, TS, DS, solids, chloride, fluoride, ammonia, nitrate, sodium, pH, colour, turbidity, conductivity, hardness, alkalinity, and calcium, including TSS that is calculated by Equation (1) are the 25 water quality parameters used as inputs for the modeling.

$$\text{Total suspended solid} = \text{Total solids} - \text{Total dissolved solids} \quad (1)$$

The unit for TSS, TS, and TDS in the equation is mg/L [24]. In addition, the statistical analysis for water quality parameters based on the raw data is shown in Table 1 and Figure 1.

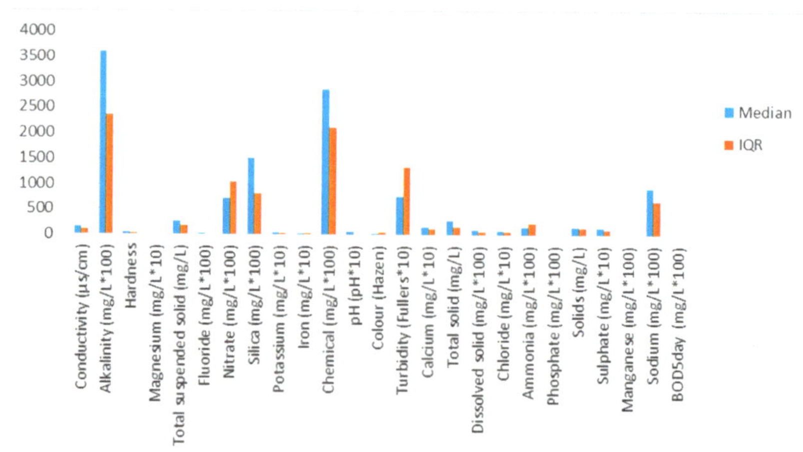

Figure 1. The statistical analysis for water quality parameters based on the raw data.

Due to the tremendous development being built along the Langat River Basin, the river has experienced several flood incidents within the basin [22,25]. Therefore, total suspended solids (mg/L), dissolved solids (mg/L × 10), and total solids (TS) were the chosen parameters to be predicted in the study.

Table 1. The statistical analysis for water quality parameters based on the raw data.

Variable	Median	IQR
Conductivity (µs/cm)	150	99
Alkalinity (mg/L × 100)	3600	2350
Hardness	44	30
Magnesium (mg/L × 10)	10	5
Total suspended solid (mg/L)	266.81	165.92
Fluoride (mg/L × 100)	30	12
Nitrate (mg/L × 100)	710	1035
Silica (mg/L × 100)	1500	800
Potassium (mg/L × 10)	46.5	21.25
Iron (mg/L × 10)	34.5	28.5
Chemical (mg/L × 100)	2850	2100
pH (pH × 10)	65	5
Colour (Hazen)	30	50
Turbidity (Fullers × 10)	750	1320
Calcium (mg/L × 10)	160	115.5
Total solid (mg/L)	284	154
Dissolved solid (mg/L × 100)	94	71
Chloride (mg/L × 10)	82	55
Ammonia (mg/L × 100)	160	228
Phosphate (mg/L × 100)	10	0
Solids (mg/L)	158	134
Sulphate (mg/L × 10)	130	97.75
Manganese (mg/L × 100)	12	11
Sodium (mg/L × 100)	920	650
BOD_{5day} (mg/L × 100)	3	6

2.2. Data Pre-Processing

The historical data were cleaned and pre-processed before being used as inputs in the modelings. According to Chen et al. [26], the practice of directly erasing the missing data is not recommended even though most researchers have applied it in their research. Therefore, in this study, the missing values from the data have been cleaned by replacing the missing values with a constant value (zero value) since the data obtained are limited. Although zero is known as a meaningless value, but according to Chollet [27], to insert missing values as zero are acceptable (i.e., with neural networks). The model will ignore the zero values since it has been trained that zero values are equal to missing data [27]. Thus, zero values have no effect on the neural network, since it cancels the corresponding weight after multiplication. However, due to limited data obtained, zero values are important to preserve the time series of the data.

Then, the data have been normalized in the range between 0 and 1 since few data have a high value which may give errors in the modeling. Next, the data have been divided into three where 70% of data was used for the training set, 15% of data was used for the testing set, and the remaining 15% for the validation set. MATLAB 2020b has been used to develop all models.

2.3. Models Development

The regression models used in the study were fine tree, medium tree, boosted tree, bagged tree, rational quadratic GPR, and lastly, exponential GPR models. The aforementioned models have been developed in MATLAB using Regression Learner Application and therefore, the models have been grouped as the regression models. Moreover, cross-validation factors were applied in the modeling and have been set to 10-folds.

Furthermore, two SVM models namely fine Gaussian SVM and medium Gaussian SVM were also applied in the study. SVM is known as kernel-based AI model as it consists of a kernel function, regression model complexity, and also regularization [28]. Both of the SVM models have been developed using Regression Learner Application. The cross-validation factors for the SVM models have also been set to 10-folds.

As for the artificial neural network (ANN) model, it was developed using Neural Network Fitting app in MATLAB and a two-layer feed forward network with two different transfer functions were used, where the sigmoid transfer function was applied in the hidden neurons while the linear transfer function was applied at the output neurons. The number of hidden neurons has been set to 1, 5, and 10, however, the ANN model with 10 hidden neurons exhibited the best performance. Thus, 10 hidden neurons have been chosen in the study. Moreover, the ANN model was trained by Levenberg-Marquardt backpropagation algorithm. The architecture of the ANN model is shown in Figure 2 while Table 2, shows the criteria selected for all models for the water quality prediction.

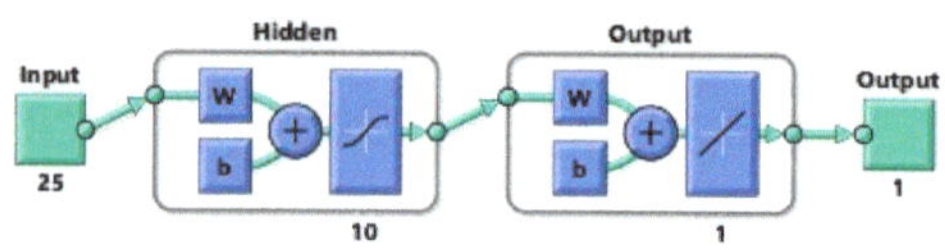

Figure 2. The architecture of the ANN model.

Table 2. The criteria selected for the water quality parameters modeling for all models.

Type of Model	Parameters	Default Value
Regression Tree -Fine Tree Model	Minimum leaf size:	4
Regression Tree -Medium Tree Model	Minimum leaf size:	12
Ensemble Tree -Boosted Tree Model	Minimum leaf size: Number of learners: Learning rate:	8 30 0.1
Ensemble Tree -Bagged Tree Model	Minimum leaf size: Number of learners:	8 30
Gaussian Process Regression -Exponential GPR	Basis function: Kernel function:	Constant Exponential
Gaussian Process Regression -Rational Quadratic GPR	Basis function: Kernel function:	Constant Rational quadratic
Artificial Neural Network	Training algorithm: Epoch:	Levenberg-Marquardt 0–100 epochs
Support Vector Machine -Fine Gaussian SVM	Kernel function: Kernel scale:	Gaussian 1.2
Support Vector Machine -Medium Gaussian SVM	Kernel function: Kernel scale:	Gaussian 4.9

2.4. Model Performance Evaluation

Root mean square error (RMSE), mean square error (MSE), and determination of coefficient (R^2) were the statistical indicators used to assess the performance of the developed models. The indicators are defined as follows [24,29]:

$$\text{MSE} = \frac{1}{n}\sum(y' - y)^2 \quad (2)$$

$$\text{RMSE} = \sqrt{\frac{1}{n}\sum(y' - y)^2} \quad (3)$$

$$R^2 = \left(\frac{n(\sum yy') - (\sum y)(\sum y')}{\sqrt{\left[n\sum y^2 - (\sum y)^2\right]\left[n\sum y'^2 - (\sum y')^2\right]}}\right)^2 \quad (4)$$

where y is the observed value, and y' is the predicted value. The n is defined as the number of data samples.

3. Results and Discussion

Table 3 shows the prediction analysis for all models in the testing dataset. Based on the table the ANN model has outperformed all models by obtaining high values of R^2 which were 0.9988, 0.9835, and 0.9880 in predicting TSS, DS, and TS, respectively. The ANN model also achieved low values of RMSE and MSE in predicting the water quality parameters. Thus, it has been shown that the ANN model has predicted the parameters accurately compared to the regression and SVM models. In the test dataset, fine Gaussian SVM and medium tree model were the worst models for predicting the parameters.

Table 3. Prediction analysis of all parameters (testing dataset).

Parameters	Type of Models	R2	MSE	RMSE
TSS (mg/L)	Fine Tree	0.8200	0.0004	0.0195
	Medium Tree	0.0000	0.0021	0.0462
	Boosted Tree	0.8100	0.0004	0.0203
	Bagged Tree	0.7000	0.0006	0.0254
	Rational Quadratic GPR	0.9600	7.6799×10^{-5}	0.0088
	Exponential GPR	0.7200	0.0006	0.0243
	Fine Gaussian SVM	0.0000	0.0021	0.0461
	Medium Gaussian SVM	0.4900	0.0011	0.0328
	Artificial Neural Network	0.9988	2.7904×10^{-5}	5.2824×10^{-3}
DS (mg/L × 100)	Fine Tree	0.5400	2.5927×10^{-6}	0.0016
	Medium Tree	0.0000	5.6248×10^{-6}	0.0024
	Boosted Tree	0.5400	2.5839×10^{-6}	0.0016
	Bagged Tree	0.4700	2.9763×10^{-6}	0.0017
	Rational Quadratic GPR	0.6200	2.1625×10^{-6}	0.0015
	Exponential GPR	0.4400	3.1423×10^{-6}	0.0018
	Fine Gaussian SVM	0.0500	5.3231×10^{-6}	0.0023
	Medium Gaussian SVM	0.2800	4.0532×10^{-6}	0.0020
	Artificial Neural Network	0.9835	6.2799×10^{-8}	2.5060×10^{-9}

Table 3. *Cont.*

Parameters	Type of Models	R2	MSE	RMSE
TS (mg/L)	Fine Tree	0.7800	0.0005	0.0213
	Medium Tree	0.0000	0.0021	0.0457
	Boosted Tree	0.8200	0.0004	0.0193
	Bagged Tree	0.5700	0.0009	0.0299
	Rational Quadratic GPR	0.9600	8.5497×10^{-5}	0.0092
	Exponential GPR	0.7500	0.0005	0.0228
	Fine Gaussian SVM	0.0000	0.0021	0.0456
	Medium Gaussian SVM	0.5400	0.0010	0.0319
	Artificial Neural Network	0.9880	0.8484×10^{-5}	5.3370×10^{-3}

The prediction analysis for all models in the overall datasets is shown in Table 4. From the table, it showed that rational quadratic GPR model has achieved the highest values of R^2 (1.00) in predicting TSS, DS, and TS. On the other hand, the rest of the models have exhibited good to fair performances in predicting the water quality parameters, where the values of $R^2 > 0.61$ for all three predictions, MSE < 0.02, and RMSE < 0.20. However, fine Gaussian SVM model has showed poor performance by having $R^2 < 0.20$, and higher values of RMSE and MSE than the other machine learning models in predicting TSS and TS. For DS prediction, medium Gaussian SVM achieved the lowest R^2 value (−0.03) and the highest RMSE and MSE values. In addition, Figure 3 shows the scatter plots for ANN model based on the respective water quality parameters.

Table 4. Prediction analysis of all parameters (overall dataset).

Parameters	Type of Models	R^2	MSE	RMSE
TSS (mg/L)	Fine Tree	0.9200	0.0017	0.0418
	Medium Tree	0.8100	0.0042	0.0649
	Boosted Tree	0.9200	0.0018	0.0429
	Bagged Tree	0.8600	0.0031	0.0556
	Rational Quadratic GPR	1.00	3.2522×10^{-7}	0.0006
	Exponential GPR	0.9500	0.0012	0.0350
	Fine Gaussian SVM	0.1600	0.0189	0.1376
	Medium Gaussian SVM	0.6100	0.0088	0.0937
	Artificial Neural Network	0.9998	4.1676×10^{-6}	0.0020
DS (mg/L × 100)	Fine Tree	0.7500	0.0040	0.0636
	Medium Tree	0.5800	0.0063	0.0826
	Boosted Tree	0.7900	0.0035	0.0590
	Bagged Tree	0.5700	0.0069	0.0832
	Rational Quadratic GPR	1.00	4.8138×10^{-8}	0.0002
	Exponential GPR	0.9000	0.0017	0.0411
	Fine Gaussian SVM	0.7500	0.0040	0.0636
	Medium Gaussian SVM	−0.0300	0.0167	0.1292
	Artificial Neural Network	0.9953	0.0001	0.0104

Table 4. *Cont.*

Parameters	Type of Models	R^2	MSE	RMSE
TS (mg/L)	Fine Tree	0.9200	0.0017	0.0409
	Medium Tree	0.8300	0.0037	0.0605
	Boosted Tree	0.9300	0.0016	0.0395
	Bagged Tree	0.8600	0.0030	0.0546
	Rational Quadratic GPR	1.00	3.118×10^{-7}	0.0006
	Exponential GPR	0.9400	0.0013	0.0358
	Fine Gaussian SVM	0.1800	0.0177	0.1331
	Medium Gaussian SVM	0.6300	0.0080	0.0895
	Artificial Neural Network	0.9970	6.5419×10^{-5}	0.0081

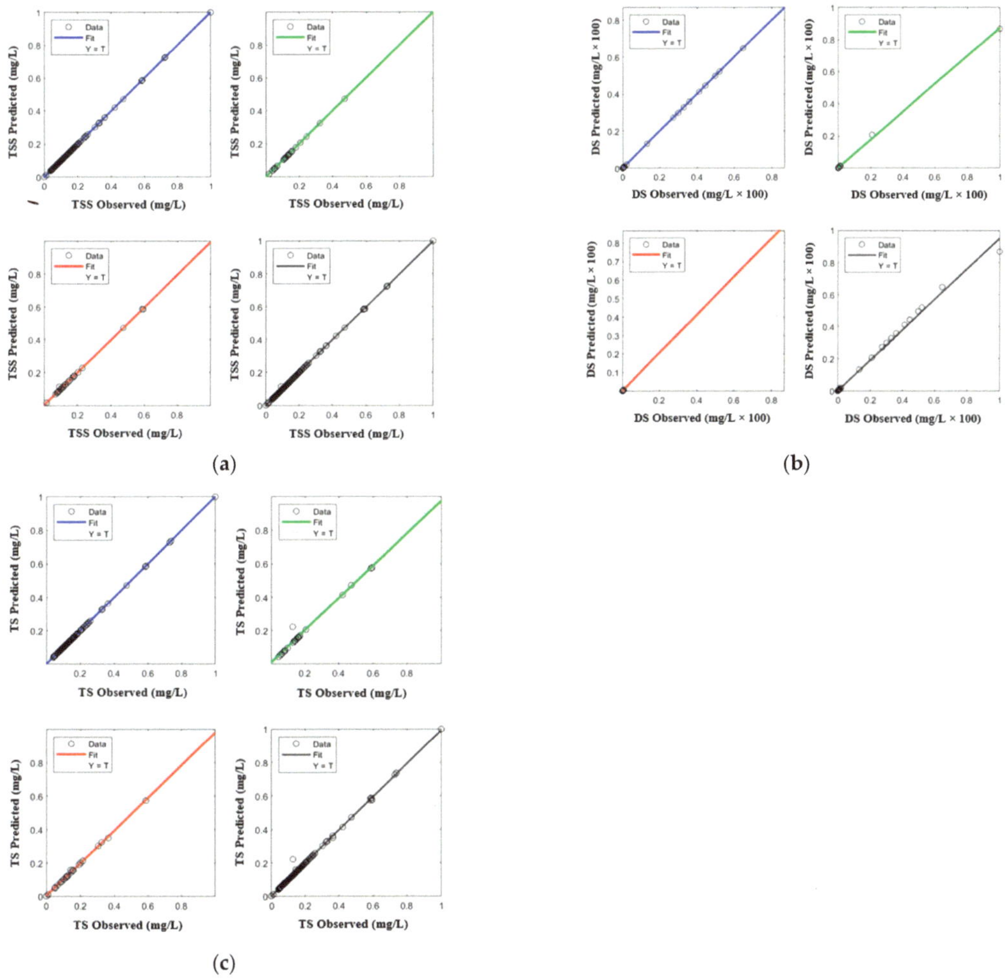

Figure 3. The scatter plots of ANN models in predicting (**a**) TSS, (**b**) DS and (**c**) TS.

Comparisons between Regression, SVM, and ANN models

Based on Tables 3 and 4, ANN model has proven to be good in predicting TS, TSS, and DS with high degree of accuracy and robustness by obtaining high values of R^2 and lower values of RMSE and MSE. Moreover, rational quadratic GPR model also exhibited similar performance, however, over-fitting might occurred during modeling since the model obtained $R^2 = 1.00$ for overall datasets in all three predictions. A research conducted by Zhou et al. [17] also used GPR models to predict water quality parameters, but it does not achieved results that is too perfect like the one obtained in this study. Similarly, the SVM models also did not performed well as it always exhibited poor performances. Past researches done by Najah et al. [16] and also several studies performed by [30] and [31] have proven that SVM model able to predict many kinds of water quality parameters. Contrary, the developed SVM models in this study did not achieved the same outcome. The reason was probably that SVM model is much suitable to be used to identify subtle patterns in a complex datasets [32] or it is because SVM model performed too well in a small datasets. Therefore, it can be concluded that ANN models were stable in forecasting the water quality parameters. In addition, fine tree, boosted tree, bagged tree, and exponential GPR models also able to be used to predict TSS, TS, and DS in Langat River.

4. Conclusions

This study has used nine different machine learning models, which consist of six regression models, two support vector machine models, and one artificial neural network model to predict dissolved solids (DS), total solids (TS), and total suspended solids (TSS) in Langat River, Malaysia. A total number of 25 water quality parameters that have been retrieved from the Department of Irrigation and Drainage (DID) have been applied as inputs in the modelling. Furthermore, ANN and rational quadratic GPR models turn out to be excellent in predicting TS, DS, and TSS as the models successfully achieved high accuracy and low errors but, the GPR model exhibited the characteristic of over-fitting since this study used small datasets. Similarly, SVM models also obtained poor accuracy in predicting the water quality parameters and over-fitting might occurred in the modeling. As for the remaining models, the models have shown fair to poor performances in predicting the water quality parameters. Therefore, ANN model was the best machine learning model in predicting DS, TSS, and TS of Langat River. Lastly, future research can be conducted to overcome this problem by changing the type of kernel function used for SVM and GPR models, or applying more complex datasets for the modeling.

Author Contributions: Conceptualization, N.N.M.R. and G.H.; methodology, N.N.M.R. and G.H.; software, N.N.M.R.; validation, N.N.M.R.; formal analysis, N.N.M.R. and G.H.; investigation, N.N.M.R. and G.H.; resources, N.N.M.R., G.H., M.M., B.M.E.E., A.O.Y.M. and M.M.K.; data curation, N.N.M.R..; writing—original draft preparation, N.N.M.R.; writing—review and editing, N.N.M.R. and G.H.; visualization, N.N.M.R., G.H., M.M., B.M.E.E., A.O.Y.M. and M.M.K.; supervision, G.H.; project administration, G.H., M.M., B.M.E.E., A.O.Y.M. and M.M.K.; funding acquisition, G.H., M.M., B.M.E.E., A.O.Y.M. and M.M.K. All authors have read and agreed to the published version of the manuscript.

Funding: This research was funded by the Fundamental Research Grant Scheme (FRGS) with the project code 20190105FRGS from the Ministry of Higher Education Malaysia, and Universiti Tenaga Nasional (UNITEN) BOLD Fund 2022, and the Deanship of Scientific Research at Umm Al-Qura University under Grant Code: (22UQU4400271DSR09).

Data Availability Statement: The data presented in this study are available on reasonable request from the corresponding author.

Acknowledgments: The authors would like to express their gratitude to the Ministry of Higher Education Malaysia for the Fundamental Research Grant Scheme (FRGS) and Universiti Tenaga Nasional (UNITEN) for the BOLD Fund, and the Deanship of Scientific Research at Umm Al-Qura University.

Conflicts of Interest: The authors declare no conflict of interest.

References

1. Varol, M. Use of water quality index and multivariate statistical methods for the evaluation of water quality of a stream affected by multiple stressors: A case study. *Environ. Pollut.* **2020**, *266*, 115417. [CrossRef]
2. Mekuria, D.M.; Kassegne, A.B.; Asfaw, S.L. Assessing pollution profiles along Little Akaki River receiving municipal and industrial wastewaters, Central Ethiopia: Implications for environmental and public health safety. *Heliyon* **2021**, *7*, e07526. [CrossRef] [PubMed]
3. Hayder, G.; Solihin, M.I.; Mustafa, H.M. Modelling of river flow using particle swarm optimized cascade-forward neural networks: A case study of Kelantan River in Malaysia. *Appl. Sci.* **2020**, *10*, 8670. [CrossRef]
4. Al-Sulttani, A.O.; Al-Mukhtar, M.; Roomi, A.B.; Farooque, A.A.; Khedher, K.M.; Yaseen, Z.M. Proposition of new ensemble data-intelligence models for surface water quality prediction. *IEEE Access* **2021**, *9*, 108527–108541. [CrossRef]
5. Jagaba, A.H.; Kutty, S.R.M.; Hayder, G.; Baloo, L.; Abubakar, S.; Ghaleb, A.A.S.; Lawal, I.M.; Noor, A.; Umaru, I.; Almahbashi, N.M.Y. Water quality hazard assessment for hand dug wells in Rafin Zurfi, Bauchi State, Nigeria. *Ain Shams Eng. J.* **2020**, *11*, 983–999. [CrossRef]
6. Fathi, E.; Zamani-Ahmadmahmoodi, R.; Zare-Bidaki, R. Water quality evaluation using water quality index and multivariate methods, Beheshtabad River, Iran. *Appl. Water Sci.* **2018**, *8*, 1–6. [CrossRef]
7. Kang, G.; Gao, J.Z.; Xie, G. Data-driven water quality analysis and prediction: A survey. In Proceedings of the 2017 IEEE Third International Conference on Big Data Computing Service and Applications (BigDataService), Redwood City, CA, USA, 6–9 April 2017; pp. 224–232.
8. Alqahtani, A.; Shah, M.I.; Aldrees, A.; Javed, M.F. Comparative Assessment of Individual and Ensemble Machine Learning Models for Efficient Analysis of River Water Quality. *Sustainability* **2022**, *14*, 1183. [CrossRef]
9. Hmoud Al-Adhaileh, M.; Waselallah Alsaade, F. Modelling and prediction of water quality by using artificial intelligence. *Sustainability* **2021**, *13*, 4259. [CrossRef]
10. Koranga, M.; Pant, P.; Kumar, T.; Pant, D.; Bhatt, A.K.; Pant, R.P. Efficient water quality prediction models based on machine learning algorithms for Nainital Lake, Uttarakhand. *Mater. Today Proc.* **2022**, *57*, 1706–1712. [CrossRef]
11. El Bilali, A.; Taleb, A. Prediction of irrigation water quality parameters using machine learning models in a semi-arid environment. *J. Saudi Soc. Agric. Sci.* **2020**, *19*, 439–451. [CrossRef]
12. Moubayed, A. Optimization Modeling and Machine Learning Techniques towards Smarter Systems and Processes. Ph.D. Thesis, The University of Western Ontario, London, ON, Canada, 2018; p. 5573.
13. Gakii, C.; Jepkoech, J. A classification model for water quality analysis using decision tree. *Eur. J. Comput. Sci. Inf. Technol.* **2019**, *7*, 1–8.
14. Niroobakhsh, M.; Musavi-Jahromi, S.H.; Manshouri, M.; Sedghi, H. Prediction of water quality parameter in Jajrood River basin: Application of multi layer perceptron (MLP) perceptron and radial basis function networks of artificial neural networks (ANNs). *Afr. J. Agric. Res.* **2012**, *7*, 4131–4139. [CrossRef]
15. Talib, A.; Amat, M.I. Prediction of chemical oxygen demand in Dondang River using artificial neural network. *Int. J. Inf. Educ. Technol.* **2012**, *2*, 259–261. [CrossRef]
16. Najah, A.; El-Shafie, A.H.A.H.; Karim, O.A.; Jaafar, O.; El-Shafie, A.H. An application of different artificial intelligences techniques for water quality prediction. *Int. J. Phys. Sci.* **2011**, *6*, 5298–5308. [CrossRef]
17. Zhao, J.; Guo, H.; Han, M.; Tang, H.; Li, X. Gaussian process regression for prediction of sulfate content in lakes of China. *J. Eng. Technol. Sci.* **2019**, *51*, 198–215. [CrossRef]
18. Dogan, E.; Sengorur, B.; Koklu, R. Modeling biological oxygen demand of the Melen River in Turkey using an artificial neural network technique. *J. Environ. Manag.* **2009**, *90*, 1229–1235. [CrossRef]
19. Basheer, A.O.; Hanafiah, M.M.; Abdulhasan, M.J. A study on water quality from Langat River, Selangor. *Acta Sci. Malays. (ASM)* **2017**, *1*, 1–4. [CrossRef]
20. Soo, E.Z.X.; Jaafar, W.Z.W.; Lai, S.H.; Othman, F.; Elshafie, A.; Islam, T.; Srivastava, P.; Hadi, H.S.O. Evaluation of bias-adjusted satellite precipitation estimations for extreme flood events in Langat river basin, Malaysia. *Hydrol. Res.* **2020**, *51*, 105–126. [CrossRef]
21. Hassim, M.; Yuzir, A.; Razali, M.N.; Ros, F.C.; Chow, M.F.; Othman, F. Comparison of rainfall interpolation methods in Langat River Basin. *IOP Conference Series: Earth Environ. Sci.* **2020**, *479*, 012018. [CrossRef]
22. Saudi, A.S.M.; Kamarudin, M.K.A.; Ridzuan, I.S.D.; Ishak, R.; Azid, A.; & Rizman, Z.I. Flood risk index pattern assessment: Case study in Langat river basin. *J. Fundam. Appl. Sci.* **2017**, *9*, 12–27. [CrossRef]
23. Rizal, N.N.M.; Hayder, G.; Yussof, S. River water quality prediction and analysis—Deep learning predictive models approach. In *Advances in Science, Engineering, and Technology (ASTI)*; (Submitted for review-in press); Springer: Berlin/Heidelberg, Germany.
24. Rizal, N.N.M.; Hayder, G.; Yusof, K.A. Water quality predictive analytics using an artificial neural network with a graphical user interface. *Water* **2022**, *14*, 1221. [CrossRef]
25. Ahmed, M.F.; Mokhtar, M.; Alam, L.; Ta, G.C.; Lee, K.E.; Khalid, R.M. Recognition of local authority for better management of drinking water at the Langat River Basin, Malaysia. *Int. J. Eng. Technol.* **2018**, *7*, 148–154. [CrossRef]
26. Chen, Y.; Song, L.; Liu, Y.; Yang, L.; Li, D. A review of the artificial neural network models for water quality prediction. *Appl. Sci.* **2020**, *10*, 5776. [CrossRef]
27. Chollet, F. *Deep Learning with Python*; Manning: Shelter Island, NY, USA, 2021.

28. Tiyasha; Tung, T.M.; Yaseen, Z.M. Asurvey on river water quality modellinh using artificial intelligence models: 2000–2020. *J. Hydrol.* **2020**, *585*, 124670. [CrossRef]
29. Ewusi, A.; Ahenkorah, I.; Aikins, D. Modelling of total dissolved solids in water supply systems using regression and supervised machine learning approaches. *Appl. Water Sci.* **2021**, *11*, 1–16. [CrossRef]
30. Li, X.; Sha, J.; Wang, Z.L. A comparative study of multiple linear regression, artificial neural network and support vector machine for the prediction of dissolved oxygen. *Hydrol. Res.* **2017**, *48*, 1214–1225. [CrossRef]
31. Abobakr Yahya, A.S.; Ahmed, A.N.; Binti Othman, F.; Ibrahim, R.K.; Afan, H.A.; El-Shafie, A.; Fai, C.M.; Hossain, M.S.; Ehteram, M.; Elshafie, A. Water quality prediction model based support vector machine model for ungauged river catchment under dual scenarios. *Water* **2019**, *11*, 1231. [CrossRef]
32. Rajaee, T.; Jafari, H. Two decades on the artificial intelligence models advancement for modeling river sediment concentration: State-of-the-art. *J. Hydrol.* **2020**, *588*, 125011. [CrossRef]

processes

Article

Extraction of Polyphenols from Unripened Coffee (Coffea Arabica) Residues and Use as a Natural Coagulant for Removing Turbidity

Diana Marcela Cuesta-Parra [1,*], Felipe Correa-Mahecha [1], Juan Pablo Rodríguez-Miranda [2], Octavio José Salcedo-Parra [3] and Edwin Rivas-Trujillo [3]

1 Direction of Chemistry and Environment, Faculty of Engineering, University of America Foundation, Bogotá 110311, Colombia; felipe.correa@profesores.uamerica.edu.co
2 Facultad de Medio Ambiente y Recursos Naturales, Universidad Distrital Francisco José de Caldas, Bogota 110711, Colombia; jprodriguezm@udistrital.edu.co
3 Facultad de Ingeniería, Universidad Distrital Francisco José de Caldas, Universidad Nacional de Colombia, Bogota 110231, Colombia; ojsalcedop@unal.edu.co (O.J.S.-P.); erivas@udistrital.edu.co (E.R.-T.)
* Correspondence: diana.cuesta@profesores.uamerica.edu.co

Abstract: The coffee agribusiness generates significant amounts of waste that becomes an environmental problem in producing countries. For example, synthetic coagulants have sustainability disadvantages. Immature coffee beans are collected together with mature beans, and their high polyphenol content makes them unsuitable for coffee production and commercialization. This paper aims to test the coagulant activity of polyphenols extracted from *Coffea arabica* residues in synthetic water samples to use them as raw material for producing a natural coagulant based on bioeconomy. It would thus allow immature coffee beans to recover, avoiding their inadequate disposition. An extract was obtained from residual green coffee beans using the ultrasound-assisted separation technique with a mixture of ethanol and water in a 1:1 ratio. The Folin–Ciocalteu method was applied for the total polyphenols quantification, resulting in a concentration of 73.54 ± 0.05 mg GAE (Gallic Acid Equivalent) per gram on a dry coffee basis (GAE/gDB). The synthetic water for the study was prepared with kaolin, showing initial turbidity of 520.90 ± 0.1 NTU (Nephelometric Turbidity Units). First, the effect of pH was determined on the coagulant activity at a fixed dose of polyphenols 2.6 mg GAE/L. Second, the dose and pH results were evaluated using a multilevel factorial design with 5.20, 3.90, 2.60, and 1.30 mg GAE/L doses and pH at 2.5, 3.0, 3.5, and 4.0. Third, the turbidity removal achieved was 99.94% at a dose of 3.9 ± 0.05 mg GAE/L and a pH of 2.5. Fourth, the result was compared with the turbidity removal of the aluminum sulfate dosed at a concentration of 3 mg/L on the same water type, with a pH variation between 5.5 and 8, obtaining 98.69% of turbidity removed. Finally, the research demonstrated that the polyphenols extracted from the residues of the *Coffea arabica* species possess a high electrochemical affinity that would allow removing turbidity by coagulation in waters at specific pH levels with similar removals to those obtained with aluminum sulfate.

Keywords: *Coffea arabica*; natural coagulant; ultrasound assisted extraction; turbidity

check for updates

Citation: Cuesta-Parra, D.M.; Correa-Mahecha, F.; Rodríguez-Miranda, J.P.; Salcedo-Parra, O.J.; Rivas-Trujillo, E. Extraction of Polyphenols from Unripened Coffee (Coffea Arabica) Residues and Use as a Natural Coagulant for Removing Turbidity. *Processes* **2022**, *10*, 1105. https://doi.org/10.3390/pr10061105

Academic Editor: Dimitris Zagklis and Georgios Bampos

Received: 29 April 2022
Accepted: 25 May 2022
Published: 1 June 2022

1. Introduction

According to the Ministry of Environment, Housing and Territorial Development in Colombia, 44% of the rural population is supplied with non-potable water, even though Colombian legal regulations require service providers to comply with water quality requirements under physicochemical and biological parameters for consumption. These provisions are not necessarily fully respected [1]. Regarding wastewater, the Superintendency of Public Services estimates that only 42.2% of the flow of water discharged to water sources in Colombia has had some treatment (Institute of Hydrology, Meteorology and Environmental Studies, 2019 (known as IDEAM in Spanish)).

Turbidity in water is due to suspended particles, which are the prime cause of deterioration in liquid quality as they adsorb different contaminants such as bacteria, nutrients, toxic compounds, and heavy metals. Thus, turbidity is considered one of the most relevant parameters for water quality assessment [2], due to the ease of implementation, low infrastructure requirements, and lower costs. Hence, coagulation and flocculation processes are the most widely used for wastewater and drinking water treatment [3,4]. Coagulation is characterized as a rapid process, which depends on water properties such as pH, temperature, and amount of dissolved and suspended particles, and involves the addition of a chemical or natural substance called coagulant, whose function is to destabilize suspended, colloidal, and dissolved matter [5].

Aluminum sulfate and aluminum polychloride (PAC) are widely used compounds in coagulating wastewater and consumption. Up to 35% of the carbon footprint of a conventional treatment water plant comes from using chemicals such as coagulants; the health and environmental effects of the mining activities for extracting bauxite, the raw material for processing aluminum, have also been reported [6]. On the one hand, the sludge generated during coagulation can reach rates between 4% and 7% concerning the treated water, implying a large amount of material available. It possesses a high aluminum content that may cause secondary pollution in the release during drying processes or the leaching when already landfilled [7] The relationship between the consumption of aluminum and the development of diseases, such as cancer, Alzheimer's, disorders of the bone system, and chromosomal modifications, has been reported [8]. Sludges from ferric coagulant treatments, on the other hand, widely used in sewage coagulation, are often implemented as additives in soil amendments; however, when applied in agricultural systems, they capture phosphorus, reducing availability in the crop and increasing the costs of fertilization, so authors have suggested the search for other methods of disposal [9].

The increasing awareness of the ecotoxicological and human health impacts of synthetic coagulants and flocculants has drawn attention to environmentally friendly approaches to water treatment, available in a growing body of research aimed at developing coagulant products from renewable, biodegradable, and non-toxic resources that boost the circular economy and reduce both the carbon footprint of coagulation and the dependence on chemicals through the partial or total substitution (Saleem & Bachmann, 2019). From a bioeconomy perspective, the production and application of polymeric coagulants of plant or microbiological origin seem to be the most viable alternatives for increasing the sustainability of water purification processes, so research on natural coagulants is a must [2,10].

As natural coagulants, there are the *Moringa oleifera* [11] and *Phaseolus vulgaris*, or the common bean proteins, the different extracts obtained from actinobacteria, starches and their derivatives, the extract tree seed known as chestnut (*Aesculus hyppocastanum*) oak or various species of the genus *Opuntia*, the plant known as nopal (Oladoja, 2015). Polysaccharides obtained from the mucilage of *Plantago major* L. have proved utility in bleaching water contaminated with dyes [12] to the same as that of other species of the genus *Plantago* and polysaccharides of *Enteromorpha* sp. (Chorophyta, macroalgae) [13]; similarly, chitosan obtained from chitin, one of the most abundant polymers on earth, is extracted from various sources such as animals, insects, and fungi; it has an activity for removing turbidity and organic matter of the water [14].

Reports also described that the polyphenols present in some plants have activity in removing dyes in synthetic waters. The ethanol or grape seed extract (*Vitis vinifera*) was evaluated as a cationic coagulant on synthetic waters with malachite green and crystal violet dyes achieving removals greater than 80% with doses of 1.5 g of extract per liter. The samples' gradual discoloration and the appearance of precipitated flocs indicate that these extracts can lead to coagulation and flocculation processes. Thus, catechin and tannic acid have been identified as the compounds responsible for removing the color [15]. Some coagulants derived from polyphenols, from plants such as *Jatropha curcas* (Abidin et al., 2013), *Acacia mearnsii* [16], and *Opuntia* spp. [17], have been tested as alternatives for turbidity removal, treatment of urban wastewater [18–20] and the removal of color in synthetic water (Lopes et al., 2019). The evaluations included the leachate treatment [21] and wastewater from the textile and dairy industry; polyphenols were also modified by gelation to generate adsorbents that remove cationic contaminants or by cationization processes that allow ampholytic coagulants [4].

Coffee, one of the most favored beverages worldwide and the leading commodity on the planet after oil, has polyphenols in each part of the fruit [22], which are related to the aroma, flavor, and color of the drink [23]; so, its residues can be used as a source of antioxidants, nutraceuticals, and preservatives of some food preparations in areas of cosmetology and water treatment [24]. Colombia is Latin America's second-largest producer; only 5% of the weight of the fresh fruit is used for the preparation of the beverage, while the remaining 95% becomes residues in the production chain [25], turning this waste into a potential source of materials for obtaining different products.

The extraction and determination of polyphenol concentrations in coffee residues are conducted by various methods. Patay and her team (2016) evaluated mature, immature beans and their pericarps of *Coffea arabica* using a 40:60 solution of isopropanol-water, then analyzed their polyphenol content by the Folin–Ciocalteu method, finding results of 2.12%, 4.14%, 1.68%, and 1.63% respectively. Prifitis and his colleagues (2018) performed ultrasound extraction of previously ground beans, although they did not mention the species used or the maturity stage: the coffee contained 42.61 mg GAE/g of dry material; the same study identified 3-chlorogenic acid (16. 61 mg/g), 4-and 5-chlorogenic acids (13.62 mg/g). The isolation of bioactive compounds present in the solid matrices of coffee by-products and green coffee has been studied, and the conventional and modern extraction techniques have been applied [26]. Among the traditional techniques is Soxhlet extraction, which has drawbacks due to the amount of solvent required, long extraction times, and in some cases, the loss of the compound to be separated [27]. New techniques include extraction with ionic solids, which uses supercritical fluids by pulsed electric fields assisted with microwaves and ultrasound, methods that improve the yield and obtain purer products [28].

2. Materials and Methods

2.1. Raw Material and Chemical Substances

Unripe green fruit of the *Coffea arabica* species, provided by the Technological Park Corporation for Coffee and Coffee Growing (Technicafé), located in the municipality of Cajibío, department of Cauca, Colombia, collected during February and March 2019 was used. The sample was kept refrigerated (7 °C) for less than a week to avoid decomposition while extraction of the polyphenols was carried out with ethanol; the quantification of phenols used the Folin–Ciocalteu reagent, sodium carbonate, and gallic acid in the calibration of the standard curve. For the jar test, the pH of the water was modified with sodium hydroxide and hydrochloric acid.

The synthetic water used for the initial turbidity control resulted from an impalpable kaolin solution. The comparison of the natural coagulant to a commercial coagulant used aluminum sulfate.

2.2. Ultrasound Assisted Extraction

The polyphenols were extracted from one sample of coffee ground following the methodology described by Cuesta and Correa (2018) with some modifications: 16 consecutive extractions were carried out using a mixture of ethanol-water (50:50) as an extraction agent at a ratio of 1:1 (solid:solution). The ground coffee and the solvent were mixed in 250 mL Schott® bottles, sonicated for 30 min in an ultrasonic bath witeg® Germany Wisd Ultrasonic cleaner Set, model WUC-D06H using a frequency of 60 kHz, afterward, the extracts were collected and vacuum filtered using a Büchner funnel; the polyphenol content of the filtrate was quantified, and this was dried using a *Heildoph® Germany G3 Hei- Vap precision* rotary evaporator at temperatures between 56 °C and 78 °C and pressures between 556 mbar and 175 mbar for solvent recovery [29].

2.3. Determination of Total Phenols of the Extract

Polyphenols extracted from the coffee fruit were quantified through the colorimetric Folin–Ciocalteu method described by Chen et al. (2015). The calibration curve was adjusted using standard solutions of gallic acid; the blank determination was prepared with distilled water, and absorbances were measured by a Jenway® *United Kingdom 6320D* at a wavelength of 765 nm. The concentration of total polyphenols is expressed in milligrams equivalent to gallic acid per liter and gram dry basis of the coffee sample (mg GAE/L and mg GAE/gDB coffee).

2.4. Sample Water

The synthetic water for the study was prepared following the methodology described by Miller and her team (2008). Indeed, 5000 mg/L of kaolin were mixed for each liter of water, and the suspension was stable for one hour, achieving turbidity of 520.90 $\pm$ 0.1 NTU and an initial pH of 7.32 $\pm$ 0.1.

2.5. Coagulation

The adaptation and coagulation procedure for the synthetic water was performed under the ASTM D 2035-90 descriptions from Liang and her team (2019) using a conventional digital flocculator. The 1000 mL samples were deposited in beakers, homogenizing them at an instantaneous mixing speed of 120 rpm. The initial turbidity and initial pH were measured, adjusting the working pH with hydrochloric acid or sodium hydroxide according to the different levels of the experimental design. Subsequently, the coagulant dose was added and homogenized for 1 min at 120 rpm. Later, the speed was reduced to 30 rpm, stirring for 20 min, after which the solutions were sedimented. Samples settled in the central part of the vessel to determine the turbidity and the final pH; as a result, the treatment was performed for each studied pH figure subjected to the same conditions but without adding a coagulant.

Turbidity was determined through the Turbiquant® Germany 1100 IR turbidimeter, following the protocol of the standard methods for the analysis of drinking water and wastewater [30]. The performance of the coagulation process was calculated, taking into account the coagulant activity as a percentage of turbidity removal and according to the following equation [3,31].

$$Coagulant\ activity\ (\%) = \frac{T_A - T_R}{T_A} \times 100\%$$

where:

T_R is the residual turbidity after treatment in Nephelometric Turbidity Units.

T_A is the turbidity of the control sample in Nephelometric Turbidity Units; it was handled under the same conditions as the test samples but without adding any coagulant.

A first test established the effect of pH on the percentage removal of turbidity at a fixed dose of 2.60 mg GAE/L, at pH of 2.0, 3.0, 4.0, 5.0, 6.0, 7.0, 8.0, 9.0, 10.0, 11.0, and 12.0, to determine the range of best coagulation performance of the extract.

Once the pH range with the best coagulant performance was known, a factorial experimental design was executed with four pH levels (2.50, 3.00, 3.50, and 4.00), and four dosage levels (5.20, 3.90, 2.60, and 1.30 mg GAE/L), with three repetitions for a total of 48 experimental tests.

Various assays removing turbidity via aluminum sulfate on synthetic water at a dosage of 3 mg/L and pH between 5.5 and 8.0 were conducted, in which improved performance of aluminum sulfate was expected based on previous studies [32,33].

2.6. Statistical Analysis

The data obtained from the turbidity removal were processed through an analysis of variance (Anova), using the data processing program SPSS Statistics 24 (IBM), under a 95% confidence level.

3. Results and Discussion

3.1. Extraction of Polyphenols

Polyphenols from immature coffee of the *Coffea arabica* species were separated using ultrasound-assisted extraction. The concentration of total phenols was 73.54 ± 0.05 mg GAE/gDB coffee, resulting in lower rates than those found in previous research projects with immature coffee beans. Cuesta and Correa (2018) obtained 129.2 ± 0.05 mg GAE/gDB coffee via extraction assisted with ultrasound at 60 kHz for 30 min; in this study, extraction with acetone was performed; however, the difference in extraction cannot be attributed only to the type of solvent since the same samples were not used in the present study. Therefore, the plant samples can have natural variations in the content of secondary metabolites caused by variability in the weather and culture (Hećimović et al., 2011). The use of alcohol mixtures as extraction agents requires applying the principles of green chemistry, which, together with ultrasound-assisted extraction and biorefinery approaches, seek to improve the environmental performance of products for the industry (Cvjetko Bubalo et al., 2018).

3.2. Effect of pH on the Removal of Turbidity

Figure 1 describes the results of the tests for removing turbidity at different pH levels and fixed doses of coagulant, 2.60 mg GAE/L to extract coffee residues. For the extract, the best removal turbidities were 95.69 ± 1.00% and 93.56 ± 1.58%, obtaining an acid pH of 2 and 3, respectively. However, between 4 and 10, a significant reduction in the coagulant activity was observed, keeping it below 40%, then there was an increase of 70.23 ± 0.70% at pH 11 and a drop to the minimum at pH 12.

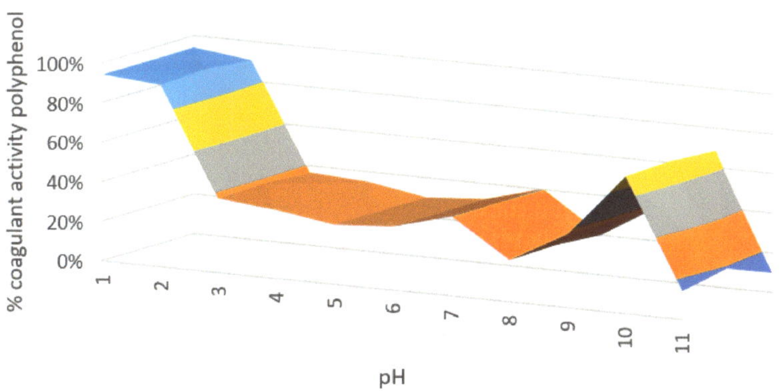

Figure 1. The pH effect on the removal of turbidity of the polyphenol extract, 2.6 mg GAE/L dose. Average levels of three determinations, n = 3. The error bars represent ± the standard deviation of the determinations.

Removal tests were conducted as per the design by taking pH, in 10 levels for understanding the removal behavior polyphenol composite. ANOVA (analysis of variance) for turbidity removal is provided in the Table 1. Each pH effect and interaction was studied through analysis of variance; the table shows the variance of factors. The F value was determined by relationship between each pH parameter and removal ratio. F value obtained as 250.8 shows the significance analysis for each pH variation separately.

Table 1. ANOVA for relationship between pH and % turbidity removal.

Source	Sum of Squares	Degrees of Freedom	Mean Square	F-Value	*p*-Value
Relationship between pH and turbidity removal	0.00101906	1	0.00101906	2.42830421	0.15359017
Intra pH and turbidity removal ratio	0.94753023	9	0.10528114	250.873741	1.0251×10^{-09}
Pure error	0.00377692	9	0.00041966		
Cor total	0.95232621	19			

Coagulant activity is influenced by pH because the active agents in coagulants generally carry anionic or cationic charges [5,31]. Korolev and Nesterov (2018) stated that the hydroxylation surface charge regulates the electrochemical behavior of kaolinite and is the most sensitive characteristic of surface charge in clays. They originate from protonation/deprotonation reactions at the edges of the particles, neutralization reactions at the surfaces, consequence of the breakage of the Si-O-Si or Al-o-Al bonds, as well as from the hydrolysis of the oxygen atoms. In their study, they found that at pH less than 3, the clay has a net positive charge; at a pH of around 3, there is a charge change; and at a pH above 3, it has a net negative charge [34,35].

Jeon et al. (2009) hypothesize that the hydroxyphenyl groups of polyphenols can interact with positively and even negatively charged contaminants through dipolar forces resulting from the electronegativity of oxygen and the loss of hydrogen ions in aqueous solutions. Thus, phenolic groups in tannins and other polyphenols cause the anionic character of the molecules (Lopes et al., 2019), whereas in polyphenols, hydroxyl groups possess a net acidic action and generate a higher level of acidity, as occurs with caffeic acid, a relatively abundant compound in coffee. Thus, the deprotonation of these groups produces stable anions due to the formation of bridge patterns and hydrogen [36], which would explain why a higher coagulating activity is achieved at low pH levels, when kaolin and polyphenols have different charges. Complexation and net charge neutralization by affinity between the pollutant and the coagulant are the main mechanisms for the precipitation of solubilized or suspended pollutants in water in the case of polyphenols (Jeon et al., 2009); however, it should be noted that acidification of the solution favors a precipitation mechanism of the clay particles (Aljerf, 2018).

In the case of *Opuntia ficus indica*, it was shown that the phenol group is responsible for the improved coagulation and flocculation activity, finding out that the optimal pH in synthetic waters with kaolin and humic acids was 10; although, there was also activity at acidic pH rates (Bouaouine et al., 2018). At pH 10, clays exhibit a high capacity for anion and cation exchange, so the cationic sites may be present on the surface and allow absorption of anions in conditions of varied pH. This phenomenon occurs due to the interactions of iron and aluminum oxides in kaolin. In another study, Bouaouine and his team (2019) used oil mill wastewaters defatted (a byproduct of olive) as bioflocculant in treating synthetic water prepared with kaolin and humic acids, concluding that the optimal removal with a dose of 100 mg/L was 92% ± 2% at a pH of 11, pH in which the present study found the activity of 70.23% ± 0.7%. The authors explained that tannins and cellulose are active functional groups in charge of flocculation. Similar findings were observed for extracts of *Opuntia ficus indica*, in which an adsorption bridge model was proposed to ionize tannins at very alkaline pH levels, establishing ionic and hydrogen bonds with the kaolin aluminum and the starches so that a contribution to the floc quality through bridges and dragging was achieved [37].

The range of initial pH at which the best performances of the aluminum sulfate were achieved corresponds to those reported by other authors (Zhao et al., 2009). Guo and his colleagues (2015) also found that the coagulant activity of sulfate aluminum had reduced when the pH assumed alkaline figures, as it happens in the present work. For acidic rates, aluminum has positive charges, so the principal coagulation mechanism occurs by charge neutralization.

A second jar test was performed to evaluate the effect of the dose, varying the concentrations of the polyphenolic extract: doses of 1.3 mg/L ± 0.05, 2.6 ± 0.05 mg/L, 3.9 ± 0.05 mg/L, 5.2 ± 0.05 mg/L were applied at pH between 2.5 and 4, similar to the best performances of the previous test. Figure 2 illustrates the results of the four coagulant doses at four different pH; the trend of the removal percentage is shown taking into account the pH used versus the concentration of added polyphenols (mg GAE/L); for pH 2.5 and 3.0, effectiveness greater than 99% is observed for all after added concentrations. The decrease in efficiency from pH 3.5 to almost 90% of removal for 3.9 mg GAE/L is evident. The pH at 4.0 confirms minimal efficiency reaching a removal percentage of less than 50% for the concentration of 2.6 mg GAE/L. These results allow concluding that the predominant efficiency of solid removal in water is visible at pH 2.5 and 3.0 for all concentrations of studied polyphenols. In Figure 2, the error bars correspond to the standard deviations of the means of three replicates per combination between polyphenol doses and pH, indicating that the 3.9 mg GAE/L and the pH 3.5 doses show better variability (standard deviation of 6.09%), followed by 2.6 mg GAE/L and pH 4.0 (3.16%), and 1.3 mg GAE/L and pH 4.0 (2.18%). The other 13 combinations present standard deviations that do not exceed 1.0%, demonstrating that the developed method provides duplicable outcomes.

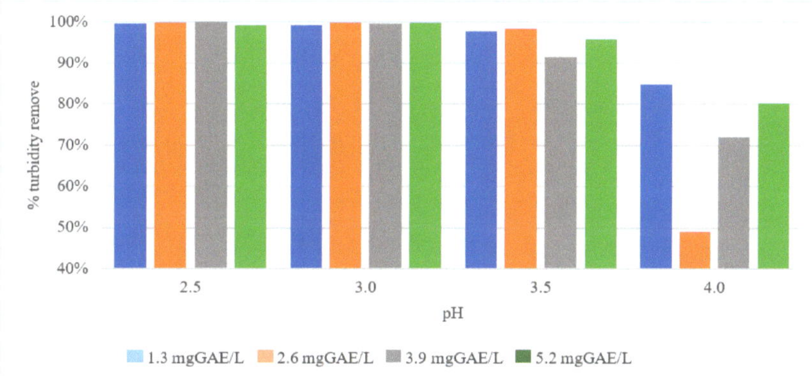

Figure 2. Relationship between coagulant dose, pH, and turbidity removal. Figure obtained with the mean removal scores of three replicates for each pH level and dose, n = 3. The error bars represent ± the standard deviation for the determinations.

On the one hand, the results specify that acidic pHs favor flocculation and coagulation, even when we used the lowest concentration of polyphenols. On the other hand, we found that the condition for producing the highest removal percentage corresponds to the combination of 3.9 mg GAE/L with pH 2.5. It is an indication that there is no dose-dependent behavior and that the most determining influence includes the pH of the sample rather than the added concentration.

Considering a p score = 0.05 with three replicates for each determination, the analysis of variance concerning the concentration of polyphenols used for the jar test trials (5.2, 3.9, 2.6, and 1.3 mg GAE/L) yielded statistically significant differences ($p < 0.0001$). At pH 4.0, a minimum rate of 45.98% for 2.6 mg GAE/L and a maximum of 87.25% for 1.3 mg GAE/L were obtained (Table 2). The results for the pH ranges studied (2.5, 3.0, 3.5, 4.0) are similar ($p < 0.00015$) to the concentration, indicating that there are differences between the four pHs studied; however, it should be noted that for pHs between 2.5 and 3.5, the percentage of removal remained above 95%, in contrast to the 80% obtained at pH 4.0.

Table 2. ANOVA for relationship between acid pH and % turbidity removal.

Source	Sum of Squares	Degrees of Freedom	Mean Square	F-Value	p-Value
Sample	0.05208757	3	0.01736252	53.038351	1.6178×10^{-12}
Polyphenol doses	0.65558436	3	0.21852812	667.551032	6.4239×10^{-29}
Relationship between pH and turbidity removal	0.18626541	9	0.02069616	63.2217978	77438×10^{-18}
Intra pH and turbidity removal ratio	0.01047545	32	0.00032736		
Cor total	0.9044128	47			

The analysis of variance for the interactions between the concentration and pH levels used for the research shows statistically significant differences ($p < 0.05$) with a maximum of 99.94% removal, pH 2.5, and 3.9 mg GAE/L. It should be noted that the results found between pH 2.5 and 3.0 were greater than 99% of removal for the four concentrations of used polyphenols (5.2, 3.9, 2.6, 1.3 mg GAE/L). These findings indicate that the polyphenols obtained from the coffee samples show improved turbidity removal at pH below 3.5, which is explained by the difference in charges between the pollutant, kaolin, and polyphenols [37].

In the case of other coagulants with polyphenolic active component, no turbidity removal studies were found at pH levels lower than 3. Bouaouine and his team (2018) evaluated the turbidity removal of *Opuntia ficus-indica* extracts using synthetic water with kaolin and humic acids. They found removals of 92% at pH of 10 to 12; the researchers also reported removals at growing pH levels between 5 and 3, reaching 88% to pH 3. The study suggests that lignin and tannin phenolic groups are responsible for the coagulant action. A later paper reported optimal level at pH 10, dose of 20 mg/L. and coagulant activity of 93% from fractionating the extract. It was concluded that the molecules responsible for the coagulant action correspond to quercetin, a flavonoid that belongs to the polyphenols, and starch. These molecules have a synergistic action in the coagulation–flocculation process at alkaline pH [37]. In the present work, the action of other components that may exert some synergistic action during the coagulation–flocculation process is not ruled out.

Vuppala et al. (2019) report removal of 99.2% and pH of 4.5 using 100 mg/L in wastewaters of the olive oil industry (olive oil mills). It is notable that they did not perform assays to pH below 4.5. Bouaouine and his team (2019) also addressed the reuse of mill defatted waters as bioflocculant in synthetic water containing kaolin. They succeeded in a removal of 92 ± 2% at pH 11 using a dose of 100 mg/L. So, tannins and starches were identified as the molecules responsible for coagulant activity; the effect was not studied at pH below 3. In this study, a coagulant activity of 70.23 ± 0.7% was reached at pH 11, indicating the need to explore in the future the removal of contaminants at alkaline pH with extracts obtained from coffee residues

Dela Justina and his team (2018) evaluated a coagulant extracted from the species *Acacia mearnssi*; subsequently cationized in wastewaters from the milk industry in a pH range between 4 and 10. The dosage used was 200 mg/L in the conducted tests, and removals greater than 90% were obtained except at pH 4, finding a relationship between pH and turbidity removal. The cationization of tannins leads to a change in the range of action of polyphenols against contaminants; thus, it is a technique to attain coagulants with better pH range [16]. Sánchez et al. (2010) evaluated the effect of pH on turbidity removal using a coagulant based on modified tannins in surface waters. They found that the coagulation process is more effective at a slightly acidic pH close to 4, removing the turbidity at 95%, which can be explained by the appearance of species with a positive charge or not soluble in water.

Hussain and his team (2019) tested the dust of the pine cones of the *Pinus gerardiana*. The coagulating activity in water with kaolin at different pH levels between 2 and 12 yielded that the best removals were 82% at pH 2 and 76% at pH 12 using doses of 0.2 mL/L of extract; however, the molecular species responsible for the coagulation effect were not identified, but the cones of this species are known for having an appreciable content of tannins and polyphenols [38].

The coagulant activity of the coffee extracts presented in this work is within the levels reported for other natural coagulants in synthetic waters with kaolin. The *Moringa oleifera* reached a removal of 98% for water with high turbidity (500 NTU) using doses of 400 mg/L of extract [39], while for *Planago ovata* removals of 95.6% are achieved at pH levels below 8 from waters with a turbidity of 250 NTU [40]. It confirms that the extracts of this agro-industrial residue would potentially remove turbidity in clay waters.

4. Conclusions

The objective of this research was to evaluate the coagulant activity of extracts of immature *Coffea arabica* residues. So, an ultrasound-assisted extraction using an ethanol-water mixture as solvent was carried out, finding that the contents of polyphenols in the extract reached 73.54 ± 0.05 mg GAE/gDB coffee, with lower figures than those reported in previous studies. The type of used solvent and the agroclimatic variables may influence the content of polyphenols and their extraction.

The pH has a significant influence on the coagulant activity of the extract with the best yields below 3 and with a local maximum pH level of 11. The increase in coagulant activity at acidic pH can be explained by the difference between the charges of kaolin (positive charges at acidic pH) and those of polyphenols (negative charges). Therefore, the coagulation mechanism can occur by charge neutralization, while at pH 11, coagulation is achieved due to the formation of ionic bonds between polyphenols and aluminum oxides on the kaolin surface.

In summary, the highest coagulant activity was 99.88% at a dose of 3.9 mg/L at pH 2.5. From statistical analysis, significant differences between pH levels and doses were found. In all treatments with pH below 3, removals over 98% were reached with no evidence of a dose-dependent effect. Finally, the comparison with the coagulant activity of aluminum sulfate showed a more notorious activity in the coffee residues, although at a different pH level.

Author Contributions: Experimentation, methodology D.M.C.-P. and F.C.-M.; conceptualization, J.P.R.-M., O.J.S.-P. and E.R.-T. All authors have read and agreed to the published version of the manuscript.

Funding: This research received no external funding.

Institutional Review Board Statement: Not applicable.

Informed Consent Statement: Not applicable.

Data Availability Statement: Data supporting reported results can be found in https://repository.uamerica.edu.co/handle/20.500.11839/6704?locale=es (accessed on 24 January 2022).

Conflicts of Interest: The authors declare no conflict of interest.

References

1. Briseño, H.; Rubiano, J. El servicio de agua potable para uso residencial en Colombia. *Rev. UDCA Actual. Divulg. Científica* **2018**, *21*, 235–242.
2. Saritha, V.; Karnena, M.K.; Dwarapureddi, B.K. "Exploring natural coagulants as impending alternatives towards sustainable water clarification"—A comparative studies of natural coagulants with alum. *J. Water Process Eng.* **2019**, *32*, 100982. [CrossRef]
3. Antov, M.G.; Vasić, V.M.; Prodanović, J.M.; Šćiban, M.B.; Đorđević, T.R.; Kukić, D.V.; Milošević, M.M. Common oak (Quercus robur) acorn as a source of natural coagulants for water turbidity removal. *Ind. Crops Prod.* **2018**, *117*, 340–346. [CrossRef]
4. Oladoja, N.A. Headway on natural polymeric coagulants in water and wastewater treatment operations. *J. Water Process Eng.* **2015**, *6*, 174–192. [CrossRef]
5. Bratby, J. *Coagulation and Flocculation in Water and Wastewater Treatment*; IWA Publishing: London, UK, 2016.
6. Saleem, M.; Bachmann, R.T. A contemporary review on plant-based coagulants for applications in water treatment. *J. Ind. Eng. Chem.* **2019**, *72*, 281–297. [CrossRef]

7. Chen, Y.S.; Ooi, C.W.; Show, P.L.; Hoe, B.C.; Chai, W.S.; Chiu, C.Y.; Wang, S.S.S.; Chang, Y.K. Removal of Ionic Dyes by Nanofiber Membrane Functionalized with Chitosan and Egg White Proteins: Membrane Preparation and Adsorption Efficiency. *Membranes* **2022**, *12*, 63. [CrossRef]
8. Choy, S.Y.; Prasad, K.N.; Wu, T.Y.; Raghunandan, M.E.; Ramanan, R.N. Performance of conventional starches as natural coagulants for turbidity removal. *Ecol. Eng.* **2016**, *94*, 352–364. [CrossRef]
9. Maroušek, J.; Stehel, V.; Vochozka, M.; Kolář, L.; Maroušková, A.; Strunecký, O.; Peterka, J.; Kopecký, M.; Shreedhar, S. Ferrous sludge from water clarification: Changes in waste management practices advisable. *J. Clean. Prod.* **2019**, *218*, 459–464. [CrossRef]
10. Sillanpää, M.; Ncibi, M.C.; Matilainen, A.; Vepsäläinen, M. Removal of natural organic matter in drinking water treatment by coagulation: A comprehensive review. *Chemosphere* **2018**, *190*, 54–71. [CrossRef]
11. Villaseñor-Basulto, D.L.; Astudillo-Sánchez, P.D.; del Real-Olvera, J.; Bandala, E.R. Wastewater treatment using Moringa oleifera Lam seeds: A review. *J. Water Process Eng.* **2018**, *23*, 151–164. [CrossRef]
12. Chaibakhsh, N.; Ahmadi, N.; Zanjanchi, M.A. Use of Plantago major L. as a natural coagulant for optimized decolorization of dye-containing wastewater. *Ind. Crops Prod.* **2014**, *61*, 169–175. [CrossRef]
13. Oladoja, N.A. Advances in the quest for substitute for synthetic organic polyelectrolytes as coagulant aid in water and wastewater treatment operations. *Sustain. Chem. Pharm.* **2016**, *3*, 47–58. [CrossRef]
14. Ang, W.L.; Mohammad, A.W.; Benamor, A.; Hilal, N. Chitosan as natural coagulant in hybrid coagulation-nanofiltration membrane process for water treatment. *J. Environ. Chem. Eng.* **2016**, *4*, 4857–4862. [CrossRef]
15. Jeon, J.R.; Kim, E.J.; Kim, Y.M.; Murugesan, K.; Kim, J.H.; Chang, Y.S. Use of grape seed and its natural polyphenol extracts as a natural organic coagulant for removal of cationic dyes. *Chemosphere* **2009**, *77*, 1090–1098. [CrossRef] [PubMed]
16. Beltrán-Heredia, J.; Sánchez-Martín, J.; Dávila-Acedo, M.A. Optimization of the synthesis of a new coagulant from a tannin extract. *J. Hazard. Mater.* **2011**, *186*, 1704–1712. [CrossRef] [PubMed]
17. Miller, S.; Fugate, E.; Oyanedel, V.; Smith, J.; Zimmerman, J. Toward Understanding the Efficacy and Mechanism of Opuntia spp. as a Natural Coagulant for Potential Application in Water Treatment. *Environ. Sci. Technol.* **2008**, *42*, 4274–4279. [CrossRef]
18. Beltrán-Heredia, J.; Sánchez-Martín, J. Municipal wastewater treatment by modified tannin flocculant agent. *Desalination* **2009**, *249*, 353–358. [CrossRef]
19. Hameed, Y.T.; Idris, A.; Hussain, S.A.; Abdullah, N. A tannin-based agent for coagulation and flocculation of municipal wastewater: Chemical composition, performance assessment compared to Polyaluminum chloride, and application in a pilot plant. *J. Environ. Manag.* **2016**, *184*, 494–503. [CrossRef]
20. Suja, F.; Hussain, S.A.; Abdullah, N.; Man, H.C.; Idris, A.; Hameed, Y.T. A tannin–based agent for coagulation and flocculation of municipal wastewater as a pretreatment for biofilm process. *J. Clean. Prod.* **2018**, *182*, 198–205.
21. Ibrahim, A.; Yaser, A.Z. Colour removal from biologically treated landfill leachate with tannin-based coagulant. *J. Environ. Chem. Eng.* **2019**, *7*, 103483. [CrossRef]
22. Panusa, A.; Petrucci, R.; Lavecchia, R.; Zuorro, A. UHPLC-PDA-ESI-TOF/MS metabolic profiling and antioxidant capacity of arabica and robusta coffee silverskin: Antioxidants vs phytotoxins. *Food Res. Int.* **2017**, *99*, 155–165. [CrossRef] [PubMed]
23. Ali, S.; Mullen, W.; Nemzer, B.; Combet, E.; Stalmach, A. Polyphenolic and Hydroxycinnamate Contents of Whole Coffee Fruits from China, India, and Mexico. *J. Agric. Food Chem.* **2013**, *61*, 5298–5309.
24. Farah, A.; Donangelo, C.M. Phenolic compounds in coffee—Minireview. *Braz. J. Plant Physiol.* **2006**, *18*, 23–36. [CrossRef]
25. Ballesteros, L.F.; Ramirez, M.J.; Orrego, C.E.; Teixeira, J.A.; Mussatto, S.I. Encapsulation of antioxidant phenolic compounds extracted from spent coffee grounds by freeze-drying and spray-drying using different coating materials. *Food Chem.* **2017**, *237*, 623–631. [CrossRef] [PubMed]
26. Galanakis, C.M. *Handbook of Coffee Processing By-Products: Sustainable Applications*; Academic Press: Cambridge, MA, USA, 2017.
27. Mourtzinos, I.; Goula, A. Polyphenols in Agricultural Byproducts and Food Waste. In *Polyphenols in Plants*; Academic Press: Cambridge, MA, USA, 2019; pp. 23–44.
28. Galanakis, C.M. Food Waste Recovery: Prospects and Opportunities. In *Sustainable Food Systems from Agriculture to Industry*; Academic Press: Cambridge, MA, USA, 2018; pp. 401–419.
29. Cuesta-Parra, D.; Correa-Mahecha, F. Obtención de fenoles a partir de granos verdes de café. *Rev. Ion* **2018**, *31*, 31–35. [CrossRef]
30. Baird, R.; Bridgewater, L. *Standard Methods for the Examination of Water and Wastewater*, 23rd ed.; American Public Health Association; American Water Works Association; Water Environment Federation: Washington, DC, USA, 2017.
31. Hussain, S.; Ghouri, A.S.; Ahmad, A. Pine cone extract as natural coagulant for purification of turbid water. *Heliyon* **2019**, *5*, e01420. [CrossRef]
32. Guo, B.; Yu, H.; Gao, B.; Rong, H.; Dong, H.; Ma, D.; Li, R.; Zhao, S. Coagulation performance and floc characteristics of aluminum sulfate with cationic polyamidine as coagulant aid for kaolin-humic acid treatment. *Colloids Surf. A Physicochem. Eng. Asp.* **2015**, *481*, 476–484. [CrossRef]
33. Zhao, H.; Liu, H.; Qu, J. Effect of pH on the aluminum salts hydrolysis during coagulation process: Formation and decomposition of polymeric aluminum species. *J. Colloid Interface Sci.* **2009**, *330*, 105–112. [CrossRef]
34. Korolev, V.A.; Nesterov, D.S. Regulation of clay particles charge for design of protective electrokinetic barriers. *J. Hazard. Mater.* **2018**, *358*, 165–170. [CrossRef]
35. Nordmark, B.A.; Przybycien, T.M.; Tilton, R.D. Effect of humic acids on the kaolin coagulation performance of Moringa oleifera proteins. *J. Environ. Chem. Eng.* **2018**, *6*, 4564–4572. [CrossRef]

36. Leopoldini, M.; Russo, N.; Toscano, M. Gas and Liquid Phase Acidity of Natural Antioxidants. *J. Agric. Food Chem.* **2006**, *54*, 3078–3085. [CrossRef] [PubMed]
37. Bouaouine, O.; Bourven, I.; Khalil, F.; Baudu, M. Reuse of olive mill wastewater as a bioflocculant for water treatment processes. *J. Clean. Prod.* **2019**, *246*, 119031. [CrossRef]
38. Taş, N.G.; Gökmen, V. Phenolic compounds in natural and roasted nuts and their skins: A brief review. *Curr. Opin. Food Sci.* **2017**, *14*, 103–109. [CrossRef]
39. Katayon, S.; Noor, M.J.M.M.; Asma, M.; Ghani, L.A.A.; Thamer, A.M.; Azni, I.; Ahmad, J.; Khor, B.C.; Suleyman, A.M. Effects of storage conditions of Moringa oleifera seeds on its performance in coagulation. *Bioresour. Technol.* **2006**, *97*, 1455–1460. [CrossRef]
40. Ramavandi, B. Treatment of water turbidity and bacteria by using a coagulant extracted from Plantago ovata. *Water Resour. Ind.* **2014**, *6*, 36–50. [CrossRef]

processes

MDPI

Review

Emerging Pollutants in Wastewater, Advanced Oxidation Processes as an Alternative Treatment and Perspectives

Alexis Rubén Bracamontes-Ruelas [1], Luis Alberto Ordaz-Díaz [2], Ana María Bailón-Salas [3], Julio César Ríos-Saucedo [4], Yolanda Reyes-Vidal [5] and Liliana Reynoso-Cuevas [6,*]

1 Departamento de Ingeniería Sustentable, Centro de Investigación en Materiales Avanzados, S.C., Calle CIMAV 110, Ejido Arroyo Seco, Durango 34147, Mexico; alexis.bracamontes@cimav.edu.mx
2 Departamento de Ingeniería en Tecnología Ambiental, Universidad Politécnica de Durango, Carretera Durango-México Km. 9.5, Durango 34306, Mexico; luis.ordaz@unipolidgo.edu.mx
3 Facultad de Ciencias Forestales y Ambientales, Universidad Juárez del Estado de Durango, Río Papaloapan, Valle del Sur, Durango 34120, Mexico; bailon_anna@hotmail.com
4 Campo Experimental Valle del Guadiana, Instituto Nacional de Investigaciones Forestales Agrícolas y Agropecuarias, Km. 5 Carretera Durango-El Mezquital, Durango 34170, Mexico; rios.julio@inifap.gob.mx
5 Cátedras CONACYT, Centro de Investigación y Desarrollo Tecnológico en Electroquímica, S.C., Parque Tecnológico Querétaro s/n, Pedro Escobedo, Querétaro 76703, Mexico; mreyes@cideteq.mx
6 Cátedras CONACYT, Centro de Investigación en Materiales Avanzados, S.C., Calle CIMAV 110, Ejido Arroyo Seco, Durango 34147, Mexico
* Correspondence: liliana.reynoso@cimav.edu.mx

Abstract: Emerging pollutants are present in wastewaters treated by conventional processes. Due to water cycle interactions, these contaminants have been reported in groundwater, surface water, and drinking waters. Since conventional processes cannot guarantee their removal or biotransformation, it is necessary to study processes that comply with complete elimination. The current literature review was conducted to describe and provide an overview of the available information about the most significant groups of emerging pollutants that could potentially be found in the wastewater and the environment. In addition, it describes the main entry and distribution pathways of emerging contaminants into the environment through the water and wastewater cycle, as well as some of the potential effects they may cause to flora, fauna, and humans. Relevant information on the SARS-CoV-2 virus and its potential spread through wastewater is included. Furthermore, it also outlines some of the Advanced Oxidation Processes (AOPs) used for the total or partial emerging pollutants removal, emphasizing the reaction mechanisms and process parameters that need to be considered. As well, some biological processes that, although slow, are effective for the biotransformation of some emerging contaminants and can be used in combination with advanced oxidation processes.

Keywords: emerging pollutants or contaminants; SARS-CoV-2; advanced oxidation processes (AOPs); mineralization; biological processes

Citation: Bracamontes-Ruelas, A.R.; Ordaz-Díaz, L.A.; Bailón-Salas, A.M.; Ríos-Saucedo, J.C.; Reyes-Vidal, Y.; Reynoso-Cuevas, L. Emerging Pollutants in Wastewater, Advanced Oxidation Processes as an Alternative Treatment and Perspectives. *Processes* **2022**, *10*, 1041. https://doi.org/10.3390/pr10051041

Academic Editors: Maria Jose Martin de Vidales, Dimitris Zagklis and Georgios Bampos

Received: 8 April 2022
Accepted: 19 May 2022
Published: 23 May 2022

1. Introduction

Conventional processes to treat wastewater, such as activated sludge and aeration lagoons, are usually designed to stabilize and remove readily biodegradable contaminants, such as nutrients and other microbiological pollutants. However, according to recent research, other constituents are present in wastewater at low concentrations (ng/L–μg/L), termed as emerging contaminants. Emerging contaminants are chemicals or substances that have the power to, directly and indirectly, harm humans, flora, and fauna [1].

Emerging pollutants can be mainly divided into the following groups: pharmaceuticals, personal care products, pesticides, and hormones; also, they can be natural or synthetic substances that are not fully regulated at a global level [2,3].

On the other hand, the list of these emerging pollutants is very long since it is increasing according to the requirements that human beings demand to improve their quality of life as the years go by [4]. So the number of organic and inorganic substances is in the millions, which has directly impacted the increase of emerging pollutants [1,5]. Moreover, since the water cycle interrelates the distribution of emerging pollutants discharged in domestic wastewater with other types of water, these emerging contaminants have also been quantified in surface water, groundwater, and drinking water [2,6,7].

Therefore, it is necessary to investigate and study the emerging pollutants based on their origin and transformation to recommend some processes that effectively treat them. Nevertheless, the conventional processes used to treat wastewater often cannot degrade some of them [8], as mentioned above.

In this sense, advanced oxidation processes (AOPs) have been of great interest in treating a wide diversity of pollutants, including emergent pollutants, present in wastewater [9]. AOPs are oxidation treatments based on the generation of potent oxidizing agents for organic compounds degradation, a clear example of which is the hydroxyl radical (HO°), which has an oxidation potential of 2.80 V. Hydroxyl radicals are equivalent to conventional oxidants such as chlorine, oxygen, and ozone [10,11]. The advantage that hydroxyl radicals show over other types of radicals is that they can react with most organic and inorganic substances [11]; therefore, AOPs promote the degradation of emerging pollutants favorably.

A general classification of AOPs that can remove emerging contaminants is shown in Figure 1. The classification was made based on that reported by Miklos et al. [10]. In addition, AOPs were classified according to the energy source (UV, chemical, and electrochemical) required to initiate the pollutant degradation reaction. Finally, it should be mentioned that the AOPs are described in more detail in the following sections.

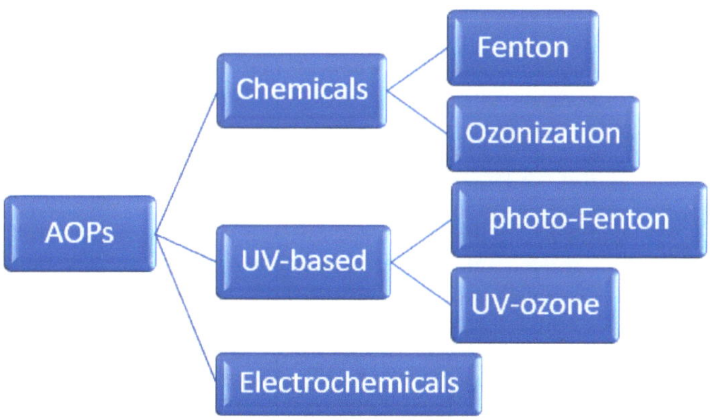

Figure 1. Most common advanced oxidation processes (AOPs) general classification, adapted and modified from [10].

In this context, the leading review objective is to suggest the consideration of the advanced oxidation process as an alternative for the removal of the emergent contaminants from wastewater due to their prevalence, hazardous, and the potential risk that they represent to the human health and the environment if their removal is not achieved until innocuous levels.

Although biological processes are currently the most widely used for wastewater treatment, several studies have shown that they have failed to adequately remove emerging pollutants [12,13]; therefore, it is necessary to attract the scientific community's attention to the need for the implementation and incorporation of other alternatives in the treatment processes, such as AOPs.

2. Emerging Pollutants and Their Relationship with the Water Cycle

Emerging contaminants are distributed throughout the water cycle, as shown in Figure 2. It should be noted that the primary inputs of emerging contaminants to surface water, groundwater, and drinking water is wastewater from households, followed by wastewater generated in industry, hospitals, livestock, and finally, wastewater produced in agriculture (Figure 2). Water from different activities converges at a certain point in the water cycle and distributes the emerging pollutants in the natural water reservoirs until they reach the drinking water (Figure 2) [14].

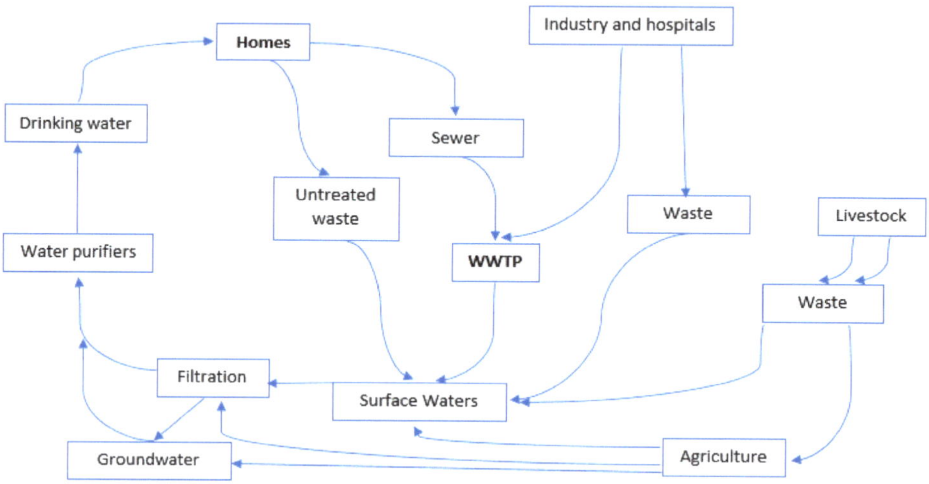

Figure 2. Distribution of emerging pollutants throughout the water cycle, adapted and modified from [14].

Emerging contaminants can cause problems for flora and fauna even in low concentrations [15]. A clear example of this is the research reported by Lee et al. [16], where the accumulation of some emerging contaminants was determined in selected fish species in the Tamsui River (Taiwan). According to the research, these fish accumulated the emerging contaminants and were then consumed by humans, causing problems to human health. Note that, to a certain extent, emerging contaminants can also reach human consumption through the trophic chain through species that have had contact with emerging contaminants and on which humans feed. In addition, it was also determined that the main source of discharge of the emerging pollutants analyzed came from products used in the home and effluents from wastewater treatment plants [16].

One of the main inputs of emerging contaminants into wastewater and the aquatic environment is through direct discharge of substances used in households, industry, hospitals, livestock, and agriculture into sewers (Figure 2). Another route is human excretion; emerging pollutants are consumed by humans and then excreted in parent compound form or as metabolites, which sometimes are not degraded or altered in wastewater treatment plants (WWTPs) [17].

The following subsections describe the significant groups of emerging contaminants.

2.1. Pharmaceuticals

Previous research on 212 medicines for human consumption has proposed that 35% of them are excreted via feces, and 64% are excreted in the urine. However, it is essential to mention that 42% of the total medicines excreted via urine are metabolites [2].

Human pharmaceuticals are considered on the official list of emerging pollutants reported by the United Nations Educational Scientific and Cultural Organization [18]. The medicines found in aqueous systems belong to the following classification groups: analgesics and anti-inflammatory drugs, antibiotics, antidepressants, antiepileptics, cardiovascular medicines, and cytostatics, among others [19,20].

Table 1 lists the toxicity level of the most harmful medicines to the aquatic environment based on the maximum mean effective concentration (EC_{50}). The most sensitive taxonomic groups were also added for each group of pharmaceutical drugs [20]. As can be seen, the group of antibiotics presents more significant toxicity in aquatic environments, according to Table 1, followed by antidepressants.

Due to their characteristics, microorganisms are the leading taxonomic group affected by antibiotics. However, this has a double effect. One is that if antibiotics reach aquatic reservoirs in adequate concentrations, the population of both pathogenic and non-pathogenic microorganisms can be reduced. Nevertheless, on the other hand, if they arrive in low concentrations, they could cause resistance of some pathogenic microorganisms to antibiotics. This factor could cause severe problems for humans and health [21,22], which makes antibiotics assigned special attention among the other groups of pharmaceuticals to be treated urgently.

Table 1. Aquatic toxicity of major drug groups, adapted and modified from [20].

Group	Extremely Toxic EC_{50} < 0.1 mgL^{-1}	Highly Toxic EC_{50} 0.1–1 mgL^{-1}	Toxic EC_{50} 1–10 mgL^{-1}	Harmful EC_{50} 10–100 mgL^{-1}	Non-Toxic EC_{50} > 100 mgL^{-1}
Analgesics			D*	D*, E*	
Antibiotics	A*	B*			
Antidepressants		D*			
Antiepileptics			C*		D*, E*
Cardiovascular medicines		D*			
Cytostatics		A*		D*, E*	

Most sensitive taxonomic groups: A* = Microorganisms; B* = Algae; C* = Cnidaria; D* = Crustaceans; E* = Fish.

Figure 3 illustrates the behavior and distribution of pharmaceutical drugs in aqueous systems, the primary and secondary production sources, and how humans can consume them. In Figure 3, wastewater generated in hospitals is indirectly included in treating human and animal diseases. However, the pharmaceutical industry and hospitals are the main generators of wastewater contaminated by these pollutants [23].

In the pharmaceutical industry, in the generation of pharmaceutical compounds, water is used at the different points of operation of the process for the washing of equipment and extraction of the medicine of interest. It is worth mentioning that the wastewater generated in pharmaceutical manufacturing (Figure 3) processes contains both the medicine in its parent form and a wide variety of toxic compounds that are harmful to humans, flora, and fauna in general [23].

Also, since pharmaceuticals are designed to increase their biological activity at low concentrations because they are used for diseases treatment in both animals and humans, their unintended presence risk in the environment and drinking water increases; thus, they should be removed from wastewater (Figure 3) and drinking water [24]. In addition, there is concern about the unintentional consumption of pharmaceuticals in drinking water, which can cause long-term adverse effects on human health [25].

To conclude this subsection, the most commonly used and consumed medicines may vary from country to country, and the level of contamination caused by their incidence in drinking water and the danger they represent to people [26]. Therefore, it is urgent to treat or remove this group of emerging contaminants (pharmaceuticals) that are present in aquatic systems.

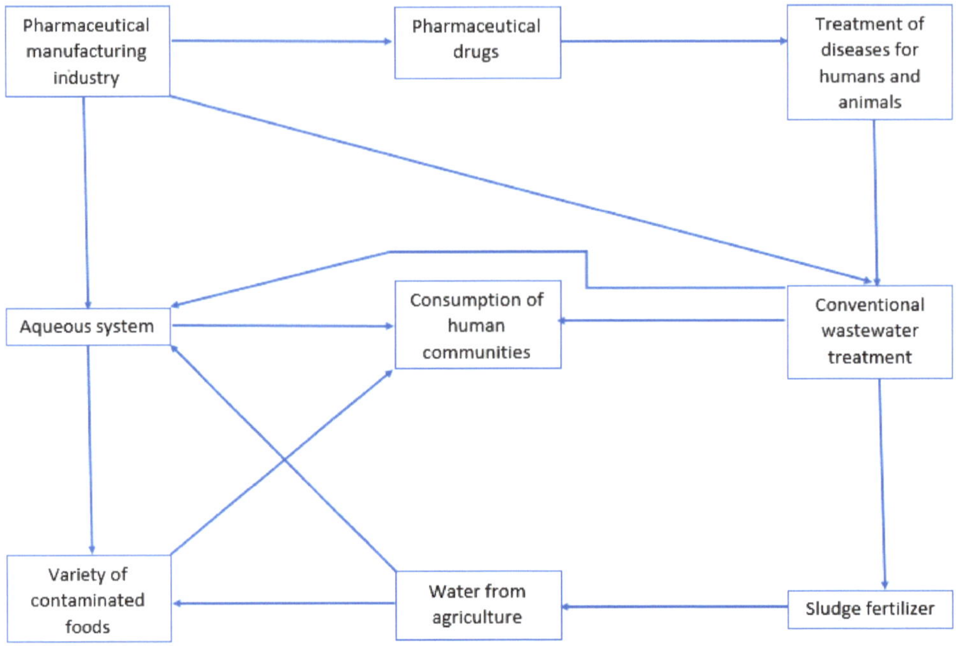

Figure 3. Pharmaceuticals distribution via aquatic routes, primary and secondary production sources, adapted from [23].

2.2. Personal Care Products

The group of emerging contaminants referred to as personal care products, also abbreviated as PCPs, can be purchased without a prescription and are commonly available in department stores and pharmacies in the health and beauty sections [27].

PCPs, like any other emerging pollutant, reach the aquatic environment through wastewater treatment plants [21]. PCPs may include mosquito repellents, surfactants, antifungal agents, antimicrobial agents, fragrances, and sunscreens, which are widely used in urban sites [21,28].

The PCPs and pharmaceuticals concentration vary according to the region and season of the year [29]. PCPs, unlike pharmaceuticals, do not have a significant change in their molecular structure, as most are designed for external use and are not metabolized by human consumption, e.g., fragrances and sunscreens [3].

Now, the PCPs toxicity will depend directly on the concentration of the emerging contaminant and the length of time the organism is exposed to the contaminant in general [30]. It is worth mentioning that plants can accumulate and metabolize PCPs, affecting them in a certain way and harming the organisms that consume them, such as humans (Figure 4) [31].

Figure 4 shows the interrelationship between wastewater contaminated by PCPs and the food chain.

It should be noted that PCPs have also been found in sludge generated in wastewater treatment plants [32]. However, there has been an increase in knowledge about PCPs commonly found in sediments, water, and biota [33]. However, there is still a lack of knowledge and information on these emerging contaminants and the possible treatments for their removal from wastewater.

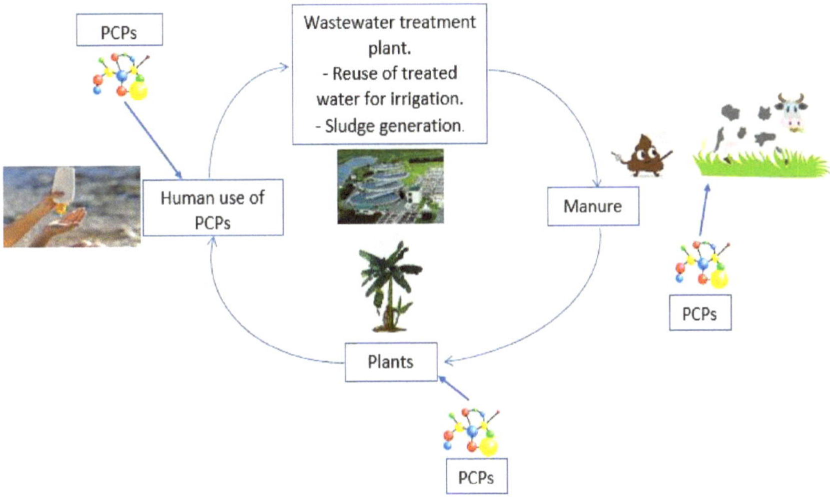

Figure 4. Distribution of personal care products (PCPs) via wastewater in the environment and the food chain, adapted and modified from [31].

2.3. Pesticides

Pesticides are a group of chemical compounds used in agriculture, mainly to regulate plant growth and eliminate insects and diseases (undesirable organisms) that may attack plants [34,35]. Pesticides can be natural or synthetic compounds [34]. Moreover, pesticides in waters can be a significant problem because they affect both ecosystems and human beings [35]. Therefore, an assessment of pesticide exposure is necessary to safeguard aquatic ecosystems [36].

The pesticides with a broad market are organophosphates, carbamates, organochlorines, neonicotinoids, triazoles, and amides [37]. Pesticides are also used in residential areas, specifically on lawns and gardens, which can cause these substances to reach the sewage system and groundwater through runoff caused by rainfall. In addition, pesticides can enter the human body through skin contact, inhalation, the eyes, or swallowing [38].

On the other hand, Rezaei Kalantary et al. [39] reported the presence of 16 pesticides in river water and drinking water; it was established that the concentration of some pesticides was very high and therefore unacceptable, as it could cause harm to people using the river water. It should be emphasized that the research presented the pesticide aldrin, which can be a danger due to the carcinogenic risk it presents to children and adolescents, a factor that promotes its long-term consumption when it is found in drinking water [39].

In addition, Mojiri et al. [40] showed that maximum pesticide concentrations in water bodies amounted to 1.21×10^5 ng/L in the Karun River in Iran, with respective bioaccumulation of pesticides in fish of up to 26.1×10^3 g/Kg, affecting both fish and the food chain.

However, the challenges in pesticide removal from wastewater are the influent composition, the pesticide molecular structure, and the pH of the pesticides contaminating the water, which varies from 0.5 to 14 [41]. Therefore, it is necessary to investigate methods that can feasibly promote their degradation and removal and counteract the abovementioned problems.

2.4. Hormones

Hormones are biochemical substances that come from the endocrine glands of humans and animals [42]. In humans, hormones are produced in the individual cells of the testes, adrenal cortex, ovaries, and placenta. The testes generate (androgen) testosterone, and the adrenal cortex aldosterone, cortisol, and dehydroepiandrosterone. The ovaries generate estrogens, generally divided into estradiol, 4-androstene-3, 17-dione, and progesterone, and finally, the placenta also generates estradiol, progesterone, and a substance called estriol [43]. Hormones are transported through the bloodstream and cause the activation of functions, including growth, reproduction, and general metabolic functions [42].

Currently, synthetic and natural hormones in water have gained a great deal of attention, attributed to the effects they can cause to organisms even though they are in low concentrations (μg/L or ng/L). In addition, hormones are known to be endocrine disruptors [44]. The arrival of endocrine disruptors in surface waters depends mainly on the efficiency, capacity, and technology of wastewater treatment plants to treat them [45].

As a result, natural and synthetic hormones reach aquatic systems through wastewater discharges from households, hospitals, agriculture, livestock, and industrial effluents [44]. For example, Lei et al. [46] found a total concentration (468 ± 27 ng/L) of estrogens such as estrone, 17β-estradiol, estriol, 17α-ethinylestradiol, and diethylstilbestrol in treated wastewater and 219 ± 23 ng/L in river water in the Beijing-Tianjin-Hebei region, China. Moreover, Aris et al. [47] concluded that the presence of synthetic estrogenic compounds, detected in various aquatic environments, has currently received much attention due to their high bioactivity, toxicity, and persistence; because they can cause health problems, such as reduced fertility and carcinogenic problems, by ingesting water and food contaminated with it [47].

In addition, they also indicated that environmental pollution caused by estrogenic compounds leads to an imbalance in the aquatic environment, where fish are the main affected since the changes generated in the characteristics of male fish cause the fish population to decrease. Similarly, it was identified that prenatal and postnatal development in women and men must have an adequate relationship between estrogens and androgens since, if not, a complete formation of the reproductive organs will not be achieved [1].

Also, Jarošová et al. [48] defined, based on 15 in vitro tests, the level of estrogenic activity (concentration) that can be safe in municipal wastewater (effluent), whereby they indicated that for extended exposure times of more than 60 days a concentration of 0.4 ng/L was safe. For exposure times of less than 60 days, a concentration of 0.5–2 ng/L was safe in terms of safe levels of estrogenic equivalents (SEQ-SSEs) to know if the effluents can be discharged into the receiving bodies.

It is worth mentioning that synthetic hormones such as ethinylestradiol (birth control pills) and drospirenone, a synthetic progestin, were the 98th most prescribed hormones in the United States [49]. Therefore, it is highly relevant to urgently quantify, treat, and remove hormones from wastewater.

2.5. SARS-CoV-2, an Important Factor in Wastewater Today

Although COVID-19 is not an emerging contaminant but an emerging disease [50], SARS-CoV-2 has also been found in wastewater from treatment plants in some countries, e.g., the USA, Australia, Spain, Japan, France, India, and the Netherlands. However, although there is no evidence that the presence of SARS-CoV-2 in wastewater is related to virus transmission, the detection of SARS-CoV-2 in wastewater may serve as an indicator to follow up on COVID-19 cases [51]. In addition, researchers such as Cao & Francis [52] have concluded a strong correlation between the SARS-CoV-2 concentration in wastewater and the cases that exist in a given community.

However, it is imperative to say that the SARS-CoV-2 virus resides in wastewater from asymptomatic, presymptomatic, and symptomatic individuals. In addition, it should be added that samples taken from different wastewaters from different sewers in Ohio in the United States of America showed different SARS-CoV-2 strain patterns due to the diversity of strains that exist today [53].

Figure 5 shows, in general, how the COVID-19 disease can be contracted through wastewater. It should be mentioned that SARS-CoV-2 can survive from hours to seven days, depending on the type of matrix in which it is found [50].

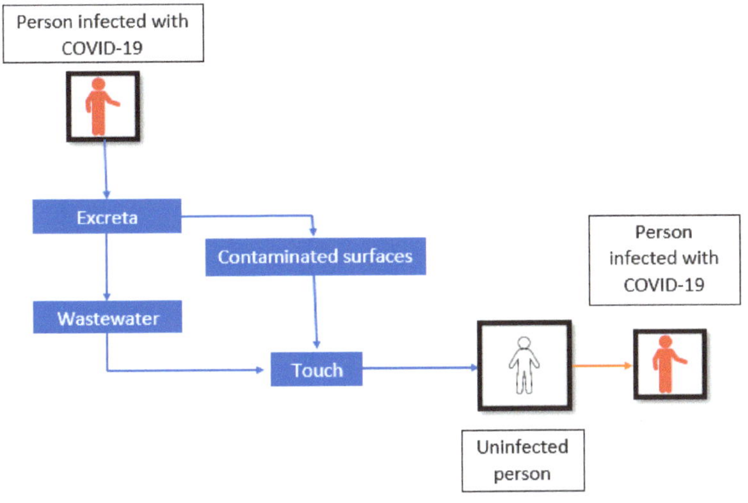

Figure 5. Possible human transmission of the COVID-19 disease through human excreta and sewage, adapted and modified from [50].

Therefore, based on the information gathered in this section, it is essential to note that SARS-CoV-2 may be a quality parameter measured in wastewater discharges soon. This may lead humans to employ new wastewater treatment methods or modifications to existing conventional methods that eliminate SARS-CoV-2.

Additionally, the SARS-CoV-2 virus is not thriving at hydrogen potential (pH) levels of 2–3 and 12–13 in a day [54]. Consequently, treatment methods that work at extreme pH can be used for SARS-CoV-2 virus removal, an example being AOPs that work in acidic ranges.

3. Emerging Pollutants Treatments in Wastewater

Some technologies applied to treat emerging pollutants are adsorption, use of membranes, advanced oxidation processes (AOPs), and biological processes. Where efficiency depends on multiple parameters, highlighting the type of technology and the pollutant concentration [55]. However, it is essential to mention that more emphasis will be placed on advanced oxidation processes (AOP) and biological treatments in this section.

3.1. Advanced Oxidation Processes (AOPs)

Advanced oxidation processes (AOPs) are recognized methods because they offer a great diversity of processes that generate hydroxyl radicals (HO°). This factor allows coupling and adapting a specific AOP based on the treatment requirements [56]. AOPs are often used when conventional wastewater treatment cannot remove or treat certain pollutants [57]. A clear example of this is some organic pollutants that show a high chemical balance and low biodegradability [58], the most recognized are emerging pollutants. In addition, it has already been shown that AOPs have high efficiency in removing trace level organic compounds compared to other water treatment technologies [59].

AOPs commonly work with an oxidizing agent, e.g., ozone (O_3) and hydrogen peroxide (H_2O_2), a catalytic agent such as Fe^{2+}, TiO_2, and other types of catalysts [60,61]. It is worth mentioning that catalysts can be modified to solve a given problem. For example, Dai et al. [61] in their research modified a catalyst to remove chlorinated volatile organic compounds.

In addition, in some cases, an external energy source such as UV radiation is included in order to degrade certain emerging pollutants that according to their molecular structure are susceptible to photocatalytic degradation [60,62]. For example, Chen et al. [62] in their research showed that some emerging contaminants present in saline wastewater tend to degrade photocatalytically. This is evidence that advanced oxidation processes can be modified and adapted depending on the molecular composition of the emerging pollutant that needs to be removed.

However, it must be ensured that the addition of external energy to the advanced oxidation process is necessary, since if the emerging contaminant of interest to be removed is not amenable to photocatalytic degradation, this would only lead to increased operating costs.

Therefore, in the following subsections, some of the most commonly used AOPs for treating emerging contaminants in wastewater are described in more detail. Emphasis will be placed on the reaction mechanisms and operating parameters (pH, catalyst dosage, oxidant dosage, among others) of the AOPs.

3.1.1. Chemicals

Fenton

AOP Fenton was named in honor of Fenton H.J.H. Fenton., who observed that tartaric acid could be oxidized by the reaction of ferrous (Fe^{2+}) salts with hydrogen peroxide (H_2O_2). Classical reactions, or Fenton reaction mechanisms, produce hydroxyl radicals (HO°) at pressure and room temperature at an acidic pH level in aqueous solution when hydrogen peroxide and ferrous ions come into contact.

Equation (1) shows how ferrous ions react with hydrogen peroxide giving hydroxyl radical as the main product. On the other hand, if there is hydrogen peroxide remaining in the process, through Equation (2), the ferric ions produced in Equation (1) will be reduced to ferrous ions again, generating protons and perhydroxyl radicals [63]. However, it should be noted that perhydroxyl radicals have less oxidative power than hydroxyl radicals [63,64]. Therefore, it is advisable to start the Fenton AOP with reagents containing ferrous ions instead of ferric ions.

It is necessary to mention that, if there is no remaining hydrogen peroxide, which can react with the ferric ions generated in Equation (1), the ferric ions generated in Equation (1) will precipitate in the form of amorphous ferric oxyhydroxides when the pH of the aqueous solution is increased from acidic to neutral [65].

$$H_2O_2 + Fe^{2+} \rightarrow Fe^{3+} + HO^\circ + OH^-,\ k_1 = 40\text{–}80\ M^{-1}s^{-1} \quad (1)$$

$$H_2O_2 + Fe^{3+} \rightarrow Fe^{2+} + HO_2{}^\circ + H^+,\ k_2 = 9.1 \times 10^{-7}\ M^{-1}s^{-1} \quad (2)$$

However, some side reactions may be negatively involved in the Fenton AOP (Equations (3)–(6)) [11]. For example, the (3)–(6) side reactions can decrease the pollutant removal performance of the Fenton AOP due to the loss of hydroxyl radicals.

$$Fe^{2+} + HO^\circ \rightarrow Fe^{3+} + OH^- \quad (3)$$

$$H_2O_2 + HO^\circ \rightarrow HO_2{}^\circ + H_2O \quad (4)$$

$$HO_2{}^\circ + HO^\circ \rightarrow O_2 + H_2O \quad (5)$$

$$HO^\circ + HO^\circ \rightarrow H_2O_2 \quad (6)$$

The appropriate pH to manage the Fenton AOP has been indicated in several articles, which should be approximately 3 [8,9]. The pH set required for the degradation of a specific pollutant may also vary based on the nature of the pollutant and the wastewater in which it is found.

The optimal dosage of hydrogen peroxide (H_2O_2) to treat wastewater needs to be determined by experimentation since a lack of hydrogen peroxide may decrease the performance of the Fenton-type AOP to treat wastewater [9]. Some publications [9,66–69] have used Equation (7) (chemical reaction) to define the theoretical amount of hydrogen peroxide required to treat the emerging contaminant.

$$C_aH_bN_cO_d + (2a + 0.5b + 2.5c - d)H_2O_2 \rightarrow aCO_2 + (2a + b + 2c - d)H_2O + cHNO_3 \quad (7)$$

In Equation (7) (2a + 0.5b + 2.5c − d) are the moles of hydrogen peroxide that are required to degrade one mole of a molecule of the type $C_aH_bN_cO_d$. It should be clarified that if the wastewater contains more contaminants and not only the emerging target pollutant to be treated, the other contaminants may decrease the removal performance of the target pollutant since Fenton AOP is a non-selective treatment process. Therefore, the following Equation (8) can be used [70,71] instead of Equation (7).

$$\text{Quantity } H_2O_2 = \frac{17}{8}\text{COD} \quad (8)$$

where the amount of hydrogen peroxide to treat the wastewater can be calculated as a function of the amount of chemical oxygen demand (COD) that is present in a certain volume of the wastewater to be treated. The concentrations of both hydrogen peroxide and COD should be in mg/L units.

It is important to mention that the load of all the pollutants present in the wastewater to be treated by the Fenton AOP is considered in the total COD of the wastewater in Equation (8).

The theoretical amount of ferrous ions, which in this case are the most common catalysts of the AOP, Fenton type, can be calculated by means of Equation (1), and the results obtained by means of Equation (7) or Equation (8), as the case may be.

Currently, Fenton-type AOP carried out in the absence of light can be classified into two major branches, which are: Homogeneous Fenton and Heterogeneous Fenton.

The difference from one method to the other is that in homogeneous Fenton, which is the process just described in more detail in this section, its catalytic process occurs throughout the liquid phase, whereas in the heterogeneous Fenton method, it occurs on the surface of the catalyst [9]. The AOP heterogeneous Fenton type emerged to solve the problems presented by the homogeneous Fenton type AOP, since some of the catalysts used are solids [72]. However, there are still more challenges for both processes.

Some examples of emerging contaminants that have been treated by the Fenton method are presented in Table 2. It should be noted that the experimentation (Table 2) was carried out in real water (wastewater) and synthetic solutions. As can be seen in Table 2, all the emerging contaminants that were treated with the AOP Fenton type, had good removal percentages in synthetic water and wastewater.

The commonly handled pH's were around the value of 3, and in what refers to the amounts of ferrous ion and hydrogen peroxide, in this case, they were very variants according to the pollutant that needed to be treated. Therefore, the AOP Fenton type, according to the information gathered, is effective in the treatment of emerging pollutants.

Table 2. Examples of emerging contaminants treated by the Fenton advanced oxidation process and process parameters.

Contaminant	Type of Water	Hydrogen Potential (pH)	Catalyst Dose and Oxidant Dose	% Removal	Reference
Amoxicillin	Synthetic	3.5	$[H_2O_2]$ = 255 mg/L $[Fe^{2+}]$ = 25 mg/L	100%	[73]
Bisphenol A	Synthetic	3	Fe^{2+}/H_2O_2 = 0.012	100% (maximum)	[74]
Estrogens *	Activated sludge waste	3	Fe^{2+}/H_2O_2 = 0.167	70–98%	[75]
Paracetamol	Synthetic	3	H_2O_2/Fe^{2+} = 2	70.37%	[76]

* Estrogens: include estrone, 17β-estradiol, estriol and 17α-ethinylestradiol.

Ozonization

Ozone has been used as an oxidant and disinfectant in water treatment. However, ozone as such (direct form) used as an oxidant is very selective and, therefore, potentially attacks groups that are rich in electrons, for example, amines and activated aromatic rings, among others [10,77]. On the other hand, ozone can be used to generate hydroxyl radicals (HO°) and indirectly carry out the oxidation of organic compounds [77] because hydroxyl radicals (HO°) are formed by the decomposition of O_3 are non-selective oxidants [78], as compared to the O_3. So it is better, from an oxidative point of view, to use ozone (O_3) to generate hydroxyl radicals because they are not selective instead of using ozone (O_3) directly to carry out the oxidation.

It is worth mentioning that in order for ozone to generate hydroxyl radicals (HO°), there must first be a considerable amount of hydroxide ions, which promote the generation of these radicals. The pH for the generation of hydroxyl radicals is a crucial factor since the pH of the wastewater to be treated must be maintained at a level approximately higher than 8 [10], but this may change depending on the nature of the contaminants to be treated and the matrix in which they are found.

The process that generates hydroxyl radicals by means of ozone mentioned above and which will be emphasized in this section, is known as indirect ozonation. The reaction mechanisms of indirect ozonation, also known as indirect reactions or hydroxyl radical pathways, are divided into initiation reactions, propagation reactions, and termination reactions [79,80].

According to Malik et al. [80], in the initiation reactions (Equations (9) and (10)), first, ozone decreases when ozone and hydroxide ions react to give a superoxide anion as a product ($O_2{}^-$) and a perhydroxyl radical ($HO_2{}^\circ$).

$$O_3 + OH^- \rightarrow O_2{}^- + HO_2{}^\circ \quad (9)$$

$$HO_2{}^\circ \leftrightarrow O_2{}^- + H^+ \quad (10)$$

Subsequently, in the propagation reactions (Equations (11)–(13)), the superoxide anion reacts with ozone to generate oxygen and $O_3{}^-$. Hydroxyl radicals (HO°) are then generated via hydrogen trioxide ($HO_3{}^\circ$). It should be noted that hydrogen trioxide ($HO_3{}^\circ$) is produced through the reaction of the protons generated in Equation (10) with the $O_3{}^-$ produced in Equation (11) [80].

$$O_3 + O_2{}^- \rightarrow O_3{}^- + O_2 \quad (11)$$

$$HO_3{}^\circ \rightarrow O_3{}^- + H^+ \quad (12)$$

$$HO_3{}^\circ \rightarrow HO^\circ + O_2 \quad (13)$$

Likewise, hydroxyl radicals react with ozone and produce $HO_4{}^\circ$ in Equation (14), which are transformed into perhydroxyl radicals ($HO_2{}^\circ$) in Equation (15) [80].

$$HO^\circ + O_3 \rightarrow HO_4{}^\circ \quad (14)$$

$$HO_4{}^\circ \leftrightarrow O_2 + HO_2{}^\circ \quad (15)$$

It should be noted that hydroxyl radicals (HO°) in the propagation reactions (Equation (16)) can also react with an organic molecule R, which can produce an interaction between the hydroxyl radical and an organic radical R° [80].

$$H_2R + HO^\circ \rightarrow HR^\circ + H_2O \quad (16)$$

Another situation that can also occur if oxygen is present is that peroxyl radicals (ROO°) are generated and start a chain reaction (Equations (17)–(19)) [80].

$$HR^\circ + O_2 \rightarrow HRO_2{}^\circ \quad (17)$$

$$HRO_2{}^{\circ} \rightarrow R + HO_2{}^{\circ} \quad (18)$$

$$HRO_2{}^{\circ} \rightarrow RO + OH^{\circ} \quad (19)$$

Now, as a final phase, it is essential to say that some organic and inorganic molecules can be involved in a reaction with HO° and produce secondary radicals, which do not produce superoxide radicals, but rather $HO_2{}^{\circ}$ y $O_2{}^{-}$. These molecules commonly terminate the chain reaction and thus ozone depletion. It is worth mentioning that carbonate and bicarbonate can also act as scavengers of hydroxyl radicals (Equations (20) and (21)) [80].

$$HO^{\circ} + CO_3^{2-} \rightarrow CO_3^{-\circ} + OH^{-} \quad (20)$$

$$HO^{\circ} + HCO_3^{-} \rightarrow HCO_3{}^{\circ} + OH^{-} \quad (21)$$

A situation that can also occur is that the hydroxyl and perhydroxyl radicals react (Equation (22)) and stop the process (Equation (22)) [80].

$$HO^{\circ} + OH_2{}^{\circ} \rightarrow O_2 + H_2O \quad (22)$$

It should be mentioned that the degree of solubility of ozone in water is temperature dependent, as well as its reaction rate [81].

However, since ozone is an unstable gas and therefore, its production must be carried out at the wastewater treatment site, the corona discharge process can be used for its generation, which is based on the principle of discharging a high voltage into a dry or cooled gas phase, which is constituted by oxygen or air, if any [82].

On the other hand, Equation (23) indicates the exposure of hydroxyl radicals (HO°) per unit of ozone used in the process, and it should be clarified that this kinetic parameter has been proposed to characterize the ozone process [78,83] and can also be used to measure the production of hydroxyl radicals with respect to the ozone consumption in the ozonation process [83].

$$R_{OH,O3} = \frac{\int \left[HO^{\circ}\right] dt}{\Delta O_3} \quad (23)$$

Table 3 shows some examples of emerging contaminants reported by Mathon et al. [84] that can be removed by direct and indirect ozonation. It should be added that some of the data presented in Table 3 are obtained by means of mathematical modeling. Emerging contaminants were added to the wastewater by means of synthetic solutions (spike), and the process was generally operated in a range of pH = 6.7–7.2, a temperature of 21 °C, and ozone dosing (O_3) of 1.6 $gO_3/gCOD$ per gram of dissolved organic carbon. It is necessary to add that, in the research, paracetamol and estrogens were considered as easily oxidized emerging pollutants, the only estrogen that was considered as a pollutant of medium oxidation difficulty was estriol in conjunction with the antibiotic ofloxacin, and metformin was considered as a pollutant of difficult oxidation. In addition, the research showed that 34 of the total emerging contaminants studied were mostly oxidized by indirect ozonation [84]. This indicates the feasibility of using indirect ozonation instead of direct ozonation, as mentioned above, simply because indirect ozonation generates hydroxyl radicals with a higher oxidative potential than ozone (O_3) [64], which is also selective.

Table 3. Examples of emerging pollutants treated by direct and indirect ozonation.

Contaminant	Type of Water	Direct Ozonation (O_3). % Removal	Indirect Ozonation (HO°). % Removal	% Large-Scale Removal	Reference
Paracetamol	Wastewater	97%	3%	ND*	[84]
Estrogens *	Wastewater	17–99%	1–83%	87–96%	[84]
Ofloxacin	Wastewater	7%	93%	80–92%	[84]
Metformin	Wastewater	1%	99%	ND*	[84]

* Estrogens: include estrone, 17β-estradiol, estriol and 17α-ethinylestradiol and ND*: is undetermined.

3.1.2. UV-Based

Photo-Fenton

The photo-Fenton process is an advanced oxidation process (AOP) that generates hydroxyl radicals (HO°) through the use of hydrogen peroxide (H_2O_2), iron salts, and UV-visible light [85]. In this case, the reactions are accelerated by light; the wavelength needed to perform the photo-Fenton AOP must be $\lambda < 580$ nm [86].

The pH level to be managed in the photo-Fenton AOP must be acidic to keep the iron in the solution [87]. Gou et al. [88], based on the information gathered in their article, the optimum pH value for operating the photo-Fenton AOP in research should be between 3 to 3.5, very similar to that of the traditional Fenton AOP. However, this may change depending on the type of contaminant to be treated and the wastewater in which it is found.

On the other hand, the literature review by Clarizia et al. [89] analyzed the feasibility of employing chelators to extend the pH range to more conveniently work the homogeneous photo-Fenton AOP and concluded that chelators increase the concentration of total organic carbon (TOC) in water, and also indicated that the biodegradability and ecotoxicity of the species formed should be evaluated. This is a field that should be studied further.

It should be clarified that the photo-Fenton AOP can also be classified into homogeneous and heterogeneous, just like the traditional Fenton AOP, and just like the photo-Fenton AOP, chelating agents can also be used in the traditional Fenton, an example of which is the research carried out by Sánchez Proaño & García Gualoto [90].

However, it is of utmost importance to mention that in both Fenton and photo-Fenton processes, problems are solved by means of heterogeneous processes or the use of chelating agents, etc., and others arise.

Now, in the reaction mechanisms of homogeneous photo-Fenton, the power of UV-visible radiation is added to the reagents used in traditional Fenton (Equations (24)–(26)) [91].

$$H_2O_2 + Fe^{2+} \rightarrow Fe^{3+} + HO^\circ + OH^- \quad (24)$$

$$Fe^{3+} + H_2O \overset{hv}{\rightarrow} Fe^{2+} + HO^\circ + H^+ \quad (25)$$

$$H_2O_2 \overset{hv}{\rightarrow} 2HO^\circ \ (\lambda < 400 \text{ nm}) \quad (26)$$

Most of the hydroxyl radicals (HO°) in the photo-Fenton AOP are generated by Equation (24) from the traditional Fenton AOP and by the photolysis of the ferric ion carried out by Equation (25). Equation (25) also indicates that oxidation by irradiation is promoted by the photoreduction of ferric ions to ferrous ions, which can react again with hydrogen peroxide if there is any remaining hydrogen peroxide [92].

As for Equation (26), hydrogen peroxide at a wavelength of less than 400 nm can also produce hydroxyl radicals in the process for the degradation of pollutants in wastewater.

The dosage of hydrogen peroxide (H_2O_2) and iron salts has to be defined experimentally as in the traditional Fenton AOP for the removal of the target pollutants. As for the theoretical doses of hydrogen peroxide and ferrous ions, they can also be calculated by means of Equations (1), (7) and (8) as in the traditional Fenton method. It should be added that the photo-Fenton AOP has been considered to disinfect water from resistant microorganisms [93].

Some examples of emerging contaminants that have been treated by photo-Fenton AOP are shown in Table 4 below. As can be seen in Table 4, the photo-Fenton process was operated at pH = 3 in the two different investigations. It should be noted that the pH used is very similar to that used in the conventional Fenton AOP since this AOP is a derivative of it. Moreover, the dosages of the catalyst (Fe^{2+}) and oxidant (H_2O_2) also varied from one investigation to another. In addition, all the removals shown amounted to 100% in both wastewater and synthetic water, which leads to establish the feasibility of this advanced oxidation process to treat and remove emerging contaminants that might be present in wastewater.

Table 4. Examples of emerging contaminants treated by the photo-Fenton advanced oxidation process (AOP) and process parameters.

Contaminant	Type of Water	Hydrogen Potential (pH)	Catalyst Dose and Oxidizer Dose	% Removal	Reference
Amoxicillin	Synthetic	3	H_2O_2/Fe^{2+} = 20	100%	[94]
Ampicillin	Synthetic	3	H_2O_2/Fe^{2+} = 20	100%	[94]
Cloxacillin	Synthetic	3	H_2O_2/Fe^{2+} = 20	100%	[94]
17α-ethinylestradiol	Wastewater with spike	3	$[Fe^{2+}]$ = 5 mg/L $[H_2O_2]$ = 4.3–15 mg/L	100%	[95]

UV/O_3

In UV-ozone AOP, the wavelength to be handled should be $\lambda < 300$ nm so that dissolved ozone can be split and a fast reaction of atomic oxygen (O) with water (H_2O) takes place, and thermally excited hydrogen peroxide (H_2O_2) is generated [10].

In the following Equations (27)–(31), the process that occurs to generate thermally excited hydrogen peroxide (H_2O_2) and hydroxyl radicals (HO°) is captured. It needs to be emphasized that there are two spin reactions allowed (Equations (27) and (28)) and that the quantum yield of the ozone decomposition is only 0.64 [96].

$$O_3 + hv \rightarrow O(^1D) + O_2(^1\Delta g) \tag{27}$$

$$O_3 + hv \rightarrow O(^3P) + O_2(3\sum_g^-) \tag{28}$$

The low quantum yield of ozone decomposition can be attributed to the recombination that takes place in the solvent (Equation (29)) [96]. As can be seen, in Equation (29), the atomic and diatomic oxygen produced in Equation (27) react again, generating ozone.

$$O(^1D) + O_2(^1\Delta g) \rightarrow O_3 \tag{29}$$

On the other hand, in order to generate thermally excited hydrogen peroxide, oxygen ($O(^1D)$) reacts rapidly with water to generate it, as denoted in Equation (30) [96].

$$O(^1D) + H_2O \rightarrow H_2O_2 \text{ (hot)} \tag{30}$$

It is worth mentioning that not all of the hydrogen peroxide generated decomposes into hydroxyl radicals (HO°) (Equation (31)) [96].

$$H_2O_2\text{(hot)} \rightarrow 2HO^{\circ} \tag{31}$$

At the end of the UV/O_3 process, the quantum yield of free hydroxyl radicals (HO°) is 0.1 [96].

Now, regarding the operating pH of the UV/O_3 AOP, Souza & Féris [97] found in their research, which focused on mineralizing caffeine in synthetic solution, that to obtain high mineralization of caffeine, the pH levels and reaction time must be kept high. The levels of pH and time used in the experimental design of this research ranged from 3 to 11 for pH and from 15 to 30 min for reaction time, respectively.

On the other hand, in another investigation by Chang et al. [98] for the removal of mefenamic acid, they established that pH plays an important role because in the first 5 min at pH levels of 4, 7, and 9, mefenamic acid remnants of 62, 53, and 50% were found. However, after 40 min of exposure, at pH levels of 4, 7, and 9, 37.5, 42.5, and 45.8% of mefenamic acid remained, so it was concluded that when the pH decreases, the removal of mefenamic acid increases, and the UV/O_3 process showed a higher removal power than the traditional ozonation process [98]. With this, as mentioned above, it is evident from the references [97,98] and their changing pH levels that pH levels can vary drastically in the UV/O_3 AOP depending on the type of emerging contaminant that needs to be treated since in one investigation the pH level should be kept high and in the other investigation, it should not. In addition, the optimum ozone dosage and radiation exposure must be calculated experimentally, in conjunction with pH, depending on the pollutants to be treated in the wastewater.

Table 5 below shows additional data that were not mentioned in the previous description of the referenced research [97,98]. Additional research is also presented [99]. The maximum removals by the UV/O_3 process ranged from 60–95%, and the water commonly used for the experimentation was synthetic water, based on the information compiled in Table 5. It should be noted that although the UV/O_3 AOP presents good removal percentages, Miklos et al. [10] point out that the AOP presents problems such as the UV lamps and the ozone generator demands large amounts of electrical energy and the efficiency of generation of hydroxyl radicals is low. Therefore, based on this information, it can be said that the UV/O_3 AOP is expensive.

Table 5. Examples of emerging contaminants treated by UV/O_3 advanced oxidation process (AOP).

Contaminant	Type of Water	% Maximum Removal	Reference
Caffeine	Synthetic	95%	[97]
Mefenamic acid	Synthetic	60–80%	[98]
Ketoprofen	Synthetic	95% (mineralization)	[99]

3.1.3. Electrochemicals

In this section, a brief description of the advanced electrochemical oxidation processes will be given since they are diverse. It is worth mentioning that some of the electrochemical AOPs, such as anodic oxidation, and the electro-Fenton process, among others, are relatively new processes that have been studied with the purpose of removing recalcitrant emerging contaminants [100].

Electrochemical AOPs are basically considered methods whose main objective is the remediation of wastewater contaminated by organic compounds (emerging pollutants) through the in situ production of reactive oxygen species such as hydroxyl radicals (HO°). One of the most studied electrochemical AOP is the electrochemical AOP of anodic oxidation; in general, in this process, a high current density (j) is applied to the anode (M) of the cell in order to produce physisorbed hydroxyl radicals on the anode surface M(HO°) [101]. The transfer of electrons, M(HO°) and M = O, or the combination of both, are part of the water cleaning process [102]. However, the oxidation capacity of electrochemical AOPs will depend on the electrochemical system used, the cell type, the electrodes, and the additional equipment used [103].

It should be added that in electrochemical AOPs, hydrodynamic parameters must be taken into account. As far as costs are concerned, the energy used to pump the water must also be considered since it could contribute to most of the energy consumption in these processes. This applies greatly if the current intensities used in the electrochemical AOP are low in order to increase the production of hydroxyl radicals (HO°) since the duration of the treatment and, therefore, the pumping time would be prolonged which would significantly increase costs [10].

On the other hand, Seibert et al. [104] concluded that, with respect to the application for industrial wastewater treatment, the costs generated by the use of electrochemical AOPs could be higher compared to those generated in conventional processes. However, if the operating conditions and parameters are optimized, they are more efficient in removing emerging pollutants.

Examples of emerging contaminants that have been removed by electrochemical AOPs in an aqueous solution are shown in Table 6. In the research reported by Brillas et al. [105], it was found that the mineralization of paracetamol in an aqueous solution is independent of pH, and it increases when the applied current and temperature are increased but decreases when the concentration of paracetamol increases more than 315 mg/L. A boron-doped diamond electrode was used in this research because anodic oxidation was performed.

Now with regard to the research of Domínguez et al. [106], it was found in the experimental design used that for the 100% oxidation of ketoprofen in synthetic solution, the most important variable was the influence of the current intensity, followed by the electrolyte concentration, the flow rate and finally the pH. In this research, pH was also considered a parameter of minor importance. The electrodes used in this research were also made of boron-doped diamond.

On the other hand, Donoso et al. [107] found that the most important factor for the removal of tartrazine in synthetic solution was also the current intensity in the anodic oxidation. Boron-doped diamond electrodes were also used in this research. It is worth mentioning that, in this case, tartrazine is an artificial colorant used in the food industry.

Therefore, based on the information gathered, electrochemical methods are efficient for removing emerging contaminants (Table 6). However, they are very expensive due to the electrical energy used to carry them out, as in the case of anodic oxidation.

Table 6. Examples of emerging contaminants treated by electrochemical advanced oxidation processes (AOP).

Contaminant	Type of Water	Important Notes	Electrochemical Method Used	% Removal	Reference
Paracetamol	Synthetic	pH independent	Anodic oxidation	100% (mineralization)	[105]
Ketoprofen	Synthetic	pH variable of minor importance	Anodic oxidation	100%	[106]
Tartrazine	Synthetic	-	Anodic oxidation	57.85–99.97%	[107]

3.2. Biological Treatments

The biotechnological use of enzymes, plants, and microorganisms such as bacteria, fungi, cyanobacteria, and microalgae have been used effectively in the bio-transformation of emerging pollutants [108].

Microorganisms, which participate in biogeochemical processes, are capable of degrading organic compounds [109] due to the ability to modify their metabolic pathways when exposed to various contaminants [110]. *Bacillus cereus*, a bacterial isolated from petroleum sludge, has been shown to be effective in the degradation of prometryn, fluometuron, and sulfamethoxazole [111]. On the other hand, microalgae have the ability to bioabsorb, bio-adsorb, and biodegrade emerging pollutants [112]. Removing parabens [113], fungicides such as triclosan, compounds related to the manufacture of plastics such as bisphenol A, 17-ethinylestradiol, a synthetic steroid, and analgesic as tramadol and diclofenac. In addition, they are capable of removing metals and heavy metals such as mercury and lead. [114]. The use of algae for the treatment of emerging contaminants is considered profitable [115], and it can be applied in developing countries since energy and economic resources are the main limitation in reducing micropollutants and bacteria in urban wastewater [114,116]. On the other hand, wetlands have been shown to be effective in reducing many emerging organic pollutants [117]. Wetlands successfully removed (81–99.9%) compounds present in plastics such as diethyl phthalate, di-isobutyl phthalate, di-n-octyl phthalate, bis(2-ehtylxexyl) phthalate, tris(1-chloro-2-propyl) [118]. Oxidoreductases, enzymes capable of degrading varied organic compounds, including phenolic compounds, have also been used [119].

However, the use of microorganisms for emerging contaminants removal is still limited. Being the specific biological processes and difficult to scale. Some contaminants exert selective pressure on some microorganisms, for example, bactericides and fungicides [120]. The case of algal-based systems has limitations such as microbiological contamination that can affect the efficiency of the system. On the other hand, it is possible to use microalgae-bacteria that form biofilms that work in synergy to improve the process [121]. The enzymes used are immobilized on various supports. However, it has been found that the enzymes are denatured, thus losing their catalytic activity [119,122]. Furthermore, the production of biomass for application at the industrial level is very low, so they have been genetically modified [123]. These strategies, called synthetic biology, allow for obtaining better results since they use the editing of sequences responsible for the degradation of compounds [124]. Considering genetically modified microorganisms as bioremediation factories [125].

Due to all these disadvantages, it has been suggested to use a combination of processes. As is the use of bioremediation, nanotechnology, and physicochemical treatments (AOPs included) to ensure that they are efficient and fast [110]. Taoufik et al. [126] state that biological processes have many advantages such as operability under mild conditions, economic feasibility, and ease of automation, such as activated sludge and oxidation ponds conventional processes used worldwide for wastewater treatment. However, these fail to remove recalcitrant or toxic contaminants such as emerging pollutants [126].

On the other hand, advanced oxidation processes present high operating costs, but meet the objective of removing recalcitrant or toxic contaminants such as emerging pollutants [126].

Therefore, Taoufik et al. [126] recommend in their research the use of hybrid processes, which employ both biological and advanced oxidation processes, in order to minimize the AOP costs by eliminating easily biodegradable pollutants through biological processes and guaranteeing the removal of emerging pollutants through advanced oxidation processes. This would be of great benefit, since most countries have conventional biological processes to treat wastewater, for example, oxidation ponds and activated sludge processes, which could be used.

4. Conclusions

Based on the information presented in this article, the following conclusions can be drawn. First, it is concluded that the main entry of emerging contaminants into the environment is through wastewater because the conventional processes commonly used to treat wastewater in different regions of the world do not have the capacity to degrade emerging contaminants in their entirety. In addition, it was also concluded that the distribution of emerging contaminants in the environment takes place primarily through the water cycle.

Now, it is of great importance to mention, according to the information gathered, that emerging contaminants affect the environment (flora, fauna, and human health) drastically, as can be seen in the subsection on pharmaceuticals, which affect several taxonomic groups in aquatic systems, such as microorganisms, algae, cnidaria, crustaceans, and fish. Also, some hormones have been reported as endocrine disruptors which can affect the poblational reproductive behavior, mainly in some aquatic species.

Moreover, emerging contaminants such as personal care products and pesticides, tend to bioaccumulate and be consumed through the food chain and the water cycle, affecting flora, fauna, and human health. In addition, pesticides have been detected in river water and drinking water at unacceptable concentrations due to their persistent, recalcitrant, and bioaccumulative characteristics; thus, they represent a global problem that must be urgently addressed.

It should be emphasized that the SARS-CoV-2 virus in the very near future could be a factor that defines the quality of water since its presence has been found in the wastewater of different countries, but this will be defined when the research on this subject is expanded, which by the way is of urgency. It should also be mentioned that the COVID-19 disease can possibly be contracted through human contact with contaminated wastewater containing the active virus (SARS-CoV-2). Therefore, it is of great importance to start looking for processes that can eliminate it, a clear example of which the advanced oxidation processes (AOP) represent.

Regarding advanced oxidation processes (AOP), it was concluded that they have good removal percentages of emerging contaminants and can be used for this purpose. However, there is still a lack of information and experimentation in real wastewater to verify the potential of these processes to remove emerging contaminants that could be present in wastewater since most of the research is carried out with synthetic solutions. On the other hand, it is ruled that as different removal parameters are added to the AOP, their cost and difficulty of operation increase, as in the case of electrochemical AOPs.

However, current research on the oxidation and removal of emerging contaminants should also focus on identifying the oxidation products that are obtained through the oxidation of the target emerging contaminants, as the oxidation products may sometimes be more toxic than the parent emerging contaminant.

Finally, it sounds tentative to suggest that hybrid processes should be formed between biological and advanced oxidation processes in order to remove emerging pollutants and address the shortcomings of such processes when used alone. This is in order to remove emerging contaminants of interest and take advantage of conventional wastewater treatment processes that are already installed in the countries, as mentioned at the end of Section 3.2 in this manuscript.

Author Contributions: Conceptualization, A.R.B.-R., Y.R.-V. and L.R.-C.; methodology, A.R.B.-R. and L.R.-C.; validation, A.R.B.-R., L.A.O.-D., A.M.B.-S. and L.R.-C.; formal analysis, A.R.B.-R., L.A.O.-D., A.M.B.-S., J.C.R.-S., Y.R.-V. and L.R.-C.; investigation, A.R.B.-R., L.A.O.-D., A.M.B.-S., J.C.R.-S., Y.R.-V. and L.R.-C.; resources, A.R.B.-R., L.A.O.-D., A.M.B.-S. and L.R.-C.; data curation, A.R.B.-R. and L.R.-C.; writing—original draft preparation, A.R.B.-R., L.A.O.-D., A.M.B.-S., J.C.R.-S., Y.R.-V. and L.R.-C.; writing—review and editing, A.R.B.-R., L.A.O.-D., A.M.B.-S., J.C.R.-S., Y.R.-V. and L.R.-C.; visualization, A.R.B.-R. and L.R.-C.; supervision, L.R.-C.; project administration, L.R.-C.; funding acquisition, Y.R.-V. and L.R.-C. All authors have read and agreed to the published version of the manuscript.

Funding: This research was funded by the National Council of Science and Technology (CONACYT) with the scholarship grant number 805457, and the project number CF-2019/102967.

Institutional Review Board Statement: Not applicable.

Informed Consent Statement: Not applicable.

Data Availability Statement: Not applicable.

Acknowledgments: To M.S.A. Luis A. Torres-Castañón from CIMAV-DGO, for his valuable technical support and collaboration.

Conflicts of Interest: The authors declare that they have no conflict of interest. The National Council of Science and Technology, which provide the scholarships grants, were not involved at any stage of the study.

Abbreviations

The following abbreviations are used in this manuscript:

AOP	Advanced Oxidation Process
EC	Emerging Contaminants or Pollutants

References

1. Ramírez-Sánchez, I.M.; Martínez-Austria, P.; Quiroz-Alfaro, M.A.; Bandala, E.R. Efectos de los estrógenos como contaminantes emergentes en la salud y el ambiente. *Tecnol. Cienc. Agua* **2015**, *6*, 31–42.
2. Tejada, C.; Quiñonez, E.; Peña, M. Contaminantes emergentes en aguas: Metabolitos de fármacos. Una revisión. *Rev. Fac. Cienc. Básicas* **2014**, *10*, 80–101. [CrossRef]
3. Rodriguez-Narvaez, O.M.; Peralta-Hernandez, J.M.; Goonetilleke, A.; Bandala, E.R. Treatment technologies for emerging contaminants in water: A review. *Chem. Eng. J.* **2017**, *323*, 361–380. [CrossRef]
4. Sánchez-Ramírez, J. Contaminantes emergentes en nuestros sistemas acuáticos: Enemigos invisibles. *Ing. Reg.* **2018**, *19*, 18–21. [CrossRef]
5. American Chemical Society. Available online: https://www.cas.org (accessed on 18 August 2021).
6. Caviedes Rubio, D.I.; Delgado, D.R.; Olaya Amaya, A. Normatividad ambiental dirigida a regular la presencia de los productos farmacéuticos residuales en ambientes acuáticos. *Rev. Juríd. Piélagus* **2017**, *16*, 121–130. [CrossRef]

7. Babuponnusami, A.; Muthukumar, K. A review on Fenton and improvements to the Fenton process for wastewater treatment. *J. Environ. Chem.* **2014**, *2*, 557–572. [CrossRef]
8. Vorontsov, A.V. Advancing Fenton and photo-Fenton wáter treatment through the catalyst design. *J. Hazard. Mater.* **2019**, *372*, 103–112. [CrossRef]
9. Wang, N.; Zheng, T.; Zhang, G.; Wang, P. A review on Fenton-like processes for organic wastewater treatment. *J. Environ. Chem.* **2016**, *4*, 762–787. [CrossRef]
10. Miklos, D.B.; Remy, C.; Jekel, M.; Linden, K.G.; Drewes, J.E.; Hübner, U. Evaluation of advanced oxidation processes for water and wastewater treatment-A critical review. *Water Res.* **2018**, *139*, 118–131. [CrossRef]
11. Bautista, P.; Mohedano, A.F.; Casas, J.A.; Zazo, J.A.; Rodriguez, J.J. An overview of the application of Fenton oxidation to industrial wastewaters treatment. *J. Chem. Technol. Biotechnol.* **2008**, *83*, 1323–1338. [CrossRef]
12. Deblonde, T.; Cossu-Leguille, C.; Hartemann, P. Emerging pollutants in wastewater: A review of the literature. *Int. J. Hyg. Environ. Health* **2011**, *214*, 442–448. [CrossRef]
13. Zhou, Y.; Meng, J.; Zhang, M.; Chen, S.; He, B.; Zhao, H.; Li, Q.; Zhang, S.; Wang, T. Which type of pollutants need to be controlled with priority in wastewater treatment plants: Traditional or emerging pollutants? *Environ. Int.* **2019**, *131*, 104982. [CrossRef]
14. Contaminación y Calidad Química Del Agua: El Problema de Los Contaminantes Emergentes. Available online: https://fnca.eu/phocadownload/P.CIENTIFICO/inf_contaminacion.pdf (accessed on 19 August 2021).
15. Saidulu, D.; Gupta, B.; Gupta, A.K.; Ghosal, P.S. A review on occurrences, eco-toxic effects, and remediation of emerging contaminants from wastewater: Special emphasis on biological treatment based hybrid systems. *J. Environ. Chem.* **2021**, *9*, 105282. [CrossRef]
16. Lee, C.C.; Hsieh, C.Y.; Chen, C.S.; Tien, C.J. Emergent contaminants in sediments and fishes from the Tamsui River (Taiwan): Their spatial-temporal distribution and risk to aquatic ecosystems and human health. *Environ. Pollut.* **2020**, *258*, 113733. [CrossRef]
17. Malmborg, J.; Magnér, J. Pharmaceutical residues in sewage sludge: Effect of sanitization and anaerobic digestion. *J. Environ. Manag.* **2015**, *153*, 1–10. [CrossRef]
18. Castillo-Zacarías, C.; Barocio, M.E.; Hidalgo-Vázquez, E.; Sosa-Hernández, J.E.; Parra-Arroyo, L.; López-Pacheco, I.Y.; Barceló, D.; Iqbal, H.N.M.; Parra-Saldívar, R. Antidepressant drugs as emerging contaminants: Occurrence in urban and non-urban waters and analytical methods for their detection. *Sci. Total Environ.* **2021**, *757*, 143722. [CrossRef]
19. Heberer, T. Occurrence, fate, and removal of pharmaceutical residues in the aquatic environment: A review of recent research data. *Toxicol. Lett.* **2002**, *131*, 5–17. [CrossRef]
20. Jones, O.A.H.; Voulvoulis, N.; Lester, J.N. Aquatic environmental assessment of the top 25 English prescription pharmaceuticals. *Water Res.* **2002**, *36*, 5013–5022. [CrossRef]
21. Stuart, M.; Lapworth, D.; Crane, E.; Hart, A. Review of risk from potential emerging contaminants in UK groundwater. *Sci. Total Environ.* **2012**, *416*, 1–21. [CrossRef]
22. Scott, L.C.; Wilson, M.J.; Esser, S.M.; Lee, N.L.; Wheeler, M.E.; Aubee, A.; Aw, T.G. Assessing visitor use impact on antibiotic resistant bacteria and antibiotic resistance genes in soil and water environments of Rocky Mountain National Park. *Sci. Total Environ.* **2021**, *785*, 147122. [CrossRef]
23. Zaied, B.K.; Rashid, M.; Nasrullah, M.; Zularisam, A.W.; Pant, D.; Singh, L. A comprehensive review on contaminants removal from pharmaceutical wastewater by electrocoagulation process. *Sci. Total Environ.* **2020**, *726*, 138095. [CrossRef] [PubMed]
24. Teixeira-Lemos, E.; Teixeira-Lemos, L.P.; Oliveira, J.; Pais do Amaral, J. Pharmaceuticals in the Environment: Focus on Drinking-Water. In *Encyclopedia of Analytical Science*, 3rd ed.; Worsfold, P., Poole, C., Townshend, A., Miró, M., Eds.; Elsevier: Amsterdam, The Netherlands, 2019; pp. 325–335. [CrossRef]
25. Mohd Nasir, F.A.; Praveena, S.M.; Aris, A.Z. Public awareness level and occurrence of pharmaceutical residues in drinking water with potential health risk: A study from Kajang (Malaysia). *Ecotoxicol. Environ. Saf.* **2019**, *185*, 109681. [CrossRef] [PubMed]
26. Kondor, A.C.; Molnár, É.; Vancsik, A.; Filep, T.; Szeberényi, J.; Szabó, L.; Maász, G.; Pirger, Z.; Weiperth, A.; Ferincz, Á.; et al. Occurrence and health risk assessment of pharmaceutically active compounds in riverbank filtrated drinking wáter. *J. Water Process Eng.* **2021**, *41*, 102039. [CrossRef]
27. Figueiredo, L.; Erny, G.l.; Santos, L.; Alves, A. Applications of molecularly imprinted polymers to the análisis and removal of personal care products: A review. *Talanta* **2016**, *146*, 754–765. [CrossRef]
28. Pal, A.; He, Y.; Jekel, M.; Reinhard, M.; Gin, K.Y.-H. Emerging contaminants of public health significance as water quality indicator compounds in the urban water cycle. *Environ. Int.* **2014**, *71*, 46–62. [CrossRef]
29. Delgado, N.; Bermeo, L.; Hoyos, D.A.; Peñuela, G.A.; Capparelli, A.; Marino, D.; Navarro, A.; Casas-Zapata, J.C. Occurrence and removal of pharmaceutical and personal care products using subsurface horizontal flow constructed wetlands. *Water Res.* **2020**, *187*, 116448. [CrossRef]
30. Pemberthy, M.D.; Padilla, Y.; Echeverri, A.; Peñuela, G.A. Monitoring pharmaceuticals and personal care products in water and fish from the Gulf of Urabá, Colombia. *Heliyon* **2020**, *6*, e04215. [CrossRef]
31. Bartrons, M.; Peñuelas, J. Pharmaceuticals and Personal-Care Products in Plants. *Trends Plant Sci.* **2017**, *22*, 194–203. [CrossRef]
32. Pérez-Lemus, N.; López-Serna, R.; Pérez-Elvira, S.I.; Barrado, E. Analytical methodologies for the determination of pharmaceuticals and personal care products (PPCPs) in sewage sludge: A critical review. *Anal. Chim. Acta* **2019**, *1083*, 19–40. [CrossRef]
33. Meng, Y.; Liu, W.; Liu, X.; Zhang, J.; Peng, M.; Zhang, T. A review on analytical methods for pharmaceutical and personal care products and their transformation products. *J. Environ. Sci.* **2021**, *101*, 260–281. [CrossRef]

34. He, B.; Ni, Y.; Jin, Y.; Fu, Z. Pesticides-induced energy metabolic disorders. *Sci. Total Environ.* **2020**, *729*, 139033. [CrossRef]
35. Hassaan, M.A.; El Nemr, A. Pesticides pollution: Classifications, human health impact, extraction and treatment techniques. *Egypt. J. Aquat. Res.* **2020**, *46*, 207–220. [CrossRef]
36. Chow, R.; Scheidegger, R.; Doppler, T.; Dietzel, A.; Fenicia, F.; Stamm, C. A review of long-term pesticide monitoring studies to assess surface water quality trends. *Water Res.* **2020**, *9*, 100064. [CrossRef]
37. Zhou, Y.; Wu, J.; Wang, B.; Duan, L.; Zhang, Y.; Zhao, W.; Wang, F.; Sui, Q.; Chen, Z.; Xu, D.; et al. Occurrence, source and ecotoxicological risk assessment of pesticides in surface water of Wujin District (northwest of Taihu Lake), China. *Environ. Pollut.* **2020**, *265*, 114953. [CrossRef]
38. Md Meftaul, I.; Venkateswarlu, K.; Dharmarajan, R.; Annamalai, P.; Megharaj, M. Pesticides in the urban environment: A potential threat that knocks at the door. *Sci. Total Environ.* **2020**, *711*, 134612. [CrossRef]
39. Rezaei Kalantary, R.; Barzegar, G.; Jorfi, S. Monitoring of pesticides in surface water, pesticides removal efficiency in drinking water treatment plant and potential health risk to consumers using Monte Carlo simulation in Behbahan City, Iran. *Chemosphere* **2022**, *286*, 131667. [CrossRef]
40. Mojiri, A.; Zhou, J.L.; Robinson, B.; Ohashi, A.; Ozaki, N.; Kindaichi, T.; Farraji, H.; Vakili, M. Pesticides in aquatic environments and their removal by adsorption methods. *Chemosphere* **2020**, *253*, 126646. [CrossRef]
41. Saleh, I.A.; Zouari, N.; Al-Ghouti, M.A. Removal of pesticides from water and wastewater: Chemical, physical and biological treatment approaches. *Environ. Technol. Innov.* **2020**, *19*, 101026. [CrossRef]
42. Sacdal, R.; Madriaga, J.; Espino, M.P. Overview of the analysis, occurrence and ecological effects of hormones in lake waters in Asia. *Environ. Res.* **2020**, *182*, 109091. [CrossRef]
43. Hormonas Esteroides. 7. Hormonas Esteroidesnivel 3—Participación Plástica y Funcional (Biopsicologia.net). Available online: https://biopsicologia.net/nivel-3-participaci%C3%B3n-pl%C3%A1stica-y-funcional/7.-hormonas-esteroides (accessed on 28 August 2021).
44. Sta Ana, K.M.; Espino, M.P. Occurrence and distribution of hormones and bisphenol A in Laguna Lake, Philippines. *Chemosphere* **2020**, *256*, 127122. [CrossRef]
45. Jálová, V.; Jarošová, B.; Bláha, L.; Giesy, J.P.; Ocelka, T.; Grabic, R.; Jurčíková, J.; Vrana, B.; Hilscherová, K. Estrogen-, androgen- and aryl hydrocarbon receptormediated activities in passive and composite samples from municipal waste and surface Waters. *Environ. Int.* **2013**, *59*, 372–383. [CrossRef] [PubMed]
46. Lei, K.; Lin, C.Y.; Zhu, Y.; Chen, W.; Pan, H.Y.; Sun, Z.; Sweetman, A.; Zhang, Q.; He, M.C. Estrogens in municipal wastewater and receiving waters in the Beijing-Tianjin-Hebei region, China: Occurrence and risk assessment of mixtures. *J. Hazard. Mater.* **2019**, *389*, 121891. [CrossRef] [PubMed]
47. Jarošová, B.; Bláha, L.; Giesy, J.P.; Hilscherová, K. What level of estrogenic activity determined by in vitro assays in municipal waste waters can be considered as safe? *Environ. Int.* **2014**, *64*, 98–109. [CrossRef] [PubMed]
48. Aris, A.Z.; Shamsuddin, A.S.; Praveena, S.M. Occurrence of 17α-ethynylestradiol (EE2) in the environment and effect on exposed biota: A review. *Environ. Int.* **2014**, *69*, 104–119. [CrossRef]
49. Ojoghoro, J.O.; Scrimshaw, M.D.; Sumpter, J.P. Steroid hormones in the aquatic environment. *Sci. Total Environ.* **2021**, *792*, 148306. [CrossRef]
50. Ahmad, J.; Ahmad, M.; Usman, A.R.A.; Al-Wabel, M.I. Prevalence of human pathogenic viruses in wastewater: A potential transmission risk as well as an effective tool for early outbreak detection for COVID-19. *J. Environ. Manag.* **2021**, *298*, 113486. [CrossRef]
51. Hu, L.; Deng, W.J.; Ying, G.G.; Hong, H. Environmental perspective of COVID-19: Atmospheric and wastewater environment in relation to pandemic. *Ecotoxicol. Environ. Saf.* **2021**, *219*, 112297. [CrossRef]
52. Cao, Y.; Francis, R. On forecasting the community-level COVID-19 cases from the concentration of SARS-CoV-2 in wastewater. *Sci. Total Environ.* **2021**, *786*, 147451. [CrossRef]
53. Ai, Y.; Davis, A.; Jones, D.; Lemeshow, S.; Tu, H.; He, F.; Ru, P.; Pan, X.; Bohrerova, Z.; Lee, J. Wastewater SARS-CoV-2 monitoring as a community-level COVID-19 trend tracker and variants in Ohio, United States. *Sci. Total Environ.* **2021**, *801*, 149757. [CrossRef]
54. Chan, K.H.; Sridhar, S.; Zhang, R.R.; Chu, H.; Fung, A.Y.; Chan, G.; Chan, J.F.; To, K.K.; Hung, I.F.; Cheng, V.C.; et al. Factors affecting stability and infectivity of SARS-CoV-2. *J. Hosp. Infect.* **2020**, *106*, 226–231. [CrossRef]
55. Rempel, A.; Gutkoski, J.P.; Nazari, M.T.; Biolchi, G.N.; Cavanhi, V.A.F.; Treichel, H.; Colla, L.M. Current advances in microalgae-based bioremediation and other technologies for emerging contaminants treatment. *Sci. Total Environ.* **2021**, *772*, 144918. [CrossRef]
56. Cheng, M.; Zeng, G.; Huang, D.; Lai, C.; Xu, P.; Zhang, C.; Liu, Y. Hydroxyl radicals based advanced oxidation processes (AOPs) for remediation of soils contaminated with organic compounds: A review. *Chem. Eng. J.* **2016**, *284*, 582–598. [CrossRef]
57. Aleksić, M.; Kušić, H.; Koprivanac, N.; Leszczynska, D.; Božić, A.L. Heterogeneous Fenton type processes for the degradation of organic dye pollutant in water—The application of zeolite assisted AOPs. *Desalination* **2010**, *257*, 22–29. [CrossRef]
58. Oller, I.; Malato, S.; Sánchez-Pérez, J.A. Combination of Advanced Oxidation Processes and biological treatments for wastewater decontamination—A review. *Sci. Total Environ.* **2011**, *409*, 4141–4166. [CrossRef]
59. Liu, Z.; Demeestere, K.; Hulle, S.V. Comparison and performance assessment of ozone-based AOPs in view of trace organic contaminants abatement in water and wastewater: A review. *J. Environ. Chem. Eng.* **2021**, *9*, 105599. [CrossRef]

60. Rubio-Clemente, A.; Chica, E.; Peñuela, G.A. Procesos de tratamiento de aguas residuales para la eliminación de contaminantes orgánicos emergentes. *Rev. Ambient. Água* **2013**, *8*, 93–103. [CrossRef]
61. Dai, Q.; Shen, K.; Deng, W.; Cai, Y.; Yan, J.; Wu, J.; Guo, L.; Liu, R.; Wang, X.; Zhan, W. HCl-Tolerant $H_XPO_4/RuO_X–CeO_2$ Catalysts for Extremely Efficient Catalytic Elimination of Chlorinated VOCs. *Environ. Sci. Technol.* **2021**, *55*, 4007–4016. [CrossRef]
62. Chen, C.X.; Yang, S.S.; Ding, J.; Wang, G.Y.; Zhong, L.; Zhao, S.Y.; Zang, Y.N.; Jiang, J.Q.; Ding, L.; Zhao, Y.; et al. Non-covalent self-assembly synthesis of AQ2S@rGO nanocomposite for the degradation of sulfadiazine under solar irradiation: The indispensable effect of chloride. *Appl. Catal. B* **2021**, *298*, 120495. [CrossRef]
63. Cai, Q.Q.; Jothinathan, L.; Deng, S.H.; Ong, S.L.; Ng, H.Y.; Hu, J.Y. Fenton- and ozone-based AOP processes for industrial effluent treatment. In *Advanced Oxidation Processes for Effluent Treatment Plants*, 1st ed.; Maulin, P.S., Ed.; Elsevier: Singapore, 2021; Volume 11, pp. 199–254. [CrossRef]
64. Legrini, O.; Oliveros, E.; Braun, A.M. Photochemical processes for water treatment. *Chem. Rev.* **1993**, *93*, 671–698. [CrossRef]
65. Pignatello, J.J.; Oliveros, E.; MacKay, A. Advanced Oxidation Processes for Organic Contaminant Destruction Based on the Fenton Reaction and Related Chemistry. *Crit. Rev. Environ. Sci. Technol.* **2006**, *36*, 1–84. [CrossRef]
66. Zhang, A.; Wang, N.; Zhou, J.; Jiang, P.; Liu, G. Heterogeneous Fenton-like catalytic removal of p-nitrophenol in water using acid-activated fly ash. *J. Hazard. Mater.* **2012**, *201–202*, 68–73. [CrossRef] [PubMed]
67. Wang, N.; Zheng, T.; Jiang, J.; Lung, W.; Miao, X.; Wang, P. Pilot-scale treatment of p-Nitrophenol wastewater by microwave-enhanced Fenton oxidation process: Effects of system parameters and kinetics study. *Chem. Eng. J.* **2014**, *239*, 351–359. [CrossRef]
68. Pastrana-Martínez, L.M.; Pereira, N.; Lima, R.; Faria, J.L.; Gomes, H.T.; Silva, A.M.T. Degradation of diphenhydramine by photo-Fenton using magnetically recoverable iron oxide nanoparticles as catalyst. *Chem. Eng. J.* **2015**, *261*, 45–52. [CrossRef]
69. López, N.; Plaza, S.; Afkhami, A.; Marco, P.; Giménez, J.; Esplugas, S. Treatment of Diphenhydramine with different AOPs including photo-Fenton at circumneutral pH. *Chem. Eng. J.* **2017**, *318*, 112–120. [CrossRef]
70. Carmona, P.B. Tratamiento de Aguas Residuales de la Industria Cosmética Mediante el Proceso Fenton y con el Sistema Fe/γ-Al_2O_3/H_2O_2. Ph.D. Thesis, Universidad Autónoma de Madrid, Madrid, Spain, 2008.
71. Taco Ugsha, M.; Mayorga Llerena, E. Aplicación del proceso Fenton en la disminución de materia orgánica en aguas residuales de la industria termoeléctrica. *Quím. Cent.* **2013**, *3*, 25–30. [CrossRef]
72. Lai, C.; Shi, X.; Li, L.; Cheng, M.; Liu, X.; Liu, S.; Li, B.; Yi, H.; Qin, L.; Zhang, M.; et al. Enhancing iron redox cycling for promoting heterogeneous Fenton performance: A review. *Sci. Total Environ.* **2021**, *775*, 145850. [CrossRef]
73. Ay, F.; Kargi, F. Advanced oxidation of amoxicillin by Fenton's reagent treatment. *J. Hazard. Mater.* **2010**, *179*, 622–627. [CrossRef]
74. Escalona, I.; Fortuny, A.; Stüber, F.; Bengoa, C.; Fabregat, A.; Font, J. Fenton coupled with nanofiltration for elimination of Bisphenol, A. *Desalination* **2014**, *345*, 77–84. [CrossRef]
75. Zhang, A.; Li, Y. Removal of steroid estrogens from waste activated sludge using Fenton oxidation: Influencing factors and degradation intermediates. *Chemosphere* **2014**, *105*, 24–30. [CrossRef]
76. Van, H.T.; Nguyen, L.H.; Hoang, T.K.; Nguyen, T.T.; Tran, T.N.H.; Nguyen, T.B.H.; Vu, X.H.; Pham, M.T.; Tran, T.P.; Pham, T.T.; et al. Heterogeneous Fenton oxidation of paracetamol in aqueous solution using iron slag as a catalyst: Degradation mechanisms and kinetics. *Environ. Technol. Innov.* **2020**, *18*, 100670. [CrossRef]
77. Zhu, S.; Dong, B.; Yu, Y.; Bu, L.; Deng, J.; Zhou, S. Heterogeneous catalysis of ozone using ordered mesoporous Fe_3O_4 for degradation of atrazine. *Chem. Eng. J.* **2017**, *328*, 527–535. [CrossRef]
78. Kwon, M.; Kye, H.; Jung, Y.; Yoon, Y.; Kang, J.W. Performance characterization and kinetic modeling of ozonation using a new method: $R_{OH,O3}$ concept. *Water Res.* **2017**, *122*, 172–182. [CrossRef]
79. Ikehata, K.; Li, Y. Ozone-Based Processes. In *Advanced Oxidation Processes for Waste Water Treatment*, 1st ed.; Ameta, S.C., Ameta, R., Eds.; Elsevier: Amsterdam, The Netherlands, 2018; Volume 5, pp. 115–134. [CrossRef]
80. Malik, S.N.; Ghosh, P.C.; Vaidya, A.N.; Mudliar, S.N. Hybrid ozonation process for industrial wastewater treatment: Principles and applications: A review. *J. Water Process Eng.* **2020**, *35*, 101193. [CrossRef]
81. Egbuikwem, P.N.; Mierzwa, J.C.; Saroj, D.P. Evaluation of aerobic biological process with post-ozonation for treatment of mixed industrial and domestic wastewater for potential reuse in agriculture. *Bioresour. Technol.* **2020**, *318*, 124200. [CrossRef]
82. Rekhate, C.V.; Srivastava, J.K. Recent advances in ozone-based advanced oxidation processes for treatment of wastewater—A review. *Chem. Eng. J. Adv.* **2020**, *3*, 100031. [CrossRef]
83. Yang, Y.; Liu, Z.; Demeestere, K.; Van Hulle, S. Ozonation in view of micropollutant removal from biologically treated landfill leachate: Removal efficiency, •OH exposure, and surrogate-based monitoring. *Chem. Eng. J.* **2021**, *410*, 128413. [CrossRef]
84. Mathon, B.; Coquery, M.; Liu, Z.; Penru, Y.; Guillon, A.; Esperanza, M.; Miège, C.; Choubert, J.M. Ozonation of 47 organic micropollutants in secondary treated municipal effluents: Direct and indirect kinetic reaction rates and modelling. *Chemosphere* **2021**, *262*, 127969. [CrossRef]
85. Polo-López, M.I.; Sánchez Pérez, J.A. Perspectives of the solar photo-Fenton process against the spreading of pathogens, antibioticresistant bacteria and genes in the environment. *Curr. Opin. Green Sustain. Chem.* **2021**, *27*, 100416. [CrossRef]
86. Oller, I.; Malato, S. Photo-Fenton applied to the removal of pharmaceutical and other pollutants of emerging concern. *Curr. Opin. Green Sustain. Chem.* **2021**, *29*, 100458. [CrossRef]
87. Cabrera-Reina, A.; Miralles-Cuevas, S.; Sánchez Pérez, J.A.; Salazar, R. Application of solar photo-Fenton in raceway pond reactors: A review. *Sci. Total Environ.* **2021**, *800*, 149653. [CrossRef]

88. Gou, Y.; Chen, P.; Yang, L.; Li, S.; Peng, L.; Song, S.; Xu, Y. Degradation of fluoroquinolones in homogeneous and heterogeneous photo-Fenton processes: A review. *Chemosphere* **2021**, *270*, 129481. [CrossRef] [PubMed]
89. Clarizia, L.; Russo, D.; Di Somma, I.; Marotta, R.; Andreozzi, R. Homogeneous photo-Fenton processes at near neutral pH: A review. *Appl. Catal. B* **2017**, *209*, 358–371. [CrossRef]
90. Sánchez Proaño, R.G.; García Gualoto, K.J. Tratamiento de aguas residuales de cargas industriales con oxidación avanzada en sistemas convencionales. *Granja* **2018**, *27*, 103–111. [CrossRef]
91. Xavier, S.; Gandhimathi, R.; Nidheesh, P.V.; Ramesh, S.T. Comparative removal of Magenta MB from aqueous solution by homogeneous and heterogeneous photo-Fenton processes. *Desalin. Water Treat.* **2015**, *57*, 12832–12841. [CrossRef]
92. Huang, Y.H.; Tsai, S.T.; Huang, Y.F.; Chen, C.Y. Degradation of commercial azo dye reactive Black B in photo/ferrioxalate system. *J. Hazard. Mater.* **2007**, *140*, 382–388. [CrossRef]
93. O'Dowd, K.; Pillai, S.C. Photo-Fenton disinfection at near neutral pH: Process, parameter optimization and recent advances. *J. Environ. Chem. Eng.* **2020**, *8*, 104063. [CrossRef]
94. Elmolla, E.S.; Chaudhuri, M. Degradation of the antibiotics amoxicillin, ampicillin and cloxacillin in aqueous solution by the photo-Fenton process. *J. Hazard. Mater.* **2009**, *172*, 1476–1481. [CrossRef]
95. Frontistis, Z.; Xekoukoulotakis, N.P.; Hapeshi, E.; Venieri, D.; Fatta-Kassinos, D.; Mantzavinos, D. Fast degradation of estrogen hormones in environmental matrices by photo-Fenton oxidation under simulated solar radiation. *Chem. Eng. J.* **2011**, *178*, 175–182. [CrossRef]
96. Von Sonntag, C. Advanced oxidation processes: Mechanistic aspects. *Water Sci. Technol.* **2008**, *58*, 1015–1021. [CrossRef]
97. Souza, F.S.; Féris, L.A. Degradation of Caffeine by Advanced Oxidative Processes: O_3 and O_3/UV. *Ozone Sci. Eng.* **2015**, *37*, 379–384. [CrossRef]
98. Chang, E.E.; Liu, T.Y.; Huang, C.P.; Liang, C.H.; Chiang, P.C. Degradation of mefenamic acid from aqueous solutions by the ozonation and O_3/UV processes. *Sep. Purif. Technol.* **2012**, *98*, 123–129. [CrossRef]
99. Illés, E.; Szabó, E.; Takács, E.; Wojnárovits, L.; Dombi, A.; Gajda-Schrantz, K. Ketoprofen removal by O_3 and O_3/UV processes: Kinetics, transformation products and ecotoxicity. *Sci. Total Environ.* **2014**, *472*, 178–184. [CrossRef]
100. Raj, R.; Tripathi, A.; Das, S.; Ghangrekar, M.M. Removal of caffeine from wastewater using electrochemical advanced oxidation process: A mini review. *CSCEE* **2021**, *4*, 100129. [CrossRef]
101. Flores Tapia, N.E.; Roman Rodríguez, M.; Coba Cabrera, R.L.; Vélez Ortiz, J.J.; Sirés Sadornil, I.; Brillas Coso, E. Procesos Electroquímicos de Oxidación Avanzados en la degradación de los ácidos trans-cinámico y trans-ferúlico. *Cienc. Digit.* **2019**, *3*, 49–60. [CrossRef]
102. Ganiyu, S.O.; Martínez-Huitle, C.A.; Oturan, M.A. Electrochemical advanced oxidation processes for wastewater treatment: Advances in formation and detection of reactive species and mechanisms. *Curr. Opin. Electrochem.* **2021**, *27*, 100678. [CrossRef]
103. Brillas, E. Recent development of electrochemical advanced oxidation of herbicides. A review on its application to wastewater treatment and soil remediation. *J. Clean. Prod.* **2021**, *290*, 125841. [CrossRef]
104. Seibert, D.; Zorzo, C.F.; Borba, F.H.; de Souza, R.M.; Quesada, H.B.; Bergamasco, R.; Baptista, A.T.; Inticher, J.J. Occurrence, statutory guideline values and removal of contaminants of emerging concern by Electrochemical Advanced Oxidation Processes: A review. *Sci. Total Environ.* **2020**, *748*, 141527. [CrossRef]
105. Brillas, E.; Sirés, I.; Arias, C.; Cabot, P.L.; Centellas, F.; Rodríguez, R.M.; Garrido, J.A. Mineralization of paracetamol in aqueous medium by anodic oxidation with a boron-doped diamond electrode. *Chemosphere* **2005**, *58*, 399–406. [CrossRef]
106. Domínguez, J.R.; González, T.; Palo, P.; Sánchez-Martín, J. Anodic oxidation of ketoprofen on boron-doped diamond (BDD) electrodes. Role of operative parameters. *Chem. Eng. J.* **2010**, *162*, 1012–1018. [CrossRef]
107. Donoso, G.; Domínguez, J.R.; González, T.; Correia, S.; Cuerda-Correa, E.M. Electrochemical and sonochemical advanced oxidation processes applied to tartrazine removal. Influence of operational conditions and aqueous matrix. *Environ. Res.* **2021**, *202*, 111517. [CrossRef]
108. Agrawal, K.; Bhatt, A.; Chaturvedi, V.; Verma, P. Bioremediation: An effective technology toward a sustainable environment via the remediation of emerging environmental pollutants. In *Emerging Technologies in Environmental Bioremediation*; Elsevier: Amsterdam, The Netherlands, 2020; pp. 165–196. [CrossRef]
109. Azaroff, A.; Monperrus, M.; Miossec, C.; Gassie, C.; Guyoneaud, R. Microbial degradation of hydrophobic emerging contaminants from marine sediment slurries (Capbreton Canyon) to pure bacterial strain. *J. Hazard. Mater.* **2021**, *402*, 123477. [CrossRef] [PubMed]
110. Meganathan, B.; Rathinavel, T.; Rangaraj, S. Trends in microbial degradation and bioremediation of emerging contaminants. *Phys. Sci. Rev.* **2021**, 000010151520210060. [CrossRef]
111. Alhefeiti, M.A.; Athamneh, K.; Vijayan, R.; Ashraf, S.S. Bioremediation of various aromatic and emerging pollutants by Bacillus cereus sp. isolated from petroleum sludge. *Water Sci. Technol.* **2021**, *83*, 1535–1547. [CrossRef] [PubMed]
112. Sutherland, D.L.; Ralph, P.J. Microalgal bioremediation of emerging contaminants-Opportunities and challenges. *Water Res.* **2019**, *164*, 114921. [CrossRef]
113. Vale, F.; Sousa, C.A.; Sousa, H.; Santos, L.; Simões, M. Parabens as emerging contaminants: Environmental persistence, current practices and treatment processes. *J. Clean. Prod.* **2022**, *347*, 131244. [CrossRef]
114. Singh, D.V.; Bhat, R.A.; Upadhyay, A.K.; Singh, R.; Singh, D.P. Microalgae in aquatic environs: A sustainable approach for remediation of heavy metals and emerging contaminants. *Environ. Technol. Innov.* **2021**, *21*, 101340. [CrossRef]

115. Gondi, R.; Kavitha, S.; Kannah, R.Y.; Karthikeyan, O.P.; Kumar, G.; Tyagi, V.K.; Banu, J.R. Algal-based system for removal of emerging pollutants from wastewater: A review. *Bioresour. Technol.* **2022**, *344*, 126245. [CrossRef]
116. Kumar, V.; Jaiswal, K.K.; Verma, M.; Vlaskin, M.S.; Nanda, M.; Chauhan, P.K.; Singh, A.; Kim, H. Algae-based sustainable approach for simultaneous removal of micropollutants, and bacteria from urban wastewater and its real-time reuse for aquaculture. *Sci. Total Environ.* **2021**, *774*, 145556. [CrossRef]
117. García, J.; García-Galán, M.J.; Day, J.W.; Boopathy, R.; White, J.R.; Wallace, S.; Hunter, R.G. A review of emerging organic contaminants (EOCs), antibiotic resistant bacteria (ARB), and antibiotic resistance genes (ARGs) in the environment: Increasing removal with wetlands and reducing environmental impacts. *Bioresour. Technol.* **2020**, *307*, 123228. [CrossRef]
118. Gikas, G.D.; Papaevangelou, V.A.; Tsihrintzis, V.A.; Antonopoulou, M.; Konstantinou, I.K. Removal of Emerging Pollutants in Horizontal Subsurface Flow and Vertical Flow Pilot-Scale Constructed Wetlands. *Processes* **2021**, *9*, 2200. [CrossRef]
119. Zdarta, J.; Jankowska, K.; Bachosz, K.; Degórska, O.; Kaźmierczak, K.; Nguyen, L.N.; Nghiem, L.D.; Jesionowski, T. Enhanced wastewater treatment by immobilized enzymes. *Curr. Pollut. Rep.* **2021**, *7*, 167–179. [CrossRef]
120. Gomes, I.B.; Maillard, J.Y.; Simões, L.C.; Simões, M. Emerging contaminants affect the microbiome of water systems—Strategies for their mitigation. *NPJ Clean Water* **2020**, *3*, 39. [CrossRef]
121. Sousa, H.; Sousa, C.A.; Simões, L.C.; Simões, M. Microalgal-based removal of contaminants of emerging concern. *J. Hazard. Mater.* **2022**, *423*, 127153. [CrossRef]
122. Al-Maqdi, K.A.; Elmerhi, N.; Athamneh, K.; Bilal, M.; Alzamly, A.; Ashraf, S.S.; Shah, I. Challenges and Recent Advances in Enzyme-Mediated Wastewater Remediation—A Review. *Nanomaterials* **2021**, *11*, 3124. [CrossRef]
123. Xiong, J.Q.; Kurade, M.B.; Jeon, B.H. Can microalgae remove pharmaceutical contaminants from water? *Trends Biotechnol.* **2018**, *36*, 30–44. [CrossRef]
124. Jaiswal, S.; Shukla, P. Alternative strategies for microbial remediation of pollutants via synthetic biology. *Front. Microbiol.* **2020**, *11*, 808. [CrossRef]
125. Bilal, M.; Iqbal, H.M. Microbial bioremediation as a robust process to mitigate pollutants of environmental concern. *Case Stud. Therm. Eng.* **2020**, *2*, 100011. [CrossRef]
126. Taoufik, N.; Boumya, W.; Achak, M.; Sillanpaa, M.; Barka, N. Comparative overview of advanced oxidation processes and biological approaches for the removal pharmaceuticals. *J. Environ. Manag.* **2021**, *288*, 112404. [CrossRef]

processes

MDPI

Article

Circular Economy Indicators for the Assessment of Waste and By-Products from the Palm Oil Sector

Pedro-Antonio Cárdenas Bejarano [1], Juan-Pablo Rodriguez-Miranda [1], Rayma Ireri Maldonado-Astudillo [2,*], Yanik Ixchel Maldonado-Astudillo [2,*] and Ricardo Salazar [3,*]

1 Facultad del Medio Ambiente y Recursos Naturales, Universidad Distrital Francisco José de Caldas, Bogotá 11021, Colombia; pacardenasb@correo.udistrital.edu.co (P.-A.C.B.); jprodriguezm@udistrital.edu.co (J.-P.R.-M.)
2 Centro de Innovación, Competitividad y Sostenibilidad (CICS), Universidad Autónoma de Guerrero, Acapulco de Juárez 39640, Mexico
3 CONACyT, Universidad Autónoma de Guerrero, Av. Lázaro Cárdenas S/N, Chilpancingo de los Bravo 39086, Mexico
* Correspondence: rimaldonado@uagro.mx (R.I.M.-A.); yimaldonado@uagro.mx (Y.I.M.-A.); rsalazarlo@conacyt.mx (R.S.)

Abstract: A circular economy (CE) promotes the reuse, reincorporation and valuation of waste and by-products under the framework of sustainable development through models and indicators that evaluate scenarios of second use and reduction in non-incorporated outputs to reduce negative externalities and pressures on the dimensions of development. A CE model applied to the transformation process of RFF in agro-industries is developed, which consists in the identification of the residue coefficients of EFB (22.48% ± 0.8), fiber (15.58% ± 0.49), husk (6.03% ± 0.66) and ash (0.55% ± 1.67). Subsequently, the valuation trends of potential second use were verified through a systematic review, which allowed the construction of the scenario of avoided costs of USD 678,721.5, a product of the total use of the outputs under bioenergy and nutrient source approaches. Finally, the RRSFM indicator was constructed, which can reach the level of 72% and a degree of improvement of 26% by 2026. In parallel, the HCRRS indicator revealed a reduction of 57.1%, 59.6% and 82.8% in emissions of t $CO_{2\text{-eq}}$ product in the comparison of scenarios for the use of residues and by-products of palm oil from agro-industries in the Casanare Department.

Keywords: circular economy; indicator; sustainable development

Citation: Bejarano, P.-A.C.; Rodriguez-Miranda, J.-P.; Maldonado-Astudillo, R.I.; Maldonado-Astudillo, Y.I.; Salazar, R. Circular Economy Indicators for the Assessment of Waste and By-Products from the Palm Oil Sector. *Processes* **2022**, *10*, 903. https://doi.org/10.3390/pr10050903

Academic Editors: Dimitris Zagklis and Georgios Bampos

Received: 31 March 2022
Accepted: 25 April 2022
Published: 3 May 2022

1. Introduction

The circular economy (CE) is an interdisciplinary approach that promotes the reuse, reincorporation and valuation of waste and by-products in order to reduce the pressures related to the dimensions of development under different strategies that recirculate energy and matter with a focus on second use in source or complementary processes [1,2]. Some authors affirm that the CE contributes to promoting sustainable development spaces [3,4], since it generates a reduction in externalities through the balanced integration of the dimensions through innovation, adaptation and environmental management [5–8], and addresses the causes that generate the disruption of sustainable development [3].

A CE incorporates different concepts [6,7], methodologies, methods, models and indicators adapted to simple or compound quantitative parameters [5,9], such as material flow, energy flow, land use and cycles of life [10,11], which allow the generation of comparative scenarios of reincorporation of second use. The foregoing is implemented at the micro-, meso- and macroeconomic scales of productive sectors [10,12], which facilitates the correction of shortcomings in the economic dimension and promotes a transition towards sustainable models.

Palm oil agro-industries are the starting point for the analysis to correct the shortcomings generated by the economic dimension, since, by providing different goods and services, negative externalities are generated that affect the dimensions of development. This is due to production failures and a low installed capacity to incorporate waste and by-products [13–15], which is aggravated by the absence of strategies that allow evaluating the benefit of incorporating outputs into the development dimensions [9,16].

This context worsens when knowing that around 23% of the waste rate and agro-industrial by-products of palm oil are reincorporated into the production process, and the remainder is abandoned on site [17,18]. This causes the contamination of bodies of water, atmospheric emissions, the alteration of the landscape, the acidification of the soil, produces leachate and affects human health [19,20]. The highlighted problem increases the magnitude and scale by evidencing the trend of the market dependent on vegetable oil products derived from palm oil [21], of which Colombia supplies and markets 4.08% of the world's production [22]. The foregoing positions Colombia as the first producing country in the Americas (above Ecuador, Honduras, Guatemala, Brazil and other countries in the region) and fourth producer worldwide (behind Indonesia, Malaysia and Thailand) [23].

The circular economy model developed in this research proposes a strategy to correct the previously highlighted deficiencies of the palm oil agro-industrial sector, since it diagnoses the source processes and values the residues and by-products under a second-use approach (use), with the purpose of establishing two quantitative indicators to estimate the incorporation of outputs through the Material Flows approach and the carbon footprint through the Other-life-cycle-based approach, which are based, applied and adjusted to the work ecosystem of the agro-industries of the Casanare Department. Additionally, the strategies developed in the model discourage the use of landfills and discharges that affect the dimensions of development; at the same time, they raise sustainable development scenarios that reduce pressures on the local environment.

2. Materials and Methods

The method of the present investigation was descriptive, since it explains and characterizes the waste and by-products from a specific case to determine the potential of second use under circular economy strategies. Likewise, a deductive method was used, since the paper analyzed data from the specific case under the first principles of approaches to valuate and use waste and by-products to propose circular economy indicators aimed at sustainable development. The methodological development used a mixed approach that was (i) qualitative, in the phases of identification, characterization and evaluation of the residues and by-products of the palm oil sector in the processing plants in the Casanare Department; and (ii) quantitative, in the assessment of exploitation approaches, the estimation of avoided costs and the projection of the model and indicators of circular economy.

2.1. Determination and Characterization of Waste and By-Products

For the construction of circular economy models and indicators, it is pertinent to characterize the source, type of waste and/or by-product, generation rate and current disposal, as developed by authors in international research, such as Rakundo et al. [24], Oliveira et al. [25], Barcelos et al. [26], Moraga et al. [9], Salguero-Puerta et al. [27] and Wang et al. [28]. The information collected can be categorized as:

A. Determination of the processing rate, measured in tons per year of fresh fruit bunch (FFB) or crude palm oil (CPO).

A.1. Average weight of FFB expressed in kg or t.
A.2. Amount of RFF required to process one ton of CPO.

B. Characterization of by-products generated in the extraction process.

B.1. Characterization of solid waste equivalent to empty clusters or coarse, fiber, shell and ash.

B.1.1. Estimation of the generation rate of solid waste and by-products, measured in t/year, kg/year, kg/RFF or t/FFB, for EFB, fiber, husk, ash and palm oil mill effluent (POME).

B.1.2. Qualify and quantify the physical and chemical properties of Organic Solid Waste—OSW (properties that can be used according to energy potential and source of nutrients).

B.2. Characterization of wastewater discharges generated in the extraction process, according to reports supported by agro-industries.

B.2.1. Physical and chemical properties of the discharge according to environmental compliance reports presented by agro-industries to the corresponding entities for the remuneration rate permit. The parameters to be measured are physical (pH and temperature) and chemical (BOD, COD, SS, SVT and fats and oil).

C. Trend analysis of production area and transformation rate.

C.1. Planted area and harvested area in hectare units (regional accumulation).

C.2. Transformation rate CPO/year and CPO/ha.

The information was collected for each agro-industry, in an interval of 5 years prior to the investigation (2015–2019), with the purpose of extracting the most pertinent and current information on the management of and approach for each residue. The information was used to calculate the production coefficients through linear or non-linear regression describing the production rate. Each data grouped in the aforementioned sections were analyzed using descriptive and inferential statistics (Pearson correlation).

Additionally, a systematic review using Google Scholar, Springer Link and ScienceDirect search engines was performed to compare with the information from the Casanare Department. For this reason, a review was performed by consulting material published within the range of 25 years (1994–2019) using the keywords waste, by-product and palm oil, given that the regional, national and international information of the last 25 years is relevant to understand the characterization and mass flow based on the technological change and growth trends of palm oil cultivation.

2.2. Valuation of Waste and By-Products of the Palm Oil Agro-Industry

Agro-industrial waste and by-products constitute a source of renewable biological resource with the potential to be incorporated to obtain bio products with high added value. Industrial development with the scheme of the exploitation and utilization of waste and by-products became a priority for business management working under the framework of green and circular economies, given the interest in them as engines of growth, innovation and job creation. The assessment and evaluation of the outputs constitutes the basis of studies related to the development of the circular economy model [29–31].

For this purpose, based on the information collected on the characterization of waste and by-products, the main trends in the valuation of palm oil agro-industrial waste and by-products were investigated. The forthcoming was developed through a systematic literature review of the last 10 years, consulting SpringerLink, ScienceDirect and Google Scholar journals, based on the methodology of Cardenas et al. [32], to identify the approaches, methods, technologies and costs associated with their respective implementation (see Table 1). In the review, the combination of the following keywords was used as the main thematic descriptors: (i) waste; (ii) valorization; and (iii) palm oil. The second phase implemented the keywords to focus on the assessment (or use approach) of waste and by-products, in terms of nutrient source, enzyme activity, biofuel and other interests. Finally, a third phase was used to identify the technological approach by which the waste and by-products are transformed, corresponding to chemical, physical and biological processes.

Table 1. Order of importance detected in relation to the valuation of palm oil residues and by-products worldwide in the database.

Phase	DT [f]	Google Scholar DD [d]	Google Scholar Q [e]	ScienceDirect DD [d]	ScienceDirect Q	SpringerLink DD	SpringerLink Q	$\bar{x}$ Q [b]	ΔQ [c]
1. DP [g]	Palm oil waste and valorization	13,400	1 [a]	1223	1	780	1	-	-
2. Leveraging Approach	Biofuel	13,200	98.5	762	62.3	337	43.2	68.1 Q3	Q4-Q3-Q2
	Nutrient source	12,500	93.3	578	47.3	361	46.3	62.3 Q3	Q4-Q2-Q2
	Enzyme activity	10,500	78.4	567	46.4	349	44.7	56.4 Q3	Q4-Q2-Q2
	Construction	8330	62.2	411	33.6	204	26.2	40.6 Q2	Q3-Q2-Q2
3. Technological approach	Technological	15,100	1	1094	1	734	1	-	
	Chemical	13,200	87.4	1047	95.7	699	95.2	92.8 Q4	Q4-Q4-Q4
	Physical	14,500	96.0	745	68.1	488	66.5	76.9 Q4	Q4-Q3-Q3
	Biological	10,200	67.5	718	65.6	425	57.9	63.7 Q3	Q3-Q3-Q3

[a] (13,200/13,400) * 100 = 98.5; [b] Average of quartile; [c] Variation of the quartile according to the database; [d] Documents detected; [e] Citation frequency index; [f] Thematic descriptors; [g] Main descriptor.

2.3. Circular Economy Model for the Reincorporation of Waste and By-Products from the Palm-Growing Sector in the Casanare Department

The implementation of environmental indicators (with emphasis on the use of waste and by-products of this research) allowed the creation of tools for the analysis of the current and future state (implementation of the model under assumed use scenarios). The assumptions were evaluated using the RRFSM indicator (Reincorporation of Waste and By-products by Material Flow) and the HCRRS indicator (Carbon Footprint of the Reincorporation of Waste and By-products).

2.3.1. RRSFM Indicator

The alternatives were proposed under a reuse, reincorporation and/or recycling scheme for waste and by-products according to the techniques and methods addressed in the Reference Framework chapter, such as:

A.1. Residual effluent as a fertigation system and an alternative water supply to reduce the capture rate.

A.2. Residual effluent as an affluent for reuse to reduce the rate of water use in the internal processes of the agribusiness.

A.3. Reuse of organic waste as a soil conditioner.

A.4. Reuse organic waste as a source of nutrients.

A.5. Reuse of organic and liquid waste as a source of energy.

After selecting the viable alternatives according to the conditions of waste and by-product production, the information was consolidated by agro-industry (Ag) in relation to: (i) Total Generation (Gt), (ii) Current Use (Ac) and (iii) Potential Use (Ap), based on the palm oil waste and by-product rates of each agro-industry according to the EFB, fiber, husk, ash and POME variables.

It should be noted that (i) the Current Use parameter corresponds to the environmental management practices implemented by agro-industries according to the alternatives investigated, and (ii) the Potential Use parameter corresponds to the total amount of the different residues and/or susceptible by-products to be used according to the current infrastructure (boilers, composting, fertigation irrigation system, among others). The valued waste and/or by-products must have ideal conditions (regulations and physical and chemical characteristics) that do not generate a negative impact due to their use.

To evaluate the RRSFM indicator at different moments (scenarios), Equations (1) to (6) were used. To determine the numerical value of the Current Moment (Ma) parameter, Equation (1) was used (modified equation of [27]), with which the coefficient (dimensionless value) of the current moment with respect to the generation of waste was calculated. This value was input to calculate the current use of each waste and by-product, according to Equation (2). These equations are exemplified through the liquid waste (LW) variable.

$$LW_{Ma} = 1 - \left(\frac{LW_{Gt} - LW_{Ac}}{LW_{Gt}}\right). \quad (1)$$

$$RRSFM_{Ma} = \frac{\sum OSW_{Ac} + LW_{Ac} + As_{Ac}}{n} \quad (2)$$

To determine the potential moment of the RRSFM indicator, the Ap variable (Current Use) was added to Equation (1) in the denominator part, as evidenced in Equation (3) (modified equation of [27]) and Equation (4), exemplified through the LW variable:

$$LW_{Mp} = 1 - \left(\frac{LW_{Gt} - LW_{Ac} - LW_{Ap}}{LW_{Gt}}\right) \quad (3)$$

$$RRSFM_{Mp} = \frac{\sum OSW_{Mp} + LW_{Mp} + As_{Mp}}{n} \quad (4)$$

Finally, the Total Use (At) parameter was determined, which defines the difference between the Total Generation (Gt) of waste and the accumulated Use (current and potential), on a scale from 0 to 1 (see Equation (5)). Since the difference is not notable when analyzing the information for each agro-industry or residue, it was necessary to use the normalization of indices (see Equation (6)), where the maximum value corresponds to 1 (assuming a 100% circular economy) and the minimum value to 0 (no circular economy assumption).

$$RRSFM = 1 - \left(\frac{Gt - Ac - Ap}{Gt}\right) \quad (5)$$

$$RRSFM_{(Normalicion)} = \frac{RRSFM - 0}{1 - 0} \quad (6)$$

The values obtained in each equation are compiled in Table 2.

Table 2. Matrix for the analysis of the RRSFM quantitative variables.

	GT			AC			Ap			At			RRSFM
Ag	OSW [a]	LW [b]	As [c]	OSW	LW	As	OSW	LW	As	OSW	LW	As	
Ag_1	OSW_{t1}	LW_{t1}	As_{t1}	OSW_{i1}	LW_{i1}	As_{i1}	OSW_{j1}	LW_{j1}	As_{j1}	OSW_{At1}	LW_{At1}	As_{At1}	Σ
Ag_2	OSW_{t2}	LW_{t2}	As_{t2}	OSW_{i2}	LW_{i2}	As_{i2}	OSW_{j2}	LW_{j2}	As_{j2}	OSW_{At2}	LW_{At2}	As_{At2}	Σ
Ag_n	OSW_{tn}	LW_{tn}	As_{tn}	OSW_{in}	LW_{in}	As_{in}	OSW_{jn}	LW_{jn}	As_{jn}	OSW_{Atn}	LWAtn	As_{Atn}	Σ

Note: [a] Organic Solid Waste (OSW) comprises EFB, fiber and husk. [b] Liquid waste (LW) refers to POME. [c] Ash (As).

2.3.2. HCRRS Indicator

A second indicator was built that estimated thecarbon footprint (CF) of the RRSFM indicator under the Other-life-cycle-based approach, in order to generate composite and articulated indicators in the circular economy model. To determine the HCRRS indicator, it was necessary to carry out a systematic review of the emission factors ($\delta_{CO_{2eq}}$) according to the type of waste and by-product generated in the RFF transformation process.

For this purpose, journals accessed through SpringerLink, ScienceDirect and Google Scholar were reviewed, based on the methodology of Cardenas et al. [32]. The main descriptors used were (i) carbon footprint and (ii) mill palm oil; RFV (EFB), POME, fiber and shell were used as complementary descriptors; and, as exclusion factors, we used (i) residues and by-products that are generated in other production phases and do not correspond to residues from the FFB processing, mainly those residues and by-products that are generated in the planting phase (see Table 3). The search period was from 2000 to 2020 of international and national publications that precisely quantified processes of the transformation of RFF or CPO in $CO_{2\text{-eq}}$.

Table 3. Order of importance detected in relation to the carbon footprint in waste and by-products of the transformation of RFF worldwide in the databases.

		Google Scholar		ScienceDirect		SpringerLink		$\bar{x}$ Q	ΔQ
Phase	**DT**	**DD**	**Q**	**DD**	**Q**	**DD**	**Q**		
1. DP	Carbon Footprint and Mill palm oil	17,000	1	1184	1	533	1	-	-
2. Waste and by-products	EFB	7420	43.65	137	11.57	Four. Five	8.44	21.2	Q2-Q1-Q1
	Fiber	5970	35.12	745	62.92	291	54.60	50.9	Q3-Q2-Q2
	Shell	4360	25.65	523	44.17	187	35.08	35.0	Q2-Q2-Q2
	POME	16,200	95.29	184	15.54	57	10.69	40.5	Q4-Q1-Q1

After identifying the emission factor ($\delta_{CO_{2eq}}$) according to the type of alternative analyzed in the RRSFM indicator, three hypothetical scenarios (moments) were constructed with variation in the rates of the reincorporation of waste and by-products to contrast under the current scenario (Ap category of the indicator RRSFM). To determine the emission for each alternative (Ea), Equations (7) and (8) were used. It is pertinent to note that the emission factor must be in units of kg $CO_{2\text{-eq}}$/t RFF.

$$\gamma = \frac{t\ Rs\ aprovechado}{t\ Rs\ generado} \tag{7}$$

$$Ea_{CO_{2eq}} = \frac{\gamma * RFF_{2019} * \delta_{CO_{2eq}}}{1000} \tag{8}$$

After determining the emission of each residue, the individual data were added to obtain the global emission of the analyzed moment. For this purpose, Equation (9) was implemented, whose result provides values in units of t CO_{2eq}.

$$HCRRS_{CO_{2\,eq}} = \sum(\vartheta_1 \partial OSW_1 \pm \vartheta_2 \partial OSW_2 \pm \vartheta_n \partial OSW_n) \tag{9}$$

3. Results and Discussion

3.1. Characterization of Waste and By-Products of the RFF Process

The production of crude palm oil can be summarized as a closed system in which the fresh fruit bunches (FFB) enter a series of processes that incorporate matter (mainly water) and energy to obtain the desired product. The RFF is selected after the evaluation in the field, which, according to the color of the fruit, is removed and transferred under electrical, mechanical or animal traction systems to the processing plant facilities.

Subsequently, the extraction begins, which separates the nuts from the peduncle or rachis (first by-product in solid state, corresponding to empty fruit bunch (EFB, also called cob or mesocarp)), by means of rotary mechanical processes [33]. The cob is composed of shells and fiber, which are equivalent to approximately 22% of the initial weight of the RFF [34]. When analyzing the information provided by the Casanare Department agro-industries, it is evident that the average weight of the cob oscillates between 20% and 24% of the initial weight of RFF; for this reason, for the present investigation, the reference value of 22% was considered.

Subsequently, the digestion or pressing process is carried out, which compresses the separated fruits to obtain palm kernel oil (PKO) or crude palm oil (CPO). Both products undergo similar processes (clarification, centrifugation, rinsing and storage); however, the thermodynamic and methodological conditions vary. In this process, the second subgroup of by-products is generated in a solid state, equivalent to walnuts, almonds, shells and fiber in a broken or compressed state, which are removed and transferred to a drying area for their respective reuse as an energy source and air conditioner ground. In the developed field review, it was evidenced that all the plants use the separation and reincorporation process as a source of combustion for boilers or soil conditioner; however, the percentages vary according to the needs of each plantation. This process is pertinent to highlight, since it produces two main products (PKO and CPO), which vary according to the source.

The subsequent process corresponds to clarification, which is made up of different threads whose purpose is to separate the liquid mixture resulting from the digestion or pressing process through gravitational processes and immiscibility of substances, mainly water and oil, aided by steam [35,36]. This process generates two relevant by-products, corresponding to liquid waste (containing sludge) and fibrous waste (suspended fibers in pressed solutions), which are treated by sedimentation processes to remove by-products. In relation to liquid waste, residual effluents are generated, and, in relation to fibrous waste, a fiber suitable for animal consumption is obtained.

It should be noted that another important source of liquid waste evidenced in the benefit plants is created through the centrifugation process, since the liquid waste prior to disposal in oxidation lagoons is subjected to mechanical systems to recover a part of the oil suspended in the wastewater. This decreases the load of suspended substances in the effluent, which favors the reincorporation of the oil into the production process and the reduction in contaminants.

It was previously mentioned that the nuts are separated from the residual fiber in the defibration and crushing process, since the interior contains the almond that is viable for obtaining other products. This process is carried out using mechanical dryers that allow the generation of a (i) palm kernel cake, (ii) palm kernel oil and (iii) shells (a by-product with potential for reuse due to its calorific value). Finally, the residue from the incineration of solid organic matter in boilers, coming from the combustion of cobs, fibers and/or husks, called ash, is found, which is equivalent to approximately 5% of the weight of the incinerated organic waste [37,38]. This product is collected in two phases: from the (i) boiler area, where the ashes are deposited, and (ii) the emission control system, which are called hard and soft ashes, respectively [39,40]. When comparing this information with the data obtained in the Casanare Department, it was revealed that the value of ash was in the range of 4.3–8% of the weight of the incinerated waste, whose percentage varies according to the combination with the different types of fuel.

In different investigations focused on the characterization of palm-oil-processing plants, the percentage or wet weight of residues and by-products after transforming 1 ton of RFF or 1 ton of CPO was quantified. Empty fruit bunches (EFB), fiber, husk, ash, POME and methane are the main residues and by-products characterized. In Tables 4 and 5, the main investigations at a global level are detailed, which quantified the output of matter after the transformation of the ton of RFF or ton of CPO, as well as the values obtained in six agro-industries of the Casanare Department.

When comparing the Reference Values (VRE) with the Values provided by the Agro-industries of the Casanare Department (VAC), a positive correlation between the data is evidenced, since the difference and proportion of waste and by-products is preserved. It is evident that the differences correspond to the type of technology and variety of the crop, which contributes to the efficiency of production and water consumption for the production of one ton of CPO.

Table 4. Quantification of the Reference Values (VRE) of waste and by-products derived from processing one ton of RFF.

Waste/By-Product Author (VRE)	EFB	Fiber	Husk	Ash	POME	Methane
Singh et al. [41] and Lik et al. [42]	23%	15%	7%	-	-	-
Abdullah and Sulaiman [34]	22%	13.50%	5.50%	-	67%	-
Abdullah and Sulaiman [34] *	14.60%	15.40%	10.40%	-	6.30%	-
Loh [39]	22%	13.50%	5.50%	-	67%	-
Loh [39] *	35%	60%	8.50%	4.60%	-	-
Vijaya et al. [43]	22%	8–16%	9–22%	2%	58.20%	10.57 m^3
Yusoff [44]	22%	13.50%	5.50%	-	67%	-
Zinatizadeh [45]	28.50%	30%	6%	0.50%	53.51%	-
Zinatizadeh [45] *		14%	6%	0.50%	53.51%	-
Kong et al. [46]	22–23%	13.5–15%	5.5–7%	-	-	-
Ali. et al. [47]	23%	15.70%	-	-	70%	-
Huailuek et al. [48]	21%	12.46%	3.95%	-	58%	-
Stichonothe et al. [49]	23%	13%	5.50%	-	86%	-
Foo and Hameed [37]	23%	14–15%	6–7%	-	-	-
Embandiri [50]	24%	14%	6%	0.42%	63%	-
Abas et al. [51]	23%	13%	5%	-	60%	-
Cenipalma [52] , Bernal [53] and Silva [54]	17.7–26.1%	11.6–15%	5–7%	-	70–80%	-
Garcia et al. [55] and Ramirez [56]	20%	13%	5%	0.53%	69%	-
Average	23.04%	16.77%	6.41%	1.43%	61.18%	10.57 m^3
Variance	3.76%	11.23%	1.87%	1.67%	17.92%	-

* Dry weight.

Table 5. Quantification of Values provided by the Agro-industries of the Casanare Department (VAC) of the waste and by-products derived from processing one ton of RFF.

Waste/By-Product Agribusiness (VAC)	EFB	Fiber	Husk	Ash	POME	Methane
A	19–23%	13–17%	7%	0.80%	80%	-
B	21–22%	15%	6%	0.50%	67%	-
C	19–25%	15%	6.40%	0.43%	63%	-
D	23%	16%	5%	0.50%	72%	-
E	20–24%	14–18%	5–7%	-	75%	-
F	22%	16%	5.80%	0.50%	60%	-
Average	22.48%	15.58%	6.03%	0.55%	69.50%	-
Variance	0.80%	0.49%	0.66%	0.15%	7.56%	-

The largest source of organic solid waste (OSW) corresponds to EFB (23.04% ± 3.76 VRE and 22.08% ± 0.8 VAC), generated in the sterilization and extraction process; its percentage of moisture content is relatively high, approximately 60% of the weight, which means that it is highly difficult to use for energy purposes. For this reason, agro-industries use dry fiber and husk as an energy source and, on occasion, sun-dried EFB, since its C/N ratio is used for soil conditioner [39].

The second by-product with the highest RSO percentage corresponds to Fiber (16.77% ± 11.23 VRE and 15.58% ± 0.49 VAC), generated in the separation and clarification process. This by-product has a low percentage of moisture and residual CPO, approximately 5% [34], and thus it is favorable to be used in incineration in boilers to obtain energy; its energy contribution corresponds to 18,795–19,060 kJ kg^{-1} [39,57,58].

The third by-product with the highest RSO contribution corresponds to Cascarilla (6.41% ± 1.87 VRE and 6.03% ± 0.66 VAC), generated in the process of the separation and breaking of the nut. This by-product has a 100% utilization rate in the agro-industries of the Casanare Department, given its calorific value of 19,500–20,750 kJ kg^{-1}, low moisture content and water absorption capacity [39,57–59].

On the other hand, the main liquid residue corresponds to the residual effluent (POME), which presents a significant variation depending on the processing plant (61.18% ± 17.92 VRE and 69.50% ± 7.56 VAC). In the literature review, it was identified that, in some processes, around 0.8–1.5 m^3 are used to process a ton of FFB, given the technology implemented by the benefit plants, and around 50% of the volume used becomes POME [39,41,49,57,60]. In relation to the Casanare processing plants, from the initial volume of water used to transform one ton of RFF, a retribution rate of 61% was estimated, which corroborates some of the abnormal values above the average reference value (VRE).

Finally, the residue with the lowest contribution corresponds to Ash (1.43% ± 1.67 VRE and 0.55% ± 0.15 VAC), derived from the combustion of organic matter, such as fiber, husk and EFB in boilers, in combination with other fuels, such as coke or coal. The variation between VRE and VAC of ash is due to a difference marked by a record of Loh [39] in dry weight, whose value is abnormal among the other records, since the humidity (remaining records) significantly decreases the concentration of ash [61,62]. In fact, it is pertinent to mention that the processes of the use of residues and organic by-products of the RFF are implemented with relative concentrations of humidity. All the mentioned data are compiled in Figure 1.

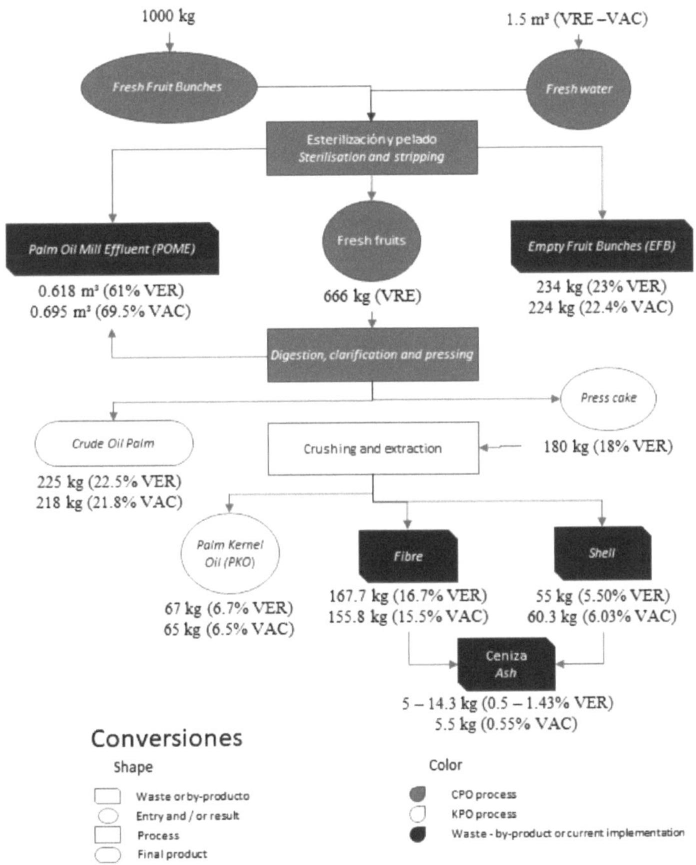

Figure 1. Diagram of the inputs and outputs of matter in the transformation process of 1 ton of RFF. Source: Adapted from Loh [39] and Abdulalla [34].

Physical and Chemical Characteristics of Palm Oil Waste and By-Products

As highlighted in the previous section, there are two groups of waste and by-products according to their physical state, solids and liquids. Solid waste is made up of EFB, husk, fiber and ash. These residues have different physical and chemical properties, such as calorific value, texture, porosity, rigidity and nutritional composition (see Table 6, which generates a variety of reuse alternatives.

The EFB (also called cob) is a lignocellulose material with a composition of 60–65% moisture, and with 1–2.5% vegetable oil impregnated by the physical separation of the rachis in the process of fruiting the sterilized bunches [63]. The EFB is the second by-product with the best proportion of potassium, manganese, phosphorus, nitrogen and zinc content, which allows visualizing implementations, such as soil conditioning, due to its nutrient contribution. On the other hand, the EFB has a high calorific value (18–19.92 MJ/kg); however, it requires prior drying processes due to its high moisture content (66–69%).

Table 6. Chemical and physical composition of the waste and by-products generated in the transformation phase of the RFF.

Waste and By-Products Chemical Composition	EFB	Fiber	Husk	Ash	POME	Author
Carbon	45.9 (%)	45.2 (%)	49.70 (%)	59,583 (%)	40.17 (%)	Abdullah et al. [62]; Loh [39]; Nalaya et al. [64]
Hydrogen	5.70 (%)	5.50 (%)	5.70 (%)	1181 (%)	5.81 (%)	Abdullah et al. [62]; Loh [39]
Nitrogen	0.8 (%)	1.10 (%)	0.40 (%)	0.088 (%)	5.26 (%)	Abdullah et al. [62]; Loh [39]
Total nitrogen	-	-	-	-	750 mg/L	Abdullah et al. [62]; Singh et al. [41]; Baharuddin et al. [65]
Sulfur	0.2 mg/L	0.23 mg/L	0.19 mg/L	-	-	Abdullah et al. [62]; Loh [39]
Oxygen	36.7 mg/L	-	-	0.981 ± 0.42 (%)	-	Abdullah et al. [62]; Loh [39]
Potassium	2.24 mg/L	1.48 mg/L	2.20 mg/L	26.3–41.0 g/kg	22,700 mg/L	Singh et al. [41]; Baharuddin et al. [65]; Vijaya et al. [57]; Tay [40]; Loh [39]
Magnesium	0.6 ± 0.2 (%)	0.49 mg/L	0.24 mg/L	20.3–22 g/kg	615 mg/L	Singh et al. [41]; Baharuddin et al. [65]; Vijaya et al. [57]; Tay [40]
Zinc	16.6 ± (2.6 mg kg^{-1})	-	-	0.1–0.3 g/kg	2.3 mg/L	Singh et al. [41]; Baharuddin et al. [65]; Vijaya et al. [57]; Tay [40]
Phosphorus	0.6 ± 0.1 (%)	0.12 mg/L	0.07 mg/L	-	180 mg/L	Baharuddin et al. [65]; Vijaya et al. [57]
Lignin	35.3 (%)	27.7 (%)	50.7 (%)	-	-	Sukiran et al. [66]
Hemicellulose	35.3 (%)	26.1 (%)	22.7 (%)	-	-	Sukiran et al. [66]
Calorific value (MJ/kg)	18–19.92	18.8–19.58	19.5–20.75	-	16.1–17.65	Soh [58]; Abdullah et al. [62]; Singh et al. [41]; Loh [67]; Loh and Choo [68]
Moisture content (%)	66–69	35–48	-	-	90–95	Soh [58]; Abdullah et al. [62]; Singh et al. [41]; Loh [67]; Loh and Choo [68]
Ash content (%)	4.6	6.1	3	-	15.2	Soh [58]; Abdullah et al. [62]; Singh et al. [41]; Loh [67]; Loh and Choo [68]
Volatile matter content (%)	87	84.9	83.4	-	77.7	Soh [58]; Abdullah et al. [62]; Singh et al. [41]; Loh [67]; Loh and Choo [68]

Table 6. *Cont.*

Waste and By-Products / Chemical Composition	EFB	Fiber	Husk	Ash	POME	Author
Biochemical Oxygen Demand	-	-	-	-	25,000 mg/L	Singh et al. [41]; Baharuddin et al. [65]; Poh et al. [14]
Chemical Oxygen Demand	-	-	-	-	50,000 mg/L	Singh et al. [41]; Baharuddin et al. [65]; Poh et al. [14]
Total solids	-	-	-	-	40,500 mg/L	Singh et al. [41]; Baharuddin et al. [65]; Poh et al. [14]
pH	6.7 ± 0.2	-	-	9.35–9.70	4.7	Singh et al. [41]; Baharuddin et al. [65]; Poh et al. [14]; Tay [40]
Fat and oil	-	-	-	-	4000 mg/L	Singh et al. [41]; Baharuddin et al. [65]; Poh et al. [14]

Note: % This value can also be determined by the difference between the content of C, H, N and S (in%) and the total of 100%, in dry weight.

Fiber contains the best proportion of nutrients compared to other organic by-products, such as nitrogen, potassium and phosphorus (see Table 6); however, it has a high content of metals, such as iron (723.2 ± 7.8), copper (24.2 ± 0.05) and aluminum (749.1 ± 14). These characteristics make it difficult for this by-product to be implemented as a soil conditioner, since it can favor the accumulation of salts and metals. For this reason, its utilization approach focuses on incineration, given its calorific value and ignition degree. The ratio to obtain the highest calorific value is from 60:40 to 90:10 in combination with husk [43,64–69].

Husk has a high content of carbon and potassium; however, it also has a low content of magnesium, nitrogen and phosphorus. Likewise, it has the best calorific power capacity among the residues and by-products derived from FFB (19.5–20.75 MJ/kg) with a low ash content (3%), which is favorable for its implementation in incineration in boilers with fiber (see Table 6). Likewise, the hull has 37% of porosity, which is beneficial for its the water absorption capacity (21–33%) [59].

Ash is a product of the OWS combustion, such as fiber, husk and mainly EFB. It has a granular and porous structure, most of it fine in size [63,64]. This by-product is not apt for incineration; however, it has a high ratio of nutrients, such as carbon, potassium and magnesium, and a low content of leachable heavy metals (less than 0.2 mg/L of cadmium, lead, copper and nickel) [70], so it is used in combination with OWS in composting processes.

On the other hand, the POME liquid residue from the grinding, centrifugation, sedimentation and other processes contains the lowest degree of carbon content (40.17%); however, it has the best composition of nutrients, such as nitrogen, potassium, phosphorus and zinc compared to other residues and by-products (see Table 6). Additionally, the said waste has a high potential calorific power (16.1–17.65 MJ/kg), derived from the anaerobic transformation of the effluent.

The concentration of BOD, COD, ST, STV and GA in the wastewater varies according to the amount of water used in the RFF transformation processes (see Table 7). The POME of the Casanare Department presents a temperature of 56.5 °C ± 1.7 °C, being a product of the sterilization process that implements the water resource at a temperature of approximately 140 °C; a pH value of 4.4; a flow rate of 22.8 m^3 ± 12.2 m^3, which varies according to the productive capacity of the processing plant; and BOD^5 and COD values above the theoretical reference average (33,225.5 ± 2564 mg/L and 63,385.3 ± 3995.7 mg/L, respectively).

Table 7. Comparison of the physical and chemical composition of the POME from the Casanare Department's agro-industries with the theoretical references.

Parameter Unit	Temperature °C	pH [a] -	Flow Rate m^3/h	BOD^5 [d] mg/L	COD [e] mg/L	SVT [f] mg/L	NKT [g] mg/L	GA [k] mg/L
Agribusiness A [h]	53–61	4.3–4.4	18.8	31,287	64,242	25,452	590	7680
Agribusiness B [h]	58	4.45	7.2	33,250	67,800	24,582	590	7500
Agro-industry C [h]	-	4.3	-	33,458	66,570	24,500	545	7770
Agro-industry D [i]	51–62	4.4	28–38	36,458	57,500	19,200	480	6800
Agribusiness E [h]	54	4.4	32	29,500	59,700	18,500	490	6850
Agribusiness F [i]	-	4.4	-	35,400	64,500	24,200	580	7650
Average	56.5	4.4	22.8	33,225.5	63,385.3	22,739.0	545.8	7375.0
Standard deviation	1.7	0.1	12.2	2564.0	3995.7	3049.1	50.0	435.1
Reference authors								
Poh et al. [71] [b]	45.8–62.1	4.4–4.5	7.8–25.4	12,520–42,630	27,840–85,267	12,000–47,667	230–780	2500–16,100
Poh et al. [71] [c]	54–66.5	4.18–4.7	5.4–65.0	32,100–56,700	67,900–87,300	41,180–47,060	525–1350	11,004–15,880
Ma and Ong [72]	80–90	4.5	-	25,000	-	-	-	8000
Ahmad et al. [73]	-	4.7	-	25,000	50,000	-	-	4000
Oswal et al. [74]	-	5	-	11,000	246,000	-	-	-
Choorit and Wisarnwan [75]	-	4.4	-	65,714	102,696	72,058	1381	9341
García [76] [j]	53	5.25	-	38,647	47,667	59,970	-	15,492
Singh et al. [41]	-	4.7	-	25,000	50,000	34,000	750	4000
Average	63.05	4.68	25.90	32,792.00	90,073.79	47,996.30	893.38	9082.14
Standard deviation	14.98	0.30	13.15	16,635.33	71,602.68	17,770.37	370.19	4340.44

Note: [a] No units for pH; [b] low harvest season; [c] peak harvest season; [d] Biochemical Oxygen Demand; [e] Chemical Oxygen Demand; [f] Total volatile solids; [g] Total Kjeldahl Nitrogen; [h] The values supported by agro-industries correspond to the 2019 period; [i] The values supported by agro-industries correspond to the 2020 period; [j] Plant reference values in the eastern zone of Colombia; [k] Fats and oils.

After the treatment of the water from POME, an effluent with dissolved substances is obtained that generates discharge restrictions, given that it exceeds the maximum values of BOD, COD, Cl, TSS and other compounds according to the current national regulations (see Table 7). However, it contains nutrient such as nitrogen, phosphorus, potassium and calcium, which are favorable for reincorporation, since they do not exceed the maximum values established for reuse purposes. When comparing the values registered by the Casanare Department's agro-industries, little discrepancy is generally reflected in relation to the reference, whose study was carried out in the eastern zone. Total nitrogen, magnesium, chlorides, sulfates and boron are above the benchmark, while total phosphorus, potassium and calcium are below it.

3.2. *Approaches and Variables Used in the Valuation of Waste and By-Products of the Palm Sector That Are Generated in the Transformation Process of RFF*

In recent years, there has been an increase in the search and development of new activities focused on the conversion of waste and by-products into energy and/or source of materials, with an emphasis on source processes to implement practices that integrate closed cycles [77]. Different academic disciplines have focused their efforts on the research and development of methodologies under closed loop schemes, which allow the incorporation of outputs, visualized as waste and/or by-products, to source processes or complementary processes. Closed loops, processes or cycles are analogies that represent the circular economy, based on material and energy flows that are likely to reincorporate outputs as a source of material or energy (in a degraded state) under valuation methodologies.

The valuation implements different methodologies appropriated by the circular economy, such as recycling, used mainly in Europe, Asia and South America, and reuse or benefit in North America [78]. Nzihou and Lifset [78] highlight that the recovery of waste and by-products refers to the treatment to obtain a benefit as a raw material or complement, or energy, in order to reduce emissions and related environmental impacts (p. 2), whose benefit can be economically estimated. Likewise, by generating closed loops, positive impacts are obtained for the social and economic components of the direct and indirect actors of the productive processes.

Valuation Approaches for Waste and By-Products of Palm Oil

Palm oil waste and by-products have a varied chemical and physical composition, such as moisture, calorific value, nutritional composition, water retention, among others, which generates a range of implementation alternatives according to the requirements of the final process. For this purpose, different methodologies are used, aided by physical, chemical and/or biological processes, to generate products derived from the transformation of waste and by-products, with the feasibility of substituting or complementing other production processes. Table 8 compiles the trends of the last 10 years in relation to the valuation of palm oil waste and by-products, with an emphasis on global trends, exploitation and technological approaches. It is important to highlight that the geographical trend in research was concentrated in Asia (73%), given the current production of CPO, supported mainly in three nations of the Asian continent, corresponding to Indonesia, Malaysia and Thailand, which generate around 66,200 thousand tons of CPO, which is equivalent to 88.16% of the CPO world production ([22], p. 19). Likewise, investigations were evidenced in the continent of America (16%) and Africa (7%), given their participation in the CPO world production, and these continents contribute almost 11.84%, equivalent to 8893 thousand tons ([22], p. 19).

Table 8. Concentrations of key trends for the valuation of residues and by-products of the palm oil processing plant.

	Geographical Area [a]						Leverage Approach [b]			Technological Focus [c]				Waste and/or By-Product [d]				
Author	**A.M**	**EU**	**AS**	**AF**	**DL**	**NS**	**EA**	**BF**	**CS**	**OI**	**PQ**	**FF**	**PB**	**EFB**	**FI**	**CA**	**CE**	**POME**
Fokam et al. [79]				x					x			x			x			
Quayson et al. [80]				x				x			x					x		x
Aprianti et al. [81]			x					x				x		x				
Foong et al. [82]					x	x		x				x		x	x	x		
Gallego et al. [83]	x							x				x		x				
Hau et al. [84]			x			x							x	x			x	x
Prasetyo et al. [85]			x							x		x				x		
Intasit et al. [86]			x				x	x			x			x				
Lemus [87]	x					x						x						x
Lam et al. [88]			x					x				x		x				
Vasu et al. [89]			x					x				x				x		x
Sinjaroonsak et al. [90]			x				x				x		x	x				
Chin et al. [91]			x					x					x	x				x
Burimsitthigul et al. [92]			x					x			x		x			x		
Dias et al. [93]	x					x				x		x				x	x	
Quintero and Torres [36]	x					x						x		x	x	x		
Shanmugarajah et al. [94]			x							x	x			x				
Zulkarnain et al. [95]			x				x				x		x	x				
Liew et al. [96]			x			x						x		x	x	x	x	
Ahmad et al. [97]			x						x			x		x				
Bukhari et al. [98]			x			x	x				x		x					x
Cheah et al. [99]			x					x		x	x		x					x
Misnon et al. [100]			x				x			x	x	x				x		
Sukira et al. [66]					x	x		x				x		x	x	x	x	
Yoo et al. [101]			x					x				x		x				
Ong et al. [102]			x						x			x					x	
Kamsani et al. [103]			x				x				x		x	x				
Idris et al. [104]			x					x					x					x
Thinagaran and Sudesh [105]			x							x	x		x					x
Junpadit et al. [106]			x				x				x		x					x
Agudelo et al. [107]	x								x			x				x	x	
Lim and Wu [108]			x			x						x	x					x

Table 8. *Cont.*

	Geographical Area [a]					Leverage Approach [b]					Technological Focus [c]			Waste and/or By-Product [d]				
Author	**A.M**	**EU**	**AS**	**AF**	**DL**	**NS**	**EA**	**BF**	**CS**	**OI**	**PQ**	**FF**	**PB**	**EFB**	**FI**	**CA**	**CE**	**POME**
Silveira et al. [109]	x						x						x		x			
Tsouko et al. [110]	x					x	x				x		x		x	x		
Fatriasari et al. [111]			x				x				x			x				
Nyakuma [112]			x					x				x		x	x	x		
Louhasakul et al. [113]			x				x	x					x					x
Neoh et al. [114]			x				x						x	x				x
Iwuagwu and Ugwuanyi [115]				x			x				x		x					x
Ishola et al. [116]			x				x				x		x	x				
Nazir et al. [117]			x				x				x	x		x				
Stemann et al. [118]			x			x		x				x		x				
Salema & Ani [119]			x					x				x		x				
Nahrul et al. [120]			x			x							x	x				x
Average	0.16	0	0.73	0.07	0.05	0.27	0.34	0.39	0.14	0.09	0.39	0.5	0.43	0.57	0.2	0.3	0.14	0.32
Standard deviation	0.37	0	0.45	0.25	0.21	0.45	0.48	0.49	0.35	0.29	0.49	0.51	0.5	0.5	0.41	0.46	0.35	0.47

Note. [a] AM = America; EU = Europe; AS = Asia; AF = Africa; DL = Different places. [b] NS = Source of nutrients; EA = Enzyme activity; BF = Biofuel; CS = Construction; OI = Other interests. [c] PQ = Chemical process; PF = Physical processes; PB = Biological process. [d] EFB = Empty Fruit Bunches; FI = Fiber; CA = Husk; CE = Ash; POME = Residual effluent from ground plant (includes sludge).

The geographical trends in relation to the exploitation approaches did not present a correlation (Pearson's chi-square test: 0.813; df: 12), which validates the fact that there is no specific exploitation approach according to the geographical area, but that there is a variety of assessment alternatives of waste and by-products.

Additionally, it was evidenced that there is no correlation between the geographical trend and technological approach (Pearson's chi-square test: 0.907; df: 15), since there is a variety of investigations with varied methodologies at different geographical scales. On the other hand, it was evidenced that there is a correlation between the exploitation approaches and the type of residue and/or by-product of FFB (Pearson's chi-square test: 0.001; df: 26), which confirms that each output of the FFB process presents a trend of use according to its physical and chemical composition. However, it was evidenced that there is no correlation between the technological approach and the type of waste and/or by-product of FFB (Pearson's chi-square test: 0.031; df: 40), since a variety of alternatives are used to transform the outputs of the RFF process according to the final need. However, there is a correlation between the exploitation approach and the technological approach (Pearson's chi-square test: 0.06; df: 22), which confirmed that there are specific methods according to the exploitation approach.

The exploitation approach is correlated to the type of waste and/or palm oil by-product, which generates trends according to the physical and chemical properties of the outputs of the RFF process. The trends of the exploitation approach are compiled into five categories, corresponding to biofuels, enzymatic activity, nutrients source, construction, and other interests, whose average in the systematic review corresponded to 39%, 35%, 15%, 7% and 4 %, respectively.

Figure 2 shows the trends and approaches that are used according to the variety of waste and/or by-products, which, in some cases, are used individually or together to obtain a complementary or substitute good or service. It is pertinent to note that the main by-product with the greatest relevance is EFB, given the abundance produced after transforming a ton of RFF (in the present investigation, the average EFB after the processing of 1 t of FFB was equivalent to 23.59% of the initial weight), followed by husk and fiber, while the main residue is POME, followed by ash (or complementary processes that use waste and by-products together with ash).

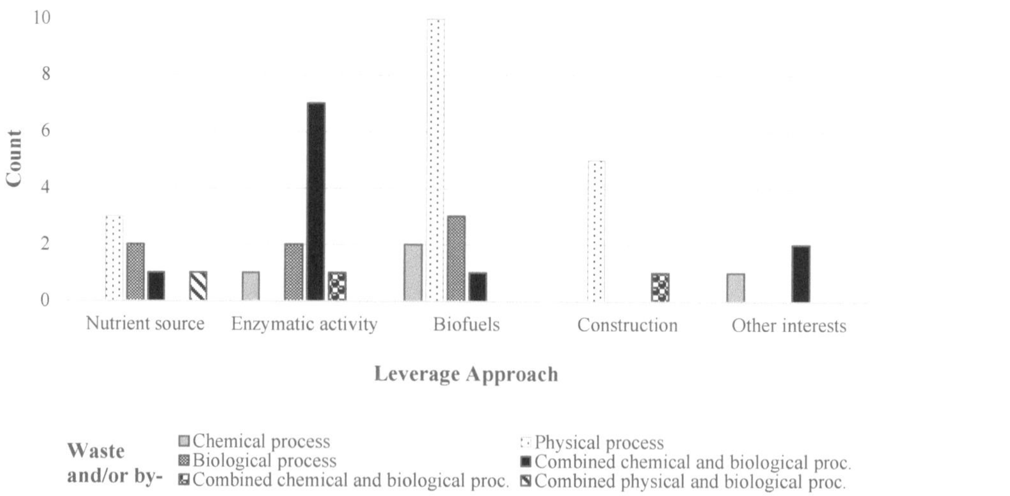

Figure 2. Concentration of exploitation approaches according to the type of waste and/or palm oil by-product.

Otherwise, Figure 3 shows the concentration of harvesting approaches for the valuation of palm oil residues and/or by-products. This concentration confirmed the non-correlation between the exploitation and technological approaches, since there is a diversity of transformation methodologies according to the exploitation trend. It is significant to highlight that the main process is the physical process (50%), followed by the biological (43%) and the chemical (39%), which were used individually or together for the transformation of the waste.

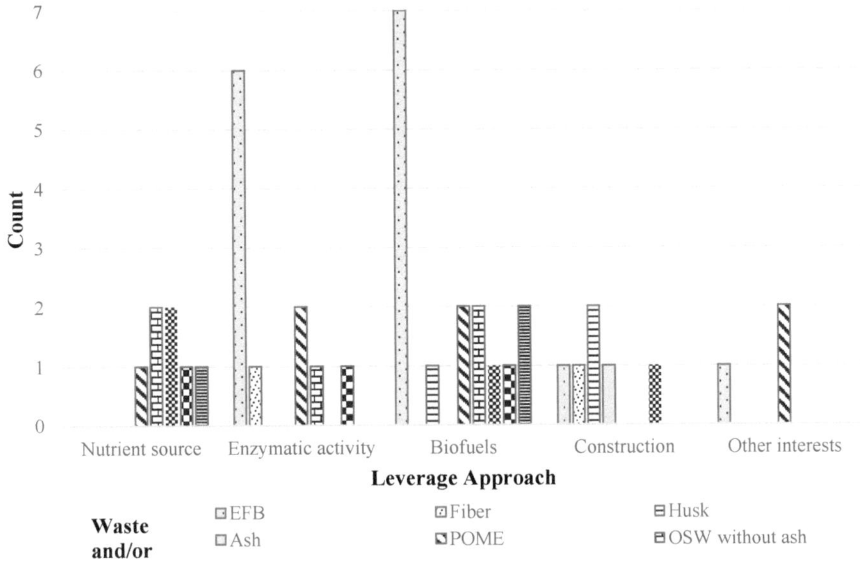

Figure 3. Concentration of technological approaches for the valuation of palm oil waste and/or by-product.

- Biofuels

The main approach for the use of the waste and/or by-product of RFF is biofuels (39% of the total review), in which organic solid waste, such as EFB, fiber, husk and liquid waste POME, is used [36,80]. The biofuels from the highlighted waste are approached from different perspectives, such as (i) complementary or substitute matter for the generation of heat in boilers; (ii) substitute or complementary material for electricity generation; and (iii) substitute material to obtain biodiesel. This is supported by physical processes, such as direct combustion, carbonization, gasification, pyrolysis, torrefaction and thermolysis, and biological processes, such as anaerobic digestion [66,88,121].

Nyakuma [112] evaluated the bioelectricity potential that can be generated by the use of EFB, fiber and husk in different proportions, obtaining a potential of 74.5–119 MWh of electricity per year depending on the efficiency of the conversion process, such as pyrolysis (65 %), gasification (50%) and direct combustion (80%). The investigation determined that, according to the calorific value of organic solid waste, the husk presents the greatest contribution, followed by fiber and EFB. Likewise, the author highlighted the complexity of transforming EFB, given the high moisture content and alkaline content, for which he recommends not implementing it as the only fuel in boilers, but combining it in different proportions with other organic solid waste. However, other authors, such as Aprianti et al. [81], suggest that EFB can be used as the only material for the combustion process, based on gasification processes (350–550 °C), since it has a composition rich in methane (22.64% vol.), carbon monoxide (29.22% vol.) and dihydrogen (3.4% vol.), which produces efficient properties in carbon conversion (95.74%) and cold gas efficiency (81.65%).

Quintero and Torres [36] identified the energy potential of palm oil residues and by-products from the Cesar Department of Colombia through the combination of different proportions of EFB, fiber and husk. In their study, they determined that the combination of fiber and husk, with 30% EFB, presented the best potential for the electrical production of KW and steam; however, it requires a greater capacity in the boiler area. Additionally, they determined that the husk had the highest calorific value among the residues and by-products, and the lowest sulfur and moisture content, which highlights its best use.

Quayson et al. [80] evaluated the production of biofuel from husk (by-product) and POME (residue) in a physical process aided by the enzymatic activity of *Aspergillus oryzae*. The husk was subjected to a carbonization process (80 °C, 24 h) to generate a structure with micropores capable of retaining impurities, while the POME was used as a source of nutrients to extract the immobilized lipase under the methanolysis method. As a result, they obtained a methyl ester with a cetane index of 48.8 and favorable cold flow properties, which generates a biodiesel suitable for use. A similar process was developed by Vasu et al. [89], for which they used a 50:50 ratio of scale and sludge from POME under pyrolysis processes.

Intasit et al. [86] carried out a similar process to obtain biodiesel from EFB and POME, aided by non-sterile solid state fermentation biological processes with *Aspergillus tubingensis* TSIP9 (a fungus present in the residual biomass of palm oil) and bioreactors. As a product, they obtained a biodiesel with a low potassium content and a high cellulose content.

Another investigation that focused on POME was carried out by Idris et al. [104], in which they used microalgae *Chlorella vulgaris* UNMACC 001 in a suspended bed to extract biodiesel. As a result, they obtained a high growth rate (0.29 day^{-1}) equivalent to 0.12 $\pm$ 0.02 g liter^{-1} day^{-1} of biomass, whose scenario is favorable for the extraction of ester with a content of 69.9% fatty acids. It is important to highlight the two previous authors, since the biological treatment of wastewater with a bioreactor is not used in the department; for this purpose, valuation goals such as these were not considered in the circular economy model.

Contrastingly, the methodology of Lam et al. [88] used EFB in conjunction with a residue from machinery operation (used oil) to generate a solid biofuel from the torrefaction method at the controlled temperature of 250 °C. As a result, they obtained a solid fuel with a high calorific power (28.0 MJ/kg) and medium carbon content (68.3%). The foregoing is pertinent to highlight, since it uses microwave technology for the torrefaction process, which is a low-cost process. Similarly, Gallego et al. [83] investigated heat variation by means of response energy gain techniques as an alternative to variable energy yield to determine the efficiency of the torrefaction process in the generation of solid biofuels. As a result, they obtained a yield of 37 t/day of biochar, a product of the transformation of 100 t/day of EFB, with an electricity potential of 0.36 MW.

- Enzymatic activity

The second approach for the use of the residues and/or by-products of FFB is the enzymatic activity (35% of the total revision), which uses organic solid residues, such as EFB, and the POME liquid residue , given their high content of carbon, nitrogen, phosphorus, potassium and cellulose. The main chemical method used for the present approach corresponds to hydrolysis, aided by biological processes of fermentation.

Iwuagwu and Ugwuanyi [115] used the POME residue as the only source of carbon and nitrogen, without a prior treatment, with concentrations of 76,000 mgL^{-1} of total solids, 15,128 mgL^{-1} of organic carbon, 114,800 mgL^{-1} of COD, 3.9 pH, among others, to obtain yeasts. For this purpose, they used fermentation at 150 rpm, 28 $\pm$ 2 °C, with bacteria typical of the wastewater treatment process, such as *Saccharomyces* sp. L31. As a result, they obtained a profitable growth of dry yeast, with a concentration of 4.42 gL^{-1}, whose amino acid content was higher than the FAO/OMS standard. It should be noted that this process generated an added value, since it presented a decomposition of 83% of the initial BOD.

A similar investigation for the generation of enzymes that used waste and/or by-products as a carbon source was developed by Silveira et al. [109], who used fiber and crushed husk, with a diameter of 1–2 mm, together with *Aspergillus niger* C BRMCTAA 82 in a submerged fermentation. As a result, they obtained a favorable yield with fiber and husk residues when compared to lipase production with other residues from other crops, given their natural content of palmitic acid and fatty acids, which represent a greater source of carbon, and consequently, a higher generation of lipase as a substitute product for industrial enzymes.

Nazir et al. [117] developed a methodology with formic acid (20% w/v) and hydrogen peroxide (10% v/v) to extract cellulose from EFB fibers, with a result of 64% (w/w) of cellulose, which registers one of the highest values of the compared benchmarks. Another investigation emphasizing cellulose extraction was carried out by Sinjaroonsak et al. [90], who extracted cellulose and xylanase from Empty Fruit Bunch (EFB) pretreated with aluminum peroxide and aided by *Streptomyces strains.*

Another relevant investigation was carried out by Zulkarnain et al. [95], who extracted biovanilla from the use of EFB, a by-product rich in lignocellulose. This process was carried out by means of alkaline hydrolysis with a result of 41% vanillic acid and 39% biovanilla.

Although the results obtained in this category are very interesting due to the different applications this method has, this approach does not present technical feasibility according to the technological conditions of the agro-industries of the Casanare Department, and, for this reason, it was not considered for the circular economy model.

- Nutrient source

The third approach for the use of the residues and/or by-products of RFF is the source of nutrients (15% of the total review), which uses organic solid residues, such as EFB, fiber and ash, and the POME liquid residue. This approach is used to obtain complementary or substitute products, given the nutritional content rich in phosphorus, nitrogen, potassium and magnesium, and low metal content, mainly aided by biological and physical processes, such as composting, bioreactors and pyrolysis. It is noteworthy to emphasize that this approach presents the union of residues and by-products of palm oil or other productive activity to generate stable products with better assimilation capacity by the substrates.

Lim and Wu [108] used POME (specifically the effluent generated in the centrifugation process) since this residue has a high content of nitrogen, phosphorus, potassium, magnesium and suspended oil, with a composition of 76% water, 12 % of residual oil and the rest in cellulose, lignin and other impurities. To generate a compost suitable for the nutritional requirements, they used a 2:1 ratio of POME and rice straw, with a duration of 4 weeks at a controlled temperature of 60 °C. As a result, they obtained a compost of 9.76 C/N, calcium (1.13 ± 0.05 g/kg), potassium (25.47 ± 0.32 g/kg), magnesium (4.87 ± 0, 19 g/kg), sodium (7.40 ± 0.03 g/kg) and phosphorus (3.62 ± 0.27 g/kg).

Hau et al. [84] used a combination of EFB, POME, ashes and residues from other production processes to generate a compost rich in carbon and nitrogen. For this purpose, they used different ratios of residues in 40-day beds with a controlled temperature, whose matter was decomposed by *Eisenia Fetida*. As a result, they obtained a superior composting performance due to their nutritional composition by combining EFB plus POME with a 1:1 ratio, containing 31.9% C, 1.08% N, 3.8% K and a ratio of 29.45 C/N. The study also generated a compost made up of EFB ash; however, it presented a low yield of carbon, nitrogen and potassium content, but a medium phosphorus content.

A study with similar results was developed by Nahrul et al. [120], in which they used different proportions of EFB with POME. To do this, they used *Eisenia Fetida* in composting beds with time periods of 40 to 45 days, with initial and final comparison of the C/N ratio. In the different proportions analyzed, they determined a reduction of 1/3 of the initial ratio, with a greater efficiency in the 50:50 ratios of EFB with POME, respectively.

Lemus [87] used the effluent from the POME treatment under facultative processes and maturation, in order to obtain a substitute product for crop fertilization. For this, the analysis of the physical and chemical components that limit or favor the productivity of t RFF/ha was carried out, such as K, Cl, N, P, S, Mg, Ca and heavy metals. Subsequently, soil irrigation was implemented for more than a year to determine the increase in foliar parameters, t RFF/ha, salinization, retention of substances in the substrate and transfer of chemical substances to the water table. As a result, a low-cost substitute product was obtained, which only involves the investment of the sprinkler irrigation system, increases leaf growth and t RFF/ha, and does not generate retention of substances in the water table. Additionally, it obtained a significant reduction in costs associated with the remuneration rate and fertilization.

- Building

The last exploitation approach corresponds to construction (14% of the total review), which uses fiber, husk and ash for its physical properties of resistance and moisture retention capacity. This approach uses the physical processes of drying at a controlled temperature to guarantee the mechanical properties of the by-product, which favors the generation of complementary products.

Fokam et al. [79] used fibers from the RFF mesocarp obtained during the nut separation process, which were previously subjected to drying (105 °C in 24 h). They used different proportions in combination with industrial cement to determine the stiffness of the mortar. As a result, they obtained a more resistant mortar, whose composition contained 3.5% residual fiber from palm oil. On the other hand, Ong et al. [102] used a comparative study between palm oil residues (husk), steel mills (steel slag) and power plants (coal bottom ash) combined with concrete to determine the stiffness (measured in Mpa). As a result, a mortar strength lower than the one corresponding to 28 days (39.44 MPa) was obtained, but higher than the one corresponding to 90 days (44.08 MPa) when using concrete with palm oil husk.

Although the implementation of palm oil waste and by-products favors the conditions for the generation of new, more resistant products, this approach was not contemplated in the circular economy model, since the installed capacity in the Department does not meet the base conditions for this implementation.

3.3. Circular Economy Model for the Reincorporation of Waste and By-Products from the Palm-Growing Sector in the Casanare Department

The circular economy model reincorporates waste and by-products from the palm oil sector under the quantitative indicators of Material Flow and Other-life-cycle-based approaches, adjusted to the approaches of Moraga et al. [9], Elia et al. [10], Rossi et al. [5] and Salguero-Puerta et al. [27]. The model was adjusted to the circular economy strategies that correspond to (i) preserving the function of products with multifunctional characteristics (strategy 1); (ii) conserving the components of the product through the reuse, recovery and reuse of parts (strategy 5); (iii) conserve embodied energy through the recovery of incineration facilities; and (iv) measure the linear economy as a reference scenario against which the model can be compared under assumptions and/or implementations of circular models (strategy 6).

This model corresponds to a specific type of measurement under the *sensu stricto aspect* (quantify aspects of quality and quantity), of material flow indicators and other-life-cycle-based approaches, which makes it a model of mixed parameters. When comparing its range of implementation with respect to Moraga et al. [9], the model has indicators of physical properties of technological cycles with an emphasis on reuse, recyclability and recovery (scope 1) and indicators that assess effects (burdens and benefits) with respect to environmental, economic and/or social concerns (scope 2). The range of implementation is adjusted at the micro level, as it focuses on a certain sector and RFF transformation process.

3.3.1. Indicator of Reincorporation of Waste and By-Products by Flow of Materials (RRSMF) of the Palm-Growing Sector of the Casanare Department

In the previous sections, the coefficients or constants for the generation of waste and by-products of palm oil in the Casanare Department were determined, which allowed the construction of the projection of the generation of waste and by-products for the period

2013–2019. The projection of the results of the year 2019 was extracted, which corresponds to the baseline for the formulation of the RRSFM indicator.

Table 9 shows the variables that make up the circular economy RRSFM indicator from the previously investigated baseline in relation to the generation of waste and transformation by-products of the RFF in the agribusinesses of the department by Casanare. The factor corresponds to the individual contribution of each agro-industry with respect to the total production. The Total Generation (Gt) variable of waste and by-products corresponds to the product of the contribution of RFF/h of each agro-industry by the waste or by-product of the year 2019. It is relevant to mention that 75% of the processing of RFF/h or CPO/h concentrates in four of the eight agro-industries or benefit plants reported for the Casanare Department.

Table 9. Total generation (Gt), current use (Ac), potential use (Ap) of palm oil residues and by-products in the Casanare Department by agribusiness or processing plant.

		Total Generation (Gt)					Current Use (Ac)				
		OSW Gt			As Gt	LW Gt	OSW Ac			As Ac	LW Ac
Cod Agro-Industry	Factor	EFB (t)	Fiber (t)	Husk (t)	Ash (t)	POME (m^3)	EFB (t)	Fiber (t)	Husk (t)	Ash (t)	POME (m^3)
Agribusiness A	0.04	300.87	208.52	80.70	7.36	930.18	75.22	104.26	40.35	5.89	0.00
Agribusiness B	0.18	1353.91	938.34	363.17	33.13	4185.80	1083.13	750.67	290.54	23.19	2511.48
Agribusiness C	0.17	1278.69	886.21	343.00	31.29	3953.26	1150.82	708.97	274.40	31.29	3162.61
Agribusiness D	0.08	601.74	417.04	161.41	14.72	1860.36	361.04	396.19	153.34	13.25	1172.02
Agribusiness E	0.1	752.17	521.30	201.76	18.40	2325.45	488.91	479.60	185.62	16.56	1627.81
Agribusiness F	0.03	225.65	156.39	60.53	5.52	697.63	135.39	109.47	42.37	2.76	0.00
Agribusiness G	0.15	1128.26	781.95	302.64	27.60	3488.17	789.78	742.85	287.51	22.08	2092.90
Agribusiness H	0.25	1880.43	1303.25	504.41	46.01	5813.61	1316.30	1238.09	479.18	36.81	3488.17
Total	1	7521.73	5213.01	2017.62	184.03	23,254.45	5400.60	4530.11	1753.31	151.82	14,054.99

		Potential Use (Ap)					RRSFM Indicator	Standardization (Scaling)
		OSW Ap			As Ap	LW Ap		
Cod Agro-Industry	Factor	EFB (t)	Fiber (t)	Husk (t)	Ash (t)	POME (m^3)		
Agribusiness A	0.04	210.61	104.26	40.35	0.88	697.63	4.62	0.924
Agribusiness B	0.18	270.78	187.67	72.63	9.94	1674.32	5	1
Agribusiness C	0.17	127.87	177.24	68.60	0.00	790.65	5	1
Agribusiness D	0.08	240.70	20.85	8.07	1.47	558.11	4.93	0.986
Agribusiness E	0.1	263.26	41.70	16.14	1.84	697.63	5	1
Agribusiness F	0.03	81.23	46.92	18.16	2.76	523.23	4.71	0.942
Agribusiness G	0.15	338.48	39.10	15.13	5.52	1395.27	5	1
Agribusiness H	0.25	564.13	65.16	25.22	9.20	2325.45	5	1
Total	1	2097.06	682.90	264.31	31.62	8662.28	-	-

The Current Use (Ac) variable of residues and by-products is not the same in all agro-industries; for this reason, the reincorporation rate varies according to each processing plant (see Figure 4). The residues and by-products with the highest average reincorporation as a second-use product are Fiber (84.12%), Scale (80%) and Ash (80%). However, the remaining value is related to the poor management used by two agro-industries, since they use only 60% ± 10% of this waste in incineration or composting processes, which makes these two plants a significant source of negative externalities.

In relation to EFB, it presents an implementation average of 65% in the plantations adjacent to the agro-industries. Agro-industry A is critical, which uses only 25% of the total generation, and the rest is available in the open. In relation to POME, only 49.12% of the total effluent is used in second use, which makes it the waste with the lowest degree of incorporation. It is important to highlight its negative impact on the dimensions of the development, since 50.88% (9199 m^3/year) are discharged into surface bodies.

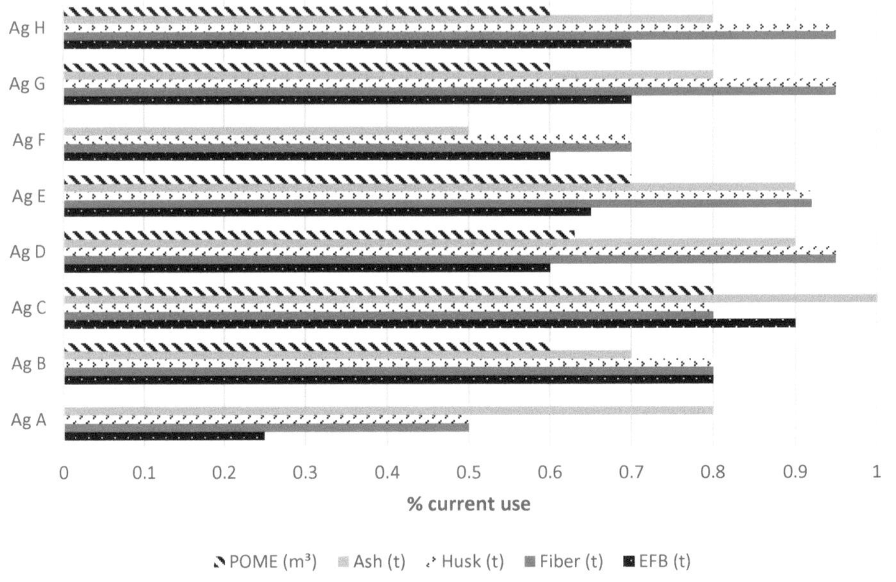

Figure 4. Percentage or degree of the current use of palm oil residues and by-products in the agro-industries in the Casanare Department.

When diagnosing the current state of use for each agro-industry, it is evident that agro-industries A and F show the least degree of progress, given their short time in industrial training. On the other hand, the remaining agro-industries show greater progress given the environmental management policies they currently employ (see Figure 5). However, all processing plants under this characterization have a high potential to improve their degree of incorporation of waste and by-products.

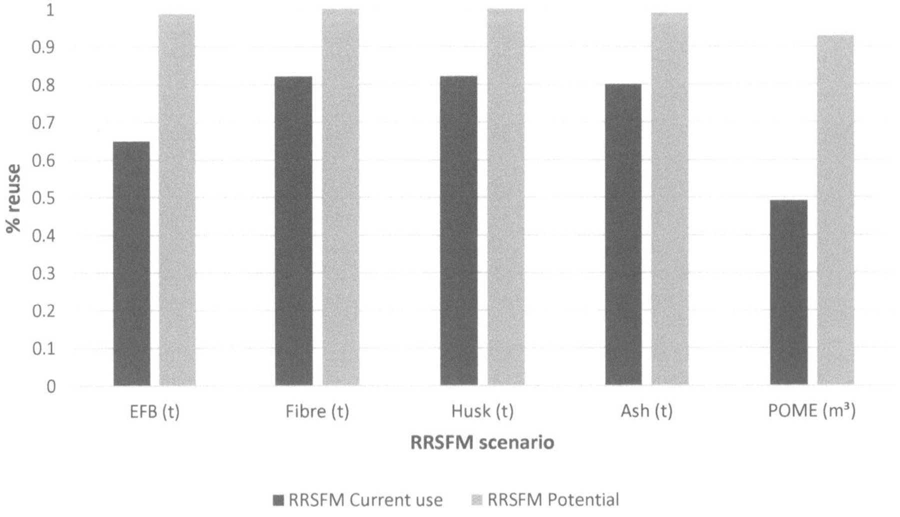

Figure 5. Comparison of the circular economy model under the current and potential scenarios of the reincorporation of palm oil waste and by-products in the Casanare Department.

The last variable that influences the quantitative model corresponds to the Potential Use (AP) under the 2026 scenario, whose value was estimated according to the installed capacity and change in territorial prospective associated with the sustainability reports and the conversations held with some environmental professionals from each processing plant. Under this variable, the waste and by-product with the highest potential degree of reuse would correspond to fiber (100%) and husk (100%), given the use in boilers that partially or totally replaces the fuel, and followed by ashes (99%) (implemented jointly with EFB and residual effluent treated for composting and soil conditioner).

It is evident that the waste with the greatest implementation difficulty under the second use framework corresponds to POME (92.87%), whose effluent presents high BOD and COD compositions that require efficient treatment systems to adjust to the maximum allowed values. However, under the methodology proposed by Lemus [87], which is coupled to the nutritional and meteorological conditions of the Casanare Department, fertigation could be carried out at a relatively low cost, and its investment can be recovered by the absence of the payment of a retributive rate and fertilization.

When processing the model under the quantitative variables, two moments of the RRSFM model were identified (see Figure 6). The first moment corresponds to the current scenario of agro-industries, which do not have a 100% use of any type of waste or by-product. This moment generates a landfill or open pit disposal of 2121 t of EFB, 682 t of fiber, 264 t of scale and 32 t of ash, and a dumping of 9199 m^3 of the residual effluent.

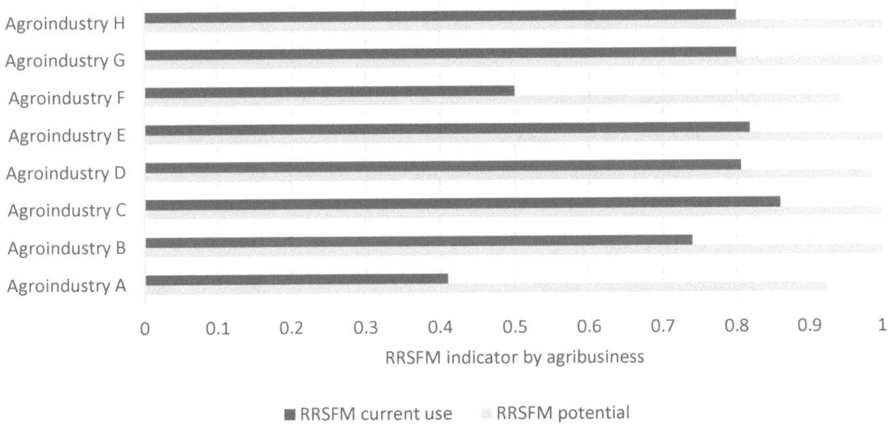

Figure 6. Current and potential RRSFM indicator of the agro-industries of the Casanare Department under the Material Flows approach.

The second moment corresponds to an optimistic scenario where the agro-industries, according to their environmental forecast for the year 2026, increase the recirculation capacity under the hypothesis of improving the physical conditions to adopt a greater degree of second use of waste and by-products. From this scenario, the reuse of EFB and ash is 99%, of fiber and husk is 100% and of POME is 92%, which is equivalent to a waste of 1.47% of the total generation of waste and by-products, mainly contributed by the low management in the incorporation of POME. However, under a more optimistic scenario, the incorporation could be increased to 100%, since the volume of unincorporated POME would correspond to 537 m^3/year, which could satisfy the water and fertilizer demand of 0.53 ha with a $\pm$2 mm layer. ha^{-1} day^{-1}.

On the other hand, it was shown that, currently, no agro-industry in the Casanare Department has an indicator above 86%. In fact, five are grouped into values in the 80–86% range, one in 51–80% and two in 0–50%. This correlates with the time of the industrial formation in the Department, in which the two agro-industries with the lowest score correspond to new processing plants established in recent years (see Figure 6). Additionally, it is pertinent to mention that the agro-industries with values greater than 80% follow the guidelines of the RSPO (Roundtable on Sustainable Palm Oil) standard of 2018.

In relation to the potential moment, an increase of 26% in the RRSFM indicator was evidenced (see Figure 6), due to the increase in the incorporation of waste and by-products. This translates into a decrease in negative externalities, improvement in environmental management and reduction in negative impacts on the dimensions of sustainable development.

3.3.2. Carbon Footprint Indicator of the Reincorporation of Waste and By-Products (HCRRS) of the Palm-Growing Sector of the Casanare Department

The model established a second indicator related to the incorporation of Material Flows under the Other-life-cycle-based approach, called Carbon Footprint of the Reincorporation of Waste and By-products (HCRRS), which corresponds to a complementary measurement suggested by Elia et al. [10] and Moraga et al. [9]. The foregoing is based on the quantification of $CO_{2\text{-eq}}$/t RFF * year 2019, emitted by the disposal of waste and by-products that emit GHGs (see Table 10).

Table 10. Palm oil waste and by-product emission factors according to the type of alternative or disposal.

Waste and By-Product	kg CO_2 eq/t RFF [a]	kgCO_2/kg EFB	kgCO_2/kg Husk + Fiber	kgCO_2/m^3 POME	Alternative or Disposal
EFB	230	1026	-	-	Open sky
EFB	1083.86	4838	-	-	Incinerated (not energy in boilers)
POME	140	-	-	0.2014	Open sky
EFB + POME	0.0231	2.51×10^{-5}	-	-	Biochar
EFB + POME	110	0.119	-	-	Composting
Fiber + Husk	941.94	-	4358	-	Incineration (energy in boiler)

[a] Adapted from Moreno et al. [122], Krishnan et al. [123], Andarini et al. [124], Subramaniam et al. [125], Rivera-Méndez et al. [126] and Stichnothe and Schuchard [127].

Moment one (M1) of the RRSFM indicator (see Figure 7) represents the negative scenario of the reuse of process outputs under the current trends corresponding to: (i) EFB disposed entirely in the open sky; (ii) fiber and husk incinerated in its entirety; and (iii) POME placed in open-air sedimentation ponds. The aforementioned scenario would emit 43,925 t CO_{2eq}, where the greatest contribution corresponds to the incineration of the fiber and husk.

Based on M1, a second moment (M2) is extracted with a change in the fuel source, in which fiber and husk incineration is replaced by a non-renewable source, such as carbon. Under this assumption, the equivalent emission would correspond to 104,857 t CO_{2eq}.

The third moment (M3) corresponds to the current scenario of the RRSFM indicator, which corresponds to: (i) the utilization of 65% of the EFB in composting and the rest in the open air; (ii) the implementation of fiber and husk of 82.14% in boilers and the rest in the open air; and (iii) the reuse of 60.4% of the effluent for composting and/or a source of nutrients, with the rest being discharged. The foregoing emits 33,860 t CO_{2eq}, where 77.3% is a product of the incineration of fiber and scale for boilers. However, this value is likely to be reduced with circular economy practices.

On the other hand, the fourth moment (M4) corresponds to the potential RRSFM scenario, whose assumption establishes (i) the total use of EFB as compost, combined in a 2:1 ratio of POME and EFB; (ii) the incorporation of 82% of fiber and husk as fuel and the rest as biochar; and (iii) the utilization of 64.47% of POME in composting and the rest used as fertigation. According to the conditions mentioned, the emissions would correspond to 30,573 t CO_{2eq}.

Finally, the fifth moment (M5) represents the scenario of non-incorporation of waste and by-products, which are incinerated in their entirety. This assumption would generate 77,107 t CO_{2eq}.

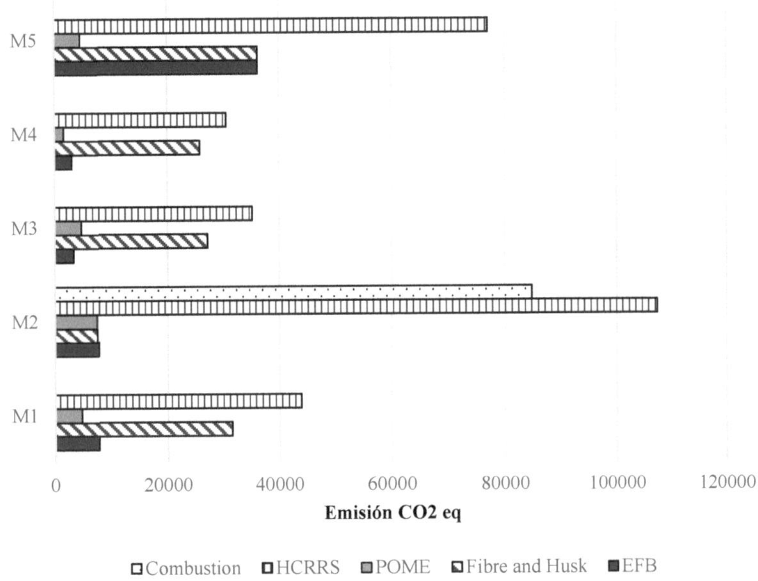

Figure 7. Estimation of the HCRRS indicator: CO_{2eq} emissions of the moments diagnosed by the RRSFM indicator.

When comparing the potential moment of the RRFSM (M4) with the other scenarios, it is evident that there is a reduction of 30.39% with respect to M1; 71.52% in relation to M2; 13.11% compared to M3; and 60.34% in relation to M5. It is prudent to highlight that the reduction would be greater if the technological capacity and business projection were inclined towards the incorporation of biodigesters and biochar.

Finally, based on Law 1819 of 2016, which established the National Carbon Tax (INAC), the cost avoided by implementing circular practices (M4) in relation to the current RRFSM would correspond to USD 18,940, which is equivalent to the other costs avoided under the analyzed circular economy model.

4. Conclusions

The circular economy model, based on the recovery of waste and by-products of the palm oil transformation process, allows visualizing comparative scenarios according to the current and potential uses (adjustable according to the projection) under two quantitative indicators. Although the indicators measure different variables (mass flow and avoided carbon dioxide emission), these two facilitate the symbiotic interpretation of environmental management practices based on sustainable development. Likewise, they illustrate the feasibility of generating different circular economy indicators that measure waste incorporation, elimination and/or prevention processes.

Author Contributions: Conceptualization, R.I.M.-A. and J.-P.R.-M.; methodology, P.-A.C.B.; formal analysis, P.-A.C.B., Y.I.M.-A. and R.S.; research, P.-A.C.B. and J.-P.R.-M.; resources, R.I.M.-A., Y.I.M.-A. and R.S. All authors have read and agreed to the published version of the manuscript.

Funding: This research received no external funding.

Institutional Review Board Statement: Not applicable.

Informed Consent Statement: Not applicable.

Data Availability Statement: Not applicable.

Conflicts of Interest: The authors declare no conflict of interest.

References

1. Barros, M.V.; Salvador, R.; de Francisco, A.C.; Piekarski, C.M. Mapping of research lines on circular economy practices in agriculture: From waste to energy. *Renew. Sustain. Energy Rev.* **2020**, *131*, 109958. [CrossRef]
2. Kirchherr, J.; Reike, D.; Hekkert, M. Conceptualizing the circular economy: An analysis of 114 definitions. *Resour. Conserv. Recycl.* **2017**, *127*, 221–232. [CrossRef]
3. Suárez-Eiroa, B.; Fernández, E.; Méndez-Martínez, G.; Soto-Oñate, D. Operational principles of circular economy for sustainable development: Linking theory and practice. *J. Clean. Prod.* **2019**, *214*, 952–961. [CrossRef]
4. Schroeder, P.; Anggraeni, K.; Weber, U. The Relevance of Circular Economy Practices to the Sustainable Development Goals. *J. Ind. Ecol.* **2018**, *23*, 77–95. [CrossRef]
5. Rossi, E.; Bertassini, A.C.; dos Santos Ferreira, C.; do Amaral, W.A.N.; Ometto, A.R. Circular Economy indicators for organizations considering Sustainability and Business Models: Plastic, textile and electro electronic cases. *J. Clean. Prod.* **2019**, *247*, 119–137. [CrossRef]
6. Velenturf, A.P.M.; Purnell, P. Principles for a sustainable circular economy. *Sustain. Prod. Consum.* **2021**, *27*, 1437–1457. [CrossRef]
7. Korhonen, J.; Honkasalo, A.; Seppälä, J. Circular Economy: The Concept and its Limitations. *Ecol. Econ.* **2018**, *143*, 37–46. [CrossRef]
8. Martínez, A.; Porcelli, M. Estudio sobre la economía circular como una alternativa sustentable frente al ocaso de la economía tradicional (primera parte). *Lex Rev. la Fac. Derecho y Cienc. Política la Univ. Alas Peru.* **2018**, *16*, 301–334. [CrossRef]
9. Moraga, G.; Huysveld, S.; Mathieux, F.; Blengini, G.; Alerts, L.; Acker, K.; Meester, S.D.; Dewulf, J. Circular economy indicators: What do they measure? *Resour. Conserv. Recycl.* **2019**, *146*, 452–461. [CrossRef]
10. Elia, V.; Grazia, M.; Tornese, F. Measuring circular economy strategies through index methods: A critical analysis. *J. Clean. Prod.* **2016**, *142*, 2741–2751. [CrossRef]
11. Prieto-Sandoval, V.; Ormazabal, M.; Jaca, C.; Viles, E. Key elements in assessing circular economy implementation in small and medium-sized enterprises. *Bus. Strateg. Environ.* **2017**, *28*, 1525–1534. [CrossRef]
12. de Oliveira, C.T.; Dantas, T.E.T.; Soares, S.R. Nano and micro level circular economy indicators: Assisting decision-makers in circularity assessments. *Sustain. Prod. Consum.* **2021**, *26*, 455–468. [CrossRef]
13. Yusuf, M. Agro-industrial Waste Materials and Their Recycled Value-Added Applications: Review. In *Handbook of Ecomaterial*; Torres, L., Vasilievna, O., Eds.; Springer: Berlin/Heidelberg, Germany, 2017; pp. 2700–2710.
14. Poh, P.E.; Wu, T.Y.; Lam, W.H.; Poon, W.C.; Lim, C.S. Palm Oil Milling Waste. In *Waste Management in the Palm Oil Industry*; Springer International Publishing: Cham, Swizterland, 2020; pp. 21–44; ISBN 978-3-030-39550-6.
15. Sadh, P.K.; Duhan, S.; Duhan, J.S. Agro-industrial wastes and their utilization using solid state fermentation: A review. *Bioresour. Bioprocess.* **2018**, *5*, 1. [CrossRef]
16. Ramos, C.; Zilber, M.A.; Silva, L.F.; Lhama, P.; Davila, F. Economía circular, innovación y sustentabilidad. *Brazilian J. Dev.* **2019**, *5*, 21261–21276. [CrossRef]
17. Rubinsin, N.; Daud, W.; Kamarudin, S.; Masdar, M.; Rosli, M.; Samsatli, S.; Tapia, J.; Ghani, W.; Lim, K. Optimization of oil palm empty fruit bunches value chain in peninsular malaysia. *Food Bioprod. Process.* **2020**, *119*, 179–194. [CrossRef]
18. Umar, M.; Urmee, T.; Jennings, P. A policy framework and industry roadmap model for sustainable oil palm biomass electricity generation in Malaysia. *Renew. Energy* **2018**, *128*, 275–284. [CrossRef]
19. Elias, M.A.; Hadibarata, T.; Sathishkumar, P. Modified oil palm industry solid waste as a potential adsorbent for lead removal. *Environ. Chem. Ecotoxicol.* **2021**, *3*, 1–7. [CrossRef]
20. Van Fan, Y.; Lee, C.T.; Lim, J.S.; Klemeš, J.J.; Kim Le, P.T. Cross-disciplinary approaches towards smart, resilient and sustainable circular economy. *J. Clean. Prod.* **2019**, *232*, 1482–1491. [CrossRef]
21. Byerlee, D.; Falcon, W.P.; Naylor, R. *The Tropical Oil Crop Revolution: Food, Feed, Fuel, and Forests*; Press, O.U., Ed.; New York Oxford University Press: New York, NY, USA, 2017.
22. USDA Oilseeds: World Markets and Trade, United States Departament of Agriculture. 2021. Available online: https://apps.fas.usda.gov/psdonline/circulars/oilseeds.pdf (accessed on 29 February 2020).
23. Ortiz, A. *La Palma de Aceite Como Ejemplo Para la Agroindustria en Colombia*; Fundación Universitaria de América: Bogotá, Colombia, 2019.
24. Rakundo, R.; Bergeron, S.; Bacoum, I.; Pelletier, N.; Doyon, M. A Methodological Approach to Designing Circular Economy Indicators for Agriculture: An Application to the Egg Sector. *Sustainability* **2021**, *13*, 8656. [CrossRef]
25. Oliveira, M.; Cocozza, A.; Zucaro, A.; Santagata, R.; Ulgiati, S. Circular economy in the agro-industry: Integrated environmental assessment of dairy products. *Renew. Sustain. Energy Rev.* **2021**, *148*, 111314. [CrossRef]
26. Barcelos, S.M.B.D.; Salvador, R.; Barros, M.V.; de Francisco, A.C.; Guedes, G. Circularity of Brazilian silk: Promoting a circular bioeconomy in the production of silk cocoons. *J. Environ. Manag.* **2021**, *296*, 113373. [CrossRef] [PubMed]
27. Salguero-Puerta, L.; Leyva-Díaz, J.C.; Cortés-García, F.; Molina-Moreno, V. Sustainability Indicators Concerning Waste Management for Implementation of the Circular Economy Model on the University of Lome (Togo) Campus. *Int. J. Environ. Res. Public Health* **2019**, *16*, 2234. [CrossRef] [PubMed]
28. Wang, N.; Guo, Z.; Meng, F.; Wang, H.; Yin, J.; Liu, Y. The circular economy and carbon footprint: A systematic accounting for typical coal-fuelled power industrial parks. *J. Clean. Prod.* **2019**, *229*, 1262–1273. [CrossRef]

29. Callejo, J.; Parra, T.; Manrique, T. *Evaluación de la Producción de los Subproductos Agroindustriales en Andalucía*; Descripción Los Subproductos: Andalucia, Spain, 2015.
30. Ferronato, N.; Rada, E.; Gorritty, M.; Cioca, L.; Ragazzi, M.; Torretta, V. Introduction of the circular economy within developing regions: A comparative analysis of advantages and opportunities for waste valorization. *J. Environ. Manag.* **2019**, *230*, 366–378. [CrossRef]
31. Maina, S.; Kachrimanidou, V.; Koutinas, A. A roadmap towards a circular and sustainable bioeconomy through waste valorization. *Curr. Opin. Green Sustain. Chem.* **2017**, *8*, 18–23. [CrossRef]
32. Cardenas, P.; Rodriguez, J.; Castillo, A. Alternatives utilization of waste: Technical and environmental considerations in the agribusiness sector. *Int. J. Mech. Prod. Eng. Res. Dev.* **2020**, *10*, 15659–15676.
33. Abdurahman, N.; Rosli, Y.; Azhari, N. Development of a membrane anaerobic system (MAS) for palm oil mill effluent (POME) treatment. *Desalination* **2011**, *226*, 208–212. [CrossRef]
34. Abdullah, N.; Sulaiman, F. The Oil Palm Wastes in Malaysia. In *Biomass Now-Sustainable Growth and Use*; Darko, M., Ed.; IntechOpen: London, UK, 2013; pp. 75–99.
35. Guimaraes, O. *Influencia del Estado de la Madurez y Condiciones de Esterilización de Los Frutos, en el Rendimiento y Calidad de Aceite Crudo de Palma Africana (Elaeis guineensis) de la Planta Extractora Oleaginosas Amazónicas S.A*; Universidad Nacional de la Amazonia: Iquitos, Peru, 2019.
36. Quintero, L.A.; Torres, C.I. Análisis de residuos sólidos de palma africana, como alternativa de aprovechamiento de energías renovables en el departamento del Cesar. *Ing. USBMed* **2019**, *10*, 8–18.
37. Foo, K.Y.; Hameed, B.H. Value-added utilization of oil palm ash: A superior recycling of the industrial agricultural waste. *J. Hazard. Mater.* **2009**, *172*, 523–531. [CrossRef]
38. Yeo, J.; How, B.; Teng, S.; Leong, W.; Ng, W.; Lim, C.; Ngan, S.; Sunarso, J.; Lam, H. Synthesis of Sustainable Circular Economy in Palm Oil Industry Using Graph-Theoretic Method. *Sustainability* **2020**, *12*, 8081. [CrossRef]
39. Loh, S.K. The potential of the Malaysian oil palm biomass as a renewable energy source. *Energy Convers. Manag.* **2017**, *141*, 285–298. [CrossRef]
40. Tay, J.H. Ash from oil-palm waste as concrete material. *J. Mater. Civ. Eng.* **1990**, *2*, 94–105. [CrossRef]
41. Singh, R.P.; Ibrahim, M.H.; Esa, N.; Iliyana, M.S. Composting of waste from palm oil mill: A sustainable waste management practice. *Rev. Environ. Sci. Bio/Technol.* **2010**, *9*, 331–334. [CrossRef]
42. Lau, H.L.N.; Choo, Y.M.; Ma, A.N.; Chuah, C.H. Quality of residual oil from palm-pressed mesocarp fiber (*Elaeis guineensis*) using supercritical CO2 with and without ethanol. *J. Am. Oil Chem. Soc.* **2006**, *83*, 893–898. [CrossRef]
43. Vijaya, S.; Ma, A.; Choo, Y.; Nik Meriam, N. Life cycle inventory of the production of crude palm oil-A gate to gate case study of 12 palm oil mills. *J. Palm Res.* **2008**, *20*, 484–494.
44. Yusoff, S. Renewable energy from palm oil e innovation on effective utilization of waste. *J. Clean. Prod.* **2006**, *14*, 87–93. [CrossRef]
45. Zinatizadeh, A.A. *Biological Treatment of Palm Oil Mill Effluent (POME) Using an Up-Flow Anaerobic Sludge Fixed Film (UASFF) Bioreactor*; University Sains Malaysia (School of Chemical Engineering): Nibong Tebal, Malaysia, 2006.
46. Kong, S.H.; Loh, S.K.; Bachmann, R.T.; Rahim, S.A.; Salimon, J. Biochar from oil palm biomass: A review of its potential and challenges. *Renew. Sustain. Energy Rev.* **2014**, *39*, 729–739. [CrossRef]
47. Ali, A.; Othman, M.; Shirai, Y.; Hassan, M. Sustainable and integrated palm oil biorefinery concept with value-addition of biomass and zero emission system. *J. Clean. Prod.* **2015**, *91*, 96–99. [CrossRef]
48. Huailuek, N.; Silalertruksa, T.; Gheewala, S. Life cycle assessment and cost-benefit analysis of palm biorefinery in Thailand for different empty fruit bunch (EFB) management scenarios. *J. Sustain. Energy Environ.* **2019**, *10*, 65–73.
49. Stichnothe, H.; Schuchardt, F. Comparison of different treatment options for palm oil production waste on a life cycle basis. *Int. J. Life Cycle Assess.* **2010**, *15*, 907–915. [CrossRef]
50. Embrandiri, A.; Ibrahim, M.; Singh, R. Palm Oil Mill Wastes Utilization; Sustainability in the Malaysian Context. *Int. J. Sci. Res. Publ.* **2013**, *24*, 389–393.
51. Abas, R.; Kamaruddin, M.; Nordin, A.; Simeh, M. A study on the Malaysian Oil Palm biomass sector-Supply and perception of palm oil millers. *Oil Palm Ind. Econ. J.* **2011**, *11*, 28–41.
52. Cenipalma. *Statistical Data Collected for Benchmarking in Oilmills by Cenipalma*; Cenipalma: Bogotá, Colombia, 2005.
53. Bernal, F.N. *El cultivo de la Palma de Aceite y su Beneficio: Guía General Para el Nuevo Palmicultor*; Fedepalma: Bogotá, Colombia, 2001; ISBN 958-96494-3-2.
54. Ponce, F.; Silva, E.; Yáñez, E.; Castillo, E. Potencial de congeneración de energía eléctrica en la agroindustria colombiana de palma de aceite: Tres estudios de caso. *Rev. Palmas* **2008**, *29*, 59–72.
55. García, J.; Cárdenas, M.; Yánez, E. Generación y uso de biomasa en plantas de beneficio de palma de aceite en Colombia. *Rev. Palmas* **2010**, *31*, 41–48.
56. Ramírez, N.; Arévalo, A.; García, J. Inventario de la biomasa disponible en plantas de beneficio para su aprovechamiento y caracterización fisicoquímica de la tusa en Colombia. *Rev. Palmas* **2015**, *36*, 41–54.
57. Subramaniam, V.; Chow, M.C.; Ma, A. Energy database of the Oil Palm. *Palm. Oil Bull. Kuala Lumpur Malays.* **2004**, *70*, 15–22.
58. Loh, S.K. Biocombustibles de segunda generación de la biomasa de palma de aceite. *Rev. Palmas* **2016**, *37*, 137–148.
59. Shafigh, P.; Jumaat, M.Z.; Mahmud, H. Mix design and mechanical properties of oil palm shell lightweight aggregate concrete: A review. *Int. J. Phys. Sci.* **2010**, *5*, 2127.

60. Cortés, C.; Cayón, D.; Aguirre, V.; Chávez, B. Respuestas de palmas de vivero a la aplicación de residuos de la planta extractora. *Rev. Palmas* **2006**, *27*, 23–32.
61. Villegas, J.M.; Avila, H. Quick-scan estimating model of Higher Heating Value of oil palm empty fruit bunches based on ash from proximate analysis data. *Rev. Ing. E Investig.* **2014**, *34*, 33–38. [CrossRef]
62. Abdullah, N.; Sulaiman, F.; Gerhauser, H. Characterisation of Oil Palm Empty Fruit Bunches for Fuel Application. *J. Phys. Sci.* **2011**, *22*, 1–24.
63. Ramírez, N.; Silva, Á.; Garzón, E.; Yáñez, E. *Caracterización y Manejo de Subproductos Del Beneficio Del Fruto de Palma de Aceite*; Fedepalma: Bogotá, Columbia, 2011.
64. Nalaya, P.; Wahid, S.; Izuan, H.E.M. Characterization of Empty Fruit Bunch Biochar Pyrolyzed at Different Temperatures with Respect to Activated Carbon and their Sorption Capacities for Pentachlorophenol. *J. Water Environ. Technol.* **2020**, *18*, 314–326. [CrossRef]
65. Baharuddin, A.; Wakisaka, M.; Shirai, Y.; Abd-Aziz, S.; Abdul Rahman, N.; Hassan, M. Co-Composting of Empty Fruit Bunches and Partially Treated Palm Oil Mill Effluents in Pilot Scale. *Int. J. Agric. Res.* **2009**, *4*, 69–78. [CrossRef]
66. Sukiran, M.A.; Abnisa, F.; Daud, W.M.A.W.; Abu Bakar, N.; Loh, S.K. A review of torrefaction of oil palm solid wastes for biofuel production. *Energy Convers. Manag.* **2017**, *149*, 101–120. [CrossRef]
67. Loh, S.K. *Oil Palm Biomass Energy Resource Data*; Malaysian Palm Oil Board: Bandar Baru Bangi, Malaysia, 2012.
68. Loh, S.K.; Choo, Y.M. Prospect, challenges and opportunities on biofuels in Malaysia. In *Advances in Biofuels*; Pokagu, R., Sarbatly, H., Eds.; Springer: Boston, MA, USA, 2013; pp. 3–14.
69. Sundram, S. The effect of Trichoderma in surface mulches supplemented with conidial drenches in the disease development of Ganoderma Basal Stem Rot in oil palm. *J. Oil Palm. Res.* **2013**, *25*, 314–339.
70. Yin, C.; Kadir, S.A.S.A.; Lim, Y.P.; Syed-Ariffin, S.N.; Zamzuri, Z. An investigation into physicochemical characteristics of ash produced from combustion of oil palm biomass waste in a boiler. *Fuel Process. Technol.* **2008**, *89*, 693–696. [CrossRef]
71. Poh, P.E.; Yong, W.; Chong, M.F. Palm oil mill effluent (POME) characteristics in high crop season and the applicability of high-rate anaerobic bioreactors for the treatment of POME. *Ind. Eng. Chem. Res.* **2010**, *49*, 11732–11740. [CrossRef]
72. Ma, A.N.; Ong, A.S.H. Pollution Control in Palm Oil Mills in Malaysia. *J. Am. Oil Chem. Soc.* **1985**, *62*, 261–266. [CrossRef]
73. Ahmad, A.L.; Ismail, S.; Bhatia, S. Water Recycling from Palm Oil Mill Effluent (POME) Using Membrane Technology. *Desalination* **2003**, *157*, 87–95. [CrossRef]
74. Oswal, N.; Sarma, P.M.; Zinjarde, S.S.; Pant, A. Palm oil mill effluent treatment by a tropical marine yeast. *Bioresour. Technol.* **2002**, *85*, 35–37. [CrossRef]
75. Choorit, W.; Wisarnwan, P. Effect of Temperature on the Anaerobic Digestion of Palm Oil Mill Effluent. *Electron. J. Biotechnol.* **2007**, *10*, 376–386. [CrossRef]
76. García, J.M. Estado actual del manejo de efluentes en Colombia. *Rev. Palmas* **1993**, *14*, 141–148.
77. Kabongo, J.D. Waste Valorization. In *Encyclopedia of Corporate Social Responsibility*; Idowu, S.O., Capaldi, N., Zu, L., Gupta, A.D., Eds.; Springer: Berlin/Heidelberg, Germany, 2013.
78. Nzihou, A.; Lifset, R. Waste Valorization, Loop-Closing, and Industrial Ecology. *J. Ind. Ecol.* **2010**, *14*, 196–199. [CrossRef]
79. Fokam, C.; Toumi, E.; Kenmeugne, B.; Wiryikfu, N.; Mevaa, L. Experimental study of the addition of oil palm mesocarp fiber on the physical and mechanical properties of fiber cement mortar composites. *SN Appl. Sci.* **2021**, *3*, 85. [CrossRef]
80. Quayson, E.; Amoah, J.; Rachmadona, N.; Morita, K.; Darkwah, L.; Hama, S.; Yoshida, A.; Kondo, A.; Ogino, C. Valorization of palm biomass waste into carbon matrices for the immobilization of recombinant Fusarium heterosporum lipase towards palm biodiesel synthesis. *Biomas Bioenergy* **2020**, *142*, 9. [CrossRef]
81. Aprianti, N.; Faizal, M.; Said, M.; Nasir, S. Valorization of Palm Empty Fruit Bunch Waste for Syngas Production Through Gasification. *J. Ecol. Eng.* **2020**, *21*, 17–26. [CrossRef]
82. Foong, S.Y.; Liew, R.K.; Yang, Y.; Cheng, Y.W.; Yuh, P.N.; Wan, W.A.; Yi, X.; Sean, C.; Vo, D.N.; Le, Q.V.; et al. Valorization of biomass waste to engineered activated biochar by microwave pyrolysis: Progress, challenges, and future directions. *Chem. Eng. J.* **2020**, *398*, 124401. [CrossRef]
83. Gallego, L.J.; Cardona, S.; Martínez, E.; Rios, L.A. Valorization of Palm-Oil Residues: Integrated Production of a Good Quality Bio-coal and Electricity via Torrefaction. *Waste Biomass Valorization* **2020**, *11*, 2273–2284. [CrossRef]
84. Hau, L.; Shamsuddin, R.; Mai, A.K.A.; Saenong, A.; Lazim, A.M.; Narasimha, M.; Low, A. Mixed Composting of Palm Oil Empty Fruit Bunch (EFB) and Palm Oil Mill Effluent (POME) with Various Organics: An Analysis on Final Macronutrient Content and Physical Properties. *Waste Biomass Valorization* **2020**, *11*, 5539–5548. [CrossRef]
85. Prasetyo, I.; Mukti, N.I.; Cahyono, R.; Prasetya, A.; Ariyanyo, T. Nanoporous Carbon Prepared from Palm Kernel Shell for CO2/CH4 Separation. *Waste Biomass Valorization* **2020**, *11*, 5599–5606. [CrossRef]
86. Intasit, R.; Cheirsilp, B.; Louhasakul, Y.; Boonsawang, P.; Chaiprapat, S.; Yeesang, J. Valorization of Palm Biomass Wastes for Biodiesel Feedstock and Clean Solid Biofuel through Non-sterile Repeated Solid-State Fermentation. *Bioresour. Technol.* **2019**, *298*, 122551. [CrossRef] [PubMed]
87. Lemus, E. Determinación del efecto ambiental y agronómico en el suelo y aguas de nivel freático por la aplicación de efluentes tratados, en sistemas de fertirrigación en cultivos de palma de aceite. In Proceedings of the XV Reunión Técnica Nacional de Palma de Aceite, Bucaramanga, Columbia, 25–27 September 2019; p. 22.

88. Lam, S.S.; Tsang, Y.F.; Yek, P.N.Y.; Liew, R.K.; Osman, M.; Peng, W.; Lee, W.H.; Park, Y.-K. Co-processing of oil palm waste and waste oil via microwave co-torrefaction: A waste reduction approach for producing solid fuel product with improved properties. *Process Saf. Environ. Prot.* **2019**, *128*, 30–35. [CrossRef]
89. Vasu, H.; Wong, C.; Vijiaretnam, N.; Chong, Y.; Thangalazhy-Gopakumar, S.; Gan, S.; Lee, L.; Ng, H. Insight into Co-pyrolysis of Palm Kernel Shell (PKS) with Palm Oil Sludge (POS): Effect on Bio-oil Yield and Properties. *Waste Biomass Valorization* **2019**, *11*, 5877–5889. [CrossRef]
90. Sinjaroonsak, S.; Chaiyaso, T.; Aran, H. Optimization of Cellulase and Xylanase Productions by Streptomyces thermocoprophilus TC13W Using Low Cost Pretreated Oil Palm Empty Fruit Bunch. *Waste Biomass Valorization* **2019**, *189*, 76–86. [CrossRef]
91. Chin, K.L.; Nurliyana, M.Y.; H'ng, P.S.; Lee, C.L.; Go, W.Z.; Khoo, P.S.; Raja, R.A.; Ashikin, S.N. Effects of Bacterial Bio-augmentation on the Methane Potential from Facultative Digestion of Palm Oil Mill Effluent and Empty Fruit Bunch. *Waste Biomass Valorization* **2019**, *11*, 3407–3418. [CrossRef]
92. Burimsitthigul, T.; Tirapanichayakul, C.; Chantrangsuwan, P.; Prasassarakich, P. Two-Stage Liquefaction of Oil Palm Shell in Alkali Supercritical Ethanol–Water for Enhanced Bio-Oil Production. *Waste Biomass Valorization* **2019**, *11*, 2745–2754. [CrossRef]
93. Dias, Y.; Souza, E.; da Costa, H.S.; Azevedo, L.C.; Penido, E.; do Amarante, C.; Marques, O.M.; Rodrigues, A. Biochar produced from Amazonian agro-industrial wastes: Properties and adsorbent potential of Cd^{2+} and Cu^{2+}. *Biochar* **2019**, *1*, 389–400. [CrossRef]
94. Shanmugarajah, B.; Chew, I.M.; Mubarak, N.M.; Choong, T.S.; Yoo, C.; Tan, K. Valorization of palm oil agro-waste into cellulose biosorbents for highly effective textile effluent remediation. *J. Clean. Prod.* **2018**, *210*, 697–709. [CrossRef]
95. Zulkarnain, A.; Bahrin, E.; Ramli, N.; Phang, L.; Abd-Aziz, S. Alkaline Hydrolysate of Oil Palm Empty Fruit Bunch as Potential Substrate for Biovanillin Production via Two-Step Bioconversion. *Waste Biomass Valorization* **2018**, *9*, 13–23. [CrossRef]
96. Liew, R.K.; Nam, W.L.; Chong, M.Y.; Phang, X.Y.; Su, M.H.; Yek, P.N.Y.; Ma, N.L.; Cheng, C.K.; Chong, C.T.; Lam, S.S. Oil palm waste: An abundant and promising feedstock for microwave pyrolysis conversion into good quality biochar with potential multi-applications. *Process Saf. Environ. Prot.* **2018**, *115*, 57–69. [CrossRef]
97. Ahmad, A.F.; Abbas, Z.; Obaiys, S.J.; Zainuddin, M.F. Effect of Untreated Fiber Loading on the Thermal, Mechanical, Dielectric, and Microwave Absorption Properties of Polycaprolactone Reinforced with Oil Palm Empty Fruit Bunch Biocomposites. *Polym. Compos.* **2018**, *39*, 1778–1787. [CrossRef]
98. Bukhari, N.; Loh, S.; Nasrin, A.; Jahim, J. Enzymatic Hydrolysate of Palm Oil Mill Effluent as Potential Substrate for Bioflocculant BM-8 Production. *Waste Biomass Valorization* **2018**, *11*, 17–29. [CrossRef]
99. Cheah, W.; Show, P.; Juan, J.; Chang, J.; Ling, T. Waste to energy: The effects of Pseudomonas sp. on Chlorella sorokiniana biomass and lipid productions in palm oil mill effluent. *Clean Technol. Environ. Policy* **2018**, *20*, 2037–2045. [CrossRef]
100. Misnon, I.I.; Mohd, N.K.; Jose, R. Conversion of Oil Palm Kernel Shell Biomass to Activated Carbon for Supercapacitor Electrode Application. *Waste Biomass Valorization* **2018**, *10*, 1731–1740. [CrossRef]
101. Yoo, H.M.; Seo, Y.C.; Park, S.W.; Kang, J.J.; Choi, H.S.; Oh, C.H. Removal Effect of Ash and Metallic Species by Washing from Empty Fruit Bunch Byproducts in Palm Mills on Pyrolytic Characteristics to Produce Bio-Crude Oil. *Waste Biomass Valorization* **2017**, *9*, 491–502. [CrossRef]
102. Ong, S.; Mo, K.; Alengaram, U.; Jumaat, M.; Ling, T. Valorization of Wastes from Power Plant, Steel-Making and Palm Oil Industries as Partial Sand Substitute in Concrete. *Waste Biomass Valorization* **2017**, *9*, 1645–1654. [CrossRef]
103. Kamsani, N.; Salleh, M.M.; Basri, S.A.; Mohamad, S.E.; Aziz, S.A.; Kamaruddin, K. Effects of Surfactant on the Enzymatic Degradation of Oil Palm Empty Fruit Bunch (OPEFB). *Waste Biomass Valorization* **2017**, *9*, 845–852. [CrossRef]
104. Idris, N.A.; Loh, S.K.; Lau, H.L.N.; Mustafa, E.M.; Vello, V.; Tan, C.; Phang, S. Cultivation of microalgae in medium containing palm oil mill effluent and its conversion into biofuel. *J. Oil Palm. Res.* **2017**, *29*, 291–299. [CrossRef]
105. Thinagaran, L.; Sudesh, K. Evaluation of Sludge Palm Oil as Feedstock and Development of Efficient Method for its Utilization to Produce Polyhydroxyalkanoate. *Waste Biomass Valorization* **2017**, *10*, 709–720. [CrossRef]
106. Junpadit, P.; Suksaroj, T.; Boonsawang, P. Transformation of Palm Oil Mill Effluent to Terpolymer Polyhydroxyalkanoate and Biodiesel Using Rummeliibacillus pycnus Strain TS8. *Waste Biomass Valorization* **2017**, *8*, 1247–1256. [CrossRef]
107. Agudelo, B.S.; Rodriguez, D.F.; Rodriguez, A.T. *Elaboración de Bloques Estructurales en Mortero Adicionado con Ceniza de Cuesco de la Palma de Aceite Africana, Para la Ciudad de Villavicencio (Meta)*; Universidad Cooperativa de Colombia: Bogotá, Colombia, 2016.
108. Lim, S.L.; Wu, T.Y. Characterization of matured vermicompost derived from valorization of palm oil mill by-product. *J. Agric. Food Chem.* **2016**, *64*, 1761–1769. [CrossRef] [PubMed]
109. Silveira, E.A.; Tardioli, P.W.; Farinas, C.S. Valorization of Palm Oil Industrial Waste as Feedstock for Lipase Production. *Biochem Biotechnol* **2016**, *179*, 558–571. [CrossRef] [PubMed]
110. Tsouko, E.; Kachrimanidou, V.; dos Santos, A.; do Nascimento Vitorino Lima, M.; Machado de Castro, A.; Guimarães, D.M.; Koutinas, A. Valorization of By-Products from Palm Oil Mills for the Production of Generic Fermentation Media for Microbial Oil Synthesis. *Appl. Microbiol. Biotechnol.* **2016**, *181*, 1241–1256. [CrossRef] [PubMed]
111. Fatriasari, W.; Anita, S.; Risanto, L. Microwave Assisted Acid Pretreatment of Oil Palm Empty Fruit Bunches (EFB) to Enhance Its Fermentable Sugar Production. *Waste Biomass Valorization* **2016**, *8*, 379–391. [CrossRef]
112. Nyakuma, B.B. Bioelectricity potential of oil palm waste in Malaysia. In *Proceedings of the 3rd International Conference Research & Education in Natural Sciences (HERTSPO 2015)*; Shkodra BENA: Shkodër, Albania, 2015; p. 6.
113. Louhasakul, Y.; Cheirsilp, B.; Prasertan, P. Valorization of Palm Oil Mill Effluent into Lipid and Cell-Bound Lipase by Marine Yeast Yarrowia lipolytica and Their Application in Biodiesel Production. *Waste Biomass Valorization* **2015**, *7*, 417–426. [CrossRef]

114. Neoh, C.; Lam, C.; Yahya, A.; Ware, I.; Ibrahim, Z. Utilization of Agro Industrial Residues from Palm Oil Industry for Production of Lignocellulolytic Enzymes by Curvularia clavata. *Waste Biomass Valorization* **2015**, *6*, 385–390. [CrossRef]
115. Iwuagwu, J.; Ugwuanyi, J. Treatment and Valorization of Palm Oil Mill Effluent through Production of Food Grade Yeast Biomass. *J. Waste Manag.* **2014**, *2015*, 439071. [CrossRef]
116. Ishola, M.M.; Taherzadeh, M.J. Effect of fungal and phosphoric acid pretreatment on ethanol production from oil palm empty fruit bunches (OPEFB). *Bioresour. Technol.* **2014**, *165*, 9–12. [CrossRef]
117. Nazir, M.; Wahjoedi, B.; Yussoff, A.W.; Abdullah, M.A. Eco-friendly extraction and characterization of cellulose from Oil palm Empty Fruit Bunches. *Bioresour. Technol.* **2013**, *8*, 2161–2172. [CrossRef]
118. Stemann, J.; Erlach, B.; Ziegler, F. Hydrothermal Carbonisation of Empty Palm Oil Fruit Bunches: Laboratory Trials, Plant Simulation, Carbon Avoidance, and Economic Feasibility. *Waste Biomass Valorization* **2013**, *4*, 441–454. [CrossRef]
119. Salema, A.; Ani, F. Pyrolysis of oil palm empty fruit bunch biomass pellets using multimode microwave irradiation. *Bioresour. Technol.* **2012**, *125*, 102–107. [CrossRef] [PubMed]
120. Nahrul, Z.; Astimar, A.; Anis, M.; Ibrahim, M.; Khalil, H.P.; Ibrahim, Z. Vermicomposting of empty fruit bunch with addition of palm oil mill efuent solid. *J. Oil Palm Res.* **2012**, *24*, 1542–1549.
121. Mekhilef, S.; Saidur, R.; Safari, A.; Mustaffa, W.E.S.B. Biomass energy in Malaysia: Current state and prospects. *Renew. Sustain. Energy Rev.* **2011**, *15*, 3360–3370. [CrossRef]
122. Moreno, J.E.; Martínez, L.A.; Belalcázar, L.C.; Rojas, N.Y. Corporate Carbon Footprint of a Palm Oil Mill. *Tecciencia* **2018**, *13*, 11–15. [CrossRef]
123. Krishnan, Y.; Bong, C.P.C.; Azman, N.F.; Zakaria, Z.; Othman, N.; Abdullah, N.; Ho, C.S.; Lee, C.T.; Hansen, S.B.; Hara, H. Co-composting of palm empty fruit bunch and palm oil mill effluent: Microbial diversity and potential mitigation of greenhouse gas emission. *J. Clean. Prod.* **2017**, *146*, 94–100. [CrossRef]
124. Andarani, P.; Nugraha, W.D. Widdya Energy balances and greenhouse gas emissions of crude palm oil production system in Indonesia (Case study: Mill P, PT X, Sumatera Island). *Int. Conf. Chem. Chem. Process Eng.* **2017**, *1823*, 020064. [CrossRef]
125. Subramaniam, V.; Hashim, Z.; Loh, S.K.; Astimar, A.A. Assessing water footprint for the oil palm supply chain- a cradle to gate study. *Agric. Water Manag.* **2020**, *237*, 106184. [CrossRef]
126. Rivera-Méndez, Y.; Rodríguez, D.; Romero, H. Carbon footprint of the production of oil palm (*Elaeis guineensis*) fresh fruit bunches in Colombia. *J. Clean. Prod.* **2017**, *149*, 743–750. [CrossRef]
127. Stichnothe, H.; Schuchardt, F. Life cycle assessment of two palm oil production system. *Biomass Bioenergy* **2011**, *35*, 3976–3984. [CrossRef]

processes

MDPI

Article

Using Periphyton Assemblage and Water Quality Variables to Assess the Ecological Recovery of an Ecologically Engineered Wetland Affected by Acid Mine Drainage after a Dry Spell

Paul Johan Oberholster [1], Yolandi Schoeman [1], Johannes Christoff Truter [2] and Anna-Maria Botha [3,*]

1 Centre for Environmental Management, University of the Free State, Bloemfontein 9300, South Africa; oberholsterpj@ufs.ac.za (P.J.O.); schoeman.yolandy@gmail.com (Y.S.)
2 Department of Paraclinical Sciences, Faculty of Veterinary Sciences, University of Pretoria, Pretoria 0028, South Africa; christoff.truter@up.ac.za
3 Department of Genetics, University of Stellenbosch, Stellenbosch 7601, South Africa
* Correspondence: ambo@sun.ac.za; Tel./Fax: +27-(21)-808-5832

Citation: Oberholster, P.J.; Schoeman, Y.; Truter, J.C.; Botha, A.-M. Using Periphyton Assemblage and Water Quality Variables to Assess the Ecological Recovery of an Ecologically Engineered Wetland Affected by Acid Mine Drainage after a Dry Spell. *Processes* **2022**, *10*, 877. https://doi.org/10.3390/pr10050877

Academic Editor: Alexandros Stefanakis

Received: 25 March 2022
Accepted: 22 April 2022
Published: 29 April 2022

Abstract: The Grootspruit valley bottom wetland in South Africa, due to the impact of acid mine drainage (AMD) from an abandoned coal mine, was severely degraded before ecologically engineered interventions, as a passive treatment process, in 2014. The surface water flow of the wetland was redirected using concrete structures to enlarge the surface area of the wetland by 9.4 ha and to optimize passive treatment. Although the ecologically engineered interventions showed an improvement in water quality after the rewetting of the enlarged wetland areas, the 2016 drought had a devastating effect on the wetland's water quality. Limited natural removal of metals and sulfate concentrations by the wetland occurred during the 2016 drought, when compared with the 2015 pre-drought conditions. This period showed higher concentrations of metals, sulfate (SO_4^{2-}), and electrical conductivity (EC) associated with the acidic surface water. Of particular interest was an observation of a substantial shift in pollutant-tolerant algae species in the ecologically engineered wetland outflow between the years 2015 and 2016. During the dry spell period of 2016, the diatoms *Gyrosigma rautenbachiae* (Cholnoky), *Craticula buderi* (Brebisson), and *Klebsormidium acidophilum* (Noris) were observed at the outflow. The latter species were not observed during the wetland surveys of 2015, before the dry spell. From late 2017 onwards, after the drought, environmental conditions started improving. In 2018, periphyton indicator species and the surface water quality were comparable to the wetland's recorded status pre-2016. The study revealed not only a regime shift, but also an ecological function loss during the drought period of 2016, followed by recovery after the dry spell. A distinct reduction in SO_4^{2-}, sodium (Na), magnesium (Mg), EC, manganese (Mn), iron (Fe), silicon (Si), aluminum (Al), and pH, following the 2016 drought, highlights the utilization of water quality variables to not only assess the passive treatment responses of an ecologically engineered wetland, but also the progress relating to ecological recovery.

Keywords: passive treatment process; valley bottom wetland; acid mine drainage; drought; eutrophication; periphyton; metal removal

1. Introduction

Wetlands are currently at risk from several sources, despite enabling legislation and the importance of wetlands. We cannot argue with the estimation that over half of the global wetlands have been lost already [1]. The anticipated and exponential increase of global mining activities remains a major concern, further threatening the conservation of wetlands. Mining activities result in significant impacts on wetlands, that range from complete wetland destruction to water quality and water flow alterations [2]. Therefore, an urgent need exists to restore and conserve wetlands due to the essentially unique ecological functions and associated ecosystem services they provide [3] as eco-technological and resource-conserving passive treatment systems [4–6]

Ecological engineering of wetlands is characterized by a complex and interrelated passive treatment process that must refer to a wide range of complementary disciplines typically aimed at restoring the lost biodiversity or providing ecosystem services in the form of water quality improvement [7]. Considering the fundamentals, in passive treatment systems that are open to environmental influences [8,9], acidity and/or metals are sequentially removed through gravity and natural ecological, physical, geochemical, and microbiological reactions [10]. Although wetland plants are regarded as the most visible aspect of many passive treatment wetland systems, they are but only one aspect of a complex and interrelated passive treatment process [10]. The passive treatment processes in wetlands are characterized by ion exchange and adsorption by plants and their substrates; bacterial and abiotic metal oxidation; precipitated metals settling; carbonate dissolution and processes associated with microbial enabled acid neutralization; and filtration and sulfate reduction and metal sulfide precipitation. The performance of such a passive system is influenced by initial water quality as an inflow determinant, site-specific conditions, minimal human intervention, and the nature of operations in specific climatic conditions [5,10–18]. According to Zedler [7], the need to provide specific and unique hydrological conditions (e.g., water quality and quantity) complicates wetland restoration because of the potential effect on microtopography and biogeochemistry from the degree of wetland wetness. Although there has been an increase in quantitative information on ecologically engineered wetlands over the past decade, existing knowledge on the impact of climate change and biological indicators post-ecological engineering in the literature is scarce. During the evaluation of restoration related wetland projects, many different variables have been applied such as vegetation and faunal components, soil analysis, and hydrological characteristics [19].

The establishment of the relationship between measured variables and the ecological function through the application of an experimental approach was proposed by Mitsch and Wilson [19] and Simenstad and Cordell [20]. However, ecosystem resilience (i.e., the ability of a particular system to recover from disturbance) evaluation may be the ultimate test for evaluating wetland restoration [21]. When wetlands optimally function as a passive treatment system, they can provide long-term, effective, and efficient treatment for many AMD sources [17,22]. In South Africa, wetlands rank among the most threatened ecosystems. According to Oberholster et al. [23] and Van Deventer et al. [24], recent studies have indicated that 65% wetland types of South African are under threat.

The Grootspruit wetland under study has been identified as a national freshwater ecosystem area and falls within a critically endangered wetland type. The Grootspruit wetland receives AMD effluent from an upstream abandoned coal mine and is ecologically engineered as a passive treatment process to improve its water quality. The formation of AMD is the result of the exposure of sulfide minerals to atmospheric, biological, and hydrological elements (such as oxygen, chemoautotrophic bacteria, and water). The resulting sulfuric acid, as an oxidation-generating agent, further imparts not only a low pH but also net acidity to water containing elevated levels of dissolved metal concentrations and sulfate, high conductivity, and low alkalinity [25]. At an increase in pH ($\geq$4.0), metal hydroxides precipitate (e.g., ferric hydroxides [$FeOH_3$] also known as "yellow boy"), with the potential to further smother aquatic biota. Dissolved metal ions can further penetrate aquatic biota membranes and cause toxicity at a lower pH range [26–28]. The effect of AMD on aquatic ecosystems is threefold, namely: (a) impacted aquatic communities impacted by AMD experience lethal pH levels and metal concentrations, which lead to a decrease in the diversity and richness of aquatic biota; (b) community assemblage is limited to resilient and tolerant organisms that can survive in these extreme AMD conditions; and (c) changes in the nutrient cycles of wetlands may cause abiotic disturbances and changes.

The autecology of periphyton in relationship with certain selected water quality variables was the selected target indicator in the current study and was subsequently applied to determine the treatment response of the wetland after a severe dry spell in 2016.

One of the worst droughts on record in the selected study area was the drought of 2016 [29]. Periphyton, in relationship with water quality variables, was selected as an indicator based on a report by Wehr and Sheath [30] that found the representation of algae, as the primary producer biomass in wetland systems, was between 30 and 50%. The presence of algae biomass is also not only a sensitive indicator of the physicochemical conditions and biological integrity of wetland systems, but it can also reflect changes in wetland water quality [31–33]. The current study's objective was to use periphyton assemblage in relationship with water quality variables to assess the responses of an ecologically engineered, critically endangered wetland, as a passive treatment system, receiving AMD before and after a severe dry spell. To the best of the authors' knowledge, the current study is the first report related to the responses of an ecologically engineered, critically endangered wetland receiving AMD before and after a dry spell.

2. Materials and Methods

2.1. Background and Description of the Study Site

The heavily impacted Grootspruit valley bottom wetland is geographically located at latitude −25.906480 and longitude 29.052827 and covers an area of 135.3 ha. The wetland area is directly impacted by AMD from an upstream underground abandoned coal mine (Figure 1). Due to AMD upstream from the unchanneled wetland, the wetland has become a channeled valley bottom wetland over space and time. The wetland is located in the Mpumalanga Upper Olifants River catchment (Quaternary Catchment B20G) of South Africa, and further forms part of a larger wetland system that lies along a Zaalklapspruit River tributary and the Grootspruit River. The Grootspruit River flows into the Wilge River, which is located approximately 35 km northwest of the town of Witbank in the Mpumalanga Province of South Africa.

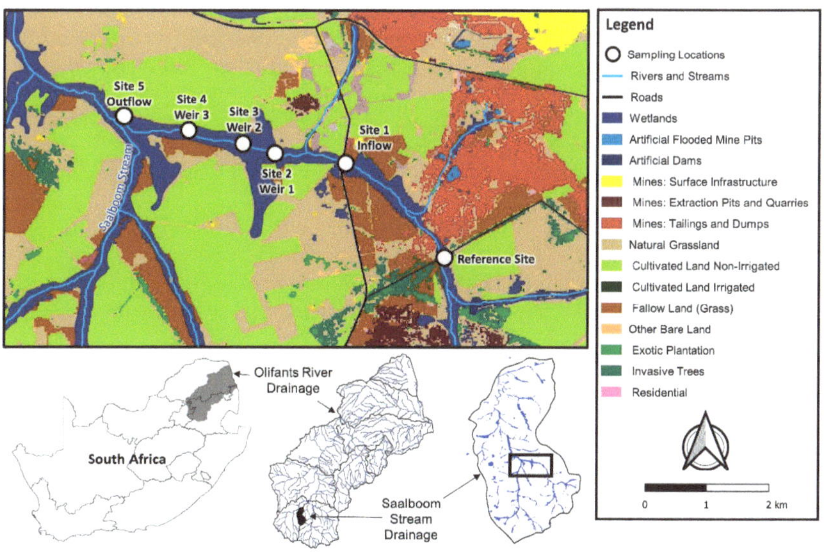

Figure 1. A map indicating the locality and extent of the Grootspruit wetland area. The map also indicates the selected sampling sites monitored over the period 2015 to 2018. The location of the study area within the Saalboom Stream quaternary catchment in the context of the larger Olifants River Basin is also indicated. The map was created with QGIS v 3.12.1 (Open Source Geospatial Foundation Project) also using the SANBI National Wetland Map Project [9] and South African National Land Cover (SANLC) 2020 datasets [34].

The Grootspruit wetland has been identified as a national freshwater ecosystem area and falls within a critically endangered wetland type. The wetland was classified as a priority wetland in a recent assessment of the conservation significance of the aquatic resources in the critical biodiversity area (CBA). The wetland falls within the Mesic Highveld Grassland Group 4 wetland vegetation group, which is further regarded as having a critically endangered threat status and has been identified as a National Freshwater Ecosystem Priority Area (NFEPA). The grass owl (*Tyto capensis*) has been observed in the wetland under study and is considered a vulnerable species for which the wetland potentially provides a habitat. The ability of the wetland to provide a water quality enhancement or ecosystem service is limited due to the channel incision caused by the decant of the abandoned coal mine upstream. Through ecological engineering, the Grootspruit wetland size was increased by 9.4 ha to allow for gravitational drainage of AMD water in 2014 [35]. The wetland enlargement process comprised the redirection of water flow using concrete structures to enlarge the wetland surface area by 9.4 ha to improve wetland water quality. The concrete structure's purpose was to change the flow of the wetland from a channeled valley bottom wetland to an unchanneled valley bottom wetland. The wetland was historically classified as an unchanneled valley bottom wetland before the upstream decanting of AMD from the abandoned coal mine. However, due to the water volume increase from the upstream decanting abandoned mine, the wetland transformed into a channeled valley bottom wetland over space and time. The transformation of the wetland caused the water quality of the wetland to deteriorate, reducing its ecosystem services. The wetland's main channels were characterized by four intervention points. The secondary incised channel was rehabilitated at three specific points (Figure 2a–f). These intervention points were strategically selected to deactivate the channelization of both the main and secondary incised wetland channels, by reducing the water velocity (<20 cm/s) through the wetland and by using man-made concrete structures to spread the flow (Figure 2). At the beginning of the ecologically engineered intervention, the water quality analysis results of Grootspruit wetland area indicated elevated sulfate, metal, and total dissolved solids (TDS) concentrations with a low average pH, all indicative of an AMD-impacted area. The wetland area was also nutrient deficient and the EC increased downstream, and hydrochemical and stable isotope contents indicated that subsurface water discharges did not support or were absent in supporting the Grootspruit wetland during the dry and wet seasons [36].

2.2. Surface Water Sampling

In situ (on-site) water quality parameters, such as pH, temperature, TDS, and EC, were measured using a handheld water quality meter (Hanna HI991300). Representative samples were taken at six specific and pre-determined sampling sites, which included the reference site 2.2 km upstream of the wetland. Samples were taken seasonally ($n = 4$; summer, winter, autumn, and spring) each year from 2015 to 2018 (Figure 1). Surface water samples were collected for chemistry analyses in triplicate in 1ℓ bottles at each wetland sampling site at a depth of ± 10 cm from the surface and kept cold on ice following the Shelton [37] sampling method. Each site's sampled water was divided further into two subsamples for specific analysis: (a) one liter was filtered through 0.45 μm pore size Whatman GF filters for dissolved nutrient analyses; (b) one liter was filtered through 1 μm Gelman glassfibre filters and preserved in nitric acid for metal analyses. Within 48 h of collection, samples were sent to an accredited chemical laboratory for analysis. The Standard Methods for the Analysis of Water and Wastewater [38] were followed for all collected water samples regarding standard chemical analysis procedures. Water quality parameters related to AMD from the literature were selected to establish the water quality of the wetland before and after the dry spell. The water samples were analyzed for the following metals and major ions: sodium (Na), calcium (Ca), magnesium (Mg), sulfate (SO_4^{2-}), silicon (Si), aluminum (Al), iron (Fe), and manganese (Mn) using an inductively coupled plasma-mass spectrometry (ICP-MS) and/or an inductively coupled plasma-optical emission spectrometer (ICP-OES). Matrix-matched standards were analyzed in parallel for quality control purposes, in addition to using the formula of Appelo and Postma [39] to determine the ionic balance. Furthermore, data were also compared with the selected value of the South African Department of Water Affairs guideline for aquatic ecosystems [40] to determine the water quality at each wetland sampling site.

Figure 2. (**a–d**). Concrete structures built in the wetland to regulate and reduce the flow. (**e**,**f**). Revegetation of the enlarged wetland area in 2015.

2.3. *Periphyton Sampling*

Periphyton was collected from submerged sand and clay (5–15 cm depth), according to the method of Hauer and Lamberti [41]. A wetland transect was placed, whereafter five samples were collected 10 m apart along the wetland transect at each sampling site. The collected samples were pooled together to form a well-mixed composite sample for each of the sampling stations and for each sampling survey over the study period of 2015–2018. The composite sample was subsequently divided into four subsamples, and comprised: (a) an unpreserved sample for benthic chlorophyll (chl-*a*) analyses; (b) an unpreserved sample for laboratory culturing and identification of doubtful filamentous algae; (c) a preserved sample for microscope soft algae identification; and (d) an unpreserved sample for the identification of diatoms. The subsample containing the soft algae was preserved in 2.5% glutaraldehyde in the field, kept cold and in a dark environment until laboratory analyses. Some sample preparation was required for the diatom samples. The organic matter of each sample site had to be cleared by means of heating in a solution of potassium dichromate and sulfuric acid. The cleared material was rinsed, diluted, and mounted in a Pleurax medium for microscopic examination.

A compound microscope (Carl Zeiss, Germany) at 1250× magnification as previously described [30,42–47] was used to identify all periphyton algae. A Sedgewick Rafter sedimentation chamber was used for sediment samples and the strip–count method [38] was used as the analysis method., Doubtful filamentous algae were cultured according to the method of Stancheva et al. [48] in the laboratory for full identification as reproductive structures are required for certain filamentous algae species identification.

Each algae's taxon was grouped according to the relative abundance, as follows: + = $\leq$ 50 (rare); ++ = 51–250 (scarce); +++ = 251–1000 (common); ++++ = 1001–5000 (abundant); and +++++ = 5001–25,000 (predominant) cells per 5 cm^2. Ice-cold absolute methanol was used to extract benthic chl-*a*. Each sample's benthic chl-*a* content was determined spectrophotometrically at 647 and 664 nm wavelengths, according to the method of Porra et al. [49].

Measuring the evenness or dominance of each algal species at each sampling site was performed by applying the Berger–Parker dominance index [50]. The dominance index comprised using actual algae cell numbers as shown in Equation (1):

$$D = N_{\max}/N \tag{1}$$

where $N_{\max}$ = the number of individuals of the most abundant species present in each sample, and N = the total number of individuals collected at each site.

The specific occurrence index by Zulkifli [51] was employed. The relative density and the application of this index established a species hierarchy in the function of their importance in the various wetland samples. The species' frequency (F) in the samples of each selected site was calculated according to Equation (2):

$$\text{SOI} = \text{Pi}/\text{P} \times 100\% \tag{2}$$

where SOI = specific occurrence index; Pi = frequency number of the species in the samples; P = total number of samples.

The periphyton species were separated into groups according to their frequency [52]:

- Absent species: not present in the samples collected;
- Rare species: present in less than 25% of the samples collected;
- Low frequency species: present in between 26 and 50% of the samples collected;
- Frequent species: present in between 51 and 90% of the samples collected;
- Permanent species: present in more than 90% of the samples collected.

2.4. Wetland Substrate Cover and Erosion

The determination of the wetland substrate cover (i.e., macrophytes) was visually according to the method of Stevenson and Bahls [53]. The degree of erosion was assessed according to the methodology of Spencer et al. [54].

2.5. Data Analysis

The Shannon and Simpson diversity indexes [55,56] were calculated based on periphyton abundance scores using Primer7 v 7.0.13 (PRIMER-e, NZ). Normality and homogeneity of variance of data were assessed using the respective Shapiro–Wilks and Levene tests. Pairwise differences between upstream and downstream water chemistry and periphyton diversity were assessed using Student's *t*-test or the Mann–Whitney U-test for non-parametric and parametric datasets, respectively. The analyses were repeated with the drought year (2016) excluded from the datasets.

Principal component analysis (PCA) was applied to assess the associations between assemblages of periphyton and water chemistry parameters across the study period and expressed as triplots [57]. The periphyton data were centered, standardized, and applied as the PCA triplot's focal plot, whereas water chemistry variables were included as supplementary variables. The Ter Braak and Smilauer method [58] was applied for the PCA triplot interpretation. A positive correlation is indicated when the angle is close to 0°, whereas an angle close to 90° indicates uncorrelated variables. Negatively correlated variables are indicated when the angle is close to 180°. Statistical analyses were performed using

Microsoft Excel, Statistica v 13 (Tibco Software, Palo Alto, CA, USA) and Canonco v 5 (Microcomputer Power, Ithaca, NY, USA).

3. Results

3.1. Dynamics of Periphyton at the Inflow of the Ecologically Engineered Wetland (2015–2018)

The periphyton biomass (chl-*a* of 47 mg m^{-2}) was measured during the dry spell in 2016 at the inflow of the wetland (Table 1). At the inflow of the wetland, periphyton mats were dominated (Berger–Parker index 0.471) by the filamentous algal species *Klebsormidium acidophilum* (Novis). The diatoms that dominated the wetland inflow sampling site during the complete study period from 2015 to 2018 included *Nitzschia clausii* (Hantzsch) (0.291); *Craticula buderi* (Brebisson) (0.278) and *Gomphonema aff. gracile* (Ehrenberg) (0.311) (Table 2).

Table 1. Survey sites (six) description and biotic and abiotic characteristics over the sampling period of 2015–2018 (n = 16).

Sites	Bottom Substrate Characteristics	In-Stream Macrophytes	Average Site Depth (cm)	Source of Impact	Bank Stability	Average Water Column Flow Velocity	Average Benthic chl-*a* mg m^{-2}	Average Thickness Layer (mm) of Bottom Hydroxide Precipitates	Turbidity (Nephelometric Unit [NTU])
Reference site	Sand, silt	*Typha capensis*	58 cm	Agriculture activities upstream	Stable	17 cm S^{-1}	13.1 mg m^{-2}	0 mm	6 NTU
Inflow (Site 1)	Sand	*Typha capensis*	22 cm	Coal mining AMD effluent	Poor	41 cm S^{-1}	37.5 mg m^{-2}	0 mm	2 NTU
Site 2	Silt, clay	*Typha capensis; Phragmites australis*	17 cm	Coal mining AMD effluent upstream	Poor	32 cm S^{-1}	16.7 mg m^{-2}	3 mm	13 NTU
Site 3		*Typha capensis; Pycreus nitidus; Kyllinga erecta*	14 cm	Coal mining AMD effluent upstream	Stable	12 cm S^{-1}	19.34 mg m^{-2}	1 mm	3 NTU
Site 4	Clay, sand	*Typha capensis; Pycreus nitidus; Fimbristylis complanata*	16 cm	Coal mining effluent upstream	Stable	11 cm S^{-1}	18.2 mg m^{-2}	1 mm	2 NTU
Outflow (Site 5)	Clay, silt	*Typha capensis; Kyllinga erecta; Fimbristylis complanata*	23 cm	Coal mining effluent upstream	Stable	23 cm S^{-1}	10.6 mg m^{-2}	0 mm	3 NTU

• Bank stability was assessed according to Spencer et al. [54]. • Instream macrophytes were identified according to Gerber et al. [59]. • Flow velocity was measured with a Magna Rod. • Turbidity was measured with a Hach 2100Q portable turbidimeter.

Table 2. Composition of the periphyton communities at the wetland area inflow and outflow over the period 2015–2018 ($n = 16$) *.

Species	Autecology of Dominant Benthic Algae	Reference Site	2015 Inflow	2015 Outflow	2016 Inflow	2016 Outflow	2017 Inflow	2017 Outflow	2018 Inflow	2018 Outflow	Frequency
Bacillariophyta											
Achnanthidium exiguum				++				+		++	33%
Amphora coffeaeformis				+				+		++	33%
Cocconeis pediculus		++								+	22%
Craticula buderi	Occurs in mine effluent characterized by moderated to elevated electrolyte content [47]		++		++++	++++	++++	+	+++		66%
Craticula cuspidate			+		+	+			+		44%
Ctenophora pulchella			+		++						22%
Cymbella kappii		+									11%
Cymbella neocistula		+		++							22%
Cymbella tumida		++									11%
Diatoma vulgaris		++		+						+	33%
Flagilaria ulna		++		++							22%
Frustulia vulgaris		+		+						++	33%

Table 2. *Cont.*

Species	Autecology of Dominant Benthic Algae	Reference Site	2015 Inflow	2015 Outflow	2016 Inflow	2016 Outflow	2017 Inflow	2017 Outflow	2018 Inflow	2018 Outflow	Frequency
Gomphonema aff. Gracile	This taxon can tolerate extremely polluted conditions and is found in abundance in mining effluent [47].		++		++++	++++	+++	+	+++		66%
Gomphonema parvulum	In general, considered to be tolerant of extremely polluted and anthropogenically modified conditions [47].		++		++++	++++	++	+	++		66%
Gomphonema pseudoaugur		++		+							22%
Gyrosigma scalproides		+									11%
Gyrosigma acuminatum	Species are found in electrolyte-rich water [47].				+++	+++	++	+	++		55%
Gyrosigma attenuatum					++		++		++		33%
Gyrosigma rautenbachiae	This taxon is associated with water anthropogenically impacted by industrial pollutants [47].		+		+++	+++		+	+		55%

Table 2. *Cont.*

Species	Autecology of Dominant Benthic Algae	Reference Site	2015 Inflow	2015 Outflow	2016 Inflow	2016 Outflow	2017 Inflow	2017 Outflow	2018 Inflow	2018 Outflow	Frequency
Melosira variance	This taxon is associated with eutrophic conditions [47].	++				+++				++++	33%
Navicula cryptotenella		++		++							22%
Navicula microcari					+			+	++		33%
Navicula pupala		++		++						+	33%
Navicula tripunctata		+++		+							22%
Nitschia clausii	This taxon is tolerant of strong pollution conditions and associated with industrial effluents [47].		++++		++++	+++	+++	++	+++	+	77%
Nitzschia communis	This taxon is associated with mining effluents [27].		+++	+	++++	++	++++	+	+		77%
Nitzschia nana				++				+		++	33%
Nitzschia sublinearis				++						++	22%
Nitzschiaintermedia		+		+							22%
Nitzschia pura			+							++	22%
Nitzschia reversa			++		++		++	+	+		55%

Table 2. *Cont.*

Species	Autecology of Dominant Benthic Algae	Reference Site	2015 Inflow	2015 Outflow	2016 Inflow	2016 Outflow	2017 Inflow	2017 Outflow	2018 Inflow	2018 Outflow	Frequency
Nitzschia umbonate	This taxon is a good indicator of eutrophic conditions [47].	++				+++++				+++	33%
Pinnularia viridiformis		++		+						+	33%
Pinnularia viridis	This taxon is known to occur in circumneutral water with electrolyte content levels that are low to moderate [47].			++				+		+++	33%
Rhopalodia gibba				++						++	22%
Synedra ulna		+++	++	++	+	+	+		+	+	88%
Tabellaria flocculosa	This taxon flourishes in circumneutral, oligotrophic, or slightly acidic water [47].			++						+++	22%
Chlorophyta											
Cladophora glomerata		++						+			22%
Closterium margaritiferum				++						+	22%

Table 2. *Cont.*

Species	Autecology of Dominant Benthic Algae	Reference Site	2015 Inflow	2015 Outflow	2016 Inflow	2016 Outflow	2017 Inflow	2017 Outflow	2018 Inflow	2018 Outflow	Frequency
Closterium peracerosum				+							11%
Closterium spinosporum				+						+	22%
Cosmarium hammeri										++	11%
Cosmarium pseudopraemorsium				++						++	22%
Microspora quadrata		+++		+			++		+		44%
Pandorina sp.	This taxon is found in meso to eutrophic types of water [46].	+++								+	22%
Scenedesmus armatus	This taxon is found in meso to eutrophic types of water [46].	++++						++		+	11%
Spirogyra Africana		++		+							11%
Spirogyra reinhardi		++								+++	11%
Staurastrum anatinum		++		++				+		+	11%
Streptophyta											
Klebsormidium acidophilum	This taxon is found in acidic water related to AMD effluent from mining activities [2].		+++	+	++++	++	+++	+	+++		77%

Table 2. *Cont.*

Species	Autecology of Dominant Benthic Algae	Reference Site	2015 Inflow	2015 Outflow	2016 Inflow	2016 Outflow	2017 Inflow	2017 Outflow	2018 Inflow	2018 Outflow	Frequency
Closterium peracerosum				+							11%
Closterium spinosporum				+						+	22%
Cosmarium hammeri										++	11%
Cosmarium pseudopraemorsium				++						++	22%
Microspora quadrata		+++		+			++		+		44%
Pandorina sp.	This taxon is found in meso to eutrophic types of water [46].	+++								+	22%
Scenedesmus armatus	This taxon is found in meso to eutrophic types of water [46].	++++						++		+	11%
Spirogyra Africana		++		+							11%
Spirogyra reinhardi		++								+++	11%
Staurastrum anatinum		++		++				+		+	11%
Streptophyta											
Klebsormidium acidophilum	This taxon is found in acidic water related to AMD effluent from mining activities [2].		+++	+	++++	++	+++	+	+++		77%

Table 2. *Cont.*

Species	Autecology of Dominant Benthic Algae	Reference Site	2015 Inflow	2015 Outflow	2016 Inflow	2016 Outflow	2017 Inflow	2017 Outflow	2018 Inflow	2018 Outflow	Frequency
Klebsormidium rivulare				+++				++		+++	33%
Mougeotia cf. *laevis*	This taxon is associated with mining effluents [60].		+		++		+	++	+	+++	66%
Zygnema cf. *cylindrospermum*				+		+					22%
Euglena sociabilis		++				+					22%
Phacus Pleuronectes		+									11%
Trachelomonas intermedia		++		++				++		+	44%
Cyanophyta											
Cylindrospermopsis raciborskii		++									11%
Merismopedia punctata		+		+							22%
Oscillatoria princeps		+++				++				++	33%
Oscillatoria tenuis		++									11%
Rhodophyta											
Batrachospermum atrum	This taxon is associated with mining effluents [61].		+++		+	++	+		++		55%

* The relative abundance of each benthic algal taxa was grouped as follows: + = ≤50 (rare); ++ = 51–250 (scarce); +++ = 251–1000 (common); ++++ = 1001–5000 (abundant); and +++++ = 5001–25,000 (predominant) cells/5 cm^2.

3.2. Dynamics of Periphyton at the Outflow of the Wetland (2015–2018)

At the outflow of the ecologically engineered wetland, between 2015 and 2016, a distinctive shift in pollution tolerant algal species was observed. During the dry spell period of 2016, the diatoms *Gyrosigma rautenbachiae* (Cholnoky), *Craticula buderi* (Brebisson), and *Klebsormidium acidophilum* (Noris) were observed at the outflow. The latter species were not observed during the wetland surveys of 2015 before the dry spell. The wetland dominant periphyton and benthic chl-*a* species in 2016 indicated a noteworthy difference from data generated during the wet season of late 2017−2018. The periphyton in the outflow of the wetland in 2015 (pre-drought conditions) and in 2018 (post-drought conditions) was dominated by green filamentous algal *Klebsormidium rivulare* (Kützing) (chl-*a* of 16.1 mg m^{-2}; 12.1 mg m^{-2}). In addition, the diatoms *Tabellaria flocculosa* (Roth) and *Nitzschia nana* (Grunow), which replaced the diatoms *Craticula buderi* (Brebisson) and *Nitzschia clausii* (Hantzsch) that were dominant during the dry spell of 2016 (Table 2), also dominated the outflow of the wetland in 2018.

Both the Shannon and Simpson diversity index values for periphyton were significantly higher downstream of the ecologically engineered Grootspruit wetland than at the inflow when the entire period of the study was collectively considered ($p = 0.022$ and 0.044, respectively) (Figures 3 and 4). Similarly, both of the aforementioned indexes indicated significantly higher diversity in the outflow than the inflow when the drought year was excluded from the analysis (Shannon: $p = 0.011$; Simpson: $p = 0.004$).

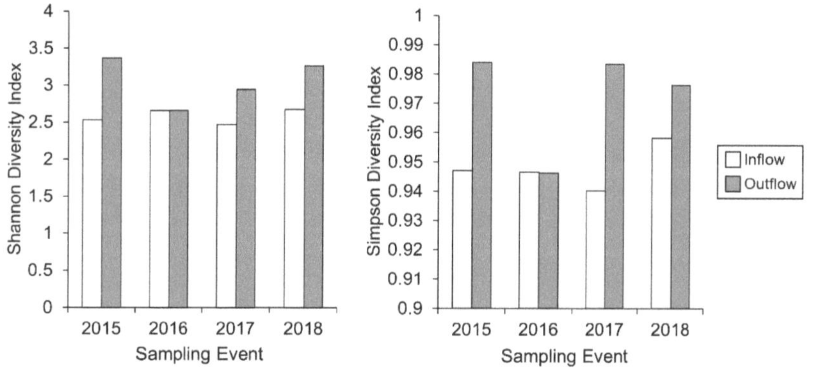

Figure 3. Shannon and Simpson diversity indexes of the periphyton at the inflow and outflow of the Grootspruit wetland over the period of 2015 to 2018.

3.3. Periphyton Dynamics after the Dry Spell of 2016

The diatoms' community structure changed as succession progressed after the drought of 2016 (Table 3). *Nitzschia umbonata* and *Melosira variance*, which are eutrophic diatom indicator species, occurred in abundance at the outflow from 2016 to 2018 (Table 2). These eutrophic indicator species were completely absent during the pre-2016 drought period of 2015.

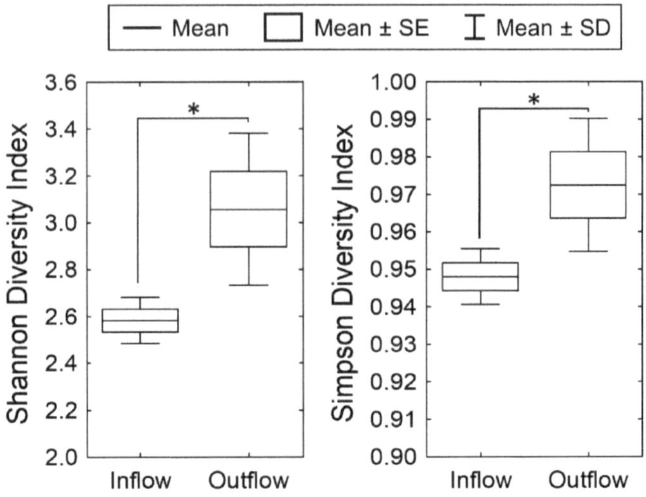

Figure 4. Shannon and Simpson diversity indexes of the periphyton at the inflow and outflow of the ecologically engineered Grootspruit wetland. The statistical analysis was performed with the drought year (2016) included and excluded. The asterisks indicate noteworthy statistical differences, $p < 0.05$.

Table 3. Average chemical and physical parameters measured during the sampling period of 2015–2018 at the inflow and outflow of the wetland (n = 16).

Parameter	Unit	Reference Site 2018	2015 Inflow	2015 Outflow	2016 Inflow	2016 Outflow	2017 Inflow	2017 Outflow	2018 Inflow	2018 Outflow
Sulfate (SO_4^{2-})	mg L^{-1}	37	1280	341	1610	1580	735	658	1380	471
Alkalinity ($CaCO_3$)	mg L^{-1}	24	8.7	141	5	6	3.8	5	2.5	44
Sodium (Na)	mg L^{-1}	11	32	26	46	43	24	22	35	28
Calcium (Ca)	mg L^{-1}	9	383		437	419	200	173	371	121
Magnesium (Mg)	mg L^{-1}	5.9	67	41	133	135	63	55	108	39
Dissolved organic carbon (DOC)	mg L^{-1}	5.2	2.2	3.7	2.5	2.8	2.8	5.6	1.9	4.3
Electrical conductivity (EC)	µS/cm	180	2357	803	2600	2450	1320	1240	2500	1060
Total phosphorus (TP)	mg L^{-1}	0.12	0.03	0.05	0.05	1.0	0.05	0.09	0.05	0.06
Total nitrogen (TN)	mg L^{-1}	1.2			0.05	0.06	0.05	0.04	0.05	0.05
Aluminium (Al)	mg L^{-1}	0.01	3.8	0.05	10.6	10.1	10.4	7.1	5.9	0.01
Iron (Fe)	mg L^{-1}	0.05	7.03	0.06	9.7	8.1	2.3	1.4	1.9	0.08
Manganese (Mn)	mg L^{-1}	0.04	14.5	0.09	17.0	16.5	11.7	6.3	13	0.02
Silicon (Si)	mg L^{-1}	0.7	6.3	4.5	8.5	6.4	7.2	3.8	9.1	2.3
pH		6.9	3.17	7.03	3.6	4.0	3.8	6.9	3.4	7.5

3.4. Surface Water Chemistry

From Table 3, it is evident that there was a major change in ionic distribution between the inflow and outflow during the drought in 2016, and during the pre-and post-drought conditions. The downstream sites from the wetland inflow to the outflow showed a significant SO_4^{2-}, decrease, whereas alkalinity increased at the outflow pre and post the 2016 drought. The SO_4^{2-} concentration of SO_4 from the wetland inflow decreased by 2%, from an average of 1610 to 1580 mg L^{-1} at the wetland outflow during the dry spell of 2016, indicating very little passive treatment capacity throughout the wetland. A 68.2% decrease in SO_4^{2-} was observed in 2018 from an average of 1480 mg L^{-1} at the inflow to 471 mg L^{-1} at the outflow sites, showing improvement in passive treatment capacity after the dry spell. The determined SO_4^{2-}, alkalinity and pH for the reference site remained comparatively unchanged during the sampling period and it was therefore not impacted by the dry spell (Table 3). The alkalinity, pH, and SO_4^{2-} concentration increased, whereas the EC decreased from an average concentration of 2500 µS cm^{-1} (by 57.8%) at the wetland inflow to 1060 µS cm^{-1} at the wetland outflow after the dry spell in 2016 (Table 3). However, the EC values were still higher in 2018 after the drought when compared with the measurements taken before the drought in 2015 (Table 3). From the surface water chemistry data, it can be concluded that there was a progressive water quality improvement from 2017 to 2018 after the drought. The metals primarily associated with mining effluent AMD (i.e., Fe, Al, and Mn) were reduced in 2018 at the wetland outflow, and the value was even below that of the reference site in the case of Mn (Table 3). The total phosphates increased at the middle and outflow sites during 2016 and 2017, but decreased in 2018 to pre-drought concentrations measured in 2015. When comparing the data generated from 2015 to 2018 with the selected value of the South African Department of Water Affairs [25] guideline for aquatic ecosystems, it was evident that, in 2016, the values for SO_4^{2-}, Al, Mn, and Fe were above the guideline values for aquatic ecosystems at the outflow. However, after the drought in 2018, the Al, Mn, and Fe values were below the South African Department of Water Affairs [40] guideline set out for aquatic ecosystems.

The DOC concentration and pH were notably higher in the wetland outflow ($p = 0.030$ and 0.004, correlatively) than in the wetland inflow, whereas Si was notably lower in the outflow ($p = 0.016$) when the drought year of 2016 was included in the analyses (Table 3). When the drought year was excluded from the analysis, the water chemistry profile was more distinct between the inflow and outflow samples. In particular, significant differences between inflow and outflow existed for the following: DOC ($p = 0.023$); pH ($p < 0.001$); calcium ($p = 0.048$); manganese ($p = 0.008$); silica ($p = 0.019$); and sulfate (SO_4^{2-}) ($p = 0.044$). The PCA triplot indicates a grouping of the four inflow samples and the outflow sample representing the 2016 drought year and association with the elements tested for, in addition to SO_4 and EC (Figure 5). The 2015 and 2016 outflow samples formed a further grouping in ordinal space and were negatively associated with the chemicals tested for and positively associated with increased pH and calcium carbonate ($CaCO_3$). Moreover, the 2017 outflow sample was located between the two aforementioned groupings, having a unique identity, as was the case with the reference site being segregated in ordinal space (Figure 5). In addition, the inflow–2016 outflow grouping was associated with periphyton indicators of extreme pollution conditions (Figure 5).

The influence of the drought year on wetland reclamation efficiency can be seen when considering differences among the inflow and outflow water chemistry profiles. When a dataset of all four sampling events (2015–2018) was assessed, three parameters, namely, pH, DOC, and Si, varied significantly among the inflow and outflow locations. Both pH and DOC were higher, indicating increased water quality. However, when the drought year was excluded from the analysis, six parameters, namely pH, DOC, Ca, Mn, Si, and SO_4^{2-}, varied significantly among the inflow and outflow locations, indicating more efficient water reclamation.

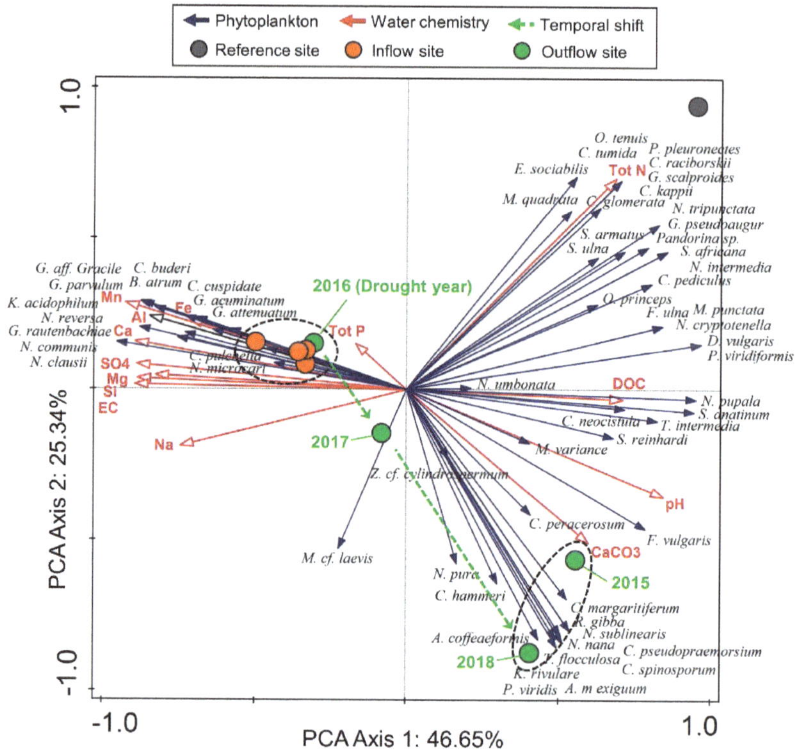

Figure 5. A principal component analysis triplot indicating the various associations between water chemistry and phytoplankton assemblage in reference to the inflow and outflow of the Grootspruit wetland and the reference site. The plot represents the entire study period (2015–2018) for the inflow and outflow locations and a reference site sampled during 2018. Phytoplankton data were applied as a focal PCA plot, and chemistry data as supplementary variables. Abbreviations: EC: electrical conductivity; Tot N: total nitrogen; DOC: dissolved organic carbon; Tot P: total phosphate.

A PCA triplot describing associations between periphyton and surface water chemistry in relation to the four sampling events and the reference site indicated a clear grouping of the 2016 drought year outflow with the inflow samples. Water quality downstream of the wetland during the drought year, therefore, corresponded to the poorer quality water released into the wetland. Although water quality and periphyton assemblage identity improved in the wetland outflow during 2017, recovery to the pre-drought state of 2015 only occurred in 2018 (Table 2, Figure 5), suggesting that a minimum of a two-year recovery period is required after a severe drought.

4. Discussion

Evidence collected from the ecologically engineered Grootspruit wetland during the 2016 drought—as one of the worst droughts on record for the Mpumalanga province, South Africa [29]—suggested that the loss of ecological functions and the resultant regime change during the drought period was a consequence of the interplay between chemical, physical, and biological processes as observed between the different sampling trips from early 2015 to 2018.

The biodiversity and autecology of the periphyton indicated a slow recovery of the wetland to its prior 2015 state, but the presence of certain diatoms, for example, *Nitzschia umbonata* and *Melosira variance*, also signaled reconfiguration to a state of eutrophication, (sampling sites 3 to 5 during 2016 to 2018). This may be partly due to the sulfate reduction indirectly affecting the nutrient kinetics of the wetland [62]. It was previously shown that sulfide produced by sulfate reduction interferes with iron-phosphate binding in wetland sediments and soils due to the iron sulfide formation [63,64], which then releases phosphates through this chemical process. In the current study, higher concentrations of phosphates were measured during the 2016 drought period at the outflow of the wetland. However, the literature [62] further shows that the quantity of released phosphates is dependent on the availability of sulfate concentrations, which were much higher in the wetland surface water outflow during 2016, in comparison with the concentrations measured in 2015 and 2018. The periphyton mats at the wetland inflow, dominated by the filamentous algal species *Klebsormidium acidophilum* (Novis) throughout the study period, were a strong indication of AMD-impacted water with low pH values and high EC [36]. Furthermore, the occurrence of the diatoms *Nitzschia clausii* (Hantzsch), *Craticula buderi* (Brebisson), and *Gomphonema aff. gracile* (Ehrenberg) at the wetland inflow during the whole study period are good indicators of mining effluence and polluted water, according to Taylor et al. [47]. The most significant changes in periphyton usually occur at a pH range of 4.7 to 5.6, which is just beyond the interval of a pH of 5.5 to 6.5 where carbonates, as a key source of both inorganic carbon and acid neutralizing capacity for photosynthesis, become rapidly depleted [65,66]. However, a proliferation of acidophilic periphyton, which increased the biomass (benthic chl-*a* concentrations) during the drought period of 2016 at certain sites, was positively correlated with a decrease in pH, as previously shown by Muller [67] and Verb and Vis [68]. This observation could possibly be ascribed to a macroinvertebrate grazing pressure decrease, or a decrease in algal competition and a nutrient cycle alteration, as suggested by Stokes [69]. In the current study, a longitudinal relationship between the water column pH increase, filamentous algae biomass (benthic chl-*a* mg m^{-2}) decrease, and a diatom diversity increase downstream from Site 2, were observed after the drought period of 2016. The lower biomass observed at Site 2 throughout the study period may be related to higher rates of oxide deposition at this specific site, since the deposition of oxide can smother periphyton and subsequently inhibit photosynthesis [70].

The occurrence of the diatoms *Nitzschia nana* (Grunow) and *Tabellaria flocculosa* (Roth) at the outflow of the wetland, which replaced the diatoms *Nitzschia clausii* (Hantzsch) and *Craticula buderi* (Brebisson) after the drought period, is a good indicator of electrolyte-poor, oligotrophic, circumneutral, or slightly acidic waters, but also moderately polluted water [47]. The occurrence of the diatoms at the outflow of the wetland serves as an indicator of the improved aquatic conditions present in the wetland in 2018.

The average chemical and physical parameters measured at the wetland outflow during the sampling period 2015–2018 also supported the two-year recovery period of the wetland after the severe drought experienced in 2016. Although a loss in some ecological functions can be ascribed to the drought occurrence, the post-2016 drought physical and chemical parameters (2018 outflow) were comparable in terms of the ecological system, and restoration performance of the services, to the 2015 outflow parameters. A distinct reduction in SO_4^{2-}, Na, Mg, EC, Al, Fe, Mn, Si, and pH, following the 2016 drought, highlighted the utilization of water quality variables to not only assess the passive treatment responses of an ecologically engineered wetland, but also the progress related to ecological recovery.

5. Conclusions

The current study demonstrated that during the prevailing drought in 2016, AMD adversely affected the ecologically engineered wetland ecosystem in two ways, namely, (a) periphyton communities were restricted to only specific tolerant and resilient organisms that were able to survive in these extreme conditions; and (b) changes in the nutrient cycles caused the wetland to become nutrient-enriched, possibly due to changes in the high concentrations of sulfate, which also affected the periphyton assemblage. A loss in ecological functions and a regime shift during the drought of 2016 in the ecologically engineered wetland system, which, in part, recovered two years after the drought, was also observed. Although water quality and periphyton assemblage identity improved in the wetland outflow during 2017, recovery to the pre-drought state of 2015 only occurred in 2018, suggesting that a minimum of a two-year recovery period is required for an ecologically engineered, passive treatment wetland system after a severe drought.

Author Contributions: Conceptualization, P.J.O. and J.C.T.; methodology, P.J.O.; software, J.C.T.; validation, P.J.O. and A.-M.B.; formal analysis, J.C.T.; investigation, P.J.O.; resources, P.J.O. and A.-M.B.; data curation, Y.S.; writing—original draft preparation, P.J.O. and Y.S.; writing—review and editing, P.J.O., J.C.T., A.-M.B. and Y.S.; visualization, A.-M.B. and J.C.T.; supervision, P.J.O. and A.-M.B.; project administration, A.-M.B.; funding acquisition, P.J.O. All authors have read and agreed to the published version of the manuscript.

Funding: The research was funded by Coaltech Research Association, grant number E2019-7.

Institutional Review Board Statement: Not applicable.

Informed Consent Statement: Not applicable.

Data Availability Statement: All relevant data presented in the article are stored according to institutional requirements and, as such, are not available online. However, all data used in this manuscript can be made available upon request to the authors.

Acknowledgments: The authors express their sincere gratitude to the co-workers for helping with the water and periphyton sampling. The authors also want to thank the anonymous referees for critically reviewing the manuscript and suggesting useful changes.

Conflicts of Interest: The authors declare no conflict of interest.

References

1. Cui, L.J. Evaluation of functions of Poyang Lake ecosystem. *Chin. J. Ecol.* **2004**, *23*, 47–51.
2. Oberholster, P.J.; Cheng, P.; Botha, A.-M.; Genthe, B. The potential of selected macroalgal species for treatment of AMD at different pH ranges in temperate regions. *Water Res.* **2014**, *60*, 82–92. [CrossRef]
3. Atazadeh, E.; Barton, A.; Shirinpour, M.; Zarghami, M.; Rajabifard, A. River management and environmental water allocation in regulated ecosystems of arid and semi-arid regions—A review. *Fundam. Appl. Limnol.* **2020**, *193*, 327–345. [CrossRef]
4. Opitz, J.; Bauer, M.; Eckert, J.; Peiffer, S.; Alte, M. Optimising Operational Reliability and Performance in Aerobic Passive Mine Water Treatment: The Multistage Westfield Pilot Plant. *Water Air Soil Pollut.* **2022**, *233*, 66. [CrossRef]
5. Kadlec, R.; Wallace, S. *Treatment Wetlands*, 2nd ed.; CRC Press: Boca Raton, FL, USA, 2009.
6. Vymazal, J. Constructed wetlands for treatment of industrial wastewaters: A review. *Ecol. Eng.* **2014**, *73*, 724–751. [CrossRef]
7. Zedler, J.B. Progress in wetland restoration ecology. *Trends Ecol. Evol.* **2000**, *15*, 402–407. [CrossRef]
8. Kadlec, R.; Knight, R.; Vymazal, J.; Brix, H.; Cooper, P.; Haberl, R. *Constructed Wetlands for Pollution Control: Processes, Performance, Design and Operation*; IWA Publishing: London, UK, 2000.
9. Mitsch, W.; Zhang, L.; Stefanik, K.; Nahlik, A.; Anderson, C.; Bernal, B.; Hernandez, M.; Song, K. Creating wetlands: Primary succession, water quality changes, and self-design over 15 years. *BioScience* **2012**, *62*, 237–250. [CrossRef]
10. Kleinmann, B.; Skousen, J.; Wildeman, T.; Hedin, B.; Nairn, B.; Gusek, J. The early development of passive treatment systems for mining-influenced water: A North American perspective. *Mine Water Environ.* **2021**, *40*, 818–830. [CrossRef]
11. Gusek, J. A periodic table of passive treatment for mining influenced water. In Proceedings of the 2009 National Meeting of the American Society of Mining and Reclamation, Billings, MT, USA, 30 May–5 June 2009; pp. 550–562. [CrossRef]
12. Nairn, R.W.; LaBar, J.A.; Strevett, K.A.; Strosnider, W.H.; Morris, D.; Neely, C.A.; Garrido, A.; Santamaria, B.; Oxenford, L.; Kauk, K.; et al. A large, multi-cell, ecologically engineered passive treatment system for ferruginous lead-zinc mine waters. In Proceedings of the International Mine Water Association Annual Meeting, Sydney, NS, Canada, 5–9 September 2010; pp. 255–258.
13. Skousen, J.; Sexstone, A.; Ziemkiewicz, P. Acid mine drainage control and treatment. In *Reclamation of Drastically Disturbed Lands*; Agronomy Monographs: Madison, WI, USA, 2000; p. 41.
14. URS. *Passive and Semi-Active Treatment of Acid Rock Drainage from Metal Mines—State of the Practice*; U.S. Army Corps of Engineers: Concord, MA, USA, 2003.
15. Watzlaf, G.R.; Schroeder, K.T.; Kairies, C. Long-term performance of alkalinity-producing passive systems for the treatment of mine drainage. In Proceedings of the 2000 National Meeting of the American Society for Surface Mining and Reclamation, Tampa, FL, USA, 11–15 June 2000; pp. 262–274. [CrossRef]
16. Wieder, R.K. *The Kentucky Wetlands Project: A Field Study to Evaluate Man-Made Wetlands for Acid Coal Mine Drainage Treatment*; Final Report to the U.S. Office of Surface Mining; Villanova University: Villanaova, PA, USA, 1992.
17. Sobolewski, A.; Sobolewski, N. Holistic design of wetlands for mine water treatment and biodiversity: A case study. *Mine Water Environ.* **2022**, *41*, 292–299. [CrossRef]

18. Skousen, J.; Zipper, C.E.; Rose, A.; Ziemkiewicz, P.F.; Nairn, R.; McDonald, L.M.; Kleinmann, R.L. Review of passive systems for acid mine drainage treatment. *Mine Water Environ.* **2017**, *36*, 133–153. [CrossRef]
19. Mitsch, W.J.; Wilson, R.F. Improving the success of wetland creation and restoration with know-how, time and self-design. *Ecol. Appl.* **1996**, *6*, 77–83. [CrossRef]
20. Simenstad, C.A.; Cordell, J.R. Ecological assessment criteria for restoring anadromous salmon habitat in Pacific Northwest estuaries. *Ecol. Eng.* **2000**, *15*, 283–302. [CrossRef]
21. Zedler, J.B.; Callaway, J. Evaluating the progress of engineered tidal wetlands. *Ecol. Eng.* **2000**, *15*, 211–225. [CrossRef]
22. Pat-Espadas, A.M.; Portales, R.L.; Amabilis-Sosa, L.E.; Gomez, G.; Vidal, G. Review of constructed wetlands for acid mine drainage treatment. *Water* **2018**, *10*, 1685. [CrossRef]
23. Oberholster, P.J.; McMillan, P.; Durgapersad, K.; Botha, A.-M.; De Klerk, A.R. The development of a wetland classification and risk assessment index (WCRAI) for non-wetland specialists for the management of natural freshwater wetland ecosystems. *Water Air Soil Pollut.* **2013**, *225*, 1833. [CrossRef]
24. Van Deventer, H.; Smith-Adao, L.; Mbona, N.; Petersen, C.; Skowno, A.; Collins, N.B.; Grenfell, M.; Job, N.; Lötter, M.; Ollis, D.; et al. *South African National Biodiversity Assessment 2018: Technical Report*; South African Inventory of Inland Aquatic Ecosystems (SAIIAE); CSIR Report Number CSIR/NRE/ECOS/IR/2018/0001/A; Council for Scientific and Industrial Research (CSIR) and South African National Biodiversity Institute (SANBI): Pretoria, South Africa, 2018; Volume 2a.
25. Hogsden, K.L.; Harding, J.S. Anthropogenic and natural sources of acidity and metals and their influence on the structure of stream food webs. *Environ. Pollut.* **2012**, *162*, 466–474. [CrossRef]
26. Jarvis, A.P.; Younger, P.L. Broadening the scope of mine water environmental impact assessment: A UK perspective. *Environ. Impact Assess. Rev.* **2000**, *20*, 85–96. [CrossRef]
27. DeNicola, D.M. A review of diatoms found in highly acidic environments. *Hydrobiologia* **2000**, *433*, 111–122. [CrossRef]
28. Van Ho, A.; Ward, D.M.; Kaplan, J. Transition metal transport in yeast. *Annu. Rev. Microbiol.* **2002**, *56*, 237–261. [CrossRef]
29. Baudoin, M.-A.; Vogel, C.; Nortje, K.; Naik, M. Living with drought in South Africa: Lessons learnt from the recent EL Nino drought period. *Int. J. Disaster Risk Reduct.* **2017**, *23*, 128–137. [CrossRef]
30. Wehr, J.D.; Sheath, R.G. Habitats of freshwater algae. In *Freshwater Algae of North America: Ecology and Classification*; Wehr, J.D., Sheath, R.G., Kociolek, J.P., Eds.; Academic Press: San Diego, CA, USA, 2003; pp. 13–29.
31. Korneva, L.G.; Mineeva, N.M. Phytoplankton composition and pigment concentrations as indicators of water quality in the Rybinsky reservoir. *Hydrobiologia* **1996**, *322*, 255–259. [CrossRef]
32. Willén, E. Phytoplankton and water quality characterization: Experiences from the Swedish large lakes Mälaren, Hjälmaren, Vätten and Vänern. *AMBIO* **2001**, *30*, 529–537. [CrossRef] [PubMed]
33. Gogoi, P.; Sinha, A.; Tayung, T.; Naskar, M.; Das Sarkar, S.; Ramteke, M.H.; Das, S.K.; Kumar, K.L.; Suresh, V.R.; Das, B.K. Unravelling the structural changes of periphyton in relation to environmental variables in a semilotic environment in the Sundarban eco-region, India. *Arab. J. Geosci.* **2021**, *14*, 2038. [CrossRef]
34. Department of Forestry, Fisheries & the Environment. *South African National Land Cover (SANLC)*; DFFE: Pretoria, South Africa, 2020.
35. Tóth, J.; Hayashi, M. The theory of basinal gravity flow of groundwater and its impacts on hydrology in Japan. *J. Groundw. Hydrol.* **2010**, *52*, 335–354. [CrossRef]
36. Oberholster, P.J.; Madlala, T.; Oberholster, P.F. *Post Restoration Monitoring of the Zaalklapspruit Wetland with Special Reference to Groundwater-Surface Water Interaction*; Coaltech Technical Report: E2019-7; Coaltech: Johannesburg, South Africa, 2019; pp. 1–21.
37. Shelton, L.R. *Field Guide for Collecting and Processing Stream Water Samples for the National Water Quality Assessment Program*; U.S. Geological Survey Open File Report 94-455; U.S. Geological Survey: Reston, VA, USA, 1994.
38. American Public Health Association (APHA). *Standard Methods for Examination of Water and Wastewater*, 20th ed.; American Public Health Association: Washington, DC, USA, 2006.
39. Appelo, C.A.J.; Postma, D. *Geochemistry, Groundwater and Pollution*, 2nd ed.; A.A. Balkema Publishers: Leiden, The Netherlands, 2005.
40. Department of Water Affairs. *South African Water Quality Guidelines: Vol. 7. Aquatic Ecosystems*, 2nd ed.; Department of Water Affairs and Forestry: Pretoria, South Africa, 1996.
41. Steinman, A.D.; Lamberti, G.A.; Leavitt, P.R. Biomass and Pigments of Benthic Algae. In *Methods in Stream Ecology*; Hauer, F.R., Lamberti, G.A., Eds.; Academic Press: Amsterdam, The Netherlands, 2006; pp. 357–377.
42. Ostenfeld, C.H.; Nygaard, G. On the phytoplankton of the Gatun Lake, Panama Canal. *Dan. Bot. Ark. Udg. Dan. Bot. Foren.* **1925**, *4*, 1–16.
43. Truter, E. *An Aid to the Identification of the Dominant and Commonly Occurring Genera of Algae Observed in Some South African Impoundments*; Department of Water Affairs, Hydrological Institute: Pretoria, South Africa, 1987.
44. Komarek, J.; Anagnostidis, K. Cyanoprokaryota: Chroococcales. In *Susswasserflora von Mitteleuropa 19/1*; Ettl, H., Gartner, H., Heynig, D., Mollenhauer, D., Eds.; Gustav Fischer: Stuttgart, Germany, 1999.
45. Komarek, J.; Anagnostidis, K. Cyanoprokaryota: Oscillatoriales. In *Susswasserflora von Mitteleuropa 19/2*; Budel, B., Gartner, G., Krienitz, L., Schagerl, M., Eds.; Elsevier: Munchen, Germany, 2005.
46. Van Vuuren, S.; Taylor, J.C.; Van Ginkel, C.; Gerber, A. *Easy Identification of the Most Common Freshwater Algae*; North-West University and Department of Water Affairs and Forestry: Pretoria, South Africa, 2006.

47. Taylor, J.C.; Harding, W.R.; Archibald, C.G.M. *An Illustrated Guide to Some Common Diatom Species from South Africa*; Water Research Commission: Pretoria, South Africa, 2007.
48. Stancheva, R.; Sheath, R.G.; Hall, J.D. Systematics of the genus Zygnema (Zygnematophyceae, Charophyta) from Californian watersheds. *J. Phycol.* **2012**, *48*, 409–422. [CrossRef] [PubMed]
49. Porra, R.J.; Thompson, W.A.; Kriedemann, P.E. Determination of accurate extinction coefficient and simultaneous equations for assaying chlorophylls a and b extracted with four different solvents: Verification of the concentration of chlorophyll standards by atomic absorption spectrometry. *Biochim. Biophys. Acta* **1989**, *975*, 384–394. [CrossRef]
50. Berger, W.H.; Parker, F.L. Diversity of planktonic foraminifera in deep-sea sediments. *Science* **1970**, *168*, 1345–1347. [CrossRef] [PubMed]
51. Zulkifli, H. Traitement des Eaux Usées par Lagunage à Haut Rendement: Structure et Dynamique des Peuplements Phytoplanctoniques (Zulkifli, H. Wastewater Treatment by High Efficiency Lagooning: Structure and Dynamics of Phytoplankton Populations). Ph.D. Thesis, Université Montpellier I, Montpellier, France, 1992.
52. Barthel, L.; De Oliveira, P.A.V.; Da Costa, R.H.R. Plankton biomass in secondary ponds treating piggery waste. *Braz. Arch. Biol. Technol.* **2008**, *51*, 1287–1298. [CrossRef]
53. Stevenson, R.J.; Bahls, L.L. Periphyton protocols. In *Rapid Bioassessment Protocols for Use in Streams and Wadeable Rivers: Periphyton, Benthic Macroinvertebrates, and Fish*, 2nd ed.; Barbour, M.T., Gerritsen, J., Snyder, B.D., Stribling, J.B., Eds.; EPA 841-B-99-002; U.S. Environmental Protection Agency, Office of Water: Washington, DC, USA, 1999.
54. Spencer, C.; Robertson, A.I.; Curtis, A. Development and testing of a rapid appraisal wetland condition index in south-eastern Australia. *J. Environ. Manag.* **1998**, *54*, 143–159. [CrossRef]
55. Shannon, C.E. A mathematical theory of communication. *Bell Syst. Tech. J.* **1948**, *27*, 379–423, 636–656. [CrossRef]
56. Simpson, E.H. Measurement of diversity. *Nature* **1949**, **163**, 688. [CrossRef]
57. Van den Brink, P.J.; Van Den Brink, N.W.; Ter Braak, J.F. Multivariate analysis of ecotoxicological data using ordination: Demonstrations of utility on the basis of various examples. *Australas. J. Ecotoxicol.* **2003**, *9*, 141–156.
58. Ter Braak, C.J.F.; Smilauer, P. *CANOCO Reference Manual and Canodraw for Windows User's Guide: Software for Canonical Community Ordination*; Microcomputer Power: New York, NY, USA, 2002; Version 4.5.
59. Gerber, A.; Cilliers, C.J.; Van Ginkel, C.; Glen, R. *Easy Identification of Aquatic Plants: A Guide for the Identification of Water Plants in and Around South African Impoundments*; Department of Water Affairs: Pretoria, South Africa, 2004.
60. Niyogi, D.K.; Lewis, W.M., Jr.; McKnight, D.M. Effects of stress from mine drainage on diversity, biomass, and function of primary producers in mountain streams. *Ecosystems* **2002**, *5*, 554–567.
61. Bray, J.P.; Broady, P.A.; Niyogi, D.K.; Harding, J.S. Periphyton communities in New Zealand streams impacted by acid mine drainage. *Mar. Freshw. Res.* **2008**, *59*, 1084–1091. [CrossRef]
62. Lasmers, L.P.M.; Tomassen, H.B.M.; Roelofs, J.G.M. Sulfate-induced eutrophication and phytotoxicity in freshwater wetlands. *Environ. Sci. Technol.* **1998**, *32*, 199–205. [CrossRef]
63. Caraco, N.; Cole, J.; Likens, G. Evidence for sulphate-controlled phosphorus release from sediments of aquatic systems. *Nature* **1989**, *341*, 316–318. [CrossRef]
64. Smolders, A.; Roelofs, J.G.M. Sulphate-mediated iron limitation and eutrophication in aquatic ecosystems. *Aquat. Bot.* **1993**, *46*, 247–253. [CrossRef]
65. Dixit, S.S.; Smol, J.P. Algal assemblages in acid-stressed lakes with particular emphasis on diatoms and chrysophytes. In *Acid Stress and Microbial Interactions*; Rao, S.S., Ed.; CRC Press: Boca Raton, FL, USA, 1989; pp. 91–134.
66. Boekken, T.; Kroglund, F.; Lindstrom, E.-A.; Carvalho, L. Acidification of rivers and lakes. In *Indicators and Methods for the Ecological Status Assessment under the Water Framework Directive: Linkage between Chemical and Biological Quality of Surface Waters*; Solimini, A.G., Cardoso, A.C., Heiskanen, A.-S., Eds.; Institute for Environment and Sustainability; European Communities: Ispra, Italy, 2006.
67. Muller, P. Effects of artificial acidification on the growth of periphyton. *Can. J. Fish. Aquat. Sci.* **1980**, *37*, 355–363. [CrossRef]
68. Verb, R.G.; Vis, M.L. Periphyton assemblages as bioindicators of mine-drainage in unglaciated Western Allegheny Plateau lotic systems. *Water Air Soil Pollut.* **2005**, *161*, 227–265. [CrossRef]
69. Stokes, W.L. *Geology of Utah. Salt Lake City, Utah: Museum of Natural History*; Utah Geological and Mineral Survey: Salt Lake City, UT, USA, 1986.
70. Anthony, M.K. Ecology of Streams Contaminated by Acid Mine Drainage Near Reefton, South Island. Master's Thesis, University of Canterbury, Christchurch, New Zealand, 1999.

processes

MDPI

Review

Oily Wastewater Treatment: Methods, Challenges, and Trends

Alexandre D'Lamare Maia de Medeiros [1,2], Cláudio José Galdino da Silva Junior [1,2], Julia Didier Pedrosa de Amorim [1,2], Italo José Batista Durval [2], Andréa Fernanda de Santana Costa [3] and Leonie Asfora Sarubbo [2,4,*]

[1] Rede Nordeste de Biotecnologia (RENORBIO), Universidade Federal Rural de Pernambuco, Rua Dom Manuel de Medeiros, s/n—Dois Irmãos, Recife CEP 52171-900, Brazil; alexandre_dlamare@outlook.com (A.D.M.d.M.); claudiocjg@gmail.com (C.J.G.d.S.J.); juliadidier@hotmail.com (J.D.P.d.A.)
[2] Instituto Avançado de Tecnologia e Inovação (IATI), Rua Potyra, n. 31, Prado, Recife CEP 50751-310, Brazil; italo.durval@gmail.com
[3] Centro de Comunicação e Design, Centro Acadêmico da Região Agreste, Universidade Federal de Pernambuco (UFPE), BR 104, Km 59, s/n—Nova Caruaru, Caruaru CEP 50670-90, Brazil; andrea.santana@ufpe.br
[4] Escola Icam Tech, Universidade Católica de Pernambuco (UNICAP), Rua do Príncipe, n. 526, Boa Vista, Recife CEP 50050-900, Brazil
* Correspondence: leonie.sarubbo@unicap.br; Tel.: +55-81-2119-4000

Abstract: The growing interest in innovations regarding the treatment of oily wastewater stems from the fact that the oil industry is the largest polluter of the environment. The harm caused by this industry is seen in all countries. Companies that produce such wastewater are responsible for its treatment prior to disposal or recycling into their production processes. As oil emulsions are difficult to manage and require different types of treatment or even combined methods, a range of environmental technologies have been proposed for oil-contaminated effluents, such as gravity separation, flotation, flocculation, biological treatment, advanced oxidation processes, and membranes. Natural materials, such as biopolymers, constitute a novel, sustainable solution with considerable potential for oily effluent separation. The present review offers an overview of the treatment of oily wastewater, describing current trends and the latest applications. This review also points to further research needs and major concerns, especially with regards to sustainability, and discusses potential biotechnological applications.

Keywords: nanocellulose; biotechnology; oily effluents; nanomaterial innovations

Citation: Medeiros, A.D.M.d.; Silva Junior, C.J.G.d.; Amorim, J.D.P.d.; Durval, I.J.B.; Costa, A.F.d.S.; Sarubbo, L.A. Oily Wastewater Treatment: Methods, Challenges, and Trends. *Processes* **2022**, *10*, 743. https://doi.org/10.3390/pr10040743

Academic Editor: Andrea Petrella

Received: 29 March 2022
Accepted: 10 April 2022
Published: 12 April 2022

1. Introduction

Large amounts of oily effluents from industries and oil spills are generated daily. Such effluents are produced in different processing stages (transport, maintenance, production, etc.) and the treatment of this large volume poses a problem for many companies [1,2]. Oily wastewater as well as oil-in-water and water-in-oil emulsions are among the main pollutants released into aquatic ecosystems [2]. These oily effluents exert enormous impacts on the environment and, consequently, society. Oily wastewater affects drinking water, groundwater, and other resources. The percolation of pollutants into the underground water supply is harmful to humans and other living beings. This process even contaminates the atmosphere (as in the case of oil burners) and the soil, exerting a negative impact on the food supply chain. Many countries have established rigid legislation to compel companies to control of their effluents, with maximum oil concentration limits within the range of 5 to 100 mg/L in wastewater discharge [3].

Free or suspended oil can be easily separated from the aqueous phase by simple physical processes. However, emulsions can contain considerable amounts of mineral oil, which is highly resistant to biochemical decomposition. Such emulsions consist of a complex mixture of water, oil, and additives, such as emulsifiers and antifoaming agents, and are chemically stabilised. Thus, proper separation is only possible through physical and/or chemical processes, which are generally expensive [1,4,5].

Harmful compounds such as phenols, petroleum hydrocarbons, and polycyclic aromatic hydrocarbons, which inhibit both plant and animal development, can be found in oily wastewater [6]. Disposal within acceptable environmental parameters remains one of the greatest challenges for environmental agencies. Thus, the treatment of these effluents has become urgent and indispensable not only for the oil industry, but for all environmental and health agencies [2].

According to Rajasulochana and Preethy (2016) [7], conventional industrial water treatment methods vary according to several factors, such as effluent composition, volume, and current environmental legislation. However, such treatments are inefficient for the removal of small oil droplets and the separation of oil emulsions. The remaining oil can cause the clogging of pipes leading to loss of useful volume in storage tanks and can even result in the need to replace equipment, leading to increased costs due to recurrent maintenance. These difficulties have stimulated the development of innovative processes and new materials to treat complex wastewater [2,8].

For the complete removal of pollutants, current research has promoted the optimisation of conventional separation processes using novel techniques and materials as well as sustainable, multidisciplinary approaches. The use of biosurfactants in demulsification processes and the use of natural sorbents for application in sorption methods are great examples of such technologies. This has resulted in major technological advances for the sector, such as advanced oxidative processes, membrane separation, as well as innovations involving biotechnological processes [3,9].

Due to the complexity of the chemical composition and physical configuration of oils in wastewater, the use of a single technology for complete remediation of free, emulsified or dispersed oil is often inefficient. A combination of technologies can be used for an efficient removal of oily pollutants, such as reverse osmosis with adsorption for oilfield-produced water, and photocatalytic reactors with ultrafiltration for oil bilge wastewater [9].

The knowledge of techniques for treating oily wastewater allows the industry to apply the best method for each specific case and evaluate possible combination approaches, aiming at a more efficient and even sustainable filtration process. Therefore, the the present review aims to describe and compare different means of treating oily effluents currently in use by the industry, demonstrate recent, innovative research, and discuss the advantages and disadvantages of new trends in the treatment of effluents contaminated mainly by oils and fats. Biotechnological methods are also explored as a sustainable alternative for the industry.

2. Oily Industrial Wastewater

Effluents contaminated with oily particles are usually associated with mechanical, automotive, and thermoelectric industries, as such effluents are a result of processes that use derivatives of fossil fuels, such as gasoline, diesel, and low-pour-point oil. These impurities can be found in different concentrations and come from different origins, including the food industry and domestic sewage (Figure 1) [1]. Effluents rich in oils and greases can be produced in different stages within the industrial setting [9,10].

This type of wastewater can cause environmental imbalances and is harmful to human health, causing diseases such as cancer as well as contaminating drinking water and groundwater resources. It may also cause mutations in plants and animals in addition to increasing chemical and biological oxygen demands in water bodies, exerting different impacts on aquatic ecosystems [3,11,12].

The oily phase in wastewater usually consists of at least four types, depending on the size and stability of the oil droplets. Free floating oil has a droplet diameter larger than 150 μm and is considered easy to remove using conventional gravitational separation methods. Dispersed oil has droplet sizes ranging from 20 to 150 μm and is relatively easy to remove using gravitational processes and/or stabilising agents [1,13]. Emulsified oil usually has droplet sizes less than 20 μm and its removal requires sophisticated processes involving the use of auxiliary techniques, such as the addition of coagulants and surfactants [10,14]. Dissolved oil has droplets less than 5 μm in size and is extremely difficult to remove, requiring the use of special chemical and physical processes [13,15,16].

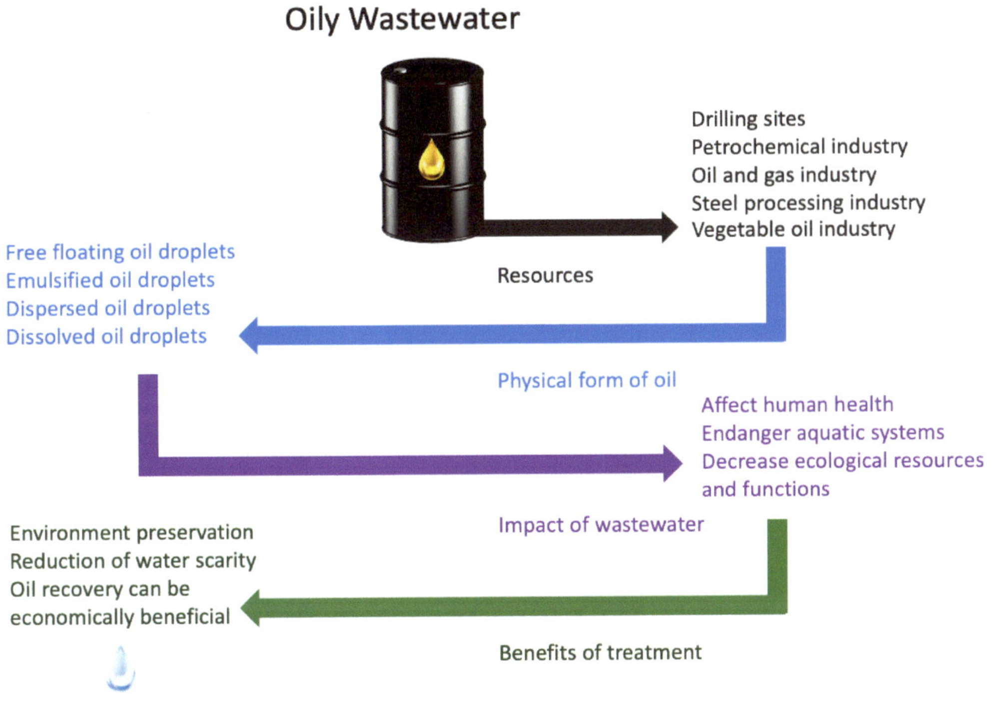

Figure 1. Resources, physical form, impact, and benefits of treatment of oily wastewater.

3. Emulsions

Contaminated wastewater consists of heavy and light hydrocarbons, oil, grease, fats, tars, lubricating oil, cutting oils, wax, etc. [17]. When oil becomes completely miscible with water, it forms an emulsion in the presence of surfactants and ions.

An emulsion is defined as a system whereby one immiscible liquid phase is dispersed as droplets (dispersed phase) in a second phase of immiscible liquid (continuous phase) [18]. The three types of emulsions are water-in-oil, oil-in-water, and multiple emulsion. A water-in-oil emulsion is formed when water droplets are dispersed throughout the continuous oil phase, as shown in Figure 2A. An oil-in-water emulsion is formed when oil droplets are dispersed throughout the continuous water phase (Figure 2B) [19]. Multiple emulsion is a complex system in which water-in-oil or oil-in-water emulsions are dispersed throughout another immiscible phase [19]. This type of emulsion includes oil-in-water-in-oil (Figure 2C) and water-in-oil-in-water (Figure 2D) [19]. Depending on the application, such emulsions can be either desirable or undesirable.

An emulsified system consists of a heterogeneous mixture of two immiscible liquids, that takes on the characteristics of a homogeneous mixture due to an external action of intense mechanical agitation or with the aid of an emulsifying agent, as demonstrated in Figure 3. When static, this system has two phases—the first (dispersed) is suspended in the form of small droplets in the second (continuous) [10].

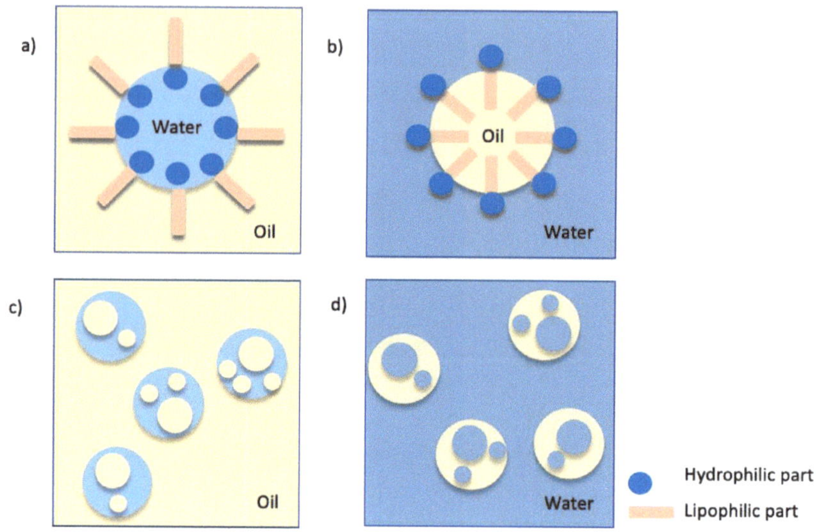

Figure 2. (**a**) Water/oil emulsion; (**b**) oil/water emulsion; (**c**) oil/water/oil emulsion; (**d**) water/oil/water emulsion.

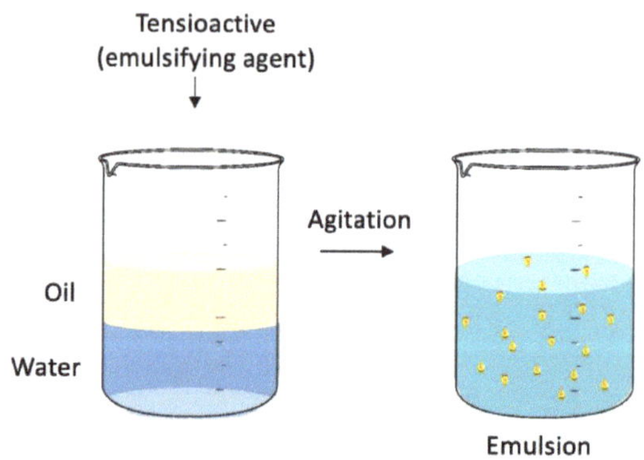

Figure 3. Emulsification schematics.

Emulsification is the process of forming an emulsion. In the oil and gas industry, emulsification problems are, for the most part, due to water-in-oil emulsions. Water-in-oil emulsion is also known as "chocolate mousse" or "mousse" among oil spill laborers [20]. Where crude oil and water are initially present in two separate phases, it is very common for turbulence and agitation through valves, pumps, and pipes to trigger the formation of an emulsion, causing major trouble for the industry [21].

The stabilisation of these emulsions can occur either chemically or physically. Chemically stabilised compounds are formed with the addition of surfactant substances (compounds derived from petroleum or produced by microorganisms that reduce surface tension or influence the surface contact between two liquids). Physically stabilised emulsions are formed without any addition of substances; stability is maintained by electrical charges inherent in the system [10,13].

Karhu et al. (2012) [22] demonstrated that it is possible to produce synthetic oil-in-water emulsions that are significantly more stable than non-synthetic emulsions. Synthetic emulsions do not lose their innate characteristics, which simplifies the study of oil removal from water. Shaw (1975) [23] states that for emulsions prepared from the homogenisation of liquid components, phase separation can be performed in a fast, simple manner.

Oil removal methods with the use of a surfactant (chemical or biological) are better suited for emulsions with a higher concentration (average of 300 ppm) [16,24–26]. This concentration is widely used in works involving the flotation of petroleum dispersed in water. However, other techniques, such as flocculation or coagulation, are often used prior to flotation [13,15,27,28]. The development of technologies for oil-in-water and water-in-oil emulsion separation has been continuous [2,29–31] and different technologies for separating oil from contaminated wastewater have been explored.

Demulsification Methods

Demulsification processes are as intricate as the formation of the emulsion itself. As mentioned above, emulsions are complex and stable, composed of a dispersed phase, a continuous phase (also called the external phase), and an emulsifying agent. Effective demulsification must, therefore, be able to disable or minimise the stability of the phases, leading to their separation [29].

Several factors affect the stability of emulsions, such as the salinity of the water, the amount of suspended and dissolved solids, pH, the presence of surfactants or emulsifiers, and the concentration of oil. Emulsions with oil concentrations higher than 5% *v/v* are generally less stable. The presence of ions, such as $MgCl_2$ and NaCl, can increase stability by forming a layer between the oil droplets and the water molecules, which leads to the immersion of the oil particles in the polar phase [32].

Daaou and Bendedouch (2012) [33] reported the effect of pH on the destabilisation of oil emulsions. The authors found that the most stable form of an oil-in-water emulsion occurs at neutral pH (pH 7). Maximum instability in the emulsion occurs when the mixture is moderately acidic (pH 5). The instability of the mixture also increases when highly alkaline (pH > 10–11), facilitating the phase separation process.

Demulsification methods are developed using three approaches: physical, chemical, and biological. In some cases, demulsification can also be divided into three operational steps: destabilisation of the oil-in-water or water-in-oil interface, oil aggregation, and gravity separation [34].

The physical method consists of separating the components through mechanical, thermal, or electrical action and usually involves high energy consumption, resulting in an expensive operation. For mechanical demulsification processes, forces are used to break the physical barrier and achieve separation of the components. Tools that can be used for the mechanical separation process include cyclones, centrifuges, and gravity settling tanks [3].

Heat is another common physical demulsification method. High temperatures of thermal demulsification affect the stiffness of the interface that promotes the coalescence of colliding droplets. Higher temperatures favour a higher collision rate between droplets, causing a reduction in the stability of the emulsion [31]. However, in addition to high costs due to requiring a large amount of energy, this method carries a high risk of accidents. The slightest carelessness in controlling the process has potential to cause serious problems [12]. Heat treatment is often carried out together with a chemical method, with the aim of increasing the efficiency of the process [31]. Microwave demulsification is another heat-based physical separation process. As this method uses microwaves, it consumes less energy than traditional heating techniques and can minimise the use of chemical demulsifiers in the separation process [35].

Electrical demulsification constitutes another physical process. The deformation of droplets and the generation of an electric field create attraction between droplets, leading to their coalescence. The electric field facilitates the rapid fusion of small droplets into larger droplets. However, adapting this technique to different emulsion types and processes remains a challenge [31].

According to Zolfaghari et al. (2016) [29], efficient chemical demulsification approaches for the separation of emulsions depend on factors such as the correct choice of chemical and its concentration. The process is performed by homogenising the components with the emulsion to ensure that the reaction takes place. In many cases, it is necessary to heat the mixture to accelerate the separation or completely break the emulsion. After these processes, adequate residence time must be allowed for the complete separation of the phases.

Various chemical agents can be applied to aid the separation of oil and water in emulsions. These agents act to neutralise the surface charge of the droplets, leading to demulsification and the agglomeration of the oil droplets into large flocs. Such agents are denominated flocculants and are generally classified into two major groups [36]. Inorganic coagulants, such as aluminium sulphate, polyaluminium chloride, and ferrous sulphates, constitute the first group and have been widely used as coagulants due to their low cost. However, large doses are required for efficient flocculation. Moreover, these coagulants are highly sensitive to pH and are inefficient for very small particles, in addition to producing large volumes of sludge that contains metals, generating problems in the disposal of this toxic material [37].

Organic polymeric flocculants constitute second group. These can have either a synthetic or natural origin. Polymeric flocculants are generally divided into four categories: non-ionic, cationic, anionic, and amphoteric polymers [38]. Despite being more expensive, polymeric flocculants are highly efficient even at low doses and generate a much smaller volume of sludge compared to inorganic coagulants. The particles formed during flocculation are generally larger and stronger, with excellent settling capacity, facilitating separation [37].

Regarding the biological approach, microorganisms and their metabolites with different biochemical structures can be used for demulsification (Table 1). These organisms have mechanisms for accessing as nourishment hydrocarbons in the emulsion [31]. The demulsifying mechanism consists of using their hydrophobic cell surfaces composed of lipids and proteins, or extracellular amphiphilic biosurfactants, to replace emulsifying molecules, acting as a reducer of interfacial tension between the emulsion phases [39]. Table 1 lists examples of biological demulsifiers.

Table 1. Biological demulsifiers.

Microorganism	Biochemical Structure	Emulsion Type	Efficiency (%)	Concentration (g/L)	Reference
Candida sphaerica	Glycolipid	Motor oil/Seawater	40	0.25	[40]
Pseudomonas aeruginosa	Glycolipid	Motor oil/Distilled water	62	8.00	[41]
Halomonas venusta	Glycoprotein	Span–Tween–Kerosene/Water	92	7.30	[39]
Bacillus subtilis	Lipopeptide	Waste crude oil/Distilled water	95	0.20	[42]
Pseudomonas aeruginosa	Glycolipid	Distilled water/n-Hexane	85	-	[43]
Bacillus mojavensis	Oxalate decarboxylase	Tween-60/Deionised water	62	0.17	[44]

Several microorganisms can be used, with different means of production and cultivation conditions. The production variables directly affect the biosurfactant structure, yield, and micelle concentration [45]. The yeast genus *Candida* [46–48] and bacteria from the genera *Pseudomonas* [49,50] and *Bacillus* [25,51] are excellent biosurfactant producers. Species of *Candida* and *Pseudomonas* are major producers of glycolipids, whereas species of *Bacillus* are major producers of lipopeptides [52].

The efficiency of a surfactant is measured by the critical micelle concentration (CMC), whereas effectiveness is related to surface and interfacial tensions. CMC is the lowest concentration of surfactant needed to form micelles. When this condition is reached, micelles are formed and no further reduction in surface tension is achieved [45]. Such properties enable a wide range of industrial applications involving emulsification, lubrication, foaming capacity, wetting capacity, solubilisation, and phase dispersion [53].

However, the cultivation of microorganisms is time-consuming and often costly [3], with unstable demulsification in some cases. Nonetheless, Bach & Gutnickn (2004) [54] state that biodemulsifiers are good alternatives to the chemically synthesised demulsifiers commonly employed in the petroleum industry. Demulsifiers of a microbiological origin have two major advantages over their chemical counterparts: low toxicity and biodegradability.

The solution for emulsions and their demulsification requires detailed knowledge of their formation as well as their stabilisation and destabilisation characteristics. Among the aforementioned methods, chemical demulsification is the most widely employed in the industry and reported in the literature. However, a more effective demulsification process can be achieved by taking advantage of the synergistic effects of the combination of one or more methods [31].

4. Conventional Treatment Methods for Oily Wastewater

Several processes have been developed to treat and purify wastewater contaminated with oil [55]. However, conventional methods can generate high operation costs and involve the use of components that may be toxic, generating secondary contaminating products [56].

Some of the most common examples of conventional methods for the treatment of oil/water effluents are evaporation, gravity separation (decantation or centrifugation), flotation, flocculation, and filtration [1,57]. When used individually, these methods have inefficient removal rates. It is therefore preferable to combine different procedures in order to maximise efficiency [2]. Furthermore, traditional methods can only be used when the oil in the effluent is not emulsified or when it is dispersed at low concentrations [58].

The installation of a water purification system and/or technologies for the treatment of wastewater can lead to considerable reduction in expenses [1]. Thus, new materials and methods have been proposed as alternatives for wastewater treatment [8].

4.1. Evaporation Separation

Separation by evaporation consists of raising the boiling temperature of the solution; as the temperature increases, the solvent, which is the substance in the mixture that has the lowest boiling point, evaporates [59]. This causes the concentration of the solute to increase considerably. In the case of oily effluents (emulsified or not), the solvent is water and the solute is the oil particles. The evaporation process usually takes place in industrial boilers [60].

Thermal and thermo-catalytic methods are also applied for oil recovery and refinement to ensure the lowest possible moisture content. The water is removed during the process in the form of steam and sent to condensers, returning to liquid form with high purity due to this simple distillation process. The water is either redirected to other sectors of the industry, such as equipment cooling, or is discarded [61]. This process is extremely effective, as it removes a high percentage of water from the effluent.

However, steam-based methods use specific equipment that requires a considerable amount of space. Moreover, accidents may occur at the high temperatures and pressures required in the process. Such methods also operate exclusively with the use of fossil fuels, such as gas or diesel, generating high energy consumption and secondary pollution [62].

4.2. Gravity Separation (Decantation or Settling Down)

The decanting process is based on the difference in density of the components in the mixture. Separation boxes can be used for this process, to remove the oil from the surface due to density difference. In some cases, the addition of chemical and/or biological surfactants is necessary to cause the oil to solidify and become denser, facilitating the separation process [1,48,63].

This is a time-consuming method, as the mixture needs to "rest" for the phases to separate. Furthermore, decantation is insufficient for separating emulsified molecules, which require a complementary method for the complete removal of oil particles [64]. In addition to decanting, centrifugation can be performed to accelerate the deposition of the solid particles through centrifugal force, thereby facilitating the separation of phases [65,66].

4.3. Flotation

The aim of the flotation method is to maintain the molecules at the top of the mixture by inserting air bubbles at the bottom, which causes the suspended particles to be dragged upwards. As there is no oil/water interaction, the molecules adhere to the bubbles via a hydrophobic interaction [1,26,67]. The density of the air is lower than the density of the particles and the bubbles ascend in the liquid promoting the occurrence of bubble-particle contact (a process denominated shock). Thus, the solute is floated to the surface. By the addition of a collector (normally a surfactant, which is recovered at the end of the process), the water at the bottom is clarified, as shown in Figure 4 [67,68].

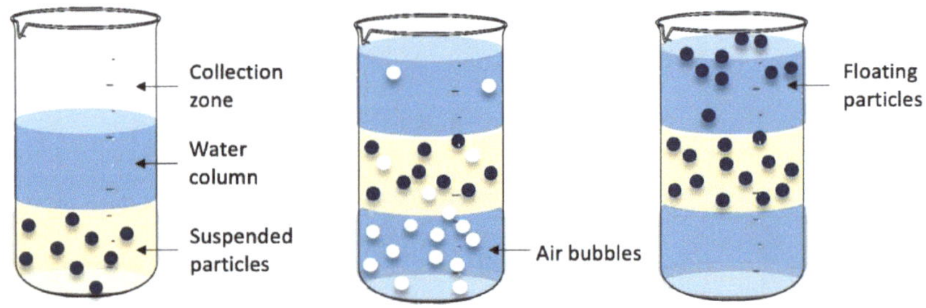

Figure 4. Steps of flotation principle.

Air flotation technology is widely used due to its remarkable separation efficiency. This is a simple process from an operational standpoint. It is also energy efficient and cost effective. This method is applicable to a wide range of oily wastewaters [69] and can be used to obtain higher separation efficiency (up to 80%) even at high loading rates with low retention times [70].

The flotation process is dependent on the surface/interfacial characteristics of the bubbles and the system of particles to be entrained. The treatment of oily wastewater differs from that of other effluents due to the small diameter of the colloidal particles, requiring the use of microbubbles. One of the main setbacks is the limitation to the quantity of microbubbles generated in a certain volume of water. This variable plays a crucial role in separating suspended material from an aqueous medium [1].

Recent research has also revealed that nanobubbles can be generated along with microbubbles in air flotation technology, thereby improving the separation performance and increasing the capture of water pollutants. Microbubble- and nanobubble-assisted flotation are promising techniques for the treatment of drinking water and wastewater, including oily effluents [71,72]. However, some features are essential to their success, such as air resistance, drag, and bubble size distribution, residence and retention times of bubbles in the pulp, solute content, particle size, gravity, surface tension, and the reagents used [67].

4.4. Coagulation (or Flocculation)

Coagulation/flocculation is a process for separating oily solutions, known for its ease of operation, economy, and low production of pollution, making it an attractive treatment option for companies. The removal of oil from wastewater through flocculation is both a chemical and physical process [73].

Coagulation occurs when the repulsive interaction between the electrical double layers is sufficiently reduced through the addition of appropriate chemicals, enabling the particles to approach each other. This process begins with destabilisation, by which the added coagulants tend to neutralise the negative charges of the layers of oily particles in suspension, reducing their electrostatic repulsion [74]. The destabilised particles then gradually begin to aggregate into large flocs. The separation of the flocs takes place through precipitation [75].

The efficiency of flocculation treatments largely depends on the conditions of the process (pH, concentration of oil, temperature, dosage, and type of chemical coagulant/flocculant). According to Yang et al. (2016) [76], when simple charge neutralisation plays a leading role, a low dosage of coagulant/flocculant is not enough to destabilise all colloidal particles. The dosage of coagulating agents, such as inorganic salts, is generally lower than that of polymeric flocculants due to the higher electron density of inorganic salts, requiring a lower dose for charge neutralisation [75].

The most frequently used coagulants/flocculants are divided into three categories according to their chemical composition: organic synthetic polymeric flocculants, inorganic salts, and natural polymeric flocculants. They all act through charge neutralisation, bridging, and sweeping action, facilitating the destabilisation and aggregation of fine oil droplets and suspended colloids. Organic synthetic polymeric flocculants, which mainly include polyacrylamide (PAM), polyacrylic acid (PAA) and their derivatives, exhibit good efficiency due to their large and diversified molecular weight, different structure possibilities (linear and branching), composition, and type of charge. As for the inorganic salts category, aluminium and iron salts are the most widely used and their effectiveness is mainly related to the existence of metal ions in oily wastewater. Finally, natural polymeric flocculants can originate from cellulose, chitosan, starch, and other polysaccharide materials. Several natural polymeric flocculants and combinations with other modified flocculants are also used to improve the coagulation/flocculation performance [75].

Charge neutralisation can occur due to a chemical reaction between positively charged coagulating agents and negatively charged colloidal particles of organic matter or by the shielding of the negatively charged sites, resulting in the precipitation of the particles. A narrow pH range (4–5.5) is considered ideal for charge neutralisation coagulation. The sweep-floc mechanism occurs in the pH 6–8 range, where conditions are suitable for the rapid formation of amorphous solids. In this step, the removal of suspended organic matter occurs by adsorption to the aluminium hydroxide ($Al(OH)_3$) precipitate, which is one of the most widely used flocculating/coagulant agents [77]. Different coagulating agents have a different ideal working pH range. An environment with a very acidic or alkaline pH is not beneficial for the treatment of wastewater with oily emulsions [75].

As mentioned above, temperature has a direct effect on coagulation methods. The optimal concentration of flocculant generally decreases as temperature increases within a suitable range. This is due to the fact that the viscosity of the carrier liquid decreases with the increase in temperature (increase in Brownian movements of colloids and, therefore, their collision frequency) so that larger particles are formed more easily. At low temperatures, the speed of floc formation is extremely slow due to the high viscosity [76].

According to Zhao et al. (2020) [75], the advantages of coagulation/flocculation lie in the fact that this technology is well established, inexpensive, and can efficiently remove oil from contaminated effluents. However, as it is a simple technique, it may be difficult to obtain desirable results when removing very small or dissolved oil particles. It is also of considerable benefit to combine this process with other treating methods, such as filtration or biological treatment, to achieve a better oil removal performance.

4.5. Filtration

Filtration is the flow of fluids through compact beds. This method consists of the mechanical separation of solid particles or particles with a larger molecular structure from a liquid suspension with the aid of a porous bed, such as a membrane, to perform separation of the phases. Filtration is often used in combination with other techniques to improve separation efficiency [57,78].

The classification of the membrane is based on the quantity of particles or residual material that it can retain. The efficiency of a filter membrane is directly related to the pore diameter. The effectiveness of the process can be measured by gravimetric tests with solid materials for coarse filters [79,80].

The conditions of the process determine the type of filter to use. A larger pore size provides a higher flow rate or lower drop in pressure. As a result, the filtration time is faster and less membrane maintenance is required due to the saturation of the pores [81,82]. The disadvantages of this method include the cost of the study to determine the ideal membrane (ultrafiltration, nanofiltration, reverse osmosis), the initial cost of implementation, and the cost of maintenance, which is required after every certain amount of filtered volume [83].

4.6. Sorption Cleaning Methods

Sorption is a general way of referring to two specific processes that occur simultaneously, i.e., adsorption, a process referring to the surface adhesion of molecules that acts according to their chemical and electrostatic interactions, and absorption, a phenomenon related to the assimilation of molecules through the sorbent mass [84].

The term "sorbent" comes from the Latin "sorbere" which means material that absorbs a liquid. This phenomenon occurs until the solid material can no longer accumulate fluid, that is, until the sorbent reaches its maximum sorption point or saturation [85].

The intermolecular forces between the fluid and the sorbent material are of great importance. They determine whether the liquid will tend to spread over the entire volume and surface of the sorbent (absorption) or minimise contact across the surface (adsorption). This mechanism occurs depending on the molecular interactions of the fluid to be sorbed with the sorbent material. Polar chemical species in the fluid have greater attractive force. In contrast, the nonpolar liquid tends to reach a state of conservation of energy, that is, when its molecules interact with each other and decrease contact with the surface [86].

To reduce the impact of oil spill events, technologies based on isolation (booms), oil collection and remediation, in addition to the use of chemical methods such as the application of dispersants, solidifiers, bioreducing agents and even the in situ burning of fuel are commonly applied [84].

A variety of sorbent materials can be effective in remediating these oil spills. Loose sorbent fibres are combined to form snares and sweeps. The particles are either used as bulk loose material or are enclosed in booms. The filling materials of booms are classified into three basic categories. The first category includes inorganic mineral products such as silica, graphite, zeolite, clay, diatomite, glass and sand. Synthetic organic products like polypropylene and polyurethane are the second category of sorbents. These have the highest oil sorption capacity, but one of their main disadvantages is their non-biodegradability. The third category of oil sorbents is of natural origin, they are biodegradable and often composed of agro-industrial residues [87,88].

Recently, great attention has been given to the use of natural sorbents, as they are biodegradable and come from renewable resources. Therefore, pollution of the environment is prevented. Examples include bark, saw dust, peat, wheat and barley straw, sugarcane bagasse and cotton. They are widely used for the control and removal of small aquatic spills [89].

5. Modern Techniques for Cleaning Oily Wastewater

The methods cited previously in this review (evaporation separation, gravity separation, flotation, coagulation/flocculation, and filtration) are considered traditional approaches for treating oily wastewater. However, these methods are often insufficient due to operational difficulties, high costs, the need for considerable factory space, the need for ex situ treatment, the generation of secondary pollutants, and low treatment efficiency [9].

Researchers have been promoting the use of novel techniques and materials for the optimisation of separation processes using multidisciplinary approaches [3]. The implementation of innovative technologies for the treatment of oily effluents provides efficient methods for the removal of these organic pollutants. In some cases, the use of combined technologies is necessary for the complete removal of pollutants [9].

5.1. Biological Treatment

Biological treatment is one of the most widely used methods for the removal of organic compounds from wastewater. These forms of treatment are classified as aerobic or anaerobic. Anaerobic systems involve less energy expenditure, can convert organic pollutants into methane (CH_4), consume fewer nutrients, and produce less pollution [90]. Aerobic biological treatments require the presence of oxygen and basic nutrients to treat concentrated wastewater, operating at a high temperature and producing a high content of pollutants due to the accelerated biodegradation kinetics. Anaerobic and aerobic systems can also be combined to treat oily wastewater without the need for any pre-treatment. This approach leads to an improvement in treatment efficiency as well as reductions in the cost and space required for implementation [91].

Some microorganisms degrade oil and ingest the oil droplets as part of their natural diet [92]. The use of microorganisms is considered very effective, economical, and sustainable for the treatment of oily effluents. Bacteria are the most widely used agents for the degradation of oil and petroleum products. These organisms function as primary degraders of oil through the production of lipase or other by-products [93]. Biosurfactants are among these metabolic by-products and assist in the degradation of biological organic compounds by increasing their solubility through emulsification, reducing the interfacial tension of the oil and forming micelles [94,95]. The biodegradation of hydrocarbons by microorganisms and their metabolites enables the conversion of hazardous substances into less toxic or non-toxic forms. This method is one of the main mechanisms by which petroleum products are removed from the environment in a simple, economical way, known as bioremediation [96,97].

5.2. Advanced Treatment Processes

Among chemical methods, advanced oxidation processes (AOPs) are widely considered for the treatment of effluents with organic residues due to their rapid oxidation reaction rates and absence of secondary pollution [98,99]. These processes are also efficient in the inactivation of pathogenic microorganisms and are therefore competitive in relation to other effluent treatment technologies [100].

AOPs involve in situ generation of highly reactive oxygen species with low selectivity, providing pathways for complete mineralisation, i.e., the conversion of micropollutants into carbon dioxide, water, and inorganic or acidic ions [101]. Supercritical water oxidation (SCWO), electrochemical catalysis, oxidation, ozonation, Fenton and photo-Fenton processes, photocatalysis, radiation, and sonolysis are examples of AOPs [102]. These processes can be divided into two main groups: photochemical, and thermal or non-photochemical methods [101]. AOPs offer high environmental compatibility in addition to being versatile, easy to apply, and safe, as such processes are carried out under ambient conditions [103,104]. Although AOPs have gained prominence for the treatment of effluents, some processes have limitations in terms of high operating costs [99].

SCWO is considered one of the most effective and promising advanced methods for the in situ and ex situ conversion of high molecular weight compounds [105]. This method involves operations with conditions above the critical point of water (374.3 °C and 22.12 mPa) [106]. SCWO consists of hydrothermal combustion that occurs in supercritical water accompanied by rapid homogeneous oxidation reaction with the components in the solution. The process starts with pressurisation of the equipment, with the addition of oxygen and a specific kind of additive so that the oxidation reaction can occur. Heating is then performed to the critical point temperature. When the reaction is completed, there is the possibility of thermal energy reutilisation, generating electric energy for industry (Figure 5). One of the great advantages of this method is that it can easily degrade toxic compounds, converting them into harmless products, such as CO_2, water, and N_2 [107,108].

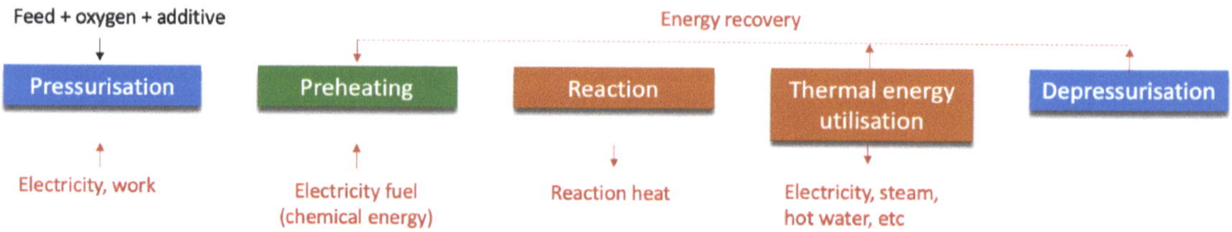

Figure 5. Basic diagram for supercritical water oxidation processes.

Despite being a very effective technique, SCWO has some disadvantages related mainly to operational and installation costs as well as the high possibility of corrosion of pipes and equipment [109]. Equipment corrosion is practically unavoidable during the process, due to the conditions necessary to carry out the reaction. Corrosive anions are formed after excessive injection of oxygen, which accelerates the corrosion rate [110]. Therefore, many SCWO plants are forced to pause or even completely interrupt their operations due to severe corrosion of system components [108]. However, several methods of salt obstruction and corrosion control are being developed to increase the useful life of the equipment, such as coating the equipment and pipes with corrosion-resistant material, pre-neutralisation, and adsorption and/or reaction in the fluidised solid phase [108,111].

Electrochemical catalytic processes have several advantages over other water treatment methods. This is a robust, easy-to-operate process, especially in cases of wastewater load fluctuations [112]. A study on the treatment of effluent from an offshore natural gas extraction platform used a Ti/Sb-SnO_2 anode modified with graphene oxide, resulting in a 58.60% reduction in chemical oxygen demand and energy efficiency of 42.63 g/Kwh [113]. Although considered one of the most efficient processes for oily water treatment, the high costs of electrodes, electrochemical catalysts, and specific coatings make it challenging to treat large volumes of effluents [90,112].

Several factors need to be considered for the appropriate use of advanced treatment processes. Efficient, inexpensive catalysts are key to the success of photocatalytic oxidation technologies [99]. For Fenton oxidation technology, an extremely acidic environmental pH of less than 3 is required for high treatment efficiency. Regarding oxidation by ozonation, optimum performance requires an alkaline environment. According to Ma et al. (2021) [99], the combination of different AOPs is necessary to circumvent the high operational costs and limited efficiency. Moreover, the improvement of the fundamentals of oxidation technology can contribute to the advancement of AOPs [99].

5.3. Membrane Separation Technology (Polymeric and Ceramic Membranes)

A membrane consists of a physical structure with specific porosity. This material is most often found in a flat and smooth macroscopic conformation and the main objective is the separation of a mixture composed of two or more components [56].

The membrane controls the passage of mass between environments, with the aid of interaction between the material to be separated and the porous surface of the membrane. Membranes occur in a wide variety of conformations and structures. The efficiency and separation rate are attributed to the membrane configuration system that is being used [57].

The use of membranes follows specific parameters according to the hydrophilicity of the material, its pore size, the flow, and the pressure supported. These are of extreme importance to maximising membrane performance for the desired application [56]. Pore size establishes the size of the particle that can pass through the membrane. This aspect is what defines classifications and specific membrane applications, as shown in Table 2.

The advantages of membrane separation processes include its lower energy demand, easy handling and maintenance, and no need to apply chemicals. The treatment process has considerable efficiency when compared to conventional techniques [56].

Table 2. Classification of membranes and respective applications.

Classification	Pore size (nm)	Retention/Removal	Reference
Microfiltration (MF)	100–5000	Suspended particles, macromolecules, fungi, and bacteria	[79]
Ultrafiltration (UF)	2–100	Proteins and viruses	[114]
Nanofiltration (NF)	0.5–2	Dissolved organic matter, heavy metals, and multivalent ions	[114]
Reverse Osmosis (RO)	0.2–1	Monovalent salts and ions (ultrapure water)	[115]

However, membrane treatment also has disadvantages. As stated by Hassan et al. (2017) [116], more than 95% of filtration membranes are produced from synthetic polymers of a fossil origin and this production requires the use of strong solvents. Therefore, this technology cannot be considered completely clean. Moreover, membranes cannot be reused, and require specific disposal. The oily effluent greatly increases their degradation time, and they can be harmful to the environment [2]. Thus, there has been growing interest in the development of novel, sustainable, biodegradable products created from natural polymers, particularly those based on nanocellulose, such as cellulose nanofibers and bacterial cellulose [117–121].

Biotechnological Filter/Strainer

Membrane filters are being developed to circumvent the disadvantages related to conventional treatment methods. Such membranes can be generally used with or without chemical additives and can be adjusted in closed, automated cycles of operation [122]. Studies also show that the efficiency of filters is greater in comparison to traditional physical methods. This is directly linked to the material used to compose this physical barrier [56].

Studies have described the use of natural polymers, due to their efficiency and sustainable characteristics [2]. Microbial cellulose is one such material. This compound has a structure chemically equivalent to plant cellulose [57,123,124]. It is produced extracellularly by aerobic bacteria in the form of a hydrated membrane, as demonstrated in Figure 6 [125]. Bacterial cellulose (BC) is a natural biomaterial that has been gaining attention demonstrating properties that are suitable for various applications [24,126].

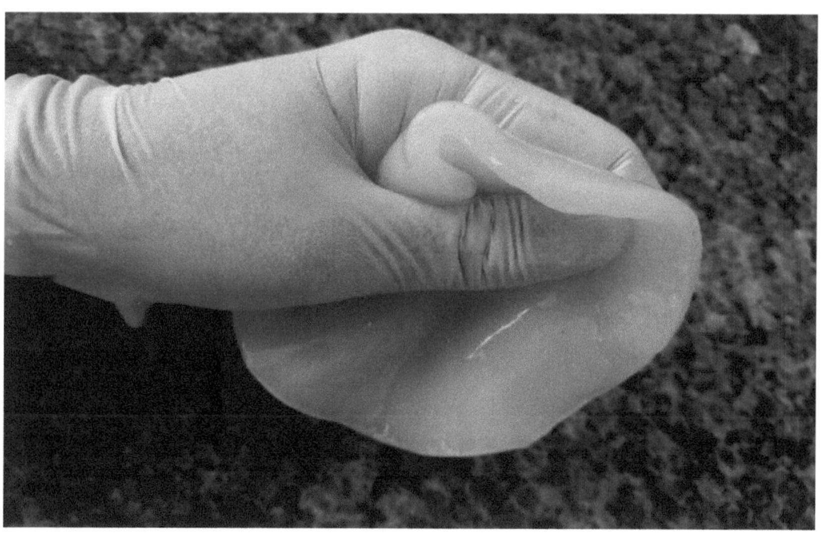

Figure 6. Bacterial cellulose membrane.

BC is a biopolymer produced by several microorganisms either in the form of a consortium [2] or alone, as in the case of bacteria of the genus *Komagataeibacter* [127]. Recent studies indicate BC to be a very promising biopolymer as an alternative to the traditional production of plant cellulose due to its wide variety of applications [57,118–120]. Donini et al. (2010) [128] state that BC has excellent mechanical properties, high crystallinity, high water retention capacity, a high degree of polymerisation, and good biodegradability [129,130]. Due to its hydrophilic character, BC increases the efficiency of the process, while keeping the final residue (water) within the environmental standards for disposal or within the parameters for reuse within the same company (Figure 7) [2].

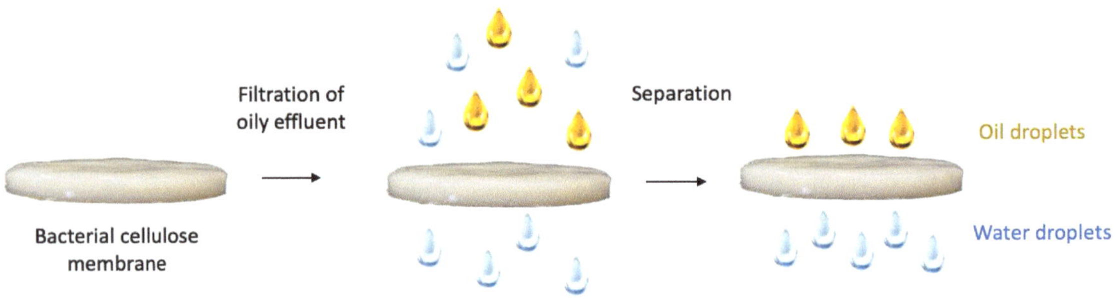

Figure 7. Schematic representation of oil-water separation process using BC as filter membrane.

A considerable advantage of BC over conventional membranes is its washability. Membranes with a reduced filtration capacity (saturated) can be removed from the filter system, washed, and reused several times without losing filtration efficiency [2].

Its nanostructure and degree of polymerisation give BC useful mechanical properties for the development of technological biomaterials [120]. Such characteristics demonstrate the advantages of using microbiological polymers, as exemplified in Table 3.

Table 3. Comparison of properties of bacterial cellulose and other polymers.

Properties	Bacterial Cellulose Matrix	Vegetable Cellulose Matrix	Polypropylene Matrix
Crystallinity degree (%)	90	62	~55
Fibre size (nm)	75	315	450
Fibre density (g cm^{-3})	1.5	0.99	0.95
Sensitivity of fibres to water	Low	High	Low
Traction force (N)	~70	47	~60
Specific deformation (%)	16	9	50
Young's modulus (GPa)	5.0	0.85	1.90
Tension force (MPa)	85	0.83	50

Galdino et al. (2020) [2] demonstrated that BC membranes are able to remove oil from synthetic effluents nearly entirely and can be washed and reused more than 20 times without losing their structural characteristics or filtering capacity. Lehtonen et al. (2021) [57] point out that control of the pressure and flow parameters is key to optimising the use of BC for better application as a filtering material on an industrial scale.

The disadvantages of this type of treatment are related to its nanometric fibres. This feature allows the bacterial cellulose membrane to remove approximately 100% of the oil present in the solution [2]. However, there is the problem of easy membrane saturation due to the relation between the membrane's low porosity and the size of the oil molecules to be filtered or adsorbed.

Several studies have been performed with the aim of modifying the BC's structure by different production or processing methods [116,131–133]. However, the lack of specific studies addressing membrane saturation and the optimization of its filtration rate remains an obstacle for the implementation of such technology in the industrial sector.

6. Critical Analysis and Future Trends

As oily wastewaters are common in nearly all industries, research on treatment methods for these toxic effluents is aimed at increasing efficiency and decreasing production costs.

The present review has cited general methods currently used by industries for the treatment of oily wastewater, and described their main characteristics. It has also pointed out the considerable potential of novel biotechnological solutions, specifically the use of bacterial cellulose as a membrane in separation processes, which, in contrast with traditional membranes, has the advantage of reusability, therefore ensuring a sustainable process.

The coagulation/flocculation method stands out among traditional techniques because it is simple, economical and can be sustainable with the use of natural polymeric flocculants in the process. However, AOPs are also quite promising because in addition to treating polluted effluents with rapid reaction rate and with the absence of secondary pollution, they are efficient in the inactivation of pathogenic microorganisms.

Many currently employed methods function for only a specific type of oily particles. In some cases, therefore, it is necessary to use a combination of more than one method for the effective treatment of more complex mixtures. The combination of conventional treatments with modern techniques can improve the performance of removing oil presenting all types of compositions and a range of particle sizes.

To be considered successful, the treatment of industrial effluents, particularly those from the oil industry, must be sustainable, biocompatible, and biodegradable. Specific studies are needed to optimise the use of certain types of treatment to enable the replacement of current methods on an industrial scale. New technologies, especially involving microbiological solutions, are expected to enable a reduction in energy costs of the treatment of wastewater as well as improve treatment and generate less waste, with lower emission of pollutants.

Author Contributions: Conceptualisation, L.A.S. and A.D.M.d.M.; validation, L.A.S., A.D.M.d.M., C.J.G.d.S.J., J.D.P.d.A. and A.F.d.S.C.; writing—original draft preparation, A.D.M.d.M., C.J.G.d.S.J. and J.D.P.d.A.; writing—review and editing, A.D.M.d.M., C.J.G.d.S.J., J.D.P.d.A., I.J.B.D. and A.F.d.S.C.; visualisation, A.D.M.d.M., A.F.d.S.C. and L.A.S.; supervision, L.A.S.; project administration, L.A.S.; funding acquisition, L.A.S. All authors have read and agreed to the published version of the manuscript.

Funding: This study was funded by Programa de Pesquisa e Desenvolvimento da Agência Nacional de Energia Elétrica (ANEEL)/Thermoelectric EPESA (Centrais Elétricas de Pernambuco S.A.)/SUAPE Ambiental and by the Brazilian fostering agencies Fundação de Apoio à Ciência e Tecnologia do Estado de Pernambuco (FACEPE), Conselho Nacional de Desenvolvimento Científico e Tecnológico (CNPq), and Coordenação de Aperfeiçoamento de Pessoal de Nível Superior (CAPES) (Grant n. Finance Code 001).

Acknowledgments: The authors are grateful to Centro de Ciências e Tecnologia of Universidade Católica de Pernambuco (UNICAP) and Instituto Avançado de Tecnologia e Inovação (IATI), Brazil.

Conflicts of Interest: The authors declare no conflict of interest.

References

1. Rocha e Silva, F.C.P.; Rocha e Silva, N.M.P.; Luna, J.M.; Rufino, R.D.; Santos, V.A.; Sarubbo, L.A. Dissolved air flotation combined to biosurfactants: A clean and efficient alternative to treat industrial oily water. *Rev. Environ. Sci. Biotechnol.* **2018**, *17*, 591–602. [CrossRef]
2. Galdino, C.J.S.; Maia, A.D.M.; Meira, H.M.; Souza, T.C.; Amorim, J.D.P.; Almeida, F.C.G.; Costa, A.F.S.; Sarubbo, L.A. Use of a bacterial cellulose filter for the removal of oil from wastewater. *Process Biochem.* **2020**, *91*, 288–296. [CrossRef]
3. Abuhasel, K.; Kchaou, M.; Alquraish, M.; Munusamy, Y.; Jeng, Y.T. Oily Wastewater Treatment: Overview of Conventional and Modern Methods, Challenges, and Future Opportunities. *Water* **2021**, *13*, 980. [CrossRef]
4. Zouboulis, A.I.; Avranas, A. Treatment of oil-in-water emulsions by coagulation and dissolved air flotation. *Colloids Surf. A Physicochem. Eng. Asp.* **2000**, *172*, 153–161. [CrossRef]
5. Demore, J.P. *Aspectos Sedimentares do Estuário da Lagoa dos Patos e Sua Interação com a Poluição por Petróleo: Subsídios Para um Plano de Contingência. Monography;* Universidade Federal do Rio Grande: Rio Grande, Brazil, 2001.
6. Njoku, K.L. Responses of accessions of *Zea mays* to crude oil pollution using growth indices and enzyme activities as markers. *Pollution* **2017**, *4*, 183–193. [CrossRef]

7. Rajasulochana, P.; Preethy, V. Comparison on efficiency of various techniques in treatment of waste and sewage water—A comprehensive review. *Resour.-Effic. Technol.* **2016**, *2*, 175–184. [CrossRef]
8. El-Gawad, H.S.A. Oil and grease removal from industrial wastewater using new utility approach. *Adv. Environ. Chem.* **2014**, *2014*, 916878. [CrossRef]
9. Adetunji, A.I.; Olaniran, A.O. Treatment of industrial oily wastewater by advanced technologies: A review. *Appl. Water Sci.* **2021**, *11*, 98. [CrossRef]
10. Zadymova, N.M.; Skvortsova, Z.N.; Traskine, V.Y.; Kulikov-Kostyushko, F.A.; Kulichikhin, V.G.; Malkin, A.Y. Rheological properties of heavy oil emulsions with different morphologies. *J. Pet. Sci. Eng.* **2017**, *149*, 522–530. [CrossRef]
11. Yu, L.; Han, M.; He, F. A review of treating oily wastewater. *Arab. J. Chem.* **2017**, *10*, 1913–1922. [CrossRef]
12. Putatunda, S.; Bhattacharya, S.; Sen, D.; Bhattacharjee, C. A review on the application of different treatment processes for emulsified oily wastewater. *Int. J. Environ. Sci. Technol.* **2019**, *16*, 2525–2536. [CrossRef]
13. Li, M.; Deng, L.; Tan, Y.; Qi, K.; Tian, X.; Yu, J.; Qin, C.; Cheng, S. Superhydrophobic/Superoleophilic Polyacrylonitrile/Ag Aerogels for the High Efficient Oil/Water Separation and Sensitive Detection of Low-Concentration Oily Sudan Dyes. *Adv. Mater. Interfaces* **2021**, *8*, 2002174. [CrossRef]
14. Chen, G.H.; He, G.H. Separation of water and oil from water-in-oil emulsion by freeze/thaw method. *Sep. Purif. Technol.* **2003**, *31*, 83–89. [CrossRef]
15. Wen, J.; Zhang, J.; Wang, Z.; Zhang, Y. Correlations between emulsification behaviors of crude oil–water systems and crude oil compositions. *J. Pet. Sci. Eng.* **2016**, *146*, 1–9. [CrossRef]
16. Luo, F.; He, L.; He, N. Simulation and experimental study of working characteristics of an improved bioreactor for degrading oily sludge. *Process Saf. Environ. Prot.* **2021**, *147*, 1201–1208. [CrossRef]
17. Srinivasan, A.; Viraraghavan, T. Oil removal from water using biomaterials. *Bioresour. Technol.* **2010**, *101*, 6594–6600. [CrossRef]
18. Lim, J.S.; Wong, S.F.; Law, M.C.; Samyudia, Y.; Dol, S.S. A review on the effects of emulsions on flow behaviours and common factors affecting the stability of emulsions. *J. Appl. Sci.* **2015**, *15*, 167–172. [CrossRef]
19. Khan, B.A.; Akhtar, N.; Khan, H.M.S.; Waseem, K.; Mahmood, T.; Rasul, A.; Iqbal, M.; Khan, H. Basics of pharmaceutical emulsions: A review. *Afr. J. Pharm. Pharmacol.* **2011**, *5*, 2715–2725. [CrossRef]
20. Wong, S.F.; Lim, J.S.; Dol, S.S. Crude oil emulsion: A review on formation, classification and stability of water-in-oil emulsions. *J. Pet. Sci. Eng.* **2015**, *135*, 498–504. [CrossRef]
21. Fingas, M. Water-in-oil emulsion formation: A review of physics and mathematical modelling. *Spill Sci. Technol. Bull.* **1995**, *2*, 55–59. [CrossRef]
22. Karhu, M.; Kuokkanen, V.; Kuokkanen, T.; Ramo, J. Bench scale electrocoagulation studies of bio oil-in-water and synthetic oil-in-water emulsions. *Sep. Purif. Technol.* **2012**, *96*, 296–305. [CrossRef]
23. Shaw, D.J. *Introduction to Colloid and Surface Chemistry*; Butterworths: London, UK, 1975.
24. Adrio, J.L.; Demain, A.L. Microbial enzymes: Tools for biotechnological processes. *Biomolecules* **2014**, *1*, 117–139. [CrossRef]
25. Durval, I.J.B.; Resende, A.H.M.; Figueiredo, M.A.; Luna, J.M.; Rufino, R.D.; Sarubbo, L.A. Studies on Biosurfactants Produced using *Bacillus cereus* Isolated from Seawater with Biotechnological Potential for Marine Oil-Spill Bioremediation. *J. Surfactants Deterg.* **2018**, *22*, 349–363. [CrossRef]
26. Rocha e Silva, N.M.P.; Almeida, F.C.G.; Silva, F.C.P.R.; Luna, J.M.; Sarubbo, L.A. Formulation of a Biodegradable Detergent for Cleaning Oily Residues Generated during Industrial Processes. *J. Surfactants Deterg.* **2020**, *23*, 1111–1123. [CrossRef]
27. Kulik, K.; Trapido, M.; Veressinina, Y.; Munter, R. Treatment of surfactant stabillized oil-in-water emulsions by means of chemical oxidation and coagulation. *Environ. Technol.* **2007**, *28*, 1345–1355. [CrossRef] [PubMed]
28. Santander, M.; Rodrigues, R.T.; Rubio, J. Modified jet flotation in oil (petroleum) emulsion/water separations. *Colloids Surf. A Physicochem. Eng. Asp.* **2011**, *375*, 237–244. [CrossRef]
29. Zolfaghari, R.; Fakhru'l-Razi, A.; Abdullah, L.C.; Elnashaie, S.S.E.H.; Pendashteh, A. Demulsification techniques of water-in-oil and oil-in-water emulsions in petroleum industry. *Sep. Purif. Technol.* **2016**, *170*, 377–407. [CrossRef]
30. Liu, J.; Wang, H.; Li, X.; Jia, W.; Zhao, Y.; Ren, S. Recyclable magnetic graphene oxide for rapid and efficient demulsification of crude oil-in-water emulsion. *Fuel* **2017**, *189*, 79–87. [CrossRef]
31. Abed, S.M.; Abdurahman, N.H.; Yunus, R.M.; Abdulbari, H.; Akbari, S. Oil emulsions and the different recent demulsification techniques in the petroleum industry—A review. *IOP Conf. Ser. Mater. Sci. Eng.* **2019**, *702*, 012060. [CrossRef]
32. Liang, H.; Esmaeili, H. Application of nanomaterials for demulsification of oily wastewater: A review study. *Environ. Technol. Innov.* **2021**, *22*, 101498. [CrossRef]
33. Daaou, M.; Bendedouch, D. Water pH and surfactant addition effects on the stability of an Algerian crude oil emulsion. *J. Saudi Chem. Soc.* **2012**, *16*, 333–337. [CrossRef]
34. Du, Y.; Si, P.; Wei, L.; Wang, Y.; Tu, Y.; Zuo, G.; Yu, B.; Zhang, X.; Ye, S. Demulsification of acidic oil-in-water emulsions driven by chitosan loaded Ti3C2Tx. *Appl. Surf. Sci.* **2019**, *476*, 878–885. [CrossRef]
35. Evdokimov, I.N.; Losev, A.P. Microwave treatment of crude oil emulsions: Effects of water content. *J. Pet. Sci. Eng.* **2014**, *115*, 24–30. [CrossRef]
36. Spinelli, L.S.; Aquino, A.S.; Pires, R.V.; Barboza, E.M.; Louvisse, A.M.T.; Lucas, E.F. Influence of polymer bases on the synergistic effects obtained from mixtures of additives in the petroleum industry: Performance and residue formation. *J. Pet. Sci. Eng.* **2007**, *58*, 111–118. [CrossRef]

37. Pérez-Calderón, J.; Santos, M.V.; Zaritzky, N. Optimal clarification of emulsified oily wastewater using a surfactant/chitosan biopolymer. *J. Environ. Chem. Eng.* **2018**, *6*, 3808–3818. [CrossRef]
38. Dao, V.H.; Cameron, N.R.; Saito, K.; Bendoraitiene, J.; Zemaitaitis, A.; Liu, G.; Ye, J.; Cheng, R.; Cheng, R.; Lan, N.T. Synthesis, properties and performance of organic polymers employed in flocculation applications. *Polym. Chem.* **2016**, *7*, 11–25. [CrossRef]
39. Cai, Q.; Zhu, Z.; Chen, B.; Zhang, B. Oil-in-water emulsion breaking marine bacteria for demulsifying oily wastewater. *Water. Res.* **2019**, *149*, 292–301. [CrossRef]
40. Luna, J.M.; Rufino, R.D.; Jara, A.M.A.T.; Brasileiro, P.P.F.; Sarubbo, L.A. Environmental applications of the biosurfactant produced by *Candida sphaerica* cultivated in low-cost substrates. *Colloids Surf. A Physicochem. Eng. Asp.* **2015**, *480*, 413–418. [CrossRef]
41. Rocha e Silva, F.C.P.; Roque, B.A.C.; Rocha e Silva, N.M.; Rufino, R.D.; Luna, J.M.; Santos, V.A.; Banat, I.M.; Sarubbo, L.A. Yeasts and bacterial biosurfactants as demulsifiers for petroleum derivative in seawater emulsions. *AMB Express* **2017**, *7*, 202. [CrossRef]
42. Yang, Z.; Zu, Y.; Zhu, J.; Jin, M.; Cui, T.; Long, X. Application of biosurfactant surfactin as a pH-switchable biodemulsifier for efficient oil recovery from waste crude oil. *Chemosphere* **2020**, *240*, 124946. [CrossRef]
43. Gidudu, B.; Chirwa, E.M.N. Biosurfactants as demulsification enhancers in bio-electrokinetic remediation of petroleum contaminated soil. *Process Saf. Environ. Prot.* **2020**, *143*, 332–339. [CrossRef]
44. Hou, N.; Wang, Q.; Sun, Y.; Li, X.; Song, Q.; Jiang, X.; Li, B.; Zhao, X.; Zang, H.; Li, D. A novel biodemulsifier of *Bacillus mojavensis* XH1–Oxalate decarboxylase with the potential for demulsification of oilfield emulsion. *J. Hazard. Mater.* **2021**, *407*, 124737. [CrossRef]
45. Rocha e Silva, N.M.; Meira, H.M.; Almeida, F.C.G.; Silva, R.C.F.S.; Almeida, D.G.; Luna, J.M.; Rufino, R.D.; Santos, V.A.; Sarubbo, L.A. Natural Surfactants and Their Applications for Heavy Oil Removal in Industry. *Sep. Purif. Rev.* **2018**, *48*, 267–281. [CrossRef]
46. Sarubbo, L.A.; Luna, J.M.; Rufino, R.D.; Brasileiro, P. Production of a low-cost biosurfactant for application in the remediation of sea water contaminated with petroleum derivates. *Chem. Eng. Trans.* **2016**, *49*, 523–528. [CrossRef]
47. Santos, D.K.F.; Meira, H.M.; Rufino, R.D.; Luna, J.M.; Sarubbo, L.A. Biosurfactant production from *Candida lipolytica* in bioreactor and evaluation of its toxicity for application as a bioremediation agent. *Process Biochem.* **2017**, *54*, 20–27. [CrossRef]
48. Silva, I.A.; Veras, B.O.; Ribeiro, B.G.; Aguiar, J.S.; Campos Guerra, J.M.; Luna, J.M.; Sarubbo, L.A. Production of cupcake-like dessert containing microbial biosurfactant as an emulsifier. *PeerJ* **2020**, *8*, e9064. [CrossRef] [PubMed]
49. Chen, C.; Sun, N.; Li, D.; Long, S.; Tang, X.; Xiao, G.; Wang, L. Optimization and characterization of biosurfactant production from kitchen waste oil using *Pseudomonas aeruginosa*. *Environ. Sci. Pollut. Res.* **2018**, *25*, 14934–14943. [CrossRef] [PubMed]
50. Silva, R.C.F.S.; Almeida, D.G.; Brasileiro, P.P.F.; Rufino, R.D.; Luna, J.M.; Sarubbo, L.A. Production, formulation and cost estimation of a commercial biosurfactant. *Biodegradation* **2018**, *30*, 191–201. [CrossRef] [PubMed]
51. Felix, A.K.N.; Martins, J.J.L.; Almeida, J.G.L.; Giro, M.E.A.; Cavalcante, K.F.; Melo, V.M.M.; Pessoa, O.D.L.; Rocha, M.V.P.; Gonçalves, L.R.B.; Aguiar, R.S.S. Purification and characterization of a biosurfactant produced by *Bacillus subtilis* in cashew apple juice and its application in the remediation of oil-contaminated soil. *Colloids Surf. B Biointerfaces* **2019**, *175*, 256–263. [CrossRef] [PubMed]
52. Markande, A.R.; Patel, D.; Varjani, S. A review on biosurfactants: Properties, applications and current developments. *Bioresour. Technol.* **2021**, *330*, 124963. [CrossRef]
53. Singh, P.; Patil, Y.; Rale, V. Biosurfactant production: Emerging trends and promising strategies. *J. Appl. Microbiol.* **2019**, *126*, 2–13. [CrossRef] [PubMed]
54. Bach, H.; Gutnick, D.L. Potential applications of bioemulsifiers in the oil industry. *Stud. Surf. Sci. Catal.* **2004**, *151*, 233–281. [CrossRef]
55. Pendergast, M.M.; Hoek, E.M.V. A review of water treatment membrane nanotechnologies. *Energy Environ. Sci.* **2011**, *4*, 1946–1971. [CrossRef]
56. Padaki, M.; Surya Murali, R.; Abdullah, M.S.; Misdan, N.; Moslehyani, A.; Kassim, M.A.; Hilal, N.; Ismail, A.F. Membrane technology enhancement in oil–water separation. A review. *Desalination* **2015**, *357*, 197–207. [CrossRef]
57. Lehtonen, J.; Chen, X.; Beaumont, M.; Hassinen, J.; Orelma, H.; Dumée, L.F.; Tardy, B.L.; Rojas, O.J. Impact of incubation conditions and post-treatment on the properties of bacterial cellulose membranes for pressure-driven filtration. *Carbohydr. Polym.* **2021**, *251*, 117073–117082. [CrossRef]
58. Hu, X.; Bekassy-Molnar, E.; Vatai, G. Study of ultrafiltration behaviour of emulsified metalworking fluids. *Desalination* **2002**, *149*, 191–197. [CrossRef]
59. Verevkin, S.P.; Zherikova, K.V.; Martynenko, E.A. Molecular versus ionic liquids: Development of a thermodynamic framework for predicting vaporization thermodynamics. *J. Mol. Liq.* **2022**, *350*, 118576. [CrossRef]
60. Aminmahalati, A.; Fazlali, A.; Safikhani, H. Study on the performance and optimization of CO boiler in the oil refinery. *Appl. Therm. Eng.* **2022**, *201*, 117790. [CrossRef]
61. Djimasbe, R.; Varfolomeev, M.A.; Al-Muntaser, A.A.; Yuan, C.; Feoktistov, D.A.; Suwaid, M.A.; Kirgizov, A.J.; Davletshin, R.R.; Zinnatullin, A.L.; Fatou, S.D. Oil dispersed nickel-based catalyst for catalytic upgrading of heavy oil using supercritical water. *Fuel* **2022**, *313*, 122702. [CrossRef]
62. Douglas, L.D.; Rivera-Gonzalez, N.; Cool, N.; Bajpayee, A.; Udayakantha, M.; Liu, G.-W.; Anita; Banerjee, S. A Materials Science Perspective of Midstream Challenges in the Utilization of Heavy Crude Oil. *ACS Omega* **2022**, *7*, 1547–1574. [CrossRef]
63. Azimi, S.C.; Shirini, F.; Pendashteh, A. Treatment of wood industry wastewater by combined coagulation–flocculation–decantation and fenton process. *Water Environ. Res.* **2020**, *93*, 433–444. [CrossRef]

64. Bach, M.; Schoenbrunn, F. Liquid–Solid Separation: De-sandin, flocculation, sedimentation and liquor filtration. In *Smelter Grade Alumina Bauxite*; Springer: Cham, Switzerland, 2022; pp. 241–267. [CrossRef]
65. Nigro, A.; Finardi, A.; Ferraro, M.M.; Manno, D.E.; Quattrini, A.; Furlan, R.; Romano, A. Selective loss of microvesicles is a major issue of the differential centrifugation isolation protocols. *Sci. Rep.* **2021**, *11*, 3589–3601. [CrossRef] [PubMed]
66. Casadei, E.; Valli, E.; Panni, F.; Donarski, J.; Gubern, J.F.; Lucci, P.; Conte, L.; Lacoste, F.; Maquet, A.; Brereton, P. Emerging trends in olive oil fraud and possible countermeasures. *Food Control.* **2021**, *124*, 107902–107915. [CrossRef]
67. Rocha e Silva, F.C.P.; Rocha e Silva, N.M.P.; Moura, A.E.; Almeida, R.G.; Luna, J.M.; Rufino, R.D.; Santos, V.A.; Sarubbo, L.A. Effect of biosurfactant addition in a pilot scale dissolved air flotation system. *Sep. Sci. Technol.* **2015**, *50*, 618–625. [CrossRef]
68. Menezes, C.T.B.; Barros, E.C.; Rufino, R.D.; Luna, J.M.; Sarubbo, L.A. Replacing Synthetic with Microbial Surfactants as Collectors in the Treatment of Aqueous Effluent Produced by Acid Mine Drainage, Using the Dissolved Air Flotation Technique. *Appl. Biochem. Biotechnol.* **2011**, *163*, 540–546. [CrossRef]
69. Wang, L.K.; Shammas, N.K.; Selke, W.A.; Aulenbach, D.B. *Flotation Technology*; Humana Press: New York, NY, USA, 2010; Volume 12, p. 680.
70. Yao, K.; Chi, Y.; Wang, F.; Yan, Y.; Ni, M.; Cen, K. The effect of microbubbles on gas-liquid mass transfer coefficient and degradation rate of COD in wastewater treatment. *Water Sci. Technol.* **2016**, *73*, 1969–1977. [CrossRef]
71. Azevedo, A.; Etchepare, R.; Rubio, J. Raw water clarification by flotation with microbubbles and nanobubbles generated with a multiphase pump. *Water Sci. Technol.* **2017**, *75*, 2342–2349. [CrossRef]
72. Brasileiro, P.P.F.; Silva, R.C.F.S.; Santos, V.A.; Sarubbo, L.A.; Benachour, M. Efficiency of microbubble production using surfactants for the treatment of oily water by flotation. *Chem. Eng. Res. Des.* **2021**, *168*, 254–263. [CrossRef]
73. Wen, X.; Du, C.; Zeng, G.; Huang, D.; Zhang, J.; Yin, L.; Tan, S.; Huang, L.; Chen, H.; Yu, G. A novel biosorbent prepared by immobilized Bacillus licheniformis for lead removal from wastewater. *Chemosphere* **2018**, *200*, 173–179. [CrossRef]
74. Wei, H.; Gao, B.; Ren, J.; Li, A.; Yang, H. Coagulation/flocculation in dewatering of sludge: A review. *Water Res.* **2018**, *143*, 608–631. [CrossRef]
75. Zhao, C.; Zhou, J.; Yan, Y.; Yang, L.; Xing, G.; Li, H.; Wu, P.; Wang, M.; Zheng, H. Application of coagulation/flocculation in oily wastewater treatment: A review. *Sci. Total Environ.* **2021**, *765*, 142795. [CrossRef] [PubMed]
76. Yang, R.; Li, H.; Huang, M.; Yang, H.; Li, A. A review on chitosan-based flocculants and their applications in water treatment. *Water Res.* **2016**, *95*, 59–89. [CrossRef] [PubMed]
77. Ahmad, A.L.; Ismail, S.; Bhatia, S. Optimization of Coagulation−Flocculation Process for Palm Oil Mill Effluent Using Response Surface Methodology. *Environ. Sci. Technol.* **2005**, *39*, 2828–2834. [CrossRef] [PubMed]
78. Silva, W.E.; Belian, M.F.; Lima, L.S.G.L.; Galembeck, A.; Alves, A.A. *BR 10 2018 009736 9 A2–Filtros à Base de Membrana de Celulose Bacteriana Modificada*; INPI: Rio de Janeiro, Brazil, 2018.
79. Anis, S.F.; Hashaikeh, R.; Hilal, N. Microfiltration membrane processes: A review of research trends over the past decade. *J. Water Process Eng.* **2019**, *32*, 100941. [CrossRef]
80. Tanudjaja, H.J.; Hejase, C.A.; Tarabara, V.V.; Fane, A.G.; Chew, J.W. Membrane-based separation for oily wastewater: A practical perspective. *Water Res.* **2019**, *156*, 347–365. [CrossRef]
81. Zheng, Y.; Yu, S.; Shuai, S.; Zhou, Q.; Cheng, Q.; Liu, M.; Gao, C. Color removal and COD reduction of biologically treated textile effluent through submerged filtration using hollow fiber nanofiltration membrane. *Desalination* **2013**, *314*, 89–95. [CrossRef]
82. Mo, X.; Ni, Y.; Liu, F. Preparation of different scale fibrous membranes and their filtration properties. *Therm. Sci.* **2021**, *25*, 1453–1459. [CrossRef]
83. Ritt, C.L.; Stassin, T.; Davenport, D.M.; Duchanois, R.M.; Nulens, I.; Yang, Z.; Ben-Zvi, A.; Segev-Mark, N.; Elimelech, M.; Tang, C.Y. The open membrane database: Synthesis, structure and performance relationships of reverse osmosis membranes. *J. Membr. Sci.* **2022**, *641*, 119927. [CrossRef]
84. Oliveira, L.M.T.M.; Saleem, J.; Bazargan, A.; Duarte, J.L.D.S.; Mckay, G.; Meili, L. Sorption as a rapidly response for oil spill accidents: A material and mechanistic approach. *J. Hazard. Mater.* **2021**, *407*, 124842. [CrossRef]
85. Ravi, Y.; Prasanthi, I.; Behera, S.; Datta, K.K.R. MIL-101(Fe) Networks Supported on Fluorinated Graphene Nanosheets as Coatings for Oil Sorption. *ACS Appl. Nano Mater.* **2022**. [CrossRef]
86. Torres, C.E.I.; Quezada, T.E.S.; Kharissova, O.V.; Kharisov, B.I.; Lafuente, M.I.G.d. Carbon-based aerogels and xerogels: Synthesis, properties, oil sorption capacities, and dft simulations. *J. Environ. Chem. Eng.* **2021**, *9*, 104886. [CrossRef]
87. Al-Majed, A.A.; Adebayo, A.R.; Hossain, M.E. A sustainable approach to controlling oil spills. *J. Environ. Manag.* **2012**, *113*, 213–227. [CrossRef] [PubMed]
88. Gheriany, I.A.E.; Saqa, F.A.E.; Amer, A.A.E.R.; Hussein, M. Oil spill sorption capacity of raw and thermally modified orange peel waste. *Alex. Eng. J.* **2020**, *59*, 925–932. [CrossRef]
89. Paulauskiene, T. Ecologically friendly ways to clean up oil spills in harbor water areas: Crude oil and diesel sorption behavior of natural sorbents. *Environ. Sci. Pollut. Res.* **2018**, *25*, 9981–9991. [CrossRef] [PubMed]
90. Varjani, S.; Joshi, R.; Srivastava, V.K.; Ngo, H.H.; Guo, W. Treatment of wastewater from petroleum industry: Current practices and perspectives. *Environ. Sci. Pollut. Res.* **2019**, *27*, 27172–27180. [CrossRef]
91. Li, J.; Sun, S.; Yan, P.; Fang, L.; Yu, Y.; Xiang, Y.; Wang, D.; Gong, Y.; Gong, Y.; Zhang, Z. Microbial communities in the functional areas of a biofilm reactor with anaerobic–aerobic process for oily wastewater treatment. *Bioresour. Technol.* **2017**, *238*, 7–15. [CrossRef]

92. Doshi, B.; Sillanpää, M.; Kalliola, S. A review of bio-based materials for oil spill treatment. *Water Res.* **2018**, *135*, 262–277. [CrossRef]
93. Habibollahi, H.; Salehzadeh, A. Isolation, optimization, and molecular characterization of a lipase producing bacterium from oil contaminated soils. *Pollution* **2018**, *4*, 119–128. [CrossRef]
94. Karlapudi, A.P.; Venkateswarulu, T.C.; Tammineedi, J.; Kanumuri, L.; Ravuru, B.K.; Dirisala, V.; Kodali, V.P. Role of biosurfactants in bioremediation of oil pollution—A review. *Petroleum* **2018**, *4*, 241–249. [CrossRef]
95. Nejad, Y.S.; Jaafarzadeh, N.; Ahmadi, M.; Abtahi, M.; Ghafari, S.; Jorfi, S. Remediation of oily sludge wastes using biosurfactant produced by bacterial isolate Pseudomonas balearica strain Z8. *J. Environ. Health Sci. Eng.* **2020**, *18*, 531–539. [CrossRef]
96. Ibrahim, M.L.; Ijah, U.J.J.; Manga, S.B.; Bilbis, L.S.; Umar, S. Production and partial characterization of biosurfactant produced by crude oil degrading bacteria. *Int. Biodeterior. Biodegrad.* **2013**, *81*, 28–34. [CrossRef]
97. Sanghamitra, P.; Mazumder, D.; Mukherjee, S. Treatment of wastewater containing oil and grease by biological method- a review. *J. Environ. Sci. Health A* **2021**, *56*, 394–412. [CrossRef] [PubMed]
98. Elmobarak, W.F.; Hameed, B.H.; Almomani, F.; Abdullah, A.Z. A Review on the Treatment of Petroleum Refinery Wastewater Using Advanced Oxidation Processes. *Catalysts* **2021**, *11*, 782. [CrossRef]
99. Ma, D.; Yi, H.; Lai, C.; Liu, X.; Huo, X.; An, Z.; Li, L.; Fu, Y.; Li, B.; Zhang, M.; et al. Critical review of advanced oxidation processes in organic wastewater treatment. *Chemosphere* **2021**, *275*, 130104. [CrossRef]
100. Garrido-Cardenas, J.A.; Esteban-García, B.; Agüera, A.; Sánchez-Pérez, J.A.; Manzano-Agugliaro, F. Wastewater Treatment by Advanced Oxidation Process and Their Worldwide Research Trends. *Int. J. Environ. Res. Public Health* **2020**, *17*, 170. [CrossRef]
101. Kanakaraju, D.; Glass, B.D.; Oelgemöller, M. Advanced oxidation process-mediated removal of pharmaceuticals from water: A review. *J. Environ. Manag.* **2018**, *219*, 189–207. [CrossRef]
102. Salimi, M.; Esrafili, A.; Gholami, M.; Jonidi Jafari, A.; Rezaei Kalantary, R.; Farzadkia, M.; Kermani, M.; Sobhi, H.R. Contaminants of emerging concern: A review of new approach in AOP technologies. *Environ. Monit. Assess.* **2017**, *189*, 414. [CrossRef]
103. Ganiyu, S.O.; Martínez-Huitle, C.A.; Oturan, M.A. Electrochemical advanced oxidation processes for wastewater treatment: Advances in formation and detection of reactive species and mechanisms. *Curr. Opin. Electrochem.* **2021**, *27*, 100678. [CrossRef]
104. Darvishmotevalli, M.; Zarei, A.; Moradnia, M.; Noorisepehr, M.; Mohammadi, H. Optimization of saline wastewater treatment using electrochemical oxidation process: Prediction by RSM method. *MethodsX* **2019**, *6*, 1101–1113. [CrossRef]
105. Duan, P.-G.; Yang, S.-K.; Xu, Y.-P.; Wang, F.; Zhao, D.; Weng, Y.-J.; Shi, X.-L. Integration of hydrothermal liquefaction and supercritical water gasification for improvement of energy recovery from algal biomass. *Energy* **2018**, *155*, 734–745. [CrossRef]
106. Krause, M.J.; Thoma, E.; Sahle-Damesessie, E.; Crone, B.; Whitehill, A.; Shields, E.; Gullett, B. Supercritical Water Oxidation as an Innovative Technology for PFAS Destruction. *J. Environ. Eng.* **2022**, *148*, 05021006. [CrossRef]
107. Brunner, G. Near and supercritical water. Part II: Oxidative processes. *J. Supercrit. Fluids* **2009**, *47*, 382–390. [CrossRef]
108. Zhang, F.; Li, Y.; Liang, Z.; Wu, T. Energy conversion and utilization in supercritical water oxidation systems: A review. *Biomass Bioenergy* **2022**, *156*, 106322. [CrossRef]
109. Tang, X.; Zheng, Y.; Liao, Z.; Wang, Y.; Yang, J.; Cai, J. A review of developments in process flow for supercritical water oxidation. *Chem. Eng. Comm.* **2020**, *208*, 1494–1510. [CrossRef]
110. Guo, S.; Xu, D.; Li, Y.; Guo, Y.; Wang, S.; Macdonald, D.D. Corrosion characteristics and mechanisms of typical Ni-based corrosion-resistant alloys in sub- and supercritical water. *J. Supercrit. Fluids* **2021**, *170*, 105138. [CrossRef]
111. Xu, D.; Huang, C.; Wang, S.; Lin, G.; Guo, Y. Salt deposition problems in supercritical water oxidation. *Chem. Eng. J.* **2015**, *279*, 1010–1022. [CrossRef]
112. Muddemann, T.; Haupt, D.; Sievers, M.; Kunz, U. Electrochemical Reactors for Wastewater Treatment. *Chem. Ing. Technol.* **2019**, *91*, 769–785. [CrossRef]
113. Pahlevani, L.; Mozdianfard, M.R.; Fallah, N. Electrochemical oxidation treatment of offshore produced water using modified Ti/Sb-SnO2 anode by graphene oxide. *J. Water Process Eng.* **2020**, *35*, 101204. [CrossRef]
114. Cassano, A.; Conidi, C.; Ruby-Figueroa, R.; Castro-Muñoz, R. Nanofiltration and tight ultrafiltration membranes for the recovery of polyphenols from agro-food by-products. *Int. J. Mol. Sci.* **2018**, *19*, 351. [CrossRef]
115. Qasim, M.; Badrelzaman, M.; Darwish, N.N.; Darwish, N.A.; Hilal, N. Reverse osmosis desalination: A state-of-the-art review. *Desalination* **2019**, *459*, 59–104. [CrossRef]
116. Hassan, E.; Hassan, M.; Abou-Zeid, R.; Berglund, L.; Oksman, K. Use of bacterial cellulose and crosslinked cellulose nanofibers membranes for removal of oil from oil-in-water emulsions. *Polymers* **2017**, *9*, 388. [CrossRef] [PubMed]
117. Costa, A.F.S.; Almeida, F.C.G.; Vinhas, G.M.; Sarubbo, L.A. Production of bacterial cellulose by *Gluconacetobacter hansenii* using corn steep liquor as nutrient sources. *Front. microbiol.* **2017**, *8*, 2027. [CrossRef] [PubMed]
118. Costa, A.F.S.; Amorim, J.D.P.; Gomes, E.A.S.; Araujo, L.M.; Sarubbo, L. Residue from the production of sugar cane: An alternative nutrient used in biocellulose production by *Gluconacetobacter Hansenii*. *Chem. Eng. Trans.* **2019**, *64*, 7–12. [CrossRef]
119. Albuquerque, R.M.B.; Meira, H.M.; Silva, I.D.; Silva, C.J.G.; Almeida, F.C.G.; Amorim, J.D.P.; Vinhas, G.M.; Costa, A.F.S.; Sarubbo, L.A. Production of a bacterial cellulose/poly(3-hydroxybutyrate) blend activated with clove essential oil for food packaging. *Polym. Polym. Compos.* **2020**, *29*, 259–270. [CrossRef]
120. Amorim, J.D.P.; Souza, K.C.; Duarte, C.R.; Duarte, I.S.; Ribeiro, F.A.S.; Silva, G.S.; Farias, P.M.A.; Stingl, A.; Costa, A.F.S.; Vinhas, G.M. Plant and bacterial nanocellulose: Production, properties and applications in medicine, food, cosmetics, electronics and engineering. a review. *Environ. Chem. Lett.* **2020**, *18*, 851–869. [CrossRef]

121. Medeiros, A.D.M.; Silva Junior, C.J.G.; Amorim, J.D.P.; Nascimento, H.A.; Converti, A.; Costa, A.F.S.; Sarubbo, L.A. Biocellulose for Treatment of Wastewaters Generated by Energy Consuming Industries: A review. *Energies* **2021**, *14*, 5066. [CrossRef]
122. Wanichapichart, P.; Kaewnopparat, S.; Buaking, K.; Puthai, W. Characterization of cellulose membranes produced by *Acetobacter xyllinum*. *Songklanakarin J. Sci. Technol.* **2002**, *24*, 855–862.
123. Brown, R.; Kuga, S. Silver labeling of the reducing ends of bacterial cellulose. *Carbohydr. Res.* **1988**, *180*, 345–350. [CrossRef]
124. Esa, F.; Tasirin, S.M.; Rahman, N.A. Overview of Bacterial Cellulose Production and Application. *Agric. Agric. Sci. Procedia* **2014**, *2*, 113–119. [CrossRef]
125. Wang, J.; Tavakoli, J.; Tang, Y. Bacterial cellulose production, properties and applications with different culture methods–A review. *Carbohydr. Polym.* **2019**, *219*, 63–76. [CrossRef] [PubMed]
126. Hussain, Z.; Sajjad, W.; Khan, T.; Wahid, F. Production of bacterial cellulose from indústrial wastes: A review. *Cellulose* **2019**, *26*, 2895–2911. [CrossRef]
127. Stasiak-różańska, L.; Płoska, J. Study on the use of microbial cellulose as a biocarrier for 1,3-dihydroxy-2-propanone and its potential application in industry. *Polymers* **2018**, *10*, 438. [CrossRef] [PubMed]
128. Donini, Í.A.N.; Salvi, D.T.B.; Fukumoto, F.K.; Lustri, W.R.; Barud, H.S.; Marchetto, R.; Messaddeq, Y.; Ribeiro, S.J.L. Biossíntese e recentes avanços na produção de celulose bacteriana. *Eclet. Quim J.* **2010**, *35*, 165–178. [CrossRef]
129. Czaja, W.K.; Young, D.J.; Kawecki, M.; Brown Jr, R.M. The future prospects of microbial cellulose in biomedical applications. *Biomacromolecules* **2007**, *8*, 1–12. [CrossRef] [PubMed]
130. Neera; Ramana, K.V.; Batra, H.V. Occurrence of cellulose-producing Gluconacetobacter spp. in fruit samples and kombucha tea, and production of the biopolymer. *Appl. Biochem. Biotechnol.* **2015**, *176*, 1162–1173. [CrossRef] [PubMed]
131. Sai, H.; Fu, R.; Xing, L.; Xiang, J.; Li, Z.; Li, F.; Zhang, T. Surface modification of bacterial cellulose aerogels' web-like skeleton for oil/water separation. *ACS Appl. Mater. Interfaces* **2015**, *7*, 7373–7738. [CrossRef]
132. Wang, Q.; Asoh, T.A.; Uyama, H. Facile fabrication of flexible bacterial cellulose/silica composite aerogel for oil/water separation. *Bull. Chem. Soc. Jpn.* **2018**, *91*, 1138–1140. [CrossRef]
133. Hou, Y.; Duan, C.; Zhu, G.; Luo, H.; Liang, S.; Jin, Y.; Zhao, N.; Xu, J. Functional bacterial cellulose membranes with 3D porous architectures: Conventional drying, tunable wettability and water/oil separation. *J. Membr. Sci.* **2019**, *591*, 117312. [CrossRef]

processes

MDPI

Article

Bioleaching and Selective Precipitation for Metal Recovery from Basic Oxygen Furnace Slag

Klemens Kremser [1,*], Sophie Thallner [2], Sabine Spiess [2], Jiri Kucera [3], Tomas Vaculovic [4], Dalibor Všianský [5], Marianne Haberbauer [2] and Georg M. Guebitz [1]

1 Department of Agrobiotechnology, IFA-Tulln, Institute of Environmental Biotechnology, University of Natural Resources and Life Sciences Vienna BOKU, Konrad-Lorenz-Straße 20, 3430 Tulln an der Donau, Austria; guebitz@boku.ac.at
2 K1-MET GmbH, Stahlstraße 14, 4020 Linz, Austria; sophie.thallner@k1-met.com (S.T.); sabine.spiess@k1-met.com (S.S.); marianne.haberbauer@k1-met.com (M.H.)
3 Department of Biochemistry, Faculty of Science, Masaryk University, Kotlarska 2, 61137 Brno, Czech Republic; jiri.kucera@sci.muni.cz
4 Department of Chemistry, Faculty of Science, Masaryk University, Kotlarska 2, 61137 Brno, Czech Republic; tomas.vaculovic@ceitec.muni.cz
5 Department of Geological Science, Faculty of Science, Masaryk University, Kotlarska 2, 61137 Brno, Czech Republic; dalibor@sci.muni.cz
* Correspondence: klemens.kremser@boku.ac.at

Abstract: Decreasing ore grades and an increasing consumption of metals has led to a shortage of important primary raw materials. Therefore, the urban mining of different deposits and anthropogenic stocks is of increasing interest. Basic oxygen furnace (BOF) slag is produced in huge quantities with the so-called Linz-Donawitz process and contains up to 5.2, 0.9, 0.1, and 0.07% of Mn, Al, Cr, and V, respectively. In the present study, sulfur-oxidizing *Acidithiobacillus thiooxidans* and iron- and sulfur-oxidizing *Acidithiobacillus ferridurans* were applied in batch and stirred tank experiments to investigate the biological extraction of metals from BOF slag. In the batch experiments, up to 96.6, 52.8, 41.6, and 29.3% of Cr, Al, Mn, and V, respectively, were recovered. The stirred tank experiments, with increasing slag concentrations from 10 to 75 g/L, resulted in higher extraction efficiencies for *A. ferridurans* and lower acid consumption. Selective metal precipitation was performed at pH values ranging between 2.5 and 5.0 to study the recovery of Mn, Al, Cr, and V from the biolixiviant. Selective precipitation of V and Cr was achieved at pH 4.0 from *A. thiooxidans* biolixiviant, while Fe and V could be selectively recovered from *A. ferridurans* biolixiviant at pH 3.0. This work revealed the potential of BOF slag as an artificial ore for urban mining and demonstrated that combining bioleaching and selective precipitation is an effective method for sustainable metal recovery.

Keywords: bioleaching; metal recovery; selective precipitation; semi-continuous setup; metallurgical slags

Citation: Kremser, K.; Thallner, S.; Spiess, S.; Kucera, J.; Vaculovic, T.; Všianský, D.; Haberbauer, M.; Guebitz, G.M. Bioleaching and Selective Precipitation for Metal Recovery from Basic Oxygen Furnace Slag. *Processes* **2022**, *10*, 576. https://doi.org/10.3390/pr10030576

Academic Editor: Avelino Núñez-Delgado

Received: 1 February 2022
Accepted: 10 March 2022
Published: 15 March 2022

1. Introduction

The use of non-renewable mineral resources is growing annually, reflecting the increase in gross domestic production (GDP) and population growth [1]. This increase leads to an estimated consumption of around 70 Gt of mineral resources per year [2]. As natural ore grades are declining and the production capacity of certain metals will consequently decrease [3–5], new alternatives for sustainable metal mining and recycling need to be developed.

Landfill mining and urban mining involve the recovery of valuable resources from all kinds of deposits and anthropogenic stocks [6]. The concept of using biological, chemical, or mechanical techniques to recover valuable metals from waste electrical and electronic equipment has already proven to be more cost-effective than original mining [7]. Furthermore, an evaluation of the economic potential of landfill mining in Europe has shown that resource recovery can become net profitable [8]. In recent years, various waste materials, such as fly ash and slag from waste incineration plants [9–11], sewage sludge [12–14], and metallurgical residues [15,16], have been investigated and revealed their potential in microbial metal recovery (bioleaching) for urban mining purposes [11,17,18]. Extremely acidophilic and chemolithotrophic bacteria, such as *Acidithiobacillus ferridurans* and *Acidithiobacillus thiooxidans*, are well known for their ability to dissolve metals directly via redoxolysis or indirectly via acidolysis and complexolysis [19,20]. The bacterial oxidation of ferrous iron and elemental sulfur generates ferric iron and sulfuric acid, facilitating the dissolution of oxide or sulfide minerals and releasing the incorporated metals.

The steelmaking process generates various by-products such as dusts, sludge, and slag. One type of steelmaking slag is the so-called Linz–Donawitz (LD) slag or, more generally, basic oxygen furnace (BOF) slag. BOF slag is produced during the downstream processing of pig iron with scrap into crude steel by injecting pure oxygen into the basic oxygen converter [21]. The main components of this slag are iron oxide, calcium oxide, and silicon dioxide, thereby making it an alkaline material. This type of slag also contains traces of potentially toxic elements, such as chromium (Cr) and vanadium (V) [22]. About 650,000 t of BOF slag is produced annually in Austria [21]. According to EUROSLAG, 18.4 Mt of steelmaking slag was produced in Europe in 2016, with a BOF slag share of 56.6% [23]. Part of the BOF slag can be recycled in the steelmaking process, but the main part is used for road construction and as an additive in the cement industry. There is no standardized regulation for the permitted application areas in Europe. BOF slag has been used as a road-building material in Austria since 1970. Based on the new legislation adopted in 2015, the conditions of its application as a road construction material have been tightened due to its V and Cr content [21,24]. The stricter regulation has led to more slag being landfilled. However, bioleaching allows for removing undesired metals from the slag and reusing it in different applications.

Many studies have mainly focused on the solubilization of valuable metals from various waste substrates using bioleaching [15,25,26], but only a few studies have addressed metal recovery from the biolixiviant [27]. Different metal recovery methods are available, such as chemical precipitation [17], biosorption [28], electrowinning [29], or bioelectrochemical systems [30]. In particular, chemical precipitation is a widely used treatment method for recovering dissolved metals [31]. Metals are removed from solutions by adjusting the pH to a value at which the solubility of the metal is minimal by adding sodium hydroxide, lime, or sodium carbonate [31]. Chemical precipitation with NaOH results in the formation of different metal hydroxides depending on the pH setting and has the advantages of cost-effectiveness and simple operation [32], which is suitable for automation, and thus, was chosen as the metal recovery method in this study.

This study aimed to evaluate the benefits of bioleaching of BOF slag from the steelmaking industry and recovery of dissolved metals by selective precipitation. Therefore, batch experiments were tested with either sulfur-oxidizing *A. thiooxidans* or iron- and sulfur-oxidizing *A. ferridurans*. Consequently, scale-up was performed in a continuously stirred tank reactor with increasing slag concentrations up to 75 g/L. A synthetic lixiviant was prepared and used for precipitation experiments to test the selective chemical precipitation of Fe, Al, Mn, V, and Cr. Combining both the processes of bioleaching and the selective precipitation of dissolved metals represents a novel biotechnological approach for urban mining of anthropogenic stocks, which was successfully demonstrated in this work.

2. Materials and Methods

2.1. Materials and Chemicals

BOF slag was obtained from an Austrian steelmaking company. The samples were dried at room temperature (22 °C), crushed in a jaw crusher (Model BB51, Retsch, Germany), and sieved through a sieve tower (Type 3D, Retsch, Germany). The fractions with particle sizes between 0.25 and 0.5 mm were used for bioleaching and control experiments. If not otherwise specified, all chemicals were of analytical grade, purchased from Sigma-Aldrich, and all solutions were prepared using deionized water.

2.2. Bacterial Strains and Growth Conditions

For bioleaching experiments, the sulfur-oxidizing bacterium *Acidithiobacillus thiooxidans* (DSM 504) and the iron- and sulfur-oxidizing *Acidithiobacillus ferridurans* (DSM 583) [33] were purchased from the German Collection of Microorganisms and Cell Culture (DSMZ, Braunschweig, Germany). *A. thiooxidans* was pre-cultivated at 30 °C and 150 rpm in a basal salt medium consisting of 3 g/L KH_2PO_4, 0.14 g/L $CaCl_2 \cdot 2H_2O$, 0.1 g/L NH_4Cl, and 0.1 g/L $MgCl_2 \cdot 6H_2O$, supplemented with 10 g/L elemental sulfur (S^0) (termed BSM-AT). The initial pH was 4.2. *A. ferridurans* was pre-cultivated at the same temperature and shaking rate in a basal salt medium containing 0.4 g/L $(NH_4)_2SO_4$, 0.4 g/L $MgSO_4 \cdot 7\,H_2O$, and 0.4 g/L KH_2PO_4, supplemented with 33.3 g/L $FeSO_4 \cdot 7\,H_2O$ (termed BSM-AF). The initial pH was adjusted to pH 1.4 with sulfuric acid.

2.3. Batch Mode Bioleaching

Bioleaching batch experiments were conducted in 250 mL Erlenmeyer flasks at 30 °C and 150 rpm for two weeks. To start the experiment, 10 mL (10% *v*/*v*) of the pre-cultivated bacterial cultures were added to 90 mL of the respective medium (BSM-AT or BSM-AF) in biological duplicates. Additionally, 1 g (10 g/L) of BOF slag (and 1 g of S^0 to the BSM-AF) were added to all the flasks. The abiotic control experiments were run with BSM-AT or BSM-AF (with 1 g of S^0). At the end of the experiments, 10 mL liquid samples were collected to determine the pH and metal concentration. The extraction efficiency was calculated as reported previously [11].

To minimize the effect of trace amounts of dissolved metals included in the culture media and inoculum, the leaching efficiency calculation was modified as follows:

$$L_{eff}[\%] = \frac{((c_L - c_M) * V_B) - (c_C * V_C)}{c_F * M_F} * 100$$

where c_L is the metal concentration in the leachate, c_M is the metal concentration in the culture media, V_B is total batch volume, c_C is the metal concentration in the inoculum, V_C is the inoculum volume, c_F metal concentration in the untreated BOF-slag, and M_F is the total mass used for the leaching experiments.

2.4. Semi-Continuous Mode Bioleaching

For scaling up, further bioleaching experiments were performed in a semi-continuous mode using a continuous stirred tank reactor (CSTR) with a working volume of 1.5 L. The CSTR consisted of 2 L glass vessels connected to an automatic pH control unit via a connected pH sensor and process control transmitter (Mettler Toledo, Vienna, Austria). Sulfuric acid (2 M) was used to control pH ($\leq$2.4) by two connected membrane pumps (ProMinent, Heidelberg, Germany). The temperature control (30 °C) and stirring (250 rpm) were provided by two hot plate stirrers (IKA, Staufen, Germany). Compressed air ($\approx$30 L/h) was used to sufficiently aerate the CSTR system (Figure 1).

Semi-continuous mode bioleaching was performed in two independent reactors containing *A. ferridurans* or *A. thiooxidans*. Prior to the bioleaching experiments, both reactors containing the respective medium (BSM-AF or BSM-AT) were inoculated with 150 mL (10% *v*/*v*) of pre-cultivated bacterial culture, achieving a sufficient cell density ($\approx 10^8$ cells/mL). First, 10 g/L of BOF slag was added to both of the bioleaching reactors, and bioleaching was performed for two weeks. After the first cycle, 1 L (2/3) of the spent medium with treated slag was removed from the reactors and replaced with respective fresh medium, 10 g/L of S^0 and 25 g/L of BOF slag. This was repeated for 50 and 75 g/L of BOF slag, resulting in a gradual increase in the converter slag and adaptation of the bacterial cultures over a total period of 8 weeks. The abiotic control experiment was performed under the same conditions using double distilled water and the same pH setpoint of 2.4.

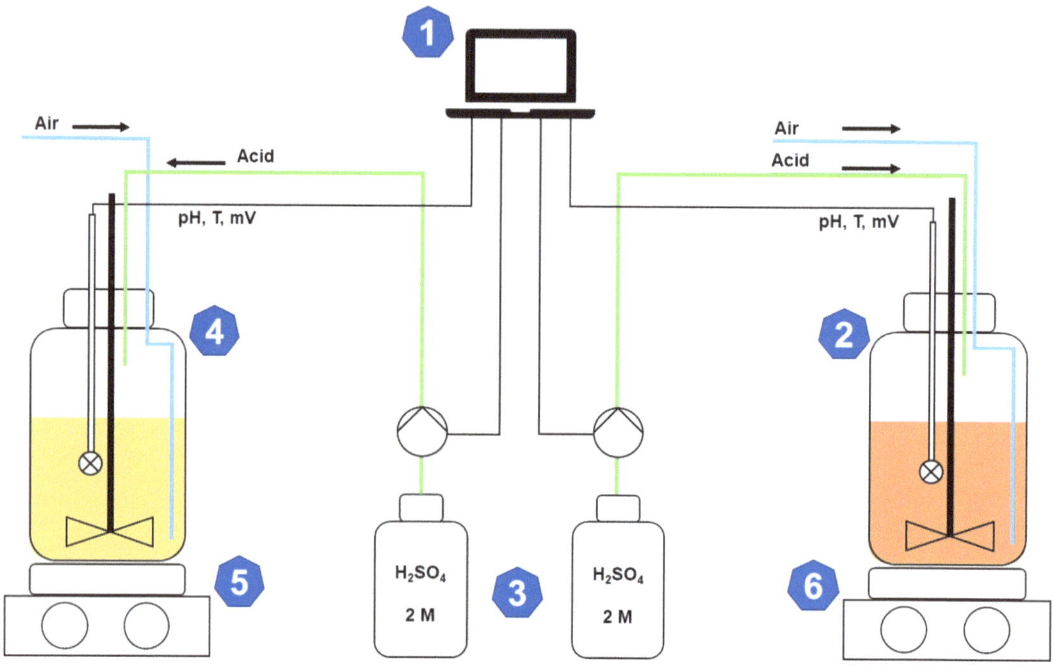

Figure 1. Continuous stirred tank reactor (CSTR) configuration. The CSTR consisted of a process control transmitter (1) controlling the pH value of two independent bioleaching cultures of *A. ferridurans* (2) and *A. thiooxidans* (4) using connected pH sensors and acid pumps (3). Temperature control and reactor stirring were done by two hot plate stirrers (5 and 6).

In all experiments, 5 mL samples were collected at the beginning (zero), middle (1 week), and end (2 weeks) of each cycle and used to measure the metal concentrations.

2.5. Chemical Analysis

The metal concentrations in the BOF slags, biolixiviants, and synthetic metal solutions were measured using inductively coupled plasma mass spectrometry (ICP-MS) as previously described [11]. Briefly, 1 g of BOF-slag was decomposed in a mixture of nitric acid and hydrogen peroxide and evaporated to dryness. Afterwards, the samples were decomposed in hydrofluoric acid, followed by leaching in perchloric acid, and evaporated to dryness. Finally, the samples were treated with hydrogen chloride and evaporated to dryness. ICP-MS was used to analyze the decomposed samples. To minimize the matrix effect and to get the best LOD, all samples were diluted with Milli Q water by a factor of 100, and a solution of Sc (400 $\mu g\ L^{-1}$) was used as an internal standard before analysis. Equally, microbial leachates were diluted with Milli Q water by a factor of 100 before ICP-MS analysis. Sc, as the internal standard (400 $\mu g\ L^{-1}$), was used to suppress the possible matrix effect. For quantification, a set of calibration solutions was prepared.

Mineral phase analysis was performed using X-ray diffraction (XRD) as previously described [11]. The samples were pulverized in a Retsch PM 100 planetary ball mill with an agate grinding jar and balls and homogenized with 10 wt.% of zincite (ZnO). Zincite was used as an internal standard for the quantification of the amorphous phase. Powder X-ray diffraction (XRD) analysis was conducted using a Panalytical X'Pert PRO MPD diffractometer with reflection geometry equipped with a cobalt tube ($\lambda K\alpha = 0.17903$ nm), Fe filter, and 1-D RMTS (X'Celerator) detector. The step size was 0.017° 2Θ, the time per step was 320 s, the angular range was 4–100° 2Θ, and the total scan duration was 247 min. The acquired data were processed using Panalytical HighScore 4 plus and Bruker AXS

Topas 4 software. For the Rietveld refinement, crystal structure models from the ICSD database were used.

2.6. Kinetic Analysis

The cell numbers were determined by counting the bacteria cells in a Neubauer improved haemocytometer with a 0.01 mm depth (BRAND GmbH, Wertheim, Germany). The pH values were measured using a Mettler Toledo S220 pH meter with a combined glass electrode.

2.7. Preparation of the Synthetic Metal Solutions

For the selective chemical precipitation of metals from the biolixiviant produced by extreme acidophiles, two simplified synthetic metal solutions were prepared according to the ICP-MS results at the end of the semi-continuous modes in the CSTR with 75 g/L of BOF slag. Therefore, the corresponding chemicals were selected and dissolved in 1 L of ultrapure water to reach the desired concentrations of metals. To distinguish between the two bioleaching mechanisms, a synthetic metal solution similar to the biolixiviant produced by *A. ferridurans* (termed SMS-AF) was prepared with ferric chloride (Fe^{3+}), and a synthetic metal solution similar to the biolixiviant produced by *A. thiooxidans* (termed SMS-AT) was prepared with ferrous sulfate (Fe^{2+}). The SMS-AF consisted of 6.72 g/L $Al_2(SO_4)_3 \cdot H_2O$, 2.4 g/L $Fe(III)Cl_3$, 0.29 g/L $CrK(SO_4)_2 \cdot 12H_2O$, 0.04 g/L V_2O_5, and 6 g/L $MnSO_4 \cdot H_2O$. The SMS-AT was composed of 3.17 g/L $Al_2(SO_4)_3 \cdot H_2O$, 17.5 g/L $FeSO_4 \cdot 7H_2O$, 0.384 g/L $CrK(SO_4)_2 \cdot 12H_2O$, 0.093 g/L V_2O_5, and 2.74 g/L $MnSO_4 \cdot H_2O$. The pH values of both the synthetic metal solutions were adjusted to pH 2.2 and 2.3 with 1 M H_2SO_4, which corresponded to the pH values at the end of both bioleaching experiments.

2.8. Selective Chemical Precipitation

Selective metal precipitation was performed in duplicate by adjusting the pH levels to 2.5, 3.0, 3.5, 4.0, 4.5, and 5.0 under a 765 Laboratory pH Meter control (Knick Elektronische Messgeräte GmbH & Co. KG, Berlin, Germany). During the experiment, 25 mL of the synthetic metal solution was stirred continuously using an IKA® C-MAG HS7 plate stirrer (IKA, Staufen, Germany) at 500 rpm, while 1 M NaOH was added until the desired pH was reached. After 45 h, the supernatant was filtered (0.2 µm), and ICP-MS analyzed the metal concentrations in the filtrate.

3. Results and Discussion

3.1. Characterization of BOF Slag

The BOF slag samples were characterized based on the metal concentrations, mineral compositions, and pH values determined according to DIN EN 15933:2012-11. The BOF slag was mainly composed of metals such as Fe, Mn, Al, Cr, and V (Table 1). Compared to other metallurgical slags [15,34,35], V, Cr, and Mn were present in much higher concentrations in the investigated BOF slag, indicating its potential to be considered a valuable secondary resource.

Table 1. The metal concentration in BOF slag determined by ICP-MS analysis.

Elements (mg/g)	The Average Values and Standard Deviations (n = 3)
Fe	225.70 ± 4.97
Mn	51.46 ± 0.95
Al	9.21 ± 0.62
Cr	1.30 ± 0.00
V	0.67 ± 0.03

The results of XRD analysis (Figure 2) show that the main phases in the BOF slag were wüstite (FeO), larnite ($Ca_2[SiO_4]$), and srebrodolskite ($Ca_2Fe_2O_5$), which constituted >75 wt.%, in agreement with earlier findings [35]. In addition to the main phases, lime (CaO), portlandite ($Ca[OH]_2$), and merwinite ($Ca_3Mg[SiO_4]_3$) were present in small amounts. The high concentration of wüstite can be explained by the fact that it is an important intermediate product formed during the blowing of the molten blast furnace iron scrap with pure oxygen. However, oxide minerals, often present in other industrial ashes and slags, have a much poorer leachability of the incorporated metals compared to sulfide minerals [11]. Most of the metals investigated are in BOF slag in oxide mineral forms, making them more accessible for reductive rather than oxidative bioleaching [36]. Nevertheless, even operationally less demanding oxidative bioleaching can extract metals significantly using ferric iron in addition to biogenic sulfuric acid.

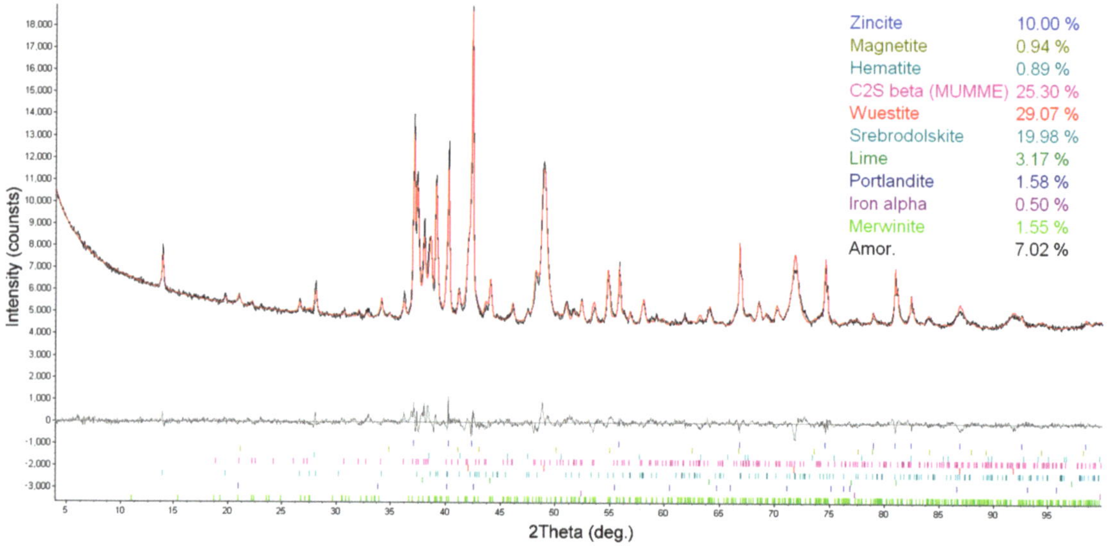

Figure 2. Results of XRD analysis of BOF slag. The main minerals are listed according to their concentration (wt.%). Zincite was used as an internal standard.

The particle size of the BOF slag ranged between 0.25 and 0.5 mm, and the determined pH value showed high alkalinity at around 12. The highly alkaline nature of these materials represents a challenging parameter for efficient bioleaching, as factors such as pulp density and high pH values have a severe impact on the extreme acidophiles employed in these processes [12,37]. On the other hand, a smaller particle size (<0.5 mm) promotes more efficient bioextraction, as more extractable metals have been reported in the fractions smaller than 4 mm in various waste ashes and slags, increasing the effective surface area for biooxidation [11,19,38].

3.2. Batch Mode Bioleaching of BOF Slag

The BOF slag was leached using two extremely acidophilic bacteria capable of oxidizing only S^0 to H_2SO_4 (*A. thiooxidans*) or oxidizing Fe^{2+} to Fe^{3+} and S^0 to H_2SO_4 (*A. ferridurans*). Both bacteria were able to extract metals more efficiently compared to the abiotic control experiments.

Sulfur-oxidizing *A. thiooxidans* leached V, Cr, Mn, and Al with efficiencies of approximately 30, 10, 37, and <1%, whereas iron- and sulfur-oxidizing *A. ferridurans* reached bioleaching efficiencies of approximately 2, 96, 42, and 53%, respectively (Figure 3). Significant differences in metal extraction were observed between the bioleaching mechanisms represented by two different *Acidithiobacillus* spp. By oxidizing Fe^{2+} to Fe^{3+}, *A. ferridurans* promoted a highly efficient Cr bioextraction of up to 100%, which may be related to its more effortless mobility and extractability under oxidative conditions compared to acid leaching [38]. Compared to other reports demonstrating the bioleaching of these waste types, this reflects one of the highest Cr recovery rates [9,36,39]. On the other hand, a negligible amount of V was recovered by bioleaching with *A. ferridurans*, contrary to previous studies [26,40]. One possible reason for this might be the higher pH value during the bioleaching experiments under Fe^{2+} oxidation compared to S^0 oxidation.

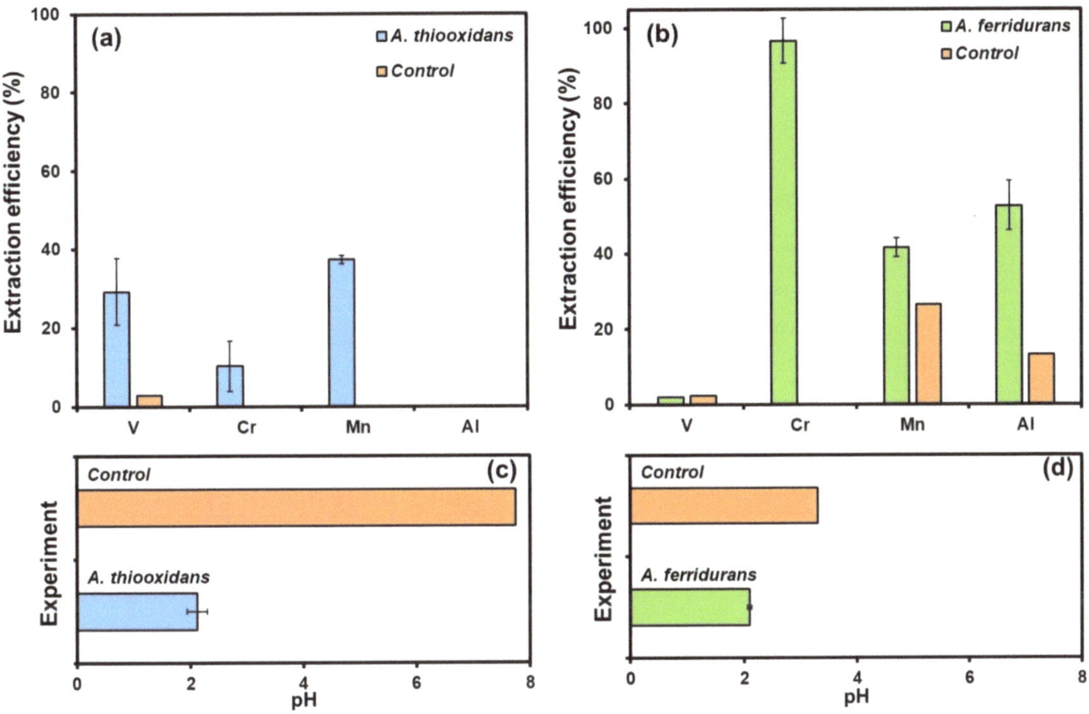

Figure 3. Bioleaching efficiency and pH values in batch modes. The metal extraction efficiency and pH values of *A. thiooxidans* (**a**,**c**) and *A. ferridurans* (**b**,**d**) and abiotic controls after two weeks. Error bars indicate the standard deviation (n = 2).

On the other hand, sulfur-oxidizing *A. thiooxidans* showed only low bioleaching efficiencies for Cr due to the absence of an oxidative environment, as this bacterial species cannot oxidize Fe^{2+} to Fe^{3+}. However, the high S^0 oxidation rate of *A. thiooxidans* corresponding to a distinct decrease in pH compared to the abiotic control (Figure 3c) led to strong acidification during bioleaching experiments, resulting in an enhanced extraction efficiency of V. The bioleaching efficiency of V tended to be within the percentages reported in previous studies [35,40]. In the case of Mn, bioleaching resulted in an extraction efficiency of approximately 40%, whereas in the control abiotic experiments, less than 1% was extracted using basal salt medium supplemented with S^0 and approximately 25% using basal salt medium supplemented with S^0 and Fe^{2+}. In investigating two different bioleaching mechanisms in batch mode, it seems that a combination of both might be suitable to dissolve various metals from BOF slag selectively. Cr and Al were more effectively removed by *A. ferridurans*, whereas the bioleaching of V occurred exclusively with *A. thiooxidans*.

3.3. Semi-Continuous Mode Bioleaching of BOF Slag

To increase the extraction efficiency and to gradually increase the slag concentration from 10 to 75 g/L, a semi-continuous mode in CSTR was operated. The bacteria were thus able to adapt to the increasing concentrations of heavy metals and the alkaline nature of the BOF slag.

At the lowest slag concentration of 10 g/L, both bacteria extracted Cr and Al with an efficiency of up to 100%, more than twice that of the abiotic controls (Figure 4). In addition, Mn and V were extracted more efficiently in both bioleaching experiments at a concentration of 10 g/L compared to the abiotic control. No further externally added acid was needed in the bioleaching processes, whereas the abiotic control required approximately 0.45 mL of 18 M sulfuric acid per gram of slag.

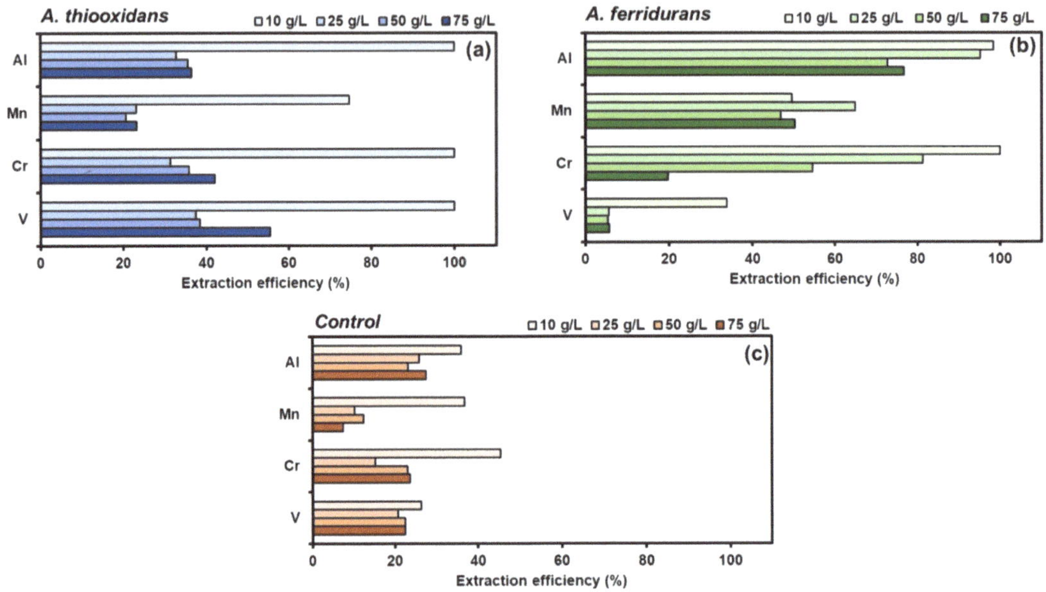

Figure 4. Bioleaching of BOF slag in a semi-continuous mode in CSTR. The metal extraction efficiency of *A. thiooxidans* (blue, **a**), *A. ferridurans* (green, **b**), abiotic control (orange, **c**) at increasing slag concentrations of 10–75 g/L.

With increasing slag concentrations, the metal extraction efficiency decreased, and the acid consumption increased for both bacteria, which is in agreement with previous studies [13,18,25]. Nevertheless, the Al and Mn bioleaching efficiencies remained consistently above 80 and 50% with *A. ferridurans*, decreasing to <40% with *A. thiooxidans* and <20% in the abiotic control. At the highest slag concentration tested (75 g/L), bioleaching with *A. thiooxidans* was more effective in extracting metals such as Cr and V, whereas bioleaching with *A. ferridurans* resulted in higher efficiency for Al and Mn, comparable to previous studies [40,41].

The acid consumption increased sharply during the bioleaching with *A. thiooxidans* at BOF slag concentrations of 50 and 75 g/L and attained higher values than *A. ferridurans* and the abiotic control experiments (Table 2). Therefore, the increased metal extractions at these slag concentrations were likely related to the external addition of sulfuric acid during the bioleaching with *A. thiooxidans*. In contrast, *A. ferridurans* required less externally added acid to maintain a low pH, probably due to the oxidation of Fe^{2+} to Fe^{3+}, which was further hydrolyzed, generating additional acidity, in addition to its higher metal tolerance [42,43]. At BOF slag concentrations between 10 and 50 g/L, *A. ferridurans* demonstrated the most promising results with (i) a bioleaching efficiency of >50% for Al, Mn, and Cr, and (ii) a lower acid consumption compared to *A. thiooxidans* and the abiotic control. Furthermore, similar results were observed regarding the selective dissolution of Mn, Cr, Al, and V at concentrations of 25–50 g/L compared to batch mode bioleaching. At 25 g/L, bioleaching with *A. ferridurans* could selectively dissolve Al and Cr, whereas the bioleaching efficiency with *A. thiooxidans* remained below 40%. The results indicate that the combined action of Fe^{3+} and biogenic H_2SO_4 led to higher acidity and even higher leaching efficiencies at elevated slag concentrations.

Table 2. Acid consumption during the CSTR bioleaching experiments compared to the chemical control for loading rates from 10 to 75 g/L.

Slag Concentration (g/L)	Acid Consumption		
	Acidithiobacillus thiooxidans	*Acidithiobacillus ferridurans*	Control
	(mL 18 M H_2SO_4/g Slag)		
10	-	-	0.44
25	0.05	-	0.27
50	0.31	0.07	0.25
75	0.49	0.19	0.23

3.4. Selective Chemical Precipitation for Metal Recovery

The selective chemical precipitation of Al, V, Cr, Mn, and Fe from the synthetic metal solutions (SMS-AT and SMS-AF) was tested by adjusting the pH from 2.5 to 5. Figure 5a shows the metals precipitation and the base consumption at each pH value of the SMS-AT. At pH 2.5, Al (4%) and Cr (2%) started to precipitate, and at pH 3, so did V (19%). At pH 4, 92% V and 75% Cr were precipitated from the solution, with 16% Al and 8% Fe. Thus, 24 mL/L 1 M NaOH was consumed. At pH 5, the entire V and Cr and 97% Al were recovered, whereas Mn (3%) and Fe (12%) remained nearly stable in the solution. These results prove that V and Cr could be selectively recovered from *A. thiooxidans* biolixiviant at pH 4, and Al at pH 5, whereas Mn and Fe remained in the solution. A previous study reported that >99% of Mn was removed at pH 11 [32]. However, it should be considered that adjusting to pH 11 is accompanied by a much higher base consumption.

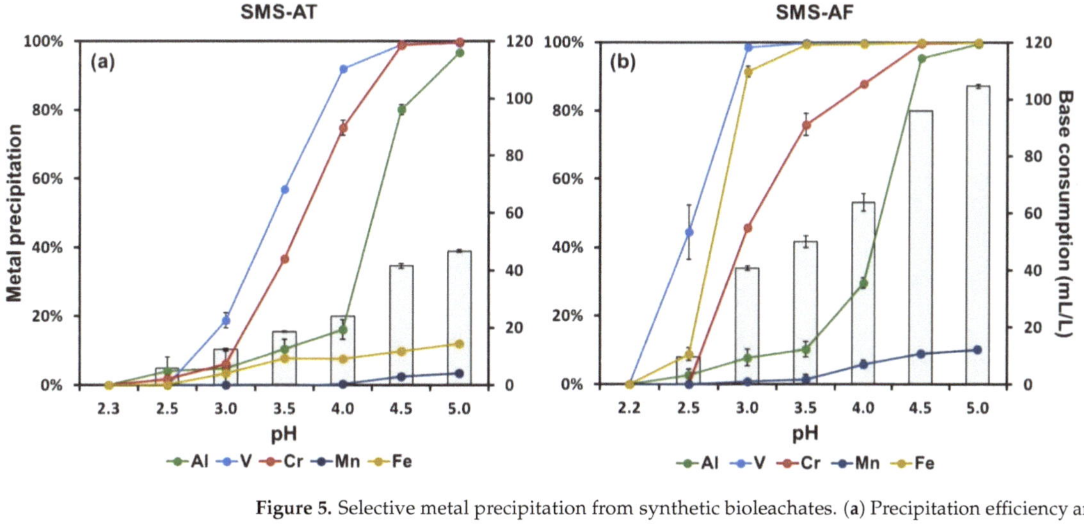

Figure 5. Selective metal precipitation from synthetic bioleachates. (**a**) Precipitation efficiency and base consumption of *A. thiooxidans* synthetic lixiviant (SMS-AT). (**b**) Precipitation efficiency and base consumption of *A. ferridurans* synthetic lixiviant (SMS-AF). Error bars indicate the standard deviation (n = 2).

Figure 5b shows the metal precipitation and the base consumption at each pH value of the SMS-AF. Significant differences were observed for Fe between the SMS-AT and SMS-AF (Figure 5a,b). At pH 3, 3% Fe from SMS-AT and 92% from SMS-AF were precipitated. This can be attributed to the oxidation state of Fe. SMS-AF was prepared with ferric chloride

(Fe^{3+}), whereas SMS-AT was prepared with ferrous sulfate (Fe^{2+}) to distinguish between the bioleaching mechanisms. Furthermore, V was precipitated from SMS-AF at pH 3 almost completely (98%), whereas only 19% V was removed from SMS-AT. In addition, Cr precipitation occurred faster in SMS-AF. At pH 3.5, 37% and 76% Cr was recovered from SMS-AT and SMS-AF. Similarly, 80% and 95% Al was precipitated from SMS-AT and SMS-AF at pH 4.5, respectively. However, the addition of base to adjust the required pH for precipitation was substantially higher for SMS-AF (to adjust to pH 5, 104.6 mL/L 1 M NaOH was consumed by SMS-AF, whereas SMS-AT consumed only 46.8 mL/L). At pH 3.5, V (100%), Fe (99%), and Cr (76%) could be selectively recovered from SMS-AF at a base consumption of 50 mL/L. At pH 4.5 and pH 5, Al was selectively precipitated at 95% and 100%, respectively. The obtained results show that it is possible to recover the selected metals from biolixiviant after slag bioleaching efficiently. Moreover, consuming large amounts of the base is unnecessary since selective metal precipitation is possible even at a lower pH than previously reported.

4. Conclusions

Novel concepts within a circular economy to recover economically critical raw materials, such as metals, are strongly required due to the depletion of natural mineral resources. This work demonstrated that chemolithotrophic acidophiles can effectively extract Al, Cr, Mn, and V from BOF slag up to a slag concentration of 75 g/L. In addition, the selective chemical precipitation of the dissolved metals from the biolixiviant using sodium hydroxide was proved to be a suitable method for metal recovery. Therefore, the combination of both methods represents a new biotechnological approach utilizing valuable secondary resources from low-grade steelmaking waste via bioleaching. Furthermore, combined bioleaching using ferric iron and biogenic sulfuric acid might lead to even higher extraction efficiency and needs to be investigated further.

Author Contributions: Conceptualization, K.K. and S.T.; methodology, K.K.; formal analysis, K.K., S.T., S.S., T.V. and D.V.; investigation, K.K., S.T. and S.S.; data curation, K.K., S.T., S.S. and D.V.; writing—original draft preparation, K.K.; writing—review and editing, S.T., G.M.G., S.S. and J.K.; visualization, K.K. and S.S.; supervision, G.M.G.; project administration, K.K. and M.H.; funding acquisition, G.M.G. and M.H. All authors have read and agreed to the published version of the manuscript.

Funding: This work was supported by the European fund for regional development, under the program Interreg V-A Austria-Czech Republic, project ATCZ183, Innovative Recycling Technology for Ashes and Slags (IRAS) and project ATCZ291, Optimization of biogenic sulfuric acid production for bioleaching applications in the waste management sector (OPTIMO).

Institutional Review Board Statement: Not applicable.

Informed Consent Statement: Not applicable.

Data Availability Statement: All data are contained in the ESI or available upon email request from the authors.

Conflicts of Interest: The authors declare no conflict of interest.

References

1. Steffen, W.; Broadgate, W.; Deutsch, L.; Gaffney, O.; Ludwig, C. The trajectory of the anthropocene: The great acceleration. *Anthr. Rev.* **2015**, *2*, 81–98. [CrossRef]
2. Wiedmann, T.O.; Schandl, H.; Lenzen, M.; Moran, D.; Suh, S.; West, J.; Kanemoto, K. The material footprint of nations. *Proc. Natl. Acad. Sci. USA* **2015**, *112*, 6271–6276. [CrossRef] [PubMed]
3. Ali, S.H.; Giurco, D.; Arndt, N.; Nickless, E.; Brown, G.; Demetriades, A.; Durrheim, R.; Enriquez, M.A.; Kinnaird, J.; Littleboy, A.; et al. Mineral supply for sustainable development requires resource governance. *Nature* **2017**, *543*, 367–372. [CrossRef] [PubMed]
4. Kerr, R.A. The coming copper peak. *Science* **2014**, *343*, 722–724. [CrossRef]
5. Mohr, S.; Giurco, D.; Retamal, M.; Mason, L.; Mudd, G. Global projection of lead-zinc supply from known resources. *Resources* **2018**, *7*, 17. [CrossRef]

6. Cossu, R.; Williams, I.D. Urban mining: Concepts, terminology, challenges. *Waste Manag.* **2015**, *45*, 1–3. [CrossRef]
7. Zeng, X.; Mathews, J.A.; Li, J. Urban Mining of E-Waste is Becoming More Cost-Effective Than Virgin Mining. *Environ. Sci. Technol.* **2018**, *52*, 4835–4841. [CrossRef]
8. Esguerra, J.L.; Laner, D.; Svensson, N.; Krook, J. Landfill mining in Europe: Assessing the economic potential of value creation from generated combustibles and fines residue. *Waste Manag.* **2021**, *126*, 221–230. [CrossRef]
9. Auerbach, R.; Ratering, S.; Bokelmann, K.; Gellermann, C.; Brämer, T.; Baumann, R.; Schnell, S. Bioleaching of valuable and hazardous metals from dry discharged incineration slag. An approach for metal recycling and pollutant elimination. *J. Environ. Manag.* **2019**, *232*, 428–437. [CrossRef]
10. Gomes, H.I.; Funari, V.; Dinelli, E.; Soavi, F. Enhanced electrodialytic bioleaching of fly ashes of municipal solid waste incineration for metal recovery. *Electrochim. Acta* **2020**, *345*, 136188. [CrossRef]
11. Kremser, K.; Thallner, S.; Strbik, D.; Spiess, S.; Kucera, J.; Vaculovic, T.; Vsiansky, D.; Haberbauer, M.; Mandl, M.; Guebitz, G.M. Leachability of metals from waste incineration residues by iron- and sulfur-oxidizing bacteria. *J. Environ. Manag.* **2021**, *280*, 111734. [CrossRef]
12. Gu, T.; Rastegar, S.O.; Mousavi, S.M.; Li, M.; Zhou, M. Advances in bioleaching for recovery of metals and bioremediation of fuel ash and sewage sludge. *Bioresour. Technol.* **2018**, *261*, 428–440. [CrossRef]
13. Chen, S.Y.; Wang, S.Y. Effects of solid content and substrate concentration on bioleaching of heavy metals from sewage sludge using Aspergillus niger. *Metals* **2019**, *9*, 994. [CrossRef]
14. Chen, S.Y.; Cheng, Y.K. Effects of sulfur dosage and inoculum size on pilot-scale thermophilic bioleaching of heavy metals from sewage sludge. *Chemosphere* **2019**, *234*, 346–355. [CrossRef]
15. Hocheng, H.; Su, C.; Jadhav, U.U. Bioleaching of metals from steel slag by Acidithiobacillus thiooxidans culture supernatant. *Chemosphere* **2014**, *117*, 652–657. [CrossRef]
16. Vestola, E.A.; Kuusenaho, M.K.; Närhi, H.M.; Tuovinen, O.H.; Puhakka, J.A.; Plumb, J.J.; Kaksonen, A.H. Acid bioleaching of solid waste materials from copper, steel and recycling industries. *Hydrometallurgy* **2010**, *103*, 74–79. [CrossRef]
17. Kremser, K.; Gerl, P.; Pellis, A.; Guebitz, G.M. A new bioleaching strategy for the selective recovery of aluminum from multi-layer beverage cans. *Waste Manag.* **2021**, *120*, 16–24. [CrossRef]
18. Kremser, K.; Thallner, S.; Schoen, H.; Weiss, S.; Hemmelmair, C.; Schnitzhofer, W.; Aldrian, A.; Guebitz, G.M. Stirred-tank and heap-bioleaching of shredder-light-fractions (SLF) by acidophilic bacteria. *Hydrometallurgy* **2020**, *193*, 105315. [CrossRef]
19. Srichandan, H.; Kumar, R.; Kumar, P. Hydrometallurgy Bioleaching approach for extraction of metal values from secondary solid wastes: A critical review. *Hydrometallurgy* **2019**, *189*, 105122. [CrossRef]
20. Sethurajan, M.; van Hullebusch, E.D.; Nancharaiah, Y.V. Biotechnology in the management and resource recovery from metal bearing solid wastes: Recent advances. *J. Environ. Manag.* **2018**, *211*, 138–153. [CrossRef]
21. Voestalpine AG LD-Schlacke Daten und Fakten. Available online: www.voestalpine.com (accessed on 12 January 2022).
22. Chaurand, P.; Rose, J.; Domas, J.; Bottero, J.Y. Speciation of Cr and V within BOF steel slag reused in road constructions. *J. Geochem. Explor.* **2006**, *88*, 10–14. [CrossRef]
23. EUROSLAG. *Euroslag Statistics*; Euroslag: Duisburg, Germany, 2016.
24. Federal Ministry for Digital and Economic Affairs. *BGBl. II Nr. 181/2015*; Federal Ministry for Digital and Economic Affairs: Vienna, Austria, 2014; pp. 1–220.
25. Xiang, Y.; Wu, P.; Zhu, N.; Zhang, T.; Liu, W.; Wu, J.; Li, P. Bioleaching of copper from waste printed circuit boards by bacterial consortium enriched from acid mine drainage. *J. Hazard. Mater.* **2010**, *184*, 812–818. [CrossRef]
26. Li, J.; Zhang, B.; Yang, M.; Lin, H. Bioleaching of vanadium by Acidithiobacillus ferrooxidans from vanadium-bearing resources: Performance and mechanisms. *J. Hazard. Mater.* **2021**, *416*, 125843. [CrossRef]
27. Ye, M.; Li, G.; Yan, P.; Ren, J.; Zheng, L.; Han, D.; Sun, S.; Huang, S.; Zhong, Y. Removal of metals from lead-zinc mine tailings using bioleaching and followed by sulfide precipitation. *Chemosphere* **2017**, *185*, 1189–1196. [CrossRef]
28. González Bermúdez, Y.; Rodríguez Rico, I.L.; Guibal, E.; Calero de Hoces, M.; Martín-Lara, M.Á. Biosorption of hexavalent chromium from aqueous solution by Sargassum muticum brown alga. Application of statistical design for process optimization. *Chem. Eng. J.* **2012**, *183*, 68–76. [CrossRef]
29. Alebrahim, M.F.; Khattab, I.A.; Cai, Q.; Sanduk, M. Practical study on the electrochemical simultaneous removal of copper and zinc from simulated binary-metallic industrial wastewater using a packed-bed cathode. *Egypt. J. Pet.* **2017**, *26*, 225–234. [CrossRef]
30. Modin, O.; Wang, X.; Wu, X.; Rauch, S.; Fedje, K.K. Bioelectrochemical recovery of Cu, Pb, Cd, and Zn from dilute solutions. *J. Hazard. Mater.* **2012**, *235–236*, 291–297. [CrossRef]
31. Zhou, P.; Huang, J.C.; Li, A.W.F.; Wei, S. Heavy metal removal from wastewater in fluidized bed reactor. *Water Res.* **1999**, *33*, 1918–1924. [CrossRef]
32. Pavlović, J.; Stopić, S.; Friedrich, B.; Kamberović, Ž. Selective removal of heavy metals from metal-bearing wastewater in a cascade line reactor. *Environ. Sci. Pollut. Res.* **2007**, *14*, 518–522. [CrossRef]
33. Moya-Beltrán, A.; Beard, S.; Rojas-Villalobos, C.; Issotta, F.; Gallardo, Y.; Ulloa, R.; Giaveno, A.; Degli Esposti, M.; Johnson, D.B.; Quatrini, R. Genomic evolution of the class Acidithiobacillia: Deep-branching Proteobacteria living in extreme acidic conditions. *ISME J.* **2021**, *15*, 3221–3238. [CrossRef]
34. Gomes, H.I.; Funari, V.; Mayes, W.M.; Rogerson, M.; Prior, T.J. Recovery of Al, Cr and V from steel slag by bioleaching: Batch and column experiments. *J. Environ. Manag.* **2018**, *222*, 30–36. [CrossRef] [PubMed]

35. Mirazimi, S.M.J.; Abbasalipour, Z.; Rashchi, F. Vanadium removal from LD converter slag using bacteria and fungi. *J. Environ. Manag.* **2015**, *153*, 144–151. [CrossRef]
36. Thallner, S.; Hemmelmair, C.; Martinek, S.; Schnitzhofer, W. Bioleaching for removal of chromium and associated metals from LD slag. *Solid State Phenom.* **2017**, *262*, 79–83. [CrossRef]
37. Fan, X.L.; Lv, S.Q.; Xia, J.L.; Nie, Z.Y.; Zhang, D.R.; Pan, X.; Liu, L.Z.; Wen, W.; Zheng, L.; Zhao, Y.D. Extraction of Al and Ce from coal fly ash by biogenic Fe3+ and H2SO4. *Chem. Eng. J.* **2019**, *370*, 1407–1424. [CrossRef]
38. Abramov, S.; He, J.; Wimmer, D.; Lemloh, M.L.; Muehe, E.M.; Gann, B.; Roehm, E.; Kirchhof, R.; Babechuk, M.G.; Schoenberg, R.; et al. Heavy metal mobility and valuable contents of processed municipal solid waste incineration residues from Southwestern Germany. *Waste Manag.* **2018**, *79*, 735–743. [CrossRef] [PubMed]
39. Pathak, A.; Srichandan, H.; Kim, D.J. Column bioleaching of metals from refinery spent catalyst by Acidithiobacillus thiooxidans: Effect of operational modifications on metal extraction, metal precipitation, and bacterial attachment. *J. Environ. Manag.* **2019**, *242*, 372–383. [CrossRef] [PubMed]
40. Dong, Y.; Lin, H.; Liu, Y.; Zhao, Y. Blank roasting and bioleaching of stone coal for vanadium recycling. *J. Clean. Prod.* **2020**, *243*, 118625. [CrossRef]
41. Mikoda, B.; Potysz, A.; Kmiecik, E. Bacterial leaching of critical metal values from Polish copper metallurgical slags using Acidithiobacillus thiooxidans. *J. Environ. Manag.* **2019**, *236*, 436–445. [CrossRef]
42. Navarro, A.C.; Von Bernath, D.; Jerez, A.C. Heavy Metal Resistance Strategies of Acidophilic Bacteria and Their Acquisition: Importance for Biomining and Bioremediation. *Biol. Res.* **2013**, *46*, 363–371. [CrossRef]
43. Asghari, I.; Mousavi, S.M. Effects of key parameters in recycling of metals from petroleum refinery waste catalysts in bioleaching process. *Rev. Environ. Sci. Biotechnol.* **2014**, *13*, 139–161. [CrossRef]

processes

Article

Comparative Metagenomics of Anaerobic Digester Communities Reveals Sulfidogenic and Methanogenic Microbial Subgroups in Conventional and Plug Flow Residential Septic Tank Systems

James Naphtali, Alexander W. Y. Chan, Faizan Saleem, Enze Li, Jacob Devries and Herb E. Schellhorn *

Department of Biology, McMaster University, 1280 Main Street West, Hamilton, ON L8S 4K1, Canada; naphtaj@mcmaster.ca (J.N.); chanw19@mcmaster.ca (A.W.Y.C.); saleef4@mcmaster.ca (F.S.); lie40@mcmaster.ca (E.L.); devrij7@mcmaster.ca (J.D.)
* Correspondence: schell@mcmaster.ca; Tel.: +1-905-525-9140 (ext. 27316)

Abstract: On-site wastewater treatment systems (OWTS) are primarily monitored using physiochemical factors, including chemical oxygen demand (COD) and residual total suspended solids (TSS), which are indirect measures of the microbial action during the anaerobic digestion process. Changes in anaerobic digester microbial communities can alter the digester performance, but this information cannot be directly obtained from traditional physicochemical indicators. The potential of metagenomic DNA sequencing as a tool for taxonomic and functional profiling of microbial communities was examined in both common conventional and plug flow-type anaerobic digesters (single-pass and recirculating). Compared to conventional digesters, plug flow-type digesters had higher relative levels of sulfate-reducing bacteria (*Desulfovibrio* spp.) and hydrogenotrophic methanogens (*Methanospirillum* spp.). In contrast, recirculating anaerobic digesters were enriched with denitrifier bacteria and hydrogenotrophic methanogens, and both were significantly correlated with physicochemical factors such as COD and TSS. Stratification of microbial communities was observed along the digester treatment process according to hydrolytic, acidogenic, acetogenic, and methanogenic subgroups. These results indicate that the high-throughput DNA sequencing may be useful as a monitoring tool to characterize the changes in bacterial communities and the functional profile due to differences in digester design in on-site systems.

Keywords: anaerobic digestion; methanogenesis; septic tank; whole-metagenome sequencing; metagenomics; bacterial communities

Citation: Naphtali, J.; Chan, A.W.Y.; Saleem, F.; Li, E.; Devries, J.; Schellhorn, H.E. Comparative Metagenomics of Anaerobic Digester Communities Reveals Sulfidogenic and Methanogenic Microbial Subgroups in Conventional and Plug Flow Residential Septic Tank Systems. *Processes* **2022**, *10*, 436. https://doi.org/10.3390/pr10030436

Academic Editors: Dimitris Zagklis and Georgios Bampos

Received: 31 January 2022
Accepted: 18 February 2022
Published: 22 February 2022

1. Introduction

Human and animal sewage wastes containing nitrogenous and phosphorous inorganic compounds such as ammonia, nitrate, and phosphate, along with organic compounds, directly contribute to increased nutrient loading in water bodies, causing eutrophication [1]. On-site anaerobic septic tank systems are a primary means for treating wastewater by reducing input waste to residual solids due to the combined metabolic activities of complex microbial communities in anaerobic digester systems [2,3]. The metabolic pathways primarily expressed by hydrolytic, acidogenic, acetogenic, and methanogenic bacterial communities drive anaerobic biomass degradation [4].

Anaerobic digestion operational efficiency can be measured using physicochemical parameters, including chemical oxygen demand (COD), total residual solids (TSS), and biogas production [5]. Observed differences in these parameters are indicators of digester performance, which, in turn, are affected by the composition of the complex digester microbial communities [5]. Influent pre-treatment methods such as microwaves and static magnetic fields can alter anaerobic digester microbial communities [6,7], and such alterations in response to pre-treatment methods and physicochemical parameters can result in enhanced biogas production and optimal anaerobic digestion [8–10]. Therefore, anaerobic digester design and operation parameters such as hydraulic retention times (HRT) and influent pre-treatment methods must be adjusted to achieve optimal functionality [11]. Though characterization of microbial communities and their metabolic potential is likely important for optimization of the anaerobic digestion process, there is little information available regarding changes in the microbial community (microbiome) in anaerobic digesters. Such information could be useful for optimizing digester performance or tracking digester failure [12,13].

Some microbial functions with respect to anaerobic digestion can be specifically enhanced by optimizing physicochemical factors. For example, a change in pH toward the acidic side can enhance the microbial hydrolysis potential [14]. However, microbial community composition analysis can be a better indicator of digester function than physicochemical parameters due to the high robustness, resilience, and ecological flexibility of the microbiome [15].

Anaerobic digester communities such as methanogens are difficult to culture or unculturable, which makes it difficult to observe the dynamic microbial changes throughout the anaerobic digestion process [16]. Advances in next-generation sequencing have resulted in culture-independent tools based on shotgun metagenomics that can potentially profile complex microbial communities [17]. Not only the community structure can be characterized, but the potential functions of certain groups (e.g., methanogenic archaea and acidogenic/hydrolytic bacteria) and their interactions can be identified [18]. While this idea has been explored for centralized municipal wastewater treatment plants (WWTPs) [2,19], there are few studies on the use of high-throughput DNA sequencing for OWTS monitoring where small installations, differing in design and mode of operation, having variable influent (e.g., from single households), are examined.

In this proof of principle study, microbial communities from two types of OWTS anaerobic digesters using conventional box-septic tanks and modified septic tanks equipped with a plug-type digester design plug flow-type were characterized using the whole-metagenome sequencing. The objectives of the study are to survey the microbial consortiums present in conventional and novel digester designs and determine how the microbial community profile and associated functional properties are influenced by digester design and mode of operation.

2. Materials and Methods

2.1. Digester Site Description

Digester samples were collected from conventional and plug flow-type recirculating and single-pass septic tank installations across Southern and Central Ontario for a total of 12 digesters from September 2018 to January 2019. All systems were located on residential rural properties and served single households. Each OWTS consisted of a digester tank connected to a downstream aerobic biofilter unit. Half of the units were equipped with an aerobic recirculating line (Supplementary Data). The four types of systems were:

- Conventional recirculating septic digesters (CR) (n = 3);
- Conventional single-pass septic digesters (CS) (n = 3);
- Plug flow-type recirculating septic digesters (PFR) (n = 3);
- Plug flow-type single-pass septic digesters (PFS) (n = 3).

For each digester, we collected sewage samples at the influent, effluent, and within the tank. In recirculating digesters, an inline valve as part of the recirculating line was used to control the proportion of aerobic effluent being pumped back into the anaerobic septic tanks. Percentages of valve openings for each recirculating digester were listed in Supplementary Data and were set by an industrial partner, Waterloo Biofilter Systems Inc. (Guelph, ON Canada). Hydraulic residence times were calculated by dividing the tank residence volume (L) and flow rate (L/day). Samples were assigned an alphanumeric code, where numbers prefixing the digester acronym represented a digester biological replicate, and numbers suffixing the digester acronym represented the sampling point, where "1" is for influent, "2" for tank, and "3" for effluent.

2.2. Digester Sample Collection

For each digester type and replicate, two effluent, tank, and influent samples were taken for chemical analysis and DNA sequencing. Effluent samples were collected from the spray nozzle feeding the aerobic biofilter unit by placing a collection vessel directly beneath the nozzle. Plug flow influent sewage was collected from the inlet opening of the digester and for tank sewage at the approximate outlet of the plug flow tube within the tank. All chemical measurements except for COD, temperature, pH, and dissolved oxygen were performed by Fleming College's Centre for Advancement of Wastewater Treatment according to protocols described in the CALA Directory of Laboratories listing (member number: 3628) and were based on Standard Methods for the Examination of Water and Wastewater, 22nd ed. [20]. Sewage sample bottles were immediately placed on ice after collection, transported within ice coolers, and stored at −80 °C before DNA extraction. Dissolved oxygen and pH were not measured on site, rather they were performed in the lab after samples were received for processing.

2.3. DNA Extraction, Sequencing, and Bioinformatics Analysis

Sewage samples were thawed to room temperature and then vacuum-filtered through sterile 0.22 μm nitrocellulose filters (Millipore, Germany) [21]. Samples were filtered until the membrane filter was completely clogged and no more liquid could pass through [22]. Filters were then folded and fit into sterile capped stock tubes pre-filled with 0.25 μL of 0.1 mm zirconium beads [23] (BioSpec Products, Bartesville, OK, USA). DNA extraction was performed by using the Norgen Biotek Soil DNA Isolation Plus Kit (Norgen Biotek Corp., Thorold, ON, Canada) following the manufacturer's protocol. Extracted DNA samples were quantified using a Qubit Fluorimeter 2.0 (Thermo Fisher Scientific, Waltham, Massachusetts) and stored at −20 °C. The shotgun metagenomic library was prepared by using the NEBNext® Ultra™ II DNA library preparation kit with an insert size of 500 bp. (New England Biolabs Inc., Whitby, Ontario). Paired-end metagenomic sequencing with a read length of 150 bp was performed using HiSeq 2500 Illumina sequencer.

Two FASTQ files (forward and reverse) were obtained for each sample. These FASTQ (quality score "Q score" was equal to or greater than Q30 files processed using Trimmomatic v0.39 [24]. The average reads per sample was 5.3 M reads (SD ± 1.6 M). Quality-filtered sequences were aligned using the DIAMOND-BLASTx algorithm against NCBI's non-redundant (nr) protein database with default BLAST parameters [25]. The BLAST output files were formatted as DIAMOND alignment archives and processed through MEGAN6 V6.18.4 for taxonomic and functional annotation [26].

2.4. Statistical Analyses

Count data were normalized using multivariate methods [27]. Relative abundance, diversity indices, rank abundance, rarefaction, and differential abundance graphs were prepared using non-normalized data. All statistical tests were performed using the "phyloseq" R package [28]. Relative abundances graphs were generated by agglomerating (aggregating low-level taxonomic groups) to either the phylum level or, in the case of Euryarchaeota abundances, at the family level. Constrained redundancy analysis (RDA) between microbial community compositions and chemical parameters and non-metric multidimensional scaling ordination methods for all taxonomic and functional samples were generated using the Bray–Curtis dissimilarity index for beta-diversity analysis. PERMANOVA tests across digester types were performed using the vegan package [29].

Comparative SIMPER analysis of Bray–Curtis dissimilarity contributions of taxa was performed using the simper function from the vegan package implemented into custom scripts created by Steinberger named simper.pretty and kruskal.pretty [30]. To identify enriched microbial groups associated with specific digesters, differential abundance pairwise comparison graphs between digester types (conventional and plug type) and between flow configurations (recirculating and single-pass) were generated using the "DESeq2" package [31,32]. Raw taxonomic counts were assessed using DESeq2 pairwise comparisons [32] of significantly enriched taxa in between digester types and flow configurations. Identified differentially abundant genera were then tested for association with chemical parameters using Spearman rank-order correlations using the R function rcorr in the "Hmisc" package and plotted with the "corrplot" package [33].

3. Results and Discussion

3.1. Whole-Community Interactions with the Physicochemical Environment

Physicochemical factors such as influent sewage load (COD, TSS), digester design and configuration, and interspecies interactions impact community compositions along the anaerobic digestion process [34]. Intra and inter-digester analysis demonstrated that differences in the microbial community across the four digester types (CR, IR, CS, and IS) were due to dissimilar taxonomic richness across the replicates ($p < 0.05$) [35]. Changes in the bacterial communities between digester groups were also significantly correlated with temperature ($p = 0.029$), NH_3 ($p = 0.007$), and pH ($p = 0.003$), suggesting that the physiochemical factors in digesters impacted microbial community composition (Figure 1 and Supplementary Data). Anaerobic digester microbial communities change in response to changes in ammonia/ammonium concentration and pH [36]. Specifically, an excessive accumulation of ammonia/ammonium can lead to inhibition of methanogenesis and favor the dominance of hydrogenotrophic microbial communities [36]. Effects of sampling point, or treatment stages (influent, tank, and effluent), were not significantly different in whole-group ($p > 0.05$) or pairwise comparison tests (Supplementary Data). Therefore, the changes in microbial communities were more likely due to digester designs and physiochemical factors than treatment stages [29,35]. In anaerobic digesters receiving plant, human, or cattle waste feed, the influent composition is one of the major driving forces of anaerobic digestion community specialization [34,37]. Although bacterial community composition in influent samples was not significantly different among the samples, additional replication may be needed to assess the effect of influent composition on the differentiation of downstream communities across digester types (Supplementary Data).

3.2. Enriched Microbial Subgroups between Digester Designs

Enrichment of microbial consortium within a community is an indicator of niche functional specializations by specific microbes [38]. Accordingly, methanogenic *Methanothrix* spp. as well as members of Proteobacteria such as *Desulfobulbus* spp., *Desulfomicrobium* spp., *Pelolinea* spp., and *Protecatella* spp. were enriched in conventional systems compared to plug flow-type systems (Figures 1 and 2 and Supplementary Data).

Plug-flow digesters are known to achieve lower COD, higher TSS removal, and methane production rates than other commonly used digester designs such as the conventional single-tube recirculating digester [39]. Consistent with this, we observed lower COD and higher TSS removal in plug-flow digesters than conventional digesters (Figure 3) possibly due to a higher solid retention time resulting in better degradation of solids and untreated sewage by the active microbial biomass (Figure 4) [40]. In addition, hydrogenotrophic methanogens, including Methanospirillaceae, Methanobacteriaceae, and Methanosarcinaceae, were enriched in plug flow-type systems (Supplementary Data). We also observed a positive correlation of TSS with the "Methanogenesis" gene subsystem (Figure 2), possibly due to the higher solid content that was available for methanogens to digest [41].

Sulfate-reducing bacteria such as *Desulfovibrio* spp. were also enriched in plug flow-type systems (Figure 2). The enrichment of hydrogenotrophic methanogens and sulfate-reducing bacterial communities may reflect elevated levels of organic acids since hydrogenotrophs are exclusive H_2-scavengers in acidic environments [42].

An inverse relationship was observed between COD, total Kjeldahl nitrogen (TKN), and NH_3 levels with abundances of *Methanothrix* spp. and *Desulfobulbus* spp. (Figure 2) suggesting that these bacterial communities may be associated with biomass degradation in the form of COD decrease and sensitivity to high nitrogen content (in the form of ammonia) [43]. However, lower COD and higher TSS removal in plug flow-type may, in addition to the action of methanogens, also be caused by uncharacterized interspecies interactions. Metagenomic studies have reported rare co-occurring phylotypes that resulted in greater abundances of niche functional features [37,38]. Further investigation may be required to reveal how interspecies and community interactions are responsible for higher waste degradation rates within the plug flow-type than in conventional septic tanks.

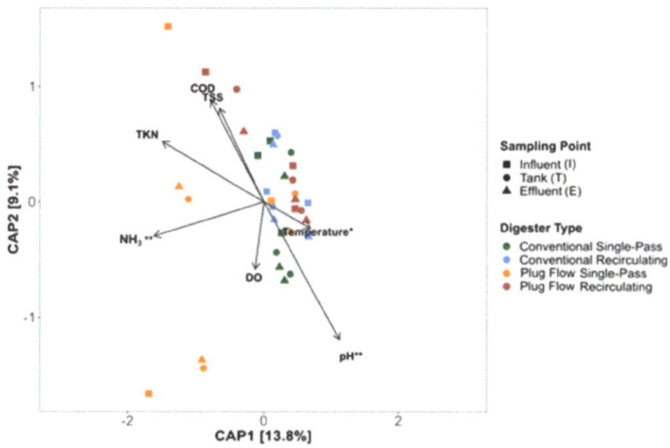

(A)

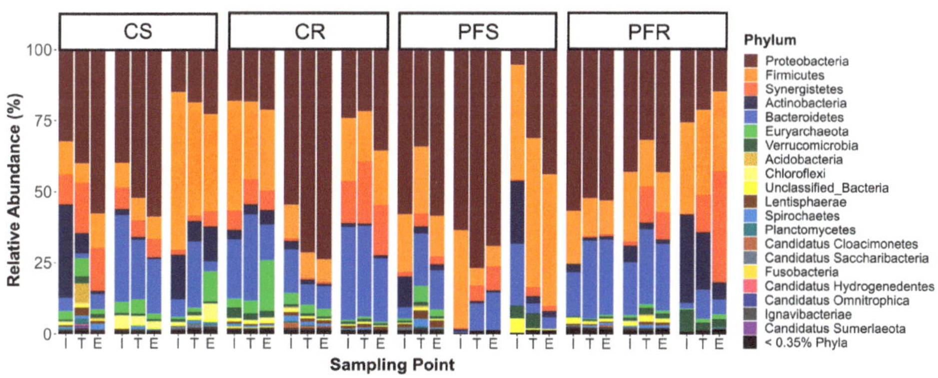

(B)

Figure 1. (**A**) Redundancy analysis of digester type and sampling point constrained against chemical parameters. Chemical parameters that significantly influenced the ordination revealed by the PERMANOVA test are denoted with asterisks (adj. $p < 0.05$), where "**" 0.01, "*" 0.05. (**B**) Relative abundance bar graphs showing the top 20 phyla for triplicate plug flow-type single-pass (PFS), plug flow-type recirculating (PFR), conventional single-pass (CS), and conventional recirculating (CR) across sampling points. The lowest abundant phyla were aggregated into a cutoff percentage labeled in black.

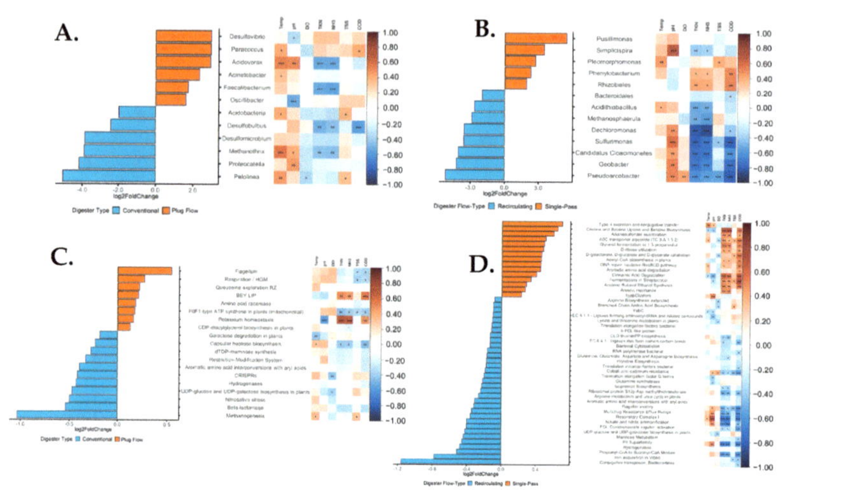

Figure 2. Pairwise comparisons with Spearman rank correlations to chemical parameters of differentially abundant genera in (**A**) conventional vs. plug flow-type, (**B**) recirculating and single-pass and differentially abundant functional genes in (**C**) conventional vs. plug flow-type, (**D**) recirculating and single-pass. Differential abundance analysis was performed using DESeq2 with an abundance cutoff of 0.1% of total agglomerated genera and subsystem Level III counts and an adjusted p-value of $p < 0.001$ for genera and $p < 0.05$ for gene counts. Significant Spearman correlations are marked with asterisks (adj. $p < 0.05$), where "***" 0.001, "**" 0.01, "*" 0.05.

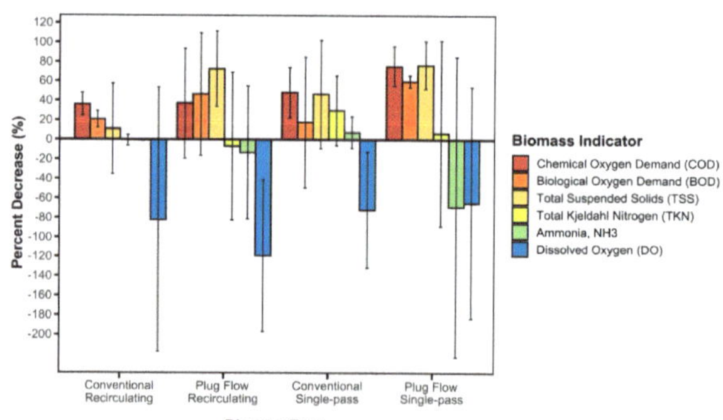

(**A**)

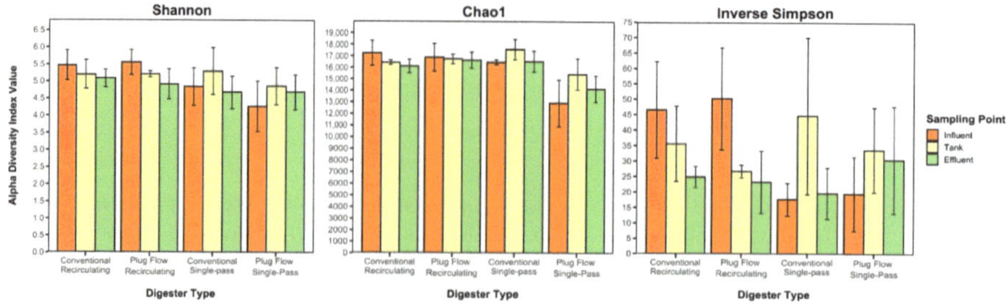

(**B**)

Figure 3. (**A**) Mean percent decrease in biomass chemical parameters across influent, tank, and effluent sampling points. Error bars indicate standard deviation. (**B**) Mean Shannon diversity, Chao1 richness, and inverse Simpson evenness indices of the four digester types over influent, tank, and effluent sewage. Error bars indicate standard error (n = 3).

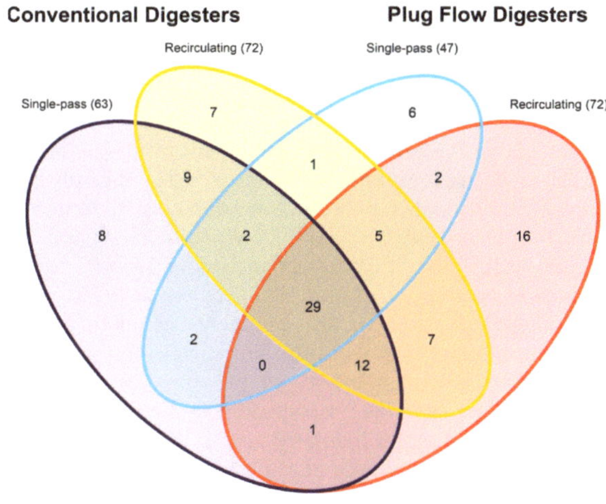

Figure 4. Shared and exclusive genera that are at least 50% prevalent across conventional single-pass (CS), conventional recirculating (CR), plug flow-type single-pass (PFS), and plug flow-type recirculating (PFR) and digesters (see Supplementary Data for list of taxa).

3.3. Enriched Microbial Subgroups among Digester Flow Configurations

Implementing downstream aerobic recirculating lines in anaerobic systems facilitates the anaerobic digestion process by replenishing nitrate for consumption by nitrate-reducing microbes (i.e., denitrifiers), preventing ammonia accumulation, and providing acidic buffering capacity by ensuring the constant use and flow of H_2 by denitrifiers [44]. Niche functional specialization was observed in recirculating systems, with recirculated digesters harboring the highest number of exclusive genera and enriched with microbial groups that were negatively correlated with TKN, NH_3, TSS levels, and COD (Figures 2 and 4 and Supplementary Data). The majority of these have roles in sulfur metabolisms and denitrification facilitated by aerobic effluent recirculation [45] (Figures 1 and 2 and Supplementary Data). The negative correlations between the recirculating-enriched bacteria and the biomass indicators (TKN, NH_3, TSS, and COD) suggested that these bacterial communities were involved in waste degradation.

Sulfur-metabolizing bacteria such as *Pseudoarcobacter*, *Sulfurimonas*, and *Geobacter* were enriched in a recirculating system. *Pseudoarcobacter* and *Sulfurimonas* are anaerobic, sulfur-oxidizing bacteria coupling denitrifying processes [46,47], while *Geobacter* spp. are sulfur-reducers, reducing elemental sulfur to sulfide [45]. The enrichment of sulfur-oxidizing denitrifiers in recirculating systems suggested higher nitrogen and propionate metabolization potential in recirculating systems, also observed from the elevated abundances of functional genes related to nitrogen metabolism and propionate oxidization (Figure 2).

In single-pass systems, compared to recirculating, sulfur-oxidizing denitrifiers and hydrogenotrophic methanogens were the main bacterial groups involved in anaerobic digestion. These results support the idea that septic tank design and flow configuration can impact the growth of specific microbial groups that facilitate nutrient removal in on-site wastewater treatment systems [47].

3.4. The Succession of Microbial Subgroups in OWTS Anaerobic Digesters

Since anaerobic digestion is a stepwise process [39,48], we focused on the succession of four functional gene subgroups in hydrolytic, acidogenic, acetogenic, and methanogenic processes in anaerobic digesters. From influent to tank to effluent, increasing relative abundances of genes in methanogenic, acidogenic, and acetogenic subgroups (glutamate, glutamine, aspartate, and asparagine metabolism under the category of amino acid and derivatives) in both conventional and plug flow-type recirculating systems (Figures 5 and 6), likely a reflection of progressive maturation of the anaerobic digestion community. Conversely, the relative abundance of hydrolytic gene subgroup (lactose and galactose uptake and use under the category of carbohydrate metabolism) decreased.

Influent sewage contained higher abundances of putative gut bacteria Lachnospiraceae and Ruminococcaceae, which were likely derived from the human sewage feed. These bacteria possess hydrolytic and acidogenic functions [39]. Effluent sulfur-reducing bacteria, nitrifying bacteria, and methanogens were significantly correlated with biomass indicators (Figure 5). Effluent-enriched sulfate-reducing deltaproteobacteria included *Desulfamplus* spp. [49], *Desulfovibrio* spp. [47], *Desulfobacterium* spp. and *Desulfobacter* spp. [50]. Elemental sulfur-reducing bacteria, including *Geobacter* spp. [45], *Desulfonatronum* spp. [51], *Desulfuromonas* spp. [52] and *Desulfobulbus* spp. [50] were also enriched in the effluent. *Methanomethylvorans* spp. are obligately methylotrophic methanogens using methanol as an electron donor [53], were also enriched in the effluent. In recirculating digesters (both conventional and plug flow-type), the alpha diversity from influent to effluent decreased slightly (Figure 3). In contrast, single-pass digesters (both conventional and plug flow-type) had the highest alpha diversity within the tank, followed by the effluent and influent (Figure 3). Effluent of recirculating digesters had higher diversity than the influent was likely due to enrichment of denitrifiers and hydrogenotrophic methanogens, in response to elevated nitrate levels replenished by the recirculating line [48].

Hydrolytic and fermentative bacteria accumulate at the early stages of anaerobic digestion, which convert complex polysaccharides, proteins, and long-chain fatty acids to volatile fatty acids, propionate, and butyrate [39,48]. At late stages, acetogens oxidize volatile and short-chain fatty acids to produce H_2 so that methanogens can use it to produce methane. Consistent with this stratification, we found gut bacteria in the influent and sulfidogenic and methanogenic microbes was enriched in the effluent. Ultimately, this may have resulted in the observed decrease in COD and TKN (Figure 5). Most of the enriched genera in effluent, such as the sulfidogenic and methanogenic microbes, were at low abundances (<5%). However, bacterial communities with lower abundance may still influence anaerobic digester functions [38]. Similarly, our results demonstrated that genes related to nitrogen metabolization, sulfur reduction, and methanogenic pathways were associated with bacteria that are in low abundance.

While this study indicates that DNA sequencing can be a useful additional tool for OWTS monitoring, additional studies are needed to identify how much replication, beyond the $n = 3$ used in this study, is needed by using power analyses to provide satisfactory information to operators and, potentially, to regulatory agencies that certify OWTS technology (Supplementary Data). In addition, as many parameters influence anaerobic digester performance, evaluating additional endpoints that are not routinely monitored by operators, particularly CO_2 and methanogenesis, may be desirable. This may, for example, establish the degree of association between biogas production and specific microbial community composition.

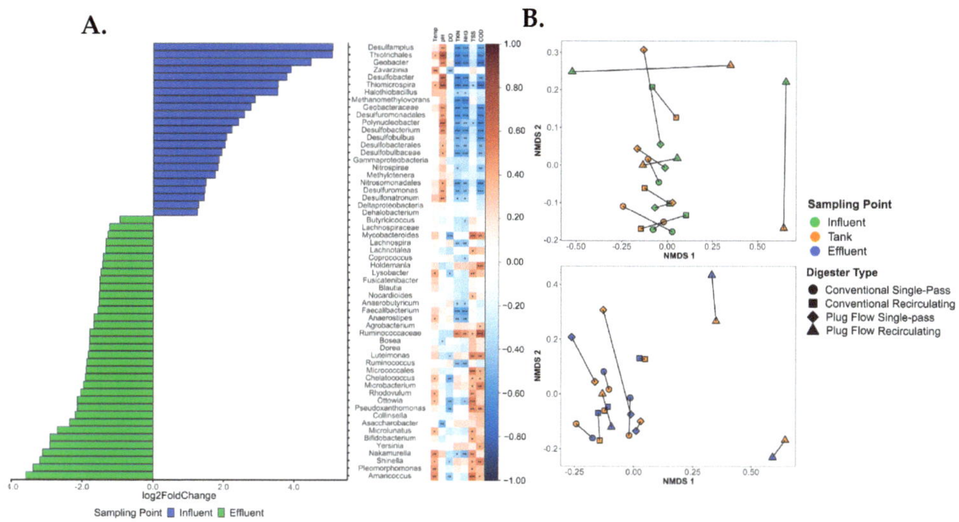

Figure 5. (**A**) Pairwise comparisons of differentially abundant genera between and Spearman rank correlations of influent and effluent sewage. Differential abundance analysis was determined using DESeq2 with an abundance cutoff at 0.01% of total agglomerated genera counts and an adjusted p-value cutoff at $p < 0.05$. Significant correlations are marked with asterisks ($p < 0.05$), where "***" 0.001, "**" 0.01, "*" 0.05. (**B**) Procrustes NMDS ordination influent vs. tank and tank vs. effluent comparisons. Significance test of the M2 procrustes statistic was performed by the R vegan function "protest" ($p < 0.05$).

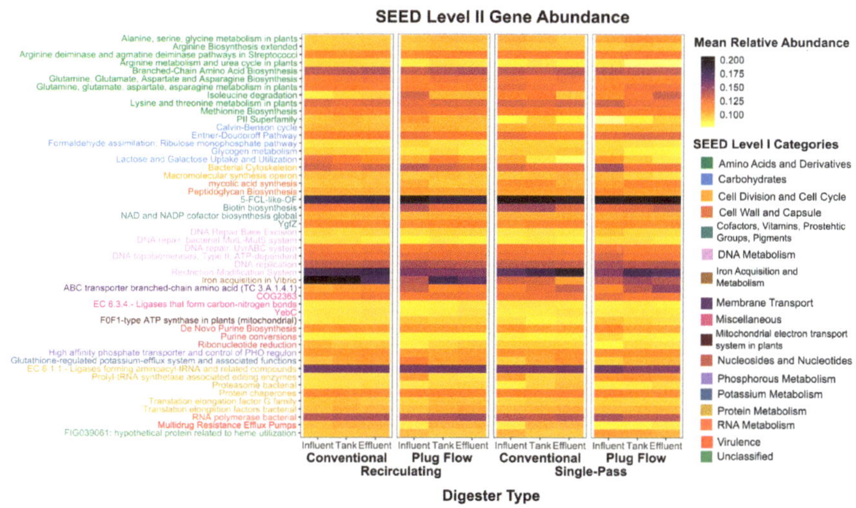

Figure 6. Top 50 most abundant functional gene subsystems (SEED Level III). Level III gene subsystems categorized alphabetically by Level I classifications are indicated in color.

4. Conclusions

- Plug flow-type anaerobic digesters exhibited a comparatively higher abundance of hydrogenotrophic methanogens and sulfate-reducing bacterial communities, while conventional digesters harbored greater propionate-oxidizing acetogen and acetoclastic methanogen populations;
- Sulfur-metabolizing bacterial communities were found to be enriched in recirculating systems, which was also supported by the increased number of nitrogen and propionate metabolization genes;
- Sulfur-oxidizing denitrifiers and hydrogenotrophic methanogens were the most abundant anaerobic digestor bacterial communities relative to recirculating systems;
- Although hydrogenotrophic methanogens and sulfur-reducing bacterial communities represented <5% of the total bacterial population, less dominant taxonomic groups may still influence anaerobic digestion processes;
- Next-generation sequencing tools represent a promising new tool, augmenting traditional technology, for monitoring anaerobic digesters in on-site waste treatment systems.

Supplementary Materials: Supporting information can be found in the electronic version of this article at https://www.mdpi.com/article/10.3390/pr10030436/s1.

Author Contributions: Conceptualization, H.E.S.; methodology, H.E.S., J.N. and A.W.Y.C.; validation, H.E.S., F.S. and E.L.; formal analysis, J.N.; writing—original draft preparation, J.N.; writing—review and editing, H.E.S., F.S., E.L. and J.D.; supervision, H.E.S. All authors have read and agreed to the published version of the manuscript.

Funding: This study was supported with funding from the NSERC Engage program (492465-15), the Ontario Water Consortium (SUB02625), and the Ontario Research Excellence fund (RE09-77). We thank Craig and Chris Jowett of Waterloo Biofilter Systems for technical advice. We also thank Waterloo Biofilter for their in-kind support of this project.

Institutional Review Board Statement: Not applicable.

Informed Consent Statement: Not applicable.

Data Availability Statement: Data will be available upon request.

Acknowledgments: We thank members of the Schellhorn lab for helpful discussions and comments on the manuscript.

Conflicts of Interest: The authors declare no conflict of interest.

References

1. Frei, R.J.; Abbott, B.W.; Dupas, R.; Gu, S.; Gruau, G.; Thomas, Z.; Kolbe, T.; Aquilina, L.; Labasque, T.; Laverman, A.; et al. Predicting Nutrient Incontinence in the Anthropocene at Watershed Scales. *Front. Environ. Sci.* **2020**, *7*, 200. [CrossRef]
2. Tran, N.H.; Reinhard, M.; Gin, K.Y.H. Occurrence and fate of emerging contaminants in municipal wastewater treatment plants from different geographical regions-a review. *Water Res.* **2018**, *133*, 182–207. [CrossRef] [PubMed]
3. Mosbaek, F.; Kjeldal, H.; Mulat, D.G.; Albertsen, M.; Ward, A.J.; Feilberg, A.; Nielsen, J.L. Identification of syntrophic acetate-oxidizing bacteria in anaerobic digesters by combined protein-based stable isotope probing and metagenomics. *Isme J.* **2016**, *10*, 2405–2418. [CrossRef] [PubMed]
4. Vanwonterghem, I.; Jensen, P.D.; Dennis, P.G.; Hugenholtz, P.; Rabaey, K.; Tyson, G.W. Deterministic processes guide long-term synchronised population dynamics in replicate anaerobic digesters. *Isme J.* **2014**, *8*, 2015–2028. [CrossRef] [PubMed]
5. Li, N.; He, J.; Yan, H.; Chen, S.S.; Dai, X.H. Pathways in bacterial and archaeal communities dictated by ammonium stress in a high solid anaerobic digester with dewatered sludge. *Bioresour. Technol.* **2017**, *241*, 95–102. [CrossRef]
6. Zhao, B.; Sha, H.; Li, J.W.; Cao, S.X.; Wang, G.; Yang, Y.Y. Static magnetic field enhanced methane production via stimulating the growth and composition of microbial community. *J. Clean. Prod.* **2020**, *271*, 122664. [CrossRef]
7. Zielinski, M.; Debowski, M.; Kazimierowicz, J. Microwave Radiation Influence on Dairy Waste Anaerobic Digestion in a Multi-Section Hybrid Anaerobic Reactor (M-SHAR). *Processes* **2021**, *9*, 1772. [CrossRef]
8. Yamada, T.; Tsuji, H.; Daimon, H. Improvement of methanogenic activity of anaerobic digestion using poly(L-lactic acid) with enhanced chemical hydrolyzability based on physicochemical parameters. *J. Environ. Manag.* **2018**, *226*, 476–483. [CrossRef]

9. Zhang, J.Y.; Lv, C.; Tong, J.; Liu, J.W.; Liu, J.B.; Yu, D.W.; Wang, Y.W.; Chen, M.X.; Wei, Y.S. Optimization and microbial community analysis of anaerobic co-digestion of food waste and sewage sludge based on microwave pretreatment. *Bioresour. Technol.* **2016**, *200*, 253–261. [CrossRef]
10. Zielinski, M.; Debowski, M.; Kazimierowicz, J. The Effect of Static Magnetic Field on Methanogenesis in the Anaerobic Digestion of Municipal Sewage Sludge. *Energies* **2021**, *14*, 590. [CrossRef]
11. Ma, S.J.; Ma, H.J.; Hu, H.D.; Ren, H.Q. Effect of mixing intensity on hydrolysis and acidification of sewage sludge in two-stage anaerobic digestion: Characteristics of dissolved organic matter and the key microorganisms. *Water Res.* **2019**, *148*, 359–367. [CrossRef] [PubMed]
12. Khafipour, A.; Jordaan, E.M.; Flores-Orozco, D.; Khafipour, E.; Levin, D.B.; Sparling, R.; Cicek, N. Response of Microbial Community to Induced Failure of Anaerobic Digesters Through Overloading With Propionic Acid Followed by Process Recovery. *Front. Bioeng. Biotechnol.* **2020**, *8*, 1434. [CrossRef] [PubMed]
13. Maurus, K.; Kremmeter, N.; Ahmed, S.; Kazda, M. High-resolution monitoring of VFA dynamics reveals process failure and exponential decrease of biogas production. *Biomass Convers. Biorefin.* **2021**. [CrossRef]
14. Lee, D.H.; Behera, S.K.; Kim, J.W.; Park, H.S. Methane production potential of leachate generated from Korean food waste recycling facilities: A lab-scale study. *Waste Manag.* **2009**, *29*, 876–882. [CrossRef]
15. Schwan, B.; Abendroth, C.; Latorre-Perez, A.; Porcar, M.; Vilanova, C.; Dornack, C. Chemically Stressed Bacterial Communities in Anaerobic Digesters Exhibit Resilience and Ecological Flexibility. *Front. Microbiol.* **2020**, *11*, 867. [CrossRef]
16. Goswami, R.; Chattopadhyay, P.; Shome, A.; Banerjee, S.N.; Chakraborty, A.K.; Mathew, A.K.; Chaudhury, S. An overview of physico-chemical mechanisms of biogas production by microbial communities: A step towards sustainable waste management. *3 Biotech* **2016**, *6*, 72. [CrossRef]
17. Luo, C.W.; Tsementzi, D.; Kyrpides, N.; Read, T.; Konstantinidis, K.T. Direct Comparisons of Illumina vs. Roche 454 Sequencing Technologies on the Same Microbial Community DNA Sample. *PLoS ONE* **2012**, *7*, e30087. [CrossRef]
18. Ju, F.; Lau, F.; Zhang, T. Linking Microbial Community, Environmental Variables, and Methanogenesis in Anaerobic Biogas Digesters of Chemically Enhanced Primary Treatment Sludge. *Environ. Sci. Technol.* **2017**, *51*, 3982–3992. [CrossRef]
19. Garcia, S.N.; Clubbs, R.L.; Stanley, J.K.; Scheffe, B.; Yelderman, J.C.; Brooks, B.W. Comparative analysis of effluent water quality from a municipal treatment plant and two on-site wastewater treatment systems. *Chemosphere* **2013**, *92*, 38–44. [CrossRef]
20. Rice, E.W.; Baird, R.B.; Eaton, A.D.; Clesceri, L.S. *Standard Methods for the Examination of Water and Wastewater*, 22th ed.; American Public Health Association, American Water Works, Water Environment Federation: Washington, DC, USA, 2012.
21. Mohiuddin, M.M.; Botts, S.R.; Paschos, A.; Schellhorn, H.E. Temporal and spatial changes in bacterial diversity in mixed use watersheds of the Great Lakes region. *J. Great Lakes Res.* **2019**, *45*, 109–118. [CrossRef]
22. Li, A.D.; Metch, J.W.; Wang, Y.L.; Garner, E.; Zhang, A.N.; Riquelme, M.V.; Vikesland, P.J.; Pruden, A.; Zhang, T. Effects of sample preservation and DNA extraction on enumeration of antibiotic resistance genes in wastewater. *Fems Microbiol. Ecol.* **2018**, *94*, fix189. [CrossRef] [PubMed]
23. Bag, S.; Saha, B.; Mehta, O.; Anbumani, D.; Kumar, N.; Dayal, M.; Pant, A.; Kumar, P.; Saxena, S.; Allin, K.H.; et al. An Improved Method for High Quality Metagenomics DNA Extraction from Human and Environmental Samples. *Sci. Rep.* **2016**, *6*, 26775. [CrossRef] [PubMed]
24. Bolger, A.M.; Lohse, M.; Usadel, B. Trimmomatic: A flexible trimmer for Illumina sequence data. *Bioinformatics* **2014**, *30*, 2114–2120. [CrossRef] [PubMed]
25. Bağcı, C.; Beier, S.; Górska, A.; Huson, D.H. Introduction to the Analysis of Environmental Sequences: Metagenomics with MEGAN. In *Evolutionary Genomics: Statistical and Computational Methods*; Anisimova, M., Ed.; Springer: New York, NY, USA, 2019; pp. 591–604.
26. Mitra, S.; Rupek, P.; Richter, D.C.; Urich, T.; Gilbert, J.A.; Meyer, F.; Wilke, A.; Huson, D.H. Functional analysis of metagenomes and metatranscriptomes using SEED and KEGG. *BMC Bioinform.* **2011**, *12*, S21. [CrossRef] [PubMed]
27. Ramette, A. Multivariate analyses in microbial ecology. *Fems Microbiol. Ecol.* **2007**, *62*, 142–160. [CrossRef]
28. McMurdie, P.J.H.S. PhyloSeq: A bioconductor package for handling and analysis of high-throughput phylogenetic sequence data. In Proceedings of the 17th Pacific Symposium on Biocomputing (PSB), Kohala Coast, HI, USA, 3–7 January 2012; pp. 235–246.
29. Oksanen, J.; Blanchet, F.G.; Kindt, R.; Legendre, P.; Minchin, P.; O'hara, R.; Simpson, G.; Solymos, P.; Stevens, M.H.H.; Wagner, H. Community Ecology Package. R Package Version 2.5-7 2013. Available online: https://rdrr.io/cran/vegan/ (accessed on 14 January 2022).
30. Steinberger, A.J. asteinberger9/seq_scripts: Release v1, Zenodo: 2018. Available online: https://zenodo.org/record/1458243#.YhSeZ5bRZPY (accessed on 14 January 2022).
31. Jonsson, V.; Osterlund, T.; Nerman, O.; Kristiansson, E. Statistical evaluation of methods for identification of differentially abundant genes in comparative metagenomics. *BMC Genom.* **2016**, *17*, 78. [CrossRef]
32. Love, M.I.; Huber, W.; Anders, S. Moderated estimation of fold change and dispersion for RNA-seq data with DESeq2. *Genome Biol.* **2014**, *15*, 550. [CrossRef]
33. Wei, T.; Simko, V.; Levy, M.; Xie, Y.; Jin, Y.; Zemla, J. Package 'corrplot'. *Statistician* **2017**, *56*, e24.
34. Zhu, X.Y.; Campanaro, S.; Treu, L.; Seshadri, R.; Ivanova, N.; Kougias, P.G.; Kyrpides, N.; Angelidaki, I. Metabolic dependencies govern microbial syntrophies during methanogenesis in an anaerobic digestion ecosystem. *Microbiome* **2020**, *8*, 22. [CrossRef]

35. Anderson, M.J. Permutational multivariate analysis of variance (PERMANOVA). *Wiley Statsref Stat. Ref. Online* **2014**, 1–15. [CrossRef]
36. Zhang, C.; Yuan, Q.; Lu, Y.H. Inhibitory effects of ammonia on syntrophic propionate oxidation in anaerobic digester sludge. *Water Res.* **2018**, *146*, 275–287. [CrossRef] [PubMed]
37. Calusinska, M.; Goux, X.; Fossepre, M.; Muller, E.E.L.; Wilmes, P.; Delfosse, P. A year of monitoring 20 mesophilic full-scale bioreactors reveals the existence of stable but different core microbiomes in bio-waste and wastewater anaerobic digestion systems. *Biotechnol. Biofuels* **2018**, *11*, 196. [CrossRef] [PubMed]
38. Rivett, D.W.; Bell, T. Abundance determines the functional role of bacterial phylotypes in complex communities. *Nat. Microbiol.* **2018**, *3*, 767–772. [CrossRef] [PubMed]
39. Dong, L.L.; Cao, G.L.; Wu, J.W.; Yang, S.S.; Ren, N.Q. Reflux of acidizing fluid for enhancing biomethane production from cattle manure in plug flow reactor. *Bioresour. Technol.* **2019**, *284*, 248–255. [CrossRef]
40. Lansing, S.; Viquez, J.; Martinez, H.; Botero, R.; Martin, J. Quantifying electricity generation and waste transformations in a low-cost, plug-flow anaerobic digestion system. *Ecol. Eng.* **2008**, *34*, 332–348. [CrossRef]
41. Vanwonterghem, I.; Jensen, P.D.; Rabaey, K.; Tyson, G.W. Temperature and solids retention time control microbial population dynamics and volatile fatty acid production in replicated anaerobic digesters. *Sci. Rep.* **2015**, *5*, 8496. [CrossRef]
42. Han, W.H.; He, P.J.; Lin, Y.C.; Shao, L.M.; Lu, F. A Methanogenic Consortium Was Active and Exhibited Long-Term Survival in an Extremely Acidified Thermophilic Bioreactor. *Front. Microbiol.* **2019**, *10*, 2757. [CrossRef]
43. Li, L.; He, Q.; Ma, Y.; Wang, X.M.; Peng, X.Y. A mesophilic anaerobic digester for treating food waste: Process stability and microbial community analysis using pyrosequencing. *Microb. Cell Factories* **2016**, *15*, 65. [CrossRef]
44. Akizuki, S.; Matsuyama, T.; Toda, T. An anaerobic-aerobic sequential batch system using simultaneous organic and nitrogen removal to treat intermittently discharged organic solid wastes. *Process Biochem.* **2016**, *51*, 1264–1273. [CrossRef]
45. Qiu, Y.Y.; Zhang, L.; Mu, X.T.; Li, G.B.; Guan, X.Q.; Hong, J.Y.; Jiang, F. Overlooked pathways of denitrification in a sulfur-based denitrification system with organic supplementation. *Water Res.* **2020**, *169*, 115084. [CrossRef]
46. Kobayashi, T.; Li, Y.Y.; Kubota, K.; Harada, H.; Maeda, T.; Yu, H.Q. Characterization of sulfide-oxidizing microbial mats developed inside a full-scale anaerobic digester employing biological desulfurization. *Appl. Microbiol. Biotechnol.* **2012**, *93*, 847–857. [CrossRef] [PubMed]
47. Saia, F.T.; Souza, T.S.O.; Duarte, R.T.D.; Pozzi, E.; Fonseca, D.; Foresti, E. Microbial community in a pilot-scale bioreactor promoting anaerobic digestion and sulfur-driven denitrification for domestic sewage treatment. *Bioprocess Biosyst. Eng.* **2016**, *39*, 341–352. [CrossRef] [PubMed]
48. Garcia-Lozano, M.; Hernandez-De Lira, I.O.; Huber, D.H.; Balagurusamy, N. Spatial Variations of Bacterial Communities of an Anaerobic Lagoon-Type Biodigester Fed with Dairy Manure. *Processes* **2019**, *7*, 408. [CrossRef]
49. Lefevre, C.T.; Trubitsyn, D.; Abreu, F.; Kolinko, S.; Jogler, C.; de Almeida, L.G.P.; de Vasconcelos, A.T.R.; Kube, M.; Reinhardt, R.; Lins, U.; et al. Comparative genomic analysis of magnetotactic bacteria from the Deltaproteobacteria provides new insights into magnetite and greigite magnetosome genes required for magnetotaxis. *Environ. Microbiol.* **2013**, *15*, 2712–2735. [CrossRef] [PubMed]
50. Liu, Z.-h.; Yin, H.; Lin, Z.; Dang, Z. Sulfate-reducing bacteria in anaerobic bioprocesses: Basic properties of pure isolates, molecular quantification, and controlling strategies. *Environ. Technol. Rev.* **2018**, *7*, 46–72. [CrossRef]
51. Zhilina, T.N.; Zavarzina, D.G.; Kuever, J.; Lysenko, A.M.; Zavarzin, G.A. Desulfonatronum cooperativum sp nov., a novel hydrogenotrophic, alkaliphilic, sulfate-reducing bacterium, from a syntrophic culture growing on acetate. *Int. J. Syst. Evol. Microbiol.* **2005**, *55*, 1001–1006. [CrossRef]
52. Badalamenti, J.P.; Summers, Z.M.; Chan, C.H.; Gralnick, J.A.; Bond, D.R. Isolation and Genomic Characterization of 'Desulfuromonas soudanensis WTL', a Metal- and Electrode-Respiring Bacterium from Anoxic Deep Subsurface Brine. *Front. Microbiol.* **2016**, *7*, 913. [CrossRef] [PubMed]
53. Lomans, B.P.; Maas, R.; Luderer, R.; den Camp, H.; Pol, A.; van der Drift, C.; Vogels, G.D. Isolation and characterization of Methanomethylovorans hollandica gen. nov., sp nov., isolated from freshwater sediment, a methylotrophic methanogen able to grow on dimethyl sulfide and methanethiol. *Appl. Environ. Microbiol.* **1999**, *65*, 3641–3650. [CrossRef]

processes

MDPI

Article

Titanium-Containing Coagulants in Wastewater Treatment Processes in the Alcohol Industry

Evgenii Kuzin [1,*], Yulia Averina [2], Andrei Kurbatov [3], Natalia Kruchinina [1] and Veniamin Boldyrev [4]

[1] Department of Industrial Ecology, D. Mendeleev University of Chemical Technology of Russia, 125047 Moscow, Russia; kruchinina.n.e@muctr.ru
[2] Department of Logistics and Economic Information, D. Mendeleev University of Chemical Technology of Russia, 125047 Moscow, Russia; averinajm@mail.ru
[3] Department of Innovation Material and Corrosion Protection, D. Mendeleev University of Chemical Technology of Russia, 125047 Moscow, Russia; andreikurbatov@yandex.ru
[4] Department of Chemistry, Bauman Moscow State Technical University, 105005 Moscow, Russia; boldyrev.v.s@bmstu.ru
* Correspondence: e.n.kuzin@muctr.ru

Abstract: As part of the work, studies of the quality of wastewater from alcohol production were carried out; the main characteristics were determined, such as the charge of dispersed particles and the content of organic compounds and fats. A comparison is made of the effectiveness of titanium salts, traditional aluminum, and iron-containing coagulants. It has been established that titanium salts are much more effective than traditional reagents, but their high cost significantly limits the scope of their application. The possibility of increasing the efficiency of traditional coagulants by adding various titanium salts in an amount of up to 10% of the mass has been proved. It has been established that the addition of titanium compounds to aluminum or iron-containing coagulants makes it possible to increase the cleaning efficiency by an average of 10–20% while a significant reduction in the dose of reagents is possible, which will have a positive effect on the cost of the cleaning process. It was found that the addition of titanium salts to a coagulant based on aluminum sulfate allows not only an increase in the cleaning efficiency but also further intensifies the processes of sedimentation and filtration of coagulation sludge, which will significantly reduce the size of the treatment equipment and reduce capital costs.

Keywords: water treatment; titanium coagulants; sedimentation

Citation: Kuzin, E.; Averina, Y.; Kurbatov, A.; Kruchinina, N.; Boldyrev, V. Titanium-Containing Coagulants in Wastewater Treatment Processes in the Alcohol Industry. *Processes* **2022**, *10*, 440. https://doi.org/10.3390/pr10030440

Academic Editor: Monika Wawrzkiewicz

Received: 22 January 2022
Accepted: 17 February 2022
Published: 22 February 2022

1. Introduction

Population growth necessitates the increase in capacity of the food industry and related industries. An increase in the volume of manufactured food products, the development of new formulations and products, and innovations in the field of food biotechnology have led to an increase in the environmental impact of enterprises. Most of the food industry's waste can be easily recycled using the biological methods (biocomposition, fermentation, etc.); the atmospheric emissions have almost no serious consequences, except for the dryers' emissions. However, treatment and disposal of the wastewater generated as a result of various food production processes remains an unresolved problem.

The processes related to the generation of large volumes of highly contaminated organic wastewater components are typical for all areas of the food industry, including the production of alcohol-containing products (alcohol, beer, and wine), the milk treatment processes, and the manufacturing of the byproducts (butter, cheese) in confectionery enterprises [1–3].

The food industry wastewater treatment is a comprehensive and extremely urgent task. The anaerobic digestion methods for highly concentrated effluents have become widespread when the total amount of dissolved and undissolved organic compounds, in terms of the COD, is more than 30,000–40,000 mg(O)/L, with the production of flammable hydrocarbons. However, this technology cannot be used for the effluents with a relatively low content of organic substances, at a COD of up to 10,000 mg(O)/L [4].

The traditional treatment method for such wastewater includes the stage of physicochemical clarification from suspended solids, undissolved organic compounds, partially emulsified fats, etc., followed by biological purification.

The coagulants (polyelectrolyte) and flocculants are used as the reagents at the stage of physicochemical purification. The most common coagulants are aluminum and iron salts. Aluminum sulfate and oxychloride are widely used in the drinking water treatment processes; however, due to their relatively high cost, they are less often used in the wastewater treatment processes. In addition, despite their high efficiency and versatility, the aluminum compounds are effective in a very limited pH range (6.0–7.5) and are subject to a strict standard for the residual aluminum content in water. Moreover, aluminum sulfate is also inefficient at low water temperatures. The iron compounds, such as ferric chloride ($FeCl_3$) and ferrous sulfate ($FeSO_4$), are quite effective reagents and have a wide applicable pH range compared to the coagulants based on aluminum compounds. However, they also have certain disadvantages, such as significant volumes of difficult-to-filter precipitate and the possible generation of complexes with organic ligands [5,6].

A new field of reactant wastewater purification is the use of coagulants based on titanium compounds. The research teams from China [7–11], Russia [12,13], and Ukraine [14] noted the high efficiency of titanium-containing coagulants as well as the absence of most disadvantages inherent in the conventional reagents. The similar results were obtained as a result of the international collaboration of institutions from France, Australia, China, Korea, and other countries [15–20].

The collaborating author team from Great Britain confirmed the high efficiency of titanium compounds in the process of removing dissolved organic substances from water [21].

The research groups from Russia proved the possible increase in the efficiency of conventional reagents based on aluminum or iron compounds by introducing a few hydrolysis products of titanium compounds into their composition (up to 10–15% by weight) that made it possible to expand the effective pH range and increase the purification efficiency and sedimentation specifications of the generated sediments [12,22].

Based on the literature review, it can be seen that data of the titanium compounds used as coagulants for the wastewater purification of the alcohol industry are not currently presented, which determines the scientific and practical novelty of the research.

The main purpose of this paper is to assess the possible use of titanium-containing coagulants in the wastewater purification processes at the alcohol industry enterprises.

2. Materials and Methods

The production of alcohol-containing products was selected as the study object. The fermentation tanks as well as separation and distillation shops are located on the site. The manufacturing department produces various types of alcoholic beverages, such as beer, wine, kvass, etc. and is standardized for all similar enterprises in the territory of the Russian Federation (over 70 entities). The production process wastewater was taken from the sewer pipes of the relevant company.

Wastewater samples were taken by averaging the 20 L capacitances and delivered to the laboratory within 4 h.

Samples of the comprehensive titanium-containing coagulants were obtained by adding hydrolysis products of titanium compounds (titanium sulfate or chloride) as well as titanium dioxide in an amount of up to 10 wt% to the conventional reagents based on aluminum and iron salts [23].

The particle size was determined by the dynamic light diffusion method using a Photocor Compact-Z device (Russia).

The zeta potential of colloidal particles was determined with a Zetasizer Nano by Malvern (Great Britain).

The pH value was determined using a digital meter pHHQ11B by HACH (Loveland, CO, USA).

The chromaticity value was determined using a DR 6000 spectrophotometer by HACH (USA).

The chemical oxygen demand was determined by the reference method with dichromate according to the procedure [24].

The biochemical oxygen demand was determined according to the applicable method (APHA 1986) [25].

The content of the suspended solids was determined by the gravimetric method using the HANNA 98703 portable turbidimeter.

The content of the fats and oils was determined gravimetrically (Soxhlet method) [26].

Titanium dioxide (TiO_2 99.8%), titanyl sulfate ($TiOSO_4 \cdot xH_2O$—99.9%), and titanium tetrachloride (99.9%, $TiCl_4$) were made by Sigma-Aldrich (St. Louis, MO, USA).

Aluminum sulfate, aluminum polyoxychloride, and ferric chloride were produced by Kemira (Helsinki, Finland).

Coagulant solutions were obtained by dissolving (magnetic stirring) a 10 g sample of powdered reagents in 100 mL of water for 60 min.

The jar test was carried out using the portable laboratory flocculator JLT 4 VELP (Italy). Wastewater samples of a given volume (500 mL) were divided into measuring cups, followed by setting it in a laboratory flocculator. With a pipet dispenser, predetermined individual dosages of preprepared complex reagents were injected into the treated water. The pH correction was adjusted using an appropriate volume of 0.05 mol/L NaOH.

During the experiments, coagulants were added at the beginning of mixing at 150 rpm for 2 min, known as the fast coagulation stage (mixing reagent with water and the formation of primary aggregates). Then, the solution underwent slow mixing at 10 rpm for 8 min, which is the flocculation and adsorption phase. This was followed by 30 min of sedimentation (quiescent settling). The samples of water after sedimentation were collected from 5 cm below the water surface.The filtration rate was determined by the passage of 100 mL of liquid through a filter with a pore size of 15 microns and by measuring the volume of the filtrate obtained in 60 s.

The clarified water was analyzed according to the methods described above.

The efficiency of treatment (Eff_{TR}) was calculated by equipment:

$$\text{Eff}_{TR} = \frac{C_{BAS} - C_{TR}}{C_{BAS}} * 100\% \tag{1}$$

C_{BAS}—concentration suspended solids (indicator value COD) before treatment, mg/L; C_{TR}—concentration suspended solids (parameter value COD) after treatment, mg/L.

3. Results and Discussion

The results of the particle sizes analysis demonstrate that the wastewater is specified by the availability of two particle peals, namely the rapid settling particles with a size of more than 10 microns and the slow settling particles with a size of about 100 nm. The zeta potential at the particle surface was (−6 mV).

The chemical composition of wastewater samples was examined at the next experimental stage (Table 1).

Table 1 shows that the priority pollutant in the water composition is the dissolved organic matter as well as the colloidal particles. The supply of such water to the urban facilities for deep biological treatment is impossible due to the prohibitively high rates of chemical and biological oxygen demand.

Among many indicators of wastewater quality (and purification efficiency) analyzed, the most common ones were selected: chemical oxygen consumption and suspended solids content. These indicators are used to correct the operation of purification facilities (dose of coagulant, flotation time, etc.) at most enterprises.

Table 1. Chemical composition of the wastewater samples.

Indicator	Unit of Measurement	Alcohol Industry Wastewater
pH		4.5
COD	mg(O)/L	6120
BOD 5	mg(O)/L	5480
Suspended solids	mg/L	2100
Chromaticity	degrees	5600
Fats	mg/L	1.2
Zeta potential	mV	−5
Particle size		130 nm, 74 μm

At the first stage, we have made a comparison of the efficiency of conventional reagents and coagulants based on the titanium compounds in the purification process for the actual alcohol industry effluents from the suspended solids (Figure 1) and the chemical oxygen demand index (COD) (Figure 2).

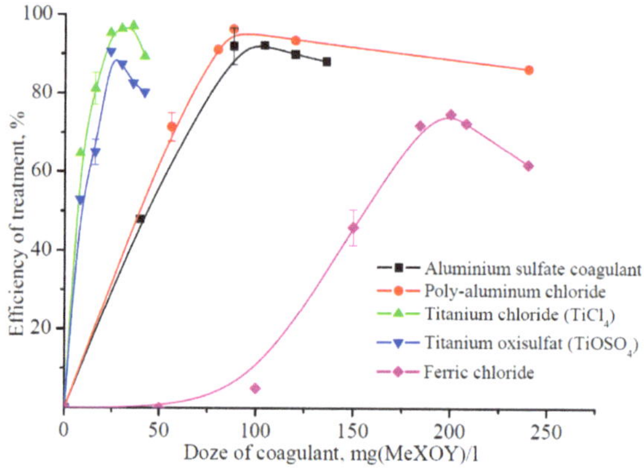

Figure 1. Dependence of the alcohol industry wastewater treatment efficiency from the suspended solids by various coagulants on the coagulant dose.

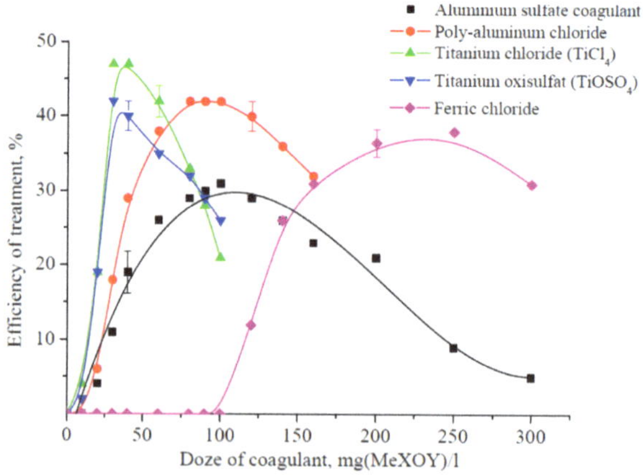

Figure 2. Dependence of the alcohol industry wastewater treatment efficiency from the organic compounds (COD index) by various coagulants.

The diagrams in Figures 1 and 2 demonstrate that the conventional coagulants based on aluminum compounds can effectively remove the suspended solids and organic compounds from water while the iron compound efficiency in this process is much lower.

The zero efficiency of ferric chloride until dose 100 mg/L can be explained by the formation of complex compounds of iron with organic ligands (for example, citrates or tartrates), which is typical for such coagulants [5,6]. The difference in effective dosages of coagulants for the COD index and suspended solids content can be explained by the cleaning mechanism. Therefore, for colloidal (suspended) particles, this is the charging neutralization and the formation of floccules, and for the COD index (content of organic substances), it is the adsorption on the surface of the hydrolysis products of titanium compounds.

The effective coagulant dose is expressed in mg (MeXOY)/L. The maximum possible treatment efficiency was achieved with the following doses of conventional reagents:

- Aluminum sulfate—88 mg (Me_XO_Y)/L;
- Aluminum oxychloride—80 mg (Me_XO_Y)/L;
- Iron chloride—200 mg (Me_XO_Y)/L.

During the purification process, the titanium compounds were not inferior to the conventional reagents in their efficiency while the minimum effective dose was, on average, 2.5–3.0 times lower and amounted to 20 and 24 mg (Me_XO_Y)/L for titanyl sulfate and titanium tetrachloride, respectively. In addition, the efficiency of titanium compounds during the process of organic compound removal, namely a decrease in the COD index, was, on average, 5–7% higher, which should have a positive effect on further biological treatment. In addition, it should be noted that when the titanium salts are used as the coagulants, the final treatment products will be titanium dioxide that has a much lower toxicity compared to the aluminum compounds [27].

The next experimental stage was to evaluate the effect of adding titanium compounds to the conventional reagents on the overall purification efficiency according to the above indicators.

The diagram in Figures 3 and 4 presents data on the effect of adding various titanium compounds on the alcohol industry wastewater treatment efficiency in terms of the content of the suspended solids and dissolved organic compounds (COD). The basic coagulant dose was 50 mg (Al_2O_3/L) for aluminum sulfate and oxychloride and 150 mg (Fe_2O_3/L) for ferric chloride (Figures 3 and 4), and the addition of titanium compounds for TiO_2 was 10 wt%.

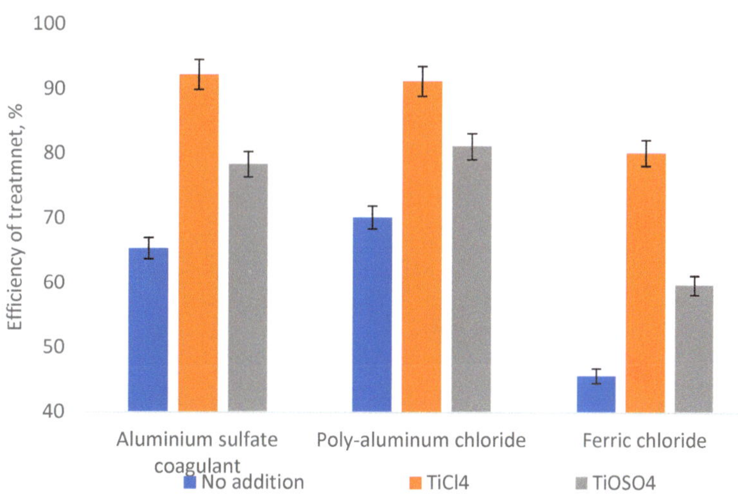

Figure 3. Efficiency of removing suspended solids from the alcohol industry wastewater by various coagulants.

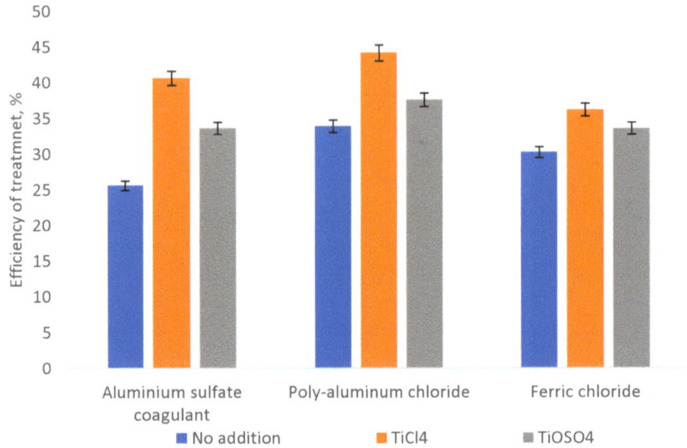

Figure 4. Efficiency of removing organic compounds (COD) from the alcohol industry wastewater by various coagulants.

The diagrams in Figures 3 and 4 demonstrate that the addition of titanium compounds in an amount of up to 10 wt% allows an increase of the wastewater treatment efficiency from the suspended solids and organic compounds. It should be noted that the greatest positive effect is made by the addition of titanium dioxide that is explained primarily by the primary nucleation processes (neutralizing coagulation) on the surface of negatively charged particles [28,29] as well as by the specific hydrolysis processes [30]. Having considered the relatively low addition (less than 10%) of titanium compounds and a decrease in the effective basic reagent dose by an average of 25–30%, the cost of the purification process can decrease by about 5–8%, depending on the wholesale price for titanium compounds.

Another feature of the food industry wastewater treatment is the low filtration rate of the generated sediments at the additional treatment stage after sedimentation. During the experiments with complex reagents, a change in the structure and shape of the resulting coagulation sludge was noted, to which the authors evaluated the filtration and sedimentation process efficiency when using the pure and complex reagents. A comparison of the precipitation filtration rate is shown in Figure 5.

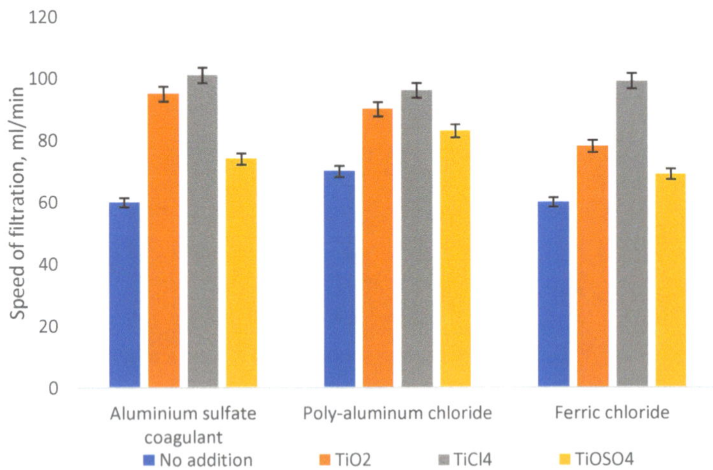

Figure 5. Filtration rate of coagulation sludge of the alcohol industry wastewater.

Based on the diagram, the availability of titanium compounds in an amount of up to 10 wt% of the conventional coagulant makes it possible to intensify the filtration processes of coagulation sludge by 20–40%. For the iron-containing coagulants, it could be increased by almost 2 times.

Based on the data obtained, the complete efficiency verification of the complex reagents required a reference standard in the form of aluminum sulfate at a dosage of 50 mg (Al_2O_3/L) with the addition of titanium dioxide (5.0 mg (TiO_2/L) as the cheapest reagent option [31]. The cost-based selection of the reagent is primarily due to the significant volumes of wastewater accepted for treatment, and, hence, the operating costs during the arrangement of the reagent facilities, the value of which will be included in the final product cost. The results of coagulation water purification in relation to all studied indicators are shown in Table 2.

Table 2. Residual concentrations and treatment efficiency.

	Initial Concentration	Aluminiumsulfate Coagulant + TiO_2		Aluminiumsulfate Coagulant	
		Concentration after Treatment	Efficiency of Treatment, %	Concentration after Treatment	Efficiency of Treatment, %
pH	4.5	-	-	-	-
COD, mg(O)/L	6120	3543.5	42.1	3965.8	35.2
BOD 5 mg(O)/L	5480	3112.6	43.2	3485.3	36.4
Suspend solids mg/L	2100	92.4	95.6	260.4	87.6
Color	5600	285.6	94.9	761.6	86.4
Fat, mg/L	1.2	0.2	87.2	0.3	72.5

Table 2 demonstrates that the use of aluminum sulfate modified with titanium compounds makes it possible to increase the alcoholic industry wastewater treatment efficiency in relation to all the studied parameters that will have a positive effect on the further deep biological treatment process.

The increased efficiency of coagulants containing titanium compounds is explained by the following phenomena:

The adsorption processes on the surface of hydrolysis products of titanium compounds. Therefore, hydroxides (orta- and metatitanic acids) and partially hydrated titanium oxide (xerogel) have a developed surface and are capable of efficient adsorption of organic pollutants during the coagulation process [28,30].

The polycondensation processes and polymerization of highly active polytitanates [15,16] have not only high coagulation activity but also a flocculating effect with the formation of gel-like structures.

The charge neutralization processes on the surface of colloidal particles of pollutants and positively charged forms of aluminum hydroxides leads to the destabilization of the colloidal system and its exit from equilibrium, with quick aggregation and precipitation of the forming structures.

4. Conclusions

According to the studies performed, we obtained information on the possible use of titanium-containing coagulants in the wastewater treatment processes at the enterprises with an alcoholic fermentation stage.

It has been proved that the coagulants based on titanium salts are significantly superior in efficiency to the conventional reagents. The effect of various added titanium compounds on the purification efficiency of the conventional reagents has been established. A positive effect of titanium compounds on the processes of removing insoluble impurities and dissolved organic substances from water has been found. It has been proved that the addition of titanium compounds in an amount of up to 10 wt% allows not only an increase in the treatment efficiency but also a significant reduction of the total coagulant demand that will significantly reduce the treatment process cost. The sludge obtained during the water treatment process, with the complex titanium-containing coagulants, has a voluminous structure that both gives off moisture and is filtered easily.

The resulting hydrolysis products of titanium compounds can be removed from water not only by settling but also by more up-to-date technologies of sediment separation, such as flotation [32,33].

The results obtained will make it possible to reduce the total coagulant demand and further expand the scope of application of titanium-containing reagents in the water treatment processes [34].

Author Contributions: V.B.—systematization of the data obtained, writing an article. E.K.—obtaining samples of complex coagulants, conducting laboratory experiments on water treatment, writing an article, discussing the results. N.K.—study of the initial composition of water, development of the experimental methodology, substantiation of the type of reagents used and dosage of agulants, discussion of the results, writing an article. A.K.—analytical control of the main indicators of water quality before and after treatment, intermediate studies of water in the process of selecting optimal dosages, discussing the results. Y.A.—statistical processing of experimental results, preparation of a literary review. All authors have read and agreed to the published version of the manuscript.

Funding: This research received no external funding.

Institutional Review Board Statement: Not applicable.

Informed Consent Statement: Not applicable.

Data Availability Statement: Not applicable.

Conflicts of Interest: The authors declare that they have no known competing financial interests or personal relationships that could have appeared to influence the work reported in this paper.

References

1. Altech Environmental Consulting Ltd. *A Review of Wastewater Management & Best Practices for Dischargers in the Food Processing Sector*; Ontario Centre for Environmental Technology Advancement: North York, ON, Canada, 2005.
2. Simate, G.S.; Cluett, J.; Iyuke, S.E.; Musapatika, E.T.; Ndlovu, S.; Walubita, L.F.; Alvarez, A.E. The treatment of brewery wastewater for reuse: State of the art. *Desalination* **2011**, *273*, 235–247. [CrossRef]
3. Farizoglu, B.; Uzuner, S. The investigation of dairy industry wastewater treatment in a biological high performance membrane system. *Biochem. Eng. J.* **2011**, *57*, 46–54. [CrossRef]
4. Falletti, L.; Conte, L.; Zaggia, A.; Battistini, T.; Garosi, D. Food Industry Wastewater Treatment Plantbased on Flotation and MBBR. *Modern Environ. Sci. Eng.* **2015**, *1*, 94–98. [CrossRef]
5. Moussas, P.; Tzoupanos, N.; Zouboulis, A. Advances in coagulation/flocculation field: Al- and Fe-based composite coagulation reagents. *Desalin. Water Treat.* **2011**, *33*, 140–146. [CrossRef]
6. Han, S.w.; Kang, L.S. Comparison of Al(III) and Fe(III) Coagulants for Improving Coagulation Effectiveness in Water Treatment June. *J. Korean Soc. Environ. Eng.* **2015**, *37*, 325–331. [CrossRef]
7. Xu, J.; Zhao, Y.; Gao, B.; Zhao, Q. Enhanced algae removal by Ti-based coagulant: Comparison with conventional Al- and Fe-based coagulants. *Environ. Sci. Pollut. Res.* **2018**, *25*, 13147–13158. [CrossRef]
8. Zhao, Y.; Gao, B.; Cao, B.; Yang, Z.; Yue, Q.; Shon, H.; Kim, J.-H. Comparison of coagulation behavior and floc characteristics of titanium tetrachloride ($TiCl_4$) and polyaluminum chloride (PACl) with surface water treatment. *Chem. Eng. J.* **2011**, *166*, 544–550. [CrossRef]
9. Zhao, Y.; Gao, B.; Zhang, G.; Qi, Q.; Wang, Y.; Phuntsho, S.; Kim, J.-H.; Shon, H.; Yue, Q.; Li, Q. Coagulation and sludge recovery using titanium tetrachloride as coagulant for real water treatment: A comparison against traditional aluminum and iron salts. *Sep. Purif. Technol.* **2014**, *130*, 19–27. [CrossRef]
10. Zhao, Y.; Gao, B.; Zhang, G.; Phuntsho, S.; Wang, Y.; Yue, Q.; Li, Q.; Shon, H. Comparative study of floc characteristics with titanium tetrachloride against conventional coagulants: Effect of coagulant dose, solution pH, shear force and break-up period. *Chem. Eng. J.* **2013**, *233*, 70–79. [CrossRef]
11. Zhao, Y.; Shon, H.; Phuntsho, S.; Gao, B. Removal of natural organic matter by titanium tetrachloride: The effect of total hardness and ionic strength. *J. Environ. Manag.* **2014**, *134*, 20–29. [CrossRef]
12. Kuzin, E.N.; Kruchinina, N.E. Titanium-containing coagulants for found rywaste water treatment. *CIS Iron Steel Rev.* **2020**, *20*, 66–69. [CrossRef]

13. Chernoberezhsky, J.M.; Minejev, D.J.; Dyagileva, A.B.; Lorenzson, A.V.; Belova, J.V. Isolation of sulphate lignin from aqueous solutions by oxo-titanium sulphate, aluminium sulphate and composite coagulant on their basis. *J. Appl. Chem. (Zhurnal Prikladnoy Khimii)* **2002**, *75*, 1730–1732.
14. Mamchenko, A.V.; Gerasimenko, N.G.; Deshko, I.I.; Pakhar', T.A. The investigation of the efficiency of coagulants based on titanium when purifying water. *J. Water Chem. Technol.* **2010**, *32*, 167–175. [CrossRef]
15. Chekli, L.; Eripret, C.; Park, S.; Tabatabai, S.; Vronska, O.; Tamburic, B.; Kim, J.; Shon, H. Coagulation performance and floc characteristics of polytitanium tetrachloride (PTC) compared with titanium tetrachloride ($TiCl_4$) and ferric chloride ($FeCl_3$) in algal turbid water. *Sep. Purif. Technol.* **2017**, *175*, 99–106. [CrossRef]
16. Galloux, J.; Chekli, L.; Phuntsho, S.; Tijing, L.; Jeong, S.; Zhao, Y.; Gao, B.; Park, S.; Shon, H. Coagulation performance and floc characteristics of polytitanium tetrachloride and titanium tetrachloride compared with ferric chloride for coal mining wastewater treatment. *Sep. Purif. Technol.* **2015**, *152*, 94–100. [CrossRef]
17. Aziz, H.A.; Rosli, M.Y.B.; Amr, S.S.A.; Hussain, S. Potential use of titanium tetrachloride as coagulant to treat semi aerobic leachate treatment. *Aust. J. Basic Appl. Sci.* **2015**, *9*, 37–44.
18. Jeon, K.-J.; Ahn, J.-H. Evaluation of titanium tetrachloride and polytitanium tetrachloride to remove phosphorus from wastewater. *Sep. Purif. Technol.* **2018**, *197*, 197–201. [CrossRef]
19. Pushpalatha, T.N.; Lokeshappa, B. The Use of Alum, Ferric Chloride and Titanium tetrachloride as Coagulants in Treating Landfill Leachate. *Int. J.Sci. Eng. Technol. Res. (IJSETR)* **2015**, *4*, 2093–2096.
20. Mashia, M.; Taguchi, Y.; Ohizumi, M.; Koyanagi, S.; Harigai, T. Removal of phosphates from wastewater by titanium(IV) sulfate with precipitation method. *Jpn. J. Water Pollut. Res.* **1985**, *8*, 668–675. [CrossRef]
21. Hussain, S.; Awad, J.; Sarkar, B.; Chow, C.W.; Duan, J.; van Leeuwen, J. Coagulation of dissolved organic matter in surface water by novel titanium (III) chloride: Mechanistic surface chemical and spectroscopic characterisation. *Sep. Purif. Technol.* **2019**, *213*, 213–223. [CrossRef]
22. Kuzin, E.N.; Krutchinina, N.E. Purification of circulating and waste water in metallurgical industry using complex coagulants. *CIS Iron Steel Rev.* **2019**, *18*, 72–75. [CrossRef]
23. Shon, H.K.; Vigneswaran, S.; Kandasamy, J.; Zareie, M.H.; Kim, J.B.; Cho, D.L. Preparation and Characterization of Titanium Dioxide (TiO_2) from Sludge produced by $TiCl_4$ Flocculation with $FeCl_3$, $Al_2(SO_4)_3$ and $Ca(OH)_2$ Coagulant Aids in Wastewater. *Sep. Sci. Technol.* **2009**, *44*, 1525–1543. [CrossRef]
24. ISO 15705:2002. *Water Quality—Determination of the Chemical Oxygen Demand Index (ST-COD)-SMALL-Scale Sealed-Tube Method*; ISO: Geneva, Switzerland, 2002.
25. American Public Health Association. *Standard Methods for the Examination of Waters and Wastewater*, 21th ed.; American Public Health Association: Washington, DC, USA, 2005; ISBN 0-87553-047-8.
26. APHA-AWWA-WPFC. *Métodos Normalizados Para el Análisis de Agua Potable y Aguas Residuales*, 17th ed.; Editorial Díaz de Santos: Madrid, Spain, 1992.
27. Zeynalov, O.A.; Kombarova, S.P.; Bagrov, D.V. On the effect of metal oxide nanoparticles on the physiology of living organisms. *Rev. Clin. Pharmacol. Drug Ther.* **2016**, *14*, 24–33. [CrossRef]
28. Shabanova, N.A.; Popov, V.V.; Sarkisov, P.D. *The Chemistry and Technology of Nanodispersed Oxides*; Learner's Guide; IKTs "Akademkniga": Moscow, Russia, 2007; p. 309.
29. Draginskiy, V.L.; Alekseeva, L.P.; Getmantsev, S.V. *Natural Water Treatment Methods and Coagulation*; Nauchnoeizdatelstvo: Moscow, Russia, 2005; p. 576.
30. Wang, T.-H.; Navarrete-López, A.M.; Li, S.; Dixon, D.A.; Gole, J.L. Hydrolysis of TiCl4: Initial Steps in the Production of TiO_2. *J. Phys. Chem. A* **2010**, *114*, 7561–7570. [CrossRef] [PubMed]
31. Kuzin, E.N.; Kruchinina, N.E.; Gromovykh, P.S.; Tyaglova, Y.V. Coagulants in the Processes of Waste Water Treatment in Dairy Complex Industry. *Chem. Sustain. Dev.* **2020**, *28*, 388–393.
32. Kolesnikov, A.V.; Savel'ev, D.S.; Kolesnikov, V.A.; Davydkova, T.V. Electroflotation extraction of highly disperse titanium dioxide TiO_2 from water solutions of electrolytes. *Glass Ceram.* **2018**, *75*, 237–241. [CrossRef]
33. Meshalkin, V.P.; Kolesnikov, A.V.; Saveliev, D.S. Analysis of physicochemical efficiency of electroflotation process for removing titanium chloride hydrolysis products from the anthropogenic effluents. *Deport Acad. Sci.* **2019**, *486*, 680–684.
34. Gan, Y.; Li, J.; Zhang, L.; Wu, B.; Huang, W.; Li, H.; Zhang, S. Potential of titanium coagulants for water and wastewater treatment: Current status and future perspectives. *Chem. Eng. J.* **2021**, *406*, 126837. [CrossRef]

processes

MDPI

Review

Recovery of Phosphorus in Wastewater in the Form of Polyphosphates: A Review

Weiran Chu [1], Yi Shi [1] and Liang Zhang [1,2,*]

[1] Key Laboratory of Industrial Biotechnology, Ministry of Education, Jiangnan University, Wuxi 214122, China; chuwr@stu.jiangnan.edu.cn (W.C.); shiyi0621@jiangnan.edu.cn (Y.S.)

[2] National Engineering Laboratory for Cereal Fermentation Technology, Jiangnan University, Wuxi 214122, China

* Correspondence: zhangl@jiangnan.edu.cn; Tel.: +86-510-85918235

Abstract: As non-renewable resource, the recovery and utilization of phosphorus from wastewater is an enduring topic. Stimulated by the advances in research on polyphosphates (polyP) as well as the development of Enhanced Biological Phosphorus Removal (EBPR) technology to achieve the efficient accumulation of polyP via polyphosphate accumulating organisms (PAOs), a novel phosphorus removal strategy is considered with promising potential for application in real wastewater treatment processes. This review mainly focuses on the mechanism of phosphorus aggregation in the form of polyP during the phosphate removal process. Further discussion about the reuse of polyP with different chain lengths is provided herein so as to suggest possible application pathways for this biosynthetic product.

Keywords: EBPR; PAOs; polyP; phosphorus reuse; wastewater

Citation: Chu, W.; Shi, Y.; Zhang, L. Recovery of Phosphorus in Wastewater in the Form of Polyphosphates: A Review. *Processes* **2022**, *10*, 144. https://doi.org/10.3390/pr10010144

Academic Editors: Dimitris Zagklis and Georgios Bampos

Received: 11 December 2021
Accepted: 4 January 2022
Published: 11 January 2022

1. Introduction

With economic and industrial development, water pollution is becoming an inevitable environmental issue and eutrophication in aqueous systems normally accompanies the excessive emission of phosphorus wastewater. To eliminate the pollution, a treatment targeting the removal of phosphate from contaminated water has been adopted since the 1950s [1,2]. In general, the technologies utilized for phosphorus removal are mainly divided into three categories, which are the chemical method [3,4], biological method [5], and chemical–biological combination method [6]. Regarding chemical phosphorus removal technology, the adsorption method is suggested as being more promising since the adsorbed phosphate could be recovered for further utilization [7]. Using ferrihydrite as the adsorbent, the maximum monolayer adsorption capacities of PO_4^{3-} was verified as 66.6 $mg \cdot g^{-1}$ [8] and the adsorbed phosphorus could then be desorbed under alkaline conditions to be reused. In comparison, the biological phosphorus removal method takes advantage of the specific metabolism properties of polyphosphate accumulate organisms (PAOs) so as to excessively absorb phosphorus from wastewater and thereafter to remove P in the form of activated sludge [9]. Alternatively, chemical–biological combination methods have been developed for enhancing the efficacy of biological treatment by introducing specific chemical substances to prevent the PAOs from losing their phosphorus accumulation capabilities, particularly in environments that are not suitable for PAOs to thrive [10]. In consideration of the corresponding properties of the above phosphorus removal technologies, decisions on the actual treatment strategies are made based on the actual water quality conditions as well as economic and environmental requirements.

Although current wastewater treatment methods can normally realize efficient phosphorus removal from the water, attention should also be paid to the provision of promising pathways for the recovery and reuse of P. Phosphorus is an element that cannot be released from the system through biological metabolism and typical wastewater treatment strategies mainly remove phosphate from water by transferring or accumulating phosphorus in a form that can be separated. The limited utilization efficacy of phosphorus-rich waste produced during the above treatment process leads to a lost resource, particularly considering that the existing phosphate reserves will be exhausted in the next 50–100 years at the current rate of extraction [11]. In this regard, strategies with the potential of removing P from the system in a form that could be further applied are receiving increasing interest. Among the available P removal strategies, Enhanced Biological Phosphorus Removal (EBPR) with PAOs as the working units is considered as the best candidate for achieving P removal and reuse simultaneously. As the drivers of this biological P removal process, PAOs generate P from their surroundings and aggregate single phosphate to form long-chain polyphosphate (polyP) [12], which is a type of polymer consisting of tens to hundreds of phosphate residues. Linked by phosphoanhydride bonds, polyP has been found in abundance in all cells in nature and it has been suggested that it is unique in its likely role in the origin and survival of species [13,14]. Recent studies also revealed that polyP could act as a storage for and generator of metabolic energy in the extracellular matrix [15] and play a role in osteoblasts for promoting cartilage mineralization [16]. In addition, polyP with different chain lengths was verified as having an impact on the activity of blood coagulation factors [17] and thus, affecting the endogenous blood clotting ability in animals [18]. The strong chelating ability of long-chain polyP could enable it to play a role in food preservation and antibacterial products [19]. Given the application potential of polyP, recovering P from wastewater via PAOs to obtain biosynthetic polyP for further utilization is considered to be a possible strategy for improving the efficacy of P reuse. Herein, reviews and a discussion are provided for the purpose of suggesting a feasible pathway to collect, recycle, and reuse phosphorus from wastewater in the form of polyP.

2. Enrichment of PolyP in PAOs during EBPR Process

EBPR is a biological phosphorus removal technology that was developed in the 1970s and is now widely used in sewage treatment plants around the world [20]. Relying on PAOs or denitrifying phosphorous accumulating bacteria (DPAOs) [21], EBPR technology can remove P by having the free P absorbed by functional microbes for the purpose of growth and accumulated within the cell in the form of polyphosphate particles [22]. The traditional EBPR process diagram is illustrated in Figure 1. Generally, by alternating anaerobic and aerobic conditions, PAOs absorb the volatile fatty acids (VFAs) in sewage during the anaerobic stage and store them as poly-β-hydroxyalkanoic acid (PHA). This stored energy substance then serves as the carbon source and provides the energy required for the phosphorus accumulation and polyphosphate synthesis in the sewage by the PAOs during the aerobic stage [23,24]. P is subsequently removed from the system when the abovementioned phosphorus-containing microorganisms are discharged as phosphorus-rich sludge from the EBPR system [25]. The content of polyphosphate can be as high as 20% of its dry weight in PAO-rich sludge [26], while it accounts for no more than 2% in normal wastewater treatment sludge. Hence, this phosphorus-rich sludge produced in the EBPR process could be considered as having the potential to be applied as the resource for phosphorus collection and further reuse. In addition, efforts could be put into introducing PAOs with enhanced capabilities for phosphate accumulation so as to have more P recycled in the form of polyP.

Although modern wastewater treatment is normally driven by biological metabolic activity, it is still an energy- and time-consuming process. As such, how to make a profit in recycling the resource from wastewater has been assessed [27,28]. Egle et al. [29] summarized the traditional EBPR system for the treatment and recovery of phosphorus in wastewater and believed that the recovery of 70% to 90% of the phosphorus in the sludge in the form of ash was the best trade-off between cost and benefit. With the limited recovery rate of the traditional phosphorus removal process, the development of strategies that recover dissolved phosphorus more efficiently in the form of polyP is considered to be beneficial with respect to the techno-economic feasibility by taking into account the application potential of this biosynthetic product.

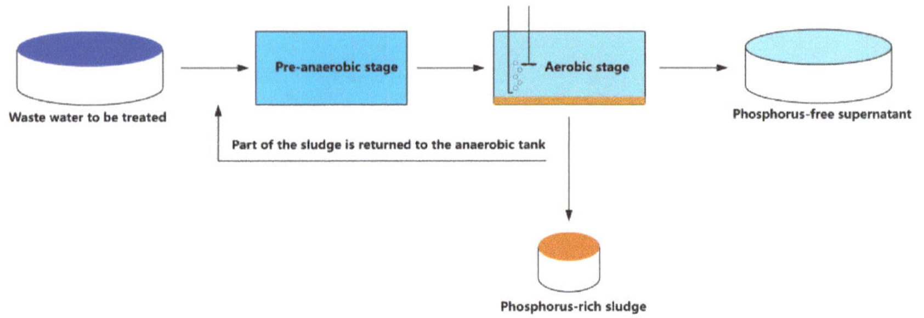

Figure 1. Diagram of traditional EBPR phosphorus removal method.

2.1. Factors Affecting PolyP Accumulation in PAOs during EBPR Process

2.1.1. Glucose Accumulating Organisms (GAOs)

In the actual sewage treatment process, the phosphorus content of the sewage is generally less than 10 mg P/L [30,31] and the concentration of P in the effluent is normally as low as 0.1 mg P/L due to having most of the phosphorus contained within the sewage influent removed during the aerobic stage [32,33]. However, the phosphorus removal efficiency of EBPR could be affected by external factors, the most influential of which is the competitive effect brought by GAOs [34]. GAOs have almost the same metabolic processes as PAOs, and the most significant difference is that the energy source of the matrix is absorbed in the aerobic phase [35]. As shown in Figure 2, GAOs decompose glycogen in the body during the anaerobic stage to provide energy and restore power to be used for organic matter absorbance and polyhydroxyalkanoates(PHA) synthesis in the cell. During the aerobic stage, this PHA is decomposed for glycogen synthesis and the maintenance of microorganism activity instead of having phosphorus accumulation. Due to the competitive relationship between GAOs and PAOs for the available VAFs, the EPBR system should be operated under conditions that are conducive to the growth of PAOs and hence, enhance the competitive advantages of PAOs over GAOs [36]. Under such conditions, the effect of GAOs on the dephosphorization efficacy would be eliminated and more phosphorus could then be enriched in the form of polyP.

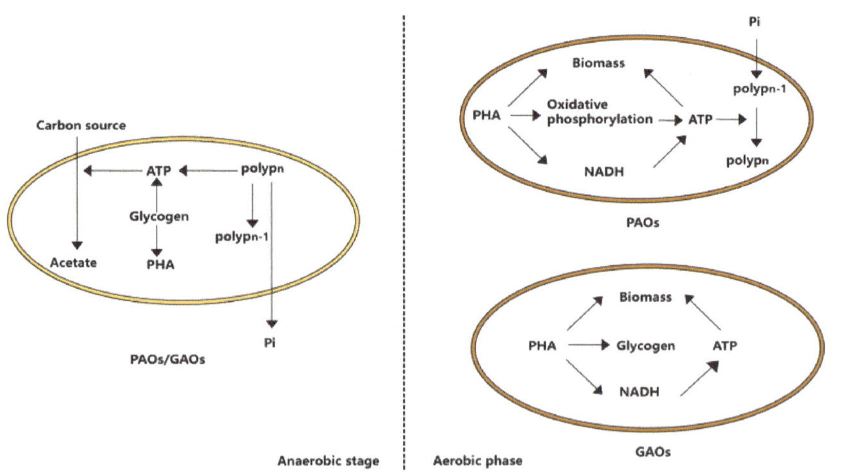

Figure 2. Anaerobic/Aerobic metabolism model of PAOs/GAOs.

2.1.2. Carbon Source

Among all environmental factors, the carbon source is considered to have the greatest impact on the efficiency of polyP synthesis in PAOs since it provides the support for PAOs to carry out all required activities. The variation in utilization rates for different carbon sources make it important to consider the type of carbon source that is available in the system as competition may occur between PAOs and GAOs if the organic substance cannot be metabolized efficiently by PAOs. To eliminate the competition, previous studies have indicated that acetate and propionate are appropriate energy sources that allow PAOs to overcome the interference brought by other competitors in the EBPR system [37–39]. When acetate is the main carbon source, the acetate permease within the PAOs enables the microbes to maintain a higher acetate adsorption rate during the phosphate accumulation metabolism (PAM) process than that in the glycogen accumulation metabolism (GAM) process performed by GAOs, even with a low acetate concentration [40]. With this source, PAOs can remove small amounts of acetic acid by maintaining superiority over acetic acid absorption activity [41]. Propionate, as another popular carbon substrate, accounts for 24 to 33% of septic tank municipal wastewater [38]. Without transferring to an intermediate, propionate can be absorbed by PAOs immediately and converted into propionyl-CoA, which can then further participate in the glycogen hydrolyzed acetyl and other metabolic processes. Coenzyme A is reduced and condensed to form PHA, which serves as a type of energy source for the formation of polyphosphate in the eutrophic stage [42]. In fact, compared to acetate, less energy and phosphorus release are required for the utilization of propionate by PAOs and hence, propionate is suggested to be the carbon source that would favor the phosphate accumulation metabolism process in PAOs [43].

In addition to the two abovementioned carbon substrates, alternative carbon sources can also be selected according to actual conditions. For instance, glucose or its polymers can replace PHA as the energy source for the absorption and polyP synthesis via PAOs. However, it should be noted that the introduction of glucose may induce glycogen metabolism activity in GAOs, resulting in a reduction in polyP production by deteriorating the EBPR process [44]. Ethanol is another economically sustainable carbon source with potential as an EBPR substrate, although its application is specific to EBPR systems composed of PAOs with acetaldehyde dehydrogenase [45]. Another reportedly valid carbon source is glycerol, which is also often used as an external carbon source for denitrification. It normally cannot induce phosphate accumulation metabolism activity in EBPR unless it is fermented into VFA in advance [46]. In addition, research also revealed that butyric acid cannot be metabolized well by GAOs at a high temperature (30 °C), thereby providing PAOs with a competitive advantage [47]. In actual sewage treatment, the production of polyP is mainly dependent on the turnover rate of PHB and polyphosphate. To obtain a larger amount of recycled polyP, the carbon substrate should be chosen for the purpose of having PAOs as the dominant microbial community as well as providing advantageous conditions for ensuring sufficient energy is available for the synthesis of polyphosphate.

2.2. Inspiration of P Recovery in the Form of PolyP by One-Step Phosphorus Removal

With the continuous in-depth research on PAOs, researchers are beginning to reflect on whether phosphorus recycling is valid and efficient within the traditional EBPR system [48]. Rather than being operated by multiple steps in EBPR, one-step phosphorus removal by sequential batch membrane bioreactors (SBMBRs) is suggested as being a more promising means by which the simultaneous removal and reuse of phosphorus may be realized [22]. Relying on the SBMBRs, polyP accumulation could be improved via genetic technology. By increasing the amount of genes encoding the Pi-specific transport system (phoU) [49] or having the overexpression of polyphosphate kinase (PPK1) [50], engineering bacteria with enhanced capabilities for Pi transport or polyP accumulation have been constructed. Until now, the overall amount of polyP synthesized during the phosphorus removal process was reported to account for about 12.7% of the dry cell weight [22], indicating the great application potential for the aims of phosphorus recovery in the form of polyP.

Without setting pre-anaerobic zones for VFA production, the abovementioned progress in constructing bacteria with the ability to accumulate phosphate under aerobic conditions suggests a simpler strategy for phosphorus removal along with effective polyP synthesis by PAOs [51].

3. PolyP Recovery from Phosphorus Removal System

Traditionally, the phosphorus-rich sludge produced from EBPR is directly used as fertilizer or dehydrated for incineration [52,53], although chemical and biological pollution accompany these operations. Instead, there may be less of an environmental impact and more economical benefits if polyP is recovered properly from the phosphorus-rich sludge. To develop an applicable polyP recovery process, the effect of the treatment on the type and chain length of polyP should be taken into consideration as the performance of polyP is largely dependent on these properties [54]. In general, the type of polyP is determined by the type of PAOs and the culture conditions, while the recovery method will affect the chain length of the obtained polyP. Herein, two recovery strategies are reviewed to provide information regarding the practical requirements for further polyP reuse.

3.1. Organic Reagent Precipitation Method

Methods for having polyP with varying chain lengths separated from microbes were preliminarily developed for the purpose of investigating its possible function involved in the microorganism metabolic process. To release polyP from the cell and remove the interference of protein, trichloroacetic acid was first applied and was followed by the continuous extraction of polyP using a combination of saturated sodium perchlorate and sodium hydroxide [55–60]. Kulaev et al. (1966) further optimized the protocol by replacing trichloroacetic acid with perchloric acid [61] and other adjustments were also performed in the following decades, depending on the specific goal of the research [62–66]. Among all the developed organic reagent precipitation methods, the use of ethanol and barium ions as the precipitation agents was reported as having the highest recovery rate [67–69].

Rather than recovering polyP with a wide range of chain lengths, research also tended to isolate polyP as an intracellular polyanion with a long phosphate backbone by developing the approach of nucleic extraction. In this method, cells are disrupted by chaotropic substances or acidic phenol beads, and the products are then purified with chloroform washing or silica gel column [70,71]. Contrasting with the previous strategy, this method is mainly targeted at polyP consisting of more than 60 phosphate residues.

3.2. High Temperature Calcination Method

Compared to the use of organic reagents to fractionate and purify polyP, the utilization of high-temperature calcination to extract the crude polyP with mixed chain lengths from sludge in wastewater treatment plants is considered to be a more economical and efficient approach, especially concerning polyP recovered from SBMBRs [29,72]. Relying on the thermal stability of polyP, organic components within the sludge or cell culture are oxidized thoroughly at a high temperature and the remaining ashes are collected for further purification to obtain polyP as the final product [72]. Generally, this method involves several processes, beginning with the recovery of PAOs from the phosphorus removal systems by having the sludge dehydrated. The obtained material is then calcined at a high temperature to remove most of the biological macromolecules, such as nucleic acid and protein. The remaining ashes are considered as the crude polyP and the pure product is collected by introducing further purification operations. In combination with the separation method described in Section 3.1, polyP with different chain lengths can be recovered thereafter.

4. Application Prospects of Biosynthetic PolyP

Until now, although polyP has been suggested as being a significant participant in various biochemical activities, such as being the activator of mitochondrial respiration regulation [57,58], the accelerating inducer of HUVEC endothelial cell tube formation [59], and the scaffold for multifunctional polyanionic protein [60], the mechanism of the biosynthesis of polyP in mammals has remained uncertain [18]. Among the various scattered functions, applying polyP as a generator and storage of metabolizable energy in the extracellular matrix (ECM) [73] in the osteoblast-led bone mineralization process [16] and supplying polyP as a functional substance for preparing procoagulants and anticoagulants in endogenous coagulation pathways [74,75] are considered as the most promising application directions of biosynthetic polyP.

4.1. Application of PolyP for Bone Reparation

PolyP has been proven to be a source of inorganic phosphate, which is required in the process of bone mineralization [76,77]. Performed by alkaline phosphatase (ALP) and mediated polyP hydrolysis (Figure 3), polyP could be applied as the activator to convert the initial transformation of amorphous calcium carbonate deposits into amorphous calcium phosphate [16]. During the process of bone or cartilage formation, the forming of blood vessels is a crucial step for tissue construction and repair. At this stage, the orientation, guidance, and ring formation of the endothelial cells need to be mediated by adenosine triphosphate (ATP) in the ECM, while polyP could be used as the back-up storage of high-energy phosphate bonds for supplying the energy required during this process. Studies have found that calcium phosphate cement (CPC) made a significant contribution to osteogenic differentiation and bone regeneration [73,78]. However, information about the effect of different salt types and different chain lengths of polyP on osteoblasts is limited. As such, further research and investigation is needed.

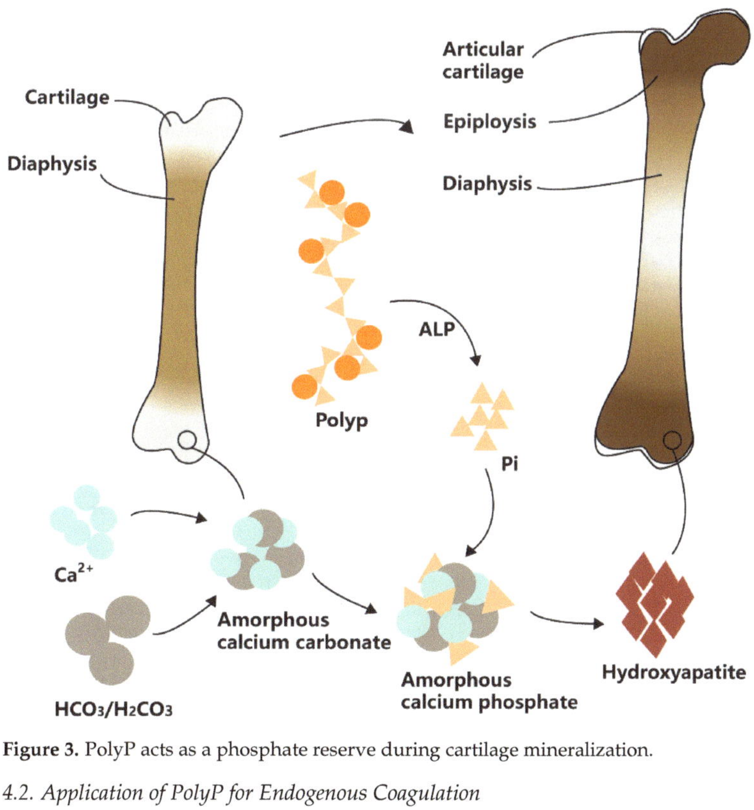

Figure 3. PolyP acts as a phosphate reserve during cartilage mineralization.

4.2. Application of PolyP for Endogenous Coagulation

Released from platelets, internal polyP was found to promote coagulation after the activation of coagulation factor XII (FXII) [74]. In addition, the introduction of external calcium-containing polyP was also suggested as having the potential to enhance the above-mentioned process [75]. When taking the chain length of polyP into consideration, the long-chain sodium polyP (45 type) revealed an inhabitation effect on blood clotting and hence, was suggested as a promising candidate for use as an anticoagulant [17]. However, despite the preliminary confirmation of the effectiveness of polyP for coagulation, it is still unclear whether there is a relationship between the properties of polyP and the resulting thrombosis effect. To fulfil this gap, research should focus on the impact of polyP with different chain lengths and salt types on coagulation and the mechanism may then be explained to some extent.

5. Conclusions

The development of new technology or optimization of current phosphorus removing strategies is an irreversible trend for realizing the effective recovery of phosphorus from wastewater. With the successful construction of engineering bacteria by introducing functional genes related with phosphorus accumulation, the use of SBMBRs to simplify the phosphorus removing process and recover the biosynthetic polyP from treatment plants is becoming promising via proper purification methods. We suggest further investigation into the mechanism of the specific influence of polyP properties on potential application areas so as to provide more fundamental information regarding the rational and practical utilization of recycled polyP.

Author Contributions: Writing—original draft preparation, W.C.; writing—review and editing, Y.S.; supervision, L.Z. All authors have read and agreed to the published version of the manuscript.

Funding: This research was supported by the National Key Research and Development Program of China (2018YFA0900300, 2021YFC2100300).

Informed Consent Statement: Informed consent was obtained from all subjects involved in the study.

Conflicts of Interest: The authors declare no conflict of interest.

References

1. Morse, G.K.; Brett, S.W.; Guy, J.A.; Lester, J.N. Review: Phosphorus removal and recovery technologies. *Sci. Total Environ.* **1998**, *212*, 69–81. [CrossRef]
2. Carpenter, S.R.; Caraco, N.F.; Correll, D.L.; Howarth, R.W.; Sharpley, A.N.; Smith, V.H. Nonpoint pollution of surface waters with phosphorus and nitrogen. *Ecol. Appl.* **1998**, *8*, 559–568. [CrossRef]
3. Ifelebuegu, A.O.; Ojo, P. Modelling the effects of ferric salt dosing for chemical phosphorus removal on the settleability of activated sludge. *J. Environ. Chem. Eng.* **2019**, *7*, 103256. [CrossRef]
4. Thistleton, J.; Berry, T.A.; Pearce, P.; Parsons, S.A. Mechanisms of Chemical Phosphorus Removal II. *Process Saf. Environ. Prot.* **2002**, *80*, 265–269. [CrossRef]
5. Mao, Y.; Xiong, R.; Gao, X.; Jiang, L.; Peng, Y.; Xue, Y. Analysis of the Status and Improvement of Microalgal Phosphorus Removal from Municipal Wastewater. *Processes* **2021**, *9*, 1486. [CrossRef]
6. Novais, S.V.; Zenero, M.D.O.; Barreto, M.S.C.; Montes, C.R.; Cerri, C.E.P. Phosphorus removal from eutrophic water using modified biochar. *Sci. Total Environ.* **2018**, *633*, 825–835. [CrossRef]
7. Chaochun, T.; Huimin, C.; Ming, L.; Xin, Y. Research progress in the use of adsorption for dephosphorization. *Ind. Water Treat.* **2015**, *37*, 1–4.
8. Ajmal, Z.; Muhmood, A.; Usman, M.; Kizito, S.; Lu, J.; Dong, R.; Wu, S. Phosphate removal from aqueous solution using iron oxides: Adsorption, desorption and regeneration characteristics. *J. Colloid Interface Sci.* **2018**, *528*, 145–155. [CrossRef]
9. Izadi, P.; Izadi, P.; Eldyasti, A. Design, operation and technology configurations for enhanced biological phosphorus removal (EBPR) process: A review. *Rev. Environ. Sci. Bio/Technol.* **2020**, *19*, 561–593. [CrossRef]
10. Ji, B.; Zhu, L.; Wang, S.; Qin, H.; Ma, Y.; Liu, Y. A novel micro-ferrous dosing strategy for enhancing biological phosphorus removal from municipal wastewater. *Sci. Total Environ.* **2020**, *704*, 135453. [CrossRef]
11. Zhang, M.; Zheng, P.; Abbas, G.; Chen, X. Partitionable-space enhanced coagulation (PEC) reactor and its working mechanism: A new prospective chemical technology for phosphorus pollution control. *Water Res.* **2014**, *49*, 426–433. [CrossRef]
12. Mino, T.; Van Loosdrecht, M.C.M.; Heijnen, J.J. Microbiology and biochemistry of the enhanced biological phosphate removal process. *Water Res.* **1998**, *32*, 3193–3207. [CrossRef]
13. Rao, N.N.; Gómez-García, M.R.; Kornberg, A. Inorganic polyphosphate: Essential for growth and survival. *Annu. Rev. Biochem.* **2009**, *78*, 605–647. [CrossRef]

14. Brown, M.R.; Kornberg, A. The long and short of it-polyphosphate, PPK and bacterial survival. *Trends Biochem. Sci.* **2008**, *33*, 284–290. [CrossRef] [PubMed]
15. Muller, W.E.G.; Schroder, H.C.; Wang, X. Inorganic Polyphosphates As Storage for and Generator of Metabolic Energy in the Extracellular Matrix. *Chem. Rev.* **2019**, *119*, 12337–12374. [CrossRef] [PubMed]
16. Wang, X.; Schroder, H.C.; Muller, W.E.G. Amorphous polyphosphate, a smart bioinspired nano-/bio-material for bone and cartilage regeneration: Towards a new paradigm in tissue engineering. *J. Mater. Chem. B* **2018**, *6*, 2385–2412. [CrossRef]
17. Yang, X.; Wan, M.; Liang, T.; Peng, M.; Chen, F. Synthetic polyphosphate inhibits endogenous coagulation and platelet aggregation in vitro. *Biomed. Rep.* **2017**, *6*, 57–62. [CrossRef]
18. Desfougeres, Y.; Saiardi, A.; Azevedo, C. Inorganic polyphosphate in mammals: Where's Wally? *Biochem. Soc. Trans.* **2020**, *48*, 95–101. [CrossRef]
19. Maier, S.K.; Scherer, S.; Loessner, M.J. Long-Chain Polyphosphate Causes Cell Lysis and Inhibits Bacillus cereus Septum Formation, Which Is Dependent on Divalent Cations. *Appl. Environ. Microbiol.* **1999**, *65*, 3942–3949. [CrossRef] [PubMed]
20. Barnard, J.L. Biological nutrient removal without the addition of chemicals. *Water Res.* **1975**, *9*, 485–490. [CrossRef]
21. Li, H.; Zhong, Y.; Huang, H.; Tan, Z.; Sun, Y.; Liu, H. Simultaneous nitrogen and phosphorus removal by interactions between phosphate accumulating organisms (PAOs) and denitrifying phosphate accumulating organisms (DPAOs) in a sequencing batch reactor. *Sci. Total Environ.* **2020**, *744*, 140852. [CrossRef] [PubMed]
22. Wang, X.; Wang, X.; Hui, K.; Wei, W.; Zhang, W.; Miao, A.; Xiao, L.; Yang, L. Highly Effective Polyphosphate Synthesis, Phosphate Removal, and Concentration Using Engineered Environmental Bacteria Based on a Simple Solo Medium-Copy Plasmid Strategy. *Environ. Sci. Technol.* **2018**, *52*, 214–222. [CrossRef] [PubMed]
23. Wang, L.; Shen, N.; Oehmen, A.; Zhou, Y. The impact of temperature on the metabolism of volatile fatty acids by polyphosphate accumulating organisms (PAOs). *Environ. Res.* **2020**, *188*, 109729. [CrossRef]
24. Ong, Y.H.; Chua, A.S.M.; Fukushima, T.; Ngoh, G.C.; Shoji, T.; Michinaka, A. High-temperature EBPR process: The performance, analysis of PAOs and GAOs and the fine-scale population study of Candidatus "Accumulibacter phosphatis". *Water Res.* **2014**, *64*, 102–112. [CrossRef] [PubMed]
25. Kristiansen, R.; Nguyen, H.T.; Saunders, A.M.; Nielsen, J.L.; Wimmer, R.; Le, V.Q.; McIlroy, S.J.; Petrovski, S.; Seviour, R.J.; Calteau, A.; et al. A metabolic model for members of the genus Tetrasphaera involved in enhanced biological phosphorus removal. *ISME J.* **2013**, *7*, 543–554. [CrossRef]
26. Metcalf & Eddy; Abu-Orf, M.; Bowden, G.; Burton, F.L.; Pfrang, W.; Stensel, H.D.; Tchobanoglous, G.; Tsuchihashi, R. AECOM. In *Wastewater Engineering: Treatment and Resource Recovery*; McGraw Hill Education: New York, NY, USA, 2014.
27. Li, B.; Udugama, I.A.; Mansouri, S.S.; Yu, W.; Baroutian, S.; Gernaey, K.V.; Young, B.R. An exploration of barriers for commercializing phosphorus recovery technologies. *J. Clean. Prod.* **2019**, *229*, 1342–1354. [CrossRef]
28. Chrispim, M.C.; Scholz, M.; Nolasco, M.A. Phosphorus recovery from municipal wastewater treatment: Critical review of challenges and opportunities for developing countries. *J. Environ. Manag.* **2019**, *248*, 109268. [CrossRef]
29. Egle, L.; Rechberger, H.; Krampe, J.; Zessner, M. Phosphorus recovery from municipal wastewater: An integrated comparative technological, environmental and economic assessment of P recovery technologies. *Sci. Total Environ.* **2016**, *571*, 522–542. [CrossRef]
30. Kim, J.-O.; Chung, J. Implementing chemical precipitation as a pretreatment for phosphorus removal in membrane bioreactor-based municipal wastewater treatment plants. *KSCE J. Civ. Eng.* **2014**, *18*, 956–963. [CrossRef]
31. Qiu, G.; Ting, Y.P. Direct phosphorus recovery from municipal wastewater via osmotic membrane bioreactor (OMBR) for wastewater treatment. *Bioresour. Technol.* **2014**, *170*, 221–229. [CrossRef]
32. Gautam, R.K.; Banerjee, S.; Gautam, P.K.; Chattopadhyaya, M. Remediation technologies for phosphate removal from wastewater: An overview. *Adv. Environ. Res.* **2014**, *36*, 177–200.
33. Gebremariam, S.Y.; Beutel, M.W.; Christian, D.; Hess, T.F. Research advances and challenges in the microbiology of enhanced biological phosphorus removal—A critical review. *Water Environ. Res.* **2011**, *83*, 195–219. [CrossRef] [PubMed]
34. Chen, J.; Wang, J.; Wang, X.; Lv, Y.; Li, D.; Hou, J.; He, X. Strengthening anoxic glycogen consumption in SNEDPR-CW as a strategy to control PAO-GAO competition under carbon limited condition. *Chemosphere* **2021**, *288*, 132617. [CrossRef]
35. Nielsen, P.H.; McIlroy, S.J.; Albertsen, M.; Nierychlo, M. Re-evaluating the microbiology of the enhanced biological phosphorus removal process. *Curr. Opin. Biotechnol.* **2019**, *57*, 111–118. [CrossRef]
36. Lopez-Vazquez, C.M.; Oehmen, A.; Hooijmans, C.M.; Brdjanovic, D.; Gijzen, H.J.; Yuan, Z.; van Loosdrecht, M.C. Modeling the PAO-GAO competition: Effects of carbon source, pH and temperature. *Water Res.* **2009**, *43*, 450–462. [CrossRef]
37. Vargas, M.; Guisasola, A.; Artigues, A.; Casas, C.; Baeza, J.A. Comparison of a nitrite-based anaerobic–anoxic EBPR system with propionate or acetate as electron donors. *Process Biochem.* **2011**, *46*, 714–720. [CrossRef]
38. Lv, X.-M.; Shao, M.-F.; Li, C.-L.; Li, J.; Xia, X.; Liu, D.-Y. Bacterial diversity and community structure of denitrifying phosphorus removal sludge in strict anaerobic/anoxic systems operated with different carbon sources. *J. Chem. Technol. Biotechnol.* **2014**, *89*, 1842–1849. [CrossRef]
39. Hsu, C.-H.; Chang, W.-C.; Chen, J.-J.; Wu, J.-S. Comparing the long-term effect of high P/COD influent on enhancement of phosphate-accumulating organisms between acetate- and propionate-fed reactors. *J. Chem. Technol. Biotechnol.* **2013**, *88*, 1071–1079. [CrossRef]

40. Schuler, A.J.; Jenkins, D. Enhanced biological phosphorus removal from wastewater by biomass with different phosphorus contents, part II: Anaerobic adenosine triphosphate utilization and acetate uptake rates. *Water Environ. Res.* **2003**, *75*, 499–511. [CrossRef] [PubMed]
41. Ahmed, Z.; Lim, B.R.; Cho, J.; Song, K.G.; Kim, K.P.; Ahn, K.H. Biological nitrogen and phosphorus removal and changes in microbial community structure in a membrane bioreactor: Effect of different carbon sources. *Water Res.* **2008**, *42*, 198–210. [CrossRef]
42. Oehmen, A.; Zeng, R.J.; Yuan, Z.; Keller, J. Anaerobic metabolism of propionate by polyphosphate-accumulating organisms in enhanced biological phosphorus removal systems. *Biotechnol. Bioeng.* **2005**, *91*, 43–53. [CrossRef]
43. Carvalheira, M.; Oehmen, A.; Carvalho, G.; Reis, M.A.M. The effect of substrate competition on the metabolism of polyphosphate accumulating organisms (PAOs). *Water Res.* **2014**, *64*, 149–159. [CrossRef]
44. Wang, Y.; Jiang, F.; Zhang, Z.; Xing, M.; Lu, Z.; Wu, M.; Yang, J.; Peng, Y. The long-term effect of carbon source on the competition between polyphosphorus accumulating organisms and glycogen accumulating organism in a continuous plug-flow anaerobic/aerobic (A/O) process. *Bioresour. Technol.* **2010**, *101*, 98–104. [CrossRef]
45. Skennerton, C.T.; Barr, J.J.; Slater, F.R.; Bond, P.L.; Tyson, G.W. Expanding our view of genomic diversity in Candidatus Accumulibacter clades. *Environ. Microbiol.* **2015**, *17*, 1574–1585. [CrossRef]
46. Yuan, Q.; Sparling, R.; Lagasse, P.; Lee, Y.; Taniguchi, D.; Oleszkiewicz, J. Enhancing biological phosphorus removal with glycerol. *Water Sci. Technol.* **2010**, *61*, 1837–1843. [CrossRef] [PubMed]
47. Wang, L.; Liu, J.; Oehmen, A.; Le, C.; Geng, Y.; Zhou, Y. Butyrate can support PAOs but not GAOs in tropical climates. *Water Res.* **2021**, *193*, 116884. [CrossRef] [PubMed]
48. Barnard, J.L.; Dunlap, P.; Steichen, M. Rethinking the mechanisms of biological phosphorus removal: Barnard et al. *Water Environ. Res.* **2017**, *89*, 2043–2054. [CrossRef]
49. Kato, J.; Yamada, K.; Muramatsu, A.; Hardoyo; Ohtake, H. Genetic improvement of Escherichia coli for enhanced biological removal of phosphate from wastewater. *Appl. Environ. Microbiol.* **1993**, *59*, 3744–3749. [CrossRef] [PubMed]
50. Hardoyo; Yamada, K.; Shinjo, H.; Kato, J.; Ohtake, H. Production and release of polyphosphate by a genetically engineered strain of Escherichia coli. *Appl. Environ. Microbiol.* **1994**, *60*, 3485–3490. [CrossRef]
51. Cokro, A.A.; Law, Y.; Williams, R.B.H.; Cao, Y.; Nielsen, P.H.; Wuertz, S. Non-denitrifying polyphosphate accumulating organisms obviate requirement for anaerobic condition. *Water Res.* **2017**, *111*, 393–403. [CrossRef]
52. Erdincler, A.; Seyhan, L. Agricultural use of municipal wastewater sludges: Phosphorus availability of biological excess phosphorus removal sludges. *Water Sci. Technol.* **2006**, *54*, 131–138. [CrossRef]
53. Kahiluoto, H.; Kuisma, M.; Ketoja, E.; Salo, T.; Heikkinen, J. Phosphorus in manure and sewage sludge more recyclable than in soluble inorganic fertilizer. *Environ. Sci Technol* **2015**, *49*, 2115–2122. [CrossRef]
54. Wang, D.; Li, Y.; Cope, H.A.; Li, X.; He, P.; Liu, C.; Li, G.; Rahman, S.M.; Tooker, N.B.; Bott, C.B.; et al. Intracellular polyphosphate length characterization in polyphosphate accumulating microorganisms (PAOs): Implications in PAO phenotypic diversity and enhanced biological phosphorus removal performance. *Water Res.* **2021**, *206*, 117726. [CrossRef] [PubMed]
55. Langen, P.; Liss, E. Über die Polyphosphate der Hele. *Naturwissenschaften* **1958**, *45*, 191. [CrossRef]
56. Langen, P.; Liss, E. Uber bildung und umsatz der polyphosphate der hefe. *Biochem. Z.* **1958**, *330*, 455–466.
57. Liss, E.; Langen, P. Zur Charakterisierung der Polyphosphate der Hefe. *Naturwissenschaften* **1959**, *46*, 151. [CrossRef]
58. Langen, P.; Liss, E. Differenzierung des Orthophosphates der Hefezelle. *Biochem. Z.* **1960**, *332*, 403–406.
59. Liss, E.; Langen, P. Uber ein hochmolekulares Polyphosphat der Hefe. *Biochem. Z.* **1960**, *333*, 193–201.
60. Liss, E.; Langen, P. Versuche zur polyphosphat-überkompensation in hefezellen nach phosphatverarmung. *Arch. Mikrobiol.* **1962**, *41*, 383–392. [CrossRef]
61. Kulaev, I.; Krasheninnikov, I.; Kokurina, N. On the Localization of Inorganic Polyphosphates and Nucleotides in Neurospora Crassa Mycelium. *Biokhimiia* **1966**, *31*, 850–859.
62. Vagabov, V.; Trilisenko, L.; Kulaev, I. Dependence of inorganic polyphosphate chain length on the orthophosphate content in the culture medium of the yeast Saccharomyces cerevisiae. *Biochem. N. Y. Engl. Transl. Biokhimiya* **2000**, *65*, 349–354.
63. Vagabov, V.; Trilisenko, L.; Kulakovskaya, E.; Kulaev, I. Study of the content of inorganic polyphosphates in Saccharomyces cerevisiae grown on different carbon sources with different O 2 concentrations in the medium. *Microbiology* **2008**, *77*, 541–546. [CrossRef]
64. Vagabov, V.M.; Trilisenko, L.V.; Kulakovskaya, T.V.; Kulaev, I.S. Effect of a carbon source on polyphosphate accumulation in Saccharomyces cerevisiae. *FEMS Yeast Res.* **2008**, *8*, 877–882. [CrossRef] [PubMed]
65. Breus, N.A.; Ryazanova, L.P.; Dmitriev, V.V.; Kulakovskaya, T.V.; Kulaev, I.S. Accumulation of phosphate and polyphosphate by Cryptococcus humicola and Saccharomyces cerevisiae in the absence of nitrogen. *FEMS Yeast Res.* **2012**, *12*, 617–624. [CrossRef]
66. Andreeva, N.; Ryazanova, L.; Dmitriev, V.; Kulakovskaya, T.; Kulaev, I. Adaptation of Saccharomyces cerevisiae to toxic manganese concentration triggers changes in inorganic polyphosphates. *FEMS Yeast Res.* **2013**, *13*, 463–470. [CrossRef]
67. Clark, J.E.; Beegen, H.; Wood, H.G. Isolation of intact chains of polyphosphate from "Propionibacterium shermanii" grown on glucose or lactate. *J. Bacteriol.* **1986**, *168*, 1212–1219. [CrossRef] [PubMed]
68. Mullan, A.; Quinn, J.P.; McGrath, J.W. A nonradioactive method for the assay of polyphosphate kinase activity and its application in the study of polyphosphate metabolism in Burkholderia cepacia. *Anal. Biochem.* **2002**, *308*, 294–299. [CrossRef]

69. Christ, J.J.; Blank, L.M. Analytical polyphosphate extraction from Saccharomyces cerevisiae. *Anal. Biochem.* **2018**, *563*, 71–78. [CrossRef] [PubMed]
70. Neef, D.W.; Kladde, M.P. Polyphosphate loss promotes SNF/SWI-and Gcn5-dependent mitotic induction of PHO5. *Mol. Cell. Biol.* **2003**, *23*, 3788–3797. [CrossRef]
71. Ault-Riché, D.; Fraley, C.D.; Tzeng, C.-M.; Kornberg, A. Novel assay reveals multiple pathways regulating stress-induced accumulations of inorganic polyphosphate in Escherichia coli. *J. Bacteriol.* **1998**, *180*, 1841–1847. [CrossRef]
72. Cyr, M.; Coutand, M.; Clastres, P. Technological and environmental behavior of sewage sludge ash (SSA) in cement-based materials. *Cem. Concr. Res.* **2007**, *37*, 1278–1289. [CrossRef]
73. Wang, S.; Xu, C.; Yu, S.; Wu, X.; Jie, Z.; Dai, H. Citric acid enhances the physical properties, cytocompatibility and osteogenesis of magnesium calcium phosphate cement. *J. Mech. Behav. Biomed. Mater.* **2019**, *94*, 42–50. [CrossRef] [PubMed]
74. Muller, F.; Mutch, N.J.; Schenk, W.A.; Smith, S.A.; Esterl, L.; Spronk, H.M.; Schmidbauer, S.; Gahl, W.A.; Morrissey, J.H.; Renne, T. Platelet polyphosphates are proinflammatory and procoagulant mediators in vivo. *Cell* **2009**, *139*, 1143–1156. [CrossRef]
75. Verhoef, J.J.; Barendrecht, A.D.; Nickel, K.F.; Dijkxhoorn, K.; Kenne, E.; Labberton, L.; McCarty, O.J.; Schiffelers, R.; Heijnen, H.F.; Hendrickx, A.P.; et al. Polyphosphate nanoparticles on the platelet surface trigger contact system activation. *Blood* **2017**, *129*, 1707–1717. [CrossRef] [PubMed]
76. Vimalraj, S. Alkaline phosphatase: Structure, expression and its function in bone mineralization. *Gene* **2020**, *754*, 144855. [CrossRef] [PubMed]
77. Murshed, M. Mechanism of bone mineralization. *Cold Spring Harb. Perspect. Med.* **2018**, *8*, a031229. [CrossRef]
78. Xia, Y.; Chen, H.; Zhang, F.; Bao, C.; Weir, M.D.; Reynolds, M.A.; Ma, J.; Gu, N.; Xu, H.H.K. Gold nanoparticles in injectable calcium phosphate cement enhance osteogenic differentiation of human dental pulp stem cells. *Nanomedicine* **2018**, *14*, 35–45. [CrossRef]

processes

MDPI

Article

Kinetics Study of Hydrothermal Degradation of PET Waste into Useful Products

Maja Čolnik [1], Darja Pečar [2], Željko Knez [1,3], Andreja Goršek [2] and Mojca Škerget [1,*]

[1] Laboratory for Separation Processes and Product Design, Faculty of Chemistry and Chemical Engineering, University of Maribor, Smetanova 17, SI-2000 Maribor, Slovenia; maja.colnik@um.si (M.Č.); zeljko.knez@um.si (Ž.K.)

[2] Laboratory for Process Systems Engineering and Sustainable Development, Faculty of Chemistry and Chemical Engineering, University of Maribor, Smetanova 17, SI-2000 Maribor, Slovenia; darja.pecar@um.si (D.P.); andreja.gorsek@um.si (A.G.)

[3] Department for Chemistry, Faculty of Medicine, University of Maribor, Taborska 8, SI-2000 Maribor, Slovenia

* Correspondence: mojca.skerget@um.si

Abstract: Kinetics of hydrothermal degradation of colorless polyethylene terephthalate (PET) waste was studied at two temperatures (300 °C and 350 °C) and reaction times from 1 to 240 min. PET waste was decomposed in subcritical water (SubCW) by hydrolysis to terephthalic acid (TPA) and ethylene glycol (EG) as the main products. This was followed by further degradation of TPA to benzoic acid by decarboxylation and degradation of EG to acetaldehyde by a dehydration reaction. Furthermore, by-products such as isophthalic acid (IPA) and 1,4-dioxane were also detected in the reaction mixture. Taking into account these most represented products, a simplified kinetic model describing the degradation of PET has been developed, considering irreversible consecutive reactions that take place as parallel in reaction mixture. The reaction rate constants (k_1–k_6) for the individual reactions were calculated and it was observed that all reactions follow first-order kinetics.

Keywords: PET waste; subcritical water; kinetics; degradation; TPA

Citation: Čolnik, M.; Pečar, D.; Knez, Ž.; Goršek, A.; Škerget, M. Kinetics Study of Hydrothermal Degradation of PET Waste into Useful Products. *Processes* **2022**, *10*, 24. https://doi.org/10.3390/pr10010024

Academic Editor: Aneta Magdziarz

Received: 6 December 2021
Accepted: 21 December 2021
Published: 23 December 2021

1. Introduction

The global production of plastic materials is still growing and exceeded 367 million metric tons in 2020 [1], making the recycling of plastic waste an urgent global commitment. Polyethylene terephthalate (PET) is the third most widely used polymer [2] in the production of packaging after polyethylene (PE) and polypropylene (PP). The primary use of PET polymer is in the textile industry, although vast amounts of this material are used to make packaging in the food, medical and cosmetic industries. PET bottle is one of the most largely consumed plastic products [3]. PET bottles are durable, clear, light, inexpensive, heat-stable, and they are usually used for beverage and food applications [4]. The global production of PET bottles reached 13.1 million tons in 2020 [5]. PET is one of the main condensation polymers, which are known to be composed of different monomers. Specifically, PET can be synthesized by esterification of monomers of terephthalic acid (TPA) and ethylene glycol (EG) or by transesterification of dimethyl terephthalate (DMT) and EG [6]. Condensation polymers have ether, ester, or amide bonds, which can be depolymerized back to monomers by solvolysis (hydrolysis, alcoholysis) under mild conditions [7]. Even though PET can be mechanically recycled several times, a large amount of waste is still generated [3,8]. Above all, recycling problems occur when mixing PET bottles by color (the significant problem present the colorants and additives [9]) and with other plastic materials (the caps of PET bottles are usually made from polyolefins and have to be recycled separately) [3]. In addition, PET also can be recycled by pyrolysis. Pyrolysis is one of the most studied methods for the chemical recycling of different polymers including PET. Due to the possibility of obtaining valuable chemicals, liquid fuel and gas, this method has received increasing attention in recent years [10–13]. In general, a high temperature

(400–800 °C) is needed for complete degradation of PET waste. Pyrolysis of PET waste mainly produces a gas phase, where more than 45% of the PET material is degraded to CO/CO_2 and the light hydrocarbons (4 wt%), where 2.8% corresponds to methane [10]. Furthermore, by pyrolysis, the residues of char decrease while the yield of gas products increases with increasing temperature and reaction time [10,14,15].

Traditionally used plastic conversion methods, such as incineration and mechanical recycling, raise some environmental concerns and are not economically attractive. In recent years, the chemical recycling of waste using supercritical fluids (SCF) has become interesting due to several advantages. Processes using SCFs represent green technology that is easy to use and has other important advantages (reducing energy consumption and emissions into the environment). One promising media is water in subcritical (SubCW) and supercritical (SCW) state, which provides a strong oxidation environment and can degrade plastic and bio-waste and produce various high value-added products [16]. These processes, if carefully optimized, can produce fuels and even monomers that can be re-used to synthesize polymers, reducing the direct demand for fossil raw materials [7]. SubCW is often used for the processing and decomposition of biomass [16–18], and currently, it is also increasingly used for chemical recycling of polymers into monomers [7,19]. Although SubCW has many beneficial properties (compared to conventional hydrolysis media), on the other hand, much work remains to be done in order to understand the reaction pathways of polymers in this media. Understanding the hydrolysis process and the degradation mechanism is crucial for scaling up the recycling process to industrial level. Due to the complex composition and different reactivity of individual components contained in waste materials, the mechanisms of reactions and the influence of process parameters are still relatively unexplored. Recycled monomers and other low molecular weight compounds are an important category of secondary raw materials used in several promising industries such as chemistry, paint industry, textile industry, and pharmacy.

PET in SubCW decomposes into the major monomers TPA and EG and other secondary products [19–21]. TPA is an organic component usually produced by homogeneous oxidation of p-xylene in the liquid phase [22]. TPA is of great industrial importance as a raw material for various widely used plastics, such as PET, polybutyl terephthalate (PBT), and bioplastics, which have been in great demand recently [23]. TPA is an important chemical and it is also widely used in the pharmaceutical industry (raw material for certain drugs) [24], paints and coatings industry (as a carrier), in arms industry (TPA smoke-for grenade filler) [25], and in the manufacture of clothing and plastic bottles [26]. Production of purified TPA reached around 84 million metric tons in 2018 and forecasts indicate that it is expected to reach 107 million metric tons by 2023 [27]. Another important carboxylic acid formed during the degradation of PET in SubCW is benzoic acid [19,23]. Benzoic acid is currently produced by partial oxidation of toluene and oxygen in the presence of manganese naphthalenes or cobalt [28]. Benzoic acid is widely used in the food industry as an antimicrobial preservative in food and beverages, especially in carbonated drinks [28]. It has the strongest antibacterial action, which is a significant cause of food spoilage. It is used in the cosmetics and personal care industries [29]. Benzoic acid is also used as a precursor of other products such as plasticizers, fungal ointments for medical use, and as a calibrating substance for bomb calorimeters [30,31]. In 2020, the global production of benzoic acid amounted to 538,770 tons and it is expected to grow to 628,350 tons by 2026 [32]. Therefore, due to the increasing use of TPA and benzoic acid, the recovery of these organic compounds from waste is highly important.

In our previous work, it was found that SubCW can be used as an efficient medium for the hydrolysis of colored and colorless PET waste. A high yield of the desired TPA product (approximately 90%) can be obtained under relatively mild hydrothermal conditions (300 °C, 95 bar, and 30 min). Nevertheless, besides TPA and EG as primary products, isophthalic acid (IPA) and 1,4-dioxane, as well as secondary products such as acetaldehyde, and benzoic acid were also identified in the reaction mixture after degradation, which indicates further degradation of primary products [19].

Although many authors have studied the hydrothermal degradation of PET waste [20,21,33] no data could be found in the literature dealing with decomposition kinetics. The aim of this work is to study the kinetics of hydrothermal degradation of colorless PET waste, to estimate kinetic data and to evaluate its degradation pathway.

2. Materials and Methods

2.1. Materials

TPA (99%), IPA (99%), 1,4-dioxane (99.8%), and acetaldehyde (99.5%) were supplied by Sigma-Aldrich (Steinheim, Germany), while benzoic acid was provided by Merck (Darmstadt, Germany). Nitrogen (N_2) (99.5%) as an inert gas was supplied by Messer (Ruše, Slovenia). Acetonitrile (ACN) (99.9%), methanol (99.9%), and trifluoroacetic acid (TFA) were obtained from Fisher Scientific (Loughborough, LE, UK), J.T. Baker (Deventer, The Netherlands), and Sigma-Aldrich. Sodium hydroxide (NaOH) and hydrochloric acid (HCl) were supplied by Sigma-Aldrich. All standards and solvents were of analytical grades.

2.2. Sample Preparation

PET colorless beverage bottles were collected, and material was prepared for experiments. First, the caps and the labels from the bottles were removed and then the material was cleaned and cut into small equal parts (1 × 1 cm).

2.3. Elemental Analysis

The colorless PET bottle waste sample was also characterized with the Perkin Elmer 2400 Series II system analyzer. The content of the main elements (carbon, hydrogen, nitrogen, and sulfur) was determined (Table 1), while oxygen was calculated by the mass balance.

Table 1. Ultimate analysis of colorless PET waste bottles.

Element	wt%
C	62.1
H	4.20
N	0.14
S	0.32
O by difference (wt%)	33.24

2.4. FTIR Analysis

Fourier transform infrared (FTIR) spectroscopy of colorless PET waste, recycled TPA, and standard of TPA were performed on an IRAffinity-1 spectrophotometer equipped with an attenuated total reflectance cell (ATR). The spectra of individual samples were recorded at wavelengths from 4000 to 400 nm against the air as the background at a resolution of 4 cm^{-1} and a total of 20 co-added scans. After all, the data were analyzed with high-performance IRsolution software.

2.5. Subcritical Water Treatment of PET

The decomposition of colorless PET waste in SubCW (material/water ratio (1/10)) was performed in a 75 mL high-temperature and high-pressure batch reactor (Parr Instruments, Moline, IL, USA). A detailed description of the experimental system (Figure 1) and procedure can be found in the literature [19]. The degradation reactions of PET were carried out at temperatures 300 and 350 °C and reaction time between 1 and 240 min. The mixture of PET and water in a weight ratio 1/10 was loaded into the batch reactor at room temperature. Before hydrolysis of PET waste, the reactor was flushed several times with nitrogen and then this gas was used to set the initial pressure in the reactor to 20 bar. The reactor was heated by a high-power heating collar controlled by a temperature controller. At the temperature of 300 and 350 °C, the pressure in the reactor was 89 ± 2 bar and 168 ± 2 bar, respectively. During the degradation of colorless PET waste, the reaction mixture was stirred constantly at 1000 min^{-1}. After a specific time, the reactor was immediately quenched with cold water, then the gas was released from the system and the products from the reactor were collected. The post-reaction mixture contained liquid and solid products. First, 1 mL of the aqueous post-reaction mixture was separated and filtered through a 45 µm filter and the filtrate was stored for further HPLC analysis. All remaining reaction mixture was treated with 4 M NaOH to form sodium salt of TPA, that is soluble in water and then filtered to remove any undegraded PET residues. Finally, TPA was precipitated with 1.5 M HCL. After filtration and drying, the concentration of formed TPA and undegraded PET waste residues in the post-reaction mixture was calculated by the following Equations (1) and (2):

$$c_{TPA} = m_{TPA}/V_{H_2O} \quad (1)$$

$$c_{PET} = m_{PET}/V_{H_2O} \quad (2)$$

where: c_{TPA} is the concentration of TPA, c_{PET} is the concentration of PET, m_{TPA} is the mass of precipitated TPA, m_{PET} is the mass of unreacted colorless PET waste and V_{H_2O} is the initial volume of distilled water before hydrolysis.

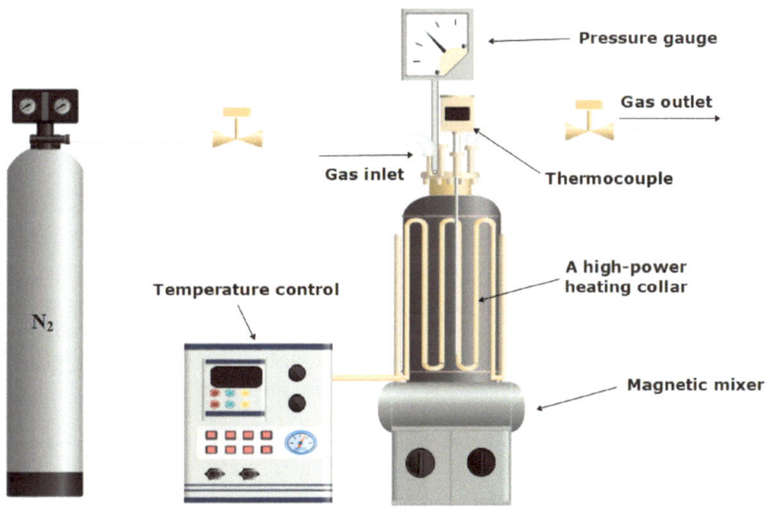

Figure 1. A batch reactor system of SubCW treatment.

2.6. HPLC Method

The concentrations of products in the aqueous phase after hydrolysis of colorless PET waste were determined using an HPLC system (model 1100, Agilent technology, Waldbronn, Germany) with an UV-Vis detector [19]. The column was an Agilent ZORBAX Eclipse XDB C18 column (4.6 × 150 mm, 3.5 µm particle diameter) and the temperature of the column was 30 °C. Mobile phase A was Milli-Q water with 0.1% TFA/ACN in a ratio of 60/40, while mobile phase B was Milli-Q water with 0.1% TFA/ACN in a ratio of 90/10. The gradient was: 0 min 90% B, 3 min 70% B, 8 min 60% B at 242 nm. The volume of initiated samples was 20 µL, while the flow rate was 1 mL/min. The identification and the quantification of the resulting products were performed using calibration curves of individual standards [19].

3. Results and Discussion

3.1. Degradation of Colorless PET Waste in SubCW

The results of our previous work [19] and data from the literature [20,21] show that the depolymerization of PET starts at 250 °C at a short reaction time, where with the reaction of hydrolysis, the main monomers, TPA and EG, are formed. At the same time, some small amounts of unreacted PET, oligomers, IPA, and 1,4-dioxane were also present in the post-reaction mixture [19,21]. With further increase in temperature to 300 °C, all PET waste degraded in the SubCW and gave the highest yield of TPA and EG [19,21]. At higher temperatures, the concentration of EG in the reaction mixture decreased with increasing temperature and time [21] and at the same time, acetaldehyde, diethylene glycol, and triethylene glycol were formed from EG via dehydration and dimerization reactions, respectively [19,21]. Similarly, the concentration of TPA started to decrease at 300 °C with the extension of reaction time (<30 min), and benzoic acid was formed due to decarboxylation reaction, where CO_2 [19] was eliminated from TPA molecules (Figure 2).

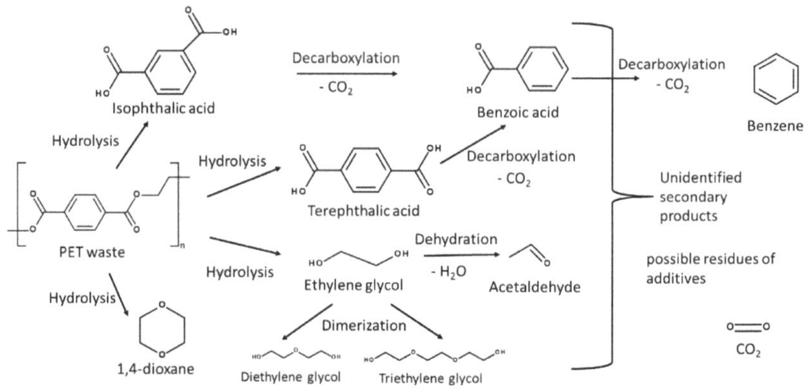

Figure 2. Reaction pathway of PET waste in SubCW.

To study the kinetics of PET degradation into valuable products, in the present work, colorless PET waste plastic was degraded in SubCW at temperatures of 300 °C and 350 °C and a reaction time of up to 240 min. The focus was on the most represented components in the aqueous phase. The concentration ($g_{product}/g_{PET}$) of main products, by-products, and secondary products are presented in Table 2.

Table 2. Concentration of degradation products ($g_{component}/g_{material}$) of colorless PET waste in SubCW.

t [1] min	$c_{(TPA)}$ g_{TPA}/g_{PET}		$c_{(benzoic\ acid)}$ $g_{benzoic\ acid}/g_{PET}$		$c_{(1,4\text{-}dioxane)}$ $g_{1,4\text{-}dioxane}/g_{PET}$		$c_{(EG)}$ g_{EG}/g_{PET}		$c_{(acetaldehyde)}$ $g_{acetaldehyde}/g_{PET}$		$c_{(IPA)}$ g_{IPA}/g_{PET}	
	300 °C	350 °C	300 °C	350 °C	300 °C	350 °C	300 °C	350 °C	300 °C	350 °C	300 °C	350 °C
1	0.73	0.71	0	0.0014	0.0075	0.0076	0.112	0.140	0.0009	0.0586	0.0021	0.0035
10	0.74	0.72	0	0.0033	0.0074	0.0075	0.224	0.162	0.0024	0.0831	0.0030	0.0043
30	0.76	0.69	0.0003	0.0063	0.0072	0.0078	0.193	0.125	0.0123	0.1405	0.0030	0.0049
60	0.75	0.64	0.0016	0.0425	0.0544	0.0156	0.129	0.086	0.0288	0.1806	0.0173	0.0145
120	0.73	0.61	0.0027	0.0502	0.0400	0.0031	0.064	0.020	0.0942	0.2689	0.0188	0.0101
180	0.70	0.56	0.0051	0.0592	0.0498	0.0013	0.016	0	0.1150	0.2753	0.0186	0.0118
240	0.68	0.44	0.0088	0.0691	0.0239	0.0004	0	0	0.1200	0.2641	0.0184	0.0121

[1] t is the reaction time at constant T and P (heating time not included).

It was found that all PET waste has decomposed already during the heating period to the desired temperature (12 min), and the main products were formed. The data in Table 2 show that the highest concentration of main products was achieved at 300 °C, namely for EG (0.22 g_{EG}/g_{PET}) at 10 min and for TPA (0.76 g_{TPA}/g_{PET}) at 30 min. The maximum concentrations of by-products (IPA and 1,4-dioxane) were also detected at the reaction temperature of 300 °C when the reaction time increased to 60 min for 1,4-dioxane and to 120 min for IPA. The concentrations of products were 0.05 $g_{1,4\text{-}dioxan}/g_{PET}$ and 0.018 g_{IPA}/g_{PET}. The secondary products (acetaldehyde and benzoic acid) were formed with hydrothermal degradation of the main products. It was found that the concentrations of these secondary products increase with increasing reaction temperature to 350 °C and with increasing reaction time. The highest concentrations of benzoic acid (0.07 $g_{benzoic\ acid}/g_{PET}$) and acetaldehyde (0.27 $g_{acetaldehyde}/g_{PET}$) were detected at 350 °C and reaction time of 240 min and 180 min, respectively. At these conditions, new unidentified products in the aqueous phase were also observed (e.g., possible residues of additives). Moreover, it was assumed that while benzoic acid is formed, part of the benzoic acid is simultaneously further degraded by the decarboxylation reaction and causes the formation of a new product—benzene [34].

In the following, based on our research and data from the literature [19,21,34], the possible degradation mechanisms of PET waste in SubCW is presented in Figure 2.

3.2. FTIR Analysis of Colorless PET Waste and Recycled TPA

The FTIR technique was used to identify the main functional groups present in the colorless PET waste before hydrolysis and in the resulting recycled TPAs after SubCW treatment. The spectrum of the colorless PET waste sample (Figure 3) contained an absorption band at 3432 cm^{-1} (OH group), at 3100–2800 cm^{-1} (aromatic and aliphatic C-H stretch), at 1720 cm^{-1} (ester carbonyl C=O stretch), and at 1453–1342 cm^{-1} (bending and wagging vibrational modes of the EG segment). The main observation of the absorption bands at 1240 cm^{-1} and 1124 cm^{-1} confirmed the terephthalate group ($OOCC_6H_4$-COO). The absorption bands were consistent with previous works [35–37].

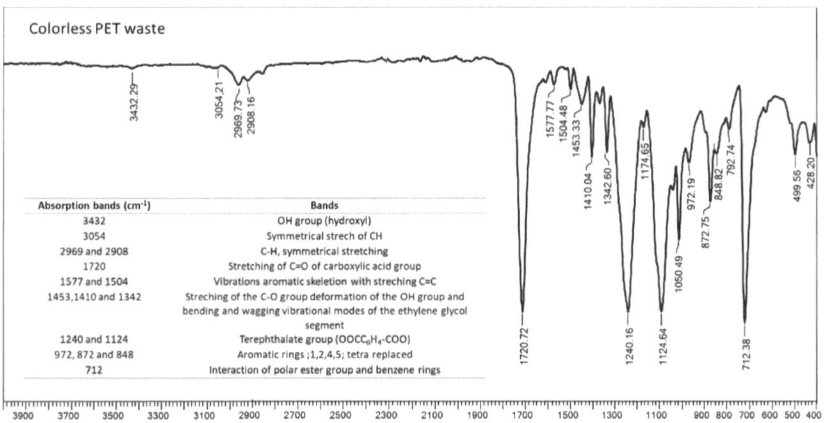

Absorption bands (cm^{-1})	Bands
3432	OH group (hydroxyl)
3054	Symmetrical strech of CH
2969 and 2908	C-H, symmetrical stretching
1720	Stretching of C=O of carboxylic acid group
1577 and 1504	Vibrations aromatic skeletion with streching C=C
1453,1410 and 1342	Streching of the C-O group deformation of the OH group and bending and wagging vibrational modes of the ethylene glycol segment
1240 and 1124	Terephthalate group ($OOCC_6H_4$-COO)
972, 872 and 848	Aromatic rings ;1,2,4,5; tetra replaced
712	Interaction of polar ester group and benzene rings

Figure 3. FTIR spectra of colorless PET waste before SubCW treatment.

FTIR analysis of a solid product obtained after hydrothermal degradation of colorless PET waste confirmed that the product contains main monomer TPA of high purity. According to Figure 4, recycled TPA presents the same characteristic absorption bands as the standard of TPA. The main IR bands of recycled TPA and TPA standard are shown at 3300–2500 cm^{-1} (O-H stretch), at 1683 cm^{-1} (C=O stretch), at 1425 cm^{-1} (O-H bend), at 1283 cm^{-1} (C-O stretch), and at 948 cm^{-1} (O-H bend). Similar results of the functional groups of TPA have been reported previously [19,20].

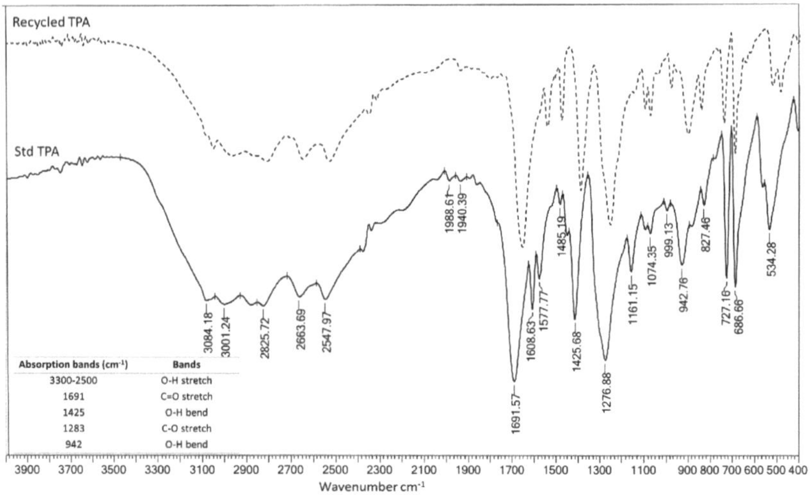

Figure 4. FTIR spectra of TPA standard and recycled TPA after hydrothermal degradation of colorless PET waste at the temperature of 350 °C and 120 min.

3.3. Kinetics of Hydrothermal Degradation of Colorless PET Waste

Due to the fact that PET waste already completely decomposes during the heating period to the desired temperature (12 min), the reaction kinetics during the heating period were also studied and degradation products were analyzed and considered for the kinetic modeling. The results show that the hydrothermal decomposition of colorless PET waste can be represented by a kinetic scheme for irreversible reactions, as presented in Figure 5. The method used in this study follows the concentration profile of PET waste (A), major PET hydrolysis products TPA (B) and EG (E), PET degradation by-products 1,4-dioxane (D) and IPA (G), and secondary products that are formed from major hydrolysis products, i.e., benzoic acid (C) and acetaldehyde (F). By prolonging the reaction time, CO_2 is cleaved from the TPA molecule [19] by decarboxylation to form the secondary product benzoic acid (C), while acetaldehyde (F) is most likely formed by dehydration of EG (Figure 6). As mentioned above, the proposed kinetic model representing the decomposition of colorless PET waste has been simplified; specifically, it does not consider unknown components that appear in the aqueous phase in the initial stage of PET degradation and most likely represent oligomers of PET polymer [21], as well as decomposition products of additives present in the waste PET plastic [7,38].

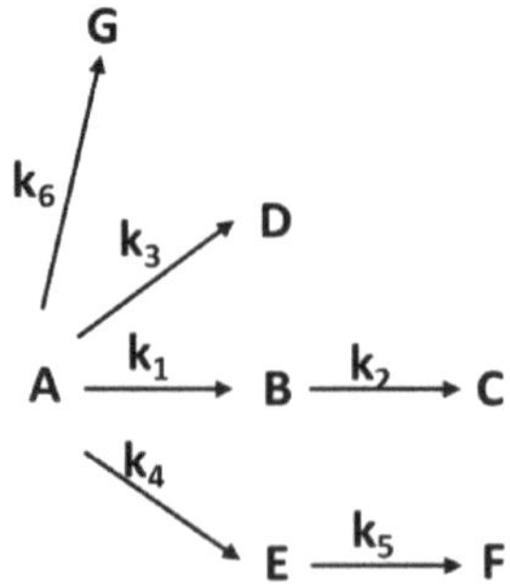

Figure 5. Scheme of proposed kinetics model of colorless PET waste degradation in SubCW. PET (A), TPA (B), benzoic acid (C), 1,4-dioxane (D), EG (E), acetaldehyde (F), IPA (G).

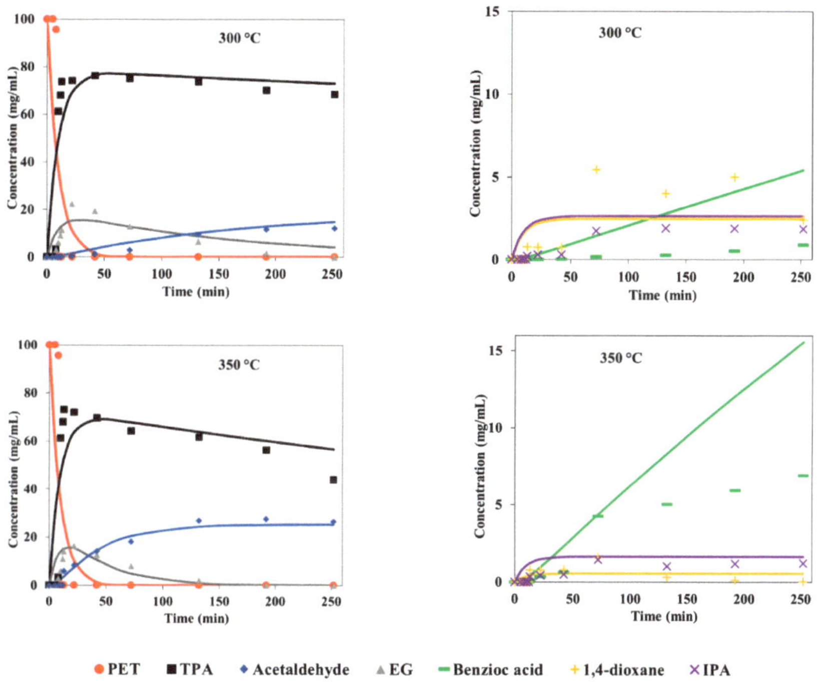

Figure 6. Colorless PET waste and its degradation products: TPA, EG, benzoic acid, 1,4-dioxane, acetaldehyde, and IPA at temperatures of 300 and 350 °C during heating period (first 12 min) and reaction at constant conditions. The symbols represent experimental measurements, while the curves represent a kinetic model.

For the kinetic modeling of the decomposition of colorless PET waste, it was assumed that the hydrolysis of PET waste follows first-order kinetics because the difference between the experimental and model data was the lowest compared to zero- and second-order kinetics. The corresponding reaction rates for all six reactions were described by the following differential Equations (3)–(9):

$$-\frac{dc_{\text{PET waste}}}{dt} = (k_1 + k_3 + k_4 + k_6\ )c_{\text{PET waste}} \tag{3}$$

$$\frac{dc_{\text{1,4-dioxane}}}{dt} = k_3 c_{\text{PET waste}} \tag{4}$$

$$\frac{dc_{\text{TPA}}}{dt} = k_1 c_{\text{PET waste}} - k_2 c_{\text{TPA}} \tag{5}$$

$$\frac{dc_{\text{benzoic acid}}}{dt} = k_2 c_{\text{TPA}} \tag{6}$$

$$\frac{dc_{EG}}{dt} = k_4 c_{\text{PET waste}} - k_5 c_{\text{EG}} \tag{7}$$

$$\frac{dc_{\text{acetaldehyde}}}{dt} = k_5 c_{\text{EG}} \tag{8}$$

$$\frac{dc_{\text{IPA}}}{dt} = k_6 c_{\text{PET waste}} \tag{9}$$

where: $c_{\text{PET waste}}$ is the concentration of PET waste, c_{TPA} the concentration of TPA, $c_{\text{benzioc acid}}$ the concentration of benzoic acid, $c_{\text{1,4-dioxane}}$ the concentration of 1,4-dioxane, c_{EG} the concentration of EG, $c_{acetaldehyde}$ the concentration of acetaldehyde and c_{IPA} the concentration of IPA; k_1, k_2, k_3, k_4, k_5 in k_6 are the reaction rate constants of individual reactions, and t is the reaction time.

Differential Equations (3)–(9) were numerically integrated using Scientist (MicroMath) software, minimizing the objective function represented by the sum of the square differences between the experimental and calculated concentrations. The calculated reaction rate constants are shown in Table 3. The standard deviation of constants k_2, k_3, and k_6 is significant due to the scattering of experimental data.

Table 3. Reaction rate constants for irreversible sequential reactions of PET waste in SubCW at different temperatures.

k (min^{-1})	300 °C	350 °C
k_1 (min^{-1})	0.07804 ± 0.00882	0.07472 ± 0.00896
k_2 (min^{-1})	0.00030 ± 0.00036	0.00101 ± 0.00044
k_3 (min^{-1})	0.00245 ± 0.00422	0.00057 ± 0.00444
k_4 (min^{-1})	0.01879 ± 0.00477	0.02615 ± 0.00545
k_5 (min^{-1})	0.00804 ± 0.00458	0.02697 ± 0.01590
k_6 (min^{-1})	0.00261 ± 0.00376	0.00169 ± 0.00392

Figure 6 shows the distribution of the PET waste and its decomposition products in dependence of the reaction time and temperature, where first 12 min present the heating period, followed by reaction at constant conditions, and comparison between experimental and model data. Results show that the concentration of PET waste significantly decreases with increasing temperature and reaction time, while TPA and EG concentrations first increase and reach the maximum at 300 °C and 42 min for TPA (76.12 mg/mL) and 22 min for EG (22.4 mg/mL). After that, the concentrations of TPA and EG start to decline. During the hydrolytic degradation reaction at 300 °C and shorter reaction times (13 min), two new by-products, i.e., 1,4-dioxane and IPA start to form, with concentrations of 0.2 mg/mL and 0.75 mg/mL, respectively. IPA is an isomer of TPA and is often added in the production of PET materials [39], while 1,4-dioxane is a cyclic ether formed as a by-product in the PET production process [40]. Corresponding reaction rate constants k_3 and k_6 at 300 °C are similar and are higher as at 350 °C, which indicates faster degradation of both compounds at higher temperature. By further prolonging the reaction time at 300 °C, the formation of benzoic acid is observed in the post-reaction mixture at a reaction time of 42 minutes and its concentration increases with time due to the decarboxylation of CO_2 from the TPA molecule, which was also confirmed by gas analysis in our previous study [19]. The highest concentration of benzoic acid is achieved at a temperature of 350 °C and reaction time of 240 min and it is 6.91 mg/mL. Further, Figure 6 also shows that at higher temperature (350 °C) the concentrations of TPA and EG decrease faster with prolongation of the reaction time as at 300 °C and, therefore, reaction rate constants k_2 and k_5, indicating the rate of formation of their degradation products benzoic acid and acetaldehyde, are higher at 350 °C. The reaction rate constant (k_1) for TPA is similar at both temperatures.

It can also be seen from Figure 6 that at both temperatures, the concentrations of products IPA, 1,4-dioxane, and acetaldehyde first increase and then begin to decline after a certain time.

The significant differences between the experimental and model results for decomposition of primary and secondary products were assigned to the formation of unidentified products. It would be expected that with further degradation of TPA, the concentration of benzoic acid would increase even more with the prolongation of the reaction time at 350 °C. However, HPLC analysis of the aqueous solution containing degradation products showed that unknown (new unidentified) degradation products began to form, which could be attributed to the degradation of benzoic acid. Lindquist and Yang [34] found that benzoic acid degraded in SubCW at 350 °C as the reaction time increased (more than 25% of benzoic acid was degraded by a reaction time of 240 min) as CO_2 was cleaved from the benzoic acid molecule, thereby causing the formation of a new benzene product [34], which was not analyzed in these study.

4. Conclusions

Kinetic study of hydrothermal degradation of PET waste to major monomers TPA and EG, and secondary products (IPA, 1-4 dioxane, acetaldehyde and benzoic acid) shows that all individual reactions follow first-order kinetics. The reaction rate constant of formation of TPA (k_1) was not significantly different at both temperatures, while the reaction rate constants k_3, and k_6, indicating formation of IPA and 1,4-dioxane, were higher at 300 °C than at 350 °C, which indicates that at higher temperatures, these compounds undergo faster further degradation. Conversely, the reaction rate constants k_2 and k_5, indicating formation of benzoic acid and acetaldehyde, were higher at 350 °C as at 300 °C, which indicates a higher degradation rate of TPA and EG at higher temperature. The maximum concentration of TPA in the aqueous phase was determined to be 300 °C and the reaction time (heating period not included) was 30 min (0.76 g_{TPA}/g_{PET}), while the highest concentration of EG was achieved at 300 °C and 10 min, and was 0.22 g_{EG}/g_{PET}. The concentration of benzoic acid, a degradation product of TPA, increased with increasing temperature and time and the highest concentration was detected at 350 °C and 240 min and was 0.69 $g_{benzoic\ acid}/g_{PET}$. Similar was observed for acetaldehyde, a degradation product of EG, where the highest concentration was achieved at 350 °C and 180 min and it was 0.27 $g_{acetaldehyde}/g_{PET}$. The concentrations of 1,4-dioxane (0.05 $g_{1,4\text{-}dioxane}/g_{PET}$) and IPA (0.018 g_{IPA}/g_{PET}) were the highest at temperature of 300 °C and reaction time of 60 min and 120 min, respectively.

Author Contributions: M.Č.: Formal analysis, Investigation, Conceptualization, Writing—original draft, Writing—review and editing. D.P.: Data curation, Formal analysis, Software, Ž.K.: Funding acquisition, Supervision, A.G. Supervision, M.Š.: Investigation, Conceptualization, Writing—review and editing, Supervision. All authors have read and agreed to the published version of the manuscript.

Funding: Special thanks to the Slovenian Research Agency (ARRS) for financial support of research program groups P2-0046: Separation processes and product design and P2-0032: Process systems engineering and sustainable development.

Informed Consent Statement: Not applicable.

Data Availability Statement: Not applicable.

Conflicts of Interest: The authors declare no conflict of interest.

References

1. Global Plastic Production 1950–2020. Available online: https://www.statista.com/statistics/282732/global-production-of-plastics-since-1950/ (accessed on 20 September 2021).
2. Allaf, R.M.; Albarahmieh, E.; Futian, M. Preparation of Sawdust-Filled Recycled-PET Composites via Solid-State Compounding. *Processes* **2020**, *8*, 100. [CrossRef]
3. Gopalakrishna, K.G.; Reddy, N. 2-Regulations on Recycling PET Bottles. In *Recycling of Polyethylene Terephthalate Bottles*; Thomas, S., Rane, A., Kanny, K., Abitha, V.K., Thomas, M.G., Eds.; William Andrew Publishing: Norwich, NY, USA, 2019; pp. 23–35, ISBN 978-0-12-811361-5.
4. Karayannidis, G.P.; Chatziavgoustis, A.P.; Achilias, D.S. Poly (Ethylene Terephthalate) Recycling and Recovery of Pure Terephthalic Acid by Alkaline Hydrolysis. *Adv. Polym. Technol.* **2002**, *21*, 250–259. [CrossRef]
5. PET Bottle Market: Global Industry Trends, Share, Size, Growth, Opportunity and Forecast 2021–2026. Available online: https://www.researchandmarkets.com/reports/5264119/pet-bottle-market-global-industry-trends-share (accessed on 26 August 2021).
6. Mandal, S.; Dey, A. 1-PET Chemistry. In *Recycling of Polyethylene Terephthalate Bottles*; Thomas, S., Rane, A., Kanny, K., VK, A., Thomas, M.G., Eds.; William Andrew Publishing: Norwich, NY, USA, 2019; pp. 1–22. ISBN 978-0-12-811361-5.
7. Queiroz, A.; Pedroso, G.B.; Kuriyama, S.N.; Fidalgo-Neto, A.A. Subcritical and Supercritical Water for Chemical Recycling of Plastic Waste. *Curr. Opin. Green Sustain. Chem.* **2020**, *25*, 100364. [CrossRef]
8. Chirayil, C.J.; Mishra, R.K.; Thomas, S. 3-Materials Recovery, Direct Reuse and Incineration of PET Bottles. In *Recycling of Polyethylene Terephthalate Bottles*; Thomas, S., Rane, A., Kanny, K., VK, A., Thomas, M.G., Eds.; William Andrew Publishing: Norwich, NY, USA, 2019; pp. 37–60, ISBN 978-0-12-811361-5.

9. Hahladakis, J.N.; Velis, C.A.; Weber, R.; Iacovidou, E.; Purnell, P. An Overview of Chemical Additives Present in Plastics: Migration, Release, Fate and Environmental Impact during Their Use, Disposal and Recycling. *J. Hazard. Mater.* **2018**, *344*, 179–199. [CrossRef] [PubMed]
10. Martín-Gullón, I.; Esperanza, M.; Font, R. Kinetic Model for the Pyrolysis and Combustion of Poly-(Ethylene Terephthalate) (PET). *J. Anal. Appl. Pyrolysis* **2001**, *58–59*, 635–650. [CrossRef]
11. Anuar Sharuddin, S.D.; Abnisa, F.; Wan Daud, W.M.A.; Aroua, M.K. A Review on Pyrolysis of Plastic Wastes. *Energy Convers. Manag.* **2016**, *115*, 308–326. [CrossRef]
12. Cheng, L.; Gu, J.; Wang, Y.; Zhang, J.; Yuan, H.; Chen, Y. Polyethylene High-Pressure Pyrolysis: Better Product Distribution and Process Mechanism Analysis. *Chem. Eng. J.* **2020**, *385*, 123866. [CrossRef]
13. Dhahak, A.; Grimmer, C.; Neumann, A.; Rüger, C.; Sklorz, M.; Streibel, T.; Zimmermann, R.; Mauviel, G.; Burkle-Vitzthum, V. Real Time Monitoring of Slow Pyrolysis of Polyethylene Terephthalate (PET) by Different Mass Spectrometric Techniques. *Waste Manage.* **2020**, *106*, 226–239. [CrossRef]
14. Singh, R.K.; Ruj, B.; Sadhukhan, A.K.; Gupta, P. Thermal Degradation of Waste Plastics under Non-Sweeping Atmosphere: Part 1: Effect of Temperature, Product Optimization, and Degradation Mechanism. *J. Environ. Manag.* **2019**, *239*, 395–406. [CrossRef]
15. Muhammad, C.; Onwudili, J.A.; Williams, P.T. Thermal Degradation of Real-World Waste Plastics and Simulated Mixed Plastics in a Two-Stage Pyrolysis–Catalysis Reactor for Fuel Production. *Energ. Fuel.* **2015**, *29*, 2601–2609. [CrossRef]
16. Knez, Ž.; Hrnčič, M.K.; Čolnik, M.; Škerget, M. Chemicals and Value Added Compounds from Biomass Using Sub- and Supercritical Water. *J. Supercrit. Fluids* **2018**, *133*, 591–602. [CrossRef]
17. Knez, Ž.; Markočič, E.; Hrnčič, M.K.; Ravber, M.; Škerget, M. High Pressure Water Reforming of Biomass for Energy and Chemicals: A Short Review. *J. Supercrit. Fluids* **2015**, *96*, 46–52. [CrossRef]
18. Pavlovič, I.; Knez, Ž.; Škerget, M. Hydrothermal Reactions of Agricultural and Food Processing Wastes in Sub- and Supercritical Water: A Review of Fundamentals, Mechanisms, and State of Research. *J. Agric. Food Chem.* **2013**, *61*, 8003–8025. [CrossRef] [PubMed]
19. Čolnik, M.; Knez, Ž.; Škerget, M. Sub- and Supercritical Water for Chemical Recycling of Polyethylene Terephthalate Waste. *Chem. Eng. Sci.* **2021**, *233*, 116389. [CrossRef]
20. Cata, A.; Miclau, M.; Ienascu, I.; Ursu, D.; Tanasie, C.; Stefanuta, M.N. Chemical Recycling of Polyethylene Terephthalate (PET) Waste Using Sub- and Supercritical Water. *Rev. Roum. Chim* **2015**, *60*, 579–585.
21. Sato, O.; Arai, K.; Shirai, M. Hydrolysis of Poly (Ethylene Terephthalate) and Poly (Ethylene 2,6-Naphthalene Dicarboxylate) Using Water at High Temperature: Effect of Proton on Low Ethylene Glycol Yield. *Catal. Today* **2006**, *111*, 297–301. [CrossRef]
22. Zhou, J.H.; Shen, G.Z.; Zhu, J.; Yuan, W.K. Terephthalic Acid Hydropurification over Pd/C Catalyst. In *Studies in Surface Science and Catalysis*; Rhee, H.-K., Nam, I.-S., Park, J.M., Eds.; New Developments and Application in Chemical Reaction Engineering; Elsevier: Amsterdam, The Netherlands, 2006; Volume 159, pp. 293–296.
23. Takebayashi, Y.; Sue, K.; Yoda, S.; Hakuta, Y.; Furuya, T. Solubility of Terephthalic Acid in Subcritical Water. *J. Chem. Eng. Data* **2012**, *57*, 1810–1816. [CrossRef]
24. Machado Cruz, R.; Boleslavská, T.; Beránek, J.; Tieger, E.; Twamley, B.; Santos-Martinez, M.J.; Dammer, O.; Tajber, L. Identification and Pharmaceutical Characterization of a New Itraconazole Terephthalic Acid Cocrystal. *Pharmaceutics* **2020**, *12*, 741. [CrossRef]
25. Muse, W.T.; Anthony, J.S.; Bergmann, J.D.; Burnett, D.C.; Crouse, C.L.; Gaviola, B.P.; Thomson, S.A. Chemical and Toxicological Evaluation of Pyrotechnically Disseminated Terephthalic Acid Smoke. *Drug. Chem. Toxicol.* **1997**, *20*, 293–302. [CrossRef]
26. Zhang, Y.M.; Sun, Y.Q.; Wang, Z.J.; Zhang, J. Degradation of Terephthalic Acid by a Newly Isolated Strain of Arthrobacter Sp.0574. *S. Afr. J. Sci.* **2013**, *109*, 1–4. [CrossRef]
27. PTA Production Capacity Globally 2023. Available online: https://www.statista.com/statistics/1065886/global-purified-terephthalic-acid-production-capacity/ (accessed on 13 September 2021).
28. Diaz-Silvarrey, L.S.; McMahon, A.; Phan, A.N. Benzoic Acid Recovery via Waste Poly (Ethylene Terephthalate) (PET) Catalytic Pyrolysis Using Sulphated Zirconia Catalyst. *J. Anal. Appl. Pyrolysis* **2018**, *134*, 621–631. [CrossRef]
29. Del Olmo, A.; Calzada, J.; Nuñez, M. Benzoic Acid and Its Derivatives as Naturally Occurring Compounds in Foods and as Additives: Uses, Exposure, and Controversy. *Crit. Rev. Food. Sci. Nutr.* **2017**, *57*, 3084–3103. [CrossRef]
30. Indrayanto, G.; Syahrani, A.; Mugihardjo; Rahman, A.; Soeharjono; Tanudjojo, W.; Susanti, S.; Yuwono, M.; Ebel, S. Benzoic Acid. In *Analytical Profiles of Drug Substances and Excipients*; Brittain, H.G., Ed.; Academic Press: Cambridge, MA, USA, 1999; Volume 26, pp. 1–46.
31. Benzoic Acid Market Size and Share, Analysis 2021–2027. Available online: https://www.gminsights.com/industry-analysis/benzoic-acid-market (accessed on 22 October 2021).
32. Benzoic Acid Global Market Volume 2015–2026. Available online: https://www.statista.com/statistics/1245227/benzoic-acid-market-volume-worldwide/ (accessed on 22 October 2021).
33. Adschiri, T.; Sato, O.; Machida, K.; Saito, N.; Arai, K. Recovery of Terepthalic Acid by Decomposition of PET in Supercritical Water. *Kagaku Kogaku Ronbunshu* **1997**, *23*, 505–511. [CrossRef]
34. Lindquist, E.; Yang, Y. Degradation of Benzoic Acid and Its Derivatives in Subcritical Water. *J. Chromatogr. A* **2011**, *1218*, 2146–2152. [CrossRef] [PubMed]
35. Palme, A.; Peterson, A.; de la Motte, H.; Theliander, H.; Brelid, H. Development of an Efficient Route for Combined Recycling of PET and Cotton from Mixed Fabrics. *Text. Cloth. Sustain.* **2017**, *3*, 1–9. [CrossRef]

36. Chen, Z.; Hay, J.N.; Jenkins, M.J. The Thermal Analysis of Poly (Ethylene Terephthalate) by FTIR Spectroscopy. *Thermochim. Acta* **2013**, *552*, 123–130. [CrossRef]
37. Manatura, K.; Samaksaman, U. Characteristics and Combustion Kinetics of Fuel Pellets Composed of Waste of Polyethylene Terephthalate and Biomass. *Glob. J. Environ. Sci. Manag.* **2021**, *7*, 625–642. [CrossRef]
38. Hansen, E. Hazardous Substances in Plastic Materials 2013. Available online: https://www.byggemiljo.no/wp-content/uploads/2014/10/72_ta3017.pdf (accessed on 10 September 2021).
39. Khaneghah, A.M.; Limbo, S.; Shoeibi, S.; Mazinani, S. HPLC Study of Migration of Terephthalic Acid and Isophthalic Acid from PET Bottles into Edible Oils. *J. Sci. Food Agric.* **2014**, *94*, 2205–2209. [CrossRef]
40. Sandy, T.; Grady Jr, L.; Meininger, S.; Boe, R. Biological Treatment of 1,4-Dioxane in Wastewater from an Integrated Polyethylene Terephthalate (PET) Manufacturing. *Proc. Water Environ. Fed.* **2001**, *2001*, 88–117. [CrossRef]

processes

MDPI

Article

Comparative Study on UV-AOPs for Efficient Continuous Flow Removal of 4-tert-Butylphenol

Saule Mergenbayeva and Stavros G. Poulopoulos *

Department of Chemical and Materials Engineering, School of Engineering and Digital Sciences, Nazarbayev University, 53 Kabanbay Batyr Ave., Nur-Sultan 010000, Kazakhstan; saule.mergenbayeva@nu.edu.kz

* Correspondence: stavros.poulopoulos@nu.edu.kz; Tel.: +7-7172-694-608

Abstract: In the present study, UV-light-driven advanced oxidation processes (AOPs) have been employed for the degradation of 4-tert-Butylphenol (4-t-BP) in water under continuous flow conditions. The effects of varying space time (10, 20, 40, 60 and 120 min) and oxidant dosage (88.3 mg/L, 176.6 mg/L and 264 mg/L) were examined. 4-t-BP degradation efficiency in the UV-induced AOPs followed the order of UV/H_2O_2 (264.9 mg/L) ≈ UV/Fe^{2+}/H_2O_2 > UV/Fe^{3+}/H_2O_2 > UV/H_2O_2 (176.6 mg/L) > UV/H_2O_2 (88.3 mg/L) > UV/Fe-TiO_2 > UV/TiO_2 > UV, while UV/Fe^{3+}/H_2O_2 was the most efficient process in terms of Total Organic Carbon (TOC) removal (at the space time of 60 min) among those tested. The combination of UV with 88.3 mg/L H_2O_2 enhanced pollutant removal from 51.29% to 93.34% after 10 min of irradiation. The presence of H_2O_2 contributed to the highest 4-t-BP and TOC removal values. Interestingly, the increase in space time from 20 to 60 min resulted in surpassing of the activity of the Fe-TiO_2 over commercial TiO_2, although it had an almost negligible positive impact on the performance of the UV/H_2O_2 system as well as H_2O_2 concentration. The results obtained showed that more than 80% of 4-t-BP could be successfully degraded by both heterogeneous and homogeneous AOPs after 60 min.

Keywords: UV-based advanced oxidation processes; continuous flow; 4-tert-Butylphenol; degradation

Citation: Mergenbayeva, S.; Poulopoulos, S.G. Comparative Study on UV-AOPs for Efficient Continuous Flow Removal of 4-tert-Butylphenol. *Processes* **2022**, *10*, 8. https://doi.org/10.3390/pr10010008

Academic Editor: Dimitris Zagklis

Received: 6 December 2021
Accepted: 14 December 2021
Published: 22 December 2021

1. Introduction

The contamination of water with various chemical compounds found in various types of water bodies [1–3] at low concentrations [4] has attracted increasing attention. These compounds have been termed as emerging pollutants (EPs) and include persistent organic compounds such as pharmaceuticals, personal care products (PCPs), and endocrine disruptive compounds (EDCs) with a high potential of bioaccumulation [5].

Among these pollutants, EDCs constitute a family of organic compounds that affect natural hormones [6]. Their widespread use has led to their release and accumulation in the aquatic environment [7,8]. For example, 4-tert-Butylphenol (4-t-BP), a representative of EDCs [9–12], has been detected in wastewater effluents and surface waters [6,13]. 4-t-BP is an alkylphenol used for the production of phenolic, polycarbonate, and epoxy resins [6]. Though detected at low concentrations in the environment, its ability to bind to the estrogen receptor [14] could result in serious damages for aquatic ecosystems and living organisms [15,16]. Due to its estrogenic activity [10], 4-t-BP has been included by the European Commission in the list of chemicals of very high concern since 2019 [17]. Thus, its toxicity [18,19] and persistence [20,21] impose the elimination of 4-t-BP in water.

Relatively few studies have been conducted to remove 4-t-BP from the aqueous environment; biodegradation and photochemical treatments have been applied mostly. However, biodegradation is not effective since long treatment times are necessary for the complete decomposition of 4-t-BP [22–24]. For instance, the presence of Spirodela polyrrhiza duckweed in environmental water samples led to 4-t-BP removals up to 56% in a 3-day cycle at 28 °C [25].

AOPs are attractive alternatives due to their low cost and environmentally friendly nature. The oxidation efficiency of these processes relies on the generation of highly reactive species such as hydroxyl radicals that have the capacity to degrade and mineralize recalcitrant organic pollutants [26–28]. Among them, UV/H_2O_2, photo-Fenton, photo-Fenton-like, and heterogeneous photocatalytic processes are gaining increasing popularity [12,13,29,30]. For instance, 4-t-BP at the concentration of 0.1 mM in water was almost completely removed within 50 min by UV/H_2O_2 [12]. Xiao et al. [31] studied the treatment of 4-t-BP under visible light heterogeneous photocatalysis, and achieved a 95% TOC reduction after 120 min.

Although a substantial amount of research has demonstrated the effectiveness of UV-induced AOPs in the removal of traditional hazardous compounds and many of emerging concern in water, most studies have been conducted in batch systems. The experimental comparison of low power 280 nm UV-photolysis, UV/H_2O_2 and UV/TiO_2 towards the degradation of 1H-benzotriazole demonstrated the superior performance of the UV/TiO_2 system in terms of degradation and mineralization [32]. Peternel et al. [33] reported about 2.5 higher mineralization efficiency for the photo-Fenton process compared with the UV/TiO_2 system for the organic reactive dye degradation in aqueous solution under 254 nm irradiation. Similarly, Martinez-Costa et al. [34] compared the performance of photo-Fenton processes and UV/H_2O_2 for the removal of the antibiotics sulfamethoxazole (SMX) and trimethoprim (TMP) in aqueous solution. It was found that UV/H_2O_2 was beneficial towards the mineralization of organic pollutant than photo-Fenton processes. Despite their promising results, lab scale batch experiments have the disadvantage of low efficiency for treating high volumes of polluted water [35]. In contrast, continuous flow systems appear to overcome such obstacles [36,37]. Silva et al. [38] studied photocatalytic degradation of SMX under simulated solar light in ultrapure water and environmental water matrices (fresh, estuarine and STP effluent). The results showed a sharp decrease in the irradiation time required for the removal of SMX under continuous flow conditions as compared with batch. Shojaeimehr et al. [39] assessed the photocatalytic degradation of diclofenac using porous carbon nitride (mp-CN) immobilized onto stainless steel (SS) plates and determined the optimum operating conditions, including catalyst loading, irradiation source and initial pollutant concentration. They concluded that the combination of continuous flow mode with the immobilized photocatalyst particles could be a promising alternative for treating contaminated water. Senthilnathan and Philip and Vishnuganth et al. [40] have also reported the effective light-driven pesticide removal using heterogeneous catalysts, N-doped TiO_2 and granular activated carbon-supported titanium dioxide (GAC-TiO_2), respectively.

In the present study, the efficiency of UV-light-driven AOPs (UV/H_2O_2, UV/Fe^{2+}/H_2O_2 and UV/Fe^{3+}/H_2O_2, UV/TiO_2 and UV/Fe-TiO_2) in terms of 4-t-BP degradation and total organic carbon removal was comparatively investigated under continuous flow conditions. To the best of our knowledge, this is the first study concerning the use of light-driven AOPs in continuous flow for the treatment of 4-t-BP in water by means of homogeneous and heterogeneous processes.

2. Materials and Methods

2.1. Chemicals and Materials

4-t-BP (99%), TiO_2-P25 (21 nm particle size, ≥99.5%), $FeCl_2$ (98%), $FeSO_4 \cdot 7H_2O$ (≥99.0%), $Fe(NO_3)_3 \cdot 9H_2O$ (≥98.0%) and methanol (MeOH) of HPLC grade were obtained from Sigma Aldrich. H_2O_2 (37.6% *w*/*w*) was obtained from Skat-Reactiv Company (Almaty, Kazakhstan). Fe-doped TiO_2 powders (Fe-TiO_2) were synthesized by the wet impregnation method with a 4% Fe:TiO_2 molar ratio [21]. Ultrapure water (resistivity 18.2 mΩ cm, Millipore) was used to prepare all solutions.

2.2. Experimental Procedures

A 1000 mL stock solution was prepared by dissolving 30 mg of 4-t-BP in ultrapure water. Depending on the system, H_2O_2, iron, TiO_2 or Fe-TiO_2 were added directly to the 4-t-BP solution before being exposed to UV light. The solution was kept under constant stirring using a digital ceramic top hotplate stirrer (Mettler Toledo). In UV/H_2O_2 processes, three different concentrations of H_2O_2 (88.3 mg/L, 176.6 mg/L and 264.9 mg/L) were used, while 5 mg/L of iron (Fe^{2+} or Fe^{3+}) was added in photo-Fenton and photo-Fenton-like processes. In UV/TiO_2 and UV/Fe-TiO_2 processes, 200 mg/L of catalyst was tested.

The experiments on the degradation of 4-t-BP were carried out in a 300 mL photoreactor equipped with a 10W UV lamp (Figure 1). The reactor was cylindrical with an external diameter of 50.8 mm and a length of 90 mm. Inlet flow rates ranged from 2.5 to 30 mL/min with the help of a peristaltic pump (Ismatec REGLO pump; IDEX Corporation, Lake Forrest, IL, USA). Ambient conditions of temperature and pressure were used, and approximately neutral pH conditions (6.7–7.1). Then, 15 mL aliquots of the treated solution were taken at different space times (10; 20; 40; 60 and 120 min) and filtered by means of 0.22 μm nylon filters. Each experiment was carried out in duplicates.

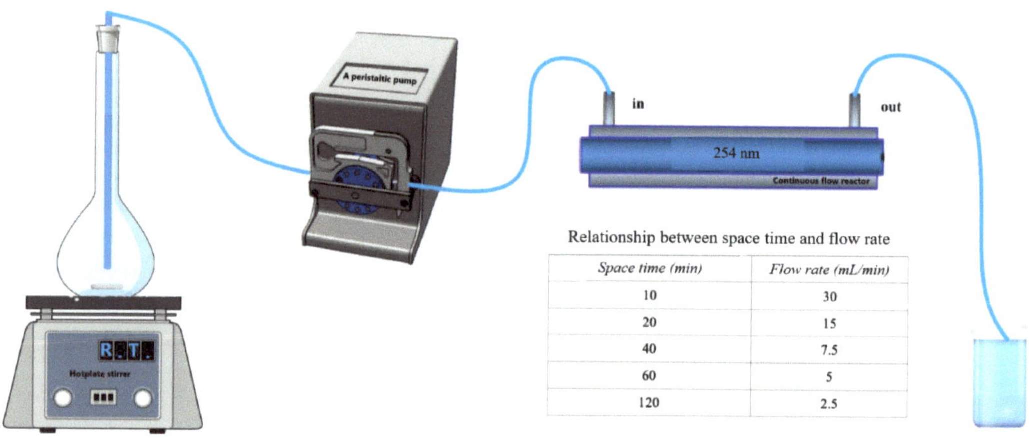

Space time (min)	Flow rate (mL/min)
10	30
20	15
40	7.5
60	5
120	2.5

Figure 1. Experimental setup for the continuous flow photodegradation of 4-t-BP.

2.3. Analytical Methods

The concentration of 4-t-BP was determined using an Agilent 1290 Infinity II high performance liquid chromatography (HPLC) unit equipped with a UV detector. The SB-C8 analytical column (2.1 × 100 mm, 1.8 μm) was operated at 25 °C. In a typical run, a sample of 1 μL was injected and the detection wavelength of 4-t-BP was set at 283 nm. A mixture of 50% MeOH and 50% ultrapure water was used as the mobile phase to elute the analytes at a flow rate of 0.4 mL/min.

TOC removal was determined by measuring the TOC in liquid samples by means of Multi N/C 3100, Analytic Jena AG Corporation (Jena, Germany). The pH of solutions was monitored using a Multi 9310 IDS meter.

3. Results and Discussion

3.1. UV-Photolysis

In order to gain insights into the performance of various UV-activated processes in continuous flow mode, a series of experiments in the absence of any catalyst and oxidant was performed by varying the space time (10, 20 40, 60 and 120 min). The degradation of 4-t-BP steadily increased from 51.3 to 89.3% with the increase in space time from 10 to 120 min (Figure 2A). On the other hand, TOC removal did not significantly increase with increasing the space time above 20 min and the values observed were considerably lower than the corresponding values for 4-t-BP degradation (Figure 2B). The increase in space time promotes both the formation of more reactive species and the contact between produced reactive species and pollutants, which, in turn, leads to higher degradation efficiencies. However, the TOC removal is not promoted to the same extent, as UV-photolysis alone cannot completely mineralize organic carbon [41–43]. For example, Wu et al. [13] achieved 60% 4-t-BP degradation after 300 min of 254 nm UV irradiation and observed the generation of several by-products, including 4-tert-butylcatechol, 4-tert-Butylphenol and 1-tert-butyl-2-methoxy-4-methylbenzene as an impurity. The results indicated that major degradation products of 4-t-BP were 4-tert-butylcatechol and 4-tert-Butylphenol dimer. Such intermediates were also detected in the UV (254 nm)/H_2O_2 system and prolongation of the process time to 16 h resulted in only 29% of TOC removal, confirming the persistent nature of generated by-products to mineralization.

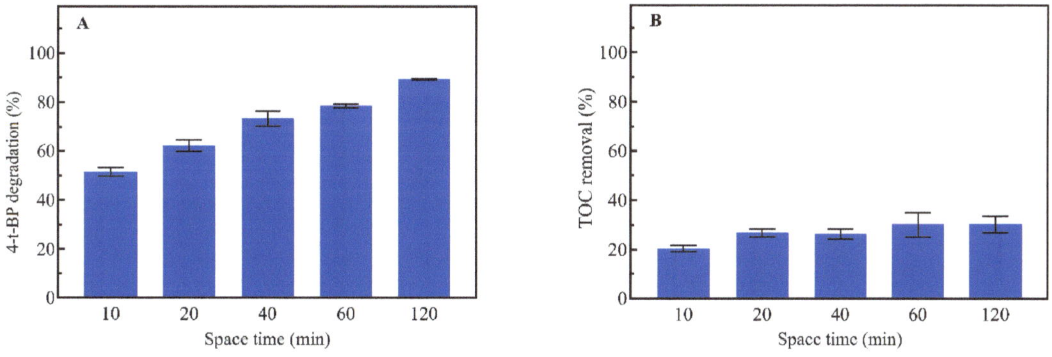

Figure 2. 4-t-BP degradation (**A**) and TOC removal (**B**) as a function of space time in the UV-photolysis.

3.2. UV/H_2O_2

Subsequently, the UV/H_2O_2 process' efficiency was examined by varying the initial concentration of hydrogen peroxide (88.3 mg/L, 176.6 mg/L and 264 mg/L) and space time values (10, 20 40, 60 and 120 min). The results obtained are shown in Figure 3 and agree with previous studies [12,21]. The UV/H_2O_2 process proceeds via direct photolysis and the formation of ·OH radicals that decompose most organic compounds without discrimination. Specifically, 93.34% of 4-t-BP degradation was observed after 10 min. This is due to the formation of sufficient amounts of hydroxyl radicals via the decomposition of H_2O_2 [44,45].

$$H_2O_2 + h\nu \rightarrow 2 \cdot OH \quad (1)$$

$$\cdot OH + \text{4-t-BP} \rightarrow \text{degradation products} \quad (2)$$

As depicted in Figure 3A, degradation efficiency was not considerably affected by H_2O_2 concentration and space time. The concentration of 176.6 mg/L of H_2O_2 seemed to be the most beneficial one in terms of 4-t-BP degradation. Excessive doses of H_2O_2 beyond a certain value may have an inhibitory effect [46,47] through the initiation of propagation reactions that form perhydroxyl radicals ($HO_2\cdot$), which are much less reactive than hydroxyl ones [48]. The detailed mechanism [49,50] is as follows (Equations (3)–(6)):

$$\cdot OH + H_2O_2 \rightarrow HO_2\cdot + H_2O \quad (3)$$

$$HO_2\cdot + H_2O_2 \rightarrow \cdot OH + H_2O + O_2 \quad (4)$$

$$HO_2\cdot + HO_2^- \rightarrow \cdot OH + HO^- + O_2 \quad (5)$$

$$\cdot OH + HO_2\cdot \rightarrow H_2O + O_2 \quad (6)$$

Similar observations were made through TOC removal measurements. As a result, the values of $[H_2O_2]_o$ = 176.6 mg/L and space time = 60 min were selected as the desired conditions for further experiments.

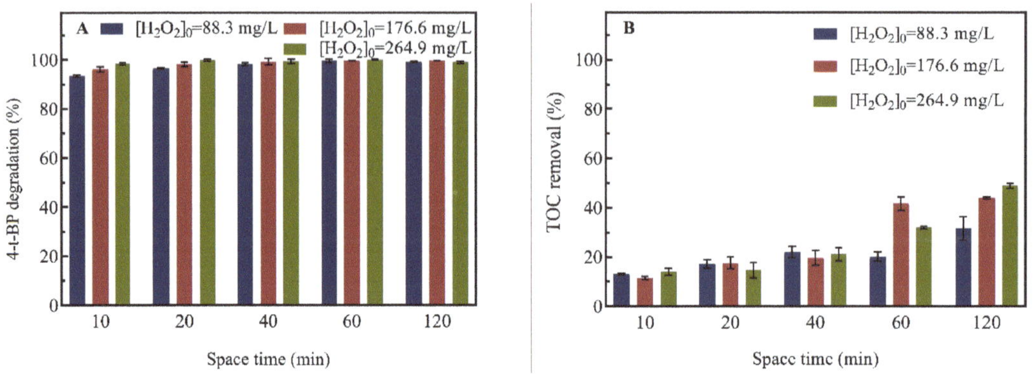

Figure 3. 4-t-BP degradation (**A**) and TOC removal (**B**) as a function of space time in the UV/H_2O_2.

The maximum TOC removal corresponding to 48.8% was achieved by applying $[H_2O_2]_o$ = 264.9 mg/L and space time = 120 min. The increase in TOC removal in the UV/H_2O_2 system in comparison to UV-photolysis could be associated with the generation of both aromatic (4-tert-butylcatechol, 4-tert-Butylphenol and 1-tert-butyl-2-methoxy-4-methylbenzene and hydroquinone) and non-aromatic products ((E)-2-Nonen-1-ol, trans-2-Decenol and (E)-2-Dodecen-1-al), where the latter are the result of a benzene ring fracturing process [13].

3.3. UV/Fe^{2+}/H_2O_2 and UV/Fe^{3+}/H_2O_2

The effect of the presence of Fe^{3+} and Fe^{2+} ions on 4-t-BP degradation was also evaluated. The initial concentration of iron ions used was 5 mg/L. The reaction between iron and H_2O_2 under UV light promotes the formation of OH [51] to effectively oxidize pollutants through Equations (7)–(14):

$$Fe^{2+} + H_2O_2 \rightarrow Fe^{3+} + \cdot OH + OH^- \quad (7)$$

$$Fe^{3+} + H_2O_2 \rightarrow Fe^{2+} + HO_2 \cdot + H^+ \quad (8)$$

$$\cdot OH + H_2O_2 \rightarrow HO_2 \cdot + H_2O \quad (9)$$

$$Fe^{2+} + \cdot OH \rightarrow Fe^{3+} + OH^- \quad (10)$$

$$Fe^{3+} + HO_2 \cdot \rightarrow Fe^{2+} + O_2 + H^+ \quad (11)$$

$$\cdot OH + \cdot OH \rightarrow H_2O_2 \quad (12)$$

$$\text{4-t-BP} + \cdot OH \rightarrow \text{Intermediates} \quad (13)$$

$$\text{Intermediates} + \cdot OH \rightarrow CO_2 + H_2O \quad (14)$$

Figure 4A shows a similar oxidation performance for Fe^{2+}/UV/H_2O_2 towards 4-t-BP degradation as in UV/H_2O_2 after 60 min of irradiation. The efficiency of Fe^{3+}/UV/H_2O_2 was slightly lower. In contrast, an enhanced TOC removal (Figure 4B) was observed for the Fe^{3+}/UV/H_2O_2 process, indicating a favorable Fe^{3+} contribution to the catalytic decomposition of H_2O_2 [52]. This can be attributed to a synergistic effect between Fe^{3+}, UV and H_2O_2. TOC removal efficiencies followed the following order: Fe^{3+}/UV/H_2O_2 > Fe^{2+}/UV/H_2O_2 > UV/H_2O_2.

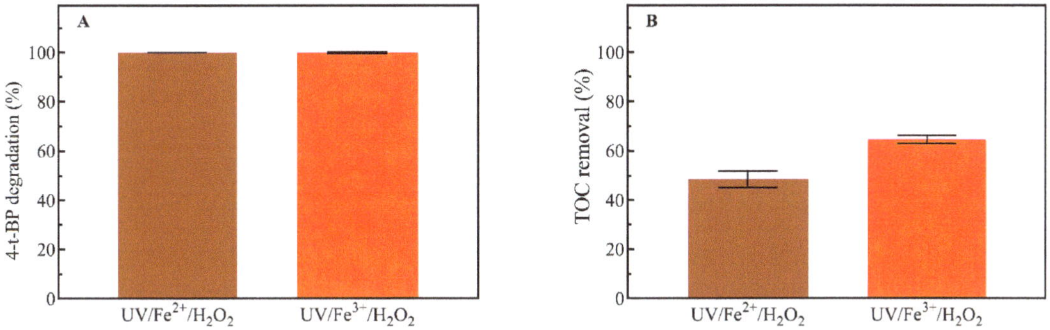

Figure 4. 4-t-BP degradation (**A**) and TOC removal (**B**) in the Fe^{2+}/UV/H_2O_2 and Fe^{3+}/UV/H_2O_2 at optimal conditions (conditions: space time = 60 min, 5 mg/L of catalyst, [H_2O_2] = 176.6 mg/L).

3.4. UV/TiO_2 and UV/Fe-TiO_2

The photocatalytic behavior of undoped TiO_2 and Fe-doped TiO_2 was examined for 4-t-BP degradation (Figure 5). The photocatalytic activity of Fe-TiO_2 relies on the dopant concentration [53–59]. Fe-TiO_2 at a mass ratio of 4% was selected as based on previously reported results [60,61].

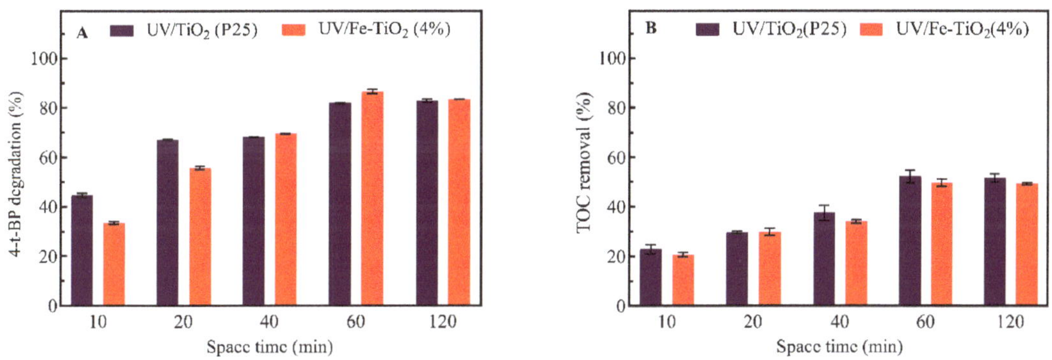

Figure 5. 4-t-BP degradation (**A**) and TOC removal (**B**) as a function of space time in the UV/TiO_2 and UV/Fe-TiO_2.

As shown in Figure 5, 4-t-BP and TOC removals were apparently increased with an increasing space time from 10 min to 60 min for both TiO_2 and Fe-TiO_2, indicating substantial impact of space time on the performance of the systems. However, increasing space time from 60 min to 120 min resulted in apparently lower 4-t-BP degradation and TOC removal. Compared to pure TiO_2, the incorporation of iron particles slightly induced the degradation of 4-t-BP, degrading 87% after 60 min instead of 82%. This can be attributed to the retardation of the electron-hole pairs recombination [62]. It is noteworthy to mention that at lower space time (10 and 20 min), the performance of TiO_2 exceeded Fe-TiO_2, reaching 44.48% and 67% of 4-t-BP degradation in comparison to 33.41% and 55.79%, respectively. Comparatively, further increases in space time to 40 min and 60 min resulted in surpassed catalytic activity of Fe-TiO_2 over TiO_2.

3.5. Comparison of UV-Induced Processes

In order to compare the performance of UV-based processes, 4-t-BP degradation and TOC removal were plotted at the space time of 60 min. As observed in Figure 6A, the highest 4-t-BP degradation was achieved in the presence of H_2O_2 with the values ranging from 99.45% to 100% and followed the order of UV/H_2O_2 (264.9 mg/L) ≈ UV/Fe^{2+}/H_2O_2 > UV/Fe^{3+}/H_2O_2 > UV/H_2O_2 (176.6 mg/L) > UV/H_2O_2 (88.3 mg/L) > UV/Fe-TiO_2 > UV/TiO_2 > UV only. Since the addition of H_2O_2 yield in a formation of higher amounts of hydroxyl radicals, the observed enhancement was expected. On the other hand, H_2O_2 concentration increase in UV/H_2O_2 system had an almost negligible effect on the degradation. As for the heterogeneous AOPs, the removal efficiency of 4-t-BP was 81.8% and 86.5% with TiO_2 and Fe-TiO_2 in the concentration of 200 mg/L, which indicates slight increased degradation of 4-t-BP than that of UV treatment only. It is worth mentioning that combination of continuous flow and UV/Fe-TiO_2 process contributed to the improvement of photo-oxidation reaction, while in a similar study [21] under batch-operated mode, the removal of 4-t-BP was higher by means of commercial TiO_2 than that of Fe-TiO_2.

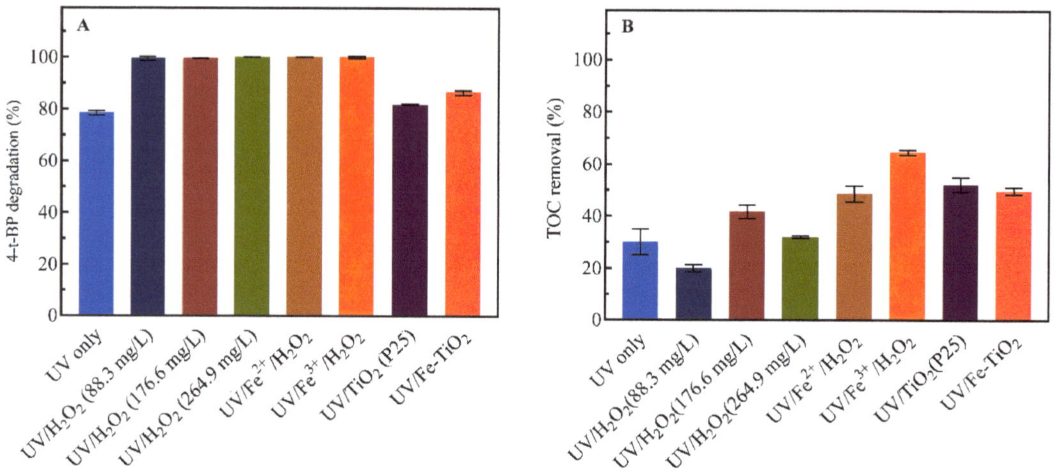

Figure 6. 4-t-BP degradation (**A**) and TOC removal (**B**) of various UV-induced AOPs at optimum space time = 60 min.

Although 4-t-BP was largely degraded by almost all applied H_2O_2 concentrations after 60 min, the TOC removal varied from 20.05 to 41.66%. UV treatment alone did not provide significant levels of TOC removal, while removals around 50% were achieved within 60 min in the presence of heterogeneous photocatalysts (Fe-TiO_2 or TiO_2). Despite the low amount of Fe^{3+} added, its combination with H_2O_2 and UV led to the highest TOC removal (65%). The rank of TOC removal efficiency followed the order of UV/Fe^{3+}/H_2O_2 > UV/TiO_2 > UV/Fe-TiO_2 > UV/Fe^{2+}/H_2O_2 > UV/H_2O_2 (176.6 mg/L) > UV/H_2O_2 (264.9 mg/L) > UV only > UV/H_2O_2 (88.3 mg/L).

4. Conclusions

In the present work, various continuous-flow UV-mediated AOPs were used for the degradation of 4-t-BP and TOC removal in water. The results showed that 93.34% 4-t-BP degradation could be achieved at relatively low space time of 10 min in the UV/H_2O_2 system with oxidant concentration of 88.3 mg/L. The presence of H_2O_2 contributes to the highest degradation efficiencies, making homogeneous AOPs (UV/H_2O_2, UV/Fe^{2+}/H_2O_2 and UV/Fe^{3+}/H_2O_2) attractive for the removal of 4-t-BP.

Generally, flow rate had a positive effect on degradation efficiency, where optimal time to achieve the highest removal for most of the studied processes was found to be 60 min. As a result, the degradation performance improvement of UV, UV/TiO_2 and UV/Fe-TiO_2 was observed with the increase in space time. Moreover, the photocatalytic activity for TOC removal ability was also evaluated. Higher TOC removal efficiencies were observed in the catalyst-assisted systems. Approximately 65% of TOC abatement in less than 60 min was achieved under UV/Fe^{3+}/H_2O_2 treatment at an optimized concentration of H_2O_2 (176.6 mg/L), probably because of the achieved synergistic effect between oxidant, catalyst, UV exposure and flow system. Overall, the findings propose attractive applications of continuous flow treatment systems for efficient water remediation contaminated with 4-t-BP.

Author Contributions: Conceptualization, S.M. and S.G.P.; methodology, S.M. and S.G.P.; validation, S.G.P.; investigation, S.M.; resources, S.G.P.; writing—original draft preparation, S.M.; writing—review and editing, S.M. and S.G.P.; supervision, S.G.P.; project administration, S.G.P.; funding acquisition, S.G.P. All authors have read and agreed to the published version of the manuscript.

Funding: This research was funded by the Nazarbayev University project "Noble metals nanocomposites hyper-activity in heterogeneous non-catalytic and catalytic reactions", Faculty development competitive research grants program for 2019–2021, Grant Number 110119FD4536.

Institutional Review Board Statement: Not applicable.

Informed Consent Statement: Not applicable.

Data Availability Statement: Not applicable.

Acknowledgments: The authors acknowledge funding support from Nazarbayev University. The technical support of the Core Facilities of Nazarbayev University is greatly acknowledged.

Conflicts of Interest: The authors declare no conflict of interest. The funders had no role in the design of the study; in the collection, analyses, or interpretation of data; in the writing of the manuscript, or in the decision to publish the results.

References

1. Heemken, O.P.; Reincke, H.; Stachel, B.; Theobald, N. The Occurrence of Xenoestrogens in the Elbe River and the North Sea. *Chemosphere* **2001**, *45*, 245–259. [CrossRef]
2. Inoue, K.; Yoshie, Y.; Kondo, S.; Yoshimura, Y.; Nakazawa, H. Determination of Phenolic Xenoestrogens in Water by Liquid Chromatography with Coulometric-Array Detection. *J. Chromatogr. A* **2002**, *946*, 291–294. [CrossRef]
3. Koh, C.-H.; Khim, J.S.; Villeneuve, D.L.; Kannan, K.; Giesy, J.P. Characterization of Trace Organic Contaminants in Marine Sediment from Yeongil Bay, Korea: 1. Instrumental Analyses. *Environ. Pollut.* **2006**, *142*, 39–47. [CrossRef] [PubMed]
4. Chang, H.-S.; Choo, K.-H.; Lee, B.; Choi, S.-J. The Methods of Identification, Analysis, and Removal of Endocrine Disrupting Compounds (EDCs) in Water. *J. Hazard. Mater.* **2009**, *172*, 1–12. [CrossRef]
5. Tang, Y.; Yin, M.; Yang, W.; Li, H.; Zhong, Y.; Mo, L.; Liang, Y.; Ma, X.; Sun, X. Emerging Pollutants in Water Environment: Occurrence, Monitoring, Fate, and Risk Assessment. *Water Environ. Res. Res. Publ. Water Environ. Fed.* **2019**, *91*, 984–991. [CrossRef]
6. Annamalai, J.; Namasivayam, V. Endocrine Disrupting Chemicals in the Atmosphere: Their Effects on Humans and Wildlife. *Environ. Int.* **2015**, *76*, 78–97. [CrossRef]
7. Acir, I.-H.; Guenther, K. Endocrine-Disrupting Metabolites of Alkylphenol Ethoxylates—A Critical Review of Analytical Methods, Environmental Occurrences, Toxicity, and Regulation. *Sci. Total Environ.* **2018**, *635*, 1530–1546. [CrossRef] [PubMed]
8. Balaguer, P.; Delfosse, V.; Grimaldi, M.; Bourguet, W. Structural and Functional Evidences for the Interactions between Nuclear Hormone Receptors and Endocrine Disruptors at Low Doses. *Comptes Rendus Biol.* **2017**, *340*, 414–420. [CrossRef] [PubMed]
9. Toyama, T.; Momotani, N.; Ogata, Y.; Miyamori, Y.; Inoue, D.; Sei, K.; Mori, K.; Kikuchi, S.; Ike, M. Isolation and Characterization of 4-tert-Butylphenol-Utilizing Sphingobium Fuliginis Strains from Phragmites Australis Rhizosphere Sediment. *Appl. Environ. Microbiol.* **2010**, *76*, 6733–6740. [CrossRef]
10. Webster, F.; Gagné, M.; Patlewicz, G.; Pradeep, P.; Trefiak, N.; Judson, R.S.; Barton-Maclaren, T.S. Predicting Estrogen Receptor Activation by a Group of Substituted Phenols: An Integrated Approach to Testing and Assessment Case Study. *Regul. Toxicol. Pharmacol.* **2019**, *106*, 278–291. [CrossRef] [PubMed]
11. Wirasnita, R.; Mori, K.; Toyama, T. Effect of Activated Carbon on Removal of Four Phenolic Endocrine-Disrupting Compounds, Bisphenol A, Bisphenol F, Bisphenol S, and 4-tert-Butylphenol in Constructed Wetlands. *Chemosphere* **2018**, *210*, 717–725. [CrossRef] [PubMed]
12. Wu, Y.; Zhu, X.; Chen, H.; Dong, W.; Zhao, J. Photodegradation of 4-tert-Butylphenol in Aqueous Solution by UV-C, UV/H_2O_2 and UV/$S_2O_8{}^{2-}$ System. *J. Environ. Sci. Health Part A* **2016**, *51*, 440–445. [CrossRef]
13. Wu, Y.; Shi, J.; Chen, H.; Zhao, J.; Dong, W. Aqueous Photodegradation of 4-tert-Butylphenol: By-Products, Degradation Pathway and Theoretical Calculation Assessment. *Sci. Total Environ.* **2016**, *566–567*, 86–92. [CrossRef]
14. Olsen, C.M.; Meussen-Elholm, E.T.M.; Hongslo, J.K.; Stenersen, J.; Tollefsen, K.-E. Estrogenic Effects of Environmental Chemicals: An Interspecies Comparison. *Comp. Biochem. Physiol. Part C Toxicol. Pharmacol.* **2005**, *141*, 267–274. [CrossRef] [PubMed]

15. Barse, A.V.; Chakrabarti, T.; Ghosh, T.; Pal, A.; Jadhao, S. One-Tenth Dose of LC50 of 4-tert-Butylphenol Causes Endocrine Disruption and Metabolic Changes in Cyprinus Carpio. *Pestic. Biochem. Physiol.* **2006**, *86*, 172–179. [CrossRef]
16. Meier, S.; Andersen, T.E.; Norberg, B.; Thorsen, A.; Taranger, G.L.; Kjesbu, O.S.; Dale, R.; Morton, H.C.; Klungsøyr, J.; Svardal, A. Effects of Alkylphenols on the Reproductive System of Atlantic Cod (Gadus Morhua). *Aquat. Toxicol.* **2007**, *81*, 207–218. [CrossRef]
17. EUR-Lex-32019D1194-EN-EUR-Lex. Available online: https://eur-lex.europa.eu/eli/dec_impl/2019/1194/oj (accessed on 4 November 2021).
18. Fan, D.; Liu, J.; Wang, L.; Yang, X.; Zhang, S.; Zhang, Y.; Shi, L. Development of Quantitative Structure–Activity Relationship Models for Predicting Chronic Toxicity of Substituted Benzenes to Daphnia Magna. *Bull. Environ. Contam. Toxicol.* **2016**, *96*, 664–670. [CrossRef]
19. Wang, L.; Liu, J.; Liu, J.; Shi, L.; Wang, Z. Application of Microcosm and Species Sensitivity Distribution Approaches in the Ecological Hazard Assessment of 4-tert-Butylphenol. *Chem. Ecol.* **2017**, *34*, 108–125. [CrossRef]
20. Janousek, R.M.; Müller, J.; Knepper, T.P. Combined Study of Source, Environmental Monitoring and Fate of Branched Alkylphenols: The Chain Length Matters. *Chemosphere* **2020**, *241*, 124950. [CrossRef] [PubMed]
21. Makhatova, A.; Ulykbanova, G.; Sadyk, S.; Sarsenbay, K.; Atabaev, T.S.; Inglezakis, V.J.; Poulopoulos, S.G. Degradation and Mineralization of 4-tert-Butylphenol in Water Using Fe-Doped TiO_2 Catalysts. *Sci. Rep.* **2019**, *9*, 19284. [CrossRef]
22. Dan, A.; Fujii, D.; Soda, S.; Machimura, T.; Ike, M. Removal of Phenol, Bisphenol A, and 4-tert-Butylphenol from Synthetic Landfill Leachate by Vertical Flow Constructed Wetlands. *Sci. Total Environ.* **2017**, *578*, 566–576. [CrossRef]
23. Nguyen, L.N.; Hai, F.I.; Yang, S.; Kang, J.; Leusch, F.D.L.; Roddick, F.; Price, W.E.; Nghiem, L.D. Removal of Pharmaceuticals, Steroid Hormones, Phytoestrogens, UV-Filters, Industrial Chemicals and Pesticides by Trametes Versicolor: Role of Biosorption and Biodegradation. *Int. Biodeterior. Biodegrad.* **2014**, *88*, 169–175. [CrossRef]
24. Nguyen, L.N.; Hai, F.I.; Price, W.E.; Kang, J.; Leusch, F.D.L.; Roddick, F.; van de Merwe, J.P.; Magram, S.F.; Nghiem, L.D. Degradation of a Broad Spectrum of Trace Organic Contaminants by an Enzymatic Membrane Reactor: Complementary Role of Membrane Retention and Enzymatic Degradation. *Int. Biodeterior. Biodegrad.* **2015**, *99*, 115–122. [CrossRef]
25. Ogata, Y.; Toyama, T.; Yu, N.; Wang, X.; Sei, K.; Ike, M. Occurrence of 4-tert-Butylphenol (4-t-BP) Biodegradation in an Aquatic Sample Caused by the Presence of Spirodela Polyrrhiza and Isolation of a 4-t-BP-Utilizing Bacterium. *Biodegradation* **2013**, *24*, 191–202. [CrossRef] [PubMed]
26. Hong, Y.; Li, C.; Zhang, G.; Meng, Y.; Yin, B.; Zhao, Y.; Shi, W. Efficient and Stable Nb_2O_5 Modified G-C_3N_4 Photocatalyst for Removal of Antibiotic Pollutant. *Chem. Eng. J.* **2016**, *299*, 74–84. [CrossRef]
27. Lee, Y.; von Gunten, U. Oxidative Transformation of Micropollutants during Municipal Wastewater Treatment: Comparison of Kinetic Aspects of Selective (Chlorine, Chlorine Dioxide, FerrateVI, and Ozone) and Non-Selective Oxidants (Hydroxyl Radical). *Water Res.* **2010**, *44*, 555–566. [CrossRef]
28. Muruganandham, M.; Suri, R.; Jafari, S.; Sillanpää, M.; Lee, G.-J.; Wu, J.; Swaminathan, M. Recent Developments in Homogeneous Advanced Oxidation Processes for Water and Wastewater Treatment. *Int. J. Photoenergy* **2014**, *2014*, 821674. [CrossRef]
29. Wu, Y.; Brigante, M.; Dong, W.; Claire, P.; Mailhot, G. Toward a Better Understanding of Fe(III)-EDDS Photochemistry: Theoretical Stability Calculation and Experimental Investigation of 4-tert-Butylphenol Degradation. *J. Phys. Chem. A* **2013**, *118*, 396–403. [CrossRef]
30. Wu, Y.; Prulho, R.; Brigante, M.; Dong, W.; Hanna, K.; Mailhot, G. Activation of Persulfate by Fe(III) Species: Implications for 4-tert-Butylphenol Degradation. *J. Hazard. Mater.* **2016**, *322*, 380–386. [CrossRef]
31. Xiao, X.; Xing, C.; He, G.; Zuo, X.; Nan, J.; Wang, L. Solvothermal Synthesis of Novel Hierarchical $Bi_4O_5I_2$ Nanoflakes with Highly Visible Light Photocatalytic Performance for the Degradation of 4-tert-Butylphenol. *Appl. Catal. B Environ.* **2014**, *148–149*, 154–163. [CrossRef]
32. Chen, Y.; Ye, J.; Li, C.; Zhou, P.; Liu, J.; Ou, H. Degradation of 1 H -Benzotriazole by UV/H_2O_2 and UV/TiO_2 : Kinetics, Mechanisms, Products and Toxicology. *Environ. Sci. Water Res. Technol.* **2018**, *4*, 1282–1294. [CrossRef]
33. Peternel, I.T.; Koprivanac, N.; Božić, A.M.L.; Kušić, H.M. Comparative Study of UV/TiO_2, UV/ZnO and Photo-Fenton Processes for the Organic Reactive Dye Degradation in Aqueous Solution. *J. Hazard. Mater.* **2007**, *148*, 477–484. [CrossRef] [PubMed]
34. Martínez-Costa, J.I.; Rivera-Utrilla, J.; Leyva-Ramos, R.; Sánchez-Polo, M.; Velo-Gala, I.; Mota, A.J. Individual and Simultaneous Degradation of the Antibiotics Sulfamethoxazole and Trimethoprim in Aqueous Solutions by Fenton, Fenton-like and Photo-Fenton Processes Using Solar and UV Radiations. *J. Photochem. Photobiol. Chem.* **2018**, *360*, 95–108. [CrossRef]
35. Chang, S.; Yang, X.; Sang, Y.; Liu, H. Highly Efficient Photocatalysts and Continuous-Flow Photocatalytic Reactors for Degradation of Organic Pollutants in Wastewater. *Chem. Asian J.* **2016**, *11*, 2352–2371. [CrossRef] [PubMed]
36. Carbonaro, S.; Sugihara, M.N.; Strathmann, T.J. Continuous-Flow Photocatalytic Treatment of Pharmaceutical Micropollutants: Activity, Inhibition, and Deactivation of TiO2 Photocatalysts in Wastewater Effluent. *Appl. Catal. B Environ.* **2013**, *129*, 1–12. [CrossRef]
37. Petala, A.; Spyrou, D.; Frontistis, Z.; Mantzavinos, D.; Kondarides, D.I. Immobilized Ag_3PO_4 Photocatalyst for Micro-Pollutants Removal in a Continuous Flow Annular Photoreactor. *Catal. Today* **2019**, *328*, 223–229. [CrossRef]
38. Silva, C.P.; Oliveira, C.; Ribeiro, A.; Osório, N.; Otero, M.; Esteves, V.I.; Lima, D.L.D. Sulfamethoxazole Exposure to Simulated Solar Radiation under Continuous Flow Mode: Degradation and Antibacterial Activity. *Chemosphere* **2020**, *238*, 124613. [CrossRef] [PubMed]

39. Shojaeimehr, T.; Tasbihi, M.; Acharjya, A.; Thomas, A.; Schomäcker, R.; Schwarze, M. Impact of Operating Conditions for the Continuous-Flow Degradation of Diclofenac with Immobilized Carbon Nitride Photocatalysts. *J. Photochem. Photobiol. Chem.* **2019**, *388*, 112182. [CrossRef]
40. Jaganathan, S.; Philip, L. Elimination of Pesticides and Their Formulation Products from Drinking Water Using Thin Film Continuous Photoreactor under Solar Radiation. *Sol. Energy* **2012**, *86*, 2735–2745. [CrossRef]
41. Piña-Pérez, Y.; Aguilar-Martínez, O.; Oros-Ruíz, S.; Gómez, R.; Tzompantzi, F. Commercial Aluminum Oxides with Different Crystalline Structures Efficient for the Mineralization of Phenolic Pollutants. *J. Photochem. Photobiol. Chem.* **2018**, *353*, 409–415. [CrossRef]
42. Chen, J.; He, Z.; Ji, Y.; Li, G.; An, T.; Choi, W. OH Radicals Determined Photocatalytic Degradation Mechanisms of Gaseous Styrene in TiO_2 System under 254 Nm versus 185 Nm Irradiation: Combined Experimental and Theoretical Studies. *Appl. Catal. B Environ.* **2019**, *257*, 117912. [CrossRef]
43. Poulopoulos, S.G.; Yerkinova, A.; Ulykbanova, G.; Inglezakis, V.J. Photocatalytic Treatment of Organic Pollutants in a Synthetic Wastewater Using UV Light and Combinations of TiO_2, H_2O_2 and Fe(III). *PLoS ONE* **2019**, *14*, e0216745. [CrossRef] [PubMed]
44. Ou, H.-S.; Ye, J.-S.; Ma, S.; Wei, C.-H.; Gao, N.-Y.; He, J. Degradation of Ciprofloxacin by UV and UV/H_2O_2 via Multiple-Wavelength Ultraviolet Light-Emitting Diodes: Effectiveness, Intermediates and Antibacterial Activity. *Chem. Eng. J.* **2016**, *289*, 391–401. [CrossRef]
45. Zhang, Y.; Zhang, J.; Xiao, Y.; Chang, V.; Lim, T. Kinetic and Mechanistic Investigation of Azathioprine Degradation in Water by UV, UV/H_2O_2 and UV/Persulfate. *Chem. Eng. J.* **2016**, *302*, 526–534. [CrossRef]
46. Afzal, T.; Isa, M.H.; Ul Mustafa, M.R. Removal of Organic Pollutants from Produced Water Using Fenton Oxidation. *E3S Web Conf.* **2018**, *34*, 02035. [CrossRef]
47. Daneshvar, N.; Salari, D.; Khataee, A.R. Photocatalytic Degradation of Azo Dye Acid Red 14 in Water on ZnO as an Alternative Catalyst to TiO_2. *J. Photochem. Photobiol. Chem.* **2004**, *162*, 317–322. [CrossRef]
48. Watts, R.J.; Sarasa, J.; Loge, F.J.; Teel, A.L. Oxidative and Reductive Pathways in Manganese-Catalyzed Fenton's Reactions. *J. Environ. Eng.* **2005**, *131*, 158–164. [CrossRef]
49. Liu, X.; Liu, Y.; Lu, S.; Wang, Z.; Wang, Y.; Zhang, G.; Guo, X.; Guo, W.; Zhang, T.; Xi, B. Degradation Difference of Ofloxacin and Levofloxacin by UV/H2O2 and UV/PS (Persulfate): Efficiency, Factors and Mechanism. *Chem. Eng. J.* **2020**, *385*, 123987. [CrossRef]
50. Samarghandi, M.R.; Dargahi, A.; Zolghadr Nasab, H.; Ghahramani, E.; Salehi, S. Degradation of Azo Dye Acid Red 14 (AR14) from Aqueous Solution Using H_2O_2/NZVI and $S_2O_8{}^{2-}$/NZVI Processes in the Presence of UV Irradiation. *Water Environ. Res. Res. Publ. Water Environ. Fed.* **2020**, *92*, 1173–1183. [CrossRef]
51. Rivas, F.J.; Beltrán, F.J.; Frades, J.; Buxeda, P. Oxidation of P-Hydroxybenzoic Acid by Fenton's Reagent. *Water Res.* **2001**, *35*, 387–396. [CrossRef]
52. Augugliaro, V.; Litter, M.; Palmisano, L.; Soria, J. The Combination of Heterogeneous Photocatalysis with Chemical and Physical Operations: A Tool for Improving the Photoprocess Performance. *J. Photochem. Photobiol. C Photochem. Rev.* **2006**, *7*, 127–144. [CrossRef]
53. Hemmati Borji, S.; Nasseri, S.; Mahvi, A.H.; Nabizadeh, R.; Javadi, A.H. Investigation of Photocatalytic Degradation of Phenol by Fe(III)-Doped TiO_2 and TiO_2 Nanoparticles. *J. Environ. Health Sci. Eng.* **2014**, *12*, 101. [CrossRef] [PubMed]
54. JImĕnez, D.H.; Triviño, G.M.H.; Villaseñor, E.A.; Mesa, E.B.; Hernández, J.A.; Zepeda, T.A.; Zapata, R.B.; Arias, A.N.A. Phenol Photocatalytic Degradation over Fe-TiO_2 Materials Synthesized by Different Methods. *Sci. Tech.* **2019**, *24*, 523–531. [CrossRef]
55. Jamil, T.S.; Gad-Allah, T.A.; Ghaly, M.Y. Parametric Study on Phenol Photocatalytic Degradation under Pure Visible and Solar Irradiations by Fe-Doped TiO_2. *Desalination Water Treat.* **2012**, *50*, 264–271. [CrossRef]
56. Moradi, V.; Ahmed, F.; Jun, M.B.G.; Blackburn, A.; Herring, R.A. Acid-Treated Fe-Doped TiO_2 as a High Performance Photocatalyst Used for Degradation of Phenol under Visible Light Irradiation. *J. Environ. Sci.* **2019**, *83*, 183–194. [CrossRef] [PubMed]
57. Nahar, M.S.; Hasegawa, K.; Kagaya, S.; Kuroda, S. Comparative Assessment of the Efficiency of Fe-Doped TiO_2 Prepared by Two Doping Methods and Photocatalytic Degradation of Phenol in Domestic Water Suspensions. *Sci. Technol. Adv. Mater.* **2007**, *8*, 286–291. [CrossRef]
58. Shawabkeh, R.; Khashman, O.; Bisharat, G. Photocatalytic Degradation of Phenol Using Fe-TiO_2 by Different Illumination Sources. *Int. J. Chem.* **2010**, *2*, 10. [CrossRef]
59. Yodsomnuk, P.; Junjeam, K.; Termtanun, M. Photoactivity of Fe and Zn-Doped TiO_2 in Phenol Degradation under Visible Light. *MATEC Web Conf.* **2018**, *192*, 03047. [CrossRef]
60. Crişan, M.; Ianculescu, A.-C.; Crișan, D.; Drăgan, N.; Todan, L.; Niţoi, I.; Oancea, P. Chapter 10—Fe-Doped TiO_2 Nanomaterials for Water Depollution. In *Nanotechnology in the Beverage Industry*; Elsevier: Amsterdam, The Netherlands, 2020; pp. 265–313. [CrossRef]
61. Sun, T.; Liu, E.; Fan, J.; Hu, X.; Wu, F.; Hou, W.; Yang, Y.; Kang, L. High Photocatalytic Activity of Hydrogen Production from Water over Fe Doped and Ag Deposited Anatase TiO_2 Catalyst Synthesized by Solvothermal Method. *Chem. Eng. J.* **2013**, *228*, 896–906. [CrossRef]
62. Moradi, H.; Eshaghi, A.; Hosseini, S.R.; Ghani, K. Fabrication of Fe-Doped TiO_2 Nanoparticles and Investigation of Photocatalytic Decolorization of Reactive Red 198 under Visible Light Irradiation. *Ultrason. Sonochem.* **2016**, *32*, 314–319. [CrossRef] [PubMed]

processes

MDPI

Review

Two-Dimensional Nanomaterials for the Removal of Pharmaceuticals from Wastewater: A Critical Review

Sergio González-Poggini, Andreas Rosenkranz and Melanie Colet-Lagrille *

Department of Chemical Engineering, Biotechnology and Materials, Faculty of Physical and Mathematical Sciences, Universidad de Chile, Beauchef 851, Santiago 8370456, Chile; sergio.gonzalez.p@ug.uchile.cl (S.G.-P.); arosenkranz@ing.uchile.cl (A.R.)

* Correspondence: mcolet@ing.uchile.cl

Abstract: The removal of pharmaceuticals from wastewater is critical due to their considerable risk on ecosystems and human health. Additionally, they are resistant to conventional chemical and biological remediation methods. Two-dimensional nanomaterials are a promising approach to face this challenge due to their combination of high surface areas, high electrical conductivities, and partially optical transparency. This review discusses the state-of-the-art concerning their use as adsorbents, oxidation catalysts or photocatalysts, and electrochemical catalysts for water treatment purposes. The bibliographic search bases upon academic databases including articles published until August 2021. Regarding adsorption, high removal capacities (>200 mg g^{-1}) and short equilibrium times (<30 min) are reported for molybdenum disulfide, metal-organic frameworks, MXenes, and graphene oxide/magnetite nanocomposites, attributed to a strong adsorbate-adsorbent chemical interaction. Concerning photocatalysis, MXenes and carbon nitride heterostructures show enhanced charge carriers separation, favoring the generation of reactive oxygen species to degrade most pharmaceuticals. Peroxymonosulfate activation via pure or photo-assisted catalytic oxidation is promising to completely degrade many compounds in less than 30 min. Future work should be focused on the exploration of greener synthesis methods, regeneration, and recycling at the end-of-life of two-dimensional materials towards their successful large-scale production and application.

Keywords: two-dimensional nanomaterials; wastewater treatment; pharmaceuticals; adsorption; photocatalytic processes; redox and electrochemical processes

Citation: González-Poggini, S.; Rosenkranz, A.; Colet-Lagrille, M. Two-Dimensional Nanomaterials for the Removal of Pharmaceuticals from Wastewater: A Critical Review. *Processes* **2021**, *9*, 2160. https://doi.org/10.3390/pr9122160

Academic Editor: Dimitris Zagklis

Received: 20 October 2021
Accepted: 24 November 2021
Published: 29 November 2021

1. Introduction

The contamination level of natural water streams has increased in recent decades due to the rapid global industrial development associated with a growing world population. This generates many inorganic contaminants (toxic heavy metals—e.g., arsenic, lead, mercury, cobalt—and toxic gases—e.g., NO_x, SO_x, CO, NH_3–), and organic contaminants including dyes, pesticides, and pharmaceuticals as well as bio-toxics [1–5]. In this scenario, the need for new, highly efficient wastewater treatment processes is compelling considering: (i) the extremely variable composition of the effluents, (ii) the refractory behaviour of several pollutants to conventional chemical and biological remediation methods (coagulation, flocculation, screening, centrifugation, flotation, aerobic and anaerobic treatments), and (iii) the low concentration at which these contaminants are present in wastewater streams.

Particularly in the case of pharmaceuticals, the latter three points are crucial in the design and development of new methods for wastewater treatment. Pharmaceuticals (and personal care products) contain diverse organic groups, such as antibiotics, hormones, and antimicrobial agents, which have a persistent toxic action on microorganisms, thus posing a considerable risk to ecosystems and human health [6–8]. They are regularly detected in natural water streams because of their non-effective elimination in treatment processes. The typical concentrations of pharmaceutical pollutants in wastewater range between

1 ng L^{-1} and 600 μg L^{-1} [9,10]. They come from varied origins, with industrial, hospital, and domestic wastewaters being the primary sources [11].

Due to the variable nature of pharmaceuticals, there is not a specific treatment that can assure their overall removal. In conventional wastewater treatment plants, carbamazepine, metoprolol, mefenamic acid, and non-steroidal anti-inflammatory drugs such as ketoprofen and diclofenac are poorly eliminated (removal efficiencies lower than 40%). The situation is similar for antibiotics such as azithromycin (removal efficiency of 45%), clarithromycin (removal efficiency of 43%), and erythromycin (removal efficiency of 30%) [12]. The conventional wastewater treatment processes used for the elimination of persistent pollutants are coagulation-flocculation, adsorption, ozonation, and biological treatment. Coagulation-flocculation processes have demonstrated to be ineffective for the elimination of most pharmaceutical micropollutants, ranging from about 0 to 50% of efficiency. Moreover, adsorption on activated carbon (which is usually employed in drinking water treatment) has proved to remove a higher percentage of pharmaceutical pollutants, with removal efficiencies between 20 and 99%, but considerably large equilibrium times [11]. Ozonation has shown high degradation efficiencies, high degradation rates, and non-selectivity. However, secondary reactions have been reported in aqueous media leading to the formation of toxic by-products with a detrimental impact on the environment [13]. Lastly, antibiotics are generally non-biodegradable, which explains the high concentrations of ciprofloxacin and tetracycline found in effluents after biological treatment. In this respect, the presence of antibiotics in water sources can contribute to the development of antimicrobial resistance, which is a global health concern since it is annually responsible for more than 23,000 deaths just in the United States [14].

Recently, nanotechnology has appeared as a promising approach to remove or degrade these pollutants in wastewater, due to the unique structural, physicochemical, and electrochemical properties shown by zero-dimensional (0D) nanoparticles, one-dimensional (1D) nanowires/nanorods, two-dimensional (2D) nanoplates/nanosheets, and three-dimensional (3D) nanostructures, as well as their functional composites [15]. Particularly, 2D nanomaterials present the thinnest structures and largest surface-to-volume ratios, which hold great promise towards faster processing times and higher treatment efficiencies than conventional chemical and biological remediation methods (please refer to Section 2 for more details).

A revision of the application of 2D materials in recent years for the removal or degradation of pharmaceuticals from wastewaters (bibliographic search made using academic databases including articles published until August 2021) suggests that they are mostly used as adsorbents, oxidation catalysts or photo-catalysts, and electrochemical catalysts [16–18].

In this regard, adsorption is one of the most widely used approaches for the removal of contaminants from wastewater due to its simplicity and cost-effectiveness. Conventional adsorbents range from granular powdered activated carbon and kaolin to chitosan, which offer high porosity, large specific surface area, and suitable functionalities for the effective capture of pollutants under optimal operating conditions (adsorbent dose, pollutant initial concentration, temperature, pH, ionic strength, mixing speed, and contact time). Recently, 2D nanomaterials, such as graphene or MXenes, have gained attention as emerging adsorbents due to their unique surface properties compared to conventional bulk adsorbents [19,20]. Particularly for the removal of antibiotics such as, tetracycline, doxycycline, and ciprofloxacin, graphene and its nanocomposites have reached removal ratios of about 100 mg of contaminant per gram of adsorbent, which can be further enhanced by modifying their degree of functionalization, or electric/magnetic properties [21]. MXenes and their nanocomposites have shown comparable removal ratios in the case of metal ions and dyes, which can be tailored by the same routes than graphene-based nanomaterials [22,23].

Heterogeneous photocatalytic processes also attract considerable attention as an effective approach for the degradation of persistent contaminants, since they represent a green route for wastewater treatment based on inexhaustible solar energy as driving force. Particularly, 2D carbon-based nanomaterials, such as graphitic carbon nitride (g-C_3N_4)

and graphene-based nanomaterials, synthesized by various routes and modified by different metals/metal oxides—such as Ag, Au, Ni, Zn, Sn, and Fe—have shown unique photoelectronic properties, high chemical stability, and good light absorption in the visible range (approximately between 450 and 460 nm) [24,25]. This type of catalysts promotes the generation of highly reactive radical species such as hydroxyl radicals ($OH^{\bullet}$), superoxide radicals ($O_2^{\bullet-}$), and holes (h^+) at their surface via photo-assisted water decomposition, which oxidize a large variety of pharmaceuticals such as ciprofloxacin, venlafaxine, sulfamethoxazole, carbamazepine, and dexamethasone. In some cases, pharmaceuticals can be even completely mineralized, as demonstrated at laboratory and pilot scale. Moreover, other type of contaminants such as dyes, petroleum hydrocarbons, and phenolic compounds have been treated with outstanding results [26–28].

Redox and electrochemical processes offer the possibility of degrading a wide range of organic compounds, such as phenols, dyes, pharmaceuticals, and personal care products, via the generation of highly reactive radical species induced through the redox or electrochemical activation of intermediate species (hydrogen peroxide (H_2O_2) or peroxymonosulfate (HSO_5^-)) on the surface of a catalyst or electrocatalyst [29,30]. Fenton and Fenton-like processes are the most widely studied and applied advanced oxidation processes (AOPs) for the degradation of these contaminants due to their simple operation, mild operating conditions, and fast formation rates of $OH^{\bullet}$ radicals compared with other AOPs such as ultrasound, ozonation, and radiolysis. In conventional Fenton processes, the Fe^{3+}/Fe^{2+} redox couple is used as intermediary for the activation of added or in-situ produced H_2O_2. However, recently other metal redox couples such as Co^{3+}/Co^{2+}, Ce^{4+}/Ce^{3+}, or Mo^{6+}/Mo^{4+} have been tested in Fenton-like processes with increased degradation rates when compared to iron [31,32]. The integration of this approach into electrochemical systems and the introduction of 2D nanomaterials for the fabrication of electrodes are expected to allow for successful scaling up of continuous operation processes, since their combination permits the rapid degradation of a diversity of contaminants due to the enhanced oxidation efficiencies established under an applied current or potential [33–36].

This article aims to summarize the key properties of 2D nanomaterials for wastewater treatment (Section 2), and their most recent applications for the removal of pharmaceuticals from wastewater via adsorption (Section 3.1), photocatalytic (Section 3.2), redox, and electrochemical processes (Section 3.3). As a result of a critical analysis of the articles revised the most promising alternatives of 2D nanomaterials for the removal of pharmaceuticals are identified, as well as the challenges they face for their future large-scale fabrication and application.

2. 2D Nanomaterials: General Aspects

Initiated by graphene discovery [37], a one-atom-thick layer of carbon atoms with honeycomb structure, the successful synthesis of 2D nanomaterials rapidly expanded from graphene derivates (graphene oxide (GO) and reduced graphene oxide (rGO)) [38] to transition metal dichalcogenides (TMDs) [39,40] and MXenes [41–44]. Their unique physicochemical, mechanical, electrical, and thermal properties suggest their usage for advanced water purification purposes such as the removal of pharmaceutical contaminants from wastewater [45,46]. In this regard, three approaches including adsorption, photocatalytic, and electrocatalytic degradation (or combinations of them), can be identified. The first approach seems to be interesting since it is easily applicable, cheap, and efficient [47]. Photocatalytic degradation using sunlight as energy source is environmentally friendly, cost-effective and efficient [48–50], while electrochemical processes (particularly combined with photocatalytic degradation) are promising to be up-scaled for continuous operation applications [4,32].

The use of 2D nanomaterials for these purposes is promising since they have a unique property combination involving high surface areas, high thermal and electrical conductivities, and partially an optical transparency. Moreover, the layer thickness of 2D nanomaterials (z-dimension) becomes adjustable by synthesis and/or mechanical/chemical delamination.

Another striking factor of 2D nanomaterials is that they have functional groups on their surface, which makes their surface chemistry intentionally tunable [43,51–53].

Regarding adsorption, single- or few-layered nanosheets provide many active sites, which can be easily accessed and contacted from both sides. Their high specific surface area and pore size are advantageous, thus enabling fast diffusion processes and surface reactions, which result in fast adsorption kinetics. The existing surface terminations on 2D nanomaterials can be used to induce specific interfacial reactions such as hydrogen bonding or π-π interactions. The overall dimensions of the used 2D nanomaterials (reduced z-dimensions) enables a fast mass transfer rate between the adsorbent and the contaminants, which results in fast equilibrium processes [47–49].

However, due to the overall purpose to remove pharmaceuticals from wastewater, the used 2D nanomaterials need to form stable dispersions in water. Their nanoscopic size induces the tendency for agglomeration, thus forming bigger clusters. This reduces not only the specific surface area, but also decreases the number of available active reaction sites and decreases the overall adsorption capacity [54,55]. In this regard, this reagglomeration tendency needs to be suppressed to ensure beneficial adsorption properties over time. The existing surface groups can be used to chemically functionalize the nanomaterials, which can help to reduce their agglomeration tendency and improve their dispersion stability. Besides chemical functionalization, the assembly of 2D and 3D structures including heterojunctions, hybrid nanocomposites, hydro- and aerogels has shown the potential to circumvent the critical agglomeration tendency [54,55]. Furthermore, the combination of different 2D nanomaterials can intentionally manipulate the specific surface area as well as the resulting interaction with the pollutants, thus increasing the respective adsorption capacity and accelerating the underlying kinetics. Similar outcomes can be achieved by doping or defect-engineering of 2D nanomaterials such as the creation of additional vacancies. Another advantage of 2D nanomaterials is that they can be easily combined with other 2D nanomaterials or 1D nanoparticles to create multifunctional hybrids. Frequently, 2D nanomaterials are self-assembled with magnetic Fe_3O_4 nanoparticles without any further crosslinking agent. Besides beneficial adsorption properties, the designed hybrid nanocomposites enable an easy solid-liquid separation, which is essential for their repetitive usage and recycling.

Regarding photocatalytic degradation, the underlying processes rely on the adsorption of light, thus exciting electrons and shifting them from the valence to the conduction band. This process produces the same amount of positively charged holes in the valence band. By adjusting the respective potential differences correctly, radical oxygen species (ROS) such as $OH^\bullet$ and $O_2^{\bullet-}$ can be formed. ROS are molecules consisting of oxygen atoms with unsaturated electron pairs, which are highly reactive and of short lifetime [24,37]. Four important species, namely superoxide anion radicals ($O_2^{\bullet-}$), hydrogen peroxide (H_2O_2), singlet oxygen ($^1O_2$) and hydroxyl radicals ($OH^\bullet$) need to be distinguished. ROS are generated in oxygen- and/or water-containing environments by complex redox reactions, while the necessary respective activation energy typically comes from the absorption of light. Superoxide anion radicals and hydrogen peroxide are generated by stepwise reduction processes or the addition of an electron to the π^* orbitals with subsequent protonation processes [37]. When further reducing hydrogen peroxide, the dissociation of the oxygen bond generates hydroxyl radicals. It is important to mention that the oxidation of water in aqueous environments induces the formation of ROS in the following order: hydroxyl radicals, hydrogen peroxide, superoxide anion radicals and singlet oxygen. In contrast, complex reduction processes of O_2 generate ROS in the respective order of superoxide anion radicals, hydrogen peroxide and hydroxyl radicals [37]. The important influence of ROS regarding photocatalytic properties has been also recently demonstrated for few- and multi-layer MXenes nanosheets. Rosales et al. showed that active TiO_2 sites on the surface of few-layer $Ti_3C_2T_x$ nanosheets, which were verified by XPS, are capable to form electron/hole pairs under UVA irradiation. As aforementioned, the electrons experience a transfer from the conduction band of TiO_2 to the nanosheets surface thus helping to separate the photogenerated carriers [56]. The existing surface terminations and MXenes outstanding electrical conductivity reduce the probability for recombination, thus notably boosting photocatalytic processes. It can be assumed that complex redox reactions (-OH surface terminations and oxygen or water) are likely to induce the generation of ROS. Few-layer $Ti_3C_2T_x$ nanosheets have shown a higher capability to generate ROS due to an increased amount of superficial TiO_2 compared to multi-layer nanosheets [57].

In this regard, the key properties of 2D nanomaterials for the successful usage in photocatalytic degradation are their specific surface area, respective bandgap energy, electron mobility and electrical conductivity [57]. Particular emphasis needs to be put on the respective bandgap engineering to match the respective energy input by visible light with the bandgap of the 2D nanomaterial [58–60]. To manipulate the bandgap and optical transparency, doping and defect engineering (vacancies) of 2D nanomaterials are well-explored approaches. In this context, doping can be realized by non-metal and metal doping with the overall purpose to improve light adsorption. Vacancies can trap electrons or holes thus manipulating the diffusion length of the respective carriers or changing the electronic structure, which potentially increases light adsorption. Electron-hole separation and migration properties can be defined by a specific interfacial design such as homo- or heterojunction interfaces, Schottky interfaces, and double heterojunctions.

In the case of pure or photo-assisted redox and electrochemical processes the same properties benefiting the photocatalytic degradation of contaminants in wastewater (namely, enhanced specific surface area, electron mobility, and charge transfer) explain the promising performance of 2D nanomaterials as oxidation catalysts and electrode materials. Particularly, titanium and carbon-based 2D nanomaterials are prospective options to conventional environmentally unfriendly and toxic catalysts and electrode materials such as cobalt and lead.

3. 2D Nanomaterials for the Removal of Pharmaceuticals in Wastewater

3.1. Adsorption Processes

2D nanomaterials have been frequently used to efficiently remove pharmaceuticals from wastewater via direct adsorption processes on the adsorbent surface suspended in the polluted liquid. In this regard, two main approaches can be identified including the direct use of as-synthesized or as-modified 2D nanomaterials or the application of binary or ternary nanocomposites.

3.1.1. Pure 2D Nanomaterials

In this group, different 2D nanomaterials including MoS_2, boron nitride (BN), GO and $Ti_3C_2T_x$ MXenes have been studied [61]. Yalei et al. fabricated flower-like, wrinkled MoS_2 nanosheets via hydrothermal processing and studied their adsorption capability for diclofenac in aqueous solutions [62]. Hydrothermal processing at 210 °C led to the formation of single-layer flower-like MoS_2 with lateral dimensions of about 1.5 μm and a specific superficial area of about 23.9 m^2 g^{-1}. The sorption behaviour of the as-fabricated flower-like, single-layer MoS_2 was studied in batch experiments depending on the interaction time (2–120 min) and diclofenac concentration (2–200 mg L^{-1}). Compared to commercial MoS_2, the flower-like MoS_2 showed notably improved removal characteristics with a faster overall removal using less material. This behaviour was connected to the structure and surface properties of the fabricated flower-like MoS_2. The maximum, theoretical sorption capacity of the flower-like MoS_2 was about 217 mg g^{-1}. This value is approximately 4 times larger than that of commercially available MoS_2 and outperforms activated carbon and rGO having removal capacities of about 63 and 59 mg g^{-1}, respectively. Regarding the sorption kinetics, a fast process with short reactions times was demonstrated thus adsorbing more than 90% of the contaminant in the first 5 min and reaching equilibrium conditions after 10 min. The experimental data were fitted with a pseudo-second order model, which implies that chemical sorption is the underlying mechanism, while the respective capacity is related to the number of active sites available.

Liu et al. optimized the synthesis of BN nanosheets using boric acid and urea as starting materials to obtain few-layered BN with a high surface area of about 900 m^2 g^{-1} and a pore volume of about 0.76 cm^3 g^{-1} [63]. Few-layer BN nanosheets were used to study their adsorption capability for estrone. The adsorption experiments were conducted in batch experiments dependent on temperature and concentration of BN. In the first 10 min of the adsorption experiments, about 80% of the initial concentration (500 μg L^{-1}) were adsorbed. After a reaction time of 2 h (equilibrium conditions), a contaminant removal of 95% was achieved (Figure 1a,b)). Materials characterization, together with the adsorption experiments, confirmed that thinner nanosheets with fewer layers provided more accessible adsorption sites, thus inducing enhanced adsorption capabilities. The data were best fitted with a pseudo-second order kinetic model and the main mechanism was identified to be chemical adsorption (further confirmed by X-ray photoelectron spectroscopy (XPS) after the adsorption experiments). The adsorption capability depended on the initial concentration of estrone and tended to increase with concentration. Data fitted with a Langmuir model verified a maximum adsorption capacity of 250 mg g^{-1} for estrone. A careful thermodynamic consideration demonstrated that the involved processes are spontaneous with negative ΔG values and entropy-driven.

Chao et al. used diboron trioxide and urea to synthesize BN nanosheets, while KOH was used to controllably induce nitrogen defects in the nanosheets structure [64]. The authors verified by X-ray diffraction (XRD) a defective structure with reduced crystallinity and weakened interlayer interaction after the use of KOH. The specific surface area was determined to be about 533 m^2 g^{-1}, with a pore size of about 2.74 nm for the defective nanosheets. Adsorption experiments for tetracycline verified an 18% increase in the maximum adsorption capacity for the defective nanosheets. The adsorption kinetics demonstrated a fast initial adsorption before reaching equilibrium conditions after 6 h. The experimental data were best fitted with a pseudo-second order model, which implies that strong intermolecular forces controlled the adsorption process. The Freundlich isotherm fitted the kinetic data in the most appropriate way, which pointed towards a multi-layer adsorption behaviour. The introduction of defects in the nanosheets changed their electronic distribution thus enhancing π-π interactions between the nanosheets and tetracycline. Chao et al. compared the adsorption capability of pristine BN with engineered BN having highly active boron vacancies and a hierarchical pore structure [65]. In this context, $ZnCl_2$, which is known to be a pore-forming agent, was utilized during the polymerization of boric oxide B_2O_3 and urea, which led to the formation of few-layered BN nanosheets with embedded Zn nanoparticles. These nanoparticles can be evaporated using high calcination temperatures, thus creating the mentioned boron vacancies and porous structure. This, in turn, helped to manipulate the electronic structure and enriched the nanosheets with active sites. The adsorption properties of these engineered nanosheets for tetracycline were studied in batch experiments as a function of the used concentration and temperature. The specific surface area was shown to increase significantly, thus reaching a value of 1104 m^2 g^{-1} for the engineered nanosheets, which may be beneficial for the involved mass transfer in adsorption processes. The adsorption capability of the created nanomaterial was determined to be about 438 mg g^{-1}, which reflects an increase by 38% compared to pristine BN. This clearly demonstrated that the precise addition of boron vacancies and creation of a porous structure helped to enhance the adsorption ability. In this regard, boron vacancies were made responsible to manipulate the electronic structure and create more active reaction sites, while the porous structure greatly increased the available surface area, both beneficially contributing towards enhanced adsorption characteristics. The obtained data were best fitted with a pseudo-second order model, which implies that the underlying mechanism relates to chemical adsorption. The authors speculated that intermolecular forces are important for the adsorption process and that the amount of adsorbed material correlated with the available active sites. The sorption isotherms were fitted with the Freundlich model, which is representative for multi-layer sorption on heterogeneous surfaces, which matches well with the characteristics of the few-layer, engineered BN. Adsorption, which was spontaneous in all cases considered, was more favourable at lower temperatures and showed negative ΔH values (exothermic reaction). Related to the respective conditions, the adsorption capability for tetracycline was shown to be pH-dependent, with an improved performance for the acidic regimes. The coexistence of other ions (especially, PO_4^{3-} and SO_4^{2-}) in solution showed an inhibition effect with decreased adsorption capabilities of about 78%. This demonstrated that electrostatic interactions are important in the overall adsorption process.

Few-layer g-C_3N_4 nanosheets were synthesized in a simple two-step thermal process based upon different precursors such as thiourea, dicyanamide or melamine to study their adsorption and photocatalytic activity towards tetracycline removal [66]. The produced nanosheets have an average thickness of 10 nm with a hierarchical morphology. Adsorption batch experiments to study the removal of tetracycline were conducted as a function of the nanomaterial loading, while keeping the temperature and pH constant throughout all experiments. g-C_3N_4 nanosheets adsorbed about 41% of the tetracycline in solution, which is 10 times higher than the value for conventional nanomaterials. Besides efficient adsorption capabilities, the nanomaterials showed excellent photocatalytic properties with about 38% of tetracycline being degraded under illumination (further revision of this sort of process is presented in Section 3.2). Adsorption and photocatalytic activity were shown to be stable over time with insignificant changes after three complete cycles. The beneficial properties of g-C_3N_4 nanosheets were related to their increased specific surface area, which increased the contact area for the adsorption of tetracycline molecules. Moreover, XPS reveals a significant content of sp^2-hybridized carbon on the surface of the nanomaterials, which implies the existence of superficial delocalized π bonds. Consequently, more tetracycline molecules were able to interact with the nanosheets surface via π-π interactions. The kinetic data of the tetracycline adsorption demonstrated a maximum adsorbed quantity under equilibrium conditions of about 40 mg g^{-1}. The kinetic data were best fitted with a pseudo-second order model, which implies that chemical adsorption is the underlying mechanism.

Zhao et al. used a microwave-assisted hydrothermal approach to synthesize 2D copper-meso-tetra(4-carboxylphenyl)porphine (Cu-TCPP) and studied their adsorption behaviour as well as photocatalytic activity for different dyes and conventional antibiotics [67]. The synthesized Cu-TCPP nanosheets showed an average size of about 500 nm, a high crystallinity, a specific surface area of 342 m^2 g^{-1}, and an average pore size of about 12 nm. The respective adsorption behaviour was studied for three different antibiotics (oxytocin, tetracycline, and norfloxacin) as a function of the adjusted pH. For oxytocin, a maximum removal rate of 47% was shown for a pH equal to 9, while the highest removal efficiency for tetracycline (72%) was found for a pH equal to 5. The adsorption kinetics verified that equilibrium conditions were reached quickly, and the overall data followed a pseudo-second order model. The maximum adsorption capacities for oxytocin and tetracycline were determined to be around 130 mg g^{-1} and 150 mg g^{-1} under equilibrium conditions. The data were fitted by Langmuir isotherms, which implies that the underlying processes for the adsorption of different antibiotics rely on the existence of different functional groups on the surface of Cu-TCPP.

Tang et al. synthesized two fluorescent, chemically and thermally stable covalent triazine frameworks via condensation reactions using aldehydes or amidines thus creating nanosheets with specific surface areas of 565 m^2 g^{-1} and an average pore size of 1.5 nm (Figure 1c,d) [68]. The nanosheets showed pronounced adsorption capabilities for different antibiotics, including nitrofurazone, nitrofurantoin, and furazolidone. The maximum adsorption capacity of 351 mg g^{-1} was found for nitrofurazone. The kinetic data were best fitted with the Langmuir model, which implies that the underlying mechanism is based upon monolayer adsorption. After having reached equilibrium conditions, the nanosheets were capable to adsorb more than 99% of the different antibiotics (Figure 1e). The underlying kinetics followed a pseudo-second order kinetic model, while the adsorption capacity correlated with the respective pore size and specific surface area.

Jiang et al. fabricated few-layered, zwitterionic covalent organic frameworks (5–7 layers) with sulfonic acids and amines evenly distributed over the nanosheets surface via solvothermal processing (specific surface area of around 187 m^2 g^{-1}) (Figure 1f) [69]. The existing functional groups induced some repulsive interactions between the layers, thus leading to self-exfoliation, which increased the number of active sites available. The adsorption kinetics were best fitted with a pseudo-second order model. For norfloxacin, an ultrafast adsorption with a 99.4% removal after 30 s was verified (Figure 1g). The speed was explained by nanochannels and zwitterionic sites available. The Langmuir isotherm fitting represented the kinetic data best, which implies that the underlying mechanisms can be best described as monolayer adsorption. The maximum adsorption capacity for norfloxacin was determined to be 824 mg g^{-1}, which was explained by the improved accessibility of active sites and their zwitterionic character.

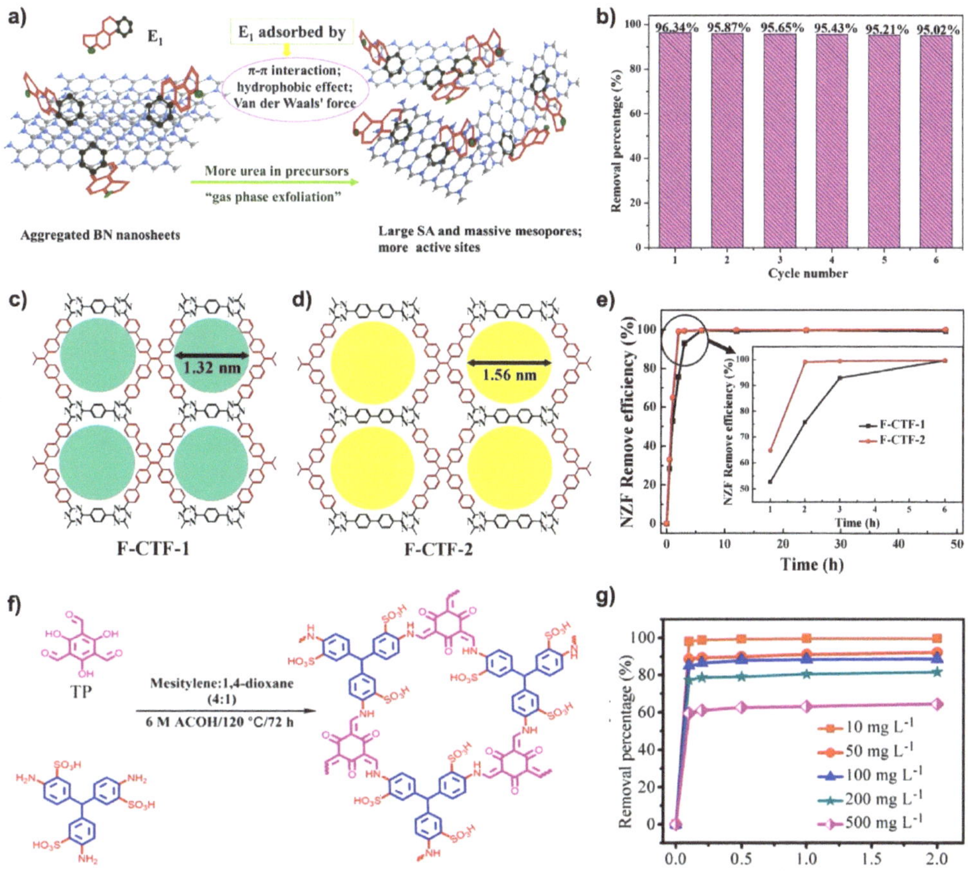

Figure 1. As-synthesized and as-modified 2D nanomaterials and their removal efficiency: (**a**,**b**) BN nanosheets and estrone removal after several cycles (reproduced with permission [63]), (**c**–**e**) 2D porous covalent triazine framework nanosheets and nitrofurazone removal by adsorbents synthesized from different precursors (reproduced with permission [68]), and (**f**,**g**) zwitterionic covalent organic frameworks and norfloxacin removal at different initial concentrations of the contaminant (reproduced with permission [69]).

Ghani et al. used sodium-intercalated, multi-layer $Ti_3C_2T_x$ nanosheets to study their adsorption behaviour and electrochemical regeneration for the removal of ciprofloxacin [70].

The sodium-intercalated nanosheets were synthesized by simply immersing multi-layer MXenes in a NaOH solution, which led to a 25% increase of the d-spacing due to the intercalated ions. The increased d-spacing was speculated to be essential for the adsorption capability due to easier accessibility for the active sites. Compared to the pristine MXenes, the sodium-intercalated MXenes showed a 2-fold enhancement in the adsorption capability. Since the superficial area was constant for both nanosheets, the improved adsorption behaviour was traced back to the increased d-spacing and more accessible -O and -OH groups of the sodium-intercalated nanosheets. The highest removal rate for ciprofloxacin was found for a pH equal to 6, while pH values above and below led to a considerable decrease of the removal. This can be well connected with the electronic states of ciprofloxacin, which largely depend on the respective pH of the solution. Kinetic data showed the adsorption of more than 80% of ciprofloxacin in the first 30 s of the experiment resembling superfast diffusion kinetics. The kinetic data were best fitted with the Elovich model, which determines chemisorption processes as the dominant processes. The post-characterization of the nanosheets after the adsorption experiments led to the assumption that the electrostatic interaction between the positively charged surface of ciprofloxacin (pH around 5.5–6) and negatively charged surface terminations may be the main underlying reason. Moreover, the successful electrochemical regeneration of the sodium-intercalated MXenes was demonstrated by immersing the loaded nanosheets in a NaOH solution.

Figure 1 summarizes the structures and performances of the pure 2D nanomaterials with the highest adsorption removal efficiencies reported, which correspond to: few-layered BN nanosheets, covalent triazine frameworks, and zwitterionic covalent organic frameworks.

3.1.2. Binary and Ternary Nanocomposites

2D nanomaterials such as MoS_2 or GO are frequently combined with ferritic Fe_3O_4 nanoparticles to create hybrid nanomaterials with a certain adsorption capability for pharmaceuticals. Wu et al. used solvothermal synthesis to create magnetic rGO/ferrite (Fe_3O_4) hybrids to study their removal performance for antibiotics from wastewater [71]. In this context, GO, Fe(III) and hydrazine hydrate (reduction of GO) were mixed under alkaline conditions and temperatures below 100 °C thus anchoring Fe_3O_4 nanoparticles on rGO. The presence of Fe_3O_4 in the nanocomposite made it magnetic, which is beneficial for the subsequent separation after wastewater treatment, while the remaining surface groups on rGO ensured a good dispersion in water. Extraction experiments for different sulfonamide antibiotics verified a concentration-dependent behaviour with increased recovery for higher amounts of nanohybrids used. The removal depended on the adjusted pH with the highest values obtained for a pH equal to 6. The excellent extraction performance of rGO/Fe_3O_4 hybrids, which outperformed rGO, Fe_3O_4 and GO/MWCNT hybrids, was traced back to the enlarged surface area of rGO and the pronounced π-π interactions between the nanohybrid and the contaminant.

Li et al. fabricated GO via a modified Hummers approach before using co-precipitation to generate magnetic GO (Fe_3O_4-GO) [72]. Subsequently, magnetic GO was surface grafted with nitrilotriacetic acid (NTA) to create magnetic, functionalized Fe_3O_4/GO nanocomposites having a specific surface area of 249 $m^2\ g^{-1}$ and an average pore diameter of 4.3 nm. The adsorption capacity of these nanocomposites for tetracycline was studied as a function of the pH, time, temperature, concentration, and ionic strength. Related to the pH value, the highest adsorption capability was verified for a pH equal to 4 with a maximum adsorption capacity of 212 mg g^{-1}. The enhanced adsorption capacity of the nanocomposite was traced back to the ionic characteristics of tetracycline at the respective pH and the available surface groups on the nanocomposite. Although showing a fast initial uptake in the first 20 min, equilibrium conditions were just reached after 600 min. The data were best fitted by a pseudo-second order model, which implies that the rate-limiting step is related to chemisorption. The adsorption mechanism was mainly related to the amino groups in the tetracycline molecule, which interacted with the π electrons of the nanocomposite as well as hydrogen bonds. Also, Li et al. fabricated homogenous Fe_3O_4/GO nanocomposites with different Fe_3O_4/GO ratios between 2.5 and 20 to study their adsorption capability for chlorpheniramine [73]. The kinetics of the chlorpheniramine adsorption showed a high initial rate, which implies a fast removal. The experimental results were fitted with a pseudo second-order model, which inferred that the adsorption occurred via chemical adsorption. Regarding the effect of the Fe_3O_4/GO ratio, an increased amount of Fe_3O_4 reduced the number of adsorption sites on GO thus decreasing the total amount of adsorbed chlorpheniramine from 300 to 138 mg g^{-1}. The authors demonstrated that the micropore volume correlated with the adsorbed chlorpheniramine concentrations, which held especially true for low Fe_3O_4/GO ratios. The adsorption behaviour was also shown to be pH dependent with the greatest removal for a pH equal to 8. This effect related to the stronger electrostatic interaction between the positive chlorpheniramine and the negative nanocomposite.

Xiao et al. fabricated ternary nanocomposites consisting of GO, MoS_2 and Fe_3O_4 nanoparticles [74]. In this regard, GO was synthesized via a modified Hummers process before producing GO-Fe_3O_4 nanocomposites via chemical solvothermal processing. The obtained nanocomposites were solution-processed to obtain ternary nanocomposites by precipitation. The fabricated ternary nanocomposites had a specific surface area of 135.14 m^2 g^{-1}, which was significantly higher than the values of the pure nanomaterials or the binary nanocomposites. The adsorption behaviour of these ternary nanocomposites was studied for different antibiotics including lecofloxacin, pazcofloxacin, and gatifloxacin. The ternary nanocomposite containing 50% MoS_2 showed the best adsorption capability for all antibiotics, thus outperforming the pure nanomaterials and binary nanocomposites. The ternary nanocomposite showed a fast adsorption behaviour and reached its maximum adsorption capacity after only 2 min. Regarding the influence of the pH, the highest adsorption was verified for a pH equal to 7, which related to the available surface terminations on GO. The underlying mechanism for the enhanced adsorption observed for the ternary nanocomposite was related to enhanced hydrogen bonding and electrostatic interactions.

Park et al. generated ternary nanocomposites consisting of supports of rGO decorated with silver and Fe_3O_4 nanoparticles via a simple nucleation and crystallization process [75]. The overall purpose of this nanocomposite was to activate peroxydisulfate (PDS), thus producing $SO_4^{\bullet-}$ radicals which can catalytically remove phenol (further revision of these processes is presented in Section 3.3.1). This, in turn, was utilized to study the degradation and catalytical removal of different pharmaceutics including acetaminophen, ibuprofen, naproxen, among others. To characterize the underlying adsorption characteristics of these ternary nanocomposites, pH-dependent batch experiments were conducted. Reference experiments verified that PDS and H_2O_2 were not capable to degrade phenol. The highest removal efficiency was shown for the ternary nanocomposite (with binary composites and pure materials as references) with k_{ini} values of 0.67, 0.46, and 0.23 h^{-1} for pH 4, 7, and 10, respectively. This effect was traced back to an increased adsorption of phenol and enhanced catalytic activity of PDS, which induced synergistic effects. The pH-dependency was explained based upon a competing reaction between $SO_4^{\bullet-}$ and $OH^{\bullet}$ radicals, which may be converted into each other depending on the predominant pH milieu. The increased k_{ini} value for a pH equal to 4 verifies the dominant role of $SO_4^{\bullet-}$ radicals in the degradation of phenol under the conditions studied. The kinetic data were fitted with a pseudo-second order model and demonstrated the greatest adsorption capacity for the ternary nanocomposites. The beneficial behaviour induced by the ternary nanocomposites was explained by the fact that the deposition of silver and Fe_3O_4 nanoparticles on rGO helped to reduce the nanosheets agglomeration. This made more reactive sites available for the adsorption of phenol. Moreover, more epoxy and hydroxyl groups were verified by XPS for the ternary nanocomposites, which can act as Lewis base sites with the possibility to chemically bond with phenolic hydroxyl groups. Subsequently, the authors studied the degradation kinetics for selective organic compounds. In the case of a pH equal to 4, a degradation of more than 90% in one hour reaction time was verified for acetaminophen, ibuprofen, 17β-estradiol, and 17α-ethinyl estradiol, while a 50–90% degradation was shown for naproxen and bisphenol A. The observed results imply that non-radical oxidation plays a dominant role in the PDS/nanocomposite system.

Zeng et al. generated via hydrothermal synthesis biochar decorated with g-MoS_2 nanosheets for the removal of tetracycline [76]. Nitrogen adsorption/desorption experiments revealed for g-MoS_2-biochar composites a specific surface area of 176.8 m^2 g^{-1} and an averaged pore diameter of about 3 nm. The produced nanocomposites exhibited a good thermal stability up to 450 °C as verified by TGA. Adsorption experiments as a function of the pH value revealed an increase of the adsorption capacity when pH increases from 2 to 4. Afterwards, it remained rather stable before dropping for pH values above 8. The beneficial effect for low pH values was correlated with the present surface charges and reduced repulsive interactions between the adsorbent and tetracycline, thus favouring adsorption. Surface charges may not be responsible for the adsorption capacity observed for medium pH values as there is only little electrostatic interaction. In this regime, it is more likely that hydrogen bonds, π-π or hydrophobic interactions contribute towards the beneficial behaviour observed. Additional ions present in the solution tended to induce a downgraded adsorption behaviour due to reduced electrostatic interaction between tetracycline and g-MoS_2. A detailed consideration of the kinetic data suggests a three-step process for the adsorption of tetracycline onto g-MoS_2 biochar composites. In an initial fast adsorption step (more than 80% adsorption in the first 4 h), the tetracycline molecules overcome the liquid resistance and are transferred to the exterior surface of the composite, which is governed by film diffusion. Subsequently, the molecules migrate from the surface to pores (intraparticle diffusion) before reaching the active sites on the interior of the composites thus being adsorbed (equilibrium conditions). Like the multiple diffusion steps necessary to reach equilibrium conditions, the underlying mechanism needs to be subdivided in different contributions including hydrogen bonding, π-π interactions, pore filling and electrostatic interactions thus inducing the beneficial adsorption ability of g-MoS_2 biochar nanocomposites for tetracycline.

González et al. generated chitin/GO nanocomposites (3:1 weight ratio) by wet-spinning to study their adsorption performance for ciprofloxacin in batch experiments by assessing the effect of the flow rate, pH value and sodium as well as calcium concentrations (water quality parameters) [77]. Scanning electron micrographs of the as-synthesized nanocomposites revealed a very broad pore size distribution with micro-, sub-micro- and nanoscopic pore sizes ranging from 10 nm up to 10 μm. A detailed analysis of the adsorption characteristics demonstrated a rather heterogeneous sorption, which may be correlated to the molecular structure of ciprofloxacin having three nitrogen atoms and two unsaturated rings. The existence of these structural features makes the interaction with functional groups on the GO nanosheets possible (electrostatic or π-π interactions). The dynamic adsorption behaviour was shown to be sensitive towards the pH adjusted, which is triggered by the respective surface charge of different surface features of ciprofloxacin. These surface charges can induce attractive or repulsive interactions with the surface terminations of GO, thus determining the adsorption behaviour.

3.1.3. Conclusions on Adsorption Processes

Adsorption is a conventional wastewater treatment process for which the use of 2D nanomaterials represents an innovation due to the enhanced surface properties they offer. In principle, their unprecedent surface area and porosity (such as in the case of MoS_2 nanosheets and covalent triazine frameworks) would explain their remarkable removal efficiencies and kinetics due to a higher amount of accessible adsorption sites as compared to conventional adsorbents (such as activated carbon). However, it has been observed that the generation of superficial delocalized π bonds during synthesis is essential for the enhancement of the adsorbent-adsorbate interaction due to the formation of strong π-π interactions (chemisorption). Furthermore, the weakening of the interlayer interaction in 2d nanomaterials such as g-C_3N_4 nanosheets and MXenes also contributes to strengthen the electrostatic surface-contaminant interaction due to the increased amount of superficial

-O and -OH groups accessible. Therefore, future synthesis approaches should be designed to produce surfaces with well distributed and accessible active sites of these types.

Enlarged surface area, π-π interactions and electrostatic interactions can also be induced by the usage of binary or ternary composites instead of pure 2D nanomaterials. In fact, rGO/Fe_3O_4 nanoparticles showed a pronounced π-π interaction with contaminants when compared to both nanomaterials separately, while $MoS_2/GO/Fe_3O_4$ composites presented an enhanced H bonding with the adsorbate and electrostatic interactions when compared to the pure nanomaterials or the binary nanocomposites. Moreover, g-MoS_2/biochar composites showed a mixture of these three adsorbate-adsorbent interactions although their kinetics were reduced when compared to rGO/Fe_3O_4 and $MoS_2/GO/Fe_3O_4$. Therefore, another interesting future research area is the development of new composites based on 2D nanomaterials.

Finally, it is worth to discuss about the recyclability of adsorbents since it is a key aspect to be considered when determining its potential for application at large scale. The regeneration of MXene-based materials is usually performed by treatment with HCl, HNO_3, or $Ca(NO_3)_2$. The regeneration mechanism is based upon the competition between the cations and the adsorbed pollutants at the surface-active sites [19,78]. In this regard, Khan et al. reported a sustained pollutant removal efficiency (over 90%) after five regeneration cycles and no significant changes in morphology [79]. Acid or alkaline solutions are used to separate ions from GO and rGO, where the pH of the solution plays an important role in the regeneration mechanism. Oxygen-containing functional groups on GO and rGO display different protonation at different pH, thus the ionization of pollutants can be easily modified by the pH in solution [21]. Kexin et al. studied the regeneration of GO/NiO nanocomposites loaded with Cr(VI) using a 0.1 M HCl solution. In their experiments, it was found that the uptake capacity slightly decreased with increasing regeneration cycles, demonstrating a steady removal efficiency (around 83%) after five cycles. Additionally, hydrophobic solutions such as ethanol and methanol can act as effective eluents for the removal of organic molecules from adsorbents. Soleimani et al. regenerated GO-cellulose nanowhiskers using ethanol, with a reduction of the efficiency about 5–10% after 3 cycles [80].

3.2. Photocatalytic Processes

Photocatalytic degradation processes are heterogeneous catalytic reactions, which typically comprise five steps: (i) mass transfer of pollutants from the bulk to the surface of the photocatalyst, (ii) adsorption of the molecules onto its surface, (iii) photocatalytic degradation reaction, (iv) desorption of products, and (v) removal of reaction product from the interphase. The working principle of photocatalytic degradation reactions by a semiconductor in an aqueous medium is based on its band structure. A semiconductor material is capable of absorbing radiation with an energy ($h\nu$) equal to or greater than its bandgap. The photons absorbed by the semiconductor induce the promotion of electrons from the valence band to its conduction band, generating holes (h^+) in the valence band. These charge carriers can either directly react with pollutants molecules adsorbed on the surface of the material or with water and oxygen molecules to form ROS. Figure 2 illustrates the possible reaction pathways that can take place after the light excitation of a photocatalyst.

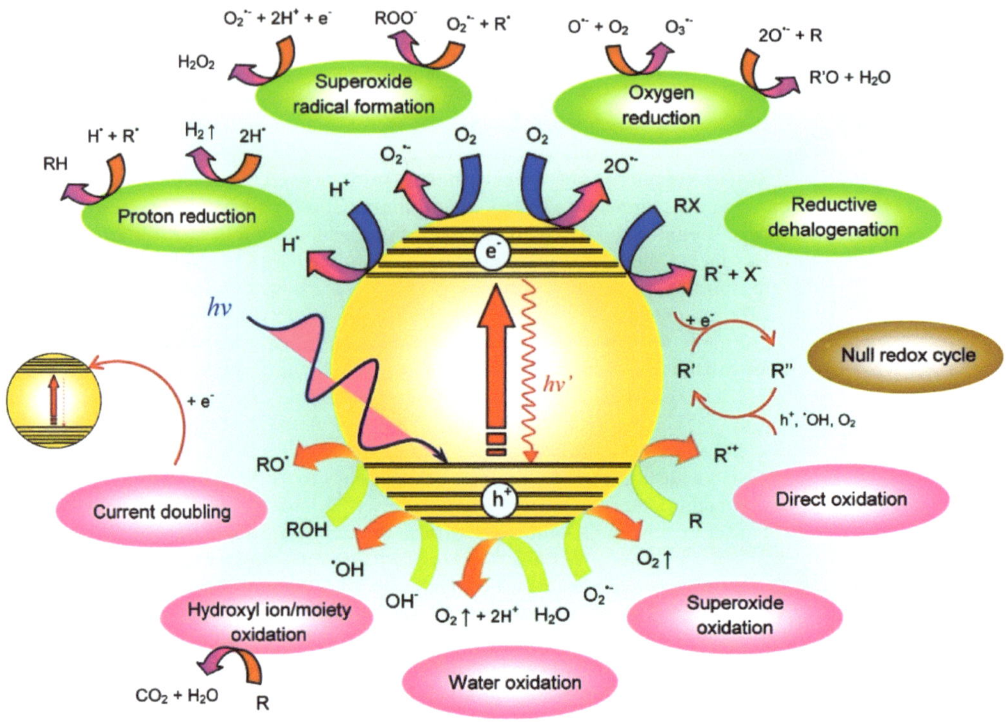

Figure 2. Possible reaction pathways arising from the excitation of photocatalysts. R denotes the general organic electron donor molecule (reproduced with permission [81]).

The efficiency of a photocatalytic degradation reaction depends on factors such as photon absorption, and photo-induced charge carrier concentration and separation. Charge carriers have the potential to recombine in the material bulk or at the surface, thus decreasing their concentration in the material and limiting the degradation efficiency. This recombination processes can occur due to a small bandgap energy, low electronic conductivity, or a high charge transfer resistance at the surface [82].

3.2.1. Graphitic Carbon Nitride-Based Materials

Graphitic carbon nitride (g-C_3N_4) has emerged as a new photocatalyst due to its bandgap energy of 2.7 eV, which can effectively absorb solar irradiation [24], and its excellent physicochemical stability. g-C_3N_4 is a metal-free semiconductor with a 2D layered structure with graphitic planes constructed from tri-*s*-trizazine units connected by planar amino groups, which can be synthesized via an one-step polymerization reaction using precursors such as cyanamide, urea, thiourea, melamine, or dicyandiamide [83]. The band structure of g-C_3N_4 consists of *p* orbitals from nitrogen forming the valence band, while the conduction band is mainly composed of the p_z orbitals of carbon [84]. Although g-C_3N_4 presents light absorption in the visible light spectrum, its performance as a photocatalyst is limited due to a reduced charge migration and a low electronic conductivity, which cause a fast recombination of the photogenerated charge carriers [24,85]. Recently, several strategies have been employed to enhance the photocatalytic performance of g-C_3N_4 including surface modification, doping, and the formation of heterostructures.

The photocatalytic activity strongly depends on the morphology, crystalline structure, and surface composition of the catalyst. This implies that materials with similar chemical composition can yield drastically different photocatalytic activity. Shi et al. prepared a supramolecular carbon nitride (MCU-C_3N_4) by dissolving melamine, cyanuric acid, and urea in dimethyl sulfoxide (DMSO), followed by a thermal treatment at 550 °C [86]. The resulting material showed a 2D morphology with a smooth surface. It was observed that after the addition of cyanuric acid and urea, a thin nanosheet structure was obtained with randomly distributed mesopores (average pore size of 6.48 nm). The bandgap energy was estimated from UV-vis diffuse reflectance spectra (UV-vis DRS) and compared with the value of pure g-C_3N_4, which was synthesized only from melamine. The bandgap energies were estimated to be 2.50 and 2.79 eV for g-C_3N_4 and MCU-C_3N_4, respectively. An increased bandgap energy and a blue-shift in MCU-C_3N_4 absorption spectra were observed, which was ascribed to the quantum confinement effect derived from the smaller size and mesoporous structures of MCU-C_3N_4. The photocatalytic activity of MCU-C_3N_4 was evaluated by the degradation of 10 mg L^{-1} of tetracycline and ciprofloxacin under visible light irradiation (420 nm cut-off filter), with a catalyst load of 0.7 g L^{-1}. MCU-C_3N_4 showed a degradation efficiency of 81.6% (80 min) and 65.80% (90 min) for tetracycline and ciprofloxacin, respectively. The reported degradation mechanism demonstrates the direct oxidation of the antibiotic molecules by h^+ and $O_2^{\bullet -}$ in the bulk.

Liu et al. demonstrated the degradation of sulfadiazine, a sulfonamide widely used as antibiotic in veterinary medicine, using a catalyst consisting of carbon nitride nanosheets with carbon vacancies (Cv-CNNs) synthesized by mixing pure carbon nitride nanosheets (CNNs) with NH_4Cl and a subsequent thermal treatment at 550 °C for 2 h [87]. The fabricated catalyst exhibited a thin few-layered sheet structure with in-plane pores of 5–7 nm. UV-vis DRS analysis revealed that Cv-CNNs presented a bandgap of 2.83 eV, slightly higher than that of the pristine CNNs (2.81 eV). Degradation experiments were performed under visible light (420 nm cut-off filter) with a catalyst load of 0.02 g L^{-1} and a sulfadiazine concentration of 10 mg L^{-1}. Sulfadiazine was completely degraded by Cv-CNNs in 90 min, in contrast to CNNs that showed an efficiency of 64.94%. This increased photocatalytic activity was attributed to an improved charge separation in Cv-CNNs, which was confirmed by photoluminescence. Furthermore, it has been reported that the formation of C vacancies led to a conversion of N exposed atoms into -NH_2 groups, which can act as hole-stabilizers and increase the charge separation in the material.

The incorporation of a different element into a semiconductor structure can tailor the electronic and optical properties and, therefore, the material photocatalytic activity [88]. A recent work by Guo et al. showcases the effects of oxygen substitution on the resulting photocatalytic performance of g-C_3N_4 [89]. Oxygen-substituted ultrathin porous g-C_3N_4 nanosheets (OCN-24-550) were prepared by a two-step hydrothermal-recalcination treatment of bulk g-C_3N_4 at 160 °C in deionized water for 24 h. The produced catalyst presented a sheet-like structure with a planar size of 2.2 μm, an average thickness of 6.8 nm with many in-plane pores, and O atoms partially incorporated into the lattice by substituting the C−N=C coordinated nitride atom. The optical bandgap was determined to be 2.54 and 2.49 eV for OCN-24-550 and pure g-C_3N_4, respectively. The photocatalytic performance of these nanosheets was studied under visible light irradiation (420 nm cut-off filter) in a 20 mg L^{-1} tetracycline solution with a catalyst load of 1 g L^{-1}. OCN-24-550 efficiently degraded 85.76% of tetracycline within 120 min, while pristine g-C_3N_4 only degraded 17.19%. The enhanced photocatalytic activity was ascribed to the synergetic effects of a larger surface area and the O substitution. This facilitated the diffusion of reactants and optimized the electronic structure by creating a local electronic polarization effect resulting in a boosted exciton dissociation and improved charge migration.

Wenjun et al. proposed an alkali metal-assisted synthesis method for the fabrication of g-C_3N_4 nanosheets with a tuneable bandgap [90]. In this regard, a novel potassium-doped porous ultrathin g-C_3N_4 (KMCN(0.05)) photocatalyst was synthesized by hydrothermal treatment of ultrathin g-C_3N_4 nanosheets in the presence of KOH (0.05 M), followed by a recalcination treatment at 550 °C for 4 h. The as-synthesized catalyst exhibited a nanosheet morphology with micro-pores having sizes between 5 and 40 nm. Moreover, the addition of potassium cations (K^+) into g-C_3N_4 nanosheets increased the pore size and specific surface area by 1.4 and 3.9 times, respectively. The calculated bandgap energy of g-C_3N_4 and KMCN(0.05) was 2.65 and 2.33 eV, respectively. The photocatalytic activity of KMCN(0.05) was evaluated under visible light irradiation using tetracycline (20 mg L^{-1}) as target pollutant and a catalyst load of 1 g L^{-1}. KMCN(0.05) demonstrated an excellent tetracycline degradation capability, exhibiting a degradation efficiency of 85.13%, while pure g-C_3N_4 showed an efficiency of only 29.60%. The authors attributed the improved photocatalytic activity to a lower charge transfer resistance, and a reduced recombination rate of the photogenerated charge carriers.

Yi et al. constructed a binary photocatalyst using few-layer $Ti_3C_2T_x$ embedded into alkalized C_3N_4 [91]. The catalyst was prepared via co-calcinating urea and $Ti_3C_2T_x$ precursors in the presence of alkali. The resulting material was composed of $Ti_3C_2T_x$ encapsulated and well-blended in alkalized C_3N_4 nanosheets. Due to the generation of NH_3 gas during urea polymerization, the etched $Ti_3C_2T_x$ lamellas were exfoliated into thinner lamellas and the interlayer spacing expanded from 1.017 to 1.249 nm. The bandgap energy of alkalized C_3N_4 was estimated to be 2.39 eV. The photocatalytic performance of the photocatalyst was tested through the degradation of tetracycline under visible light irradiation (420 nm cut-off filter) with a concentration of 20 mg L^{-1} of tetracycline and 0.2 g L^{-1} of catalyst. Alkalized C_3N_4 showed degradation efficiency of 77%, which was significantly higher compared to pure C_3N_4 (10%). The authors associated this enhanced photocatalytic activity with the formation of a Schottky junction, which prevented charge recombination of charge carriers with $Ti_3C_2T_x$ acting as an electron acceptor. When reusing the catalyst for 4 times, no changes were observed between each cycle of tetracycline removal. Additionally, radical scavenging experiments indicated that h^+ was the most significant active species during the degradation of tetracycline.

3.2.2. Bismuth Oxyhalides-Based Materials

In recent decades, bismuth oxyhalides (BiOX with X = Cl, Br, I) have gained great attention due to their photocatalytic activity towards the removal of contaminants in wastewaters. BiOX possess a layered structure composed of a $[Bi_2O_2]^{2+}$ layer and two halogen ions layers. BiOX photocatalysts are chemically inert, non-hazardous, and corrosion-resistant in aqueous environments. These catalysts absorb visible light radiation having bandgap energies of 3.22, 2.64, and 1.77, for BiOCl, BiOBr, and BiOI, respectively [92]. The valence band of BiOX crystals is composed of p orbitals of X, while the conduction band is comprised of Bi 6p orbitals [93]. Previous studies have reported the degradation of several pharmaceutical molecules using BiOX materials such as 171α-ethinyl estradiol, estriol, carbamazepine, and levofloxacin [92,94–96]. However, the degradation performance of BiOX is still restricted by a poor light harvesting ability and a reduced charge transfer efficiency related to its low conductivity.

Yingying et al. reported the synthesis of BiOBr nanosheets employing L-cysteine as a directing agent (Cyst-BiOBr) by a hydrothermal method at 180 °C for 24 h. [97]. Cyst-BiOBr nanosheets showed an even distribution with a size range of 0.05 to 0.5 μm, an interplanar lattice distance of 0.277 nm, a high surface area and oriented growth in the (110) direction. It was found that Cyst-BiOBr exhibited a bandgap energy of 2.30 eV, which was smaller compared to the bandgap energy of BiOBr nanosheets synthesized using the amino acids arginine and glycine (2.51 and 2.62 eV, respectively). The reduced bandgap of Cyst-BiOBr was attributed to its smaller particle size, which altered the band structure. The introduction of cysteine into the precursor solution promoted the complexation of Bi^{3+} with the amino acid, which controlled the release of these ions to form BiOBr crystallites when adding Br^- to the solution. The evaluation of the photocatalytic activity of Cyst-BiOBr was performed through the degradation of cefazolin, a first-generation cephalosporin, under visible light irradiation (420 nm cut-off filter) with a catalyst load of 0.2 g L^{-1} and a cefazolin concentration of 10 mg L^{-1}. After 3 h of irradiation, a degradation efficiency of 87% was attained.

The studies of Chu-Ya et al. provided an insight on the impact of thickness on the photocatalytic activity of bismuth oxybromide $Bi_{24}O_{31}Br_{10}$ nanosheets (BOB) [98]. These nanosheets were synthesized via hydrothermal method at 160 °C for 12 h, controlling the thickness by adjusting the concentration of NH_4Br during synthesis. The resulting material showed a uniform 2D sheet structure, with nanosheets size between 0.5 and 1.0 μm and different thicknesses, noted as BOB-S (40 nm), BOB-M (85 nm) and BOB-L (130 nm). The estimated optical bandgap energies were 2.30, 2.33, and 2.35 eV for BOB-S, BOB-M, and BOB-L, respectively. The photocatalytic performance was tested by the degradation of tetracycline under visible light (420 nm cut-off filter) with a catalyst load of 0.34 g L^{-1} and 10 mg L^{-1} of tetracycline. After a reaction time of 90 min, a degradation efficiency of 95, 70, and 45% was achieved for BOB-S, BOB-M, and BOB-L, respectively. The increment in the degradation rate for BOB-S was ascribed to its higher charge transfer efficiency, which was supported by electrochemical impedance spectroscopy analyses, and a higher concentration of h^+ due to an increased valence band potential with an energy top of −0.06 eV (in contrast to −0.51 and −0.42 eV for BOB-M and BOB-S, respectively).

Wang et al. synthesized $BiOCl_{0.75}I_{0.25}$ nanospheres via precipitation method, followed by calcination at 200 °C for 2 h. $BiOCl_{0.75}I_{0.25}$ showed a flower-like hierarchical nanosphere structure consisting in globular nanosheets about 0.5 μm in size [99]. Compared to the pure materials, $BiOCl_{0.75}I_{0.25}$ exhibited a smaller size and a bandgap energy of 2.12 eV, which was lower than BiOCl (3.31 eV) but higher compared to BiOI (1.73 eV). The degradation of acetaminophen, an analgesic and antipyretic drug, was performed under visible light employing a reactor with a catalyst load of 0.6 g L^{-1} and acetaminophen concentration of 20 mg L^{-1}. After 3 h of constant stirring, a degradation efficiency of 80% was accomplished, which was superior to that obtained for pure BiOCl (34%) and BiOI (54%). The increase was attributed to a greater generation of $O_2^{\bullet-}$.

The photocatalytic performance of a semiconductor can also be enhanced by creating heterostructures. Heterostructures are composed of combinations of different semiconductors in a multilayer arrangement, which can induce an efficient charge separation, enhance light absorption, and generate band alignments that favour the separation of electrons and holes [82]. In this regard, novel Bi_5O_7Br/BiOBr 2D/3D heterojunctions were prepared by Lulu et al. by a one-step hydrolysis [100]. The fabricated material displayed a morphology consisting of Bi_5O_7Br nanosheets well dispersed on the surface of BiOBr nanospheres. The photocatalytic activity was studied by the degradation of carbamazepine under visible light irradiation using a catalyst load of 0.5 mg mL^{-1} and a carbamazepine concentration of 10 mg L^{-1}. Pristine BiOBr and Bi_5O_7Br showed a moderated activity after 90 min of irradiation, reaching a 25.8 and 40% degradation efficiency, respectively. In contrast, Bi_5O_7Br/BiOBr achieved a degradation efficiency of 92%. This behaviour was ascribed to an increase of the charge carriers separation efficiency within the material, while h^+ was the dominant active species in the degradation mechanism.

A recent study by Haibo et al. revealed the effects of coupling BiOX with metallic Bi and carbon [101]. Ultrathin Bi/C/BiOX (X = Cl, Br) nanosheets were prepared by a solvothermal method followed by a calcination with Ar protection. This process resulted in microspheres with a diameter of 2 μm, which were decorated with several nanosheets with a thickness of 30 and 40 nm for BiOCl and BiOBr, respectively. Tetracycline degradation experiments were performed under visible light (420 nm cut-off filter) with a catalyst load of 0.2 g L^{-1} and a tetracycline concentration of 10 mg L^{-1}. Bi/C/BiOCl and Bi/C/BiOBr demonstrated a degradation efficiency of 80 and 56.3%, respectively, while the pure catalysts BiOCl and BiOBr degraded only 45.3 and 24%, respectively. The enhanced photocatalytic activity of Bi/C/BiOCl was explained by synergetic contributions originating from the surface plasmon resonance effect of Bi and the presence of oxygen vacancies.

Guodong et al. developed a novel $Bi_4Ti_3O_{12}$/BiOBr p-n junction. $Bi_4Ti_3O_{12}$ was synthesized via a solid state (diffusion) method followed by an in-situ growth of BiOBr [102]. The obtained $Bi_4Ti_3O_{12}$/BiOBr heterojunction consisted of $Bi_4Ti_3O_{12}$ particles evenly decorated with round BiOBr nanosheets (diameter of 100–200 nm). The photocatalytic activity of $Bi_4Ti_3O_{12}/BiOBr$ was studied by the degradation of ciprofloxacin and tetracycline with concentrations of 20 mg L^{-1} under visible light (400 nm cut-off filter). After a reaction time of 210 min, $Bi_4Ti_3O_{12}/BiOBr$ achieved a degradation efficiency of 96 and 97% for ciprofloxacin and tetracycline, respectively. This value was more than 1.5 times increased compared to the degradation efficiency of pristine $Bi_4Ti_3O_{12}$ and BiOBr. This enhanced performance was traced back to the increased photoactivity of the p-n junction structure, which decreased the recombination rate of the charge carriers.

3.2.3. TiO_2-Based Materials

TiO_2 is a widely studied semiconductor with a bandgap energy around 3.0 to 3.2 eV. The band structure of TiO_2 is comprised of O 2*p* orbitals forming the valence band and a conduction band composed of Ti 3*d* orbitals [103]. Since a semiconductor can absorb photons with energies equal or greater than its bandgap, TiO_2 can only absorb the light in the UV region, which accounts for only 5% of the solar spectrum. Moreover, TiO_2 suffers from recombination of charge carriers, which prohibits surface reactions to occur. Coupling TiO_2 with other semiconductors has been proven an efficient way to reduce the charge carriers recombination rate and to extend the light absorption edges of the material [82]. In this regard, Lin et al. synthesized TiO_2 nanofibers enfolded with BN nanosheets (TiO_2-BN) via electrospinning [104]. XPS analysis of BN nanosheets and TiO_2 nanofibers indicated the existence of B-O-Ti bonds. The optical properties of the fabricated catalyst showed that the absorption of visible light was improved with the addition of BN nanosheets, which could be a result of an energy rearrangement stemming from the formation of B-O-Ti bonds. The photodegradation of ibuprofen (5 mg L^{-1}) by TiO_2-BN was tested under visible light irradiation with a catalyst load of 0.2 g L^{-1}. TiO_2-5 wt.% BN exhibited a 100% degradation efficiency after 2 h under light irradiation, while pure TiO_2 showed an efficiency of 60%. Furthermore, the photocatalytic degradation increased with increasing BN content. This behaviour was attributed to an improved charge carriers separation produced by BN nanosheets wrapping onto TiO_2. Moreover, negatively charged BN nanosheets increased the dispersion of holes.

Xu et al. reported the fabrication of $CF/TiO_2/Bi_2WO_6$ nanostructured heterojunctions by coating carbon fibres (CF) with TiO_2 nanorods by a dip-coating hydrothermal technique, followed by an in-situ growth of Bi_2WO_6 nanosheets [105]. The fabricated catalyst consisted of carbon fibres coated by TiO_2 nanorods (diameter of about 200 nm), which were decorated with $BiWO_6$ nanosheets with diameters ranging from 350 to 500 nm and a thickness of about 15 nm. $CF/TiO_2/Bi_2WO_6$ exhibited a long-wavelength absorption edge at 440 nm and photocatalytic activity towards the degradation of tetracycline. Degradation experiments using 10 mg L^{-1} of tetracycline and 3 g L^{-1} of catalyst under solar light irradiation showed that $CF/TiO_2/Bi_2WO_6$ can degrade 95.1% of tetracycline in 60 min, a significantly higher efficiency compared to CF/TiO_2 (29.6%). Additionally, radical trapping experiments revealed that $O_2^{\bullet -}$ and h^+ can be regarded as the main active species during the degradation of tetracycline.

Leong et al. synthesized $Ni(OH)_2$-modified coral-like rutile TiO_2 ($Ni(OH)_2$-TiO_2) via hydrolysis of $TiCl_4$ followed by hydrothermal crystallization [106]. The synthesized catalysts showed a coral-like loops morphology consisting of TiO_2 nanorod aggregates decorated with $Ni(OH)_2$ nanosheets clusters. UV-vis DRS revealed that pure rutile TiO_2 exhibits a light absorption edge at 410 nm with the deposition of $Ni(OH)_2$ generating additional light absorption in the range from 600 to 800 nm, which was ascribed to *d-d* transitions of Ni(II). Tetracycline photodegradation experiments were performed to assess the photocatalytic activity of $Ni(OH)_2$-TiO_2 under visible light irradiation (420 nm cut-off filter) using a solution with 10 g L^{-1} catalyst and 100 mg L^{-1} tetracycline. After 2 h of irradiation, a 74% of tetracycline was degraded by $Ni(OH)_2$-TiO_2, which was slightly higher than pristine TiO_2 (65%). The greater efficiency was connected with a higher adsorption ability induced by $Ni(OH)_2$ incorporation and the extended light absorption range exhibited by $Ni(OH)_2$-TiO_2.

3.2.4. Conclusions on Photocatalytic Processes

Compared to adsorption, for which processes focus mainly on the interaction between the pollutant molecules and the material surface, photocatalytic processes comprise charge separation/transfer mechanisms and the formation of ROS, which ultimately interact with pollutants. To develop effective photocatalysts, the design of the band structure is crucial for the generation of charge carries (h^+ and electrons) with adequate energy levels that allow for their interaction with the target pollutants. In this respect, bandgap energies are commonly reported in the literature, but detailed band structure characterizations are required to properly study charge transfer and photocatalytic degradation mechanisms. These studies might include band positions and composition, as well as the study of potential mid-gap states.

It has been demonstrated that the bandgap energy can be modulated by controlling the morphology and size of two-dimensional nanomaterials. Photocatalysts Cyst-BiOBr (2.30 eV) and BOB-S (2.30 eV) exhibited smaller bandgap energies compared to control catalysts produced by alternative methods. Even though these catalysts exhibit bandgap energies that allows them to absorb a large spectrum of light, the degradation rates exhibited during photocatalytic degradation experiments (27.17 and 31.05 mg pollutant (g catalyst min)$^{-1}$) were significantly lower than that calculated for catalysts with larger bandgaps such as Cv-CNNs (555.6 mg pollutant (g catalyst min)$^{-1}$). In addition to bandgap energy, charge separation and charge transfer have a great influence on the degradation rate, since the higher rate exhibited by Cv-CNNs was attributed to the incorporation of carbon vacancies, which increase electronic conductivity and improves charge separation.

Moreover, the synthesis of heterostructures (binary and tertiary systems) has proven to enhance charge separation, producing photocatalysts with higher degradation activities. In this regard, $Ti_3C_2T_x$ embedded into alkalized C_3N_4 ($Ti_3C_2T_x/C_3N_4$) exhibited a degradation rate of 128.33 mg tetracycline (g catalyst min)$^{-1}$, nearly 8 times higher than that of other g-C_3N_4-based materials such as MCU-C_3N_4. This high degradation rate is ascribed to the presence of $Ti_3C_2T_x$, which act as an electron acceptor and increase the charge separation. Similarly, Bi/C/BiOCl displayed a degradation rate of 266.67 mg tetracycline (g catalyst min)$^{-1}$, related to synergetic contributions from surface plasmon resonance effects of Bi and the presence of oxygen vacancies in the structure.

With respect to heterostructure engineering, the fabrication of binary and ternary photocatalysts can be considered as an effective modification method and further research should be done on coupling two-dimensional materials with metal nanoparticles, inorganic semiconductors, and conducting materials with the overall aim of enhancing charge carriers separation.

3.3. *Redox and Electrochemical Processes*

Heterogeneous catalysis based on redox and electrochemical processes has been widely applied for the degradation of organic contaminants, particularly pharmaceuticals, in water [33–36,107]. Conventionally, highly oxidizing hydroxyl ($OH^\bullet$) and superoxide ($O_2^{\bullet-}$) ROS are produced via direct activation of H_2O_2 at the surface of a catalyst containing redox metallic active sites, such as Fe^{3+}/Fe^{2+} or Co^{3+}/Co^{2+} (Fenton or Fenton-like reactions), or via electrochemical activation of oxygen and water molecules at the surface of metal-based electrocatalysts (electrochemical processes). Additionally, photo-electrocatalytic generation of electron-holes (h^+) can take place together with the generation of ROS, if photoactive semiconductors are used as electrodes. More recently, sulfate radical ($SO_4^{\bullet-}$)-based advanced oxidation processes via peroxymonosulfate (HSO_5^-) activation on the surface of catalysts containing transition metals have gained attention to remove toxic and refractory pollutants from water due to the significant oxidation activity of this radical.

The degradation efficiency of all these processes strongly depends on the properties of the catalytic materials employed. Particularly, an increase in the catalyst (support) surface area, catalyst active centers dispersion, and charge transfer properties have a significant impact on the materials performance, which is reflected by superior degradation kinetics of the pollutants. 2D nanomaterials offer these enhanced properties, as discussed in Section 2, and recently have been started to be explored as catalysts and/or catalyst supports for redox and (photo-)electrochemical degradation of organic pollutants with promising results.

In this section, the most recent works on peroxymonosulfate activation, Fenton and Fenton-like reactions, and (photo-)electrochemical processes using 2D catalysts are discussed. This includes a description of the catalysts main properties, their degradation efficiency, optimal operating conditions (temperature, pH, catalyst dosage, pollutant concentration, activated species concentration), and catalysts stability.

3.3.1. Peroxymonosulfate (PMS) Activation

Ma et al. prepared hollow structure cobalt hydroxide (h-$Co(OH)_2$) through a modified solvothermal-hydrothermal process using $Co(NO_3)_2 \cdot 6H_2O$ dissolved in isopropanol and glycerine (solvothermal stage) followed by redissolution of the precipitate obtained in water (hydrothermal stage). They compared its effectiveness for PMS activation and ibuprofen degradation with that of neat $Co(OH)_2$ prepared via a conventional precipitation method (p-$Co(OH)_2$) [108]. Unlike p-$Co(OH)_2$ which generally exhibits a broken and uneven flat nanostructure (100–500 nm in size), h-$Co(OH)_2$ represented hollow microscale spheres (about 1 μm diameter) with a thin nanosheets shell. In this regard, it was observed that p-$Co(OH)_2$ and h-$Co(OH)_2$ correspond to the β-$Co(OH)_2$ and α-$Co(OH)_2$ phase, respectively. Both presented Co^{3+} on its surface due to oxidation of Co^{2+}. The specific surface area and total pore volume of p-$Co(OH)_2$ were 23.9 $m^2\ g^{-1}$ and 0.088 $cm^3\ g^{-1}$, respectively, while these for h-$Co(OH)_2$ were 132.36 $m^2\ g^{-1}$ and 0.57 $cm^3\ g^{-1}$, respectively. Control tests for ibuprofen (initial concentration: 10 μmol L^{-1}) degradation indicated that almost no removal of the contaminant was reached by contact with PMS (0.2 mmol L^{-1} at pH equal to 7.0) in the absence of catalyst. However, the addition of p-$Co(OH)_2$ or h-$Co(OH)_2$ to the previous solution (0.2 g L^{-1}) promoted a fast (within about 10 min) and efficient degradation of the contaminant (83.6 and 98.6%, respectively, following a pseudo-first order kinetic model). This was promoted by the effective PSM activation on the surface of these catalysts considering that the adsorption of ibuprofen was negligible (<0.2%). The better performance of h-$Co(OH)_2$ over p-$Co(OH)_2$ in terms of degradation efficiency was explained by its higher specific surface area and pore volume, as well as its different crystalline phase, which would enhance the interaction of PMS with the catalyst. The formation of superficial $Co(OH)^+$ has been reported as the rate-limiting step for radical production, and also it is the key cobalt species to activate PMS. PMS can also consume Co^{3+} to regenerate Co^{2+}, which is a critical process to maintain the chain reactions at relative low cobalt concentrations. The generated sulfate radicals ($SO_4^{\bullet-}$), as well as $OH^{\bullet}$ radicals, attack the organic molecules for their degradation or mineralization (CO_2 + H_2O as products). When h-$Co(OH)_2$ was used as catalyst 6.3% of the ibuprofen decomposition can be attributed to $OH^{\bullet}$ radicals, while 75.7% can be attributed to $SO_4^{\bullet-}$ radicals. After the degradation process a dissolution of 2.36% of cobalt was observed for this catalyst.

On the search of cobalt-free PMS activators due to the high toxicity and scarcity of this metal, Ding et al. prepared an α-Fe_2O_3/MXene nanocomposite (FM) through hydrothermal processing using $FeCl_3 \cdot 6H_2O$ aqueous solutions of concentrations 0.5 M, 1 M, and 2 M (FM-1, FM-2, and FM-3, respectively). They compared its effectiveness for PMS activation and salicylic acid degradation with that of pure MXenes and α-Fe_2O_3 prepared by the same hydrofluoric acid etching and hydrothermal processing, respectively [109]. XRD verified the characteristic peaks of MXene and α-Fe_2O_3 in the composite, with the latter gradually intensified as the content of α-Fe_2O_3 increased. Infrared spectroscopy revealed the interfacial contact between MXenes and α-Fe_2O_3, which indicated that the iron oxide particles were successfully grafted onto MXene nanosheets. SEM and TEM micrographs demonstrated that α-Fe_2O_3 nanoparticles with smaller size were well-dispersed on the surface of MXene nanosheets, which hold especially true for the FM-2 nanocomposite, while some agglomerations were found for the FM-3 nanocomposite. ICP-OES showed that the FM-1, FM-2, and FM-3 nanocomposites contained 8.9, 17.1, and 26.6 wt.% of α-Fe_2O_3, respectively, while XPS indicated that Fe^{3+} is the predominant oxidation state of iron on the fresh catalyst surface. The BET specific surface areas of as-prepared MXene, α-Fe_2O_3, and the FM-2 nanocomposite were 19.3, 22.4, and 37.2 m^2 g^{-1}, respectively. Control tests for the degradation of salicylic acid (initial concentration: 20 mg L^{-1}) indicated that almost no removal of the contaminant was reached by contact with PMS (200 mg L^{-1} at pH equal to 7.0) in the absence of catalyst or in the presence of as-prepared MXenes (0.2 g L^{-1}). The maximum degradation was obtained for the FM-2 nanocomposite (0.2 g L^{-1}), which allowed for a decomposition of 96.7% of salicylic acid in 120 min (following a pseudo-first order kinetic model). This was promoted by the effective PSM activation on the surface of this catalyst with negligible adsorption of salicylic acid (<2.5%). The superior activity of FM nanocomposites over pure α-Fe_2O_3 was cross-correlated with the smaller size of α-Fe_2O_3 on the MXenes surface, which supressed the nanoparticles agglomeration, as well as the higher specific surface areas of the nanocomposites. However, an excess of α-Fe_2O_3 on the surface of MXene (as in the case of FM-3) resulted in a decreased degradation rate, probably due to the overlap of the active sites. The negative effect that an increase in the catalyst dosage (over 0.3 g L^{-1}) has on the degradation rate of salicylic acid can be explained by the accelerated consumption of PMS for the generation of $SO_5^{\bullet-}$ radicals, leading to lower availability of PMS to produce $SO_4^{\bullet-}$ radicals. The $SO_5^{\bullet-}$ radicals possess a weaker oxidation capacity than $SO_4^{\bullet-}$ radicals, which results in a decreased degradation rate. The results of quenching analyses demonstrated that the principal radical species controlling the oxidation reaction was $SO_4^{\bullet-}$. It was observed that the degradation capacity of the FM-2 catalyst decreased from 97.2 to 87.9% after five cycles, which was related to the adsorption of intermediate products on the surface of the catalyst and leaching of active constituents.

Similarly, looking for the replacement of cobalt as catalyst material, Yang et al. synthesized a Cu-incorporated Mo_2C (Cu-Mo_2C) catalyst with an exposed (002) facet and different copper contents (1 to 10 wt.%) through a one-step calcination method using aqueous solutions of ammonium molybdate tetrahydrate ($(NH_4)_6Mo_7O_{24}\cdot 4H_2O$) and cupric acetate monohydrate ($Cu(C_2H_3O_2)_2\cdot H_2O$). They studied its effectiveness for PMS activation and antibiotics degradation [110]. The XRD patterns of the Cu-Mo_2C catalysts showed the characteristic peaks of the hexagonal β-Mo_2C phase and the metallic copper phase, while XPS verified the incorporation of Cu on the Mo_2C surface. SEM and TEM images present the pristine Mo_2C as highly aggregated hexagonal nanosheets with a thickness of 28.7 nm, while the incorporation of copper reduced the particle size and promoted the particles disaggregation. HRTEM-SAED indicated that the hexagonal 5% Cu-Mo_2C had a pure crystalline structure with a (002) lattice plane. HAADF-EDS confirmed that copper was evenly distributed on the surface of the Mo_2C nanosheets. Control tests for tetracycline (initial concentration: 40 ppm) degradation indicated that a low removal of the contaminant (around 18% in 60 min) was reached by contact with PMS (0.3 g L^{-1} at pH equal to 5.0) in the absence of catalyst or in the presence of pristine Mo_2C, Co_3O_4, and α-Fe_2O_3 (34, 28, and 46%, respectively, in 60 min when 0.2 g L^{-1} of catalyst was added). A degradation efficiency of almost 100% in only 20 min was obtained in the presence of 5% Cu-Mo_2C catalyst (0.2 g L^{-1}), which was promoted by the effective PSM activation on the surface of this catalyst (Figure 3a,b). A further copper addition (>5 wt.%) decreased the catalyst efficiency probably due its agglomeration on the surface of the Mo_2C nanosheets. Degradation kinetics verified a pseudo-first order kinetic model for all cases studied. The results of quenching analyses demonstrated that the principal radical species controlling the oxidation reaction was $SO_4^{\bullet-}$, followed by $OH^{\bullet}$ and $O_2^{\bullet-}$ radicals. Additionally, electrochemical impedance spectroscopy showed that the incorporated copper might facilitate the interfacial charge transfer, which is beneficial for the valence transition and PMS activation. It was observed that the degradation capacity of the 5% Cu-Mo_2C catalyst decreased from 100% to 91% after six cycles, which was associated with leaching of Cu (0.78 wt.%) and Mo (0.12 wt.%) after the stability experiment.

3.3.2. Fenton and Fenton-like Reactions

Nie et al. prepared ultrathin iron-cobalt oxide nanosheets (CoFe-ONSs) through a surfactant-aided co-reduction process using $Fe(NO_3)_3 \cdot 9H_2O$, $Co(NO_3)_2 \cdot 6H_2O$, and cetyl trimethyl ammonium bromide (CTAB) aqueous solutions with $NaBH_4$ as reducing agent. They investigated their effectiveness for H_2O_2 activation and tetracycline degradation, and compared with that of $CoFe_2O_4$ nanoparticles [111]. SEM micrographs of the CoFe-ONSs revealed hierarchical structures assembled by ultrathin nanosheets with wrinkles, while TEM confirmed the interconnection between the nanosheets and lateral sizes of the nanosheets of about 300–500 nm. The XRD patterns of the CoFe-ONSs were associated to the standard structure of $CoFe_2O_4$, while EDS verified that the major elements are Fe, Co, and O, which were evenly distributed in the material structure. XPS confirmed that superficial Fe and Co presented oxidation states of +2 and +3. AFM demonstrated that the height of nanosheets was 4.6 nm. The specific surface area of CoFe-ONSs was 144.1 $m^2 g^{-1}$, while the pore volume accounted for 0.638 $cm^3 g^{-1}$ with an average pore size of 16.6 nm. Control tests for tetracycline (initial concentration: 50 mg L^{-1}) degradation indicated that a low contaminant removal (around 16.8% in 50 min) was obtained by contact with H_2O_2 (20 mM at pH equal to 7.0) in the absence of catalyst or in the presence of $CoFe_2O_4$ nanoparticles (around 59% in 50 min when 0.3 g L^{-1} of catalyst was added). A degradation efficiency of 83.5% within 50 min was reached for the CoFe-ONSs catalyst, which was promoted by synergistic effects induced by the nanostructures and H_2O_2 (H_2O_2 direct degradation + adsorption of the contaminant on the CoFe-ONSs surface + H_2O_2 activation on the surface of the catalyst). The degradation kinetics showed a pseudo-first order kinetic model for all cases studied. The better performance of CoFe-ONSs over $CoFe_2O_4$ was attributed to its distinctive hierarchical structure with interconnected wrinkled nanosheets, which contributed with an extensive specific surface area and an ultrathin nature offering numerous active sites and fast electron transport that benefit the contaminant adsorption and its degradation via Fenton-like reaction. The results of quenching analyses demonstrated that the principal radical species controlling the oxidation reaction was $OH^{\bullet}$. The efficiency of tetracycline degradation after five succeeding cycles slightly decreased (from 83.5 to 74.4%) probably due to intermediate products occupying active sites and leaching of the active metals from CoFe-ONSs (0.045 mg Fe L^{-1} and 0.016 mg Co L^{-1} after each cycle).

On the search of an enhanced iron-based catalyst for Fenton and Fenton-like reactions, Xing et al. prepared a 2D π-d conjugated metal-organic framework (MOF) Fe_3(hexaiminotriphenylene)$_2$ ($Fe_3(HITP)_2$) through an ammonia-assisted precipitation method using $FeCl_3$ and HITP·6HCl aqueous solutions. They studied its efficiency for H_2O_2 activation and tetracycline degradation under dark conditions and under visible light illumination, and compared with that of Fe_2O_3 and Fe_3O_4 in dark [112]. XRD of $Fe_3(HITP)_2$ revealed a poor crystallinity, but high purity. TEM indicated that $Fe_3(HITP)_2$ displayed an irregular nanosheet structure with a smooth surface and hexagonal pores (MOF honeycomb arrangement), while HAADF-STEM verified a homogeneous distribution of Fe, N, and C within the nanosheets (an average iron content of 19.3 wt.% as measured by ICP and XPS). XPS, XAFS, FTIR, and Raman confirmed the formation of π-d conjugated 2D structures with $Fe\text{-}N_4$ active sites and mixed iron oxidation states (+2 and +3). Finally, the BET specific surface area of $Fe_3(HITP)_2$ was 260.2 $m^2 g^{-1}$. By UV-vis diffuse reflectance, it was verified that $Fe_3(HITP)_2$ absorbed light over the entire UV-vis range and part of the infrared range. Control tests for tetracycline (initial concentration: 20 mg L^{-1}) degradation indicated that its adsorption on the $Fe_3(HITP)_2$ surface (0.4 g L^{-1} at pH equal to 4.5) under dark was negligible (3% in 30 min), while low degradation (17.3% in 30 min) was reached in the presence of H_2O_2 (20 mM at pH equal to 4.5). The degradation efficiency did not improve under visible light illumination in the presence of the catalyst (13% in 30 min) or in the presence of H_2O_2 (2.7% in 30 min), respectively. However, in the presence of $Fe_3(HITP)_2$ and H_2O_2 under visible illumination, a degradation efficiency of around 96.7% was reached in 30 min, which suggests a synergistic effect of the catalyst, H_2O_2, and visible light in a photo-Fenton-like reaction (Figure 3c,d). The $Fe_3(HITP)_2$ and H_2O_2 system under visible light illumination outperformed Fe_2O_3 (31.5% of degradation in 30 min) and Fe_3O_4 (41.8% of degradation in 30 min), which was connected to the catalyst remarkable high specific surface area, high density of Fe-N_4 active sites, and superior light absorption. The results of quenching analyses demonstrated that the principal radical species controlling the oxidation reaction was $O_2^{\bullet-}$, rather than $OH^{\bullet}$ radicals observed in traditional Fenton or Fenton-like reactions. Electron paramagnetic resonance spectroscopy results suggest that light illumination promotes the dynamic cycle Fe^{2+}/Fe^{3+}. DFT calculations indicated that $Fe_3^{3+}(HITP)_2$ and $Fe_3^{2+}(HITP)_2$ possess narrow band gaps of 0.37 and 0.33 eV, respectively. No decay of the degradation efficiency was observed after 8 cycles, even though 0.037 mg L^{-1} of iron were leached from the catalyst after each cycle, demonstrating its high stability.

3.3.3. Electrochemical Processes

Mi et al. prepared stable reduced graphene oxide (rGO)/Ce-WO_3 nanosheets to modify carbon felt (CF) as an electro-Fenton cathode through a hydrothermal method using aqueous solutions of rGO synthesized by the Hummers method, $Na_2WO_3 \cdot 2H_2O$, and $CeCl_3 \cdot 7H_2O$ with variable mass ratio of rGO and Ce-WO_3. Subsequently, they compared their effectiveness for H_2O_2 production and activation, and ciprofloxacin degradation [113]. SEM micrographs of CF/rGO/Ce-WO_3 cathodes showed nanosheets randomly loaded on the rough surface of CF fibres, while TEM confirmed the plate-like morphologies of rGO/Ce-WO_3 nanoparticles with diameters of about 8–12 nm with Ce-WO_3 nanosheets distributed on the surface of rGO. HR-TEM showed that rGO/Ce-WO_3 nanoparticles presented good crystallinity with WO_3 planes clearly exposed and CeO_2 crystallites highly dispersed. XPS revealed the presence of C, O, W, and Ce, with the addition of RGO on the surface of CF greatly increasing the degree of oxidation, abundance of superficial active oxygen species, and Ce^{4+}, Ce^{3+}, W^{6+}, and W^{5+} coexisting on the surface. The specific surface area of rGO/Ce-WO_3 was 69.71 m^2 g^{-1}, while its pore volume and average pore size were 0.120 cm^3 g^{-1} and 14.084 nm, respectively. EIS tests showed that CF/rGO/Ce-WO_3 electrodes presented lower electron transfer impedance than CF/Ce-WO_3. This indicates that rGO could promoted the electron transfer between the electrode surface and the solution. Degradation tests were carried out at room temperature with an electrolyte solution containing 50 mg L^{-1} of ciprofloxacin, 0.05 mol L^{-1} of Na_2SO_4, and 0.1 mmol L^{-1} of $FeSO_4 \cdot 7H_2O$ at a pH equal to 3.0. The anode (counter electrode) was a platinum plate of 1×0.5 cm^2, and the cathode was the CF/rGO/Ce-WO_3 fabricated using a $6 \times 3 \times 0.5$ cm^3 3D carbon felt. Oxygen was bubbled into the solution at a flow rate of 100 mL min^{-1} up to 30 min prior to each degradation experiment. It was determined that the optimal rGO/Ce-WO_3 mass ratio, Ce/W molar ratio, and (RGO/Ce-WO_3)/CF loading amount in the cathode were 1:3, 0.05, and 4.5 wt.%, respectively, which exhibited a degradation efficiency of 100% in 60 min at an applied current of 400 mA. This high degradation efficiency can be explained by the effective generation of H_2O_2 and radical species on the high specific surface of the electrode, while its activity was enhanced by addition of Ce as a dopant. The results of quenching analyses suggest that oxygen (O_2) is reduced in the presence of protons (H^+) on the surface of the cathode to produce H_2O_2 via a two-electron transfer process. This gives rise to the generation of $O_2^{\bullet-}$ radicals, which can produce further H_2O_2 (via $OH^{\bullet}$ and $OOH^{\bullet}$ radicals) or oxidize the organic pollutant (Figure 3e,f). Therefore, ciprofloxacin degradation occurs via $OH^{\bullet}$ radicals generated by H_2O_2 through a Fenton-like process and $O_2^{\bullet-}$ radicals generated via O_2 reduction and Fenton-like processes. The degradation of ciprofloxacin marginally reduced (by 5.1%) after five cycles due to mechanical wearing of the cathode coating.

Tang et al. prepared g-C_3N_4 nanosheets supported on TiO_2 nanotube arrays (TCNs) as photo-electrocatalytic anodes through direct calcination of melamine (0.2, 0.5 and 1 g in 15 mL of ethyl alcohol) onto TiO_2 nanotube arrays (TNAs). They investigated their effectiveness for the degradation of tetracycline compared with bare TNAs anodes [114].

SEM and TEM micrographs showed well-ordered and highly dense, pure TNAs with tube diameters between 100 and 125 nm, while the added g-C_3N_4 nanosheets were uniformly anchored on the TNAs. The tube walls of TiO_2 nanotubes tend to thicken with increasing g-C_3N_4 content due to their attachment to the tube surfaces. XRD identified anatase TiO_2 and increasing peaks intensities for g-C_3N_4 with increasing amount of melamine in the precursor solution. EDX confirmed that g-C_3N_4 was evenly distributed over the entire TNAs. XPS verified the interaction between electron-rich nitrogen in g-C_3N_4 and the electron deficient Ti^{4+} in TiO_2, which would benefit charge separation and transfer at their interface. UV-visible diffuse reflection spectroscopy revealed a stronger light absorption of TCNs compared with pure TNAs and g-C_3N_4, while photoluminescence measurements showed that TCN-2 (TCNs prepared with 0.5 g of melamine) presented the most stable separation of photogenerated charge carriers. Degradation tests were carried out in an aqueous electrolyte containing 10 mg L^{-1} of tetracycline and 0.1 M of Na_2SO_4. The anode was the as-prepared TNAs or TCNs photoelectrode with an active area of 4 cm^2, while a platinum plate was used as cathode (counter electrode) and an Ag/AgCl electrode was used as reference. The system was kept under dark conditions for 30 min to reach adsorption-desorption equilibrium prior to photo-electrochemical measurements under simulated solar light irradiation. The photo-electrochemical degradation process showed a higher efficiency than electrochemical and photocatalytic processes for both TNAs and TCNs anodes. In the case of TCN-2 electrodes, a degradation efficiency of 95% within 60 min was reached when a potential of 1.0 V vs. Ag/AgCl was applied. The results of quenching analyses demonstrate that the electron holes (h^+), generated under simulated solar light irradiation at the TiO_2/g-C_3N_4 heterojunction are the main radical species acting in the degradation of tetracycline. The $OH^{\bullet}$ and $O_2^{\bullet -}$ radicals produced at the cathode showed a lower degradation effect than h^+, although it was not negligible. Finally, tests under continuous flow (0.84 mL min^{-1}) were conducted using a real water source electrolyte reaching a degradation of tetracycline of 80% in 60 min with unchanged morphology and structure of the TCN-2 anode.

Recently, Wang et al. prepared 2D Ni-MOF derived NiO@C nanosheet arrays modified Ti/TiO_2NTs/PbO_2 anodes through a solvothermal process using $NiCl_2 \cdot 6H_2O$ and 1,4-dicarboxybenzene acid dissolved in a solution containing water, ethanol and N,N-dimethylformamide, in which TiO_2NTs were immersed. This process was followed by a carbonization stage of the produced electrode in highly pure N_2 for the effective generation of the NiO@C nanosheet arrays. Finally, β-PbO_2 active layers were electrodeposited on Ti/TiO_2NTs/NiO@C, Ti/TiO_2NTs, and bare Ti plate electrodes using an electrolyte containing $Pb(NO_3)_2$ and HNO_3, and different electrodeposition times (15, 20, 30 and 45 min) [115]. SEM images showed well-ordered and highly dense TiO_2NTs with a tube diameter of around 140 nm. The tubes were covered by highly dispersed NiO@C petal-like nanosheets with an average thickness of 30 nm. PbO_2-electroplated Ti/TiO_2NTs/NiO@C electrodes showed smaller metal oxide particles, which were uniformly dispersed on their surface and increased the availability of active sites. XRD demonstrated the presence of anatase TiO_2, cubic NiO, and β-PbO_2 phases as well as smaller grain sizes for the latter in the case of Ti/TiO_2NTs/NiO@C/PbO_2 electrodes. XPS confirmed Ni oxidation states of 0 and +2, which implies the presence of Ni metal and NiO in the structure. The electrochemical performance was tested using Ti/TiO_2NTs/NiO@C/PbO_2, Ti/TiO_2NTs/PbO_2 or bare Ti/PbO_2 plate electrodes as anode, a platinum plate as cathode (counter electrode), and KCl-saturated calomel electrode as reference, immersed in a 2 mol L^{-1} H_2SO_4 solution at 60 °C. It was observed that the Ti/TiO_2NTs/NiO@C/PbO_2 electrode with electrodeposition time of 20 min showed the largest electrochemical active area and the fastest electron transfer rate. In addition, the electrode showed: (i) a high oxygen evolution potential that favours the formation of more reactive oxygen species (particularly, $OH^{\bullet}$ radicals), and (ii) a service life larger than 40 h (in accelerated life tests). Both aspects are desirable for the degradation of organic pollutants. The degradation tests were carried out in an electrolyte aqueous solution containing 10 mg L^{-1} of salicylic acid and 0.1 mol L^{-1} of Na_2SO_4

at a pH equal to 5. The prepared electrodes were tested as anode and a titanium plate (2 cm^2) was used as cathode. A current density of 70 mA cm^{-2} was applied for 3 h. The Ti/TiO_2NTs/NiO@C/PbO_2 electrode showed the highest degradation rate (pseudo-first order kinetics) due to its enhanced electrochemical performance, reaching a removal of around 96.29% of salicylic acid even after 8 cycles. The degradation of salicylic acid was attributed to indirect electrochemical oxidation via mediating $OH^{\bullet}$ radicals generated at the Ti/TiO_2NTs/NiO@C/PbO_2 electrode surface.

Figure 3 summarizes the structure and performance of the 2D nanomaterials with the highest degradation efficiencies reported for PMS activation (Mo_2C nanosheets), Fenton and Fenton-like reactions (2D π-d conjugated metal organic framework $Fe_3(HITP)_2$), and electrochemical processes (rGO-Ce/WO_3 nanosheets).

3.3.4. Conclusions on Redox and Electrochemical Processes

PMS activation and Fenton or Fenton-like reactions are redox processes which greatly depend on the generation of active species such as $SO_4^{\bullet -}$ or ROS ($OH^{\bullet}$, $O_2^{\bullet -}$) for the degradation of pharmaceuticals. In this regard, the results reported in Sections 3.3.1 and 3.3.2 indicate that 2D nanomaterials are promising catalysts for these processes due to their enhanced surface area and charge transfer properties. Additionally, as observed for Cu-Mo_2C in PMS activation and $Fe_3(HITP)_2$ in Fenton-like reactions, high disaggregation and density of active sites is crucial for obtaining high degradation rates and efficiencies. Therefore, the improvement of the synthesis methods towards intelligent materials design appears as an important aspect to be considered in future studies for their successful application in wastewater treatment.

Together with structure and surface design of 2D nanomaterials to obtain increased surface areas and highly dispersed active centres, the ability to control their optical transparency via bandgap engineering is an effective method to enhance their degradation properties. As observed for $Fe_3(HITP)_2$ in Fenton-like reactions and TCNs in electrochemical processes, the photogenerated electron holes (h^+) can also act as oxidation agents together with ROS, showing even higher degradation yields. Considering that PMS activation seems to be the most effective degradation method of pharmaceuticals in wastewater due to the lower degradation times and the variety of degraded compounds reported, it is thought that the development of new 2D nanomaterials for PMS activation with enhanced optical properties is a promising research area.

Finally, electrochemical processes offer a more straightforward path for the design of continuous operation devices than conventional heterogeneous catalysis methods (PMS activation and Fenton reactions), which is attractive in terms of large-scale wastewater treatment. Electrochemical cells allow for the ad-hoc and facile control of the reduction and oxidation kinetics, which is important in real operations where wastewater composition can be variable over time. This combined with the use of improved photocatalytic electrode materials (as shown for TCNs) and the integration of PMS electro-activation can also be an interesting area for future research.

It is worth to highlight that the replacement of scarce and/or toxic metals, such as cobalt and lead, as part of catalyst or electrocatalyst is of great importance for the success of the methods reported in Section 3.3. The environmental and health impact that these technologies might have at the end of their life is an aspect to be studied to promote real, sustainable alternatives with a positive impact on environment. In this regard, recycling processes of catalysts or electrocatalysts should be developed aiming to reduce the impact of waste worldwide. This aspect is critical considering the instability reported for several materials after a small number of cycles (up to 8 cycles), which should be further reduced in the future.

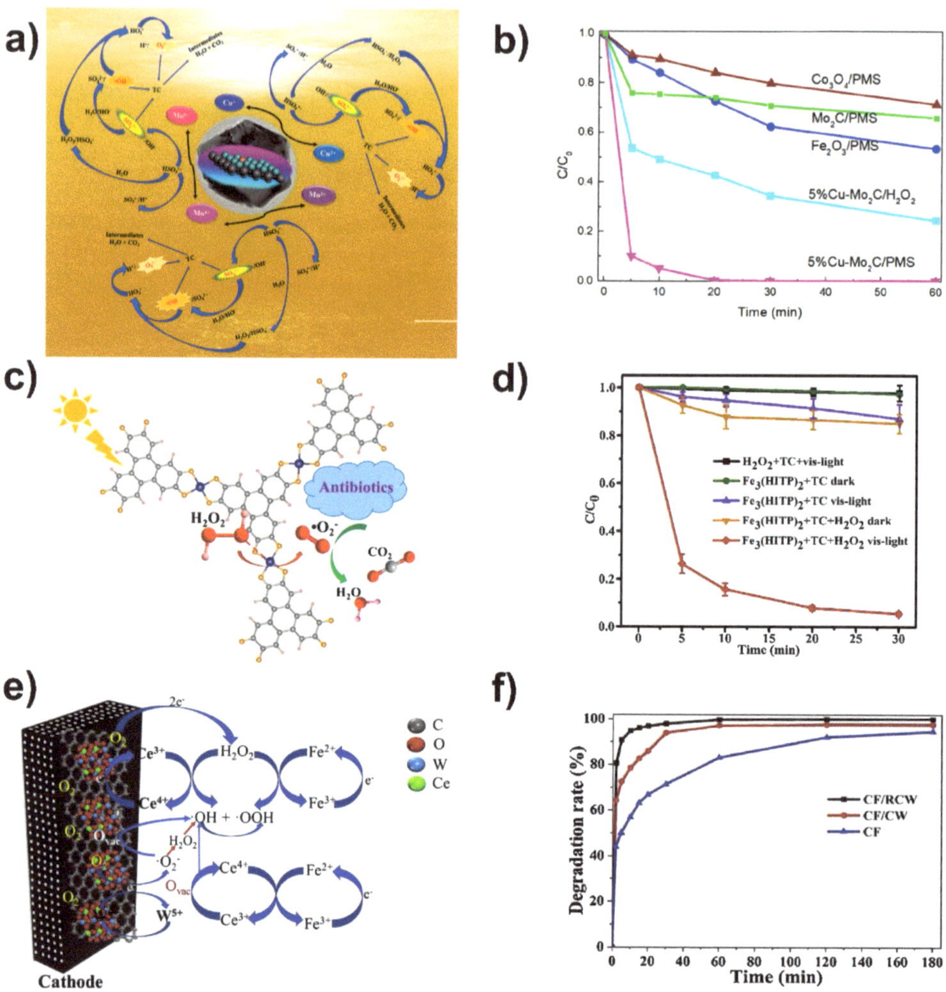

Figure 3. 2D nanomaterials for redox and electrochemical processes and their removal efficiency: (**a**,**b**) hexagonal Mo_2C nanosheets with (002) facets and Cu incorporation for tetracycline removal via PMS activation (reproduced with permission [110]), (**c**,**d**) 2D π-d conjugated metal-organic framework $Fe_3(HITP)_2$ and tetracycline removal via photo-Fenton like reaction (reproduced with permission [112]), and (**e**,**f**) rGO-Ce/WO_3 nanosheets modified carbon felt cathode and ciprofloxacin removal via electro-Fenton process (reproduced with permission [113]).

4. Overview and Outlook

In this review, the state-of-the-art regarding the ability of 2D nanomaterials to remove pharmaceutical pollutants from wastewater is summarized and critically discussed. In this regard, adsorption, photocatalytic, and redox/electrochemical processes have been identified as main approaches. Current trends in 2D nanomaterials development for wastewater treatment tend to the research of materials composed of earth-abundant elements (e.g., C, Bi, Ti, Mo, and Fe), the design of rational material-pollutant interactions, and the fabrication of diverse morphologies for a superior removal or degradation of pollutants.

Table 1 summarizes the performance of pure 2D nanomaterials and binary/ternary nanocomposites as adsorbents of pharmaceutical contaminants in wastewater. The results suggest that materials based on MoS_2, MOFs, and MXenes are the most promising for the removal of many compounds, showing a fast uptake within a few minutes or even seconds, short equilibrium times, and maximum adsorption capacities between 200 and 800 mg g^{-1}. The exceptional adsorption capacity of these nanomaterials is associated with a superior specific surface area (in the range of 10 to 100 m^2 g^{-1}), and a surface structure which offers an increased number of active sites (such as -OH and -O) that promote chemisorption of the pollutants. Moreover, GO/Fe_3O_4 binary nanocomposites demonstrated fast uptakes within 10 min, equilibrium times of 30 min, and maximum adsorption capacities around 250 mg g^{-1} due to pronounced π-π interactions between the nanohybrid and the contaminant.

Table 2 summarizes the reviewed studies regarding photocatalytic processes. To compare different material performances, the degradation rate normalized by the catalyst load is used as an indicator for the degradation capacity. This parameter is found suitable since it considers the initial concentration, efficiency, reaction time, and catalyst load. In this context, the fabrication of heterostructures appears as a promising strategy to obtain highly active photocatalysts, allowing for the enhancement of visible light absorption, thus increasing the charge carriers separation and reducing charge recombination. $Ti_3C_2T_x$ embedded into alkalized C_3N_4 ($Ti_3C_2T_x/C_3N_4$) exhibited a degradation rate of 128.33 mg tetracycline (g catalyst min)$^{-1}$, which is nearly 8 times higher than that of other g-C_3N_4-based materials such as MCU-C_3N_4. This enhanced photodegradation activity was ascribed to its binary structure, which prevents the recombination of charge carriers, while $Ti_3C_2T_x$ acted as an electron acceptor. Similarly, the high photocatalytic activity achieved by Bi/C/BiOCl (266.67 mg tetracycline (g catalyst min)$^{-1}$) was connected to synergetic contributions from surface plasmon resonance effects of Bi and the presence of oxygen vacancies. Additionally, carbon nitride nanosheets catalysts with carbon vacancies (Cv-CNNs) exhibited a degradation rate of 555.56 mg sulfadiazine (g catalyst min)$^{-1}$, which was explained by an improved charge separation generated by the incorporated carbon vacancies.

Table 3 summarizes the performance of the 2D nanomaterials used for redox and electrochemical degradation of pharmaceuticals in wastewater. The results suggest that oxidation processes based on peroxymonosulfate activation are the most promising method for degradation, reaching degradation rates of nearly 100% in less than 30 min. This is associated with the high activity of $SO_4^{\bullet -}$ radicals, as well as their capacity to enhance the generation of $OH^{\bullet}$ radicals in water, which seems to be favoured by metal carbide nanomaterials such as Cu-MoC_2 nanosheets. Furthermore, this nanomaterial showed a strong ability to degrade many contaminants under natural pH and temperature conditions in different water sources. In the case of Fenton-like processes, it is worth to highlight the performance of the metal-organic framework $Fe_3(HITP)_2$ and its ability to absorb visible light, which enhanced the degradation rate for a wide range of contaminants when compared with conventional iron-based Fenton catalysts.

Table 1. 2D nanomaterials reported for the removal of pharmaceuticals in wastewater via adsorption processes.

Pure 2D Nanomaterials							
Material	Pollutant	Specific area/m^2 g^{-1}	Concentration (pollutant/adsorbent)	Wastewater characteristics	Optimal adsorption characteristics	Maximum adsorption capacity	Ref.
Flower-like single-layer MoS_2	diclofenac	23.9	10 mg L^{-1}/ 0.2–0.8 g L^{-1} (optimal: 0.5 g L^{-1})	Synthetic, temperature: 25 °C	>90% (in 5 min)/ equilibrium time: 10 min	217 mg g^{-1}	[62]
Few-layer boron-nitride nanosheets	estrone	900	500 µg L^{-1}/ 0.05 g L^{-1}	Synthetic, temperature: 25 °C	>80% (in 10 min)/ equilibrium time: 2 h	249 mg g^{-1}	[63]

Table 1. *Cont.*

Pure 2D Nanomaterials							
Defective boron-nitride nanosheets	tetracycline	533	200–250 mg L^{-1} (optimal: 250 mg L^{-1})/ 0.2–2 g L^{-1} (optimal: 2 g L^{-1})	Synthetic, temperature: 25 °C	Fast uptake within 50 min/ equilibrium time: 6 h	1101 mg g^{-1}	[64]
Defective boron-nitride nanosheets with boron vacancies	tetracycline	1104	200 mg L^{-1}/ 0.2 g L^{-1}	Synthetic, pH = 3.2, temperature: 25 °C	Fast uptake within 50 min/ equilibrium time: 3 h	438 mg g^{-1}	[65]
Few-layer g-C_3N_4	tetracycline	178	20 mg L^{-1}/ 0.2 g L^{-1}	Synthetic, pH = 6.5–7.5, temperature: 25 °C	>75% (in 60 min)/ adsorption + photocatalytic process	40 mg g^{-1}	[66]
Copper-porphyrin	tetracycline	342	10 mg L^{-1}/ 0.2 g L^{-1}	Synthetic, pH = 5	>85% (in 6 h)/ equilibrium time: 30 min/ adsorption + photocatalytic process	150 mg g^{-1}	[67]
Covalent triazine framework nano-sheets	nitrofurazone	565	50 ppm/ 0.5 g L^{-1}	Synthetic, pH = 6	Fast uptake within 60 min/ equilibrium time: 2 h	351 mg g^{-1}	[68]
Few-layer, zwitter-ionic COFs	norfloxacin	187	10–500 mg L^{-1} (optimal: 10 mg L^{-1})/ 0.5 g L^{-1}	Synthetic, pH = 8	99.4% (in 30 s)/ equilibrium time: 1 min	824 mg g^{-1}	[69]
Multi-layer $Ti_3C_2T_x$	ciprofloxacin	---	10 mg L^{-1}/ 0.4 g L^{-1}	Synthetic, pH = 5.5, temperature: 25 °C	>80% (in 30 s)/ equilibrium time: few minutes	208 mg g^{-1}	[70]
Binary and ternary nanocomposites							
Material	Pollutant	Specific area/m^2 g^{-1}	Concentration (pollutant/adsorbent)	Wastewater characteristics	Optimal adsorption characteristics	Maximum adsorption capacity	Ref.
rGO/Fe_3O_4	sulfonamide antibi-otics	---	trace amounts/ 0.5–3 g L^{-1} (optimal: 2 g L^{-1})	Tap water, sewage	Fast uptake within 10 min/ equilibrium time: 15 min	---	[71]
GO/Fe_3O_4	tetracycline	249	30 and 50 mg L^{-1} (optimal: 50 mg L^{-1})/ 9.6 g L^{-1}	Synthetic, pH = 2–11, temperature: 25 °C	Fast uptake within 20 min/ equilibrium time: 10 h	212 mg g^{-1}	[72]
GO/Fe_3O_4 (suspen-sions)	chlorpheniramine	221	120 mg L^{-1}/ ---	Synthetic, pH = 6	Fast uptake within 5 min/ equilibrium time: 60 min	300 mg g^{-1}	[73]

Table 1. *Cont.*

Binary and ternary nanocomposites							
MoS_2/GO/ Fe_3O_4	lecofloxacin, paz-cofloxacin, gati-floxacin	135	trace amounts/ 0.05–0.25 g L^{-1} (optimal: 0.2 g L^{-1})	Synthetic, pH = 0.5−4	Fast uptake within 2 min/ equilibrium time: 4 min	5 mg g^{-1}	[74]
rGO/Ag/ Fe_3O_4	acetaminophen, ibupro-fen, naproxen	- - -	- - -/ 0.1 g L^{-1}	Synthetic, pH = 4–10, temperature: 25 °C	>99% of acetaminophen (in 1 h)/equilibrium time: 6 h	- - -	[75]
g-MoS_2/biochar	tetracycline	176.8	100 mg L^{-1}/ 0.4 g mL^{-1}	Synthetic, pH = 2–11, temperature: 25 °C	>80% (in 4 h)/ equilibrium time: 15 h	249 mg g^{-1}	[76]
chitin/GO	ciprofloxacin	- - -	4–850 mg L^{-1}/ 5 g L^{-1}	Synthetic, pH = 6.3, temperature: 25 °C	Fast uptake within 100 min/ equilibrium time: 6 h	270 mg g^{-1}	[77]

Table 2. 2D nanomaterials reported for the degradation of pharmaceuticals in wastewater via photocatalytic processes.

Material	Band gap/eV	Pollutant	Specific area/m^2 g^{-1}	Concentration (pollutant/catalyst)	Wastewater characteristics	Maximum degrada-tion	Degradation rate / $\frac{mg\ pollutant}{g\ catalyst-min}$	Ref.
Graphitic Carbon Nitride-Based Materials								
MCU-C_3N_4	2.79	tetracycline, ciprofloxacin	122.53	10 mg L^{-1}/ 0.7 g L^{-1}	Synthetic	81.60% of tetracycline (in 80 min) 65.80% of ciprofloxacin (in 90 min)	14.57 10.44	[86]
Cv-CNNs	2.83	sulfadiazine	77.72	10 mg L^{-1}/ 0.02 g L^{-1}	Synthetic, temperature: 25 °C	100% (in 90 min)	555.56	[87]
OCN-24-550	2.54	oxytetracycline	126	20 mg L^{-1}/ 1 g L^{-1}	Synthetic	85.76% (in 120 min)	14.29	[89]
KMCN(0.05)	2.33	tetracycline	50.19	20 mg L^{-1}/ 1 g L^{-1}	Synthetic, temperature: ambient	85.13% (in 60 min)	28.38	[90]
$Ti_3C_2T_x/C_3N_4$	- - -	tetracycline	37.20	20 mg L^{-1}/ 0.2 g L^{-1}	Synthetic	77% (in 60 min)	128.33	[91]
Bismuth oxyhalides-based materials								
Material	Band gap/eV	Pollutant	Specific area/m^2 g^{-1}	Concentration (pollutant/catalyst)	Wastewater characteristic	Maximum degrada-tion	Degradation rate / $\frac{mg\ pollutant}{g\ catalyst-min}$	Ref.
Cyst-BiOBr	2.3	cefazolin	11.38	10 mg L^{-1}/ 0.2 g L^{-1}	Synthetic, pH = 6.7,tem-perature: 20 °C	87% (in 180 min)	24.17	[97]
$Bi_{24}O_{31}Br_{10}$	2.3	tetracycline	8.57	10 mg L^{-1}/ 0.34 g L^{-1}	Synthetic, temperature: ambient	95% (in 90 min)	31.05	[98]

Table 2. *Cont.*

Material	Band gap/eV	Pollutant	Specific area/m² g⁻¹	Concentration (pollutant/catalyst)	Wastewater characteristic	Maximum degradation	Degradation rate / $\frac{mg\ pollutant}{g\ catalyst-min}$	Ref.
Bismuth oxyhalides-based materials								
$BiOCl_xI_y$	2.12	acetaminophen	34.93	20 mg L^{-1}/ 0.6 g L^{-1}	Synthetic, pH = 7.36, temperature: 25 °C	80% (in 180 min)	14.81	[99]
$Bi_5O_7Br/BiOBr$	- - -	carbamazepine	- - -	10 mg L^{-1}/ 0.5 g L^{-1}	Synthetic, temperature: 25 °C	92% (in 90 min)	20.44	[100]
Bi/C/BiOCl	- - -	tetracycline	57.80	10 mg L^{-1}/ 0.2 g L^{-1}	Synthetic, temperature: 25 °C	80% (in 15 min)	266.67	[101]
$Bi_4Ti_3O_{12}/BiOBr$	- - -	ciprofloxacin, tetracycline	- - -	20 mg L^{-1}/ - - -	Synthetic, temperature: ambient	96% of ciprofloxacin (in 210 min)97% of tetracycline (in 210 min)	- - -	[102]
TiO_2-based materials								
Material	Band gap/ eV	Pollutant	Specific area/m² g⁻¹	Concentration (pollutant/catalyst)	Wastewater characteristic	Maximum degradation	Degradation rate / $\frac{mg\ pollutant}{g\ catalyst-min}$	Ref.
TiO_2-BN	- - -	ibuprofen	49.60	5 mg L^{-1}/ 0.2 g L^{-1}	Synthetic	100% (in 120 min)	20.83	[104]
$CF/TiO_2/Bi_2WO_6$	- - -	tetracycline	- - -	10 mg L^{-1}/ 3 g L^{-1}	Synthetic, temperature: 20 °C	95.10% (in 60 min)	5.28	[105]
$Ni(OH)_2$-TiO_2	- - -	tetracycline	96.60	100 mg L^{-1}/ 10 g L^{-1}	Synthetic	74% (in 120 min)	6.17	[106]

Table 3. 2D nanomaterials reported for the degradation of pharmaceuticals in wastewater via redox and electrochemical processes.

Material	Pollutant	Specific area/m² g⁻¹	Concentration (pollutant/catalyst)	Wastewater characteristics	Maximum degradation	Ref.
Peroxymonosulfate Activation						
h-$Co(OH)_2$	ibuprofen	132.36	10 μmol L^{-1}/ 0.2 g L^{-1}	0.2 mmol L^{-1} PMS, pH = 3–11 (optimal: 7), room temperature	~98.6% (in 10 min)	[108]
α-Fe_2O_3/Mxene	salicylic acid	37.2	10–100 mg L^{-1} (optimal: 20 mg L^{-1})/ 0.1–1 g L^{-1} (optimal: 0.2 g L^{-1})	0.1–0.8 g L^{-1} PMS (optimal: 0.2 g L^{-1}), pH = 3–9 (optimal: 7), temperature: 25 °C	~96.7% (in 120 min)	[109]
Cu-Mo_2C	Tetracycline [1]	—	10–50 ppm (optimal: 40 ppm)/ 0.05–0.40 g L^{-1} (optimal: 0.2 g L^{-1})	0–0.5 g L^{-1} PMS (optimal: 0.3 g L^{-1}), pH = 3–11 (optimal: 5), temperature: 15–45 °C (optimal: 25 °C) [2]	~100% (in 20 min)	[110]

Table 3. *Cont.*

Fenton and Fenton-like reactions						
Material	Pollutant	Specific area/m^2 g^{-1}	Concentration (pollutant/catalyst)	Wastewater characteristics	Maximum degradation	Ref.
CoFe-ONSs	tetracycline	144.1	40–70 mg L^{-1} (optimal: 50 mg L^{-1})/ 0.2–1.2 g L^{-1} (optimal: 0.3 g L^{-1})	5–40 mM H_2O_2 (optimal: 20 mM), pH = 3–8 (optimal: 7), temperature: 15–35 °C (optimal: 25 °C) [3]	~83.5% (in 50 min)	[111]
$Fe_3(HITP)_2$	Tetracycline [4]	260.2	20 mg L^{-1}/ 0.1–0.4 g L^{-1} (optimal: 0.4 g L^{-1})	4–40 mM H_2O_2 (optimal: 20 mM), pH = 3–9 (optimal: 4.5), room temperature, under visible light illumination	~96.7% (in 30 min)	[112]
Electrochemical processes						
Material	Pollutant	Specific area/m^2 g^{-1}	Concentration (pollutant)	Wastewater characteristics	Maximum degradation	Ref.
CF/RGO-Ce/WO_3	ciprofloxacin	69.71	50 mg L^{-1}	0.05 mol L^{-1} Na_2SO_4, 0.05–0.30 mmol L^{-1} $FeSO_4{\cdot}7H_2O$ (optimal: 0.1 mmol L^{-1}), pH = 3.0, applied current: 400 mA room temperature	~100% (in 60 min)	[113]
TiO_2NTs/g-C_3N_4	Tetracycline [5]	—	10 mg L^{-1}	0.1 mol L^{-1} Na_2SO_4, applied voltage: 0.2–2.5 V vs. Ag/AgCl (optimal: 1.0 V), under solar light illumination [6]	~95% (in 60 min)	[114]
Ti/TiO_2NTs/NiO@C/PbO_2	salicylic acid	—	10–90 mg L^{-1} (optimal: 30 mg L^{-1})	0.025–0.3 mol L^{-1} Na_2SO_4 (optimal: 0.05–0.1 mol L^{-1}), pH = 3–11 (optimal: 3–5), applied current: 30–120 mA cm^{-2} (optimal: 70 mA cm^{-2})	~96.29% (in 180 min)	[115]

[1]. Additionally, other organics were almost completely degraded in about 60 min. [2]. Additionally, the effects on the degradation rate of different anions, cations and natural water sources were investigated. No significant drawbacks were observed, reaching complete or almost complete degradation in about 60 min. [3]. Additionally, the effects on the degradation rate of different anions and natural water sources were investigated. Suppression of tetracycline degradation was observed in 50 min, probably due to the consumption of active species by the added chemicals. [4]. Additionally, other organics were highly degraded in about 45 min. [5]. Additionally, other organics were highly degraded in about 60 min. [6]. Additionally, the effect on the degradation rate of a natural water source was investigated. No significant drawbacks were observed, reaching 80% of degradation in about 60 min.

Although large scale application of 2D nanomaterials for the removal of pollutants in wastewater is highly desired, most of the studies to date wereperformed at laboratory scale. Moreover, several challenges need to be addressed to achieve full-scale treatment systems, including large-scale fabrication, devices and automatization design, regeneration systems, and management of secondary residues (when produced). In this regard, the potential up-scaling must be considered when developing novel synthesis routes. For instance, it has been identified that synthesis routes such as liquid-phase oxidation, chemical vapor deposition, and oxidative exfoliation reduction are well suited for large scale of graphene, based upon environmental, economic, and social indicators [116]. In the case of MXenes, Naguib et al. reported a generic method for large-scale delamination of Mxenes (V_2CT_x and Ti_3CNT_x) using tetrabutylammonium hydroxide (TBAOH), choline hydroxide, or n-butylamine as an organic base. The authors reported a successful up-scaling of the process from few milliliters to a liter batch without significant changes of MXenes surface chemistry and overall quality [117]. Despite these recent reports, large scale production of 2D nanomaterials remains in its infancy and further research is required to take the processes to pilot scale before full-scale applications are feasible. In this regard, the study of alternative methods that could accelerate the transition from batch experiments to continuous production is key for the success of the methods revised in this work in the short term.

2D nanomaterials have been demonstrated to be well suited to degrade and reduce pharmaceutical pollutants in wastewater. Although, despite extensive research conducted, they still encounter several challenges to be addressed in the future:

- It is critical to replace toxic reagents during the synthesis of these materials (e.g., cyanide and hydrazine), which turn into secondary residues at the end of the fabrication process. In this regard, several studies highlight the importance of green synthesis methods such as the replacement of HF, an etchant agent widely used during $Ti_3C_2T_x$ MXene synthesis, by less toxic chemicals. For instance, Limbu et al. reported a green protocol for the room temperature synthesis of reduced $Ti_3C_2T_x$ MXene via a simple L-ascorbic acid treatment [118].
- Toxicity studies on humans and environmental impact assessments need to be addressed, in the light of studies performed on graphene, GO, and rGO reporting that these materials can elicit toxic effects both in-vitro and in-vivo [119]. Furthermore, life cycle analyses should be considered in the development of every so-called sustainable material to establish the real impact of its fabrication and use on the environment [120].
- More research needs to be conducted on the regeneration and reuse of 2D nanomaterials for wastewater treatment to assure and improve sustainability. Moreover, a material with a high selectivity for the target pollutant is required for real wastewater applications. Real wastewaters contain different chemicals like organic and inorganic molecules, as well as counter ions, which can lead to a reduced efficiency than that reported at the laboratory scale employing one-component solutions.

Author Contributions: Conceptualization, M.C.-L. and S.G.-P.; formal analysis, M.C.-L., A.R. and S.G.-P.; investigation, M.C.-L., A.R. and S.G.-P.; writing—original draft preparation, M.C.-L., A.R. and S.G.-P.; writing—review and editing, M.C.-L., A.R. and S.G.-P. All authors have read and agreed to the published version of the manuscript.

Funding: This research received no external funding.

Institutional Review Board Statement: Not applicable.

Informed Consent Statement: Not applicable.

Conflicts of Interest: The authors declare no conflict of interest.

References

1. Ghosh, A.; Nayak, A.K.; Pal, A. Nano-Particle-Mediated Wastewater Treatment: A Review. *Curr. Pollut. Rep.* **2017**, *3*, 17–30. [CrossRef]
2. Zunita, M. Graphene oxide-based nanofiltration for Hg removal from wastewater: A mini review. *Membranes* **2021**, *11*, 269. [CrossRef]
3. Kommu, A.; Singh, J.K. A review on graphene-based materials for removal of toxic pollutants from wastewater. *Soft Mater.* **2020**, *18*, 297–322. [CrossRef]

4. Sirés, I.; Brillas, E.; Oturan, M.A.; Rodrigo, M.A.; Panizza, M. Electrochemical advanced oxidation processes: Today and tomorrow. A review. *Environ. Sci. Pollut. Res.* **2014**, *21*, 8336–8367. [CrossRef] [PubMed]
5. Ibrahim, Y.; Kassab, A.; Eid, K.; Abdullah, A.M.; Ozoemena, K.I.; Elzatahry, A. Unveiling fabrication and environmental remediation of mxene-based nanoarchitectures in toxic metals removal from wastewater: Strategy and mechanism. *Nanomaterials* **2020**, *10*, 885. [CrossRef] [PubMed]
6. Chen, H.; Jing, L.; Teng, Y.; Wang, J. Multimedia fate modeling and risk assessment of antibiotics in a water-scarce megacity. *J. Hazard. Mater.* **2018**, *348*, 75–83. [CrossRef]
7. Aus der Beek, T.; Weber, F.A.; Bergmann, A.; Hickmann, S.; Ebert, I.; Hein, A.; Küster, A. Pharmaceuticals in the environment-Global occurrences and perspectives. *Environ. Toxicol. Chem.* **2016**, *35*, 823–835. [CrossRef]
8. Liu, J.L.; Wong, M.H. Pharmaceuticals and personal care products (PPCPs): A review on environmental contamination in China. *Environ. Int.* **2013**, *59*, 208–224. [CrossRef]
9. Lindsey, M.E.; Meyer, M.; Thurman, E.M. Analysis of Trace Levels of Sulfonamide and Tetracycline Antimicrobials in Groundwater and Surface Water Using Solid-Phase Extraction and Liquid Chromatography/Mass Spectrometry. *Anal. Chem.* **2001**, *73*, 4640–4646. [CrossRef]
10. Santos, J.L.; Aparicio, I.; Callejón, M.; Alonso, E. Occurrence of pharmaceutically active compounds during 1-year period in wastewaters from four wastewater treatment plants in Seville (Spain). *J. Hazard. Mater.* **2009**, *164*, 1509–1516. [CrossRef]
11. Luo, Y.; Guo, W.; Ngo, H.H.; Nghiem, L.D.; Hai, F.I.; Zhang, J.; Liang, S.; Wang, X.C. A review on the occurrence of micropollutants in the aquatic environment and their fate and removal during wastewater treatment. *Sci. Total Environ.* **2014**, *473–474*, 619–641. [CrossRef] [PubMed]
12. Monteil, H.; Péchaud, Y.; Oturan, N.; Oturan, M.A. A review on efficiency and cost effectiveness of electro- and bio-electro-Fenton processes: Application to the treatment of pharmaceutical pollutants in water. *Chem. Eng. J.* **2019**, *376*, 119577. [CrossRef]
13. Kharel, S.; Stapf, M.; Miehe, U.; Ekblad, M.; Cimbritz, M.; Falås, P.; Nilsson, J.; Sehlén, R.; Bester, K. Ozone dose dependent formation and removal of ozonation products of pharmaceuticals in pilot and full-scale municipal wastewater treatment plants. *Sci. Total Environ.* **2020**, *731*, 139064. [CrossRef] [PubMed]
14. Morris, D.; Harris, S.; Morris, C.; Commins, E.; Cormican, M. *Hospital Effluent: Impact on the Microbial Environment and Risk to Human Health*; Report No. 162; Environmental Protection Agency: Wexford, Ireland, 2016.
15. Zeng, M.; Chen, M.; Huang, D.; Lei, S.; Zhang, X.; Wang, L.; Cheng, Z. Engineered two-dimensional nanomaterials: An emerging paradigm for water purification and monitoring. *Mater. Horiz.* **2021**, *8*, 758–802. [CrossRef]
16. Ihsanullah, I. Boron nitride-based materials for water purification: Progress and outlook. *Chemosphere* **2021**, *263*, 127970. [CrossRef]
17. Yang, Y.; Wu, M.; Zhu, X.; Xu, H.; Ma, S.; Zhi, Y.; Xia, H.; Liu, X.; Pan, J.; Tang, J.-Y.; et al. 2020 Roadmap on two-dimensional nanomaterials for environmental catalysis. *Chin. Chem. Lett.* **2019**, *30*, 2065–2088. [CrossRef]
18. Liu, Y.; Zeng, X.; Hu, X.; Hu, J.; Zhang, X. Two-dimensional nanomaterials for photocatalytic water disinfection: Recent progress and future challenges. *J. Chem. Technol. Biotechnol.* **2019**, *94*, 22–37. [CrossRef]
19. Jeon, M.; Jun, B.M.; Kim, S.; Jang, M.; Park, C.M.; Snyder, S.A.; Yoon, Y. A review on MXene-based nanomaterials as adsorbents in aqueous solution. *Chemosphere* **2020**, *261*, 127781. [CrossRef]
20. Velusamy, S.; Roy, A.; Sundaram, S.; Kumar Mallick, T. A Review on Heavy Metal Ions and Containing Dyes Removal Through Graphene Oxide-Based Adsorption Strategies for Textile Wastewater Treatment. *Chem. Rec.* **2021**, *21*, 1570–1610. [CrossRef]
21. Faysal Hossain, M.D.; Akther, N.; Zhou, Y. Recent advancements in graphene adsorbents for wastewater treatment: Current status and challenges. *Chin. Chem. Lett.* **2020**, *31*, 2525–2538. [CrossRef]
22. Jun, B.M.; Kim, S.; Heo, J.; Park, C.M.; Her, N.; Jang, M.; Huang, Y.; Han, J.; Yoon, Y. Review of MXenes as new nanomaterials for energy storage/delivery and selected environmental applications. *Nano Res.* **2019**, *12*, 471–487. [CrossRef]
23. Yaqub, A.; Shafiq, Q.; Khan, A.R.; Husnain, S.M.; Shahzad, F. Recent advances in the adsorptive remediation of wastewater using two-dimensional transition metal carbides (MXenes): A review. *New J. Chem.* **2021**, *45*, 9721–9742. [CrossRef]
24. Zhang, S.; Gu, P.; Ma, R.; Luo, C.; Wen, T.; Zhao, G.; Cheng, W.; Wang, X. Recent developments in fabrication and structure regulation of visible-light-driven g-C_3N_4-based photocatalysts towards water purification: A critical review. *Catal. Today* **2019**, *335*, 65–77. [CrossRef]
25. Lopes, J.L.; Martins, M.J.; Nogueira, H.I.S.; Estrada, A.C.; Trindade, T. Carbon-based heterogeneous photocatalysts for water cleaning technologies: A review. *Environ. Chem. Lett.* **2021**, *19*, 643–668. [CrossRef]
26. Antonopoulou, M.; Kosma, C.; Albanis, T.; Konstantinou, I. An overview of homogeneous and heterogeneous photocatalysis applications for the removal of pharmaceutical compounds from real or synthetic hospital wastewaters under lab or pilot scale. *Sci. Total Environ.* **2021**, *765*, 144163. [CrossRef]
27. Ren, G.; Han, H.; Wang, Y.; Liu, S.; Zhao, J.; Meng, X.; Li, Z. Recent advances of photocatalytic application in water treatment: A review. *Nanomaterials* **2021**, *11*, 1804. [CrossRef]
28. Shurbaji, S.; Huong, P.T.; Altahtamouni, T.M. Review on the visible light photocatalysis for the decomposition of ciprofloxacin, norfloxacin, tetracyclines, and sulfonamides antibiotics in wastewater. *Catalysts* **2021**, *11*, 437. [CrossRef]
29. Atalay, S.; Ersöz, G. Advanced Oxidation Processes. In *Novel Catalysts in Advanced Oxidation of Organic Pollutants*; Springer: Cham, Switzerland, 2016; ISBN 9781789848908.
30. Chen, G.; Yu, Y.; Liang, L.; Duan, X.; Li, R.; Lu, X.; Yan, B.; Li, N.; Wang, S. Remediation of antibiotic wastewater by coupled photocatalytic and persulfate oxidation system: A critical review. *J. Hazard. Mater.* **2021**, *408*, 124461. [CrossRef]

31. Liu, Y.; Zhao, Y.; Wang, J. Fenton/Fenton-like processes with in-situ production of hydrogen peroxide/hydroxyl radical for degradation of emerging contaminants: Advances and prospects. *J. Hazard. Mater.* **2021**, *404*, 124191. [CrossRef]
32. Wang, J.; Zhuan, R. Degradation of antibiotics by advanced oxidation processes: An overview. *Sci. Total Environ.* **2020**, *701*, 135023. [CrossRef]
33. Zhang, J.; Zhou, Y.; Yao, B.; Yang, J.; Zhi, D. Current progress in electrochemical anodic-oxidation of pharmaceuticals: Mechanisms, influencing factors, and new technique. *J. Hazard. Mater.* **2021**, *418*, 126313. [CrossRef]
34. Bampos, G.; Petala, A.; Frontistis, Z. Recent Trends in Pharmaceuticals Removal from Water Using Electrochemical Oxidation Processes. *Environments* **2021**, *8*, 85. [CrossRef]
35. Dao, K.C.; Yang, C.-C.; Chen, K.-F.; Tsai, Y.-P. Recent Trends in Removal Pharmaceuticals and Personal Care Products by Electrochemical Oxidation and Combined Systems. *Water* **2020**, *12*, 1043. [CrossRef]
36. Liu, L.; Chen, Z.; Zhang, J.; Shan, D.; Wu, Y.; Bai, L.; Wang, B. Treatment of industrial dye wastewater and pharmaceutical residue wastewater by advanced oxidation processes and its combination with nanocatalysts: A review. *J. Water Process Eng.* **2021**, *42*, 102122. [CrossRef]
37. Novoselov, K.S.; Geim, A.K.; Morozov, S.V.; Jiang, D.; Zhang, Y.; Dubonos, S.V.; Grigorieva, I.V.; Firsov, A.A. Electric Field Effect in Atomically Thin Carbon Films. *Science* **2004**, *306*, 666–669. [CrossRef] [PubMed]
38. Marcano, D.C.; Kosynkin, D.V.; Berlin, J.M.; Sinitskii, A.; Sun, Z.; Slesarev, A.; Alemany, L.B.; Lu, W.; Tour, J.M. Improved Synthesis of Graphene Oxide. *ACS Nano* **2010**, *4*, 4806–4814. [CrossRef] [PubMed]
39. Manzeli, S.; Ovchinnikov, D.; Pasquier, D.; Yazyev, O.V.; Kis, A. 2D transition metal dichalcogenides. *Nat. Rev. Mater.* **2017**, *2*, 17033. [CrossRef]
40. Rosenkranz, A.; Costa, H.L.; Baykara, M.Z.; Martini, A. Synergetic effects of surface texturing and solid lubricants to tailor friction and wear—A review. *Tribol. Int.* **2021**, *155*, 106792. [CrossRef]
41. Anasori, B.; Lukatskaya, M.R.; Gogotsi, Y. 2D metal carbides and nitrides (MXenes) for energy storage. *Nat. Rev. Mater.* **2017**, *2*, 16098. [CrossRef]
42. Naguib, M.; Barsoum, M.W.; Gogotsi, Y. Ten Years of Progress in the Synthesis and Development of MXenes. *Adv. Mater.* **2021**, *2103393*, 1–10. [CrossRef]
43. Wyatt, B.C.; Rosenkranz, A.; Anasori, B. 2D MXenes: Tunable Mechanical and Tribological Properties. *Adv. Mater.* **2021**, *33*, 2007973. [CrossRef] [PubMed]
44. Ihsanullah, I. MXenes (two-dimensional metal carbides) as emerging nanomaterials for water purification: Progress, challenges and prospects. *Chem. Eng. J.* **2020**, *388*, 124340. [CrossRef]
45. Rasool, K.; Pandey, R.P.; Rasheed, P.A.; Buczek, S.; Gogotsi, Y.; Mahmoud, K.A. Water treatment and environmental remediation applications of two-dimensional metal carbides (MXenes). *Mater. Today* **2019**, *30*, 80–102. [CrossRef]
46. Dervin, S.; Dionysiou, D.D.; Pillai, S.C. 2D nanostructures for water purification: Graphene and beyond. *Nanoscale* **2016**, *8*, 15115–15131. [CrossRef] [PubMed]
47. Minale, M.; Gu, Z.; Guadie, A.; Kabtamu, D.M.; Li, Y.; Wang, X. Application of graphene-based materials for removal of tetracyclines using adsorption and photocatalytic-degradation: A review. *J. Environ. Manag.* **2020**, *276*, 111310. [CrossRef]
48. Li, M.; Liu, Y.; Zeng, G.; Liu, N.; Liu, S. Graphene and graphene-based nanocomposites used for antibiotics removal in water treatment: A review. *Chemosphere* **2019**, *226*, 360–380. [CrossRef]
49. Lu, L.; Yang, Q.; Xu, Q.; Sun, Y.; Tang, S.; Tang, X.; Liang, H.; Yu, Y. Two-dimensional materials beyond graphene for the detection and removal of antibiotics: A critical review. *Crit. Rev. Environ. Sci. Technol.* **2021**, *51*, 1–32. [CrossRef]
50. Yang, X.; Chen, Z.; Zhao, W.; Liu, C.; Qian, X.; Zhang, M.; Wei, G.; Khan, E.; Hau Ng, Y.; Sik Ok, Y. Recent advances in photodegradation of antibiotic residues in water. *Chem. Eng. J.* **2021**, *405*, 126806. [CrossRef]
51. Jakubczak, M.; Szuplewska, A.; Rozmysłowska-Wojciechowska, A.; Rosenkranz, A.; Jastrzębska, A.M. Novel 2D MBenes—Synthesis, Structure, and Biotechnological Potential. *Adv. Funct. Mater.* **2021**, *31*, 2103048. [CrossRef]
52. Papageorgiou, D.G.; Kinloch, I.A.; Young, R.J. Mechanical properties of graphene and graphene-based nanocomposites. *Prog. Mater. Sci.* **2017**, *90*, 75–127. [CrossRef]
53. Compton, O.C.; Cranford, S.W.; Putz, K.W.; An, Z.; Brinson, L.C.; Buehler, M.J.; Nguyen, S.T. Tuning the Mechanical Properties of Graphene Oxide Paper and Its Associated Polymer Nanocomposites by Controlling Cooperative Intersheet Hydrogen Bonding. *ACS Nano* **2012**, *6*, 2008–2019. [CrossRef] [PubMed]
54. Wu, Z.; Shang, T.; Deng, Y.; Tao, Y.; Yang, Q. The Assembly of MXenes from 2D to 3D. *Adv. Sci.* **2020**, *7*, 1903077. [CrossRef]
55. Quero, F.; Rosenkranz, A. Mechanical Performance of Binary and Ternary Hybrid MXene/Nanocellulose Hydro- and Aerogels—A Critical Review. *Adv. Mater. Interfaces* **2021**, *8*, 2100952. [CrossRef]
56. Ye, S.; Yan, M.; Tan, X.; Liang, J.; Zeng, G.; Wu, H.; Song, B.; Zhou, C.; Yang, Y.; Wang, H. Facile assembled biochar-based nanocomposite with improved graphitization for efficient photocatalytic activity driven by visible light. *Appl. Catal. B Environ.* **2019**, *250*, 78–88. [CrossRef]
57. Rosales, M.; Garcia, A.; Fuenzalida, V.M.; Espinoza-González, R.; Song, G.; Wang, B.; Yu, J.; Gracia, F.; Rosenkranz, A. Unprecedented arsenic photo-oxidation behavior of few- and multi-layer $Ti_3C_2T_x$ nano-sheets. *Appl. Mater. Today* **2020**, *20*, 100769. [CrossRef]
58. Chaves, A.; Azadani, J.G.; Alsalman, H.; da Costa, D.R.; Frisenda, R.; Chaves, A.J.; Song, S.H.; Kim, Y.D.; He, D.; Zhou, J.; et al. Bandgap engineering of two-dimensional semiconductor materials. *Jpn. 2D Mater. Appl.* **2020**, *4*, 29. [CrossRef]

59. Lu, J.; Wu, J.; Carvalho, A.; Ziletti, A.; Liu, H.; Tan, J.; Chen, Y.; Castro Neto, A.H.; Özyilmaz, B.; Sow, C.H. Bandgap Engineering of Phosphorene by Laser Oxidation toward Functional 2D Materials. *ACS Nano* **2015**, *9*, 10411–10421. [CrossRef]
60. Ba, K.; Jiang, W.; Cheng, J.; Bao, J.; Xuan, N.; Sun, Y.; Liu, B.; Xie, A.; Wu, S.; Sun, Z. Chemical and Bandgap Engineering in Monolayer Hexagonal Boron Nitride. *Sci. Rep.* **2017**, *7*, 45584. [CrossRef]
61. Zhao, P.; Jian, M.; Zhang, Q.; Xu, R.; Liu, R.; Zhang, X.; Liu, H. A new paradigm of ultrathin 2D nanomaterial adsorbents in aqueous media: Graphene and GO, MoS_2, MXenes, and 2D MOFs. *J. Mater. Chem. A* **2019**, *7*, 16598–16621. [CrossRef]
62. Zhang, Y.; Yin, Z.; Dai, C.; Zhou, X.; Chen, W. Interfacial thermodynamics and kinetics of sorption of diclofenac on prepared high performance flower-like MoS_2. *J. Colloid Interface Sci.* **2016**, *481*, 210–219. [CrossRef]
63. Liu, G.; Zhang, Z.; Yan, C.; Wang, Y.; Ma, X.; Gao, P.; Feng, Y. Adsorption of estrone with few-layered boron nitride nanosheets: Kinetics, thermodynamics and mechanism. *Chemosphere* **2018**, *207*, 534–542. [CrossRef] [PubMed]
64. Chao, Y.; Zhang, J.; Li, H.; Wu, P.; Li, X.; Chang, H.; He, J.; Wu, H.; Li, H.; Zhu, W. Synthesis of boron nitride nanosheets with N-defects for efficient tetracycline antibiotics adsorptive removal. *Chem. Eng. J.* **2020**, *387*, 124138. [CrossRef]
65. Chao, Y.; Tang, B.; Luo, J.; Wu, P.; Tao, D.; Chang, H.; Chu, X.; Huang, Y.; Li, H.; Zhu, W. Hierarchical porous boron nitride with boron vacancies for improved adsorption performance to antibiotics. *J. Colloid Interface Sci.* **2021**, *584*, 154–163. [CrossRef] [PubMed]
66. Tian, C.; Zhao, H.; Sun, H.; Xiao, K.; Keung Wong, P. Enhanced adsorption and photocatalytic activities of ultrathin graphitic carbon nitride nanosheets: Kinetics and mechanism. *Chem. Eng. J.* **2020**, *381*, 122760. [CrossRef]
67. Zhao, S.; Li, S.; Zhao, Z.; Su, Y.; Long, Y.; Zheng, Z.; Cui, D.; Liu, Y.; Wang, C.; Zhang, X.; et al. Microwave-assisted hydrothermal assembly of 2D copper-porphyrin metal-organic frameworks for the removal of dyes and antibiotics from water. *Environ. Sci. Pollut. Res.* **2020**, *27*, 39186–39197. [CrossRef]
68. Tang, Y.; Huang, H.; Xue, W.; Chang, Y.; Li, Y.; Guo, X.; Zhong, C. Rigidifying induced fluorescence enhancement in 2D porous covalent triazine framework nanosheets for the simultaneously luminous detection and adsorption removal of antibiotics. *Chem. Eng. J.* **2020**, *384*, 123382. [CrossRef]
69. Jiang, W.; Cui, W.; Liang, R.; Qiu, J. Zwitterionic surface charge regulation in ionic covalent organic nanosheets: Synergistic adsorption of fluoroquinolone antibiotics. *Chem. Eng. J.* **2021**, *417*, 128034. [CrossRef]
70. Ghani, A.A.; Shahzad, A.; Moztahida, M.; Tahir, K.; Jeon, H.; Kim, B.; Lee, D.S. Adsorption and electrochemical regeneration of intercalated $Ti_3C_2T_x$ MXene for the removal of ciprofloxacin from wastewater. *Chem. Eng. J.* **2021**, *421*, 127780. [CrossRef]
71. Wu, J.; Zhao, H.; Chen, R.; Pham-Huy, C.; Hui, X.; He, H. Adsorptive removal of trace sulfonamide antibiotics by water-dispersible magnetic reduced graphene oxide-ferrite hybrids from wastewater. *J. Chromatogr. B* **2016**, *1029–1030*, 106–112. [CrossRef]
72. Li, M.; Liu, Y.; Zeng, G.; Liu, S.; Hu, X.; Shu, D.; Jiang, L.; Tan, X.; Cai, X.; Yan, Z. Tetracycline absorbed onto nitrilotriacetic acid-functionalized magnetic graphene oxide: Influencing factors and uptake mechanism. *J. Colloid Interface Sci.* **2017**, *485*, 269–279. [CrossRef]
73. Li, C.; Chen, C.-H.; Chen, W. Different influences of nanopore dimension and pH between chlorpheniramine adsorptions on graphene oxide-iron oxide suspension and particle. *Chem. Eng. J.* **2017**, *307*, 447–455. [CrossRef]
74. Xiao, R.; Wang, S.; Ibrahim, M.H.; Abdu, H.I.; Shan, D.; Chen, J.; Lu, X. Three-dimensional hierarchical frameworks based on molybdenum disulfide-graphene oxide-supported magnetic nanoparticles for enrichment fluoroquinolone antibiotics in water. *J. Chromatogr. A* **2019**, *1593*, 1–8. [CrossRef]
75. Park, C.M.; Heo, J.; Wang, D.; Su, C.; Yoon, Y. Heterogeneous activation of persulfate by reduced graphene oxide–elemental silver/magnetite nanohybrids for the oxidative degradation of pharmaceuticals and endocrine disrupting compounds in water. *Appl. Catal. B Environ.* **2018**, *225*, 91–99. [CrossRef]
76. Zeng, Z.; Ye, S.; Wu, H.; Xiao, R.; Zeng, G.; Liang, J.; Zhang, C.; Yu, J.; Fang, Y.; Song, B. Research on the sustainable efficacy of g-MoS_2 decorated biochar nanocomposites for removing tetracycline hydrochloride from antibiotic-polluted aqueous solution. *Sci. Total Environ.* **2019**, *648*, 206–217. [CrossRef] [PubMed]
77. González, J.A.; Bafico, J.G.; Villanueva, M.E.; Giorgieri, S.A.; Copello, G.J. Continuous flow adsorption of ciprofloxacin by using a nanostructured chitin/graphene oxide hybrid material. *Carbohydr. Polym.* **2018**, *188*, 213–220. [CrossRef]
78. Shahzad, A.; Nawaz, M.; Moztahida, M.; Jang, J.; Tahir, K.; Kim, J.; Lim, Y.; Vassiliadis, V.S.; Woo, S.H.; Lee, D.S. $Ti_3C_2T_x$ MXene core-shell spheres for ultrahigh removal of mercuric ions. *Chem. Eng. J.* **2019**, *368*, 400–408. [CrossRef]
79. Khan, A.R.; Husnain, S.M.; Shahzad, F.; Mujtaba-ul-Hassan, S.; Mehmood, M.; Ahmad, J.; Mehran, M.T.; Rahman, S. Two-dimensional transition metal carbide ($Ti_3C_2T_x$) as an efficient adsorbent to remove cesium (Cs^+). *Dalt. Trans.* **2019**, *48*, 11803–11812. [CrossRef] [PubMed]
80. Zhang, K.; Li, H.; Xu, X.; Yu, H. Synthesis of reduced graphene oxide/NiO nanocomposites for the removal of Cr(VI) from aqueous water by adsorption. *Microporous Mesoporous Mater.* **2018**, *255*, 7–14. [CrossRef]
81. Teoh, W.Y.; Scott, J.A.; Amal, R. Progress in Heterogeneous Photocatalysis: From Classical Radical Chemistry to Engineering Nanomaterials and Solar Reactors. *J. Phys. Chem. Lett.* **2012**, *3*, 629–639. [CrossRef]
82. Tahir, M.; Tasleem, S.; Tahir, B. Recent development in band engineering of binary semiconductor materials for solar driven photocatalytic hydrogen production. *Int. J. Hydrogen Energy* **2020**, *45*, 15985–16038. [CrossRef]
83. Zhao, Z.; Sun, Y.; Dong, F. Graphitic carbon nitride based nanocomposites: A review. *Nanoscale* **2015**, *7*, 15–37. [CrossRef] [PubMed]

84. Wang, X.; Maeda, K.; Thomas, A.; Takanabe, K.; Xin, G.; Carlsson, J.M.; Domen, K.; Antonietti, M. A metal-free polymeric photocatalyst for hydrogen production from water under visible light. *Nat. Mater.* **2009**, *8*, 76–80. [CrossRef] [PubMed]
85. Wang, X.L.; Yang, H.G. Facile fabrication of high-yield graphitic carbon nitride with a large surface area using bifunctional urea for enhanced photocatalytic performance. *Appl. Catal. B Environ.* **2017**, *205*, 624–630. [CrossRef]
86. Shi, Y.; Huang, J.; Zeng, G.; Cheng, W.; Yu, H.; Gu, Y.; Shi, L.; Yi, K. Stable, metal-free, visible-light-driven photocatalyst for efficient removal of pollutants: Mechanism of action. *J. Colloid Interface Sci.* **2018**, *531*, 433–443. [CrossRef] [PubMed]
87. Liu, M.; Zhang, D.; Han, J.; Liu, C.; Ding, Y.; Wang, Z.; Wang, A. Adsorption enhanced photocatalytic degradation sulfadiazine antibiotic using porous carbon nitride nanosheets with carbon vacancies. *Chem. Eng. J.* **2020**, *382*, 123017. [CrossRef]
88. Jian, J.; Jiang, G.; van de Krol, R.; Wei, B.; Wang, H. Recent advances in rational engineering of multinary semiconductors for photoelectrochemical hydrogen generation. *Nano Energy* **2018**, *51*, 457–480. [CrossRef]
89. Guo, H.; Niu, C.-G.; Feng, C.-Y.; Liang, C.; Zhang, L.; Wen, X.-J.; Yang, Y.; Liu, H.-Y.; Li, L.; Lin, L.-S. Steering exciton dissociation and charge migration in green synthetic oxygen-substituted ultrathin porous graphitic carbon nitride for boosted photocatalytic reactive oxygen species generation. *Chem. Eng. J.* **2020**, *385*, 123919. [CrossRef]
90. Wang, W.; Xu, P.; Chen, M.; Zeng, G.; Zhang, C.; Zhou, C.; Yang, Y.; Huang, D.; Lai, C.; Cheng, M.; et al. Alkali Metal-Assisted Synthesis of Graphite Carbon Nitride with Tunable Band-Gap for Enhanced Visible-Light-Driven Photocatalytic Performance. *ACS Sustain. Chem. Eng.* **2018**, *6*, 15503–15516. [CrossRef]
91. Yi, X.; Yuan, J.; Tang, H.; Du, Y.; Hassan, B.; Yin, K.; Chen, Y.; Liu, X. Embedding few-layer $Ti_3C_2T_x$ into alkalized g-C3N4 nanosheets for efficient photocatalytic degradation. *J. Colloid Interface Sci.* **2020**, *571*, 297–306. [CrossRef]
92. Ahern, J.C.; Fairchild, R.; Thomas, J.S.; Carr, J.; Patterson, H.H. Characterization of BiOX compounds as photocatalysts for the degradation of pharmaceuticals in water. *Appl. Catal. B Environ.* **2015**, *179*, 229–238. [CrossRef]
93. Sharma, K.; Dutta, V.; Sharma, S.; Raizada, P.; Hosseini-Bandegharaei, A.; Thakur, P.; Singh, P. Recent advances in enhanced photocatalytic activity of bismuth oxyhalides for efficient photocatalysis of organic pollutants in water: A review. *J. Ind. Eng. Chem.* **2019**, *78*, 1–20. [CrossRef]
94. Tang, L.; Lv, Z.; Xue, Y.; Xu, L.; Qiu, W.; Zheng, C.; Chen, W.; Wu, M. MIL-53(Fe) incorporated in the lamellar BiOBr: Promoting the visible-light catalytic capability on the degradation of rhodamine B and carbamazepine. *Chem. Eng. J.* **2019**, *374*, 975–982. [CrossRef]
95. Gao, X.; Peng, W.; Tang, G.; Guo, Q.; Luo, Y. Highly efficient and visible-light-driven BiOCl for photocatalytic degradation of carbamazepine. *J. Alloys Compd.* **2018**, *757*, 455–465. [CrossRef]
96. Xu, J.; Wang, Y.; Niu, J.; Chen, M. Facile construction of BiOBr/BiOCOOH p-n heterojunction photocatalysts with improved visible-light-driven photocatalytic performance. *Sep. Purif. Technol.* **2019**, *225*, 24–32. [CrossRef]
97. Chen, Y.; Tian, H.; Zhu, W.; Zhang, X.; Li, R.; Chen, C.; Huang, Y. L-Cysteine directing synthesis of BiOBr nanosheets for efficient cefazolin photodegradation: The pivotal role of thiol. *J. Hazard. Mater.* **2021**, *414*, 125544. [CrossRef] [PubMed]
98. Wang, C.-Y.; Zhang, X.; Qiu, H.-B.; Huang, G.-X.; Yu, H.-Q. $Bi_{24}O_{31}Br_{10}$ nanosheets with controllable thickness for visible-light-driven catalytic degradation of tetracycline hydrochloride. *Appl. Catal. B Environ.* **2017**, *205*, 615–623. [CrossRef]
99. Wang, X.; Bi, W.; Zhai, P.; Wang, X.; Li, H.; Mailhot, G.; Dong, W. Adsorption and photocatalytic degradation of pharmaceuticals by $BiOClxI_y$ nanospheres in aqueous solution. *Appl. Surf. Sci.* **2016**, *360*, 240–251. [CrossRef]
100. Zhang, L.; Yue, X.; Liu, J.; Feng, J.; Zhang, X.; Zhang, C.; Li, R.; Fan, C. Facile synthesis of Bi_5O_7Br/BiOBr 2D/3D heterojunction as efficient visible-light-driven photocatalyst for pharmaceutical organic degradation. *Sep. Purif. Technol.* **2020**, *231*, 115917. [CrossRef]
101. Li, H.; Long, B.; Ye, K.-H.; Cai, Y.; He, X.; Lan, Y.; Yang, Z.; Ji, H. A recyclable photocatalytic tea-bag-like device model based on ultrathin Bi/C/BiOX (X = Cl, Br) nanosheets. *Appl. Surf. Sci.* **2020**, *515*, 145967. [CrossRef]
102. Shen, G.; Pu, Y.; Sun, R.; Shi, Y.; Cui, Y.; Jing, P. Enhanced visible light photocatalytic performance of a novel heterostructured $Bi_4Ti_3O_{12}$/BiOBr photocatalyst. *New J. Chem.* **2019**, *43*, 12932–12940. [CrossRef]
103. Sivula, K.; van de Krol, R. Semiconducting materials for photoelectrochemical energy conversion. *Nat. Rev. Mater.* **2016**, *1*, 15010. [CrossRef]
104. Xu, P.; Shen, X.; Luo, L.; Shi, Z.; Liu, Z.; Chen, Z.; Zhu, M.; Zhang, L. Preparation of TiO_2/Bi_2W_6 nanostructured heterojunctions on carbon fibers as a weaveable visible-light photocatalyst/photoelectrode. *Environ. Sci. Nano* **2018**, *5*, 327–337. [CrossRef]
105. Leong, S.; Li, D.; Hapgood, K.; Zhang, X.; Wang, H. $Ni(OH)_2$ decorated rutile TiO_2 for efficient removal of tetracycline from wastewater. *Appl. Catal. B Environ.* **2016**, *198*, 224–233. [CrossRef]
106. Lin, L.; Jiang, W.; Nasr, M.; Bechelany, M.; Miele, P.; Wang, H.; Xu, P. Enhanced visible light photocatalysis by TiO_2–BN enabled electrospinning of nanofibers for pharmaceutical degradation and wastewater treatment. *Photochem. Photobiol. Sci.* **2019**, *18*, 2921–2930. [CrossRef]
107. Taoufik, N.; Boumya, W.; Achak, M.; Sillanpää, M.; Barka, N. Comparative overview of advanced oxidation processes and biological approaches for the removal pharmaceuticals. *J. Environ. Manag.* **2021**, *288*, 112404. [CrossRef]
108. Ma, M.; Chen, L.; Zhao, J.; Liu, W.; Ji, H. Efficient activation of peroxymonosulfate by hollow cobalt hydroxide for degradation of ibuprofen and theoretical study. *Chin. Chem. Lett.* **2019**, *30*, 2191–2195. [CrossRef]
109. Ding, M.; Chen, W.; Xu, H.; Shen, Z.; Lin, T.; Hu, K.; Lu, C.h.; Xie, Z. Novel α-Fe_2O_3/MXene nanocomposite as heterogeneous activator of peroxymonosulfate for the degradation of salicylic acid. *J. Hazard. Mater.* **2020**, *382*, 121064. [CrossRef]

110. Yang, L.; Chen, H.; Jia, F.; Peng, W.; Tian, X.; Xia, L.; Wu, X.; Song, S. Emerging Hexagonal Mo_2C Nanosheet with (002) Facet Exposure and Cu Incorporation for Peroxymonosulfate Activation Toward Antibiotic Degradation. *ACS Appl. Mater. Interfaces* **2021**, *13*, 14342–14354. [CrossRef]
111. Nie, M.; Li, Y.; Li, L.; He, J.; Hong, P.; Zhang, K.; Cai, X.; Kong, L.; Liu, J. Ultrathin iron-cobalt oxide nanosheets with enhanced H_2O_2 activation performance for efficient degradation of tetracycline. *Appl. Surf. Sci.* **2021**, *535*, 147655. [CrossRef]
112. Xing, D.; Cui, Z.; Liu, Y.; Wang, Z.; Wang, P.; Zheng, Z.; Cheng, H.; Dai, Y.; Huang, B. Two-dimensional π–d conjugated metal–organic framework Fe_3(hexaiminotriphenylene)$_2$ as a photo-Fenton like catalyst for highly efficient degradation of antibiotics. *Appl. Catal. B Environ.* **2021**, *290*, 120029. [CrossRef]
113. Mi, X.; Han, J.; Sun, Y.; Li, Y.; Hu, W.; Zhan, S. Enhanced catalytic degradation by using RGO-Ce/WO_3 nanosheets modified CF as electro-Fenton cathode: Influence factors, reaction mechanism and pathways. *J. Hazard. Mater.* **2019**, *367*, 365–374. [CrossRef]
114. Tang, H.; Shang, Q.; Tang, Y.; Yi, X.; Wei, Y.; Yin, K.; Liu, M.; Liu, C. Static and continuous flow photoelectrocatalytic treatment of antibiotic wastewater over mesh of TiO_2 nanotubes implanted with g-C_3N_4 nanosheets. *J. Hazard. Mater.* **2020**, *384*, 121248. [CrossRef]
115. Wang, J.; Xu, M.; Liang, X.; Zhang, Y.; Yang, D.; Pan, L.; Fang, W.; Zhu, C.; Wang, F. Development of a novel 2D Ni-MOF derived NiO@C nanosheet arrays modified Ti/TiO2NTs/PbO_2 electrode for efficient electrochemical degradation of salicylic acid wastewater. *Sep. Purif. Technol.* **2021**, *263*, 118368. [CrossRef]
116. Lee, X.J.; Hiew, B.Y.Z.; Lai, K.C.; Lee, L.Y.; Gan, S.; Thangalazhy-Gopakumar, S.; Rigby, S. Review on graphene and its derivatives: Synthesis methods and potential industrial implementation. *J. Taiwan Inst. Chem. Eng.* **2019**, *98*, 163–180. [CrossRef]
117. Naguib, M.; Unocic, R.R.; Armstrong, B.L.; Nanda, J. Large-scale delamination of multi-layers transition metal carbides and carbonitrides "MXenes". *Dalt. Trans.* **2015**, *44*, 9353–9358. [CrossRef]
118. Limbu, T.B.; Chitara, B.; Orlando, J.D.; Garcia Cervantes, M.Y.; Kumari, S.; Li, Q.; Tang, Y.; Yan, F. Green synthesis of reduced $Ti_3C_2T_X$ MXene nanosheets with enhanced conductivity, oxidation stability, and SERS activity. *J. Mater. Chem. C* **2020**, *8*, 4722–4731. [CrossRef]
119. Guo, X.; Mei, N. Assessment of the toxic potential of graphene family nanomaterials. *J. Food Drug Anal.* **2014**, *22*, 105–115. [CrossRef] [PubMed]
120. Cossutta, M.; Vretenar, V.; Centeno, T.A.; Kotrusz, P.; McKechnie, J.; Pickering, S.J. A comparative life cycle assessment of graphene and activated carbon in a supercapacitor application. *J. Clean. Prod.* **2020**, *242*, 118468. [CrossRef]

processes

Article

Optimization, Kinetics, Thermodynamic and Arrhenius Model of the Removal of Ciprofloxacin by Internal Electrolysis with Fe-Cu and Fe-C Materials

Tra Huong Do [1,*], Xuan Linh Ha [2], Thi Tu Anh Duong [1], Phuong Chi Nguyen [1], Ngoc Bich Hoang [3] and Thi Kim Ngan Tran [3,4,*]

1 Chemistry Faculty, Thai Nguyen University of Education, Thai Nguyen 250000, Vietnam; anhttd@tnue.edu.vn (T.T.A.D.); phuongchi0715@gmail.com (P.C.N.)
2 International School, Thai Nguyen University, Thai Nguyen 250000, Vietnam; haxuanlinh@tnu.edu.vn
3 Institute of Environmental Technology and Sustainable Development, Nguyen Tat Thanh University, Ho Chi Minh 700000, Vietnam; bichhn@ntt.edu.vn
4 Faculty of Food and Environmental Engineering, Nguyen Tat Thanh University, Ho Chi Minh 700000, Vietnam
* Correspondence: huongdt.chem@tnue.edu.vn (T.H.D.); nganttk@ntt.edu.vn (T.K.N.T.); Tel.: +84-392-073-898 (T.H.D.); +84-765-712-086 (T.K.N.T.)

Abstract: The ciprofloxacin (CIP) removal ability of a Fe-Cu electrolytic material was examined with respect to pH (2–9), time (15–150 min), shaking speed (100–250 rpm), material mass (0.2–3 g/L), temperature (298, 308, 323) and initial CIP concentration (30–200 mg/L). The Fe-Cu electrolytic materials were fabricated by the chemical plating method, and Fe-C materials were mechanically mixed from iron powder and graphite. The results show that at a pH value of 3, shaking time 120 min, shaking speed 250 rpm, a mass of Fe-Cu, Fe-C material of 2 g/L and initial CIP concentration of 203.79 mg/L, the CIP removal efficiency of Fe-Cu material reached 90.25% and that of Fe-C material was 85.12%. The removal of CIP on Fe-Cu and Fe-C materials follows pseudo-first-order kinetics. The activation energy of CIP removal of Fe-Cu material is 14.93 KJ/mol and of Fe-C material is 16.87 KJ/mol. The positive ΔH proves that CIP removal is endothermic. A negative entropy of 0.239 kJ/mol and 0.235 kJ/mol (which is near zero and is also relatively positive) indicated the rapid removal of the CIP molecules into the removed products.

Keywords: internal microelectrolysis; Fe-Cu and Fe-C; remove; ciprofloxacin; aqueous solution

Citation: Do, T.H.; Ha, X.L.; Duong, T.T.A.; Nguyen, P.C.; Hoang, N.B.; Tran, T.K.N. Optimization, Kinetics, Thermodynamic and Arrhenius Model of the Removal of Ciprofloxacin by Internal Electrolysis with Fe-Cu and Fe-C Materials. *Processes* **2021**, *9*, 2110. https://doi.org/10.3390/pr9122110

Academic Editors: Dimitris Zagklis and Georgios Bampos

Received: 11 October 2021
Accepted: 16 November 2021
Published: 24 November 2021

1. Introduction

Antibiotic resistance, a serious problem to global health, is currently increasing to dangerous levels, especially in developing countries [1,2]. Since antibiotic residue is one of the reasons that cause antibiotic resistance, significant efforts have been devoted to developing techniques for the removal of antibiotic residue from water environments. Techniques that hold potential in the removal of antibiotics are diverse and might include adsorption [3], photocatalytic removal or advanced oxidation [4], water-soluble polymers [5–7]. Internal micro electrolysis is a technique that has been gaining popularity recently in the pretreatment process of wastewater, especially industrial wastewater. This method has been frequently applied to treat wastewater containing persistent organic substances with high concentrations of pollutants, usually in wastewater containing textile dyes [8–10], pharmaceuticals [11,12], paper chemicals [13], coal gasification [14], Cu-EDTA complexes [15], nitrogen compounds from sewage [16], electroplating and mechanical industrial discharges [17–21], oily substances [22] and TNT and RDX explosives [23].

The process of internal electrolysis commences when two materials, having different electrode potentials, contact and form a pair of microelectrodes. For Fe-C, Fe-Cu systems, iron plays the role of anode and copper or carbon is the cathode, which is similar to the pair of micro-batteries in metal corrosion. With a pair of micro-batteries with a potential of about 1.2 V, a small current of about μA appears, acting as a redox agent in the removal reaction of organic compounds adsorbed on the electrode surface. Due to such a principle, the process of Fe-C and Fe-Cu microelectrolysis is also called internal microelectrolysis. Therefore, it is possible to dissolve iron without using an external current by establishing microbatteries in the form of Fe-C or Fe-Cu composite materials, which is an important advantage in the internal engineering of wastewater pretreatment electrolysis [8–27]. The reactions that take place during electrolysis are as follows:

$$\text{Reaction at the anode (oxidation): } Fe \rightarrow Fe^{2+} + 2e \tag{1}$$

$$\text{Reaction at the cathode (reduction): } 2H^{+} + 2e \rightarrow 2[H] \rightarrow H_2 \tag{2}$$

When the oxygen is presented: acid conditions [24]:

$$O_2 + 4H^{+} + 4e \rightarrow 2O + 4[H] \rightarrow 2H_2O \tag{3}$$

In addition, the removal of organic compounds in the internal electrolysis method is also thought to be based on the reduction of iron at valence, the reduction of [H], the oxidation of ·O, the formation of complexes, the colloidal accumulation of iron ions and the adsorption of iron hydroxide [25].

According to Yang [25], Fe-C or Fe-Cu materials embedded in the electrolyte will form corrosive batteries with electronegative metal corroded from the anodic reaction: $Me \rightarrow Me^{n+} + ne$. Corresponding to the anodic process is the cathode depolarization process on the more electropositive part, such as releasing H_2 gas in an acidic environment, depolarizing dissolved oxygen or other substances or ions. In the presence of Fe^{2+} and H_2O_2 in the electrolyte, there can also be a Fenton reaction to form ·OH radicals. If in the solution the organic substances RX (organochlorine compound) and RNO_2 (aromatic nitrocyclic compound) are components capable of accepting electrons on the anode surface (Fe) to transfer to the cathode, they are deduced according to dechlorination and deaminoation reactions. Then the pollutant will be converted into non-toxic or less toxic intermediate products, which are easier to biodegrade. Therefore, in terms of chemical and physicochemical natures, the process of internal electrolysis will be different from the process of non-valent iron in the treatment of pollutant waste (producing A-sized internal electrolyte and producing radicals, OH).

Based on the above-mentioned advantages and analysis, the method of internal electrolysis is regarded as an efficient and affordable technology for decomposing antibiotics in aqueous media. There have been many studies on Fe-Cu and Fe-C materials; however, research on CIP removal on these two materials is still limited, and no author has fully studied the dynamic process, thermodynamics or activation energy of CIP removal of Fe-Cu and Fe-C. In order to improve the efficiency of antibiotic treatment in this paper, we studied the influence of factors, such as pH, treatment time, material weight, shaking speed, concentration and temperature, on the yield, study of kinetics, thermodynamics and Arrhenius model in the removal of ciprofloxacin in aqueous solution of Fe-C and Fe-Cu electrolytic materials.

2. Materials and Methods

2.1. Producing Fe-C and Fe-Cu Materials

2.1.1. Producing Fe-C Materials

A mixture was formed with the following ingredients: 95% Fe, 3% graphite and 2% bentonite binder additive. The material was dried at 80–105 °C for 2 h, then underwent mass sintering at 500–600 °C for 4 h, followed by natural cooling. The material was then stored in a desiccator for further studies [26].

2.1.2. Producing Fe-Cu materials

Fe powder with a size of less than 50 μm and purity of 99.9% (PA, China) was soaked in 30% NaOH solution for 10 min to remove grease and clean the surface. The surface was activated by treatment in 7.4% HCl for 3 min. A diluted HCl solution was prepared with 37% HCl solution. The material was then washed several times with water, dried at 105 °C for 2 h, then allowed to cool and stored in a sealed glass jar. The Fe-Cu sample was fabricated by chemical plating in 5% $CuSO_4$ solution (wt%). To be specific, a total of 100 g of Fe powder was added to 1 L of 5% $CuSO_4$ solution for a period of 2 min. The mixture was then washed several times with water and dried at 105 °C for 3 h under N_2. The material was then stored in a desiccator for further study [27].

2.1.3. Investigation of Surface Characteristics, Structure and Chemical Composition of Fe-C and Fe-Cu Electrolytic Materials

The surface morphological characteristics of Fe-C materials and Fe-Cu were determined using scanning electron microscopy (SEM). The material structure was analyzed by X-ray diffraction (XRD). The material composition was analyzed by energy dispersive spectroscopy (EDX). The specific surface area of the materials was analyzed by N_2 adsorption-desorption isotherm (BET). The results are shown in documents [26,27].

2.2. *Ciprofloxacin*

Ciprofloxacin (CIP) is an antibiotic of the 4 quinolone group of the 2nd generation of the Fluoroquinolone antibiotic system [28].

Molecular formula: $C_{17}H_{18}FN_3O_3$.

Molar Mass: 331.346.

The UV-Vis spectrum of CIP in aqueous solution is shown in Figure 1.

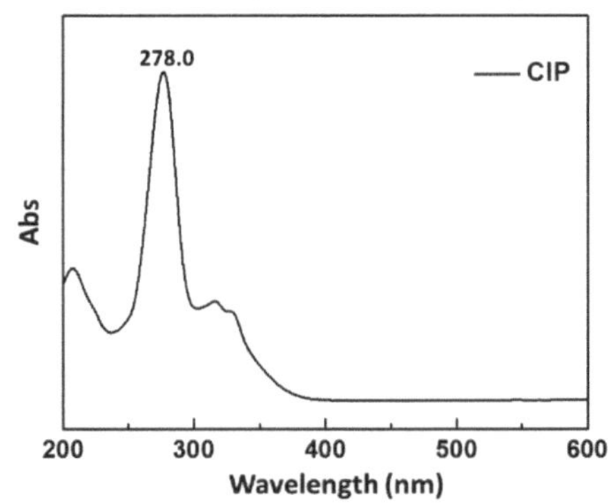

Figure 1. UV-Vis Spectrum of CIP, concentration 10 mg/L, pH = 3.

2.3. *CIP Removal*

The experimental factors, such as solution pH, time, the weight of Fe-Cu and Fe-C materials, shaking speed, initial CIP concentration and heat, were surveyed to evaluate the process of CIP removal. To ensure repeatability, each experiment was performed at least 3 times under the same conditions. The results are presented as the average of the 3 experiments. Briefly, a certain amount of Fe-Cu and Fe-C materials was introduced into each 100 mL Erlenmeyer flask, followed by the addition of a certain amount of CIP solution. A total of 1 mol/L NaOH and 0.1 mol/L HCl were used to adjust the pH of the CIP solution. The flasks were shaken on a shaker with conditions varying depending on the experiment Table 1).

Table 1. Parameters in the experiments.

Experiment/Parameters	pH	Time (min)	Material Mass (g)	Shaking Speed (rpm)	CIP Initial Concentration (mg/L)	Temperature (K)
Effect of pH	2 to 9	120	0.1	250	50	298
Effect of time	3	15–180	0.1	250	50	298
Effect of mass of initial materials	3	120	0.01–0.15	250	50	298
Effect of shaking speed	3	120	0.1	100–250	50	298
Effect of CIP initial concentration	3	120	0.1	250	50	298
Effect of temperature	3	120	0.1	250	100	298, 308, 323

The concentrations of CIP before and after treatment with Fe-Cu and Fe-C materials were determined by ultraviolet-visible spectroscopy measured on Hitachi UH5300 machine at University of Medicine and Pharmacy—University of Science and Technology Thai Nguyen. They were measured with a wavelength range from 190 to 1100 nm, a scanning speed from 10 to 6000 nm/s, a wavelength accuracy of ±0.3 nm and noise <0.0001 nm.

The CIP removal efficiency was calculated according to the following formula:

$$H\% = \frac{(C_0 - C_t)}{C_0} \times 100\% \quad (4)$$

where C_0 is the initial concentration of CIP solution before removal (mg/L), C_t is the concentration of CIP solution after removal (mg/L) and H is the removal efficiency (%).

The Arrhenius formula is:

$$\ln k = -\frac{E_a}{RT} + \ln A_o \quad (5)$$

where k is the reaction rate constant.

E_a (KJ/mol) is the Arrhenius activation energy.

A_o is the factor that has the same value and has the same units as k.

The enthalpy change of the CIP removal process could be identified as the E_a-value, which could be calculated by the following equation with the assumption that 1 mol of CIP molecules was degraded [29].

$$\Delta H^0 = E_a + RT \quad (6)$$

After obtaining E_a, the entropy change of the reaction could be calculated from the following equation, in which the K_B (Boltzmann constant) and h (Plank constant) are known.

$$\ln(k/T) = \ln(K_B/h) + \Delta S^0/RT - (\Delta H^0/RT) \quad (7)$$

Plotting ln(k/T) against ($\Delta H^0/RT$) yielded an intercept value that is interpreted as ΔS^0 and is used for the calculation of the activation Gibbs free energy as follows [29]:

$$\Delta G^0 = \Delta H^0 - T\Delta S \quad (8)$$

3. Results and Discussion

3.1. Effect of Initial pH on CIP Removal

From Equations (1) and (2), it is shown that the pH value has a great influence on the reaction rate and the redox's ability to generate [H]. When the pH becomes more acidic, the amount of H^+ provided for the reaction is enough or in excess and can make the rate of electrolysis faster. As the initial pH value becomes lower, the [H] concentration becomes higher. Moreover, in the presence of O_2, the cathode reduction of the internal electrolysis reaction can also occur according to the following reaction:

$$O_2 + 4H^+ + 4e \rightarrow 2O\cdot + 4[H] \rightarrow 2H_2O;\ E_0\ (O_2/H_2O) = 1.23\ V \quad (9)$$

Thus, more H^+ will produce more [H] and O, leading to the higher redox capacity of CIP and achieving better CIP treatment efficiency.

On the other hand, the reactions of Fe in solutions with different pH can be represented by the following equations [25]:

$$Fe \rightarrow Fe^{2+} + 2e \tag{10}$$

$$Fe + 2H^+ \rightarrow Fe^{2+} + H_2 \tag{11}$$

$$Fe^{2+} \rightarrow Fe^{3+} + e \tag{12}$$

According to the Nernst Equation, the reducing ability of Fe^{2+}/Fe will increase as the pH decreases. The initial pH value also affects the rate of corrosion reactions of Fe-C; Fe-Cu materials to form Fe^{2+}, Fe^{3+}, $Fe(OH)_2$ and $Fe(OH)_3$. In a more acidic environment than Fe^{2+}, Fe^{3+} is easily formed, but it is difficult to precipitate $Fe(OH)_2$ and $Fe(OH)_3$. On the contrary, when the pH is high, the acidity decreases and the presence of dissolved oxygen will easily form $Fe(OH)_2$ and $Fe(OH)_3$, and the concentration will be gradually increased with the reaction time. Iron hydroxides are also indirect agents for the partial removal of CIP, as well as intermediate compounds of the treatment process by adsorption, coagulation and precipitation.

As shown in Figure 2, CIP removal under treatment with Fe-C and Fe-Cu materials decreases with increasing pH value. With Fe-Cu materials, the results show that the CIP removal efficiency increases when the pH increases from 2 to 3, eventually peaking at pH = 3 with a removal efficiency of 93.97%. In the increasing pH range from 3 to 9, the efficiency decreases again; at pH = 9, the removal efficiency was only 56.29%. Similar to Fe-Cu materials, Fe-C materials peak when investigating pHs in the range from 2 to 9, the highest removal efficiency was 88.90%, achieved at pH = 3, when the pH gradually approached the basic environment, the removal efficiency decreased gradually. This is explained by the reduced concentration of Fe^{2+} and Fe^{3+} ions as the pH increases, leading to a decrease in CIP removal by oxidation-reduction, chemical reactions and flocculation. However, due to the adsorption of activated carbon in Fe-C material [25], the CIP removal efficiency of this material is reduced compared to that of Fe-Cu at pH values ≤ 3. As a result, we chose pH = 3 as the optimal pH value for CIP treatment of both Fe-Cu and Fe-C materials.

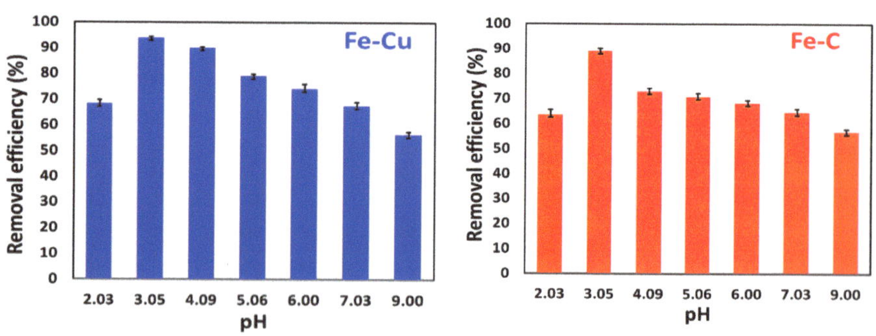

Figure 2. Effect of initial pH on CIP removal under Fe-Cu and Fe-C treatment.

3.2. *Effect of Reaction Time on CIP Removal*

Figure 3 shows that at pH = 3, when investigating at a time interval of 15–120 min, the CIP removal efficiency of Fe-Cu and Fe-C both increased rapidly. For Fe-Cu material, the CIP removal efficiency increased from 64.52% to 94.39%, higher than that of Fe-C material, which increased from 52.75% to 91.68%. During the next period of 120–180 min, the removal efficiency of both materials increased quite slowly (about 1%/30 min). This can be explained as follows: as the reaction time increases, the concentration of [H], Fe^{2+} and Fe^{3+} generated in the electrochemical reaction increases and the CIF content decomposed more. When the reaction time is too long, the CIP removal rate of the internal electrolytic material will be slower, possibly because a large amount of H^+ is consumed along with the increase in the reaction time, leading to a decrease in [H] and increased oxygen content in the aqueous material, which increased the amount of OH^- participating in the reaction with H^+, thereby reducing [H] [27]. Therefore, the time of 120 min was chosen as the optimal time for CIP processing. Based on the results in Figure 4, it can be seen that at the same initial investigation conditions, Fe-Cu materials have a better CIP removal efficiency than Fe-C materials.

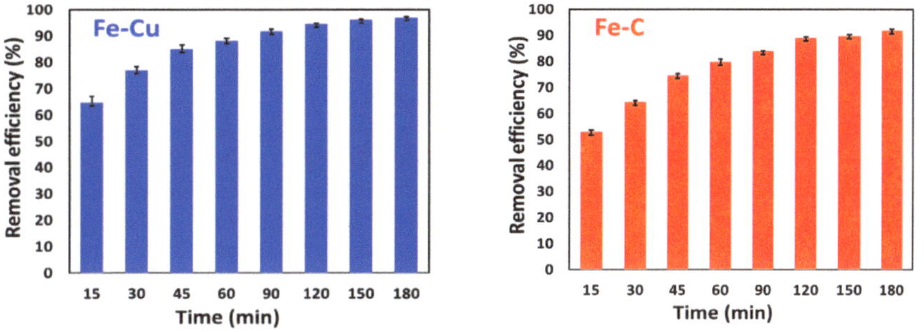

Figure 3. Effect of reaction time on CIP removal under Fe-Cu and Fe-C treatment.

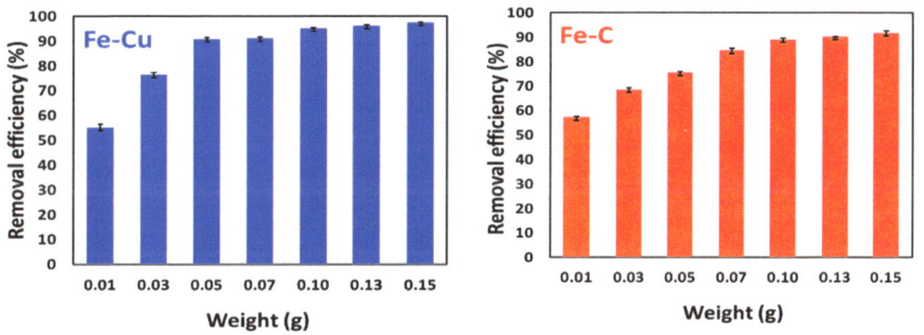

Figure 4. Effect of material mass on CIP removal under Fe-Cu and Fe-C treatment.

3.3. Effect of Material Mass on CIP Removal

When the mass of Fe-Cu increased from 0.01 to 0.10 g, the CIP removal efficiency increased sharply (54.97–95.01%). When the material weight increased from 0.10 to 0.15 g, the CIP removal efficiency increased more slowly and insignificantly (about 1%), which is similar to Fe-C materials. However, under the same initial material mass conditions, the CIP removal treated under Fe-Cu materials is always higher than that of Fe-C materials. This can be explained as follows: when the amount of Fe-C and Fe-Cu increased, more microbatteries were formed, which will lead to an increase in current efficiency during internal electrolysis. Therefore, the CIP removal efficiency increased. However, the sufficiently large quantity of iron existing in the material can inhibit electrode reactions. Therefore, the removal of contaminants should resort to the action of noncovalent iron [30], which holds a weaker removal capacity than the internal electrolysis [31]. Therefore, we choose the optimal mass of Fe-C and Fe-Cu materials for the CIP removal to be 0.1 g or 2 g/L to conduct further studies.

3.4. Effect of Shaking Speed

The shaking speed increases the dissolved oxygen content of the solution and the possibility of pollutant diffusion to the Fe-C and Fe-Cu electrode contact surfaces, as well as the rapid dispersion of the treated products at the electrodes polar into the solution. However, under a low pH, the dissolved oxygen content is less than in an alkaline medium. The influence of shaking speed on CIP removal efficiency can be explained by the following reasons:

(1) The contact ability between the liquid phase (CIP solution)–solid phase (Cu and C cathode): the higher the shaking speed is, the larger the amount of CIP solution in contact with the cathode and vice versa. In addition, the faster shaking speed also increases the transport speed of the CIP solution and intermediate removal products to participate in the reaction between the (liquid) phase and the solid phase.

(2) Increasing the dissolved oxygen content in the solution: the higher the shaking speed is, the greater the dissolved oxygen content in the reaction solution is. In other words, oxygen in the air readily dissolves into the reaction solution under conditions of high shaking speeds. The process of internal electrolysis is the process of self-corrosion of iron at the anode (sacrificial electrode); then if there is oxygen at the cathode, the oxygen will be reduced and combined with the H^+ available in the acidic environment to produce H_2O_2. This agent then meets the nascent Fe^{2+}, where the Fenton reaction occurs in a pH of 2 to 3 to generate ·OH radicals, those ·OH radicals are capable of strongly oxidizing CIP. Therefore, CIP is not only reduced on the surface of the Cu and C cathodes but also enhanced by the Fenton mechanism in an acidic environment during internal electrolysis.

Figure 5 shows that when the shaking speed is higher, the CIP removal efficiency of Fe-Cu and Fe-C materials also increases. Therefore, we chose a shaking speed of 250 rpm as the optimal speed for the next studies.

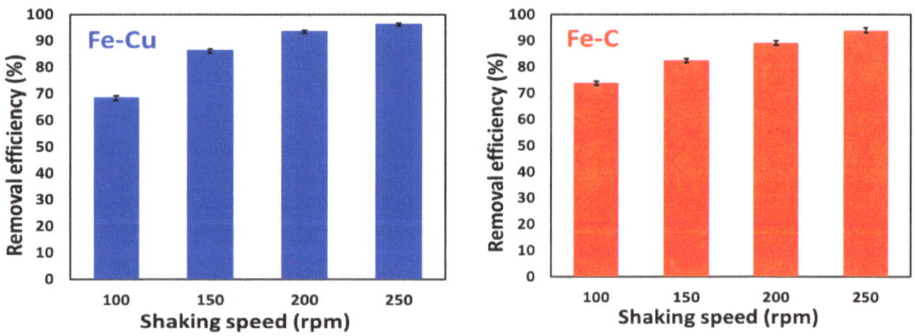

Figure 5. Effect of shaking speed on CIP removal under Fe-Cu and Fe-C treatment.

3.5. Effect of Initial Concentration CIP

As shown in Figure 6, when the initial CIP concentration is 33.44 mg/L, the CIP removal efficiency of both Fe-Cu and Fe- C materials is greater than 95.5%. When increasing the CIP concentration from 51.19 to 203.79 mg/L, the CIP removal efficiency of Fe-Cu and Fe-C materials both decreased gradually. With an initial CIP concentration of 203.79 mg/L, the CIP removal efficiency of Fe-Cu materials was 90.25% and Fe-C materials was 85.12%. This can be explained as follows: CIP removal depends on oxidation reactions by free radicals, mainly ·HO (Figure 6). At low CIP concentrations, the reaction rate will be less constrained by concentration. Furthermore, the lifetime of ·HO is too short to participate in the reaction with CIP. As the initial CIP concentration is gradually increased, the probability of a reaction between CIP and HO* increases, so the removal efficiency of CIP increases. However, at a high concentration of CIP solution, the mass of the material remains the same, the amount of ·HO formed in the internal electrolysis method decreases, so the CIP removal efficiency decreases [9]. However, the results from Figure 6 also show that Fe-Cu and Fe-C materials still give high CIP removal efficiency when the concentration is large.

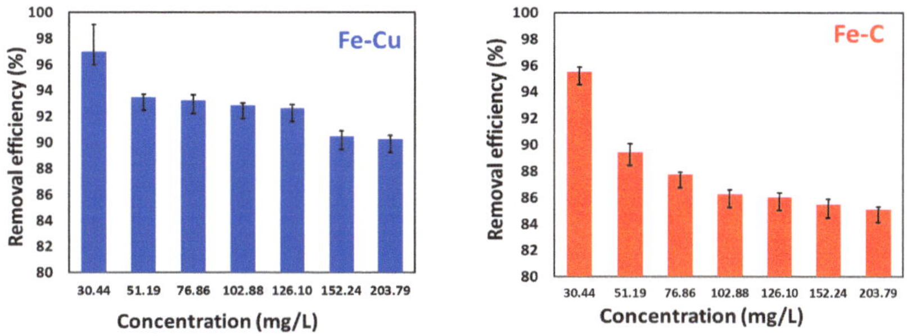

Figure 6. Effect of initial concentration CIP on CIP removal under Fe-Cu and Fe-C treatment.

3.6. *Effect of Reaction Temperature on CIP Removal*

The survey results show that, for both Fe-Cu and Fe-C materials, when the temperature increases, the CIP removal efficiency increases. This can be explained as follows: as the temperature increases, the reaction rate increases, the ·HO radical concentration increases, so the CIP removal efficiency increases. Figure 7 shows that the CIP removal reaction is endothermic.

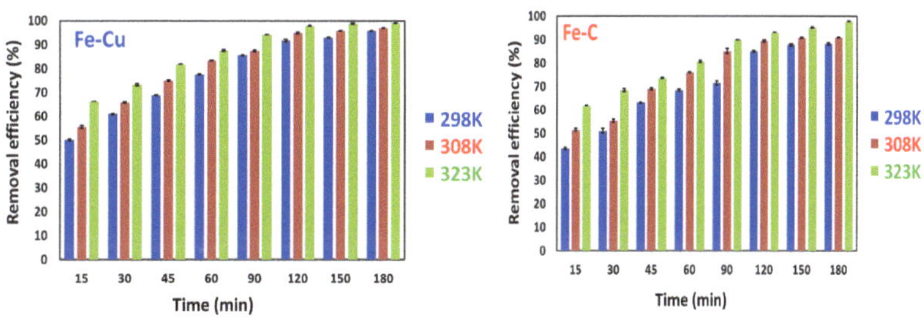

Figure 7. Effect of reaction temperature on CIP removal under Fe-Cu and Fe-C treatment.

3.7. *Kinetics of CIP Removal Treated by Fe-C and Fe-Cu Materials*

For the results of the influence of time on the CIP removal efficiency, we conducted a kinetic investigation of the CIP removal process according to the apparent kinetic equations of first, second and third as follows:

$$\text{The first apparent kinematics equation: } \ln C_t = -\, k_1.t + A_1 \quad (13)$$

$$\text{The second apparent kinematics equation: } 1/C_t = k_2.t + A_2 \quad (14)$$

$$\text{The third apparent kinematics equation: } 1/Ct^2 = 2k_3.t + A_3 \quad (15)$$

where k_1, k_2 and k_3 are the first, second and third-order reaction rate constants, respectively, and A_1, A_2, and A_3 are constants.

The results of the first, second and third order kinetic equations of CIP removal in aqueous solution of Fe-Cu and Fe-C materials are shown in Figure 8.

The results of Table 2 show that the CIP removal process at 298, 308 and 323 K of both Fe-Cu and Fe-C materials follows the apparent first-order kinetic model because the correlation coefficient of R^2 is high (>0.95), compared to the second (>0.8094)) and third-order (>0.5879) kinetic models.

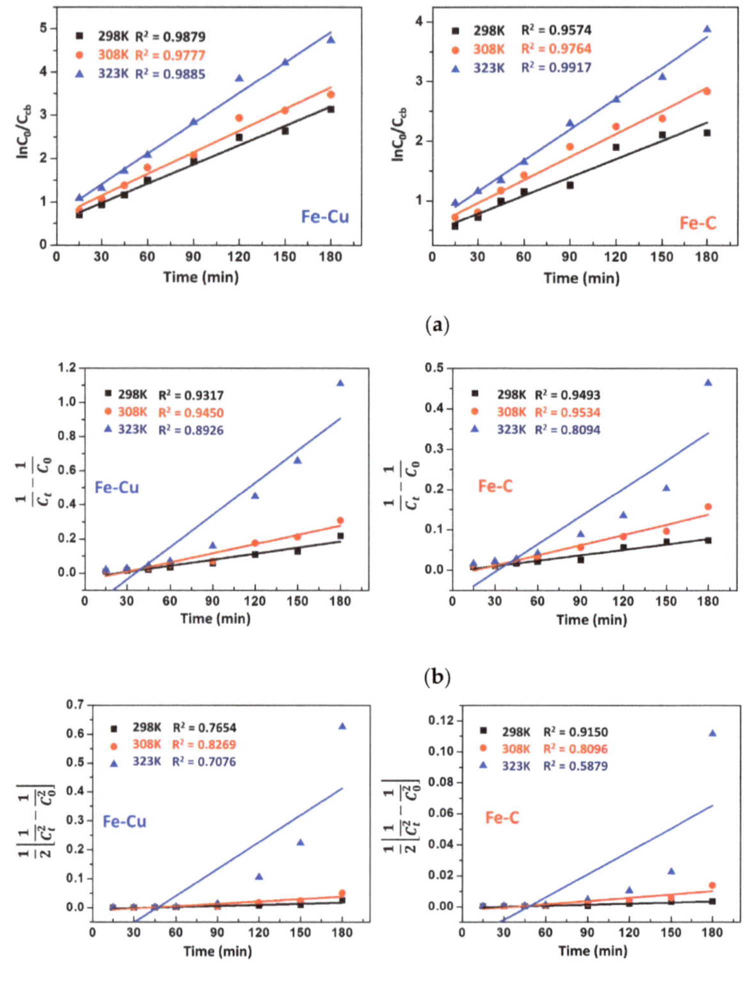

Figure 8. (**a**) The apparent first-order kinetic equation of Fe-Cu and Fe-C materials at 298, 308 and 323 K; (**b**) The apparent second-order kinetic equation of Fe-Cu and Fe-C materials at 298, 308 and 323 K; (**c**) The apparent third-order kinetic equation of Fe-Cu and Fe-C materials at 298, 308 and 323 K.

The CIP removal reaction rate constants of Fe-Cu and Fe-C materials both decrease with increasing temperature. This once again confirms that the CIP removal reaction of the two materials is endothermic.

At the same temperature, the CIP removal reaction rate constant of Fe-Cu material is larger than that of Fe-C material. This is consistent with the previous survey results; Fe-Cu materials showed higher CIP removal efficiency than Fe-C materials.

The results of calculating the amount of E_a activation are based on the relationship between lnk and 1/T (Figure 9a)

Table 2. The calculation results of first-order, second-order and third-order CIP removal reaction rate constants and linear regression coefficients of Fe-Cu and Fe-C materials at different temperatures.

T (K)	Fe-Cu		Fe-C	
First-Order Kinetic Equation Removal of CIP				
	k (min^{-1})	R^2	k (min^{-1})	R^2
298	0.0148	0.9879	0.0102	0.9574
308	0.0166	0.9777	0.0129	0.9764
323	0.0234	0.9885	0.0173	0.9917
Second-order kinetic equation removal of CIP				
	k ($min^{-1}·L·mg^{-1}$)	R^2	k ($min^{-1}·L·mg^{-1}$)	R^2
298	0.0012	0.9317	0.0004	0.9493
308	0.0018	0.9450	0.0008	0.9534
323	0.0063	0.8926	0.0023	0.8094
Third-order kinetic equation removal of CIP				
	k ($min^{-1}·(L·mg)^{-2}$)	R^2	k ($min^{-1}·(L·mg)^{-2}$)	R^2
298	0.0001	0.7654	0.00002	0.9150
308	0.0003	0.8269	0.00007	0.8096
323	0.0031	0.7076	0.0005	0.5879

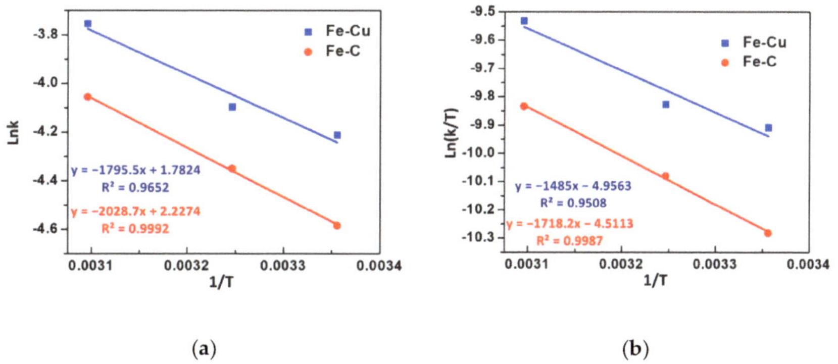

Figure 9. (**a**) Relationship between lnk and 1/T; (**b**) Relationship between ln(k/T) and 1/T.

As shown in Table 3, the activation energy of CIP removal of Fe-C materials is higher than that of Fe-Cu materials, showing that Fe-Cu materials decompose CIP better than Fe-C materials. This can be explained as follows: microscopic bimetallic electrolytic materials prepared by deposition, plating the second transition metals on the iron surface can greatly enhance the reduction of pollutants [32,33]. Previous studies showed that transition metals, such as Ni, Pd, Cu, Co, Au and Ru, could enhance the catalytic reactivity of Fe^0 [34]. The catalytic mechanisms of bimetallic internal electrolytic materials could be elaborated via two main explanations. First, bimetallic internal electrolysis and transition metal additives might promote the generation of atomic hydrogen ([H]abs), which might be absorbed on the surface of the material and induce indirect reduction [35,36]. Second, surface additives (i.e., metal transitions) might accelerate Fe^0 oxidation by the formation of pairs of electrochemical batteries, which in turn causes a direct reduction on the active site catalytic activity by accepting electrons [37,38].

Table 3. Activation energy of CIP removal under Fe-Cu and Fe-C treatment.

Material	Ea (kJ/mol)
Fe-Cu	14.93
Fe-C	16.87

Xu et al. (2008) [39] used Fe-Cu electrolytic materials—nitrobenzene (100 mg/L)—to treat an aqueous solution at the optimum pH range of 3–4, the weight of the internal electrolyte material of 50 g/L, treatment time of 90 min, achieving a nitrobenzene removal efficiency of 90%. The authors have fabricated internal electrolytic materials by two methods. In the first method, iron scrap (CT3 scrap) with a surface area of 0.3–0.4 m^2/g, washed with soap to remove oil, was used. The catalyst was pure copper foil with a thickness of 0.12 mm and was cut into 3 × 1 cm thin pieces before the experiments, mixed in the ratio Fe:Cu = 10:1 (mass). In the second method, scrap iron (CT3 scrap) with a surface area of 0.3–0.4 m^2/g, washed with soap to remove oil, was plated in a $CuSO_4$ solution with a concentration of 0.1 mol/L. For the Fe-Cu electrolytic material by the chemical plating method, after 60 min of treatment, the nitrobenzene removal efficiency reached 95%. This shows that the Fe-Cu electrolytic material fabricated by the chemical plating method are able to catalyze and create better reactions than internal electrolytic materials made by the mechanical mixing method of a mixture of scrap Fe and copper foil.

3.8. Thermodynamics of Remove CIP

The thermodynamic results of CIP removal are shown in Figure 9b and Table 4.

Table 4. Thermodynamic parameters for the removal of CIP by materials Fe-Cu and Fe-C.

T (K)	ΔH^0 (KJ/mol)		ΔS^0 (KJ/mol)		ΔG^0 (KJ/mol)	
	Fe-Cu	Fe-C	Fe-Cu	Fe-C	Fe-Cu	Fe-C
298	2.493	2.494			73.639	72.538
308	2.576	2.578	−0.239	−0.235	76.110	74.972
323	2.700	2.702			79.816	78.622

The positive ΔH proves that CIP removal is endothermic. A negative entropy of −0.239 and −0.235 KJ/mol (which is near zero and is also relatively positive) indicates the rapid removal of the CIP molecules into the removal products [38].

4. Conclusions

The ciprofloxacin removal ability of Fe-Cu electrolytic materials and Fe-C materials was explored with respect to a number of experimental parameters. Fe-Cu was fabricated by the chemical plating method, and Fe-C materials were fabricated mechanically from iron powder and graphite. The considered conditions were pH, time, the mass of material, shaking speed, initial concentration of CIP and temperature. The results show that at a pH value of 3, shaking time of 120 min, shaking speed of 250 rpm, the mass of Fe-Cu and Fe-C material of 2 g/L, initial CIP concentration of 203.79 mg/L, resulted in a CIP removal efficiency of the Fe-Cu material of 90.25% and that of the Fe-C material of 85.12%. The removal of CIP of Fe-Cu and Fe-C materials follows pseudo-first-order kinetics. The activation energy of CIP removal of Fe-Cu material is 14.93 KJ/mol, that of Fe-C material is 16.87 KJ/mol. The positive ΔH proves that CIP removal is endothermic. A negative entropy of 0.239 and 0.235 kJ/mol (which is near zero and is also relatively positive) shows the rapid removal of the CIP molecules into the removal products.

Under the same reaction conditions, the CIP removal efficiency of Fe-Cu materials is higher than that of Fe-C materials. The results of using Fe-Cu and Fe-C internal electrolytic materials in CIP removal show that the material is capable of decomposing CIP at large concentrations with high efficiency.

From the above results, it is possible to combine Fe-Cu and Fe-C materials with biological methods to treat antibiotics in water environments.

Author Contributions: Writing—original draft preparation, T.H.D. and T.K.N.T.; data curation, X.L.H. and P.C.N.; conceptualization, P.C.N. and N.B.H.; methodology, N.B.H. and X.L.H.; formal analysis, T.T.A.D. and T.H.D.; writing—review and editing T.H.D., T.T.A.D. and T.K.N.T. All authors have read and agreed to the published version of the manuscript.

Funding: This research received no external funding.

Institutional Review Board Statement: Not applicable.

Informed Consent Statement: Not applicable.

Data Availability Statement: All the data is available within the manuscript.

Conflicts of Interest: The authors declare no conflict of interest.

References

1. Huertas, E.; Salgot, M.; Hollender, J.; Weber, S.; Dott, W.; Khan, S.; Schäfer, A.; Messalem, R.; Bis, B.; Aharoni, A.; et al. Key objectives for water reuse concepts. *Desalination* **2008**, *218*, 120–131. [CrossRef]
2. Tertigas, D.; Barber, G. Determining the Rate of Development of Antibiotic Resistance to Streptomycin and Doxycycline in Escherichia coli. *Sci. McMaster Undergrad. Sci. J.* **2019**, *3*, 2–7. [CrossRef]
3. Grimes, K.L.; Dunphy, L.J.; Loudermilk, E.M.; Melara, A.J.; Kolling, G.L.; Papin, J.A.; Colosi, L.M. Evaluating the efficacy of an algae-based treatment to mitigate elicitation of antibiotic resistance. *Chemosphere* **2019**, *237*, 124421. [CrossRef] [PubMed]
4. Karthikeyan, S.; Gupta, V.K.; Boopathy, R.; Titus, A.; Sekaran, G. A new approach for the remove of high concentration of aromatic amine by heterocatalytic Fenton oxidation: Kinetic and spectroscopic studies. *J. Mol. Liq.* **2012**, *173*, 153–163. [CrossRef]
5. Palacio, D.A.; Urbano, B.F.; Rivas, B.L. Water-soluble polymers with the ability to remove amoxicillin as emerging pollutant from water. *Environ. Technol. Innov.* **2021**, *23*, 101589. [CrossRef]
6. Rivas, B.L.; Oñate, P.; Palacio, D.A. Removal of oxytetracycline by polymers. an overview. *J. Chil. Chem. Soc.* **2020**, *54*, 4. [CrossRef]
7. Palacio, D.A.; Leiton, L.M.; Urbano, B.F.; Rivas, B.L. Tetracycline removal by polyelectrolyte copolymers in conjunction with ultrafiltration membranes through liquid-phase polymer-based retention. *Environ. Res.* **2020**, *182*, 109014. [CrossRef]
8. Zhang, X.B.; Dong, W.Y.; Sun, F.Y.; Yang, W.; Dong, J. Remove efficiency and mechanism of azo dye RR2 by a novel ozone aerated internal micro-electrolysis filte. *J. Hazard. Mater.* **2014**, *15*, 77–87.
9. Li, P.; Liu, Z.; Wang, X.; Guo, Y.; Wang, L. Enhanced decolorization of methyl orange in aqueous solution using iron-carbon micro-electrolysis activation of sodium persulfate. *Chemosphere* **2017**, *180*, 100–107. [CrossRef]
10. Yang, Z.; Ma, Y.; Liu, Y.; Li, Q.; Zhou, Z.; Ren, Z. Remove of organic pollutants in near-neutral pH solution by Fe-C micro-electrolysis system. *Chem. Eng. J.* **2017**, *315*, 403–414. [CrossRef]
11. Xu, X.; Cheng, Y.; Zhang, T.; Ji, F.; Xu, X. Treatment of pharmaceutical wastewater using interior microelectrolysis/Fenton oxidation-coagulation and biological remove. *Chemosphere* **2016**, *152*, 23–30. [CrossRef] [PubMed]
12. Zhang, L.; Yue, Q.; Yang, K.; Zhao, P.; Gao, B. Analysis of extracellular polymeric substances (EPS) and ciprofloxacin-degrading microbial community in the combined Fe-C micro-electrolysis-UBAF process for the elimination of high-level ciprofloxacin. *Chemosphere* **2018**, *193*, 645–654. [CrossRef] [PubMed]
13. Liu, M.; Wang, L.; Xiao, X.; He, Z. Fe/C micro electrolysis and Fenton oxidation process for the removal of recalcitrant colored pollutants from mid-stage pulping effluent. *J. Bioresour. Bioprod.* **2018**, *3*, 118–122. [CrossRef] [PubMed]
14. Ma, W.; Han, Y.; Xu, C.; Han, H.; Ma, W.; Zhu, H.; Li, K.; Wang, D. Enhanced remove of phenolic compounds in coal gasification wastewater by a novel integration of micro-electrolysis with biological reactor (MEBR) under the micro-oxygen condition. *Bioresour. Technol.* **2018**, *251*, 303–310. [CrossRef] [PubMed]
15. Chen, R.H.; Chai, L.Y.; Wang, Y.Y.; Liu, H.; Shu, Y.D.; Zhao, J. Remove of organic wastewater containing Cu-EDTA by Fe-C micro-electrolysis. *Trans. Nonferr. Met. Soc. China* **2012**, *22*, 983–990. [CrossRef]
16. Lim, E.-T.; Jeong, G.-T.; Bhang, S.-H.; Park, S.-H.; Park, D.-H. Evaluation of pilot-scale modified A2O processes for the removal of nitrogen compounds from sewage. *Bioresour. Technol.* **2009**, *100*, 6149–6154. [CrossRef]
17. Wang, Y.P.; Wang, L.J.; Peng, P.Y.; Lu, T.H. Treatment of naphthalene derivatives with iron-carbon micro-electrolysis. *Trans. Nonferr. Met. Soc. China* **2006**, *16*, 1442–1447. [CrossRef]
18. Jin, Y.-Z.; Zhang, Y.-F.; Li, W. Micro-electrolysis technology for industrial wastewater treatment. *J. Environ. Sci.* **2003**, *13*, 334–338.
19. Zhu, Y.; Fang, Z.; Xia, Z. Study on the reaction materials for micro-electrolysis treatment of wastewater. *Membr. Sci. Technol.* **2001**, *4*, 56–60.
20. Fan, J.-H.; Ma, L.-M. The pretreatment by the Fe–Cu process for enhancing biological degradability of the mixed wastewater. *J. Hazard. Mater.* **2009**, *164*, 1392–1397. [CrossRef]
21. Zhu, Q.; Guo, S.; Guo, C.; Dai, D.; Jiao, X.; Ma, T.; Chen, J. Stability of Fe-C Micro-Electrolysis and Biological Process in Treating Ultra-High Concentration Organic Wastewater. *Chem. Eng. J.* **2014**, *255*, 535–540. [CrossRef]
22. Li, G.; Guo, S.; Li, F. Treatment of oil field produced water by anaerobic process coupled with micro-electrolysis. *J. Environ. Sci.* **2010**, *22*, 1875–1882. [CrossRef]

23. Oh, S.Y.; Chiu, P.C.; Kim, B.J.; Cha, D.K. Enhancing Fenton oxidation of TNT and RDX through pretreatment with zero-valent iron. *Water Res.* **2003**, *37*, 4275–4283. [CrossRef]
24. Fan, L.; Ni, J.; Wu, Y.; Zhang, Y. Treatment of bromoamine acid wastewater using combined process of micro-electrolysis and biological aerobic filter. *J. Hazard. Mater.* **2009**, *162*, 1204–1210. [CrossRef] [PubMed]
25. Lin, H.; Lin, Y.; Liu, L. Treatment of dinitrodiazophenol production wastewater by Fe/C and Fe/Cu internal electrolysis and the COD removal kinetics. *J. Taiwan Inst. Chem. Eng.* **2016**, *58*, 148–154. [CrossRef]
26. Huong, D.T.; Nguyen, V.T.; Ha, X.L.; Nguyen Thi, H.L.; Duong, T.T.; Nguyen, D.C.; Nguyen Thi, H.T. Enhanced Remove of Phenolic Compounds in Coal Gasification Wastewater by Methods of Microelectrolysis Fe-C and Anaerobic-Anoxic- Oxic Moving Bed Biofilm Reactor(A2O-MBBR). *Processes* **2020**, *8*, 1258. [CrossRef]
27. Huong, D.T.; Van Tu, N.; Anh, D.T.T.; Tien, N.A.; Ngan, T.T.K.; Van Tan, L. Removal of Phenol from Aqueous Solution Using Internal Microelectrolysis with Fe-Cu: Optimization and Application on Real Coking Wastewater. *Processes* **2021**, *9*, 720. [CrossRef]
28. Campoli-Richards, D.M.; Monk, J.P.; Price, A.; Benfield, P.; Todd, P.A.; Ward, A. Ciprofloxacin: A Review of its Antibacterial Activity, Pharmacokinetic Properties and Therapeutic Use. *Drug Eval.* **1988**, *35*, 373–447. [CrossRef] [PubMed]
29. Rezaei, M.; Nezamzadeh-Ejhieha, A. The ZnO-NiO nano-composite: A brief characterization, kinetic and thermodynamic study and study the Arrhenius model on the sulfasalazine photoremove. *Int. J. Hydrogen Energy* **2020**, *45*, 24749–24764. [CrossRef]
30. Zhang, X.; Lin, Y.M.; Shan, X.Q.; Chen, Z.L. Remove of 2, 4, 6 -trinitrotoluene (TNT) from explosive wastewater using nanoscale zero-valent iron. *Chem. Eng. J.* **2010**, *158*, 566–570. [CrossRef]
31. Yuan, S.H.; Kang, J.; Wu, X.H.; Wang, L.L.; Chen, J.; Lu, X.H. Development of a novel internal electrolysis system by iron connected with carbon: Treatment of nitroaromatic compounds and case of engineering application. *J. Environ. Eng.* **2010**, *136*, 975–982. [CrossRef]
32. Koutsospyros, A.; Pavlov, J.; Fawcett, J.; Strickland, D.; Smolinski, B.; Braida, W. Remove of high energetic and insensitive munitions compounds by Fe/Cu bimetal reduction. *J. Hazard. Mater.* **2012**, *219*, 154–181.
33. Lai, B.; Zhang, Y.; Chen, Z.; Yang, P.; Zhou, Y.; Wang, J. Removal of p-nitrophenol (PNP) in aqueous solution by the micron-scale iron-copper (Fe/Cu) bimetallic particles. *Appl. Catal. B Environ.* **2014**, *144*, 154–830. [CrossRef]
34. Cwiertny, D.M.; Bransfield, S.J.; Livi, K.J.; Fairbrother, D.H.; Roberts, A.L. Exploring the Influence of Granular Iron Additives on 1,1,1- Trichloroethane Reduction. *Environ. Sci. Technol.* **2006**, *40*, 6837–6843. [CrossRef]
35. Lin, C.J.; Lo, S.L.; Liou, Y.H. Dechlorination of trichloroethylene in aqueous solution by noble metal-modified iron. *J. Hazard. Mater.* **2004**, *116*, 219–228. [CrossRef]
36. Bransfield, S.J.; Cwiertny, D.M.; Livi, K.; Fairbrother, D.H. Influence of transition metal additives and temperature on the rate of organohalide reduction by granular iron: Implications for reaction mechanisms. *Appl. Catal. B Environ.* **2007**, *76*, 348–356. [CrossRef]
37. Xu, W.Y.; Gao, T.Y. Dechlorination of carbon tetrachloride by the catalyzed Fe-Cu process. *J. Environ. Sci.* **2007**, *19*, 792–799. [CrossRef]
38. Ma, L.M.; Ding, Z.G.; Gao, T.Y.; Zhou, R.F.; Xu, W.Y.; Liu, J. Discoloration of methylene blue and wastewater from a plant by a Fe/Cu bimetallic system. *Chemosphere* **2004**, *55*, 1207–1212. [CrossRef]
39. Wenying, X.; Ping, L.; Jinhong, F. Reduction of nitrobenzene by the catalyzed Fe/Cu process. *J. Environ. Sci.* **2008**, *20*, 915–992.

processes

MDPI

Article

Characteristics of Hydrochar and Liquid Products Obtained by Hydrothermal Carbonization and Wet Torrefaction of Poultry Litter in Mixture with Wood Sawdust

Rafail Isemin [1,*], Natalia Muratova [1], Sergey Kuzmin [1], Dmitry Klimov [1], Vadim Kokh-Tatarenko [1], Alexander Mikhalev [1], Oleg Milovanov [1], Antoine Dalibard [2], Olayinka Ahmed Ibitowa [3], Manuel Nowotny [3], Mathieu Brulé [4], Fouzi Tabet [5] and Bernd Rogge [6]

1 Biocenter, Tambov State Technical University, Sovetskaya St. 106, 392000 Tambov, Russia; muratova-nat@rambler.ru (N.M.); tregulyaj@mail.ru (S.K.); marcelldm@mail.ru (D.K.); vadim6873@yandex.ru (V.K.-T.); malex_v@list.ru (A.M.); milowanov.o@yandex.ru (O.M.)

2 Fraunhofer-Institut für Grenzflächen-und Bioverfahrenstechnik IGB, Fraunhofer-Gesellschaft, Nobelstraße 12, 70569 Stuttgart, Germany; antoine.dalibard@igb.fraunhofer.de

3 COAST-Zentrum für Umwelt-und Nachhaltigkeitsforschung, Universität Oldenburg, Ammerländer Heerstr. 114-118, 26129 Oldenburg, Germany; olayinka.ahmed.ibitowa@uni-oldenburg.de (O.A.I.); manuel.nowotny@uol.de (M.N.)

4 Department of Natural Resources Management and Agricultural Engineering, Agricultural University of Athens (AUA), Iera Odos 75, 11855 Athens, Greece; mathieubrule82@gmail.com

5 Opti'Tech, Schletterstrasse 12, 04107 Leipzig, Germany; fouzi.tabet@gmail.com

6 Inter Center GmbH, Allee der Kosmonauten 32 B, 12681 Berlin, Germany; br@intercentre.de

* Correspondence: penergy@list.ru; Tel.: +7-475-263-0446

Citation: Isemin, R.; Muratova, N.; Kuzmin, S.; Klimov, D.; Kokh-Tatarenko, V.; Mikhalev, A.; Milovanov, O.; Dalibard, A.; Ibitowa, O.A.; Nowotny, M.; et al. Characteristics of Hydrochar and Liquid Products Obtained by Hydrothermal Carbonization and Wet Torrefaction of Poultry Litter in Mixture with Wood Sawdust. *Processes* **2021**, *9*, 2082. https://doi.org/10.3390/pr9112082

Academic Editors: Dimitris Zagklis and Georgios Bampos

Received: 22 October 2021
Accepted: 17 November 2021
Published: 20 November 2021

Abstract: Poultry farms with floor-standing poultry generate large amounts of poultry litter waste. The direct application of this waste as an organic fertilizer does not ensure sustainable and cost-efficient utilization of all waste fractions, and can also be linked to environmental hazards. Therefore, the development of new technologies is required for processing poultry litter into a safe product with higher added value. In this work, the characteristics of activated carbon derived from hydrochar, along with the liquid products obtained from hydrothermal carbonization (HTC) and the wet torrefaction (WT) of poultry litter, were investigated. Poultry litter (PL) was applied in a mixture with sawdust (SD) in the following ratios: 1:0 (PL/SD 1:0), 1:1 (PL/SD 1:1), 1:2 (PL/SD 1:2), and 2:1 (PL/SD 2:1). WT processing took place in an innovative fluidized bed system in a superheated steam medium with low overpressure (less than 0.07 MPa) at 300 °C and 350 °C for 30–45 min. Conventional HTC processing was performed in a water medium at 220 °C for 1–4 h. The hydrochar produced in the experiments was activated with steam for 1 h at 450–750 °C. The porosity characteristics of activated hydrochar were measured, including pore size, pore volume, and specific surface area, in view of potential industrial applications as an adsorbent. Additionally, the contents of 5-hydroxymethylfurfural (HMF), as high-value product, were determined in the liquid products obtained from HTC processing, as well as in the condensate obtained after WT processing. Specific surface areas of the activated hydrochars may still be too low for application as adsorbent material. Hence, its use as a biofertilizer and soil improver should be preferred. Interestingly, the liquid fraction obtained from the innovative WT process displayed a significantly higher 5-HMF content compared to the conventional HTC process.

Keywords: hydrothermal carbonization; steam torrefaction; biomass; waste treatment; activated carbon; fluidized bed

1. Introduction

In various regions of Russia, and abroad, the lack of technologies to ensure efficient disposal of biowaste, generated by agricultural, food, timber, and wood processing industries, results in environmental hazards and emissions.

An innovative pathway applicable for a wide range of biowaste and organic waste is the process of hydrothermal carbonization (HTC), taking place at a temperature range of 180–250 °C and an overpressure of 2–10 MPa [1,2]. This process generates a solid product, hydrochar, which is similar in properties to brown coal. The hydrochar yield is in the range 35–60%, depending on process conditions. Hydrochar can be further converted into high-value products, such as synthesis gas and liquid fuel, soil improver, or sorbents [3].

In a study investigating the use of HTC for municipal solid waste processing [4], a hydrochar yield of 58% of the initial raw material was obtained. During the process, the heating value of the biomass increased 1.41-fold based on weight-heat content, or 9-fold based on volumetric heat content. Furthermore, the HTC process can be successfully applied for the processing of medical waste and plastic waste [5], since organic chlorine in polyvinyl chloride wastes can be converted into an inorganic form during the process. The efficiency of waste dechlorination increased up to 83–90% as a result of the HTC co-treatment of PVC medical waste together with lignin, or wood chips, as co-substrate. Furthermore, HTC has been successfully used for PVC processing [6] at 180–260 °C, during which active dehydrochlorination was observed. Chlorinated hydrocarbons, including chlorophenols, could only be identified at trace levels in the final products, and polychlorinated dibenzodioxins, in addition to dibenzofurans, were not detected. Hence, the HTC of municipal solid waste may be applicable as an environmentally friendly process for the removal of toxic organic products. Additionally, the effects of HTC on the distribution of heavy metals, and subsequent pyrolysis of sewage sludge, were studied in comparison with the direct pyrolysis of sewage sludge [7]. Under this configuration, HTC reduced the content of acid-soluble and recoverable fractions of heavy metals; reducing the risk of pollution by heavy metal compounds in the environment as heavy metals remained in an inactive form in the inorganic fraction of hydrochar.

Hydrochar generated by HTC can be used in high-value applications, such as adsorption processes [8,9] and the synthesis of activated carbon [10–12]. HTC produces hydrochar with higher amounts of oxygen functional groups and low levels of aromatization, which facilitates further chemical activation [10,13]. In [10], hydrochar obtained from starch and eucalyptus sawdust was exposed to chemical activation using potassium hydroxide. The researchers obtained high-quality activated carbon with high pore surface (2700 m^2/g) and hydrogen absorption capacity (16.4 mmol/g) [10]. In another study, activated carbon, obtained by the chemical activation of hydrochar with potassium hydroxide, yielded CO_2 absorption rates of up to 4.8 mmol/g at 298 K and 1 atm [11]. It is reported that activated carbon generated from hydrochar originating from rye straw, displayed high pore surface (2200 m^2/g) and CO_2 adsorption capacity (20 μmol/g) [14]. Furthermore, chemical activation of rice husk hydrochar with H_3PO_4 also yielded high specific pore surface areas (2610 m^2/g) [15], while chemical activation with KOH yielded an even higher specific pore surface (3362 m^2/g), and therefore produced higher adsorption capacities [16].

Poultry litter is among the most dangerous categories of biowaste. This is especially alarming considering the risks posed to people, and the environment in surrounding areas, in connection with the commissioning of large poultry farms which are engaged in the production of meat. The environmental hazards caused relate to the daily inflow of dozens of tons of droppings, along with the long-term accumulation of this waste in unauthorized storage facilities at the farming sites which do not possess sufficient arable land for spreading the biowaste. For example, a typical poultry farm hosting 40,000 laying hens, or 10 million broilers, respectively, generates between 35,000 and 83,000 tons of poultry litter, in addition to over 400,000 m^3 of wastewater, annually, with high contents in organic components. This cycle results in potentially hazardous materials that pose a threat to public health and the environment. According to the World Health Organization, droppings and sewage from poultry enterprises can be a factor in the transfer of pathogens for more than 100 infectious and invasive diseases, including zoonoses. In addition, organic waste provides a favorable environment for the development, and long-term survival, of pathogenic microflora and can contain increased amounts of pollutants, including heavy metals, pesticides, antibiotics, radioactive substances, weed seeds, and other contaminants. Therefore, it is extremely difficult to ensure the environmental safety of these wastes, as existing technologies for disinfecting poultry droppings are unreliable, ineffective, and very expensive.

On the other hand, opportunities exist for the development of innovative pathways converting poultry litter into high-value products. Nitrogen-containing hydrochar, which can be obtained by the HTC of poultry litter, may act as an adsorbent for electrolytes. This provides a promising candidate as a sorption material for water purification from heavy metals (for example, Pb^{2+}, Cd^{2+}, Cu^{2+}, U^{6+}) or from organic pollutants (for example, dyes) due to the abundance of surface functional groups [13,17]. This broadens the prospective uses of hydrochar obtained by the HTC of nitrogen-containing biowaste, such as manure and poultry litter, as a raw material for the production of highly efficient sorbents.

Nevertheless, the HTC process, when is carried out using state-of-the-art technologies, is a low productivity batch process, which should be carried out at elevated pressure. These harsh operating conditions, along with low performances, may significantly complicate and increase the cost of waste processing.

Wet torrefaction (WT) is sometimes assimilated to HTC, and involves a similar process carried out at lower operational pressures, thereupon reducing processing costs. However, a significant drawback from the reduced pressure is the longer processing time required for biomass treatment. This issue can be resolved through the integration of a fluidized bed design into the WT process. In order to optimize performance, the WT process can be performed under the specific operating conditions of vapothermal carbonization in a superheated steam environment. Earlier [18], we performed preliminary studies on the WT of ground biomass in a fluidized bed in an environment of superheated steam. These previous studies have shown that the duration of the WT process in a fluidized bed can be reduced to 15–20 min, versus 4–12 h when carrying out WT by means of state-of-the art technologies. The advantage of WT is the possibility of obtaining hydrochar, the characteristics of which are comparable to the hydrochar obtained by conventional HTC, while applying mild processing conditions, in particular lower overpressure [19,20]. Superheated steam can be used as a medium for drying biofuels and wood [21]. The use of superheated steam in the heat treatment of biomass is considered a promising alternative to traditional technologies, as superheated steam is an inert medium (at temperatures below 600 °C) [22] with interesting thermodynamic properties, allowing for high heat and mass transfers [23].

The aim of this work is to compare the characteristics of the activated hydrochar (AH) obtained by physical steam activation, as well as liquid products obtained by processing a mixture of poultry litter (PL) and sawdust (SD) with HTC and WT processes. WT is performed through an innovative time-efficient process comprising a fluidized bed operated with superheated steam as a fluidization medium. The key parameters investigated in this study were on the pore characteristics and specific surface area of the activated hydrochar, which were important parameters for its application as an absorbent material. The concentrations of organic compounds (TOC) and 5-HMF in the liquid products were also measured in view of the recovery of 5-HMF as a high-value compound from the liquid fraction.

The yields of non-condensable gases obtained from the innovative WT process have already been reported in previous studies [24,25].

2. Experimental Setup and Experimental Technique

Figure 1 presents the pilot plant for the wet torrefaction (WT) of biowaste in a fluidized bed and an environment of superheated steam. The plant consists of the following components: feedstock input bunker (1), reactor for carbonization in a fluidized bed (2), cyclone separating gases from hydrochar particles exiting the reactor (3), hydrochar collection bunker (4). The apparatus also comprises a boiler for steam generation, a superheater, and a steam cooler (not shown). The reactor (2) for wet torrefaction is connected to gas distribution inlets supplying superheated steam under the fluidized bed. The side walls of the reactor were heated electrically to maintain the required temperature in the reactor. A detailed description of the operation of the pilot reactor for WT has been provided in [25].

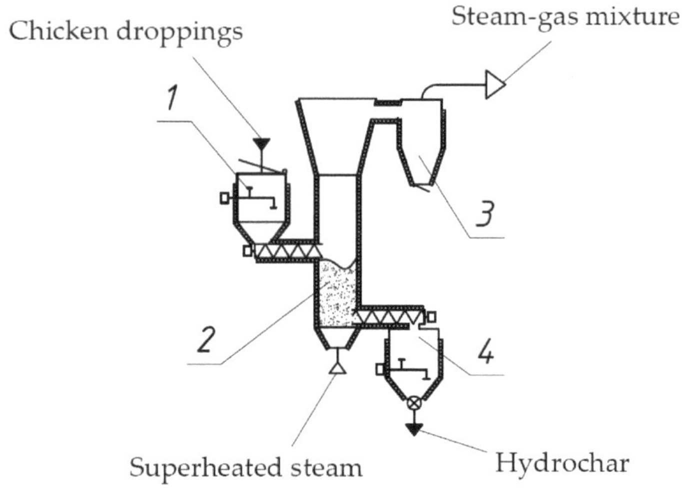

Figure 1. Diagram of the vapothermal carbonization apparatus of biowaste in a fluidized bed in a superheated steam environment: (**1**) feedstock bunker for biomass; (**2**) reactor; (**3**) cyclone; (**4**) product bunker for hydrochar recovery.

Before initiating the experiment, roughly 6 L of hydrochar from a previous batch was poured into the wet torrefaction reactor. Subsequently, 2 kg of raw material, which would be carbonized, was poured into the feedstock bunker. The pre-heating of the reactor was performed while the apparatus was purged with nitrogen until the operating temperature was reached. Subsequently, nitrogen input was stopped. Saturated steam was generated and supplied by the superheater, then fed into the gas inlets. The reactor temperature was set for the desired operating conditions (310–330 °C or 350 °C). Subsequently, the biomass feedstock was fed into the reactor for wet torrefaction, and loading was completed after 15–17 min. Since the initiation of feedstock feeding, non-condensable gases (CO_2, CO, H_2 and CH_4) were continuously sampled downstream of the condenser and characterized with a gas analyzer (Vario Plus Industrial Syngas analyzer). After the concentration of these substances decreased to the initial levels preceding the start of the biomass supply to the reactor, the experiment was stopped, and the generated hydrochar was discharged into the product bunker. The duration of the WT process ranged from 30 to 40 min and decreased with an increase in the process temperature from 300 °C to 350 °C. The entire unit was then purged with cold nitrogen and the cooled hydrochar was unloaded from the bunker for further analyses.

HTC of the poultry litter (PL) in mixture with sawdust (SL) was performed in a 250 mL stainless steel autoclave with a PTEF inlet, and subsequent physical activation of the obtained hydrochar was performed at higher temperatures in a steam medium. Each HTC experiment contained a total mass of 100 g and was carried out at 220 °C for 1, 4 and 8 h, respectively. The sealed autoclave was placed into a heated convection oven. After completing the reaction, the autoclave was taken out of the oven and placed in a fume hood at room temperature to cool the reaction products. Vacuum filtration was applied to separate liquid products from the hydrochar. The separated hydrochar was washed with 1 L of deionized water and dried overnight at 120 °C. Subsequent activation tests were carried out with steam at 450 and 750 °C for 1 h. To this effect, 1 g of hydrochar was filled into a ceramic container and placed into a tube furnace. Steam generated by the evaporation of the deionized water under a nitrogen atmosphere was flushed into a round bottom flask at the end of the tube furnace where it made contact with the hydrochar. A PT100 thermo-element, connected with a temperature controller, measured the temperature in the middle of the furnace. After the reaction, the furnace was cooled down to 80 °C and the obtained activated hydrochar was washed with 500 mL deionized water and finally dried for 12 h at 120 °C.

Subsequent physical activation of the WT and HTC hydrochars was performed at higher temperatures (450 and 750 °C) in a steam medium. This process was completed in order to obtain activated hydrochars with high pore volumes and specific surface areas, with their potential application as adsorbents in industrial and environmental processes.

The elemental analysis was performed with a EuroEA 3000 Elemental Analyzer, and ash contents were determined according to ASTM D3174-02. Surface characterization was done by infrared spectroscopy and solid-state NMR. IR spectra were measured under ATR conditions (diamond ATR crystal) with a Tensor 27 (Bruker, USA) spectrometer and OPUS software for spectra interpretation. After background recording, each sample was measured with 25 scans. Next, 13C and 1H solid state NMR spectra were obtained using an Avance III HD NMR spectrometer (Bruker BioSpin) with a cryomagnet having a nominal field of 11.75 T. Resonance frequencies were 125.86 MHz for 13C and 500.46 MHz for 1H. A standard (double resonance) 4.0 mm WVT probe (Bruker), which can be operated at spinning speed up to 15 kHz, was used for the experiments. At 13C, for 1H, MAS NMR spectra were referenced to solid 1,1,1,3,3,3-hexamethyl-2,2-bis (trimethylsilyl) trisilane, with resonance signals occurring at 6.9 ppm (13C) and 0.1 ppm (1H) relative to the primary reference TMS (0 ppm).

Solid-state spectra were recorded using a rotor-synchronized Hahn echo pulse sequence (p/2–t–p–t–acquire; t: rotor period) and standard cross-polarization (CP) pulse sequence, respectively. A total of 13C {1H} CP MAS NMR was used with a flip-back pulse on the 1H spins following the data acquisition process. Recycling delays amounted to 15–120 s for the Hahn echo experiments and 15–60 s for the CP experiment, respectively. For the 1H NMR and 13C Hahn echo experiments, 16, and up to 2112 scans, were collected. The sample was recorded within 1 day. The studied samples were operated to set the Hartman–Hahn match under the MAS conditions with the match optimized for a spinning speed of 14 kHz. Data processing was performed with TopSpin 3.5 software (Bruker).

The specific surface areas were evaluated by nitrogen sorption with a TRISTAR II 3020 (Micromeritics GmbH, Germany) followed by BET analysis. After drying the samples for 12 h under a vacuum at 150 °C, the free space was measured with helium prior to the recording of the isotherms. Adsorption and desorption data were collected at a pressure range of p/p0 = 0.005–0.95 (36 points, randomly chosen by measurement mode) and p/p0 = 0.95–0.1 (19 points, linear distribution), respectively. Specific surface areas were calculated using the BET formalism in accordance with the Rouquerol criteria.

Thermal stability was evaluated via thermogravimetric analysis (TGA) with a TGA 4000 (Perkin Elmer Corporation, USA) setup, including a furnace with balance, temperature sensor, heating/cooling elements, Al_2O_3 crucible, and Pyris software. Before the measurement, the apparatus was purged with nitrogen for 30 min at a rate of 40 mL min^{-1}, and the crucible was baked out. Measurements were carried out under a nitrogen atmosphere. The temperature was varied as follows: hold for 1 min at 30 °C, ramp from 30 °C to 750 °C with 10 °C min^{-1}, and, finally, hold for 60 min at 750 °C.

The separated liquid products of HTC were stored in a fridge before undergoing further analyses. The determination of 5-HMF amounts were completed using the Merck quick test.

3. Results and Discussion

The analysis results of the HTC and WT hydrochar after activation are presented in Table 1.

Table 1. Characteristics of activated hydrochars obtained after WT or HTC followed by physical steam activation.

Process	Feedstock Mixture	Processing Conditions	Activation Conditions	Ash Content %	Carbon Content %	SSA, m^2/g	Pore Volume, cm^3/g	Pore Diameter, nm
WT	PL/SD (2:1)	40 min/ 300 °C	-	41	34	13	0.035	11
			1 h/450 °C	53	41	13	0.015	8
			1 h/750 °C	58	37	460	0.258	2
WT	PL/SD (1:1)	40 min/ 300 °C	-	29	50	5	0.017	12
			1 h/450 °C	37	46	10	0.021	8
			1 h/750 °C	53	44	300	0.185	2
WT	PL/SD (1:2)	40 min/ 300 °C	-	20	60	7	0.027	10
			1 h/450 °C	28	60	29	0.032	4
			1 h/750 °C	34	56	447	0.229	2
WT	PL/SD (1:0)	30 min/ 350 °C	-	39	37	14	0.044	13
			1 h/450 °C	51	41	15	0.001	6
			1 h/750 °C	57	34	226	0.148	3
WT	PL/SD (1:1)	30 min/ 350 °C	-	31	49	10	0.031	12
			1 h/450 °C	44	43	13	0.018	6
			1 h/750 °C	52	46	298	0.159	2
WT	PL/SD (2:1)	30 min/ 350 °C	-	19	75	5	0.014	11
			1 h/450 °C	25	61	35	0.013	7
			1 h/750 °C	40	55	470	0.252	2
WT	PL/SD (1:2)	30 min/ 350 °C	-	19	59	5	0.020	8
			1 h/450 °C	24	67	7	0.021	2
			1 h/750 °C	35	55	491	0.261	2
HTC	PL/SD (1:0)	1 h/220 °C	1 h/450 °C	49	44	14	0.003	9
			1 h/750 °C	54	38	118	0.048	5
		4 h/220 °C	1 h/450 °C	51	47	36	0.017	5
			1 h/750 °C	60	35	125	0.030	5
HTC	PL/SD (1:1)	1 h/220 °C	1 h/450 °C	24	58	125	0.017	27
			1 h/750 °C	19	72	325	0.101	4
		4 h/220 °C	1 h/450 °C	28	63	167	0.003	2
			1 h/750 °C	21	68	445	0.033	3

According to data shown in Table 1, the characteristics of activated the hydrochar (AH) obtained by WT were influenced by feedstock composition and the processing temperature. Thus, with an increase in the proportion of sawdust (SD) in the mixture, from 33% to 66%, the carbon content in AH increased from 34% to 60% at a process temperature of 300 °C. At a process temperature of 350 °C, the composition of the initial feedstock had a lower effect on the characteristics of AH, whose carbon contents ranged from 49% to 75%, with the maximum carbon content observed in AH obtained from a mixture containing 67% of poultry litter and 33% sawdust (ratio 2:1).

Physical activation in a steam environment at 750 °C significantly increased the pore volumes and specific surface areas (SSA) of AH, while reducing the pore sizes. Moreover, AH characteristics were affected by the composition of the initial biomass, with higher proportions of PL in the feedstock mixture resulting in high pore volumes and specific surface areas of 0.258–0.252 cm^3/g and 460–470 m^2/g, operating at temperatures of 300 and 350 °C, respectively. However, the application of PL alone resulted in lower values, revealing interactions caused by the addition of SD in the feedstock, which resulted in higher porosities and specific surface areas.

The characteristics of the AH obtained by HTC were influenced by both the composition of initial feedstock, and the activation temperature. Furthermore, considering the higher activation temperature of 750 °C, the characteristics of AH for the investigated parameters (pore volume, specific surface area and pore diameter) were similar to AH obtained using the innovative WT process. These results show that HTC could be replaced with WT processing under mild conditions (lower overpressure), obtaining AH of similar characteristics after subsequent physical activation.

The carbon content in AH obtained by WT and HTC of PL, after physical activation in a steam environment at 450–750 °C, was 1.7–1.8 and 1.7–2.2-fold higher than in the biochar obtained by another research group who investigated the pyrolysis of PL at 400–600 °C, respectively [26].

SSA of AH produced by WT and HTC, and activated with steam at 450 °C, was 3.5–6.7-fold higher than in biochar obtained by the pyrolysis of PL at 400–600 °C [26].

An increase of the hydrochar activation temperature to 750 °C, obtained by WT or HTC from PL as sole feedstock without sawdust, results in SSA of AH up to 125–236 m^2/g. Furthermore, the addition of a share of SD in the feedstock results in significantly higher SSA of up to 447–491 m^2/g. Nevertheless, the physical steam activation at 450–750 °C does not yield high SSA values above 3000 m^2/g, which can be obtained by chemical activation processes [24].

In a previous study [19], we have shown that hydrochar obtained from a mixture of PL and SD before and after ac7tivation by pyrolysis, contains large amount of nutrients that make activated biochar a valuable fertilizer: potassium 30.7–57.7 mg/g, magnesium 9.7–19.2 mg/g, phosphorus 20.3–39.7 mg/g. Furthermore, in addition to nutrient supply as a biofertilizer, hydrochar may have additional benefits as a soil improver, due to its porosity and nutrient sorption properties, increasing water and nutrient retention [24,26].

An additional benefit of converting PL into raw or activated hydrochar is the removal of the pathogenic microflora which is contained in large amounts in raw PL. Hence, pyrolysis processes result in the efficient disinfection of PL [27]. Following pyrolysis, the pH of PL decreased to neutral values (pH—6.2–6.8), which are favorable for plant development. The total nitrogen content in poultry litter amounts to 8–9%, of which 10–12% is in an easily degradable water-soluble form. The presence of large amounts of readily available nitrogen creates a danger to the development of plants. Pyrolysis leads to a sharp decrease in the content of water-soluble and easily decomposable forms of nitrogen, decreased by 0.67% in weight, and the removal of unpleasant odors.

Table 2 shows the results of the analysis of the liquid products of WT and HTC.

According to the results shown in Table 2, the method of thermal feedstock treatment (WT or HTC) does not significantly affect the content of total organic carbon in the liquid phase.

The content of 5-HMF in the liquid phase after WT decreases along with increased proportions of poultry litter in the mixture, from 0.74–0.8 g/L at a PL share of 33%, down to 0.32–0.42 g/L at a PL share of 66%.

An increase in WT process temperature resulted in decreased 5-HMF concentrations. However, the concentrations of 5-HMF in the liquid products after WT were 13–27-fold higher than after HTC. The higher 5-HMF concentrations in the liquid products of WT might be caused by very short contact times of biomass particles with superheated steam, not exceeding 2 s. These short contact times might prevent the decomposition of 5-HMF. This might occur in the HTC process as the liquid phase remains in contact with biomass during the entire reaction time of 4 h.

The results also show that sawdust (SD) is more suitable than poultry litter (PL) for the production of 5-HMF by wet torrefaction (WT) in a fluidized bed in a superheated steam environment.

Table 2. Composition of the liquid products of WT and HTC.

Method	Substrate	Reaction Condition	TOC, g/L	5-HMF, g/L
WT	PL/SD (2:1)	300 °C	14	0.42
WT	PL/SD (1:1)	300 °C	10	0.54
WT	PL/SD (1:2)	300 °C	13	0.8
WT	PL/SD (1:0)	350 °C	14	0.3
WT	PL/SD (1:1)	350 °C	9	0.19
WT	PL/SD (2:1)	350 °C	13	0.32
WT	PL/SD (1:2)	350 °C	12	0.74
HTC	PL/SD (1:0)	4 h/220 °C	12	0.02
HTC	PL/SD (1:1)	4 h/220 °C	9	0.02
HTC	PL/SD (2:1)	4 h/220 °C	10	0.08
HTC	PL/SD (1:2)	4 h/220 °C	8	0.06
HTC	PL/SD (0:1)	4 h/220 °C	7	6.15

Cellulose is considered the main raw material for the production of HMF in the near future, since it is the most affordable non-food polysaccharide [28]. In [29], the process of obtaining 5-HMF from sugarcane bagasse by HTC was studied in a water medium at temperatures of 200–300 °C. Increased HMF yields were observed when the temperature increased from 200 °C to 270 °C. Above 270 °C, the yield of 5-HMF decreased, and the compound was undetected at 300 °C (>10 min reaction time). The maximum yield of 5-HMF obtained in the process reached 3.09 wt%. This 5-HMF yield is much higher than the yields obtained in our study, even while applying the innovative fluidized bed WT technology. This might be due to the fact that we used poultry litter biowaste as a main substrate, which contains lower amounts of cellulose. Indeed, HTC of sawdust (SD) as sole substrate yielded 76–307-old higher HMF contents than the HTC of the mixtures of poultry litter (PL) and sawdust (SD).

In large-scale industrial production of 5-HMF, reactor performance may be largely determined by the contact time of biomass and the water medium [28]. High-temperature (≥170 °C) coal-water dehydration methods imply shorter contact times between the biomass and the water medium (water or steam) in the reactor (amounting to seconds, or fractions of a second) and might be very promising for large-scale production. However, the possibility of using the process of wet torrefaction (WT) in a fluidized bed in a superheated water vapor environment for the production of 5-HMF, requires additional research. In particular, important prerequisites for future full-scale applications of the WT process were: the integration of a superheated steam generation process, a heat recovery and recycling system connected to the re-condensation of the liquid products and nitrogen and water recycling or a recovery system.

4. Conclusions

Studies have shown that both methods of heat treatment (WT and HTC) produce activated hydrochars with similar characteristics, following physical steam activation at 450–750 °C. Interestingly, in the innovative WT process operated as a fluidized bed in an environment of superheated steam, biomass treatment can be completed within 30–40 min, while the conventional HTC process may last for 1–4 h.

The obtained activated hydrochars had specific pore surface areas (SSA) up to 491 m^2/g, yet higher SSA may be achieved by chemical activation methods. Further studies are needed to investigate the potential application of activated hydrochars as biofertilizers and soil improvers. Due to the low SSA, and the presence of nutrients, the obtained activated hydrochar may be more suitable for agricultural applications, rather than as industrial adsorbents. In this configuration, biomass treatment by WT/HTC, and the subsequent physical steam activation, alleviates the drawbacks arising from direct poultry litter application to the soil, since pathogens and excess nitrogen were removed, while the transportability, storability, and soil-improving properties were improved.

The mixture of poultry litter and sawdust is not the optimal feedstock for 5-HMF production due to low cellulose contents, as highlighted by the much higher 5-HMF yield obtained after processing sawdust as sole feedstock. Still, the valorization of the 5-HMF contained in the liquid products of WT/HTC might improve the economics of the biowaste treatment processes.

In this regard, the innovative fluidized bed WT process yielded 13–27-fold higher 5-HMF contents compared with the conventional HTC process. Hence, it would be interesting to further investigate the WT process in view of 5-HMF production from diverse feedstocks.

More research is required to prove that the WT process, in a fluidized bed in a superheated steam medium, can be considered a promising technology for biowaste treatment and 5-HMF production. This research would focus on the investigation of diverse feedstock and their mixtures, as well as the optimization of energy efficiency during the process through efficient insulation and the addition of heat exchangers, along with the efficient recovery or recycling of water and nitrogen flows.

Author Contributions: Conceptualization, R.I.; methodology, O.A.I., S.K. and A.D.; software, N.M.; validation, D.K., O.M. and M.N.; formal analysis, M.B.; investigation, A.D., O.A.I. and V.K.-T.; resources, A.M.; data curation, F.T.; writing—original draft preparation, R.I.; writing—review and editing, M.B. and F.T.; visualization, V.K.-T.; supervision, R.I.; project administration, R.I. and B.R.; funding acquisition, R.I. and M.N. All authors have read and agreed to the published version of the manuscript.

Funding: This research was funded by the Fund for Assistance to Small Innovative Enterprises in Science and Technology, Russia (contract No. 410GR/49827 dated 24 October 2019) and Bundesministerium für Wirtschaft und Energie (BMWi), Germany (contract № ZF 4703301SA9).

Institutional Review Board Statement: Not applicable.

Informed Consent Statement: Not applicable.

Conflicts of Interest: The authors declare no conflict of interest.

References

1. Kumar, M.; Olajire Oyedun, A.; Kumar, A. A review on the current status of various hydrothermal technologies on biomass feedstock. *Renew. Sustain. Energy Rev.* **2018**, *81*, 1742–1770. [CrossRef]
2. Wang, T.; Zhai, Y.; Zhu, Y.; Li, C.; Zeng, G.; Zhu, Y. A review of the hydrothermal carbonization of biomass waste for hydrochar formation: Process conditions, fundamentals, and physicochemical properties. *Renew. Sustain. Energy Rev.* **2018**, *90*, 223–247. [CrossRef]
3. Lin, Y.; Ma, X.; Peng, X.; Yu, Z.; Fang, S.; Lin, Y.; Fan, Y. Combustion, pyrolysis and char CO_2-gasification characteristics of hydrothermal carbonization solid fuel from municipal solid wastes. *Fuel* **2016**, *181*, 905–915. [CrossRef]
4. Lu, L.; Namioka, T.; Yoshikawa, K. Effects of hydrothermal treatment on characteristics and combustion behaviors of municipal solid wastes. *Appl. Energy* **2011**, *88*, 3659–3664. [CrossRef]
5. Shen, Y.; Yu, S.; Ge, S.; Chen, X.; Ge, X.; Chen, M. Hydrothermal carbonization of medical wastes and lignocellulosic biomass for solid fuel pro-duction from lab-scale to pilot-scale. *Energy* **2017**, *118*, 312–323. [CrossRef]
6. Poerschmann, J.; Weiner, B.; Woszidlo, S.; Koehler, R.; Kopinke, F.-D. Hydrothermal carbonization of poly (vinyl chloride). *Chemosphere* **2015**, *119*, 682–689. [CrossRef] [PubMed]
7. Liu, T.; Liu, Z.; Zhehg, O.; Lang, Q.; Xia, Y.; Peng, N.; Cai, C. Effect of hydrothermal carbonization on migration and environmental risk of heavy metals in sewage sludge during pyrolysis. *Bioresour. Technol.* **2018**, *247*, 282–290. [CrossRef] [PubMed]

8. Liu, Z.; Zhang, F.S. Removal of lead from water using biochars prepared from hydrothermal liquefaction of biomass. *J. Hazard. Mater.* **2009**, *167*, 933–939. [CrossRef]
9. Liu, Z.; Zhang, F.S.; Wu, J. Characterization and application of chars produced from pinewood pyrolysis and hydrothermal treatment. *Fuel* **2010**, *89*, 510–514. [CrossRef]
10. Sevilla, M.; Fuertes, A.; Mokaya, R. High density hydrogen storage in superactivated carbons from hydrothermally carbonized renewable organic materials. *Energy Environmental. Sci.* **2011**, *4*, 1400–1410. [CrossRef]
11. Sevilla, M.; Fuertes, A.B. Sustainable porous carbons with a superior performance for CO_2 capture. Energy Environmental. *Science* **2011**, *4*, 1765–1771.
12. Sevilla, M.; Maciá-Agulló, J.A.; Fuertes, A.B. Hydrothermal carbonization of biomass as a route for the sequestration of CO_2: Chemical and structural properties of the carbonized products. *Biomass Bioenergy* **2011**, *35*, 3152–3159. [CrossRef]
13. Sevilla, M.; Fuertes, A. The production of carbon materials by hydro-thermal carbonization of cellulose. *Carbon* **2009**, *47*, 2281–2289. [CrossRef]
14. Falco, C.; Marco-Lozar, J.; Salinas-Torres, D.; Morallon, E.; Cazor-la-Amoros, D.; Titirici, M.; Lozano-Castello, D. Tailoring the porosity of chem-ically activated hydrothermal carbons: Influence of the precursor and hydro-thermal carbonization temperature. *Carbon* **2013**, *62*, 346–355. [CrossRef]
15. Wang, L.; Guo, Y.; Zou, B.; Rong, C.; Ma, X.; Qu, Y.; Li, Y.; Wang, Z. High surface area porous carbons prepared from hydrochars by phosphoric acid activation. *Bioresour. Technol.* **2011**, *102*, 1947–1950. [CrossRef] [PubMed]
16. Zhang, Z.; Qu, Y.; Guo, Y.; Wang, Z.; Wang, X. A novel route for preparation of high-performance porous carbons from hydrochars by KOH activation. *Colloids Surf. A Physicochem. Eng. Asp.* **2014**, *447*, 183–187.
17. Straten, J.W.; Schleker, P.; Krasowska, M.; Veroutis, E.; Granwehr, J.; Auer, A.A.; Hetaba, W.; Becker, S.; Schlegl, R.; Heumann, S. Nitrogen-functionalized hydrothermal carbon materials by using urotropine as the nitro-gen precursor. *Chem. A Eur. J.* **2018**, *24*, 12298–12317. [CrossRef] [PubMed]
18. Isemin, R.L.; Mikhalev, A.V.; Muratova, N.S.; Kogh-Tatarenko, V.S.; Teplitskii, Y.S.; Buchilko, E.K.; Greben'kov, A.Z.; Pitsukha, E.A. Im-proving the Efficiency of Biowaste Torrefaction. *Therm. Eng.* **2019**, *66*, 521–526. [CrossRef]
19. Arauzo, P.J.; Maziarka, P.A.; Olszewski, M.P.; Isemin, R.L.; Muratova, N.S.; Ronsse, F.; Kruse, A. Valorization of the poultry litter through wet torrefaction and different activation treatments. *Sci. Total. Environ.* **2020**, *732*, 139288. [CrossRef]
20. Ghanim, B.; Pandey, D.; Kwapinski, W.; Leahy, J. Hydrothermal carbonisation of poultry litter: Effects of treatment temperature and residence time on yields and chemical properties of hydrochars. *Bioresour. Technol.* **2016**, *216*, 373–380. [CrossRef] [PubMed]
21. Kovensky, V.; Borodulya, V.; Teplitskii, Y.; Palchenok, G.; Slizhuk, D. Simulation of biofuel drying with superheated steam in a fluidized bed. *Eng. Phys. J.* **2010**, *83*, 715–719.
22. Fernandez, E.; Santamaria, L.; Amutio, M.; Artetxe, M.; Arregi, A.; Lopez, G.; Bilbao, J.; Olazar, M. Role of temperature in the biomass steam pyrolysis in a conical spouted bed reactor. *Energy* **2022**, *238*, 122053. [CrossRef]
23. Zhang, D.; Chen, X.; Qi, Z.; Wang, H.; Yang, R.; Lin, W.; Li, J.; Zhou, W.; Ronsse, F. Superheated steam as carrier gas and the sole heat source to enhance biomass torrefaction. *Bioresour. Technol.* **2021**, *331*, 124955. [CrossRef] [PubMed]
24. Isemin, R.; Melezhyk, A.; Kuzmin, S.; Nebyvayev, A.; Muratova, N.; Mikhalev, A.; Milovanov, O.; Teplitskii, Y.; Buchilko, E.; Pitsukha, E.; et al. Production of Activated Carbon from Biochar Obtained by Wet Torrefaction of Chicken Manure as Sole Feedstock, and in Mixture with Sawdust in a Fluidized Bed Powered with Superheated Steam. *Chem. Eng. Trans.* **2021**, *86*, 1519–1524.
25. Isemin, R.; Marias, F.; Muratova, N.; Kuzmin, S.; Klimov, D.; Mikhalev, A.; Milovanov, O.; Brulé, M.; Tabet, F. Wet Torrefaction of Poultry Litter in a Pilot Unit: A Numerical Assessment of the Process Parameters. *Processes* **2021**, *9*, 1835. [CrossRef]
26. Hadroug, S.; Jellali, S.; Leahy, J.J.; Kwapinska, M.; Jeguirim, M.; Hamdi, H.; Kwapinski, W. Pyrolysis Process as a Sustainable Management Option of Poultry Manure: Characterization of the Derived Biochars and Assessment of their Nutrient Release Capacities. *Water* **2019**, *11*, 2271. [CrossRef]
27. Isemin, R.L.; Mikhalev, A.V.; Milovanov, O.Y.; Stepantsova, L.V.; Solopov, V.A. Development of the Technology of Poultry Litter Treatment into Organic Fertilizer through Its Decontamination by the Low Temperature Pyrolysis Method. *Int. J. Chem. Eng. Appl.* **2017**, *3*, 184–188. [CrossRef]
28. Chernyshev, V.M.; Kravchenko, O.A.; Ananikov, V.P. Conversion of plant biomass to furan derivatives and sustainable access to the new generation of polymers, functional materials and fuels. *Russ. Chem. Rev.* **2017**, *86*, 357–387. [CrossRef]
29. Iryani, D.A.; Kumagai, S.; Nonaka, M.; Sasaki, K.; Hirajima, T. Production of 5-hydroxymethyl Furfural from Sugarcane Bagasse under Hot Compressed Water. *Procedia Earth Planet. Sci.* **2013**, *6*, 441–447. [CrossRef]

processes

MDPI

Review

Biological Indicators for Fecal Pollution Detection and Source Tracking: A Review

Enze Li, Faizan Saleem, Thomas A. Edge and Herb E. Schellhorn *

Department of Biology, McMaster University, Hamilton, ON L8S 4L8, Canada; lie40@mcmaster.ca (E.L.); saleef4@mcmaster.ca (F.S.); edget2@mcmaster.ca (T.A.E.)
* Correspondence: schell@mcmaster.ca

Abstract: Fecal pollution, commonly detected in untreated or less treated sewage, is associated with health risks (e.g., waterborne diseases and antibiotic resistance dissemination), ecological issues (e.g., release of harmful gases in fecal sludge composting, proliferative bacterial/algal growth due to high nutrient loads) and economy losses (e.g., reduced aqua farm harvesting). Therefore, the discharge of untreated domestic sewage to the environment and its agricultural reuse are growing concerns. The goals of fecal pollution detection include fecal waste source tracking and identifying the presence of pathogens, therefore assessing potential health risks. This review summarizes available biological fecal indicators focusing on host specificity, degree of association with fecal pollution, environmental persistence, and quantification methods in fecal pollution assessment. The development of practical tools is a crucial requirement for the implementation of mitigation strategies that may help confine the types of host-specific pathogens and determine the source control point, such as sourcing fecal wastes from point sources and nonpoint sources. Emerging multidisciplinary bacterial enumeration platforms are also discussed, including individual working mechanisms, applications, advantages, and limitations.

Keywords: fecal pollution; fecal source tracking; fecal indicator; water disinfection; bioengineering; bacterial enumeration

Citation: Li, E.; Saleem, F.; Edge, T.A.; Schellhorn, H.E. Biological Indicators for Fecal Pollution Detection and Source Tracking: A Review. *Processes* **2021**, *9*, 2058. https://doi.org/10.3390/pr9112058

Academic Editor: Dimitris Zagklis

Received: 24 October 2021
Accepted: 15 November 2021
Published: 17 November 2021

1. Introduction

Fecal pollution may cause public health and ecological issues, as it may be associated with the release of pathogens, antibiotic-resistant organisms, high nutrient loads, and the emission of harmful gases [1–4]. Commonly affected areas include soil, sediment, farmland, wastewater treatment plants (WWTPs), and many water bodies such as drinking water sources, recreational water, and groundwater [2,4–7]. Fecal pollution can occur from untreated sewage. In developing and undeveloped countries, untreated sewage is sometimes discharged directly into surface water bodies [8,9]. Polluted water may infiltrate into the ground, rendering groundwater no longer safe for drinking and agricultural purposes [8–10]. According to the WHO, at least 2 billion people worldwide used an unsecured drinking water source polluted with feces in 2017 [11], causing diseases like diarrhea, typhoid, cholera, and dysentery. In 2016, diarrheal diseases ranked as the 8th leading cause of mortality among all ages (more than 1.6 million deaths) and the 5th leading cause of death among children under 5 years old (around 0.45 million deaths, mainly in South Asia and Africa) across 195 countries [12].

Pathogens of different types and infectious doses may differ based on their existing environments, exposure conditions, host susceptibility, and seasonality [13–15]. For example, swimmers exposed to fecal-polluted recreational waters usually develop mild to no symptoms. In contrast, people who cook with fecal-polluted drinking water tend to have acute and severe symptoms. Children under 5 years old and seniors over 70 years old are statistically more vulnerable to waterborne diseases due to reduced immunity [11,16]. Viral pathogens and protozoan parasites may have relatively lower infectious doses than bacterial pathogens, making them a more serious threat [17]. After seasonal rainfalls, short periods of elevated fecal pollution are frequently reported, suggesting a seasonality of waterborne disease outbreaks [14,15]. Table 1 summarizes recent epidemics of waterborne and foodborne diseases that occurred within the last decade. These diseases, however, often have mild or even no early symptoms. Even when symptoms are severe, it remains challenging to attribute diseases to specific fecal pollution exposure [3]. These frequently reported outbreaks signify the importance of the need for fecal pollution monitoring.

Table 1. Examples of outbreaks of waterborne and foodborne diseases from 2011 to 2021.

Leading Pathogen	Diseases	Year	Country	No. of Cases	Citation
Salmonella spp.	Salmonellosis	2010	China	324	[18]
	Salmonellosis	2010	USA	695	[19]
	Gastroenteritis	2013	USA	634	[20]
Cryptosporidium	Cryptosporidiosis	2010	Canada	12	[21]
	Cryptosporidiosis	2011	Australia	1141	[22]
Shigella sonnei	Shigellosis	2012	China	134	[23]
	Shigellosis	2012	Turkey	4239	[24]
E. coli O157:H7	Diarrhea/Vomiting	2012	Canada	5	[25]
	Diarrhea/Vomiting	2012	Japan	115	[26]
Campylobacter	Campylobacteriosis	2016	UK	69	[27]
	Campylobacteriosis	2016	New Zelaland	967	[28]
Hepatitis A virus	Hepatitis A	2016	USA	292	[29]
	Jaundice	2016	India	233	[30]
Rotavirus G2P	Gastroenteritis	2017	Germany	32	[16]
Rotavirus G3P	Gastroenteritis	2017	Australia	2319	[31]
Adenovirus	Respiratory illness	2018	USA	168	[32]
	Fever	2019	China	27	[33]
Norovirus	Diarrhea/Vomiting	2018	South Korea	199	[34]
	Gastroenteritis	2020	Brazil	176	[35]

Another health concern is the undesired dissemination of antibiotic-resistant bacteria (ARB) and antibiotic resistance genes (ARGs) to the environment. Antibiotic residues reach the sewage systems via feces, urine, and body remains [36], which explains why WWTPs are often a hotspot for mixing antibiotics, ARB and ARGs and for their spread into the receiving environment [36]. Elevated ARB and ARGs are frequently measured in areas with increased exposure to municipal wastewater [2,37–39], resulting in a correlation between fecal pollution and antibiotic resistance dissemination. Although antibiotic concentrations in sewage are much lower than their therapeutic dosage in humans or animals, they may still be high enough to affect susceptible environmental bacteria and pose a selective pressure for the evolution of resistant bacteria [2,40]. Multidrug-resistant bacterial pathogens may cause difficult-to-treat or even untreatable infections because conventional antibiotics may not be effective. Moreover, ARGs can spread among non-resistant bacteria by horizontal gene transfer.

Ecological issues associated with fecal pollution are noticeable as well. Fecal wastes are rich in organic matters and nutrients such as ammonium, sulfate, phosphate and potassium [41]. Therefore, their discharge into the environment can be seen as nutrient loads, which can alleviate nutrient limitation and stimulate nutrient recycling, but sometimes causes eutrophication and the emission of harmful gases such as NH_3, H_2S, N_2O and CH_4 [42,43]. Birds can also heavily impact water systems. For example, cormorant defecation in shallow-water ecosystems can lead to algal bloom formation [44]. Defecation by a small cormorant population may not have a significant impact, but a group living in large colonies makes their feces quantitatively relevant as a nutrient input. Similarly, fish can also contribute to the hyper-bloom of cyanobacteria through intensive consumption of herbivorous zooplankton, although some fish such as tilapia can digest cyanobacteria completely [45].

Fecal sludge is widely used as a fertilizer with or without treatment in agriculture due to its high nutrient loads. Fecal sludge application may increase soil fertility, especially regarding carbon retention, but retention of bioavailable phosphorus and nitrogen can be low [46]. Furthermore, untreated fecal sludge poses risks to the crops because of pathogens, heavy metals, and toxic levels of ammonia and nitrite; additionally, the intensive fermentation activities within untreated fecal sludge can result in hypoxia in the rhizosphere [47,48]. To avoid these risks, treatment of fecal sludge or wastewater is required before its agricultural reuse. Composting, the aerobic and thermophilic mineralization of organic matters, is a standard fecal sludge treatment [49]. It effectively eliminates enteric pathogens and parasite eggs due to increased internal temperatures within the fecal sludge. Moreover, mineralized nutrients have improved plant uptake. However, the emission of biogas contaminants produced during composting, including sulfide, ammonia and greenhouse gases (e.g., methane, carbon dioxide and nitrous oxides), is inevitable [41,43,50]. Meanwhile, manure compost can potentially disseminate AGRs and ARB from livestock to soil/crops and finally to humans [51].

This review summarizes five fecal indicator bacteria and four fecal indicator viruses, focusing on their host specificity, degree of association with fecal pollution, environmental persistence, and quantification methods in fecal pollution assessment, together with individual enumeration methods (Figure 1), aiming to serve as a resource for researchers and engineers to evaluate indicators and methodologies for a specific application.

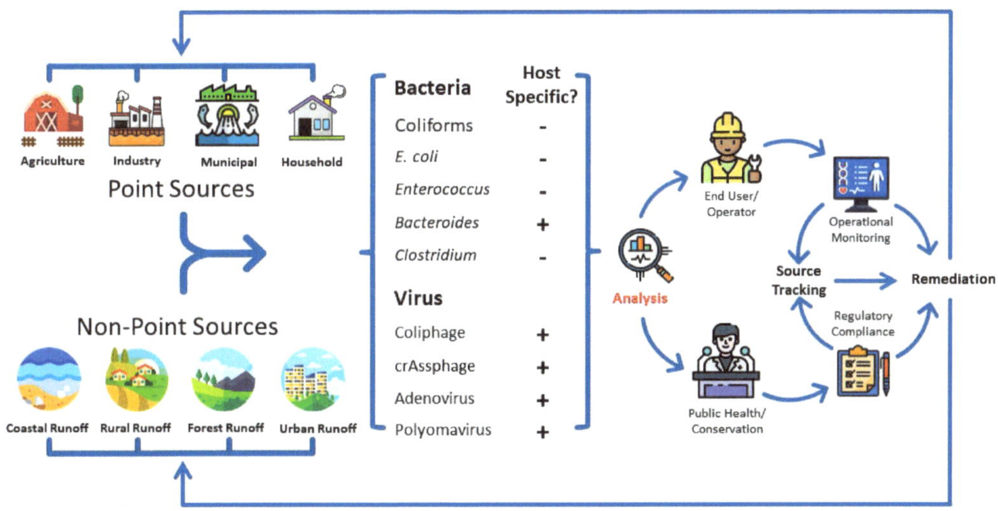

Figure 1. Relationship between the use of selected microbial markers, potential monitoring sites, and diagnostic information by process operators and regulators. Use of these biological markers can facilitate actionable remedial measures by public health officials.

2. Fecal Pollution Detection Using Biological Markers

Fecal pollutants causing water quality problems may come from point sources and/or nonpoint sources. Point source pollution is fecal pollution events resulting from a single identifiable source, such as discharge from a WWTP or a leak from a septic tank. In contrast, nonpoint source pollution results from diffuse sources, such as runoff from residential areas and farmlands. Fecal waste from diffuse sources can be a persistent and cryptic input to the environment that may be present at low concentrations and difficult to trace compared to fecal wastes from point sources because diffuse sources can be a mixture of fresh and aged fecal wastes, and therefore require more complicated methods for source tracking [52]. According to the United States Environmental Protection Agency (U.S. EPA), fecal wastes from diffuse sources have become the leading cause of water quality problems in the U.S. [53]. Accurate identification of the source not only helps confine the types of host-specific pathogens but also determines the source control point and remediation strategies. There are several available indicators for fecal pollution detection or fecal source tracking purposes. Here we provide updates on five bacterial and four viral indicators recommended by the WHO, U.S. EPA and other studies.

2.1. Bacterial Markers

One major purpose of fecal pollution detection is to identify the presence of pathogens to identify the fecal waste source and potential health risks. With the realization that waterborne transmission of bacterial diseases was a public health concern in the 18th century, governments established sanitary systems to monitor drinking water safety and build water treatment facilities [54]. A question arises—should one monitor waterborne pathogens directly or employ proxy indicators? There are many bacterial, protozoan, and viral enteric pathogens that can cause diseases. Their methods of detection are often laborious, time-consuming, and require specialized expertise and biosafe laboratory settings. Pathogens in environmental water are sporadic and sparse, making it difficult or costly to measure these pathogens [3]. In addition, waterborne pathogens in the environment are less dangerous at low concentrations compared to those in GI tracts, but their laboratory enumeration often involves concentration from large volumes of water samples; therefore, these concentrated pathogens may cause severe human diseases during their measurement [1,13]. Another reason we should monitor fecal indicators is that there may be unknown waterborne pathogens that will be overlooked if we only monitor known pathogens. Therefore, detection of fecal pollution-associated pathogens is a multifaceted challenge.

Water quality monitoring to detect fecal pollution can employ biological fecal indicators to proxy many potential pathogens. A typical fecal indicator should fulfill the following criteria [1,3,55]:

(1) It should be detected only in feces-polluted environments and should not be detected in non-fecal-polluted environments; i.e., it should have a high association with fecal pollution with little or no background noise;
(2) It should be detectable in the same host as the pathogens, and its abundance should reflect pathogen abundance both inside and outside the host; i.e., it should have tight correlations with pathogens/diseases;
(3) It should be detectable in the host intestine, feces, and also fecal-polluted environmental waters so that it can be assessed using inexpensive and simple methods; i.e., it should have good detection sensitivity;
(4) It should not multiply outside the host intestine but should be viable or in a detectable form for some time comparable to fecal pathogens;
(5) It should be safe to measure for field and lab workers, not causing severe human diseases.

The value of common fecal indicator bacteria (FIB) is that they occur widely in the feces of birds, mammals, and humans, posing a robust way to detect fecal pollution. However, lack of host specificity is also a problem when determining the specific source of fecal pollution, i.e., fecal pollution source tracking. Advances in the field of microbial source tracking have now provided new fecal indicators that can be used to identify specific fecal pollution sources (e.g., human sewage). If a proxy is used to detect human fecal pollution, ideally it should be exclusively of human sources and not be found in animals. Here we introduce well-validated bacterial, viral and molecular indicators, emphasizing their association with pathogens, their presence in fecal wastes, their host specificity, their persistence in the environment, and their pathogenicity.

Host specificity is one of the most critical criteria for any fecal indicator because it determines whether a given fecal indicator can be used to identify the specific sources of fecal pollution, i.e., fecal source tracking [56]. Accurate source tracking helps water quality regulators to locate pollution sources, such as fecal pollution from a nearby WWTP or an animal farm. Host specificity can be readily assessed by the proportion of the false-positive results detected in nontarget fecal pollution sources [57]. Fecal indicators with absolute human specificity must be detected only in humans and must not be detectable in any animals, whereas those of poor human specificity may also be detected in animals, plants, or environments without association with fecal pollution. Livestock in a farm/slaughterhouse [58,59] and wild animals in forests/natural reserves [60] can also be sources of fecal pollution and pathogens. Thus, fecal indicators specific to one animal host are worth noticing because they help identify the species of animal hosts and the possible source locations.

Sensitivity is another critical criterion which is evaluated by the positive proportions in target fecal pollution sources. Fecal indicators of perfect sensitivity should be detectable in all target fecal pollution sources, whereas those of poor sensitivity may be detectable in a minor proportion of target fecal pollution sources in situations where pollution is high. Sensitivity measurement is, of course, affected by fecal indicator concentrations. Even fecal indicators of high sensitivity may be undetectable in target samples if they are diluted or have undergone significant losses through sample processing steps, causing false negatives.

2.1.1. Coliforms

Coliform bacteria, or total coliforms or simply coliforms, are a group of Gram-negative, facultatively anaerobic, non-spore-forming and rod-shaped bacteria that can ferment lactose to acid and gas at ~37 °C within 48 h, constituting ~10% of gut microorganisms [61]. Traditionally, total coliforms included four closely related genera within the family Enterobacteriaceae: *Escherichia*, *Enterobacter*, *Klebsiella* and *Citrobacter* [62]. But today, some species of over 19 bacterial genera (e.g., *Hafnia*, *Serratia* and *Raoultella*) can be classified as coliforms based on the phenotypic characteristics above [63]. Total coliforms are used to evaluate water treatment efficacy in fecal wastes removal and assess the integrity of water distribution systems [64] because they mainly originate from the GI tracts of humans and warm-blooded animals (i.e., mammals and birds), and thus, their presence in water can indicate possible fecal pollution. In 1975, the U.S. EPA issued the National Primary Drinking Water Regulations, which allowed a maximum of 5% drinking water samples positive for total coliforms tested in a month [65]. Then, in 1989, the U.S. EPA published the first Total Coliform Rule, which supplemented a Maximum Contaminant Level Goal of zero total coliforms in nationwide drinking waters [66]. To date, total coliforms are still routinely monitored in drinking water.

The recognition of free-living coliforms in non-enteric environments, such as freshwater, soil and vegetation, that do not have a history associated with fecal pollution [67,68] has cast doubt on the exclusive association between total coliforms and fecal pollution. For example, *Klebsiella variicola* is a member of total coliforms but is typically a plant endosymbiont [68]. Thus, not all members of total coliforms are suitable indicators of fecal pollution, and presence of total coliforms in drinking water does not necessarily indicate fecal pollution [64]. To overcome this issue, fecal coliforms, which are defined as coliforms originating from the intestines of warm-blooded animals capable of fermenting lactose at 44.5–45.5 °C, are proposed as a monitoring target [69]. As a subgroup of total coliforms, fecal coliforms have better associations with fecal pollution because the non-fecal-origin part is excluded [70]. Fecal coliforms detected in aquafarm harvesting areas such as an oyster bed are indicative of fecal pollution, causing ecological issues and economic losses [71]. Note that neither total coliform nor fecal coliform is a natural taxon but a method-based bacteria category for convenience. We now understand that even within the category of fecal coliforms, some are not exclusively of fecal origin, hence the transition to *Escherichia coli* as a better fecal indicator. Among fecal coliforms, *E. coli* is the dominant species isolated from water (>95%) and is the most frequently used indicator [72]. Figure 2 describes the relationships among total coliforms, fecal coliforms and *E. coli*.

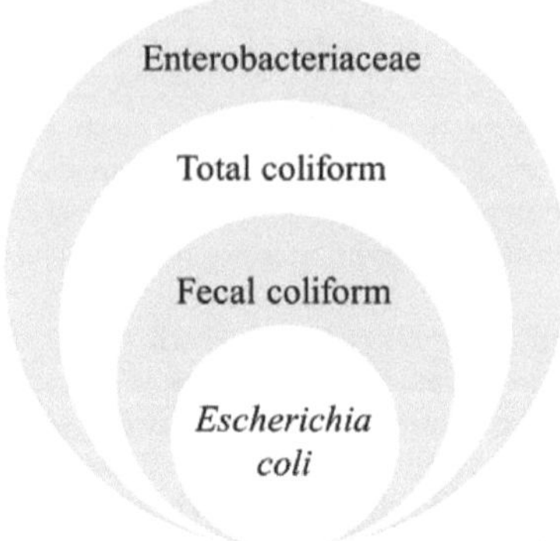

Figure 2. Relationships of coliforms and *Escherichia coli*.

Since coliforms can be cultured with ease, there are culture-based and culture-independent methods to detect and enumerate coliforms in water. Frequently used culture-based enumeration methods include the (i) classical plate count on violet-red-bile-lactose agar medium [73] or chromogenic media based on the activities of β-D-galactosidase (of coliforms) and β-glucuronidase (of *E. coli*) [74], (ii) most probable number procedure in lauryl tryptose broth [75] and IDEXX systems, (iii) membrane filtration in selective M-Endo medium [76] and (iv) lateral-flow immune-chromatographic assay using antibodies recognizing coliform genera [77]. Culture-independent methods include (i) real-time PCR targeting the *lacZ* gene of coliforms [78,79], (ii) fluorescence-based DNA microarray probing *E. coli* [80], (iii) flow cytometry [81] and (iv) next-generation sequencing or metagenomics that profile all microorganisms, including total/fecal coliforms [82].

Except for a few strains such as *E. coli* O157:H7, fecal coliforms are generally not considered to be pathogenic. However, drinking water free of coliforms or not exceeding coliform regulatory thresholds is not necessarily safe to drink. Seven waterborne disease outbreaks, affecting more than 3000 people across 5 American states between 1991 and 1992, occurred in drinking water not exceeding the 1989 Maximum Contaminant Levels [83]. Protozoal parasites *Giardia* and *Cryptosporidium* were identified as the etiological agents. Therefore, coliforms should not be the sole indicator for successful drinking water quality assessment.

2.1.2. *Escherichia coli*

Escherichia coli is used as a fecal indicator as part of drinking water regulations [54]. The U.S. EPA added *E. coli* to the water monitoring and assessment list in 1985 [84]. According to an early investigation by the U.S. EPA, *E. coli* has the highest correlation to swimming-associated diseases at freshwater beaches on the east coast of the U.S. [85]. First, it shows a strong association with fecal pollution, as it is found in all feces of warm-blooded animals and a few reptiles at high concentrations [54,86], and remains at detectable levels in waters subject to animal fecal pollution, such as sewage, rivers, lakes, groundwater and coastal areas [87,88]. In addition, *E. coli* is a facultative anaerobe and is easy to culture, so it can be readily measured using both culture-based and culture-independent methods. Second, *E. coli* is highly associated with a fecal origin and is less likely to be present in natural environments free of animal feces. It was believed to be of exclusively fecal origin when selected as a fecal indicator, which, however, is no longer accurate with the recognition of "naturalized" *E. coli*, which persists and may replicate in non-enteric/non-fecal environments such as soil, sediment, beaches and aquatic plants [89,90]. "Naturalized"

E. coli may not come from a recent fecal input, though it is still believed to have diverged from *E. coli* of fecal origins.

It is important to distinguish between fecal pollution caused by human wastes and that caused by animals, because human fecal pollution is generally associated with high health risks from human-specific pathogens and high loads of antibiotics and ARBs. One major limitation of *E. coli* as a fecal indicator lies in its poor specificity to human feces. Despite the substantial genetic diversity of *E. coli* fecal isolates, epidemiological studies have identified weak associations between a particular *E. coli* clone and a given host [91,92]. For example, a B1 subgroup clone that exhibits the *hly* virulence gene is unique to animals [91], and a B2 subgroup VIII, O81 serotype clone is specific to humans [93]. *E. coli* strains showing stringent host specificity may be specialized candidates for fecal source tracking indicators. Still, most *E. coli* strains discovered in human fecal samples are not specific to humans, causing difficulties in fecal pollution source tracking.

E. coli was initially believed to have poor survival and to be unlikely to multiply after being released into the environment through deposition of fecal wastes because of the environment being so different from animal gastrointestinal tracts and because of environmental stresses such as insufficient organic matter [94], sunlight inactivation and high salinity [95]. However, prolonged *E. coli* survival can be found at low temperatures [96], and enhanced *E. coli* replication is seen at higher temperatures, where relatively sufficient nutrients are likely to support its regrowth [97]. Lake sediments rich in organic contents also favor the survival of *E. coli* at lower temperatures [98]. When grown as biofilms, *E. coli* has a significantly higher tolerance against chlorine than planktonic *E. coli* [99]. In conclusion, the persistence of *E. coli* is greatly affected by environmental conditions.

Changes in climate, such as an increase in temperature and rainfall, may cause fluctuations of *E. coli* counts in waters [100–102]. A rainfall event can result in up to a three-fold increase of *E. coli* counts in recreational waters compared to before rainfall, and this increase can persist for up to 12 h [102,103]. After a rainfall event, increases in *E. coli* counts may occur due to the "first flush" from stormwater drainage or agricultural waste runoff coming along with the flow [103,104]. Furthermore, bacteria including *E. coli* tend to form flocs, adhere to rocks and other particles, then settle down to the sediment of freshwater ecosystems. A heavy rain or storm event may resuspend these bacterial cells from sediments, leading to a 30–37 times higher bacterial load in water [105,106]. Thus, climate and meteorological parameters should be monitored before and during water sample collection for *E. coli* testing to help interpret data.

Most strains of *E. coli* are commensal or symbiotic to their animal hosts in the gastrointestinal tracks [107]. For example, indigenous gut *E. coli* strains can be highly competitive for proline, and their depletion of proline helps inhibit the colonization of pathogenic *E. coli* strain O157:H7 [108]. These non-pathogenic strains are considered safe fecal indicators (categorized as biosafety Level I organisms in the U.S.) and can be enumerated by similar methods, as described for coliforms in environmental waters. In addition, there are specialized methods that only detect certain *E. coli* strains, such as ATCC 47,076, in drinking water [109]. Pathogenic *E. coli* strains, including the aforementioned O157, a Shiga toxin-producing *E. coli* (STEC) serotype categorized as a biosafety Level II organism, caused 5441 infections across the United States in 1996 [110]. Our previous study [111] identified a nine base-pair conserved DNA sequence insertion located in the *ybiX* gene that is unique to O157 strains and can be used for O157 quantification by qPCR. Pathogenic *E. coli* strains may have better correlations with diseases, and therefore, their enumeration can be a means to assess potential health risks to the public.

2.1.3. Enterococcus

Enterococci share traits with *E. coli* that are useful for environmental monitoring. First, they both are facultative anaerobic bacteria detected abundantly in the GI tracts of humans and many animals and are shed in feces at high concentrations. Furthermore, both are easy to culture, both have generally low pathogenicity (at Biosafety Level 1, except for a few species), both have abundances highly associated with GI diseases, and both are subsequently common indicators for water quality monitoring, fecal pollution detection and health risk assessment by culture-based and culture-independent methods [54]. Apart from in humans, enterococci have been detected in insects (e.g., drosophila), domestic animals (e.g., cattle and cats), plants (e.g., forage crops) [112–114], as well as non-enteric environments (e.g., soil and beach sand) [54]. These discoveries indicated a limited host specificity of enterococci and possible origins other than feces.

Both *E. coli* and enterococci have long been used as FIB because of their association with waterborne diseases [115]. Two epidemiological studies recognized enterococci as good indicators of GI diseases among swimmers at both freshwater and marine beaches [116,117]. Enterococci have been the sole recommended FIB by U.S. EPA for marine waters since 1986 [118] because they have higher survival in salt water than coliforms (including *E. coli*) and are thought to be more tightly associated with human pathogens [119,120]. However, using enterococci as fecal indicators has been criticized as well because enterococci are not exclusively of fecal origin and have limited host specificity, as aforementioned. These characteristics may hinder accurate fecal pollution source tracking because non-fecal-origin enterococci may not be associated with health risks, and animal-specific enterococci-associated pathogens may have variable survival strategies in the environment and health risks from human-specific pathogens [121].

The persistence of enterococci in the non-enteric environment is robust, as enterococci can survive across a wide range of temperatures (from 10 to 45 °C) [54], are intrinsically resistant to multiple antibiotics (e.g., ampicillin, cephalosporin and kanamycin) [122], and can survive high salinity (6.5% w/w NaCl) and alkaline (up to pH 10) [123]. *Enterococcus faecium* and *Enterococcus faecalis* are generally the two most abundant species of *Enterococcus* found in human fecal microbiota [124]. They generally have a longer persistence in marine water than *E. coli* [125,126]. They tend to adhere to biotic or abiotic surfaces to form biofilms, protecting them from environmental stresses such as temperature fluctuations, UV radiation, predation, and wave action [127]. Heavy rain or a storm can disrupt these biofilms and release the enterococci into the water column [128]. In environmental waters, adverse conditions, such as solar radiation, starvation, and low temperatures, may decrease enterococci culturability and induce them to enter the VBNC state [129,130]. VBNC cells are unlikely to be detected via culture-based methods.

Epidemiological studies support an association between enterococci from human fecal sources (e.g., urban sewage) and waterborne diseases in recreational waters [116,131,132]. Higher enterococci densities are associated with an increasing occurrence of GI illnesses among beachgoers at beaches impacted by treated sewage effluent [116]. In addition, human pathogens such as *Giardia* and *Salmonella* are found in higher densities and positively correlated with enterococci concentrations in recreational waters impacted by sewage effluent [131,132]. Along with human fecal pollution, exposure to animal feces poses health threats. Globally, one-third of childrens' deaths due to GI illnesses in 2015 were caused by pathogens that are found in animal feces [133]. It is predicted that enterococci densities as low as 35 CFUs/100 mL from cattle manure pose health risks [121]. Thus, enterococci levels are indicative of general health risks from both human and animal fecal pollution. However, without associated source tracking information, it is difficult to unambiguously identify the fecal hosts/sources of enterococci.

2.1.4. *Clostridium perfringens*

Clostridium perfringens is a Gram-positive, spore-forming, biofilm-forming, anaerobic and pathogenic species (at biosafety Level II) associated with feces [134]. It is among the first fecal indicators in aquatic environments [135]. *C. perfringens* can be found in the gut of fish and warm-blooded animals, such as pigs and ducks [136–138], as well as in some non-enteric environments, such as soil, streams and marine sediments [139,140]. Despite having limited host specificity, *C. perfringens* can be present in higher abundances in human feces than animal feces. As a pathogenic species, its presence itself indicates possible health risks to humans and animals.

Species of the genus *Clostridium* are obligate anaerobes and cannot multiply in environmental waters. Spores of *C. perfringens* are coated by a thick peptidoglycan cortex structure that provides resistance against environmental stresses such as chlorination, heat and UV radiation and resistance against toxins in industrial wastes [141]; thus, *C. perfringens* spores can survive longer in polluted waters than other FIB, especially in freshwaters [141,142]. The inability to multiply in non-enteric environments and the prolonged environmental viability of its spores make *C. perfringens* a better indicator of long-term or accumulative fecal pollution. Criticisms against the usage of *C. perfringens* also focus on its prolonged viability and wide distribution in aquatic environments. Its spores can be detected in places far from the pollution input site, indicative of either remote or old fecal pollution [143]. Besides, its abundance does not always correlate with abundances of other FIB [139]. Hence, combined usage of *C. perfringens* and other FIB may better predict pathogens.

Detection and quantification of *C. perfringens* can be conducted by culture-based and culture-independent methods. It can be isolated from water samples by the membrane filtration and Rapid Fung double tube methods [144], then cultivated in anaerobic chambers at 44–46 °C [145]. Available media include mCP medium, tryptose-sulfite-cycloserine agar, and tryptone-sulfite-neomycin agar [146,147]. Rapid and culture-independent enumeration methods include qPCR targeting the *cpn60* gene [148], DNA sequencing targeting 16S rDNA genes or the whole genome, and the newly developed Saltatory Rolling Circle Amplification targeting the *cpa* gene, which is conducted at a constant temperature and does not require a thermocycler or gel electrophoresis [149].

2.1.5. Bacteroides

Species of genus *Bacteroides* are frequently used fecal indicators for source tracking purposes. They are low-pathogenic indicators (biosafety Level I) with promising host specificity compared to facultative anaerobic FIB [150]. *Bacteroides* are obligate anaerobes of gut origins, unlikely to multiply in a non-enteric environment, and are present at significantly higher densities in the GI tracts and feces of humans than coliforms or enterococci [57]. *Bacteroides* can be found in warm-blooded animal hosts such as humans, dogs, pigs, and also fish, which are cold-blooded [151–153]. Unlike coliforms or enterococci, *Bacteroides* spp. isolated from humans and some animal hosts are unique to one or a group of hosts [154]. Table 2 shows examples of host-specific 16S rDNA sequences of these *Bacteroides* spp. that can be promising molecular markers to identify fecal pollution from humans and animal hosts [155,156] by culture-independent methods, such as qPCR.

Table 2. Host-specific 16S rDNA fragments of *Bacteroides* spp. used to discriminate fecal pollution from human and animal hosts.

Host	Primer Set	Sequence (5′ → 3′)	Citation
Human	HF183F/Bac708R	F: ATCATGAGTTCACATGTCCG R: CAATCGGAGTTCTTCGTG	[157]
	HF134F/HF654R	F: GCCGTCTACTCTTGGCC R: CCTGCCTCTACTGTACTC	[157]
	HuBac566F/HuBac 692R	F: GGGTTTAAAGGGAGCGTAGG R: CTACACCACGAATTCCGCCT	[158]
	BacHF/BacHR	F: CTTGGCCAGCCTTCTGAAAG R: CCCCATCGTCTACCGAAAATAC	[159]
Cattle	CF128F/Bac708R	F: CCAACYTTCCCGWTACTC R: See above	[157]
	CS406F/Bac581R	F: GAAGGATGAAGGTTCTATGGATTGT R: CGCTCCCTTTAAACCCAATAAA	[160]
	CS621F/Bac725R	F: AACCACAGCCCGCGATT R: CAATCGGAGTTCTTCGTGATATCTA	[160]

Table 2. *Cont.*

Host	Primer Set	Sequence (5′ → 3′)	Citation
Swine	Bac41F/PS183R	F: TACAGGCTTAACACATGCAAGTCG R: CTCATACGGTATTAATCCGCCTTT	[160]
	PigBac1F/PigBac1R	F: CGGGTTGTAAACTGCTTTTATGAAG R: CGCTCCCTTTAAACCCAATAAA	[160]
	PF163F/Bac708R	F: GCGGATTAATACCGTATGA R: See above	[161]
Dog	BacCan-545F/Bac725R	F: GGAGCGCAGACGGGTTTT R: See above	[155]
Chicken	C160F-HU/Bac265R-HU	F: AAGGGAGATTAATACCCGATGATG R: CCGTTACCCCGCCTACTAC	[156]
Duck	Bac366F-HU/Duck474R-HU	F: TTGGTCAATGGGCGGAAG R: GCACATTCCCACACGTGAGA	[156]

Unlike facultative anaerobic FIB, the isolation and cultivation of *Bacteroides* cells require anoxic chambers, which is costly, time-consuming and labor-intensive [162]. *Bacteroides* can grow at 37 °C in many selective media, such as Brain Heart Infusion medium, Bacteroides Bile Esculin agar, and Blood Hemin Vitamin K medium containing antibiotics [162–164]. In most applications, *Bacteroides* grows into visible colonies on plates within 24 h, and colonies will be large enough to enumerate in 36 and 48 h. Some *Bacteroides* species can be identified to the species level within four hours by a few biochemical tests, such as the API ZYM system that detects activities of *Bacteroides*-specific enzymes [165].

The use of biological molecular markers bypasses the difficulties of anaerobic cultivation and enables the routine detection of *Bacteroides* spp. The aforementioned host-specific 16S rDNA markers can be used to detect and quantify *Bacteroides* in environmental samples and thus enable source tracking of the fecal pollution. Among human-specific *Bacteroides* 16S rDNA sequences, the region to be amplified by the primer set HF183F (Human Fecal 183 forward primer) and Bac708R (reverse primer) ranks among the most recognized markers to assess human fecal pollution and to discriminate it from animal fecal pollution because of its high sensitivity and specificity to humans [166]. Although HF183 fluorescent probe-based assays may sometimes cross-react with animal feces, its concentration in human feces can be 1000-fold more, or higher, than in animal feces [167]. In 2012, the U.S. EPA recommended an illness-based recreational water quality threshold of 32 illnesses per 1000 primary contact recreators using enterococci or *E. coli* [118]. In 2020, as *Bacteroides* HF183 becomes widely used, a threshold of 525 HF183 copies/100 mL recreational water was proposed by Boehm and Soller [168]. Apart from 16S rDNA sequences, there are other *Bacteroides* human molecular markers, such as genes encoding a hypothetical protein of *Bacteroidales*-like cell surfaces [169], which showed a >97% human specificity.

Obligate anaerobes such as *Bacteroides* cannot grow or multiply in environmental waters and have shorter survival periods than aerobes, facultative anaerobes, or viruses [57]. This property can provide a time progression of fecal pollution in environmental waters after *Bacteroides* spp. are released [170]. Molecular markers of both viable and non-viable cells can be detected using qPCR and DNA sequencing, so that they are less affected by cell states [171]. For example, the portion of culturable enterococci and *E. coli* in unfiltered freshwater can be reduced by 90% within 1.2 h and 1.4 h in sunlight, whereas it takes more than 6 h for the same reduction for HF183 and human polyomavirus [167].

2.2. Viral Markers

Fecal pollution detection using conventional FIB has several limitations. For general FIB of poor host specificity (e.g., *E. coli* and enterococci), they do not discriminate between human and animal fecal sources and therefore are inappropriate for source tracking [172]. Human-specific fecal indicators, mainly genetic markers of *Bacteroides* spp., are capable of fecal pollution source tracking from both point and non-point sources [150] but are not always representative of enteric viral pathogens or protozoa. This is because viruses and parasite protozoa react to environmental stresses and wastewater treatments differently from bacteria [173]. There are outbreaks caused by waterborne viral pathogens or protozoa when FIB are not present or are below the regulated levels in water samples, as aforementioned [83]. One reason could be that FIB quantification, especially by culture-based methods, can provide false negative results if cells are in the VBNC state [174]. Therefore, it is important to employ fecal indicator viruses in water quality assessment for their higher correlation with enteric viral pathogens. These viral indicators can be either of fecal origin, such as coliphage and crAssphage, whose hosts are gut bacteria, or highly associated with human activities, such as pepper mild mottle virus, a plant virus which is commonly found in dietary pepper products [175].

2.2.1. Coliphage

In human guts, the viral community (termed human gut virome) is dominated by bacteriophages [176]. Coliphages are bacteriophages that specifically infect coliforms and are not pathogenic to humans. They are generally found in the gut and share similar morphologies, sizes, structures and resistance against disinfection methods with enteric viruses [177,178]. For example, coliphages can be resistant to environmental stresses, such as hydrostatic pressure (less than $1-\log_{10}$ reduction at 600 MPa for 1 h), thermal inactivation ($4-\log_{10}$ reduction at 72 °C for 20 min) and high or low pHs (no significant $\log_{10}$ reductions between pH 4 to 11 for 3 h) [179,180]. Therefore, coliphages can act as indicators for enteric viruses and human fecal pollution in rivers, drinking water and underground water [181]. Based on the infection site, coliphages are classified as somatic coliphages, which are DNA phages that infect coliform cells through their attachment to the outer cell wall, and male-specific (or F+) coliphages, which can be DNA or RNA phages that infect male (F+) host cells through the fertility (F) pili [61]. Apart from *E. coli*, somatic coliphages can infect other pathogenic gut bacteria, such as *Shigella* spp. and the multi-drug resistant *Klebsiella pneumoniae* [182].

Two culture-based methods recommended by the U.S. EPA for enumerating somatic coliphages in freshwater and groundwater are the spot-test assay for their presence/absence and the single-agar-layer plate assay to quantify the plaque-forming units (PFU) [183,184]. The former requires 24 h, including 18 h for incubation and 6 h for the spot test, whereas the latter requires 16 h from sample processing to plaque enumeration. Hence, both assays do not allow for rapid detection. Another Bluephage method relies on the potential of *E. coli* strains overexpressing β-glucuronidase when facing coliphage-mediated bacterial lysis. The substrate resulting from the overexpressed β-glucuronidase is chromogenic and produces a blue color that indicates coliphage density [185]. This method features an excellent sensitivity (as low as 1 PFU/100 mL), an excellent specificity ensured by the coliphage, and a rapid procedure within 6.5 h (2.5 h for pre-growing host *E. coli* strains, 4 h for the assay) [186]. Thus, the Bluephage method can provide a rapid and sensitive estimation of fecal pollution in water samples.

F+ RNA coliphages are divided into four genogroups (I–IV) based on genomic organization and serological cross-reactivity. Higher abundances of genogroups II and III are detected in rivers, streams and groundwater affected by municipal sewage [187], whereas higher abundances of genogroups I and IV are detected in waters contaminated by animal farms or bovine wastewaters [188]. Therefore, genogroups II and III are identified as specific to human fecal pollution, whereas I and V are identified as specific to animal fecal pollution. F+ RNA coliphages can be enumerated by a modified Bluephage method, which selects F+ (male) *E. coli* (e.g., strains CB14 and CB16) as the host [189]. Furthermore, *E. coli* strain CB-390 allows for simultaneous detection of somatic and F+ RNA coliphages [190]. Since somatic and F+ coliphages have varying environmental persistence, the simultaneous detection of both can result in a more accurate estimation of total fecal pollution. For source tracking purposes, however, separate enumeration of somatic and F+ coliphages is more suitable, because somatic coliphages may be $4-\log_{10}$

more abundant than F+ coliphages due to enhanced environmental persistence, which can mask the population of F+ coliphages [191].

2.2.2. CrAssphage

CrAssphage is a double-stranded DNA bacteriophage that infects *Bacteroides intestinalis*. Unlike most bacteriophages, the discovery of crAssphage was not based on biological experiments but was based on bioinformatics. In 2014, its genome was assembled for the first time by the de novo cross-assembly method from 12 previously published human fecal viromes [176]. It is named after the software *crAss*. This name is unconventional because it does not indicate any phylogeny, taxonomy or host information. So far, crAssphage shows no homology to any known virus. In 2018, crAssphage was isolated from human feces for the first time, and its host, *Bacteroides intestinalis*, a human gut symbiont, was identified [192].

The presence of crAssphage is common among human fecal viromes and fecal metagenomes, as it is present in 342 out of 466 datasets surveyed (73%) [176]. In addition, it is highly abundant in the human gut, accounting for 90% of human fecal viromes and 1.68% of total human fecal metagenomes, making it the most abundant phage in the human gut. crAssphage was thought to be of exclusively human fecal origin, but recently it has been found in cats [56] and monkeys [193]. The crAssphage relatives detected in monkeys are distantly related to the published human crAssphage genome, suggesting a long evolutionarily divergent history in humans and primates. Nevertheless, crAssphage is highly specific to humans, showing a specificity up to 98% [56].

A known host is required for any virus to be cultured. When the crAssphage genome was first identified, its host was predicted by in silico approaches, including OFR homology searches (blastn), CRISPR spacer identification, and co-occurrence profiling of the crAssphage genome and 151 fecal bacterial metagenomes [176]. Among the predicted hosts, *Bacteroides* was believed to be the most likely host. In 2018, a phage that dominated 96% of total viral reads of the *Bacteroides intestinalis* strain APC919/174 plate was isolated and found to be related to a known but previously uncultured crAssphage genome. That was the first successful isolation of crAssphage.

Quantification of crAssphage in environmental waters is mainly by culture-independent methods, such as metagenomics and qPCR. Primers in PCR/qPCR targeting predicted coding regions of the crAssphage genome have been designed and validated [172]. The two most frequently used primer sets CPQ_056 and CPQ_064 (Table 3) exhibit the highest sensitivity and specificity, meaning that PCR assays using these primer sets can obtain (1) PCR products of correct sizes among human feces/sewage samples; (2) no PCR product among animal fecal samples; (3) PCR products of correct sizes among diluted human feces/sewage samples; or (4) no spurious PCR products, such as primer-primer dimers. A number of subsequent studies have validated the amplification capability of the two primer sets [194–198].

Table 3. Available primers for qPCR detection of crAssphage, HAdV and HPyV.

Target Virus	qPCR Assay	Primer or Probe Sequence (5′ → 3′)	Target Gene	Citation
CrAssphage	CPQ_056	F: CAGAAGTACAAACTCCTAAAAAACGTAGAG R: GATGACCAATAAACAAGCCATTAGC Probe: [FAM]-AATAACGATTTACGTGATGTAAC-[MGB]	Predicted coding region	[172]
	CPQ_064	F: TGTATAGATGCTGCTGCAACTGTACTC R: CGTTGTTTTCATCTTTATCTTGTCCAT Probe: [FAM]-CTGAAATTGTTCATAAGCAA-[MGB]	Predicted coding region	[172]
HAdV	VTB-1-HAdVF	F: GCCTGGGGAACAAGTTCAGA R: GCGTAAAGCGCACTTTGTAAG Probe: [Quasar]-CAGTCGCTGYGACCTGTCTGTGGTT-[BHQ2]	Hexon	[199]
	VTB-2-HAdVC	F: GAGACGTACTTCAGCCTGAAT R: GATGAACCGCAGCGTCAA Probe: [FAM]-CCTACGCACGACGTGACCACAGA-[BHQ1]	Hexon	[199]

Table 3. *Cont.*

Target Virus	qPCR Assay	Primer or Probe Sequence (5′ → 3′)	Target Gene	Citation
HPyV	BKPyV and JCPyV	SM2: AGTCTTTAGGGTCTTCTACCTTT P6: GGTGCCAACCTATGGAACAG KGJ3 probe: [FAM]-TCATCACTGGCAAACAT-[MGB]	Conserved T-antigen	[200]
	HPyV-2	F: ATGTTTGCCAGTGATGATGAAAA R: GGAAAGTCTTTAGGGTCTTCTACCTTT Probe: [FAM]-AGGATCCCAACACTCTACCCCACCTAAAAAGA-[MGB]	JVGP5	[201]

So far, the following features make crAssphage a fecal indicator virus suitable for human fecal pollution detection and source tracking. First, it is ubiquitous and abundant in human feces and sewage (5.28–7.28 $\log_{10}$ copies/100 mL) [194]; it is rarely detected in environments free of human fecal pollution [56]; and it has concentrations strongly correlated with sewage ARGs [2], pathogenic viruses [198,202] and conventional FIB [197]. In addition, it is not pathogenic to humans [193] and has molecular markers that can be detected by rapid, culture-independent and cost-effective methods [172], although it is unclear whether crAssphage can replicate itself outside human intestines but within *Bacteroides* cells.

2.2.3. Human Adenovirus (HAdV)

Adenovirus was first isolated in 1953 from human adenoids, hence the name [203]. Adenovirus is not a taxonomic unit and represents a group of viruses from six genera of the family Adenoviridae: *Atadenovirus*, *Aviadenovirus*, *Ichtadenovirus*, *Mastadenovirus*, *Siadenovirus* and *Testadenovirus* [60]. Although adenoviruses share a similar morphology, their hosts and genomic organizations differ from genus to genus. For example, hosts of mastadenoviruses are exclusively mammals, whereas hosts of aviadenoviruses are exclusively birds. In addition, mastadenoviruses can synthesize unique proteins V and IX; the former is responsible for redistributing viral components into or out of the host cell nucleus, while the latter is a transcriptional activator [204,205]. So far, adenovirus has been detected in species of almost every vertebrate class, such as birds, mammals, reptiles, amphibians and fish, but has not been detected in invertebrates [60]. All human-specific adenoviruses (HAdV) belong to the genus *Mastadenovirus*.

HAdV is highly specific to humans. As it targets a conserved region of the *hexon* gene, HAdV is not detected by qPCR in samples that do not contain human feces, meaning there are no false-positive cases, whereas somatic coliphages, *Bacteroides* HF183 and crAssphage CPQ_056 are detected in the feces of cows, chickens and cats [56,206]. Except for HAdV, animal-specific adenoviruses can be used to identify animal fecal pollutions as well. Bovine adenovirus is found to be specific to cows [58], and porcine adenovirus is found to be specific to pigs [207]. Hence, despite the broad host spectrum of the adenovirus group, specific adenoviruses have excellent host specificity, enabling source tracking of human and animal feces.

HAdV in groundwater has a stable persistence and may survive for long periods. For example, the low decay rates in groundwater (0.0076 $\log_{10}$/d at 4 °C and 0.0279 $\log_{10}$/d at 20 °C) result in a long natural T_{90} (time required for $1-\log_{10}$ reduction) of 131 d at 4 °C and 36 d at 20 °C [208]. In WWTPs, HAdV is highly resistant to UV disinfection and is frequently detected in the effluent and receiving environmental waters [209,210]. Among four enteric viral pathogens, echovirus, coxsackievirus, poliovirus and HAdV type 2, a UV dose of 27 mW/cm^2 sufficiently brought 3−log reductions to the first three viruses, whereas it is 119 mW/cm^2 for HAdV type 2 [211]. One possible reason is that adenoviruses have a double-stranded DNA genome, which is more stable than RNA or single-stranded DNA genomes. Additionally, HAdV uses human cellular enzymes to repair its DNA damage [212]. In this case, HAdV is not truly resistant to UV radiation but is actively able to repair the damage to DNA brought by UV.

HAdV is typically found at much lower concentrations than *E. coli*, crAssphage, or *Bacteroides*, often at levels close to or lower than its detection limit. For example, HAdV can be detected in sewage samples at ~10^5 gene copies/L, and only 10^2–10^3 gene copies/L in rivers, urban streams and estuarine waters [209]. Such low concentrations in environmental waters can be a challenge for HAdV plaque assays, because it is estimated that 1–10^3 virions are required for 1 PFU to form depending on the virus [213]. Hence, environmental water samples should be concentrated for HAdV plaque assay by membrane filtration or other methods [214]. In culture, HAdV may take up to 1 week to infect, reproduce within and lyse human cells to generate visible PFUs [215]. The required time for a HAdV plaque assay is not practical for routine water quality monitoring or rapid decision making. Additionally, HAdV is pathogenic to humans, causing upper respiratory symptoms and fever, though it causes mild or asymptomatic infection in most cases [216].

Culture-independent methods to enumerate HAdV from environmental waters include regular PCR, nested PCR, qPCR, integrated cell culture qPCR, and flow cytometry [210,217]. qPCR assays using a nested primer set and a degenerate primer set targeting the *hexon* gene (Table 3) indicate excellent the host specificity of HAdV and bovine adenovirus (no false positives), with an overall sensitivity of 78% and 73%, respectively [58]. Additionally, fluorescent antibodies targeting two HAdV proteins, *Hexon* and *E1A*, can be used in fluorescence-activated cell sorting for HAdV rapid detection with excellent linearity between fluorescence intensity and HAdV concentration ($r^2 > 0.9$) and sensitivity in the range of 1–10^4 PFU per assay [210].

2.2.4. Human Polyomavirus (HPyV)

HPyVs are a group of human enteric viruses that belong to the genera *Alpha-*, *Beta-* and *Delta-Polyomavirus* within the family *Polyomaviridae*, with circular double-stranded DNA genomes [218]. First isolated in 1971 from human brain tissue, there are 14 distinct HPyV species identified so far [218,219]; some are human feces-specific (e.g., HPyV-6 and HPyV-12) and some are human urine-specific (e.g., HPyV-9, BKPyV and JCPyV). These HPyVs are potential source tracking indicators due to their human specificity, environmental prevalence and resilience. Animal-specific PyVs, such as bovine, porcine, and ovine PyVs, are specific to cattle, pigs, or sheep, so they can be used for the identification and source tracking of animal fecal pollution [220].

HPyVs are more specific to humans when compared to FIB and are less likely to be detected in non-human fecal samples. For example, qPCR assays targeting HPyV (the T-antigen gene), *Bacteroides* HF183 and crAssphage (CPQ_56) among 359 animal feces samples exhibit a human specificity of 99.72%, 96.65% and 98.61%, respectively [56]. Despite the excellent human specificity, HPyV has variable sensitivity based on sample types. For example, urine-specific HPyV is detected in 6 out of 26 urine samples from healthy individuals, ranging from 6.61×10^2 to 1.2×10^7 copies/mL, showing a limited sensitivity of 23.1% [200]. In WWTPs, however, HPyV is detected in 36/37 influent and effluent wastewater samples, showing a high sensitivity of 97.3% [209].

HPyV is primarily excreted in human urine and feces; hence, its presence in environmental waters reflects anthropogenic pollution. BKPyV and JCPyV, specific to human urine, are the first HPyV identified in environmental samples (from sewage) [221]. Since then, increasing HPyV species have been detected in many environmental waters. HPyV is prevalent in rivers, urban streams, recreational beaches and drinking water [209,222,223]. WWTPs are believed to be the key point for HPyV releasing to open water bodies. Though unable to multiply, JCPyV was reported to have a T_{90} as long as 64 d in sewage at 20 °C [224]. Viruses adsorbed onto particle surfaces may have prolonged survival relative to free viruses, possibly because particles reduce contact between viruses and disinfectants [225]. In that case, HPyVs excreted in human feces are expected to have longer environmental persistence than HPyVs being excreted in the urine.

Like HAdVs, HPyVs are present at low concentrations in environmental waters, and therefore HPyV enumeration often involves the concentration of large volumes of water samples. Simple membrane filtration, using a 0.22 or 0.45 μm filter, fails to concentrate HAdV or HPyV particles, as their size is much smaller than the pore size of regular filters. Salt and acidic pH may increase viral adsorption to the filter because viruses in ambient waters are negatively charged [226]. Hence, current membrane filtration for virus concentration often requires pH adjustment of the water sample, for example, acidification using HCl if a nitrocellulose membrane is used, to obtain a higher HPyV recovery without introducing PCR inhibitors [200,227].

Concentrated HPyVs can be quantified by plaque assay using human fetal glial cells or tumor cells, immunofluorescence, microarray and PCR/qPCR [56,228,229]. qPCR targeting the conserved T-antigen of BKPyV and JCPyV is the most extensively used molecular method for HPyV detection in environmental waters (Table 3), with a sensitivity (i.e., limit of detection) as low as 10 copies per reaction [200]. HPyVs detected by immunofluorescence assay are generally $2-\log_{10}$ less than HPyVs detected by qPCR, because only infectious HPyVs can be detected by immunofluorescence assay or plaque assay, whereas both infections and non-infectious HPyVs can be detected by qPCR [228]. For some samples, it is important to analyze the infectivity of viruses, and culture-based methods remain the gold standard. Infectious HPyVs do not necessarily cause diseases, as HPyVs are generally not pathogenic and are categorized as Biosafety Level II.

Compared to bacterial indicators, HPyV better correlates with viral pathogens in the environmental waters. For example, norovirus, a human enteric pathogen, is generally not routinely monitored partially because of its pathogenicity and RNA genome. Norovirus cannot be efficiently predicted by FIB because it has longer environmental persistence and may be present in high concentrations in non-disinfected wastewater effluent and receiving waters, such as rivers, urban streams and estuarine water, without causing outbreaks [209,230]. Among these waters, HPyV is not only prevalent but also significantly correlates ($p < 0.05$) with norovirus (determined by reverse transcription qPCR). In recreational coastal beaches impacted by non-point source fecal pollution, HPyV levels may not be significantly associated with levels of *Bacteroides*, enterococci, total coliforms or fecal coliforms, but can be significantly associated with levels of HAdV [222]. The poor associations between fecal indicator viruses and FIB strengthen the need to employ fecal indicator viruses for water quality monitoring, as FIB are often poorly correlated with viral pathogens, whereas HAdVs and HPyVs themselves are human enteric pathogens, sharing similar environmental surviving strategies with other human enteric viral pathogens, and are therefore of greater potential to indicate health risks.

3. Emerging Bioengineering Enumeration Methods

All current biological methods have limitations. Culture-based methods (e.g., plate counting and membrane filtration) are labor- and time-consuming and often lead to delayed results that are insufficient to catch sudden bacterial outbreaks or to support proactive actions. It is possible that by the time fecal pollution is detected in drinking water using culture-based methods, the investigated water may have already been distributed and consumed. In addition, culture-based methods provide information on a fraction of the total bacterial population, because non-culturable bacteria or VBNC cells are not included. A limitation for molecular methods such as qPCR and metagenomics is that they cannot distinguish DNA of living and dead cells, i.e., they cannot exclude the interference of relic DNA. It is estimated that relic DNA can persist for weeks to years in soil, account for 40% of total DNA being sequenced, and have inflated the observed richness of prokaryotes and fungi by 55% [231]. Sample types (e.g., water, soil, plant) and DNA extraction methods also impact the DNA recovery. Multi-disciplinary platforms integrated with cutting-edge techniques in the fields of biology, engineering, chemistry and computer science may be powerful alternatives or supplements to current methods to overcome (at least some of) the existing limitations.

3.1. A Time-Lapse Coherent Imaging Platform

A lens-free imaging platform capable of early detection and species classification of living FIB on agar plates is a promising new technology [232], though manual membrane filtration of water samples and inoculation of concentrated indicator cells are required. When an inoculated agar plate is placed in the incubator, the imaging system scans the plate every 30 min, simultaneously with the bacterial incubation. Captured plate images are sent to the integrated software, where periodically captured plate images are digitally stitched together in chronological order to form time-lapse image stacks. These stacks are reconstructed to reveal static impurities (e.g., dust, air bubble and light speckle) and growing bacterial colonies on the agar surface.

Fifteen pure culture agar plates of three total coliform bacteria (*E. coli*, *Klebsiella aerogenes*, and *Klebsiella pneumoniae*) are blindly tested on this platform. For validation of colony recognition sensitivity and precision, colonies on these plates are also manually enumerated. A colony recognition sensitivity of 80% is achieved when *E. coli*, *K. aerogenes* and *K. pneumoniae* are incubated for ~6.8 h, ~8.8 h, and ~6.0 h, respectively. When incubation continues to 12 h, a sensitivity higher than 95% is achieved for all three coliforms. After 6 h incubation, precision for all three coliforms is close to 100%. A species classification accuracy of ~80% is achieved when *E. coli*, *K. aerogenes* and *K. pneumoniae* are incubated for ~8.0 h, ~12.0 h, and ~7.6 h, respectively. For *E. coli* and *K. pneumoniae*, higher species classification accuracies (~97.2% and ~98.5%) are achieved when incubated for 12 h. Colony recognition sensitivity, colony recognition precision, and species classification accuracy are positively correlated with the incubation period.

This platform enables a rapid (87 s to scan a 60-mm diameter petri dish), high-resolution (with a linewidth scanning resolution of ~3.5 μm) and automatic plate image capture for every 30 min (interval adjustable). Such a high scanning resolution enables early recognition of emerging colonies that are invisible to naked eyes. Compared to EPA-approved culture-based methods (e.g., 1103.1 and 1604 methods), this platform is more time- and labor-saving. In the blind testing, a whole test (bacterial incubation, colony recognition and species classification) can be completed within 12 h, with an excellent limit of detection as low as 1 CFU/L, at a low cost of only $0.6, including the cost of preparing an agar plate. However, this platform is not fully automatic because manual sample processing is still required.

3.2. An Online Culture-Independent Imaging Platform

Drinking water requires faster and more reliable quality monitoring than water for other purposes because it is directly consumed by humans. Though water utilities monitor water quality regularly, if culture-based methods are applied, results come out 1–3 d delayed and do not provide information of unculturable bacteria or VBNC cells. To resolve these problems, an automated, online and culture-independent platform capable of enumerating total bacteria within 10 min has been designed [233]. This platform is installed within drinking water pipeline systems, where drinking water flows through a flow cell under the pressure of the water source or an inlet pump. To initiate a measurement, the flow cell becomes sealed to trap 6 μL of water sample inside. The stagnant trapped water sample is scanned by a lens-based imaging system at different focuses, i.e., it undergoes a 3D scanning. Captured images are sent to the platform-integrated software for analysis. Then the flow cell reopens and gets flushed, preparing for the next measurement. By combining images of different focuses (i.e., depths), a 3D model of the water sample can be established from which particles will be recognized, together with their 59 optical parameters, such as area, length, contrast and granularity. Then, the pre-trained software classifies bacteria and abiotic particles based on the 59 optical parameters.

In blind testing using monotype suspensions containing either bacterial or abiotic particles, a classification accuracy of >80% is achieved for all bacterial suspensions, and less than 8% of abiotic particles are mistakenly classified as cells. The average classification accuracy for monotype suspensions is 90%. In the blind testing using mixed suspensions containing both bacteria and abiotic particles, the overall classification accuracy is 78%. It is worth noticing that for mixed suspensions, due to lack of manual curation, false negative (i.e., bacteria being classified as abiotic particles) and false-positive cases (abiotic particles being classified as bacteria) can both occur but cannot be distinguished [234]. Based on the small sampling volume (6 μL), this platform has a theoretical lower limit of detection of 1.6×10^2 particles/mL. Water samples with lower particle concentrations may be associated with high uncertainty due to sample inhomogeneity. On the other hand, an upper limit of 5×10^6 particles/mL is also determined, because water samples with higher particle concentrations may not be detected due to strong background light scattering, and heavy water pollution will significantly shorten the lifespan of a flow cell.

This platform enables a fully automatic, rapid and high-frequency (10-min time resolution) monitoring of bacteria and abiotic particles, making it possible to detect sudden water quality changes and classify them as biotic or abiotic. This platform does not require chemical supplies or produce any hazardous wastes. Acquired data can be wirelessly transferred to a data server in real time (i.e., online). The only required maintenance is to replace the flow cell every few months. As a culture-independent platform without species preference, it is suitable to estimate the total bacterial biomass. However, dead or inactive bacteria (e.g., those inactivated by UV treatment), which are no longer of health risk, can still be detected, as long as their cells remain intact, with the same or similar optical parameters as living cells. In addition, it does not apply to enumerating virus or protozoa, which are relatively either too small or too big compared to water chemical particles.

3.3. An On-Site Microflow Cytometry Free of Fluorescence Staining

Flow cytometry can be used to enumerate bacteria and viruses in water samples, such as coliforms in drinking water and HAdV in sewage effluents [81,210]. These methods have limited on-site use because fluorescence staining is often involved, which requires a bulky cytometer, expensive reagents, and complicated sample preparation. The development of a microfluid-based portable microflow cytometer enables rapid and on-site *E. coli* enumeration without fluorescence staining [235]. A major part of this platform is a four-layer panel; between the middle two layers is a 50-μm-wide microfluidic channel. The water sample and sheath fluid are injected into the channel by a syringe pump. The sheath fluid is injected at a much higher rate than the water sample so that cells in the water sample are forced to pass through the channel one by one unidirectionally towards the outlet. An input light beam is perpendicular to the channel, where light is differentially scattered by cells, particles or water. An on-chip lens system beneath the panel records the side-scattered light signals to separate *E. coli* cells and particles.

This microflow cytometer is tested using two suspensions, one made of pure *E. coli* cells diluted in PBS buffer and the other made of both *E. coli* cells and polystyrene beads with diameters of 2 and 4 μm to mimic particles from environmental waters. For a pure *E. coli* suspension, the sample-to-sheath flow-rate ratio is found to be negatively correlated with detection efficiency. When *E. coli* is injected at 50 μL/h with a constant sheath flow of 313 μL/h, a high detection efficiency (97%) can be achieved. When the sample flow rates increase to 100 and 150 μL/h, the detection efficiencies decrease to 63% and 45%, respectively. Regardless of sample flow rates, the high signal/noise ratios (9.7 to 11) are supposed to reduce the possibility of false counting. For mixed suspension, the presence of 2 and 4 μm diameter polystyrene beads does not affect the detection of *E. coli* cells, because the site-scatter signal intensity is proportional to particle size, and therefore, the beads have significantly higher side-scatter signal intensities than *E. coli* cells. Furthermore, even the 2 and 4 μm beads have distinct intensities and can be distinguished. Therefore, the presence of particles larger than 2 μm is unlikely to affect the detection of *E. coli* by the platform.

This microflow cytometer features a rapid (50 s for a measurement), less costly (no fluorescence staining required) and on-site (portable equipment) enumeration of *E. coli* in water samples, with excellent cell detection efficiencies at low sample-to-sheath flow rate ratios. The high signal/noise ratios decrease false cell counting from the background noise, and particles of similar or larger sizes of *E. coli* do not interfere with cell detection. However, interference may arise when particles smaller than 2 μm are present; in that case, cells require other staining techniques, such as fluorescence staining, to be distinguished.

One major limitation of this platform lies in its limited throughput, as higher sample injection rates would decrease detection efficiencies. Additionally, this platform is not suitable for long-term operation, because the throughput decreases over time due to cell precipitation in the syringe during measurements. In addition, *E. coli* and water have close refractive indexes, so a pre-concentration is required to increase cell density by 3–6 orders of magnitude; otherwise, site-scatter signals produced by *E. coli* cannot be sufficiently separated from the background noise.

4. Conclusions

Traditional water monitoring relies on *E. coli*, enterococci and total coliform testing, which are well-validated and widely used methods. These are routinely monitored in drinking water or recreational water because they are associated with waterborne illnesses and can be enumerated with ease. However, their limited host specificity hampers accurate source tracking of fecal pollution from nonpoint sources. It is particularly important to identify human sources of fecal pollution, as they usually present high health risks. Source tracking indicators, such as certain species/strains of *Bacteroides*, human adenovirus and crAssphage, have better host specificity and therefore can be used for human fecal pollution source tracking. Whether using fecal indicator bacteria or viruses, a single fecal indicator is unlikely to predict all pathogens or diseases; hence, multiple indicators are recommended to be used in combination to assess the water quality and health risks throughout. These additional tools can augment traditional monitoring and serve as a toolbox for researchers and engineers to determine appropriate combinations based on their needs.

The enumeration of fecal indicators by culture-based methods is relatively inexpensive and easy to perform. Among them, plate counting is the most common for bacterial indicators. In addition, cell-culture-based methods remain the gold standard to assess the infectivity of viral indicators/pathogens. However, culture-based methods are time-consuming, often leaving insufficient time for the water quality regulatory agencies to take proactive actions. In addition, unculturable, difficult-to-culture and viable-but-not-culturable fecal indicators are unlikely to be enumerated by culture-based methods. Thus, culture-based methods tend to underestimate the health risks. Molecular methods, such as qPCR and digital PCR, on the other hand, are rapid, of good replicability and have broader applicability to various fecal indicators. One limitation is that molecular methods cannot distinguish detected genetic materials from living or dead cells; therefore, they tend to overestimate the health risks, because dead or inactive indicators/pathogens cannot be excluded.

Emerging multidisciplinary bacterial enumeration platforms often incorporate cutting-edge techniques in biology, engineering and computer science, aiming to simplify procedures, prompt automation, improve detection sensitivity, and reduce the requirements of time, cost and labor. The first platform discussed here enables automatic colony plate counting and species classification; the second platform is suitable for high-frequency and long-term quality monitoring of drinking water, and the third platform introduces an on-site flow cytometry based on microfluidics without fluorescence staining. These platforms currently have individual limitations, but still can be powerful alternatives or supplements to current biological methods.

Author Contributions: Writing—original draft preparation, E.L., F.S.; writing—review and editing, H.E.S., T.A.E.; investigation, H.E.S., T.A.E.; supervision, H.E.S. All authors have read and agreed to the published version of the manuscript.

Funding: We gratefully acknowledge funding supporting environmental studies provided by the Ontario Research Excellence Fund (Grant # RE09-077, Ontario Ministry of Economic Development, Job Creation and Trade), the Ontario Ministry of Environment, Conservation and Parks (MECP), and the Natural Sciences and Engineering Research Council (NSERC) of Canada (Grant # ALLRP 554507-20).

Institutional Review Board Statement: Not applicable.

Informed Consent Statement: Not applicable.

Data Availability Statement: Not applicable.

Acknowledgments: We wish to thank our municipal and provincial partners, including the Ministry of the Environment, Conservation and Parks (MECP) and Toronto Public Health, for helpful discussions in support of our monitoring studies.

Conflicts of Interest: The authors declare no conflict of interest.

References

1. Glassmeyer, S.T.; Furlong, E.T.; Kolpin, D.W.; Cahill, J.D.; Zaugg, S.D.; Werner, S.L.; Meyer, M.T.; Kryak, D.D. Transport of chemical and microbial compounds from known wastewater discharges: Potential for use as indicators of human fecal contamination. *Environ. Sci. Technol.* **2005**, *39*, 5157–5169. [CrossRef]
2. Karkman, A.; Pärnänen, K.; Larsson, D.J. Fecal pollution can explain antibiotic resistance gene abundances in anthropogenically impacted environments. *Nat. Commun.* **2019**, *10*, 80. [CrossRef]
3. Cabral, J.P. Water microbiology. Bacterial pathogens and water. *Int. J. Environ. Res. Public Health* **2010**, *7*, 3657–3703. [CrossRef] [PubMed]
4. Paruch, L.; Paruch, A.M.; Eiken, H.G.; Sorheim, R. Faecal pollution affects abundance and diversity of aquatic microbial community in anthropo-zoogenically influenced lotic ecosystems. *Sci. Rep.* **2019**, *9*, 19469. [CrossRef] [PubMed]
5. Anderson, K.L.; Whitlock, J.E.; Harwood, V.J. Persistence and differential survival of fecal indicator bacteria in subtropical waters and sediments. *Appl. Environ. Microbiol.* **2005**, *71*, 3041–3048. [CrossRef]
6. Porter, K.D.H.; Quilliam, R.S.; Reaney, S.M.; Oliver, D.M. High resolution characterisation of *E. coli* proliferation profiles in livestock faeces. *Waste Manag.* **2019**, *87*, 537–545. [CrossRef] [PubMed]
7. He, Q.; Qiu, S.; Jiang, Y.; Wu, Z.; Liu, Z. Land-use change caused microbial pollution in a karst underground river, Chongqing, China. *Environ. Earth Sci.* **2016**, *75*, 709. [CrossRef]
8. Rehman, F.; Subyani, A.M.; Cheema, T.; Harbi, H.M.; Azeem, T.; Naseem, A.A.; Ullah, M.F.; Riaz, O.; Ur Rehman, S. Contribution to the comparative study of treated and untreated wastewater: A case study (Wadi Fatima and Al Misk Lake) in Saudi Arabia. *Arab. J. Geosci.* **2020**, *13*, 842. [CrossRef]
9. Abd-Elhamid, H.F.; Abd-Elmoneem, S.M.; Abdelaal, G.M.; Zelenakova, M.; Vranayova, Z.; Abd-Elaty, I. Investigating and Managing the Impact of Using Untreated Wastewater for Irrigation on the Groundwater Quality in Arid and Semi-Arid Regions. *Int. J. Environ. Res. Public Health* **2021**, *18*, 7485. [CrossRef]
10. Kuroda, K.; Murakami, M.; Oguma, K.; Muramatsu, Y.; Takada, H.; Taldzawa, S. Assessment of Groundwater Pollution in Tokyo Using PPCPs as Sewage Markers. *Environ. Sci. Technol.* **2012**, *46*, 1455–1464. [CrossRef]
11. World Health Organization & the United Nations Children's Fund. *Progress on Drinking Water, Sanitation and Hygiene: 2017 Update and SDG Baselines*; WHO: Geneva, Switzerland, 2017; Licence: CC BY-NC-SA 3.0 IGO.
12. Troeger, C.; Blacker, B.F.; Khalil, I.A.; Rao, P.C.; Cao, S.; Zimsen, S.R.; Albertson, S.B.; Stanaway, J.D.; Deshpande, A.; Abebe, Z. Estimates of the global, regional, and national morbidity, mortality, and aetiologies of diarrhoea in 195 countries: A systematic analysis for the Global Burden of Disease Study 2016. *Lancet Infect. Dis.* **2018**, *18*, 1211–1228. [CrossRef]
13. Fuhrmeister, E.R.; Ercumen, A.; Pickering, A.J.; Jeanis, K.M.; Ahmed, M.; Brown, S.; Arnold, B.F.; Hubbard, A.E.; Alam, M.; Sen, D.; et al. Predictors of Enteric Pathogens in the Domestic Environment from Human and Animal Sources in Rural Bangladesh. *Environ. Sci. Technol.* **2019**, *53*, 10023–10033. [CrossRef]
14. Wanjugi, P.; Sivaganesan, M.; Korajkic, A.; McMinn, B.; Kelty, C.A.; Rhodes, E.; Cyterski, M.; Zepp, R.; Oshima, K.; Stachler, E.; et al. Incidence of somatic and F+ coliphage in Great Lake Basin recreational waters. *Water Res.* **2018**, *140*, 200–210. [CrossRef]
15. Jagai, J.S.; Castronovo, D.A.; Monchak, J.; Naumova, E.N. Seasonality of cryptosporidiosis: A meta-analysis approach. *Environ. Res.* **2009**, *109*, 465–478. [CrossRef]
16. Niendorf, S.; Ebner, W.; Marques, A.M.; Bierbaum, S.; Babikir, R.; Huzly, D.; Maaßen, S.; Grundmann, H.; Panning, M. Rotavirus outbreak among adults in a university hospital in Germany. *J. Clin. Virol.* **2020**, *129*, 104532. [CrossRef] [PubMed]
17. Owusu-Ansah, E.d.-G.J.; Sampson, A.; Amponsah, S.K.; Abaidoo, R.C.; Dalsgaard, A.; Hald, T. Probabilistic quantitative microbial risk assessment model of norovirus from wastewater irrigated vegetables in Ghana using genome copies and fecal indicator ratio conversion for estimating exposure dose. *Sci. Total Environ.* **2017**, *601*, 1712–1719. [CrossRef] [PubMed]
18. Wei, S.H.; Huang, A.S.; Liao, Y.S.; Liu, Y.L.; Chiou, C.S. A large outbreak of salmonellosis associated with sandwiches contaminated with multiple bacterial pathogens purchased via an online shopping service. *Foodborne Pathog. Dis.* **2014**, *11*, 230–233. [CrossRef] [PubMed]
19. Minnesota Department of Health. Salmonellosis. 2010. Available online: https://www.health.state.mn.us/diseases/reportable/dcn/sum10/salmonellosis.html (accessed on 15 August 2021).
20. U.S. Centers for Disease Control and Prevention. Multistate Outbreak of Multidrug-Resistant Salmonella Heidelberg Infections Linked to Foster Farms Brand Chicken (Final Update). Available online: https://www.cdc.gov/salmonella/heidelberg-10-13/ (accessed on 21 August 2021).
21. Hopkins, J.; Hague, H.; Hudgin, G.; Ross, L.; Moore, D. An Outbreak of Cryptosporidium at a Recreational Water Park in Niagara Region, Canada. *J. Environ. Health* **2013**, *75*, 28–33.

22. Waldron, L.S.; Ferrari, B.C.; Cheung-Kwok-Sang, C.; Beggs, P.J.; Stephens, N.; Power, M.L. Molecular epidemiology and spatial distribution of a waterborne cryptosporidiosis outbreak in Australia. *Appl. Environ. Microbiol.* **2011**, *77*, 7766–7771. [CrossRef]
23. Chen, T.; Leung, R.K.; Zhou, Z.; Liu, R.; Zhang, X.; Zhang, L. Investigation of key interventions for shigellosis outbreak control in China. *PLoS ONE* **2014**, *9*, e95006. [CrossRef]
24. Topal, S.; Çelik, H.; Yilmaz, Ş.; Özoğlu, E.; Karaoğlanoğlu, O.; Temel, F.; Şirin, H. Outbreak of *Shigella sonnei* infection in Terme City, Turkey, September 2012. *Turk Hij. Deney. Biyol. Derg.* **2019**, *76*, 141–148. [CrossRef]
25. Gill, A.; Huszczynski, G. Enumeration of *Escherichia coli* O157:H7 in Outbreak-Associated Beef Patties. *J. Food Prot.* **2016**, *79*, 1266–1268. [CrossRef]
26. Asia Pacific Emerging Infections Network, University of Washington, Seattle, USA. Japan: *E. coli* O157:H7 Outbreak, Pickled Cabbage. 2012. Available online: https://blogs.uw.edu/apecein/2012/08/31/japan-e-coli-o157h7-outbreak-pickled-cabbage/ (accessed on 11 August 2021).
27. Kenyon, J.; Inns, T.; Aird, H.; Swift, C.; Astbury, J.; Forester, E.; Decraene, V. Campylobacter outbreak associated with raw drinking milk, North West England, 2016. *Epidemiol. Infect.* **2020**, *148*, e13. [CrossRef] [PubMed]
28. Institute of Environmental Science and Research Funded by Ministry of Health, New Zealand. Annual Summary of Outbreaks in New Zealand. 2016. Available online: https://surv.esr.cri.nz/surveillance/annual_outbreak.php?we_objectID=4741 (accessed on 2 September 2021).
29. Viray, M.; Hofmeister, M.; Johnston, D.; Krishnasamy, V.; Nichols, C.; Foster, M.; Balajadia, R.; Wise, M.; Manuzak, A.; Lin, Y. Public health investigation and response to a hepatitis A outbreak from imported scallops consumed raw—Hawaii, 2016. *Epidemiol. Infect.* **2019**, *147*, E28. [CrossRef] [PubMed]
30. Rakesh, P.; Mainu, T.C.R.; Raj, A.; Babu, D.; Rajiv, M.; Mohandas, K.S.; Das, A.; Balasubramanian, A. Investigating a community wide outbreak of hepatitis A in Kerala, India. *J. Fam. Med. Prim. Care* **2018**, *7*, 1537.
31. Maguire, J.E.; Glasgow, K.; Glass, K.; Roczo-Farkas, S.; Bines, J.E.; Sheppeard, V.; Macartney, K.; Quinn, H.E. Rotavirus epidemiology and monovalent rotavirus vaccine effectiveness in Australia: 2010–2017. *Pediatrics* **2019**, *144*, e20191024. [CrossRef]
32. Kujawski, S.A.; Lu, X.; Schneider, E.; Blythe, D.; Boktor, S.; Farrehi, J.; Haupt, T.; McBride, D.; Stephens, E.; Sakthivel, S.K.; et al. Outbreaks of Adenovirus-associated Respiratory Illness on 5 College Campuses in the United States, 2018–2019. *Clin. Infect. Dis.* **2021**, *72*, 1992–1999. [CrossRef]
33. Dai, M.; Wu, Y.; Tan, H.; Deng, J.; Hou, M.; Peng, W.; Chen, G.; Li, Y.; Li, H.; Pan, P. Cross-infection of adenovirus among medical staff: A warning from the intensive care unit in a tertiary care teaching hospital in China. *Int. J. Infect. Dis.* **2020**, *98*, 390–397. [CrossRef]
34. Belluz, J. The Dreaded Norovirus Has Hit the Winter Olympics. at Least 199 People Are Sick. *Vox*. 2018. Available online: https://www.vox.com/science-and-health/2018/2/6/16980028/winter-olympics-norovirus (accessed on 15 September 2021).
35. Fumian, T.M.; Ferreira, F.C.; de Andrade, J.d.S.R.; Canal, N.; Silva Gomes, G.; Teixeira, L.B.; Miagostovich, M.P. Norovirus Foodborne Outbreak Associated with the Consumption of Ice Pop, Southern Brazil, 2020. *Food Environ. Virol.* **2021**, 1–7. [CrossRef]
36. Rizzo, L.; Manaia, C.; Merlin, C.; Schwartz, T.; Dagot, C.; Ploy, M.; Michael, I.; Fatta-Kassinos, D. Urban wastewater treatment plants as hotspots for antibiotic resistant bacteria and genes spread into the environment: A review. *Sci. Total Environ.* **2013**, *447*, 345–360. [CrossRef]
37. Palacios, O.A.; Contreras, C.A.; Muñoz-Castellanos, L.N.; González-Rangel, M.O.; Rubio-Arias, H.; Palacios-Espinosa, A.; Nevárez-Moorillón, G.V. Monitoring of indicator and multidrug resistant bacteria in agricultural soils under different irrigation patterns. *Agric. Water Manag.* **2017**, *184*, 19–27. [CrossRef]
38. Proia, L.; Anzil, A.; Subirats, J.; Borrego, C.; Farre, M.; Llorca, M.; Balcazar, J.L.; Servais, P. Antibiotic resistance along an urban river impacted by treated wastewaters. *Sci. Total Environ.* **2018**, *628–629*, 453–466. [CrossRef] [PubMed]
39. Liu, C.; Li, X.; Zheng, S.; Kai, Z.; Jin, T.; Shi, R.; Huang, H.; Zheng, X. Effects of wastewater treatment and manure application on the dissemination of antimicrobial resistance around swine feedlots. *J. Clean. Prod.* **2021**, *280*, 123794. [CrossRef]
40. Luczkiewicz, A.; Jankowska, K.; Fudala-Ksiazek, S.; Olanczuk-Neyman, K. Antimicrobial resistance of fecal indicators in municipal wastewater treatment plant. *Water Res.* **2010**, *44*, 5089–5097. [CrossRef]
41. Andreev, N.; Ronteltap, M.; Boincean, B.; Wernli, M.; Zubcov, E.; Bagrin, N.; Borodin, N.; Lens, P.N.L. Lactic acid fermentation of human urine to improve its fertilizing value and reduce odour emissions. *J. Environ. Manag.* **2017**, *198*, 63–69. [CrossRef]
42. Shen, Y.; Ren, L.; Li, G.; Chen, T.; Guo, R. Influence of aeration on CH_4, N^2O and NH_3 emissions during aerobic composting of a chicken manure and high C/N waste mixture. *Waste Manag.* **2011**, *31*, 33–38. [CrossRef]
43. Han, Z.; Sun, D.; Wang, H.; Li, R.; Bao, Z.; Qi, F. Effects of ambient temperature and aeration frequency on emissions of ammonia and greenhouse gases from a sewage sludge aerobic composting plant. *Bioresour. Technol.* **2018**, *270*, 457–466. [CrossRef]
44. Petkuviene, J.; Vaiciute, D.; Katarzyte, M.; Gecaite, I.; Rossato, G.; Vybernaite-Lubiene, I.; Bartoli, M. Feces from piscivorous and herbivorous birds stimulate differentially phytoplankton growth. *Water* **2019**, *11*, 2567. [CrossRef]
45. Wang, Y.; Gu, X.; Zeng, Q.; Mao, Z.; Wang, W. Contrasting response of a plankton community to two filter-feeding fish and their feces: An in situ enclosure experiment. *Aquaculture* **2016**, *465*, 330–340. [CrossRef]
46. Börjesson, G.; Kätterer, T. Soil fertility effects of repeated application of sewage sludge in two 30-year-old field experiments. *Nutr. Cycl. Agroecosyst.* **2018**, *112*, 369–385. [CrossRef]
47. Pradhan, S.K.; Cofie, O.; Nikiema, J.; Heinonen-Tanski, H. Fecal sludge derived products as fertilizer for lettuce cultivation in urban agriculture. *Sustainability* **2019**, *11*, 7101. [CrossRef]

48. Bartelme, R.P.; Oyserman, B.O.; Blom, J.E.; Sepulveda-Villet, O.J.; Newton, R.J. Stripping Away the Soil: Plant Growth Promoting Microbiology Opportunities in Aquaponics. *Front. Microbiol.* **2018**, *9*, 8. [CrossRef] [PubMed]
49. Pradhan, S.K.; Nikiema, J.; Cofie, O.O.; Heinonen-Tanski, H.; Drechsel, P. Fecal sludge-derived pellet fertilizer in maize cultivation. *J. Water Sanit. Hyg. Dev.* **2016**, *6*, 474–481. [CrossRef]
50. Hashemi, S.; Han, M. Optimizing source-separated feces degradation and fertility using nitrifying microorganisms. *J. Environ. Manag.* **2018**, *206*, 540–546. [CrossRef]
51. Yoshizawa, N.; Usui, M.; Fukuda, A.; Asai, T.; Higuchi, H.; Okamoto, E.; Seki, K.; Takada, H.; Tamura, Y. Manure compost is a potential source of tetracycline-resistant *Escherichia coli* and tetracycline resistance genes in Japanese farms. *Antibiotics* **2020**, *9*, 76. [CrossRef]
52. Korajkic, A.; McMinn, B.R.; Harwood, V.J. Relationships between microbial indicators and pathogens in recreational water settings. *Int. J. Environ. Res. Public Health* **2018**, *15*, 2842. [CrossRef]
53. U.S. Environmental Protection Agency. Basic Information about Nonpoint Source (NPS) Pollution. 2020. Available online: https://www.epa.gov/nps/basic-information-about-nonpoint-source-nps-pollution (accessed on 5 August 2021).
54. Edberg, S.C.; Rice, E.W.; Karlin, R.J.; Allen, M.J. *Escherichia coli*: The best biological drinking water indicator for public health protection. *Symp. Ser. Soc. Appl. Microbiol.* **2000**, *88*, 106S–116S. [CrossRef]
55. Rochelle-Newall, E.; Nguyen, T.M.; Le, T.P.; Sengtaheuanghoung, O.; Ribolzi, O. A short review of fecal indicator bacteria in tropical aquatic ecosystems: Knowledge gaps and future directions. *Front. Microbiol.* **2015**, *6*, 308. [CrossRef]
56. Ahmed, W.; Gyawali, P.; Feng, S.; McLellan, S.L. Host Specificity and Sensitivity of Established and Novel Sewage-Associated Marker Genes in Human and Nonhuman Fecal Samples. *Appl. Environ. Microbiol.* **2019**, *85*, e00641-19. [CrossRef]
57. Tran, N.H.; Gin, K.Y.; Ngo, H.H. Fecal pollution source tracking toolbox for identification, evaluation and characterization of fecal contamination in receiving urban surface waters and groundwater. *Sci. Total Environ.* **2015**, *538*, 38–57. [CrossRef]
58. Ahmed, W.; Goonetilleke, A.; Gardner, T. Human and bovine adenoviruses for the detection of source-specific fecal pollution in coastal waters in Australia. *Water Res.* **2010**, *44*, 4662–4673. [CrossRef]
59. Hundesa, A.; Bofill-Mas, S.; Maluquer de Motes, C.; Rodriguez-Manzano, J.; Bach, A.; Casas, M.; Girones, R. Development of a quantitative PCR assay for the quantitation of bovine polyomavirus as a microbial source-tracking tool. *J. Virol. Methods* **2010**, *163*, 385–389. [CrossRef] [PubMed]
60. Harrach, B.; Tarján, Z.L.; Benkő, M. Adenoviruses across the animal kingdom: A walk in the zoo. *FEBS Lett.* **2019**, *593*, 3660–3673. [CrossRef] [PubMed]
61. Robinson, R.K. *Encyclopedia of Food Microbiology*; Academic Press: New York, NY, USA, 2014.
62. Masiello, S.N.; Martin, N.H.; Trmcic, A.; Wiedmann, M.; Boor, K.J. Identification and characterization of psychrotolerant coliform bacteria isolated from pasteurized fluid milk. *J. Dairy Sci.* **2016**, *99*, 130–140. [CrossRef]
63. Martin, N.H.; Trmčić, A.; Hsieh, T.-H.; Boor, K.J.; Wiedmann, M. The evolving role of coliforms as indicators of unhygienic processing conditions in dairy foods. *Front. Microbiol.* **2016**, *7*, 1549. [CrossRef]
64. Craun, G.F.; Berger, P.S.; Calderon, R.L. Coliform bacteria and waterborne disease outbreaks. *J.-Am. Water Work. Assoc.* **1997**, *89*, 96–104. [CrossRef]
65. U.S. Environmental Protection Agency. *National Interim Primary Drinking Water Regulations*; U.S. Environmental Protection Agency: Washington, DC, USA, 1976.
66. U.S. Environmental Protection Agency. *National Primary Drinking Water Regulations Total Coliforms Final Rule*; U.S. Environmental Protection Agency: Washington, DC, USA, 1989.
67. Das, S.K.; Das, S.K. Influence of phosphorus and organic matter on microbial transformation of arsenic. *Environ. Technol. Innov.* **2020**, *19*, 100930. [CrossRef]
68. Rosenblueth, M.; Martinez, L.; Silva, J.; Martinez-Romero, E. Klebsiella variicola, a novel species with clinical and plant-associated isolates. *Syst. Appl. Microbiol.* **2004**, *27*, 27–35. [CrossRef]
69. Kabler, P.W.; Clark, H.F. Coliform group and fecal coliform organisms as indicators of pollution in drinking water. *J.-Am. Water Work. Assoc.* **1960**, *52*, 1577–1579. [CrossRef]
70. *Coliform Bacteria in Drinking Water Supplies*; New York State Department of Health, Center for Environmental Health: New York, NY, USA, 2017.
71. Florini, S.; Shahsavari, E.; Ngo, T.; Aburto-Medina, A.; Smith, D.J.; Ball, A.S. Factors Influencing the Concentration of Fecal Coliforms in Oysters in the River Blackwater Estuary, UK. *Water* **2020**, *12*, 1086. [CrossRef]
72. Bartram, J.; Ballance, R. *Water Quality Monitoring: A Practical Guide to the Design and Implementation of Freshwater Quality Studies and Monitoring Programmes*; CRC Press: Boca Raton, FL, USA, 1996.
73. Leclercq, A.; Wanegue, C.; Baylac, P. Comparison of fecal coliform agar and violet red bile lactose agar for fecal coliform enumeration in foods. *Appl. Environ. Microbiol.* **2002**, *68*, 1631–1638. [CrossRef] [PubMed]
74. Alonso, J.; Soriano, A.; Amoros, I.; Ferrus, M. Quantitative determination of *E. coli*, and fecal coliforms in water using a chromogenic medium. *J. Environ. Sci. Health Part A* **1998**, *33*, 1229–1248. [CrossRef]
75. Evans, T.M.; LeChevallier, M.W.; Waarvick, C.E.; Seidler, R.J. Coliform species recovered from untreated surface water and drinking water by the membrane filter, standard, and modified most-probable-number techniques. *Appl. Environ. Microbiol.* **1981**, *41*, 657–663. [CrossRef]

76. Dunling, W.; Fiessel, W. Evaluation of media for simultaneous enumeration of total coliform and *Escherichia coli* in drinking water supplies by membrane filtration techniques. *J. Environ. Sci.* **2008**, *20*, 273–277.
77. Tominaga, T. Rapid quantification of coliforms in ready-to-eat foods using lateral-flow immunochromatographic assay. *J. Food Saf.* **2020**, *40*, e12835. [CrossRef]
78. Hu, S.; Yu, Y.; Li, R.; Xia, X.; Xiao, X.; Li, X. Real-time TaqMan PCR for rapid detection and quantification of coliforms in chilled meat. *Food Anal. Methods* **2016**, *9*, 813–822. [CrossRef]
79. Martin, M.C.; Martinez, N.; del Rio, B.; Ladero, V.; Fernandez, M.; Alvarez, M.A. A novel real-time polymerase chain reaction-based method for the detection and quantification of lactose-fermenting Enterobacteriaceae in the dairy and other food industries. *J. Dairy Sci.* **2010**, *93*, 860–867. [CrossRef]
80. Li, X.; Harwood, V.J.; Nayak, B.; Staley, C.; Sadowsky, M.J.; Weidhaas, J. A novel microbial source tracking microarray for pathogen detection and fecal source identification in environmental systems. *Environ. Sci. Technol.* **2015**, *49*, 7319–7329. [CrossRef] [PubMed]
81. Cheswick, R.; Cartmell, E.; Lee, S.; Upton, A.; Weir, P.; Moore, G.; Nocker, A.; Jefferson, B.; Jarvis, P. Comparing flow cytometry with culture-based methods for microbial monitoring and as a diagnostic tool for assessing drinking water treatment processes. *Environ. Int.* **2019**, *130*, 104893. [CrossRef]
82. Xia, Y.; Li, A.-D.; Deng, Y.; Jiang, X.-T.; Li, L.-G.; Zhang, T. MinION nanopore sequencing enables correlation between resistome phenotype and genotype of coliform bacteria in municipal sewage. *Front. Microbiol.* **2017**, *8*, 2105. [CrossRef]
83. Moore, A.C.; Herwaldt, B.L.; Craun, G.F.; Calderon, R.L.; Highsmith, A.K.; Juranek, D.D. Waterborne disease in the United States, 1991 and 1992. *J.-Am. Water Work. Assoc.* **1994**, *86*, 87–97. [CrossRef]
84. U.S. Environmental Protection Agency. *Test Methods for Escherichia coli and Enterococci in Water by the Membrane Filter Procedure*; (Method #1103.1); U.S. Environmental Protection Agency: Washington, DC, USA, 1985.
85. U.S. Environmental Protection Agency. *Ambient Water Quality Criteria for Bacteria—1986*; U.S. Environmental Protection Agency: Washington, DC, USA, 1986.
86. Ishii, S.; Sadowsky, M.J. *Escherichia coli* in the Environment: Implications for Water Quality and Human Health. *Microbes. Environ.* **2008**, *23*, 101–108. [CrossRef] [PubMed]
87. Souid, F.; Agoubi, B.; Hamdi, M.; Telahigue, F.; Kharroubi, A. Groundwater chemical and fecal contamination assessment of the Jerba unconfined aquifer, southeast of Tunisia. *Arab. J. Geosci.* **2017**, *10*, 231. [CrossRef]
88. Dheenan, P.S.; Jha, D.K.; Das, A.K.; Vinithkumar, N.V.; Devi, M.P.; Kirubagaran, R. Geographic information systems and multivariate analysis to evaluate fecal bacterial pollution in coastal waters of Andaman, India. *Environ. Pollut.* **2016**, *214*, 45–53. [CrossRef] [PubMed]
89. Byappanahalli, M.N.; Whitman, R.L.; Shively, D.A.; Sadowsky, M.J.; Ishii, S. Population structure, persistence, and seasonality of autochthonous *Escherichia coli* in temperate, coastal forest soil from a Great Lakes watershed. *Environ. Microbiol.* **2006**, *8*, 504–513. [CrossRef] [PubMed]
90. Power, M.L.; Littlefield-Wyer, J.; Gordon, D.M.; Veal, D.A.; Slade, M.B. Phenotypic and genotypic characterization of encapsulated *Escherichia coli* isolated from blooms in two Australian lakes. *Environ. Microbiol.* **2005**, *7*, 631–640. [CrossRef]
91. Escobar-Paramo, P.; Le Menac'h, A.; Le Gall, T.; Amorin, C.; Gouriou, S.; Picard, B.; Skurnik, D.; Denamur, E. Identification of forces shaping the commensal *Escherichia coli* genetic structure by comparing animal and human isolates. *Environ. Microbiol.* **2006**, *8*, 1975–1984. [CrossRef]
92. Gordon, D.M.; Cowling, A. The distribution and genetic structure of *Escherichia coli* in Australian vertebrates: Host and geographic effects. *Microbiology* **2003**, *149*, 3575–3586. [CrossRef]
93. Clermont, O.; Lescat, M.; O'Brien, C.L.; Gordon, D.M.; Tenaillon, O.; Denamur, E. Evidence for a human-specific *Escherichia coli* clone. *Environ. Microbiol.* **2008**, *10*, 1000–1006. [CrossRef]
94. Williams, A.; Avery, L.; Killham, K.; Jones, D.L. Persistence of *Escherichia coli* O157 on farm surfaces under different environmental conditions. *J. Appl. Microbiol.* **2005**, *98*, 1075–1083. [CrossRef]
95. Sinton, L.W.; Hall, C.H.; Lynch, P.A.; Davies-Colley, R.J. Sunlight inactivation of fecal indicator bacteria and bacteriophages from waste stabilization pond effluent in fresh and saline waters. *Appl. Environ. Microbiol.* **2002**, *68*, 1122–1131. [CrossRef]
96. Brennan, F.P.; Grant, J.; Botting, C.H.; O'Flaherty, V.; Richards, K.G.; Abram, F. Insights into the low-temperature adaptation and nutritional flexibility of a soil-persistent *Escherichia coli*. *FEMS Microbiol. Ecol.* **2013**, *84*, 75–85. [CrossRef]
97. Vital, M.; Hammes, F.; Egli, T. Competition of *Escherichia coli* O157 with a drinking water bacterial community at low nutrient concentrations. *Water Res.* **2012**, *46*, 6279–6290. [CrossRef]
98. Haller, L.; Poté, J.; Loizeau, J.-L.; Wildi, W. Distribution and survival of faecal indicator bacteria in the sediments of the Bay of Vidy, Lake Geneva, Switzerland. *Ecol. Indic.* **2009**, *9*, 540–547. [CrossRef]
99. Schwering, M.; Song, J.; Louie, M.; Turner, R.J.; Ceri, H. Multi-species biofilms defined from drinking water microorganisms provide increased protection against chlorine disinfection. *Biofouling* **2013**, *29*, 917–928. [CrossRef] [PubMed]
100. Mellefont, L.A.; McMeekin, T.A.; Ross, T. Performance evaluation of a model describing the effects of temperature, water activity, pH and lactic acid concentration on the growth of *Escherichia coli*. *Int. J. Food Microbiol.* **2003**, *82*, 45–58. [CrossRef]
101. Abia, A.L.; Ubomba-Jaswa, E.; Momba, M.N. Impact of seasonal variation on *Escherichia coli* concentrations in the riverbed sediments in the Apies River, South Africa. *Sci. Total Environ.* **2015**, *537*, 462–469. [CrossRef] [PubMed]

102. Stocker, M.D.; Pachepsky, Y.A.; Hill, R.L.; Shelton, D.R. Depth-Dependent Survival of *Escherichia coli* and Enterococci in Soil after Manure Application and Simulated Rainfall. *Appl. Environ. Microbiol.* **2015**, *81*, 4801–4808. [CrossRef]
103. Kleinheinz, G.T.; McDermott, C.M.; Hughes, S.; Brown, A. Effects of rainfall on *E. coli* concentrations at Door County, Wisconsin beaches. *Int. J. Microbiol.* **2009**, *2009*, 876050. [CrossRef]
104. Mugler, C.; Ribolzi, O.; Viguier, M.; Janeau, J.L.; Jarde, E.; Latsachack, K.; Henry-Des-Tureaux, T.; Thammahacksa, C.; Valentin, C.; Sengtaheuanghoung, O.; et al. Experimental and modelling evidence of splash effects on manure borne *Escherichia coli* washoff. *Environ. Sci. Pollut. Res. Int.* **2021**, *28*, 33009–33020. [CrossRef]
105. Stumpf, C.H.; Piehler, M.F.; Thompson, S.; Noble, R.T. Loading of fecal indicator bacteria in North Carolina tidal creek headwaters: Hydrographic patterns and terrestrial runoff relationships. *Water Res.* **2010**, *44*, 4704–4715. [CrossRef]
106. Abia, A.L.; James, C.; Ubomba-Jaswa, E.; Benteke Momba, M.N. Microbial Remobilisation on Riverbed Sediment Disturbance in Experimental Flumes and a Human-Impacted River: Implication for Water Resource Management and Public Health in Developing Sub-Saharan African Countries. *Int. J. Environ. Res. Public Health* **2017**, *14*, 306. [CrossRef]
107. Ducarmon, Q.R.; Zwittink, R.D.; Hornung, B.V.H.; van Schaik, W.; Young, V.B.; Kuijper, E.J. Gut Microbiota and Colonization Resistance against Bacterial Enteric Infection. *Microbiol. Mol. Biol. Rev.* **2019**, *83*, e00007-19. [CrossRef] [PubMed]
108. Momose, Y.; Hirayama, K.; Itoh, K. Competition for proline between indigenous *Escherichia coli* and *E. coli* O157: H7 in gnotobiotic mice associated with infant intestinal microbiota and its contribution to the colonization resistance against *E. coli* O157: H7. *Antonie Van Leeuwenhoek* **2008**, *94*, 165–171. [CrossRef] [PubMed]
109. Panhwar, S.; Aftab, A.; Keerio, H.A.; Sarmadivaleh, M.; Tamer, U. A Novel Approach for Real-Time Enumeration of *Escherichia coli* ATCC 47076 in Water through High Multi-Functional Engineered Nano-Dispersible Electrode. *J. Electrochem. Soc.* **2021**, *168*, 037514. [CrossRef]
110. Centers for Disease Control and Prevention. *National Shiga Toxin-Producing Escherichia coli (STEC) Surveillance Annual Report, 2016*; CDC: Atlanta, GA, USA, 2018.
111. Wong, S.Y.; Paschos, A.; Gupta, R.S.; Schellhorn, H.E. Insertion/deletion-based approach for the detection of *Escherichia coli* O157: H7 in freshwater environments. *Environ. Sci. Technol.* **2014**, *48*, 11462–11470. [CrossRef]
112. Cox, C.R.; Gilmore, M.S. Native microbial colonization of Drosophila melanogaster and its use as a model of Enterococcus faecalis pathogenesis. *Infect. Immun.* **2007**, *75*, 1565–1576. [CrossRef]
113. Ghosh, A.; Kukanich, K.; Brown, C.E.; Zurek, L. Resident Cats in Small Animal Veterinary Hospitals Carry Multi-Drug Resistant Enterococci and are Likely Involved in Cross-Contamination of the Hospital Environment. *Front. Microbiol.* **2012**, *3*, 62. [CrossRef]
114. Cai, Y. Identification and characterization of Enterococcus species isolated from forage crops and their influence on silage fermentation. *J. Dairy Sci.* **1999**, *82*, 2466–2471. [CrossRef]
115. Napier, M.D.; Haugland, R.; Poole, C.; Dufour, A.P.; Stewart, J.R.; Weber, D.J.; Varma, M.; Lavender, J.S.; Wade, T.J. Exposure to human-associated fecal indicators and self-reported illness among swimmers at recreational beaches: A cohort study. *Environ. Health* **2017**, *16*, 103. [CrossRef]
116. Wade, T.J.; Sams, E.; Brenner, K.P.; Haugland, R.; Chern, E.; Beach, M.; Wymer, L.; Rankin, C.C.; Love, D.; Li, Q. Rapidly measured indicators of recreational water quality and swimming-associated illness at marine beaches: A prospective cohort study. *Environ. Health* **2010**, *9*, 66. [CrossRef] [PubMed]
117. Wade, T.J.; Calderon, R.L.; Brenner, K.P.; Sams, E.; Beach, M.; Haugland, R.; Wymer, L.; Dufour, A.P. High sensitivity of children to swimming-associated gastrointestinal illness: Results using a rapid assay of recreational water quality. *Epidemiology* **2008**, *19*, 375–383. [CrossRef]
118. U.S. Environmental Protection Agency. *Recreational Water Quality Criteria*; U.S. Environmental Protection Agency: Washington, DC, USA, 2012.
119. Wymer, L.J.; Brenner, K.P.; Martinson, J.W.; Stutts, W.R.; Schaub, S.; Dufour, A.P. *The EMPACT Beaches Project: Results from a Study on Microbiological Monitoring in Recreational Waters*; USEPA: Washington, DC, USA, 2005.
120. Health and Welfare Canada. *Guidelines for Canadian Recreational Water Quality*, 3rd ed.; Healthy Environments and Consumer Safety Branch, Health Canada: Ottawa, ON, Canada, 2012.
121. Soller, J.A.; Schoen, M.E.; Bartrand, T.; Ravenscroft, J.E.; Ashbolt, N.J. Estimated human health risks from exposure to recreational waters impacted by human and non-human sources of faecal contamination. *Water Res.* **2010**, *44*, 4674–4691. [CrossRef] [PubMed]
122. Miller, W.R.; Munita, J.M.; Arias, C.A. Mechanisms of antibiotic resistance in enterococci. *Expert Rev. Anti-Infect.* **2014**, *12*, 1221–1236. [CrossRef] [PubMed]
123. Cheng, W.; Chen, J.-C. Effect of cultivation broth pH, temperature and NaCl concentration on virulence of an Enterococcus-like bacterium to the giant freshwater prawn Macrobrachium rosenbergii. *Dis. Aquat. Org.* **1999**, *36*, 233–237. [CrossRef]
124. Dubin, K.; Pamer, E.G. Enterococci and their interactions with the intestinal microbiome. *Microbiol. Spectr.* **2017**, *5*, 5–6. [CrossRef]
125. Roslev, P.; Iversen, L.; Sonderbo, H.L.; Iversen, N.; Bastholm, S. Uptake and persistence of human associated Enterococcus in the mussel Mytilus edulis: Relevance for faecal pollution source tracking. *J. Appl. Microbiol.* **2009**, *107*, 944–953. [CrossRef]
126. Nishiyama, M.; Iguchi, A.; Suzuki, Y. Identification of Enterococcus faecium and Enterococcus faecalis as vanC-type vancomycin-resistant enterococci (VRE) from sewage and river water in the provincial city of Miyazaki, Japan. *J. Environ. Sci. Health Part A* **2015**, *50*, 16–25. [CrossRef]
127. Piggot, A.M.; Klaus, J.S.; Johnson, S.; Phillips, M.C.; Solo-Gabriele, H.M. Relationship between enterococcal levels and sediment biofilms at recreational beaches in South Florida. *Appl. Environ. Microbiol.* **2012**, *78*, 5973–5982. [CrossRef]

128. Wiegner, T.N.; Mead, L.H.; Molloy, S.L. A comparison of water quality between low-and high-flow river conditions in a tropical estuary, Hilo Bay, Hawaii. *Estuaries Coasts* **2013**, *36*, 319–333. [CrossRef]
129. Deller, S.; Mascher, F.; Platzer, S.; Reinthaler, F.F.; Marth, E. Effect of solar radiation on survival of indicator bacteria in bathing waters. *Cent. Eur. J. Public Health* **2006**, *14*, 133–137. [CrossRef]
130. Heim, S.; Del Mar Lleo, M.; Bonato, B.; Guzman, C.A.; Canepari, P. The viable but nonculturable state and starvation are different stress responses of Enterococcus faecalis, as determined by proteome analysis. *J. Bacteriol.* **2002**, *184*, 6739–6745. [CrossRef]
131. Dorevitch, S.; Doi, M.; Hsu, F.C.; Lin, K.T.; Roberts, J.D.; Liu, L.C.; Gladding, R.; Vannoy, E.; Li, H.; Javor, M.; et al. A comparison of rapid and conventional measures of indicator bacteria as predictors of waterborne protozoan pathogen presence and density. *J. Environ. Monit.* **2011**, *13*, 2427–2435. [CrossRef]
132. Mansilha, C.R.; Coelho, C.A.; Reinas, A.; Moutinho, A.; Ferreira, S.; Pizarro, C.; Tavares, A. Salmonella: The forgotten pathogen: Health hazards of compliance with European Bathing Water Legislation. *Mar. Pollut. Bull.* **2010**, *60*, 819–826. [CrossRef]
133. Wang, H.; Naghavi, M.; Allen, C.; Barber, R.M.; Bhutta, Z.A.; Carter, A.; Casey, D.C.; Charlson, F.J.; Chen, A.Z.; Coates, M.M. Global, regional, and national life expectancy, all-cause mortality, and cause-specific mortality for 249 causes of death, 1980–2015: A systematic analysis for the Global Burden of Disease Study 2015. *Lancet* **2016**, *388*, 1459–1544. [CrossRef]
134. Koo, B.S.; Hwang, E.H.; Kim, G.; Park, J.Y.; Oh, H.; Lim, K.S.; Kang, P.; Lee, H.Y.; Jeong, K.J.; Mo, I.; et al. Prevalence and characterization of Clostridium perfringens isolated from feces of captive cynomolgus monkeys (*Macaca fascicularis*). *Anaerobe* **2020**, *64*, 102236. [CrossRef]
135. Matches, J.R.; Liston, J.; Curran, D. Clostridium perfringens in the environment. *Appl. Microbiol.* **1974**, *28*, 655–660. [CrossRef] [PubMed]
136. Sabry, M.; Abd El-Moein, K.; Hamza, E.; Abdel Kader, F. Occurrence of *Clostridium perfringens* Types A, E, and C in Fresh Fish and Its Public Health Significance. *J. Food Prot.* **2016**, *79*, 994–1000. [CrossRef] [PubMed]
137. Scott, A.; Tien, Y.C.; Drury, C.F.; Reynolds, W.D.; Topp, E. Enrichment of antibiotic resistance genes in soil receiving composts derived from swine manure, yard wastes, or food wastes, and evidence for multiyear persistence of swine *Clostridium* spp. *Can. J. Microbiol.* **2018**, *64*, 201–208. [CrossRef] [PubMed]
138. Milton, A.A.P.; Sanjukta, R.; Gogoi, A.P.; Momin, K.M.; Priya, G.B.; Das, S.; Ghatak, S.; Sen, A.; Kandpal, B.K. Prevalence, molecular typing and antibiotic resistance of *Clostridium perfringens* in free range ducks in Northeast India. *Anaerobe* **2020**, *64*, 102242. [CrossRef]
139. Fujioka, R.S. Monitoring coastal marine waters for spore-forming bacteria of faecal and soil origin to determine point from non-point source pollution. *Water Sci. Technol.* **2001**, *44*, 181–188. [CrossRef] [PubMed]
140. Skanavis, C.; Yanko, W.A. Clostridium perfringens as a potential indicator for the presence of sewage solids in marine sediments. *Mar. Pollut. Bull.* **2001**, *42*, 31–35. [CrossRef]
141. Paredes-Sabja, D.; Sarker, N.; Setlow, B.; Setlow, P.; Sarker, M.R. Roles of DacB and spm proteins in clostridium perfringens spore resistance to moist heat, chemicals, and UV radiation. *Appl. Environ. Microbiol.* **2008**, *74*, 3730–3738. [CrossRef] [PubMed]
142. Wang, G.; Paredes-Sabja, D.; Sarker, M.; Green, C.; Setlow, P.; Li, Y.q. Effects of wet heat treatment on the germination of individual spores of C lostridium perfringens. *J. Appl. Microbiol.* **2012**, *113*, 824–836. [CrossRef] [PubMed]
143. Savichtcheva, O.; Okabe, S. Alternative indicators of fecal pollution: Relations with pathogens and conventional indicators, current methodologies for direct pathogen monitoring and future application perspectives. *Water Res.* **2006**, *40*, 2463–2476. [CrossRef]
144. Vijayavel, K.; Kashian, D. Evaluation of Clostridium perfringens as a tracer of sewage contamination in sediments by two enumeration methods. *Environ. Monit. Assess.* **2014**, *186*, 5617–5624. [CrossRef]
145. Sartory, D. Membrane filtration enumeration of faecal clostridia and *Clostridium perfringens* in water. *Water Res.* **1986**, *20*, 1255–1260. [CrossRef]
146. Sartory, D.P.; Field, M.; Curbishley, S.M.; Pritchard, A.M. Evaluation of two media for the membrane filtration enumeration of *Clostridium perfringens* from water. *Lett. Appl. Microbiol.* **1998**, *27*, 323–327. [CrossRef]
147. Marshall, R.S.; Steenbergen, J.F.; McClung, L. Rapid technique for the enumeration of *Clostridium perfringens*. *Appl. Microbiol.* **1965**, *13*, 559–563. [CrossRef]
148. Karpowicz, E.; Novinscak, A.; Barlocher, F.; Filion, M. qPCR quantification and genetic characterization of *Clostridium perfringens* populations in biosolids composted for 2 years. *J. Appl. Microbiol.* **2010**, *108*, 571–581. [CrossRef]
149. Milton, A.A.P.; Momin, K.M.; Priya, G.B.; Ghatak, S.; Gandhale, P.N.; Angappan, M.; Das, S.; Sen, A. A novel in situ methodology for visual detection of *Clostridium perfringens* in pork harnessing saltatory rolling circle amplification. *Anaerobe* **2021**, *69*, 102324. [CrossRef]
150. Bernhard, A.E.; Field, K.G. Identification of nonpoint sources of fecal pollution in coastal waters by using host-specific 16S ribosomal DNA genetic markers from fecal anaerobes. *Appl. Environ. Microbiol.* **2000**, *66*, 1587–1594. [CrossRef] [PubMed]
151. Pilla, R.; Suchodolski, J.S. The Role of the Canine Gut Microbiome and Metabolome in Health and Gastrointestinal Disease. *Front. Vet. Sci.* **2019**, *6*, 498. [CrossRef]
152. Bergamaschi, M.; Tiezzi, F.; Howard, J.; Huang, Y.J.; Gray, K.A.; Schillebeeckx, C.; McNulty, N.P.; Maltecca, C. Gut microbiome composition differences among breeds impact feed efficiency in swine. *Microbiome* **2020**, *8*, 110. [CrossRef] [PubMed]
153. Kabiri, L.; Alum, A.; Rock, C.; McLain, J.E.; Abbaszadegan, M. Isolation of Bacteroides from fish and human fecal samples for identification of unique molecular markers. *Can. J. Microbiol.* **2013**, *59*, 771–777. [CrossRef] [PubMed]

154. Ko, H.Y.; Cho, K.; Park, S.; Kim, J.H.; Kang, J.H.; Jeong, Y.S.; Choi, J.D.; Sin, Y.; Lee, C.; Ko, G. Host-Specific Bacteroides Markers-Based Microbial Source Tracking in Aquaculture Areas. *Microbes Environ.* **2018**, *33*, 151–161. [CrossRef]
155. Kildare, B.J.; Leutenegger, C.M.; McSwain, B.S.; Bambic, D.G.; Rajal, V.B.; Wuertz, S. 16S rRNA-based assays for quantitative detection of universal, human-, cow-, and dog-specific fecal Bacteroidales: A Bayesian approach. *Water Res.* **2007**, *41*, 3701–3715. [CrossRef]
156. Kobayashi, A.; Sano, D.; Hatori, J.; Ishii, S.; Okabe, S. Chicken- and duck-associated *Bacteroides-Prevotella* genetic markers for detecting fecal contamination in environmental water. *Appl. Microbiol. Biotechnol.* **2013**, *97*, 7427–7437. [CrossRef]
157. Bernhard, A.E.; Field, K.G. A PCR assay To discriminate human and ruminant feces on the basis of host differences in *Bacteroides-Prevotella* genes encoding 16S rRNA. *Appl. Environ. Microbiol.* **2000**, *66*, 4571–4574. [CrossRef]
158. Layton, A.; McKay, L.; Williams, D.; Garrett, V.; Gentry, R.; Sayler, G. Development of Bacteroides 16S rRNA gene TaqMan-based real-time PCR assays for estimation of total, human, and bovine fecal pollution in water. *Appl. Environ. Microbiol.* **2006**, *72*, 4214–4224. [CrossRef]
159. Reischer, G.H.; Kasper, D.C.; Steinborn, R.; Farnleitner, A.H.; Mach, R.L. A quantitative real-time PCR assay for the highly sensitive and specific detection of human faecal influence in spring water from a large alpine catchment area. *Lett. Appl. Microbiol.* **2007**, *44*, 351–356. [CrossRef]
160. Okabe, S.; Okayama, N.; Savichtcheva, O.; Ito, T. Quantification of host-specific *Bacteroides-Prevotella* 16S rRNA genetic markers for assessment of fecal pollution in freshwater. *Appl. Microbiol. Biotechnol.* **2007**, *74*, 890–901. [CrossRef]
161. Dick, L.K.; Bernhard, A.E.; Brodeur, T.J.; Santo Domingo, J.W.; Simpson, J.M.; Walters, S.P.; Field, K.G. Host distributions of uncultivated fecal *Bacteroidales* bacteria reveal genetic markers for fecal source identification. *Appl. Environ. Microbiol.* **2005**, *71*, 3184–3191. [CrossRef]
162. Bacic, M.K.; Smith, C.J. Laboratory maintenance and cultivation of bacteroides species. *Curr. Protoc. Microbiol.* **2008**, *9*, 13C-1. [CrossRef]
163. Livingston, S.J.; Kominos, S.D.; Yee, R.B. New medium for selection and presumptive identification of the *Bacteroides fragilis* group. *J. Clin. Microbiol.* **1978**, *7*, 448–453. [CrossRef] [PubMed]
164. Ho, P.L.; Ho, L.Y.; Yau, C.Y.; Tong, M.K.; Chow, K.H. A Novel Selective Medium for Isolation of *Bacteroides fragilis* from Clinical Specimens. *J. Clin. Microbiol.* **2017**, *55*, 384–390. [CrossRef] [PubMed]
165. Jenkins, S.A.; Drucker, D.B.; Keaney, M.G.; Ganguli, L.A. Evaluation of the RAPID ID 32A system for the identification of *Bacteroides fragilis* and related organisms. *J. Appl. Bacteriol.* **1991**, *71*, 360–365. [CrossRef] [PubMed]
166. Ahmed, W.; Hughes, B.; Harwood, V.J. Current status of marker genes of *Bacteroides* and related taxa for identifying sewage pollution in environmental waters. *Water* **2016**, *8*, 231. [CrossRef]
167. Kirs, M.; Caffaro-Filho, R.A.; Wong, M.; Harwood, V.J.; Moravcik, P.; Fujioka, R.S. Human-associated *Bacteroides* spp. and human polyomaviruses as microbial source tracking markers in Hawaii. *Appl. Environ. Microbiol.* **2016**, *82*, 6757–6767. [CrossRef]
168. Boehm, A.B.; Soller, J.A. Refined ambient water quality thresholds for human-associated fecal indicator HF183 for recreational waters with and without co-occurring gull fecal contamination. *Microb. Risk Anal.* **2020**, *16*, 100139. [CrossRef]
169. Shanks, O.C.; Kelty, C.A.; Sivaganesan, M.; Varma, M.; Haugland, R.A. Quantitative PCR for genetic markers of human fecal pollution. *Appl. Environ. Microbiol.* **2009**, *75*, 5507–5513. [CrossRef]
170. Toledo-Hernandez, C.; Ryu, H.; Gonzalez-Nieves, J.; Huertas, E.; Toranzos, G.A.; Santo Domingo, J.W. Tracking the primary sources of fecal pollution in a tropical watershed in a one-year study. *Appl. Environ. Microbiol.* **2013**, *79*, 1689–1696. [CrossRef] [PubMed]
171. Zimmer-Faust, A.G.; Thulsiraj, V.; Marambio-Jones, C.; Cao, Y.; Griffith, J.F.; Holden, P.A.; Jay, J.A. Effect of freshwater sediment characteristics on the persistence of fecal indicator bacteria and genetic markers within a Southern California watershed. *Water Res.* **2017**, *119*, 1–11. [CrossRef] [PubMed]
172. Stachler, E.; Kelty, C.; Sivaganesan, M.; Li, X.; Bibby, K.; Shanks, O.C. Quantitative CrAssphage PCR Assays for Human Fecal Pollution Measurement. *Environ. Sci. Technol.* **2017**, *51*, 9146–9154. [CrossRef] [PubMed]
173. Okoh, A.I.; Sibanda, T.; Gusha, S.S. Inadequately treated wastewater as a source of human enteric viruses in the environment. *Int. J. Environ. Res. Public Health* **2010**, *7*, 2620–2637. [CrossRef]
174. Liu, Y.; Gilchrist, A.; Zhang, J.; Li, X.F. Detection of viable but nonculturable *Escherichia coli* O157:H7 bacteria in drinking water and river water. *Appl. Environ. Microbiol.* **2008**, *74*, 1502–1507. [CrossRef] [PubMed]
175. Kitajima, M.; Sassi, H.P.; Torrey, J.R. Pepper mild mottle virus as a water quality indicator. *NPJ Clean Water* **2018**, *1*, 19. [CrossRef]
176. Dutilh, B.E.; Cassman, N.; McNair, K.; Sanchez, S.E.; Silva, G.G.; Boling, L.; Barr, J.J.; Speth, D.R.; Seguritan, V.; Aziz, R.K.; et al. A highly abundant bacteriophage discovered in the unknown sequences of human faecal metagenomes. *Nat. Commun.* **2014**, *5*, 4498. [CrossRef]
177. McMinn, B.R.; Huff, E.M.; Rhodes, E.R.; Korajkic, A. Concentration and quantification of somatic and F+ coliphages from recreational waters. *J. Virol. Methods* **2017**, *249*, 58–65. [CrossRef]
178. McMinn, B.R.; Rhodes, E.R.; Huff, E.M.; Korajkic, A. Decay of infectious adenovirus and coliphages in freshwater habitats is differentially affected by ambient sunlight and the presence of indigenous protozoa communities. *Virol. J.* **2020**, *17*, 1. [CrossRef]
179. Tomat, D.; Balagué, C.; Casabonne, C.; Verdini, R.; Quiberoni, A. Resistance of foodborne pathogen coliphages to thermal and physicochemical treatments applied in food manufacture. *Innov. Food Sci. Emerg. Technol.* **2015**, *30*, 184–191. [CrossRef]

180. Guan, D.; Kniel, K.; Calci, K.R.; Hicks, D.T.; Pivarnik, L.F.; Hoover, D.G. Response of four types of coliphages to high hydrostatic pressure. *Food Microbiol.* **2006**, *23*, 546–551. [CrossRef] [PubMed]
181. Vergara, G.; Goh, S.; Rezaeinejad, S.; Chang, S.; Sobsey, M.; Gin, K. Evaluation of FRNA coliphages as indicators of human enteric viruses in a tropical urban freshwater catchment. *Water Res.* **2015**, *79*, 39–47. [CrossRef] [PubMed]
182. Muniesa, M.; Mocé-Llivina, L.; Katayama, H.; Jofre, J. Bacterial host strains that support replication of somatic coliphages. *Antonie Van Leeuwenhoek* **2003**, *83*, 305–315. [CrossRef]
183. U.S. Environmental Protection Agency. *Male-Specific (F+) and Somatic Coliphage in Water by Two-Step Enrichment Procedure*; U.S. Environmental Protection Agency: Wahington, DC, USA, 2001.
184. U.S. Environmental Protection Agency. *Method 1602: Male-Specific (F+) and Somatic Coliphage in Water by Single Agar Layer (SAL) Procedure*; U.S. Environmental Protection Agency: Wahington, DC, USA, 2001.
185. Muniesa, M.; Ballesté, E.; Imamovic, L.; Pascual-Benito, M.; Toribio-Avedillo, D.; Lucena, F.; Blanch, A.; Jofre, J. Bluephage: A rapid method for the detection of somatic coliphages used as indicators of fecal pollution in water. *Water Res.* **2018**, *128*, 10–19. [CrossRef]
186. Mendez, J.; Toribio-Avedillo, D.; Mangas-Casas, R.; Martinez-Gonzalez, J. Bluephage, a method for efficient detection of somatic coliphages in one hundred milliliter water samples. *Sci. Rep.* **2020**, *10*, 2977. [CrossRef]
187. Haramoto, E.; Yamada, K.; Nishida, K. Prevalence of protozoa, viruses, coliphages and indicator bacteria in groundwater and river water in the Kathmandu Valley, Nepal. *Trans. R. Soc. Trop. Med. Hyg.* **2011**, *105*, 711–716. [CrossRef]
188. Lee, J.E.; Lee, H.; Cho, Y.H.; Hur, H.G.; Ko, G. F+ RNA coliphage-based microbial source tracking in water resources of South Korea. *Sci. Total Environ.* **2011**, *412*, 127–131. [CrossRef]
189. Toribio-Avedillo, D.; Martin-Diaz, J.; Blanco-Picazo, P.; Blanch, A.R.; Muniesa, M. F-specific coliphage detection by the Bluephage method. *Water Res.* **2020**, *184*, 116215. [CrossRef]
190. Agullo-Barcelo, M.; Galofre, B.; Sala, L.; Garcia-Aljaro, C.; Lucena, F.; Jofre, J. Simultaneous detection of somatic and F-specific coliphages in different settings by *Escherichia coli* strain CB390. *FEMS Microbiol. Lett.* **2016**, *363*. [CrossRef]
191. Benjamin-Chung, J.; Arnold, B.F.; Wade, T.J.; Schiff, K.; Griffith, J.F.; Dufour, A.P.; Weisberg, S.B.; Colford, J.M., Jr. Coliphages and Gastrointestinal Illness in Recreational Waters: Pooled Analysis of Six Coastal Beach Cohorts. *Epidemiology* **2017**, *28*, 644–652. [CrossRef]
192. Shkoporov, A.N.; Khokhlova, E.V.; Fitzgerald, C.B.; Stockdale, S.R.; Draper, L.A.; Ross, R.P.; Hill, C. ΦCrAss001 represents the most abundant bacteriophage family in the human gut and infects *Bacteroides* intestinalis. *Nat. Commun.* **2018**, *9*, 4781. [CrossRef]
193. Edwards, R.A.; Vega, A.A.; Norman, H.M.; Ohaeri, M.; Levi, K.; Dinsdale, E.A.; Cinek, O.; Aziz, R.K.; McNair, K.; Barr, J.J.; et al. Global phylogeography and ancient evolution of the widespread human gut virus crAssphage. *Nat. Microbiol.* **2019**, *4*, 1727–1736. [CrossRef]
194. Kongprajug, A.; Mongkolsuk, S.; Sirikanchana, K. CrAssphage as a potential human sewage marker for microbial source tracking in Southeast Asia. *Environ. Sci. Technol. Lett.* **2019**, *6*, 159–164. [CrossRef]
195. Stachler, E.; Crank, K.; Bibby, K. Co-occurrence of crAssphage with antibiotic resistance genes in an impacted urban watershed. *Environ. Sci. Technol. Lett.* **2019**, *6*, 216–221. [CrossRef]
196. Ahmed, W.; Payyappat, S.; Cassidy, M.; Besley, C. A duplex PCR assay for the simultaneous quantification of *Bacteroides* HF183 and crAssphage CPQ_056 marker genes in untreated sewage and stormwater. *Environ. Int.* **2019**, *126*, 252–259. [CrossRef] [PubMed]
197. Stachler, E.; Akyon, B.; de Carvalho, N.A.; Ference, C.; Bibby, K. Correlation of crAssphage qPCR Markers with Culturable and Molecular Indicators of Human Fecal Pollution in an Impacted Urban Watershed. *Environ. Sci. Technol.* **2018**, *52*, 7505–7512. [CrossRef] [PubMed]
198. Wu, Z.; Greaves, J.; Arp, L.; Stone, D.; Bibby, K. Comparative fate of CrAssphage with culturable and molecular fecal pollution indicators during activated sludge wastewater treatment. *Environ. Int.* **2020**, *136*, 105452. [CrossRef]
199. Wolf, S.; Hewitt, J.; Greening, G.E. Viral multiplex quantitative PCR assays for tracking sources of fecal contamination. *Appl. Environ. Microbiol.* **2010**, *76*, 1388–1394. [CrossRef] [PubMed]
200. McQuaig, S.M.; Scott, T.M.; Lukasik, J.O.; Paul, J.H.; Harwood, V.J. Quantification of human polyomaviruses JC virus and BK virus by TaqMan quantitative PCR and comparison to other water quality indicators in water and fecal samples. *Appl. Environ. Microbiol.* **2009**, *75*, 3379–3388. [CrossRef] [PubMed]
201. Pal, A.; Sirota, L.; Maudru, T.; Peden, K.; Lewis, A.M., Jr. Real-time, quantitative PCR assays for the detection of virus-specific DNA in samples with mixed populations of polyomaviruses. *J. Virol. Methods* **2006**, *135*, 32–42. [CrossRef] [PubMed]
202. Ahmed, W.; Zhang, Q.; Kozak, S.; Beale, D.; Gyawali, P.; Sadowsky, M.J.; Simpson, S. Comparative decay of sewage-associated marker genes in beach water and sediment in a subtropical region. *Water Res.* **2019**, *149*, 511–521. [CrossRef] [PubMed]
203. Rowe, W.P.; Huebner, R.J.; Gilmore, L.K.; Parrott, R.H.; Ward, T.G. Isolation of a cytopathogenic agent from human adenoids undergoing spontaneous degeneration in tissue culture. *Proc. Soc. Exp. Biol. Med.* **1953**, *84*, 570–573. [CrossRef]
204. Parks, R.J. Adenovirus protein IX: A new look at an old protein. *Mol. Ther.* **2005**, *11*, 19–25. [CrossRef]
205. Matthews, D.A. Adenovirus protein V induces redistribution of nucleolin and B23 from nucleolus to cytoplasm. *J. Virol.* **2001**, *75*, 1031–1038. [CrossRef]

206. Noble, R.T.; Allen, S.M.; Blackwood, A.D.; Chu, W.; Jiang, S.C.; Lovelace, G.L.; Sobsey, M.D.; Stewart, J.R.; Wait, D.A. Use of viral pathogens and indicators to differentiate between human and non-human fecal contamination in a microbial source tracking comparison study. *J. Water Health* **2003**, *1*, 195–207. [CrossRef]
207. Hundesa, A.; De Motes, C.M.; Albinana-Gimenez, N.; Rodriguez-Manzano, J.; Bofill-Mas, S.; Sunen, E.; Girones, R.R. Development of a qPCR assay for the quantification of porcine adenoviruses as an MST tool for swine fecal contamination in the environment. *J. Virol. Methods* **2009**, *158*, 130–135. [CrossRef]
208. Ogorzaly, L.; Bertrand, I.; Paris, M.; Maul, A.; Gantzer, C. Occurrence, survival, and persistence of human adenoviruses and F-specific RNA phages in raw groundwater. *Appl. Environ. Microbiol.* **2010**, *76*, 8019–8025. [CrossRef] [PubMed]
209. Hewitt, J.; Greening, G.E.; Leonard, M.; Lewis, G.D. Evaluation of human adenovirus and human polyomavirus as indicators of human sewage contamination in the aquatic environment. *Water Res.* **2013**, *47*, 6750–6761. [CrossRef]
210. Li, D.; He, M.; Jiang, S.C. Detection of infectious adenoviruses in environmental waters by fluorescence-activated cell sorting assay. *Appl. Environ. Microbiol.* **2010**, *76*, 1442–1448. [CrossRef]
211. Gerba, C.P.; Gramos, D.M.; Nwachuku, N. Comparative inactivation of enteroviruses and adenovirus 2 by UV light. *Appl. Environ. Microbiol.* **2002**, *68*, 5167–5169. [CrossRef]
212. Eischeid, A.C.; Thurston, J.A.; Linden, K.G. UV disinfection of adenovirus: Present state of the research and future directions. *Crit. Rev. Environ. Sci. Technol.* **2011**, *41*, 1375–1396. [CrossRef]
213. Flint, S.J.; Enquist, L.W.; Racaniello, V.R.; Skalka, A.M. *Principles of Virology: Molecular Biology, Pathogenesis, and Control*; ASM Press: Washington, DC, USA, 2000.
214. Sassoubre, L.M.; Love, D.C.; Silverman, A.I.; Nelson, K.L.; Boehm, A.B. Comparison of enterovirus and adenovirus concentration and enumeration methods in seawater from Southern California, USA and Baja Malibu, Mexico. *J. Water Health* **2012**, *10*, 419–430. [CrossRef] [PubMed]
215. Lipp, E.K.; Lukasik, J.; Rose, J.B. Human enteric viruses and parasites in the marine environment. *Methods Microbiol.* **2001**, *30*, 559–588.
216. Lachiewicz, A.; Cianciolo, R.; Miller, M.; Derebail, V. Adenovirus causing fever, upper respiratory infection, and allograft nephritis complicated by persistent asymptomatic viremia. *Transpl. Infect. Dis.* **2014**, *16*, 648–652. [CrossRef] [PubMed]
217. Demoliner, M.; Gularte, J.S.; Girardi, V.; Eisen, A.K.A.; de Souza, F.G.; Staggemeier, R.; Henzel, A.; Spilki, F.R. Microbial Source Tracking in Small Farms: Use of Different Methods for Adenovirus Detection. *Water Air Soil Pollut.* **2021**, *232*, 63. [CrossRef]
218. Schoch, C.L.; Ciufo, S.; Domrachev, M.; Hotton, C.L.; Kannan, S.; Khovanskaya, R.; Leipe, D.; Mcveigh, R.; O'Neill, K.; Robbertse, B. NCBI Taxonomy: A comprehensive update on curation, resources and tools. *Database* **2020**, *2020*, baaa062. [CrossRef] [PubMed]
219. Padgett, B.; Zurhein, G.; Walker, D.; Eckroade, R.; Dessel, B. Cultivation of papova-like virus from human brain with progressive multifocal leucoencephalopathy. *Lancet* **1971**, *297*, 1257–1260. [CrossRef]
220. Bofill-Mas, S.; Rusiñol, M.; Fernandez-Cassi, X.; Carratalà, A.; Hundesa, A.; Girones, R. Quantification of human and animal viruses to differentiate the origin of the fecal contamination present in environmental samples. *BioMed Res. Int.* **2013**, *2013*, 192089. [CrossRef]
221. Bofill-Mas, S.; Pina, S.; Girones, R. Documenting the epidemiologic patterns of polyomaviruses in human populations by studying their presence in urban sewage. *Appl. Environ. Microbiol.* **2000**, *66*, 238–245. [CrossRef]
222. McQuaig, S.; Griffith, J.; Harwood, V.J. Association of fecal indicator bacteria with human viruses and microbial source tracking markers at coastal beaches impacted by nonpoint source pollution. *Appl. Environ. Microbiol.* **2012**, *78*, 6423–6432. [CrossRef]
223. Albinana-Gimenez, N.; Clemente-Casares, P.; Calgua, B.; Huguet, J.M.; Courtois, S.; Girones, R. Comparison of methods for concentrating human adenoviruses, polyomavirus JC and noroviruses in source waters and drinking water using quantitative PCR. *J. Virol. Methods* **2009**, *158*, 104–109. [CrossRef]
224. Bofill-Mas, S.; Albinana-Gimenez, N.; Clemente-Casares, P.; Hundesa, A.; Rodriguez-Manzano, J.; Allard, A.; Calvo, M.; Girones, R. Quantification and stability of human adenoviruses and polyomavirus JCPyV in wastewater matrices. *Appl. Environ. Microbiol.* **2006**, *72*, 7894–7896. [CrossRef] [PubMed]
225. Templeton, M.R.; Andrews, R.C.; Hofmann, R. Particle-associated viruses in water: Impacts on disinfection processes. *Crit. Rev. Environ. Sci. Technol.* **2008**, *38*, 137–164. [CrossRef]
226. Haramoto, E.; Katayama, H. Application of acidic elution to virus concentration using electropositive filters. *Food Environ. Virol.* **2013**, *5*, 77–80. [CrossRef]
227. Ahmed, W.; Harwood, V.; Gyawali, P.; Sidhu, J.P.; Toze, S. Comparison of concentration methods for quantitative detection of sewage-associated viral markers in environmental waters. *Appl. Environ. Microbiol.* **2015**, *81*, 2042–2049. [CrossRef]
228. Calgua, B.; Barardi, C.R.M.; Bofill-Mas, S.; Rodriguez-Manzano, J.; Girones, R. Detection and quantitation of infectious human adenoviruses and JC polyomaviruses in water by immunofluorescence assay. *J. Virol. Methods* **2011**, *171*, 1–7. [CrossRef]
229. Wong, M.V.; Hashsham, S.A.; Gulari, E.; Rouillard, J.-M.; Aw, T.G.; Rose, J.B. Detection and characterization of human pathogenic viruses circulating in community wastewater using multi target microarrays and polymerase chain reaction. *J. Water Health* **2013**, *11*, 659–670. [CrossRef] [PubMed]
230. Hewitt, J.; Leonard, M.; Greening, G.E.; Lewis, G.D. Influence of wastewater treatment process and the population size on human virus profiles in wastewater. *Water Res.* **2011**, *45*, 6267–6276. [CrossRef] [PubMed]

231. Carini, P.; Marsden, P.J.; Leff, J.W.; Morgan, E.E.; Strickland, M.S.; Fierer, N. Relic DNA is abundant in soil and obscures estimates of soil microbial diversity. *Nat. Microbiol.* **2016**, *2*, 16242. [CrossRef]
232. Wang, H.; Koydemir, H.C.; Qiu, Y.; Bai, B.; Zhang, Y.; Jin, Y.; Tok, S.; Yilmaz, E.C.; Gumustekin, E.; Rivenson, Y. Early detection and classification of live bacteria using time-lapse coherent imaging and deep learning. *Light Sci. Appl.* **2020**, *9*, 118. [CrossRef]
233. Højris, B.; Christensen, S.C.B.; Albrechtsen, H.-J.; Smith, C.; Dahlqvist, M. A novel, optical, on-line bacteria sensor for monitoring drinking water quality. *Sci. Rep.* **2016**, *6*, 23935. [CrossRef] [PubMed]
234. Simões, J.; Yang, Z.; Dong, T. An ultrasensitive fluorimetric sensor for pre-screening of water microbial contamination risk. *Spectrochim. Acta Part A Mol. Biomol. Spectrosc.* **2021**, *258*, 119805. [CrossRef] [PubMed]
235. Guo, T.; Wei, Y.; Xu, C.; Watts, B.R.; Zhang, Z.; Fang, Q.; Zhang, H.; Selvaganapathy, P.R.; Deen, M.J. Counting of *Escherichia coli* by a microflow cytometer based on a photonic–microfluidic integrated device. *Electrophoresis* **2015**, *36*, 298–304. [CrossRef] [PubMed]

processes MDPI

Review

Wastewater Treatment Using Constructed Wetland: Current Trends and Future Potential

Ikrema Hassan [1,*], Saidur R. Chowdhury [2,3], Perdana K. Prihartato [4] and Shaikh A Razzak [5,6]

1 Department of Civil Engineering, College of Engineering, Taibah University, Medina 42353, Saudi Arabia
2 Department of Civil Engineering, College of Engineering, Prince Mohammad Bin Fahd University, Dhahran 34754, Saudi Arabia; schowdhury1@pmu.edu.sa
3 SC Environmental Solutions, Greystone Walk Dr, Toronto, ON M1K 5J3, Canada; saidurc@gmail.com
4 Environmental Scientist, Saudi Arabia Oil Company, Dhahran 34464, Saudi Arabia; perdana.karim@gmail.com
5 Department of Chemical Engineering, King Fahd University of Petroleum & Minerals, Dhahran 31261, Saudi Arabia; srazzak@kfupm.edu.sa
6 Interdisciplinary Research Center for Membranes and Water Security, King Fahd University of Petroleum & Minerals, Dhahran 31261, Saudi Arabia
* Correspondence: imohammed@taibahu.edu.sa; Tel.: +966-565-333-757

Abstract: Constructed wetlands (CW) is an environmentally friendly technique for removing pollutants from wastewater and has been applied to municipal wastewater, petroleum refinery wastewater, agriculture drainage, acid mine drainage, etc. The past decade has seen a remarkable number of innovations in the exponentially growing field of microbiology. This manuscript covers a critical review of key aspects of CW, such as various types of CW, the contaminants and their removal mechanisms, degradation pathways, challenges and opportunities, materials, applications, and theory with a focus on recent advances in the last three decades. In addition, an attempt has been taken to project future advances in the field of CW and facilitate these advances by framing key unsolved problems in CW. Guidelines are prepared for the fast-growing CW field through the standardization of key design aspects. This review covers the evaluation of the current state-of-the-art of CW technology and provides definitions and performance metric nomenclature in an effort to unify the fast-growing CW community. It also contains an outlook on the emerging trends in CW and proposes future research and development directions.

Keywords: constructed wetland; wastewater; plants; microorganisms; remediation; degradation

Citation: Hassan, I.; Chowdhury, S.R.; Prihartato, P.K.; Razzak, S.A. Wastewater Treatment Using Constructed Wetland: Current Trends and Future Potential. *Processes* **2021**, *9*, 1917. https://doi.org/10.3390/pr9111917

Academic Editor: Dimitris Zagklis

Received: 7 September 2021
Accepted: 20 October 2021
Published: 27 October 2021

1. Introduction

In the last decade, environmental awareness has increased, and the treatment of environmental pollution and contamination has become the main agenda of concerned governmental bodies around the world. Usually, a suitable environmental remediation method for a certain type of waste is selected based on the effectiveness of the degradation process and the cost of the method [1–3]. More importantly, the environmental impact of the selected method is of special concern because in some remediation methods the daughter product of the degradation process is more toxic than the original contaminant. Scientists and researchers believe that there is no single universal remediation method suitable for all types of contaminants and all sources; instead, an effective remediation program may involve the collective implementation of two or more methods [1,2]. Wetlands are one of the effective remediation technologies that now attract the environmentalist for the treatment of wastewater contaminants [2]. Mother Nature, through natural wetlands, removes contaminants from water resources through different natural processes including biodegradation, sorption, phytostabilization, phytoextraction, and rhizofiltration [2,4–7]. Constructed wetlands are developed to mimic the natural processes in removing/degrading contaminants in wastewater [8,9]. Municipal wastewater, industrial wastewater (particularly petroleum refinery and sour-water treatment wastewater products), agriculture wastewater, stormwater, textile wastewater, landfill leachates, mining drainage, etc., are examples of contaminated wastewater that can be treated using wetlands [10–12]. Constructed wetlands use natural processes in degrading contamination; therefore, it is an environmentally friendly remediation method with minimum adverse environmental impact [13]. CW is designed to treat wastewater from different sources similar to common effluent treatment plants [14].

Wetlands are generally defined by the saturation of soil for a long enough period for anaerobic conditions to develop. There are various types of wetlands including natural fresh and salt-water wetlands and constructed wetlands [2]. Wetlands constructed for remediation of contaminants integrate complicated inclusive processes that involve water, substrate (soil), plants, animals, microorganisms, and the environment. Constructed wetlands implemented various remediation methods including biodegradation, phytoremediation, and natural attenuation [15]. The main processes that occur in wetlands include physical processes such as filtration and sedimentation, chemical processes such as adsorption and precipitation, and biological processes such as biodegradation and plant assimilation [16].

The vast majority of the wetlands are characterized by a high density of vascular plants. High-density vegetation results in slowing the water flow, creating microenvironments, and providing sorption sites for contaminants and attachments sites for microorganisms [5]. As the plants die and fall in the water, the parts of the plants above the water create additional sorption and exchange sites. In addition, plant debris is a suitable source of organic carbon and nutrients (nitrogen and phosphorous) for microorganism metabolism. The oxygen concentration is very low in wetlands due to soil saturation conditions. Therefore, the vegetation in wetlands is limited to species (vascular plants) that can grow at low oxygen concentrations [5].

Microorganisms play a dominant role in the contaminant degradation process and in the transformation of the contamination between the environment and the plant. Usually, a microbial consortium is involved in contaminant degradation and the transformation processes [9]. There are specific degradation pathways for each microorganism to degrade a specific contaminant. The success of the process of contaminant degradation relies on the existence of microorganisms that are required for the degradation process and suitable environmental conditions [17,18].

Constructed wetlands have been used all around the world for the treatment of wastewater. For instance, CWs have been used in Europe since the second half of the last century; the first country to use CWs in Europe was Germany [16]. Many other countries including the United Kingdom, Austria, Slovenia, Switzerland, and Denmark have CWs in operation. Some countries in Africa have been using CWs, such as South Africa, Tanzania, Kenya, and Seychelles. Some statistics show that the cost of subsurface CWs in Africa for wastewater treatment is around US $5 per person compared to mechanical wastewater treatment (i.e., activated sludge system), which costs around US $50 per person [16,17]. A recent study showed that the total cost (including operation, maintenance, and disposal of sludge) of wastewater treatment varies between €0.30 to €0.88 per m^3 [19]. In China there are more than 400 CWs [20].

Constructed wetlands are characterized by very slow water flow and shallow water depth. The slow water flow results in a long retention time that facilitates the settlement of the sediment and increases the contact time between the wastewater and the components of the wetland [21]. Thus, in the design, it is critical to choose a suitable flow velocity to provide enough time for microorganisms to degrade the contaminants. The components of the constructed wetland affect the degradation processes. For instance, the number of sorption sites in the soil and the substrate control the bioavailability and degradation process [21]. There are two schools of thought concerning bioavailability; some scientists believe that bacteria can degrade contaminants without the need for the contaminants to be desorbed [21]. On the other hand, some researchers believe that the contaminants need to be desorbed first before bacteria can degrade them. Some types of bacterial strains are capable of producing biosurfactants to facilitate the desorption process [22]. Therefore, for constructed wetlands, understanding the capabilities of the available microorganism consortium is essential for a successful treatment process.

The main functions of CWs are to improve wastewater quality by degradation or sorption, control the flood by creating storage for the storm rainfall and surface runoff, and recycle nutrients [4,8,11]. However, the values of wetlands include providing recreational areas and providing suitable data for research and education. The advantages of CWs include (i) the low construction cost compared to other remediation methods, (ii) an environmentally friendly approach that is viewed with favor by the public, (iii) relatively low maintenance and operation costs, (iv) high flexibility in the landscape design to provide habitat for wildlife and organisms, and (v) wetlands facilitate the reuse of treated water. Although they have great advantages, there are some limitations of CWs that include (i) requiring large land areas compared to the other remediation methods, (ii) the inconsistency of the treatment compared to the other wastewater treatment methods, (iii) not suitable to treat if the discharge should meet specific standards, (iv) ammonia and pesticides have detrimental effects on the plants and microorganisms, and (v) low tolerance of near-complete drying conditions [9,17]. The local regulations and bylaws directly affect the selection of the type of flow in constructed wetlands. Some locations have stringent laws that prohibit surface flow, and the designers have to choose the vertical flow in these cases [23].

2. Wetland Treatment Systems

Generally, constructed wetlands can be classified according to different criteria such as hydrology (surface-flow and subsurface-flow), macrophyte types (free-floating, emergent, and submerged), and flow path (horizontal or vertical) [24]. As mentioned before, there are many types of constructed wetlands including surface flow (SF) wetlands, subsurface flow (SSF) wetlands, and hybrid systems, which encompass surface and subsurface flow wetlands [25]. The hybrid system is a multistage system in which the treatment is performed in different units that are designed for specific functions. For instance, for wastewater treatment, some units are designed to promote aerobic reactions, while other units are designed for anaerobic conditions. A wetland equipped with an air pump connected to a subsurface network of air distribution pipes is called aerated CW [26]. The air bubbles introduced by the air pump can increase the oxygen transfer rate in horizontal flow (HF) or vertical flow (VF) type wetlands and effectively creates aerobic conditions. The estimated oxygen consumption rate in CW could be 250 g of $O_2/m^2/d$ with air flow rate and distribution to be $\geq 0.6\ m^3/m^2/h$ and 30 cm × 30 cm, respectively. Mechanically aerated wetlands can provide higher oxygen transfer rates (more than $1\ m^3/m^2/h$). The aerated wetland has improved nitrification/denitrification capacity as well as better biogeochemical conditions in the wetland subsurface and the plants [26,27].

In the literature, scientists and engineers use various terms to describe wetlands. For instance, in free-surface flow (FSF) the surface of the wastewater flow is above the soil (substrate in the CWs) [2–4]. Alternatively, some scientists use the term SF surface flow to denote the FSF. Additionally, the term SSF is used for sub-surface flow system in which the wastewater flows horizontally or vertically such that the water level is below the ground surface. The abbreviation HSSF is used for horizontal subsurface flow or alternatively SSHF subsurface horizontal flow [2,4,5,13]. The terms VSSF and SSVF are used for the vertical subsurface flow or subsurface vertical flow, respectively. Sometimes the abbreviation of the constructed wetlands CWs is added to the above-mentioned terms. For example, CWs is added to HF to become HFCWs [1,22]. In an attempt to unify the terms used in the literature, this article will use the following abbreviations: SF for surface flow, VSSF for vertical subsurface flow, and HSSF for horizontal subsurface flow.

2.1. Surface Flow (SF)

Surface flow wetlands (SF) is where the surface of the wastewater flow is above the soil (substrate) [28]. The SF is composed of a substrate (clay and sand soil) to support the roots of the plants and water control structures that maintain a shallow depth of water above the soil (see Figure 1) [4,5]. The bottom of the soil is sealed to maintain water within the system. SF wetlands can be used to treat wastewater while supporting wildlife habitats. In the SF, the concentration of oxygen varies with depth [29]. Depending on the water level, the top layers (near the water surface) have a high oxygen concentration, while the bottom layers have a very low (or zero) oxygen concentration [2]. Therefore, the top layers support aerobic processes (nitrification), and the bottom layer supports anaerobic (denitrification) processes. SF can be used to treat stormwater, agricultural runoff, and mining wastewater. SF wetlands require a large area compared with the subsurface flow wetlands. The advantages of the surface flow wetlands are that their construction and operational cost are lower than the subsurface flow wetlands, and it is simple technology compared with the subsurface flow wetlands. The disadvantages include odor problems and more area required compared to subsurface flow CWs and other remediation methods [29].

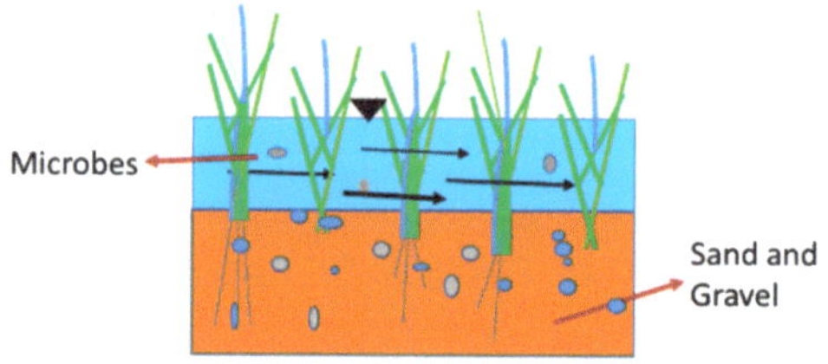

Figure 1. Schematic layout of free surface flow.

2.2. Subsurface Flow (SSF)

Subsurface flow (SSF) wetland substrate is composed of porous soil (rock and gravel) sealed from the bottom [9,13]. In the SSF, the water level should be below the soil level all the time (see Figure 2). The flow path in the SSF wetland is either a horizontal or vertical flow path. SSF wetlands are suitable for the treatment of wastewater, provided that the solid concentration in the wastewater is low. For high solid concentration in the wastewater, sedimentation tanks or basins are used to remove the solids from the wastewater [25,30,31]. The main advantages of SSF wetlands include tolerance of cold weather, fewer odor problems compared to SF wetlands, greater sorption and exchange sites compared to SF wetlands, and efficient use of the land compared to SF wetlands. SSF wetlands can be constructed below public parks because the water level is below the ground surface. The limitations of SSF wetlands include their higher cost compared to SF wetlands, used for small flows, and clogging of the pores [26,27].

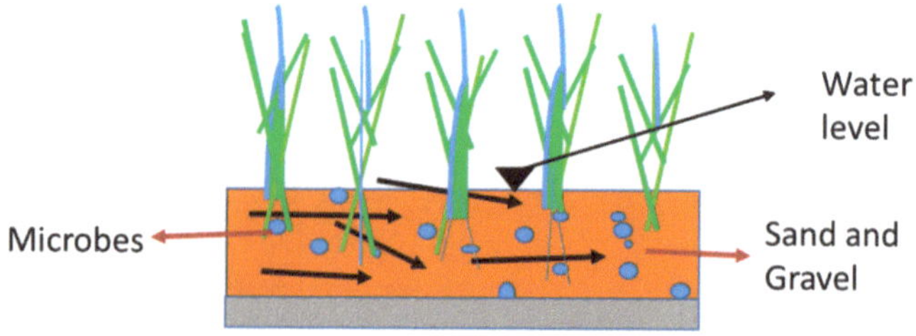

Figure 2. Schematic layout of horizontal subsurface flow.

The complex processes that contribute to purifying the wastewater in wetlands include sorption, filtration through the substrate, sedimentation, settlement, ion exchange, plant sorption, and microorganism oxidation [2,4,5]. Constructed wetlands with the subsurface flow have been used to treat heavy oil-produced water containing asphalt, gelatinoids, wax, polymers, surfactant, and other artificial products. The constructed wetland has been approved to be successful in treating heavy oil-produced water. There is no proper design guideline that specifies the filtration rate; as such, some studies used 0.52 m day^{-1} [25]. Previous research showed that the removal of COD is highly affected by DO concentration. The anaerobic conditions and high organic loading rate can reduce the efficiency of COD removal. There are two types of flows in SSF: the first is the horizontal subsurface flow (HSSF), and the second type is vertical subsurface flow (VSSF) [32].

2.2.1. Horizontal Subsurface Flow Constructed Wetland (HSSF)

The horizontal subsurface flow (HSSF) is composed of a channel located below the ground surface [9,13]. The channel is filled with selected types and sizes of gravel and sand. The channel is planted with a suitable plant. There are certain considerations that should be taken into account when designing the channel. For instance, the water flow in HSSF is designed such that the level of the water is below the ground surface. Solids should also be removed from the wastewater before entering the channel to avoid clogging [13]. A sedimentation tank or filter are usually used to remove the solid particles from the wastewater. The cross-sectional area of the HSSF plays a dominant role in the efficiency of the decontamination process. The oxygen level in the HSSF should be enough to promote the growth and metabolism of aerobic bacteria as well as plants. Many mechanisms facilitate oxygen transfer to the subsurface environment. Plant roots contribute to the increase in oxygen in the environment near the roots [5,9,13].

2.2.2. Vertical Subsurface-Flow Constructed Wetlands (VSSF)

Some constructed wetlands are designed to have vertical water flow. In such a case, it is called vertical subsurface-flow constructed wetlands (VSSF) [30]. Similar to HSSF, the flow of VSSF should be passed through filter beds before entering the CW channels (See Figure 3). The flow in VSSF is either by gravity or by pumps. Certain considerations should be taken into account in the design of VSSF. The time required for the wastewater to percolate through the vertical channels should be calculated. In most cases, the flow in VSSF is controlled such that there is a time interval between the loading events, and intermittence loading is adopted [33]. The plant roots transfer some oxygen to the subsurface, and dryness periods allow oxygen diffusion to subsurface. Thus, the oxygen level in VSSF is high, which promotes the growth of aerobic bacteria. Therefore, VSSF enhances aerobic degradation and is found to be suitable for nitrification and is more aerobic than HSSF. Previous studies found that VSSF promotes a wide range of degradation and decontaminations processes including, biodegradation, adsorption, precipitation, and filtration [34]. It was also found that the VSSF is successful in removing the organic matter and suspended solids. The VSSF capital cost is less than the HSSF capital cost because the VSSF needs a smaller area size compared to HSSF [24].

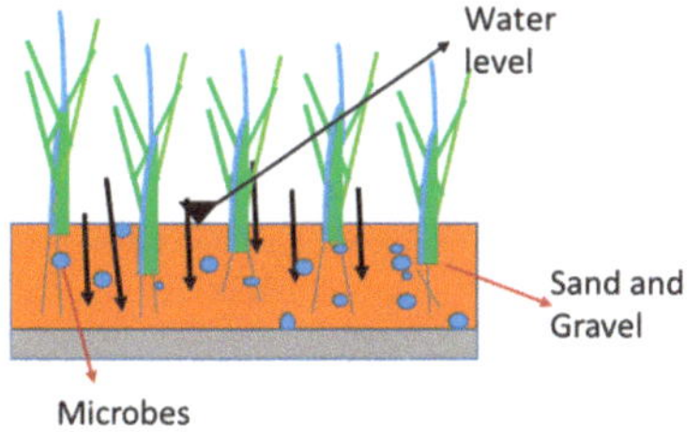

Figure 3. Schematic layout of vertical subsurface flow constructed wetlands.

2.3. Constructed Wetlands Design, Construction, Operation and Maintenance

2.3.1. Design

(a) General design criteria

The design of wetlands contains many components that influence the performance of wetlands. The performance of CW depends on the location, climate, weather, type of wastewater, type of runoff, and expected variations during the year [9,32,34]. In general, constructed wetlands are designed to simulate the natural wetlands with special attention to the system that leads to the enhancement of the wastewater quality. Researchers, engineers, and scientists recommended that the design of CWs should be simple. It is highly recommended to use native and available media as well as environmentally friendly technologies [9]. The CWs should be designed in such a way that results in low maintenance and minimizes the use of pumps. The CW design guidelines include the detention time, size, length, width, depth, etc. More space and a greener environment are required to build CWs than other technologies [32]. The CWs can be designed on a small scale to serve a single family or can be designed on a large scale to serve a community [32]. Special consideration should be given to design parameters such as geometry of the CW, influent quality, media type, microorganisms and plant species, water depth, and hydraulic retention time. The geometry is basically dependent on the type of the CW. For instance, for surface flow CW, the basin slope should be between 3:1 and 5:1. Impervious media should be used at the bottom of the CW. The hydraulic retention time should be between 2 and 3 days. The quality of the influent includes the biological oxygen demand (BOD) and total suspended solids (TSS). For the BOD and TSS, loading could be between 20 and 30 mg/L or 45 and 50 kg/ha per day. On the other hand, for subsurface flow, the BOD and TSS should be around 7–16 g/m^2 per day and 20 g/m^2 per day, respectively. The recommended minimum length is 15 m, and the maximum width is 61 m. The hydraulic conductivity varies in the longitudinal and transversal directions. A previous study showed that the hydraulic conductivity subsurface flow of CW varies between 1400 and 2800 m/day, and another study recommended 1000 m/day for the first 30% of the length. However, for ageing systems, clogging of the soil pores is expected, which can result in reduced porosity of the system. Pervious studies show, for ageing systems, that the hydraulic conductivity is between 30 and 107 m/d in the outlet zone [35–38].

For a single family, the general consideration that should be taken in the design of CWs includes removal of trees and grass, construction of a fence around the selected area to stop compaction, and use of inorganic fine-grained soil for backfilling. The size of CWs should be 1 m^2 for 30 L (1 m^2/30 L) influent or the detention time of 5 days. Provision should be taken that the length should be double the width. The wastewater can be estimated at 210 L/capita/day, and the number of people can be estimated as two persons per bedroom. Therefore, for a two-bedroom household, the area of CWs should be 28 m^2 with dimensions of 7.5 m × 3.75 m [32]. In a previous study, a vertical flow CW of surface area of 24.4 m^2 and depth of 1.3 m was divided into two equal sizes (3.5 × 3.5 × 1.5 m) and used to treat wastewater from two-family residence of eight persons. The system includes two settling tanks, a vault tank, zeolite tank, effluent collection tank, and pump to feed the CW with 150 L wastewater every 3 h. The system was designed to treat 150 L/capita/day. Coarse gravel (20–40 mm diameter) and fine gravel (2–10 mm diameter) were used as the porous media. The zeolite tank was used to remove ammonia and phosphorus. Removal efficiencies were 96.4% for BOD, 94.4% for COD, 90.8% for TKN, 92.8% for ammonia, 61.6% for OP, and 69.8% for TP [39]. A previous study used HSSF CW for a single family of four people. The system consisted of two settling tanks (each 1.3 × 1.3 × 1.2 m, width × length × depth), CW (10 × 2.5 × 0.5 m), and a zeolite tank (1.6 × 1.3 m). The mean BOD loading was about 120–150 mg/L, the hydraulic residence time was 7 days, and the flow rate was 0.6 m^3/d. The mean BOD effluent concentration was about 30 mg/L indicating 87% removal [40].

A septic tank is required to filter the solid or scum before the CWs. It is recommended to load the CWs by gravity (gravity inflow); however, pumping can be used to load the

CWs but avoids large doses [5]. The water depth in the cells should be 600 mm or less to ensure enough oxygen concentration in CWs to support microorganism metabolism. Polyvinylchloride (PVC), high-density polyethylene (HDPE), or polypropylene (PPE) of 1.0 mm may be used to prevent leakage from the CWs [5,24,32]. The liner should be free from pinholes and damage, and provision should be taken during the insulation to prevent punctures of the lining. The CWs should be filled with gravel of different sizes. Generally, the CWs can be divided into three sections that should be filled with gravel, including the inlet and outlet (the far edges), the middle bottom layer, and the middle top layer. Large gravel size of 38 to 76 mm should be placed near the inlet and outlet (the edges), medium gravel size of 12.7 to 25.4 mm should be placed at a depth of 18 inches in the middle bottom layer, and a small size of 9.5 to 12.7 mm should be placed to form the top middle layer. The design of inflow and outflow is crucial for efficient CWs [5,24,32]. The following equation can be used to calculate the amount of water in the system:

$$Q_{in} - Q_{out} + P - ET = dv/dt \quad (1)$$

where Q_{in} = influent wastewater flow (m^3/s); Q_{out} = effluent wastewater flow (m^3/s); P= precipitation (m^3/s); ET= evapotranspiration (m^3/s); V= volume of water (m^3); t = time.

(b) Vertical flow (VF): the inflow in VF is required to be passed through a filtration system to prevent clogging of the subsurface flow system [13]. Materials with high sorption sites, such as charcoal and clay, may be used in the filtration system, which can support the growth of pathogens [30]. Therefore, the selection of suitable materials, such as gravel and slag, that reduces the growth of microorganisms can contribute positively to the quality of the effluent and can control the infection. On the other hand, temperature has a significant effect on the biological treatment of wastewater. Therefore, the temperature is a very important parameter in the design of CWs. Low temperature can result in slowing down the degradation process, and as a result, the retention time should be increased in the design. For example, for an influent of 112 kg/ha, a detention time of 11 days is required for a temperature around 5 °C, and a detention time of 5 days is required in the summer at higher temperatures [30]. The detention time is controlled by changing the number of cells in operation and/or by changing the water depth.

(c) Horizontal flow (HF): the flow in HF can be modelled using Darcy's Law as follows

$$q = k.\ i.\ A \quad (2)$$

where: q = average daily flow rate (m^3/d); k = hydraulic conductivity (m/d); i = hydraulic gradient (m/m); A = cross sectional area (m^2).

The HF CW system should rely on gravity flow; reduce the amount of filling and cutting, and consider the time required for the system to be fully functional [41]. In addition, the design should be performed based on the hydraulic and organic design criteria; however, other parameters should be measured such as nutrient removal and pathogen reduction. In HF, the water flow should be designed to maintain water flow levels above the surface of the bed depending on the type of CW [24,42,43]. In the HF the degradation in the top surface layer should be designed to support aerobic degradation, while the deep layers should be designed to support anaerobic degradation [41].

2.3.2. Construction

Wetlands should be constructed to meet regulations and high standards of reliability and safety. Special consideration and attention should be paid to the compaction and elevations [41]. Proper compaction is required because compaction controls the infiltration process, especially in the case of SSF. The elevation is equally important to maintain gravity flow. Specifications should be prepared based on site conditions and include regulatory body requirements [32]. In SSF, a sedimentation or filtration tank with suitable retention time is needed for the removal of solid waste to avoid clogging of the subsurface hydraulic system. Usually, the sedimentation tank is placed to receive the influent. The loading of the system should also be adequate. Table 1 contains information about the design of different types of CWs.

Table 1. Design, operation, and maintenance of CWs.

Type of Constructed Wetland	Design	Operation	Maintenance
Surface Flow (SF)	Simple, requires a large land area, uses gravity flow	Low operation cost, simple operation procedure, high evapotranspiration, low temperature, affects the microbial activity	Greatly affected by temperature flocculation, odor and mosquito problems, low maintenance cost
Surface Subsurface flow (SSF)	Complex, needs less area than SF, needs sedimentation tank, needs pumping	Provides more sorption sites than SF, relatively higher operation cost, flow should be uniform with low solid concentration, transpiration only	Greater cold tolerance, less odor and pests. Clogging problems, higher maintenance cost
Vertical Subsurface flow (VSF)	Complex, needs less area than SSF, needs sedimentation pond, needs pumping	Provides more sorption sites than SF, relatively higher operation cost, flow should be uniform with low solid concentration, transpiration only	Greater cold tolerance, less odor and pests. Clogging problems, higher maintenance cost

2.3.3. Operation

After construction, the wetland should be tested for the water level. The channeling and erosion, if any, should be fixed in the trial period [44]. It is preferred to allow a full growing season before introducing the wastewater that needs to be treated. The water level should be maintained as per the design because if the wetland were to dry, remobilization of contaminants may occur. Table 1 contains information about the operation of different types of CWs [44,45].

2.3.4. Maintenance

A plan and schedule for regular maintenance should be prepared and followed strictly to ensure that the wetland performance is up to the required level [45,46]. It is very important to ensure that there is enough contact time between the waste to be treated and the components of the wetlands (microorganism, soil, and plant). A suitable environment for microorganisms and plant growth should be maintained all the time to achieve optimum degradation of contaminants [24,45,46]. Table 1 contains information about the maintenance of different types of CWs.

3. Types of Contaminants Treated in CW

Over the last two decades, the bioremediation technology of CWs has been shown to be an efficient natural process method for wastewater treatment. CWs have been proved to be effective in the treatment of domestic, municipal, agricultural, and industrial wastewaters [29]. CWs, a promising green technology, are also very effective in treating organic-rich wastewater [47], toxic metal contaminated water (e.g., Cr (VI), Se) [48], pharmaceuticals and personal care products (PPCPs) [23], oil refinery wastewater, diesel [41], and agricultural drainage water. In the following subsections of the paper, the use of constructed wetlands in removal of different contaminants is discussed.

3.1. Domestic and Municipal Wastewater Contaminants

Expansion of urban populations has resulted in an increase in the footprint of domestic water supply and sewerage, giving a rise to greater quantities of wastewater. Properly managed municipal wastewater treatment would result in reducing surface water contamination problems [4,5,24]. The process of wastewater treatment not only conserves valuable water resources, but it also takes advantage of the nutrients contained in sewage to grow crops. Wastewater is increasingly viewed as an additional source that can provide a new source of good quality water. However, in many parts of the world, especially in under-developed countries, the operation cost of the wastewater treatment plant is an issue [5]. On the other hand, the applications of CWs for the treatment of domestic and municipal wastewater have been rapidly increasing throughout the world because of their unique advantages of lower operational and maintenance costs. Different types of contaminants (such as nitrogen, phosphorus, organics, solids, metals, and coliforms) can be removed from wastewater by CW treatment methods. This method includes a complex inter-connected system of plants, media, bulk water, and biomass population. The removal of biological oxygen demand (BOD), chemical oxygen demand (COD), total suspended solids (TSS), nitrogen and phosphorous pollutants, heavy metals, and different organic matter (e.g., amino acids, fatty acids, proteins, carbohydrates, lipids) by CW systems have gained substantial attention in recent years [25,49]. In subsurface flow wetlands, the removal of contaminants depends on a diverse range of co-existing physical, chemical, and biological routes depending on numerous environmental and operational factors.

Nitrogen, one of the principal pollutants in wastewater, exists in wastewater in both organic and inorganic forms. The removal of nitrogen through nitrification and denitrification can be achieved using constructed wetlands. The design should carefully consider the parameters that affect the processes such as pH, temperature, dissolved oxygen concentration, and ammonia concentration. Other factors that affect the nitrification and denitrification processes include the plant species and the wetland type [50].

Organic nitrogen presents in amino acids (a form of peptide chain that makes proteins), urea (dispose of ammonia by mammals when amino acids are used for energy production), uric acids (produced by birds and insects), and purine and pyrimidines (involved in DNA making) [25]. The inorganic forms (e.g., ammonium (NH_4^+), nitrite (NO_2^-), nitrate (NO_3^- nitrous oxide (N_2O), and dissolved elemental nitrogen or nitrogen gas (N_2) as well as free ammonia (NH_3)) can be removed in subsurface flow wetland systems, and the removal processes includes biological (i.e., ammonification, nitrification, denitrification, plant uptake, biomass assimilation, dissimilatory nitrate reduction) and physicochemical routes (e.g., ammonia volatilization, and adsorption) [25,51,52]. The organic compounds can be reduced aerobically and anaerobically. The common interactions are as follows:

Aerobic organic degradation:

$$C_6H_{12}O_6 + 6O_2 \rightarrow 6CO_2 + 6H_2O \quad (3)$$

Anaerobic organic degradation:

$$C_6H_{12}O_6 \rightarrow 2CH3CHOHCOOH \text{ (Lactic acid)} \quad (4)$$

$$C_6H_{12}O_6 \rightarrow 2CH_3CH_2OH \text{ (Ethanol)} + 2CO_2 \quad (5)$$

The removal routes of nitrogen and organics in CW systems are critically dependent on different environmental parameters, such as pH, temperature, oxygen and operational strategies, i.e., presence of organic carbon, hydraulic load, feeding mode, retention time, pollutant loading, recirculation, and plant harvesting [31,53,54].

Phosphorus removal in the CW systems involves the immobilization/removal of phosphorus through chemical precipitation, bacterial action, plant uptake and related physico-chemical pathways, and the adsorptive capacity of the gravel layer. Some studies reported removal of BOD, COD, NH_4-N and Total Nitrogen (TN) to be as 77%, 78%, 99%, and 100%, respectively [55]. Different substrate media (such as rice husk, organic wood-mulch, zeolite, aggregates, alum sludge, slag, peat, maert, compost, and shale) were employed to increase the performance of contaminant removal from wastewater by different CW methods. In addition, the common media substrates, such as sand, fine to coarse gravel, limestone, zeolite, calcite, native soil, basic oxygen furnace slag, activated carbon, lightweight aggregates, etc., are used in VF, FWS (free water surface), and HF constructed wetland systems to remove TSS, TN, TP, SO_4^{2-}, COD, BOD, and organic matters from domestic and municipal wastewater [56]. Moreover, free-floating, submerged, emergent plants, and also microalgae can treat P (>90% of removal) in CWs. Plant species, hydraulic retention time, temperature, type of CWs, effluent concentration, and seasonal changes can influence the removal efficiency of P in CWS [57].

3.2. Toxic Metals/Metaloids Contaminated Water

Treating industrial wastewater, particularly metal-containing wastewater in wetlands, has received increased attention [58]. Wetland soils characterized by their reduced condition and high organic matter content could accumulate heavy metals. Soil and its overlying water, and vegetation, play a major role to remove the metals within wetlands. Dissolved or particulate forms of metals are introduced into wetland systems through the wastewater. The free metal ions are the most biologically available fraction, and sediments in the wetland systems can serve as sinks and reservoirs for metals [59].

Heavy metals are one of the major pollutants in industrial wastewater and gain special concern to treat in easiest and less expensive methods [58,59]. Nowadays, the removal of heavy metals using CWs shows promising results and high efficiency. The heavy metal removal performance of the wetland treatment system depends on the inflow water quality and the metal removal mechanisms. Metal mass removal ranges from 27% to 99% within the wetland. It has been reported that an average of 7.6 kg/ha/year, or 54% of influent metals loadings, could be sequestered within the internal wetland compartments [60,61].

Table 2 presents the heavy metal removal efficiency from different types of wastewaters using different types of plants. Previous studies have assessed the removal of some heavy metals including Cu and Ni from constructed wetlands [49]. For example, the heavy metal removal performance by a laboratory-scale wetland system treating storm runoff has been studied. It was found that the heavy metal removal efficiency in CW ranged from 81.7% to 91.8% with the rates of 36.6–372.7 mg/m^2/day for Cu, 75.8–95.3% and 30.8–387 mg/m^2/day for Pb, and 82.8–90.4% and 33.6–362.1 mg/m^2/day for Zn, respectively [49,60–64]. Heavy metals such as Cu, Cd, Zn, Pb, Ni, and Co could be readily removed by constructed wetland systems, and the metal uptake seems to be influenced by the types of media used and the types of wastewaters to be treated. The extent of different removal mechanisms might vary depending on the speciation of inflow metals and physicochemical water parameters (e.g., redox, pH, and types of wetland systems).

Table 2. Heavy metals removal by CW treatment.

Metals	Types of Waste	Media Used in CW	Plants	Type of WC	Removal Efficiency (%)	Sources
Cu	Domestic wastewater	Native sediments	Macrophyte Myriophyllum spicatum	Not mentioned	>84	[60]
Zn	Industrial Wastewater	Sediments	Macrophyte	Free water surface wetland	55	[62]
Ni	Industrial Wastewater	Sediments	Macrophyte	Free water surface wetland	69	[62]
Fe	Industrial Wastewater	Compost-based and gravel-based	Macrophyte	Free water surface wetland	83	[62]
Pb	Domestic wastewater	Gravel-filled subsurface-flow	Typha latifolia	Surface flow	>75	[49]
Cd	Domestic wastewater	Gravel-filled subsurface-flow	Typha latifolia	Surface flow	>80	[49]
B	Mine tailings waste	Ashes-sewage sludge mixture with gravel sand	Macrophyte	Not mentioned	>70	[64]
Cr	Tannery Wastewater	Sediments	T. Latifolia and P. australis	Not mentioned	82	[65]
Al	Domestic wastewater	Gravel-filled horizontal subsurface-flow	Macrophyte Myriophyllum spicatum	Not mentioned	>84	[60]
Co	Domestic wastewater	Gravel-filled subsurface-flow	Typha latifolia	Surface flow	>70	[49]
As	Mine tailings waste	Ashes-sewage sludge mixture with gravel sand	Macrophyte	Not mentioned	>70	[64]
Se	Oil refinery wastewater	Sand, peat moss	Algal species	Not mentioned	96%	[48]

3.3. Industrial Wastewater

The discharge of industrial wastewater into the aquatic environment has been one of the primary threats to the ecosystem. Industrial wastewater management faces high technical challenges to achieve its effective treatment for the delivery of a high-quality treated effluent [66]. CWs, especially in developing countries, have been proposed to offer a low-cost and low maintenance treatment alternative for industrial effluent. The feasibility of treating industrial wastewater within a pilot-scale wetland in Argentina was studied. It was found that the average metal removal efficiencies were 83%, 82%, 69%, and 55% for Fe, Cr, Ni, and Zn, respectively [49,61–65].

Previous study reported significant removal of chemical and microbial parameters such as COD (68%), BOD (71%), ammoniacal nitrogen (66%), total phosphorus (21.5%), and total nitrogen (68.5%), and fecal coliforms (47.2%) were achieved in Duckweed-microalgae constructed wetland (DM-CW) [47]. Constructed wetlands (CWs) using *S. grossus* plat could be used to remove diesel from wastewater [41]. The removal of total suspended solids and chemical oxygen demand (COD) from a 0.1% diesel concentration of wastewater were 100 and 75.4%, respectively.

3.4. Oilfield Produced Wastewater

The treatment and production processes used in the oil and gas industry have a significant impact on the environment and contribute to the contamination of natural ecosystems including surface water [67]. The oil exploration and production processes result in a substantial amount of highly contaminated water. The wastewater from oilfields can contain toxic organic and inorganic pollutants. The wastewater from oilfields is usually stored in large ponds in remote areas. The wastewater from the oilfield can contain various components, including groundwater, water, and chemical compounds used in the extraction process. Previous studies showed that the quality of oilfield wastewater can be improved using a suitable type of CW [68]. Many plants, characterized by high tolerance to saltwater, can be used to decontaminate oilfields wastewater, including *acacia nilotica, cotton, Brassica napus,* and *ricinus communis* [29].

A recent study showed that CW can be used to decontaminate/treat oilfield wastewater. The results showed that the contaminant concentration was reduced from 500 mg/L to 30 mg/L [69]. In addition, the results showed that a substantial amount between 30 and 50% of water was lost in the process due to the evapotranspiration process. It has been proved that CWs can completely remove nitrogen and phosphorous from petroleum-contaminated wastewater. The decontamination of petroleum wastewater is known to be a companied with a 10–12% increase in the pH of the water flow.

Petroleum refinery or oilfield produced wastewater can be treated in vertical surface flow constructed wetland (VSFCW) systems vegetated with *Eichhornia crassipes* (water hyacinth) to meet the compliance permissible limit set by Federal Environmental Protection Agency (FEPA) and WHO for discharge into surface waters [28]. Bio-treatment of petroleum wastewater in vertical surface flow constructed wetland resulted in the removal efficiencies of 91.5% turbidity, 94.6% BOD_5, 80.2% COD, 92.6% TPH, 9 mg 0.4% oil and grease, 94% cadmium, 92.5% lead, 93% chromium, 94.8% iron, 92.2% nickel, and 57.7% chloride [28].

3.5. Oil Refinery Wastewater

Large volumes of wastewater during extraction processes are produced as a result of production, storage, distribution, and processing of petroleum [70]. Oil refinery wastewater is also generated by accidents due to spills from water/fuel mixtures and leaks from underground storage or water [70]. Oil exploration and exploitation activities generate a variety of toxic wastewater consisting of different organic compounds and inorganic and heavy metals, thus leading to pollution of the terrestrial and aquatic ecosystems [71]. Organic contaminants produced by oil refinery industrial activity include BTEX (benzene, toluene, ethylbenzene, and xylene), PAH, linear alkylbenzene sulfonates chlorinated hydrocarbons, benzene and methyl tert-butyl ether (MTBE), aliphatic and aromatic petroleum hydrocarbons consisting of methyl-tetrabutyl ether, phenol, 2,3,5,6-tetramethylphenol, naphthalene, xylene, tetradecane, 4-chloro-3-methylphenol, and 3-tert-butylphenol at different concentrations [72]. Table 3 shows the plants that can be used in CW for oil refinery wastewater.

Table 3. Major contaminant removal from oil refinery wastewater in different wetland systems.

Parameters	Methods	Plants	Removal Efficiency (%)	Sources
Electrical conductivity	VSF-CW	C. *alternifolius* and C. *dactylon*, Eichhornia crassipes	>75	[29]
Turbidity	VSF-CW	Eichhornia crassipes, Cyperus alternifolius and *Cynodon dactylon* (L.) Pers.	50–90	[29]
BOD	VSF-CW	Eichhornia crassipes	94.6	[28,49]
COD	VSF-CW	Eichhornia crassipes; P. *australis*, Typha angustifolia and T. *latifolia*	80	[28,49]
Oil and grease	VSF-CW	Eichhornia crassipes	58–94	[29]
Total petroleum hydrocarbon (TPH),	HF-CW, VF-CW	Scirpus grossus (bulrush), *Salix* spp., *Scirpus* spp., *Juncus* spp. and P. *australis*	>75	[29,73]
Heavy metals (e.g., cadmium, lead, copper, nickel, zinc, iron and chromium)	VSF-CW	Eichhornia crassipes, Typha latifolia	48–92	[29,74]
Nitrate-nitrogen, sulphate and chloride	VSF-CW	Cyperus alternifolius, Eichhornia crassipes, Typha Orientalis	>58	[29,75]

A previous study reported that refined wastewater has higher toxicity than crude oil due to the addition of metals and their speciation during the refining processes [28]. Many heavy metals such as Fe, Cr, Cd, Pb, Hg, Ag, Mo, Cu, Zn, and As from oil exploration and exploitation activities can be found in petroleum wastewater [71]. Petroleum refinery secondary wastewater can be treated by vertical surface flow constructed wetland (VSF-CW) vegetated with Eichhornia crassipes [28]. The wastewater having relatively elevated levels of turbidity (18.30 NTU), BOD (20.40 mg/L), COD (86 mg/L), TPH (16.6 mg/L), oil and grease (18.4 mg/L), heavy metals (cadmium (0.034 mg/L), lead (0.12 mg/L), chromium (0.47 mg/L), iron (1.54 mg/L), nickel (0.09 mg/L)), and chloride (1412 mg/L) can be treated with the 90% to 95% of removal efficiency by vertical surface flow constructed wetland (VSF-CW). Plants, with the natural ability to uptake metals, can remove heavy metals in CWs by microbiota uptake [76], plant uptake, as well as adsorption onto media and sediments in the system [77]. In CWs, additional processes such as biosorption, bioaccumulation, redox transformation, dissimilatory sulphate reduction, and precipitation as insoluble salts can remove heavy metals from wastewater [48,49,51]. A previous study showed that Cu can be removed using the calcined sludge; therefore, treating different waste at CW can enhance removal efficiency [78].

3.6. Pharmaceuticals and Personal Care Products (PPCPs)

Over the last few decades, the production of a wide array of pharmaceuticals and personal care products (PPCPs) has been seen in the world due to unprecedented advancements in the medicine and cosmetics industry that revolutionized healthcare and beautification facilities [79]. Biotechnology, a cost-effective and environmentally sound approach, has gained great interest that uses aquatic macrophytes and microalgae for wastewater treatment [23,34,80]. Phytoremediation in CW is a well-established and efficient method for decontamination of PPCPs from the environment. Scientists and engineers from several countries have paid attention to the potential of aquatic macrophytes to treat and recycle pollutants from municipal and industrial wastewater.

The phytoremediation process in CW, one of the methods used to remove PPCPs, can act as a primary, secondary, or tertiary treatment step [80]. A previous study reported that caffeine (>99.9%), acetaminophen (99.98%), IBU (99.6%), naproxen (99.4%), and triclosan (98%) were up-taken in CW treatment [23]. Another study showed more than 95% removal efficiencies for caffeine, carboxy-ibuprofen, hydrocinnamic acid, hydroxy-ibuprofen, methyl dihydrojasmonate, oxybenzone, salicylic acid, and IBU, while 70–90% of naproxen, diclofenac, galaxolide, and tonalide. Treatment of water contaminated with

PPCPs such diclofenac, IBU, naproxen, tonalide, and bisphenol using as HSSF planted with *P. australis* is promising, and the removal was around 90% [43]. The removal can be attributed to biodegradation as well as biosorption. Various studies show evidence that plant species including *Berulaerecta, Irispseudacorus, Juncus effusus, P. ustralis,* and *Typha latiflolia* can be used to remove/degrade 90% of IBU and iohexol from wastewater. On the other hand, it was proven that CWs planted with *Scirpus validus* can reduce 74% of carbamazepine and 98% of naproxen from wastewater in hydroponic conditions [81,82]. Different types of plants such as *Phragmites australis, Typha Angustifolia, P. australis, Spathiphyllum, Oryza sativa, Capsicum annuum, Scirpus validus, Populus nigra, Typha angustifolia, P. australis, Cucumis sativus*, etc., were used in HSSF, VF, and SSF type CWs to remove PPCPs (e.g., Caffeine, Ibuprofen, Naproxen, Chlortetracycline, Diclofenac, Carbamazepine, Tonalide and Ketoprofen etc.). The removal of these PPCPs was found to be in the range of 70% to 95% [20,51,52].

In a recent study, five endocrine disrupting chemicals (EDCs), such as Bisphenol A (BPA), Nonylphenol (NP), Nonylphenol monoethoxylate (NP1EO), Nonylphenol diethoxylate (NP2EO), and Triclosan (TCS), were treated in HSF and VF CWs system. The removal range was found to be 48% to 99%. The removal mechanisms of these EDCs were biodegradation under aerobic or/and anaerobic conditions as well as adsorption onto the organic matter present in the CW substrate. Compared to the VF CW system, HF CW demonstrated low efficiency in removing BPA. On the other hand, higher removal efficiencies were achieved for NP1EO and NP2EO in the HF CW system [83].

All PPCPs can be considered as persistent organic pollutants (POPs). In addition to PPCPs, other chemicals such as aldrin, chlordane, dieldrin, endrin, heptachlor, mirex, toxaphene, hexachlorobenzene (HCB), polychlorinated biphenyls (PCB) DDT, and polycyclic aromatic hydrocarbons (PAHs) such as polychlorinated dibenzo-p-dioxins/dibenzofurans were treated with subsurface flow systems (SSF) and free surface flow types constructed wetland [84,85]. Phytoremediation process in CW can remove most POPs, and nearly 100% efficiency in removing organic compounds from wastewater may be achieved in such systems. The plants used in CW to treat different POPs were *Acorus calamus* L., Carex rostrate Stokes, *Phragmites australis, Schoenoplectus lacustris* (L.) *Palla, and Typha latifolia* L., *Nymphaea alba* L., *Nuphar lutea Myriophyllum spicatum Lemna minor* L., *Spirodela polyrhiza* L., and *Eichhornia crassipes* [85]. The removal capacity by phytoremediation process in CW were found to be 66% to 100% when free water surface systems (FWS), subsurface flow systems (SSF), or vertical subsurface flow (VFS) constructed wetlands were used [85].

3.7. Agricultural Wastewater

Wastewater discharge into natural systems, rivers, lakes, and groundwater can cause a detrimental impact on the future use or function of these natural aquatic systems. Rainfalls, floods, and overflow can carry wastewater from livestock and spread contamination [86]. The wastewater from livestock processes needs to be treated to reduce the nutrients concentration, organic matters, and suspended solids [86]. The SF constructed wetlands support wildlife habitats. On the other hand, the SSF prevents the spread of mosquitoes in CWs, reduces the bad odor associated with CWs, and results in avoiding human contacts [87].

It has been proven that CW can be used to treat the BOD, TSS, and facial coliforms from dairy parlour wastewater. CW planted with different macrophytes show 75% to 85% removal of TSS from agricultural wastewater. The retention time is directly proportional to the efficiency of the removal process of dissolved humic substance. Results showed that 7-d retention time resulted in 47% to 75% removal of nitrogen, 37% to 74% removal of phosphorous, and 99% fecal coliform removal. The selection of suitable plant species is crucial in the degradation process. For instance, CWs planted with *Juncus, Scirpus Typha* and mixed spices resulted in 90%, 92%, 94%, 91%, and 99% removal of BOD_5, TTS, NH_3-N, TKN, and facial coliforms, respectively [88].

Some other operational challenges include the removal of pesticides. Constructed wetlands can be used to remove fungicides, insecticides, and herbicides from wastewater mixed with agriculture wastewater. Recent research concluded that the deposit load has an important effect on the removal of pesticides from the wastewater [89].

A recent study discussed the removal efficiency of boscalid from rinsing water produced during cleaning of pesticide spraying equipment using two pilot-scale horizontal subsurface flow (HSSF) constructed wetlands (CWs) [90]. The CWs were planted with common reeds (*Phragnites australis*) in fine gravel (code name FG-R) and the other cobbles (CO-R). The results showed that the removal in both systems ranged from 49% to 100%, indicating the efficiency of HSSF systems in treating boscalid-polluted agricultural equipment rinsing water in the agricultural area.

Another study investigated the removal of pesticides using horizontal subsurface flow (HSSF) CW systems that were operated continuously for a long period of time (one year). The CWs (containing either *Phragmites australis* or *Typha latifolia*) were able to remove up to 73.7% and 58.4% of pesticides in the system, respectively. High concentrations of terbuthylazine were detected in roots, leaves, and shoots of both plants of the CWs [91]. In another study, three horizontal subsurface flow (HSSF) constructed wetlands (CWs) planted with *Phragmites australis* in fine gravel layers were used to treat fluopyram from rinsing water produced during the cleaning of pesticide spraying equipment. Higher fluopyram removal efficiency (>96%) was achieved in the HSSF system. The fluopyram removal mechanisms were achieved through microbial biodegradation, adsorption on porous media, and bioaccumulation processes [92].

3.8. Compost and Landfill Leachates

The leachate from landfill facilities can be harmful and can be a real threat to the environment. CWs can be used as one of the alternatives to treat landfill leachate. For instance, HSSF CW planted with *Vetiver* was used to treat leachate from composting facility in Isfahan, Iran [42]. The study period was five months, during which the results of leachate treated with CWs were compared to leachate that was not treated. The loading rate used for the HSSF CW was 24 L/d. The results showed that 75% of the BOD_5 was removed. Although the removal of COD was around 53%, it was considered as an improvement compared to the removal without the CWs, which was around 26%. The removal of NH_3-N, NO_3-N, and TN was around 70%, 74%, and 74% respectively. This study showed that a sub-surface horizontal flow constructed wetland planted with *Vetiver* has the potential to be used as a leachate pre-treatment or treatment method to treat highly contaminated composting leachate [42].

3.9. Mine Drainage

Mining activities use a considerable amount of water in extracting valuable parts from ores. The process resulted in a highly toxic solution of acid mine drainage (AMD). The AMD is carcinogenic and has a detrimental effect on the environment. Nevertheless, CW can be used to treat AMD and reduce the toxicity in the effluent. CW is an environmentally friendly process that encompasses various processes including biological, chemical, biogeochemical, geochemical, and physical processes to reduce contamination in the wastewater. Heavy metal compounds are the main component of AMD, which is characterized by low pH [93]. The mechanisms that are responsible for heavy metal removal from AMD include sedimentation, sorption, accumulation in the plant, and precipitation. Heavy metals accumulation in the plant represents part of the removal, but not a high percentage. It has been proven that a significant amount of heavy metals is removed from AMD in CW by sedimentation [93].

3.10. Storm Water Runoff

Storm runoff is a highly variable and intermittent feedstock that can be treated by constructed wetlands. There has been increasing interest in the use of constructed wetlands to remediate storm runoff, which often contains high levels of metals, mainly including Cu, Cd, Zn, and Pb [24,58,59]. Previous work investigated the heavy metal removal performance by a laboratory-scale wetland system treating storm runoff. The removal efficiencies and rates for metals monitored ranged from 81.7% to 91.8% and 36.6–372.7 mg/m^2/day for Cu, 75.8–95.3% and 30.8–387 mg/m^2/day for Pb, and 82.8–90.4% and 33.6–362.1 mg/m^2/day for Zn, respectively. Stormwater runoff where systems receive low hydraulic and constituent loads can be treated by constructed wetland systems (CWs). The CWs remove contaminants, nutrients, and solids from runoff water [49,74].

3.11. Microplastic on CWs

Synthetic organic polymer plastics have been heavily used in different purposes such as household and personal care products, medical applications, clothing, agricultural (e.g., mulching,) engineering materials, infrastructure, and land use practices media [94,95]. The micro and nano plastics are invisible, and humans are exposed to these products through the use of personal care products containing microplastics [96,97]. The extensive use of plastics products (which are very slow biodegradable materials) in human everyday life has resulted in posing a risk to humans and the ecosystem. The distribution and transport of microplastics (MPs) in CWs depend on morphological features and the apertures of the substrate, soil texture, and vegetation coverage [98]. Moreover, macroinvertebrates were used in horizontal subsurface flow constructed wetlands (as tertiary treatment) to investigate the distribution of MPs in CW and found the potential role of macroinvertebrates in MPs distribution and removal [99]. Recent study showed that wetland plants, substrate composition, hydraulic loading rate, hydraulic retention time (HRT), and biofilm in CWs can influence the migration, transformation, and removal behaviors of MPs [100–102]. Multi-stage constructed wetlands (CWs) played a crucial role in treating the MPs (>89% of removal) in water. Plant species such as *P. australis*, *Lolium perenne* (garden cress), *Sphagnum palustre* (peat moss), *Vigna radiata* (mung bean), *Spirodela polyrhiza* (great duckweed), and *Allium cepa* (garden onion) were used in CWs to reduce MPs or control MPs migration [103]. The transformation of toxic MP compounds into nontoxic or less toxic forms through accumulation by roots or immobilization within plants species is one of the mechanisms to control MPs migration and distribution in water [20,103]. Previous studies showed that mangroves can uptake considerable amounts of microplastics [93,97–99]. The retention of microplastic in the mangrove depends on the paling pattern and density, roots of the mangrove, as well as the dead plant debris and branches [94]. Another study showed that the uptake of microplastics by plants depends on particle characteristics such as size and charge, as well as plant features [103,104]. It was found that microplastic particles are attracted to the soil organic carbon resulting in an increased stability of microplastic in the soil sediments [105,106]. A recent study presented evidence of the accumulation and retention of microplastic particles in freshwater reeds [107]. The average concentration of microplastics on reeds samples was found be around 4.9 items/individual [107]. Very few studies are available in this area in the current literature; therefore, the numbers here are not confirmed by other studies [107]. A recent study showed that microplastic fibers contributed positively on the growth of plant community, and this was attributed to the reduction in soil bulk density and increase in aeration due to the existence of the microfibers [108]. Another study showed plastic particles can contribute to an increase in plant biomass [103]. Some studies reported negative toxic effects of microplastics on plant germination and root elongation [103,109]. On the other hand, contradicting results have been reported on the effect of the plastic particles on plant photosynthesis [103].

4. Microorganisms Used in the Wetland

The design of wetland components includes the substrate (the soil matrix), the hydrology (water flow and water control structures), and the plants. However, important components of wetlands such as communities of microbes and invertebrates are developed naturally [51,110]. Bacteria play a dominant role in contaminant degradation and in the transformation of contaminants between the environment and the plant. Different types of heterotrophic and autotrophic bacteria contribute to contaminant degradation. Heterotrophic bacteria utilize organic carbon for their metabolism; therefore, they are active in the degradation of organic pollutants such as petroleum hydrocarbon. On the other hand, autotrophic bacteria use carbon dioxide for cell growth [111].

In aerobic digestion, the organic pollutant serves as electron donor and oxygen serves as electron acceptor. In this process, the pollutant is decomposed to CO_2 through a certain pathway, which involves the production of daughter products of the original contaminant [23,73,112]. It is very important to mention that in some cases some daughter products of these processes are more toxic than the original pollutant. Therefore, it is very important to ensure that degradation happens at the end of the processes (i.e., produce CO_2 and H_2O). The shallow surface of the surface flow constructed wetlands and the top layers of the subsurface flow constructed wetlands provide a suitable environment (i.e., oxygen concentration) for aerobic degradation of contaminants [23].

In anaerobic digestion, the organic pollutant serves as electron donor and CO_2 serves as electron acceptor. Some other chemical compounds serve as electron acceptors including nitrate, nitrite, sulphate, and carbonate. In denitrification, nitrate and nitrite serve as electron acceptors, and the organic pollutant serves as the electron donor [13,23]. Table 4 shows various types of bacteria that contribute to nitrification and denitrification processes

Table 4. Bacterial strains involved in nitrification and denitrification in constructed wetlands.

	Bacteria	Porous Media	Process	Reference
1	*Pseudomonas falva* WD-3	Brown soil (0–4 mm diameter), Sludge (5–10 mm diameter) and Gravel (40–50 mm diameter)	Denitrification	[113]
2	*Alcaligenes faecalis strain WT14*	Sediments	Heterotrophic nitrification and aerobic denitrification	[114]
3	*Albidiferax*	Rough sand (1–2 mm diameter), garvel (10–20 mm diameter) and Gravel (30–50 mm diameter)	Nitrification	[115]
4	*Candidatus Nitrosotenuis*	Quartz sand (5–10 mm diameter), cativated carbon (2–5 mm diameter) and Cobble (15–30 mm diameter)	Nitrification	[116]
5	*Nitrosomonas*	Quartz sand (5–10 mm diameter), cativated carbon (2–5 mm diameter) and Cobble (15–30 mm diameter)	Nitrification	[116]
6	*Nitrosopumilus*	Quartz sand (5–10 mm diameter), cativated carbon (2–5 mm diameter) and Cobble (15–30 mm diameter)	Nitrification	[116]

The metabolism of each type of bacteria is optimum under specific environmental conditions including pH, temperature, oxygen concentration, and sunlight. In general, pH around 6–7 is suitable for most types of bacteria. A temperature around 28–30 is suitable for most types of bacteria. However, thermophilic bacterial strains survive under very high temperatures, such as *Sulfolobus Solfataricus, Bacillus licheniformis,* and *Thermomonas hydrothermalis* [110].

Temperature has an effect on the microbial activity in the wetlands; therefore, the temperature can directly affect the success of the treatment process. Low temperatures slow down the microbial activities, while very high temperatures have detrimental effects on the microorganisms. Sorption and sedimentation processes are also affected by the temperature. Temperature directly affects the concentration of dissolved oxygen in the water and the oxidation-reduction potential. High temperatures cause low dissolved oxygen environments. Temperatures between 20 and 30 °C are suitable for the growth of plants as well as bacterial growth [11,23].

There are several aspects concerning microorganisms in enhancing the outcome of constructed wetlands through prolonging the survival of the microorganisms or improving bacterial viability and persistence in the wastewater environment. This can be divided into two main factors: the wetland environment and the characteristics of microorganisms. Survival and persistence of bacteria in the wetlands are affected by changes in soil pH, nutrients, electron acceptors, osmotic stress, temperature (cold or hot weather), UV exposure, and chemicals [117]. In addition to the wetland environment, the characteristics of the microorganisms play a significant role in their existence; for example, some bacteria form biofilms that protect themselves from external stresses. Another mechanism for bacteria to survive is to produce spores [118]. In the event of severe weather and nutrient deprivation, bacteria will die, eventually producing endospores which have a very hard shell and protect them [119]. Typically, under extremely poor living conditions, endospores are in a state of dormancy (sleeping condition); once the environmental conditions improve, the spore will germinate and outgrow [1].

Alga contributes to removal of contaminants in constructed wetlands. For instance, *Chlorella vulgaris* and *Scenedesmus obliquus* strains can remove selenium from river water when treated with a wetlands system by volatilization and accumulation [48]. Plants and microorganisms can contribute to the volatilization of selenium. The volatilization of selenium has been approved to be dependent on nutrients (N and P). Previous studies showed that the volatilization of selenium is more sensitive to increase in P than the increase in N [120].

5. Phytoremediation and Bacteria Interaction

Phytoremediation in constructed wetlands includes phytoextraction, phytovolatilization, phytodegradation, rhizodegradation, and phytostabilization [15]. Different plants, as well as phytoplankton, are used in wetlands [121]. The most common plants used in constructed wetlands are *P. australis* (35.29%), *T. latifolia* (17.65%), *J. effusus* (8.82%), *P. arundinacea* (7.84%), and *Z. latifolia* (5.88%). Other plants also are used in the wetlands [1,79,81]. Cattails (*Typha latifolua*) and reed (*Phragmites australis*) are common on most wetlands. Some of these plants are vascular and some are non-vascular plants. The main role of the plants in the wetlands is to provide sorption sites and exchange mediums for microorganisms and the pollutant [121]. The plants provide an organic carbon source for microorganisms and slow down the flow of water, therefore enhancing the sedimentation and pollutant settlement. Moreover, plants provide means for soil stabilization, and the roots increase the soil permeability. The plants are important elements in the aesthetic value of the wetland. Not all plants can tolerate high concentrations of wastewater or pollutants [122]. The selection is made based on which plant can grow better. For instance, *Acorus calamus* tolerates an acidic environment, *Iris versicolor* can tolerate high nutrient concentrations, *Scripus validus* is effective in contaminant removal, and *Scirpus pungens* can remove heavy metals [123].

The contaminant concentration in the wetlands can affect the growth of the plants. In some cases, the above-water parts of the plants are not affected by the contaminant, but the length of the roots is affected [4,5,24]. This can be attributed to the fact that the roots are in direct contact with the contaminated water. Depending on plant tolerance to the specific contaminant, the stem height, as well as the root length, can be affected with physiological toxicity at certain contaminant concentrations [73,121].

5.1. Conventional Contaminant Removal Mechanisms

The conventional processes in the constructed wetlands include physical, chemical, geochemical, and biological processes. Figure 4 shows the removal mechanisms of different wastewaters in CW. The biodegradation processes are initiated as the contaminated water flows through various components of the wetlands including substrate (gravel and/or sand), plants, and microorganisms [4,5,24]. The pollutants are removed/degraded by different mechanisms such as filtration and sedimentation of suspended particles [124]. Other mechanisms also contribute to the removal of pollutants such as sorption, photolysis, precipitation, volatilization, and plant uptake [124–126]. The mechanisms by which plants can degrade contaminants include metabolic and hydraulic processes. The plants grow rapidly and transfer the contaminant from the environment around the roots to the leaves. The indigenous microorganisms in the vicinity of the roots play a dominant role in the removal process. Microorganisms facilitate the following processes: phytotransformation, rhizosphere bioremediation, phytostabilization, phytoextraction, or rizofiltration [122].

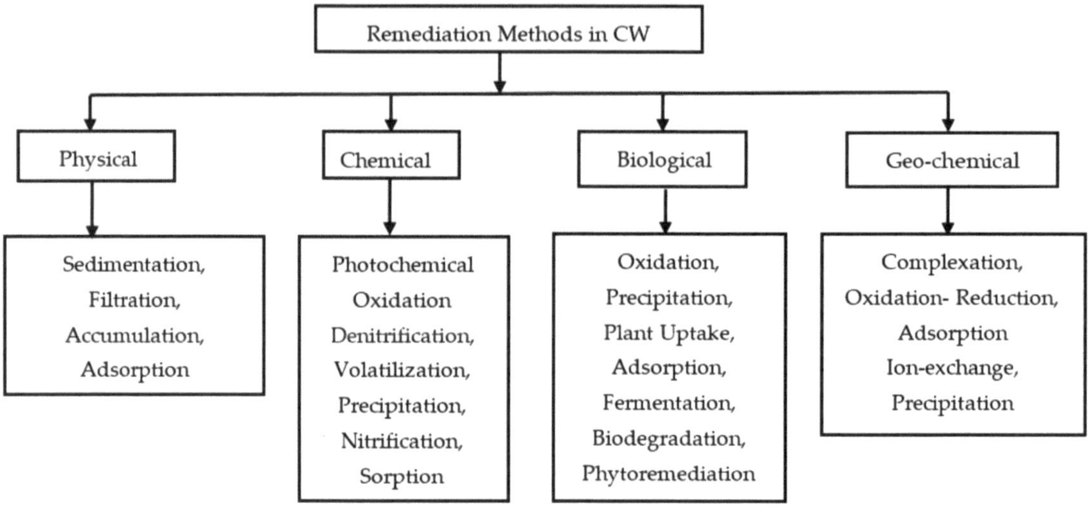

Figure 4. Remediation mechanisms in CW for different wastewaters.

Various plants show not only high potential in removing organic contaminants from wastewater but also transforming the contaminant into harmless by-products [122]. The pollutants are degraded due to the process of releasing certain enzymes, roots exudates, and also the accumulation of organic carbon in the soil. For example, rizofiltration is the process through which the dissolved heavy metals are removed from the water to the roots and the to the plant leaves [82]. On the other hand, organic matters are degraded/ removed from wastewaters through various mechanisms that contribute positively to decreasing the concentration of pollutants. Volatilization, photochemical oxidation, sedimentation, sorption, microbial degradation, fermentation, and aerobic and anaerobic respiration are mechanisms that are involved in the degradation of organic pollutants [76,82]. Nitrification denitrification, plant uptake, volatilization, and adsorption are the mechanisms through which the concentrations of various types of nitrogen, organic and inorganic, are reduced [112]. Phosphorous, a major component of the storm runoff and agriculture wastewater, is removed from wastewater through adsorption, precipitation, and plant uptake [112].

5.2. Kinetics of Biotreatment in CW

Researchers have investigated the use of various kinetics modeling approaches including first-order kinetics, the constant stirred tank reactor (CSTR), Monod kinetics, and Monod-CSTR kinetics to describe the degradation process in the CW. For instance, a previous study investigated the removal rates of BOD_5, COD, TPH, and NO_3-N from wastewater using CW planted with *Eichhornia crassipes* [29]. The results showed that the removal constant rates R^2 were around 0.89 to 0.99 for BOD_5, COD, NO_3-N, and TPH [127]. For BOD_5 removal, in vegetated VSF-CW, the estimated area-based first-order (K-C) and the first-order (K-C *) removal rate constants of BOD_5 (k_A) were 0.12 and 0.16 md^{-1}, respectively [29,52]. The area-based first-order (K-C)-CSTR removal rate constant is 0.57 m day^{-1} while the Monod-CSTR kinetics maximum removal rate (K_{max}) and multiple Monod-CSTR kinetics (K_{max}) were 3.27 and 3.47 gm^{-2} d^{-1}, respectively, in the vegetated VSF-CW. For the COD removal in the vegetated VSF-CW, the k_A (first-order K-C) and k_A (first-order K-C *) for COD were 0.07 and 0.11 $mday^{-1}$, respectively, while the k_A (first-order CSTR flow), K_{max} (Monod-CSTR), and K_{max} (multiple Monod-CSTR) were 0.21 m day^{-1}, 4.12 gm^{-2} day^{-1}, and 4.35 gm^{-2} day^{-1}, respectively. Previous studies showed that the first-order

CSTR model is not precise enough to describe the BOD_5 and NO_3-N removal. On the other hand, Monod CSTR model showed high match between the model and the test results [52].

The contaminant removal can be described by the mathematical model ($J = k.C$). The rate (k) is dependent on area of wetland, volume of flow, and temperature. The rate coefficient represents the contaminant degradation speed [52]. The relation between the degradation and constructed wetland inflow and out flow is described by the second equation in Table 5. The water flow in the constructed wetland cannot be described by Plug Flow (PF) nor Continuous Stirred Tank Reactor (CSTR). The best model found to describe the constructed wet lands is the Tank-in-Series (TIS) model, which it takes into consideration background concentration and the time effect on the process [127]. A single equation can be formed using a combination of non-ideal flow and background concentration to describe the contaminant removal in the wetlands [52]. This equation can be denoted by P-k-C^*, where P is the pollutant weathering, k is the first-order kinetic, and C^* is the non-zero background concentration [127]. Monod kinetics can be combined with CSTR to form a model describing the relation between the inlet and outlet concentrations through the half saturation constant of the limiting substrate and maximum pollutant rates. Multiple Monod kinetics, which assumes more than one substrate that limits the rate of contaminant degradation, can be used to express the reactions in the constructed wetlands. Table 5 below contains kinetics equations that can be used in the design of constructed wetlands [127].

Table 5. Kinetics equations that are used in CWs.

Equation	Parameters	Description
$J = k \cdot C$	J: Is the contaminant removal per unit area $g \cdot m^{-2} \cdot d^{-1}$ C: Contaminant concentration ($g \cdot m^{-3}$) k: Rate coefficient $m \cdot d^{-1}$	-
$\frac{C_{out}}{C_{in}} = e^{(-\frac{k_A}{q})}$	q: is the hydraulic loading rate (m/d) kA: the decomposition constant in $m \cdot d^{-1}$	First-order equation
$\frac{C_{out}}{C_{in}} = e^{(-k_v t)}$	Kv: Account for time $Cout$: effluent concentration Cin: influent concentration	First-order equation (uses the hydraulic residence time (HRT,t)
$(C - C^*)/(C_i - C^*) = e^{(-kt)}$	C: effluent concentration Ci: influent concentration C^*: background concentration	Combination of non-ideal flow and background concentration (Pollutant weathering, first-order kinetics and non-zero background concentration)
$\frac{C_{in}-C_{out}}{\tau} = -K_{max}\left(\frac{C_{out}}{C_{half}} + C_{out}\right)$	τ: hydraulic retention time $Chalf$: half saturation constant of limiting substrate $Kmax$: maximum pollutant removal rates	Combination of Continuous Stirred Tank Reactor (CSTR) and Monod kinetics
$K_3 = \frac{q(C_{in}-C_{out})(C_{out1}+C_{half1})(C_{out2}+C_{half2})}{C_{out1}C_{out2}}$	$K3$: maximum areal pollutant removal rate $K3$, $g \cdot m^{-2} \cdot d^{-1}$ $Chalf1$, $Chalf2$: half saturation constant of limiting substrates $Cout1$, $Cout2$: the outlet concentrations of limiting substrates	-

6. Economical Evaluation and Challenges in the Field

In general, there are variations in the construction cost of SSF CW compared to the construction cost of SF CW [2,4]. However, there are common items that should be considered in the calculation of the cost of both approaches. The cost can be divided into two stages: the acquisition and design stage and the construction and commissioning stage [5,19]. The first stage includes land acquisition, geotechnical investigation, and system design. The construction and commissioning stage includes site mobilization (site office, fence and access road), construction activities (plastic liners, filtration, media, vegetation, flow system, hydraulic structures), supervision fees, and commissioning (trial period) [24]. Special consideration should be taken in the cost regarding the availability of water and additional area if required.

The cost of CW is mainly dependent on the specification and the design components, which vary from one system to another. The costs of media, plant species (vegetation), hydraulic structures, pumps, and filtration systems are considered as the main design components that contribute to the majority of the total cost [19]. The selection of suitable components for the wastewater to be treated is crucial for the maintenance and operating costs. Economical CW systems should have low maintenance and operation cost. In most cases, the selection of remediation method is made based on the efficiency of the method (output/cost) [128]. The operation and maintenance costs include the personnel need to run the facility, the energy cost, facility infrastructure maintenance, plant replacement and disposal, cleaning, pumps and tools replacement, and maintenance quality control (sampling and analysis cost) [24].

The challenges facing constructed wetlands include methane production. It has been observed that an increase in the concentration of CH_4 in the air is associated with constructed wetlands used to treat sewer wastewater [126]. Methane is considered one of the important contributors (greenhouse gases) to ozone destruction. There are specific types of plants that when used in constructed wetlands produce high CH_4, for instance, *Z. latifolia*, *J. effusus*, and *P. australis*. On the other hand, some plants including *P. arundinacea*, *C. papyrus*, and *T. latifolia* produce less CH_4 in the air compared to the aforementioned types. Not only the types of plants affect the production of CH_4 but also the intensity and the combination of the plants have a significant effect on the emission of CH_4 [44]. Potential for development and challenges in biogas (e.g., CH_4) as an energy resource is one of the alternative options to reduce greenhouse gas emissions from the constructed wetland [129]. Wetland biomass as a cooking fuel can offset some of the demand for unsustainable and unhealthy solid fuels (such as charcoals). The potential to employ proven small-scale biogas technologies in CW can be utilized as an energy source in the community. Depending on the plant type selected, wetland biomass can yield energy for a 200-person community ranging from about 12–150 GJ/year [129]. Thus, biomass replacing wood can reduce stress on ecosystems. Moreover, biomass can also replace manure as a fuel source; this organic matter can be used as fertilizer in agricultural activities, thus offsetting some of the needs for artificial fertilizers [129].

7. Conclusions, Innovations, and New Directions

CWs have unique characteristics, as they are not only a simple and environmentally friendly treatment method but also continuously evolving to contribute to the global ecosystem. CWs are a home for native and migratory wildlife and create a balance between the land development and the need for green cover [130]. The new innovations and directions in the CW are as follows:

(I) Potential use of wetlands to sustain food source

Recent discussion in the scientific community and in engineering societies landed on some potential uses for CWs in the near future. CWs can be utilized in agriculture schemes. CWs can result in increasing fertility of the soil, creating a belt to protect the agriculture schemes from storm runoff, and providing shade from wind. It is highly recommended for the stakeholder representing governments to increase the uses of CWs.

The agricultural schemes use fertilizers (phosphate), and an excess amount of these fertilizers is usually flashed with storm runoff and contaminates the nearby water bodies. High phosphorus concentration promotes the growth of algae. Thus, covering the surface of water bodies results in low oxygen concentration and decreases the quality of water. The watershed conservation authorities in many parts of the world are responsible for management of the storm runoff, which in many cases is associated with agriculture wastewater. These governmental bodies strive to develop regulations to protect the public interest in water bodies. However, the framers are usually in conflict with these government bodies regarding the regulation and operation of the agricultural schemes. The runoff from agriculture schemes contributes negatively to the quality of the receiving water bodies. The negative effect in the receiving waters can occur as change in salinity or excess nutrients. Constructed wetlands can be used as a receiving water body that contributes to reducing the nutrients in the runoff and decreases the erosion effect. CWs can provide a shade for the agricultural schemes and minimize the adverse effect of the foods. Nevertheless, CWs host birds and can be used as a source of timber and wood.

(II) Wetlands as mitigation and adaptation to the climate change

In the last decade, the greenhouse gas effect in climate change has gained increasing interest in the scientific community. Constructed wetlands and natural wetlands can play a dominant role in reducing the adverse effect of the greenhouse gases. The plants can contribute and utilize the carbon, which is a major components of greenhouse gases. On the other hand, recent studies showed that certain microorganisms can be used in carbon fixation. CWs, as a part of the global ecosystem, can also be affected by the phenomena resulting from greenhouse gases such as sea level rise and the groundwater level. It has been recognized that the CWs can store large amount of carbon; however, CWs can release CO_2, CH_4, and NO when subjected to disturbance. Therefore, it is essential for a sustainable environment to keep the CWs running and increase the area covered with the CWs.

(III) Wetland Wildlife Considerations

The CWs support a diversity of microorganisms, birds, and plants. The environment in CWs is suitable for other wildlife species. Microorganisms can be introduced to the CWs or can form colonies naturally. The vegetation in the CWs plus the microorganism colonies form a suitable habitat for a wide range of species. Various types of plants and animals have the capability to grow in contaminated water, including air-breathing invertebrates such as water beetles, bugs, water snails, etc. These species can contribute to the ecological cycles.

(IV) Biomass and Renewable Energy Source

Constructed wetlands (CWs) are a biological technology, require optimum maintenance, and are effective in treating wastewater under the ambient environment. Biomass, as a source of renewable energy, harvested from CW can be used as a renewable energy source. The CW-treated effluent can also be used in the agricultural field. For example, plants such as *Phragmites* spp., *Typha* spp., A. donax, and *C. papyrus* used in CW can yield about 1500 g to 6000 g/m^2 of dry mass per square meter (g/m^2) [129]. Methane yields vary from about 170 to 360 L of methane per kilogram of dry mass (LN/kg). They further reported that an average of 110 MJ through direct combustion or 1660 L of methane from biogas can be generated from 1 m^2 of CW planted with A. donax. Thus, bioenergy production from CW can reduce reliance on fossil fuels and has the potential to offset energy and irrigation needs in different regions. The microbial fuel cell (MFC), which is used to generate energy from wastewater, is characterized by anaerobic conditions at the anode and aerobic environment at the vicinity of the anode. Similar conditions naturally exist in CW, where a high concentration of oxygen presents at the water surface and anoxic conditions at the bottom of the CW. Therefore, MFC can be integrated into CW to promote the use of CW as powerful tool for treatment of wastewater and generation of energy [131,132]. The basic concept of the MCF is the use of anerobic digestion of high-strength wastewater to produce biogas. The fact that CWs require low energy for operation and low maintenance costs, plus the integration of MFC, could make the CW the best choice for the authorities and policy makers [132,133]. From the current state, it is obvious that the technology has the potential to provide sustainable wastewater management as well as a bioenergy source without creating any burden on water resources.

(V) The removal of POPs and microplastics on CW

The current knowledge is limited about the accumulation and retention of microplastics in plant. There is a potential to advance the understanding of the interactions between micro- and nanoplastics on plants and macrophytes. The is a gap in knowledge that needs to be addressed including the discussion about plant exposure conditions, the uptake mechanisms, and the toxic effect of the microplastic on the plants. It was reported that 83–89% of MPs are removed using integrated vertical flow CW systems, and 27–60% are removed by surface flow constructed wetlands; however, more studies are required to address the gap of knowledge [100].

POPs such as polychlorinated biphenyls (PCBs) and polycyclic aromatic hydrocarbons (PAHs) can be removed in CW systems. The removal capacity by phytoremediation processes in CW were found to be between 66 to 100%. A recent study showed that five endocrine disrupting chemicals (EDCs) were removed with different percentages of 48% to 99% in HSF and VF CW systems [83]. The removal mechanisms include plant uptake, adsorption to the bottom sediments (in the SF system) or substrate of the beds (in the SSF systems), as well as adsorption to suspended solids, followed by sedimentation. Although POPs can be removed in CW, there are some limitations that need to be addressed including the effect of changes of temperature from season to season, and the uptake pathways need to be investigated to determine the contribution of each uptake mechanism.

Author Contributions: Conceptualization, I.H. and S.R.C.; methodology, I.H. and P.K.P.; formal analysis, S.A.R. and I.H.; investigation, I.H. and S.R.C.; resources, S.A.R.; data curation, I.H. and S.R.C.; writing—original draft preparation, I.H. and S.R.C.; writing—review and editing, I.H., S.R.C., P.K.P. and S.A.R.; visualization, P.K.P. and S.A.R.; supervision, S.A.R.; funding acquisition, S.A.R. All authors have read and agreed to the published version of the manuscript.

Funding: The support received from Interdisciplinary Research Center for Membranes and Water Security, King Fahd University of Petroleum & Minerals, Dhahran, Saudi Arabia, with funding grant and financial support for this work through project No. INMW2105.

Data Availability Statement: The data presented in this study are available in all tables and figures of this article.

Acknowledgments: Authors would like to acknowledge the support received from Interdisciplinary Research Center for Membranes and Water Security, King Fahd University of Petroleum & Minerals, Dhahran, Saudi Arabia, with funding grant and financial support for this work through project No. INMW2105. Authors highly appreciate for the continuous support and contribution received from the Department of Civil Engineering at Prince Mohammad Bin Fahd University and Taibah University during the collection and preparation of all necessary data and information for this research.

Conflicts of Interest: The authors declare no conflict of interest, and the funders had no role in the decision to publish the results.

References

1. Santos, M.; Melo, V.F.; Serrat, B.M.; Bonfleur, E.; Araújo, E.M.; Cherobim, V.F. Hybrid technologies for remediation of highly Pb contaminated soil: Sewage sludge application and phytoremediation. *Int. J. Phytoremediat.* **2021**, *23*, 328–335. [CrossRef]
2. Omondi, D.O.; Navalia, A.C. *Constructed Wetlands in Wastewater Treatment and Challenges of Emerging Resistant Genes Filtration and Reloading*; IntechOpen: London, UK, 2020.
3. Ji, Z.; Tang, W.; Pei, Y. Constructed wetland substrates: A review on development, function mechanisms, and application in contaminants removal. *Chemosphere* **2021**, *286*, 131564. [CrossRef]
4. Donde, O.O. Wastewater Management Techniques: A Review of Advancement on the Appropriate Wastewater Treatment Principles for Sustainability. *Environ. Manag. Sustain. Dev.* **2017**, *6*, 40–58. [CrossRef]
5. Thorslund, J.; Jarsjo, J.; Jaramillo, F.; Jawitz, J.W.; Manzoni, S.; Basu, N.B.; Chalov, S.R.; Cohen, M.J.; Creed, I.F.; Goldenberg, R.; et al. Wetlands as large-scale nature-based solutions: Status and challenges for research, engineering and management. *Ecol. Eng.* **2017**, *108*, 489–497. [CrossRef]
6. Mustafa, H.M.; Hayder, G. Recent studies on applications of aquatic weed plants in phytoremediation of wastewater: A review article. *Ain Shams Eng. J.* **2021**, *12*, 355–365. [CrossRef]
7. Can an Integrated Constructed Wetland in Norfolk Reduce Nutrient Concentrations and Promote In Situ Bird Species Richness? Available online: https://link.springer.com/article/10.1007/s13157-019-01247-7 (accessed on 21 September 2021).
8. Almuktar, S.A.A.N.; Abed, S.N.; Scholz, M. Wetlands for wastewater treatment and subsequent recycling of treated effluent: A review. *Environ. Sci. Pollut. Res.* **2018**, *25*, 23595–23623. [CrossRef]

9. Brovelli, A.; Carranza-Diaz, O.; Rossi, L.; Barry, D.A. Design methodology accounting for the effects of porous medium heterogeneity on hydraulic residence time and biodegradation in horizontal subsurface flow constructed wetlands. *Ecol. Eng.* **2011**, *37*, 758–770. [CrossRef]
10. Chen, J.; Liu, Y.-S.; Su, H.-C.; Ying, G.-G.; Liu, F.; Liu, S.-S.; He, L.-Y.; Chen, Z.-F.; Yang, Y.-Q.; Chen, F.-R. Removal of antibiotics and antibiotic resistance genes in rural wastewater by an integrated constructed wetland. *Environ. Sci. Pollut. Res.* **2015**, *22*, 1794–1803. [CrossRef]
11. Christofilopoulos, S.; Kaliakatsos, A.; Triantafyllou, K.; Gounaki, I.; Venieri, D.; Kalogerakis, N. Evaluation of a constructed wetland for wastewater treatment: Addressing emerging organic contaminants and antibiotic resistant bacteria. *New Biotechnol.* **2019**, *52*, 94–103. [CrossRef]
12. Vymazal, J.; Zhao, Y.; Mander, Ü. Recent research challenges in constructed wetlands for wastewater treatment: A review. *Ecol. Eng.* **2021**, *169*, 106318. [CrossRef]
13. Dan, T.H.; Quang, L.N.; Chiem, N.H.; Brix, H. Treatment of high-strength wastewater in tropical constructed wetlands planted with Sesbania sesban: Horizontal subsurface flow versus vertical downflow. *Ecol. Eng.* **2011**, *37*, 711–720. [CrossRef]
14. Padalkar, A.V. Rakesh Kumar Common Effluent Treatment Plant (CETP): Reliability Analysis and Performance Evaluation. *Water Sci. Eng.* **2018**, *11*, 205–213. [CrossRef]
15. Truu, J.; Truu, M.; Espenberg, M.; Nõlvak, H.; Juhanson, J. Phytoremediation and Plant-Assisted Bioremediation in Soil and Treatment Wetlands: A Review. *Open Biotechnol. J.* **2015**, *9*, 85–92. [CrossRef]
16. Barik, D. *Energy from Toxic Organic Waste for Heat and Power Generation*, 1st ed.; Woodhead Publishing: Sawston, UK, 2018. Available online: https://www.elsevier.com/books/energy-from-toxic-organic-waste-for-heat-and-power-generation/barik/978-0-08-102528-4 (accessed on 29 August 2021).
17. Azubuike, C.C.; Chikere, C.B.; Okpokwasili, G.C. Bioremediation techniques–classification based on site of application: Principles, advantages, limitations and prospects. *World J. Microbiol. Biotechnol.* **2016**, *32*, 1–18. [CrossRef] [PubMed]
18. Bruch, I.; Fritsche, J.; Bänninger, D.; Alewell, U.; Sendelov, M.; Hürlimann, H.; Hasselbach, R.; Alewell, C. Improving the treatment efficiency of constructed wetlands with zeolite-containing filter sands. *Bioresour. Technol.* **2011**, *102*, 937–941. [CrossRef]
19. Pajares, E.M.; Valero, L.G.; Sánchez, I.M.R. Cost of Urban Wastewater Treatment and Ecotaxes: Evidence from Municipalities in Southern Europe. *Water* **2019**, *11*, 423. [CrossRef]
20. Rezania, S.; Park, J.; Rupani, P.F.; Darajeh, N.; Xu, X.; Shahrokhishahraki, R. Phytoremediation potential and control of Phragmites australis as a green phytomass: An overview. *Environ. Sci. Pollut. Res.* **2019**, *26*, 7428–7441. [CrossRef] [PubMed]
21. Ortega-Calvo, J.-J.; Harmsen, J.; Parsons, J.R.; Semple, K.T.; Aitken, M.D.; Ajao, C.; Eadsforth, C.; Galay-Burgos, M.; Naidu, R.; Oliver, R.; et al. From Bioavailability Science to Regulation of Organic Chemicals. *Environ. Sci. Technol.* **2015**, *49*, 10255–10264. [CrossRef] [PubMed]
22. Biosurfactants-Types, Sources and Applications. Available online: https://scialert.net/abstract/?doi=jm.2015.181.192 (accessed on 29 August 2021).
23. Nguyen, P.M.; Afzal, M.; Ullah, I.; Shahid, N.; Baqar, M.; Arslan, M. Removal of pharmaceuticals and personal care products using constructed wetlands: Effective plant-bacteria synergism may enhance degradation efficiency. *Environ. Sci. Pollut. Res.* **2019**, *26*, 21109–21126. [CrossRef]
24. Vymazal, J. Constructed Wetlands for Wastewater Treatment: Five Decades of Experience. *Environ. Sci. Technol.* **2011**, *45*, 61–69. [CrossRef]
25. Saeed, T.; Sun, G. A review on nitrogen and organics removal mechanisms in subsurface flow constructed wetlands: Dependency on environmental parameters, operating conditions and supporting media. *J. Environ. Manag.* **2012**, *112*, 429–448. [CrossRef]
26. Boog, J.; Nivala, J.; Aubron, T.; Wallace, S.; Sullivan, C.; Van Afferden, M.; Müller, R.A. Treatment Wetland Aeration without Electricity? Lessons Learned from the First Experiment Using a Wind-Driven Air Pump. *Water* **2016**, *8*, 502. [CrossRef]
27. Gunter, L.; Gabriela, D.; Jaime, N.; Anacleto, R.S. *Wetland Technology: Practical Information on the Design and Application of Treatment Wetlands*; Langergraber, G., Dotro, G., Nivala, J., Rizzo, A., Stein, O.R., Eds.; IWA Publishing: London, UK, 2020.
28. Adewuyi, G.O.; Olowu, R. Assessment of Oil and Grease, Total Petroleum Hydrocarbons and Some Heavy Metals in Surface and Groundwater within the Vicinity of NNPC Oil Depot in Apata, Ibadan Metropolis, Nigeria. *Int. J. Res. Rev. Appl. Sci.* **2012**, *13*, 166–174.
29. Agarry, S.E.; Oghenejoboh, K.M.; Latinwo, G.K.; Owabor, C.N. Biotreatment of petroleum refinery wastewater in vertical surface-flow constructed wetland vegetated with Eichhornia crassipes: Lab-scale experimental and kinetic modelling. *Environ. Technol.* **2018**, *41*, 1793–1813. [CrossRef]
30. Perdana, M.C.; Sutanto, H.B.; Prihatmo, G. Vertical Subsurface Flow (VSSF) constructed wetland for domestic wastewater treatment. *IOP Conf. Ser. Earth Environ. Sci.* **2018**, *148*, 012025. [CrossRef]
31. Zhao, Y.J.; Hui, Z.; Chao, X.; Nie, E.; Li, H.J.; He, J.; Zheng, Z. Efficiency of two-stage combinations of subsurface vertical down-flow and up-flow constructed wetland systems for treating variation in influent C/N ratios of domestic wastewater. *Ecol. Eng.* **2011**, *37*, 1546–1554. [CrossRef]
32. Nivala, J.; Headley, T.; Wallace, S.; Bernhard, K.; Brix, H.; van Afferden, M.; Müller, R.A. Comparative analysis of constructed wetlands: The design and construction of the ecotechnology research facility in Langenreichenbach, Germany. *Ecol. Eng.* **2013**, *61*, 527–543. [CrossRef]

33. Hoffmann, A.A.; Montgomery, B.L.; Popovici, J.; Iturbeormaetxe, I.; Johnson, P.H.; Muzzi, F.; Greenfield, M.; Durkan, M.; Leong, Y.S.; Dong, Y.; et al. Successful establishment of Wolbachia in Aedes populations to suppress dengue transmission. *Nature* **2011**, *476*, 454–457. [CrossRef]
34. Hijosa-Valsero, M.; Matamoros, V.; Sidrach-Cardona, R.; Martín-Villacorta, J.; Bécares, E.; Bayona, J.M. Comprehensive assessment of the design configuration of constructed wetlands for the removal of pharmaceuticals and personal care products from urban wastewaters. *Water Res.* **2010**, *44*, 3669–3678. [CrossRef] [PubMed]
35. Critical Design Parameters for Constructed Wetlands Natural Wastewater Treatment Systems. Available online: https://www.natsci.upit.ro/issues/2017/volume-6-issue-12/critical-design-parameters-for-constructed-wetlands-natural-wastewater-treatment-systems/ (accessed on 2 October 2021).
36. Constructed Wetlands Treatment of Municipal Wastewaters. Document Display NEPIS US EPA. Available online: https://nepis.epa.gov/Exe/ZyNET.exe/200053ND.TXT?ZyActionD=ZyDocument&Client=EPA&Index=2000+Thru+2005&Docs=&Query=&Time=&EndTime=&SearchMethod=1&TocRestrict=n&Toc=&TocEntry=&QField=&QFieldYear=&QFieldMonth=&QFieldDay=&IntQFieldOp=0&ExtQFieldOp=0&XmlQuery=&File=D%3A%5Czyfiles%5CIndex%20Data%5C00thru05%5CTxt%5C00000001%5C200053ND.txt&User=ANONYMOUS&Password=anonymous&SortMethod=h%7C-&MaximumDocument 1&FuzzyDegree=0&ImageQuality=r75g8/r75g8/x150y150g16/i425&Display=hpfr&DefSeekPage=x&SearchBack=ZyActionL&Back=ZyActionS&BackDesc=Results%20page&MaximumPages=1&ZyEntry=1&SeekPage=x&ZyPURL (accessed on 22 September 2021).
37. Ávila, C.; Nivala, J.; Olsson, L.; Kassa, K.; Headley, T.; Mueller, R.A.; Bayona, J.M.; García, J. Emerging organic contaminants in vertical subsurface flow constructed wetlands: Influence of media size, loading frequency and use of active aeration. *Sci. Total Environ.* **2014**, *494–495*, 211–217. [CrossRef] [PubMed]
38. Baptestini, G.C.F.; Matos, A.T.; Martinez, M.A.; Borges, A.C. Hydraulic Conductivity Variability in Horizontal Subsurface Flow Constructed Wetlands. *Eng. Agríc.* **2017**, *37*, 333–342. [CrossRef]
39. Gikas, G.D.; Tsihrintzis, V.A. A small-size vertical flow constructed wetland for on-site treatment of household wastewater. *Ecol. Eng.* **2012**, *44*, 337–343. [CrossRef]
40. Gikas, G.D.; Tsihrintzis, V.A. On-site treatment of domestic wastewater using a small-scale horizontal subsurface flow constructed wetland. *Water Sci. Technol.* **2010**, *62*, 603–614. [CrossRef]
41. Al-Baldawi, I.; Abdullah, S.R.S.; Anuar, N.; Suja, F.; Idris, M. Performance assessment of pilot horizontal sub-surface flow constructed wetlands for removal of diesel from wastewater by Scirpus grossus. *Water Sci. Technol.* **2013**, *68*, 2271–2278. [CrossRef]
42. Bakhshoodeh, R.; Alavi, N.; Majlesi, M.; Paydary, P. Compost leachate treatment by a pilot-scale subsurface horizontal flow constructed wetland. *Ecol. Eng.* **2017**, *105*, 7–14. [CrossRef]
43. Ávila, C.; Pedescoll, A.; Matamoros, V.; Bayona, J.M.; García, J. Capacity of a horizontal subsurface flow constructed wetland system for the removal of emerging pollutants: An injection experiment. *Chemosphere* **2010**, *81*, 1137–1142. [CrossRef]
44. Wu, H.; Zhang, J.; Ngo, H.H.; Guo, W.; Hu, Z.; Liang, S.; Fan, J.; Liu, H. A review on the sustainability of constructed wetlands for wastewater treatment: Design and operation. *Bioresour. Technol.* **2015**, *175*, 594–601. [CrossRef] [PubMed]
45. Beharrell, M. Chapter 19—Operation and Maintenance for Constructed Wetlands. In *Wetlands Ecosystems in Asia*; Wong, M.H., Ed.; Developments in Ecosystems; Elsevier: Amsterdam, The Netherlands, 2004; Volume 1, pp. 347–359.
46. Gunes, K.; Tuncsiper, B.; Masi, F.; Ayaz, S.; Leszczynska, D.; Hecan, N.F.; Ahmad, H. Construction and maintenance cost analyzing of constructed wetland systems. *Water Pract. Technol.* **2011**, *6*, wpt2011043. [CrossRef]
47. Bouali, M.; Zrafi, I.; Mouna, F.; Bakhrouf, A. Pilot study of constructed wetlands for tertiary wastewater treatment using duckweed and immobilized microalgae. *Afr. J. Microbiol. Res.* **2012**, *6*, 6066–6074. [CrossRef]
48. Huang, J.-C.; Suárez, M.C.; Yang, S.I.; Lin, Z.-Q.; Terry, N. Development of a Constructed Wetland Water Treatment System for Selenium Removal: Incorporation of an Algal Treatment Component. *Environ. Sci. Technol.* **2013**, *47*, 10518–10525. [CrossRef] [PubMed]
49. Lim, P.E.; Tay, M.G.; Mak, K.Y.; Mohamed, N. The effect of heavy metals on nitrogen and oxygen demand removal in constructed wetlands. *Sci. Total Environ.* **2003**, *301*, 13–21. [CrossRef]
50. Zhi, W.; Yuan, L.; Ji, G.; He, C. Enhanced Long-Term Nitrogen Removal and Its Quantitative Molecular Mechanism in Tidal Flow Constructed Wetlands. *Environ. Sci. Technol.* **2015**, *49*, 4575–4583. [CrossRef] [PubMed]
51. Schmidt, I.; Sliekers, O.; Schmid, M.; Cirpus, I.; Strous, M.; Bock, E.; Kuenen, J.G.; Jetten, M.S.M. Aerobic and anaerobic ammonia oxidizing bacteria â€" competitors or natural partners? *FEMS Microbiol. Ecol.* **2002**, *39*, 175–181. [CrossRef] [PubMed]
52. Saeed, T.; Sun, G. Kinetic modelling of nitrogen and organics removal in vertical and horizontal flow wetlands. *Water Res.* **2011**, *45*, 3137–3152. [CrossRef]
53. Ong, S.-A.; Uchiyama, K.; Inadama, D.; Ishida, Y.; Yamagiwa, K. Performance evaluation of laboratory scale up-flow constructed wetlands with different designs and emergent plants. *Bioresour. Technol.* **2010**, *101*, 7239–7244. [CrossRef]
54. Lavrova, S.; Koumanova, B. Influence of recirculation in a lab-scale vertical flow constructed wetland on the treatment efficiency of landfill leachate. *Bioresour. Technol.* **2010**, *101*, 1756–1761. [CrossRef]
55. Tee, H.-C.; Lim, P.-E.; Seng, C.-E.; Nawi, M.-A.M. Newly developed baffled subsurface-flow constructed wetland for the enhancement of nitrogen removal. *Bioresour. Technol.* **2012**, *104*, 235–242. [CrossRef]

56. Kumar, S.; Dutta, V. Efficiency of Constructed Wetland Microcosms (CWMs) for the Treatment of Domestic Wastewater Using Aquatic Macrophytes. In *Environmental Biotechnology: For Sustainable Future*; Sobti, R.C., Arora, N.K., Kothari, R., Eds.; Springer: Singapore, 2019; pp. 287–307.
57. Rezania, S.; Kamyab, H.; Rupani, P.F.; Park, J.; Nawrot, N.; Wojciechowska, E.; Yadav, K.K.; Ghahroud, M.L.; Mohammadi, A.A.; Thirugnana, S.T.; et al. Recent advances on the removal of phosphorus in aquatic plant-based systems. *Environ. Technol. Innov.* **2021**, *24*, 101933. [CrossRef]
58. Liu, M.; Li, X.; He, Y.; Li, H. Aquatic toxicity of heavy metal-containing wastewater effluent treated using vertical flow constructed wetlands. *Sci. Total Environ.* **2020**, *727*, 138616. [CrossRef] [PubMed]
59. Zhang, C.; Yu, Z.; Zeng, G.; Jiang, M.; Yang, Z.; Cui, F.; Zhu, M.; Shen, L.; Hu, L. Effects of sediment geochemical properties on heavy metal bioavailability. *Environ. Int.* **2014**, *73*, 270–281. [CrossRef]
60. Lesage, E.; Mundia, C.; Rousseau, D.P.L.; Van De Moortel, A.M.K.; Du Laing, G.; Meers, E.; Tack, F.M.G.; De Pauw, N.; Verloo, M.G. Sorption of Co, Cu, Ni and Zn from industrial effluents by the submerged aquatic macrophyte *Myriophyllum spicatum* L. *Ecol. Eng.* **2007**, *30*, 320–325. [CrossRef]
61. Removal of Metals in Constructed Wetlands: Review. Practice Periodical of Hazardous, Toxic, and Radioactive Waste Management. Available online: https://ascelibrary.org/doi/abs/10.1061/%28ASCE%291090-025X%282008%2912%3A2%2896%29 (accessed on 21 September 2021).
62. Hadad, H.R.; Maine, M.A.; Bonetto, C.A. Macrophyte growth in a pilot-scale constructed wetland for industrial wastewater treatment. *Chemosphere* **2006**, *63*, 1744–1753. [CrossRef]
63. Huang, Y.-C.; Rao, A.; Huang, S.-J.; Chang, C.-Y.; Drechsler, M.; Knaus, J.; Chan, J.C.C.; Raiteri, P.; Gale, J.D.; Gebauer, D. Uncovering the Role of Bicarbonate in Calcium Carbonate Formation at Near-Neutral pH. *Angew. Chem. Int. Ed.* **2021**, *60*, 16707–16713. [CrossRef] [PubMed]
64. Stoltz, E.; Greger, M. Influences of wetland plants on weathered acidic mine tailings. *Environ. Pollut.* **2006**, *144*, 689–694. [CrossRef]
65. Calheiros, C.S.C.; Rangel, A.O.S.S.; Castro, P.M.L. Constructed wetland systems vegetated with different plants applied to the treatment of tannery wastewater. *Water Res.* **2007**, *41*, 1790–1798. [CrossRef]
66. Stefanakis, A.; Akratos, C.S.; Tsihrintzis, V.A. *Vertical Flow Constructed Wetlands—Eco-Engineering Systems for Wastewater and Sludge Treatment*; Elsevier: Amsterdam, The Netherlands, 2014; p. 392.
67. Stefanakis, A.I.; Prigent, S.; Breuer, R. Integrated Produced Water Management in a Desert Oilfield Using Wetland Technology and Innovative Reuse Practices. In *Constructed Wetlands for Industrial Wastewater Treatment*; John Wiley & Sons, Ltd.: Hoboken, NJ, USA, 2018; pp. 23–42.
68. Biological Treatment of Petrochemical Wastewater. Available online: https://www.intechopen.com/chapters/62888 (accessed on 22 September 2021).
69. Stefanakis, A.; Akratos, C.S.; Tsihrintzis, V.A. Chapter 3—VFCW Types. In *Vertical Flow Constructed Wetlands*; Stefanakis, A., Akratos, C.S., Tsihrintzis, V.A., Eds.; Elsevier: Boston, MA, USA, 2014; pp. 27–38.
70. Jain, M.; Majumder, A.; Ghosal, P.S.; Gupta, A.K. A review on treatment of petroleum refinery and petrochemical plant wastewater: A special emphasis on constructed wetlands. *J. Environ. Manag.* **2020**, *272*, 111057. [CrossRef] [PubMed]
71. Mustapha, H.I. *Treatment of Petroleum Refinery Wastewater with Constructed Wetlands*; CRC Press: London, UK, 2018; ISBN 978-0-429-45092-1.
72. Saien, J.; Shahrezaei, F. Organic Pollutants Removal from Petroleum Refinery Wastewater with Nanotitania Photocatalyst and UV Light Emission. *Int. J. Photoenergy* **2012**, *2012*, e703074. [CrossRef]
73. Khan, S.; Afzal, M.; Iqbal, S.; Khan, Q.M. Plant-bacteria partnerships for the remediation of hydrocarbon contaminated soils. *Chemosphere* **2013**, *90*, 1317–1332. [CrossRef] [PubMed]
74. Qasaimeh, A.; AlSharie, H.; Masoud, T. A Review on Constructed Wetlands Components and Heavy Metal Removal from Wastewater. *J. Environ. Prot.* **2015**, *6*, 710–718. [CrossRef]
75. Shahin, S.A.; Mossad, M.; Fouad, M. Evaluation of copper removal efficiency using water treatment sludge. *Water Sci. Eng.* **2019**, *12*, 37–44. [CrossRef]
76. Haberl, H.; Wackernagel, M.; Krausmann, F.; Erb, K.-H.; Monfreda, C. Ecological footprints and human appropriation of net primary production: A comparison. *Land Use Policy* **2004**, *21*, 279–288. [CrossRef]
77. Aslam, M.M.; Malik, M.; Baig, M.A.; Qazi, I.A.; Iqbal, J. Treatment performances of compost-based and gravel-based vertical flow wetlands operated identically for refinery wastewater treatment in Pakistan. *Ecol. Eng.* **2007**, *30*, 34–42. [CrossRef]
78. Tony, M.A.; Purcell, P.J.; Zhao, Y. Oil refinery wastewater treatment using physicochemical, Fenton and Photo-Fenton oxidation processes. *J. Environ. Sci. Health Part A* **2012**, *47*, 435–440. [CrossRef]
79. Carvalho, P.N.; Basto, M.C.P.; Almeida, C.M.R.; Brix, H. A review of plant-pharmaceutical interactions: From uptake and effects in crop plants to phytoremediation in constructed wetlands. *Environ. Sci. Pollut. Res.* **2014**, *21*, 11729–11763. [CrossRef]
80. Verlicchi, P.; Zambello, E. How efficient are constructed wetlands in removing pharmaceuticals from untreated and treated urban wastewaters? A review. *Sci. Total Environ.* **2014**, *470–471*, 1281–1306. [CrossRef]
81. Ferreira, A.R.; Ribeiro, A.; Couto, N. Remediation of Pharmaceutical and Personal Care Products (PPCPs) in Constructed Wetlands: Applicability and New Perspectives. In *Phytoremediation: Management of Environmental Contaminants*; Ansari, A.A., Gill, S.S., Lanza, G.R., Newman, L., Eds.; Springer International Publishing: Cham, Switzerland, 2017; Volume 5, pp. 277–292.

82. Zhang, D.Q.; Hua, T.; Gersberg, R.M.; Zhu, J.; Ng, W.J.; Tan, S.K. Carbamazepine and naproxen: Fate in wetland mesocosms planted with Scirpus validus. *Chemosphere* **2013**, *91*, 14–21. [CrossRef]
83. Papaevangelou, V.A.; Gikas, G.D.; Tsihrintzis, V.A.; Antonopoulou, M.; Konstantinou, I.K. Removal of Endocrine Disrupting Chemicals in HSF and VF pilot-scale constructed wetlands. *Chem. Eng. J.* **2016**, *294*, 146–156. [CrossRef]
84. Wojciechowska, E. Removal of persistent organic pollutants from landfill leachates treated in three constructed wetland systems. *Water Sci. Technol.* **2013**, *68*, 1164–1172. [CrossRef] [PubMed]
85. Polińska, W.; Kotowska, U.; Kiejza, D.; Karpińska, J. Insights into the Use of Phytoremediation Processes for the Removal of Organic Micropollutants from Water and Wastewater; A Review. *Water* **2021**, *13*, 2065. [CrossRef]
86. Nan, X.; Lavrnić, S.; Toscano, A. Potential of constructed wetland treatment systems for agricultural wastewater reuse under the EU framework. *J. Environ. Manag.* **2020**, *275*, 111219. [CrossRef]
87. Wang, M.; Zhang, D.; Dong, J.; Tan, S.K. Application of constructed wetlands for treating agricultural runoff and agro-industrial wastewater: A review. *Hydrobiologia* **2018**, *805*, 1–31. [CrossRef]
88. Cronk, J.K. Constructed wetlands to treat wastewater from dairy and swine operations: A review. *Agric. Ecosyst. Environ.* **1996**, *58*, 97–114. [CrossRef]
89. Ignatowicz, K. Removal of Pesticides from Wastewater by the Use of Constructed Wetlands. *J. Ecol. Eng.* **2020**, *21*, 219–223. [CrossRef]
90. Papaevangelou, V.A.; Gikas, G.D.; Vryzas, Z.; Tsihrintzis, V.A. Treatment of agricultural equipment rinsing water containing a fungicide in pilot-scale horizontal subsurface flow constructed wetlands. *Ecol. Eng.* **2017**, *101*, 193–200. [CrossRef]
91. Gikas, G.D.; Pérez-Villanueva, M.E.; Tsioras, M.; Alexoudis, C.; Pérez-Rojas, G.; Masís-Mora, M.; Lizano-Fallas, V.; Rodríguez-Rodríguez, C.E.; Vryzas, Z.; Tsihrintzis, V.A. Low-cost approaches for the removal of terbuthylazine from agricultural wastewater: Constructed wetlands and biopurification system. *Chem. Eng. J.* **2018**, *335*, 647–656. [CrossRef]
92. Parlakidis, P.; Mavropoulos, T.; Vryzas, Z.; Gikas, G.D. Fluopyram removal from agricultural equipment rinsing water using HSF pilot-scale constructed wetlands. *Environ. Sci. Pollut. Res.* **2021**, 1–13. [CrossRef]
93. Pat-Espadas, A.M.; Portales, R.L.; Amabilis-Sosa, L.E.; Gómez, G.; Vidal, G. Review of Constructed Wetlands for Acid Mine Drainage Treatment. *Water* **2018**, *10*, 1685. [CrossRef]
94. Kumar, R.; Sharma, P.; Bandyopadhyay, S. Evidence of microplastics in wetlands: Extraction and quantification in Freshwater and coastal ecosystems. *J. Water Process. Eng.* **2021**, *40*, 101966. [CrossRef]
95. Horton, A.A.; Walton, A.; Spurgeon, D.J.; Lahive, E.; Svendsen, C. Microplastics in freshwater and terrestrial environments: Evaluating the current understanding to identify the knowledge gaps and future research priorities. *Sci. Total Environ.* **2017**, *586*, 127–141. [CrossRef] [PubMed]
96. Duis, K.; Coors, A. Microplastics in the aquatic and terrestrial environment: Sources (with a specific focus on personal care products), fate and effects. *Environ. Sci. Eur.* **2016**, *28*, 1–25. [CrossRef] [PubMed]
97. Helcoski, R.; Yonkos, L.T.; Sanchez, A.; Baldwin, A.H. Wetland soil microplastics are negatively related to vegetation cover and stem density. *Environ. Pollut.* **2020**, *256*, 113391. [CrossRef]
98. Chen, Y.; Li, T.; Hu, H.; Ao, H.; Xiong, X.; Shi, H.; Wu, C. Transport and fate of microplastics in constructed wetlands: A microcosm study. *J. Hazard. Mater.* **2021**, *415*, 125615. [CrossRef] [PubMed]
99. Wang, Q.; Hernández-Crespo, C.; Santoni, M.; Van Hulle, S.; Rousseau, D.P.L. Horizontal subsurface flow constructed wetlands as tertiary treatment: Can they be an efficient barrier for microplastics pollution? *Sci. Total Environ.* **2020**, *721*, 137785. [CrossRef] [PubMed]
100. Zhou, X.; Zhao, Y.; Pang, G.; Jia, X.; Song, Y.; Guo, A.; Wang, A.; Zhang, S.; Ji, M. Microplastic abundance, characteristics and removal in large-scale multi-stage constructed wetlands for effluent polishing in northern China. *Chem. Eng. J.* **2021**, *430*, 132752. [CrossRef]
101. Zou, Y.; Ye, C.; Pan, Y. Abundance and characteristics of microplastics in municipal wastewater treatment plant effluent: A case study of Guangzhou, China. *Environ. Sci. Pollut. Res.* **2021**, *28*, 11572–11585. [CrossRef] [PubMed]
102. Long, Z.; Pan, Z.; Wang, W.; Ren, J.; Yu, X.; Lin, L.; Lin, H.; Chen, H.; Jin, X. Microplastic abundance, characteristics, and removal in wastewater treatment plants in a coastal city of China. *Water Res.* **2019**, *155*, 255–265. [CrossRef]
103. Mateos-Cárdenas, A.; van Pelt, F.N.A.M.; O'Halloran, J.; Jansen, M.A.K. Adsorption, uptake and toxicity of micro- and nanoplastics: Effects on terrestrial plants and aquatic macrophytes. *Environ. Pollut.* **2021**, *284*, 117183. [CrossRef] [PubMed]
104. Li, J.; Zhang, H.; Zhang, K.; Yang, R.; Li, R.; Li, Y. Characterization, source, and retention of microplastic in sandy beaches and mangrove wetlands of the Qinzhou Bay, China. *Mar. Pollut. Bull.* **2018**, *136*, 401–406. [CrossRef]
105. Nor, N.H.M.; Obbard, J.P. Microplastics in Singapore's coastal mangrove ecosystems. *Mar. Pollut. Bull.* **2014**, *79*, 278–283. [CrossRef]
106. Li, R.; Yu, L.; Chai, M.; Wu, H.; Zhu, X. The distribution, characteristics and ecological risks of microplastics in the mangroves of Southern China. *Sci. Total Environ.* **2020**, *708*, 135025. [CrossRef] [PubMed]
107. Yin, L.; Wen, X.; Huang, D.; Zeng, G.; Deng, R.; Liu, R.; Zhou, Z.; Tao, J.; Xiao, R.; Pan, H. Microplastics retention by reeds in freshwater environment. *Sci. Total Environ.* **2021**, *790*, 148200. [CrossRef] [PubMed]
108. Lozano, Y.M.; Rillig, M.C. Effects of Microplastic Fibers and Drought on Plant Communities. *Environ. Sci. Technol.* **2020**, *54*, 6166–6173. [CrossRef]

109. Pflugmacher, S.; Sulek, A.; Mader, H.; Heo, J.; Noh, J.H.; Penttinen, O.-P.; Kim, Y.; Kim, S.; Esterhuizen, M. The Influence of New and Artificial Aged Microplastic and Leachates on the Germination of *Lepidium sativum* L. *Plants* **2020**, *9*, 339. [CrossRef]
110. Mohammad, B.T.; Al Daghistani, H.I.; Jaouani, A.; Abdel-Latif, S.; Kennes, C. Isolation and Characterization of Thermophilic Bacteria from Jordanian Hot Springs: Bacillus licheniformis and Thermomonas hydrothermalis Isolates as Potential Producers of Thermostable Enzymes. *Int. J. Microbiol.* **2017**, *2017*, e6943952. [CrossRef] [PubMed]
111. Stefanakis, A.I. Constructed Wetlands for Sustainable Wastewater Treatment in Hot and Arid Climates: Opportunities, Challenges and Case Studies in the Middle East. *Water* **2020**, *12*, 1665. [CrossRef]
112. Rani, M.U.; Appaiah, A. Optimization of culture conditions for bacterial cellulose production from Gluconacetobacter hansenii UAC09. *Ann. Microbiol.* **2011**, *61*, 781–787. [CrossRef]
113. Tang, M.; Zhang, F.; Yao, S.; Liu, Y.; Chen, J. Application ofPseudomonas flavaWD-3 for sewage treatment in constructed wetland in winter. *Environ. Technol.* **2014**, *36*, 1205–1211. [CrossRef]
114. Chen, J.; Xu, J.; Zhang, S.; Liu, F.; Peng, J.; Peng, Y.; Wu, J. Nitrogen removal characteristics of a novel heterotrophic nitrification and aerobic denitrification bacteria, Alcaligenes faecalis strain WT14. *J. Environ. Manag.* **2021**, *282*, 111961. [CrossRef]
115. Wang, P.; Zhang, H.; Zuo, J.; Zhao, D.; Zou, X.; Zhu, Z.; Jeelani, N.; Leng, X.; An, S. A Hardy Plant Facilitates Nitrogen Removal via Microbial Communities in Subsurface Flow Constructed Wetlands in Winter. *Sci. Rep.* **2016**, *6*, 33600. [CrossRef] [PubMed]
116. Wang, L.; Pang, Q.; Peng, F.; Zhang, A.; Zhou, Y.; Lian, J.; Zhang, Y.; Yang, F.; Zhu, Y.; Ding, C.; et al. Response Characteristics of Nitrifying Bacteria and Archaea Community Involved in Nitrogen Removal and Bioelectricity Generation in Integrated Tidal Flow Constructed Wetland-Microbial Fuel Cell. *Front. Microbiol.* **2020**, *11*, 1385. [CrossRef]
117. Locatelli, A.; Spor, A.; Jolivet, C.; Piveteau, P.; Hartmann, A. Biotic and Abiotic Soil Properties Influence Survival of Listeria monocytogenes in Soil. *PLoS ONE* **2013**, *8*, e75969. [CrossRef]
118. Maldonado, L.A.; Quintana, E.T. Unexpected Properties of Micromonosporae from Marine Origin. *Adv. Microbiol.* **2015**, *05*, 452–456. [CrossRef]
119. McKenney, P.T.; Driks, A.; Eichenberger, P. The Bacillus subtilis endospore: Assembly and functions of the multilayered coat. *Nat. Rev. Microbiol.* **2013**, *11*, 33–44. [CrossRef] [PubMed]
120. Gupta, M.; Gupta, S. An Overview of Selenium Uptake, Metabolism, and Toxicity in Plants. *Front. Plant Sci.* **2017**, *7*, 2074. [CrossRef]
121. Wang, J.; Liu, X.; Zhang, X.; Liang, X.; Zhang, W. Growth Response and Phytoremediation Ability of Reed for Diesel Contaminant. *Procedia Environ. Sci.* **2011**, *8*, 68–74. [CrossRef]
122. Zhang, B.Y.; Zheng, J.S.; Sharp, R.G. Phytoremediation in Engineered Wetlands: Mechanisms and Applications. *Procedia Environ. Sci.* **2010**, *2*, 1315–1325. [CrossRef]
123. Moshiri, G.A. *Constructed Wetlands for Water Quality Improvement*; CRC Press: Boca Raton, FL, USA, 2020.
124. Shelef, O.; Gross, A.; Rachmilevitch, S. Role of Plants in a Constructed Wetland: Current and New Perspectives. *Water* **2013**, *5*, 405–419. [CrossRef]
125. Zhang, C.-B.; Liu, W.-L.; Wang, J.; Ge, Y.; Ge, Y.; Chang, S.X.; Chang, J. Effects of monocot and dicot types and species richness in mesocosm constructed wetlands on removal of pollutants from wastewater. *Bioresour. Technol.* **2011**, *102*, 10260–10265. [CrossRef] [PubMed]
126. Vymazal, J.; Březinová, T. Accumulation of heavy metals in aboveground biomass of Phragmites australis in horizontal flow constructed wetlands for wastewater treatment: A review. *Chem. Eng. J.* **2016**, *290*, 232–242. [CrossRef]
127. Gajewska, M.; Skrzypiec, K. Kinetics of nitrogen removal processes in constructed wetlands. *E3S Web Conf.* **2018**, *26*, 1. [CrossRef]
128. Allen, E.W. Process water treatment in Canada's oil sands industry: I. Target pollutants and treatment objectives. *J. Environ. Eng. Sci.* **2008**, *7*, 123–138. [CrossRef]
129. Avellán, T.; Gremillion, P. Constructed wetlands for resource recovery in developing countries. *Renew. Sustain. Energy Rev.* **2019**, *99*, 42–57. [CrossRef]
130. Zhi, W.; Ji, G. Constructed wetlands, 1991–2011: A review of research development, current trends, and future directions. *Sci. Total Environ.* **2012**, *441*, 19–27. [CrossRef]
131. Wang, Y.; Zhao, Y.; Xu, L.; Wang, W.; Doherty, L.; Tang, C.; Ren, B.; Zhao, J. Constructed wetland integrated microbial fuel cell system: Looking back, moving forward. *Water Sci. Technol.* **2017**, *76*, 471–477. [CrossRef] [PubMed]
132. Srivastava, P.; Yadav, A.K.; Mishra, B.K. The effects of microbial fuel cell integration into constructed wetland on the performance of constructed wetland. *Bioresour. Technol.* **2015**, *195*, 223–230. [CrossRef] [PubMed]
133. Doherty, L.; Zhao, X.; Zhao, Y.; Wang, W. The effects of electrode spacing and flow direction on the performance of microbial fuel cell-constructed wetland. *Ecol. Eng.* **2015**, *79*, 8–14. [CrossRef]

processes

MDPI

Article

Energy Flexibility Chances for the Wastewater Treatment Plant of the Benchmark Simulation Model 1

George Skouteris [1], Mario Alejandro Parra Ramirez [1], Sebastian Felix Reinecke [1,*] and Uwe Hampel [1,2]

[1] Helmholtz-Zentrum Dresden-Rossendorf, Institute of Fluid Dynamics, 01328 Dresden, Germany; g.skouteris@hzdr.de (G.S.); m.parra-ramirez@hzdr.de (M.A.P.R.); u.hampel@hzdr.de (U.H.)
[2] Chair of Imaging Techniques in Energy and Process Engineering, Technical University of Dresden, 01062 Dresden, Germany
* Correspondence: s.reinecke@hzdr.de

Abstract: Future energy systems must mainly generate electricity from renewable resources. To deal with the fluctuating availability of wind and solar power, new versatile electricity markets and sustainable solutions concentrating on energy flexibility are needed. In this research, we investigated the potential of energy flexibility achieved through demand-side response for the wastewater treatment plant of the Benchmark Simulation Model 1. First, seven control strategies were simulated and assessed. Next, the flexibility calls were identified, two energy flexibility scenarios were defined and incorporated into the model, and the control strategies were evaluated anew. In this research, the effluent ammonia concentration needed to be maintained within the limits for as long as possible. Strategy 5, which controlled ammonia in Tank 5 at a low value and did not control any nitrate in Tank 2, of Scenario 1, which was characterized by an undetermined on/off aeration cycle, was then found to be the best. Although this control strategy led to high total energy consumption, the percentage of time during which aeration was nearly suspended was one of the highest. This work proposes a methodology that will be useful to plant operators who should soon reduce energy consumption during spikes in electricity prices.

Keywords: energy flexibility; wastewater treatment plant; Benchmark Simulation Model 1; control strategies; aeration; water quality

Citation: Skouteris, G.; Parra Ramirez, M.A.; Reinecke, S.F.; Hampel, U. Energy Flexibility Chances for the Wastewater Treatment Plant of the Benchmark Simulation Model 1. *Processes* **2021**, *9*, 1854. https://doi.org/10.3390/pr9101854

Academic Editor: Dimitris Zagklis

Received: 20 August 2021
Accepted: 16 October 2021
Published: 19 October 2021

1. Introduction

To achieve the goals of the Paris Agreement, use of fossil energy must soon be abandoned. Hence, renewable energy production from wind and solar resources must significantly increase, particularly for countries like Germany with only low hydropower capacity and a decided nuclear power phase-out. However, because of the intermittency of wind and solar power, their prioritization will lead to increasing fluctuations in electricity production, and the use of renewable resources in electric grids will increase their uncertainty and variability. Therefore, electricity markets, due to this increasing share of fluctuating renewable energies, are currently going through a phase of agitating transition. In the future, electric grids will have to rely on storage and reserve capacities as well as demand-side resources to address this problem in supply and demand, or their continued operation may be challenged. New smart applications that will manage to adjust quickly to changes of demand and supply should therefore be considered [1–5].

The German electricity market is Europe's largest, with an annual power consumption of around 530 TWh and a generation capacity of 184 GW [6]. In 2015, renewable energy contributed 195.9 TWh (about 30%) towards electricity generation, which, compared to 2005, corresponds to an increase of 213% [2]. Therefore, Germany, which is committed to increasing the contribution of renewable energy to 80% of the total by 2050 [1], will soon need its energy sector to be reorganized. Sustainable solutions for integration of renewable energy have already begun focusing on the potential for energy flexibility as well as storage facilities. Any deficits or surpluses attributed to the unsustainable nature of the renewable energy production will have to be balanced by energy flexibility measures [1,5,7]. Flexibility refers to modifications to the energy consumption pattern in terms of time and quantity [1]. Flexibility in electricity systems is offered on either the supply or demand side, as well as by the electricity grids by spreading load changes over regions or markets. However, not every infrastructure or technology can provide the same kind of flexibility. While one difference is the time during which flexibility is provided, another important difference is whether positive flexibility, i.e., additional generation, or negative flexibility, i.e., less generation, can be offered—from the demand-side management side, positive flexibility comprises reduction in demand, whereas negative flexibility comprises the opposite [8].

In principle, the two major ways of providing energy flexibility are energy storage and the demand-side response [9]. The demand-side response is an emerging approach to improve asset management in the electricity sector where network capacity is acquired from demand-side actors like major commercial customers who reduce their electricity demand at peak times either by using their own electricity generation, or by shifting electricity consumption outside peak time periods [10]. Load-shedding, which curtails system load to match generation supply [11], and load-shifting, which shifts the demand in the peak hours to the off-peak hours reducing peak demand at specific time periods [12], can be used for temporal energy flexibility.

However, the energy sector cannot be separated from the water sector as the two strongly interrelate, and both are equally essential in modern life. In order to provide and distribute drinking water and to treat and discharge wastewater, substantial amounts of energy are consumed [13]. Indeed, 2–3% of the world's electrical energy is used for water supply and sanitation purposes, and 1–18% of electrical energy in urban areas is used to treat and transport water and wastewater [14]. Wastewater treatment (WWT) consumes approximately 1% of a country's total electricity, and electricity for a plant's operation corresponds up to 50% of its operational costs [3]. On the other hand, water generates power and delivers or recovers energy for use—44% of total global water withdrawals are used for energy production [13,15]. As such, the efficient and sustainable management of combined water and energy systems has already become a priority [14]. In particular, WWT plants seem to be able to provide ancillary services for electric grids through their consumption and production processes. Those with anaerobic digestion have already been found to contribute to energy flexibility solutions due to their electrical consumers, energy production via combined heat and power units, and gas storage units [5,7,13,16].

This work sought to investigate energy flexibility potential in the WWT plant of the Benchmark Simulation Model 1 (BSM1); this plant lacks any energy production through anaerobic treatment of sludge, so the present work solely concentrated on energy flexibility options relating to electricity consumption. All energy consumed, e.g., for aeration, pumping, and mixing, is a cost factor for the operation of the model; however, what mostly determines the cost is the aeration energy, which is the energy consumed to provide oxygen to the aerobic processes, and the pumping energy for the recycling of the plant. Out of the two, aeration energy is the more important [5,17,18]. Aeration is the most electricity-consuming process in a WWT plant, accounting for 40–75% of total electricity demand [3]. Aeration is one of the key components of the activated sludge process, and it is one of the most important aspects of modeling WWT plants [19]. Aeration takes place in large, engineered tanks, in which bacteria need aerobic conditions to convert ammonia to nitrate. The aeration tank, as the largest electricity consumer, is one of the basic modules of a WWT plant that can provide reserve control either through intermittent operation or with pure oxygen [16]. Towards this end, sector-coupling and its synergies, e.g., use of electrical energy surplus from the grid to produce hydrogen and oxygen with an electrolyzer, with hydrogen being used for long-term storage and oxygen for enhancement of the purification processes, may also help [20]. Nevertheless, compliance with effluent quality standards is the main objective of WWT plant operators and it is unlikely that approaches for energy flexibilization reach full-scale applications if they greatly affect treatment efficiency. In the case of aeration control, the main risk is increased ammonia load in the effluent due to insufficient nitrification.

In this work, we aimed at dealing with price spikes (high peak prices) under real-time pricing tariffs. During real-time pricing, electricity consumers are charged prices that vary over short time intervals, typically hourly, and are quoted one day or less in advance to reflect contemporaneous marginal supply costs, where prices spikes are related to periods of low renewable energy production and the need for load-shedding as a demand-side response [7,21]. Therefore, two energy flexibility scenarios concentrating on load-shedding opportunities within the European Power Exchange spot (day-ahead) electricity market (EPEX SPOT) were investigated under both the default and another six more advanced control strategies of the BSM1. After that, all useful information was extracted and all the necessary comparisons were made.

2. Methods and Approach

2.1. Benchmark Simulation Model 1 and User-Defined Control Strategies under Study

The BSM1 is a useful simulation platform for testing and comparing control strategies in WWT processes. It defines a plant layout, influent loads, test procedures, and evaluation criteria. The plant consists of a five-compartment activated sludge reactor, i.e., two anoxic reactors (Tanks 1 and 2) followed by three aerobic reactors (Tanks 3, 4, and 5), and one 10-layer secondary settler. The BSM1 can also be seen as a figure in [22]. Therefore, the plant combines nitrification (aerobic zone) with pre-denitrification (anoxic zone) for biological nitrogen removal [22–24]. Sludge in the secondary settler is recycled to the first anoxic bioreactor (returned activated sludge) in order to maintain the microbiological population; however, some sludge in excess is pumped out of the settler to keep the sludge concentration constant. In addition, some of the mixed liquor is recycled to the first anoxic bioreactor (internal recycle) to enhance nitrogen removal [25]. The International Water Association (IWA)'s Activated Sludge Model 1 (ASM1) as defined by [26] was selected to model the biological reaction, and the double exponential settling velocity model of [27] was selected as a representation of the settling process [18,28,29].

Two reference configurations have already been defined for the BSM1, which are the open-loop configuration where no control is involved and the closed-loop configuration with a defined and active default control strategy added to it [30]. The closed-loop configuration involves two feedback control loops, where the first aims to maintain nitrate concentration in the second anoxic (unaerated) tank (Tank 2) at a set point value of 1 g_N m^{-3}, and the second aims to maintain dissolved oxygen (DO) concentration in the third aeration tank (Tank 5) at a set point value of 2 g_{O2} m^{-3}—this is the model's default control strategy. Both default controllers that are used are proportional–integral controllers [22,24,25]. However, it should be emphasized that the default control strategy described above is not the best strategy available, and it is only provided as a test case for the user. This strategy only uses two control handles, the internal recirculation flowrate (for nitrate control), and the oxygen transfer rate in last tank (for DO control). However, there are approximately 30 individual control handles that can be manipulated within the defined benchmark plant and are thus available for the implementation of user-defined control strategies, dramatically increasing the plant's flexibility [22,30].

The influent dynamics are defined for three different weather conditions: dry, rain, and storm [22]. Each of the weather files contains 14 days of influent data at 15 min intervals and defines the influent, which is assumed to be primary effluent. In general, these files were developed based on real data and represent expected diurnal variations in the influent flowrates, chemical oxygen demand (COD) concentrations, and nitrogen concentrations—the dry weather file depicts what is normal diurnal variations [18,30]. To calculate performance, the system (closed-loop configuration) is first run to steady state by applying a defined constant influent file over a 150-day period of stabilization with no noise on the measurements. Dynamic simulation then occurs by first applying 14 days of the dry influent weather file followed by 14 days of the dry, rain, or storm influent file depending on which weather conditions the user selects. The performance of the benchmark is then evaluated for the last 7 days of dynamic data [22,24,29].

To carry out simulations, dry weather conditions were selected, a commercial software (SIMBA#) was used, and the methodology mentioned above was applied. To this end, the following steps were executed: (i) the software ran at steady state for 150 days and stored the file; (ii) it loaded the last file, ran with data given in the dry weather influent file for 14 days (0–14), and stored the file; (iii) it loaded the last file, ran with data given in the dry influent file for 7 days (0–7), and stored the file; (iv) it loaded the last file, ran with data given in the dry weather influent file for 7 days (7–14), and stored the file; and finally (v) it calculated mean values of effluent-related parameters for the last 7 days (7–14). However, ASM1 was slightly modified to meet the needs of the software.

In this work, the default control strategy, Strategy 1, as defined above, was first simulated. Subsequently, additional basic control loops, as given by [31], were added, and six new user-defined strategies, Strategies 2–7, were created and simulated (see Table 1). The control loops that were additionally added aimed at either controlling the DO concentration in the first and the second aerobic tank by manipulating the respective mass transfer coefficient k_La values based on DO concentration measurements in each reactor, or by controlling ammonia in the last aerated tank by manipulating the DO concentration set point, i.e., a primary ammonia controller, together with three secondary DO controllers in a cascade control configuration. In regard to ammonia-related strategies, both a strict set point for ammonia concentration and a looser one were tested. Finally, Strategies 2–7 appeared in two versions; one included a nitrate controller, whereas the other included no such controller (see Table 2). Information on the scenarios relating to energy flexibility solutions is given later.

Table 1. Controllers in the studied strategies.

Controllers	Description
DO	DO concentration control by manipulating k_La through aeration
NO_3-N	Nitrate nitrogen concentration control by manipulating internal recirculation flowrate
NH_4-N	Ammonia concentration control by manipulating the DO set point in the DO loop (cascade control)

Table 2. Location and set points of the controllers in the different control strategies.

Strategy	DO		NO_3-N		NH_4-N	
	Location	SP (mg_{O2} L^{-1})	Location	SP (mg_N L^{-1})	Location	SP (mg_N L^{-1})
Strategy 1	Tank 5	2	Tank 1	1	—	
Strategy 2	Tanks 3, 4, and 5	2	Tank 1	1	—	
Strategy 3	Tanks 3, 4, and 5	2	—	—	—	
Strategy 4	Tanks 3, 4, and 5	0–4	Tank 1	1	Tank 5	1
Strategy 5	Tanks 3, 4, and 5	0–4	—	—	Tank 5	1
Strategy 6	Tanks 3, 4, and 5	0–4	Tank 1	1	Tank 5	3.5
Strategy 7	Tanks 3, 4, and 5	0–4	—	—	Tank 5	3.5

SP: set point, DO: dissolved oxygen.

2.2. Energy Market in Germany and Analysis of Energy Consumption Data

The different electricity markets include the forward/future market, the day-ahead market, the intraday market, and the balancing market. The forward/future markets run from years before to the day before delivery. In the day-ahead market, electricity is traded one day before delivery, whereas in the intraday market, electricity is traded on the delivery day. Finally, the balancing market, which is the last in a sequence of electricity markets, occurs just minutes before the energy delivery [32–36]. The European Market Coupling, of which EPEX SPOT is a major contributor, allows for electricity to flow freely across borders. As such, it is distributed more efficiently all over Europe because flows follow price signals. EPEX SPOT, which was our focus of interest, operates the most liquid day-ahead and intraday markets in Europe. Both markets fulfill different purposes, and they are indispensable links of the energy value chain [37].

In regard to this work, the most important market was the day-ahead. In Europe, day-ahead markets are also known as spot markets, or markets where electricity is traded for immediate use; however, they are practically forward/future markets, as they deal with electricity and reserves for future use. However, the term forward/future market is mostly used for markets that take place before the day-ahead market [33]. The day-ahead market allows electric energy trading one day ahead of energy delivery. It is cleared around noon the day before to the day on which energy is to be delivered. Clearing at noon one day before means 12 h anticipation regarding the first delivery hour, and 35 h anticipation regarding the last delivery hour. Such anticipation is needed by some energy producers, like nuclear plants or coal plants, due to their technical limitations on operational flexibility [32]. In more detail, the trading is driven by its participants. A buyer needs to assess the amount of energy they need in order to fulfil their customer's requirements for the following day, and the cost of its purchase price hour by hour. The seller also submits the quantity they can deliver next day, and at what price level on an hourly basis. The deadline for the members to submit the price and the quantity for which they seek to make an agreement is 12:00 noon Central Europe Time (CET). These bids are inserted into a complex algorithm to calculate the clearing price. This process takes around 42 min. From 12:00 midnight CET next day, the sellers deliver the power at the contracted rate [38].

However, the day-ahead market has to be complemented by the intraday and balancing markets to overcome unplanned events, e.g., unexpected power plant outages, and changing weather conditions, e.g., better wind forecasts, thus allowing market participants to correct for shifts in their day-ahead nominations [33,39]. In Germany, the intraday markets, typically organized by power exchanges, take the form of continuous trading allowing purchasing and selling of electricity throughout the whole day, up to a few minutes before delivery. On the intraday market, market participants trade continuously, 24 h a day, with delivery on the same day. Electricity can be traded up to 5 min before delivery, and through hourly, half-hourly, or quarter-hourly contracts. As this allows for a high level of flexibility, members use the intraday market to make last-minute adjustments and to balance their positions closer to real time, something already mentioned above. However, intraday prices have not been studied as intensively as their day-ahead counterparts, and research typically focuses only on short-term forecasts, so it assumes knowledge of the day-ahead prices. The final balancing of the demand and supply is achieved through the balancing markets, which are controlled by system operators and aim at securing the system's stability [37,39]. More detail on the German Energy Market is given by the Federal Ministry of Economic Affairs and Energy in [40].

In this work, as we concentrated on the German electricity market, we used day-ahead electricity prices as provided by the EPEX SPOT for the calculations. At the time of writing of this document, EPEX SPOT data were available until the end of 2020. For this work, we needed to know electricity prices for at least 14 consecutive days, 24 h a day, as the software used a 14 day data set on a loop for the whole duration of simulations. As it was decided for this work to concentrate on the most recent data, we selected a set of data in autumn 2020 with the values lying between 21 September 2020 and 4 October 2020—this is also shown in Figure 1. During this time, the average electricity price was equal to 39.99 € MWh^{-1}, with 50% of the values being equal to or lower than the average.

To carry out this research, the educated guess was made that one fourth of the time under study should have been subject to energy flexibility. The 75th percentile for the selected set of data was calculated to determine the corresponding price limit for purchase of electricity, and it was found to be equal to 51.15 € MWh^{-1}. The time above the limit comprised 19 flexibility calls ranging from 1 hour to 17 consecutive hours. To explain further, a price of 51.15 € MWh^{-1} means that a plant operator can unrestrictedly buy electricity any time it is being sold at a price below or equal to 51.15 € MWh^{-1}, whereas flexible solutions must be identified each time electricity is being sold at a price over 51.15 € MWh^{-1}. During these times, options like shutting down the plant, switching off items of equipment, or reducing DO concentration in aeration tanks can be taken into consideration [22].

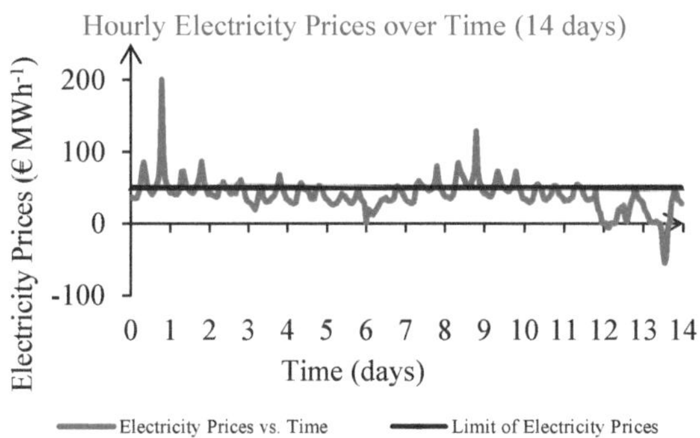

Figure 1. Day-ahead electricity prices over time.

In this work, two energy flexibility scenarios plus the default scenario without energy flexibility were tested per strategy—these scenarios took effect only during the flexibility calls, while the rest of the time the aeration was as per the applied strategy. Both scenarios were selected bearing in mind that the nitrifying bacteria, to the best of our knowledge, would not be endangered at any time. Therefore, the evaluation of the simulation results in terms of effluent water quality was carried out based on ammonia concentration. Scenario 1 recommended shutting off aeration in all three aeration tanks; however, any time ammonia concentration would be over 4 $mg_N\ L^{-1}$, the aeration was reinstated. Scenario 2 recommended intermittent aeration, which was subject to a 1 h on/off aeration cycle during which aeration was suspended for the first 30 min and reinstated for the remaining 30 min. Scenario 1 was then characterized by an undetermined on/off aeration cycle that depended on changes of the effluent quality, whereas Scenario 2 depended on a well-determined aeration scheme, which was unaffected by any changes of treated water quality. However, for both scenarios, air flowrate was never equal to zero during the aeration shut-off periods, as a minimum very low air flowrate was maintained to ensure enough mixing without having to use mechanical mixers. This minimum air flowrate corresponded to a k_La-value of 20 d^{-1} according to guidelines provided by IWA [30]. An overview of the studied flexibility scenarios can be found in Table 3 below.

Table 3. Overview of studied flexibility scenarios.

	Scenario 1	**Scenario 2**
Flexibilization option	Complete aeration shut-off	Intermittent aeration (one hour cycle with 30 min off and 30 min on)
Minimum air flow during shut-off	$k_La \geq 20\ d^{-1}$	$k_La \geq 20\ d^{-1}$
Affected tanks of BSM1	Tanks 3, 4, and 5	Tanks 3, 4, and 5
Condition regarding effluent quality	Effluent ammonia concentration below 4 $mg_N\ L^{-1}$	No condition

3. Simulations and Discussion

As a first step, the control strategies had to be evaluated against ammonia concentrations in the effluent. In this approach, simulations where carried out for each of the strategies described in Tables 1 and 2 without any consideration for energy prices (no-energy-flexibility scenario). Simulations were then conducted for the seven control strategies with the integration of energy prices according to Scenario 1 and Scenario 2. The resulting time series from the simulations for two of the seven strategies, Strategy 1 and Strategy 5, are shown in Figures 2 and 3 below, respectively. Further analysis of the performance of the strategies in the different scenarios is presented in Section 3.3.

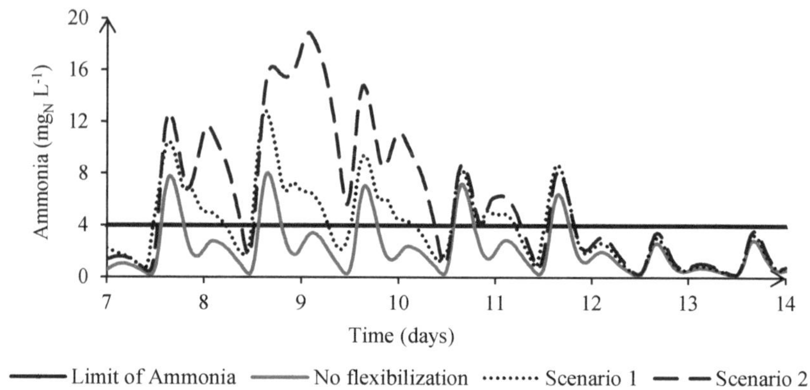

Figure 2. Course of ammonia concentration over time for Strategy 1.

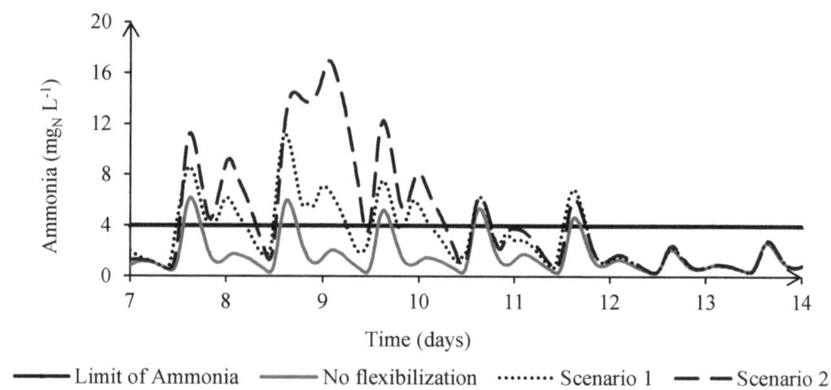

Figure 3. Course of ammonia concentration over time for Strategy 5.

Based on Figure 2 above, the time series for the no-flexibilization scenario indicated that the presented control strategies were unable to maintain the ammonia concentration in the effluent below the critical value as defined for the BSM1 model (4 mg_N L^{-1}) throughout the whole time period of 7 days. The same was true for the remaining strategies. The fact that this also applied to the default strategy (strategy 1) indicates that the BSM1 model has been designed so that water quality in the effluent does not meet the required criteria all the time.

Taking into account the results described above, a comparison between the strategies in relation to their flexibility capacity and treatment performance was carried out for each scenario. The percentage of time during which the concentration limit for ammonia in the effluent was not met was quantified, along with the energy consumption as the sum of aeration energy and pumping energy. In this work, the pumping energy was regarded as the sum of the energy consumed for the removal of the excess sludge for the internal recycle and for the external one. Results are summarized in the following sections.

3.1. Control Strategies: No-Energy-Flexibility Scenario

The parameters for the control strategy evaluation for the scenario without energy flexibilization are presented in Table 4 below.

Table 4. No-flexibility scenario: parameters.

Strategy	Percentage of Time over the Ammonia Concentration Limit	Energy Consumption (MWh over 7 Days)
1	15.9	27,466.7
2	14.8	26,299.3
3	14.1	27,280.8
4	12.4	30,319.5
5	10.3	32,895.7
6	26.9	24,774.8
7	26.2	26,043.6

As seen in the table above, addition of the nitrate control loop to a strategy slightly deteriorated performance in terms of removal of ammonia. In other words, when comparing two strategies between which the only difference was the existence or not of the nitrate control loop, the one that lacked it ensured a larger percentage of time during which ammonia concentration in the effluent was below 4 mg_N L^{-1}. After running the simulations, lower nitrate concentrations in the effluent were recorded when the nitrate control loop was used, which could have been a result of poorer nitrification. This could have produced higher ammonia concentrations in the effluent, which subsequently increased the length of time during which concentration was above the critical concentration. After all, the role of the nitrate control loop is not related to nitrification—rather, its role is to improve the denitrification that is taking place in the anoxic zone.

Strategy 5, which controlled ammonia in Tank 5 at 1 mg_N L^{-1} and lacked any nitrate control loop, had the best performance of all strategies. On the other hand, Strategy 6, which controlled ammonia in Tank 5 at the much higher value of 3.5 mg_N L^{-1} and included the nitrate control loop, was the worst. This means that the selection of the set point in ammonia-based control strategies was very important. At this point, it is also worth mentioning that the three water quality parameters of interest that are not strongly or not at all related to nitrogen, that is COD (concentration limit in the effluent: 100 g_{COD} m^{-3}), total suspended solids (TSS) (concentration limit in the effluent: 30 g_{SS} m^{-3}), and 5-day biochemical oxygen demand (BOD_5) (concentration limit in the effluent: 10 g_{BOD5} m^{-3}) [22], always remained below the limit for all strategies, making ammonia concentration the critical parameter that required continued attention.

However, in terms of total energy consumption, results were different. As already mentioned, energy in this work was assumed to be the sum of electricity that was consumed for pumping and aeration—the formulas and the coefficients used for the calculation are provided in [22]. In this case, upon the comparison of two strategies for which the only difference was the existence or not of the nitrate control loop, the one including the control loop consumed less energy. This was expected as the nitrate control loop kept alternating the flowrate of the internal recycle and adjusting it to the values required. On the other hand, when the nitrate control loop was excluded, the flowrate remained constant regardless of the needs of the WWT plant. This time, Strategy 6, which controlled ammonia in Tank 5 at 3.5 mg_N L^{-1}, and included a nitrate control loop, appeared to have the best performance of all strategies. On the other hand, Strategy 5, which controlled ammonia in Tank 5 at the much lower value of 1 mg_N L^{-1} and lacked the nitrate control loop, was the worst. From the above, it can be said that the percentage of time over the limit and the energy consumption rate are inversely related. However, the percentage of time over the limit for Strategy 5 was 2.6 times less than that in Strategy 6, and the energy consumed was 1.3 imes higher accordingly. To this end, Strategy 5 was better than Strategy 6; however, the final decision always depends on what the plant operators consider to be more important.

3.2. Energy Flexibility Scenarios

3.2.1. Scenario 1: Undetermined On/Off Aeration Cycle

Scenario 1 ordered that air would be shut off in all three aeration tanks each time the electricity price was equal to or higher than 51.15 € MWh^{-1} and the ammonia concentration at the exit of the aerobic zone was below 4 mg_N L^{-1}—any time the concentration was over 4 mg_N L^{-1}, aeration did not stop or, if had already been interrupted, it was reinstated. As for the analysis above, the evaluation parameters are summarized in Table 5.

Table 5. Scenario 1: parameters.

Strategy	Percentage of Time over the Ammonia Concentration Limit	Energy Consumption (MWh over 7 Days)	Percentage of Time of No Aeration
1	44.91	25,930.5	10.57
2	45.06	25,074.7	10.12
3	45.06	25,806.7	10.12
4	36.53	30,828.1	12.50
5	35.01	32,436.1	12.35
6	50.75	26,641.4	10.12
7	48.20	27,487.7	10.12

Similarly to the case where the no-energy-flexibility scenario was applied, Strategy 5 both had the best performance with respect to percentage of time during which ammonia concentration in the effluent exceeded the limit and consumed the largest amount of energy over time. Even though Strategy 6 was no longer the strategy that consumed the lowest amount of total energy, it still remained the best among the ammonia-related control strategies. In terms of percentage of time during which aeration was shut off, this strategy was characterized by the second longest period during which aeration could remain suspended, behind only Strategy 4, for which difference with Strategy 5 was the inclusion of the nitrate controller. To this end, there seems to be some contradiction, as the strategy that maintained the air blowers switched off for more time than most of remaining strategies also had the highest energy consumption rate. However, as seen in Figure 4, which shows the course of the sum of the k_La values in all three aeration tanks for Strategies 3, 5, and 7, which were selected because all of them excluded the nitrate controller and their comparison was feasible, this occurred due to nature of the control strategy. In regard to this strategy, it is true that the air system was shut down for a long time, but for the rest of time, DO concentration values could freely soar up to 4 mg_{O2} L^{-1} each time that was required, significantly increasing the k_La values and consequently the energy consumption. However, based on the priorities set, Strategy 5 was still the best strategy in terms of performance. This strategy ensured the shortest period of time during which ammonia concentration surpassed the limit, which was the highest priority, and, at the same time, it ensured a quite long period of time during which aeration remained shut off, which was the second most important condition, as the idea was to buy as little electricity as possible during the times this was expensive.

3.2.2. Scenario 2: Predetermined On/Off Aeration Cycle

Scenario 2 ordered that air would be shut off in all three aeration tanks each time the electricity price was equal to or higher than 51.15 €, according to the determined aeration cycle of 30 min on/30 min off. This means that for a fixed 12.5% of the 25% of the time that was subject to energy flexibility, the aeration would be suspended. Like the analysis before, the evaluation parameters are summarized in Table 6.

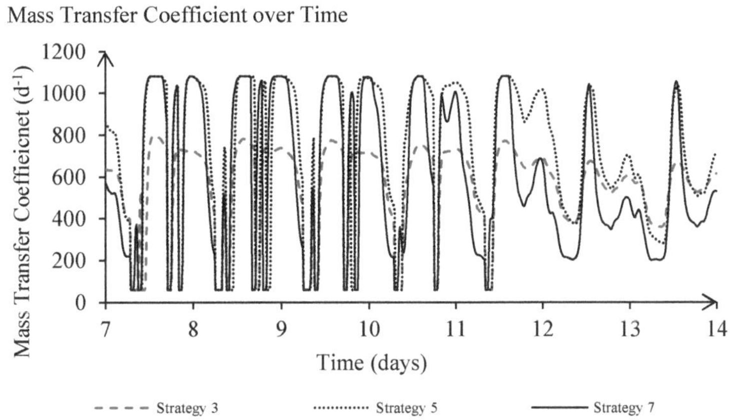

Figure 4. Mass transfer coefficient over time.

Table 6. Scenario 2: parameters.

Strategy	% of Time over the Ammonia Concentration Limit	Energy Consumption (MWh over 7 Days)
1	66.47	25,367.5
2	60.48	23,838.9
3	59.88	24,309.7
4	43.41	29,369.6
5	41.32	30,923.2
6	58.53	25,930.8
7	57.63	26,531.3

Similarly to before, Strategy 5 once again had the best performance in terms of ammonia removal at the exit of the plant. Strategy 5 was also the best strategy with respect to the percentage of time during which the air blowers remained switched off; however, in this scenario, this made no difference as all strategies were characterized by the same percentage. Finally, the total energy consumed by Strategy 5 was again the largest, with Strategy 6 being the one that consumed the least among the ammonia-related strategies, which was in line with Scenario 1.

3.3. Aggregated Results and Final Discussion

Figure 5, containing the aggregated results for the potential flexibility options studied in this work, is shown below.

It is clear from Figure 5a that Scenario 1 removed more ammonia than Scenario 2. All control strategies for Scenario 2 ended up with a higher percentage of time over the ammonia concentration limit than their respective strategies for Scenario 1. This means that the undetermined cycle of Scenario 1, which was governed by effluent quality restrictions, complied with the regulations for a longer period of time than the predetermined cycle of Scenario 2, which was based on a fixed on/off aeration cycle.

Strategy 5 always had the best performance regardless of the scenario. However, as Strategy 5 for Scenario 1 maintained the ammonia concentration in the effluent within limits for a longer time period than any other strategy in both scenarios, this strategy seemed to be the most attractive. Of course, that strategy was characterized by the highest energy consumption overall (see Figure 5b); however, the total energy consumed was of low priority to this work and instead, the important consideration was the amount of energy that had to be bought during the flexibility calls. Even though the time during which aeration was shut off was slightly less (12.35%) than the 12.5% that was ensured by all strategies for Scenario 2 and Strategy 4 for Scenario 1 (see Figure 5c), the fact that the latter had longer time periods during which the ammonia concentration was over the limit helped to determine Strategy 5 for Scenario 1 as the best possible option. From the above, it is clear that the control of ammonia at the exit of Tank 5 was critical. However, despite the fact that Strategies 4 and 5 for both scenarios managed to control the system best, Strategies 6 and 7 significantly failed to achieve this, in spite of all four having aimed at controlling ammonia in the same place. This means that it was not the strategy itself that determined the performance, but the selected set point. Therefore, the stricter the set point, the better the ammonia removal, but, on the other hand, the stricter the set point, the larger the amount of total energy consumed.

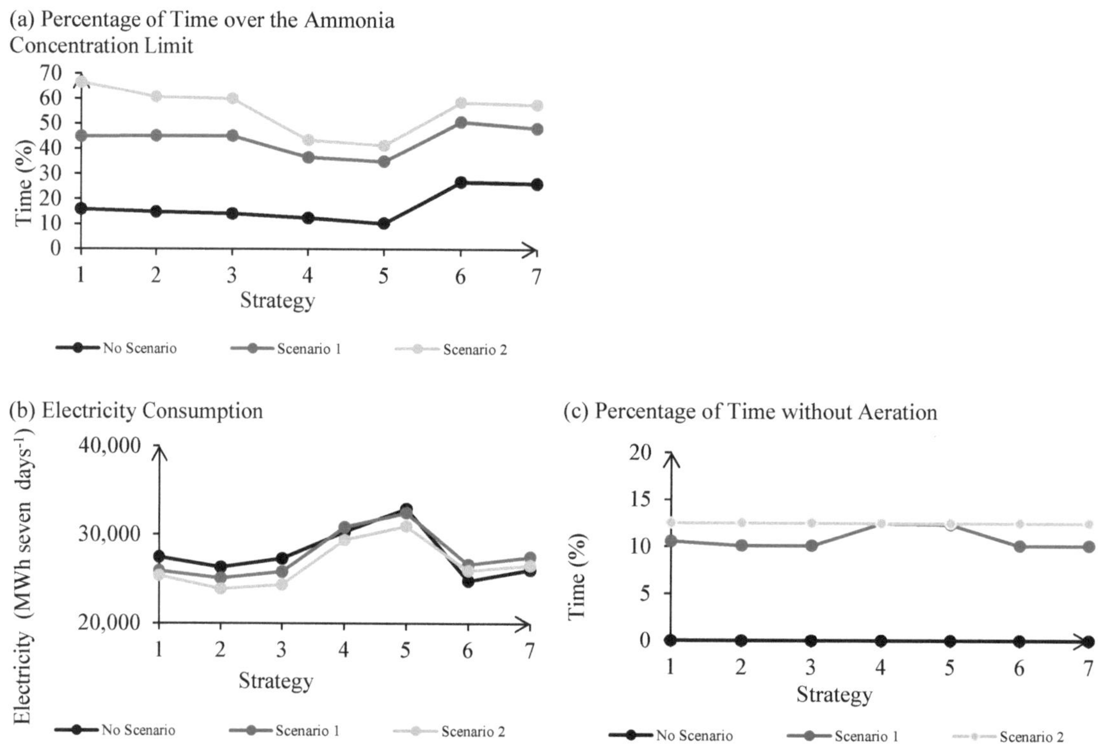

Figure 5. No scenario, Scenario 1, and Scenario 2: aggregated results for the percentage of time above ammonia concentration limit (**a**), electricity consumption (**b**) and percentage of time without aeration (**c**).

Now, if priority had been given to the amount of time during which aeration had been suspended, Strategy 5 for Scenario 1 would have been replaced by Strategy 4 for Scenario 1 as the best option, meaning that Scenario 1 always outmatched Scenario 2. This is due to the fact that all strategies for Scenario 2, plus Strategy 4 for Scenario 1, had the same percentage of time during which aeration was shut off (12.5%); however, Strategy 4 for Scenario 1 had a shorter period of time during which the ammonia concentration at the exit surpassed the limit. Although Strategy 4 had a slightly higher total energy consumption for Scenario 1 than for Scenario 2, the maximum time during which equipment was switched off for Scenario 2 was half an hour, whereas for Scenario 1, that time was equal to three hours. In general, the more time the equipment is switched off, the greater the possibility that the bacterial culture is affected. Even though the model showed that three consecutive hours of almost zero air provision had no detrimental effect on the bacterial culture, as ammonia removal was able to reinstate when air blowers were switched on again, we recommend that this also be confirmed experimentally.

Even though this is a preliminary piece of work in the field, it is quite useful, as it provides the reader with a methodology for dealing with demand-side energy flexibility in wastewater treatment plants. This work managed to evaluate a number of different control strategies against potential energy flexibility options and opened up new possibilities for more options to follow. Therefore, this methodology can soon be applied to real wastewater treatment plants. In any case, more scenarios must be tested and evaluated in the near future, and concepts like on-site energy storage and the wastewater treatment plant as a micro-grid, together with on-site energy production, either related to anaerobic digestion or not, must be considered. However, these preliminary results appear to be quite promising, and the methodology used can be easily adopted by plant operators who are interested not only in reducing energy consumption when it is expensive, but also in enhancing the water sustainability credentials of their plants.

4. Conclusions

The definite future prioritization of weather-dependent renewable energy resources in the energy sector will unavoidably lead to fluctuations in electricity production. Because of this insecurity in production, all energy producers and consumers are due to explore energy flexibility schemes in the near future. WWT plants are rather energy-intensive, so potential energy flexibility solutions are imperative so that their operators can cope with high peak electricity prices successfully.

In this work, we sought ways of providing on-site energy flexibility through demand-side response for the wastewater treatment plant of the BSM1. The plant was first tested under seven different control strategies, including the default strategy along with six additional more advanced strategies which controlled DO in the whole aerobic zone, either with or without an ammonia primary controller. This analysis, which was the reference point for this work, was free from any energy flexibility scenario. Next, two energy flexibility scenarios were put in place and all strategies were evaluated against the new operating conditions.

Results showed that when the no-energy-flexibility scenario was employed, the ammonia concentration in the effluent was at times higher than the upper limit, so there was a percentage of time during which treated water was not suitable for discharge. Introduction of the flexibility scenarios in the model further diminished this percentage of time; however, Scenario 1, which was based on an undetermined on/off aeration cycle, outperformed Scenario 2, which was based on a predetermined one. Strategy 5 for Scenario 1, which controlled ammonia in Tank 5 at a rather strict set point, was found to be the best option, although the most expensive. However, in this research, the amount of energy that had to be bought during the flexibility calls was more critical than the overall energy consumed, so Strategy 5 for Scenario 1 still remained the best choice as it had a slightly larger percentage of time during which the air blowers were switched off.

Author Contributions: Conceptualization: S.F.R. and U.H.; Methodology: G.S. and M.A.P.R.; Software: G.S. and M.A.P.R.; Validation: G.S. and M.A.P.R.; Formal Analysis: G.S.; Investigation: G.S. and M.A.P.R.; Resources: U.H.; Data Curation: G.S.; Writing—Original Draft Preparation: G.S., Writing—Review and Editing: G.S. and M.A.P.R., Visualization: G.S. and M.A.P.R.; Supervision: S.F.R. and U.H.; Project Administration: S.F.R., Funding Acquisition: S.F.R. and U.H. All authors have read and agreed to the published version of the manuscript.

Funding: This work is partly funded by Initiative and Networking Fund of the Helmholtz Association in the frame of the Clean Water Technology Lab CLEWATEC—a Helmholtz Innovation Lab.

Institutional Review Board Statement: Not applicable.

Informed Consent Statement: Not applicable.

Conflicts of Interest: The authors declare no conflict of interest. Also, the funders had no role in the design of the study; in the collection, analyses, or interpretation of data; in the writing of the manuscript, or in the decision to publish the results.

References

1. Schäfer, M.; Hobus, I.; Schmitt, T.G. Energetic flexibility on wastewater treatment plants. *Water Sci. Technol.* **2017**, *76*, 1225–1233. [CrossRef]
2. Bublitz, A.; Keles, D.; Zimmermann, F.; Fraunholz, C.; Fichtner, W. *A Survey on Electricity Market Design: Insights from Theory and Real-World Implementations of Capacity Remuneration Mechanisms*; Working Paper Series in Production and Energy; Karlsruhe Institute of Technology (KIT); Institute for Industrial Production (IIP): Karlsruhe, Germany, 2018; No. 27.
3. Brok, N.B.; Munk-Nielsen, T.; Madsen, H.; Stentoft, P.A. Flexible control of wastewater aeration for cost-efficient, sustainable treatment, IFAC Workshop on Control of Smart Grid and Renewable Energy Systems. *IFAC PapersOnLine* **2019**, *52*, 494–499. [CrossRef]
4. Fallahi, Z.; Henze, G.P. Interactive buildings: A review. *Sustainability* **2019**, *11*, 3988. [CrossRef]
5. Schäfer, M. Short-term flexibility for energy grids provided by wastewater treatment plants with anaerobic sludge digestion. *Water Sci. Technol.* **2020**, *81*, 1388–1397. [CrossRef] [PubMed]
6. Chen, S.; Härdle, W.K.; López Cabrera, B. *Regularization Approach for Network Modeling of German Energy Market*; IRTG 1792 Discussion Paper 2018-017; Humboldt University of Berlin: Berlin, Germany, 2018.
7. Seier, M.; Schebek, L. Model-based investigation of residual load smoothing through dynamic electricity purchase: The case of wastewater treatment plants in Germany. *Appl. Energy* **2017**, *205*, 210–224. [CrossRef]
8. Bertsch, J.; Growitsch, C.; Lorenczik, S.; Nagl, S. *Do We Need an Additional Flexibility Market in the Electricity System?—A System-Economic Analysis for EUROPE*; Paper based on the study: Flexibility options in European Electricity Markets in high RES-E Scenarios, for the International Energy Agency; Institute of Energy Economics, University of Cologne: Köln, Germany, 2013.
9. Yilmaz, S.; Xu, X.; Cabrera, D.; Chanez, C.; Cuony, P.; Patel, M.K. Analysis of demand-side response preferences regarding electricity tariffs and direct load control: Key findings from a Swiss survey. *Energy* **2020**, *212*, 118712. [CrossRef]
10. Langendahl, P.-A.; Roby, H.; Potter, S.; Cook, M. Smoothing peaks and troughs: Intermediary practices to promote demand side response in smart grids. *Energy Res. Soc. Sci.* **2019**, *58*, 101277. [CrossRef]
11. Ford, J.J.; Bevrani, H.; Ledwich, G. Adaptive load shedding and regional protection. *Int. J. Electr. Power Energy Syst.* **2009**, *31*, 611–618. [CrossRef]
12. Balakumar, P.; Sathiya, S. Demand side management in smart grid using load shifting technique. In Proceedings of the 2017 International Conference on Electrical, Instrumentation and Communication Engineering (ICEICE2017), Karur, India, 27–28 April 2017.
13. Schäfer, M.; Gretzschel, O.; Steinmetz, H. The possible roles of wastewater treatment plants in sector coupling. *Energies* **2020**, *13*, 2088. [CrossRef]
14. Aymerich, I.; Rieger, L.; Sobhani, R.; Rosso, D.; Corominas, L.L. The difference between energy consumption and energy cost: Modelling energy tariff structures for water resource recovery facilities. *Water Res.* **2015**, *81*, 113–123. [CrossRef]
15. Larsen, M.A.D.; Drews, M. Water use in electricity generation for water-energy nexus analyses: The European case. *Sci. Total Environ.* **2019**, *651*, 2044–2058. [CrossRef] [PubMed]
16. Schäfer, M.; Gretzschel, O.; Schmitt, T.G.; Knerr, H. Wastewater treatment plants as system service provider for renewable energy storage and control energy in virtual power plants—A potential analysis. *Energy Procedia* **2015**, *73*, 87–93. [CrossRef]
17. Revolar, S.; Vega, P.; Vilanova, R.; Francisco, M. Optimal control of wastewater treatment plants using economic-oriented model predictive dynamic strategies. *Appl. Sci.* **2017**, *7*, 813. [CrossRef]
18. Husin, M.H.; Rahmat, M.F.; Wahab, N.A.; Sabri, M.F.M.; Suhaili, S. Proportional-integral ammonium-based aeration control for activated sludge process. In Proceedings of the 2020 13th International UNIMAS Engineering Conference (EnCon), Kota Samarahan, Malaysia, 27–28 October 2020.
19. Schraa, O.; Rieger, L.; Alex, J. Development of a model for activated sludge aeration systems: Linking air supply, distribution, and demand. *Water Sci. Technol.* **2017**, *75*, 552–560. [CrossRef]
20. Gretzschel, M.O.; Schäfer, M.; Steinmetz, H.; Pick, E.; Kanitz, E.; Krieger, S. Advanced wastewater treatment to eliminate organic micropollutants in wastewater treatment plants in combination with energy-efficient electrolysis at WWTP Mainz. *Energies* **2020**, *13*, 3599. [CrossRef]
21. Barbose, G.; Goldman, C. *A Survey of Utility Experience with Real Time Pricing*; LBNL-54238; Lawrence Berkley National Laboratory: Berkeley, CA, USA, 2004.
22. Alex, J.; Benedetti, L.; Copp, J.; Gernaey, K.V.; Jeppsson, U.; Nopens, I.; Pons, M.N.; Steyer, J.P.; Vanrolleghem, P. *Benchmark Simulation Model No. 1 (BSM1)*; University of Lund: Lund, Sweden, 2008.
23. Liu, H.; Yoo, C. Cascade control of effluent nitrate and ammonium in an activated sludge process. *Desalination Water Treat.* **2015**, *57*, 21253–21263. [CrossRef]
24. Santín, I.; Pedret, C.; Vilanova, R. Control strategies for ammonia violations removal in BSM1 for dry, rain and storm weather conditions. In Proceedings of the 23rd Mediterranean Conference on Control and Automation (MED), Paper ThAT 2.3, Torremolinos, Spain, 16–19 June 2015.
25. Mulas, M.; de Araùjo, A.C.B.; Baratti, R.; Skogestad, S. Optimized control structure for a wastewater treatment benchmark. In Proceedings of the 9th International Symposium on Dynamics and Control of Process Systems (DYCOPS 2010), Leuven, Belgium, 5–7 July 2010.

26. Henze, M.; Grady, C.P.L., Jr.; Gujer, W.; Marais, G.V.R.; Matsuo, T. *Activated Sludge No 1, Scientific and Technical Reports No 1*; International Association on Water Pollution Research and Control: London, UK, 1987.
27. Takács, I.; Patry, G.G.; Nolasco, D. A dynamic model of the clarification-thickening process. *Water Res.* **1991**, *25*, 1263–1271. [CrossRef]
28. Vanrolleghem, P.A.; Gillot, S. Robustness and economic measures as control benchmark performance criteria. *Water Sci. Technol.* **2002**, *45*, 117–126. [CrossRef]
29. Stare, A.; Vrečko, D.; Hvala, N.; Strmčnik, S. Comparison of control strategies for nitrogen removal in an activated sludge process in terms of operating costs: A simulation study. *Water Res.* **2007**, *41*, 2004–2014. [CrossRef] [PubMed]
30. Gernaey, K.V.; Jeppsson, U.; Vanrolleghem, P.A.; Copp, J.B. *Benchmarking of Control Strategies for Wastewater Treatment Plants*; Scientific and Technical Report, No. 23, IWA Task Group on Benchmarking of Control Strategies for Wastewater Treatment Plants; IWA Publishing: London, UK, 2014.
31. Gernaey, K.V.; Vrecko, D.; Rosen, C.; Jeppsson, U. BSM1 versus BSM1_LT: Is the control strategy performance ranking maintained? In Proceedings of the 7th International IWA Symposium on Systems Analysis and Integrated Assessment in Water Management, Washington, DC, USA, 7–9 May 2007.
32. Morales, J.M.; Conejo, A.J.; Madsen, H.; Pinson, P.; Zugno, M. Balancing Markets. In *Integrating Renewables in Electricity Markets, Operational Problems*; Springer: New York, NY, USA, 2014; Chapter 4; pp. 101–136.
33. KU Leuven Energy Institute. *EI Fact Sheet: The Current Electricity Market Design in Europe*; EI-FACT SHEET 2015-01; KU Leuven Energy Institute: Leuven, Belgium, 2015.
34. Erbach, G. *Understanding Electricity Markets in the EU*; Members' Research Service PE 593.519; European Parliamentary Research Service: Brussels, Belgium, 2016.
35. Van der Veen, R.A.C.; Hakvoort, R.A. The electricity balancing market: Exploring the design challenge. *Util. Policy* **2016**, *43*, 186–194. [CrossRef]
36. Mazzi, N.; Pinson, P. Wind power in electricity markets and the value of forecasting. In *Renewable Energy Forecasting, From Models to Applications*; Woodhead Publishing Series in Energy; Woodhead Publishing: Sawston, UK, 2017; pp. 259–278.
37. Home | EPEX SPOT. Available online: https://www.epexspot.com/en (accessed on 15 April 2021).
38. Khoshrou, A.; Pauwels, E.J.; Dorsman, A.B. The evolution of electricity price on the German day-ahead market before and after the energy switch (CEVI). *Renew. Energy* **2019**, *134*, 1–13. [CrossRef]
39. Maciejowska, K.; Nitka, W.; Weron, T. Day-ahead vs. intraday-Forecasting the price spread to maximize economic benefits. *Energies* **2019**, *12*, 631. [CrossRef]
40. BMWi. Available online: https://www.bmwi.de/Navigation/DE/Home/home.html (accessed on 15 April 2021).

processes

Article

A Comparative Study of Improvement of Phycoremediation Using a Consortium of Microalgae in Municipal Wastewater Treatment Pond Systems as an Alternative Solution to Africa's Sanitation Challenges

Paul J. Oberholster [1], Maronel Steyn [2] and Anna-Maria Botha [3,*]

1 Centre for Environmental Management, University of the Free State, Bloemfontein 9300, South Africa; OberholsterPJ@ufs.ac.za
2 CSIR, Smart Places Unit, Water Centre, P.O. Box 320, Stellenbosch 7599, South Africa; msteyn@csir.co.za
3 Genetics Department, University of Stellenbosch, Private Bag X1, Matieland, Stellenbosch 7601, South Africa
* Correspondence: ambo@sun.ac.za

Abstract: The reuse of wastewater has been observed as a viable option to cope with increasing water stress in Africa. The present case studies evaluated the optimization of the process of phycoremediation as an alternative low-cost green treatment technology in two municipality wastewater treatment pond systems that make up the largest number of domestic sewage treatment systems on the African continent. A consortium of specific microalgae (*Chlorella vulgaris* and *Chlorella protothecoides*) was used to improve the treatment capacity of domestic wastewater at two operational municipality wastewater pond systems under different environmental conditions in South Africa. Pre- and post-phycoremediation optimization through mass inoculation of a consortium of microalgae, over a period of one year under different environmental conditions, were compared. It was evident that the higher reduction of total phosphates (74.4%) in the effluent, after treatment with a consortium of microalgae at the Motetema pond system, was possibly related to (1) the dominance of the algal taxa *C. protothecoides* (52%), and to a lesser extent *C. vulgaris* (36%), (2) more cloudless days, (3) higher air temperature, and (4) a higher domestic wastewater strength. In the case of the Brandwag pond treatment system, the higher reduction of total nitrogen can possibly be related to the dominance of *C. vulgaris*, different weather conditions, and lower domestic wastewater strength. The nutrient reduction data from the current study clearly presented compelling evidence in terms of the feasibility for use of this technology in developing countries to reduce nutrient loads from domestic wastewater effluent.

Keywords: consortia of microalgae; sustainable development goals; phycoremediation; pond treatment systems; domestic wastewater; Africa

Citation: Oberholster, P.J.; Steyn, M.; Botha, A.-M. A Comparative Study of Improvement of Phycoremediation Using a Consortium of Microalgae in Municipal Wastewater Treatment Pond Systems as an Alternative Solution to Africa's Sanitation Challenges. *Processes* **2021**, *9*, 1677. https://doi.org/10.3390/pr9091677

Academic Editor: Dimitris Zagklis

Received: 25 July 2021
Accepted: 9 September 2021
Published: 18 September 2021

1. Introduction

Attention towards addressing sanitation and wastewater issues in Africa has grown in recent years [1]. While the aim of improving access to sanitation is included both in the Millennium Development Goals (MDGs) (2000–2015) and the Sustainable Development Goals (SDGs) (2015–2030), it is gaining prominence in the SDGs. Although overall service coverage increased, the target was missed by a margin of almost 700 million people, many living in sub-Saharan Africa. The new SDGs were endorsed by the UN General Assembly in September 2015, which included a commitment to provide universal access to sanitation and drinking water services by 2030 [2]. Population and urban growth are interlinked and the foremost challenge facing Africa's sanitation provision. Rapid urbanization poses major infrastructure, economic, environmental, and social problems. Africa's urban population will double by 2030, and the difficulties African cities currently face in providing sanitation will be exacerbated. Rates of urbanization in Africa are the highest in the world (5.8% in sub-Saharan Africa) [1].

The capacity of insufficient wastewater treatment ponds (WWTPs) to cope with increasing wastewater loads due to an increase in population numbers is among the technical, social, economic, and environmental challenges facing governments in Africa. Another key challenge is to cope with variation in pollution loads caused by uncontrolled discharges into the sewage network (e.g., industrial discharge). Furthermore, high operational and maintenance costs lead to inappropriate sludge disposal and odor. Power cuts also play a major role in cases where energy is needed for the treatment process. Financial problems were observed to impact all the selected African countries, which negatively affects the operation, maintenance, or upgrade of WWTPs [3]. Release of insufficiently treated wastewater into the environment is also observed when treatment plants are dysfunctional or temporarily disconnected, which is common in countries such as Ghana and South Africa [3,4]. Workers and managers in charge of treatment plants also lack the skills to maintain them and are not motivated to maintain treatment plants due to low wages.

In a comparative study on wastewater practices of seven African countries, namely Algeria, Burkina Faso, Egypt, Ghana, Morocco, Senegal, and Tunisia [3], it was evident that at least seven out of ten WWTPs are either waste stabilization ponds or activated sludge. Table 1 shows an overview of the most used sanitation infrastructure in these counties. In all the selected countries, both activated sludge (AS) and waste stabilization pond systems presented 68−100% of all implemented units still in operation. AS systems are applied mostly by private entities (e.g., industry) in Ghana, while pond treatment systems are preferred by public entities. In Burkina Faso, only pond systems are used [3]. However, in the other five countries, both AS and pond systems have been used. Overall, waste stabilization pond systems are the most used treatment technology in Africa.

Table 1. Overview of the insufficient wastewater treatment plants (WWTPs) in operation at the seven selected countries (Algeria, Burkina Faso, Egypt, Ghana, Morocco, Senegal, and Tunisia) [3].

Country	Total Number	First Most Used Technology	Second Most Used Technology	Third Most Used Technology	Feed Flow Rate (m^3/h)
Burkina Faso	2	100% pond systems	N/A	N/A	96 (for the largest pond system)
Ghana	19 [1] (4 ponds under construction)	42% are pond systems	26% are activated sludge (AS) or aerated tanks	16% are anaerobic digesters	1–25
Senegal	9	56% are ponds	44% are AS	N/A	N/A
Algeria	123 (96 under construction, of which 60 are AS and 36 are pond systems)	55% are ponds	45% are AS	N/A	8–2750
Egypt	>99	Between 65% and 85% are AS	About 10% are pond systems	Others	–
Morocco	>100	>77% are pond systems	5% are AS	Trickling filters	12–4914
Tunisia	109	82% are AS	13% are pond systems	Trickling filters and wetlands	4–3250

[1] Among the WWTPs under construction, 60 are AS and 36 are pond systems.

Conventional domestic wastewater treatment systems (including AS reactors and trickling filters) [5] are labor- and energy-intensive, often requiring expensive infrastructure and regular maintenance, leaving the municipalities of rural communities in developing countries unable to treat the domestic inflow [6]. This led to the need and emergence of cost-effective wastewater treatment systems (also referred to as ecotechnologies), which require less infrastructure and use little or no energy [7]. Among these ecotechnologies are wastewater stabilization pond (WSP) systems [7]. WSP systems are regarded as the wastewater treatment method of choice worldwide [8]. In cooler climates, including Europe, WSP systems are used in small rural communities of around 2000 people [9]. However, in warmer climate conditions (including Africa and Asia), WSP systems are used for larger communities of up to a million citizens [10]. WSP systems use a series of different pond types, namely anaerobic ponds, facultative ponds, oxidative ponds, and aerobic ponds, each creating its own specific microbial, chemical, and biochemical characteristics, based on the oxygen level and nutrient content of the respective environments [11,12]. Each pond is dedicated to addressing a specific aspect of wastewater through the metabolic activities of the algae and microorganisms, while the organic and chemical content of the wastewater is metabolized and removed from the liquid via the cycling of nutrients [13].

Sustainable development and the global move towards a more circular use of resources where waste is reduced and resources are recycled, have driven a paradigm shift within the scientific community with regard to wastewater solutions [14]. Furthermore, nutrient recovery from domestic wastewater extends beyond direct economic benefits to human health and also environmental benefits, reducing eutrophication [14].

Phycoremediation is a biological cleanup technology involving the use of micro- or macro-algae for the biological transformation of contaminants, including nutrients such as inorganic and organic carbon, phosphorus, nitrogen, sulfates, and heavy metals [12,15,16]. During the process of phycoremediation, algae utilize nitrogen, phosphorus, carbon, and other salts from the wastewater for their growth [17]. This cleanup process has been a research subject for many decades [18]. Although a vast amount of literature related to phycoremediation and domestic wastewater treatment is available from developing countries, little is known of this technology on the African continent.

There are only a few known published case studies in Africa from the literature on phycoremediation and domestic wastewater treatment. One of these case studies is the Belmont Valley integrated algae pond system (IAPS) using natural occurring algae in the algae raceway. The IAPS is a derivation from the Oswald designed algal integrated wastewater pond system. The IAPS technology was introduced to South Africa in 1996 and a pilot plant was designed and commissioned at the Belmont Valley WWTW in Grahamstown. The IAPS primary treatment of domestic wastewater makes use of an advanced facultative pond which houses an in-pond anaerobic digester, followed by high-rate algae oxidation ponds with paddlewheel mixing. The IAPS technology makes use of gravity, solar energy, and natural occurring algae in the high-rated algae ponds [19]. Other places in Africa where IAPS has been used are Morocco and Zimbabwe [20]. However, in the case study by Oberholster et al. [12], the authors used specific phycoremediation to improve the treatment capacity of domestic wastewater in existing pond systems in South Africa, by mass culturing and inoculation of a consortium of selected algae.

The objective of the current study was to present comparative data of the nutrient treatment capacity of two municipal wastewater pond systems in South Africa, after implementing the phycoremediation technology process in maturation ponds by mass inoculation of a consortium of selected microalgae. Data were recorded before the technology was implemented at both plants and monitored after phycoremediation was introduced at the two wastewater treatment plants. The latter two municipal systems treat $\leq$0.5 and 4.5 mL day^{-1} of domestic wastewater and are located in two different locations with vastly different climate conditions (coastal and inland).

2. Materials and Methods

2.1. Background

The improvement of the process of phycoremediation in the two municipal pond treatment systems utilized a specific consortium of microalgal species (*Chlorella vulgaris* and *Chlorella protothecoides*) to remove nutrients through mass inoculation. Renuka [21] stated that *Chlorella* spp. is one of the most explored microalga genera in relation with nutrient removal in different types of wastewater. The microalgae scaling-up from laboratory cultures to the outdoor photobioreactors was performed according to Oberholster et al. [12]. The intention was to implement a self-sustaining system that can be operated independently of electricity or expensive chemicals and can be effectively maintained within financial and capacity constraints of developing countries [12].

The Motetema WWTP system (25°6′3.87″ S and 29°28′6.78″ E) is located inland in the Limpopo province of South Africa, in a temperate climatic zone, and treats 4.5 mL day^{-1} of domestic wastewater. This WWTP system consists of twelve earth ponds organized in two series of six each, parallel to one another. Of the twelve ponds, only six ponds are operated at a time, while the other six ponds are dried for sludge removal [12]. The Brandwag WWTP system, on the other hand, has a coastal location and is near the small town of Brandwag in the Western Cape, South Africa (34.0493° S and 22.0573° E). This WWTP system consists of a series of seven earth ponds, namely anaerobic ponds, primary facultative ponds, secondary facultative ponds, and maturation ponds, that naturally overflow gravitationally from one pond to another. The above-mentioned pond treatment system treats 0.5 mL day^{-1} of domestic effluent and is categorized as a micro-WWTP system (Figure 1).

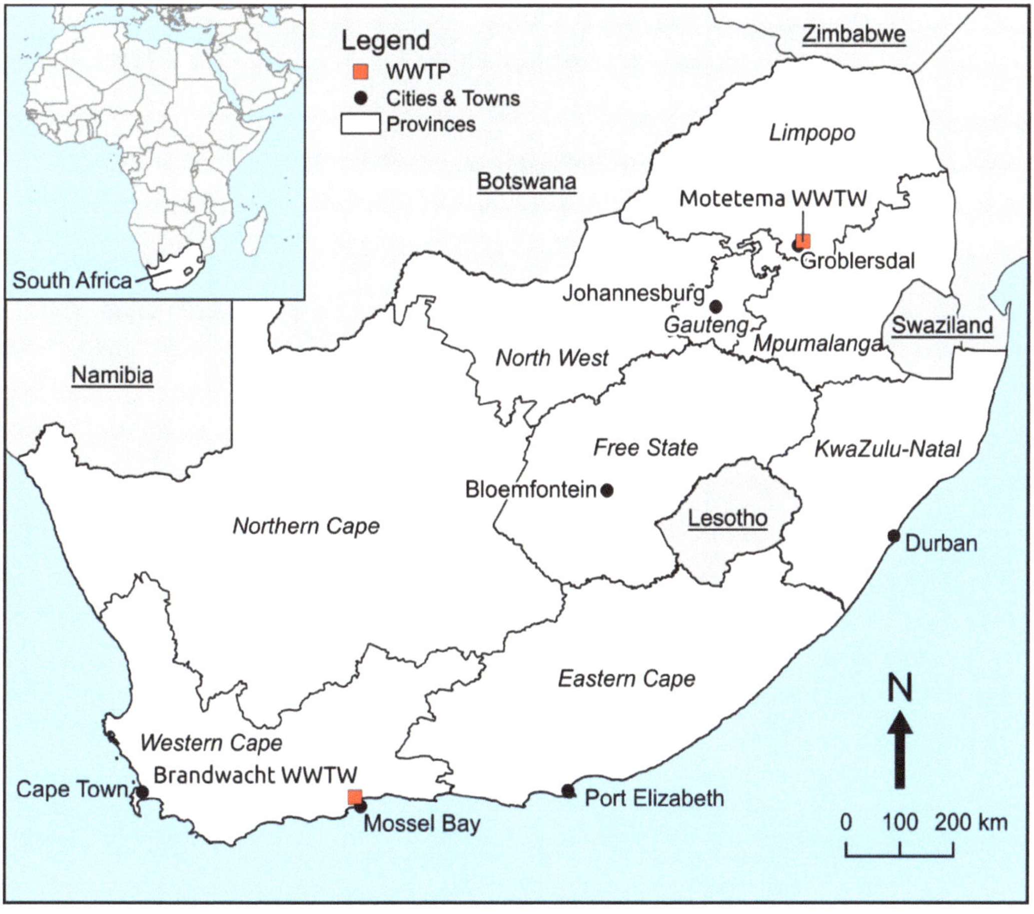

Figure 1. The locations of the Motetema and Brandwag wastewater treatment pond (WWTP) systems in South Africa.

Five semi-transparent photobioreactor tanks with a capacity of 5000 L each, diameter of 1800 mm, and height of 2040 mm, were installed at the Motetema wastewater pond system (Figure 2A), and three at the Brandwag wastewater pond system (Figure 2B), also with a capacity of 5000 L each [12,22].

(**A**) (**B**)

Figure 2. Photobioreactors for the outdoor mass culturing of the consortium of specific microalgae at the Motetema (**A**) and Brandwag (**B**) municipal WWTP systems.

The microalgae strains, *Chlorella vulgaris* (Beijerinck, ATCC: 30821) and *Chlorella protothecoides* (Kruger, ATCC: 30411), were acquired from American Type Culture Collection (ATCC) and cultured as described by the ATCC protocol and used as starting culture. In each of these outdoor bioreactors, a volume of 25 L of the two algal species (50:50%) was added to 1000 L of dechlorinated tap water and 10 g of synthetic fertilizer (10 phosphorus:20 nitrogen:10 potassium) for the start-up culture in the outdoor photobioreactors. This culture was upscaled to 5000 L in each photobioreactor and released in the selected different oxidation ponds on a 3 to 4 weekly basis, depending on the season. The photobioreactors' algae dry weights (wt.) during the different culturing phases were the following: lag phase (average dry wt. 1.12 ($\pm$0.13) mg L^{-1}), exponential phase (average dry wt. 1.69 ($\pm$0.15) mg L^{-1}), and in the stationary phase (1.18 ($\pm$0.13) mg L^{-1}) for the Motetema WWTP outdoor photobioreactors. In the Brandwag WWTP outdoor photobioreactors, it was: lag phase (average dry wt. 1.01 ($\pm$0.18) mg L^{-1}), exponential phase (average dry wt. 1.55 ($\pm$0.11) mg L^{-1}), and in the stationary phase (1.10 ($\pm$0.9) mg L^{-1}). Solar illuminance reached 1251 μmol m^{-2} s^{-1} during midday in the Motetema WWTP outdoor photobioreactors and 1077 μmol m^{-2} s^{-1} in the Brandwag WWTP outdoor photobioreactors. Temperature ranges from 29 to 31 °C during midday in the Motetema WWTP outdoor bioreactors and 27 to 28 °C in the Brandwag WWTP outdoor bioreactors. The pH in the Motetema WWTP outdoor photobioreactors was 9.1, and 8.8 in the Brandwag WWTP outdoor photobioreactors during midday (12h00–13h00). All the above data were generated in duplicate during the startup of the mass algae culturing in the photobioreactors. The cultured microalgae were release from the photobioreactors when they reached a chlorophyll-a value of 250 mg L^{-1} to prevent overshadowing and suspension in the photobioreactors [23]. The protocol of Porra et al. [24] was followed for the determination of suspended chl-*a* (μg L^{-1}) in the photobioreactors. The volume of cultured algae in the photobioreactors was stirred manually every four days with an oar by the wastewater plant operator. The total volume of cultured algae being released in the maturation ponds at one time (every three or four weeks) was 15,000 L at the Motetema pond treatment system and 12,000 L at the Brandwag pond treatment system. Of the 5000 L in each of the outdoor photobioreactors, only 4000 L was released at each mass inoculation, while 1000 L was left in the photobioreactor for subculturing. The pond systems are based on natural overflow from one pond to another, using no electricity, and are gravity-fed. Residence times for both the Motetema and Brandwag pond treatment systems were between 26 and 29 days depending on environmental conditions, for example, rain conditions. The average depth of the maturation ponds in both systems was 1.5 m. Although the photobioreactors were connected to all the maturation ponds by a piping network, only the first three maturation ponds, after the facultative pond, were inoculated with the consortium of microalgae in both systems. Chemical data presented in the current study were related to the last two ponds, which include the effluents of the last pond over a period of one year.

2.2. Physical and Chemical Sampling Analyses

Water samples were taken before and after treatment (one year later) at the outflow of the last two maturation ponds at each of the case study locations and analyzed according to Oberholster et al. [12]. In short, the first water samples were taken two times ($n = 2$) before any occurrence of mass microalgae inoculation at the two pond treatment systems, while the last water samples were taken three times ($n = 3$), four weeks after the last mass microalgae inoculation over the period of one year. On-site water quality parameters, such as temperature, pH, and electrical conductivity (EC), were measured using a Hanna HI991300 handheld water quality meter. The sampled water from each sampling location was divided into two subsamples for the following analyses: (a) one liter for dissolved nutrient analyses, and (b) one liter which was kept for the analyses of chemical oxygen demand (COD) and sulfates. Samples were kept on ice and sent to an accredited chemical laboratory for analysis within 48 h of collection. The standard chemical analysis procedures for all the water samples followed the approved analytical methods detailed in "Standard Methods for the Analysis of Water and Wastewater" [25]. Data of the effluent of the last maturation pond of each of the pond systems were compared with the South African wastewater discharge standards [6]. The following methods were used for the analyses: American Public Health Association (APHA) 4500-N: total nitrogen; APHA 5310-B: total organic carbon; APHA 5220-D/HACH Method 8000: total COD; APHA 4500-P: total phosphorus; APHA 2540-D: suspended solids; APHA 4500, APHA 4500-SO_4 G: sulfate; APHA 4500-NH_3 H: ammonia; APHA 4500-PO_4 G: phosphate. Wastewater samples before and after microalgae treatment of the different ponds were filtered through 0.22 µm pore size Whatman GF filters to separate the microalgae from the treated water for the determination of nitrogen and phosphorus uptake by the microalgae.

2.3. Phytoplankton Sampling and Analyses

Phytoplankton sampling was performed at the outflow of the last two maturation ponds of each of the case studies before and one year after treatment. The water sample was divided into two subsamples: (a) preserved for microscope soft algae identification and (b) unpreserved for diatom identification. The soft algae subsample was fixed in 2.5% glutaraldehyde in the field and kept cold and in the dark until the laboratory analysis. Diatom samples of each sampling site were cleared of organic matter by heating in a potassium dichromate and sulfuric acid solution and the cleared material was rinsed, diluted, and mounted in a Pleurax medium for microscopic examination. All algae were identified using a compound microscope (Carl Zeis, Jena, Germany) at a 1250× magnification [26–29]. Samples were sedimented in a Sedgewick–Rafter sedimentation chamber and analyzed using the strip–count method [30]. The Berger–Parker dominance index [31] was used to measure the evenness or dominance of each algal species at each sampling pond using actual algae cell numbers:

$$D = N_{\max}/N \tag{1}$$

where $N_{\max}$ = the number of individuals of the most abundant species present in each sample, and N = the total number of individuals collected at each pond site.

2.4. Weather Condition at Each WWTP Location

Weather data were generated from World Weather Online [32] to determine the influence of climate conditions on the treatment performance of the consortium of algae.

2.5. Statistical Analyses

Water quality data were captured in Microsoft Excel Spreadsheets for both wastewater treatment works. Simple error bar graphs for the water quality parameters were created in SigmaPlot (Version 14), and statistical analysis was performed using SigmaStat 14. The Mann–Whitney Rank Sum test was used to determine statistical significance for each parameter in each of the ponds for both wastewater treatment plants. Where data was normally distributed, the *t*-test was performed to determine significance. In all tests, the level of significance was adopted at = 0.05.

3. Results and Discussion

Chinnasamy et al. [33] envisaged that the use of a consortium of algae can be a promising alternative to increase the efficiency of wastewater treatment, but that further research is needed. They, however, speculated that competition between native and non-native algal strains may be a problematic factor for more effective treatment [33]. Furthermore, they also stated that environmental conditions such as climate and different types of wastewater may hamper the effective treatment of the consortia of algal strains [33]. Nevertheless, a study by Ruiz-Martinez et al. [34] showed that after a 42-day culturing period in WWTP effluent, a mixture of the microalgae *Chlorococcales* and cyanobacteria isolated from a submerged anaerobic bioreactor achieved reduction efficiencies forNH_4-N and PO_4-P of 67.2% and 97% respectively.. Silva-Benavides and Torzillo [35] compared the removal efficiency of *Chlorella* and a *Chlorella–Planktothrix* co-culture grown in municipal wastewater and showed that co-culture of *Chlorella–Planktothrix* removed the highest nitrogen concentration (80%) over the two-day exposure period. In 2015, Renuka et al. [21] presented a comprehensive review on different microalgal consortia in various types of wastewater. Nonetheless, for all the above studies, a consortium of microalgae was used to test for their nutrient removal in batch flask cultures, small high-rated algal ponds, small algal race ways, or outdoor photobioreactors [21,36]. The scaling-up of microalgal cultures to outdoor conditions presents additional challenges, for example variable irradiance, temperature, and rainfall [37]. These challenges are not observed in the constant environment experience by small-scale laboratory cultures. To the authors' knowledge, the current study is the first report on a comparative study using two existing municipality pond treatment systems to treat domestic wastewater by improving the process of phycoremediation using a specific consortium of algae and mass inoculation over a period of one year. Table 2 summarizes the data, showing the percentage reduction after treatment as well as the standard deviation. Figures 3 and 4 shows the standard error and statistical significance of Brandwag wastewater treatment works and Motetema wastewater treatment works, respectively.

Table 2. Median of physical and chemical parameters measured before (day 0) and after treatment (one year later) with a specific consortium of microalgae at the Motetema (MT) wastewater pond system and Brandwag wastewater pond system.

Parameters	Motetema Wastewater Pond System						Brandwag Wastewater Pond System					
	Before Treatment Pond 6 (STDEV)	After Treatment Pond 6 (STDEV)	Reduction after Algae Treatment (%)	Before Treatment Pond 7 (Outflow) (STDEV)	After Treatment Pond 7 (Outflow) (STDEV)	Reduction after Algae Treatment (%)	Before Treatment Pond 5 (STDEV)	After treatment Pond 5 (STDEV)	Reduction after Algae Treatment (%)	Before Treatment Pond 6 (Outflow) (STDEV)	After Treatment Pond 6 (Outflow) (STDEV)	Reduction after Algae Treatment (%)
Total nitrogen (mgL)	50 (3)	24 (7)	52.00	41 (5)	11 (4)	73.1	70 (11)	45 (7)	43.0	28 (7)	17 (4)	35.4
Total organic carbon (mgL)	53 (10)	32 (12)	39.62	50 (4)	23 (3)	54.0	195 (18)	69 (9)	69.0	52 (11)	42 (8)	22.2
Total chemical oxygen demand (mgL)	140 (26)	145 (46)	−3.57	122 (72)	114 (42)	6.6	572 (83)	147 (23)	75.0	235 (39)	97 (11)	60.0
Total phosphorous (mgL)	17 (2)	19 (9)	11.76	12 (2)	6 (1)	50.0	9.5 (4)	2.7 (0.9)	74.5	9 (3)	2.2 (0.3)	74.4
Sulfate as SO_4 dissolved (mgL)	195 (90)	73 (20)	62.56	184 (85)	78 (24)	58.0	81 (10)	113 (21)	−45.0	172 (13)	127 (11)	26.9
Ortho Phosphate as P (mgL)	12 (1)	2.36 (0.79)	80.33	8 (3)	1.36 (0.7)	83.0	5.8 (1.1)	2.1 (0.6)	77.0	3.7 (0.8)	0.74 (0.4)	87.0
Ammonia as N (mgL)	20 (13)	9 (4)	55.00	19 (2)	0.1 (0.85)	99.4	48 (11)	32 (7)	43.0	27 (9)	23 (8)	16.6
Electrical conductivity (mSm)	114 (10)	185 (33)	−62.28	118 (2)	160 (22)	−36.0	117 (11)	125 (21)	−7.1	112 (10)	128 (9)	−16.0
pH (Lab) 20 °C	8.1 (0.15)	8.4 (0.06)	−3.70	8.2 (0.06)	9.1 (0.49)	10.9	7.7 (0.03)	8.8 (0.03)	−14.1	8.2 (0.02)	8.7 (0.18)	−6.1

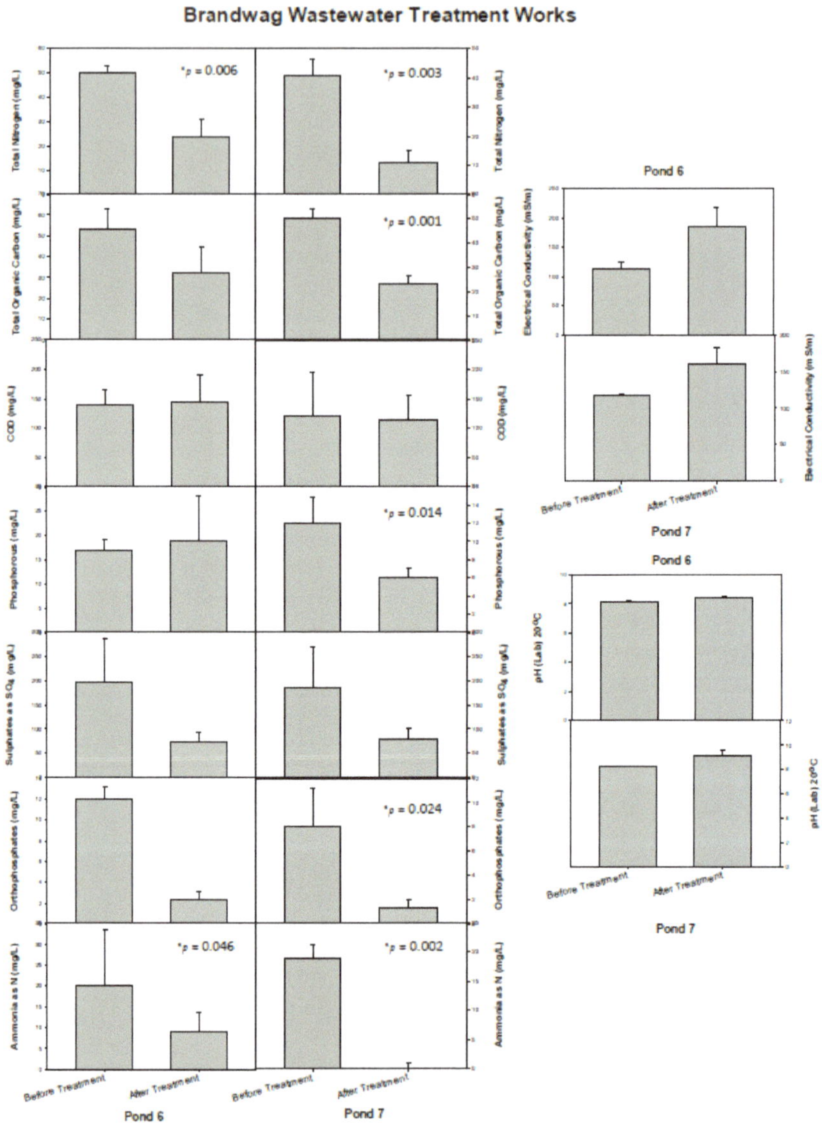

Figure 3. Water quality results from Brandwag wastewater pond treatment before and after phycoremediation treatment. * *p*-values shown for those parameters that were statistically significantly different before and after treatment.

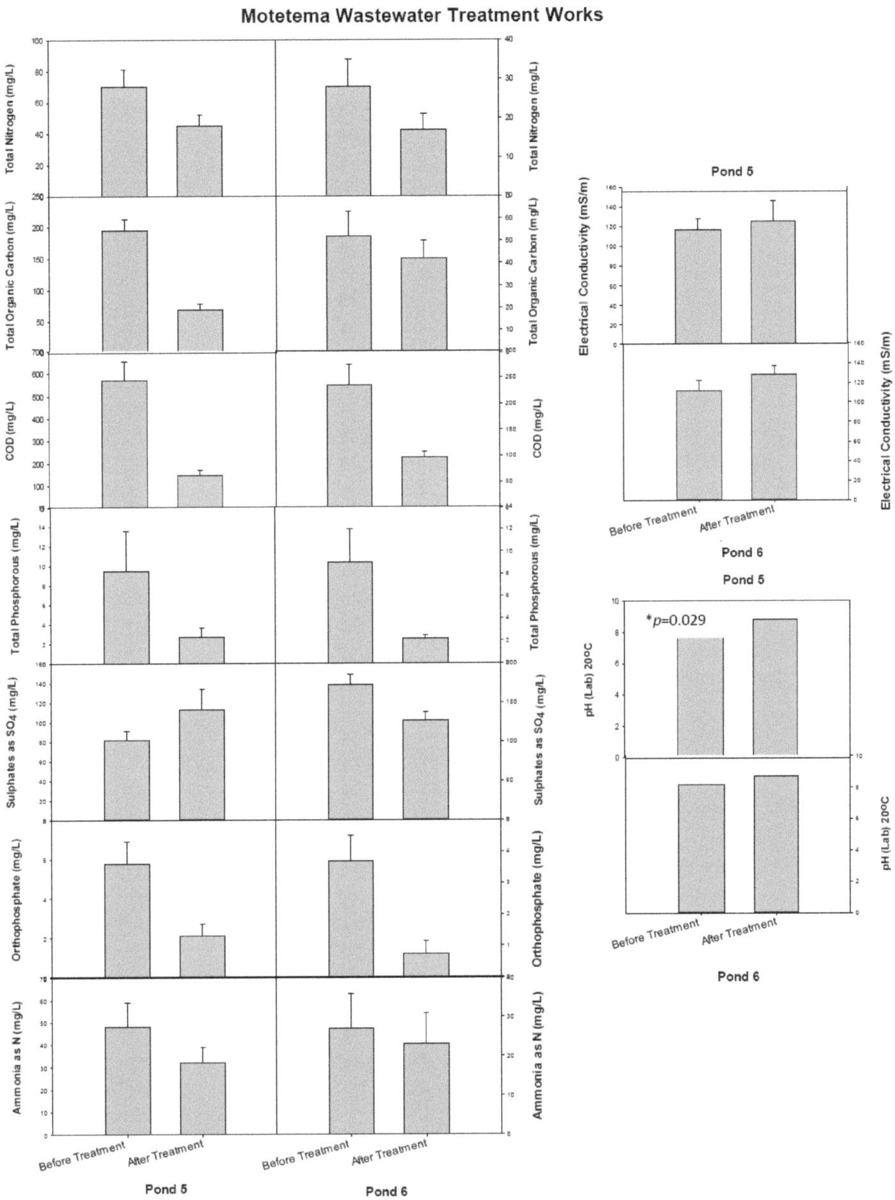

Figure 4. Water quality results from Motetema wastewater pond treatment plants before and after phycoremediation treatment. * *p*-values shown for those parameters that were statistically significantly different before and after treatment.

In the current study, the average pH of both pond treatment systems increased after treatment with the specific algae consortium. The average pH was determined from sampling data two times ($n = 2$) before any occurrence of mass microalgae inoculation at the two pond treatment systems, and three times ($n = 3$) one year after the mass microalgae inoculation. The increase in pH values after specific algae treatment suggested a higher photosynthetic rate, drawing more dissolved CO_2 from the water (Table 2). The pH values in the two pond treatment systems during the study period did not reach a value of 11, which is required for the precipitation of phosphorus, indicating that the microalgae were the main mechanism for the removal of phosphorus in the two treatment pond systems [38] (Table 2). The highest reduction of total nitrogen (73.1%) occurred in the last maturation pond of the Brandwag pond treatment system from pre- to post-treatment. Nevertheless, the highest removal of total phosphorus (74.4%) from pre- to post-specific mass algae treatment occurred in the effluent of the last maturated pond at the Motetema pond system. The uppermost removal of ammonia (99.4%) after specific algae mass treatment occurred at the effluent of the last maturated pond of the Brandwag pond system (Table 2). The N:P ratios before mass algae treatment in the final effluent of the Brandwag and Motetema pond systems were 3.4:1 and 3.5:1. After one year of mass microalgae treatment at the Brandwag and Motetema pond systems, the N:P ratios were 1.8:1 and 10:1. Although both the systems displayed a reduction of unfiltered COD after specific algae treatment in the last pond (Motetema pond treatment system 92 mgL^{-1} (60% reduction); Brandwag pond treatment system 114 mgL^{-1} (6.6% reduction)), they were unable to reduce unfiltered COD levels to meet the South African effluent discharge standard of 75 mgL^{-1} (Table 2). However, a large portion of this residual COD was possibly related to the microalgae biomass that increased the chemical demand of dissolved oxygen levels [12]. The treatment performance of microalgae is strongly related to three primary nutrients: C, N, and P. The C (C as COD):N:P ratios before mass algae treatment in the final effluent at the Brandwag and Motetema pond systems were 1.4:4:15 and 1:9:30 [39]. After one year of mass microalgae treatment of the Brandwag and Motetema pond systems, the C:N:P ratios in the final effluent were 1.1:12:22 and 1.4:7:63. Comparing the water quality data of the final effluent of both systems with the South African general standards for wastewater effluent, it was evident that the following standard target concentrations were not met after one year's treatment with specific microalgae: ammonia (23 mgL^{-1}) in the final effluent of the Motetema pond treatment system (South African general effluent standard 6 mgL^{-1}), total nitrogen (17 mgL^{-1}) in the final effluent of the Motetema pond treatment system (South African general effluent standard 15 mgL^{-1}), and EC (160 mSm^{-1}) in the final effluent of the Brandwag pond treatment system (South African general effluent standard 150 mSm^{-1}) (Table 2). Overall, a higher reduction of all the selected parameters was observed in the last maturation pond of both pond treatment systems, except for EC values that increased from the second to last maturation pond to the last maturation pond (Table 2). To reach the latter nutrient target values, it is suggested to increase the number of algae photobioreactors.

Mahapatra et al. [40] reported on the treatment efficiency of a natural algae-based sewage treatment plant in India of 67.65 million liters per day treatment capacity, with a residence time of 14.3 days. Their study showed a moderate performance, with the removal of total COD (60%) and filterable COD (50%), as sewage travels from the inlet to the outlet [40]. Furthermore, nitrogen content showed sharp variations with total Kjeldahl nitrogen removal of 36%, ammonium removal efficiency of 18%, nitrate (NO_3-N) removal efficiency of 22%, and nitrite removal efficiency of 57.8%. The predominant algae classes were euglenoids (in facultative pond) and chlorophyceae in the maturation ponds. However, in the current study, the resident times of the two pond treatment systems were much longer with a much lower treatment capacity per day, as in the case of the study by Mahapatra et al. [40]. Furthermore, Mahapatra et al. [40] compared data generated from the inlet to the outlet with treatment of natural algae. In the current study, the authors evaluated the treatment capacity of the last two maturation ponds before and after mass treatment with a specific consortium of algae before and one year after treatment.

Before treatment with the specific consortium of algae (day 0), the dominant taxa consisted of 40% *Microcystis aeruginosa* and 24% *Micractinium pussillum* in the last maturation pond of the Brandwag pond treatment system (Figure 5A). Although *M. aeruginosa* is one of the most common species in freshwater ecosystems, it is not common in pond treatment systems [41,42]. However, after one year of algae inoculation, the algae assemblage changed, with 45% *C. vulgaris* as the dominant taxa, followed by *C. protothecoides* with 37%.

It must be mentioned that a natural strain of *C. vulgaris* occurred in relatively low numbers at the last pond of the Brandwag pond treatment system before the inoculation of the consortium of microalgae. These findings suggest that the combination of both *C. vulgaris* and *C. protothecoides* was able to suppress the dominance of *M. aeruginosa* after inoculation with the consortium of specific microalgae (Figure 5B).

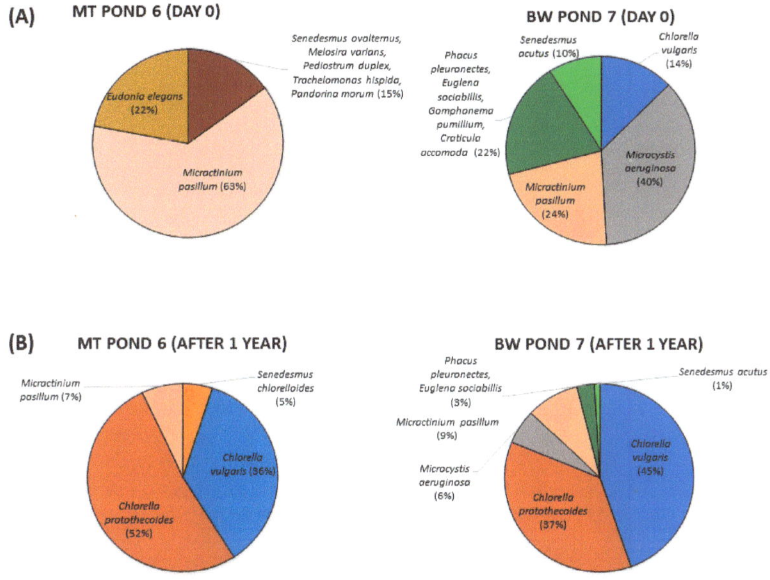

Figure 5. (**A**) Distribution of natural algae species in the last maturation pond before mass inoculation of the consortium of specific microalgae species (day 0) at the Motetema (MT) and Brandwag (BW) municipality WWTP systems. (**B**) Distribution of algae species in the last maturation pond 1 year after inoculation of the consortium of specific microalgae species at the Motetema (MT) and Brandwag (BW) municipality WWTP system.

At the Motetema pond treatment system, before the inoculation of the consortium of algae, *M. pussillum* was the dominant taxa (63%) in the last pond (Figure 5A), followed by the green algal *Eudorina elegans*. One year later, after the inoculation with the consortium of specific algae, the last pond was dominated by *C. protothecoides* (52%) and *C. vulgaris* (36%) (Figure 5B). No natural *Chlorella* spp. was observed at the Motetema pond treatment system before the inoculation, as in the case of the Brandwag pond treatment system.

Almomani and Örmeci [43] showed that the microalgae *C. vulgaris* can grow and perform well at 20 and 36 °C in 500 mL sterilized reactors. Zhang et al. [44] showed that *Chlorella* spp. can grow in surface water temperatures in the temperature range of 5–30 °C, while Shi and Shi [45] reported significant growth of *C. protothecoides* at 28 °C. However, Shi and Shi [45] showed that a lower algae growth was observed at 35 °C. A study by González-Camejo et al. [36] used two outdoor photobioreactors to evaluate the effects of ambient temperature on an indigenous microalgae-nitrifying bacteria culture dominated by *Chlorella*. The authors conducted four experiments over different seasons, maintaining the temperature-controlled photobioreactor at 25 °C, while the temperature in the non-temperature-controlled photobioreactors was allowed to vary with the surrounding environmental conditions [36]. The authors reported that temperature in the range of 15–30 °C had no significant effects on the microalgae cultivation performance; however, when the temperature rose to 30–35 °C, the microalgae viability was significantly reduced.

In the current study, the monthly air temperature did not increase above 31 °C at the two localities of the two treatment plants.

The different observations in the treatment capacity at the two pond treatment systems under study can possibly be related to the characteristics of the domestic wastewater as well as environmental conditions, for example weather patterns. The strength of wastewater refers to its organic matter content, where the higher the organic matter content, the 'stronger' the wastewater, which is measured by its biological oxygen demand (BOD_5) or COD value [46]. From the current study, it was evident that there was a large difference in the COD values between the two pond treatment systems before the treatment with a specific consortium of microalgae. The Brandwag WWTP system had a value of 140 mgL^{-1} at Pond 6 and the Motetema WWTP system had a value of 576 mgL^{-1} at Pond 5 (Table 2). The latter observation indicates that the community of the town of Brandwag had high water consumption that caused a weaker wastewater inflow into the Brandwag WWTP [46], which may have played a role in the results of the current study.

Borowitzka [37] stated that the number of sunny days and degree of cloud cover can determine the outcome of intensive outdoor algae culture systems. In the current study, the Motetema pond treatment system experienced more cloudless days than in the case of the Brandwag WWTP (Figure 6C,D). Cassidy [47] reported that the optimal growth temperature for *C. vulgaris* was between 25 and 30 °C, while Latala [48] observed poor or no growth at temperatures between 5 and 10 °C. Even though the minimum air temperatures measured at both localities (11 °C at the Brandwag pond treatment system and 4 °C at Motetema pond treatment system) were not optimal for the growth of *C. vulgaris*, a previous study by Oberholster et al. [12] showed that the temperature at the Motetema pond treatment system was not a major concern during the winter months, since these low air temperatures occur sporadically for two or three days during the winter season (Figure 6A,B). However, during the winter season, the microalgae culturing in the photobioreactors for the mass microalgae inoculation into the maturation ponds shift from a four-week cycle in the summer to a five-week cycle to reach the targeted chlorophyll-a value of 250 μgL^{-1} before inoculation at both pond treatment localities. The latter shift was possibly due to slower growth by the cultured microalgae in the photobioreactors at lower air temperatures (Figure 6A,B).

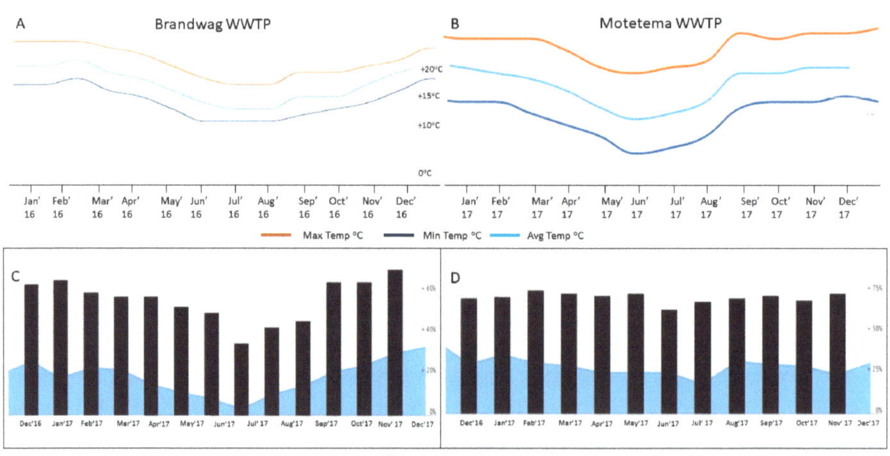

Figure 6. (**A**) Maximum, minimum, and average air temperature at the Brandwag municipality WWTP system during 2017. (**B**) Maximum, minimum, and average air temperature at the Motetema municipality WWTP system during 2016. (**C**) The percentages of cloud cover and humidity at the Brandwag municipality WWTP system in 2017. (**D**) The percentages of cloud cover and humidity at the Motetema municipality WWTP system in 2017.

It was evident from the data in the current study that the higher reduction of total phosphates (74.4%) in the effluent, after treatment with a consortium of microalgae at the Motetema WWTP system, was possibly related to some of the following factors: (1) the dominance of the algal taxa *C. protothecoides* (52%), and to a lesser extent, *C. vulgaris* (36%) in the microalgae consortium, (2) more cloudless days, (3) higher air temperature, and (4) a higher domestic wastewater strength. In the case of the Brandwag WWTP system, the higher reduction of total nitrogen can possibly be related to the dominance of *C. vulgaris* in the microalgae consortium, less cloudless days, lower air temperature, and a lower domestic wastewater strength. According to Mara [46], when comparing with other treatment methods, waste stabilization ponds are the most important method for wastewater treatment in developing countries, since space is available, and temperature is suitable for the process.

4. Conclusions

The algal diversity in wastewater stabilization systems is greatly influenced by the design parameters, environmental conditions, such as temperature, solar radiation, and domestic wastewater strength, as well as operation conditions. However, it was evident that more cloudless days, higher air temperature, and domestic wastewater strength may have played a role in the treatment capacity and dominance of certain algal species of the pond treatment systems of the two case studies, after mass inoculation with a specific consortium of microalgae. From the case studies presented, it was evident that improvement of phycoremediation through specific microalgae consortiums can play a major role in the removal of nutrients from domestic wastewater of WSP systems in Africa. The construction of the whole phycoremediation treatment systems (five photobioreactors, piping and installation), treating around 4.5 mL day^{-1} at a cost of 13,800 USD, make it a feasible option for developing countries to improve their existing pond treatment systems to reduce eutrophic conditions in receiving waterbodies. The implementation of phycoremediation in maturation treatment pond systems in African countries could also effectively minimize greenhouse effects, since the algae mass culturing in maturation ponds is a carbon-absorbing process, which can be used by municipalities in the carbon trading market and to provide a subsidy to reward their contributions to environmental protection. Currently, further research is in progress for using the algae biomass in the circular economy as a biofertilizer in African countries with a dominant agricultural economy.

Author Contributions: Conceptualization, P.J.O. and M.S.; methodology, M.S.; software, A.-M.B.; validation, P.J.O., M.S. and A.-M.B.; formal analysis, A.-M.B.; investigation, P.J.O.; resources, M.S.; data curation, A.-M.B.; writing—original draft preparation, A.-M.B.; writing—review and editing, P.J.O. and M.S.; visualization, A.-M.B.; supervision, P.J.O. and A.-M.B.; project administration, M.S.; funding acquisition, M.S. All authors have read and agreed to the published version of the manuscript.

Funding: This research contract PECG1/08/2019/141 was funded by the Climate Finance and Environment Division (PECG1) of the African Development Bank (AfDB).

Institutional Review Board Statement: Not applicable.

Informed Consent Statement: Not applicable.

Data Availability Statement: Weather data is available from https://weathersa.co.za.

Acknowledgments: The authors express their gratitude to the African Development Bank for funding the current study. We also want to thank the municipalities involved and providing resources for the sampling of the Brandwag WWTP and Motetema WWTP.

Conflicts of Interest: The authors declare no conflict of interest.

References

1. Oberholster, P.; Munyayi, R.C.; Turton, A.; Du Plessis, A. Overview. In *Sanitation and Wastewater Atlas of Africa*; AfDB: Abidjan, Ivory Coast; UNEP: Nairobi, Kenya; GRID-Arendal: Arendal, Norway, 2020; Volume 283, pp. 9–35, ISBN 978-82-7701-194-3.
2. Weststrate, J.; Dijkstra, G.; Eshuis, J.; Gianoli, A.; Rusca, M. The sustainable development goal on water and sanitation: Learning from the Millennium Development Goals. *Soc. Indic. Res.* **2019**, *143*, 797–800. [CrossRef]
3. Nikiema, J.; Figoli, A.; Weissenbacher, N.; Langergraber, G.; Marrot, B.; Moulin, P. Wastewater treatment practices in Africa—Experiences from seven countries. *SSP* **2013**, *14*, 26–34. Available online: https://hdl.handle.net/10568/40210 (accessed on 5 July 2010).
4. Oberholster, P.J.; Botha, A.-M.; Chamier, J.; De Klerk, A.J. Longitudinal trends in water chemistry and phytoplankton assemblage downstream of the Riverview WWTP in the Upper Olifants River. *Ecohydrol. Hydrobiol.* **2013**, *13*, 41–51. [CrossRef]
5. Sasongko, N.A.; Noguchi, R.; Ito, J.; Demura, M.; Ichikawa, S.; Nakajima, M.; Watanabe, M.M. Engineering Study of a Pilot Scale Process Plant for Microalgae-Oil Production Utilizing Municipal Wastewater and Flue Gases: Fukushima Pilot Plant. *Energies* **2018**, *11*, 1693. [CrossRef]
6. South Africa, Department of Water Affairs. *Revision of General Authorisations in Terms of Section 39 of the National Water Act, 1998 (Act No. 36 of 1998), No. 665 of 2013*; Department of Water Affairs: Pretoria, South Africa, 2013.
7. Gray, N.F. (Ed.) How nature deals with waste. In *Biology of Wastewater Treatment*, 2nd ed.; Series on Environmental Science and Management; Imperial College Press: London, UK, 2004; Volume 4, pp. 1–131. [CrossRef]
8. Mohammed, B. Design and performance evaluation of a wastewater treatment unit. *AU JT* **2006**, *9*, 193–198.
9. Bucksteeg, K. German experiences with sewage treatment ponds. *Water Sci. Technol.* **1987**, *19*, 17–23. [CrossRef]
10. Barceló, D.; Petrovic, M. Waste water treatment and reuse in the Mediterranean region. In *The Handbook of Environmental Chemistry*; Springer: Berlin/Heidelberg, Germany, 2011; Volume 14.
11. Bitton, G. Formula handbook for environmental engineers and scientists. In *Environmental Science and Technology*; A Wiley-Interscience Series of Texts and Monographs; Schoor, J.L., Zehnder, A., Eds.; John Wiley & Sons: Hoboken, NJ, USA, 1998; Volume 117.
12. Oberholster, P.J.; Cheng, P.-H.; Genthe, B.; Steyn, M. The environmental feasibility of low-cost algae-based sewage treatment as a climate change adaption measure in rural areas of SADC countries. *J. Appl. Phycol.* **2019**, *31*, 355–639. [CrossRef]
13. Anderson, C.R.; Condron, L.M.; Clough, T.J.; Fiers, M.; Stewart, A.; Hill, R.A.; Sherlock, R.R. Biochar induced soil microbial community change: Implications for biogeochemical cycling of carbon, nitrogen and phosphorus. *Pedobiologia* **2011**, *54*, 309–320. [CrossRef]
14. Otoo, M.; Drechsel, P. (Eds.) *Resource Recovery from Waste: Business Models for Energy, Nutrient and Water Reuse in Low- and Middle-Income Countries*; Routledge—Earthscan: Oxon, UK, 2018. Available online: https://hdl.handle.net/10568/93011 (accessed on 5 July 2010).
15. Kang, D.; Kim, K. Real Wastewater Treatment Using a Moving Bed and Wastewater-Borne Algal–Bacterial Consortia with a Short Hydraulic Retention Time. *Processes* **2021**, *9*, 116. [CrossRef]
16. Dębowski, M.; Rusanowska, P.; Zieliński, M.; Dudek, M.; Romanowska-Duda, Z. Biomass Production and Nutrient Removal by Chlorella vulgaris from Anaerobic Digestion Effluents. *Energies* **2018**, *11*, 1654. [CrossRef]
17. Mohsenpour, S.F.; Hennige, S.; Willoughby, A.A.; Adeloye, A.; Gutierrez, T. Integrating micro-algae into wastewater treatment: A review. *Sci. Total Environ.* **2021**, *752*, 142–168. [CrossRef]
18. Dębowski, M.; Zieliński, M.; Kazimierowicz, J.; Kujawska, N.; Talbierz, S. Microalgae Cultivation Technologies as an Opportunity for Bioenergetic System Development—Advantages and Limitations. *Sustainability* **2020**, *12*, 9980. [CrossRef]
19. Mambo, P.M.; Westensee, D.K.; Zuma, B.M.; Cowan, A.K. The Belmont Valley integrated algae pond system in retrospect. *Water SA* **2014**, *40*, 385–394. [CrossRef]
20. Oswald, W.J. Ponds in the 21st century. *Water Sci. Technol.* **1995**, *31*, 1–8. [CrossRef]
21. Renuka, N.; Sood, A.; Prasanna, R.; Ahluwalia, A.S. Phycoremediation of wastewaters: A synergistic approach using microalgae for bioremediation and biomass generation. *Int. J. Environ. Sci. Technol.* **2015**, *12*, 1443–1460. [CrossRef]
22. Van den Berg, M.F.; Botha, A.M.; Bierman, A.; Oberholster, P. Assessing domestic wastewater effluent with a battery of bioassays after treatment with a specific consortium of microalgae and different flocculation methods. *Water Air Soil Pollut.* **2020**, *231*, 257. [CrossRef]
23. Masaraure, R. The Development of an Approach for the Production and Use of Algae to Treat Urban Wastewater. 2016. Available online: https://repository.up.ac.za/bitstream/handle/2263/57260/Masaraure_Development_2016.pdf?isAllowed=y&sequence=1 (accessed on 5 July 2010).
24. Porra, R.J.; Thompson, W.A.; Kriedemann, P.E. Determination of accurate extinction coefficients and simultaneous equations for assaying chlorophylls *a* and *b* extracted with four different solvents: Verification of the concentration of chlorophyll standards by atomic absorption spectrometry. *Biochim. Biophys. Acta* **1989**, *975*, 384–394. [CrossRef]
25. American Public Health Association; American Water Works Association; Water Pollution Control Federation. *Standard Methods for the Examination of Water and Wastewater*, 19th ed.; APHA, AWWA, and WPCF: Washington, DC, USA, 1992.
26. Taylor, J.C.; Harding, W.R.; Archibald, C.G.M. *An Illustrated Guide to Some Common Diatom Species from South Africa*; WRC Report, No. TT 282/07; Water Research Commission: Pretoria, South Africa, 2007; pp. 1–178.

27. Truter, E. *An Aid to the Identification of the Dominant and Commonly Occurring Genera of Algae Observed in Some South African Impoundments*; Department of Water Affairs, Hydrological Research Institute: Pretoria, South Africa, 1987. Available online: https://www.dws.gov.za/iwqs/reports/tr/TR_135_Truter_1987_An_aid_to_the_identification_of_algae_in_South_African_impoundments_s.pdf (accessed on 5 July 2010).
28. Van Vuuren, S.; Taylor, J.; Gerber, A.; Van Ginkel, C. *Easy Identification of the Most Common Freshwater Algae: A Guide for the Identification of Microscopic Algae in South African Freshwaters*; North-West University and Department of Water Affairs and Forestry: Pretoria, South Africa, 2006. Available online: https://www.dws.gov.za/iwqs/eutrophication/NEMP/Janse_van_Vuuren_2006_Easy_identification_of_the_most_common_freshwater_algae.pdf (accessed on 5 July 2010).
29. Wehr, J.D.; Sheath, R.G. Freshwater habitats of algae. In *Freshwater Algae of North America: Ecology and Classification*; Wehr, J.D., Sheath, R.G., Eds.; Academic Press: Cambridge, MA, USA, 2003; pp. 11–57.
30. American Public Health Association. *Standard Methods for Examination of Water and Wastewater*, 20th ed.; American Public Health Association, American Water Works Association, Water Environment Federation: Washington, DC, USA, 2006.
31. Berger, W.H.; Parker, F.L. Diversity of planktonic foraminifera in deep-sea sediments. *Science* **1970**, *168*, 1345–1347. [CrossRef]
32. World Weather Online. 2021. Available online: https://www.worldweatheronline.com/ (accessed on 17 July 2021).
33. Chinnasamy, S.; Bhatnagar, A.; Claxton, R.; Das, K.C. Biomass and bioenergy production potential of microalgae consortium in open and closed bioreactors using untreated carpet industry effluent as growth medium. *Bioresour. Technol.* **2010**, *101*, 3097–3105. [CrossRef]
34. Ruiz-Martinez, A.; Garcia, N.M.; Romero, I.; Seco, A.; Ferrer, J. Microalgae cultivation in wastewater: Nutrient removal from anaerobic membrane bioreactor effluent. *Bioresour. Technol.* **2012**, *126*, 247–253. [CrossRef]
35. Silva-Benavides, A.M.; Torzillo, G. Nitrogen and phosphorus removal through laboratory batch cultures of microalgae Chlorella vulgaris and cyanobacterium Planktothrix isothrix grown as monoalgal and as co-cultures. *J. Appl. Phycol.* **2011**, *24*, 267–276. [CrossRef]
36. González-Camejo, J.; Aparicio, S.; Ruano, M.V.; Borrás, L.; Barat, R.; Ferrer, J. Effect of ambient temperature variations on an indigenous microalgae-nitrifying bacteria culture dominated by *Chlorella*. *Bioresour. Technol.* **2019**, *290*, 121788. [CrossRef]
37. Borowitzka, M.A. Culturing microalgae in outdoor ponds. In *Algal Culturing Techniques*; Andersen, R.A., Ed.; Academic Press: Cambridge, MA, USA, 2005; Chapter 14, pp. 205–206.
38. Larsdotter, K. Wastewater treatment with microalgae—A literature review. *Vatten* **2006**, *62*, 31–38. Available online: https://www.tidskriftenvatten.se/wp-content/uploads/2017/04/48_article_2125.pdf (accessed on 5 July 2010).
39. Falkowski, P.G.; Raven, J.A. *Aquatic Photosynthesis*, 2nd ed.; Princeton University Press: Princeton, NJ, USA, 2007.
40. Mahapatra, D.M.; Chanakya, H.N.; Ramachandra, T.V. Treatment efficacy of algae-based sewage treatment plants. *Environ. Monit. Assess.* **2013**, *185*, 7145–7164. [CrossRef] [PubMed]
41. Ndlela, L.L.; Oberholster, P.J.; Van Wyk, J.H.; Cheng, P.H. An overview of cyanobacterial bloom occurrences and research in Africa over the last decade. *Harmful Algae* **2016**, *60*, 11–26. [CrossRef]
42. Vasconcelos, V.M. Toxic cyanobacteria (blue−green−algae) in Portuguese fresh-waters. *Arch. Hydrobiol.* **1994**, *130*, 449–451. [CrossRef]
43. AlMomani, F.; Örmeci, B. Assessment of algae-based wastewater treatment in hot climate regions: Treatment performance and kinetics. *Process Saf. Environ. Prot.* **2020**, *141*, 140–149. [CrossRef]
44. Zhang, Q.; Zhan, J.-J.; Hong, Y. The effects of temperature on the growth, lipid accumulation and nutrient removal characteristics of Chlorella sp. HQ. Desalin. *Water Treat.* **2016**, *57*, 10403–10408. [CrossRef]
45. Shi, C.; Shi, X. Characterization of three genes encoding the subunits of light-independent protochlorophylide reductase in Chlorella protothecoides CS-41. *Biotechnol. Prog.* **2016**, *22*, 1050–1055. [CrossRef]
46. Mara, D. *Domestic Wastewater Treatment in Developing Countries*; Earthscan: London, UK, 2003. Available online: https://www.pseau.org/outils/ouvrages/earthscan_ltd_domestic_wastewater_treatment_in_developing_countries_2003.pdf (accessed on 10 February 2005).
47. Cassidy, K.O. Evaluating Algal Growth at Different Temperature. Master's Thesis, University of Kentucky, Lexington, KY, USA, 2011. Available online: https://uknowledge.uky.edu/bae_etds/3 (accessed on 22 August 2021).
48. Latala, A. Effects of salinity, temperature and light on the growth and morphology of green planktonic algae. *Oceanologia* **1991**, *31*, 119–138. Available online: http://www.iopan.gda.pl/oceanologia/OC_31/OC_31_119-138.pdf (accessed on 22 August 2021).

processes

MDPI

Article

Improving the Treatment Efficiency and Lowering the Operating Costs of Electrochemical Advanced Oxidation Processes

Thorben Muddemann [1,*,†], Rieke Neuber [2,*,†], Dennis Haupt [3], Tobias Graßl [2], Mohammad Issa [3], Fabian Bienen [4], Marius Enstrup [1], Jonatan Möller [1], Thorsten Matthée [2], Michael Sievers [3] and Ulrich Kunz [1]

1 Institute of Chemical and Electrochemical Process Engineering, Clausthal University of Technology, 38678 Clausthal-Zellerfeld, Germany; enstrup@icvt.tu-clausthal.de (M.E.); jonatan.moeller@tu-clausthal.de (J.M.); kunz@icvt.tu-clausthal.de (U.K.)
2 CONDIAS GmbH, 25524 Itzehoe, Germany; grassl@condias.de (T.G.); matthee@condias.de (T.M.)
3 CUTEC Research Center for Environmental Technologies, Clausthal University of Technology, 38678 Clausthal-Zellerfeld, Germany; dennis.haupt@cutec.de (D.H.); mohammad.issa@cutec.de (M.I.); michael.sievers@cutec.de (M.S.)
4 Institute of Engineering Thermodynamics, German Aerospace Center, 70569 Stuttgart, Germany; fabian.bienen@dlr.de
* Correspondence: thorben.muddemann@tu-clausthal.de (T.M.); neuber@condias.de (R.N.)
† These authors contributed equally to this work.

Abstract: Electrochemical advanced oxidation processes (EAOP®) are promising technologies for the decentralized treatment of water and will be important elements in achieving a circular economy. To overcome the drawback of the high operational expenses of EAOP® systems, two novel reactors based on a next-generation boron-doped diamond (BDD) anode and a stainless steel cathode or a hydrogen-peroxide-generating gas diffusion electrode (GDE) are presented. This reactor design ensures the long-term stability of BDD anodes. The application potential of the novel reactors is evaluated with artificial wastewater containing phenol (COD of 2000 mg L^{-1}); the reactors are compared to each other and to ozone and peroxone systems. The investigations show that the BDD anode can be optimized for a service life of up to 18 years, reducing the costs for EAOP® significantly. The process comparison shows a degradation efficiency for the BDD–GDE system of up to 135% in comparison to the BDD–stainless steel electrode combination, showing only 75%, 14%, and 8% of the energy consumption of the BDD–stainless steel, ozonation, and peroxonation systems, respectively. Treatment efficiencies of nearly 100% are achieved with both novel electrolysis reactors. Due to the current density adaptation and the GDE integration, which result in energy savings as well as the improvements that significantly extend the lifetime of the BDD electrode, less resources and raw materials are consumed for the power generation and electrode manufacturing processes.

Keywords: boron-doped diamond; EAOP®; electrochemical cell; gas diffusion electrode; hydrogen peroxide; ozonation; peroxone; wastewater treatment

Citation: Muddemann, T.; Neuber, R.; Haupt, D.; Graßl, T.; Issa, M.; Bienen, F.; Enstrup, M.; Möller, J.; Matthée, T.; Sievers, M.; et al. Improving the Treatment Efficiency and Lowering the Operating Costs of Electrochemical Advanced Oxidation Processes. *Processes* **2021**, *9*, 1482. https://doi.org/10.3390/pr9091482

Academic Editors: Dimitris Zagklis and Georgios Bampos

Received: 30 July 2021
Accepted: 19 August 2021
Published: 24 August 2021

1. Introduction

Electrochemical advanced oxidation processes (EAOP®) are promising key elements for achieving a circular economy. EAOP® provide versatile, efficient, and clean methods for wastewater treatment through the use of electrons and their ability to oxidize wastewater pollutants either directly or indirectly [1]. The electron transfer between the organic compound and the electrode occurs at the electrode surface with the direct electrochemical oxidation. In contrast, indirect electrochemical oxidation involves intermediates that perform the electron exchange between the organics and the electrode in the bulk of the electrolyte [1]. "Active" and "non-active" anodes are used for oxidation, whereby "active" anodes with low oxygen overpotentials (OVP), e.g., dimensionally stable anodes or Pt, favor the selective oxidation of organic species and are preferred for the generation of reactive chlorine species [2,3]. In contrast, anodes with high OVP show "non-active" behavior and enable complete organic oxidation. The ideal anodes for this purpose are made of SnO_2, PbO_2, Ti_4O_7, or boron-doped diamond (BDD) [4]. BDD anodes have high corrosion and electrochemical stability and a wide potential window for water discharge, resulting in high yields of hydroxyl radicals (·OH, Equation (1) [5]); therefore, reactors with BDD electrodes producing ·OH are capable of mineralizing nearly every organic compound [2,6,7] and were, therefore, chosen for this study.

In addition, electrochemical oxidation processes, especially those involving diamond electrodes, are easy to control and can be supplied with energy from both fossil and renewable energy sources [8].

Although BDD-based EAOP® systems are highly efficient, the high cost of the BDD electrode material [1,9,10], the operating costs, and the formation of toxic byproducts [11,12] hinder their integration in many applications. Usually, BDDs are used in combination with a H_2 evolution reaction (HER) at a stainless steel cathode (Equation (2)):

$$H_2O + BDD \rightleftharpoons BDD(\cdot OH)_{ads} + H^+ + e^- \quad (1)$$

$$2H_2O + 2e^- \rightleftharpoons H_2 + 2OH^- \quad (2)$$

This combination has the disadvantage of the H_2 evolution, which does not contribute to the water treatment performance, and might lead to the formation of foam during treatment [13]. This results in higher energy consumption owing to the gas isolation characteristics [12], a more complex process handling, and hinders the upscaling of the cells [3]. These disadvantages can be overcome by using a gas diffusion electrode (GDE). These electrodes produce hydrogen-peroxide (H_2O_2), using the oxygen reduction reaction for the generation of the oxidizing agents H_2O_2 and $HO_2{}^-$ (Equations (3) and (4)). H_2O_2 is one of the most versatile chemical compounds due to its specific reactivity over a wide pH range [14]. It has the advantage of being able to decompose into environmentally friendly and non-toxic products [15–17].

$$O_2 + 2H^+ + 2e^- \rightleftharpoons H_2O_2 \quad (3)$$

$$O_2 + H_2O + 2e^- \rightleftharpoons HO_2^- + OH^- \quad (4)$$

GDEs are 3D electrode systems that form a three-phase boundary layer, whereby the reaction takes place between the solid catalyst material, the gaseous oxygen, and the liquid phase of the electrolyte (Figure 1b) [18]. GDEs overcome the gaseous mass transfer limitations of 2D cathodes [19]. The usage of a GDE results in lower cell voltages owing to the lower cathodic potential of the oxygen reduction compared to the water reduction, whereby the gas evolution is avoided. Furthermore, such GDEs are inexpensive (approx. 1000 € m^{-2}).

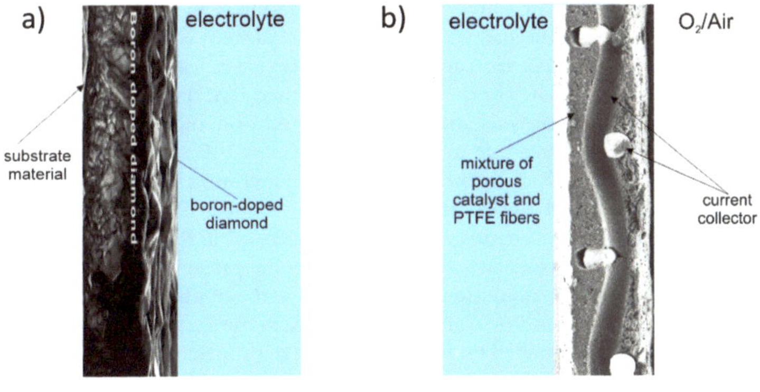

Figure 1. Scanning electron microscope images of a (**a**) boron-doped diamond anode and (**b**) a carbon-based H_2O_2-producing gas diffusion electrode on a Ni carrier.

BDD anodes are also multi-layer systems, as shown in Figure 1a. The BDD coating is built on top of a suitable carrier, exhibiting a high overpotential for the oxidation of water, which enables the direct oxidation of contaminants due to the weakly adsorbed ·OH [9]. The price range for such anodes in the literature is 12,000–18,000 € m^{-2} [9,10].

The first aim of this study is to reduce the anodic diamond wear off, enabling a long BDD service life and lower operational costs, in particular. The second aim is to combine the direct oxidation behavior of a BDD anode with the indirect oxidation behavior of cathodic-generated H_2O_2 to enable both higher degradation efficiency and lower energy consumption.

This is also crucial for the future-oriented efficient coupling of EAOP® with regenerative energy (RE) [8] for sustainable, autonomous, and low carbon emission applications [8]. Whether coupled with a wind turbine, photovoltaic element, or low-energy renewable sources, e.g., a microbial fuel cell, the aim must be to make better use of the collected energy by using an optimized reactor in combination with an optimized operation [20].

The basis of this study is an artificial wastewater (WW) containing phenol, which is a well-known model pollutant that is used to simulate the WW caused by pharma production in this study. In the literature, the electrochemical treatment of phenolic WW has been investigated by many authors [17] and the electrochemical treatment systems have been summarized. Luo et al. used a H_2O_2-producing GDE (10 cm^2; Vulcan XC-72-based) and its electroactivation of ·OH to treat 50 mg L^{-1} phenol in 0.2 mol L^{-1} Na_2SO_4. With the best operation point, they reached a TOC mineralization efficiency of 85% [21]. In-cell and ex-cell Fenton experiments were performed by Agladze et al. for the treatment of 282 ppm phenol in Na_2SO_4 and NaCl electrolytes with 90–98% treatment efficiency. The lifetime of the GDEs was the main problem in the investigations, as the GDEs destructed during the experiments (up to 2 h) [22]. Huang et al. also used in situ generation and activation of H_2O_2 with a metal–carbon catalyst for the treatment of 40 mg L^{-1} phenol. They achieved a similar removal efficiency rate of 87%, but the metal–carbon catalyst released iron during operation [23]. A similar investigation with a MnO_2–graphite electrode was performed by Abbas et al. [24], with activation using electro-Fenton. After 6 h, it was found that 88% of the phenol was removed by applying 0.08 kA m^{-2} at 60 °C with a current efficiency (CE) of approximately 4%. Higher efficiencies were shown by Ma et al. with ferric sulfate and potassium permanganate adsorbed onto active bentonite in a slurry-bed electrolytic reactor with graphite electrodes. They presented a 99% removal rate for 0.52 g L^{-1} phenol with the H_2O_2-producing and -activating process [25].

Other authors investigated the anode side of the electrolysis cell to treat phenolic WW. Shi et al. investigated a nanostructured, boron-doped diamond (NBDD) anode for the degradation of 0.04 M phenol in an acid milieu. They showed that the NBDD had a larger effective surface area of 31% in comparison to their standard BDD, which revealed a TOC removal of approx. 50% in comparison to 40% (both at 1 kA m^{-2}) [26]. PbO_2-based anodes are also known for ·OH generation. Gui et al. showed an average degradation efficiency for 100 mg L^{-1} phenol of 91.1% with this type of electrode [27]. Unfortunately, PbO_2 anodes are known to release toxic Pb^{2+} ions and have lower activity rates in comparison to BDD anodes [2]. Li et al. stated that the anodic property affects the pathway of phenol electrolysis and that ·OH-generating anodes are superior to other anode systems. They showed that ·OH-generating Ti/SnO_2-Sb degraded the TOC of an 80 mL phenolic solution to 99%, whereby non-radical-generating anodes such as Ti/RuO_2 and a Pt anode showed much lower efficiencies [28].

The presented studies described in the literature focused only on one electrode for the treatment of phenolic WW. In the present study, we prepare and use an improved BDD in combination with a GDE for the treatment of WW. Furthermore, the system does not depend on the pH value, as with the Fenton process. The aim is to degrade up to 99% or more of phenol to avoid toxic byproducts [29] and the comparison with a state-of-the-art electrolysis system with a H_2-generating cathode in an also novel reactor design and with ozonation and peroxone processes [30].

The oxidation of phenol involves several steps and intermediates before the degradation results in CO_2 and water as the final products [31]. The intermediates are precursors of methyl radicals, which lead to the etching of the diamond layer [32]. For this reason, the system studied here is also suitable for investigating how long a diamond electrode can be used in WW treatment before this consumable material has to be replaced at the end of its lifetime.

2. Materials and Methods

2.1. Artificial Wastewater

In this study, an artificial WW containing phenol was used to simulate the WW from a pharmaceutical manufacturer to ensure reproducibility. The same composition of 9 g L^{-1} Na_2SO_4 ($\geq$99%, p.a., article number: 8560.1) and 0.85 g L^{-1} phenol ($\geq$99.5%, p.a; article number: 0040.1) in demineralized water (Behropur B10 from behr Labor-Technik, Germany) was subjected to the electrolysis setups and the ozonation and peroxone processes. H_2O_2 at a concentration of 30 wt.% (article number: 9681.1) was used for the peroxone experiments. The chemicals were purchased from Carl Roth GmbH & Co., KG (Karlsruhe, Germany).

Na_2SO_4 was selected as the conductive salt for the artificial WW. The addition of conductive salts is necessary to operate the cell within an acceptable and realistic voltage, temperature, and energy regime. Na_2SO_4 was used to enable a better comparison with the results found in the literature. Halogen salts or acids were not used to prevent the generation of other hazardous substances, e.g., adsorbable organic halides by Cl^- oxidation [3]. It is noted that the oxidation of sulfate ($SO_4{}^{2-}$) to persulfate ($S_2O_8{}^{2-}$) is possible with BDD electrodes, due to $\cdot$OH formation and the high overpotential.

This side reaction was of minor relevance, as $SO_4{}^{-}\cdot$ readily reacts with organic compounds. At low current densities, the probability for a recombination of sulfate radical to persulfate is decreased. Independently in the absence of a transition metal, UV radiation, and temperatures >60 °C, the activation of persulfate to perform oxidation reactions of organic compounds is neglectable and was not an influencing factor for the performed degradation experiments [4,33–35].

2.2. Electrolysis Setups

Both electrolysis cells were built from stainless steel using a filter press design (SSZ100/ES and SSZ100/GDE by CONDIAS GmbH, Itzehoe, Germany; Figure 2). A BDD anode (DIACHEM® by CONDIAS) was coupled with a stainless steel cathode and a carbon-based GDE (Printex L6 on Ag-plated Ni mesh; Covestro AG, Leverkusen, Germany; details about the manufacturing process are given in the Supplementary Information Figure S1) using cell designs a and b, respectively. The Printex catalyst was chosen for the GDE due to its high yield of H_2O_2 in acid and alkaline environments in comparison to other pure carbon catalysts [36,37]. The DIACHEM® electrode was coated with 15 µm as a multi-layer on tantalum substrate (thickness of 2 mm). The production using the hot-filament-activated chemical vapor deposition (HF-CVD) process [38] and the related pre-treatment was performed in several steps especially adapted to the material characteristics in order to manufacture the DIACHEM® coating. The main improvement was the triple multi-layer coating.

In design a, one end plate was used as the active cathode, whereby the electrolyte flows from the bottom to the top to ensure homogeneous working electrodes. In design b, the GDE was integrated into the endplate to ensure the electrolyte, air, and power inputs pass through one cell element [39]. These requirements were achieved using a structured flow field milled to the steel body; therefore, in design b, two feed streams must pass the steel body. The PTFE frame has two functions—one is to seal the GDE and the other is to route the electrolyte to the electrochemical reaction zone. Regardless of the cathodic half-shell, the cell design includes important features for effective and efficient operation of the DIACHEM® electrodes for an EAOP® treatment. The special features are the homogeneous electrolyte flow through the flow channels of the inlet and outlet and the bifunctionality of the stainless steel part that combines the cathode with the electrolyte distribution, while we followed the concept of "the fewer the cell parts, the better the cell design, the lower the risk of malfunction". In addition, the anode edges are protected from the influence of the electrolytes, the electric field, and their influence on electrode wear by the flat gasket design. The cells were fed upwards with WW from a tank (18 L and 9 L WW volume for cell designs a and b, respectively) with a constant velocity of 0.298 ± 0.019 m s^{-1} through a centrifugal pump (WPDC-06.7L-10M-24-VP, Rotek Handels GmbH, Hagenbrunn, Austria). The flow areas for cell designs a and b were 180 and 140 mm^2, respectively. Synthetic air (Linde, Germany) was used for the GDE with a 3 × stoichiometric excess of O_2 in the air (127, 76, and 26 mL min^{-1} for 0.5, 0.3, and 0.1 kA m^{-2}, respectively) and the pressure in the air compartment of the cell was adjusted to approx. 40 mbar. The power was applied with a TDK Lambda power supply (type Genesys, Tokyo, Japan). The voltage, current, temperature, and pH were monitored using multimeters (NI 9239 by NI, Austin, TX, USA), a PT100 thermometer (type HSRTD by Omega Engineering, Inc., Norwalk, CT, USA), and a pH probe (tecLine 201020 by JUMO GmbH & Co. KG, Fulda, Germany), respectively.

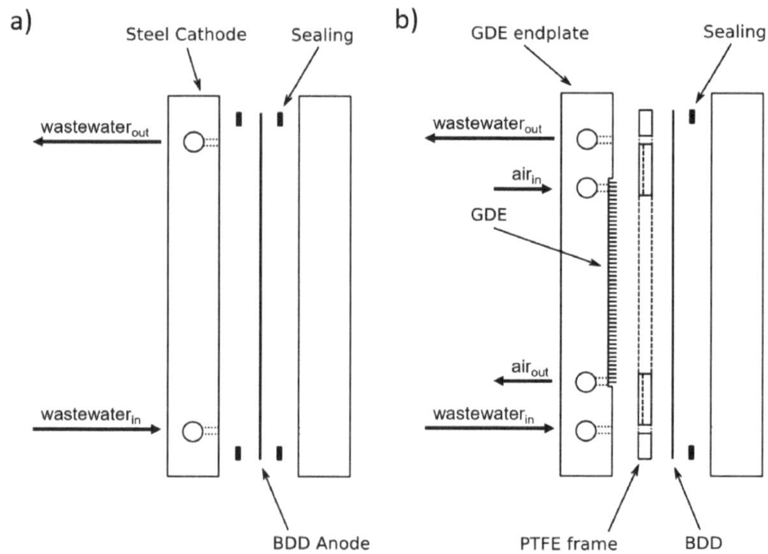

Figure 2. The investigated boron-doped diamond-based electrolysis cells with (**a**) a steel cathode and an active electrode area of 82.6 cm^2. (**b**) A gas diffusion electrode integrated into a steel body with an active electrode area of 41.3 cm^2. A sealing with a PTFE frame was used to ensure similar electrode distances of 2 mm. The process streams are sketched. For more details see the Figures S3 and S4 in the Supplementary Information.

2.3. *Ozone and Peroxone Process*

An O_3 generator and a concentration measurement device from Anseros Klaus Nonnenmacher GmbH (Tübingen, Germany, type Generator COM and Ozomat GM) were used. A reactor and a O_3 destructor used to destroy unreacted O_3 completed the ozone process. A schematical drawing of this setup is given in [30]. It was also used for the peroxone process in this work, whereby a dosing pump (KNF Neuberger GmbH, SIMDOS 02, Freiburg, Germany) was added to enable a continuous dosing of the H_2O_2 solution into the reactor, forming superoxide radicals from O_3 and H_2O_2. Experiments were conducted using various power inputs of the O_3 generator (25% = 0.51 mgO_3 s^{-1}, 50% = 0.63 mgO_3 s^{-1}, 75% = 0.74 mgO_3 s^{-1}, 100% = 0.88 mgO_3 s^{-1}) and in combination with H_2O_2 dosing, with an O_3 power input of 25% + 0.03 mL min^{-1} or + 0.07 mL min^{-1} H_2O_2 solution.

2.4. Analytical Methods and Performance Data

Photometrical cuvette tests (Macherey-Nagel GmbH & Co. KG CSB 40, 300, 600, and 1500, Düren, Germany) were used to determine the chemical oxygen demand (COD). It is known that H_2O_2 negatively affects COD tests, resulting in overdetermination; therefore, two methods were applied to ensure correct values. The method according proposed by Lee et al. was used to correct the measured values with correlation lines [40] for the GDE electrolysis experiments. The correlation lines for the investigated WW are given in the Supplementary Information (Figure S4). The method according proposed by Issa et al. [41] was used for the peroxone experiments. H_2O_2 concentrations were determined using iodometry [42]. To calculate the specific energy consumption per mass O_3 (E_{sc,O_3}), the power consumption (P_{O_3-gen}) and the gaseous volume flow rate ($\dot{V}_{gas}$) of the O_3 generator were used, as well as the concentration of O_3 (c_{O_3}) according to Equation (5). The specific O_3 demands per mass COD were calculated (Equation (6)), whereby the following additional parameters were used, namely the time (t), differential between the initial and the final COD concentration (Δc_{COD}), and treated wastewater volume (V_{WW}). Finally, the specific energy consumption per mass COD was calculated using Equation (7):

$$E_{sc,O_3}\left[\frac{\text{kWh}}{\text{kg}_{O_3}}\right] = \frac{P_{O_3-gen}[\text{kW}]}{\dot{V}_{gas}\left[\frac{\text{L}}{\text{h}}\right] \cdot c_{O_3}\left[\frac{\text{kg}}{\text{L}}\right]} \tag{5}$$

$$\frac{m_{O_3}}{m_{COD}}\left[\frac{\text{kg}_{O_3}}{\text{kg}_{COD}}\right] = \frac{\dot{V}_{gas}\left[\frac{\text{L}}{\text{h}}\right] \cdot c_{O_3}\left[\frac{\text{kg}}{\text{L}}\right] \cdot t[\text{h}]}{\Delta c_{COD}\left[\frac{\text{kg}}{\text{L}}\right] \cdot V_{WW}[\text{L}]} \tag{6}$$

$$E_{sc,O_3,COD} = \frac{P_{O_3-gen}[\text{kW}] \cdot t[\text{h}]}{\Delta c_{COD}\left[\frac{\text{kg}}{\text{L}}\right] \cdot V_{WW}[\text{L}]} \tag{7}$$

During the electrolysis tests, the current densities (j) used were adjusted to the prevailing COD concentration. The j for mineralization without mass transfer limitation and the ideal treatment curve were calculated with the model proposed by Panizza et al. [43].

The specific energy consumption per volume ($E_{sc,el.,Vol.}$) and per kg_{COD} ($E_{sc,el.,COD}$) of the electrolysis setups were calculated according to Equations (8) and (9) with the cell voltage U, the current density j, the active electrode area A, the volume of the wastewater V_{WW} and the Δc_{COD}:

$$E_{sc,el.,Vol.} = \frac{\int_{t=0}^{t}\left(U\,[\text{V}] \cdot j\left[\frac{\text{kA}}{\text{m}^2}\right] \cdot A\left[\text{m}^2\right]\right)dt}{V_{WW}\,[m^3]} \tag{8}$$

$$E_{sc,el.,COD} = \frac{\int_{t=0}^{t}\left(U\,[\text{V}] \cdot j\left[\frac{\text{kA}}{\text{m}^2}\right] \cdot A\left[\text{m}^2\right]\right)dt}{\Delta c_{COD}\left[\frac{\text{kg}}{\text{L}}\right] \cdot V_{WW}[\text{L}]} \tag{9}$$

For determination of the diamond coating thickness, the beta backscattering methodology was used (FISCHERSCOPE® MMS® PC2 in combination with ^{147}Pm beta radiation source; Helmut Fischer GmbH, Sindelfingen, Germany). The measurements were performed on 225 points (9 lines across the sample's longitudinal axis, each with 25 measurement points). The high number of measurements, hence the high resolution of the data, were possible due to the custom-made, self-implemented automatic measurement.

The detection of surface changes of the diamond coating was performed with a color 3D laser scanning microscope (LSM; VK-9700K, Keyence Corporation, Osaka, Japan). For the image analysis, VK Analyzer Version 2.5.0.1 was used.

The operational expenses (OPEX) for different electrode lifetimes were calculated according to Equation (10) with the price M, the estimated electrode lifetime ($LT_{estimated}$), and the annual WW volume flow rate ($\dot{V}_{WW}$):

$$OPEX_{Nb\ or\ Ta}\left[\frac{€}{m^3}\right] = \frac{M_{substrate}\,[€] + M_{BDD\ coating}\,[€]}{LT_{estimated}\,[a]\cdot \dot{V}_{WW}\left[\frac{m^3}{a}\right]} \tag{10}$$

For a better comparison, the $OPEX_{ratio}$ (Equation (11)) was considered.

$$OPEX_{ratio} = \frac{OPEX_{Nb}}{OPEX_{Ta}} \tag{11}$$

The OPEX for the energy demand was calculated based on the results shown in Section 3.2 and Equation (12) with the electric energy (P) and the energy costs (C). Here, the OPEX ratio ($OPEX_{ratio,energy}$) was also considered for a better comparison (Equation (13)):

$$OPEX_{energy,\ \ steel\ or\ GDE}\left[\frac{€}{m^3}\right] = P\left[\frac{kWh}{m^3}\right]\cdot C\left[\frac{€}{kWh}\right] \tag{12}$$

$$OPEX_{ratio,\ energy} = \frac{OPEX_{energy,\ steel}}{OPEX_{energy,\ GDE}} \tag{13}$$

3. Results

3.1. BDD Coating Thickness

To investigate the degradation of the BDD coating in order to estimate the lifetime of the novel Ta-based BDD (BDD-Ta), the WW treatment with cell type a was carried out 77 times to achieve a total runtime of 12,222 h (Figure S5 shows the COD degradation in dependence of time). Figure 3 shows LSM micrographs before and after the test cycle. The j was set to 0.5 kA m^{-2} in the first 24 h, to 0.3 kA m^{-2} between 24 and 96 h, and to 0.1 kA m^{-2} for the remaining time of each batch. For more experimental details, please refer to Section 3.2.

Figure 3. Laser scanning microscope photographs of the Ta-based boron-doped diamond anode (**a**) before initial operation and (**b**) after 12,222 h of operation.

The test conditions during treatment lead to corrosive reactions to the BDD surface due to intermediates of the phenol degradation [31]. This leads to a change of the morphology appearance, the crystal edges becoming rougher, and the diamond wearing off [32]. During the test period, the diamond coating thickness was measured, with the results given in Figure 4. It is obvious that the treatment conditions reduced the coating thickness by 4.8% during the test.

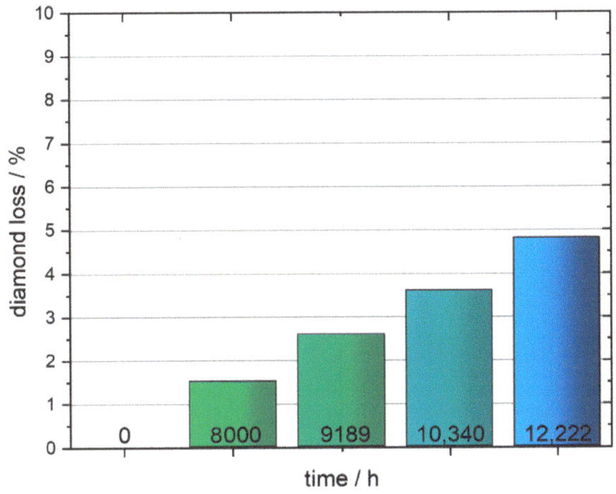

Figure 4. Evolution of the diamond loss during phenolic wastewater treatment over 12,222 h.

Linear prediction was performed based on the known results of an additional long-term test, which was also aimed at diamond loss due to the described reactions (see Figure S6 in the Supplementary Information). Here, the diamond loss exhibits a linear progression over sufficient data points over a coating thickness range of 10 to 6 µm. Additional information about the calculations is given in the Supplementary Information. Other prediction options that are theoretically possible due to the low data density in Figure 4 (e.g., exponential progression) were, therefore, discarded. Consequently, the prediction of the lifetime of DIACHEM® was performed using linear extrapolation from 15 to 6 µm based on the described durability test, resulting in a lifetime of approx. 18 years. The estimated and also the measured service life clearly exceeds the values in the literature, e.g., 100–200 h for Ti-based BDD anodes [44].

3.2. *Treatment Performance of the Electrolysis and the Ozone and Peroxone Processes*

Figure 5a shows the results of the electrolyzer processes treating WW in cell type a with a stainless steel cathode and in cell type b with a H_2O_2-producing GDE in combination with a BDD anode. These results are compared to the ideal mineralization.

All electrolysis-based experiments were conducted over 7 days and at least three times to elucidate statistical errors. To avoid mass transfer limitations due to insufficient amounts of organics, j was set between 0.5 and 0.1 kA m^{-2} according to Figure 5a. With cell design a, degradation of the organics was achieved, which was congruent with the ideal model progression according to Panizza et al. This was achieved via a homogeneous flow of electrolyte over the anode, sufficiently high turbulence inside the cell, as well as optimized j. The electrolysis in combination with a GDE cathode showed accelerated degradation. While the COD values with H_2O_2 interference were similar to those with the stainless steel cathode, the corrected values showed accelerated degradation of up to 35% within the first 48 h. In the COD range below about 30 mg L^{-1}, both processes performed similarly due to the low residual concentration. Both processes reached a concentration of approx. 10 mg L^{-1} on the 6th day, which remained almost constant during day 7. A mineralization rate of >99.5% after 6 days was achieved with both processes. During purification, H_2O_2 concentrations of 20.34 ± 1.09 mmol L^{-1} at 0.5 kA m^{-2} after 24 h, of 7.54 ± 0.92 mmol L^{-1} at 0.3 kA m^{-2} on day 4, and of 5.40 ± 0.76 mmol L^{-1} at 0.1 kA m^{-2} on day 7 were measured. This is an advantage of the implemented operation with an adjustment of the j, since the residual H_2O_2 concentration after the purification process is low, e.g., in comparison to Agladze et al., who achieved 113–340 mmol L^{-1} [22].

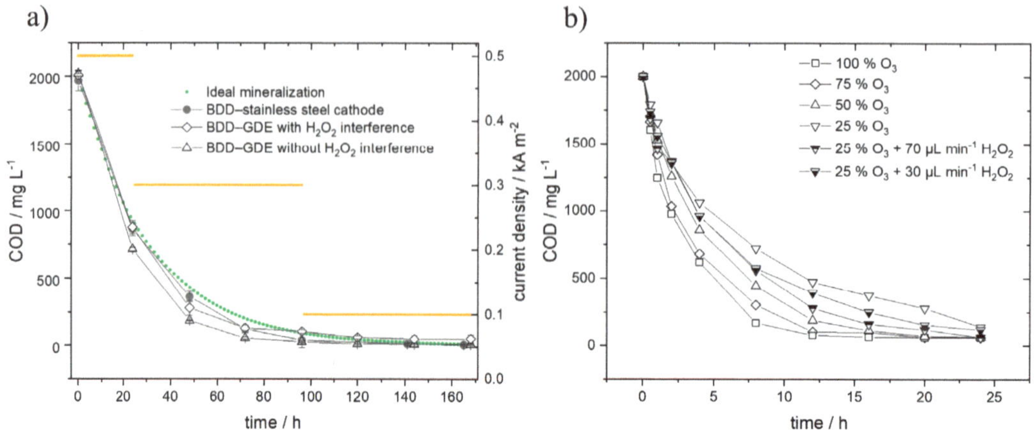

Figure 5. (**a**) COD concentrations of 18 L (cell type a) and 9 L (cell type b) during the 160 h experiments with two different electrode combinations. BDD, boron-doped diamond anode; GDE, gas diffusion electrode. The selected and adjusted current density values are shown on the 2nd *y*-axis. (**b**) COD degradation with O_3 and O_3 + H_2O_2 of 2 L wastewater during 24 h experiments. The varied power settings of the O_3 generator and the H_2O_2 dosing flow rate for the peroxone process are given in the legend above. It is noted that different treatment times were chosen due to the different system designs.

The measured concentrations represent an equilibrium between the formation reaction (Equation (4)), H_2O_2 oxidation at the anode (Equation (14)), the reaction with the hydroxyl radicals formed on the anode side (Equation (15)), self-decomposition (Equation (16)), and the reaction with the organics [45].

$$H_2O_2 \rightarrow H_2O + O_2 + 2e^- \quad (14)$$

$$H_2O_2 + \cdot OH \rightarrow H_2O + \cdot O_2H \quad (15)$$

$$2H_2O_2 \rightarrow 2H_2O + O_2 \quad (16)$$

Table 1 shows the specific energy demands after 6 days and after reaching a COD concentration of 99 mg L^{-1} [46], which is the permitted discharge concentration for WW from pharmaceutical productions according to German law (if a circular economy is not yet being pursued).

Table 1. Energy demands of the investigated electrolysis setups for COD removal at room temperature.

Electrode Configuration	Energy Demand Over 6 Days		Energy Demand Reaching Concentration of Discharge	
	$E_{sc,el.,vol.}$ /kWh m^{-3}	$E_{sc,el.,COD}$ /kWh $kgCOD^{-1}$	$E_{sc,el.,vol.}$ /kWh m^{-3}	$E_{sc,el.,COD}$ /kWh $kgCOD^{-1}$
BDD–stainless steel cathode	87.57 ± 3.91	43.79 ± 1.95	78.89 ± 3.84	39.45 ± 1.92
BDD–GDE	80.46 ± 0.73	40.23 ± 0.37	58.84 ± 1.18	29.42 ± 0.59

Taking the same treatment time into account, the usage of the H_2O_2-generating GDE instead of the stainless steel cathode with HER results in energy savings of approx. 9%. This can be attributed to the higher reduction potential at the cathode (reduction of oxygen to hydrogen peroxide compared to water reduction to hydrogen) on the one hand and to a significant reduction of the cathodic overpotential caused by the application of highly specific catalysts on the other hand. This is also the case when considering the full 7 day cleanup cycle (92.24 ± 4.00 kWh m^{-3} or 46.12 ± 2.00 kWh $kgCOD^{-1}$ versus 83.75 ± 0.72 kWh m^{-3} or 41.88 ± 0.36 kWh $kgCOD^{-1}$). If the accelerated degradation of the organics due to the BDD–GDE combination is considered, the differences in energy consumption are more significant. Here, cell design b showed an accelerated degradation of up to 35% compared to cell design a. This is explained by the higher concentration of oxidizing agents and the combination of $\cdot OH$ and H_2O_2 enabling the formation of additional oxidation species by consecutive reactions, resulting in more efficient decomposition of organic pollutants [47].

For the ozone process, the WW was treated with different O_3 inputs and H_2O_2 was additionally dosed for the peroxone process (Figure 5b). As expected, a higher O_3 input leads to a faster decrease in the COD, although only partially to a lower final value. At power input rates of 100%, 75%, and 50%, final COD values of about 60 mg L^{-1} were achieved; however, this value is reached after different test times. At power input rates of 100%, 75%, and 50% the minimum COD value were reached after about 12, 20, and 24 h, respectively. The final value of the lowest power input was about 150 mg L^{-1}. Column 5 in Table 2 shows the specific O_3 demand per COD during the 24 h treatment.

Table 2. Influence of ozone and peroxone inputs on wastewater treatment after 24 h experiments at room temperature. $P_{O3\text{-}gen}$ denotes the power consumption for the selected power settings of the O_3 generator. E_{sc,O_3}, E_{sc,H_2O_2}, and $E_{sc,O_3,COD}$ denote the specific energy per mass O_3, per mass H_2O_2, and per mass COD, respectively. The mass O_3 demand in dependence to the mass COD is given by $\frac{m_{O_3}}{m_{COD}}$. The energy demand for the production of the used H_2O_2 is estimated as 15 kWh $kgH_2O_2{}^{-1}$ according to [30].

Power Setting (O_3 Generator) and Optional H_2O_2 Dosing (Peroxone) /%	$P_{O_3-gen.}$ /W	E_{sc,O_3} /kWh $kgO_3{}^{-1}$	E_{sc,H_2O_2} /kWh $kgH_2O_2{}^{-1}$	$\frac{m_{O_3}}{m_{COD}}$ /mgO_3 $mgCOD^{-1}$	$E_{sc,O_3,COD}$ /kWh $kgCOD^{-1}$
Idle	17.6	0	-	-	-
25%	30.4	16.5	-	11.9	196.3
50%	47	20.6	-	14.1	291.5
75%	61.5	22.9	-	16.6	380.8
100%	76	24.1	-	19.5	470.0
25% + 0.03 mL min^{-1}	-	16.5	15	11.74 *	349.0
25% + 0.07 mL min^{-1}	-	16.5	15	11.39 *	539.0

* The impact of H_2O_2 is considered.

It shows that a lower power input to the O_3 generator and thus a lower O_3 concentration in the reactor lead to a lower specific energy demand for the removal of phenol. The mass transport of the O_3 limits the phenol degradation rate significantly. The minimum COD value is reached after about 24 h with a power input of 25% and a H_2O_2 dosing of 70 μL min^{-1}, similar to the ozone process with a power input of 50%. Consequently, a small amount of H_2O_2 significantly accelerates the degradation of phenol due to the formation of superoxide radicals (Equations (17) and (18)):

$$H_2O_2 \rightleftarrows HO_2^- + H^+ \tag{17}$$

$$HO_2^- + O_3 \rightarrow \Delta O_2H + O_3^- \tag{18}$$

With a power input of 25% and a H_2O_2 dosing of 30 μL min^{-1}, the final value is higher (120 mg L^{-1}). It can be observed that the addition of H_2O_2 increases the specific energy demand compared to the ozone process with a power input of 25%, but again decreases in terms of the degradation rate compared to the ozone process with a power input of 50%. Furthermore, the solubility of O_3 in water is considerably lower than that of H_2O_2 [48]. Consequently, O_3 tends to outgas in comparison to H_2O_2 and thus leaves the reactor without colliding with phenol. In contrast, the ozone process with a power input of 75%, exhibits both a lower specific energy demand as well as a greater decrease in the COD value than the peroxonation process with a power input of 25% and H_2O_2 addition of 70 μL min^{-1}.

The results of the experiments show that up to 97% mineralization can be achieved using ozonation or peroxonation, requiring 4.9 to 13.4 times more energy than the electrolysis process with BDD–GDE and 4.5 to 12.3 more than the BDD–stainless steel process.

3.3. Operational Expense of Ta-BDD-Based EAOP®

With the objective of determining the effects of the innovative materials in a real application, a case study was designed based on the reported test conditions. This approach was taken to determine the advantages of the material improvements and to investigate the industrial applications using the EAOP®, as it is a known fact that the material costs for tantalum substrates are factor 4–5 higher than for niobium. The assumptions for the case study were:

- 2.5 m^3 h^{-1} WW from pharmaceutical production;
- COD degradation from 2 g L^{-1} to 0.02 g L^{-1};
- 7, 12, or 18 year lifetime for tantalum-based DIACHEM® electrodes, based on the BDD coating thickness investigation carried out in this study;
- 2 year lifetime for niobium-based DIACHEM® electrodes as the reference material, which is the previous standard material for BDD electrodes, based on our industrial experience, which is consistent with Comninellis et al. [49];
- 0.1 € kWh^{-1} energy costs;
- Specific charge of degradation with cell BDD–stainless steel process is equal to specific charge of degradation with cell BDD–GDE process;
- Adapted current density from 0.5 kA m^{-2} to 0.1 kA m^{-2} as performed during the investigations.

DIACHEM® electrodes are consumable due to the described wear mechanism, meaning they have to be replaced at the end of their lifetime. In addition to energy costs, the costs for replacement electrodes are part of the OPEX of an industrial wastewater treatment plant. Different cell configurations were considered in the case study. In order to evaluate the influence of electrode replacement on OPEX, the cell equipped with the BDD of niobium and stainless steel cathodes was compared with the cell equipped with BDD made of tantalum and stainless steel cathodes. In order to evaluate the influence of the cathode type, a cell equipped with BDD anodes with stainless steel cathodes was compared with a cell equipped with BDD anodes and GDE cathodes.

The OPEX values of the different scenarios were determined mainly for OPEX of electrode replacement related to the increased lifetime of the BDD electrodes due to the investigated material improvements (Figure 6).

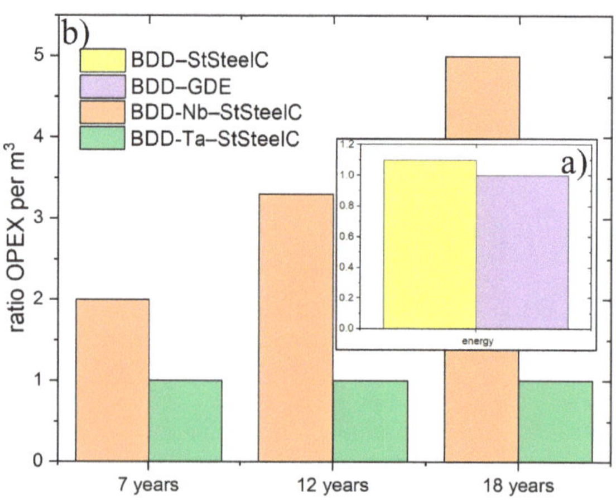

Figure 6. Operational expenses for the case study related to (**a**) energy and (**b**) electrode replacement costs for EAOP® cell configurations. StSeelC, BDD-Nb and BDD-Ta denotes stainless steel cathode, boron-doped diamond on niobium substrate and boron–doped diamond on tantalum substrate, respectively.

The improvements were mainly due to the novel electrode based on the tantalum substrate and the specific coating instead of the niobium-based BDD electrode, in addition to the adapted process parameter and the specific edge-protecting cell design. The purchase costs for the substrate material and the costs for the electrode production were based on real offers from suppliers and effort calculations of a real BDD production with HF-CVD processes.

The OPEX comparison for energy costs shows that the values for the cell with the GDE cathode instead of the stainless steel cathode were 10% better. This advantage is independent of the lifetime of the DIACHEM® anode; therefore, the impact of electrode lifetime was not considered at this point.

The results of the case study show the economic benefits for an industrial application due to the DIACHEM® material innovations. By changing the substrate material, implementing the adapted coating, and using this material in a cell like type a, which was designed to give the longest lifetime for DIACHEM® through the protection of electrode edges, operating cost advantages can be achieved, despite the higher purchase costs for the substrate. The OPEX values for replacement electrodes can be reduced by up to a factor of 5 compared to niobium-based BDD electrodes, depending on the electrode lifetime.

4. Conclusions and Outlook

Two electrochemical reactors based on a BDD anode in combination with a stainless steel cathode (type a) or a GDE (type b) were designed and compared with each other for the treatment of artificial pharmaceutical wastewater and against the ozonation and peroxonation processes. The usage of a BDD–GDE system achieves a degradation efficiency of up to 135% in comparison with the BDD–stainless steel system. In addition, the BDD–GDE system does only require 75%, 14%, and 8% of the energy demands of the BDD–stainless steel, ozonation, and peroxonation systems, respectively.

The treatment efficiency rates of nearly 100% for both investigated electrolysis systems allow lower specific energy consumption rates compared to most of the investigated electrochemical processes in the literature (Table 3).

Table 3. Comparison of electrochemical phenolic degradation and specific energy consumption results from our study with those from the literature. The main working electrodes cited in the literature are denoted as "C" for cathode and "A" for anode, while details about the investigated electrolysis process are specified in the "special feature" column. Furthermore, c_{Phenol}, CE, and TOC denote the concentration of phenols, current efficiency, and total organic carbon, respectively.

Main Electrode	Special Feature	c_{Phenol} [mg L^{-1}]	Na_2SO_4 [g L^{-1}]	Mineralization Rate [%]	CE [%]	Specific Energy per Mass COD or TOC [kWh kgCOD/TOC^{-1}]	Ref.
C	Generated and activated H_2O_2	50	28.4	85 (TOC)	-	-	[21]
C	In-cell and ex-cell Fenton	282	1	90 (COD)	-	56.3	[22]
C	Generated and activated H_2O_2	40	7.1	87 (TOC)	-	-	[23]
C	Electrochemical catalytic treatment	520	4	99 (COD)	-	-	[25]
A	Nanostructured BDD	9153	49 g L^{-1} H_2SO_4	50 (TOC)			[26]
A	Modified PbO_2-based anode	100	14.2	91.1 (phenol)	-	-	[27]
A	PbO_2 anode	50	7.1	94 (TOC)	1.8	3759	[50]
A	TiO_2 with activated carbon fiber	100	10.7	83.3 (phenol)	-	-	[51]
A	BDD anode	850	9	>99 (COD)	≈100	43.79	This study
A+C	BDD anode + H_2O_2 GDE	850	9	>99 (COD)	≈100	40.23	This study

High levels of mineralization probably make the issue of transformation products obsolete and should be investigated in further studies to allow the water to be reused or ecologically discharged. Since traditional BDD anodes are considered a cost driver for EAOP® applications in the literature, the long-term stability of the newly developed tantalum-based BDD was determined. The diamond wear-off of this next-generation anode allows the assumption that the BDD has a long-term stability of up to 18 years when process parameters and cell design are adapted to this specific anode material. Based on this finding, a case study was conducted to demonstrate that the significantly increased long-term stability of the tantalum-based BDD lowers the operational costs as compared to the previous standard material niobium for industrial applications, as the OPEX for replacement electrode is reduced by up to 80%.

Due to the significantly higher electrode costs for BDDs compared to GDEs and the "lack of well-functioning and stable" [44] anodes described in literature, the long-term stability of the anode was assessed in this study. In future investigations, the long-term performance of the GDE should also be evaluated in cell type b for use periods >1 year. Regarding the operational parameters, the differential pressure in the air compartment and the air flow rate should be optimized for further energy savings [42]. To further reduce the carbon emissions and costs of EAOPs and to ensure they operate in more eco-sustainable manners, BDD-based treatment processes were coupled with RE in previous studies, with these results compared to the current study in Table 4. It is obvious that the E_{sc} competes with the literature by using the investigated electrode combination and the current adaption operation. Future studies should investigate how cell types a and b can be powered by RE. While the mentioned studies show that BDD anodes can be operated in a battery-buffered manner or directly coupled with a photovoltaic or wind turbine, the effects of load fluctuations on the lifetime of a GDE are unknown. As such, it may be important to consider the cost/benefit ratio for interconnected buffer storage.

While the ozone and peroxone processes operate volumetrically, the electrolysis processes require larger electrode areas for upscaling to ensure proper mass transfer at the anode surface; therefore, a cell with a greater electrode area needs to be developed to enable easy upscaling in addition to numbering up.

The novel electrolysis with Tantalum-based BDD and GDE is highly suitable for wastewater treatment fulfilling the requirements of a sustainable system for a circular economy.

Table 4. Comparison of the specific energy values per volume and COD of the electrolysis systems with those of the BDD-based electrochemical oxidation systems from literature, which were driven by regenerative energy. Here, ww, CE, PV, and WT denote the concentration wastewater, current efficiency, photovoltaic energy, and wind turbine, respectively.

Main Electrode	Regenerative Energy	Type of ww	Mineralization [%]	$E_{sc,el.,COD.}$ [kWh kgCOD^{-1}]	$E_{sc,el.,vol.}$ [kWh m^{-3}]	Ref.
BDD	PV	Industrial	99 (TOC)	-	896	[52]
BDD	PV	lignosulfonate	90 (TOC)	-	100	[20]
BDD	PV	Soil washing fluid spiked with diesel	57–76 (TOC)	2300	-	[53]
BDD	WT	herbicides	>99 (TOC)	-	18.75 kAh m^{-3}	[54]
BDD	-	Phenolic	>99 (COD)	43.79	87.57	This study
BDD + H_2O_2-GDE	-	Phenolic	>99 (COD)	40.23	80.46	This study

Supplementary Materials: The following are available online at https://www.mdpi.com/article/10.3390/pr9091482/s1: Figure S1: Schematic sequence of GDE production by calandering. The calandered GDEs were coated on one side of the substrate mesh. Figure S2: Exploded view of cell design a with a stainless steel cathode and a boron-doped diamond anode. Figure S3: The structured steel plate (a) and the PTFE frame (b) in cell design b. Figure S4: Correlation between the measured COD values, including an overdetermination due to H_2O_2 interference, and the real COD values for phenol in 9 g L^{-1} Na_2SO_4 solution. Different H_2O_2 concentrations were investigated, which are given in the figure legend above. Table S1: Equations of the linear fits from Figure S4. The coefficient of determination (R^2) was 0.9985for the given equations. Figure S5: COD degradation in dependence of time during 77 batch treatments, with a total runtime of 12,222 h with cell type a. Figure S6: Durability test of a BDD anode. The diamond loss was between 6 and 10 μm. The coefficient of determination (R^2) of the linear regression was 0.994. The conditions for the accelerated corrosion test were in accordance with those published in [31,32] for diamond etching due to high current densities ($j >> 1$ kA m^{-2}) and the degradation of an electrolyte of short-chain CH_3-containing molecules.

Author Contributions: T.M. (Thorben Muddemann) and R.N., conceptualization, methodology, investigation, validation, data curation, visualization, writing—original draft preparation, writing—review and editing. D.H., investigation, writing—review and editing. T.G., writing—review and editing. M.I., investigation. F.B., investigation. M.E., investigation. J.M., investigation. T.M. (Thorsten Matthée), conceptualization, writing—review and editing. M.S. and U.K., supervision, project administration, funding acquisition, writing—review and editing. All authors have read and agreed to the published version of the manuscript.

Funding: This research was funded by Federal Ministry of Education and Research (Bundesministerium für Bildung und Forschung), BMBF, contract numbers 03XP0107G and 03XP0107B.

Institutional Review Board Statement: Not applicable.

Informed Consent Statement: Not applicable.

Data Availability Statement: Not applicable.

Acknowledgments: The authors thank the Federal Ministry of Education and Research (Bundesministerium für Bildung und Forschung), BMBF, Germany, for funding this study.

Conflicts of Interest: The authors declare no conflict of interest. The funders had no role in the design of the study; in the collection, analyses, or interpretation of data; in the writing of the manuscript, or in the decision to publish the results.

References

1. Martínez-Huitle, C.A.; Panizza, M. Electrochemical oxidation of organic pollutants for wastewater treatment. *Curr. Opin. Electrochem.* **2018**, *11*, 62–71. [CrossRef]
2. Särkkä, H.; Bhatnagar, A.; Sillanpää, M. Recent developments of electro-oxidation in water treatment—A review. *J. Electroanal. Chem.* **2015**, *754*, 46–56. [CrossRef]
3. Ganiyu, S.O.; Martínez-Huitle, C.A.; Rodrigo, M.A. Renewable energies driven electrochemical wastewater/soil decontamination technologies: A critical review of fundamental concepts and applications. *Appl. Catal. Environ.* **2020**, *270*, 118857. [CrossRef]
4. Ganiyu, S.O.; Martínez-Huitle, C.A.; Oturan, M.A. Electrochemical advanced oxidation processes for wastewater treatment: Advances in formation and detection of reactive species and mechanisms. *Curr. Opin. Electrochem.* **2020**, *27*, 100678. [CrossRef]
5. Jiang, H.; Dang, C.; Liu, W.; Wang, T. Radical attack and mineralization mechanisms on electrochemical oxidation of *p*-substituted phenols at boron-doped diamond anodes. *Chemosphere* **2020**, *248*, 126033. [CrossRef] [PubMed]
6. Muddemann, T.; Haupt, D.; Sievers, M.; Kunz, U. Elektrochemische Reaktoren für die Wasserbehandlung. *Chem. Ing. Tech.* **2019**, *91*, 769–785. [CrossRef]
7. Simon, R.G.; Stöckl, M.; Becker, D.; Steinkamp, A.-D.; Abt, C.; Jungfer, C.; Weidlich, C.; Track, T.; Mangold, K.-M. Current to clean water—Electrochemical solutions for groundwater, water, and wastewater treatment. *Chem. Ing. Tech.* **2018**, *90*, 1832–1854. [CrossRef]
8. Ganiyu, S.O.; Martínez-Huitle, C.A. The use of renewable energies driving electrochemical technologies for environmental applications. *Curr. Opin. Electrochem.* **2020**, *22*, 211–220. [CrossRef]
9. Radjenovic, J.; Sedlak, D.L. Challenges and opportunities for electrochemical processes as next-generation technologies for the treatment of contaminated water. *Environ. Sci. Technol.* **2015**, *49*, 11292–11302. [CrossRef] [PubMed]
10. Barndõk, H.; Hermosilla, D.; Negro, C.; Blanco, A.M. Comparison and predesign cost assessment of different advanced oxidation processes for the treatment of 1,4-dioxane-containing wastewater from the chemical industry. *ACS Sustain. Chem. Eng.* **2018**, *6*, 5888–5894. [CrossRef]
11. He, Y.; Lin, H.; Guo, Z.; Zhang, W.; Li, H.; Huang, W. Recent developments and advances in boron-doped diamond electrodes for electrochemical oxidation of organic pollutants. *Sep. Purif. Technol.* **2018**, *212*, 802–821. [CrossRef]
12. Muddemann, T.; Bulan, A.; Sievers, M.; Kunz, U. Avoidance of chlorine formation during electrolysis at boron-doped diamond anodes in highly sodium chloride containing and organic-polluted wastewater. *J. Electrochem. Soc.* **2018**, *165*, J3281–J3287. [CrossRef]
13. Muddemann, T.; Haupt, D.; Sievers, M.; Kunz, U. Electrochemical reactors for wastewater treatment. *Chem. Biol. Eng. Rev.* **2019**, *6*, 142–156. [CrossRef]
14. Campos-Martin, J.M.; Blanco-Brieva, G.; Fierro, J.L.G. Hydrogen peroxide synthesis: An outlook beyond the anthraquinone process. *Angew. Chem. Int. Ed.* **2006**, *45*, 6962–6984. [CrossRef]
15. Giomo, M.; Buso, A.; Fier, P.; Sandona, G.; Boye, B.; Farnia, G. A small-scale pilot plant using an oxygen-reducing gas-diffusion electrode for hydrogen peroxide electrosynthesis. *Electrochim. Acta* **2008**, *54*, 808–815. [CrossRef]
16. Elvers, B. *Ullmann's Encyclopedia of Industrial Chemistry*; Chemie: Berlin/Heidelberg, Germany, 1991.

17. Busca, G.; Berardinelli, S.; Resini, C.; Arrighi, L. Technologies for the removal of phenol from fluid streams: A short review of recent developments. *J. Hazard. Mater.* **2008**, *160*, 265–288. [CrossRef]
18. Moussallem, I.; Jörissen, J.; Kunz, U.; Pinnow, S.; Turek, T. Chlor-alkali electrolysis with oxygen depolarized cathodes: History, present status and future prospects. *J. Appl. Electrochem.* **2008**, *38*, 1177–1194. [CrossRef]
19. Cornejo, O.M.; Murrieta, M.F.; Castañeda, L.F.; Nava, J.L. Characterization of the reaction environment in flow reactors fitted with BDD electrodes for use in electrochemical advanced oxidation processes: A critical review. *Electrochim. Acta* **2019**, *331*, 135373. [CrossRef]
20. Dominguez-Ramos, A.; Aldaco, R.; Irabien, A. Photovoltaic solar electrochemical oxidation (PSEO) for treatment of lignosulfonate wastewater. *J. Chem. Technol. Biotechnol.* **2010**, *85*, 821–830. [CrossRef]
21. Luo, H.; Li, C.; Wu, C.; Zheng, W.; Dong, X. Electrochemical degradation of phenol by in situ electro-generated and electro-activated hydrogen peroxide using an improved gas diffusion cathode. *Electrochim. Acta* **2015**, *186*, 486–493. [CrossRef]
22. Agladze, G.R.; Tsurtsumia, G.S.; Jung, B.-I.; Kim, J.-S.; Gorelishvili, G. The "In-cell" and "Ex-cell" Fenton treatment of phenol, 4-chlorophenol and aniline. *J. Appl. Electrochem.* **2007**, *37*, 385–393. [CrossRef]
23. Huang, B.-C.; Jiang, J.; Wang, W.-K.; Li, W.-W.; Zhang, F.; Jiang, H.; Yu, H.-Q. Electrochemically catalytic degradation of phenol with hydrogen peroxide in situ generated and activated by a municipal sludge-derived catalyst. *ACS Sustain. Chem. Eng.* **2018**, *6*, 5540–5546. [CrossRef]
24. Abbas, Z.I.; Abbas, A. Oxidative degradation of phenolic wastewater by electro-fenton process using MnO_2-graphite electrode. *J. Environ. Chem. Eng.* **2019**, *7*, 103108. [CrossRef]
25. Ma, H.; Zhang, X.; Ma, Q.; Wang, B. Electrochemical catalytic treatment of phenol wastewater. *J. Hazard. Mater.* **2009**, *165*, 475–480. [CrossRef] [PubMed]
26. Shi, L.; Xu, F.; Gao, J.; Yuen, M.; Sun, S.; Xu, J.; Jia, K.; Zuo, D. Nanostructured boron-doped diamond electrode for degradation of the simulation wastewater of phenol. *Diam. Relat. Mater.* **2020**, *109*, 108098. [CrossRef]
27. Gui, L.; Chen, Z.; Chen, B.; Song, Y.; Yu, Q.; Zhu, W.; Hu, Q.; Liu, Y.; Zheng, Z.; Ze, L.; et al. Preparation and characterization of ZnO/PEG-Co(II)-PbO_2 nanocomposite electrode and an investigation of the electrocatalytic degradation of phenol. *J. Hazard. Mater.* **2020**, *399*, 123018. [CrossRef]
28. Li, X.-Y.; Cui, Y.-H.; Feng, Y.; Xie, Z.-M.; Gu, J.-D. Reaction pathways and mechanisms of the electrochemical degradation of phenol on different electrodes. *Water Res.* **2005**, *39*, 1972–1981. [CrossRef]
29. Wu, Z.; Zhou, M. Partial degradation of phenol by advanced electrochemical oxidation process. *Environ. Sci. Technol.* **2001**, *35*, 2698–2703. [CrossRef] [PubMed]
30. Haupt, D.; Muddemann, T.; Kunz, U.; Sievers, M. Evaluation of a new electrochemical concept for vacuum toilet wastewater treatment—Comparison with ozonation and peroxone processes. *Electrochem. Commun.* **2019**, *101*, 115–119. [CrossRef]
31. Suzuki, H.; Araki, S.; Yamamoto, H. Evaluation of advanced oxidation processes (AOP) using O_3, UV, and TiO_2 for the degradation of phenol in water. *J. Water Process. Eng.* **2015**, *7*, 54–60. [CrossRef]
32. Panizza, M.; Sineé, G.; Duo, I.; Ouattara, L.; Comninellis, C. Electrochemical polishing of boron-doped diamond in organic media. *Electrochem. Solid State Lett.* **2003**, *6*, D17. [CrossRef]
33. de Araújo, D.M.; Sáez, C.; Cañizares, P.; Rodrigo, M.A.; Martínez-Huitle, C.A. Improving the catalytic effect of peroxodisulfate and peroxodiphosphate electrochemically generated at diamond electrode by activation with light irradiation. *Chemosphere* **2018**, *207*, 774–780. [CrossRef]
34. Serrano, K.G. A critical review on the electrochemical production and use of peroxo-compounds. *Curr. Opin. Electrochem.* **2020**, *27*, 100679. [CrossRef]
35. Liu, H.; Bruton, T.A.; Doyle, F.M.; Sedlak, D.L. In situ chemical oxidation of contaminated groundwater by persulfate: Decomposition by Fe(III) and Mn(IV) containing oxides and aquifer materials. *Environ. Sci. Technol.* **2014**, *48*, 10330–10336. [CrossRef] [PubMed]
36. Assumpção, M.; De Souza, R.; Rascio, D.; Silva, J.C.; Calegaro, M.; Gaubeur, I.; Paixao, T.; Hammer, P.; Lanza, M.; Santos, M. A comparative study of the electrogeneration of hydrogen peroxide using Vulcan and Printex carbon supports. *Carbon* **2011**, *49*, 2842–2851. [CrossRef]
37. Barros, W.; Wei, Q.; Zhang, G.; Sun, S.; Lanza, M.; Tavares, A.C. Oxygen reduction to hydrogen peroxide on Fe_3O_4 nanoparticles supported on Printex carbon and Graphene. *Electrochim. Acta* **2015**, *162*, 263–270. [CrossRef]
38. Fryda, M.; Matthée, T.; Mulcahy, S.; Hampel, A.; Schäfer, L.; Tröster, I. Fabrication and application of Diachem electrodes. *Diam. Relat. Mater.* **2003**, *12*, 1950–1956. [CrossRef]
39. Neuber, R.; Grassl, T.; Matthée, T. Elektrochemische Zelle. EP3816322, 29 October 2020.
40. Lee, E.; Lee, H.; Kim, Y.K.; Sohn, K.; Lee, K. Hydrogen peroxide interference in chemical oxygen demand during ozone based advanced oxidation of anaerobically digested livestock wastewater. *Int. J. Environ. Sci. Technol.* **2011**, *8*, 381–388. [CrossRef]
41. Issa, M.; Muddemann, T.; Haupt, D.; Kunz, U.; Sievers, M. Simple catalytical approach for removal of analytical interferences caused by hydrogen peroxide in a standard chemical oxygen demand test. *J. Environ. Eng.* **2021**. [CrossRef]
42. Muddemann, T.; Haupt, D.R.; Sievers, M.; Kunz, U. Improved operating parameters for hydrogen peroxide-generating gas diffusion electrodes. *Chem. Ing. Tech.* **2020**, *92*, 505–512. [CrossRef]
43. Panizza, M.; Brillas, E.; Comninellis, C. Application of boron-doped diamond electrodes for wastewater treatment. *J. Environ. Eng. Manag.* **2008**, *18*, 139–153.

44. Moradi, M.; Vasseghian, Y.; Khataee, A.; Kobya, M.; Arabzade, H.; Dragoi, E.-N. Service life and stability of electrodes applied in electrochemical advanced oxidation processes: A comprehensive review. *J. Ind. Eng. Chem.* **2020**, *87*, 18–39. [CrossRef]
45. Ginovska, B.; Dupuis, M.; Camaioni, D.M. The H_2O_2+OH→HO_2+H_2O reaction in aqueous solution from a charge-dependent continuum model of solvation. *J. Chem. Phys.* **2008**, *129*, 014506. [CrossRef]
46. Verordnung über Anforderungen an das Einleiten von Abwasser in Gewässer (Abwasserverordnung-AbwV). Available online: https://www.gesetze-im-internet.de/abwv/AbwV.pdf (accessed on 29 July 2021).
47. Sievers, M. Advanced Oxidation Processes. In *Treatise of Water Science*; Wilderer, P., Ed.; Elsevier: Oxford, UK, 2011.
48. Haynes, W.M. *CRC Handbook of Chemistry and Physics: A Ready-Reference Book of Chemical and Physical Data*, 97th ed.; CRC Press: London, UK, 2017.
49. Kraft, A. Doped diamond: A compact review on a new, versatile electrode material. *Int. J. Electrochem. Sci.* **2007**, *2*, 355–385.
50. Duan, X.; Ma, F.; Yuan, Z.; Chang, L.; Jin, X. Electrochemical degradation of phenol in aqueous solution using PbO_2 anode. *J. Taiwan Inst. Chem. Eng.* **2012**, *44*, 95–102. [CrossRef]
51. Jin, P.; Chang, R.; Liu, D.; Zhao, K.; Zhang, L.; Ouyang, Y. Phenol degradation in an electrochemical system with TiO_2/activated carbon fiber as electrode. *J. Environ. Chem. Eng.* **2014**, *2*, 1040–1047. [CrossRef]
52. García-García, A.; Martínez-Miranda, V.; Martínez-Cienfuegos, I.G.; Almazán-Sánchez, P.T.; Castañeda-Juárez, M.; Linares-Hernández, I. Industrial wastewater treatment by electrocoagulation-electrooxidation processes powered by solar cells. *Fuel* **2015**, *149*, 46–54. [CrossRef]
53. Henrique, J.; De Andrade, D.C.; Neto, E.L.B.; Da Silva, D.R.; Dos Santos, E.V. Solar-powered BDD-electrolysis remediation of soil washing fluid spiked with diesel. *J. Chem. Technol. Biotechnol.* **2019**, *94*, 2999–3006. [CrossRef]
54. Souza, F.; Lanza, M.; Llanos, J.; Sáez, C.; Rodrigo, M.A.; Cañizares, P. A wind-powered BDD electrochemical oxidation process for the removal of herbicides. *J. Environ. Manag.* **2015**, *158*, 36–39. [CrossRef] [PubMed]

processes

MDPI

Article

Dark Fermentation of Sweet Sorghum Stalks, Cheese Whey and Cow Manure Mixture: Effect of pH, Pretreatment and Organic Load

Margarita Andreas Dareioti [1], Aikaterini Ioannis Vavouraki [2], Konstantina Tsigkou [1], Constantina Zafiri [3] and Michael Kornaros [1,*]

[1] Laboratory of Biochemical Engineering & Environmental Technology (LBEET), Department of Chemical Engineering, University of Patras, 1 Karatheodori Str., University Campus, 26504 Patras, Greece; m.dareioti@gmail.com (M.A.D.), ktsigkou@chemeng.upatras.gr (K.T.)

[2] School of Mineral Resources Engineering, Technical University of Crete, 73100 Chania, Greece; avavouraki@isc.tuc.gr

[3] Green Technologies Ltd., 5 Ellinos Stratiotou Str., 26223 Patras, Greece; nzafeiri@tee.gr

* Correspondence: kornaros@chemeng.upatras.gr; Tel.: +30-2610997418

Citation: Dareioti, M.A.; Vavouraki, A.I.; Tsigkou, K.; Zafiri, C.; Kornaros, M. Dark Fermentation of Sweet Sorghum Stalks, Cheese Whey and Cow Manure Mixture: Effect of pH, Pretreatment and Organic Load. *Processes* **2021**, *9*, 1017. https://doi.org/10.3390/pr9061017

Academic Editor: Jean-Louis Lanoiselle

Received: 5 May 2021
Accepted: 7 June 2021
Published: 9 June 2021

Abstract: The aim of this study was to determine the optimal conditions for dark fermentation using agro-industrial liquid wastewaters mixed with sweet sorghum stalks (i.e., 55% sorghum, 40% cheese whey, and 5% liquid cow manure). Batch experiments were performed to investigate the effect of controlled pH (5.0, 5.5, 6.0, 6.5) on the production of bio-hydrogen and volatile fatty acids. According to the obtained results, the maximum hydrogen yield of 0.52 mol H_2/mol eq. glucose was measured at pH 5.5 accompanied by the highest volatile fatty acids production, whereas similar hydrogen productivity was also observed at pH 6.0 and 6.5. The use of heat-treated anaerobic sludge as inoculum had a positive impact on bio-hydrogen production, exhibiting an increased yield of 1.09 mol H_2/mol eq. glucose. On the other hand, the pretreated (ensiled) sorghum, instead of a fresh one, led to a lower hydrogen production, while the organic load decrease did not affect the process performance. In all experiments, the main fermentation end-products were volatile fatty acids (i.e., acetic, propionic, butyric), ethanol and lactic acid.

Keywords: agro-industrial wastes; dark fermentation; cheese whey; ensiled sweet sorghum; cow manure; pH effect; organic load effect; pretreatment effect

1. Introduction

Renewable energy production has attracted great interest from the international community during the last decades, with biomass being one of the oldest and most promising energy sources. To that end, organic wastes and by-products such as animal manures, wastewaters, energy crops, agricultural and agro-industrial residues have been extensively investigated as potential feedstocks for efficient production of biofuels. Towards a circular economy, the valorization of agro-industrial wastes, livestock wastes and by-products, and the generation of renewable energy from wastes among others, are in line with sustainable development. Anaerobic digestion (AD) is considered as an alternative waste management process that complies with the European circular economy concept [1].

Cheese whey (CW), for example, is the main by-product of cheese manufacturing containing significant amounts of carbohydrates (4–5%), mainly lactose, soluble proteins (0.6–0.8%), lipids (0.4–0.5%), lactic acid less than 1% and salts that may range from 1% up to 3% [2]. Due to its high organic content (up to 70 g Chemical Oxygen Demand (COD)/L), CW disposal contributes to environmental pollution with lactose being mainly responsible for its high COD. The small size and the seasonal production, along with the geographic dispersion of the producers, make difficult and effective management for added-value applications. Among hydrogen- and biogas-producing agro-industrial wastes, CW appears to be one of the most promising feedstocks [3]. Additionally, cow manure is one of the most polluting agro-industrial by-products. Large amounts of liquid cow manure (LCM) are usually poorly managed, while decomposing livestock manure is considered to be a targeted environmental pollutant due to its high organic matter, nutrients (i.e., nitrogen and phosphorous concentrations), methane and ammonia emissions, and also pathogens.

Sweet sorghum was mainly used as food and animal feed, while it is considered as feedstock for bioenergy production due to high concentrations of fermentable free sugars [4,5]. Sweet sorghum stalks have been examined for biogas (methane) production [6,7] as it is one of several plant species that has been identified as a potential "energy crop" for the production of liquid biofuels and biogas, due to its high crop yields and biodegradability [8,9]. Sweet sorghum's biomass is composed mainly of cellulose, hemicellulose, and lignin, while its fraction of soluble sugars is rich in glucose and sucrose [10]. Fermentable soluble sugars are the primary source for the production of bioethanol and biogas, whereas second-generation biofuels (biogas included) may be produced through the exploitation of their content.

Ensiling is an efficient approach for sweet sorghum fermentation due to residual nutrients, sugars, and low cost. During the ensiling process, the lactic acid bacteria and water-soluble carbohydrates prevail, achieving good quality silage fermentation [11]. A lignocellulosic feedstock is characterized by complex structural composition, while usually pretreatment is required in order to remove the rigid lignin, exposing cellulose and hemicellulose. Alkali pretreatment, among others, enhances dissolution of hemicellulose and, thus, delignification of the biomass is achieved, maximizing carbohydrates solubilization [12,13].

Among conversion technologies for energy exploitation of biomass, biological processing focusing particularly on biogas production [14]. The latter has economic and environmental advantages compared to chemical (thermal) technologies. With the aim to increase both process and equipment performance, in terms of produced biogas, multiple waste streams of organic substrates can be mixed to generate a rather homogeneous feedstock for anaerobic digesters. To this end, co-digestion of different types of organic substrates, such as agro-industrial wastewaters [15] combined with energy crops [16], have been increasingly applied in order to enhance biogas production. At the same time, co-digestion of multiple feedstocks may contribute to overcoming a number of problems either of managerial types, such as seasonal availability of wastes, or, most importantly, of the microbial process itself, such as nutrient imbalance and presence of inhibiting compounds, among other factors.

In particular, the two-stage AD of organic substrates aiming at simultaneous hydrogen and methane production has been reported to provide higher process efficiency and energy recoveries as compared to the conventional one-stage AD [17]. Anaerobic acidogenesis or dark fermentation is known as the first step in the AD of soluble organic materials to hydrogen and CO_2. Carbohydrates are one of the main energy booster nutrients and sources of hydrogen that fermentation processes require. Therefore wastes, energy crops, and/or carbohydrate-rich agricultural residues can be considered as potential substrates for hydrogen production [18].

Under fermentative conditions, carbohydrates degradation is accompanied by the production of hydrogen and metabolic products, mainly volatile fatty acids (VFAs) (i.e., acetic, propionic, butyric acid), lactic acid, and ethanol. The biochemical pathway(s) followed, and consequently, the production of hydrogen and fermentation end-products are highly dependent on the conditions of the process, such as pH [19–21], among others. A number of studies point to a pH value of 5.0 to 6.0 for maximum hydrogen yields [19,22,23].

According to observations and theoretical considerations reported in the literature, homoacetogenesis and propionic acid production would occur predominantly at pH 5.0 or higher. Homoacetogenesis is unlikely to occur at pH 5.0 or lower due to the fact that the production of acetic acid acts as an uncoupler under such conditions [24]. The production of methane is strongly related to hydrogen-consumption, under acidic conditions [25]. According to Lay [22], heat-treated inoculum may also change the microbial community during acidogenesis by developing hydrogen-producing bacteria. Thermal treatment is considered as a pretreatment process for biogas production enhancement during the AD process since organic matter degradation and methane production are significantly increased [26,27]. Several studies have been conducted for the inoculum thermal pretreatment efficiency using a variety of substrates [28].

Referring to published studies, waste streams of animal origin such as liquid swine or buffalo manure have been tested for hydrogen production via dark fermentation, combined mainly with a carbon source such as glucose, potato and pumpkin waste. More specifically, for the case of liquid swine manure and glucose at pH 5.0, a yield of 1.63 mol H_2/mol glucose was achieved [29], while a mixture of potato waste, pumpkin waste and buffalo manure at pH 7.5 led to 135.6 cm^3 H_2/VS [30]. The dark fermentation of cheese whey could result in various yields, strongly depended on the pH value and the physicochemical characteristics of the substrate, presenting a range of 0–1.5 mol H_2/mol lactose [31]. Concerning the dark fermentation of sweet sorghum and the expected hydrogen yields, a maximum of 2.6 mol H_2/mol C6 sugar was achieved after alkaline pretreatment, while sweet sorghum leaves reached 213.14 L H_2/kg of fermentable sugars [18]. To our knowledge, there are no studies involving such an agro-industrial waste mixture as the substrate for the process of dark fermentation.

The aim of this study was to investigate the effect of pH during dark fermentation of an agro-waste mixture consisting of CW and LCM with sweet sorghum stalks on hydrogen production and end-products (i.e., VFAs, lactic acid, ethanol) distribution. The pH values tested ranged from 5.0 to 6.5 with a 0.5 increment and were maintained constant throughout the fermentation process. This study also assessed experimentally the influence of heat-treated sludge as inoculum, the organic loading of the initial substrate, and the role of ensiled sorghum (ES) on the obtained hydrogen yield.

2. Materials and Methods

2.1. Substrates and Inoculum

The tested waste mixture in this study consisted of 55% fresh sweet sorghum stalks (FS) (suspension of 8% *w/w* FS dry matter in water), 40% CW, and 5% LCM (*v/v/v*). This study, in fact, is a sequel to a previous study in which equal organic loading was used, consisting of 55% olive mill wastewater (OMW), 40% CW, and 5% LCM [21]. The reason for using sweet sorghum in this study was to replace OMW, which exhibits seasonal availability (October to February) during a year-round operation of a biogas plant [15], and compare the two system performances as OMW consists not only of carbohydrates but also of fats/oils, which could lead to lower hydrogen production (per g VS) [32]. Additionally, anaerobic co-digestion of ES sorghum, LCM and CW [12] has already exhibited many advantages such as process stability, methane yield improvement and process feasibility due to the seasonal production of these kinds of wastewaters, as well as nutrients addition minimization.

The raw agro-wastes, i.e., CW and LCM, used in this study were collected from small local plants in the area of Patras (Western Greece). In particular, CW was provided from a cheese factory producing mainly white "feta" cheese with daily production of approximately 30 m^3 of CW, while LCM was collected from a dairy farm breeding 230 cows. Following collection, both wastes were immediately stored in the freezer at −18 °C until subsequent use throughout the experimentation period. Sweet sorghum (*Sorghum bicolor L. Moench var. Keller*) was cultivated at the University of Patras facilities according to European regulation (EC 2092/91) and was harvested in early October. Geo-coordinates of the fields were 38°17′51.14 ″ N, 21°48′2.28 ″ E. After harvesting, the sweet sorghum was chopped into pieces (of particle size 1–3 cm), dried at 55 °C, and then ground into 1 mm particle size with a kitchen blender and sieved to powder of <315 μm diameter. A mechanical shredding step is essential to reduce particle size and increase the surface area of the sweet sorghum prior to fermentation. Acclimatized anaerobic culture seed sludge, which was obtained from a lab-scale anaerobic acidogenic continuous stirred tank reactor (CSTR) that was being fed with a mixture of 55% OMW, 40% CW, and 5% LCM (v/v/v) was used as inoculum. Prior to batch experiments, centrifugation (4500 rpm) of the inoculum was performed in order to remove the soluble part.

2.2. Experimental Conditions

All experiments were conducted in a double-walled, cylindrical stainless steel (INOX 316) fermentor under controlled temperature and stirring rate conditions. The temperature was kept constant at 37 °C (mesophilic conditions) via a thermocouple controller and the stirring rate was set at 70 rpm by a geared motor drive unit installed on the top of the fermentor, ensuring thus homogeneous mixing and facilitating rapid diffusion of the produced hydrogen. The working volume of the fermentor was adjusted to 900 mL. The amount of anaerobic sludge used as inoculum was 20% (v/v) of the working volume, while the remaining consisted of the tested waste mixture. The pH of the mixed liquor was kept constant during the experiments via automatic control (using a HACH PID-controller) by adding NaOH or HCl solution (6 N) drop-wise as constant regulated pH values during dark fermentation have exhibited higher hydrogen productivity comparing to unregulated experiments [33]. Anaerobic conditions were imposed on the culture by sparging with nitrogen gas at the beginning of each experiment.

Anaerobic batch experiments were performed in duplicate to study the acidogenesis of an agro-waste mixture of FS, CW, and LCM (in a ratio of 55:40:5, v/v/v) at different constant pH values, namely pH 5.0, 5.5, 6.0 and 6.5. When the optimum pH value was selected, based on the obtained results, additional anaerobic batch experiments were carried out to assess and compare their performance in terms of bio-hydrogen yield and acidified end-products. The following experiments were thus carried out using: (a) heat-treated enriched anaerobic sludge (100 °C for 20 min) [28] instead of the acclimatized anaerobic culture seed sludge (to assess the effect of inoculum), (b) half initial mixed substrate concentration, i.e., 17.15 g total carbohydrates/L (to assess the effect of initial organic loading) and (c) pretreated sorghum (ensiled), instead of a fresh one. The ES pretreatment was performed at a constant temperature of 80 °C for 120 min with the addition of alkaline solution 1.0% NaOH and 1.0% KOH (w/w) (optimum conditions for the pretreatment of ensiled sorghum, ES) [12]. The reactor configuration and the sets of experiments are described in detail in Figure 1.

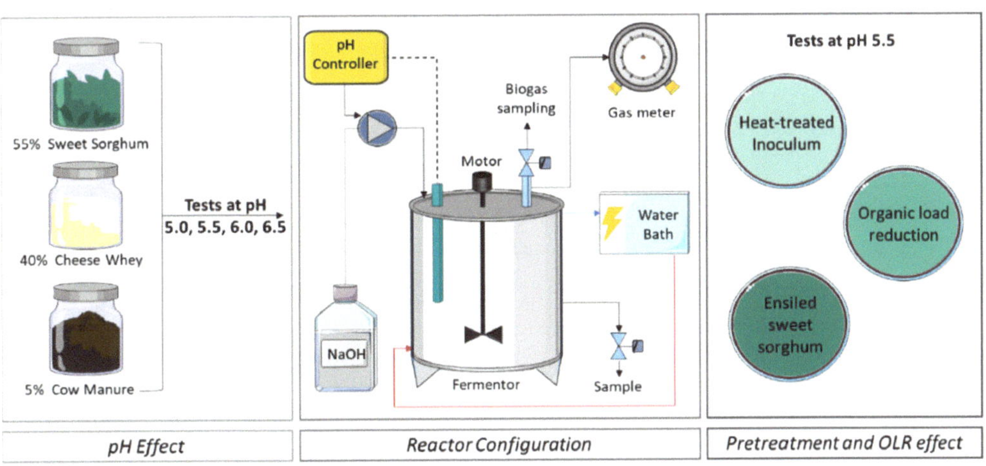

Figure 1. Reactor configuration and sets of experiments.

2.3. Analytical Methods

Detailed physicochemical characterization was performed in all raw agro-waste streams (in triplicate) and fermented effluents (fermentation end-products, in duplicate).

The off-line pH measurements were carried out using an electrode (Thermo Scientific Orion 3-star benchtop pH meter. Total Suspended Solids (TSS), Volatile Suspended Solids (VSS), Total Solids (TS), Volatile Solids (VS), total and soluble COD, Biochemical Oxygen Demand (BOD_5), Total Kjeldahl Nitrogen (TKN), ammonium nitrogen, total and ortho—phosphates and alkalinity were determined according to *Standard Methods for the examination of Water and Wastewater* [34]. Concerning the TSS/VSS measurement and the soluble characteristics (soluble organic carbon (SOC), lactic acid, VFAs, etc.), the insoluble residue was separated from the supernatant by filtration (Whatman® glass microfiber filters Grade GF/F (GE Healthcare Life sciences, Chicago, Illinois, USA)). Carbohydrates determination was accomplished after colorimetric measurement of a colored sugar derivative, which was produced through the addition of L-tryptophan, sulfuric and boric acid [35]. Total organic carbon (TOC) and SOC were analyzed with a Carbon TOC-V analyzer (Shimadzu, Kyoto, Japan). Composition analysis of VFAs, ethanol, biogas and lactic acid were measured by Gas (Agilent, CA, USA) and Ion (Dionex, CA, USA) Chromatography as described in detail by Dareioti et al. [21]. Briefly, for the VFAs and ethanol determination, a capillary column (DB-FFAP, 30 m in length, 0.25 mm I.D. and 0.25 mm film) and a flame ionization detector were used, while for the biogas composition a capillary column (HP-PLOT/Q, 30 m in length, 0.53 mm I.D. and 40 mm packing film) and a thermal conductivity detector were used. Lactic acid was measured using an analytical column (AS19 length 4 × 250 mm and 7.5 mm I.D.), a guard column (4 × 50 mm length and 12 mm I.D.) and an electron conductivity detector.

Cellulose, hemicellulose and lignin content of dried sweet sorghum were measured according to the NREL method [36] and high-performance liquid chromatography (HPLC) analysis [37]. Structural carbohydrates (i.e., glucose, xylose, arabinose) were determined, and the cellulose and hemicellulose contents estimation followed (Equations (1) and (2)):

$$\text{Cellulose } (\%\ \text{VS}) = \text{Glucose } (\%\ \text{VS})/1.11 \tag{1}$$

$$\text{Hemicellulose } (\%\ \text{VS}) = [\text{Xylose } (\%\ \text{VS}) + \text{Arabinose } (\%\ \text{VS})]/1.13 \tag{2}$$

where: 1.11 represents the ratio of the molecular weights of glucose to glucan (180/162), whereas 1.13 represents the ratio of the molecular weights of xylose and arabinose to xylan (150/132).

Total biogas and hydrogen volume, produced during each experiment, was converted at standard temperature and pressure conditions (STP) (0 °C and 1 atm).

3. Results and Discussion

3.1. Chemical Composition of Each Substrate

Prior to experiments, chemical composition characterization of each substrate was performed. Table 1 presents the average values obtained during the CW and LCM characterization, whereas the chemical composition of FS and ES, after drying and milling, is given in Table 2. In particular, CW can be described as a high organic load waste stream, mainly due to its carbohydrates (lactose) content, whereas LCM was characterized by a high concentration of VFAs and nitrogen. The physicochemical characteristics of both substrates are similar to other studies [12,38]. Sweet sorghum mainly consisted of soluble and insoluble polysaccharides (cellulose and hemicellulose) and high lignin content. During the ensiling procedure, soluble carbohydrates are converted mainly to lactic acid and ethanol by fermentative bacteria [39]. Thus, the amount of soluble carbohydrates in the ES was low, whereas lactic acid and ethanol were present.

Table 1. Chemical composition of CW and LCM.

Parameter	Units	CW	LCM
pH	-	6.17 ± 0.03	7.26 ± 0.18
TSS	g/L	38.00 ± 0.11	48.60 ± 0.24
VSS	g/L	34.60 ± 0.14	35.80 ± 0.53
TS	g/L	69.40 ± 0.34	52.40 ± 0.93
VS	g/L	60.17 ± 0.25	35.83 ± 0.64
Total COD	g/L	75.00 ± 0.29	62.50 ± 2.12
Soluble COD	g/L	66.73 ± 0.17	23.02 ± 0.26
TOC	g/L	31.45 ± 0.64	24.19 ± 0.16
Total carbohydrates [1]	g/L	64.40 ± 0.90	8.12 ± 0.31
Soluble carbohydrates [1]	g/L	62.80 ± 1.04	1.20 ± 0.01
TKN	g/L	0.84 ± 0.08	3.36 ± 0.00
Ammonium-N	g/L	0.10 ± 0.02	2.39 ± 0.02
Alkalinity	g $CaCO_3$/L	0.50 ± 0.00	12.38 ± 0.32
TVFA	g/L	0.07 ± 0.00	7.24 ± 0.02

[1] in equivalent glucose (eq. glucose).

Table 2. Chemical composition of FS and ES.

Parameter	FS	ES
pH	5.6 ± 0.1	4.6 ± 0.2
Moisture (% wet weight)	76.7 ± 0.1	77.7 ± 0.3
TS (% wet weight)	23.3 ± 1.0	22.3 ± 0.2
VS (%DM)	94.5 ± 0.2	94.4 ± 0.1
Ash (%DM)	5.5 ± 0.3	5.6 ± 0.8
TOC (%DM)	59.8 ± 0.1	51.0 ± 1.3
Total carbohydrates [1] (%DM)	81.0 ± 3.1	36.9 ± 1.0
Soluble carbohydrates [1] (%DM)	49.0 ± 0.3	2.9 ± 0.1
Cellulose (%DM)	23.0 ± 0.4	27.9 ± 2.1
Hemicellulose (%DM)	11.0 ± 0.7	12.5 ± 1.2
Lignin (%DM)	8.0 ± 0.4	11.9 ± 3.1
Total Nitrogen, TKN (%DM)	0.1 ± 0.0	0.2 ± 0.0
Proteins (%DM)	0.7 ± 0.1	1.1 ± 0.1
Lactic Acid (%DM)	N.D.	1.2 ± 0.1
Ethanol (%DM)	N.D.	8.9 ± 0.1

[1] in equivalent glucose (eq. glucose); DM: Dry Matter; N.D.: Not Detected.

3.2. Effect of pH

Batch experiments were carried out using a mixture of agro-wastes at a ratio of 55% FS, 40% CW and 5% LCM (*v/v/v*) in order to assess the effect of pH on bio-hydrogen production and the composition of dark fermentation end-products. The pH of the agro-industrial waste mixture was controlled at 5.0, 5.5, 6.0 and 6.5. At regular intervals (4 ± 2 h), samples were withdrawn for composition analysis, i.e., determination of carbohydrates, lactic acid, VFAs, ethanol, etc. Qualitative and quantitative composition analysis of produced biogas was performed throughout the course of each experiment. The consumption of total carbohydrates was high in all pH values tested (Figure 2a), with a maximum carbohydrates' degradation of 77% at pH 6.5. However, the soluble carbohydrates consumption was similar (~95–96%) at all pH values (data not shown), suggesting that the microorganisms' ability to consume carbohydrates was not altered within this pH range. The dominant soluble end-products were VFAs (i.e., acetic, propionic, butyric), ethanol and lactic acid in all pH values tested (Figure 2b). Both acetic and butyric acid are well-known metabolites in the anaerobic hydrogen fermentation of carbohydrates, whereas propionic acid is considered an undesirable product of hydrogen fermentation [40]. Acetic and butyric acid were the most abundant end-products at each pH (except pH 5.0), whereas the concentrations of propionic acid and ethanol were substantially lower (<2.3 g/L and 0.9 g/L, respectively). Limited amounts (<0.25 g/L) of *i*-butyric, valeric, *i*-valeric and caproic were detected in all tests. Accumulation of lactic acid, as an intermediate product, was observed at pH 5.0 (8.86 g lactic acid/L), whereas at higher pH it was metabolized during the process and thus not detected at the end of each batch test (Figure 2b). Lactate accumulation at pH 5.0 accompanied by low or no H_2 production has been also observed during the dark fermentation of municipal solid wastes using mixed culture as inoculum [41,42]. As depicted in Figure 2b, the distribution of main soluble end-products was similar at all pH values, apart from pH 5.0. Hence, at these pH values, biogas and hydrogen productivity was also approximately at the same level (Figure 2c). Figure 2a presents additional results at each pH tested, including organic carbon and solids' degradation.

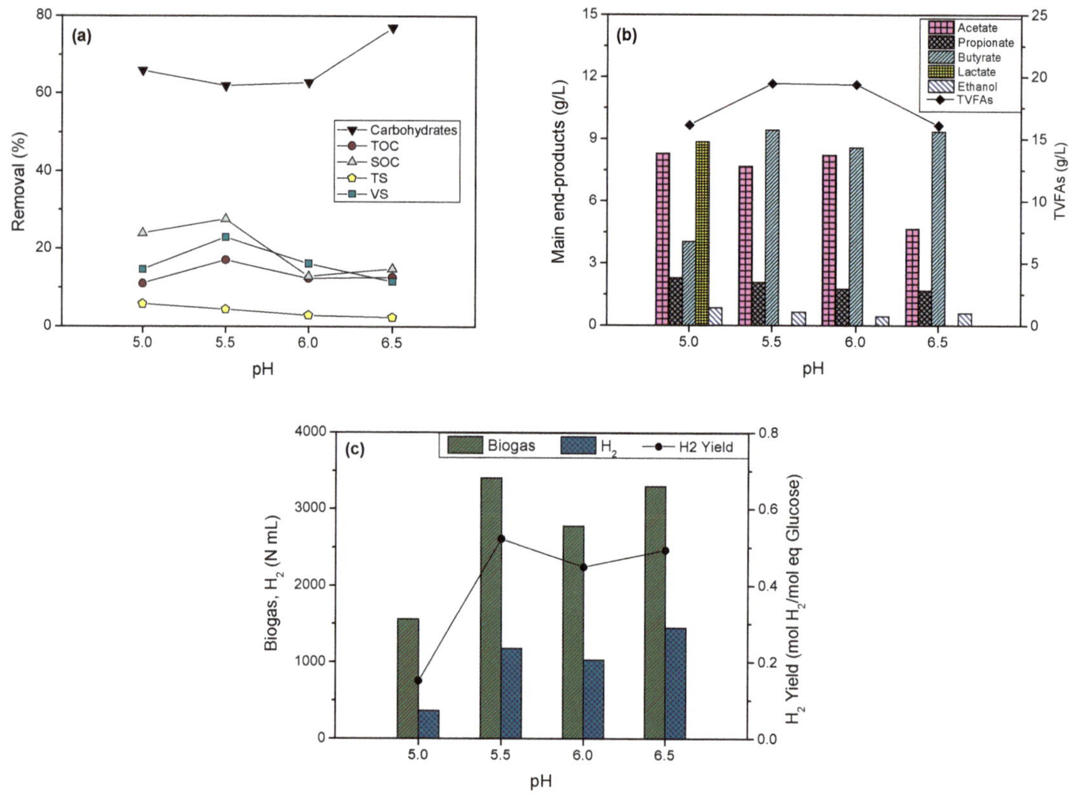

Figure 2. Effect of pH on (**a**) carbohydrates, TOC, SOC, TS, VS removal; (**b**) main soluble end-products; (**c**) biogas and hydrogen production and hydrogen yield.

As shown, the highest degradation of TOC was obtained at pH 5.5 (17.1%), as well as VS degradation (22.9%). Taken together, these data show that the pH value of 5.5 is the most suitable in terms of hydrolysis. Hydrogen yield (calculated as mol H_2/ mol carbohydrate consumed) is a good indicator for the effectiveness of hydrogen production and represents the capability of microorganisms to convert carbohydrates into hydrogen gas. Figure 2c presents the net hydrogen yield (mol H_2/mol eq. glucose of total consumed carbohydrates) at each pH tested. The maximum hydrogen yield of 0.52 mol H_2/mol eq. glucose was observed at pH 5.5, which is considered satisfactory since similar tests with untreated seed sludge generally exhibit low H_2 productivity (<1 mol H_2/mol glucose) [31]. The results indicate that pH had indeed an effect on hydrogen yield and, thus, pH adjustment during dark fermentation of the studied agro-industrial mixture may determine the type of anaerobic fermentation pathway followed during the anaerobic bio-hydrogen process.

In literature, many studies have been presented about the effect of initial and controlled pH on hydrogen production using many different substrates, including the substrates used in this article, as mono substrates. Many of them are in line with the results of the present study, where the optimum pH concerning the H_2 yield appeared to be 5.5, with similar behavior up to pH 6.5. For instance, in the study of Van Ginkel et al. [23], the highest rate of hydrogen production (74.7 mL $H_2/L_R \cdot h$) using sucrose and a heat-treated inoculum was observed at pH 5.5. Similarly, the maximum hydrogen content of 26.9% and production rate of 31.8 mL H_2/h were observed at pH 5.5 by Hernández and Rodríguez [43] who studied pig manure fermentation in an anaerobic batch reactor. Fang and Liu [19] studied the conversion of glucose to hydrogen at different pH values (4.0 to 7.0) and suggested that pH 5.5 was the optimum with the highest hydrogen content (64.2%) and yield of 2.1 ± 0.1 mol H_2/mol glucose. However, Davila-Vazquez et al. [3] studied the effect of initial pH (3.88–8.12) on hydrogen production using three different substrates, namely lactose, CW powder and glucose. pH 7.5 was found as the optimum for lactose and glucose, whereas pH 6.0 was suggested for the CW powder. According to Elbeshbishy et al. [20], one of the most critical factors of dark fermentation is pH due to the end-products determination, the microbial community structure, and the catalysis of intracellular metabolism. In order to enhance H_2 production and avoid inhibition, pH maintenance with an optimum range is of great importance. On the other hand, as various inhibitors of dark fermentation have been identified, co-fermentation of several feedstocks could control or even minimize inhibition by factors such as nutrients, organic load [42] or toxic compounds [44].

Figure 3 displays the batch experimental results at the optimum pH 5.5. Figure 3a presents the concentration of total and soluble carbohydrates over the fermentation process, indicating that the achieved degradation was 62.8% and 96.1%, respectively. 90% of total carbohydrates' degradation was accounted for by the degradation of soluble ones, whereas the rest (10%) was due to hydrolysis and degradation of particulate carbohydrates. The carbohydrate fermentation contributed to the increase of VFAs, lactic acid, and ethanol concentrations (Figure 3b). The most abundant metabolic products were lactic and acetic acid during the first 10 h of the batch experiment. Significant accumulation of butyric acid (~9.5 g/L) was observed after a period of 10 h, as a result of lactic acid bioconversion, which was also accompanied by simultaneous production of hydrogen (Figure 3c). The produced biogas consisted exclusively of hydrogen and carbon dioxide. No methane was observed in any of the batch experiments that were conducted although LCM was used in the mixture. The complete absence of methane could be attributed to a number of reasons with the most valid ones being the use at a very low percentage (5% *v/v*) of a rather recalcitrant to biomethanization substrate (LCM), fermented under acidic conditions. Herein, it can be mentioned that pH can exert remarkable effects on end-products distribution. For instance, at the lowest pH tested in this study (5.0) accumulation of lactic acid, without further bioconversion, was observed due to kinetic limitations in the reactions converting lactic acid to butyric acid and subsequently to hydrogen production.

Taking into account the results obtained in this work and the concept of biorefinery, in which the best compromise between hydrogen and VFAs production has to be considered in an integrated way aiming to maximize energy production in the overall system comprising of an acidogenic and methanogenic reactor, the pH 5.5 was considered as optimum for further work. Moreover, the use of pH 5.5 implies the use of a lower amount of alkaline solution for pH control, leading to lower cost and minimum potential inhibition to the subsequent step of methanogenesis due to sodium cations' presence compared to operation at higher pH values [45].

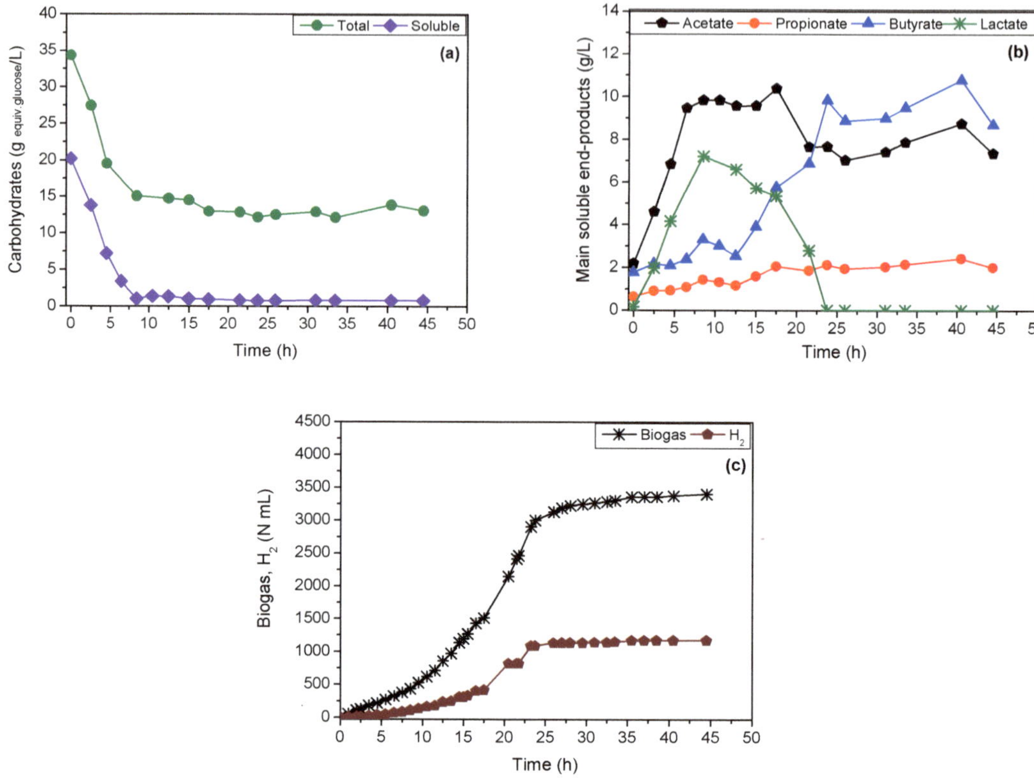

Figure 3. (**a**) Consumption of carbohydrates; (**b**) evolution of main VFAs and lactic acid and (**c**) gaseous products (biogas and hydrogen) during batch acidogenesis of agro-waste mixture (in ratio of 55:40:5), at pH 5.5.

3.3. Kinetic Analysis

It is well known that lactic acid is produced from glucose via three metabolic pathways, i.e., the homofermentative (Equation (3)), the heterofermentative (Equation (4)) and the bifidum pathway (Equation (5)). In all three pathways, the hydrogen balance is zero, i.e., no hydrogen is consumed nor produced.

Lactic acid, as a major intermediate soluble product, may also decompose to acetic and propionic acid (Equation (6)) [46] or to butyric acid accompanied by simultaneous production of hydrogen via the following metabolic pathway (Equation (7)) [21,47].

In the batch experiments conducted in this study, lactic acid depletion was accompanied by a simultaneous consumption of acetic acid and hydrogen. According to Bhat and Barker [48], in the *C. kluyveri* fermentation, butyric acid may be formed by condensation of two moles of acetic acid with two moles of hydrogen (Equation (8)). Dareioti et al. [21] reported briefly, using mass balance calculations, the bioconversion of acetic acid and hydrogen to butyric acid.

$$C_6H_{12}O_6 \rightarrow 2CH_3CH(OH)COOH \quad (3)$$

$$C_6H_{12}O_6 \rightarrow CH_3CH(OH)COOH + CH_3CH_2OH + CO_2 \quad (4)$$

$$2C_6H_{12}O_6 \rightarrow 3CH_3COOH + 2CH_3CH(OH)COOH \quad (5)$$

$$3CH_3CH(OH)COOH \rightarrow 2CH_3CH_2COOH + CH_3COOH + CO_2 + H_2O \quad (6)$$

$$2CH_3CH(OH)COOH \rightarrow CH_3CH_2CH_2COOH + 2CO_2 + 2H_2 \quad (7)$$

$$2CH_3COOH + 2H_2 \rightarrow CH_3CH_2CH_2COOH + 2H_2O \quad (8)$$

Taking into account the transformation pathways (Equations (3)–(8)), the experimental data obtained at the optimum pH (5.5) of this study were used to calculate mass balances in order to estimate the pathways that result in hydrogen production under these conditions (Table 3). In order to facilitate calculations, they were carried out before and after the peak detection of lactic acid (8.5th h). Firstly, the contribution of each pathway (Equations (3)–(5)) to the production of intermediates was investigated (0–8.5 h). Ethanol formation was attributed to glucose degradation via the heterofermentative pathway (Equation (4)) with simultaneous lactic acid production (Table 3, Reaction (2)). On the other hand, the produced acetic acid was attributed to the bifidum pathway (Equation (5)), whereas the rest glucose was decomposed to lactic acid via the homofermentative pathway (Equation (3)). Moreover, in the first period of the experiment (0–8.5 h), a slight lactic acid depletion was observed. The propionic acid formation was attributed to lactic acid bioconversion with simultaneous acetic acid production, according to Equation (6) (Table 3, Reaction (4)). Butyric acid production was correlated with lactic acid degradation with simultaneous hydrogen production (Table 3, Reaction (5)) and also, with acetic acid and hydrogen depletion (Table 3, Reaction (6)). An error of 6.48% was calculated in the mass balances of lactic acid accumulation, which is attributed to the complexity of microbial populations present in the mixed culture used in our experiments.

Table 3. Mass balances for hydrogen production at the optimum pH 5.5.

Reactions for Period: 0–8.5 h	Equation	Calculated Values (mmoles)						
		Glucose	Lactic	Ethanol	Butyric	Acetic	Propionic	H_2
1. $C_6H_{12}O_6 \rightarrow 2CH_3CH(OH)COOH$	(3)	−6.76	+13.52					
2. $C_6H_{12}O_6 \rightarrow CH_3CH(OH)COOH + CH_3CH_2OH + CO_2$	(4)	−9.04	+9.04	+9.04				
3. $2C_6H_{12}O_6 \rightarrow 3CH_3COOH + 2CH_3CH(OH)COOH$	(5)	−75.25	+75.25			+112.87		
4. $3CH_3CH(OH)COOH \rightarrow 2CH_3CH_2COOH + CH_3COOH + CO_2 + H_2O$	(6)		−13.08			+4.36	+8.72	
5. $2CH_3CH(OH)COOH \rightarrow CH_3CH_2CH_2COOH + 2CO_2 + 2H_2$	(7)		−16.40		+8.20			+16.40
6. $2CH_3COOH + 2H_2 \rightarrow CH_3CH_2CH_2COOH + 2H_2O$	(8)				+5.80	−11.60		−11.60
Net calculated values		−91.05	+68.33	+9.04	+14.08	+105.63	+8.72	+4.80
Experimental values		−91.05	+64.17	+9.04	+14.08	+105.63	+8.72	+4.78
Error (%)		0.00	6.48	0.00	0.00	0.00	0.00	0.42
Reactions for period: 8.5–44.5 h								
7. $C_6H_{12}O_6 \rightarrow 2CH_3CH(OH)COOH$	(3)	−10.67	+21.34					
8 $C_6H_{12}O_6 \rightarrow CH_3CH(OH)COOH + CH_3CH_2OH + CO_2$	(4)	−0.70	+0.70	+0.70				
9. $2C_6H_{12}O_6 \rightarrow 3CH_3COOH + 2CH_3CH(OH)COOH$	(5)	−0.00	+0.00			+0.00		
10. $3CH_3CH(OH)COOH \rightarrow 2CH_3CH_2COOH + CH_3COOH + CO_2 + H_2O$	(6)		−8.60			+2.87	+5.74	
11. $2CH_3CH(OH)COOH \rightarrow CH_3CH_2CH_2COOH + 2CO_2 + 2H_2$	(7)		−69.18		+34.59			+69.18
12. $2CH_3COOH + 2H_2 \rightarrow CH_3CH_2CH_2COOH + 2H_2O$	(8)				+10.95	−21.90		−21.90
Net calculated values		−11.37	−55.74	+0.70	+45.54	−19.03	+5.74	+47.28
Experimental values		−11.37	−55.74	+0.70	+45.54	−19.80	+5.74	+47.55
Error (%)		0.00	0.00	0.00	0.00	3.89	0.00	0.57

+: production; −: consumption.

A calculation was then carried out at the second phase of the experiment (8.5–44.5 h), where lactic acid bioconversion with simultaneous hydrogen production mainly took place. The continued degradation of carbohydrates after the 8.5th h till the end of the experiment was calculated in the same way as described in the first period, in order to estimate the further lactic and acetic acid production (Table 3, Reactions (7)–(9)).

Equation (6) was used in order to correlate the propionic acid production via lactic acid consumption (Table 3, Reaction (10)), whereas the rest of lactic acid was decomposed to butyric acid and hydrogen, according to Equation (7) (Table 3, Reaction (11)). Finally, an additional calculation was carried out regarding acetic acid depletion with the corresponding hydrogen towards the rest of butyric acid production (Table 3, Reaction (12)). According to Table 3, marginal errors of 3.89% and 0.57% were estimated in the mass balances of acetic acid and hydrogen, respectively, which however, can be partially attributed to experimental errors and also to biomass production, which was not encountered in all calculations.

3.4. Heat-Treated Inoculum

In an attempt to estimate the effect of inoculum pretreatment on hydrogen productivity, a batch experiment was performed at the optimum pH 5.5, using heat-treated enriched anaerobic sludge (Figure 4a–c), which could potentially enhance the sludge's *Clostridia* characteristics [22]. The total and soluble carbohydrates consumption during the course of this experiment is presented in Figure 5a, reaching up to 58.9% and 94.9%, respectively, which is almost similar to the ones obtained with untreated inoculum. The main metabolic end-product observed was butyric acid (Figure 4b), which was produced from the bioconversion of lactic acid, while the concentration of acetic acid was substantially lower than in the previous batch experiments. The pretreatment of inoculum prior to fermentation led to different metabolic pathways via killing or inactivating one or more bacterial types presented in a typical anaerobic mixed culture. Such types usually refer to non-spore-forming bacteria, while *Clostridium*, *Bacillus* (spore-forming bacteria) and acetic/propionic/lactic acid bacteria are able to survive. Hydrogenotrophic methanogens as well as *Enterobacter* or *Citrobacter*, which are hydrogen-producers are also eliminated [49]. In particular, a significant accumulation of lactic acid (13.7 g lactic acid/L) was realized during the first 30 h of the experiment, whereas acetic acid production of 1.6 g/L was also observed. This indicates a change in the main metabolic pathway followed towards homofermentation. The lactic acid concentration was almost two-fold the one measured in the previous experiments which resulted in higher production of butyric acid with simultaneous hydrogen production (Figure 4c). The experimental results are in accordance with the aforementioned information about the microbial changes.

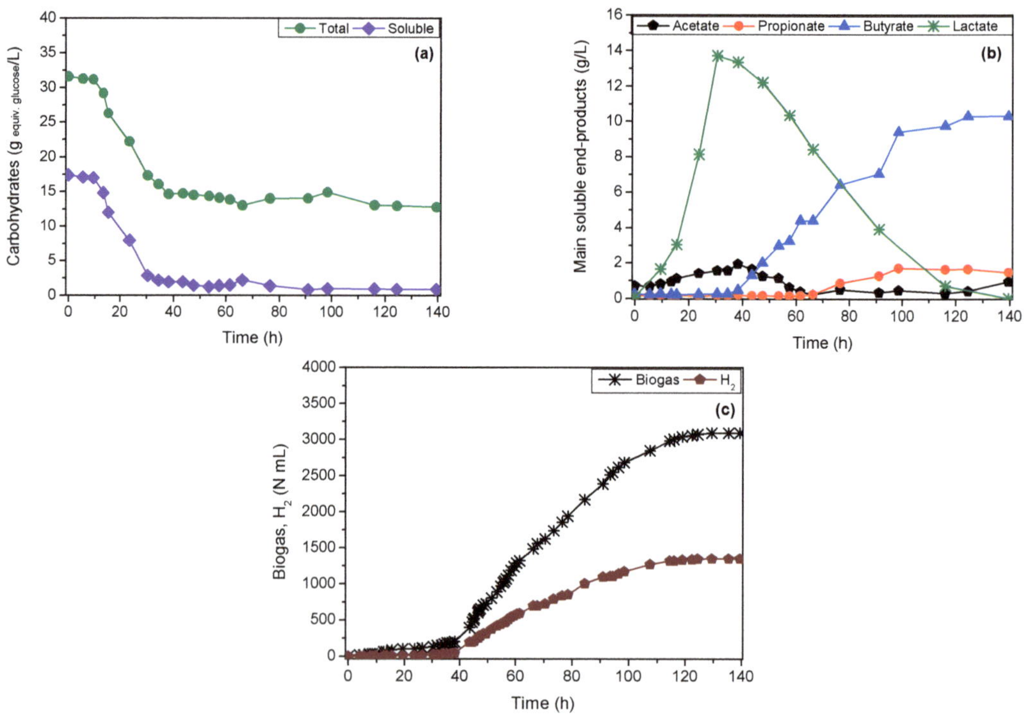

Figure 4. (**a**) Consumption of carbohydrates; (**b**) evolution of main VFAs and lactic acid and (**c**) gaseous products (biogas and hydrogen) using heat pre-treated enriched anaerobic sludge, at pH 5.5.

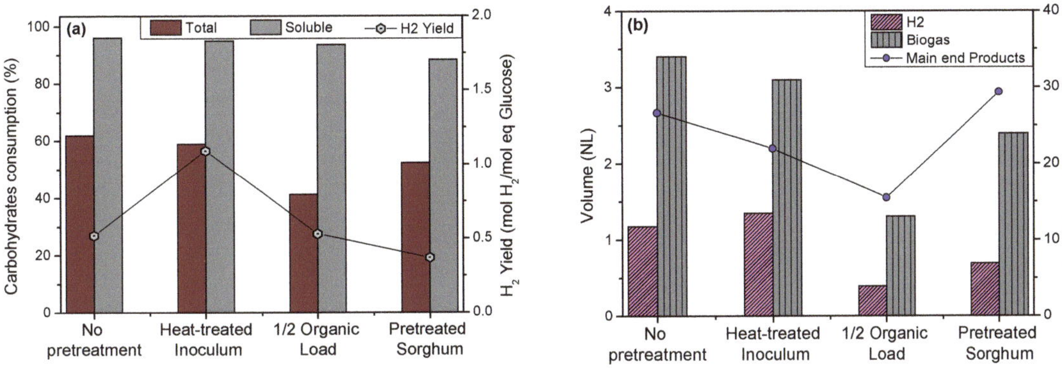

Figure 5. (**a**) Carbohydrates consumption and H_2 yield; (**b**) biogas, H_2 and soluble main end-products accumulation at pH 5.5 for all tested factors.

In Table 4 the mass balance calculations, obtained in the same way as described in the previous batch, are reported using the experimental data of batch with heat-treated inoculum. The calculations were also carried out before and after the peak detection of lactic acid (30.5th h). As previously mentioned, the main metabolic pathway of glucose degradation was homofermentation (Table 4, Reaction (1)). Consequently, a higher concentration of lactic acid was observed compared to the previous batch experiments with untreated inoculum. Moreover, in the first period (0–30.5 h) no lactic acid bioconversion was observed, which occurred only in the second phase (30.5–139.5 h). Using mass balance calculations, a negligible error of 1.74% was obtained from the bioconversion of lactic acid. In the second period of the experiment (30.5–139.5 h) similar calculations were conducted, revealing higher hydrogen production, due to the higher lactic acid amount, compared to the experiment with untreated sludge that was carried out at pH 5.5. The heat-treated inoculum resulted in equal carbohydrate consumption compared to the test at pH 5.5 with untreated inoculum. However, even if the produced main end-products were decreased, the H_2 yield increase because of the different metabolic pathways (Figure 5).

Clostridium species are considered to be the dominant microorganisms responsible for butyric acid fermentation [50]. It has been widely reported that heat-treatment is one of the most promising established methods able to eradicate bacteria in inoculums, that have an adverse effect on hydrogen fermentation and to accelerate thus an enrichment of microbial population with hydrogen-producing bacteria, such as *Clostridium* species, which are highly tolerant to extreme environments due to their spore-forming ability [49,51]. Indeed, the overall hydrogen yield estimated in this test was increased (1.09 mol H_2/mol eq. glucose) compared to the previous experiments where untreated inoculum was used (0.52 mol H_2/mol eq. glucose).

This is in agreement with previous studies mentioned by Wong et al. [28], where untreated and heat-treated inocula of the same origin were compared regarding their efficiency in hydrogen production. More specifically, for most of the reported tests, the H_2 yield of the heat-treated inoculum was usually two to four times higher compared to the untreated inoculum tests. On the contrary, only a few studies mention lower H_2 yields after the heat-pretreatment of the inoculum (compared to the corresponding yield of untreated inoculum). Indicatively, according to Dessi et al. [51], the H_2 efficiency decreased almost four times after the heat shock of the inoculum, for the thermophilic batch dark fermentation of xylose. In a recent study by Słupek et al. [52], it was shown that higher concentrations of hydrogen and methane were obtained during the dark fermentation process under mesophilic conditions than in the process under thermophilic conditions.

Table 4. Mass balances for hydrogen production using heat-treated inoculum at pH 5.5.

Reactions for Period: 0–8.5 h	Equation	Calculated Values (mmoles)						
		Glucose	Lactic	Ethanol	Butyric	Acetic	Propionic	H_2
1. $C_6H_{12}O_6 \rightarrow 2CH_3CH(OH)COOH$	(3)	−42.02	+84.04					
2. $C_6H_{12}O_6 \rightarrow CH_3CH(OH)COOH + CH_3CH_2OH + CO_2$	(4)	−0.96	+0.96	+0.96				
3. $2C_6H_{12}O_6 \rightarrow 3CH_3COOH + 2CH_3CH(OH)COOH$	(5)	−8.96	+8.96			+13.44		
4. $3CH_3CH(OH)COOH \rightarrow 2CH_3CH_2COOH + CH_3COOH + CO_2 + H_2O$	(6)		−0.00			+0.00	+0.00	
5. $2CH_3CH(OH)COOH \rightarrow CH_3CH_2CH_2COOH + 2CO_2 + 2H_2$	(7)		−0.00		+0.00			+0.00
6. $2CH_3COOH + 2H_2 \rightarrow CH_3CH_2CH_2COOH + 2H_2O$	(8)				+0.00	−0.00		−0.00
Net calculated values		−51.94	+93.96	+0.96	+0.00	+13.44	+0.00	+0.00
Experimental values		−51.94	+92.35	+0.96	+0.00	+13.44	+0.00	+0.00
Error (%)		0.00	1.74	0.00	0.00	0.00	0.00	0.00
Reactions for period: 8.5–44.5 h								
7. $C_6H_{12}O_6 \rightarrow 2CH_3CH(OH)COOH$	(3)	−9.31	+18.62					
8 $C_6H_{12}O_6 \rightarrow CH_3CH(OH)COOH + CH_3CH_2OH + CO_2$	(4)	−0.00	+0.00	+0.00				
9. $2C_6H_{12}O_6 \rightarrow 3CH_3COOH + 2CH_3CH(OH)COOH$	(5)	−1.94	+1.94			+2.91		
10. $3CH_3CH(OH)COOH \rightarrow 2CH_3CH_2COOH + CH_3COOH + CO_2 + H_2O$	(6)		−13.46			+4.49	+8.97	
11. $2CH_3CH(OH)COOH \rightarrow CH_3CH_2CH_2COOH + 2CO_2 + 2H_2$	(7)		−85.46		+42.73			+85.46
12. $2CH_3COOH + 2H_2 \rightarrow CH_3CH_2CH_2COOH + 2H_2O$	(8)				+10.21	−20.42		−20.42
Net calculated values		−11.25	−78.36	+0.00	+52.94	−13.02	+8.97	+65.04
Experimental values		−11.25	−78.36	+0.00	+52.94	−13.76	+8.97	+60.12
Error (%)		0.00	0.00	0.00	0.00	5.38	0.00	8.18

+: production; −: consumption.

3.5. *Initial Organic Load*

A batch experiment with half the initial organic load (17.15 g total carbohydrates/L) was conducted at the optimum pH 5.5 in order to assess substrate inhibition phenomena to hydrogen and VFAs production. The evolution of carbohydrates consumption, soluble main end-products, and biogas production are presented in Figure S1 (Supplementary Material). Based on our results, no effect of the initial organic load was identified, as the hydrogen yield (0.53 mol H_2/mol eq. glucose) was almost the same, compared to the previously conducted batch experiment at pH 5.5 (0.52 mol H_2/mol eq. glucose) (Figure 5a). The hydrogen yield was calculated in this work for initial soluble carbohydrates concentration (20.2 g carbohydrates/L, as shown in Figure 5a). Such behavior usually indicates that, under conditions, a higher organic load may become inhibitory due to acid production or increased hydrogen partial pressure. However, the aforementioned results do not lead to a clear conclusion as the system performance also depends on other factors such as the inoculum origin, the type, and the composition of the substrate [20]. Hence, avoidance of such inhibitory conditions is considered mandatory for the achievement of maximum hydrogen conversion efficiency and production rate at high substrate concentrations. Few studies have suggested that higher hydrogen molar yield was found at low organic load [3], whereas high initial substrate concentration caused high initial hydrogen production, increased hydrogen partial pressure and acid toxicity or pH inhibition. Slezak et al. [53] found the highest hydrogen yield (76.1 mL/g VS) using kitchen wastes at an organic load of 14.3 g VS/L, after several tests ranging from 4.1 to 48.2 g initial VS/L, while Rangel et al. [54] attained the maximum hydrogen yield treating residual biomass at an initial organic load of 10 g VS/L and an inoculum to substrate ratio of 1. Based on the results of the current study, it is considered that the inhibitory/toxic thresholds are specific to each system, depending on the type of substrate and inoculum used, while it seems that for the current tested organic load the inhibitory threshold was not reached.

3.6. *Effect of Ensiled Sorghum*

The effect of using sweet sorghum obtained following an ensiling procedure, instead of a fresh one, on hydrogen production was also assessed in this study. The experiment was conducted at constant controlled pH of 5.5. Figure S2a illustrates the total and soluble carbohydrates bioconversion to soluble (Figure S2b) and gaseous (Figure S2c) end-products during the fermentation (Supplementary Material). 52.2% removal of total carbohydrates and 88.2% of soluble ones were observed during this experiment as can be seen in Figure 5a. The distribution of end-products was quite similar to the experiment with FS. The most abundant metabolic end-products were acetic and butyric acid, whereas lactic acid was initially accumulated (reaching 4.9 g/L) and then metabolized to butyric acid. However, hydrogen production in this experiment (Figure 5b) was lower (694.5 mL) than the H_2 production obtained in the previous similar experiment conducted with FS, most probably due to the significantly lower content of carbohydrates (Table 2). As a result, the calculated hydrogen yield was thus also lower (0.37 mol H_2/mol eq. glucose) compared to the one obtained with FS (Figure 5b).

4. Conclusions

Batch experiments were carried out in order to investigate the effect of pH on hydrogen production and the distribution of end-products. The maximum hydrogen yield (0.52 mol H_2/mol eq. glucose) was observed at pH 5.5 accompanied by the highest VFAs production, although similar hydrogen productivities were also realized in pH 6.0 and 6.5. No substrate inhibition was identified, as the half organic load test exhibited the same yield. Heat-treated inoculum resulted in twofold hydrogen yield (1.09 mol H_2/mol eq. glucose) compared to the use of typical inoculum most probably due to partial inactivation or death of hydrogen-consuming microbial populations. On the other hand, utilization of ES instead of fresh one led to limited hydrogen yield, leading to the conclusion that the inoculum heat-treatment and fresh sorghum use are of great importance for the maximization of hydrogen yields.

Supplementary Materials: The following are available online at https://www.mdpi.com/article/10.3390/pr9061017/s1, Figure S1: (a) Consumption of carbohydrates; (b) evolution of main volatile fatty acids and lactic acid; (c) gaseous products (biogas and hydrogen) using half initial organic load, at pH 5.5, Figure S2: (a) Consumption of carbohydrates; (b) evolution of main volatile fatty acids and lactic acid (c) gaseous products (biogas and hydrogen) using pre-treated ensiled sorghum, at pH 5.5.

Author Contributions: Conceptualization, M.A.D.; methodology, M.A.D. and A.I.V.; validation, M.A.D., A.I.V. and K.T.; formal analysis, M.A.D.; investigation, M.A.D., A.I.V. and K.T.; resources, C.Z. and M.K.; data curation, M.A.D. and K.T.; writing—original draft preparation, M.A.D. and A.I.V.; writing—review and editing, A.I.V., K.T. and M.K.; visualization, K.T.; supervision, C.Z. and M.K.; project administration, M.K.; funding acquisition, M.K. All authors have read and agreed to the published version of the manuscript.

Funding: This research was co-funded by the European Union (European Social Fund—ESF) and Greek national funds through the Operational Program "Education and Lifelong Learning" of the National Strategic Reference Framework (NSRF)—Research Funding Program: Heracleitus II. Investing in knowledge society through the European Social Fund.

Institutional Review Board Statement: Not applicable.

Informed Consent Statement: Not applicable.

Data Availability Statement: Data is contained within the article, in Section 3.

Conflicts of Interest: The authors declare no conflict of interest.

References

1. Pires, A.; Martinho, G. Waste hierarchy index for circular economy in waste management. *Waste Manag.* **2019**, *95*, 298–305. [CrossRef]
2. Gelegenis, J.; Georgakakis, D.; Angelidaki, I.; Mavris, V. Optimization of biogas production by co-digesting whey with diluted poultry manure. *Renew. Energy* **2007**, *32*, 2147–2160. [CrossRef]
3. Davila-Vazquez, G.; Alatriste-Mondragón, F.; de León-Rodríguez, A.; Razo-Flores, E. Fermentative hydrogen production in batch experiments using lactose, cheese whey and glucose: Influence of initial substrate concentration and pH. *Int. J. Hydrogen Energy* **2008**, *33*, 4989–4997. [CrossRef]
4. Hu, S.; Wu, L.; Persson, S.; Peng, L.; Feng, S. Sweet sorghum and Miscanthus: Two potential dedicated bioenergy crops in China. *J. Integr. Agric.* **2017**, *16*, 1236–1243. [CrossRef]
5. Appiah-Nkansah, N.B.; Li, J.; Rooney, W.; Wang, D. A review of sweet sorghum as a viable renewable bioenergy crop and its techno-economic analysis. *Renew. Energy* **2019**, *143*, 1121–1132. [CrossRef]

6. Matsakas, L.; Rova, U.; Christakopoulos, P. Evaluation of dried sweet sorghum stalks as raw material for methane production. *Biomed Res. Int.* **2014**, *2014*. [CrossRef] [PubMed]
7. Głąb, L.; Sowiński, J.; Chmielewska, J.; Prask, H.; Fugol, M.; Szlachta, J. Comparison of the energy efficiency of methane and ethanol production from sweet sorghum (*Sorghum bicolor* (L.) Moench) with a variety of feedstock management technologies. *Biomass Bioenergy* **2019**, *129*, 105332. [CrossRef]
8. Claassen, P.A.M.; de Vrije, T.; Budde, M.A.W.; Koukios, E.G.; Gylnos, A.; Reczey, K. Biological hydrogen production from sweet sorghum by thermophilic bacteria. In Proceedings of the 2nd World Conference on Biomass for Energy, Industry and Climate Protection, Rome, Italy, 10–14 May 2004; pp. 1522–1525.
9. Ntaikou, I.; Gavala, H.N.; Kornaros, M.; Lyberatos, G. Hydrogen production from sugars and sweet sorghum biomass using Ruminococcus albus. *Int. J. Hydrogen Energy* **2008**, *33*, 1153–1163. [CrossRef]
10. Panagiotopoulos, I.A.; Bakker, R.R.; De Vrije, T.; Koukios, E.G.; Claassen, P.A.M. Pretreatment of sweet sorghum bagasse for hydrogen production by Caldicellulosiruptor saccharolyticus. *Int. J. Hydrogen Energy* **2010**, *35*, 7738–7747. [CrossRef]
11. Miaoyin, D.; Qiaoqiao, L.; Xu, F.; Wang, S.; Chen, J.; Li, W. Effects of microbial inoculants on the fermentation characteristics and microbial communities of sweet sorghum bagasse silage. *Sci. Rep.* **2020**, *10*, 837.
12. Dareioti, M.A.; Kornaros, M. Anaerobic mesophilic co-digestion of ensiled sorghum, cheese whey and liquid cow manure in a two-stage CSTR system: Effect of hydraulic retention time. *Bioresour. Technol.* **2015**, *175*, 553–562. [CrossRef]
13. Yu, N.; Tan, L.; Sun, Z.-Y.; Nishimura, H.; Takei, S.; Tang, Y.-Q.; Kida, K. Bioethanol from sugarcane bagasse: Focused on optimum of lignin content and reduction of enzyme addition. *Waste Manag.* **2018**, *76*, 404–413. [CrossRef]
14. Kumar, A.; Samadder, S.R. A review on technological options of waste to energy for effective management of municipal solid waste. *Waste Manag.* **2017**, *69*, 407–422. [CrossRef]
15. Dareioti, M.A.; Dokianakis, S.N.; Stamatelatou, K.; Zafiri, C.; Kornaros, M. Biogas production from anaerobic co-digestion of agroindustrial wastewaters under mesophilic conditions in a two-stage process. *Desalination* **2009**, *248*, 891–906. [CrossRef]
16. Giuliano, A.; Bolzonella, D.; Pavan, P.; Cavinato, C.; Cecchi, F. Co-digestion of livestock effluents, energy crops and agro-waste: Feeding and process optimization in mesophilic and thermophilic conditions. *Bioresour. Technol.* **2013**, *128*, 612–618. [CrossRef]
17. Tsigkou, K.; Tsafrakidou, P.; Kopsahelis, A.; Zagklis, D.; Zafiri, C.; Kornaros, M. Used disposable nappies and expired food products valorisation through one- & two-stage anaerobic co-digestion. *Renew. Energy* **2020**, *147*. [CrossRef]
18. Bundhoo, Z.M.A. Potential of bio-hydrogen production from dark fermentation of crop residues: A review. *Int. J. Hydrogen Energy* **2019**, *44*, 17346–17362. [CrossRef]
19. Fang, H.H.P.; Liu, H. Effect of pH on hydrogen production from glucose by a mixed culture. *Bioresour. Technol.* **2002**, *82*, 87–93. [CrossRef]
20. Elbeshbishy, E.; Dhar, B.R.; Nakhla, G.; Lee, H.S. A critical review on inhibition of dark biohydrogen fermentation. *Renew. Sustain. Energy Rev.* **2017**, *79*, 656–668. [CrossRef]
21. Dareioti, M.A.; Vavouraki, A.I.; Kornaros, M. Effect of pH on the anaerobic acidogenesis of agroindustrial wastewaters for maximization of bio-hydrogen production: A lab-scale evaluation using batch tests. *Bioresour. Technol.* **2014**, *162*, 218–227. [CrossRef] [PubMed]
22. Lay, J.J. Modeling and optimization of anaerobic digested sludge converting starch to hydrogen. *Biotechnol. Bioeng.* **2000**, *68*, 269–278. [CrossRef]
23. Ginkel, S.; Van Sung, S.; Lay, J.-J. Biohydrogen Production as a Function of pH and Substrate Concentration. *Environ. Sci. Technol.* **2001**, *35*, 4726–4730. [CrossRef] [PubMed]
24. Rogers, P.; Gottschalk, G. Biochemistry and regulation of acid and solvent production in clostridia. In *The Clostridia and Biotechnology*; Woods, D.R., Ed.; Butterworth-Heinemann: Stoneham, MA, USA, 1993; pp. 25–50.
25. Wainaina, S.; Lukitawesa; Kumar Awasthi, M.; Taherzadeh, M.J. Bioengineering of anaerobic digestion for volatile fatty acids, hydrogen or methane production: A critical review. *Bioengineered* **2019**, *10*, 437–458. [CrossRef] [PubMed]
26. Liu, X.; Wang, W.; Gao, X.; Zhou, Y.; Shen, R. Effect of thermal pretreatment on the physical and chemical properties of municipal biomass waste. *Waste Manag.* **2012**, *32*, 249–255. [CrossRef] [PubMed]
27. Pilli, S.; Yan, S.; Tyagi, R.D.; Surampalli, R.Y. Thermal pretreatment of sewage sludge to enhance anaerobic digestion: A review. *Crit. Rev. Environ. Sci. Technol.* **2015**, *45*, 669–702. [CrossRef]
28. Wong, Y.M.; Wu, T.Y.; Juan, J.C. A review of sustainable hydrogen production using seed sludge via dark fermentation. *Renew. Sustain. Energy Rev.* **2014**, *34*, 471–482. [CrossRef]
29. Wu, X.; Zhu, J.; Dong, C.; Miller, C.; Li, Y.; Wang, L.; Yao, W. Continuous biohydrogen production from liquid swine manure supplemented with glucose using an anaerobic sequencing batch reactor. *Int. J. Hydrogen Energy* **2009**, *34*, 6636–6645. [CrossRef]
30. Ghimire, A.; Frunzo, L.; Pontoni, L.; d'Antonio, G.; Lens, P.N.L.; Esposito, G.; Pirozzi, F. Dark fermentation of complex waste biomass for biohydrogen production by pretreated thermophilic anaerobic digestate. *J. Environ. Manag.* **2015**, *152*, 43–48. [CrossRef]
31. De Gioannis, G.; Friargiu, M.; Massi, E.; Muntoni, A.; Polettini, A.; Pomi, R.; Spiga, D. Biohydrogen production from dark fermentation of cheese whey: Influence of pH. *Int. J. Hydrogen Energy* **2014**, *39*, 20930–20941. [CrossRef]
32. Alibardi, L.; Cossu, R. Effects of carbohydrate, protein and lipid content of organic waste on hydrogen production and fermentation products. *Waste Manag.* **2016**, *47*, 69–77. [CrossRef]

33. Penniston, J.; Gueguim Kana, E.B. Impact of medium pH regulation on biohydrogen production in dark fermentation process using suspended and immobilized microbial cells. *Biotechnol. Biotechnol. Equip.* **2018**, *32*, 204–212. [CrossRef]
34. APHA. Standard Methods for the Examination of Water and Wastewater. In *Standard Methods*, 22nd ed.; Byrd Prepess: Springfield, WA, USA, 2012; ISBN 9780875532356.
35. Joseffson, B. Rapid spectrophotometric determination of total carbohydrates. In *Methods of Seawater Analysis*; Grasshoff, K., Ehrhardt, M., Kremling, K., Eds.; Verlag Chemie International: Weinheim, Germany; Deerfield Beach, FL, USA; Basel, Switzerland, 1983; pp. 340–342. ISBN 3527259988.
36. Sluiter, A.; Hames, B.; Ruiz, R.; Scarlata, C.; Sluiter, J.; Templeton, D.; Crocker, D. NREL/TP-510-42618 analytical procedure—Determination of structural carbohydrates and lignin in Biomass. *Lab. Anal. Proced.* **2008**, *1*, 1–16.
37. Vavouraki, A.I.; Volioti, V.; Kornaros, M.E. Optimization of thermo-chemical pretreatment and enzymatic hydrolysis of kitchen wastes. *Waste Manag.* **2014**, *34*, 167–173. [CrossRef] [PubMed]
38. Flores-Mendoza, A.P.; Hernández-García, H.; Cocotle-Ronzón, Y.; Hernandez-Martinez, E. Methanogenesis of raw cheese whey: pH and substrate–inoculum ratio evaluation at mesophyll temperature range. *J. Chem. Technol. Biotechnol.* **2020**, *95*, 1946–1952. [CrossRef]
39. Gallagher, D.; Parker, D.; Allen, D.J.; Tsesmetzis, N. Dynamic bacterial and fungal microbiomes during sweet sorghum ensiling impact bioethanol production. *Bioresour. Technol.* **2018**, *264*, 163–173. [CrossRef]
40. Castelló, E.; Ferraz-Junior, A.D.N.; Andreani, C.; del Pilar Anzola-Rojas, M.; Borzacconi, L.; Buitrón, G.; Carrillo-Reyes, J.; Gomes, S.D.; Maintinguer, S.I.; Moreno-Andrade, I. Stability problems in the hydrogen production by dark fermentation: Possible causes and solutions. *Renew. Sustain. Energy Rev.* **2020**, *119*, 109602. [CrossRef]
41. Ghimire, A.; Sposito, F.; Frunzo, L.; Trably, E.; Escudié, R.; Pirozzi, F.; Lens, P.N.L.; Esposito, G. Effects of operational parameters on dark fermentative hydrogen production from biodegradable complex waste biomass. *Waste Manag.* **2016**, *50*, 55–64. [CrossRef]
42. Tsigkou, K.; Tsafrakidou, P.; Athanasopoulou, S.; Zafiri, C.; Kornaros, M. Effect of pH on the Anaerobic Fermentation of Fruit/Vegetables and Disposable Nappies Hydrolysate for Bio-hydrogen Production. *Waste Biomass Valorization* **2019**, *11*, 539–551. [CrossRef]
43. Hernández, M.; Rodríguez, M. Hydrogen production by anaerobic digestion of pig manure: Effect of operating conditions. *Renew. Energy* **2013**, *53*, 187–192. [CrossRef]
44. Vavouraki, A.I.; Zakoura, M.V.; Dareioti, M.A.; Kornaros, M. Biodegradation of Polyphenolic Compounds from Olive Mill Wastewaters (OMW) during Two-stage anaerobic Co-digestion of Agro-industrial mixtures. *Waste Biomass Valorization* **2019**, *11*, 5783–5791. [CrossRef]
45. Zagklis, D.; Papadionysiou, M.; Tsigkou, K.; Tsafrakidou, P.; Zafiri, C.; Kornaros, M. Effect of pH on the Economic Potential of Dark Fermentation Products from Used Disposable Nappies and Expired Food Products. *Appl. Sci.* **2021**, *11*, 4099. [CrossRef]
46. Antonopoulou, G.; Gavala, H.N.; Skiadas, I.V.; Angelopoulos, K.; Lyberatos, G. Biofuels generation from sweet sorghum: Fermentative hydrogen production and anaerobic digestion of the remaining biomass. *Bioresour. Technol.* **2008**, *99*, 110–119. [CrossRef]
47. De Gioannis, G.; Friargiu, M.; Massi, E.; Muntoni, A.; Polettini, A.; Pomi, R. Effect of pH value on biohydrogen production from cheese whey. In Proceedings of the 4th International Symposium on Energy from Biomass and Waste, Venice, Italy, 12–15 November 2012; pp. 1–12.
48. Bhat, J.V.; Barker, H.A. Clostridium lacto-acetophilum Nov. Spec. and the Role of Acetic Acid in the Butyric Acid Fermentation of Lactate. *J. Bacteriol.* **1947**, *54*, 381–391. [CrossRef]
49. Łukajtis, R.; Hołowacz, I.; Kucharska, K.; Glinka, M.; Rybarczyk, P.; Przyjazny, A.; Kamiński, M. Hydrogen production from biomass using dark fermentation. *Renew. Sustain. Energy Rev.* **2018**, *91*, 665–694. [CrossRef]
50. Kumar, A.; Jianzheng, J.; Li, Y.; Baral, Y.N.; Ai, B. A review on bio-butyric acid production and its optimization. *Int. J. Agric. Biol.* **2014**, *16*, 1019–1024.
51. Dessì, P.; Porca, E.; Frunzo, L.; Lakaniemi, A.; Collins, G.; Esposito, G.; Lens, P.N.L. Inoculum pretreatment differentially affects the active microbial community performing mesophilic and thermophilic dark fermentation of xylose. *Int. J. Hydrogen Energy* **2018**, *43*, 9233–9245. [CrossRef]
52. Słupek, E.; Makoś, P.; Kucharska, K.; Gębicki, J. Mesophilic and thermophilic dark fermentation course analysis using sensor matrices and chromatographic techniques. *Chem. Pap.* **2020**, *74*, 1573–1582. [CrossRef]
53. Slezak, R.; Grzelak, J.; Krzystek, L.; Ledakowicz, S. The effect of initial organic load of the kitchen waste on the production of VFA and H2 in dark fermentation. *Waste Manag.* **2017**, *68*, 610–617. [CrossRef] [PubMed]
54. Rangela, C.; Sastoqueb, J.; Calderonb, J.; Mosquerab, J.; Velasquezd, P.; Cabezabc, I.; Acevedobd, P. Hydrogen Production by Dark Fermentation Process: Effect of Initial Organic Load. *Chem. Eng.* **2020**, *79*. [CrossRef]

MDPI
St. Alban-Anlage 66
4052 Basel
Switzerland
www.mdpi.com

Processes Editorial Office
E-mail: processes@mdpi.com
www.mdpi.com/journal/processes

www.ingramcontent.com/pod-product-compliance
Lightning Source LLC
Chambersburg PA
CBHW041432210926
51850CB00013B/3413
9783725829583